CONTENTS

INDUSTRIAL VENTILATION

A Manual of Recommended Practice *for Design*

28th Edition

ACGIH®

Signature Publications

Previous Editions

1st Edition — 1951
2nd Edition — 1952
3rd Edition — 1954
4th Edition — 1956
5th Edition — 1958
6th Edition — 1960
7th Edition — 1962
8th Edition — 1964
9th Edition — 1966
10th Edition — 1968
11th Edition — 1970
12th Edition — 1972
13th Edition — 1974
14th Edition — 1976
15th Edition — 1978
16th Edition — 1980
17th Edition — 1982
18th Edition — 1984
19th Edition — 1986
20th Edition — 1988
21st Edition — 1992
22nd Edition — 1995
23rd Edition — Metric — 1998
24th Edition — 2001
25th Edition — 2004
26th Edition — 2007
27th Edition — 2010

ISBN: 978-1-607260-57-8

ACGIH®
Kemper Woods Center
1330 Kemper Meadow Drive
Cincinnati, Ohio 45240-4148
Telephone: 513-742-2020 Fax: 513-742-3355
Email: Publishing@acgih.org
http://www.acgih.org

FOREWORD

The Manual has undergone many changes in the past few issues. A major reorganization into two manuals ("Design" and "Operation and Maintenance"), the more intense science and derivation of equations and the details of high temperature operation are a few of many improvements that have been made going forward from the 26th Edition.

Now in the 28th Edition we take another step forward by incorporating metric (SI) values to augment the improved text and information. The term 'hybrid' is a common word at the time of this writing. It applies to energy savings in automobiles, design of golf clubs, and many other items to reflect the process of change from one product or process to a new, improved version. We mentioned hybrid because this Manual is a step toward the compatibility of terminology and mathematics.

ACGIH® did publish a Metric Manual in 1998 as a duplicate of the 23rd Edition with its traditional IP values. At that time, many of the problems of adapting worldwide mathematics and terminology became evident. Some nations or regions used Pascals for the unit of pressure. Others used 'mm H_2O' (millimeters of water) that corresponded to the 'in. H_2O' (inches of water) used in the United States for their units (see Appendix A9 for a discussion of pressure units in the SI system in this Manual).

In addition, there were variances of duct sizes worldwide. There were European standards for duct sizes and many were similar to the sizes encountered in previous IP Manuals and used by North American contractors (SMACNA) and suppliers. But duct sizes were also varied worldwide and this made it difficult to settle on one set of standards.

In the 14 years since that SI Manual was published, there still has not been a consensus for these units worldwide in the ventilation industry. Other items such as Air/Cloth ratio (a value representing the velocity of air through filtering media) have some accepted ranges in the international industry but are measured in feet/minute. At the same time, a value in SI units would not be comparable but the Manual now adds some of these values to start a dialogue.

For the most part, values in SI units are shown alongside the familiar IP values and enclosed in brackets [ ]. In many cases, sample problems are solved in SI units also. SI values have also been included in Chapter 9 tables, some problem solutions and calculation sheets. In addition, many of the figures are duplicated in SI units and all of the VS plates in Chapter 13 have IP and SI values.

This is a transition but also hopefully a first step to an international understanding of the materials of construction and methods and terminology. Practitioners who are familiar with duct sizes and specified fans in IP units will now be more familiar with materials and equipment furnished in other parts of the world.

The new Manual does revise some values in tables and solutions to better reflect the level of detail and calculations needed to design a ventilation system. In almost every case in the measurement of industrial ventilation systems, there is seldom a need to determine pressure (IP units) in the hundredths of an inch of water gauge (0.01" wg). If not measured, then why go the extra level of detail when calculating the system's need?

In response to that, some values (duct area, losses per foot of duct, etc.) have fewer decimal places in tables. Static pressure in system design (IP units) is now calculated to one decimal place (tenths of an inch – water gauge). That means some rows on the calculation sheet show zero values because of rounding but these are counteracted by rows that may have been rounded up. When checked with pervious Manuals, problem solutions checked out almost exactly and certainly close enough to specify a fan and motor.

The hope is that the 28th Edition will begin more international use of this Manual as a resource for system design. It is also hoped that feedback from our users will provide good data for future editions to focus on items such as duct sizes to refine the data even further.

Metric (SI) to English (IP) Conversions

This issue is being encountered more and more in the international nature of business and regulatory standards. There have been guidelines published and this Manual uses the U.S. Department of Defense Document SD-10 (published December 2003) for its nomenclature and presentation.

There are some key definitions to be considered going forward in the Manual:

Metric Units (SI): A system of basic measures defined by the International Symbol of Units on "Le Systeme International d' Unites (SI)". These units are described in IEEE/ASTM SI 10.

Inch-Pound Units (IP): The standards as previously adopted in the United States and some other parts of the world. There are still some units that have been adopted internationally in some areas.

Soft Conversion: The process of changing a measurement from inch-pound (IP) units to equivalent metric (SI) units with acceptable measurement tolerances without changing the physical configuration of the item. For example, one pound = 453.592 grams.

Hard Conversion: The process of changing a measurement in inch-pound units (IP) to metric units (SI), which necessitates physical configuration changes of the item outside those permitted by established measurement tolerances. For example, one pound = 454 grams.

This Manual uses hard conversions for many of the standards in Chapter 13 and other parts where there are problem solutions. Soft conversion values are used in tables such as Flash Point Temperatures and other areas where exact values were published previously. This will allow for ease of measurement and verification and will provide a more suitable and comparable solution to problems.

In many cases, where hard conversions provide different input data (for example, the air volume specified for a particular piece of equipment), problem solutions will obviously give different results. This should be expected – duct sizes are different, ranges between duct sizes are different and in many cases, the SI solution will be more accurate and provide less energy. This has to be weighed against the varied duct sizes and the problems with fabrication.

INDUSTRIAL VENTILATION COMMITTEE

ACKNOWLEDGMENTS

Industrial Ventilation is a true Committee effort. It brings into focus useful practical ventilation data from all parts of the world in one source. The Committee membership of industrial ventilation engineers and industrial hygienists represents a diversity of experience and interests that ensures a well-rounded cooperative effort.

From the First Edition in 1951, this effort has been successful as witnessed by the acceptance of the "Ventilation Manual" throughout industry, by governmental agencies, and as a worldwide reference and text.

We are grateful for the faith and firm foundation provided by past Committees and members listed below as well as numerous consultants and contributors. Almost every solution in Chapter 13 has been provided by industry in its quest to improve plant ventilation. Special acknowledgment is made to the Division of Occupational Health, Michigan Department of Health, for contributing their original field manual, which was the basis of the First Edition, and to Mr. Knowlton J. Caplan who supervised the preparation of the Manual.

To many other individuals and agencies who have made specific contributions and have provided support, suggestions, and constructive criticism, our special thanks.

INDUSTRIAL VENTILATION COMMITTEE

Previous Members

G.M. Adams, 2004-2008
A.G. Apol, 1984-2002
H. Ayer, 1962-1966
R.E. Bales, 1954-1960
J. Baliff, 1950-1956; Chair, 1954-1956
J.C. Barrett, 1956-1976; Chair 1960-1968
J.L. Beltran, 1964-1966
D. Bonn, Consultant, 1958-1968
D.J. Burton, 1988-1990
K.J. Caplan, 1974-1978; Consultant, 1980-1986
A.B. Cecala, 1998-1999
G. Carlton, 1999-2002
W.M. Cleary, 1976-2006; Chair, 1978-1984
M. Davidson, 1995-1998
R. Dayringer, 2004-present
L. Dickie, 1984-1994; Consultant, 1968-1984
T.N. Do, 1995-2000
N. Donovan, Editorial Consultant, 1950-2008
D.L. Edwards, 2003-present
B. Feiner, 1956-1968
M. Flynn, 1989-1995
M. Franklin, 1991-1994; 1998-2001
S.E. Guffey, 1984-present
J.F. Hale, 2004-present
G.M. Hama, 1950-1984; Chair, 1956-1960
R.P. Hibbard, 1968-1994
R.T. Hughes, 1976-present; Chair, 1989-2001
G.Q. Johnson, 2001-2008
H.S. Jordan, 1960-1962
J. Kane, Consultant, 1950-1952
J. Kayse, Consultant, 1956-1958
J.F. Keppler, 1950-1954; 1958-1960
G.W. Knutson, 1986-2012
G. Lanham, 1998-present, Vice Chair, 2008-present
J.J. Loeffler, 1980-1995; Chair, 1984-1989
J. Lumsden, 1962-1968
J.R. Lynch, 1966-1976
K.R. Mead, 1995-2001
G. Michaelson, 1958-1960
K.M. Morse, 1950-1951; Chair, 1950-1951
R.T. Page, 1954-1956
K.M. Paulson, 1991-present; Vice Chair, 1996-2008
O.P. Petrey, Consultant, 1978-1999
G.S. Rajhans, 1976-1995; Vice Chair, 1994-1995; Chair, 2002-present
K.E. Robinson, 1950-1954; Chair, 1952-1954
A. Salazar, 1952-1954
E.L. Schall, 1956-1958
M.M. Schuman, 1962-1964; Chair, 1968-1978
J.C. Soet, 1950-1960
J.L. Topmiller, 2004-present
A.L. Twombly, 1987-2001
J. Willis, Consultant, 1952-1956
R. Wolle, 1966-1974
A.W. Woody, 1998-present
J.A. Wunderle, 1960-1964

DEFINITIONS

Aerosol: An assemblage of small particles, solid or liquid, suspended in air. The diameter of the particles may vary from 100 microns down to 0.01 micron or less, e.g., dust, fog, smoke.

Air Cleaner: A device designed for the purpose of removing atmospheric airborne impurities such as dusts, gases, mists, vapors, fumes, and smoke. (Air cleaners include air washers, air filters, electrostatic precipitators, and charcoal filters.)

Air Filter: An air-cleaning device that removes light particulate loadings from normal atmospheric air before introduction into the building. Usual range: loadings up to 3 grains per thousand cubic feet (0.003 grains per cubic foot). Note: Atmospheric air in heavy industrial areas and in-plant air in many industries have higher loadings than this, and dust collectors are then indicated for proper air cleaning.

Air Horsepower: The theoretical horsepower required to drive a fan if there were no losses in the fan; that is, if its efficiency were 100 percent.

Aspect Ratio: The ratio of the width to the length; AR = W/L.

Aspect Ratio of an Elbow: The width (W) along the axis of the bend divided by depth (D) in the plane of the bend; AR = W/D.

Blast Gate: Sliding damper.

Blow (throw): In air distribution, the distance an air stream travels from an outlet to a position at which air motion along the axis reduces to a velocity of 50 fpm. For unit heaters, the distance an air stream travels from a heater without a perceptible rise due to temperature difference and loss of velocity.

Brake Horsepower: The horsepower actually required to drive a fan. This includes the energy losses in the fan and can be determined only by actual test of the fan. (This does not include the drive losses between motor and fan.)

Capture Velocity: The air velocity at any point in front of the hood or at the hood opening necessary to overcome opposing air currents and capture the contaminated air at that point by causing it to flow into the hood.

Comfort Zone (Average): The range of effective temperatures over which the majority (50% or more) of adults feel comfortable.

Convection: The motion resulting in a fluid from the differences in density and the action of gravity. In heat transmission this meaning has been extended to include both forced and natural motion or circulation.

Deflagration: A propagation of a combustion zone that occurs at a velocity that is less than the speed of sound in the unreacted medium.

Density: The ratio of the mass of a specimen of a substance to the volume of the specimen. The mass of a unit volume of a substance. When weight can be used without confusion, as synonymous with mass, density is the weight of a unit volume of a substance.

Density Factor: The ratio of actual air density to density of standard air. The product of the density factor and the density of standard air (0.075 lb/ft^3) will give the actual air density in pounds per cubic foot; Density = df × 0.075 lb/ft^3 (the density of standard air).

Dust: Small solid particles created by the breaking up of larger particles by processes, i.e., crushing, grinding, drilling, explosions, etc. Dust particles already in existence in a mixture of materials may escape into the air through such operations as shoveling, conveying, screening, sweeping, etc.

Dust Collector: An air-cleaning device to remove heavy particulate loadings from exhaust systems. Usual range of particulate loading: 0.003 grains per cubic foot or higher.

Entry Loss: Loss in pressure caused by air flowing into a duct or hood (inches H_2O).

Fumes: Small, solid particles formed by the condensation of vapors of solid materials.

Gases: Formless fluids that tend to occupy an entire space uniformly at ordinary temperatures and pressures.

Hard Conversion: In a hard conversion a new rationalized number is created that is convenient to work with and remember. Example 48 × 25 = 1200 mm

Hood: A shaped inlet designed to capture contaminated air and conduct it into the exhaust duct system.

Hood Flow Coefficient: The ratio of flow caused by a given hood static pressure compared to the theoretical flow that would result if the static pressure could be converted to velocity pressure with 100 percent efficiency. *NOTE: This was defined as Coefficient of Entry in previous editons.*

Humidity, Absolute: The weight of water vapor per unit volume, pounds per cubic foot or grams per cubic centimeter.

Humidity, Relative: The ratio of the actual partial pressure of the water vapor in a space to the saturation pressure of pure water at the same temperature.

Inch of Water: A unit of pressure equal to the pressure exerted by a column of liquid water one inch high at a standard temperature.

Lower Explosive Limit: The lower limit of flammability or explosibility of a gas or vapor at ordinary ambient temperatures expressed in percent of the gas or vapor in air by volume. This limit is assumed constant for temperatures up to 250 F. Above these temperatures, it should be decreased by a factor of 0.7 since explosibility increases with higher temperatures.

Manometer: An instrument for measuring pressure; essentially a U-tube partially filled with a liquid, usually water, mercury or a light oil, so constructed that the amount of displacement of the liquid indicates the pressure being exerted on the instrument.

Micron: A unit of length, the thousandth part of 1 mm or the millionth of a meter (approximately 1/25,000 of an inch).

Minimum Design Duct Velocity: Minimum air velocity required to move the particulates in the air stream (fpm).

Mists: Small droplets of materials that are ordinarily liquid at normal temperature and pressure.

Plenum: Pressure equalizing chamber.

Plug Flow: The velocity of a fluid is assumed to be constant across any cross section of the duct perpendicular to the axis of the duct.

Pressure, Static: The potential pressure exerted in all directions by a fluid at rest. For a fluid in motion, it is measured in a direction normal to the direction of flow. Usually expressed in inches water gauge when dealing with air. (The tendency to either burst or collapse the pipe.)

Pressure, Total: The algebraic sum of the velocity pressure and the static pressure (with due regard to sign).

Pressure, Vapor: The pressure exerted by a vapor. If a vapor is kept in confinement over its liquid so that the vapor can accumulate above the liquid, the temperature being held constant, the vapor pressure approaches a fixed limit called the maximum or saturated vapor pressure, dependent only on the temperature and the liquid. The term vapor pressure is sometimes used as synonymous with saturated vapor pressure.

Pressure, Velocity: The kinetic pressure in the direction of flow necessary to cause a fluid at rest to flow at a given velocity. Usually expressed in inches water gauge.

Radiation, Thermal (Heat): The transmission of energy by means of electromagnetic waves of very long wavelength. Radiant energy of any wavelength may, when absorbed, become thermal energy and result in an increase in the temperature of the absorbing body.

Replacement Air: A ventilation term used to indicate the volume of controlled outdoor air supplied to a building to replace air being exhausted.

Slot Velocity: Linear flow rate of contaminated air through a slot, fpm.

Smoke: An air suspension (aerosol) of particles, usually but not necessarily solid, often originating in a solid nucleus, formed from combustion or sublimation.

Soft Conversion: An inch-pound measurement or value is converted to its exact or nearly exact metric equivalent. Example 48 × 25.4 = 1219.4 mm

Specific Gravity: The ratio of the mass of a unit volume of a substance to the mass of the same volume of a standard substance at a standard temperature. Water at 39.2 F is the standard substance usually referred to. For gases, dry air, at the same temperature and pressure as the gas, is often taken as the standard substance.

Standard Air: Dry air at 70 F and 29.92 (in Hg) barometer. This is substantially equivalent to 0.075 lb/ft^3. Specific heat of dry air = 0.24 BTU/lb/F.

Temperature, Effective: An arbitrary index that combines into a single value the effect of temperature, humidity, and air movement on the sensation of warmth or cold felt by the human body. The numerical value is that of the temperature of still, saturated air that would induce an identical sensation.

Temperature, Wet-Bulb: Thermodynamic wet-bulb temperature is the temperature at which liquid or solid water, by evaporating into air, can bring the air to saturation adiabatically at the same temperature. Wet-bulb temperature (without qualification) is the temperature indicated by a wet-bulb psychrometer constructed and used according to specifications.

Threshold Limit Values (TLVs®): The values for airborne toxic materials that are to be used as guides in the control of health hazards and represent time-weighted concentrations to which nearly all workers may be exposed for 8 hours per day over extended periods of time without adverse effects (see Appendix).

Transport (Conveying) Velocity: See Minimum Design Duct Velocity.

Turn-Down Ratio: The degree to which the operating performance of a system can be reduced to satisfy part-load conditions. Usually expressed as a ratio; for example, 30:1 means the minimum operation point is 1/30th of full load.

Vapor: The gaseous form of substances that are normally in the solid or liquid state and that can be changed to those states either by increasing the pressure or decreasing the temperature.

ABBREVIATIONS

A . . . . . area
acfm . . . . . actual flow rate in cubic feet/min
AH . . . . . air horsepower
am^3/s . . . . . actual flow rate in cubic meters/sec
AR . . . . . aspect ratio
A_s . . . . . slot area
B . . . . . barometric pressure
bhp . . . . . brake horsepower
bhp_a . . . . . brake horsepower, actual
bhp_s . . . . . brake horsepower, standard air
BTU . . . . . British Thermal Unit
BTUH . . . . . BTU per hour
C . . . . . degree, Celsius
C_e . . . . . hood flow coefficient
CLR . . . . . centerline radius
D . . . . . diameter
df . . . . . overall density factor
df_e . . . . . elevation density factor
df_p . . . . . pressure density factor
df_t . . . . . temperature density factor
df_m . . . . . moisture density factor
dscf . . . . . dry standard cubic feet
dscfm . . . . . dry standard cubic feet per minute
ET . . . . . effective temperature
f . . . . . Moody diagram friction coefficient
F . . . . . degree, Fahrenheit
F'_d . . . . . loss per unit length (duct)
F_h . . . . . hood entry loss coefficient
F_{el} . . . . . elbow loss coefficient
F_{en} . . . . . entry loss coefficient
fpm . . . . . feet per minute
fps . . . . . feet per second
F_s . . . . . slot loss coefficient
ft^2 . . . . . square foot
ft^3 . . . . . cubic foot
g . . . . . gravitational force, ft/sec/sec
gpm . . . . . gallons per minute
gr . . . . . grains
h . . . . . total heat (enthalpy)
H . . . . . height
h_h . . . . . hood entry loss
h_e . . . . . overall hood entry loss
h_{el} . . . . . elbow loss
h_{en} . . . . . entry loss
h_d . . . . . loss in straight duct run
HEPA . . . . . high-efficiency particulate air filters
hp . . . . . horsepower
hr . . . . . hour
h_s . . . . . slot or opening entry loss
HV . . . . . humid volume (ft^3 mix/lbm dry air)
HVAC . . . . . heating, ventilation, and air conditioning
in . . . . . inch
in^2 . . . . . square inch
"wg . . . . . inches water gauge
L . . . . . length
lb . . . . . pound
lbm . . . . . pound mass
LEL . . . . . lower explosive limit
$\dot{m}$ . . . . . mass flow rate
m^2 . . . . . meters squared
m^3 . . . . . cubic meters
ME . . . . . mechanical efficiency
mg . . . . . milligram
min . . . . . minute
mm . . . . . millimeter
mm wg . . . . . millimeters water gauge
MRT . . . . . mean radiant temperature
m/s . . . . . meters per sec
MW . . . . . molecular weight
ρ . . . . . density of air in lb/ft^3
ppm . . . . . parts per million
psi . . . . . pounds per square inch
PWR . . . . . power
Q . . . . . flow rate in cfm
Q_{corr} . . . . . corrected flow rate at a junction
R . . . . . degree, Rankin
RH . . . . . relative humidity
rpm . . . . . revolutions per minute
s . . . . . second
scfm . . . . . standard cubic feet per minute
sfpm . . . . . surface feet per minute
SG . . . . . specific gravity
SP . . . . . static pressure
SP_{gov} . . . . . higher static pressure at junction of 2 ducts
SP_h . . . . . hood static pressure
SP_s . . . . . SP, system handling standard air
STP . . . . . standard temperature and pressure
T . . . . . temperature
TLV® . . . . . Threshold Limit Value
TP . . . . . total pressure

V ....velocity, fpm
V_d ....duct velocity
VP ....velocity pressure
VP_d ....duct velocity pressure
VP_r ....resultant velocity pressure
VP_s ....slot velocity pressure
V_s ....slot velocity
V_t ....duct transport velocity
W ....watt
$\dot{w}$ ....work
ω moisture content (lbm H_2O/lbm dry air)
z ....elevation in feet above sea level

Chapter 1

EXPOSURE ASSESSMENT

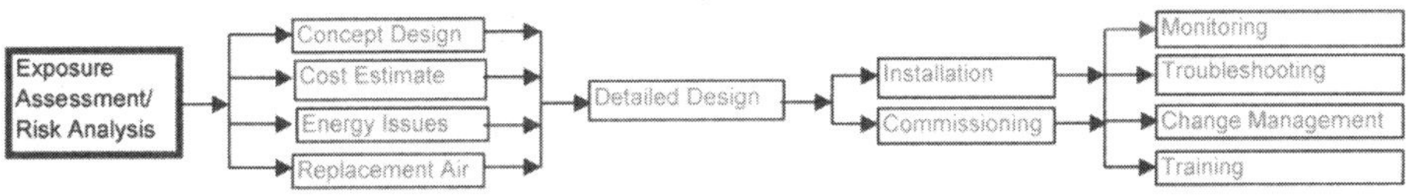

NOTE: Equations with notation followed by (IP) are designated for inch-pound system only; Equations followed by (SI) are designated for metric use only. If Equation bears neither, then it applies to both systems.

1.1 INTRODUCTION

Adverse health effects can occur when employees are exposed to occupational hazards. These adverse effects may occur immediately after exposure (such as the effects of carbon monoxide), or after a long latency period (such as the effects of asbestos). Exposure assessment involves the tasks of qualitatively or quantitatively evaluating the frequency, duration, magnitude, and route of exposure to occupational hazards. This assessment should be based on knowledge in the disciplines of industrial hygiene, toxicology, and epidemiology. The purpose of the assessment is to prevent hazardous exposures and any resulting adverse health effects. Industrial hygienists possess skills specific to conducting exposure assessments. The order of practice in industrial hygiene is hazard 1) anticipation, 2) recognition, 3) evaluation, and 4) control.

This places exposure assessment (evaluation) as the third step in the industrial hygiene procedure and control as the final step. When considering industrial ventilation systems as a solution to occupational exposures, a three part methodology should be considered:

1) Evaluate whether the process generates potential chemical and/or physical hazards (Section 1.3),
2) Determine if employees are potentially exposed to the hazards (Sections 1.2 – 1.5),
3) Determine if exhaust ventilation as an engineering control is the preferred method of hazard control (Section 1.8).

Due to the initial and long-term capital expenditures required to implement control systems, the installation of an exhaust ventilation system should only occur if other easier and less costly methods of control are not feasible. Personal protective equipment (PPE), such as respirators, should not be considered at this stage. PPE is rarely an easier or less costly control method when compared to engineering controls.

The method of answering the three basic questions will vary based on whether the process currently exists or is under proposal. However, both scenarios require a thorough process review conducted with the input of an experienced occupational safety and health professional. Review will typically include the following steps:

1) Identify potential hazardous chemicals and physical agents. Review the corresponding physical, chemical and toxicological properties and applicable exposure criteria.
2) Research the documented exposure levels and necessary control approaches for similar operations or processes. These can be either internal or external to a specific facility.
3) Evaluate the process using a process management approach, investigating worst case scenarios and control approaches necessary to reduce the potential for adverse health effect.
4) Evaluate the process from the mindset of the traditional industrial hygiene hierarchy-of-controls[1.1] (see Section 1.8), examining the potential for hazard substitution or significant reduction.
5) Identify the applicable design requirements specified in Federal, state or local standards and codes holding regulatory authority over the industry, facility and/or process.
6) Where exposure control through exhaust ventilation is necessary, identify applicable design approaches for the process under evaluation.

1.2 HAZARDS OF THE OPERATION

The first task is to identify (anticipate and recognize) all potential hazards involved in the process or operation. Hazards are numerous and can include workers' exposure to vapors, gases, liquids, fumes, dusts, noise, heat, explosive environments, oxygen-deficient atmospheres, heat, cold, vibrations, and ergonomic concerns. Control of airborne hazards are the focus of this Manual.

The obvious airborne hazards include individual products, chemicals, etc. that are directly involved in the operation or process. Other hazards include chemical compounds and/or by-products that may form during a reaction or intermediate step. Also, some products subjected to heat or moisture may release contaminants that are hazards. Combined exposures to more than one contaminant should also be considered, especially when the two or more contaminants affect the same biological system or organ.

1.3 IDENTIFY THE INHERENT HAZARDS

Inherent hazards are physical and chemical properties of the materials and fall into three broad categories: Health, Flammability, and Reactivity. Reducing the inherent hazard of a material takes a deliberate change such as substituting a less hazardous material (i.e., less toxic or less flammable) or modifying the form of the material (i.e., larger particle sizes, lower volatility solvents). However, these options may not be possible due to the required chemical/physical properties of the product. The first part of controlling the risk is to understand the inherent hazards of the processing materials.

1.3.1 Health Hazards. In the workplace, there is potential for an employee exposure to contaminants through injection, ingestion, or respiration of airborne contaminants or through skin contact. Toxicologists and industrial hygienists set exposure limits based on results of epidemiological studies, an assessment of the chemical structure of the molecule and results from animal testing and clinical trials when available. They consider the timeframe of concern (acute, chronic) and the part of the body affected. If available data are sufficient to set a specific numerical limit, an Occupational Exposure Limit (OEL) is established. OELs are generally expressed as either parts per volume of air (i.e., thousand, million, billion), or an airborne mass concentration in milligrams, micrograms, or nanograms of contaminant per cubic meter of air, over a set period of time. The set time period is used to weight the airborne mass concentration, providing the Time Weighted Average (TWA). Normally a TWA is for 8 hours, however, a TWA can also be for 15 minutes, known as a Short-Term Exposure Limit (STEL). STELs are provided to supplement TWA OELs for substances where there are both chronic and acute health effects. Ceiling limits are not TWAs, as they are instantaneous concentrations not to be exceeded for any length of time. See Section 1.12 for more details of exposure levels and testing procedures.

Some materials are regulated by government agencies to ensure physical, chemical, biological or radiological safety. OELs and general safety standards contribute to a safe work environment. While some of these OELs and standards are enforceable by OSHA, others are recommendations based primarily on the potential for health effects. For some agents, OELs have not been established. Occupational Exposure Control Bands (OECB) are another way of determining an acceptable level of control for these hazards. In some cases, ventilation by itself is not sufficient to protect an employee from a hazard. In the pharmaceutical industry where drugs such as cytotoxins and hormones cause health effects at extremely low doses, an OECB[1.2] may be chosen to provide a relative measure of the inherent hazard. The pharmaceutical industry has been using OECBs successfully for years. Only the top OECBs (Band 1) can be effectively controlled without additional measures. OELs and OECBs are numbers that may be difficult to comprehend from a physical standpoint (Table 1-1).

Air sampling equipment is readily available for analyzing a number of airborne health hazards. However, choosing the correct monitoring and analytical procedure, calibrating equipment and conducting monitoring in a manner that provides meaningful, accurate, and significant results can be difficult. If the facility does not employ an industrial hygiene, safety or plant engineering staff capable of performing personnel or process monitoring, an industrial hygiene consultant should be contacted. The American Board of Industrial Hygiene (ABIH) certifies industrial hygiene professionals in a range of industrial hygiene practices with the designation certified industrial hygienist (CIH). A list of board certified industrial hygienists can be found at *www.abih.org*.

1.3.2 Flammability Hazards. Organic molecules and some inorganic molecules have the potential to oxidize very rapidly, generating large amounts of combustion gases in a small timeframe. If this rapid combustion were to occur in process equipment it could burst or rupture from over-pressurization. A deflagration propagates the combustion zone at less than the speed of sound. A detonation propagates faster than the speed of sound and cannot be controlled. Similar to the Fire Triangle (ignition plus oxygen plus fuel), the Explosion Pentagon lists five conditions for a deflagration to occur:

1) ignition source
2) fuel
3) oxygen or other oxidizer
4) mixing, and
5) confinement.

The dry environment of some operations can also increase the risk of static electric discharge. The inherent flammability and combustibility hazards described must be known to prevent a potential deflagration or explosion in accordance with National Fire Protection Association (NFPA) standards (Table 1-2). Any combustible material, and some that would normally be considered non-combustible, can burn rapidly when finely divided into very small particles, fibers, fines, chips, chunks, flakes, or a mixture of these forms. Such a dust, when suspended in the appropriate concentration, may deflagrate and even become explosive. The force from a combustible dust explosion is capable of causing death and severe eye injury to employees as well as catastrophic damage to property.

Types of combustible dusts include, but are not limited to:

- Metal dusts (e.g., aluminum, magnesium, etc.)
- Wood dusts
- Plastic dusts
- Organic dusts (e.g., biosolids, sugar, grain, paper, soap, and dried blood)
- Textile dusts

Industries that may handle combustible dusts include, but are not limited to:

- Agricultural
- Chemical
- Forest and furniture making
- Metal processing
- Paper
- Pharmaceutical
- Recycling (e.g., metal, paper and plastic)
- Textile
- Wastewater treatment

Since the 1980s, hundreds of dust-related fires and explosions have resulted in the deaths or severe injury of hundreds of employees as well as costly economic and emotional impacts. Nearly every state in the United States has been impacted by such fires and explosions. Failure to properly assess the hazards and risks associated with such events can yield catastrophic results.

In many combustible dust incidents, employees and managers were unaware of the potential for and serious nature of

TABLE 1-1. Visualizing the Potent Compounds Containment Challenge

Potency Bands	Potency Bands: International Program on Chemical Safety	Occupational Exposure Limits Expressed in:			ALLOWABLE DUST PARTICLES TO BE ≤ OEL		COMPARE TO CLEAN ROOM STANDARDS See Note below.		
		PARTICLE DIAMETER, (micrometers)			100	10	0.5	Federal Standard 209E	ISO 146441
					# particles / ft^3 [# particles / m^3]	# particles / ft^3 [# particles / m^3]	# particles / ft^3 [# particles / m^3]	# particles / ft^3 [# particles / m^3]	# particles / ft^3 [# particles / m^3]
		10^{-3}g/m^3 (milligrams)	10^{-6}g/m^3 (micrograms)	10^{-9}g/m^3 (nanograms)					
Band 1 (Low Toxicity)	Band A (Not harmful, not irritating, low pharmaceutical activity)	10	10,000	10,000,000	283 [10,000]	283,000 [10,000,000]	2,270,000,000 [80130 × 10^6]		
Band 2 (Intermediate Toxicity)		1	1,000	1,000,000	28.3 [1,000]	28,300 [1,000,000]	227,000,000 [80130 × 10^5]		
	Band B (Harmful, may be irritant and/or moderate pharmacological effect)	0.1	100	100,000	2.83 [100]	2,830 [100,000]	22,700,000 [80130 × 10^4]		
Band 3 (Potent)	Band C (Moderate toxic and/or high pharmacological activity)	0.01	10	10,000	2.83 per 10 ft^3 [100 / 10 m^3]	283 [10,000]	2,270,000 [80130 × 10^3]		Class 9: 997,000 [35194000]
	Band D (Toxic, may be corrosive, sensitizing or genotoxic and/or very high pharmaceutical activity. Often termed potent.)	0.001	1	1,000	2.83 per 100 ft^3 [100 / 100 m^3]	28.3 [1,000]	227,000 [80130 × 10^2]	Class 100,000	Class 8: 99,700 [3519400]
Band 4 (Highly Potent)	Band E (Extremely toxic, may be corrosive, sensitizing or genotoxic and/or extremely high pharmaceutical activity. Often referred to as potent.)	0.0001	0.1	100	2.83 per 1,000 ft^3 [100 / 10^3 m^3]	2.83 [100]	22,700 [801300]	Class 10,000	Class 7: 9,970 [351940]
		0.00001	0.01	10	2.83 per 1,000,000 ft^3 [100 / 10^6 m^3]	0.283 [10]	2,270 [80130]	Class 1,000	Class 6: 997 [35194]
		0.000001	0.001	1	2.83 per 1,000,000,000 ft^3 [100 / 10^9 m^3]	0.0283 [1]	227 [8013]	Class 100	Class 5: 99.7 [3519]
		0.0000001	0.0001	0.1	2.83 per 1,000,000,000,000 ft^3 [100 / 10^{12} m^3]	0.00283 [0.1]	22.7 [801]	Class 10	Class 4: 9.97 [352]

NOTE: This list introduces the range of containment tools, but it cannot possibly describe all possible permutations with unit operations and improvements in capability not yet published. Work with equipment vendors and reliable containment testing protocols to ensure the desired capability.

TABLE 1-2. Deflagration Conditions (see Section 1.3.2)

Dust (any finely divided solid material, < 420 microns in diameter) (decreasing particle size increases the potential for deflagration)

- K_{st} – deflagration index for dusts in bar-m/s [KP_a-m/s] (size explosion vent area)
- P_{max} – maximum pressure developed in an unvented vessel in bar [KP_a] (set design pressure for equipment designed for explosion containment)
- Dust Deflagration Hazard Classes:

Hazard Class	K_{St}	P_{max}
St-1	200 [20,000]	10 [1,000]
St-2	200–300 [20,000–30,000]	10 [1,000]
St-3	> 300 [> 30,000]	12 [1,200]

- MIE – Minimum Ignition Energy – minimum amount of energy release in a combustible mixture that can cause flame propagation in millijoules (mJ; relative risk of static ignition) (< 25 mJ a threshold of concern)
- Flammable Limits – minimum (LFL) and maximum (LOC or Limiting Oxidant Concentration) concentrations in a gaseous oxidizer that will propagate a flame (set inerting limits)
- Resistivity – ability of solids to hold a charge of static electricity (> 10^9 ohm • m – threshold of ability to hold a charge)

Liquid (increasing vapor pressure increases the potential for deflagration)

- Flashpoint – minimum temperature at which a liquid evolves vapor in sufficient concentration to form an ignitable mixture at the surface of the liquid
- Liquid Deflagration Hazard Classes

Hazard Class	Flash Pt	Boiling Pt
Flammable IA	< 73 F [< 23 C]	< 100 F [< 38 C]
Flammable IB	< 73 F [< 23 C]	> 100 F [> 38 C]
Flammable IC	> 73 F [> 23 C], < 100F [< 38 C]	—
Combustible II	> 100 F [> 38 C], < 140 F [< 60 C]	—
Combustible IIIA	> 140 F [> 60 C], < 200 F [< 93 C]	—
Combustible IIIB	> 200 F [> 93 C]	—

- MIE – Minimum Ignition Energy – minimum amount of energy release in a combustible mixture that can cause flame propagation in mJ (relative risk of static ignition) (most flammable liquids < 1 mJ)·
- Flammable Limits – minimum (LFL) and maximum (UFL) concentrations in a gaseous oxidizer that will propagate a flame
- Conductivity – ability to allow the flow of static electric charge (conductive liquid > 10^4 picoSiemens (pS)/m; semiconductive > 10^2, < 10^4 pS/m; non-conductive < 50 pS/m)

dust explosions. In one case, six employees were killed and more than 30 injured in Kinston, North Carolina, due to the buildup of combustible dust above a suspended ceiling in an otherwise clean manufacturing facility.[1.3] Additionally, such hazards often are not clearly identified on the material safety data sheets (MSDSs) of combustible substances.

In an effort to increase the awareness of the hazards posed by combustible dust fires and explosions, the U.S. Chemical Safety Board (CSB) has issued numerous reports regarding such events. On March 11, 2008, the federal Occupational Safety and Health Administration (OSHA) issued an instruction, (CPL-03-00-008), Combustible Dust National Emphasis Program (Reissued), that contains policies and procedures for inspecting workplaces that create or handle combustible dusts. Additionally, the National Fire Protection Association (NFPA) has released several standards that should be consulted to aid in the recognition and control of hazards posed by combustible dusts. All employers are encouraged to review available CSB, OSHA, and NFPA publications and resources in an effort to properly protect their employees and facilities from the devastating impacts of combustible dust fires and explosions.

1.3.3 Reactivity Hazards. Runaway reactions can be caused by materials that are readily capable of water reaction, detonation, explosive decomposition, polymerization, or self-reaction at normal temperature and pressure. Oxidizers are another category of physical hazard. Consideration should be given to possible conditions that can impact the design of ventilation systems, especially when venting closed chemical processes.

1.3.4 Physical Hazards. Other hazards can exist when installing a new process or when altering existing plant operations. These can include (but are not limited to) noise, vibrations, heat, skin contact with contaminants, and excessive moisture. In some cases, ventilation can be used to deal with these hazards, but often other corrective actions must be taken.

1.3.5 Regulatory Issues Pertaining to Hazards. Many chemicals have specific handling requirements in environmental or occupational health and safety regulations. Government agencies issue environmental permits for air emissions, wastewater discharges and hazardous/solid waste disposal. The Drug Enforcement Agency has regulatory authority for certain controlled substances. Check with environmental, health, and safety resources to determine what requirements apply to the project.

1.4 POTENTIAL EXPOSURE DURING NORMAL EQUIPMENT OPERATION

Once the potential hazards have been identified, the next

step is to identify exposure criteria related to the individual hazards. Some exposure criteria are not guidelines, but are legal standards and regulations that require adherence. The majority of the legal standards and regulations relating to occupational exposures in the United States are established by the Occupational Safety and Health Administration (OSHA) in the U.S. Department of Labor (USDOL).[1.4] These standards and regulations cover a multitude of occupational safety and health concerns. OSHA establishes specific standards for exposure to chemical and physical hazards called Permissible Exposure Limits (PELs).[1.5] These are the legal limits usually encountered when conducting occupational hygiene assessments in work environments. When the hazard is one of flammable materials or explosive vapors, OSHA has adopted the criteria developed by the National Fire Protection Association (NFPA) by reference.

A number of states have established agreements with OSHA to conduct safety and health inspections within their own states. These agreement "state plan states" are required to establish standards that are at least as stringent as the Federal OSHA standards, but may be more so. If the state where the process or operation occurs is one of the state plan states, then the standards of that state apply. Under the auspices of Section 5 of the Toxic Substances Control Act (TSCA), the U.S. Environmental Protection Agency (EPA) may also establish a New Chemical Exposure Limit (NCEL) for new chemical substances covered under the authority of TSCA. The NCEL is determined based on information provided as part of TSCA's premanufacture notice (PMN) application process and is issued as a TSCA Section 5(e) Consent Order. In addition to the exposure limit, the comprehensive NCELs provisions, (modeled after OSHA's PEL program), include requirements addressing performance criteria for sampling and analytical methods, periodic monitoring, respiratory protection, and recordkeeping. USEPA generally extends these Section 5(e) order requirements to other manufacturers and processors of the same chemical substances via a Section 5(a)(2) Significant New Use Rule (SNUR).

On the non-regulatory side, one source of guidelines is the Recommended Exposure Limits (RELs) published by the National Institute for Occupational Safety and Health (NIOSH). These RELs provide additional information regarding the adequacy of a current PEL or for establishing a new PEL. The RELs also suggest physical and biological exposure assessments. The RELs are published under the authority of the Occupational Safety and Health Act of 1970 and the Federal Mine Safety and Health Act of 1977. In addition, NIOSH recommends appropriate preventive measures to reduce or eliminate the identified adverse health and safety effects of these hazards. To formulate these recommendations, NIOSH evaluates all known and available medical, biological, engineering, chemical, trade, and other information relevant to the hazard. These recommendations are then published and transmitted to OSHA and the Mine Safety and Health Administration (MSHA) for use in promulgating legal standards.

A second non-regulatory source of guidelines from a non-governmental corporation is the ACGIH® *TLVs® and BEIs®* book. The majority of these exposure criteria have a corresponding PEL and in most cases, the ACGIH® TLV® and OSHA PEL are the same. However, as new medical and toxicological data are generated, the TLVs® can respond more quickly than the PELs (which require public and legal hearings before they can change). Consequently, the TLV® is often more current.

Other sources of guidance to consider are the Workplace Environmental Exposure Levels (WEELs) published by the American Industrial Hygiene Association (AIHA) and the exposure values for potential and confirmed carcinogen exposures set by the International Agency for Research on Cancer (IARC). Individual companies may also develop their own internal exposure guidelines based on their knowledge of a product or its manufacturing process.

The decision regarding which exposure guidelines to follow is not always obvious. Some organizations seek to follow the most stringent guidelines while others may choose to use some fraction of an identified guideline. (The chosen fraction is intended to provide a margin of safety in ensuring exposures will remain within the guideline limits at all times.) Such decisions should be made in consultation with an experienced occupational safety and health professional. The OSHA PELs (the state OSHA PELs in state plan states) are the legally binding limits. Additional review by a legal professional may also be desired. Consideration should be given to the potential for future revisions of exposure guidance. The guidance generating organization will often announce pending changes in advance to allow opportunity for feedback and planning. Industry trade associations can also be a valuable source of information regarding pending changes to regulations that could potentially affect particular industries.

1.5 POTENTIAL EXPOSURE OTHER THAN DURING NORMAL OPERATION

Exposures do not only occur during the normal operation of a piece of equipment. There are many other times exposure can occur such as startup, shutdown, charging, discharging, quality sampling, cleaning, and maintenance. A thorough knowledge of the process and what tasks the operators actually perform during the various unit operations provides the best exposure assessment. Some of the more common conditions that cause exposures include:

1) Energy Added to Process Step: The greater the energy input, the greater the potential for contaminants to escape the unit operation.
    a) Elevation change – i.e., material dropping by gravity from one level to another mixes with air in the equipment. When the material is stopped at the lower level, fines (small airborne particulates) are expelled.

b) Rotary or reciprocating motion – i.e., mills, drills, grinders, etc., input high energy with rotary motion reducing particle size and dispersing particulates.

c) Pressurization from external process – i.e., compressed air or nitrogen and pressure pneumatic conveying can fluidize powders and push them out of any available equipment opening or crack.

d) Liquids – i.e., atomization or rapid depressurization, aeration from open falling liquids (like a waterfall), elevating the temperature and vapor pressure of volatile liquids, and water hose cleanup create very small liquid aerosols due to the energy input and the surface tension of water. If dealing with potent or highly potent compounds in liquids, these seemingly small sources become important.

2) Manual Intervention: During these tasks, operators and mechanics come in close contact with the product. The operator technique and time pressures can lead to significant contaminant generation, close to the operator's breathing zone. Examples include:

a) Dispensing – i.e., scooping from one drum to another to weigh ingredients creates dust at the scoop and in both drums.

b) Sampling – i.e., grab samples taken for quality purposes.

c) Cleaning equipment – i.e., Wash-down and wipe-down are close contact tasks.

d) Maintenance – i.e., even when cleaned, product held up in equipment crevasses can cause exposure during disassembly.

i) Cleaning – i.e., using compressed air to blow down inaccessible places moves contaminants everywhere. The contaminants then need to be removed by manual washing, wiping or vacuuming.

3) Problem Materials: Understand material characteristics throughout the range of process conditions. For instance, opening up material transfer lines or ventilation ducts to poke out plugs and similar maintenance activities can lead to airborne exposures. The process project team should be able to completely describe material properties and avoid employee exposures. Example material characteristics include:

a) Liquid properties – viscosity, surface tension, vapor pressure throughout the range of operating conditions, corrosiveness, flammability, products of thermal decomposition, etc.

b) Cleaning agents – solvents, detergents, pH, potential for causing dermatitis.

c) Powder properties – angle of repose, shear, compression, hygroscopicity (ability to absorb water), friability, flammability, deflagration, melting point, etc.

4) Waste Streams: Process air emissions, wastewaters from process operations and equipment cleaning and waste disposal are all exposure opportunities.

5) Process Upsets: Process safety studies are needed to ensure prevention or protection or both for these general hazards. If relief devices are used, they must discharge to a safe location. Operation of a relief device would disperse materials over a wide area. If toxic compounds are used, consider an alternate process safety strategy that contains the overpressure. Also consider:

a) Create overpressure or vacuum – i.e., review to see if compressed gasses or liquids are supplied to relatively weak equipment.

b) Dust or flammable liquid deflagrations – i.e., consider if ignition sources and static electricity are controlled (grounding and bonding) and the right electrical classification is in place as well as flammable liquids handling practices and equipment.

c) Runaway reaction – i.e., review to ensure reaction conditions or sequencing are well understood.

1.6 POTENTIAL SOURCE IDENTIFICATION

Familiarity with how these airborne contaminants are generated and how specific process operators could be exposed is strongly encouraged. The exposure assessment process should be thorough, and should evaluate various activities including start-up, shutdown, ongoing adjustments, changeover, normal operations, maintenance, cleaning, product sampling, etc. Many of these operations require the operator to interact with the process and may greatly influence the operator's overall exposure. This type of exposure evaluation is called Task Based Exposure monitoring and is important in determining where controls are necessary, and when evaluating new or modified control systems. See Table 1-3 for an example of a task evaluation conducted in advance of designing a new operation. The evaluation example is designed to identify activities with exposure generating potential while unloading hazardous material from a bulk container known as a "super-sack." In addition to identifying suspected exposure generating activities and their anticipated frequencies, the Table identifies possible exposure control strategies to protect the worker from these potential exposures.

1.7 ASSESSING THE EXPOSURE

After the potentially hazardous chemical and physical exposures are identified, toxicological or occupational health references should be consulted for guidance on the level of hazard associated with each constituent. An industrial hygienist or other occupational health professional should be involved in this phase of health risk identification and evaluation.

One important source of toxicological data is the Material

TABLE 1-3. Example Task Based Exposure Assessment (Reprinted with permission from Procter & Gamble)

EXPOSURE SOURCE	FREQUENCY	ENGINEERING CONTROLS	ADMINISTRATIVE CONTROLS & PPE
1 Hazardous material on outer surface of supersack	Infrequently but possible		• Ensure that the vendor provides clean supersacks • Wear PPE during cleaning • Use vacuum cleaning for spills
2 Spills due to supersack getting damaged during transport	Infrequently but possible	• Portable vacuum cleaner with HEPA filter readily available in warehouse	• Emergency spill cleanup procedure should be in place
3 Spills on top of the surge bin and spills to the floor while dumping supersack	Whenever dumping a supersack	• Dump hopper has hole sized for supersack spout	• Supersack spout should be long enough so that it is inserted into the dump hopper • Follow proper dumping procedure • Wear PPE while dumping supersacks
4 Dusting from surge bin during operation	Whenever running	• Bin under negative pressure from dust control • High face velocity at dump hopper opening	
5 Empty supersacks handling	Whenever dumping a supersack		• Deflate supersack into dump hole, and put it in a scrap bin or plastic bag for disposal • Wear PPE while discarding supersacks
6 Dust/spillage from bin cleanout	1–2x/yr	• Provide vacuum cleaning capability • Minimize maintenance by ensuring internal clean design of hopper; Hopper angle > 60 deg; Discharge opening large enough for easy flow	• Wear PPE while cleaning spills • Use vacuum cleaning for cleanup
7 Damaged flexible connections	Infrequently but possible	• Robust design (tube type)· • Material w/ good flexibility (neoprene or equivalent)	• Wear PPE when repairing flexibles
8 Dusting from LIW due to air being displaced while filling the LIW hopper	Whenever running	• Provide aspirating duct for LIW hopper to vent to dust controlled equipment	
9 Dusting from dump hopper opening	Whenever running	• Provide adequate face velocity	
10 Airborne dust from large spill		• Dedicated room for hazardous material dumping and metering with wall fan to maintain 1 m/s face velocity across the door. This will prevent exposures outside the hazardous handling area.	• Wear PPE when cleaning up spills • Use vacuum cleaning for spill cleanup
11 Dusting from transfer conveyors	Whenever running	• Conveyor under negative pressure from dust control	
12 Dust/spillage from belt conveyor entry for inspection, cleanout and maintenance	1x/shift to1x/mo	• Provide face velocity of 3.3 fpm [1 m/s] at access doors • Provide transparent access doors for easy inspection • Material handling design improvements need to minimize reason for housing entry	• Wear PPE whenever entering the equipment

Safety Data Sheet (MSDS). The Hazard Communication Regulations promulgated by OSHA[1.6] require employers to have an employee-accessible MSDS for every potentially hazardous chemical that an employee may encounter. In addition to containing toxicity exposure data, the MSDS will often contain information on likely routes of exposure and physical properties that assist in evaluation of the hazard. It is important to note, however, that the reliability of information received on an MSDS is never certain and additional resources should be consulted in order to complete the review. One common problem with the MSDS is located in Section Two. Section Two is the place on the MSDS where the hazardous materials are listed. Many chemical suppliers have been known to list non-traditional chemical nomenclature, chemical synonyms, chemical family names, etc. in this section. This can become confusing. However, the Chemical Abstracts Service (CAS)[1.7] number is also listed in this section, and the CAS number can be cross-referenced to the correct specific chemical name.

Numerous other sources of literature exist for use by occupational health professionals when gathering toxicological backgrounds on potential exposures. A partial list of helpful resources includes: 1) The American Conference of Governmental Industrial Hygienists' (ACGIH®) *Documentation of the TLVs® and BEIs®*; 2) National Institute for Occupational Safety and Health (NIOSH) Criteria Documents; 3) *Chemical Hazards of the Workplace*, published by Proctor and Hughes; 4) Chemical Reviews by the National Toxicology Program, published by the National Institutes of Health (NIH); 5) *Dangerous Properties of Industrial Materials*, by Irving R. Sax; 6) *Patty's Industrial Hygiene*, by George Clayton; 7) *Occupational Diseases – A Guide to Their Recognition*, published by the U.S. Department of Health, Education and Welfare; and 8) *NIOSH Pocket Guide to Chemical Hazards*, published by the U.S. Department of Health and Human Services. Finally, an Internet search can lead to useful information although the user is strongly urged to avoid reference sources with uncertain credibility and to be careful of publishing dates.

1.8 HIERARCHY OF EXPOSURE CONTROL OPTIONS

The practice of Industrial Hygiene (IH) is not the scope of this manual. However, "Hierarchy of Exposure Control" is one of the IH tenets with which the plant leadership group (i.e., plant engineering, operations team, etc.) should be familiar. This concept categorizes the IH approaches used to strategically control potentially hazardous exposure. While the number/grouping of strategy levels and some of the terminology may vary from reference to reference, this approach is generally consistent throughout the IH profession. In this manual, we will consider five strategy levels. Often, more than one strategy is used at the same time to protect employee health. The selection depends upon the risk that must be managed, the availability of effective control technology and the cost to implement and maintain that technology. In some cases, regulating authorities will determine control strategy. The strategy levels, in their order of recommended consideration/implementation are:

1) Elimination/Substitution: This strategy level removes the hazardous exposure by eliminating the contaminant or exchanging it with a less harmful substitution. (Example: Substituting a less toxic abrasive for silica in an abrasive blasting activity.) Toxic, reactive, and flammable materials are good candidates for this approach. Many times there are alternate chemicals that can be used to reduce worker risk.
2) Process Modification: This approach modifies the work process to eliminate or reduce the hazardous exposure. (Example: Reducing the temperature in a mixing vat to reduce the amount of vapor generation.) Note that the best time to influence process design is during the conceptual phase of the project, before firm project funding commitments have been made. Other examples include:
    a) Changing the physical attributes of the chemicals/materials received; powders with low dust (fines) and sodium hypochlorite liquid in place of chlorine gas.
    b) Modifying receiving methods; delivery of materials in tote boxes that can be sealed during transfer to process instead of bags that have to be handled by employees.
    c) Isolation; locate operations involving highly hazardous materials in rooms with limited employee access and separate ventilation systems.
3) Engineering Controls: Design features incorporated into the work process in such a way that the hazardous materials are contained within the process equipment or captured and eliminated from the work environment prior to personnel exposure. (Example: Local exhaust ventilation installed to capture welding fumes in a production welding operation.)
4) Administrative Control Procedures: Includes identifying procedural, not equipment based, ways to limit an individual's time in the area where the exposures are occurring. This technique is used often to manage heat stress, radiation and ergonomic issues. (Example: Rotating work assignments at a hazard-generating task. Note that administrative controls do not control contaminant exposures and are hard to manage. They rely on staffing behavior and require regular tracking.)
5) Personal Protective Equipment (PPE): Includes protective equipment for eyes, face, head, and extremities, protective clothing, respiratory devices, protective shields, etc., worn by an individual while performing exposure-generating tasks. This is the lowest ranked exposure control method. For some hazards and industries, specif-

ic PPE is identified by regulation. In other circumstances, regulations may prohibit the selection of specific PPE (i.e., respirators) unless effective engineering controls are proven not feasible, or while they are being instituted.(1.6) (Example: Respiratory protection for asbestos abatement activities.) Note that PPE can be effective if stringently managed and has a low initial capital cost. Over time, though, the cost of continuing PPE use outpaces that of instituting and maintaining engineering controls. Additionally, the protection provided by PPE is based on selection, use and training, and since use of some PPE is uncomfortable and often hot, it can be difficult to wear for long periods.

1.9 COMMON AIRBORNE HAZARDS

After a determination is made that ventilation systems are the best option for control of the hazards, the source control, conveyance to and specification of an air control device depend on the physical nature of the hazard. If contaminant (hazard) passes through an air-cleaning device into the environment, it is then a pollutant. These pollutants are categorized by size (measured in microns) and their physical nature in the system at the measured conditions as discussed in Section 1.10. For example a liquid pollutant may go through phase changes from liquid to vapor and back to liquid within the confines of the system duct and hood. Each category of material represents its own particular problems of capture and control. Details for the designs of these systems and specifications for equipment are included in Chapters 2 and 5.

1.10 AIRBORNE CONTAMINANTS

Airborne contaminants are generated in a variety of ways. Understanding how workplace contaminants are generated aids in understanding when an industrial ventilation system is required or if an existing system is providing adequate control. The major workplace application of industrial ventilation is in control of employee exposure to airborne particles and vapors. There are a variety of particulate types and control techniques vary depending on the individual process and type of particulate.

In general, the sources of particulate contaminants can be found at the point of the operation, i.e., at the point of cutting or grinding, at the point where a chemical reaction occurs, at points where heat is applied, at points where materials are transferred, and at other locations. At times, the contaminant generation sources may be less obvious, such as drying areas, material storage areas, vaporization of contained liquids, process leaks, etc. In most circumstances, a thorough evaluation of the entire process, including maintenance activities, should reveal the sources of airborne contaminants. When source identification appears elusive, the combination of direct reading instruments and a map or layout of the production area can be used to create a contaminant concentration contour map. Tracking the contours to their epicenter will usually lead to identification of the contamination release point and thus, the point of maximum effectiveness for local exhaust ventilation.

Particles are classified by size and the typical unit of measurement is the micrometer (µm) or micron (µ), which is one millionth of a meter (10^{-6} m). Examples of dimensions defined in microns are provided below, and in Table 1-4:

- Sheet of copy paper: 100 µ thick
- Hair: 50 to 70 µ diameter
- Visible particles: > 10 µ (depending on lighting conditions)
- Fumes: < 1 µ

1.10.1 Particulates. A particulate is defined as a solid or semi-liquid particle from mineral, chemical, or organic materials that can remain suspended in the air due to its small size. Normally airborne particulates are below 500 µ. If the particulate is defined as 'smoke' it generally will include particles less than 1µ in diameter. Smoke is defined as an aerosol mixture usually formed by combustion processes such as burning of wood, tobacco, oil, or coal.

Solid particles with diameters less than 100 microns can easily be moved by air currents. Particles that are less than 100 microns (inhalable) can enter a person's respiratory tract through the nose, mouth and upper airways. Most inhalable particulates are deposited on the mucous membranes before reaching the thoracic and respirable regions. Thus, larger particles generally affect the upper airways (i.e., an acid mist). Particles that are smaller than 10 microns can penetrate deep into the respiratory tract and can damage areas of the inner lung through a number of biological or physical mechanisms (i.e., respirable crystalline silica). Particles ≤ 4 µ (respirable) can penetrate to the gas exchange region (alveolar sacs) potentially damaging this fragile structure. Sources of particulate from processes include:

1) Expulsion of Fine Particulate: Sudden compaction of a falling mass of particles. Compaction of a mass of falling dusty material occurs when it impacts the floor or pours into a container, such as a bin or a tote. The sudden compaction expels air and fine dust from the container. Particularly when a large mass of dusty materials enters a container, the air inside the container is displaced. This displaced air carries fine particulate with it as it escapes from all available openings of the container (Figure 1-1).
2) Mechanical Compression of Products Carrying Fine Particles: The compression of a bulky product carrying fine particulate also expels dusty air out of all available openings into the surrounding areas. This mechanism occurs when squeezing the air out of bags containing powders, such as deflating a sack of flour to seal it. Some processes squeeze the air out of the product with compression rolls that cause a rapid expulsion of dust.

TABLE 1-4. Particle Size Ranges and Classifications for Aerosols[1,8]

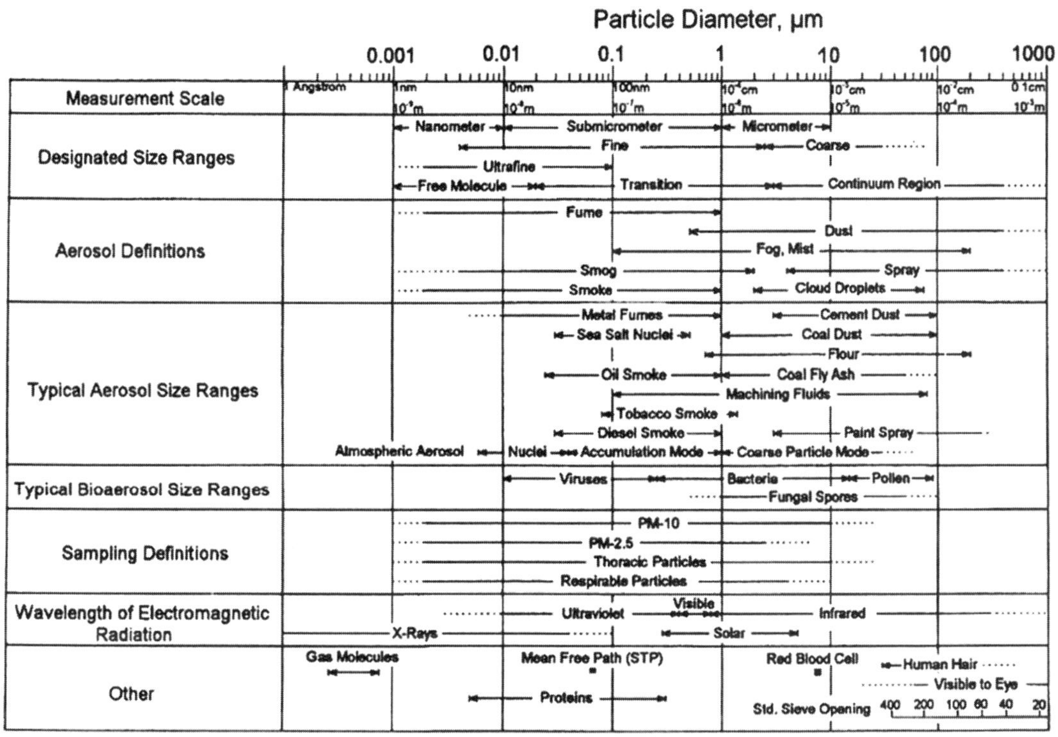

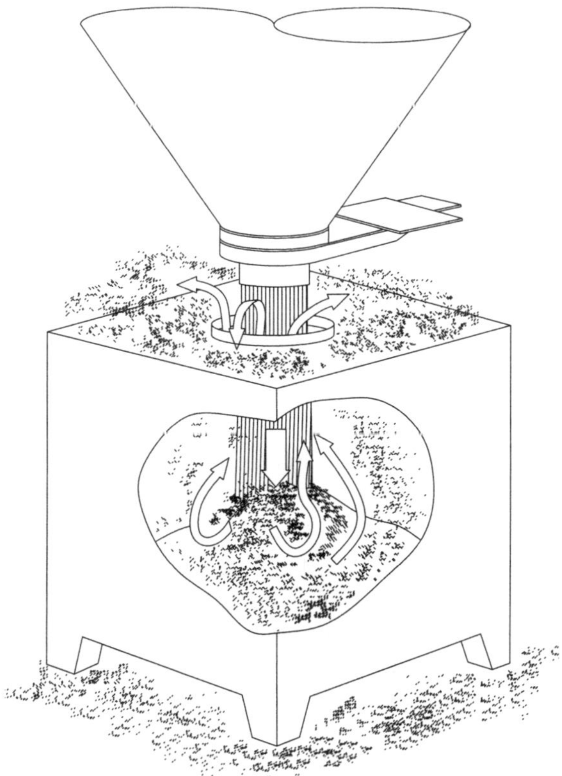

FIGURE 1-1. Displaced air containing fine particulates (reprinted with permission from Procter & Gamble)

Figure 1-2 illustrates one common compression dust source.

3) When materials (e.g., powders, grains, pellets, etc.) fall from a conveying system into a container or onto a floor, two dust-generating mechanisms are present. The sudden compaction of the material expels dusty air. This generates the greatest quantity of dust of the two mechanisms. The second mechanism generates dust when a material falls through the surrounding air. The action of the material falling through the air causes fine particles to be stripped from the material stream and entrained in facility air currents. Figure 1-3 shows an example of these two mechanisms of dust generation.

4) Pressurized Air Leaks: Bulk solid powders can be transported for considerable distances, both horizontally and vertically, by pneumatic conveying systems. Powder and air move through the pipeline, much like liquid. However, if a positive-pressure pneumatic conveying system is not tightly sealed or if the pipeline leaks, it can cause a serious dust problem. Because of the high velocity and pressure of the air inside a pneumatic conveying system, a powerful air "jet" can be formed at the point of the leak. The air jet can transport fine dusts to a place away from the source. In addition, the relief of pneumatic cylinders or pressure relief devices will also create air "jets" and move dusty air into surrounding areas.

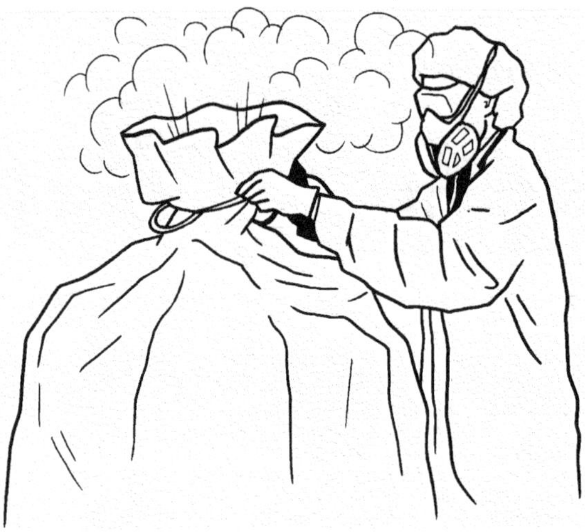

FIGURE 1-2. Dust expulsion by mechanical compression (reprinted with permission from Procter & Gamble)

FIGURE 1-3. Dust generated from falling particles (reprinted from *Hemeon's Plant & Process Ventilation, Third Edition*, with permission from D. Jeff Burton)

5) Vibration: Vibration is another mechanism that generates dust. In a belt conveyor system, the vibration of the belt propels some particles airborne or knocks them off the belt into the housing or onto the floor as spillage.
6) Machine Actions Such as Grinding or Cutting: When large particles are ground into smaller particles, high velocity dust can be generated as the disintegrating particles are thrown by the high velocity of the grinder. Usually, the fine dust travels in the air stream created by the high-speed grinding wheel.
7) Crushing spilled powders into smaller particles.

1.10.2 Liquid Aerosols. Pollutants can also be found in liquid form. When particle sizes are between 0.1 and 200 μ these aerosols are frequently called 'mist' or 'fog.' Generally they are suspended liquid droplets formed by condensation of water vapor or atomization of liquids. This may include condensation from a chemical process or coolants applied to the surface of machining operations. Collection of liquid aerosols is normally provided by dry filtration media in the form of a mist collector, centrifugal mechanical collectors or filter pads.

Normal sources of aerosol hazards include:

1) Aerosols from atomization/spraying: Aerosols are generated by means of rapid depressurization when the liquid, under pressure, is passed through a low pressure nozzle. Atomization produces smaller sizes of aerosols than spraying. When the liquid product under pressure is passing through the spray nozzle and depressurized, aerosols are produced in the cone shaped zone in front of the nozzle. Normally higher pressure will result in smaller aerosol diameters. Spray painting operations are common generators of both aerosols and organic vapors. Aerosols generated are usually > 1 μ, and limited somewhat by use of high volume low-pressure (HVLP) equipment. Unintended generation of aerosol spray may also occur during high-pressure drop filling nozzles on a packing line and the operation of relief valves.
2) Atomization produces smaller sizes of aerosol than spraying. Aerosols from splashes/splatters: Aerosols can be generated when a liquid drop or a liquid rod hit dry surfaces (splatters) or liquid surfaces (splashes). When a liquid drop falls onto a dry, hard surface from a given height, it spreads out on the surface and forces the edge of the liquid "sheet" to expand. As the edge becomes thinner, some small liquid droplets can be released, and these subsequently collapse into smaller aerosols. The greater the height of the fall, the more aerosols are produced. An example of a common mechanical process causing an occupational exposure to vapors and aerosols is a machining process where a coolant/lubricant is pumped onto a cutting tool. The flow of the coolant and mechanical activity (shearing action) of the tool combine to generate a coolant/lubricant aerosol (oil mist) and vapors, from the heat generated by the tools causing a phase change.
3) Aerosols from bubbles bursting: Bubbles can be formed when liquids fall from a high level to a lower level. This fall forces air into the liquid and creates bubbles. More specifically, bubbles can be made by the following two mechanisms:
   a) Bubbles made by aeration of liquid. Bottle filling operation is an example of this mechanism.
   b) Bubbles made by aeration from clean-up methods.

One additional aerosol mechanism is condensation of

vapors. A common example is atmospheric fog. Fog is formed when the air temperature has dropped below the dew point for a vapor causing tiny droplets to form.

Aerosolization is the process where aerosols are formed by the violent mixtures of liquid and air. There are several factors that can influence the amount and characteristics of the process:

1) Mechanical agitation increases the formation of liquid aerosols. The more violent the agitation, the greater the energy that enters into the liquid system to break surface tension. Designing the process to minimize agitation will significantly reduce the aerosols formed.
2) Pressurized processes produce more aerosols. For example, high pressure filling nozzles can be converted to a low-pressure design to minimize aerosol formation.
3) Temperature also has an effect on aerosolization. In some cases, the amount of aerosol doubles as the liquid temperature increases from 60 F to 90 F [16 C to 32 C].
4) Aerosolization processes can be modified by the addition of materials that reduce or increase surface tension. By reducing surface tension using surfactants, aerosolization decreases; increasing surface tension using hydrophobic particles will promote aerosolization.

1.10.3 Fumes. Fumes are generally very small particles (< 1 micron) formed by the condensation from gases of volatilized molten metals. Welding is the most common process that produces a metallic fume. However, other processes that can cause significant exposure to metallic fumes include metal melting, smelting, brazing, silver soldering, and other foundry activities. Also, many fumes oxidize during the volatilization and condensation process, i.e., Zinc Oxide fume. Fumes can flocculate or coalesce into larger particles, but they can also be so small that they pass through the fabric of a baghouse. In those cases, an Electrostatic Precipitator may be needed to collect them.

1.10.4 Vapors. Vapors are gases at room temperatures. Commonly they are formed by the evaporation of volatile liquids such as organic solvents. Many vapors also have a low odor threshold and can initially be detected at relatively low concentrations. To remove vapors from an air stream they must either be cooled sufficiently to be condensed or adsorbed onto activated material such as charcoal or be thermally destroyed in an oxidizer. Examples of vapor generating mechanisms include:

- Evaporation from a pail or tank of volatile liquid with a low vapor pressure (i.e., acetone used for cleaning) at ambient conditions.
- Volatilization of resins during formation of polymer products (i.e., styrene from fiberglass chop and gel coat spray activities).
- Drying of solvent carriers used to deposit the contained solid pigments (i.e., automotive spray painting).
- Operating a tank containing a volatile liquid (i.e., paint dip operations or furniture stripping).
- Spills of a volatile liquid during open handling (i.e., paint mixing operations).
- Operation of a relief valve or sampling ports (i.e., chemical/petroleum manufacturing).
- Evolution of an acid gas from an open surface tank operation (i.e., hydrochloric or nitric acid from pickling tanks).
- Volatilization of isocyanates from polyurethane application (i.e., methylene bisphenyl isocyanate (MDI) from spray-on truck bed liners or foaming operations).
- Evolution of chlorinated hydrocarbons from open surface tanks (i.e., degreasing operations).
- Evolution of sterilants from disinfection/sterilization equipment (i.e., ethylene oxide, etc.).
- Evolution of organic vapors from cardboard or particle board production (i.e., formaldehyde).
- Evolution of organic vapors from foundry core making activity (i.e., phenol – formaldehyde).
- Evolution of a metallic vapor from a liquid metal spill (i.e., mercury from a broken thermometer).
- Evolution of vapors as coolant is heated near the surface of a metal cutting tool.

1.11 WORKPLACE AIR QUALITY ASSESSMENT ISSUES

The exposure guidelines discussed earlier are applicable to those individuals whose work tasks lead to their exposure to one or more identified hazards. Care must also be taken not to expose individuals who are not involved in the contaminant-generating process. One example with potential for such exposures is an office environment next to or near the production area. Non-production areas (offices, lunchroom, meeting rooms, etc.) should have a positive pressure relative to the production area. Normally, the production area will have exhaust systems requiring replacement air and the desired pressure relationship can be achieved by oversupplying makeup air to the office areas. This oversupply will transfer into the production areas as a source of shop replacement air. Where grills are intended to facilitate air transfer, take caution to assure correct airflow from the "clean" to the "contaminated" areas under the full range of expected operational conditions. Careful separation between office HVAC air intakes and contaminated air discharging from the exhaust ventilation systems in the production area is critical to minimize contaminant entrainment into the office HVAC system. Chapters 2, 5, 8 and 10 address the design considerations necessary to prevent contaminant reentry.

While architectural issues are outside the scope of this manual, be aware that uncontrolled infiltration between spaces can be a major contaminant pathway. Examples of such pathways include dropped ceilings and plenum spaces, vertical pipe/electrical chases, inactive ventilation systems, poorly sealed walls, doors, and windows, and under-floor chases and pits.

1.12 EXPOSURE MONITORING

The sampling protocol should include both personal and area monitoring and wherever possible, should establish exposure baselines for later comparison to concentration measurements made after system installation or modification. This comparison will show the effectiveness of the ventilation system and may possibly be used to document that the personal exposures are within established guidelines. Post installation exposures that approximate non-detectable concentrations are the optimum. However, in many cases, designing the ventilation system to achieve "zero" exposure is not warranted. The actual acceptable levels will vary depending upon the process and the contaminant toxicity. One practice used by ventilation system designers is to seek to reduce exposures to some fraction (as low as 20%) of the desired occupational exposure limit (TLV®, PEL, REL, etc.).

Other considerations for the use of area monitoring include the ability to identify when and where contaminants migrate into adjacent areas (including nearby outdoor air intakes). In addition to being a potential indoor air quality concern for nearby office areas, some migrating contaminants might react with chemicals or activities in adjacent areas to result in a more serious hazard. For example, if welding or cutting operations are conducted in the presence of chlorinated hydrocarbons, (such as the type used in some solvents) hazardous concentrations of phosgene and hydrogen chloride can be produced. When hydrogen chloride is produced, one very noticeable concern is the resulting extreme metal corrosion.

1.12.1 Personal Monitoring. Personal monitoring for air contaminants is also known as "breathing zone" monitoring. In fact, the basis for establishing compliance or non-compliance with TLVs® and PELs is determined by conducting monitoring in the employee's breathing zone. The breathing zone is defined as a hemisphere forward of an employee's shoulders with a radius of approximately 6 to 9 inches [152–229 mm]. During breathing zone monitoring, the air sampling device (normally a filter cassette or absorbent tube) is usually attached to the employee's clothing below the mouth/nose. However, one example of placing the sampling device in another area involves evaluating welding fume exposure. Then it is important to position the air-sampling device inside the welding helmet to obtain accurate exposure information.

1.12.2 TWA Monitoring. Most of the regulated air contaminants list a maximum amount of contaminant that can be detected in the employee's breathing zone based on a Time Weighted Average (TWA). The TWA is defined as the employee's average airborne exposure in any 8-hour work shift. When conducting TWA sampling it is important to eliminate errors associated with fluctuations in exposure by sampling as much of the 8-hour shift as possible. Every effort should be made to include all segments of the work shift that result in the highest level of employee exposure. These segments may occur during routine set-up, change over, end of shift clean up or intermittent operations, etc. Observations of the operation and operator interviews are required in determining when high exposures may occur. Also, since an 8-hour sample is only a snapshot of the employee's exposure it is important to determine if the air sampling was conducted on a typical day. Several things can result in altered sampling results including a worker new to the operation, reduced or increased work load, intermittent work operations, windows and shipping doors being open or closed (summer/winter conditions), alternate product use, etc.

1.12.3 STEL Monitoring. Some of the regulated air contaminants list a maximum amount of contaminant that can be detected in the employee's breathing zone based on a Short-Term Exposure Limit (STEL). The STEL is defined as a 15-minute time-weighted average exposure. When conducting STEL monitoring, it is important to obtain a sample that represents the highest 15-minute exposure that occurs during the workday. Before conducting STEL monitoring it is very important to observe the operation and interview the operator to determine when breathing zone sampling should be conducted.[1.9] Usually it is necessary to conduct repeated monitoring during specific operation(s) that are believed to cause abnormally high employee exposure. The STEL requires a 15-minute sample. If the operation takes only 10 minutes the air-sampling pump should run for another five minutes. Otherwise the last five minutes are time-weight averaged as "zero" exposure.

1.12.4 Ceiling Exposure Monitoring. A few of the regulated air contaminants list a maximum amount of contaminant that can be detected in the employee's breathing zone as a ceiling.[1.11] For some ceiling limits the employee's exposure is never to exceed the listed concentration at any time during the workday. Other ceiling limits list acceptable maximum peaks above the ceiling for a maximum time. To conduct ceiling monitoring it is necessary to observe the operation and interview the operator concerning intermittent operations that may cause elevated exposures. It is also important, whenever possible, to conduct ceiling monitoring using direct reading (instantaneous) monitoring equipment. When direct reading equipment is not available, air sampling can be conducted for the smallest volume of air necessary to provide credible results. Air sampling volumetric flow rates and minimum sampling time information can normally be obtained by contacting a certified industrial hygiene laboratory.

1.12.5 Engineering Monitoring. Employee Time Weighted Average (TWA) exposure monitoring assists in determining when new or improved engineering controls are required to reduce operator exposures. However, they do not show where the exposure originates. Engineering samples attempt to determine the contribution of specific operations to the operator exposures. These samples may either be employee breathing zone or area monitoring that is conducted during a specific task(s). Often this monitoring practice is referred to as "task based" monitoring. When conducted in the worker's breathing zone, it could represent STEL or ceiling monitoring if the air contaminants involved had corresponding STEL or ceiling exposure limits. Regardless of whether the sampling is conducted on an employee or in an area, multiple samples may be required in order to obtain a statistically meaningful result. There are three general types of engineering samples:

a) Fractional or Task Based Monitoring: The air sample is collected only during specific operations. If the task is broken down into steps, the exposure from each step can be determined. Direct reading instrumentation is useful in this type of sample. However, air sampling pumps and collection media can also be used for this process. When the samples are collected on media, the media can be frequently changed to isolate the individual task contributions. Repeated monitoring and controlled conditions (same operation, operator, part produced, length of sample time) may be required to obtain a meaningful result. Task based monitoring can be a useful commissioning tool and help determine system effectiveness when conducted pre- and post-system installation or upgrade.

b) Emission Rate Monitoring: Contaminant generation rates are often very difficult to determine. A series of specific samples can assist in estimating the generation rates. For example, pouring hot metal into some sand molds generates carbon monoxide (CO). If the mold is isolated and placed in a capture hood, the amount of CO generated can be determined by measuring the concentration in the exhaust and, knowing the flow rate, a generation rate can be calculated.

c) Area Monitoring: Stationary area samples at potential emission points can indicate the significance of a source. The observed concentrations are often higher, because they are located close to the source, and must be properly interpreted. This method of sampling may also be a useful commissioning tool when conducted pre- and post-system change. However, for results to be conclusive, care needs to be taken to sample the same area under the same condition. For instance, samples have to be obtained in the same area, doors and windows closed or open, fans on or off, same process, same procedures, same production rate, etc.

As with all sampling procedures, it is important that the sample method, placement, duration, etc. are well documented and the samples labeled as engineering samples. Failure to do so may lead to future misinterpretation of the measured concentrations as a representation of actual work exposures.

1.12.6 Video Use. Another engineering sampling tool is the use of video exposure monitoring (VEM). This procedure combines a direct-reading instrument and data logger with a video camera to allow a detailed evaluation of the worker's interaction with the process. Both the data logger and the video camera clocks are synchronized prior to sampling so that periods of higher exposure concentration may be analyzed in detail to pinpoint the source of exposure.[1.10] This approach facilitates precise exhaust hood designs that should minimize unnecessary interference with the work process.

1.12.7 Monitoring Equipment Calibration. Whenever air monitoring is conducted it is critical that the monitoring equipment be calibrated. When a direct reading instrument is used it should be calibrated before use according to the manufacturer's instructions. This will normally include providing the monitor with clean air (setting a zero point) then using a manufacturer supplied standard gas for single point calibration of the monitor. When monitoring is conducted using a sampling pump, the pump should be calibrated either pre- or pre- and post-use. A variety of calibration equipment can be used for this purpose. Examples of equipment available for calibration include a bubble meter, precision rotameters, or a dry calibrator. The calibration must be conducted with the selected sampling media in line.

1.12.8 Selecting a Laboratory for Processing Monitoring Results. Whenever industrial hygiene (IH) monitoring is conducted, accurate analysis of the sample media or bulk samples are required. AIHA provides a system for accreditation of industrial hygiene laboratories. To achieve accreditation, the laboratory undergoes a rigorous inspection. To maintain accreditation the inspection is repeated at the end of each three-year cycle. Also, quarterly proficiency and analytical testing (PAT) samples are analyzed. These samples test the ability of the laboratory to accurately analyze a variety of categories of materials. These categories include: metals, silica, asbestos, solvents, beryllium, formaldehyde, etc. A list of industrial hygiene accredited laboratories is maintained on the AIHA website at *www.aiha.org*. Certified Industrial Hygiene Laboratories provide chemical analysis via approved analytical procedures. However, occasionally a sampling and/or analytical procedure is found to have limited collection efficiency, and results in under reporting the actual contaminant concentration. Thus, some chemical analytical procedures have been, and will continue to be, modified or changed to provide a more accurate picture of reality. For some air contaminants, it may be important to select an industrial hygiene and consultation laboratory based on the sampling and analytical methods they use. Examples include hexavalent chromium and isocyanate sample collection, transfer, storage, and analytical procedures having been modified to better capture, stabilize, preserve, and detect these reactive materials.

1.12.9 Monitoring for Air Contaminants in Confined Spaces. Monitoring air contaminants in a confined space is used to determine if:

1) the space has a respirable atmosphere and is safe to enter, or

2) the internal atmosphere continues to be respirable while employees are in the space performing their work activity.

The most common method of obtaining air contaminant information about a confined space is the use of a multi-gas meter. These meters are available from a variety of manufacturers and normally come equipped to test for three to five contaminants. The most commonly monitored contaminants include oxygen, combustibles, hydrogen sulfide, carbon monoxide and organic vapors. However, there are a variety of sensors or detectors available that can often be substituted or ordered separately for monitoring special conditions. When conducting monitoring in a confined space it is extremely important to follow all manufacturer's instructions on proper instrument calibration and appropriate safety procedures before placing the multi-gas meter in the confined space. The multi-gas meter's sensors must be calibrated before use. Normally they are calibrated with methane gas. If the combustible gas in the confined space is something other than methane, the meter reading will not be correct. If the actual gas in the confined space is known, a conversion chart supplied by the gas meter manufacturer can be used to convert the methane reading to the actual reading. It is important not to use the methane reading when the meter has been calibrated with methane, and a different combustible gas is present in the monitored space. A false low reading could be produced, thereby potentially endangering occupants of the monitored space.

Another issue with the use of multi-gas meters is the false sense of security a low combustible gas reading may convey to the user. A combustible gas reading is provided in percent of the lower explosive level (LEL), which is greatly different than the scale used to measure the toxic limit for gases (typically in parts per million). Thus, even at 1% to 2% of the LEL reading, a toxic and possibly Immediately Dangerous to Life and Health (IDLH) atmosphere may be present, depending on the gas present.

When monitoring confined spaces, it is extremely important to monitor from outside the space. If possible, an intrinsically safe meter equipped with information logging and/or an alarm should be introduced into the space before anyone enters. A better practice is to use a probe attached to the meter that can be extended through a small sampling port into the space to be monitored. In an atmosphere where the concentration of combustible gas is above the UEL, opening a large hole could allow enough ambient air into the confined space to move the combustible concentration into the explosive range. When using a probe, sufficient time must be allowed for the meter's pump to pull the sampled air through the length of the probe to the sensor.

Since multi-gas meters rely on sufficient oxygen to detect combustible gases, oxygen monitoring is conducted first. In an atmosphere that is above the upper explosive limit (UEL) and oxygen deficient, the operator may observe a quick spike in the meter's combustible gas reading, followed by a rapid decrease in readings to near zero.

1.13 LEGAL AND CODE REQUIREMENTS

The design and installation of an industrial ventilation system may be impacted by legal or code requirements governing many other aspects of building design such as fire prevention and suppression, electrical design, smoke management, and more. In addition to the required standards mentioned above, state and municipal authorities may require their own building codes to be followed. When multiple codes apply, it is necessary to review all of the applicable codes and then design to meet the most stringent requirements. It should also be remembered that these codes prescribe minimum standards. Engineering analysis and professional judgment may require that these minimum design standards be exceeded in order to adequately reduce the potential for loss of life or property. Some of the most common building codes and standards are discussed further below.

1.13.1 NFPA. The National Fire Protection Association (NFPA) develops, publishes, and disseminates an extensive list of consensus codes and standards that are intended to minimize the risks associated with fire and other hazards. OSHA has adopted many of these as consensus standards. Other Federal, state and local codes also reference NFPA codes and standards. Within the U.S., the proper design and installation of every mechanical ventilation system will be affected by one or more of the 300 NFPA codes and standards. These codes and standards can impact design attributes for nearly every aspect of the system design, from the minimum flow rates, to the duct thickness, to the type of material used for the fan blades. NFPA codes are voluminous. They can often be found in public libraries or they may be purchased from NFPA through the Internet. Be aware that NFPA frequently uses the subjective term "adequate ventilation" without further definition of what is adequate to prevent fire and explosion. Thus, the task of determining adequate ventilation is imposed on the reader.

1.13.2 Building Codes. At the state and municipal levels, governments often enforce their own building codes. Traditionally, these codes were adopted (with or without modification) from one of three model codes: The Uniform Building Code (UBC)[(1.11)] [International Congress of Building Officials, ICBO],[(1.12)] the National (formerly Basic) Building Code (NBC) [Building Officials and Code Administrators, BOCA],[(1.13)] and the Standard Building Code (SBC) [Southern Building Code Congress International, SBCCI].[(1.14)] More recently, representatives from ICBO, BOCA and SBCCI collectively created the International Code Council (ICC). The role of the ICC is to develop a single set of comprehensive and coordinated national model construction codes without regional limitations. While the new ICC codes are slowly being adopted into state and municipal building codes, not all jurisdictions have adopted them. Since many jurisdictions modify the standardized codes to meet their jurisdictional preferences, individuals seeking copies of relevant building codes are encouraged to pursue such copies through their local code authority.

1.13.3 State and Municipal Fire Codes. Each jurisdiction may have its own fire codes, which may reference NFPA codes, but they may also have some codes that are specific to their state or region. It is, therefore, important to check with the local building authority to determine the local code requirements.

1.13.4 Other Code Requirements. There may be other codes, standards, covenants, etc., that apply to the installation of a new system. For example, some business parks have covenants that do not permit certain operations. Others prohibit the use of discharge stacks above a certain height that in turn could create contaminant migration into windows or outdoor air intake. Operations at government owned facilities will often have their own design and performance stnadards. Many organizations have minimum process ventilation requirements established by corporate standards.

1.13.5 Emission Requirements. The U.S. Environmental Protection Agency (USEPA), state environmental agencies and local air pollution agencies have regulations that must be addressed when contaminants are released into the atmosphere. These regulations may require the installation of air cleaning devices that will have an effect on the industrial ventilation system. They may also have requirements regarding fugitive emissions that are discharged from the building via open doors, windows and general ventilation systems. These regulations may mandate the use of local exhaust ventilation with air cleaning devices. The permitting and testing process associated with these environmental regulations can be a daunting task with long delays and large financial repercussions if done inappropriately. It is recommended that an individual familiar with local and national environmental air regulations be consulted prior to the design or source-altering modification of industrial ventilation systems.

1.13.6 Air Emission Surveys. For existing and newly installed ventilation systems, the emission rate of the airborne contaminants discharged into the atmosphere may require exhaust stream monitoring or indirect monitoring through concentration measurements at the property line. These measurements can be compared with regulatory thresholds or past measurements for compliance purposes and may also serve to establish a baseline for future measurements. Estimations may be calculated knowing the quantity of chemicals consumed within a known period of time.

1.13.7 Permits. Regulatory agencies may require installation and operation permits for a new or altered ventilation system. Often the installation permits must be obtained before construction can begin and before the operation or process is started up. Operating permits are obtained after the operation has begun, while operating under the installation permit. Often the installation permit requires emission monitoring and states a specific timeframe for the testing. Many operating permits will also require periodic retesting to verify compliance with applicable environmental regulations. Consult local and state regulatory agencies before planning a new ventilation system or revision to an existing system to determine permit requirements.

Emission estimates are usually required as part of the permit application process. Specification of well-designed local exhaust systems that capture contaminants close to the source for subsequent treatment or material reclamation can increase the likelihood of permit success while providing a financial incentive to the operation. The permitting process can be complex, time-consuming, and have a great impact on construction and start-up timeframes. Permit requirements need to be addressed by a knowledgeable professional early in the design stage.

1.14 SETTING AN EXPOSURE CONTROL STRATEGY

Providing process containment has many possible benefits. A simple containment, such as an enclosed hood, is a well-documented method of enhancing contaminant control and reducing the air-flow requirements for achieving contaminant control. Enclosing machining centers has proven an effective method of reducing oil mist exposures. Decisions on more elaborate process containment strategies should consider the many potential benefits of process containment. Table 1-5 lists the containment strategies commonly used in order of increasing capability, a rangc of dcmonstratcd capability, and somc of the basic pro and con issues to consider. All strategies must consider plant and process safety considerations, and document the successful proof of containment with standardized test protocols. In addition, the strategy for removal of emissions from the process has other potential benefits including:

- Reduced potential for product cross contamination;
- Reduced reconciliation issues for controlled substances;
- Improved process reliability;
  - Reduced product losses
  - Reduced non-productive housekeeping time
  - Reduced impact of problem materials that are very fine dusts or slippery materials, even when not potent compounds
- Reduced risk of dust or vapor deflagration, or fire or both;
- Reduced operating costs to prevent worker exposure [E.g., price of purchasing personal protective equipment (PPE)];
  - Inefficiencies while wearing PPE – donning/

TABLE 1-5. Containment Tools to Reduce Exposures

Containment Category and description	Approximate range of capability (Very unit operation and operator technique dependent)	Advantages	Disadvantages
1 - Dilution ventilation & no engineering controls Supply and exhaust large volumes of air (typically > 10 air changes/hr) through the manufacturing suite to dilute airborne emissions below health limits.	Sometimes B and A (See Table 1-1)	- Disperses widespread emission sources such as vapors - No equipment to add to standard unit operation equipment	- Does not control at source, spreads emissions around the room requiring substantial cleanup efforts - Air velocities too low to control most particulate contaminants - High cost for once through conditioned HVAC air
2 - Local exhaust ventilation (LEV) Hoods or enclosures on process equipment that exhaust air at the emission sources to collection equipment and away from the operator's breathing zone.	Bands A and B	- Capture emissions at their sources with well designed hoods - Reduced room HVAC air volumes	- Contaminants could be pulled into product if LEV applied incorrectly - Product pulled into LEV must be discarded - Additional systems to operate - Operators must be trained in correct use
3 - Downflow booths Small room or enclosure with low velocity (100 ft/min) downward airflow to push contaminants away from the operator's breathing zone.	Bands B, C, and D	- Emissions pushed away from operator's breathing zone - Useful for manual operations for which a more contained approach is not feasible	- Emissions land on the floor – cleanup issue - Operator technique can interfere - Additional systems to operate - Operators must be trained in correct use
4 - Closed process design All steps of the process are sealed with little chance of release to the suite. Examples include: automated dispensing, vertically stacked process, transfers with Intermediate Bulk Containers and Active/Passive containment valves and multiple unit operations in one housing.	Bands B, C, and D	- Emission sources within the physical confines of the unit operation - No easy path for external contamination to enter the product - Can separate product part of equipment from technical part of the equipment to limit size of manufacturing space - Need for housekeeping is reduced - Reduced need for HVAC air volumes	- Higher capital cost for equipment - Process flexibility may be limited - Some unit operations cannot easily be CIP'd so some equipment entry is required for cleaning
5 - Isolators Specialized rigid or flexible enclosures built around equipment, often with a dedicated HEPA air filtration system. Manipulations of the process through built in gloves. Materials passed in and out through airlock-like chambers or other devices.	Bands D and E	- Emission sources within the physical confines of the unit operation - No easy path for external contamination to enter the product - Can separate product part of equipment from technical part of the equipment to limit size of manufacturing space - Need for housekeeping is reduced - Reduced need for HVAC air volumes	- Ergonomic limitations to accommodate different size people make process difficult to operate and must be resolved to be feasible - Higher capital cost for custom designed equipment - Process flexibility may be limited - Some unit operations cannot easily be CIP'd so some equipment entry is required for cleaning
6 - Robotics The intent is no manual manipulation in the interior of the isolator because a robotic device does all operations. Similar pass in and pass out issues as for an isolator.	Band E	- No manual intervention with the process - Emission sources within the physical confines of the unit operation - No easy path for external contamination to enter the product - Can separate product part of equipment from technical part of the equipment to limit size of manufacturing space - Need for housekeeping is reduced - Reduced need for HVAC air volumes	- Not demonstrated on a large scale - Design requires finding ways to safely decontaminate the robotic equipment for maintenance personnel to work on it - Higher capital cost for custom designed equipment - Process flexibility may be limited - Some unit operations cannot easily be CIP'd so some equipment entry is required for cleaning

NOTE: This list introduces the range of containment tools but does not describe all possible permutations with unit operations. The Bands listed in column 2 are the same as used in Table 1-1.

doffing, communication;

- Meeting PPE regulatory requirements – fit test, medical certification, etc.;
- Costs for continuous stream of PPE to support process; and
- Management effort to enforce PPE procedures – OSHA requires management to enforce correct procedures.

1.14.1 Exposure Control Strategy Documentation. Controlling processing risk is a function of both the inherent hazards of the materials used, handled or processed and the exposures provided by the process. At the beginning of the project, clarify both the general manufacturing practice (GMP) requirements for processing conditions to make the desired product to specifications and non-GMP requirements for controlling exposures. For each process the process project team should develop an exposure assessment and control strategy. The required complexity of the exposure control strategy depends upon the complexity and inherent hazards of the process, and could include:

- A complete description of the process, product, product-flow and identified employee exposure mechanisms.
- The chemical and physical hazards associated with the employee exposure (dusts, fumes mists, product handling, noise, equipment use, etc).
- The applicable regulatory and/or recommended Occupational Exposure Limits.
- Any actual or referenced air sampling data.
- The facilities exposure control strategy, including general work practices, equipment maintenance, and spill cleanup methods.
- The maintenance and monitoring plan for installed exposure control equipment.
- The recommended air-monitoring program.
- Any recommended PPE.
- Process specific recommended Safe Work Practices.
- Any hazard specific recommended employee medical monitoring program.

1.15 VENTILATION SYSTEM WORKER SAFETY AND HEALTH ISSUES

Employees installing, performing maintenance on or testing ventilation systems can be exposed to a variety of hazards depending on the system.

1.15.1 Toxic Materials. Many exhaust ventilation systems are installed to control employee exposure to toxic materials, and to transport these materials. Thus, worker exposure to hazardous materials can occur during work on ventilation systems. Before employees work on ventilation systems that may contain hazardous materials, the material safety data sheets (MSDSs) and other available data associated with the hazardous materials should be reviewed and appropriate PPE provided. Typical hazards and corresponding PPE include:

- Inhalation (respiratory protection);
- Ingestion (hygiene practices including washing before eating, drinking, smoking or applying cosmetics);
- Dermal or skin contact (some materials can cross the skin barrier and cause a significant increase in the body burden of that toxic material);
- Spread of the toxic material (provision of work clothing and change facilities can reduce the spread of a toxic material out of the workplace or to unwanted areas in the workplace);
- Falling objects (hard hat, bump cap);
- Eye injury (safety glasses, goggles or face shield);
- Noise (earplugs or muffs with an adequate noise reduction ratio (NRR));
- Cuts and abrasions (gloves, gauntlet arm protection); and
- Liquid splash (chemical apron, or disposable suit as appropriate).

1.15.2 Fall Protection. When ventilation system work must be performed from elevated platforms, use of adequate fall protection equipment and procedures is required. This equipment may include approved full torso safety harness with a shock-absorbing lanyard. The lanyard must be fastened to a support capable of arresting a fall.

1.15.3 Machine Guarding. Guarding is required to protect employees from nip-points at belts and pulleys that can cause severe compression or amputation injuries. When employees work on ventilation systems they should:

- Not remove machine guards (i.e., pulley or sheave belt fan guards on fan drives) unless the equipment has been locked out both electrically and mechanically.
- Watch for guards that are missing or only partially enclose the mechanical pinch point.
- Watch out for the pinch point at ventilation system door openings. When opening access doors on operating equipment under negative pressure, the vacuum in the equipment can slam the door shut on an unwary worker, or pull unprotected hands or arms across sharp metal openings.
- Make sure that hands and equipment are clear of the rotary valve before jogging it to clean the net pocket in the rotor when cleaning.

1.15.4 Lockout. To prevent injury by energized mechanical equipment all personnel must follow lockout requirements provided in:

- OSHA Standard 29 CFR 1910.147, and
- American National Standards Institute (ANSI) Standard Z244.1.

Lockout procedures should ensure that all energy sources be deactivated. The energy sources can be electrical, pneumatic, hydraulic, and mechanical (gravity force from full buckets on a bucket elevator running backwards, exhaust fans rotating because pressure differential between the inside and outside of the building).

1.15.5 Combustible Dusts. Any combustible material, and some that would normally be considered non-combustible, can burn rapidly when finely divided into very small particles, fibers, fines, chips, chunks, flakes, or a mixture of these forms. Such a dust, when suspended in the appropriate concentration, may deflagrate and even become explosive. The force from a combustible dust explosion is capable of causing death and severe injury to employees as well as catastrophic damage to property.

Types of combustible dusts include, but are not limited to:

- Metal dusts (e.g., aluminum, magnesium, etc.)
- Wood dusts
- Plastic dusts
- Organic dusts (e.g., sugar, grain, paper, soap, and dried blood)
- Textile dusts

Industries that may handle combustible dusts include, but are not limited to:

- Agricultural
- Chemical
- Forest and furniture making
- Metal processing
- Paper
- Pharmaceutical
- Recycling (e.g., metal, paper and plastic)
- Textile
- Wastewater treatment

Since the 1980s, hundreds of dust-related fires and explosions have resulted in the deaths or severe injury of hundreds of employees as well as costly economic and emotional impacts. Nearly every state in the United States has been impacted by such fires and explosions. Failure to properly assess the hazards and risks associated with such events can yield catastrophic results.

In many combustible dust incidents, employees and managers were unaware of the potential for and serious nature of dust explosions. In one case, six employees were killed and more than 30 injured in Kinston, North Carolina, due to the buildup of combustible dust above a suspended ceiling in an otherwise clean manufacturing facility.[1.15] Additionally, such hazards often are not clearly identified on the material safety data sheets (MSDSs) of combustible substances.

In an effort to increase the awareness of the hazards posed by combustible dust fires and explosions, the U.S. Chemical Safety Board (CSB) has issued numerous reports regarding such events. On March 11, 2008, the federal Occupational Safety and Health Administration (OSHA) issued an instruction, CPL-03-00-008, Combustible Dust National Emphasis Program (Reissued), which contains policies and procedures for inspecting workplaces that create or handle combustible dusts. Additionally, the National Fire Protection Association (NFPA) has released several standards that should be consulted to aid in the recognition and control of hazards posed by combustible dusts. All employers are encouraged to review available CSB, OSHA, and NFPA publications and resources in an effort to properly protect their employees and facilities from the devastating impacts of combustible dust fires and explosions.

REFERENCES

1.1 Occupational Safety and Health Administration: OSHA Technical Manual, OSHA TED 1-0.15A, OSHA (1999).

1.2 National Institute for Occupational Safety and Health: FAQs About Control Banding, NIOSH (April 1, 2005).

1.3 U.S. Chemical Safety Board: West Pharmaceutical Services Dust Explosion and Fire (January 29, 2003).

1.4 U.S. Department of Labor: Occupational Safety and Health Administration (OSHA) Code of Federal Regulation 29 CFR 1910 (revised June 20, 1996).

1.5 U.S. Department of Labor: Occupational Safety and Health Administration (OSHA) 29 CFR 1910.1000 through 1910.1099 (revised June 20, 1996).

1.6 U.S. Department of Labor: Occupational Safety and Health Administration (OSHA) Code of Federal Regulations, 29 CFR 1910.1200 (1976).

1.7 American Chemical Society: Chemical Abstract Service, a division of the American Chemical Society (revised June 20, 1996).

1.8 Hinds, W.C.: Aerosol Technology. Wiley and Sons, Inc., New York, NY (1999).

1.9 U.S. Department of Labor: Occupational Safety and Health Administration (OSHA) Code of Federal Regulations, 29 CFR 1910.1000 Z-2 (June 20, 1996).

1.10 National Institute for Occupational Safety and Health: Analyzing Workplace Exposure Using Direct Reading Instruments and Video Exposure Monitoring Techniques. NIOSH, Cincinnati, Ohio, U.S. Department of Health and Human Services, Public

Health Service, Centers for Disease Control, DHHS (NIOSH) Publication No. 92-104 (1992).

1.11 Uniform Building Code (UBC) (1997).

1.12 International Congress of Building Officials (ICBO): Serving states west of the Mississippi (1997).

1.13 Building Code Officials and Administrators International (BOCA): Serving the north central and northeast United States, Country Club Hills, IL (1999).

1.14 Southern Building Code Congress International (SBCCI): Serving the south central and southeast United States, Birmingham, AL (1999).

1.15 U.S. Chemical Safety Board: West Pharmaceutical Services Dust Explosion and Fire (January 29, 2003).

Chapter 2

PRELIMINARY DESIGN

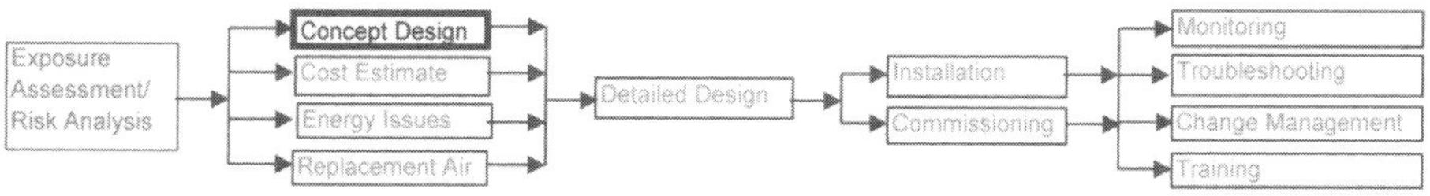

NOTE: Equations with notation followed by (IP) are designated for inch-pound system only; Equations followed by (SI) are designated for metric use only. If Equation bears neither, then it applies to both systems.

2.1 INTRODUCTION

The health hazard potential of an airborne substance is characterized by the Threshold Limit Value (TLV®) (see Chapter 1). The TLVs® refer to airborne concentrations of chemical substances and represent conditions under which it is believed that nearly all workers may be repeatedly exposed, day after day, over a working lifetime without adverse health effects. The time-weighted average (TLV-TWA) is defined as the time-weighted average concentration for a conventional 8-hour workday and a 40-hour workweek to which it is believed that nearly all workers may be repeatedly exposed for a lifetime without adverse effects. Exposures for work shifts lasting longer than an 8-hour day or 40 hours a week must be evaluated by a knowledgeable industrial hygienist. The TLV-TWA is usually used to determine potential workplace health hazards. TLV® values are published by the American Conference of Governmental Industrial Hygienists (ACGIH®). Annual revisions are made as more evidence accrues on the toxicity of the substance. Appendix A of this Manual provides the current TLV® list for chemical substances as of the date of publication and annual values are published separately by ACGIH®.

Regulatory bodies such as national, provincial, state, etc. governments may use other occupational exposure limits (OEL). For example, the United States uses the Occupational Safety and Health Administration's (OSHA) Permissible Exposure Limit (PEL). Although it is not always the case, the TLV® is usually more stringent than the regulatory OEL. When there is a difference between the TLV® and regulatory OEL, the OEL must be met and the TLV® is considered state of the practice. Good design practice points toward using the most stringent.

Occasionally, industrial ventilation systems are designed to improve housekeeping or for fire protection. Refer to standards organizations such as the National Fire Protection Association (NFPA) information on combustible dusts or flammable vapor limits.

Ventilation systems used in industrial plants are of two generic types. The SUPPLY system is used to furnish air, usually tempered, to a work space. The EXHAUST system is used to remove the contaminants generated by an operation in order to maintain a healthy work environment.

A complete ventilation program must consider both the supply and the exhaust systems. If the overall quantity of air exhausted from a workspace is greater than the quantity of outdoor air supplied to the space, the plant interior will experience a lower pressure than the local atmospheric pressure. This may be desirable when using a dilution ventilation system to control or isolate contaminants in a specific area of the overall plant. Often this condition occurs simply because local exhaust systems are installed and consideration is not given to the corresponding supply air systems (see Chapter 10). Air will then enter the plant in an uncontrolled manner through cracks, walls, windows, and doorways. This typically results in 1) employee discomfort in winter months for those working near the plant perimeter, 2) exhaust system performance degradation, possibly leading to loss of contaminant control and a potential health hazard, and 3) higher heating and cooling costs.

Supply systems are used for two purposes: to create a comfortable environment in the plant (the Heating, Ventilating and Air Conditioning (HVAC) System); and to replace air exhausted from the plant (REPLACEMENT System). Many times, supply and exhaust systems are coupled, as in dilution control systems (see Chapter 4).

A well-designed supply air system will consist of an air inlet section, filters, heating and/or cooling equipment, a fan, ducts, and register/grilles for distributing the air within the workspace. The filters, heating and/or cooling equipment, and fan are often combined into a complete unit called an air house or air supply unit. If part of the air supplied by a system is recirculated, a RETURN system is used to bring the air back to the air handling units.

Exhaust ventilation systems are classified in two groups: the GENERAL exhaust system and the LOCAL exhaust system. The general exhaust system can be used for heat control and/or removal of contaminants generated in a space by flushing out the space with large quantities of air. When used for heat control, the air may sometimes be tempered and recycled. When used for contaminant control (the dilution system), enough outdoor air must be mixed with the contaminant so that the average concentration in the worker's breathing zone is reduced to a safe level. The contaminated air is then typically discharged to the atmosphere. A supply system is usually used in conjunction with a general exhaust system to replace the air exhausted.

Dilution ventilation systems are normally used for contaminant control only when local exhaust is impractical because large quantities of tempered replacement air are required to offset the air exhausted can lead to high operating costs. Chapter 4 describes the basic features of general ventilation systems and their application to contaminant and fire hazard control.

Local exhaust ventilation systems operate on the principle of capturing a contaminant at or near its source. It is the preferred method of control because it is more effective and the smaller exhaust flow rate results in lower equipment and energy costs.

Local exhaust systems are comprised of up to four basic elements: 1) the hood(s), 2) the duct system (including the exhaust stack and/or recirculation duct), 3) the air cleaning device, and 4) the fan. The purpose of the hood is to collect the contaminant generated in an air stream directed toward the hood. A duct system must then transport the contaminated air to the air cleaning device, if present, or to the fan. In the air cleaner, the contaminant is removed from the air stream. The fan must overcome all resistance due to friction, hood entry, and fittings in the system while producing the design flow rate. The duct on the fan outlet usually discharges the air to the atmosphere in such a way that it will not be re-entrained by the replacement and/or HVAC systems. In some situations, the cleaned air is returned to the plant for the controlled buildings and nearby buildings. Chapter 10, Section 10.8 and Chapter 11, Section 11.3 discuss exhaust air recirculation.

This chapter focuses on the preliminary design aspects of exhaust ventilation systems, but the principles described also apply to supply systems.

2.2 PROJECT GOALS AND SUCCESS CRITERIA

Because the design and installation of a local exhaust ventilation system involves approval by many outside agencies and potential interface with varied plant processes and departments, management must select a team that is responsive to identifying the end user(s) and their needs. In most cases, the opportunity for error or problems can occur at the interface between company and regulator, between company and contractor or among project team members themselves. Communication and organization can be the key to a successful installation and team members must have or develop these skills to bring success to the project (see Chapter 5, Figure 5-1).

2.2.1 Small Projects or Small Organizations and Success Criteria. It is recognized that simple projects and smaller operations may not need a team or special organization to complete the installation of a ventilation system. However, even the smallest system has requirements to meet safety and environmental regulations. Because of these regulatory aspects, all projects should have minimum organization and documentation.

All projects usually begin with the identification of the problem to be solved. The organization must also keep a focus on any requirements of proof of performance for completion (see Chapter 8 of *Industrial Ventilation: A Manual of Recommended Practice for Operation and Maintenance* [the O&M Manual]). This may be informal through normal plant or department communications and usually would be in response to the following issues:

1) Addition of new process that requires ventilation controls;
2) Change to existing process that requires additions or revisions to existing systems;
3) Measured or perceived safety and health issues that can be improved with ventilation;
4) Response to plant labor committees to improve ventilation for worker comfort or safety;
5) Needed improvements to poor design that render the present system ineffective or a waste of energy; and/or
6) Failure of the present system to meet required emission levels.

In response to the needs for system installation or improvement, management mobilizes the necessary plant resources. In a small plant this may be just the plant manager or plant engineer working in conjunction with outside contractors or engineers. Instructions for installation may be given verbally or with a few sketches and followed with a formal or informal proposal. Parallel to this effort, commercial requirements such as funding, cost controls and budget management would be required.

Even the smallest systems require a review on the impact to existing plant resources such as electrical power, floor space and maintenance staff. This can include some preliminary engineering from vendors or engineering staff to provide a concept of the design. Production disruptions and adjustments will also be identified and planned. In general, every project would be organized in phases:

1) Feasibility and concept design – The idea from the process is studied and verified and defined for the new project; sizes for equipment are estimated and preliminary estimates are made to see if the project is feasible. The issues of Proof of Performance and Commissioning could be included at this stage.
2) Definition and funding – The design is refined so that scope can be written for instructions to designers or for design/build firms; more detailed (±20%) estimates are made so that funding can be acquired; possible work required to start permit process is determined.
3) Detailed design – This should be done either by in-house, independent or design/build engineering staffs with enough detail to evaluate the process impacts and detailed cost issues.
4) Construction – This can begin during some of the detailed design phase after design approval and securing of required permits.
5) Startup and Commissioning – This is final phase where ownership of the project is transferred from the construction organization to the final owner.

2.2.2 Larger Projects and the Keys to Success. As systems become larger and more complicated or have implications for meeting regulations (or both), the need for more formal organization and document controls becomes mandatory. This includes the documentation of the basis for design, the conceptual and detailed design drawings and verification of effectiveness of the system. In addition, maintenance and service records for the completed installation need to be kept. This is required for a proper transfer of ownership from the project team to the actual operators of the system (see Chapter 2 of the O&M Manual).

The same goals and phases also govern smaller projects but the organization may be less formal for small projects. However, communications, especially among team members, should also be documented.

Organization starts with the identification of the person or persons responsible to receive the system and possibly responsible for its operation. Maintenance and production may be assigned responsibility for the upkeep of the exhaust system and replacement parts, and must be kept informed throughout the design process.

The first step is a simple document to define the expectations or success criteria for the operation of the system. The expectations become the directive to determine all further effort. In its simplest form, this document notes existing problems or shortcomings to be resolved and can be in long hand or outline form. These concerns could include exceeding OSHA limits, poor performance of existing control technology, etc. Other advisory groups such as The American Conference of Governmental Industrial Hygienists (ACGIH®), The American Industrial Hygiene Association (AIHA) and NFPA can provide supplemental data for occupational health and safety exposure limits. In some cases, it may be good to refer to previous projects and what had been considered successful project completions. These can include those within the company but also may include other industry success stories. Evaluating unsuccessful projects may also provide insight into deficiencies to avoid. This may be done by looking at a Design Basis to reference the closure requirements. Looking at the end of project requirements can better define what needs to be accomplished at the beginning of a new installation.

The document would then identify measurable goals (dust exposure, emission levels, bringing a process on line by a certain date, etc.) so that plant management, design engineers and contractors stay focused on the system requirements. It would also identify the benefits (cost or energy savings, avoidance of fines, etc.) so that the clear intent of the installation is maintained. Note that this may require some study and review. Any system, no matter how small, that includes responsibility to regulatory agencies and has impact on worker health and safety should include this important first step.

Evaluation should also include an assessment of potential risks (see Chapter 1). This includes evaluation of potential risks for worker exposure to the dusts, mists, fumes, vapor or heat from the process. These risks primarily include inhalation but may include other exposures such as skin absorption. This document also may include input from manufacturers of new or existing equipment or processes to be controlled to see if there are alternative methods to reduce exposure. Before the scope is defined, it must be determined that all practical means have been investigated to remove or reduce the pollutants at their source before adding controls.

Any restrictions on the IV system or controlled process should be listed. This may include access to equipment that is hindered by hoods or ergonomic considerations (e.g., as workers need to reach into enclosures or over other restrictions from the system). Maintenance and construction worker access and safety must also be identified (see O&M Manual, Chapter 5).

During this early stage the need for environmental permits must be addressed. In many states this process can take more than a year and potentially delay the construction or start-up. If sufficient resources are not available within the company, then a permit specialist (consultant or law firm) should be contacted early in the project schedule.

The Clean Air Act Amendments of 1990 have changed many of the requirements, especially with issues such as Maximum Achievable Control Technology (MACT) Standards (Title III), permits (Title V), non-attainment areas, permits to install and permits to operate, etc. In many cases, preliminary estimates are done with regard to engineering data (emission factors, air volume, stack heights and locations, etc.) that must be accomplished even before organizing a project team. Similarly there may also be a need to determine if studies are needed for Prevention of Significant Deterioration (PSD) permits or other provisions that may be required. These early reviews may actually provide opportunity to save on the installation by considering alternate processes or materials to eliminate or reduce the need for pollution controls.

2.3 LARGE PROJECT TEAM ORGANIZATION

After project goals have been determined and a person is designated to receive and own the finished project, the task of organizing within the plant begins. Again, the size of the project may determine the experience needed to proceed.

At a minimum, representatives from Process, Purchasing, Maintenance and Plant Engineering are required (Figure 2-1). In smaller operations, one or two persons may hold all these positions. In addition, there probably are requirements for approval by regulatory agencies (requiring stack testing for emissions or industrial hygiene testing for OSHA issues). These health and safety reviews by plant professionals or consultants should be included for any system that has an impact on the plant environment.

While workstation operators do not have to be part of the larger design team, they must be consulted regularly during the design process and may have suggestions that may make the project work smoother. For example, mockups of hood and enclosure designs can define operability problems that can be addressed during the design phase.

2.4 TEAM RESPONSIBILITY MATRIX (TRM)

At this point an outline of responsibilities and team members should be developed (Figure 2-1). This outline is called a "Team Responsibility Matrix" or "TRM", and can also include the requirements of outside resources such as consultants, e.g., IV design specialists or special service companies (industrial hygiene firms, etc.). At the same time, the Project Closure Document (Figure 2-2) should be completed to determine the persons responsible for final acceptance of the project. Some preliminary work can also be accomplished for the Commissioning process, such as a list of proofs of performance. These could include items such as required filter bag life, emission levels from the collector, TLV® near operator station,

Maintenance - m Safety - s Operations - o Plant Engineering - pe Electrical Controls - c Quality - q Plant Manager - pm Purchasing - p Environmental - e	Team Review Members	Scope Definition Review (Input to Design Basis)	Design Basis Review (Input to Design Team)	Design Review (Input to Construction)	Approval Plant Manager
Plant Issues					
Environmental Regulations & Permits	e,s,pm				
Health & Safety Regulations	s,pm				
Fire Protection Regulations	s,pm				
Process Modifications (P&IDs) (Process Heat and Material Balance)	o,pe,q,pm				
Layout of Plant Equipment & LEV Isometric	pe,o				
Location of Control Equipment	pe,o				
Plant Rules & Regulations	s,o,pe,p				
Future System Expansion	o,pm,e				
Energy Requirements	o,pe,pm				

Design Issues

A. Hoods					
Ergonomics of Hood	o,pe,s,m				
Machine Access Requirements	o,s,m				
Materials of Hood Construction	pe,m				
Safety Interlocks	o,s,pe,c				
Lighting Requirements	o,s,pe,c				
Air Volume Requirements & Basis for Design	e,pe				
B. Duct					
Duct Construction Materials	pe,s,e,o				
Flanges/Welded/Spiral	pe,e				
Access Doors	pe,e				
Elbows/Fittings	pe,e				
Transport Velocities	me				
Duct Sizing	pe,e				
Blast Gates/Orifice Plates	pe,e				
Methods of Support	pe,e				
C. Fan					
Type of Fan	pe,e				
Location (Clean Air Side, etc.)	pe,e				
Construction Material	pe,o,e				
Specifications	pe,e				
Special Temperature Requirements	pe,o,e				
Future Changes to System	o,s,pm,pe				
Safety Factor	o,pm,q,pe				

FIGURE 2-1. Sample Team Responsibility Matrix

Class/Rotation/Arrangement	pe				
Traps/Drains	pe				
Stack Design	e				
Noise	s,pe				
Vibration	s,pe				
D. Air Pollution Control Device					
Type of Device	e,pe,m				
Auxiliary Connections	e,pe,m				
Handling of Collected Materials	e,pe,m				
Energy Considerations	o,pe,pm				
Construction Materials	pe				
Specifications	e,pe				
Special Temperature Requirements	o				
Future Changes to System	o,pm,e				
Safety Factor	o,pm,e				
Guarantees	pm,e				
Life Cycle Costs	pm,p,e				
Air/Cloth Ratio for Filters	e				
E. Replacement Air (Air Volume exhausted by LEVs in Operation Area)					
Number and Type of Units	m,pe,e				
Fuel Type	pe,e,pm				
Air Volume	e,pm,pe				
Temperature Rise	pe,e,pm				
Controls	pe,c,s				
Insurance Requirements	p,pe,s				
Location	pe,pm				
Duct Design	pe				
Supports	pe				

Project Management Issues

Cost Estimates	o,pm,p				
Management of Scope Changes	pm,p				
Drawing Standards	pe				
Document Control & Distribution	p,pe				
Transfer of Ownership	(all)				
Management of Safety	s,pm				
Schedule	p,pm,o				

FIGURE 2-1 (Cont.). Sample Team Responsibility Matrix

Project #	Project Title:
	Target Completion Date:

	Sign & Date (All signatures required to close PCD)	
Project Manager		1. Project Manager must complete this page before issuance of Design Basis. 2. Checklist items may be completed at any point thereafter. 3. Final closure approvals may be obtained once the checklist is complete. Checklist item exceptions indicated by N/A on the checklist.
Receivers		
Operations (O)		
Maintenance (M)		
Environmental (E)		
Plant Manager (PM)		
Safety (S)		
Plant Engineer (PE)		
Quality (Q)		
Electrical Controls (C)		
Purchasing (P)		

Extension Requests		
Revision Date	**Reason** (Attach list of tasks needed to complete)	**Plant Manager (Sign and Date)**
1.		
2.		
3.		

Final Closure Approvals	**Sign & Date**	**Signatures indicate agreement that project is completed and can be closed.**
Operations Manager		
Plant Engineer		

FIGURE 2-2. Sample Project Closure Document (PCD)

etc. Typical plant personnel to be included are shown on the form but may be expanded based on particular project needs. See Chapter 2 of the O&M Manual for a complete discussion on commissioning and system evaluation.

The purpose of the TRM is to ensure that the proper resources are used to determine the plant and project needs before the design begins. The boxes on the form would contain the names of the individuals responsible for the input to the Design Basis (instructions to the Design Team) and the project. The individuals would initial opposite their name to indicate that the information has been given to the Project Manager for issue. These same individuals would initial in the remaining boxes after issuance of the Design Basis and the construction package (instructions to the contractors and/or bidders). This minimizes delays and scope changes as the project proceeds. It also avoids late input from outside sources that could impede the project timing and success.

This places a lot of pressure on the Project Manager to ensure that proper individuals are contacted at the beginning of the project. For example, a maintenance foreman may have experience with particular types of electrical controls and that input would be important information for the Project Team.

2.5 PROJECT TEAM SAFETY

A prime consideration when beginning these projects is safety. This includes the safety of the audit process since readings of pollutant and energy outputs may be required and extends both to workers and outside testing and engineering firms. The data required for the design of air pollution control or industrial ventilation systems may not be normal measurements taken in the process. Special plant precautions may be required to manage the safe gathering of information.

For instance, many air pollution control systems require scaffolding to perform source emissions acceptance tests. Initial testing, adjusting and balancing technicians may need cherry picker trucks to access sub-main ducts located over some processes. The contract must be written to inform them of the safety needs such as respiratory and fall protection.

2.5.1 Process and Equipment Safety Studies. Similarly, the attachment of air pollution control devices to existing processes may have impacts on the processes themselves. Process safety reviews may be necessary to evaluate the impact of the system additions. For instance, the purchasing department may need to locate sources and storage facilities for treatment chemicals, or filter media wet collectors may require additional permitting to discharge into the industrial waste treatment plant or the sanitary sewer. In addition, personnel safety or fire and explosion studies may be required based on the nature of the project.

2.6 DOCUMENT CONTROL

Smaller projects may have very little in the way of drawings, specifications or design calculations. Because ventilation projects may have regulatory or safety implications, there should be some record of the system design and maintenance requirements. Small systems may eventually need expansion and the more information available to the engineer, the better the opportunity for a successful project.

The control of project documents begins immediately. For larger projects this includes minutes of planning meetings, meetings with contractors and consultants and the exchange of information such as "Scope of Work" and bid proposals. The document control may be as simple as an "engineering and correspondence" file kept by the Plant Engineer or Project Manager. On large projects the distribution of documents and management of communications may be handled by a project clerk. Proper document control can mean the difference between a successful project and a disaster.

Document control can also serve to keep the project focused. Often a project is expanded as other plant needs are addressed. This is not always a negative thing since pollution control projects often can be the opportunity to improve plant efficiencies and reduce operating costs. Document control can be used to manage the input for scope definition (scope creep) and resultant project costs.

Document control can also be invaluable for the avoidance and settlement of project disputes. Many project problems can be attributed to the lack of communications. This can include the correct definition of the scope but also extends to a realistic definition of expectations. System guarantees and requirements usually become the focal point if a system does not meet performance standards. Communication of the expectations, acceptance by engineering firms or contractors, and management of the information are required to gain solutions to project disputes.

Document control also extends to plan and specification review and the expectations of the process. Many times a system is designed by a consultant or contractor and there is no clear understanding of the review and approval process. The plant is asked to review complicated engineering controls and equipment to install the system. A determination must be made as to who is qualified to conduct plan and specification review and who has ultimate responsibility for approval. There must be a communication from the project team regarding other implied approvals. This includes issues such as consistency between the architectural, structural and mechanical drawings, interferences on drawings that may be missed and who is responsible for back charges, etc.

In addition corporate headquarters or some other authority outside the plant may have final review authority. If that is the case, they should be involved throughout the project and not just at the end. In general, there are no clear rules that always apply in this area, but communication of expectations reduces the chances for disputes and allows all parties to consider risks and costs during the bid process.

2.7 PROJECT TEAM ORGANIZATION, SELECTION AND SKILLS

The size and complexity of the project has some bearing on the selection of the Project Manager in the scope development and project execution phases. Smaller jobs may only require the Plant Engineer to serve in a part-time role but complex installations may require full-time leadership and responsibility.

This could also be reflected in the size of the support staff. It is best to keep continuity between the development of concepts and delivery of the final completed project. The final receiver of the completed installation should be involved as early as possible so that expectations are listed and the coordination of operator training can be accomplished. Similarly, maintenance hand-off is critical to the continued operation of the system and should be included through the development of the scope definition phase.

The organization, even on small projects, should be defined explicitly. During these temporary assignments, there may be role changes that may not be compatible with normal plant or company organizations. The importance of being able to manage under these conditions is central to the success of the project and must be supported and maintained by plant and company management. It is important to include anyone who may have the ability to change or delay the project after it is organized. For environmental projects this would especially apply to health and safety staff. The time to include them is before the project has proceeded to design or construction.

At the same time, the project organization may result in the inclusion of personnel not normally familiar with the disciplines and schedule requirements of a complex installation. Care must be taken to properly train all members to be a team asset. At a minimum this training should include: 1) cost management; 2) schedule control; and 3) communications skills.

2.8 RESPONSIBILITY FOR FINAL APPROVAL OF BUDGET, TECHNICAL MERIT AND REGULATORY ISSUES

After building the project organization the responsibilities inside the group are determined. The primary purpose of the project team is to manage the installation so that it can become the property of the plant. One of the first determinations to be made is under what conditions will the installation be accepted by the plant. This acceptance may require more than one set of conditions. The installation is usually impacted by regulations that can include: improvements to plant ambient air conditions, safety requirements, requirement to meet emission regulations, and installed equipment to meet plant and regulatory safety requirements.

As mentioned earlier, the approval process can also include plant contacts reviewing complicated engineering drawings, calculations and specifications. This can include a tacit approval of periphery items such as physical dimensions or connections to plant equipment. These approvals may have cost impacts. For example, a plant project team may be asked for the selection between two alternate control schemes that have cost, technical and regulatory implications. Members can be well versed in plant and process operations but may not possess the technical expertise to approve these issues. At that point, another outside resource may need to be considered.

Ultimately the Project Manager is responsible to management and must sign off on all decisions (Figure 2-1). A team member may be designated for review of certain aspects of the installation but final approval must come from the Project Manager. The important factor is ensuring that the Project Manager is not inundated with minutia such as review of each dimension on a drawing. The Project Manager's time should be devoted to larger problems such as ensuring that the most efficient and effective system is installed. At the same time, good communications between the Project Manager and all team members would ensure that decisions are made and reviewed in a timely manner so as not to impact the schedule.

2.9 COMMUNICATION OF PLANT (AND PROJECT) REQUIREMENTS

With the completion of the organization, areas of responsibilities, and document control issues, the team turns outward to communicate the project's requirements to the person or company responsible for the system design and to the final user. In simplified terms, a project design can be considered to have three distinct steps to complete: Conceptual Design, Design Definition and Detailed Design.

2.9.1 Project Feasibility and Conceptual Design. During this phase, a plan is developed to define feasibility and preliminary design; the minimum requirements from this part of the project are to have the following information:

1) Concept Description
2) Clearly stated objectives (i.e., exposures below "x", reduction of environmental emissions by "y", etc.)
3) Equipment list
4) Process Flow Diagram
5) Heat and Material Balance
6) Process and Instrumentation Diagrams (P&IDs)
7) List of existing equipment (e.g., fans, motors)
8) Instrument list
9) Milestone schedule
10) Preliminary cost estimate
11) Available utilities
12) Building codes and regulatory requirements, e.g., environmental permits (changes to existing permits, increase in emissions, etc.), relevant fire codes
13) Equipment layout, floor space, site location

14) Studies list (fire protection, safety, etc.)

15) Proofs of performance for existing systems

16) Time critical dates (e.g., scheduled maintenance shutdown, production schedules)

The final documents would be the compilation of all of the above information in a suitable format for presentation to the owner of the system (plant and company management). The above information may have been accumulated using company resources or studies and outside resources may also have been used for this task.

The level of accuracy at this stage must be sufficient to identify major problems that may impact the final installation and its cost. It is also noted that some option analysis may not be resolved until the project reaches Scope Definition but hopefully everything is defined before proceeding with Detailed Design.

2.9.2 Design Definition – Defining and Communicating the Scope. After the completion of the conceptual design, this second phase would be developed. For small systems, this may be simply done with a few lines of description. In more complicated projects, a formal Design Basis becomes the method of communication. This can be accomplished only after all studies are complete and all design options have been determined. Basically all of the above information would have been "frozen" in place in order to proceed. In addition, long lead time items such as permits and equipment purchases would be implemented or ordered and cost estimates would be refined and more detailed.

The Design Basis (Figure 2-3) is authored by the project team and simply is a detailed set of instructions to the design team. Figure 2-3 should be imported onto the project team's company letterhead and expanded as needed. This also includes a list of the expected deliverables at the final detail design phase (including project goals). The issuance of the Design Basis may include other review requirements where other parties review the conceptual design before proceeding with detailed design. This second review is used for more complicated projects where other company and outside resources may be needed to refine the concepts. For example, the Project Team may initially review alternate control schemes with each cost and schedule implication considered or it may choose to pass them on for further review by corporate representatives. These efforts obviously take longer but may actually reduce overall project schedule (and cost) by reducing confusion during the design definition and detailed design phases.

Since the Design Basis document comes from the Project Team, the first decision would be who shall author, publish and review the document. Again, project size may influence the need for text and standards input but the team can use this opportunity to communicate their particular requirements for completion and acceptance. The Design Basis should be signed by all team members before submission to the selected design manager, firm or team. It also becomes the attachment for scope definition for competitive design bids if that is the direction taken.

At a minimum, the Design Basis must include the expectations for the project; any applicable standards that must be met and proofs of performance. These may include regulatory requirements: ANSI, NFPA, plant or local standards, safety, and delivery requirements (drawing methods and detail, schedule, etc.). A sample Design Basis is shown in Figure 2-4. Information and headings may be changed to reflect the details and requirements of each project. At this point, the owner should have an idea of the annual energy budget and O&M costs.

The Design Basis is then given to the chosen design firm or individual and becomes the document for management of detailed design. The completed Design Basis can also be used as scope instructions to Design/Build contractors (see Sections 2.10 and 2.12).

The extent of the detailed design is also to be determined during this issue. For example one owner may want all of the engineering completed as one package. In addition to the ventilation design there may be a requirement for the design of the electrical power and control, foundations, structural tie-ins to the owner's building, etc. Other companies may have in-house resources already in place for these services. They may already have an electrical design or contracting firm that does all electrical power, controls and/or energy management in the plant. In those cases, it may be better to have this work performed outside the ventilation design contract as long as information is freely transmitted between all parties.

Similarly, other organizational issues can be determined at the issuance of the Design Basis including the distribution of drawings and review methods for approval of designs and contractor prints. It can also lay out very specific limits of responsibility such as the requirements of the engineer or design/build contractor to review certified prints from vendors and make sure that foundations match anchor bolt layouts of equipment. The more information included in the Design Basis the less opportunity there is for dispute and project cost overruns.

2.9.3 Detailed Design. This final phase is the one most identified with the project. It is the final set of instructions to the installer. Details of design considerations for all of the major components of systems are included in Chapters 4, 5, 6, 7, 8 and 10. In addition, the calculating methods for system sizing are included in Chapter 9. This phase also includes the final review set of plans and specifications that the Plant Management sees before construction bidding. In the case of Design/Build contracts it represents the document deliverables for the installation.

At a minimum, this phase should include enough detail to clearly communicate the final system to be installed. Drawings must be to the detail level requested in Design Basis and may

Design Basis		
Project:	**Plant:**	
Date:	**Location:**	
A. Project Description		
B. Scope of Design Basis		
C. Attached or Reference Material:		
D. Minimum System Requirements:		
E. Regulatory Requirements:		
F. Equipment Requirements (LEV System Requirements and Structural Issues):		
G. Plant Safety Requirements:		
H. Duct Specification Requirements:		
I. Interface with Other Engineering Disciplines:		
J. Power and Controls:		
K. Transfer of Ownership to Plant:		
L. Technical Documentation:		
M. Technology Transfer and Training to Operators:		
Title	**Signature**	**Date**
Project Manager		
Team Member		
Team Member		
Team Member		
Team Member		

Use this figure as a guide. Companies are encouraged to reprint and rearrange this figure on their own letterhead.

FIGURE 2-3. Sample Design Basis form

Project: Sand System Ventilation **Date**: 1/26/05
Plant: Local Foundry
Location: Cincinnati, OH

A. **Project Description:** Provide complete engineered solution to the emissions from a Green Sand processing operation as shown on Plant Layout drawing # 100. System will exhaust dust from all emission points and convey to a new baghouse to be located outdoors behind the plant. System will meet all requirements for emissions as required by local and applicable federal laws and guarantee a minimum bag life of one year at full volume.

B. **Scope of Design Basis**: System engineering will include all hoods and enclosures (including supports), selection of airflow requirements, duct design and routing and selection of control device, fan, motor and drive. Electrical and compressed air connections to the equipment to be furnished by others.

C. **Attached or Referenced Materials**: Plant layout drawings showing location of equipment to be ventilated, process requirements of sand (flow rate, moisture content and temperature), applicable emission limits, OSHA in-plant dust levels required, Plant Safety Rules.

D. **Minimum System Requirements**: System to be guaranteed to collect dust at all enclosures to meet the requirements of USEPA Method 204 for capture as well as all applicable codes for operation and emission. This will be done while maintaining minimum transport velocities to keep material from building up in all ducts. Design to consider moisture from several sources in the sand process and must operate under all weather conditions in the plant. Expected winter temperature in the sand system area can be 50 F [10 C].

E. **Regulatory Requirements**: System will fall under the MACT standards as published in December 2002 and meet all of these requirements as well as applicable emission limits in effect at this date. No further requirements for future changes are anticipated at this time.

F. **Equipment Requirements (LEV System Requirements and Structural Issues)**: New baghouse to be a baghouse with cloth bags at an air/cloth ratio to guarantee a bag life of one-year while delivering full design volume and capture at all hoods. Bags will have top access for changing and maintenance and include screw conveyor and rotary valve(s). Maximum rpm = 20. Fan(s) to be backward-inclined wheel design, Class III or heavier based on volume, temperature and pressure calculated. Fan will discharge into a free standing stack complete with test ports and access for emission testing. Fan to be belt driven for up to 150 HP [745 watts]. Fan will be direct drive if connected power is over 150 HP [745 watts]. Fan to operate on a stable point of fan curve with operating pressure less than 90% of maximum pressure at rated speed. Fan to be equipped with access door, removable wheel, and drain. Motors to be high efficiency. Equipment to be painted per plant specifications. All structural design to meet applicable codes and seismic requirements.

G. **Plant Safety Requirements**: Design to meet all plant safety requirements and attached rules. Designer to furnish company safety manual for approval before commencing and evidence of drug testing and related verification.

H. **Duct Specification Requirements**: New system to meet all requirements for thickness and stiffening per SMACNA round and square duct standards. All fittings to have entry angle of a maximum of 45 degrees with 30 degrees preferred. Cleanout doors to be located on maximum 20 foot [6.1 m] centers and at all elbows and hoods. System to be designed with blast gates for air balancing. All elbows to have R/D of 2.0 minimum with 2.5 preferred.

I. **Interface With Other Engineering Disciplines**: Designer responsible for coordination with in-house structural (Contact name:_________) and electrical engineering departments (Contact Name: _______) for the design of structural supports, concrete foundations and power and control of system (see Section J).

J. **Power and Controls**: In-house engineering staff will design all electrical power and controls for support of the new system. Designer will coordinate with Electrical Engineer and provide sequence of operation, horsepower and other requirements for design of the system.

K. **Transfer of Ownership to Plant**: This Design Basis is for design services only. Design will have reviews with plant project team at 50% completion and 100% completion. After approval of design, information package

FIGURE 2-4. Sample Design Basis

Project: Sand System Ventilation **Date**: 1/26/03

(drawings, specifications for purchased equipment and the names of three suitable vendors for all equipment purchased) will be in electronic format with paper copies. Drawings to be stamped by Professional Engineer.

L. Technical Documentation: Designer to furnish calculations for determination of all air volumes as well as ACGIH® calculation sheets for the selection of duct sizes, velocities and static pressure requirements of the fan. All drawings to be done in CAD per plant drawing standards.

M. Technology Transfer and Training to Operators: Designer to provide training program for the installation of final system including requirements from installing contractor for installation manuals and personnel training.

Signatures: ______________________ (Project Manager)
______________________ (Team Member)
______________________ (Team Member)
______________________ (Team Member)
______________________ (Team Member)

FIGURE 2-4 (Cont.). Sample Design Basis

have company drafting standards included. Normally the contract would require the completion of "as-built" drawings and the turnover of electronic copies for the plant's files. The level of detail may extend from single line drawings with few dimensions to extensive double line drawings that show details for shop fabrication. Since the cost differential between the two can be extensive, it is important that the Design Basis communicate the expectations of the project and plant management.

Specifications may be included on the drawings or added as a separate document per various industry and company standards. There are advantages to both methods. On smaller projects the inclusion of information on the drawings keeps one single source of information for future reference since specification books may be stored in different locations from the drawings. Larger and more complicated projects may be better served by the use of specification text packages that may have clearer information for transfer to the contractor.

At the completion of detailed design, a more defined construction schedule usually can be determined and should be included, especially on Design/Build projects. At the same time, certified vendor prints and cut sheets should be included in the package.

2.10 DESIGN/BUILD, IN-HOUSE DESIGN OR OUTSIDE CONSULTANT

After the project team has begun with the development of the Design Basis, a decision should be made regarding the method of design completion. Each of the three methods listed above has its advantages and disadvantages and the choice may reflect the preferences of the plant management as much as the project team. This may also define the level of instructions in the Design Basis.

Detailed information may be required for consultants especially if there is a bid process among firms that may not have done previous work in the plant. The Design Basis becomes a bid document (either for design firms or for design/build firms) and will be reviewed by all unsuccessful bidders. It may contain proprietary process information, security and secrecy of process and/or planning and so this may be an important issue. Again, document control is important during early organization especially for the issuance of Secrecy Agreements and return of information from unsuccessful bidders.

When selecting a method, the plant management and Project Team must be realistic in its ability to manage the efforts of any of the three types. In-house resources are easier to control with respect to confidential plant process but may not have the depth of expertise to consider different control methods, new technology or alternate designs. The project may need the abilities of experienced design/build or consulting firms to ensure that the project meets all requirements. If in-house staffing is selected then the project would encounter many of the same issues listed below for design-construct, including review of their capabilities. It should be noted that the more organizations involved in the process, whether in-house or outside, the more requirements for information hand-offs and reviews.

If in-house design is not selected then the next decision is between the design-construct method and design/build (turnkey) method. The former uses a detailed engineering package to convey information to the construction contractor (builder). Sometimes this is known as "plan and spec." The builder may be a general contractor, a specialty contractor (mechanical or sheet metal firm, for example) or a combination.

2.11 DESIGN-CONSTRUCT METHOD (SEPARATE RESPONSIBILITIES FOR ENGINEERING AND INSTALLATION)

The Design-Construct engineering package would contain sufficient drawings, specifications, logic drawings and other materials to convey the requirements of the system and the physical dimensions to contractors for bidding and installation. The drawings may be stamped as required by the project or the regulatory agencies. This would require a Professional Engineer for the design.

2.11.1 Selection of Engineering Firm. If the choice is to proceed with Design-Construct, then the selection of the engineering firm is obviously the next important issue. If the Project Team is in place, they may choose to pre-qualify one or more firms for a presentation of experience and capabilities. These firms could be specialty companies who design only industrial ventilation systems or departments in large multi-disciplined firms. If a detailed Design Basis has been developed to hand over to engineering bidders, the selection process can move more easily because the definition of scope is usually clearer. State and national professional engineering societies also have guidelines for the selection of firms. They focus on experience and quality of work.

Any firm must be able to provide references on similar projects. A common problem is the use of HVAC companies for the design of exhaust systems. The requirements for these two system types are different even though both involve the movement of air. [IV is a specialized subset of HVAC design.] An HVAC engineer would normally be experienced in the design of building mechanical systems, supply duct systems, chillers and air handlers. They may not possess the required skills to design an industrial ventilation system that uses specialty collection devices, material handling fans, heavy gauge duct and involves special issues like minimum transport velocities and hood design. At the same time, an industrial ventilation firm may be unable to consider all of the special requirements of a complicated air conditioning installation.The fundamental difference between IV and HVAC is the manner in which air is distributed. To mitigate stratification, HVAC systems are designed to mix the room air with the incoming supply air. IV systems avoid mixing clean and contaminated air in the worker's breathing zone. In addition, the process may have environmental conditions that need to be maintained and/or there may be emissions from the process that affect the environment.

An important consideration is always cost. The Project Team must consider all aspects of cost analysis. An engineering firm with higher hourly rates may actually cost less if their total hours are less. Similarly an experienced firm may be able to provide a design that is less expensive or more efficient even if the engineering costs are higher.

The major consideration should be life cycle costs that include initial capital costs but also consider the operating costs over the life of the system (see Chapter 12). A low initial cost installation or design by an inexperienced engineer may burden the plant with high power and maintenance costs for 20 years or more.

Similar considerations must be made for the use of in-house engineering. Larger companies now use their in-house staff to manage specialty consultants and design firms rather than performing the engineering themselves. This keeps overhead lower by not keeping large staff between projects. Where in-house staff is used, special care must be taken to keep them educated on new techniques and collector, hood and fan technology to ensure the system is efficiently designed.

During the selection process for the consulting or in-house engineer, a realistic schedule must be communicated. This includes milestone dates (beginning and end of engineering, construction, commissioning, etc.). The regulatory issues such as permit application and approval must also be considered in the schedule.

In addition, all of the disciplines must be determined and division of responsibility made. For example, a design may include civil engineers for foundations and roof loads, electrical engineers for power and controls, mechanical engineers for the air-moving systems, chemical engineers for process safety reviews and structural engineers for duct and collector supports. The needs may extend to permit application specialists and construction managers. The plant may want to keep one or more of these disciplines in-house even though a specialist handles the main project engineering. Electrical engineering may already be handled by a firm familiar with knowledge of locations of all plant equipment and load capacities.

After the screening process for an engineering firm is complete, a decision must be made as to the method of payment. Very large projects may be paid on a fee basis where the payment is a percentage of the total construction cost. Projects that fall into the size range of most industrial ventilation systems usually would be performed on a fixed price or Time and Material (T&M) basis.

A fixed price proposal is usually the best method for the User because it is easier to manage budgets. To ask an engineering firm to bid this way, the Design Basis and schedule must be very explicit and there must be clear methods described for scope changes and their management.

Some firms may have a tendency to bid low while asking for change orders as they occur. Other firms may build some factor of safety into the price and seldom ask for scope changes unless they are significant. When checking references it is important to ascertain the scope management history of the firms bidding the project. As mentioned earlier, T&M rates can be very misleading. A company with lower rates can actually have higher total costs because hours are higher (cost = hours × rate).

It is especially difficult to choose engineering firms for blanket order arrangements strictly on rates. If forced to be most competitive on rates, less experienced personnel may be substituted or qualifications can be inflated to 'play the rate game.' It is always best to choose engineering firms on qualifications, efficiency and experience and not on initial costs. The premium paid for less experienced engineering is paid over the life of the project. Also, if a particular engineer or group has the knowledge and experience for a project or technology, they may need to be specified by name in the negotiation of the contract.

Important issues with regard to selection of installing contractors and the management of the construction are included in Chapter 1 of the O&M Manual.

2.12 DESIGN/BUILD (TURNKEY) METHOD – SINGLE SOURCE OF RESPONSIBILITY

This method can be reflected in many different types of partnerships and can include:

1) An engineering firm as the prime contractor partnering with an installer to provide a turnkey project;
2) The installer as prime contractor using their own in-house design staff;
3) The installer as prime contractor using an engineering firm or other resources for the design; or
4) Partnerships or joint ventures between design and installation firms to provide turnkey installations.

As with Design-Construct there are inherent advantages and disadvantages. When the owner defines the project, the Design Basis can be issued as a set of instructions to the design team. If a Design/Build approach is selected then the Design Basis can be issued directly to the turnkey bidders. When proposals are returned, the owner can be assured that everyone is bidding to the same general scope and project requirements.

However, each design/build contractor would be given flexibility in their presentation to take the contract based on their own preliminary design. This could include different methods of hooding, different air volumes, different collection devices, etc. This puts the premium on the experience and capability of the design/build contractor. Since the contractor would be absorbing the risk of the design and delivery, they would have to provide a system with enough surety of design to complete the project profitably but not be so safe in their choices that they make their price too high. Where specific control approaches (i.e., hood-type) or mechanical performance specifications (i.e., minimum capture velocities) are required by the owner, they should be clearly stated in the Design Basis to not leave these particular items to the discrepancy of the turnkey contractor's design team.

2.13 PROJECT TEAM AND SYSTEM EVALUATION

This range of acceptability may have to be evaluated by the Project Team for either method. It is no different than the selection of the correct engineering firm when pre-screening for a design-construct project but there are a few differences. The turnkey method does relieve the Project Team of potential burden of ruling on disputes between designer and installer if there is a system failure on a design-construct project. In these latter cases, the Team must determine if it was a design or installation flaw (or both) to assign back charges and move the project to completion.

In Design-Construct, the engineer would design to the standards in the Design Basis. In every case it costs very little more for an engineer to design with enough factor of safety to ensure they would always have a successful installation with no fear of errors or omissions. For example, the costs to design a system with 20,000 acfm [9.44 am^3/s] are marginally more than that required to design the same system with 10,000 acfm [4.7 am^3/s]. Moreover, the 20,000 acfm system would, in most cases, work with more margin of safety than the smaller one. Unfortunately, the owner pays for this safety factor for the remaining life of the project. They pay in higher installation costs and higher operating costs.

A Design/Build proposal forces the bidder to consider their own risk for performance and factor of safety. Since the bidding would be competitive they must build their expertise and experience into their price. However, the owner must now make his purchase decision based on the review of many proposals that may have varied design parameters. One design build firm may propose 10,000 acfm [4.7 am^3/s] and the other may propose 20,000 acfm [9.44 am^3/s] for the same process. The Project Team must now decide which is correct (and possibly ignore the price implications). They must also decide if the company giving the lower price and smaller system can provide the guarantee if there is non-performance. Life cycle cost analysis by the User will help determine which proposal is viable in the long term. See Chapter 12, Section 12.4, Cost Comparison Methods.

2.14 PROJECT RISK AND NON-PERFORMANCE

The implications of non-performance go beyond the obvious. For instance, a system that cannot meet guarantees of emission levels may delay the start of a major process installation at the plant with an impact far in excess of the ventilation system's cost. Similarly, the system may actually work and perform to standards set but may require inordinate amounts of maintenance and other resources to keep running. These factors may not have been included in any performance guarantee.

Using the Design-Construct method (separate design and installing firms) opens possibilities for conflicts as the installation progresses. Drawings furnished by the engineer may be inaccurate or incomplete and this may provide opportunity for the installer to recover extra costs associated with these errors or omissions. In the United States Supreme Court case *United States v. Spearin*,[2.1] it was determined that the owner warranted that the engineering package (drawings and specifications) was accurate and sufficient enough to build the project. (Note that this Manual is not intended as a law reference and that court rulings can be altered at any time by review and appeal.)

From that case, the Court's decision produced what is now called the Spearin Doctrine.[2.1] Under that doctrine, the contractor can recover from the party who supplies the plans and specifications (usually the owner) the costs for delays and added costs due to errors or omissions. In response to this doctrine, many owners include provisions in the bidding and contract documents to lessen its effects. These may include requiring the contractor to assume responsibilities for final checks of the drawings or some other methods. As can be expected this can become a complicated issue on some large or risky projects that may have the potential for cost and design disputes. Because the dispute may eventually happen between consultant and owner, it is important in the selection process to have agreement on all responsibility issues before design begins.

The Project Team must be informed in order to make clear and concise decisions and may need legal input on complicated installations. The idea that using design/build methods relieves the owner of any responsibility is short-sighted. The contractor may give a guarantee but the financial strength and quality of the guarantor must be determined. At the same time, precise expectations must be communicated to the contractor. This may include the requirements to meet emission levels, levels of dust in the plant, bag life for fabric filters, maximum pressure drops, etc. Certain financial and time aspects may be tied to the performance as long as these are clearly stated during the bidding process so the contractors can include these risks in their bids.

It is just as important to make these requirements realistic and enforceable. Requiring extremely low dust levels in the plant may not be possible because ambient levels from nearby areas may already be higher than the guarantee request. At the same time issues such as housekeeping, material handling methods or some other factors may be completely outside of the control and scope of the ventilation system. Even though the owner may get a guarantee from a consultant or installer, the reality may be that this guarantee can never be enforced.

2.14.1 Communication of Risk. Any time a contract is entered into between the owner and outside suppliers, risks in the delivery of that contract will exist. The Project Team must make an assessment of these risks and determine how much of the risk should be shared by the other parties. Systems that have a history of simple and predictable operation may not have much to consider for risk costs or contingencies. All risks must be considered in systems that are attached to new processes or involve new technologies. It is always best to communicate these risks to all parties before contracts are signed so that a plan is in place in case systems do not meet the requirements.

These communications include, but are not limited to:

1) What are the expectations of the system at start-up (emissions, in-plant dust levels, bag life, pressure drop, etc.)?
2) What are the expectations of the system during normal operation (are contingencies necessary for an accidental spill, fire or explosion, i.e., a purge, a full shutoff of one or both of the supply and exhaust systems)?
3) Does "risk free" imply excess costs to ensure compliance?
4) What outside influences can affect the guarantee of the system?
5) How should risk factors be conveyed to vendors (engineers and product suppliers)?
6) Who absorbs risks: engineer? equipment supplier? contractor?
7) Should design follow or comply with published guidelines or some other recommendations?
8) What system is in place to mediate the conflicts between parties if there is a failure to meet guarantee?

Any system provided by the three methods discussed in Section 2.10 must have the same goals. They must meet all of the regulatory, process and safety requirements of the project at the lowest life-cycle cost. These lowest costs are never really known even after the project is installed and running to specification; nevertheless, as the project proceeds and develops, the Project Team would be required to use its experience, training, outside resources and judgment to make the best decisions to meet the goals.

2.14.2 Communicating Proof of Performance. Proof of Performance is the defining requirement for any installation. This guarantee could be limited to meeting the intent of the Design Basis (i.e., provide an air volume of "x" acfm for a particular process with a minimum transport velocity of "y" fpm in all duct branches), meeting other guidelines such as ACGIH® recommendations for hood design, or even meeting all applicable codes and regulations such as in-plant dust levels or emissions. It should be noted that hood design recommendations provided by ACGIH®, ASHRAE or similar resources may be stated in a range (i.e., control velocity of 150 to 250 fpm or something similar. This would apply whether in IP or SI units.). If a proof of performance is based on these references, then it would be necessary to focus on values within those published ranges.

Though it is easiest to demand a proof of performance based on regulatory levels such as Permissible Exposure Limits (PELs) or emission limits, it is important to determine the factors that can be controlled by the system and the project including background ambient levels in the plant before operation of the system. One method may be to take area exposure levels in the plant at key locations near the new system. Readings could be taken before the system operates and after the system is commissioned. If background exposure levels before system operation (from other sources outside the new system) already exceed the guarantee, there could be a case that the new system could never meet these requirements.

See Chapter 2 of the O&M Manual for details of the system design, installation and project teams when proceeding with the Commissioning process and verifying Proof of Performance.

2.15 USING PLANT PERSONNEL AS PROJECT RESOURCES

Ultimately the system is received and used by plant personnel. After final acceptance, there is usually a person designated as the receiver of the completed project. This may be the Plant Manager, Operations Manager or Maintenance Manager. In addition, someone would be designated as responsible for the ongoing operation and maintenance of the system.

There have been many documented cases where successful installations meeting all startup guarantees are altered, removed or even sabotaged after the contractor leaves the site. This may happen because they do not represent workable solutions to meet the production, access or maintenance requirements of the people ultimately required to use them. Whether it is cardboard mounted over hood openings, replacement air systems diverted, hoods removed or fans turned off, the results are the same. A system purchased and installed with the best of intentions and at significant cost is left idle or debilitated and does not meet its intended goals.

When the Project Team includes operations and maintenance personnel, project goals are easier to manage into the commissioning phase because there is "buy-in" from the end user with input in early decisions. Even for small projects, the experience of the operator would help ensure that the system would be used and maintained. Ability to reach around hoods, removal and replacement of guards, safety interlocks and other issues can be most effectively addressed by including the input of those actually using the equipment. This information should be gathered using a questionnaire format or at least through interviews with written comments.

Similarly, maintenance implications can also make or break an installation. This not only includes maintenance access to the process being ventilated but also extends to the ventilation system hardware. Collectors should be selected for easy access bag removal and replacement. Fans should be properly fitted with machine guards. Motors and controls should be specified to match existing capabilities and training.

Access to duct and equipment should include work platforms and proper ladders or stairs to get materials and equipment to high maintenance areas. Many times a large system may involve more maintenance or operating personnel to handle the new issues of collector operations and dust removal. Planning for these needs while the project is still in the development and installation phase can save training costs and avoid possible safety issues. Changes to plant operations may include new requirements for fall protection or other safety requirements.

2.16 INTERFACE BETWEEN THE PLANT AND PROJECT

The plant must be prepared for major new construction. This includes the extra contractor traffic on site, security for plant entry, enforcement of plant and fire safety regulations and inclusion of other rules for use of plant facilities such as receiving docks, restrooms and cafeterias. Construction requirements must be coordinated with production, shipping and other plant needs to ensure that there is a minimum of interference among these efforts.

Normally the Project Manager may need to be involved in securing building permits or other permits for construction and operation. Some permits may require long lead times and must be built into the project schedules right at the beginning.

The installation of the system would also impact the plant in other ways. The auxiliary equipment requirements for the system itself is the most obvious way. Unless new facilities are added to handle the needs of the system, the plant's electrical power, compressed air, water, sewage or other systems may be extended to their limits or require intermittent shut-down periods during certain phases of the installation process. This should be considered in the ventilation design and suitable plans should be included for expansion of these systems.

Similarly, any new exhaust system should include consideration for replacement air. If the plant is in balance before the project, it may just require a supply volume equal to the exhaust. At the same time, the Project Manager may want to cure some previous under-design of replacement air by adding more supply to newer projects. In any case, the placement of the supply air may have effects on adjacent areas not normally considered in the project.

Project planning and design definition may need to consider a rebalance of the supply and exhaust air in the whole plant. Details for the installation of these systems are included in Chapter 10 (Supply Air Systems).

2.17 IMPACT OF NEW SYSTEMS ON PLANT OPERATION

While the positive impacts of the local exhaust systems on the plant environment can be easy to identify, there may be other unpredicted influences on the operation of the new system. Because the process itself may now be more enclosed to provide capture and control, there may be heat build-up. This may translate to higher duct and system temperatures. It also may cause formations of different chemicals in the exhaust gas streams or change the dew point or acid dew point.

At the same time, the local exhaust system may now include long runs of hot duct through the plant and there can be condensation issues that had not been accounted for in design. For systems involving heat and moisture in the gas stream, it is important that the Project Team consider these other effects on the plant environment as well as the plant's effect on the local exhaust system.

Frequently, other issues may arise when a different local exhaust system is installed. The local exhaust system must meet its stated goals but also may cause other issues to be addressed as it is being installed.

Final success criteria for the Preliminary Design Phase is the definition of a project that meets all of the regulatory, safety and operations needs of the plant. It is then feasible to move to Detailed Design Phase.

REFERENCE

2.1 *United States v. Spearin*, 248 U.S. 132 (1918).

Chapter 3

PRINCIPLES OF VENTILATION

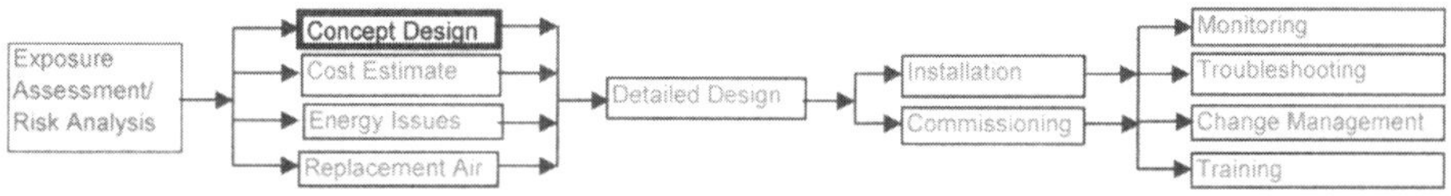

NOTE: Equations with notation followed by (IP) are designated for inch-pound system only; Equations followed by (SI) are designated for metric use only. If Equation bears neither, then it applies to both systems.

3.1 INTRODUCTION

The importance of clean uncontaminated air in the industrial work environment is well known. Modern industry with its complexity of operations and processes uses an increasing number of chemical compounds and substances, many of which are highly toxic. The use of such materials may result in particulates, gases, vapors, and/or mists in the workroom air in concentrations that exceed safe levels. Heat stress can also result in unsafe or uncomfortable work environments. Effectively designed ventilation offers a solution to these problems. Ventilation can also serve to control odor, moisture, and other undesirable environmental conditions.

The application of ventilation to solve the problems of worker exposure and general plant hygiene involves a process of technical solutions to problems of air and particulate movement.[3.1] In general, these are problems of fluid (mass) movement as well as calculations of energy transfer. Because the formulae for these problems can get involved with advanced mathematics and physics, simpler methods have been developed to reduce the solution of complex equations by using more manageable forms. These methods are based on the same basic laws that are of prominent use in the physical sciences: Conservation of Mass and Conservation of Energy. To solve the problems of industrial ventilation (how much air to apply to a hood designed to protect a worker, how much horsepower is required for a fan, how large a duct should be to convey a design airflow, etc.), these basic principles apply.

Chapters 5 through 9 provide calculation methods for design of components for an Industrial Ventilation System. This Chapter will provide most of the basics for the development of those methods as well as examples of how the Laws of Physics are derived for easier use in this text. It must be restated that the methods use simpler algebraic solutions and generate a level of accuracy acceptable for general use in the industry. It is understood that more accurate requirements for the laboratory and research may also require more rigorous mathematic solutions.

In addition, the design of the industrial ventilation system is based on steady state/steady flow conditions (where conditions at any point in the system do not change with time). That condition rarely exists for any period of time but this restriction will be the basis of how these formulae and equations are written for the Manual.

Table 3-1 and inside the back cover provide the very basic mathematic definitions and relationships required for the remaining chapters as well as metric equivalents. A familiarity with basic scientific notation and definitions is a minimum requirement for the solution of industrial ventilation problems. Before investigating the design chapters in this Manual, please be familiarized with these basic definitions. They will be referenced in the following sections and chapters.

TABLE 3-1. Common Physical Constants

QUANTITY	SYMBOL	PHYSICAL DIMENSIONS	MAGNITUDE (IP UNITS)	MAGNITUDE (SI UNITS)
SPECIFIC HEAT OF AIR	C_P	$\frac{\text{ENERGY}}{m-T}$	$0.24\ \frac{BTU}{lb_m\ F}$	$1.005\ \frac{kJ}{kg-k}$
SPECIFIC WEIGHT OF WATER	∂		$62.4\ \frac{lb_f}{ft^3}$	$1000\ \frac{kg}{m^3}$
STANDARD AIR	ρ_{std}	$\frac{m}{L^3}$	$0.075\ \frac{lb_m}{ft^3}$	$1.204\ \frac{kg}{m^3}$
	T_{std}	T	70 F, 530 R	21.2 C, 294 K
	P_{std}	$\frac{F}{L^2}$	14.7 psi, 407 in wg	101, 384 kPa
	ω	$\frac{m-\text{water}}{m-\text{dry air}}$	0	0
SP IN EXIT PLANE OF DUCT	SP_{exit}	$\frac{F}{L^2}$	0 in. wg	0
EARTH'S GRAVITATIONAL FIELD	g	$\frac{L}{t^2}$	$32.2\ \frac{ft}{s^2}$	$9.8\ m/s^2$

3.2 CONSERVATION OF MASS

Both conservation principles will be written for a fixed volume section (referred to as a "duct segment" in the calculation sheet, but here termed a "control volume" (c.v.) that must be explicitly defined, usually by a drawing. (Various terms may or may not be applicable depending upon the c.v. being analyzed.) The general physical law for Conservation of Mass states that the rate of mass flow into a c.v. (by all flow streams) equals the rate at which mass leaves the c.v. (by all flow streams). Symbolically, for steady flow, this can be written as:

$$\sum \dot{m}_{in} = \sum \dot{m}_{out} \quad [3.1]$$

where $\dot{m}$ is defined as the mass flow rate (pounds per minute) [kg/s].

Note: This principle is "general" in the sense that it contains no physical constants and hence is equally valid for (and applicable to) all fluids (air, water vapor, gas, etc).

EXAMPLE PROBLEM 1 (Conservation of Mass)

In Figure 3-1, two airstreams are combined through a junction or fitting and a single flow exits:

$$\dot{m}_1 + \dot{m}_2 = \dot{m}_3$$

Using the definition for mass (see inside back cover):

$$\rho_1 V_1 A_1 + \rho_2 V_2 A_2 = \rho_3 V_3 A_3$$

where:

ρ = density (pounds per cubic foot) [kg/m³]
V = velocity (feet per minute) [m/s]
A = area (square feet) [m²]

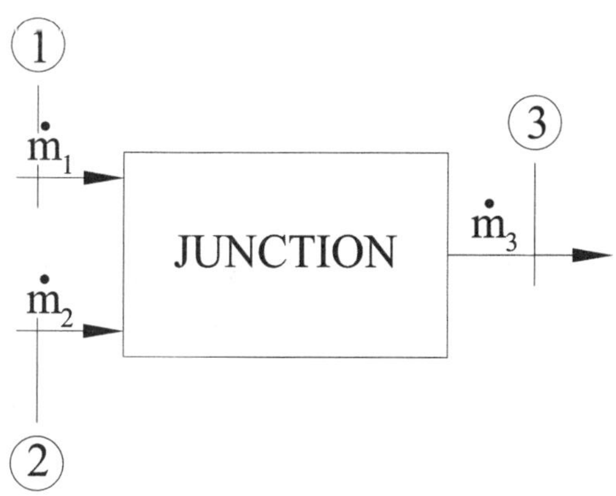

FIGURE 3-1. Conservation of Mass in a duct junction

For standard air (see Section 3.9 for definition) then

$$\rho_1 = \rho_2 = \rho_3 = \rho_{std} \text{ and so}$$

$$V_1A_1 + V_2A_2 = V_3A_3$$

Since "Q" (Volumetric Flow Rate or "Volume" is defined as V*A then:

$$Q_1 + Q_2 = Q_3$$

The Conservation of Mass as applied across a heater is shown in Figure 3-1a. In this case, there is a change in density as the air is heated but the mass rate of flow of air going into the heater and out of the heater are identical (conserved) whereas volumetric flow rates (Q) will change. Thus in this case:

$$\dot{m}_1 = \dot{m}_2$$

If the air going into the heater is assumed to be standard and then heated to a new condition with a density of ρ_2 then the equation can be stated as:

$$\rho_{std} V_1 A_1 = \rho_2 V_2 A_2 \quad [3.2a]$$

Applying the definition of Q (= VA) [3.2b]

Then:

$$Q_2 = \frac{\rho_{std}}{\rho_2} Q_1 \quad \text{or} \quad Q_2 = \frac{Q_1}{df_2} \quad [3.3]$$

where $$df = \frac{\rho_{act}}{\rho_{std}} \quad [3.4]$$

Note that this shows the relationship between standard and actual air conditions when the density is known. However, it does not consider the mass of moisture when the air stream contains water vapor.

Figure 3-2 depicts a system with standard air ($\dot{m}_{air}$) and moisture ($\dot{m}_{H_2O}$) entering and a mixture of the two leaving. The mixture would be a value in ACFM considering the density and actual flow change of the two streams. And so:

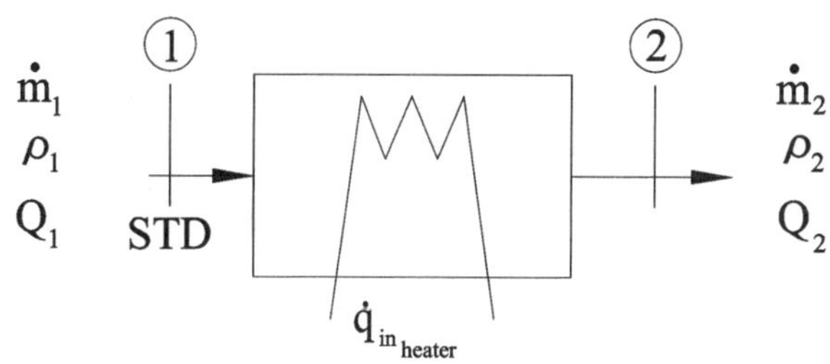

FIGURE 3-1a. Conservation of Mass in a heater

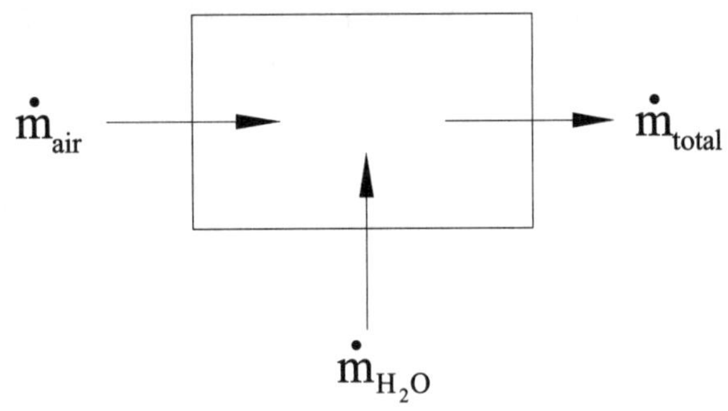

FIGURE 3-2. Conservation of Mass

$$(\dot{m}_{air}) + (\dot{m}_{H_2O}) = (\dot{m}_{total})$$

$$(\dot{m}_{total}) = (\dot{m}_{air})\left(1+\left[\frac{m_{H_2O}}{m_{air}}\right]\right)$$

Where $\frac{m_{H_2O}}{m_{air}}$ is defined as "ω" (pounds of water per pounds of dry air) [kg H_2O/kg – dry air].

And so:

$$(\dot{m}_{total}) = (\dot{m}_{air})(1+\omega)$$

The exiting stream will have a new volume (Q_{act}) and a new density (ρ_{act}).

$$(\dot{m}_{total}) = (Q_{act})(\rho_{act}) = (Q_{std})(\rho_{std})(1+\omega)$$

$$Q_{act} = Q_{std}\frac{1}{\left(\frac{\rho_{act}}{\rho_{std}}\right)}(1+\omega)$$

$$Q_{act} = Q_{std}\left(\frac{1}{df_{act}}\right)(1+\omega) \qquad [3.5]$$

This is the equation relationship between standard and non-standard air and is shown as Equation 3 on the calculation sheet.

3.3 CONSERVATION OF ENERGY

Conservation of Energy in a ventilation system is the basis for the equations and formulae to calculate losses in duct sections. It is also used to determine the work required by the fan to move the air in a system. These principles are governed by the First Law of Thermodynamics. They have been "simplified" for application to ventilation problems but the principles still guide the overall procedure involved in design and the calculation sheet. In contrast to applications of the Law of Conservation of Mass, energy cannot only be transferred into or out of the c.v. by both air streams but also by non-flow means: by thermal input (heat source or heat exchanger, for example) ($\dot{q}_{in}$) or by mechanical input (work provided by a fan) (w_{in}). Refer to Table 3-1 and inside the back cover for basic definitions and uses of all symbols. In basic mathematical terms the equation for Conservation of Energy can be written:

$$\sum(\dot{m}e)_{in} + \dot{q}_{in} + w_{in} = \sum(\dot{m}e)_{out} \qquad [3.6]$$

where $$e \equiv \frac{SP}{\rho} + \frac{V^2}{2g_c} + u$$

This definition of "e" basically includes the energy being conveyed by air streams into or out of a control volume (c.v.) and is divided into the following three components:

Potential Energy Component = $\frac{SP}{\rho}$

Kinetic Energy Component = $\frac{V^2}{2g_c}$

Internal Energy Component = u

EXAMPLE PROBLEM 2 (Conservation of Energy)

Assume that no heat is added or removed from a system and no work is performed on the system by the fan ($q_{in} = w_{in} = 0$ in Equation 3.6) and density (ρ) is constant. And so:

$$\sum(\dot{m}e)_{in} = \sum(\dot{m}e)_{out}$$

$$\dot{m}_1\left(\frac{SP_1}{\rho}+\frac{V_1^2}{2g_c}+u_1\right) = \dot{m}_2\left(\frac{SP_2}{\rho}+\frac{V_2^2}{2g_c}+u_2\right)$$

$\dot{m} = \rho Q$ **and** $Q_1 = Q_2$

Substituting in the equation and using the definitions on the inside back cover of the Manual, the new relationship for the duct system can be defined as:

$$SP_1 + VP_1 = SP_2 + VP_2 + \rho(u_2 - u_1) = \qquad [3.7]$$
$$SP_2 + VP_2 + \sum losses_{1-2}$$

3.4 SYSTEM PRESSURES (STATIC, VELOCITY, TOTAL)

There are three different but mathematically related pressures associated with a moving air stream. The measurement of flow and pressure as well as the ability to predict flow and pressure conditions through calculations is the basis for design of industrial ventilation systems. The conditions and orientation of pressure in a duct is shown in Figure 3-3.

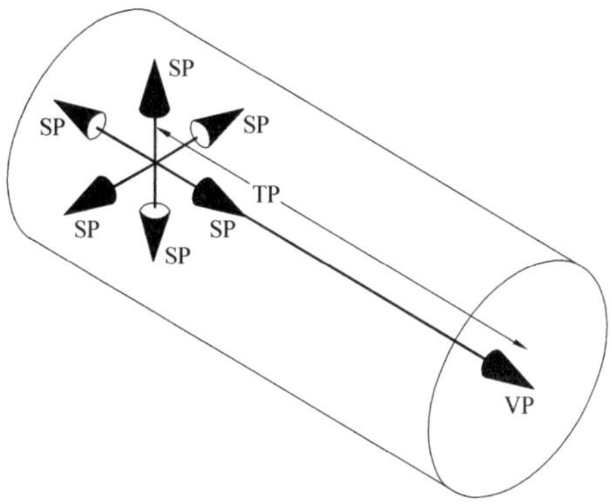

FIGURE 3-3. SP, VP, and TP at a point

Static Pressure (SP) is defined as the pressure in the duct that tends to burst or collapse the duct. In industrial ventilation system (IVS) design it is usually measured with a water manometer and units are "wg (inches water gauge or inches of water) in IP system or Pascals [Pa] in SI system. SP can be positive or negative with respect to the local atmospheric pressure, but must be measured perpendicular to the airflow. The holes in the side of a Pitot tube (see Appendix C) or a small hole carefully drilled to avoid internal burrs that disturb the airflow (never punched) into the side of a duct will yield SP. SP does not vary laterally across a duct but does decrease in the direction of flow in a duct with constant diameter.

Velocity Pressure (VP) is the representation of kinetic energy of an air stream and is defined as:

$$VP = \frac{\rho V^2}{2g_c} \quad [3.8]$$

When the units of measurement are changed to fit IVS standards and density factor is considered, the formula can be rewritten in the following form:

$$VP = \left(\frac{V}{4005}\right)^2 df \quad [3.8a]\ IP$$

$$[VP = \left(\frac{V}{1.29}\right)^2 df] \quad [3.8a]\ SI$$

When solving for velocity (when VP is known) the formula can be algebraically rearranged to:

$$Velocity = (4005)\sqrt{\frac{VP}{df}} \quad [3.8b]\ IP$$

$$[V = 1.29\sqrt{\frac{VP}{df}}] \quad [3.8b]\ SI$$

VP cannot be directly measured in a duct system but is determined by subtracting the measured SP from the measured TP (both of which are obtainable using the proper field equipment). This subtraction can be obtained mathematically by using Equation 3.9 or by proper connection of the measuring device to a manometer (see Figure 3-5 and discussion). Like SP, the VP is calculated in ("wg) [Pa] for purposes of this text.

Total Pressure (TP) is defined as the algebraic sum of the static and velocity pressures or:

$$TP = SP + VP \quad [3.9]$$

It can be measured in a duct or hood system by placing the probe directly in the path of flow. This will capture both the VP and SP components since static pressure is equal in all directions. Air or any other fluid will always flow from a region of higher TP to a region of lower TP in the absence of work addition (a fan). TP can be positive or negative with respect to atmospheric pressure and is a measure of the energy content of the air stream, always dropping as the flow proceeds downstream through a duct. The only place it will increase in magnitude in an open duct system is across the fan due to the external energy input.

TP can be measured with an impact tube pointing directly upstream and connected to a manometer. It will vary across a duct due to the change of velocity across its cross section and therefore single readings of TP will not be representative of the energy content. Appendix C illustrates procedures for measurement of SP, VP and TP in a duct system.

NOTE: It then can also be stated that the change in Total Pressure (SP + VP) from point "1" to point "2" in the above Example Problem 2 is the sum of the losses encountered between those two points. In Figure 3-8, these are the losses encountered because of the friction in the duct plus the elbow. This is a key basis for the determination of losses in a system and will be used for system calculations in Chapter 9.

The significance of these pressures can be illustrated as follows in a non-flow situation. Assume a duct segment with both ends being sealed and then pressurized to an SP of 0.1 pounds per square inch (psi) [+690 Pa] above the atmospheric pressure as shown in Figure 3-4 (+0.1 psi). If a small hole were drilled into the duct wall and connected to one side of a U-tube manometer, the reading would be approximately +2.77 "wg (= +0.1 psi) [+690 Pa]. Note the way the manometer to the left is deflected. If the water in the side of the manometer exposed to the atmosphere is higher than the water level in the side connected to the duct, then the pressure read by the gauge is positive (greater than atmospheric). Because there is no velocity, the velocity pressure value is 0.0 "wg [0 Pa] and SP = TP (since TP = SP + VP).

A probe that faces the flow is called an impact tube and will measure TP. In this example, a manometer connected to an impact tube (the one on the right) will also read 2.77 "wg [690 Pa]. Finally, if one side of a manometer were connected to the

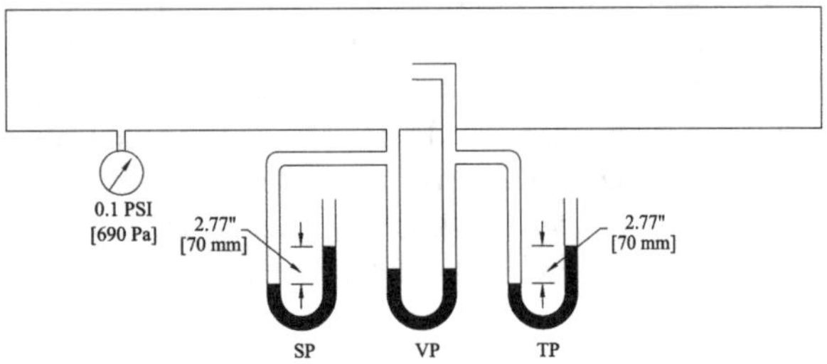

FIGURE 3-4. Measurement of SP, VP, and TP in a pressurized duct

impact tube and the other side were connected to the static pressure opening (the center one), the manometer would read the difference between the two pressures. Since VP = TP + SP, a manometer so connected would read VP directly. In this example, there is no flow and hence VP = 0.0 [0 Pa] as indicated by the lack of manometer deflection.

If the duct ends were removed and a fan placed midway in the duct, the situation would change to the one shown in Figure 3-5. Upstream of the fan, SP and TP are negative (less than atmospheric). This is called the suction side. Downstream of the fan, both SP and TP are positive. This is called the pressure side. Regardless of which side of the fan is considered, VP is always positive (because it represents the kinetic energy of a moving air stream and that cannot be negative). Note that the direction in which the manometers are deflected shows whether SP and TP are positive or negative with respect to the local atmospheric pressure.

3.5 SYSTEM LOSS COEFFICIENTS

To calculate the effects of different conditions in an industrial ventilation system, Equation 3.7 must be made manageable so that the required work for the fan and system resistance (losses) can be calculated. These system losses (changes in SP or ΔSP) are mostly due to friction encountered in the system and can be divided into the following categories:

- duct wall friction
- friction due to hood configuration
- elbows (turning of the air in the duct system)
- branch or 'wye' fittings (turning of the air in the combining streams)
- contractions (air is squeezed through a smaller duct or opening)

In addition, there are other losses in the system such as those encountered going through a filter bag or scrubber but those are determined by other methods (and are usually supplied by the manufacturer). In the above five bullet points the change to SP (ΔSP) is related to the attendant VP (Kinetic Energy) through dimensionless Coefficients (specified as "F") i.e., ΔSP = (F)(VP). This loss is also defined as "h" (with a subscript) as in h_{el} is the loss through an elbow. The coefficients for the different system components (elbows, hoods, etc.) are available in existing charts and tables (in Chapter 9) and are based on empirical data and usage.

To illustrate the use of coefficients, the following examples consider conditions encountered.

EXAMPLE PROBLEM 3 (Hood with No Resistance)

This condition cannot exist because losses are incurred with even the most efficient hood design. However, if such a hood could be constructed (as shown in Figure 3-6), Equation 3.7 would be stated as:

$$SP_o + VP_o = SP_h + VP_h$$

In this case, VP_h is also the VP_d, where subscript h refers to hood and subscript d refers to duct. Therefore, the equation can be rewritten:

$$SP_o + VP_o = SP_h + VP_d$$

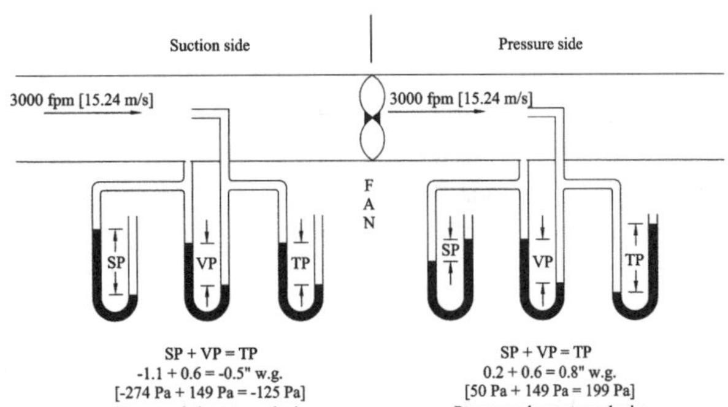

FIGURE 3-5. SP, VP, and TP at points in a ventilation system

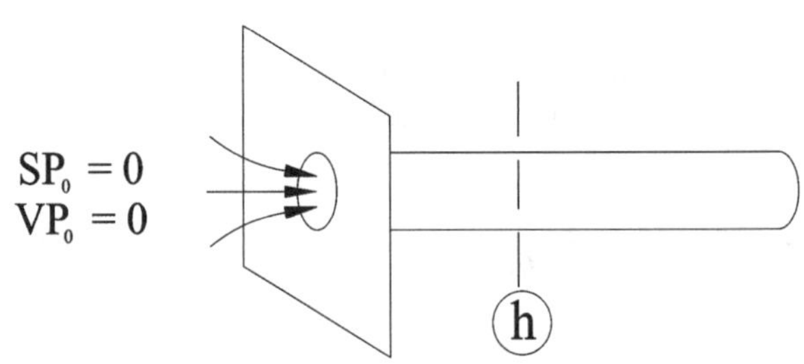

FIGURE 3-6. Simple exhaust hood

Simply stated, at some distance from the face of any hood both SP and VP are zero (no air movement and pressure is at ambient conditions). However, after entry into a no-loss hood, the SP and VP would equal each other as indicated by the new conditions at Point 1.

Equation 3.7 could then be stated:

$$0 + 0 = SP_h + VP_d$$

or:

$$SP_h = -VP_d$$

Another way to describe this loss using the factor designation of F_a is:

$$SP_h = F_a VP_d$$

Where $F_a = 1$ and is called the 'acceleration' or 'Bernoulli' coefficient.

Note that this is not a loss due to acceleration but merely trading energy equally between two forms (potential and kinetic).

EXAMPLE PROBLEM 4 (Hood Loss)

When Figure 3-6 is applied to any other hood (with losses) then the conditions through Equation 3.7 can be stated as

$$SP_o + VP_o = SP_h + VP_d + (F_h \times VP_d)$$

$$SP_h = -(F_a + F_h)\, VP_d \qquad \textbf{[3.10]}$$

Since $SP_o = VP_o = 0$ and $F_a = 1$, then:

$$SP_h = -(1 + F_h)\, VP_d \qquad \textbf{[3.10a]}$$

or

$$SP_h = -(VP_d + h_h) \qquad \textbf{[3.11]}$$

where: $h_h = F_h(VP_d)$ **[3.12]**

And h_h is called the hood entry loss. The hood SP (SP_h) would be the sum of the hood loss (h_h) and the energy transfer as air moves from stillness outside the hood to the energy as it travels at the velocity in the duct (F_aVP_d = $1VP_d$). The coefficients are combined normally as shown in Equation 3.10a.

Coefficients for the application of hood design (F_h) are included in Chapters 6 and 13. Equation 3.10 is the basis for determination of the Hood Static Pressure of a *simple* hood. (There may be additional losses in a *compound* hood where slots are used for air distribution) (see Chapter 6). In summary, the SP downstream of the hood is negative (less than atmospheric) due to two effects:

1) Energy exchange in the air stream from potential to kinetic ($F_aVP_d = 1VP_d$)*
2) Hood entry losses ($F_h \times VP_d$); this is a function of the shape of the hood and its inherent inefficiencies.

**NOTE: This is the same as the loss calculated in Example Problem 3 and is not "acceleration." This is an energy transfer.*

An alternate method of describing hood entry losses is by the "Hood Flow Coefficient" (C_e). This was previously called "Hood Loss Coefficient." It is defined as the square root of the ratio of duct velocity pressure to hood static suction, or

$$C_e = \sqrt{\frac{VP}{|SP_h|}} \qquad \textbf{[3.13]}$$

If there were no hood losses (Example Problem 3), then SP = VP and C_e = 1.00. However, as hoods always have some inefficiency, C_e is always less than 1.00. An important feature of C_e is that it is a constant for any given hood. It can, therefore, be used to determine the flow rate if the hood static suction is known.

$$Q \equiv VA$$

$$V = 4005\sqrt{\frac{VP}{df}} \qquad \text{(IP Units)}$$

Therefore: $Q = (4005\sqrt{\frac{VP_d}{df}})(A)$

From Equation 3.10: $\sqrt{VP_d} = C_e(\sqrt{SP_h})$

Therefore: $$Q = 4005(C_e)(\sqrt{\frac{SP_h}{df}})(A) \qquad \textbf{[3.14] IP}$$

$$[Q = 1.29(C_e)(\sqrt{\frac{SP_h}{df}})(A)] \qquad \textbf{[3.14] SI}$$

Note that for Standard Air (see Section 3.9) df = 1. If the value for C_e is known then a quick calculation can be made to determine flow into the hood without the use of a complete duct traverse (see Appendix C). By knowing C_e and a measurement of SP_h, the flow rate of a hood can be quickly estimated and corrective action can be taken if the calculated flow rate does not agree with the design flow rate.

This can be a useful tool for troubleshooting systems that may have lost airflow. C_e is an elusive value when designing and measuring a system. Values in this text are estimates for standard hood designs. Field conditions may alter designs and the actual value for C_e would be measured at start-up. The start-up value would be used for comparison rather than using the estimate from the design calculations.

EXAMPLE PROBLEM 5 (Straight Duct Losses)

The losses along a length of straight duct are somewhat more complicated. Unlike hoods where the only contributing factors are the shape and the energy transfer of air as it moves into the hood, the losses in a straight duct depend on (are a function of):

- Velocity of the air moving in the duct (V)
- Density of the air (ρ)
- Length of the duct (L)
- Diameter of the duct (D)
- Roughness of the walls of the duct (ε)
- Viscosity of the air (μ)

Mathematically this would be described as:

$$\Delta SP = \text{Func}(V, \rho, L, D, \varepsilon, \mu)$$

When all of the like coefficients are combined in the respective equations, the terms can be combined into the following dimensionless quantities (this is done to reduce the number of variables involved to a manageable size):

$$\frac{\Delta SP}{VP} = \text{Func}(Re, \varepsilon, \frac{L}{D})$$

where 'Re' is the Reynolds Number (see inside back cover) and is a measure of turbulence in the duct. Note that here and for all other cases, the ΔSP and VP will always be related by a loss coefficient ("F") and in this case the loss coefficient for straight duct (f_d) is a function of two items – Reynolds Number (Re) and the roughness of the duct (ε). In general, for a straight duct the equation is stated:

$$\Delta SP = f_d(\frac{L}{D})(VP_d) \quad [3.15]$$

Since duct lengths are in feet [meters] and duct diameters are measured in inches [mm], the equation can be refined further:

$$\Delta SP = h_d = \frac{f_d}{D(in)} \frac{12\ in}{1\ ft}(L(ft)(VP_d)) = \frac{12 f_d}{D}(L)(VP_d)$$

If we define $F'_d = \frac{12 f_d}{D} \equiv \frac{F_d}{L}$ where F'_d is a loss coefficient per unit length (feet) (IP only)

Then the duct pressure loss (h_d) is:

$$h_d = F'_d(L)(VP_d) = F_d(VP_d) \quad [3.16]$$

Where F'_d is determined by the empirical relationship:

$$F'_d = \frac{aV^b}{Q^c}(\frac{1}{ft}) \quad [3.17]\ IP$$

$$[F'_d = \frac{aV^b}{Q^c}(\frac{1}{m})] \quad [3.17]\ SI$$

The original values for friction or loss coefficients (sometimes also called factors) (as a function of Re and roughness) were provided on the Moody Diagram.(3.2) These were values used for the Equivalent Foot Method of system loss calculations and used through the 18th Edition of this Manual. After the change to the Velocity Pressure Method, work by Loeffler(3.3) provided values to calculate F'_d. These are included in table and graph form in Chapter 9, Table 9-4 and Figures 9-b (IP) and 9-c (IP).

The values shown in Table 3-2 are substituted in Equation 3.15 to calculate F'_d. Beginning with the 25th Edition of this Manual, the values were further changed to combine all metal duct as one value. This was substituted because most metal duct is coated with contaminants soon after operation begins and differences are relatively small when comparing. The values used in Chapter 9 for system calculations are the more conservative values shown in Table 3-2 for "Other Sheet Metal Duct," i.e., a = 0.0307 [0.0155], b = 0.533, c = 0.612. Values for flexible duct should still be used when encountered. Note that SI values are in [ ] in Table 3-2.

The losses for all of the other components of a duct system (elbows, branch entries) can be stated in a similar manner:

Loss through an elbow = $\Delta SP_{el} = h_{el} = F_{el}(VP_d)$

Loss through a branch (at entry only) =
$\Delta SP_{en} = h_{en} = F_{en}(VP_d)$

Where h_{el} and h_{en} are the elbow and entry pressure losses, respectively.

For a contraction (decrease in diameter of a straight duct), the loss is dependent on the degree of abruptness of the contraction as well as the energy required for the increase in velocity through the fitting. When duct expands, there is a regain

TABLE 3-2. Correlation Equation Constants (F'_d)

Duct Material	a	b	c
Aluminum, black iron, stainless steel	0.0425 [0.0162]	0.465	0.602
Other sheet metal and plastic duct	0.0307 [0.0155]	0.533	0.612
Flexible duct, fabric wires covered	0.0311 [0.0186]	0.604	0.639

coefficient that applies (see Chapter 5).

These loss coefficients have been determined either under laboratory or field conditions and are presented in Chapters 5 and 9. Coefficients for various hoods designed to meet special conditions or processes are shown in Chapters 6 and 13. All of these are based on the same principle of defining a loss coefficient (F) that relates the loss (ΔSP or "h") and the Velocity Pressure (VP).

3.6 THE FAN IN THE SYSTEM

The equations for Conservation of Mass and Energy also apply to the work provided by the fan. Equation 3.6 can be rewritten for conditions in and out of the fan as follows (assuming no heat loss or gain in the fan) (Figure 3-7):

$$\dot{m}_1\left(\frac{SP_1}{\rho}+\frac{V_1^2}{2g_c}+u_1\right)+\dot{W}_{in(fan)}=\dot{m}_2\left(\frac{SP_2}{\rho}+\frac{V_2^2}{2g_c}+u_2\right)$$

To calculate the work and losses in the fan (combining the values for VP and those shown on inside back cover):

$$\dot{W}_{in(fan)(act)}=Q[(SP+VP)_2-(SP+VP)_1]+\dot{m}(u_2-u_1)$$

$$= Q[\Delta TP] + m\ (U_2 - U_1)$$

Where Fan Loss = $\dot{m}(u_2-u_1)=\dot{m}\Delta u$

Since there is no way to evaluate directly, the actual fan work is assessed by using an efficiency value (η) where:

$$\dot{W}_{in(fan)act}\cong\frac{Q(\Delta TP)}{\eta} \qquad [3.18]$$

and when the ΔVP = 0 in the fan, the equation can be stated:

$$\dot{W}_{in(fan)act}\cong\frac{Q(\Delta SP)}{\eta} \qquad [3.19]$$

3.7 APPLYING THE FAN TO THE SYSTEM (SYSTEM CURVE)

The previous section considered the work provided by the fan and its efficiency overcoming the system resistances. In addition, Section 3.5 and its examples showed the method to determine and use Coefficients to predict system losses (ΔSP) for the components found. Restating Equation 3.19 the total system losses can be defined for a simple system (Figure 3-8).

$$\dot{W}_{in(fan)act}(\eta)=Q(\Delta SP)_{fan}=Q\sum losses_{(1-4)}$$

$$\left[F_a+F_h+F_{el}+F_{d(1-2)}+F_{d(3-4)}\right](VP)=\Delta SP_{sys}$$

$$(\text{constant})(VP) = \Delta SP_{sys}$$

Using Equation 3.7 to relate VP and Velocity (and also Volume) then:

$$\Delta SP_{sys} = (K)(Q^2) \qquad [3.20]$$

Equation 3.20 can then be plotted as a relationship between the volume (Q) conveyed through the system and the amount of pressure required by the fan to overcome the resistance in the system (System Static Pressure). This relationship is shown as the System Curve in Figure 3-9.

With the relationship defined by Equation 3.20, the value for "K" is basically the inclusion of all of the loss coefficients (F) in the system and is based on the information from the calculation sheet (see Chapter 9 for calculation methods). The more resistance, the higher the value for "K" and the steeper the system curve (see Curve K_1Q_2 in Figure 3-9). If the system has less internal resistance then Curve K_2Q_2 would be more representative of the curve. Another way to interpret the relationships of the two curves would be to state that for identical pressures, Curve K_2 provides more airflow than K_1. Any change to the internal losses of the system will cause the curve to move to either extreme. In Chapter 5 there is a discussion of the significance of the System Curve and its use in determining the design operating point of the installed system. The intersection of the System Curve and the Fan Curve will be the predicted operating point of the system.

Note that more complicated systems may result in more complicated system curves. For example, the losses through a filter may actually be more linear (ΔSP_{fan} = (K)(Q)) than follow Equation 3.20. In cases where the filter losses are a large

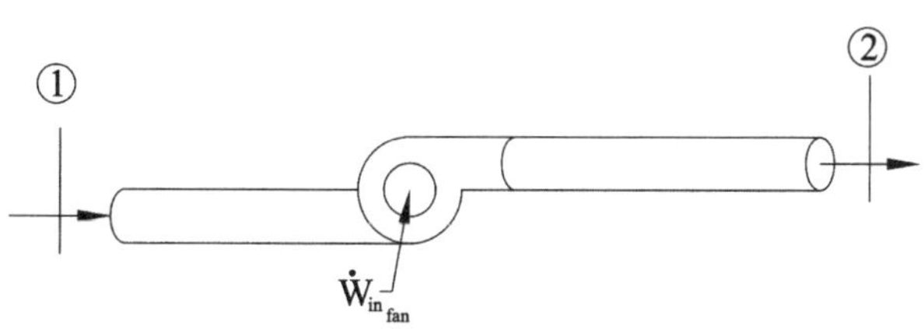

FIGURE 3-7. Fan work example

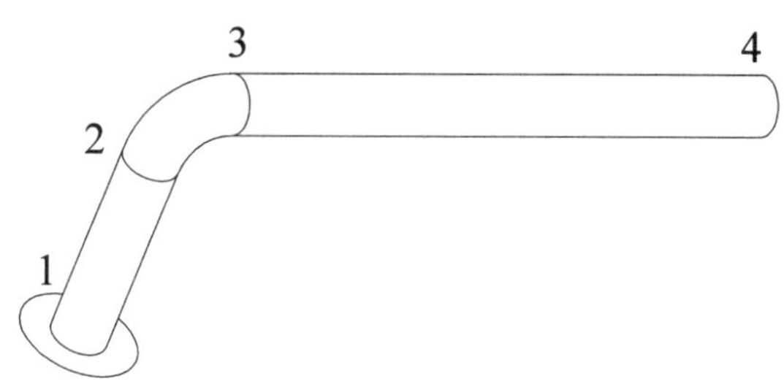

FIGURE 3-8. Simple duct system

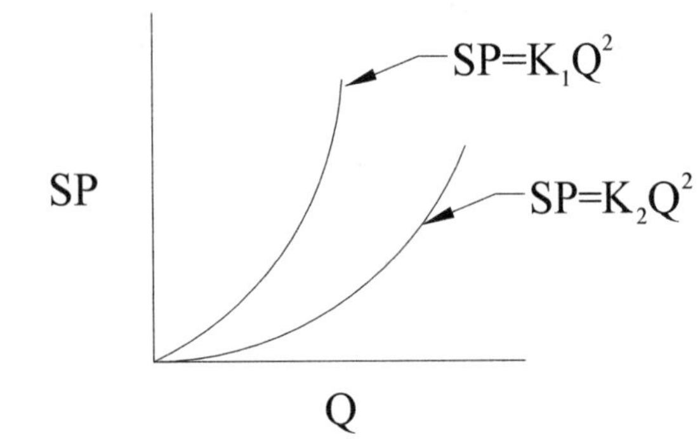

FIGURE 3-9. System curve

proportion of the total system losses, a value for K may actually include linear and exponential constants.

3.8 TRACKING PRESSURE VARIATIONS THROUGH A SIMPLE SYSTEM

The application of the design principles will be demonstrated by an analysis of the simple system shown in Figure 3-10. The normally vertical exhaust stack is shown horizontal to facilitate graphing the variation of static, total, and velocity pressures. In the example, the grinder wheel hood requires 300 acfm (Q) [142 m³/s] and the duct diameter (d) is constant at 3.5 inches [89 mm] (0.0668 ft² [0.006 m²] area). This yields a duct velocity of (300 acfm/0.0668 ft² =) 4491 feet per minute [22.81 m/s] and a VP of 1.26 "wg [313 Pa] (Equation 3.8a). The details for calculating these values are also included in Chapters 5 and 9.

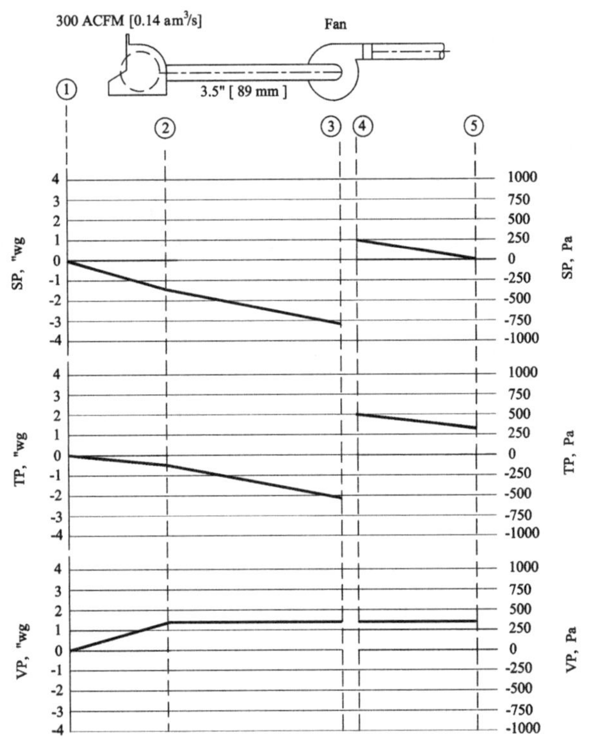

FIGURE 3-10. Variation of SP, VP, and TP through a ventilation system

In the example, the graphical relationship among TP, VP and SP is maintained per Equation 3.8 (TP = SP + VP). All pressures are at zero value some distance from the face of the hood. To get air induced into the face of the hood there is work required by the fan. The ΔSP of the hood is the combination of the resistance due to the shape of the hood plus the change of the potential energy of the air at rest to the velocity (kinetic energy) now achieved in the hood and duct.

A Grinder Hood with tapered takeoff has a value for F_h of 0.4 (see VS-80-10 in Chapter 13). Using Equation 3.10a, the value for the SP_h is calculated as -(1+ 0.4)(1.26) = -1.76 "wg [(1.4)(313) = 438 Pa] and is shown at Point "2" on the Static Pressure plot. Velocity Pressure was already calculated as +1.26 "wg [313 Pa] so TP at Point 2 is calculated as (-1.76 +1.26 =) -0.5 "wg [-438 + 313 = -125 Pa].

As the air and dust proceed toward the fan, additional friction and static pressure loss is accumulated. This is shown on the static pressure graph as the slanting line ending at Point "3". The difference between the value at Point "2" and Point "3" for static pressure (ΔSP) is calculated from Equation 3.16. Velocity Pressure is constant so there is a corresponding change in Total Pressure also for this segment.

There is similar resistance encountered in the straight duct leaving the fan (Segment 4-5). The static pressure requirements for this segment would also be calculated using Equation 3.16. Note this equation does not differ for air under negative pressure (before the fan) or positive pressure (after the fan).

Finally, the work required by the Fan is calculated by using Equation 3.16 or 3.17. Knowing the volume (Q), fan efficiency from the manufacturer, the difference between the negative value for TP (or SP) at the fan inlet and the positive number at the outlet, the work can be determined. Chapter 7 details fan energy and horsepower requirements for system installations.

3.9 ASSUMED CONDITIONS (STANDARD AIR)

As mentioned previously, two basic principles of fluid mechanics govern the flow of air in industrial ventilation systems: conservation of mass and conservation of energy (see Sections 3.2 and 3.3). These are essential physical laws that state that all mass and all energy must be completely accounted for. Coverage of fluid mechanics is not in the purview of this Manual; reference to any standard fluid mechanics textbook will show the derivation of these principles. However, it is important to know what simplifying assumptions are made for industrial ventilation systems in this Chapter and included in the principles discussed below. They include:

1) Air is assumed to be at 70 F [21 C] and heat transfer effects are neglected. If the temperature inside the duct is significantly different from the air temperature surrounding the duct, heat transfer will occur and may need to be considered in more complicated designs. Significant heat transfer can lead to changes in the duct air temperature and hence in the volumetric flow rate (but not mass flow). Under normal conditions this will have negligible effect on the operation and measurement in the system.
2) Compressibility effects are neglected. If the overall pressure drop from the start of the system to the fan is greater than about -20 "wg [-5 kPa], then the density will change by about 5% and the volumetric flow rate will also change. Standard conditions consider no effects of high negative pressures in the system.
3) The air is assumed to be dry. Water vapor in the air stream will lower the air density and correction for this effect, if present, should be made. Chapter 5 describes the necessary psychrometric analysis required when encountering systems with significant moisture (Dew Point > 80 F) [27 C].
4) Elevation is assumed to be at sea level where no atmospheric effects are encountered in the system operation or measurement.
5) The weight and volume of the contaminant in the air stream is ignored. This is permissible for the contaminant concentrations in typical exhaust ventilation systems. For high concentrations of solids (> 70 gr/dscf) [>1.29 x 10^{-4} kg/nm³] or significant amounts of gases other than air, corrections for this effect should be included.

Please note that Standard Air Conditions are seldom actually achieved and that the cumulative effects of small deviations from the accepted conditions (70 F, Sea Level, 80 F Dew Point, SP > -20 "wg) [21 C, Sea Level, 27 C Dew Point, SP > -5 kPa] can cause problems in measurement and design.

SCFM in the context of this Manual allows for Temperature up to 100 F [38 C], Dew Point up to 80 F [27 C] and Elevation up to 1000' ASL [305 m ASL]. Each of those variations can result in a density change of 5%.

A more restrictive definition of standard air is stated as "dscfm" or dry standard cubic feet per minute. This term is used for emissions standards as well as psychrometric base. The definition is for air at 70 F [21 C], no moisture, no duct pressure and at sea level, rather than above allowable ranges.

EXAMPLE PROBLEM 6 (Density of Standard Air) (IP Units)

Air under standard conditions has a density (ρ) of 0.075 lbm/ft³ [1.204 kg/m³]. This value can be calculated using the Ideal Gas Law Equation:

$$P = \rho\, R_g\, T \qquad [3.21]$$

The Ideal Gas Law or Equation of State describes the interrelationship between Pressure (P), Density (ρ) and Temperature (T) for gases. These are related through a constant (R_g) that is unique for each gas. The constant is calculated using the universal Gas Constant (R_u) and the molecular weight (M) of the gas through the following:

$$R_g = \frac{R_u}{M} \qquad [3.22]$$

where: $R_u = 1545.4 \frac{\text{ft} - \text{lbf}}{\text{mole}-°\text{R}}$ and $M_{air} = 28.941 \frac{\text{lbm}}{\text{mole}}$

$$R_g = \frac{1545.4}{28.941} = 53.34 \frac{\text{ft} - \text{lbf}}{\text{lbm}-°\text{R}} \quad \text{for air (IP)}$$

Solving for the density of Standard Air with Equation 3.21 (Temperature in °R = °F + 460 and psi is converted to lbf/ft² to reconcile units):

$$\rho = \frac{P}{R_g T} = \frac{(14.7)(144 \frac{\text{in}^2}{\text{ft}^2})}{(53.35)(530)} = 0.075 \frac{\text{lbm}}{\text{ft}^3}$$

In SI units the derivation becomes:

$$R_u = 8.3145 \frac{\text{joules}}{\text{mole}-°\text{K}}$$

$$1 \text{ joule} = 1000 \text{ pa} - \text{m}^3$$

$$R_g = \frac{8{,}314.5 \frac{\text{Pa}-\text{m}^3}{\text{mole}-°\text{K}}}{28.941 \frac{\text{kg}}{\text{mole}}} = 287.5 \frac{\text{Pa}-\text{m}^3}{\text{kg}-°\text{K}}$$

$$\rho = \frac{P}{R_g T} \frac{101{,}384 \text{ Pa}}{\left(287.5 \frac{\text{Pa}-\text{m}^3}{\text{kg}-°\text{K}}\right)(293\text{K})} = 1.204 \frac{\text{kg}}{\text{m}^3}$$

This same formula can be used to solve for density of any gas, knowing temperature and pressure conditions and value of R_g. Note: In contrast to conservation principles, this "law" is specific (contains material constants) and, therefore, does not have as wide a range of applicability (i.e., don't try to use it for water).

3.10 ASSUMED CONDITIONS (NON-STANDARD AIR)

Most systems will not operate under "Standard" conditions as defined in Section 3.6. Formulae for losses in the system will be less as the density of the air or other gas decreases in the system. Conservation of Mass and Energy still apply and Equation 3.4 is in effect in systems where changes in external heat are determined.

The basic formulae as determined for SP losses still apply. The relationship between Velocity and Velocity Pressure is shown in Equations 3.8, 3.8a and 3.8b. As VP changes with density, then losses will also vary. The Coefficients as defined in Section 3.5 are constant and based on the physical shapes and characteristics of each piece. But losses will vary when the density of the air being conveyed also changes. This is apparent as the calculation sheet is employed for system design.

Note that the same formulae apply for Standard Air by inserting the value of density factor (df) equal to 1. The Ideal Gas Law (see Section 3.9) can also be used to determine density (and density factor) for air and other gases under different conditions. These conditions can include:

- Pressure (altitude above sea level where system is located and absolute pressure inside the duct especially under conditions of high negative pressure caused by the fan)
- Temperature of the air stream
- Moisture in the air stream

There are also other cases where the changes in conditions affect operation of an industrial ventilation system. Air itself has the capacity to hold and distribute heat. For most conditions this can be predicted by use of the same equations for Conservation of Energy (see Section 3.3).

EXAMPLE PROBLEM 7 (Heat and ΔT)

Figure 3-11 shows a simple heater unit with air flowing in and out. By Conservation of Mass Laws, the mass rate of air flowing into and out of the heater is constant. If an assumption is made that there is no appreciable change in pressure across the heater (the magnitude of heater energy (thermal) input will overwhelm ΔVP or ΔSP across the heater) then Equation 3.4 can be rewritten to:

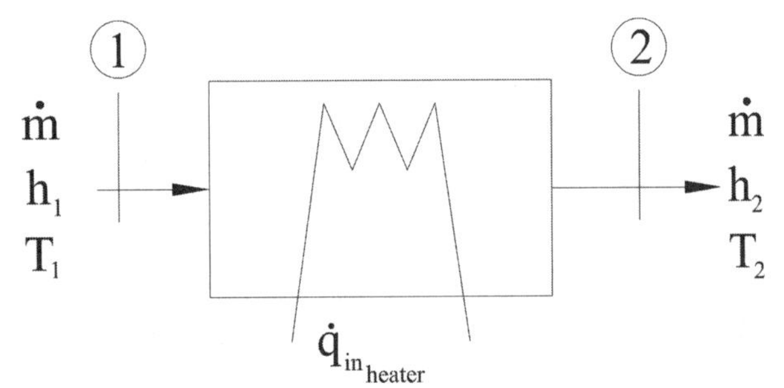

FIGURE 3-11. Energy gained by air through a heater

$$\dot{m}_1\left(\frac{SP}{\rho}+u\right)_1+\dot{q}_{in(heater)}=\dot{m}_2\left(\frac{SP}{\rho}+u\right)_2$$

For an ideal gas (air) this would calculate the heat added to the air stream. With $\dot{m}_1 = \dot{m}_2$ and the definition of enthalpy (see inside back cover), this calculates as:

$$\dot{q}_{in(heater)} = \dot{m}(h_2 - h_1) \qquad \text{and}$$

$$\dot{q}_{in(heater)} = \dot{m}C_p(T_2 - T_1)$$

Where C_p is the Constant of Specific Heat (for air this value is shown in Table 3-1).

3.11 DENSITY AND DENSITY FACTOR

Almost all systems must consider the effects of density during operation. In past Manuals, the basic rule was not to consider density in calculation when the following criteria were met: temperature below 100 F [38 C], dew point below 80 F [26.7 C], pressure inside duct > -20 "wg [-5 kPa] and elevation below 1000 feet [305 m] above sea level.

Each of these factors by themselves has only a 5% effect on final air volume. If, however, all of the conditions are in place, then the effects would be: (0.95)(0.95)(0.95)(0.95) = 0.814 or a misrepresentation of volume of almost 20%.

It is recommended that density be considered for almost all systems. The use of Standard Air (df = 1, see Section 3.10) should be relegated only to the simplest of systems (SSP < 12 "wg) where no moisture or heat is added and the plant location is near sea level. Failure to follow the guidelines may lead to an underestimation of fan requirements.

Density (as a function of temperature and pressure) can be calculated through the Ideal Gas Law. Example Problem 8 shows the method to calculate density of air at elevated temperature but similar calculations for duct pressure can be done by inserting P_{act} through the Ideal Gas Law (see Example Problem 6). The calculated density is then compared to the density of standard air to calculate density factor. Note: The values are dimensionless and identical in SI and IP units.

The calculated value df will change the duct VP through Equation 3.8a. This then changes the system resistance as the loss coefficients (F) are multiplied by VP. It also changes the volume of air. This cascading effect of air conditions and their effect on system design makes it extremely important that proper procedures be followed. These procedures for design continue in the following chapters.

EXAMPLE PROBLEM 8 (df and Temperature)

Temperature

The same method from Example Problem 6 can be used

to calculate the density and df of air at an elevated temperature T_{act}. This value can also be calculated using the Ideal Gas Law Equation (Equation 3.21). The constant was calculated in Example Problem 6 and remains the same for air at all conditions (IP units):

$$R_g = \frac{1545.4}{28.941} = 53.34 \frac{ft-lbf}{lbm-R} \quad \text{for air}$$

Solving for the density of the heated air with Equation 3.21 (Temperature in R = F + 460 F = T_{act} + 460 F and psi is converted to lbf/ft^2 to reconcile units):

$$\rho = \frac{P}{R_g T} = \frac{(14.7)(144\frac{in^2}{ft^2})}{(53.35)(T_{act}+460)} = \rho_{act}\frac{lbm}{ft^3}$$

Note that with all other factors in the formula being the same (P, R_g) this is the inverse ratio of the temperatures in Rankin and so formula for density of air at elevated temperature would be:

$$\rho_{act} = \rho_{std}\left(\frac{T_{std}}{T_{act}}\right) \quad \textbf{[3.23]}$$

and so:

$$df_t = \frac{\rho_{act}}{\rho_{std}} = \left(\frac{T_{std}}{T_{act}}\right) \quad \textbf{[3.24]}$$

where all temperatures are stated in degrees Rankin (R). In SI units, the base Temperature is 293 K and actual Temperature is (T + 273) K. Equation 3.24 would apply.

Formulas for pressure effects (either due to the absolute pressure in the duct or the elevation of the plant above sea level can be derived from the same Ideal Gas Equation (and considering the units of "wg) to the following:

Elevation

$$df_e = [1-((6.73)(10^{-6})(z))]^{5.258} \quad \textbf{[3.25] IP}$$

$$[df_e = [1-((22.08)(10^{-6})(z))]^{5.258}] \quad \textbf{[3.25] SI}$$

where z = elevation of the system in ft [m] above sea level.

The issue of absolute pressure in the duct can be more complicated. As system conditions are calculated for each duct segment, the absolute pressure is technically not known until all other factors are computed. This can be tedious and will have a minor effect on most duct segments. By convention the absolute pressure effect will only be considered in the last segment before the fan so that proper specification can be made. In systems where extreme accuracy is required, absolute duct pressure may need to be considered in all segments.

Duct Pressure

$$df_p = \frac{407 + SP_{duct}}{407} \quad \textbf{[3.26] IP}$$

$$[df_p = \frac{101384 + SP_{duct}}{101384}] \quad \textbf{[3.26] SI}$$

where duct pressure is stated in "wg [Pa].

All Density Factors Considered

The df in the duct would then be the product of all density effects:

$$df = (df_t)(df_m)(df_e)(df_p) \quad \textbf{[3.27]}$$

The calculation of the actual SP and VP in an industrial ventilation system requires the knowledge of the air conditions within the duct system. As air decreases in density it will have fewer molecules in contact with surfaces of the duct system and will require less work for conveyance. The methods in Chapter 9 consider these conditions by keeping all loss coefficients (F) constant under all conditions and changing the VP to reflect the calculation of losses in any component. And so with

$$\Delta SP_{std} = F\ (VP)$$

and VP considers df (Equation 3.8a) then

$$\Delta SP_{act} = h = (\Delta SP_{std})(df)$$

Simply stated, the loss in a section component or complete system is a function of its loss coefficient (F) and df. The values for df as a function of temperature (df_t), moisture (df_m), elevation (df_e) and absolute pressure (df_p) are also given in Chapter 9. When considering Standard Air (air defined as being 70 F, containing no moisture and at sea level) a value of df =1 can be inserted in any equation.

REFERENCES

3.1 Boyers, A.: Private Communication to G. Lanham (April 2005).

3.2 Moody, L.F.: Friction Factors for Pipe Flow. ASME Trans. 66:672 (1944).

3.3 Loeffler, J.J.: Simplified Equations for HVAC Duct Friction Factors. ASHRAE Journal, p. 76 (January 1980).

Chapter 4

GENERAL INDUSTRIAL VENTILATION

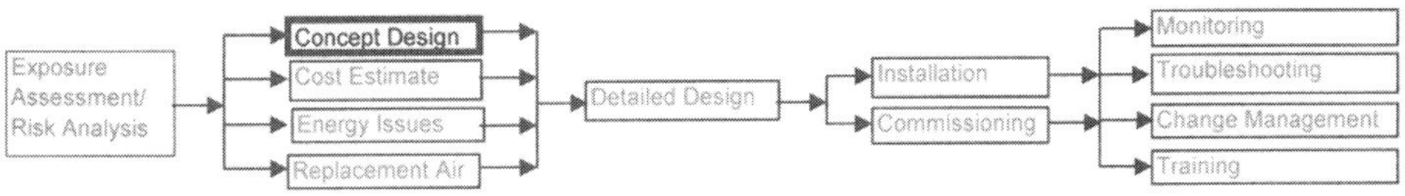

NOTE: Equations with notation followed by (IP) are designated for inch-pound system only; Equations followed by (SI) are designated for metric use only. If Equation bears neither, then it applies to both systems.

4.1 INTRODUCTION

"General industrial ventilation" is a broad term that refers to the supply and exhaust of air with respect to an area, room, or building. It can be divided further into specific functions as follows:

1) *Dilution Ventilation* is the dilution of contaminated air with uncontaminated air for the purpose of controlling potential airborne health hazards, fire and explosive conditions, odors, and nuisance type contaminants. Dilution ventilation can also include the control of airborne contaminants (vapors, gases, and particulates) generated within tight buildings.

   Dilution ventilation is not as satisfactory for health hazard control as is local exhaust ventilation. Circumstances may be found in which dilution ventilation provides an adequate amount of control more economically than a local exhaust system. One should be careful, however, not to base the economic considerations entirely upon the first cost of the system since dilution ventilation frequently exhausts large amounts of heat from a building, which may greatly increase the energy cost of the operation.(4.1)

2) *Heat Control Ventilation* is the control of indoor atmospheric conditions associated with hot industrial environments such as are found in foundries, laundries, bakeries, etc., for the purpose of preventing acute discomfort or injury.

4.1.1 Generation of Gases and Vapors. Contaminant gases may be released from storage cylinders deliberately or inadvertently as part of a process or they may be created by chemical reactions in a process.

Vapors also can be released from storage containers deliberately or inadvertently and they also are sometimes created by chemical reactions in a process. Because they are due to evaporation of a liquid, their generation rate increases directly with surface area exposed to the air and even at a greater rate with an increase in temperature.

The concentration in the space above a liquid in a closed container in which liquid remains is equal to the vapor pressure at that temperature divided by the pressure of all the gases and vapors in that space. If the container space contains a mixture of air and the vapor at atmospheric pressure, then the concentration of the vapor will be the vapor pressure divided by the atmospheric pressure. For example, if the vapor pressure at 70 F [21 C] is 203.5 "w.g. [380 mmHg], then at an atmospheric pressure of 407 "w.g. [760 mmHg], the saturation concentration of the vapor would be 203.5/407 [380/760] (i.e., 50% or 500,000 ppm).

If air/vapor mixture is withdrawn from that space and replaced by uncontaminated air, the space concentration will fall below the saturation concentration. The concentration will gradually rise back to saturation concentration as long as some liquid remains in the container. That space concentration will exponentially approach saturation concentration with time. That is, the change in concentration with time will become slower and slower as the concentration increases.

Evaporation rate is directly related to concentration in the air relative to the saturation concentration. At any time, some molecules are condensed to liquid state while others are escaping the surface by evaporation. At lower concentrations, the fraction evaporating is much greater than the fraction condensing and the concentration increases rapidly. At higher concentrations, the two rates become more and more equal and the concentration increases correspondingly slower. At equilibrium, the two processes are equal and the concentration does not change at all or bounces back and forth within a narrow range.

If the temperature changes, the vapor pressure changes markedly. The vapor pressure can never quite reach zero (frozen into a solid) or atmospheric pressure (100% concentration in the space above the liquid). As the vapor pressure changes with temperature, the saturation concentration changes proportionally. The space concentration in a closed container can be predicted very accurately from its temperature.

The time required to re-establish the saturation concentration will fall sharply if the liquid surface is agitated.

Outside of a closed container (e.g., a puddle of liquid), one might think that the generate rate would be very high since the concentration in the air is unlikely to achieve more than a small fraction of the saturation concentration because of air exchange with the outdoors. However, there is a thin layer of air that adheres to any surface, including a puddle of liquid thus limiting the exchange to the room air. The concentration in that layer will generally be much higher than the air even a few millimeters above it. Hence, the apparent concentration gradient is much lower.

However, room air currents flowing over the puddle increases the exchange rate with the thin layer, greatly increasing the generation rate. That is why fanning a puddle of liquids with moderate vapor pressures (e.g., ethanol) will greatly increase their rate of evaporation (or cooling). Hence, blowing dilution air over a liquid surface will greatly increase evaporation rate.

Since evaporation occurs from surfaces, generate rate due to evaporation is directly proportional to the total surface area exposed to room air. In this regard, it is vital to understand that agitating a surface increases its surface area (and breaks up the thin layer of air). If the agitation is violent enough to produce particles of liquid, the surface area is vastly increased and the generation rate increases dramatically.

For a mixture of evaporating liquids, the rate of evaporation of each one is affected by the other components. The concentration in the air of each will be governed by its vapor pressure. However, if the liquid evaporates completely, then the concentration of each is proportional to the number of moles created from the mass of each as it is evaporated.

The amount of generation per surface area is extremely difficult to predict even for quiet, undisturbed surfaces. For a surface disturbed by room air currents or agitation, it can be determined for an exact set of conditions empirically but cannot be modeled so that that knowledge can be extrapolated to predict generation rates for a different set of conditions.

The change in air concentration of a contamination can be dramatic if there is a large spill; in minutes the concentration can jump from barely measurable to deadly if the chemical is both volatile and highly toxic. Even if it is not notably volatile or toxic, the chemical can evaporate to dangerous air concentrations if it is spread over a large surface area or if it is released in a relatively small confined space.

Ironically, air drawn across the surface of an evaporating liquid strips away the saturated layer of vapor accelerating the evaporation rate. Thus, it is very difficult to estimate evaporation rates when general ventilation systems are drawing air over a spill or through a confined space (see Section 4.4) that still contains the evaporating liquid because the evaporation rate (G) itself will be a function of the ventilation rate.

4.2 DILUTION VENTILATION PRINCIPLES

The principles of dilution ventilation system design are as follows:

1) Select from available data the amount of air required for satisfactory dilution of the contaminant. The values tabulated in Table 4-1 assume perfect distribution and dilution of the air and solvent vapors. These values must be multiplied by the selected Mixing Factor (m_i) (see Section 4.3.1).
2) Locate the exhaust openings near the sources of contamination, if possible, in order to obtain the benefit of "spot ventilation."
3) Locate the air supply and exhaust outlets such that the air passes through the zone of contamination. The operator should remain between the air supply and the source of the contaminant.
4) Replace exhausted air by use of a supply air system. This supply or replacement air should be heated or possibly cooled to satisfy the temperature requirements of the space. Dilution ventilation systems usually handle large flows of air by means of low pressure fans (see Figures 4-11 and 4-12). Adequate quantities of supply air must be provided if the system is to operate satisfactorily.
5) Avoid re-entry of the exhausted air by discharging the exhaust high above the roof line or by assuring that no window, outdoor supply air intakes, or other such openings are located near the exhaust discharge.
6) Minimize any opportunity for flexible duct to "kink" or become twisted. Use rigid duct where possible (see Figure 4-13).

4.3 DILUTION VENTILATION FOR HEALTH

The use of dilution ventilation for health hazards has four limiting factors: 1) the quantity of contaminant generated must not be too great or the airflow rate necessary for dilution will be impractical; 2) workers must be far enough away from the contaminant source or the generation of contaminant must be in sufficiently low concentrations so that workers will not have an exposure in excess of the established TLV® (TLVs® should be used as guidelines only and not as absolute criteria for a safe and acceptable workplace); 3) the toxicity of the contaminant must be low (substances of unknown toxicity should be treated as highly toxic until proven otherwise); and 4) the generation of contaminants must be reasonably uniform.

Dilution ventilation is used most often to control the vapors from organic liquids with a TLV® of 100 ppm or higher. In order to successfully apply the principles of dilution to such a problem, factual data are needed on the rate of vapor generation or on the rate of liquid evaporation. Usually such data can be obtained from the plant if adequate records on material consumption are kept.

4.3.1 General Dilution Ventilation Equation. The ventilation rate needed to maintain a constant concentration at a uniform generation rate is derived by starting with a fundamental material balance (assuming no contaminant in the air supply).

Rate of Accumulation = Rate of Generation – Rate of Removal

or

$$V_r\,dC_g = G\,dt - Q'C_g\,dt \qquad [4.1]$$

where: V_r = volume of room
G = rate of generation of contaminant
Q' = effective volumetric flow rate
C_g = concentration of gas or vapor at time t
t = time

At a steady state, $dC_g = 0$

$$G\,dt = Q'C_g\,dt$$

$$\int_{t_1}^{t_2} G dt = \int_{t_1}^{t_2} Q'C_g dt$$

At a constant concentration (C_g) and uniform generation rate (G), then

$$G(t_2 - t_1) = Q'C_g\,(t_2 - t_1)$$

$$Q' = \frac{G}{C_g} \qquad [4.2]$$

Due to incomplete mixing, a mixing factor (m_i) is introduced to the rate of ventilation; thus:

$$Q' = \frac{Q}{m_i} \qquad [4.3]$$

TABLE 4-1. Dilution Air Volumes for Vapors

The following values are tabulated using the TLV® values shown in parentheses (parts per million). TLV® values are subject to revision if further research or experience indicates the need. If the TLV® value has changed, the dilution air requirements must be recalculated. The values on the table must be multiplied by the evaporation rate (pts/min) [l/min] to yield the effective ventilation rate (Q). (See Equation 4.5 (IP and SI))

Liquid (TLV® in ppm)*	ft^3 [m^3] of Air (STP) Required for Dilution to TLV®** Per Pint [liter] Evaporation
Acetone (500)	11,025 [695]
n-Amyl acetate (50)	54,400 [3,427]
Benzene (0.5)	NOT RECOMMENDED
n-Butanol (butyl alcohol) (C 50)	88,000 [5,594]
n-Butyl acetate (150)	20,400 [1,285]
Butyl Cellosolve (2-butoxyethanol) (20)	NOT RECOMMENDED
Carbon disulfide (10)	NOT RECOMMENDED
Carbon tetrachloride (5)	NOT RECOMMENDED
Cellosolve (2-ethoxyethanol) (5)	NOT RECOMMENDED
Cellosolve acetate (2-ethoxyethyl acetate) (5)	NOT RECOMMENDED
Chloroform (10)	NOT RECOMMENDED
1-2 Dichloroethane (ethylene dichloride) (10)	NOT RECOMMENDED
1-2 Dichloroethylene (200)	26,900 [1,694]
1,4 Dioxane (20)	NOT RECOMMENDED
Ethyl acetate (400)	10,300 [649]
Ethyl alcohol (1000)	6,900 [435]
Ethyl ether (400)	9,630 [606]
Gasoline (300)	REQUIRES SPECIAL CONSIDERATION
Isoamyl alcohol (100)	37,200 [2,344]
Isopropyl alcohol (400)	13,200 [832]
Isopropyl ether (250)	11,400 [718]
Methyl acetate (200)	25,000 [1,575]
Methyl alcohol (200)	49,100 [3,039]
Methyl n-butyl ketone (5)	NOT RECOMMENDED
Methyl Cellosolve (2-methoxyethanol) (0.1)	NOT RECOMMENDED
Methyl Cellosolve acetate (2-methoxyethyl acetate) (0.1)	NOT RECOMMENDED
Methyl chloroform (350)	11,390 [736]
Methyl ethyl ketone (200)	22,500 [1,418]
Methyl isobutyl ketone (50)	64,600 [4,070]
Methyl propyl ketone (200)	19,900 [1,254]
Naphtha (coal tar)	REQUIRES SPECIAL CONSIDERATION
Naphtha VM & P (300)	REQUIRES SPECIAL CONSIDERATION
Nitrobenzene (1)	NOT RECOMMENDED
n-Propyl acetate (200)	17,500 [1,103]
Stoddard solvent (100)	30,000–35,000 [1,890–2,205]
1,1,2,2-Tetrachloroethane (1)	NOT RECOMMENDED
Tetrachloroethylene (perchloroethylene) (25)	159,400 [10,042]
Toluene (50)	75,700 [4,769]
Trichloroethylene (50)	90,000 [5,670]
Xylene (100)	33,000 [2,075]

*See Threshold Limit Values in Appendix A.

**The tabulated dilution air quantities must be multiplied by the selected m_i value.

where: Q = actual ventilation rate, acfm [am³/s]
Q′ = effective ventilation rate, acfm [am³/s]
m_i = a factor to allow for incomplete mixing

Equation 4.2 then becomes:

$$Q = \left(\frac{G}{C_g}\right) m_i \quad [4.4]$$

This Mixing Factor (m_i) is based on several considerations:

1) The distribution of supply air introduced into the room or space being ventilated (Figure 4-1) and how well it mixes with room air.
2) The toxicity of the solvent. Although TLVs® are only guidelines for toxicity levels and TLVs® and toxicity are not synonymous, the following guidelines have been suggested for increasing the appropriate m_i value:

   Slightly toxic material: TLV® > 500 ppm
   Moderately toxic material: TLV® ≤ 100–500 ppm
   Highly toxic material: TLV® < 100 ppm

3) The judgment of any other circumstances that an industrial hygienist determines to be important based on experience and the specific workspace. Included in these criteria are such considerations as:
   a) Duration of the process, operational cycle, and normal locations of workers relative to sources of contamination.
   b) Location and number of points of generation of the contaminant in the workroom or area.
   c) Seasonal changes in the amount of natural ventilation.
   d) Operational effectiveness of the mechanical air moving devices.
   e) Other circumstances which may affect the concentration of hazardous material in the breathing zone of the workers.

The Mixing Factor (m_i) selected, depending on the above considerations, ranges from 1 to 10.

4.3.2 Calculating Dilution Ventilation for Steady State Concentration. The concentration of a gas or vapor at a steady state can be expressed by the material balance equation shown in Equation 4.2:

$$Q' = \frac{G}{C_g}$$

Therefore, the rate of flow of uncontaminated air required to maintain the atmospheric concentration of a hazardous material at an acceptable level can be easily calculated if the generation rate can be determined. Usually, the acceptable concentration (C_g) expressed in parts per million (ppm) is considered to be the Threshold Limit Value (TLV®). For liquid solvents, the rate of generation is

$$G = \frac{\text{CONSTANT} \times SG \times ER}{MW}$$

where:

G = generation rate, acfm [am³/s]
CONSTANT = 403 (the volume in ft³ that 1 pt of liquid, when vaporized, will occupy at STP, ft³/pt) [24 (the volume that 1 liter of liquid, when vaporized, will occupy at STP, m³/l)]
SG = Specific gravity of volatile liquid
ER = evaporation rate of liquid, pts/min [l/s]
MW = molecular weight of liquid

Thus, $Q' = G/C_g$ can be expressed as

$$Q' = \frac{403 \times 10^6 \times SG \times ER}{MW \times C_g} \quad [4.5]\ \text{IP}$$

$$[Q' = \frac{24 \times 10^6 \times SG \times ER}{MW \times C_g}] \quad [4.5]\ \text{SI}$$

EXAMPLE PROBLEM 1 (Dilution Airflow with Constant Evaporation of Contaminant)

Methyl chloroform is lost by evaporation from a tank at a rate of 1.5 pints [0.71 l] per 60 minutes [3,600 s]. What is the effective ventilation rate (Q′) and the actual ventilation rate (Q) required to maintain the vapor concentration at the TLV®?

TLV = 350 ppm, SG = 1.32, MW – 133.4, Assume m_i = 5

Assuming perfect dilution, the effective ventilation rate (Q′) is

$$Q' = \frac{(403)(10^6)(1.32)(1.5/60)}{(133.4)(350)} = 284 \text{ acfm}$$

$$[Q' = \frac{24\,(10^6)(1.32)(0.71/3600)}{(133.4)(350)} = 0.13 \text{ am}^3/\text{s}]$$

Due to incomplete mixing the actual ventilation rate (Q) is

$$Q = \frac{(403)(10^6)(1.32)(1.5/60)\,(5)}{(133.4)(350)} = 1424 \text{ acfm}$$

$$[Q = \frac{24\,(10^6)(1.32)(0.71/3600)\,(5)}{(133.4)(350)} = 0.67 \text{ am}^3/\text{s}]$$

4.3.3 Contaminant Concentration Buildup **(Figure 4-2).** The concentration of a contaminant can be calculated after any time interval. Rearranging the differential material balance (Equation 4.1) results in

$$\frac{dC_g}{G - Q'C_g} = \frac{dt}{V_r}$$

FIGURE 4-1

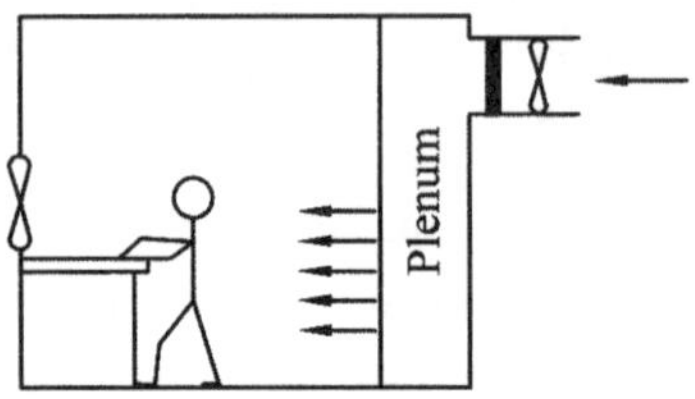

BEST AIR INLET
Best exhaust
$m_i = 1.0$ minimum

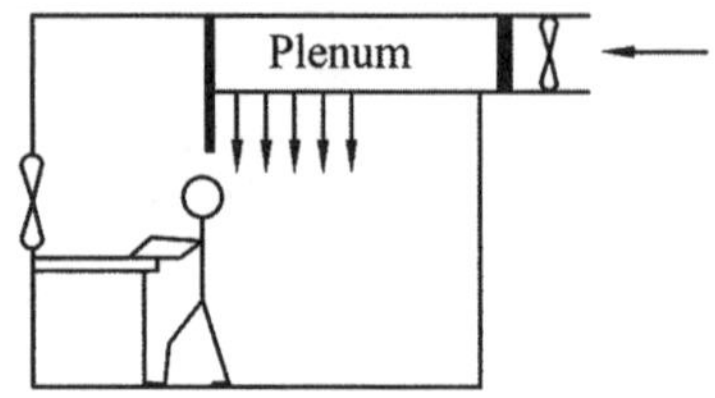

BEST AIR INLET
Best exhaust
$m_i = 1.0$ minimum

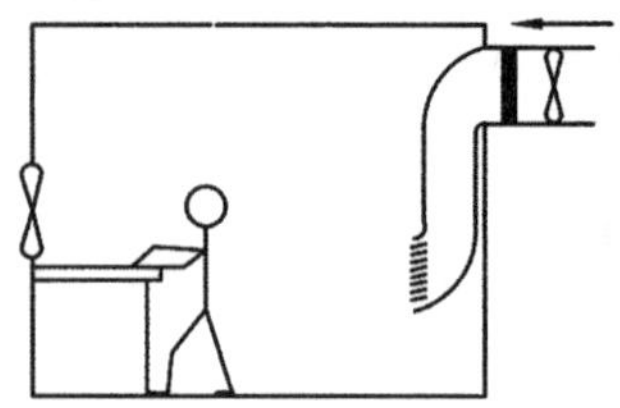

BEST AIR INLET
Best exhaust
$m_i = 1.0$ minimum

FAIR AIR INLET
Best exhaust
$m_i = 2.5$ minimum

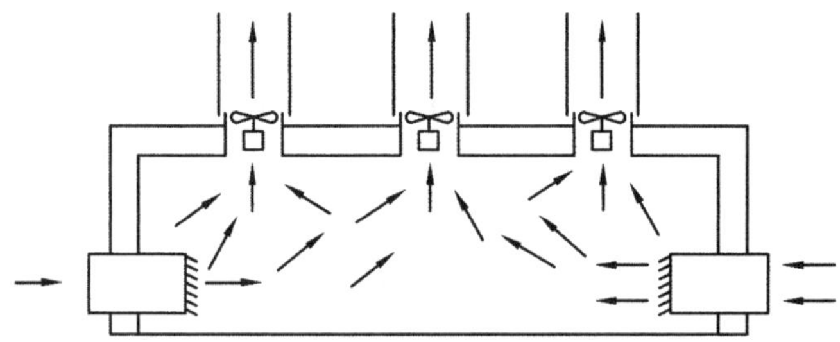

FAIR
$m_i = 2$ to 5
Ref. 4.2

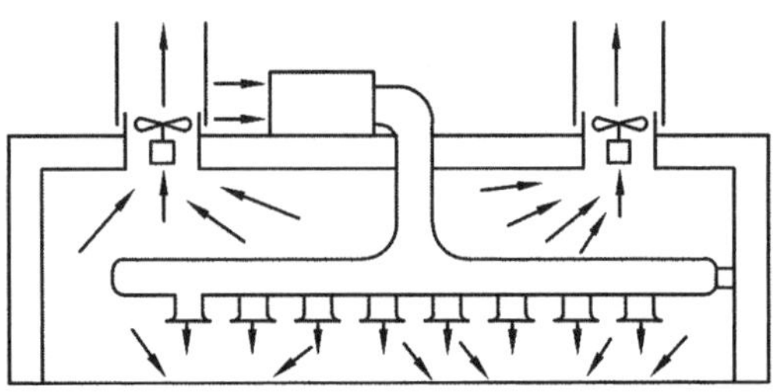

GOOD
$m_i = 1.5$ to 2
Ref. 4.2

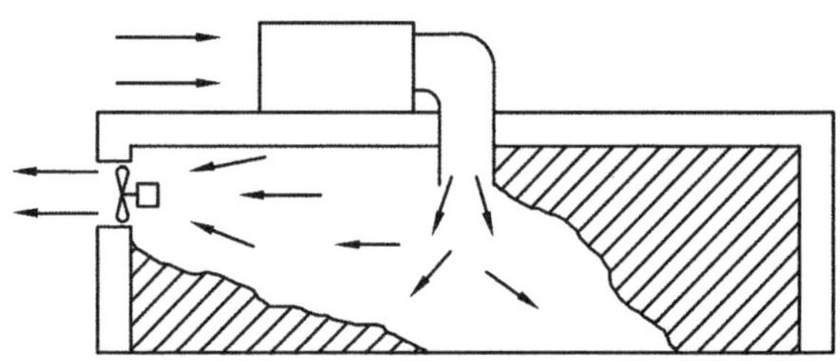

POOR
$m_i = 5$ to 10
Ref. 4.2

NOTE: The Mixing Factors (m_i) listed here consider only the inlet and the exhaust locations and are subjective. To select the Mixing Factor (m_i) used in the equation, the number and location of the employees, the source of the contaminant, and the toxicity of the contaminant must also be considered.

ACGIH®	TITLE MIXING FACTORS SUGGESTED FOR INLET AND EXHAUST LOCATIONS	FIGURE 4-1 DATE 1-12

CHECK CODES, REGULATIONS, AND LAWS (LOCAL, STATE, AND NATIONAL) TO ENSURE THAT DESIGN IS COMPLIANT.

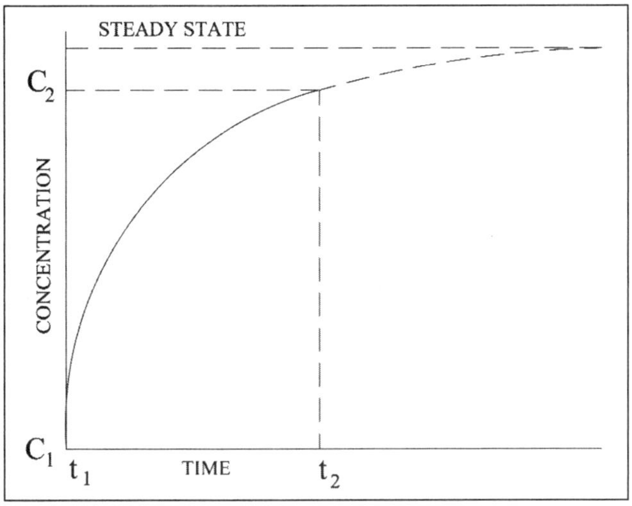

FIGURE 4-2. Contaminant concentration buildup

which can be integrated to yield

$$\ln\left(\frac{G - Q'C_{g_2}}{G - Q'C_{g_1}}\right) = -\frac{Q'(t_2 - t_1)}{V_r} \quad [4.6]$$

where subscript 1 refers to the initial condition and subscript 2 refers to the final condition. If it is desired to calculate the time required to reach a given concentration, rearranging gives $t_2 - t_1$, or Δt.

$$\Delta t = -\frac{V_r}{Q'}\left[\ln\left(\frac{G - Q'C_{g_2}}{G - Q'C_{g_1}}\right)\right] \quad [4.7]$$

If $C_{g_1} = 0$, then the equation becomes

$$\Delta t = -\frac{V_r}{Q'}\left[\ln\left(\frac{G - Q'C_{g_2}}{G}\right)\right] \quad [4.8]$$

NOTE: The concentration C_{g_2} is ppm or parts/10^6 (e.g., if C_{g_2} = 200 ppm, enter C_{g_2} as 200/10^6).

If it is desired to determine the concentration level (C_{g_2}) after a certain time interval, $t_2 - t_1$ or Δt, and if $C_{g_1} = 0$, then the equation becomes

$$C_{g_2} = \frac{G\left[1 - e^{-\left(\frac{Q'\Delta t}{V_r}\right)}\right]}{Q'} \quad [4.9]$$

NOTE: To convert C_{g_2} to ppm, multiply the answer by 10^6.

EXAMPLE PROBLEM 2 (Time to Reach a Concentration with Constant Evaporation of Contaminant)

[IP] Methyl chloroform is being generated under the following conditions: G = 1.2 acfm, Q = 6000 acfm, V_r = 100,000 ft^3, C_{g_1} = 0, C_{g_2} = 200 ppm or 200÷10^6, m_i = 3.

$$\Delta t = \frac{V_r}{Q'}\left[\ln\left(\frac{G - Q'Cg_2}{G}\right)\right]$$

$Q' = Q/m_i = 6000/3 = 2000$ acfm

$$\Delta t = \frac{100{,}000}{2000}\left[\ln\left(\frac{1.2 - 2000\,(200 \div 10^6)}{1.2}\right)\right]$$

$$= 20.3 \text{ minutes}$$

Using the same value, what will the concentration be after 60 minutes?

$$Cg_2 = G\left[\frac{1 - e^{-\left(\frac{Q'\Delta t}{V_r}\right)}}{Q'}\right] \times 10^6$$

$$Cg_2 = 1.2\left[\frac{1 - e^{-\left(\frac{2000 \times 60}{100{,}000}\right)}}{2000}\right] \times 10^6 = 419 \text{ ppm}$$

[SI] Methyl chloroform is being generated under the following conditions: G = 0.0005663 am^3/s, Q = 2.831 am^3/s, V_r = 2832 m^3, C_{g_1} = 0, C_{g_2} = 200 ppm or 200÷10^6, m_i = 3.

$$\Delta t = \frac{V_r}{-Q'}\left[\ln\left(\frac{G - Q'Cg_2}{G}\right)\right]$$

$Q' = Q/m_i = 2.831/3 = 0.944$

$$\Delta t = \frac{-2832}{.944}\left[\ln\left(\frac{.0005663 - (0.944)(200/10^6)}{.0005663}\right)\right]$$

$$= 1216 \text{ seconds}$$

Using the same value, what will the concentration be after 60 minutes [3600 seconds]?

$$Cg_2 = G\left[\frac{1 - e^{-\left(\frac{Q'\Delta t}{V_r}\right)}}{Q}\right] \times 10^6$$

$$Cg_2 = .0005663\left[\frac{1 - e^{-\left(\frac{.994 \times 3600}{2832}\right)}}{.994}\right] \times 10^6 = 409 \text{ ppm}$$

***4.3.4 Rate of Purging* (Figure 4-3).** Where a quantity of air is contaminated but where further contamination or generation has ceased, the rate of decrease of concentration over a period of time is as follows:

$$V\,dC_g = -Q'C_g\,dt$$

$$\int_{C_1}^{C_2} \frac{dC_g}{C_g} = -\frac{Q'}{C_g}\int_{t_1}^{t_2} dt$$

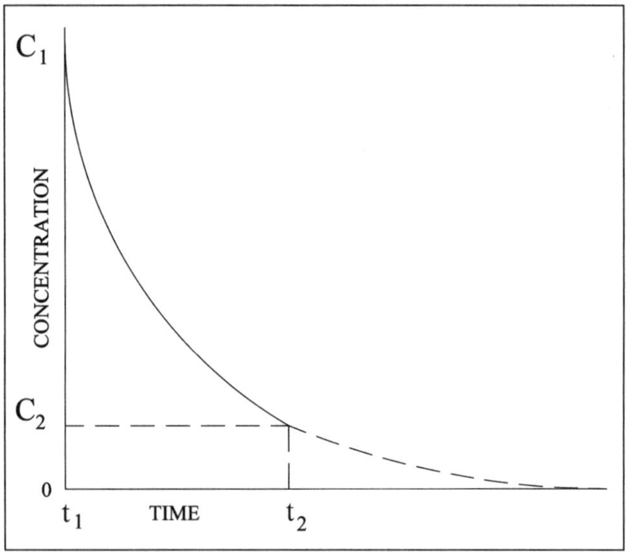

FIGURE 4-3. Rate of purging

$$\ln\left(\frac{C_{g_2}}{C_{g_1}}\right) = -\frac{Q'}{V}(t_2 - t_1)$$

or,

$$C_{g_2} = C_{g_1} e^{-\left[\frac{Q'(t_2 - t_1)}{V}\right]} \qquad \textbf{[4.10]}$$

EXAMPLE PROBLEM 3 (Dilution of Contaminant Concentration after Removal of Source) (IP Units)

In the room of the example in Section 4.3.3, assume that ventilation continues at the same rate (Q' = 2000 acfm), but that the contaminating process is interrupted. How much time is required to reduce the concentration from 100 (C_{g_1}) to 25 (C_{g_2}) ppm?

$$t_1 - t_2 = -\frac{V}{Q'}\ln\left(\frac{C_{g_2}}{C_{g_1}}\right) = 69.3 \text{ min}$$

In the problem above, if the concentration (C_{g_1}) at t_1 is 100 ppm, what will concentration (C_{g_1}) be after 60 minutes (Δt)?

$$C_{g_2} = C_{g_1} e^{\left(-\frac{Q'\Delta t}{V}\right)} = 30.1 \text{ ppm}$$

$$Cg_2 = 100e^{-\left(\frac{2000\times 60}{100{,}000}\right)} = 30.1 \text{ ppm}$$

Solution in SI units is same:

$$\left[Cg_2 = 100e^{-\left(\frac{944\times 3600}{2832}\right)} = 30.1 \text{ ppm}\right]$$

4.3.5 Cyclic Operations. It is not unusual for contaminant generation to be cyclic, especially in batch operations. If the cycles are short compared to the total buildup time, the concentrations may be only marginally higher and lower than they would have been if the same amount of contaminant were released at a constant rate (see Figure 4-4). However, if the cycles are very long compared to the buildup period, then the peaks and valleys diverge substantially from the continuous exponential curve. In investigating the 8-hour average exposure, these deviations are of little concern because the average will be the same as long as the buildup period is less than eight hours.

For short-term averages, cycles can become very important, especially in the event that the cycles are longer than the measurement period. For example, if the peak-to-peak time in Figure 4-5 is one hour long, then a 15-minute average taken at the peak could be much higher than the 15-minute average taken at the trough.

4.4 CONFINED SPACE VENTILATION

When considering whether a space should be considered a confined space for purposes of ventilation design, a functional definition is useful. A confined space is a location: 1) where concentrations of hazardous air contaminants can build up rapidly due to its relatively small size and limited air exchange with the outside, and 2) a person will enter the space. Using this definition, a 10 ft × 10 ft utility room housing Chlorine cylinders is a confined space if the cylinders are leaking or there is a significant risk they could leak. Likewise, a 7 ft deep trench in which workers are standing at the bottom and spraying epoxy could be considered a confined space. The space is even more hazardous if egress is difficult, whether due to the small size of the opening, the need to climb a ladder to escape, or any other reason. Mixing tanks with entry ports just large enough for a worker to enter and exit via a ladder are classic examples of the latter.

Monitoring, purging, and ventilation are the keys to protecting workers from airborne contaminants in confined spaces. However, ventilation is not sufficient by itself to keep confined space entries safe. The OSHA Confined Space Standard (29 CFR 1910.146) is an excellent basis for a standard operating procedure. Every part of that procedure is important. Although this is a ventilation text, monitoring is discussed since it is affected by ventilation practices. Likewise, ventilation apparatus can impede rescue operations, and must be considered.

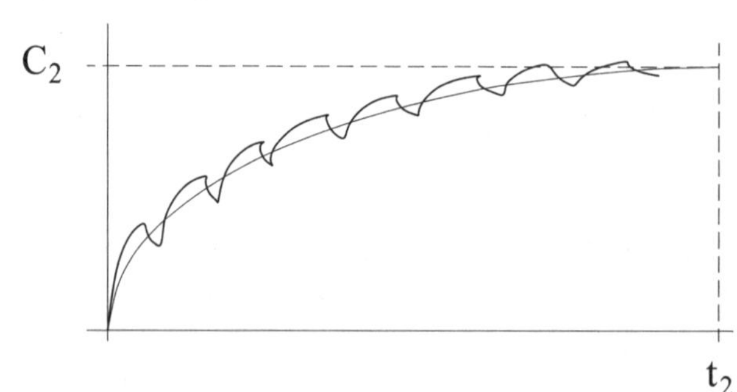

FIGURE 4-4. Cyclic generation, short cycles

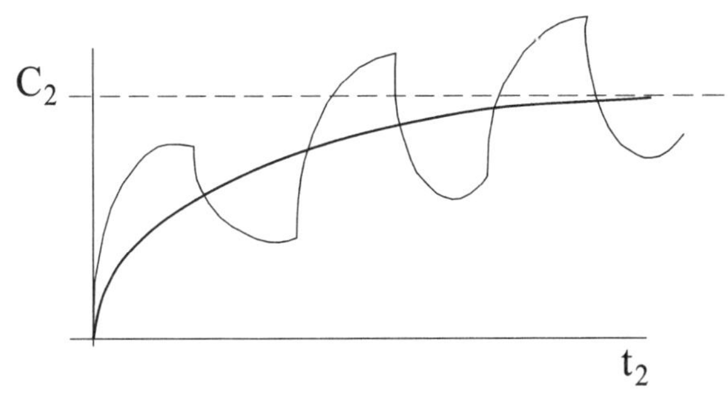

FIGURE 4-5. Cyclic generation, long cycles

An important value to compute is the time required to achieve seven room air changes (t_{7RC})

$$t_{7RC} = 7\left(\frac{Vol_{room}}{Q}\right) \quad [4.11]$$

The time computed using Equation 4.11 is a useful benchmark and is a reasonable estimate of a minimum time for safe entry for most situations if no contaminant is being generated and the supply air is clean.

If concentrations reach a significant non-zero concentration in the exhaust air and do not fall lower for a substantial length of time, then steady-state conditions have been achieved. The time-dependent terms in Equation 4.1 become zero and this is expressed by:

$$C_{exh} = \frac{G}{Q} + C_{sup} \quad [4.12]$$

Where:

C_{exh} = Concentration of exhaust gas (ppm)
G = Generation Rate (acfm) [am³/s]
Q = Exhaust Flow Rate (acfm) [am³/s]
C_{sup} = Vapor concentration in supply air (ppm)

If monitoring confirms that C_{sup} is zero, then Generation Rate (G) must be greater than zero, which means that contaminant is still evolving within the confined space. If the concentration is unacceptable, increasing the airflow to a higher level should reduce the concentration linearly:

$$Q_{new} = Q_{original}\left(\frac{C}{C_t}\right) \quad [4.13]$$

Where:

C = Original concentration of exhaust gas (ppm)
Q = Exhaust Flow Rate (acfm) [am³/s]
C_t = Target concentration of exhaust gas (ppm)

4.4.1 Reducing Generation Rates. If an existing system inadequately protects workers, it is possible to reduce the exposures to acceptable levels by:

1) Draining liquids completely.
2) Minimizing the potential for spills and leaks from the work.
3) Installing local exhaust hoods within the space to collect contaminants released by the work (see Chapter 6, Design Issues – Hoods), especially welding effluent.
4) Locating the exhaust and supply air intake and discharge points (see sections below) so that airflow patterns are effective in mixing the contaminated air with the ducted air. (see Section 4.4.3). If the initial concentration is very high or if the room containing the confined space is relatively small or poorly ventilated, it may be advisable to start with negative pressure ventilation and switch to positive pressure.
5) If experience or logic indicates that some sections of the space will have concentrations higher than the average for the total volume, move the flexible duct to that location for some fraction of the initial purge period. The fraction should depend on the size of the poorly ventilated area compared to the whole and to the severity of the difference in concentrations.
6) Locating clean air intake points where there are no contaminants.
7) Purging the space with at least the minimum level of airflow and for at least the minimum time period determined from relevant experience. Do not allow entry until after the minimum time has passed *and* measured concentrations are below target levels. Both requirements should be met, not just one.
8) Continue ventilating the space with at least the minimum level of airflow determined from relevant experience.
9) If the concentration is still unacceptably high after an unacceptably long period of time, try increasing the airflow volume (Q) to the point that the highest concentration within the space is lower than the target concentration (C_t).

4.4.2 Providing Clean Supply Air. The air blown or drawn into a confined space should be free of hazardous contaminants. In general:

1) Locate fresh air intakes for blowers well away from potential sources of contamination, including the air pouring out of the confined space if air is blown into the space (see Figure 4-6).
2) Remove sources of exposures from the vessel's access hole if air will be drawn into that opening. If that cannot be done, blow air into the vessel and ensure that the fan intake draws from clean air.

When blowing air into a space, it is necessary only to place the fan in an area having non-contaminated air (clean area) or connect inlet ductwork to it and place its open end to a clean

FIGURE 4-6. Avoid locating fan near opening

area (Figure 4-10). This can be problematic if the fan is placed on top of the purged vessel since the air escaping from it will contaminate the area. If the fan is placed on or near the vessel, it is especially important to run inlet ductwork as far as necessary to reach a clean area.

When exhausting air from a space, the inlet port of the space becomes the path of clean air. If the fan is placed near the port (Figure 4-6), outlet ductwork should be employed to carry the contaminated air to a space that is either uninhabited or is large enough and has enough ventilation that excessive concentrations will not build up due to the exhausted vessel air.

In cases where a gasoline-powered fan is employed (Figure 4-7), the effluent from the engine exhaust is a serious hazard. The engine exhaust should not be near the air intake of the blower for positive pressure systems or near the vessel inlet port for negative pressure systems.

FIGURE 4-7. Gasoline powered fan

The range of concentrations within a ventilated space is strongly influenced by the mixing factor (m_i) (see Section 4.3.1), a parameter that describes how efficiently the contaminated air is mixed within the clean air as it passes through the space. It is extremely difficult to predict the mixing factor for a volume without extensive sampling data taken over time. Mixing factors normally apply to pinpoint locations within the space and not to the entire space.

The range of concentrations within a space can be very large. This is especially true for the typical case where the contaminant is not spread uniformly throughout the space. The less diffuse the source, the greater the concentration gradients within the space. Likewise, if the worker is very close to the source, the mixing factor will be extremely high.

Mixing factors (m_i) typically range from low (< 0.2) to high (> 5) within the same space. A lower range is sometimes achievable by judicious placement of the exhaust points and supply points in relationship to the sources. Blowing air into a space at high velocities will produce better mixing because of attendant high kinetic energies, but the same high velocities can produce worker complaints, especially if the supply air is relatively cool. Furthermore, sometimes it is better to have poor mixing for locations where little contaminant is released. Even in areas with strong sources, high velocity supply air can blow the contaminant into adjacent clean areas.

4.4.3 Blowing into Versus Exhausting from the Space. In most cases, airflow should be blown into a confined space rather than exhausting from it. This is generally advisable because air drawn into the confined space is likely not to mix well with the contaminated air in the space. This will increase the mixing factor and the required airflow rate (Q) and/or purge time. It also is important to blow air into the vessel if a significant source is near the opening to the vessel (e.g., exhaust pipe from a vehicle).

As shown in Figure 4-8, the air exhausted from the room has little kinetic energy to stir the surrounding air. Note that even the pressure system can still have poor mixing in parts of the space (Figure 4-9).

Three important exceptions where exhaust design shown in Figure 4-8 is preferred are:

1) During initial purging of highly contaminated spaces if a worker must be near the port where the air would be exhausted.

2) During initial purging of highly contaminated spaces if the room with the vessel is small enough and poorly ventilated enough to become significantly contaminated by the vessel effluent.

3) If a tent or other shelter is installed above the vessel's exhaust opening. The air concentration inside the shelter will be about the same as the air discharging from the confined space through it if positive pressure ventilation is employed.

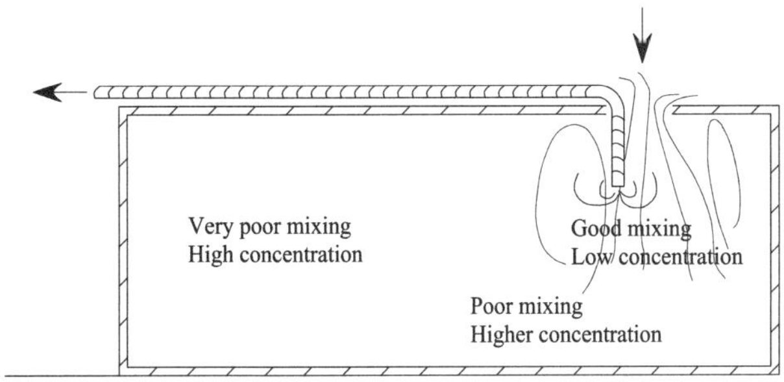

FIGURE 4-8. Concentrated exhaust and poor mixing

In any case, air should be exhausted from the space until the effluent levels are low enough to discharge to the room. The exhaust air should be released into a room that is either unoccupied or is large enough or well ventilated enough to prevent excessive concentration buildup. If those conditions cannot be met then use an exhaust system to clear the area.

4.5 MIXTURES — DILUTION VENTILATION FOR HEALTH

In many cases, the parent liquid for which dilution ventilation rates are being designed will consist of a mixture of solvents. The common procedure used in such instances is as follows.

When two or more hazardous substances are present, their combined effect, rather than that of either individually, should be given primary consideration. *In the absence of information to the contrary, the effects of the different hazards should be considered as additive.* That is, if the sum of the following fractions,

$$\frac{C_{g_1}}{TLV_1} + \frac{C_{g_2}}{TLV_2} + \ldots + \frac{C_{g_n}}{TLV_n} \qquad [4.14]$$

exceeds unity, then the threshold limit of the mixture should be considered as being exceeded. "C_g" indicates the observed atmospheric concentration and "TLV®" indicates the corresponding threshold limit. In the absence of information to the contrary, the dilution ventilation should, therefore, be calculated on the basis that the effect of the different hazards is additive. The air quantity required to dilute each component of the mixture to the required safe concentration is calculated, and the *sum* of the air quantities is used as the required dilution ventilation for the mixture.

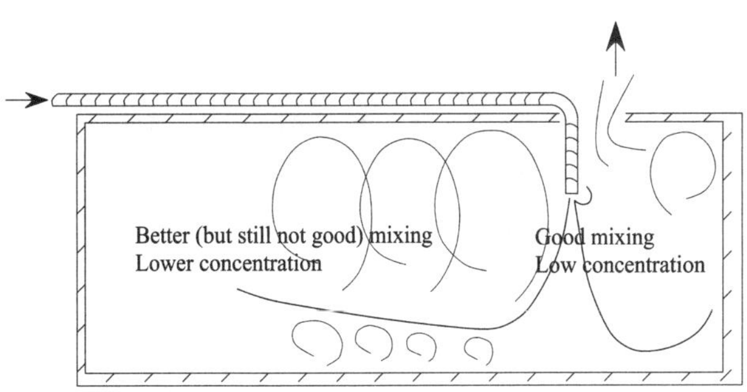

FIGURE 4-9. Concentrated supply and better mixing

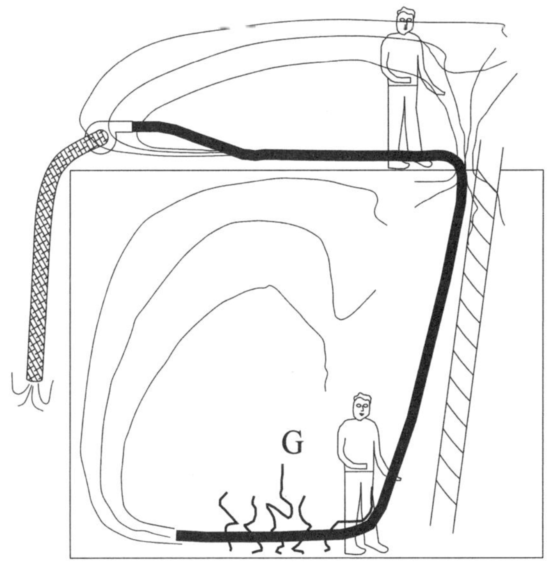

FIGURE 4-10. Blowing air into a confined space

Exceptions to the above rule may be made when there is good reason to believe that the chief effects of the different harmful substances are not additive but independent, as when purely local effects on different organs of the body are produced by the various components of the mixture. In such cases, the threshold limit ordinarily is exceeded only when at least one member of the series itself has a value exceeding unity, e.g.,

$$\frac{C_{g_1}}{TLV_1} \text{ or } \frac{C_{g_2}}{TLV_2}$$

Therefore, where two or more hazardous substances are present and it is known that the effects of the different substances are not additive but act independently on the different organs of the body, the required dilution ventilation for each component of the mixture should be calculated and the highest acfm [am³/s] obtained should be used as the dilution ventilation rate.

EXAMPLE PROBLEM 4 (Dilution Airflow with Constant Evaporation of Two Contaminants)

A cleaning and gluing operation is being performed; methyl ethyl ketone (MEK) and toluene are both being released. Both have narcotic properties, and the effects are considered additive. Air samples disclose concentrations of 150 ppm MEK and 50 ppm toluene. Using Equation 4.14, the sum of the fractions [(150/200) + (50/50) = 1.75] is greater than unity, and the TLV® of the mixture is exceeded. The volumetric flow rate at standard conditions required for dilution of the mixture to the TLV® would be as follows:

Assume 2 pints of each are being released every 60 min. Select an m_i value of 4 for MEK and an m_i value of 5 for toluene; SG for MEK = 0.805, for toluene = 0.866; MW for MEK = 72.1, for toluene = 92.13.

$$\text{Q for MEK} = \frac{(403)(0.805)(10^6)(4)(2/60)}{72.1 \times 200} = 3000 \text{ acfm}$$

$$[\text{Q for MEK} = \frac{(24)(0.805)(10^6)(4)(.94/3600)}{72.1 \times 200} = 1.41 \text{ am}^3\text{/s}]$$

$$\text{Q for toluene} = \frac{(403)(0.866)(10^6)(5)(2/60)}{92.13 \times 50} = 12{,}627 \text{ acfm}$$

$$[\text{Q for toluene} = \frac{(24)(0.866)(10^6)(5)(.946/3600)}{93.13 \times 50} =$$

$$5.92 \text{ am}^3\text{/s}]$$

Q for mixture = 3000 + 12,627 = 15,627 acfm

[Q for mixture = 1.41 + 5.92 = 7.33 $\text{am}^3\text{/s}$]

4.6 DILUTION VENTILATION FOR FIRE AND EXPLOSION (IP Units Only)

Another function of dilution ventilation is to reduce the concentration of vapors within an enclosure to below the lower explosive limit. It should be stressed that this concept is never applied in cases where workers are exposed to the vapor. In such instances, dilution rates for health hazard control are always applied. The reason for this will be apparent when comparing TLVs® and lower explosive limits (LELs).

The TLV® of xylene is 100 ppm. The LEL of xylene is a 1% content ratio or 10,000 ppm. An atmosphere of xylene safeguarded against fire and explosion usually will be kept below 25% of the LEL or 2500 ppm. Exposure to such an atmosphere may cause severe illness or death. However, in baking and drying ovens, in enclosed air drying spaces, within ventilation ducts, etc., dilution ventilation for fire and explosion is used to keep the vapor concentration to below the LEL.

Equation 4.5 [IP] can be modified to yield air quantities to dilute below the LEL. By substituting LEL for TLV®:

$$Q = \frac{(403)(\text{SG liquid})(100)(\text{ER})(S_f)}{(\text{MW liquid})(\text{LEL})(B)} \text{ (for Standard Air)} \quad \textbf{[4.15] IP}$$

NOTES: 1. *Since LEL is expressed in percent (parts per 100) rather than ppm (parts per million as for the TLV®), the coefficient of 1,000,000 becomes 100.*

2. *S_f is a safety coefficient that depends on the percentage of the LEL necessary for safe conditions. In most ovens and drying enclosures, it has been found desirable to maintain vapor concentrations at not more than 25% of the LEL at all times in all parts of the oven. In properly ventilated continuous ovens, an S_f coefficient of 4 (25% of the LEL) is used. In batch ovens, with good air distribution, the existence of peak drying rates requires an S_f coefficient of 10 or 12 to maintain safe concentrations at all times. In non-recirculating or improperly ventilated batch or continuous ovens, larger S_f coefficients may be necessary.*

3. *B is a constant that takes into account the fact that the lower explosive limit of a solvent vapor or air mixture decreases at elevated temperatures. B = 1 for temperatures up to 250 F; B = 0.7 for temperatures above 250 F.*

EXAMPLE PROBLEM 5 — Dilution Airflow to Avoid Explosive Mixture with Constant Evaporation of Solvent (IP Units Only)

A batch of enamel dipped shelves is baked in a recirculating oven at 350 F for 60 minutes. Volatiles in the enamel applied to the shelves consist of two pints of xylene. What oven ventilation rate, in acfm, is required to dilute the xylene vapor concentration within the oven to a safe limit at all times?

For xylene, the LEL = 1.0%; SG = 0.88; MW = 106; S_f = 10; B = 0.7. From Equation 4.15 [IP]:

$$Q = \frac{(403)(0.88)(2/60)(100)(10)}{(106)(1.0)(0.7)} = 159 \text{ scfm}$$

Since the above equation is at standard conditions, the airflow rate must be converted from 70 F to 350 F (operating conditions):

$$Q_A = (\text{acfm}_{\text{stp}}) \text{ (Ratio of Absolute Temperature)}$$

$$= (\text{acfm}_{\text{stp}}) \frac{(460 \text{ F} + 350 \text{ F})}{(460 \text{ F} + 70 \text{ F})}$$

$$Q_A = 159\left(\frac{810}{530}\right)$$

$$= 243 \text{ acfm}$$

EXAMPLE PROBLEM 6 (Dilution Airflow to Avoid Explosive Mixture with Varying Evaporation of Solvent)

In many circumstances, solvent evaporation rate is non-uniform due to the process temperature or the manner of solvent use.

A 6 ft diameter muller is used for mixing resin sand on a 10-minute cycle. Each batch consists of 400 pounds of sand, 19 pounds of resin, and 8 pints of ethyl alcohol (the ethyl alcohol evaporates in the first two minutes). What ventilation rate is required?

For ethyl alcohol, LEL = 3.28%; SG = 0.789; MW = 46.07; S_f = 4; B = 1

$$Q = \frac{(403)(0.789)(8/2)(100)(4)}{(46.07)(3.28)(1)} = 3367 \text{ acfm}$$

Another source of data is the National Fire Protection Association's *Standards for Class A Ovens and Furnaces*.[4.3] This contains a more complete list of solvents and their properties. In addition, it lists and describes a number of safeguards and interlocks that must always be considered in connection with fire dilution ventilation. See also Reference 4.4.

4.7 FIRE DILUTION VENTILATION FOR MIXTURES

It is common practice to regard the entire mixture as consisting of the components requiring the highest amount of dilution per unit liquid volume and to calculate the required air quantity on that basis. (This component would be the one with the highest value for SG/(MW)(LEL).)

4.8 VENTILATION FOR HEAT CONTROL

Ventilation for heat control in a hot industrial environment is a specific application of general industrial ventilation. The primary function of the ventilation system is to prevent the acute discomfort, heat-induced illness, and possible injury of those working in or generally occupying a designated hot industrial environment. Heat-induced occupational illnesses, injuries, or reduced productivity may occur in situations where the total heat load may exceed the defenses of the body and result in a heat stress situation. It follows, therefore, that a heat control ventilation system or other engineering control method must follow a physiological evaluation in terms of potential heat stress for the occupant in the hot industrial environment.

Due to the complexity of conducting a physiological evaluation, the criteria presented here are limited to general considerations. It is strongly recommended, therefore, that the NIOSH Publication No. 86-113, *Criteria for a Recommended Standard, Occupational Exposure to Hot Environments*,[4.5] be reviewed thoroughly in the process of developing the heat control ventilation system.

The development of a ventilation system for a hot industrial environment usually includes the control of the ventilation airflow rate, velocity, temperature, humidity, and airflow path through the space in question. This may require inclusion of certain phases of mechanical air-conditioning engineering design which is outside the scope of this Manual. The necessary engineering design criteria that may be required are available in appropriate publications of the American Society of Heating, Refrigerating and Air-Conditioning Engineers (ASHRAE) handbook series.

4.9 HEAT BALANCE AND EXCHANGE

An essential requirement for continued normal body function is that the deep body core temperature be maintained within the acceptable range of about 98.6 F [37 C] ± 1.8 F [1 C]. To achieve this, body temperature equilibrium requires a constant exchange of heat between the body and the environment. The rate and amount of the heat exchange are governed by the fundamental laws of thermodynamics of heat exchange between objects. The amount of heat that must be exchanged is a function of 1) the total heat produced by the body (metabolic heat), which may range from about 1 kilocalorie (kcal) per kilogram (kg) of body weight per hour (1.16 watts) at rest to 5 kcal/kg body weight/hour (7 watts) for moderately hard industrial work; and 2) the heat gained, if any, from the environment. The rate of heat exchange with the environment is a function of air temperature and humidity, skin temperature, air velocity, evaporation of sweat, radiant temperature, and type, amount, and characteristics of the clothing worn, among other factors. Respiratory heat loss is of little consequence in human defenses against heat stress.

The basic heat balance equation is:

$$\Delta S = (M - W) \pm C \pm R - E \qquad [4.16]$$

where: ΔS = change in body heat content

(M–W) = total metabolism – external work performed

C = convective heat exchange

R = radiative heat exchange

E = evaporative heat loss

To solve the equation, measurement of metabolic heat production, air temperature, air water vapor pressure, wind velocity, and mean radiant temperature are required.

The major modes of heat exchange between man and the environment are conduction, convection, radiation, and evaporation.

4.9.1 Conduction. Other than for brief periods of body contact with hot tools, equipment, floors, etc., which may cause burns, conduction plays a minor role in industrial heat stress. Because of the typically small areas of contact between either body surfaces or its clothing and hot or cold objects, heat exchange by thermal conduction is usually not evaluated in a heat balance equation for humans. The effect of heat exchange by thermal conduction in human thermal regulation is important only when large areas of the body are in contact with surfaces that are at temperatures different from average skin temperature (nominally 95 F [35 C]), as when someone is prone or supine for long periods. It is important also when even small body areas are in contact with objects that provide steep thermal gradients for heat transfer, as when someone is standing on very cold or very hot surfaces.

FIGURE 4-11. Centrifugal fan

The equations for calculating heat exchange by convection, radiation, and evaporation are available in Standard International (SI) units, metric units, and English units. In SI units, heat exchange is in watts per square meter of body surface (W/m^2). The heat exchange equations are available in metric and English units for both the semi-nude individual and the worker wearing conventional long-sleeved work shirt and trousers. The values are in kcal/h or British thermal units per hour (BTU/h) for the "standard worker" defined as one who weighs 154 lbs [70 kg] and has a body surface area of 19.4 ft^2 [1.8 m].

4.9.2 Convection. The rate of convective heat exchange between the skin of a person and the ambient air immediately surrounding the skin is a function of the difference in temperature between the ambient air (T_a), the mean weighted skin temperature (T_{sk}) and the rate of air movement over the skin (V_a). This relationship is stated algebraically for the "standard worker" wearing the customary one layer work clothing ensemble as:

$$C = 0.65\,V_a^{0.6}\,(T_a - T_{sk}) \quad \textbf{[4.17] IP}$$

$$[C = 7\,V_a^{0.6}\,(T_a - T_{sk})] \quad \textbf{[4.17] SI}$$

where: C = convective heat exchange, BTU/h [k_{cal}]

V_a = air velocity, fpm [m/s]

T_a = air temperature, F [C]

T_{sk} = mean weighted skin temperature, usually assumed to be 95 F [35 C]

When $T_a > 95$ F [35 C], there will be a gain in body heat from the ambient air by convection. When $T_a < 95$ F [35 C], heat will be lost from the body to the ambient air by convection.

4.9.3 Radiation. Infrared radiant heat exchange between the exposed surfaces of a person's skin and clothing varies as a function of the difference between the fourth power of the absolute temperature of the exposed surfaces and that of the surface of the radiant source or sink, the exposed areas and their emissivities. Heat is gained by thermal radiation if the facing surface is warmer than the average temperature of the exposed skin and clothing and *vice versa*. A practical approximation for infrared radiant heat exchange for a person wearing conventional clothing is:

$$R = 15.0\,(T_W - T_{sk}) \quad \textbf{[4.18] IP}$$

$$[R = 6.8\,(T_W - T_{sk})] \quad \textbf{[4.18] SI}$$

where: R = radiant heat exchange, BTU/h [k_{cal}]

T_w = mean radiant temperature, F [C]

T_{sk} = mean weighted skin temperature

4.9.4 Evaporation. The evaporation of water (sweat) or other liquids from the skin or clothing surfaces results in a heat loss from the body. Evaporative heat loss for humans is a function of airflow over the skin and clothing surfaces, the water vapor partial pressure gradient between the skin surface and the surrounding air, the area from which water or other liquids are evaporating and mass transfer coefficients at their surfaces.

$$E = 2.4\,V_a^{0.6}\,(\rho_{sk} - \rho_a) \quad \textbf{[4.19] IP}$$

$$[E = 14.4\,V_a^{0.6}\,(\rho_{sk} - \rho_a)] \quad \textbf{[4.19] SI}$$

where: E = evaporative heat loss, BTU/h [k_{cal}]

V_a = air velocity, fpm [m/s]

ρ_a = water vapor pressure of ambient air, mmHg

ρ_{sk} = water vapor pressure on the skin, assumed to be 42 mmHg at a 95 F [35 C] skin temperature

FIGURE 4-12. Axial flow fan

4.10 ADAPTIVE MECHANISM OF THE BODY

Even people in generally good health can adjust physio-

FIGURE 4-13. Ribbed flexible duct

logically to thermal stress only over a narrow range of environmental conditions. Unrestricted blood flow to the skin, an unimpeded flow of dry, cool air over the skin surface and sweating are prime defenses in heat stress. Although heat produced by muscle activity reduces the impact of cold stress, it can add substantially to the total challenge during heat stress. Diminished health status, medications, limited prior thermal exposure, among other factors, increase danger to thermal stresses.

The reflex control of blood flow is the body's most effective and important first line of defense in facing either cold or heat stress. Reducing blood flow to the skin of the hands, feet, fingers, and toes is an important measure for reducing heat loss in a cold environment. Blood flow to the skin, however, increases many-fold during heat stress. Its effect is to increase rates of heat distribution in the body and maximize conductive, convective, radiative, and evaporative heat losses to the environment (Figure 4-14). Its cost is often to reduce perfusion of other organs, especially the brain, and reduce systemic arterial blood pressure leading to reduced consciousness, collapse, heat exhaustion, and other heat-induced illnesses.

Reflex sweating during the physical activities of exercise, work, and/or heat stress often brings large volumes of body water and electrolytes (salts) to the skin surface. Heat is lost when the water in sweat evaporates. Whether the electrolytes remain on the skin surface or are deposited in clothing, they are nonetheless permanently lost to the body. The electrolyte content of a typical American diet usually provides adequate electrolyte replacement for these losses. Electrolyte replacement fluids, however, may be necessary for people on salt-restricted diets and those who commonly sustain periods of prolonged and profuse sweating. It is essential for everyone that the lost body water and electrolytes are replaced in the same volume and proportion as lost in sweat. Muscle spasms, cramps, gastrointestinal disturbances, and general malaise, among other signs and symptoms, commonly develop when they are not.

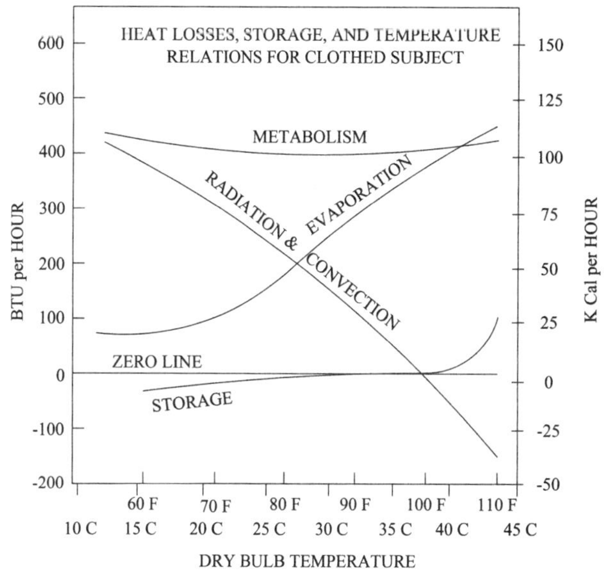

FIGURE 4-14. Heat losses, storage, and temperature relations

4.11 ACCLIMATIZATION

People in general good health normally develop heat acclimatization in a week or so after intermittently working or exercising in a hot environment. Its effect is to improve the comfort and safety of the heat exposure. It occurs because of an increase in total circulating blood volume, an improved ability to maintain systemic arterial blood pressure during heat stress, a developed ability to produce larger volumes of more dilute sweat, the rate of production of which is more precisely matched to the heat load. Heat acclimatization rapidly diminishes even after a day or so of discontinued activity in the heat — most is lost after about a week.

4.12 ACUTE HEAT DISORDERS

A variety of heat disorders can be distinguished clinically when individuals are exposed to excessive heat. A brief description of these disorders follows.

4.12.1 Heatstroke (also called "Sunstroke"). Heat stroke is a **life-threatening** condition which, without exception, demands immediate emergency medical care and hospitalization. Before medical care arrives, move the person to a shaded area, check for other injuries, ensure there is an unobstructed airway, remove or loosen clothing, and flood the body surface with free-flowing, tepid (not cold) water. Vigorous fanning helps cooling. Heat stroke develops when body heat gains from exercise, work, and/or a hot environment overwhelm normal thermoregulatory defenses. Characteristically, sweating has ceased, the skin is hot and dry, and deep body temperature is above about 104 F [40 C]. The person may be either diaphoretic, semiconscious, unconscious or agitated, delirious, and in convulsions. Demand medical care even if consciousness returns — lethal effects may develop in the next 24 to 72 hours.

4.12.2 Heat Exhaustion (also Called "Exercise-induced Heat Exhaustion," "Heat Syncope"). Heat exhaustion most commonly occurs in people who are not heat acclimatized and who are in poor physical condition, obese, inappropriately dressed for a heat stress and exercising, or working energetically in the heat at unaccustomed and/or demanding tasks. It is characterized by lightheadedness, dizziness, vision disturbances, nausea, vague flu-like symptoms, tinnitus, weakness, and occasionally, collapse. The person's deep body temperature is typically in a normal range or only slightly elevated; the skin is moist and cool but may be reddened by its high rate of blood flow. Heat exhaustion develops when there is reflex demand for blood flow to the skin to dissipate body heat and a simultaneous reflex demand for blood flow to exercising muscles to meet metabolic needs of increased activity. These peripheral distributions of blood volume reduce systemic arterial pressure and brain blood flow, causing most of the symptoms of heat exhaustion. Rest in a cool environment where there is freely flowing, dry air usually remediates symptoms quickly. Although heat exhaustion is debilitating and uncomfortable, it is not often a long-term health threat. There are considerable dangers, of course, for anyone operating machinery when consciousness is impaired because of heat exhaustion or for any other reason.

4.12.3 Heat Cramps ("Muscle Cramps") and Heat Rash ("Prickly Heat," "Miliaria Rubia"). Spontaneous, involuntary, painful, and prolonged muscle contractions commonly occur in otherwise healthy people when both body water and electrolyte levels have not been restored after extended periods of heavy sweating during exercise and/or heat stress. Full recovery can be expected in about 24 hours with the use of electrolyte replacement fluids and rest. Heat rash is an acute, inflammatory skin disease characterized by small red, itchy or tingling lesions, commonly in areas of skin folds or where there is abrasive clothing. It commonly disappears when these areas are kept dry, unabraded and open to free flowing, dry air.

4.13 ASSESSMENT OF HEAT STRESS AND HEAT STRAIN

Heat Stress is defined by environmental measurements of air temperature, humidity, airflow rate, the level of radiant heat exchange, and evaluation of a person's metabolic heat production rate from exercise and/or work. Heat stress is the load on thermoregulation. *Heat Strain* is defined as the cost to each person facing heat stress. Although all people working at the same intensity in the same environment face the same level of heat stress, each is under a unique level of heat strain. Almost any environmental thermal exposure will be comfortable and safe for some, but endangering, even lethal to others. Because disabilities, danger, and death arise directly from heat strain no measure of heat stress is a reliable indicator of a particular person's heat strain, or the safety of the exposure.

4.13.1 Evaluation of Heat Stress. Dry-bulb air temperature (DB: so-called "dry-bulb" temperature) is measured by calibrated thermometers, thermistors, thermocouples, and similar temperature-sensing devices which themselves do not produce heat and which are protected from the effects of thermal conduction, evaporation, condensation, and radiant heat sources and sinks. Relative humidity is evaluated psychrometrically as a function of the steady state difference between "dry-bulb" temperature and that indicated by the temperature of a sensor covered with a freely evaporating, water-saturated cotton wick. Such a measure reports "NWB" (natural wet-bulb temperature) when the wetted sensor is affected only by prevailing air movement, and "WB" (when it is exposed to forced convection). Free air movement is measured with an unobstructed anemometer. Infrared radiant "heat transfer" is typically measured by a temperature sensor at the center of a 6-inch, hollow, copper sphere painted flat ("matte") black. Such a measure reports "GT" (globe temperature) (Figure 4-15). A person's metabolic heat production is usually evaluated from an estimated level of average physical activity (Table 4-2).

Although there are a number of different indices for evaluating heat stress, none is reliable as a sole indicator of heat strain for a specific person. "Dry-bulb" temperature is the least valuable measure of heat stress because it provides no information about ambient relative humidity, or heat exchange by convec-

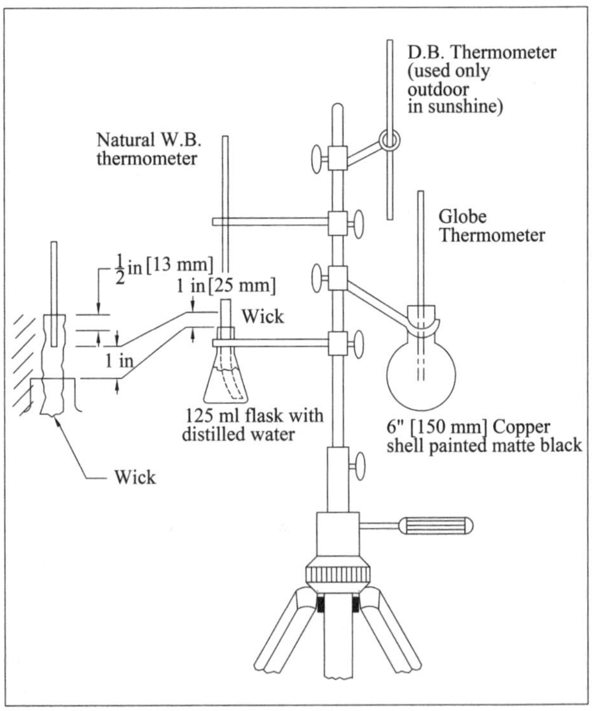

FIGURE 4-15. Determination of wet-bulb globe temperature

TABLE 4-2. Estimating Energy Consumed by Task/Work Performed

A. Body position and movement	Btu/hr [kcal/min]*	
Sitting	0.07 [0.3]	
Standing	0.14 [0.6]	
Walking	0.05 – 0.71 [2.0 – 3.0]	
Walking uphill	Add 0.06/ft rise [0.8 /meter rise]	
B. Type of work kcal/min	**Btu/hr** [Average kcal/min]	**Range**
Hand work – light	0.10 [0.4]	0.05 – 0.28 [0.2 – 1.2]
Hand work – heavy	0.22 [0.9]	
Work one arm – light	0.24 [1.0]	0.17 – 0.60 [0.7 – 2.5]
Work one arm – heavy	0.40 [1.7]	
Work both arms – light	0.36 [1.5]	0.24 – 0.83 [1.0 – 3.5]
Work both arms – heavy	0.56 [2.5]	
Work whole body – light	0.83 [3.5]	0.50 – 3.58 [2.5 – 15.0]
Work whole body – moderate	1.19 [5.0]	
Work whole body – heavy	1.67 [7.0]	
Work whole body – very heavy	2.14 [9.0]	
C. Basal metabolism	0.24 [1.0]	
D. Sample calculation**		
Assembling work with heavy hand tools		
1. Standing	0.14 [0.6]	
2. Two-arm work	0.83 [3.5]	
3. Basal metabolism	0.24 [1.0]	
TOTAL	1.24 Btu/hr [5.1 kcal/min]	

*For standard worker of 154 lbs [70 kg] weight and body surface of 19.4 ft^2 [1.8 m^2].

**Example of measuring metabolic heat production of a worker when performing initial screening.

tion or radiation, and gives no estimate of the metabolic heat production. "Wet-Bulb, Globe Temperature" (WBGT) is often used as an index of heat stress. When there is a source of radiant heat transfer (solar radiation, hot surfaces of machinery):

$$WBGT = 0.7\ T_{nwb} + 0.2\ T_g + 0.1\ T_a \quad \textbf{[4.20]}$$

where: T_{nwb} = natural wet-bulb temperature
T_g = globe temperature
T_a = ambient temperature

When radiant heat transfer is negligible:

$$WBGT = 0.7\ T_{nwb} + 0.3\ T_g \quad \textbf{[4.21]}$$

WBGT evaluates more factors contributing to heat stress than does the measure of DB alone. It does not, however, effectively evaluate the importance of mass and energy transfer from human skin by convection which is essential for the removal of heat from the skin surface and the formation of water vapor from secreted sweat. Nor does WBGT evaluate the importance of metabolic heat production in heat stress. Under many environmental conditions, heat produced by metabolism is the predominant, sometimes lethal, stressor.

4.13.2 Evaluation of Heat Strain. The incidence and severity of heat strain will vary greatly among people, even though all are exposed to the same level of heat stress. Paying attention to the early signs and symptoms of heat strain is the best first line of defense against debilitating heat-induced discomfort and injuries. It is dangerous, inappropriate, and irresponsible to consider a heat stress as safe for all when some exposed to it show heat strain signs and symptoms, while others do not. Acute heat strain is indicated by:

Visible Sweating: Thermoregulatory reflexes normally fine-tune with precision the rate of sweating to the rate at which body heat must be lost to maintain homeostasis. Normally, there is no liquid water on the skin surface in a tolerable heat stress because water brought to the skin surface by sweating readily forms invisible water vapor in the process of evaporative cooling. Although an all too common occurrence in the workplace, liquid sweat either on the skin surface, or soaked into clothing, is a sure sign of heat strain. It indicates the level of sweating required to keep body temperature in a normal range cannot be matched by the rate of water evaporation from the skin surface to the environment. It is necessary either to increase the airflow rate over skin and clothing surfaces, lower ambient temperature and relative humidity, reduce radiant heat gain, and/or reduce metabolic heat production if progressive heat disabilities are to be avoided. Visible sweating is an indisputable sign of heat strain.

Discontinued Sweating: A hot, dry skin for someone exposed to heat stress is a dangerous sign. It indicates either suppression of sweating, as perhaps by prescription, or even over-the-counter medications, or an entry level into heat stroke. The appearance of a hot, dry skin for someone in a heat stress demands immediate attention and corrective actions.

Elevated Heart Rate: Short term increases in heart rate are normal for episodic increases in work load. In a heat stress, however, a sustained heart rate greater than 160/min for those younger than about 35 years, or 140/min for those who are older, is a sign of heat strain.

Elevated Deep Body Temperature: A sustained deep body temperature greater than 100.4 F [38 C] is a sign of heat strain in someone exposed to heat stress.

Decreased Systemic Arterial Blood Pressure: A fall in blood pressure of more than about 40 Torr in about 3.5 minutes for someone working in a heat stress indicates a heat-induced disability. Reduced consciousness, feeling of weakness, vision disturbances, and other signs and symptoms are likely to follow.

Personal Discomfort: Heat strain may be indicated by people exposed to heat stress by severe and sudden fatigue, nausea, dizziness, lightheadedness, or fainting. Others may complain of irritability, mental confusion, clumsiness for otherwise competently executed skills, forgetfulness, general malaise and the development of sometimes vague, flu-like symptoms, and paradoxical chills and shivering.

Infrequent Urination: Urinating less frequently than normal and the voiding of a small volume of dark-colored urine is a sign of whole body dehydration. Such dehydration compromises the body's ability to maintain a large enough circulating blood volume so that normal blood pressure is maintained in the face of the combined stressors of exercise and heat exposure. People who work or exercise in the heat need to develop the habit of drinking adequate volumes of water at frequent enough intervals to maintain the same patterns of urination they have when not heat stressed. Those who sweat heavily for long periods should also evaluate with their physicians whether electrolyte replacement fluids are needed.

4.14 WORKER PROTECTION

There is improved safety, comfort, and productivity when those working in the heat are:

1) In generally good physical condition, not obese, heat acclimatized, and experienced in the heat stressing job. They also need to know how to select clothing and maintain whole body hydration and electrolyte levels to provide the greatest comfort and safety.
2) In areas that are well-ventilated and shielded from infrared radiant heat sources.
3) Knowledgeable about the effects of their medications affecting cardiovascular and peripheral vascular function, blood pressure control, body temperature maintenance, sweat gland activity, metabolic effects, and levels of attention or consciousness.
4) Appropriately supervised when there is a history of abuse or recovery from abuse of alcohol or other intoxicants.
5) Provided accurate verbal and written instructions, frequent training programs, and other information about heat stress and strain.
6) Able to recognize the signs and symptoms of heat strain in themselves and others exposed to heat stress and know the appropriate steps for their remediation (Figures 4-16 and 4-17).

4.15 VENTILATION CONTROL

The control method presented here is limited to a general

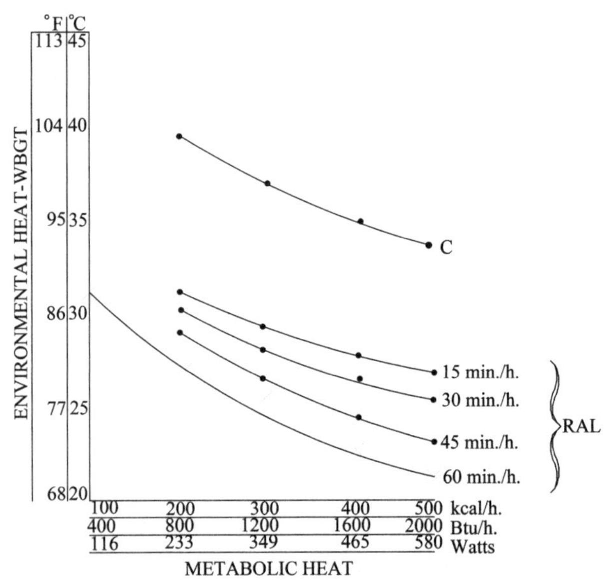

C = Ceiling Limit
RAL = Recommended Alert Limit

*For "Standard worker" of 70 kg (154 lbs) body weight and 1.8 m^2 (19.4 ft^2) body surface

FIGURE 4-16. Recommended heat stress alert limits (unacclimatized workers)

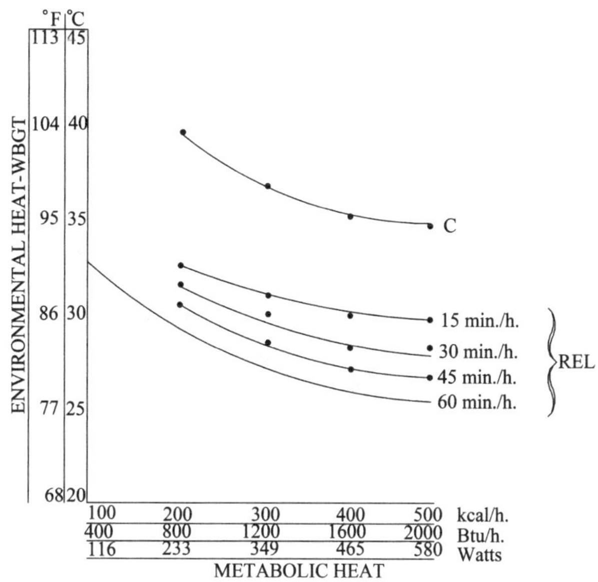

C = Ceiling Limit
REL = Recommended Exposure Limit
*For "Standard worker" of 70 kg (154 lbs) body weight and 1.8 m^2 (19.4 ft^2) body surface

FIGURE 4-17. Recommended heat stress exposure limits (acclimatized workers)

engineering approach. Due to the complexity of evaluating a potential heat stress producing situation, it is essential that the accepted industrial hygiene method of recognition, evaluation, and control be utilized to its fullest extent. In addition to the usual time limited exposures, it may be necessary to specify additional protection which may include insulation, baffles, shields, partitions, personal protective equipment, administrative control, and other measures to prevent possible heat stress. Ventilation control measures may require a source of cooler replacement air, an evaporative or mechanically cooled source, a velocity cooling method, or any combination thereof. Specific guidelines, texts, and other publications or sources should be reviewed for the necessary data to develop the ventilation system.

4.16 VENTILATION SYSTEMS

Exhaust ventilation can be used to remove excessive heat and/or humidity if a replacement source of cooler air is available. If it is possible to enclose the heat source, such as in the case of ovens or certain furnaces, a gravity or forced air stack may be all that is necessary to prevent excessive heat from entering the workroom. If a partial enclosure or local hood is indicated, control velocities, as shown in Chapters 6 and 13, can be estimated from the volume of air to be exhausted.

Many operations do not lend themselves to local exhaust. General ventilation may be the only alternative. To determine the required general ventilation, the designer must estimate the acceptable temperature or humidity rise. The first step in determining the required volumetric flow is to determine the sensible and latent heat load. Next, determine the volumetric flow to dissipate the sensible heat and the volumetric flow to dissipate the latent heat. The required general ventilation is the larger of the two volumetric flows.

The sensible heat rise can be determined by the following:

$$H_s = Q_s \times \rho \times c_p \times \Delta T \times (60\ \text{min/hr}) \qquad \textbf{[4.22] IP}$$

$$[H_s = Q_s \times \rho \times c_p \times \Delta T \times (60\ \text{s/min})] \qquad \textbf{[4.22] SI}$$

where: H_s = Sensible heat gain, BTU/hr
Q_s = Volumetric flow for sensible heat, acfm [am³/s]
ρ = Density of the air, lbm/ft³ [kg/m³]
c_p = Specific heat of the air, BTU/lbm F [kJ /kg-deg C]
ΔT = Change in temperature, F [C]

For air, c_p = 0.24 BTU/lbm-F [1.0 w-s/kg-C] and ρ = 0.075 lbm/ft³ [1.20 kg/m³].
Consequently, the equation becomes:

$$H_s = 1.08 \times Q_s \times \Delta T$$

$$[H_s = 20 \times Q_s \times \Delta T]$$

or

$$Q_s = H_s \div (1.08 \times \Delta T) \qquad \textbf{[4.23] IP}$$

$$[Q_s = H_s \div (20 \times \Delta T)] \qquad \textbf{[4.23] SI}$$

In order to use this equation, it is necessary to first estimate the heat load. This will include solar radiation, people, lights, and motors as well as other particular sources of heat. Of these, solar radiation, lights, and motors are all completely sensible. The people heat load is part sensible and part latent. In the case of hot processes that give off both sensible and latent heat, it will be necessary to estimate the amounts or percentages of each. In using the above equation for sensible heat, one must decide the amount of temperature rise that will be permitted. Thus, in a locality where 90 F [32 C] outdoor ω dry-bulb may be expected, if it is desired that the inside temperature not exceed 100 F [38 C], or a 10 F [6 C] rise, a certain airflow rate will be necessary. If an inside temperature of 95 F [35 C] is required, the airflow rate will be doubled.

For latent heat load, the procedure is similar although more difficult. If the total amount of water vapor is known, the heat load can be estimated from the latent heat of vaporization, 970 BTU/lb (IP units). In a manner similar to the sensible heat calculations, the latent heat gain can be approximated by:

$$H_l = Q_l \times \rho \times c_l \times \Delta\omega \times (60\ \text{min/hr}) \times (1\ \text{lb/7000 grains}) \qquad \textbf{[4.24] IP}$$

where: H_l = Latent heat gain, BTU/hr
Q_l = Volumetric flow for latent heat, acfm
ρ = Density of the air, lbm/ft³
c_l = Latent heat of vaporization, BTU/lbm
$\Delta\omega$ = Change in absolute humidity of the air, grains-water/lbm-dry air

For air, c_l is approximately 970 BTU/lb and ρ = 0.075 lbm/ft³. Consequently, the equation becomes:

$$H_l = 0.62 \times Q_l \times \Delta\omega$$

or

$$Q_l = H_l \div (0.62 \times \Delta\omega) \quad \textbf{[4.25] IP}$$

If the rate of moisture released ($\dot{m}$ in pounds per hour) is known, then:

$$\dot{m} = Q_l \times \rho \times \Delta\omega \times (1\ \text{lb}/7000\ \text{gr}) \times (60\ \text{min/hr})$$
$$= Q_l \times \rho \times \Delta\omega \div (0.00857)$$

or

$$Q_1 = \dot{m} \div (\rho \times \Delta\omega \times 0.00857) \quad \textbf{[4.26] IP}$$

The term "grains-water per pound-air difference" is taken from the psychrometric chart or tables, and represents the difference in moisture content of the outdoor air and the conditions acceptable to the engineer designing the exhaust system. The air quantities calculated from the above two equations **should not** be added to arrive at the required quantity. Rather, the higher quantity should be used since both sensible and latent heat are absorbed simultaneously. Furthermore, in the majority of cases, the sensible heat load far exceeds the latent heat load so the design can be calculated only on the basis of sensible heat.

The ventilation should be designed to flow through the hot environment in a manner that will control the excess heat by removing it from that environment. Figures 4-18 and 4-19 illustrate this principle.

In SI units, the procedure is similar. If the total amount of water vapor is known, the heat load can be estimated from the latent heat of vaporization, 2.256×10^6 W-s/kg. In a manner similar to the sensible heat calculations, the latent heat gain can be approximated by:

$$H_1 = Q_1 \times \rho \times c_1 \times \Delta\omega \div (60\ \text{s/min})$$

where:

H_l = Latent heat gain, W

Q_l = Volumetric flow for latent heat, m³/s

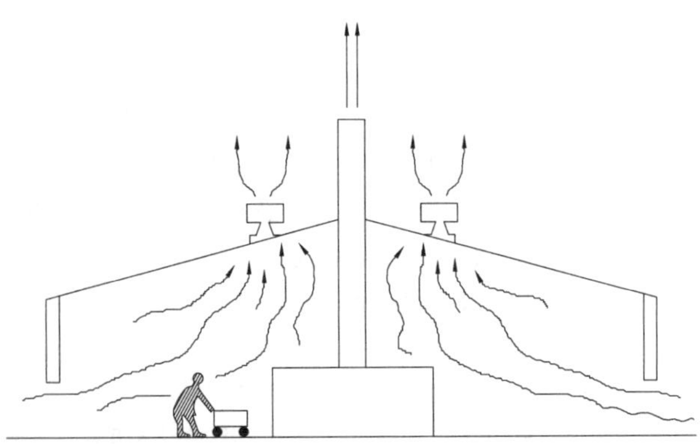

FIGURE 4-18. Good natural ventilation and circulation

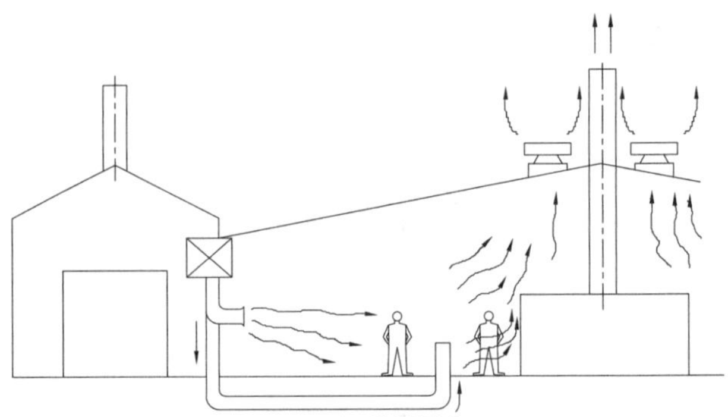

FIGURE 4-19. Good mechanically supplied ventilation

ρ = Density of the air, kg/m³

c_l = Latent heat of vaporization, W-s/kg

$\Delta\omega$ = Change in absolute humidity of the air, kg-water/kg-dry air

For air, c_l is approximately 2.256×10^6 W-s/kg and ρ = 1.2 kg/m³. Consequently, the equation becomes

$$H_l = 45 \times 10^3 \times Q_l \times \Delta\omega$$

or

$$Q_l = \frac{H_l}{45 \times 10^3 \times \Delta\omega} \quad \textbf{[4.27] SI}$$

If the rate of moisture released ($\dot{m}$ in kilograms per second) is known, then

$$\dot{m} = Q_l \times \rho \times \Delta\omega$$

or

$$Q_l = \frac{\dot{m}}{\rho \times \Delta\omega} \quad \textbf{[4.28] SI}$$

4.17 VELOCITY COOLING

If the air dry-bulb or wet-bulb temperatures are lower than 95 – 100 F [35 – 38 C], the worker may be cooled by convection or evaporation. When the dry-bulb temperature is higher than 95 – 100 F [35 – 38 C], increased air velocity may add heat to the worker by convection; if the wet-bulb temperature is high also, evaporative heat loss may not increase proportionately, and the net result will be an increase in the worker's heat burden. Many designers consider that supply air temperature should not exceed 80 F for practical heat relief.

Current practice indicates that air velocities in Table 4-3 can be used successfully for direct cooling of workers. For best results provide directional control of the air supply (Figure 4-20) to accommodate daily and seasonal variations in heat exposure and supply air temperature.

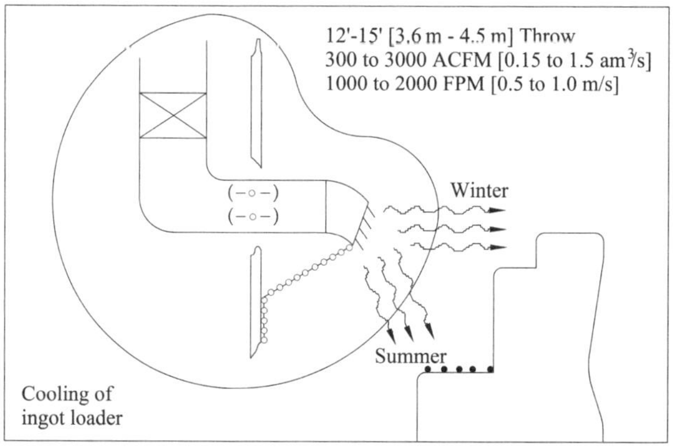

FIGURE 4-20. Spot cooling with volume and directional control

4.18 RADIANT HEAT CONTROL

Since radiant heat is a form of heat energy which needs no medium for its transfer, radiant heat cannot be controlled by any of the above means. Painting or coating the surface of hot bodies with materials having low radiation emission characteristics is one method of reducing radiation.

For materials such as molten masses of metal or glass that cannot be controlled directly, radiation shields are effective. These shields can consist of metal plates, screens, or other material interposed between the source of radiant heat and the workers. Shielding reduces the radiant heat load by reflecting the major portion of the incident radiant heat away from the operator and by re-emitting to the operator only a portion of that radiant heat which has been absorbed. Table 4-4 indicates the percentage of both reflection and emission of radiant heat associated with some common shielding materials. Additional ventilation will control the sensible heat load but will have only a minimal effect, if any, upon the radiant heat load (Figure 4-21).

4.19 PROTECTIVE SUITS FOR SHORT EXPOSURES

For brief exposures to very high temperatures, insulated aluminized suits and other protective clothing may be worn. These suits reduce the rate of heat gain by the body but provide no means of removing body heat; therefore, only short exposures may be tolerated.

4.20 RESPIRATORY HEAT EXCHANGERS

For brief exposure to air of good quality but high temperature, a heat exchanger on a half-mask respirator face piece is available. This device will bring air into the respiratory passages at a tolerable temperature but will not remove contaminants nor furnish oxygen in poor atmospheres.

4.21 REFRIGERATED SUITS

Where individuals must move about, cold air may be blown

TABLE 4-3. Acceptable Comfort Air Motion at the Worker

	Air Velocity, fpm* [m/s*]
Continuous Exposure	
Air conditioned space	50 – 75 [0.25 – 0.38]
Fixed work station, general ventilation or spot cooling: Sitting	75 – 125 [0.38 – 0.63]
Standing	100 – 200 [0.50 – 1.00]
Intermittent Exposure, Spot Cooling or Relief Stations	
Light heat loads and activity	1000 – 2000 [5 – 10]
Moderate heat loads and activity	2000 – 3000 [10 – 15]
High heat loads and activity	3000 – 4000 [15 – 20]

*Note: Velocities greater than 1000 fpm [5 m/s] may seriously disrupt the performance of nearby local exhaust systems. Care must be taken to direct air motion to prevent such interference.

TABLE 4-4. Relative Efficiencies of Common Shielding Materials

Surface of Shielding	Reflection of Radiant Heat Incident Upon Surface	Emission of Radiant Heat from Surface
Aluminum, bright	95	5
Zinc, bright	90	10
Aluminum, oxidized	84	16
Zinc, oxidized	73	27
Aluminum paint, new, clean	65	35
Aluminum paint, dull, dirty	40	60
Iron, sheet, smooth	45	55
Iron, sheet, oxidized	35	65
Brick	20	80
Lacquer, black	10	90
Lacquer, white	10	90
Asbestos board	6	94
Lacquer, flat black	3	97

into a suit or hood worn as a portable enclosure. The usual refrigeration methods may be used with insulated tubing to the suit. It may be difficult, however, to deliver air at a sufficiently low temperature. If compressed air is available, cold air may be delivered from a vortex tube worn on the suit. Suits of this type are commercially available.

4.22 ENCLOSURES

In certain hot industries, such as in steel mills, it is unnecessary and impractical to attempt to control the heat from the process. If the operation is such that remote control is possible,

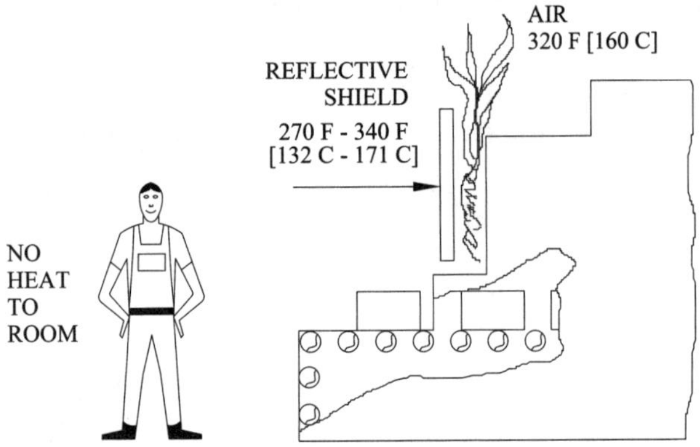

FIGURE 4-21. Heat shielding

an air conditioned booth or cab can be utilized to keep the operator reasonably comfortable in an otherwise intolerable atmosphere.

4.23 INSULATION

If the source of heat is a surface giving rise to convection, insulation at the surface will reduce this form of heat transfer. Insulation by itself, however, will not usually be sufficient if the temperature is very high or if the heat content is high.

REFERENCES

4.1 American Industrial Hygiene Association: The Occupational Environment: Its Evaluation, Control & Management, Second Edition (2003).

4.2 Air Force: AFOSH Standard 161.2 (1977).

4.3 National Fire Protection Association 8G Standard for Ovens and Furnaces (2007).

4.4 Feiner, B.; Kingsley, L.: Ventilation of Industrial Ovens. Air Conditioning, Heating and Ventilating, pp. 82-89 (December 1956).

4.5 U.S. Department of Health and Human Services, PHS, CDC, NIOSH: Occupational Exposure to Hot Environments, Revised Criteria (1986).

Chapter 5

DESIGN ISSUES – SYSTEMS

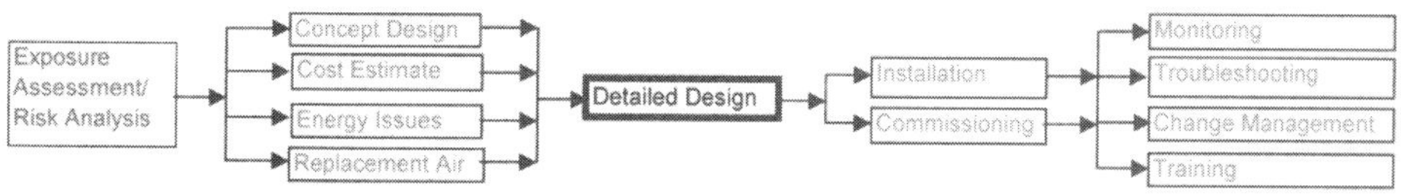

NOTE: Equations with notation followed by (IP) are designated for inch-pound system only; Equations followed by (SI) are designated for metric use only. If Equation bears neither, then it applies to both systems.

5.1 ADMINISTRATION OF INDUSTRIAL VENTILATION SYSTEM DESIGN

5.1.1 Design Organization and Administration. Successful industrial ventilation systems installations are not limited to the hardware placed in the plant. They include the proper communication of ideas, responsibilities, expectations and verification among all parties on the project. Because most industrial ventilation systems can include regulatory requirements (OSHA, EPA or State emission limits, etc.) there must be proper document control and execution to complement the actual installation.

A method to implement controls and provide the cross-checks and proper design decisions is the use of a Project Team concept for the installation of local exhaust ventilation systems. Figure 5-1 shows the flow chart for the information and communications between the owner and the parties responsible for design and installation of the project. Because the installation of these systems also impacts regulatory responsibilities and legal issues (guarantees for proof of performance, OSHA exposures, environmental emission, etc.), the need for communication and proper transfer and control of information is more critical than usual plant projects.

The size of the Project Team may vary based on the critical nature of the materials controlled, the size of the project, and the size of the company. In some cases, a large company may furnish all of the services in the diagram. Smaller companies, or those not having the specific skills required for air-cleaning installations, may use outside resources for some or all of the services. In either case, it is important that the Project Team documents the decision process and the communications relative to it (electronic and hard copies). Methods and guidelines for the organization of Project Teams and methods of design management are included in Chapter 2.

5.1.2 Tools for Design Communications. There must be a method to ensure the implementation of the system design requirements. Good project organization can provide the tools to give good communication tools and chains of command and responsibility. Many of the project requirements are determined in the Preliminary Design Phase and at the same time problems and pitfalls in a schedule and installation can be identified. The final product of this Preliminary Design Phase is the Design Basis. In effect, this is a set of expectations and instructions to the Design Team.

The Design Basis should include information to define the successful completion of the project. Elements can include the specification of the types of standards for material and equipment to be selected. It can also include the requirements for passing all applicable tests for emissions and OSHA limits for exposure in the project area or at operator stations as well as special plant or company reliability requirements such as production levels or operating time. It also would include the requirements for training and operation of the system.

After completion of the design and issuance of a design package to the selected contractor (usually through a bidding process), the contract must then be completed and verified for final acceptance by the owner of the system. Until this transfer of responsibility is made, the system remains in the hands of the Project Team and Project Manager. Details and forms for this process are included in Chapter 2 of the O&M Manual (Commissioning and Proof of Performance). It should be used for systems of all sizes. Smaller systems may only require "as-built" drawings and maintenance manuals for purchased equipment as additional documentation. Even the smallest system should document details for design including ACGIH® calculation sheets, methods for selection of hood volumes, fan curves and other related information. This information is then available for use in the event of future alterations or troubleshooting (see Chapter 5 of the O&M Manual).

The Design Basis should also indicate the owner and the format for receiving the information through the commissioning process. Drawings may be furnished in electronic as well as traditional printed formats. There also may be requirements to input the specifications and maintenance requirements into plant management software. Results of compliance stack tests will need to be communicated to appropriate plant and regula-

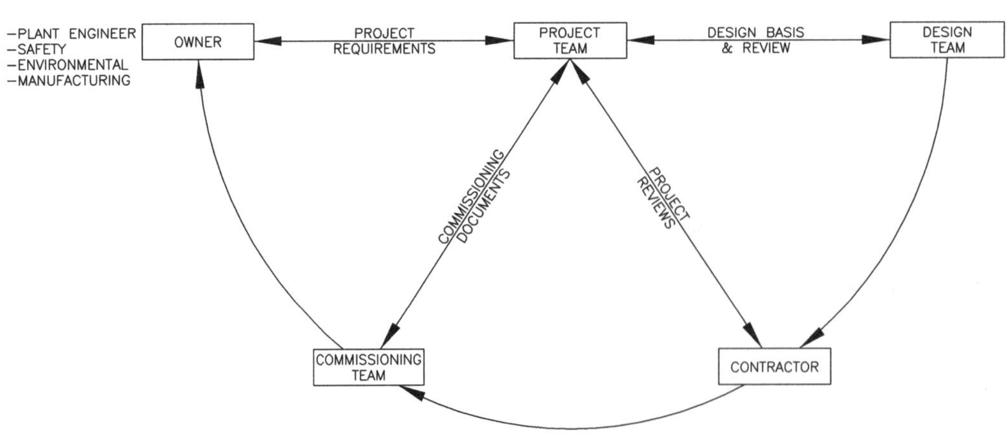

FIGURE 5-1. Organizational flow chart

tory personnel. All of these items will need to be considered to close the project properly.

As with all transfer of information, the requirements for acceptance by the Owner must be clear before the project is bid so that vendors and contractors have included all in their costs and prices. Similarly, plant personnel responsible for other in-house closure requirements must be informed well before the project is completed to ensure that they have sufficient time to review and approve the transfer of ownership. It may be prudent to start at the end of the project and determine the requirements for commissioning and then work forward to a point where design can meet the final needs of the system. This includes detailed information with regard to issues for proper completion of the project and who is responsible for the ownership of the system after installation.

If the Calculation Sheet is the basic design document, then the Design Basis, Team Responsibility Matrix, Project Closure Documents and Commissioning Documents (checklists and proof of performance) should be considered the basic project management documents. All of these should be in the project file as the installation is completed and successfully handed off to the owner. It would be best if all of these elements could be in place as the Project Team is organized so that the goals of the project are known and listed well before design begins.

5.1.3 Detail Design Administrative Process (Industrial Ventilation Systems and Local Exhaust Ventilation Systems). This Chapter focuses on the design considerations for hood and duct exhaust systems. The designation for these types of systems is Industrial Ventilation Systems. As a subset of Industrial Ventilation Systems there may be General Exhaust, Local Exhaust Ventilation, Process Air and Supply Air systems. For any industrial ventilation system to be constructed properly, it must be designed with the correct combination of hoods, duct systems (duct runs and sizes), and exhaust stacks. This information must be communicated to the party responsible for construction. This could be done through verbal or written communications. However, since these systems require knowledge of plant processes, regulatory agency requirements and the needs and requirements of other outside parties, clearly written instructions should be used even for the simplest system.

After the system project has been defined and a Design Basis (see Chapter 2) has been issued, the next normal phase is "detailed design." Company personnel, consulting engineers, or the engineering staff of design/build contractors may accomplish this phase. Normally, the Design Basis is furnished by the Owner and should be tailored to fit the needs of the particular party responsible for actually installing the system. It should be the specification to the design team, which means it describes how the system should be conceived and designed and defines the desired outcome of the project.

If company personnel design the system, the Design Basis instructions may not be as formal as they should be if outside consultants or design/build contractors are responsible for the detailed design. For the latter, it is crucial that there be a contract and clear agreement on expectations among all parties. The Design Basis may be used during the request for engineering quotations as a scope document. Later, it can be used as the project management document during the detailed design phase and continue as a guide for Commissioning.

The final deliverables for this phase normally should be detailed drawings and specifications as well as a Commissioning Plan. Just as the Design Basis serves as instructions to the designer, the engineering "package" provides the instructions to the installer. It should include the requirements for construction, possibly the methods of installation, and instructions for commissioning.

5.1.4 Drawings and Level of Detail. Communication of the design intent is usually made via permanent records such as drawings, specifications and written scope documents. These can be in paper and/or electronic format. Verbal communications that are not recorded should be avoided during the design phase. In addition, the level of design detail usually would be determined during the Design Basis phase. Drawings may vary from basic single line sketches to extremely detailed computer aided design (CAD) drawings that include isometric views and scale models. Likewise, the specifications and other instructions may simply be typed documents included as notes on the drawings or they may be part of pre-programmed engineering specification software.

The level and presentation of the detail of the design should consider the needs and level of sophistication and experience of the person reading the instructions. Single line sketches may be suitable for experienced fabricators and installers. Less experienced installers may need all dimensions displayed and locations for flanges, weld symbols and other details shown. The Project Manager and Team must know and communicate the requirements for the level of detail through the Design Basis. For that reason, it is very helpful to know who may be bidding the installation early in the project.

A very detailed package may allow for more bidders but may be needlessly expensive if all bidders are experienced. More detail takes more time and increases engineering costs. A more detailed package usually requires more field measurement, more views on the drawings, and more total drawings. This usually will put the designer at more risk of back charges and require more time spent taking field dimensions and checking drawings. On the other hand, while it is true that less detailed drawings are usually less expensive they may allow for interpretations by the installer that can result in back charges or misunderstandings as the system is put in place. In addition, less detailed drawings may result in more risk to the installer and may be reflected in higher construction bids.

Experienced designers may be able to use CAD techniques or templates to reduce drawing time and lasers and other resources are now available for more detailed field measurements. Hence, less detailed drawings may include all the dimensions necessary for installation and eliminate only frivolous duplications.

An example of dimensioning is shown in Figure 5-2. One set of dimensions shows only the distance between pieces of equipment. It allows some flexibility by the installer to choose duct lengths and flange locations to suit their installation techniques but still meet the requirements of the design.

Figure 5-3 shows every piece dimensioned in detail. This may be necessary on some projects where there are specific connection requirements, special duct routing is required to ensure clearance from predicted obstructions, or all the duct segments are fabricated off-site and the installation is on a tight schedule. As stated earlier, such detail also may be necessary if the installers have limited experience and the design intent must be correspondingly more explicit. However, it is important not to require unnecessary precision in installation. The example shown in Figure 5-3 may actually cost the project extra money if companies are held to exact dimensions as displayed. This drawing assumes that the duct designer knows the best and most cost effective location of all pieces and flanges. In reality, it may be different for each fabricator and installer. However, the design should balance the system (Chapter 9). If the contractor takes too many liberties with the design, the system may not be balanced and the design flows may not be achieved.

Some specific items that usually require more detailed design for all types of installers include structural supports for duct and hoods, location of fire suppression nozzles, and other features required to meet codes and regulations. Some of these details may have to be accomplished after final installation and inspection, especially National Fire Protection Association (NFPA) requirements.

Since accuracy is the important issue, the level of detail is determined by the needs of the end user. In the case of systems installed either in a new plant or retrofit to existing conditions, the size of local exhaust ventilation system ducts usually makes it difficult to route the ducts around all interferences. This is especially the case when attempting to re-route large duct to avoid small process devices or facility piping or lighting. Accurate field measurements will locate these but the project must also consider moving small interferences. This is especially important if the re-routing will require many offsets and elbows, which can drive up the installation cost and the static pressure (and power) requirements at the fan.

5.2 DESIGN OPTIONS FOR INDUSTRIAL VENTILATION SYSTEMS

The following information contains recommendations and experiences of good engineering practice. However, Codes, Regulations, and personal experiences with particular materials and construction may be more restrictive. In all cases, the most restrictive code or specification should supersede any recommendation included below.

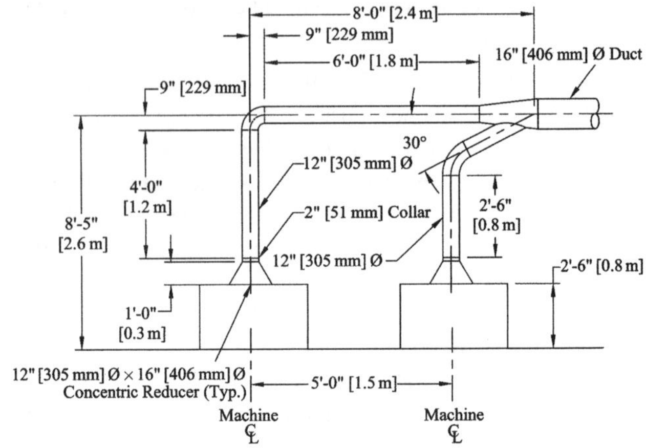

FIGURE 5-3. Drawing with detailed dimensions

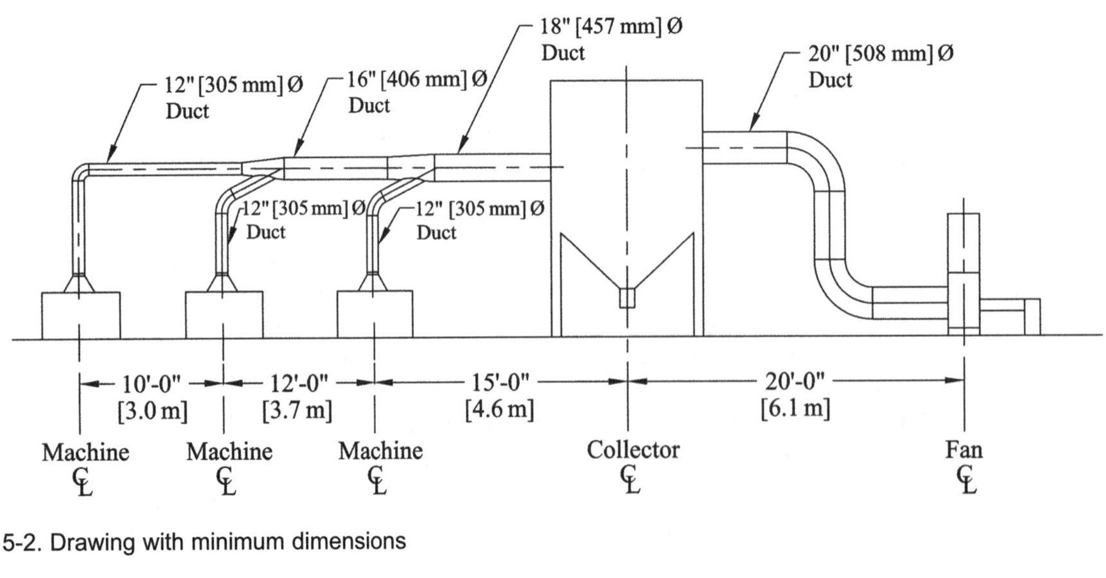

FIGURE 5-2. Drawing with minimum dimensions

5.2.1 Basic System Types — Dilution versus Local Exhaust Ventilation Design. The primary purpose of an industrial ventilation system is to maintain a safe level of airborne workplace contaminants by controlling them and removing them from the worker's environment. The method and equipment must be selected for the specific process, work flow, and worker tasks involved. Generally, the size and type of the equipment is based on the process and ergonomics, the size of hoods and duct (if used) and is based on an optimal tradeoff of reliability, operating cost and initial cost.

For worker protection there are basically two types of ventilation systems: "Dilution" (also called "General") and "Local Exhaust." Dilution ventilation mixes large amounts of clean air with contaminated air to keep concentrations below allowable limits (Figure 5-4). The design information for the installation of dilution or general ventilation systems is included in Chapter 4. This chapter will focus on the design considerations for local exhaust ventilation systems with calculation methods included in Chapter 9. Normally, dilution design is used to control the potential for fire or explosive conditions or to dilute odor. Dilution ventilation also can include the control of airborne contaminants (e.g., vapors, gases, and particulates), but should be limited to relatively less toxic contaminants that meet the following criteria:

a) The airflow necessary for dilution of contaminants must be replaced by supply or replacement air. This requires consideration of filtering, tempering, delivery methods and cost of the second system;
b) Workers must not be too close to contaminant source (i.e., within arm's length), especially if the generation rate is high or the toxicity is significant;
c) The toxicity of the contaminant must be low; and
d) The evolution rate of contaminants must be reasonably uniform.

While dilution systems mix clean air with contaminated air, local exhaust ventilation systems capture contaminants at the generation points and remove the contaminants from the workplace through a duct system (Figure 5-5). In addition, local exhaust ventilation systems also create a path for exhaust streams of materials from plant processes, improving their efficiency.

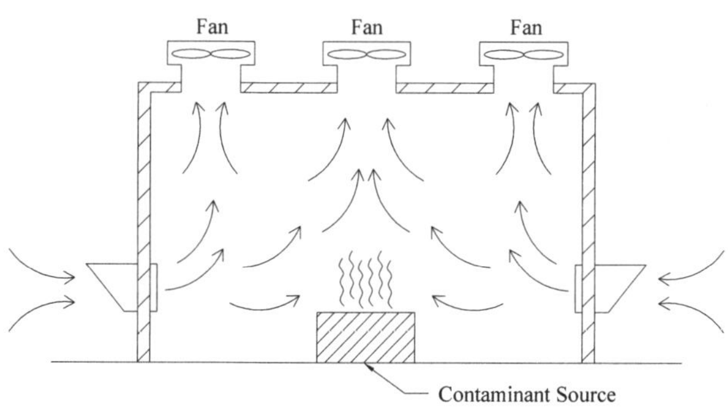

FIGURE 5-4. Dilution or general ventilation

5.2.2 Direct Discharge of Emissions to Atmosphere versus Air-Cleaning Device. In some cases, exhaust air with low levels of contaminants can be discharged directly to the atmosphere outside the workplace. This would be based on the considerations that:

a) No government regulations prohibit it;
b) Levels are predictable and verifiable;
c) Other nuisances like odors are not sent into the atmosphere; and
d) the discharge of the contaminants does not cause a neighborhood nuisance.

Higher potential emission levels and toxic contaminants often should be removed from the air stream by appropriate air-cleaning devices. Whether high or low toxicity, air discharged outside the plant must conform to both federal and local emission standards and not cause a neighborhood nuisance. In situations where the contaminant levels and toxicity are very low, it may also be possible to clean contaminants and return the cleaned air to the work areas. Details for the selection and design of Air-Cleaning Devices are included in Chapter 8 and an explanation of when and how air can be recirculated is included in Chapter 10.

The requirements for Air-Cleaning Devices are normally determined by regulations at federal, state or local levels. Before beginning the design process, a determination must be made concerning the use of air-cleaning devices and required efficiencies or discharge limits.

5.2.3 Local Exhaust Ventilation System Orientation. No two local exhaust ventilation systems are exactly alike, which becomes clear when viewing plan view drawings of systems. Hoods may be scattered throughout the plant or be located close together. The method of connecting the hood, air-cleaning device, and fan can be very different from system to system. For example, all hoods may be located at one end of the plant and the duct system connected through a long tapered main duct (Figure 5-2). Another designer may choose to locate the collector in the center of the system and include shorter runs of duct to the edges of the layout (Figure 5-5). A third design may include an individual air-cleaner and fan at each hood (Figure 5-6). In addition, there may be variations and combinations of all methods.

Each type of system configuration has its own advantages and disadvantages. At the same time, a designer may be confined in the design style by architectural considerations or the limitations of the physical space where equipment is located. For example, there may be only one possible location for the air-cleaning device. Very early in the design process, even as the Project Team is being chosen, an audit should determine any options for the physical location of equipment to be venti-

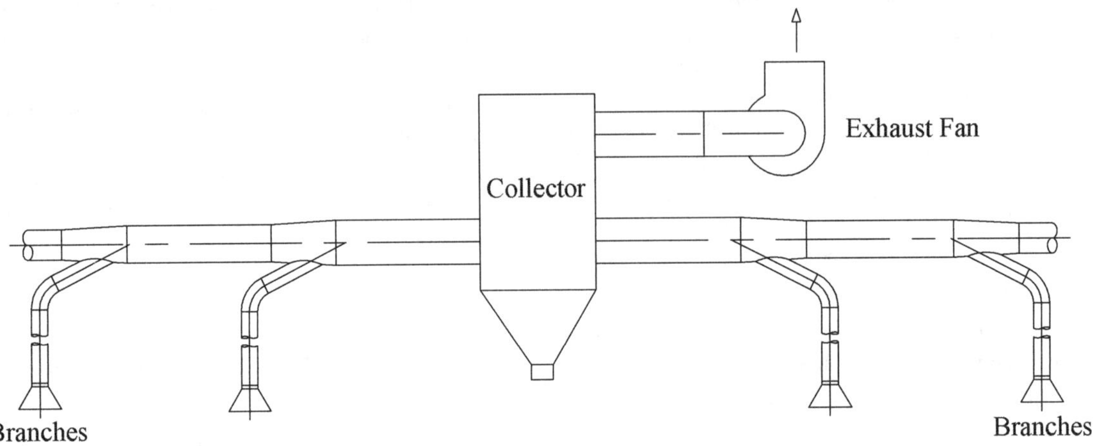

FIGURE 5-5. Local exhaust ventilation system

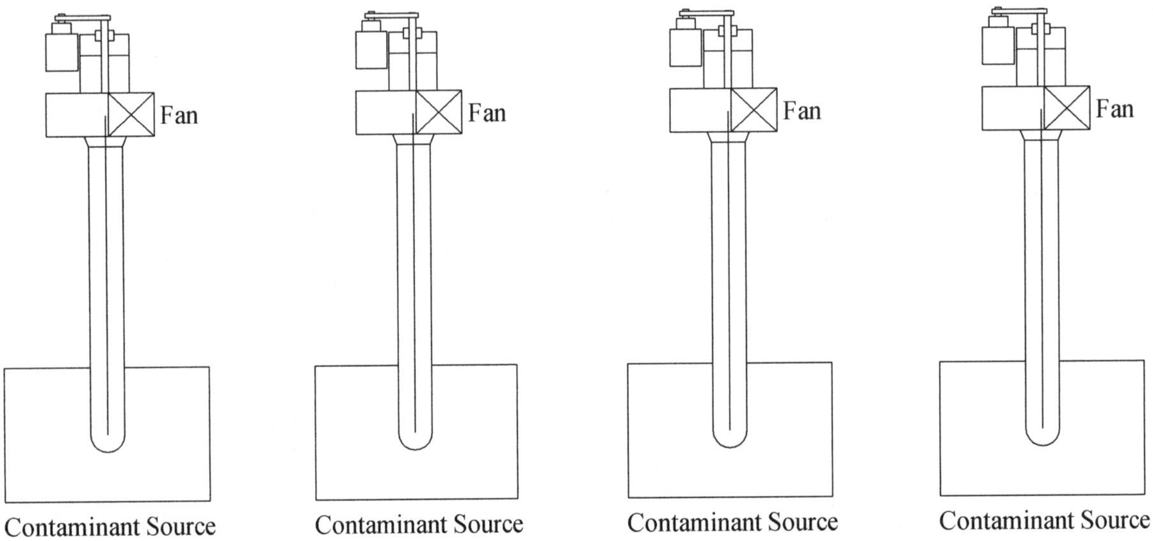

FIGURE 5-6. On-line design (single fan and/or collector for single or small group of contaminant sources)

lated as well as options for the locations of collection equipment. Alternative options may be determined by available property, but also may include restrictions due to the location of exhaust stacks, electrical power sources, soil or building structural conditions, or access for removal of collected pollutants. In some cases, lease or purchase agreements may include requirements for noise at the property line or hiding of heavy equipment from street view.

The Design Basis may include these restrictions or recommendations but many times the actual locations are being determined as the detailed design phase proceeds.

5.3 DESIGN PROCEDURES

5.3.1 Introduction. The duct system that connects the hoods, air-cleaning device(s), and fan must be properly designed. This process is much more involved than merely connecting pieces of duct. If the system is not carefully designed in a manner that reliably ensures that all required flow rates will be realized, adequate contaminant control may not be achieved. In addition, minimum transport velocities must be maintained in all branches and main ducts at all times during operation if the system is handling particulate matter. Duct systems require large amounts of air to convey relatively small amounts of contaminant. For that reason, they are one of the least efficient items in the plant or process. Careful design can provide the required system goals utilizing the least amount of power and initial cost. In addition, the designer must consider reliability, maintenance, and equipment life.

Detailed calculation procedures as listed in Chapter 9 are used to determine the duct sizes and the fan operating point (system flow rate and pressure) required by the system. Chapters 7 and 9 describe how to select a fan based on these results.

5.3.2 Preliminary Steps in the Design Process. With almost all design efforts, proper organization of data and information will simplify the process. In order to coordinate design efforts with all personnel involved (including the equipment or process operator as well as maintenance, health, safety, fire, and environmental personnel), the designer should have, at a minimum, the following data available at the start of the design process:

1) A layout of the operations, workroom, building (if necessary), etc. The available location(s) for the air-cleaning device and fan should be determined. An important aspect that must be considered at this time is the location of the final system exhaust point (where the air exits the system – usually a stack or fan discharge). It is extremely important that the discharged air not re-enter the workspace, either through openings in the building perimeter or through replacement air unit intakes. Key calculations for the proper selection of the emission point(s) design (stacks, etc.) are included in Section 5.12.

2) A line sketch of the duct system layout, including plan and elevation dimensions, fan location, air-cleaning device location, etc. Number, letter, or otherwise identify each branch and section of main duct on the line sketch for convenience (Figure 5-7). Types of systems as referenced in Section 5.3.2 show different configurations with respect to location of the fan in the system. Most systems, especially when handling particulate, will locate the fan on the clean air side of the collection device. Other considerations may force the location of the fan before the collector. If possible, locate the system fan close to pieces of equipment with high static pressure losses. This will facilitate balancing and may result in lower operating costs.

   Locating the fan (and air-cleaning device) in the center of the system (Figure 5-5) may yield a smaller system static pressure requirement.

3) Use "hard" (sheet metal or solid plastic, etc.) duct whenever possible and keep flexible duct lengths as straight and short as possible. Flexible duct is susceptible to sagging and excessive bending, which increases static pressure losses, these additional losses usually cannot be predicted accurately. Even if flexible duct could be mounted in straight sections, without sags or bends, its pressure losses per foot for straight sections can be more than twice the values of metal duct.

4) A design or sketch of the desired hood for each operation with direction and elevation of outlet for duct connection. Hood sketches can be in isometric or plan and elevation views. Enough detail must be included to determine the anticipated opening sizes, location and size of slots and other factors that will determine air volumes and hood static pressures.

5) Information about the details of the operation(s), specifically toxicity, worker access/use, physical and chemical characteristics, required flow rate at hoods or enclosures, minimum required duct (transport) velocity (see Section 5.3.5), hood entry losses, and required capture velocities at the hood face. Special attention should be given to room air turbulence (cross ducts, supply air delivery, and other disturbing air movement) and incompatibilities between dusts, fumes or vapors that might be intermixed in the exhaust system to assure that they do not result in fire or explosion hazards, destructive corrosion or a toxic mixture. If any mixture is incompatible, separate ventilation systems or appropriate air-cleaning devices should be provided. Please consult all latest appropriate standards including ASTM E 2012-06 (Standard Guide for the Preparation of a Binary Chemical Compatibility Chart) for selection of materials of construction as well as the design routing of ventilation systems.

6) Information relevant to the process such as temperature, moisture content and elevation (above sea level) should be provided for each hood and duct branch.

7) The method and location of the replacement air distribution devices as they affect each hood's performance. The type and location of supply fixtures can dramatically affect contaminant control by creating undesirable turbulence at the hood (see Chapter 10). Perforated plenums or perforated duct may provide better replacement air distribution with fewer adverse effects on hood performance.

5.3.3 Calculation Methods to Optimize Design. The actual design procedure is a continuing process and does not end with the initial system calculations. Calculations and evaluation may need to be repeated several times including 1) during the original conceptual design, 2) during final drive speed specification from "as-built" drawings, and 3) when providing a tool for the air balance technician. In addition, the designer must not consider this merely a simple method to size ducts and fan. It should also be used to identify ducts with very high velocities that could wear prematurely, and to analyze the branches with the highest pressure drop so changes can be made to the design to reduce system static pressure. For example, a small branch duct in a large volume system may represent the highest static pressure loss. By increasing the flow at the hood, making the duct larger and reducing the friction loss-

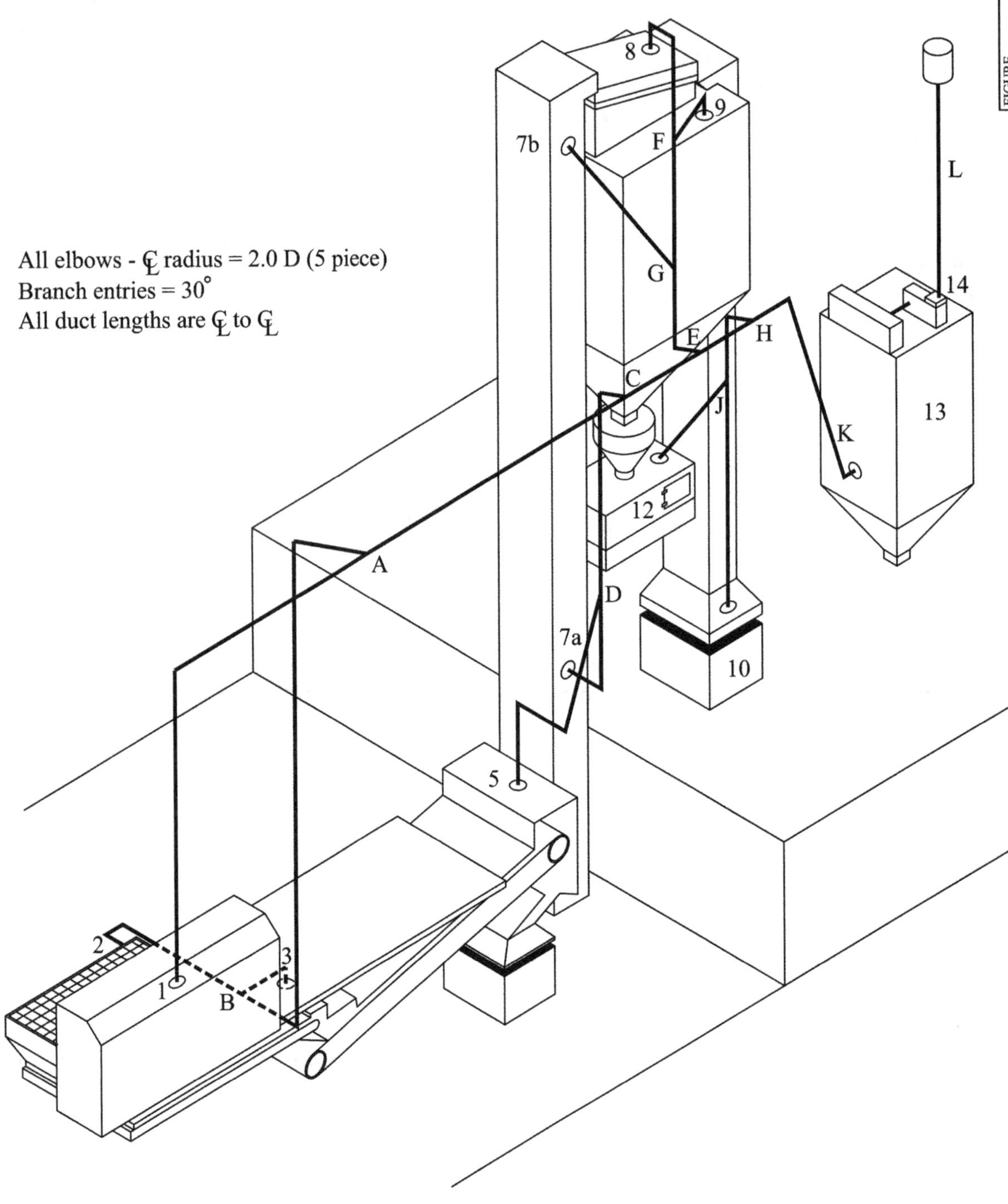

TITLE
SINGLE LINE ISOMETRIC SKETCH OF LOCAL EXHAUST VENTILATION SYSTEM

FIGURE 5-7

DATE 1-12

CHECK CODES, REGULATIONS, AND LAWS (LOCAL, STATE, AND NATIONAL) TO ENSURE THAT DESIGN IS COMPLIANT.

es in the duct, the overall system pressure may go down with a small total increase in total flow. This can result in lower system horsepower requirement.

Similarly, the system design usually only considers the conditions at initial start-up and installation. After the system is in use, it will lose some effectiveness as dust covers the duct interior wall (changing friction losses) and fan impellers and collectors show wear and dust buildup. The designer must consider the conditions during the operating life of the system. For instance, where volumetric flow, face velocities or transport velocities are selected from a range of values in the Tables, the upper end of the range should be considered if the system cannot be easily shut down for routine maintenance.

5.3.4 Design Calculations to Estimate System Performance. The calculation methods are used primarily to engineer the system (determine duct sizes, estimate static pressure requirements for fan selection, etc.). However the data can also be used to predict a range of operation that can be used to support field analysis of systems. Static pressures calculated at branches using the methods in Chapter 9 can be used as a start point to predict possible findings when troubleshooting systems. Note that calculation sheet data are for system design only and this will not duplicate the actual conditions. Hood losses, actual duct losses after material coats the inside walls, and other fabrication influences such as grinding of welds, etc., will all impact the actual results. Most of the estimated values for losses in system components are just that, estimates, even though based on laboratory and other field research, there is very little chance of exact duplication in most systems.

5.3.5 Selection of Duct Velocities. In systems that are intended to carry particulate, a minimum conveying velocity is necessary to ensure that the particulate will not settle in the duct. Conversely, when a system handling 'clean' air is installed in a quiet area, it may be necessary to keep velocities low to avoid excessive duct noise. When axial flow fans are used and no material is present to settle in duct, velocities of 1,000 [5.08 m/s] to 1,500 fpm [7.62 m/s] could be preferred. In a gas or vapor exhaust system installed in a typical factory environment where none of these restrictions apply, the velocity may be selected to yield the lowest annual operating cost.

To determine the optimum economic velocity, the system must first be designed at an assumed velocity and the total initial costs of duct material, fabrication and installation estimated. Optional duct and operating costs can be determined for other duct velocities for comparison. This optimum economic velocity will normally range from under 2,000 fpm [10.16 m/s] to over 4,000 fpm [20.32 m/s]. Lengthy expected service periods and system operating times tend to lower the optimum velocity while high interest rates and duct costs tend to raise the optimum. In general, a velocity of 2,500 [12.70 m/s] to 3,000 fpm [15.24 m/s] will not result in equivalent total annual costs much in excess of the true optimum. *NOTE: The transport velocity requirements of the material handled will supersede the economic aspects of duct sizing.*

The type of material being transported in the duct dictates the minimum velocity. For systems handling particulate, a minimum design velocity (Table 5-1) is required to prevent settling and plugging of the duct.[5.1] On the other hand, excessively high velocities are wasteful of power and may cause rapid abrasion and destruction of ducts. Minimum recommended design velocities are higher than theoretical and experimental values to protect against practical contingencies.

The following comments should be considered when using Table 5-1:

1) Plugging or closing one or more branches will reduce the total flow rate in the system and correspondingly will reduce the velocities in at least some sections of the duct system.
2) Damage to ducts by denting, for example, will increase the resistance and decrease the flow rate and velocity in the damaged portion of the system.
3) Leakage of outside air into ducts will increase flow rate and velocity downstream of the leak but will decrease airflow upstream and in other parts of the system.
4) Corrosion or erosion of the fan wheel, accumulation on the fan wheel, or slipping of a fan drive belt will reduce flow rates and velocities.
5) Velocities must be adequate to pick up or re-entrain dust that has settled due to improper operation of the local exhaust ventilation system.

The designer is cautioned that for some conditions such as sticky materials, condensing conditions in the presence of dust, strong electrostatic effects, etc., velocity alone may not be sufficient to prevent plugging and other special measures may be necessary.

5.4 DISTRIBUTION OF AIRFLOW IN DUCT SYSTEMS

A simple exhaust system consists of a hood, duct segments, and special fittings leading to and from an exhaust fan. A complex system is merely an arrangement of several simple exhaust systems connected to a common duct and one or more fans. Therefore, when designing a system of multiple hoods and branches, the same design methods apply. In a multiple branch system, however, it is also necessary to provide a means of distributing airflow properly among the branches. This can be accomplished either by a "balanced" design or by the use of blast gates or orifice plates.

Air will always take the path of least resistance. A natural balance at each junction will occur; that is, the exhaust flow rate will distribute itself according to the pressure losses of the merging flow paths. The designer must provide distribution such that the design airflow at each hood will not fall below the minimums listed in Chapter 6 and/or 13. To do so, the designer must ensure that all flow paths (ducts) entering a junction

TABLE 5-1. Range of Minimum Duct Design Velocities

Nature of Contaminant	Examples	Design Velocity
Vapors, gases, smoke		Any desired velocity (economic optimum velocity usually 1000–2000 fpm) [5.08–10.16 m/s]
Fumes, metal smokes	Welding	2000–2500 fpm [10.16–12.70 m/s]
Very fine light dust	Cotton lint, wood flour, litho powder	2500–3000 fpm [12.7–15.24 m/s]
Dry dusts and powders	Fine rubber dust, Bakelite molding powder dust, jute lint, cotton dust, shavings (light), soap dust, leather shavings	3000–3500 fpm [15.24–17.78 m/s]
Average industrial dust	Grinding dust, buffing lint (dry), wool jute dust (shaker waste), coffee beans, shoe dust, granite dust, silica flour, general material handling, brick cutting, clay dust, foundry (general), limestone dust, packaging and weighing asbestos dust in textile industries	3500–4000 fpm [17.78–20.32 m/s]
Heavy dusts	Sawdust (heavy and wet), metal turnings, foundry tumbling barrels and shake-out, sand blast dust, wood blocks, hog waste, brass turnings, cast iron boring dust, lead dust	4000–4500 fpm [20.32–22.86 m/s]
Heavy or moist dusts	Lead dusts with small chips, moist cement dust, buffing lint (sticky), quick-lime dust	4500 fpm [22.86 m/s] and up

will have equal calculated static pressure requirements.

To accomplish this, the designer has a choice of a balanced design (no gates or orifice plates) or balancing with blast gates or orifice plates. The object of both methods is the same: to obtain the desired flow rate at each hood in the system while maintaining transport velocity in all duct sections.

5.4.1 Balance by Design versus Blast Gate/Orifice Plate Methods. The two methods, labeled Balance by Design Method and Blast Gate/Orifice Plate Method, are outlined below. Table 5-2 shows some relative advantages and disadvantages of the two methods. Orifice plates are essentially fixed blast gates and have many of the same advantages and disadvantages of each method. The method of calculating orifice plate openings can be found in other texts with varying results. The location of blast gates and orifice plates are dependent on the location within the duct system (near elbows and hoods or other disturbances), so care must be taken to keep as much straight duct as possible (at least five duct diameters preferred) before and after their location to get predictable results. The losses due to blast gates (as a function of insertion depth) are difficult to predict because of the different blade shapes and clearances. Data may be made available from the damper manufacturer, however, installation usually requires field adjustment.

5.4.2 Balance by Design Procedure. This procedure provides for achievement of desired airflow (a "balanced" system) without the use of blast gates or orifice plates. It is often called the "Static Pressure Balance Method." The designer calculates the pressure loss of each duct segment for an exhaust hood to the junction with the next branch based on hood design data, fittings, and total duct length. At each junction, the static pressure (SP) for each parallel path of airflow must be the same. Where the ratio of the value of the governing SP to the lower SP is greater than 1.2, redesign of the branch with the lower pressure loss should be considered. This may include a change of duct size, selection of different fittings, and/or modifications to the hood design. Chapter 9 details the calculation method for this procedure.

The Balance by Design Method would usually be selected where highly toxic materials are controlled or to safeguard against tampering with blast gates (and consequently subjecting personnel to potentially excessive exposures), when excessive abrasion of the dampers is a major concern. The Balance by Design Method is highly recommended for systems that exhaust explosives, radioactive dusts and biological materials to minimize the possibility of accumulations in the system caused by a blast gate or orifice plate obstruction.

5.4.3 Blast Gate/Orifice Plate Procedure. This procedure depends on the use of blast gates and/or orifice plates located in branches or mains to provide the restrictions to balance static pressures. Blast gates (also sometimes called "cutoffs") or dampers must be adjusted after installation in order to achieve the desired flow at each hood.

Data and pressure loss calculations involved are the same as for the "Balance by Design" procedure; however, the duct sizes, fittings and flow rates are not adjusted. The blast gates are set after installation to provide the required static pressures to deliver the design flow rates. A change in any of the blast gate settings will change the flow rates in all of the other branches. Readjusting the blast gates during the system balancing process

TABLE 5-2. Relative Advantages and Disadvantages of Blast Gates versus "Balance by Design" Methods

Balance by Design Method	Blast Gate/Orifice Plate Method
1. Flow rates cannot be changed easily by workers or at the whim of the operator.	1. Flow rates may be changed relatively easily. Such changes are desirable where pickup of unnecessary quantities of material may affect the process.
2. There is little degree of flexibility for future equipment changes or additions. The duct is "tailor made" for the job.	2. Depending on the fan and motor selected, there is somewhat greater flexibility for future changes or additions.
3. The choice of exhaust flow rates for a new operation may be incorrect. In such cases, some duct revisions may be necessary.	3. Correcting improperly estimated exhaust flow rates is relatively easy within certain ranges.
4. No unusual erosion or accumulation problems will occur.	4. Partially closed blast gates may cause erosion thereby changing resistance or causing particulate accumulation.
5. Duct will not plug if velocities are chosen correctly.	5. Duct may plug if blast gate insertion depth has been adjusted improperly.
6. Total flow rate may be greater than design due to higher air requirements.	6. Balance may be achieved with design flow rate; however, the net energy required may be greater than the Balance by Design Method.
7. The system must be installed exactly as designed, with all obstructions cleared and length of runs accurately determined.	7. Moderate variations in duct layout are possible.
8. Small ducts chosen for static pressure balance may be required to operate at high velocities causing premature wear.	8. Operators can change blast gate settings possibly putting the system out of balance.

sometimes can result in increases to the actual fan static pressure and increased fan power requirements. Calculation methods for the employment of these balancing devices are included in Chapter 9. At each junction, the flow rates of two joining ducts are achieved by blast gate adjustments that result in the desired static pressure balance. Similarly, orifice plate opening sizes may be changed to reflect actual requirements at start-up or when system revisions are made. However, orifice plate design usually infers a more permanent installation because there is less chance of operator adjustment.

It should be noted that the Blast Gate/Orifice Plate Method would theoretically require less total flow and horsepower in the system because volume increases to balance pressures at branches will not occur. With this method, the static pressure needed to balance the branch will be the difference between the calculated static pressures in the joining branches. In practice, many balancers iteratively increase the insertion depth while balancing. This can result in higher system static pressures and greater energy use than increasing the volumetric flow by the Balance by Design procedure. See Chapter 4 of the O&M Manual for a discussion of balancing methods and techniques to reduce the total static pressure in a system balanced by blast gates.

Sometimes it is a practice to design systems on the assumption that only a fraction of the total number of hoods will be used at a time and the flow to the branches not used will be shut off with dampers or blast gates. For tapered system designs (see Section 5.5) where particulate is transported, this practice may lead to plugging in the main duct due to settled particulate. This procedure is not recommended unless minimum transport velocity can be assured in all ducts during any variation of closed blast gates. It is better to design these systems with individual branch lines all converging very close to the fan inlet to minimize the lengths of duct mains. Some NFPA Standards prohibit intermittent use of blast gates as shut off valves.

5.5 LOCAL EXHAUST VENTILATION SYSTEM TYPES

5.5.1 Tapered Main versus Plenum Design. There are two general classes of duct system designs: Tapered Main Systems and Plenum Systems. The duct in a tapered main system gradually gets larger as flows are merged together, thus keeping duct velocities nearly constant. If the system transports particulate (dust, mist or condensable vapors), the tapered system maintains the minimum velocity required to prevent settling. The duct in a plenum system is generally larger than that in a tapered system and the velocity in it is usually low. Particulate in the air stream can settle out in the large ducts. Certain mist and coolant control systems are designed this way to encourage settling of droplets in the duct. Figures 5-2 and 5-5 illustrate a tapered system while Figure 5-8 illustrates a plenum system. Regardless of which system is used, following proper procedure provides a workable system design.

Plenum systems differ from the tapered main designs illustrated earlier. In a tapered main system, the minimum transport velocity is maintained in all horizontal and vertical ducts. In a plenum exhaust system (Figure 5-8), minimum transport velocities are maintained only in the branch ducts to prevent

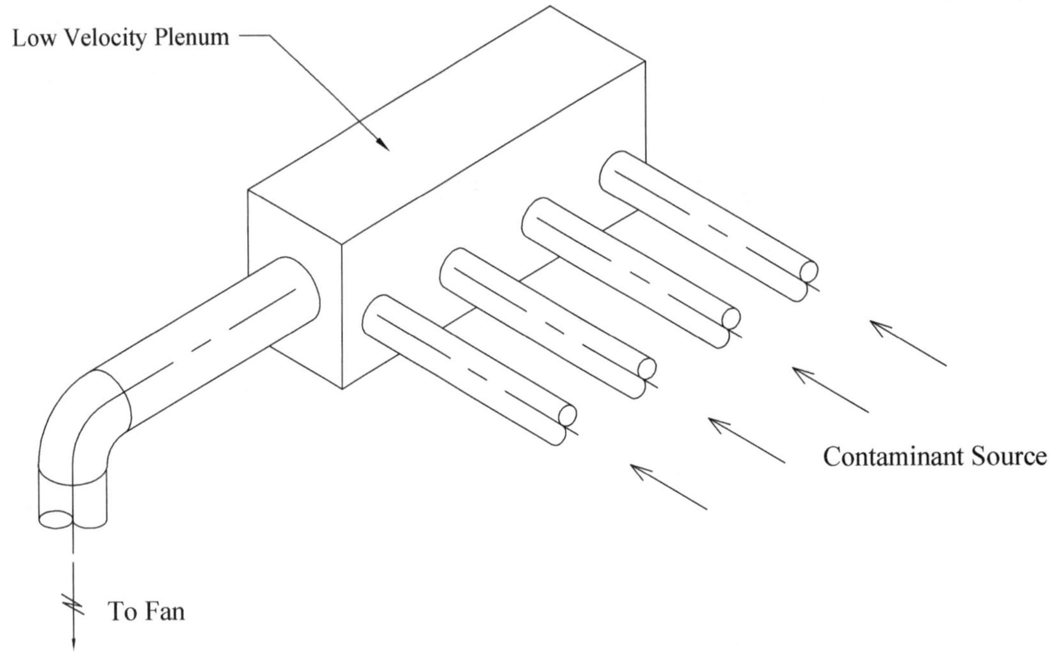

FIGURE 5-8. Plenum duct system

settling of particulate matter. The main duct (plenum) is oversized and design velocities decrease below minimum transport velocity values (many times below 1000 fpm [5.08 m/s]). The function of this plenum is to provide a low-pressure loss path for airflow from the various branches to the air-cleaner or the fan. This helps to maintain balanced exhaust in all of the branches and often minimizes operating power.

5.5.2 Plenum Design Advantages and Disadvantages. In most cases, tapered main systems are used for local exhaust ventilation system designs. However, plenum systems offer some advantages when handling of mists or transport velocities are not an issue.(5.2) The methods have varying success based on the materials being collected. Advantages of the plenum exhaust system include the following:

1) Branch ducts can be added, removed or relocated at any convenient point along the main duct limited only by the total airflow and pressure available at the fan. *(NOTE: Systems may need to be rebalanced every time a line change is made.)* Some systems are designed to automatically adjust to change in the number of active exhaust points. For example, a static pressure controller could maintain the static pressure by changing a variable frequency drive on the exhaust fan. Up to the capacity of the system, increasing the number of hoods increases the volumetric flow through the system.
2) Branch ducts can be closed off and the flow rate in the entire system reduced as long as minimum transport velocities are maintained in the remaining branches.
3) The main duct can act as a preliminary separator (settling chamber) for large particulate matter or liquids and refuse material that might be undesirable in the air-cleaner or fan. It is important to allow for removal of this collected material during the operation of the system through drains, drag conveyors, etc.

Limitations of the plenum design include the following:

1) Sticky, linty materials tend to clog the main duct. Buffing dust and lint are subject to this limitation and the plenum design is not recommended for these materials.
2) Materials that are subject to direct or spontaneous combustion must be handled with care. Some wood dust or oil mist has been handled successfully in systems of this type but buffing dust and lint are not recommended. Explosive dusts such as magnesium, aluminum, titanium or grain dusts cannot be handled in systems of this type. Applicable NFPA and other codes may require tapered main systems and minimum transport velocities in all ducts.

5.5.3 Plenum System Design Considerations. Control flow rates, hoods and duct sizes for all branches are calculated in the same manner as with tapered duct systems and shown in Chapter 9. The branch segment with the greatest pressure loss will govern the static pressure required in the main duct and fan. Other branches will be designed to operate at this static pressure (Balance by Design Method) or locking dampers can be used to adjust their pressure loss to the same static pressure as the governing branch.

Where the main plenum is relatively short or where the air-cleaners or fans can be spaced along the duct, static pressure losses due to airflow in the main plenum can be ignored. For extremely long plenums, it is necessary to calculate the static pressure loss along the main in a manner similar to that used in the balanced and blast gate methods of Chapter 9. Design plenum velocities are usually less than 50% of the branch velocity design duct velocities and can be as low as 1000 fpm [5.08 m/s]. Note that lower plenum velocities will result in larger sized plenums and possibly higher initial installation costs. Duct connections to air-cleaners, fans and discharge to outdoors are handled in the normal manner with consideration to minimum transport velocity.

Various types of plenum exhaust systems are used in industry (Figure 5-9). They include both self-cleaning and manual-cleaning designs. Self-cleaning types include pear-shaped designs that incorporate a drag chain conveyor in the bottom of the duct. This is used to convey the dust to a chute, tote box, hopper or other enclosure for disposal. Another self-cleaning design uses a rectangular main with a belt conveyor. In these types, the conveyors may be run continuously or on periodic cycles to empty the main duct before considerable buildup and clogging occur. A third type of self-cleaning design utilizes a standard conveying main duct system to remove the collected material from a hopper bottom main duct above. Such a system is usually run continuously to avoid clogging of the pneumatic air circuit. Manual-cleaning designs may be built into the floor or may be large enclosures behind the equipment to be ventilated. Experience indicates that these should be generously oversized, particularly the under floor designs, to permit added future exhaust capacity as well as convenient housekeeping intervals.

5.5.4 Tapered Main Design Considerations. The tapered main system is the standard design method for most local exhaust ventilation systems. A properly sized tapered main system will provide relatively constant velocities throughout the duct network. If these velocities meet the minimum requirements of transport velocity (see Section 5.3.5) particulate can be transported to the collection device. However, the flow of any gas stream through a duct system can result in eddies and places of high turbulence, particularly at elbows and junctions of two branch ducts. Higher minimum velocities may be specified where dropout of material is especially dangerous (flammable and toxic materials). This is especially the case for extremely long runs of duct or sections where there are several fittings in close proximity.

The more streamlined the system (longer radius elbows, small angled branch entries, efficiently designed hoods, etc.) the less horsepower is normally required. This can come at a higher initial price but the cost of operating horsepower lasts through the life of the system (sometimes 20 years or more). The designer should be cautioned to the effects of using cheaper but less energy-efficient parts in the system design.

5.6 SYSTEM REDESIGN

Many ventilation systems are changed after installation (processes are changed, operations are relocated, equipment is added to or removed from the production floor, etc.). When such changes occur, the effect of the proposed change(s) to the ventilation system must be calculated. Often, systems are changed without adequate design, resulting in catastrophic changes to some hood flow rates. The result is that worker safety and health are jeopardized. The O&M Manual considers operations and techniques applied to existing systems already designed. Chapter 8 in the O&M Manual provides guidelines and methods to be considered when systems are redesigned or field changes are made. It is important to note that the same techniques and calculation methods employed for the original system design also apply to the revisions to systems after installation.

5.7 SYSTEM COMPONENTS

After the basic shape and orientation of the system (see Section 5.2.3) is determined the focus changes to the individual component design. Local exhaust ventilation systems are comprised of four basic elements: hood(s), duct system (including the exhaust duct, discharge stack and/or recirculation duct), air-cleaning device and fan. Details for the design and specification of these components are included in Chapters 6 through 9 of this Manual.

The hood collects contaminant generated by a process or operation in an air or other gas stream. These contaminants may be particulate (solid and/or liquid) or gaseous. A duct system transports the contaminated air to the air-cleaning device, if present, and to the fan. The air-cleaner removes the contaminant from the air stream. The fan must overcome all the losses due to friction in the hoods, duct system and collection device while producing the required flow rate. The outlet duct from the fan usually discharges the air to the atmosphere in such a way that it will not be brought directly back into the plant (re-entrained) by the replacement and/or HVAC systems. In some situations, the cleaned air is recirculated to the plant (see Chapter 10).

5.8 HOODS

The type of hood (also sometimes referred to as "enclosure" or "receiver") to be used will depend on the physical characteristics of the process equipment, the contaminant generation mechanism and the operator/equipment interface. Hoods may be of a wide range of physical configurations but can be grouped into three general categories: enclosing, capturing or exterior and receiving hoods. Calculation methods for the determination of airflow and static resistance of hoods is included in Chapter 6. In addition, there are examples of many types of hoods for specific processes included in Chapter 13.

In Chapter 13, hoods are classified by process and use and include a drawing number preceded by "VS" for "Ventilation

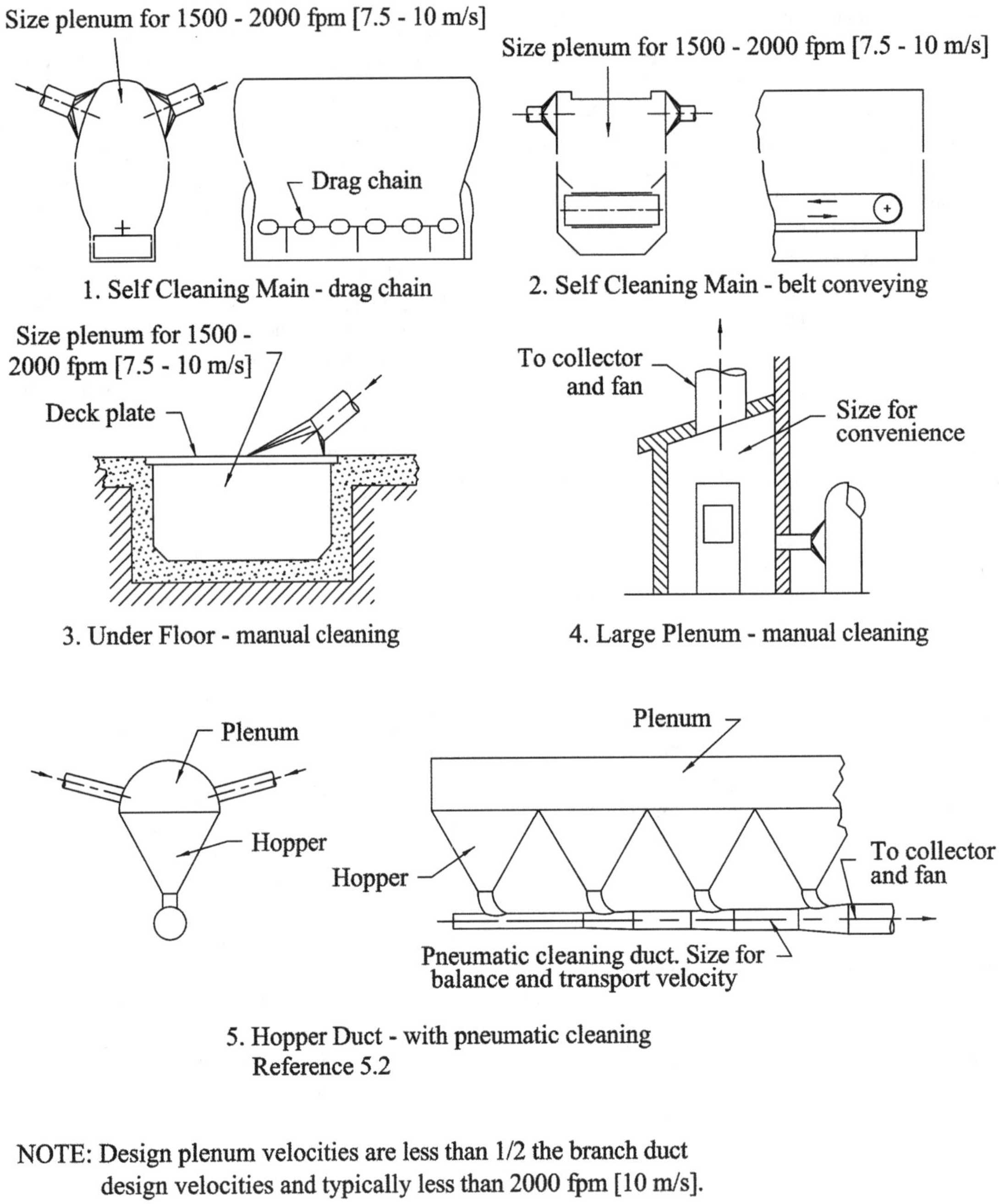

FIGURE 5-9. Types of plenum duct designs

Sheet." For example, VS-35-20 is located in Chapter 13 and shows a particular design of hood used for glove box ventilation. Information on this VS plate includes minimum flow and expected resistance (static pressure) from that particular type of hood.

5.9 DUCT SYSTEMS

5.9.1 Duct Design Issues. After the hood design and locations have been determined, they are connected through a duct system to the collection device and/or fan. The method of sizing duct systems is described in detail in Chapter 9.

If a ventilation system is to operate efficiently and reliably, careful attention must be given to its design. The process is much more involved than merely connecting hoods and pieces of duct to a fan. If the system is not carefully planned in a manner that inherently ensures that the design flow rates will be realized, contaminant control may not be achieved. These methods should be used even for the simplest installation. Records of the design should be maintained for reference in case there are future revisions.

In addition, the designer must consider initial capital costs, reliability, maintenance, and equipment life. There are a number of publications available that the designer should utilize. Chapter 9 provides detailed discussion of the design factors for all major ventilation system components. Detailed procedures for system design are included. Concept design criteria for over 150 specific industrial processes are also provided in Chapter 13. Other organizations providing specific ventilation system component information include Sheet Metal and Air Conditioning Contractor's National Association (SMACNA)[5.3,5.4] for sheet metal duct and Air Movement and Control Association (AMCA)[5.5] for fans.

5.10 FANS AND BLOWERS

To move air in a local exhaust ventilation system, energy is required to overcome the system losses. These losses are caused by the restrictions of the duct system, filter resistance and other factors. A powered air-moving device such as a fan or an ejector will provide this energy. Selection of an air-moving device can be a complex task and the designer is encouraged to take advantage of all available information from applicable trade associations as well as from individual manufacturers. Chapter 7 discusses the characteristics and design considerations for the selection of the correct type of fan for the local exhaust ventilation system.

Air moving devices can be divided into two basic classifications: ejectors and fans. Ejectors are sometimes used when it is not desirable to have contaminated air such as corrosive, flammable, explosive, hot or sticky materials pass directly through the air moving device. They are extremely inefficient and generally have high noise levels but may be necessary for special conditions such as handling long continuous strands of paper trim.

Fans (also called "blowers") can be divided into three basic groups: axial, centrifugal and special designs. As a general rule, axial fans are used for flow rates at lower resistances and centrifugal fans are used for flow rates at higher resistances. Axial fans in most cases are used for clean air applications although there are special designs that can handle air streams with minimal amounts of particulate.

5.11 AIR-CLEANING DEVICES

Often dusts, fumes and toxic or corrosive gases should not be discharged directly to the atmosphere. To meet most regulations for air emissions, an air-cleaning device or other form of collection device will be required to separate (or render harmless) the contaminants from the air stream. The contaminants or emissions can be in many physical forms including gas, liquid or solid or combinations of all three. They may also include vapors of water or acids in the gas stream that require special considerations. Each exhaust system handling such materials should be provided with an adequate air-cleaner as outlined in Chapters 6 and 8 of the O&M Manual.

The nature of the materials being collected, the required efficiencies and the temperatures of the air (or gas) stream will determine the collection methods required. Chapter 8 discusses most available technologies in detail. The air-cleaning device must be designed with reliable operating parameters. Many installations also require emissions monitoring or proof of continual operation by measuring direct or surrogate conditions in the system. This has taken the emphasis from proof of performance just at start-up and replaced it with more conservative selections.

In addition, maintenance and operating costs must be considered for the correct selection. In general, the system can be operated through many cycles of start-up and shut down. The air-cleaner must operate in stable conditions through these cycles. It must be accessible for maintenance and one must also consider if operation will be required even if there are problems with the device. The latter would require a design with "off-line" access so maintenance or repairs can be performed while the unit is operating.

Before the air cleaning device is selected, it is most important to know these maintenance and access requirements as well as the physical characteristics of the air stream. Other issues include the physical size of the equipment and how it will be installed in the plant as well as the methods of removing the collected contaminants.

Ultimately, the device must perform reliably and provide the efficiencies required to meet local/state and federal regulations. These requirements are normally listed in the Design Basis and the commissioning documents. This may include requirements for outlet loading or an overall efficiency rating for the unit itself. Before any information can be included in the Design Basis, careful research must be done to determine the correct application for the air control device and the guarantees needed from vendors to have a successful installation. These contractual guarantees may also extend past the initial installation and include maintenance and replacement parts (filter bags, etc.) for a period of time.

For example, one vendor may select an air-cleaning device that is smaller and will meet all requirements at start-up. But operation over the life of the unit may result in higher pressure drop (horsepower), or require more changes of bags or more maintenance to keep operating at required efficiencies. See Chapter 12 for information on system cost considerations. Life cycle costs must include the requirements for electric power costs as well as on-going operating costs when making the best selection of the air-cleaning device. Focusing on initial cost only may result in a financial burden borne for the remaining life of the system.

The designer will also need to consider the change in pressure drop (over time) of the collection device in many cases. If a system is started with clean bags and is not seeded with a pre-coat, then filter ΔP (pressure drop across the bag media expressed in "wg [Pa]) may be extremely low and initial flows may be higher than design. This can have a negative effect on the operation of the system because the higher velocities through the media can embed particles in spaces between the media fibers and retard effective cleaning. In addition, the system may be connected to a process where high flows have a negative impact. Similarly, a high initial flow may give false flow readings as the system is started and balanced.

Pre-coating of bags may be the best solution to reduce the impact of high fluctuations in ΔP. Another method would be to add artificial resistance to the fan by employing an outlet damper and feedback circuit to provide a constant inlet static pressure to the dust collector. The use of a Variable Frequency Drive (VFD) is another possible solution but has higher initial costs. (Note: If a VFD or inlet fan damper is used for volume control, the requirement will still remain for minimum transport velocities in the duct system.) The designer will need to consider energy usage and other issues, but the design must always be able to provide the design flow at the maximum pressure drop encountered (i.e., baghouse at maximum ΔP).

5.12 DISCHARGE STACKS

The final component of the ventilation system is the exhaust stack, an extension of the exhaust duct above the roof or grade. Assuming all exhaust emission levels are met and maintained, there are still two prime design considerations for the placement of an exhaust stack for a local exhaust ventilation system. First, the air exhausted should escape the building envelope so it does not return directly into building air intakes. Second, once it has escaped the building envelope, the stack should provide sufficient dispersion so that the plume does not cause an unacceptable situation when it reaches the ground.

The exhaust stack should incorporate a "stack cap" to prevent entry of precipitation and ice. (In addition, the fan should incorporate a drain port so that moisture does not settle in its housing and cause problems at start-up.) If the exhaust stack design includes horizontal runs the duct should be slightly inclined toward a drain point. Large heavy vertical exhaust stacks should not be supported directly by the fan.

When placing an exhaust stack on the roof of a building, the designer must consider several factors. The most important is the pattern of the air as it passes the building. Even in the case of a simple building design with a perpendicular wind, the airflow patterns over the building can be complex to analyze. Figure 5-10 shows the complex interaction between the building and the wind. A stagnation zone forms on the upwind wall.

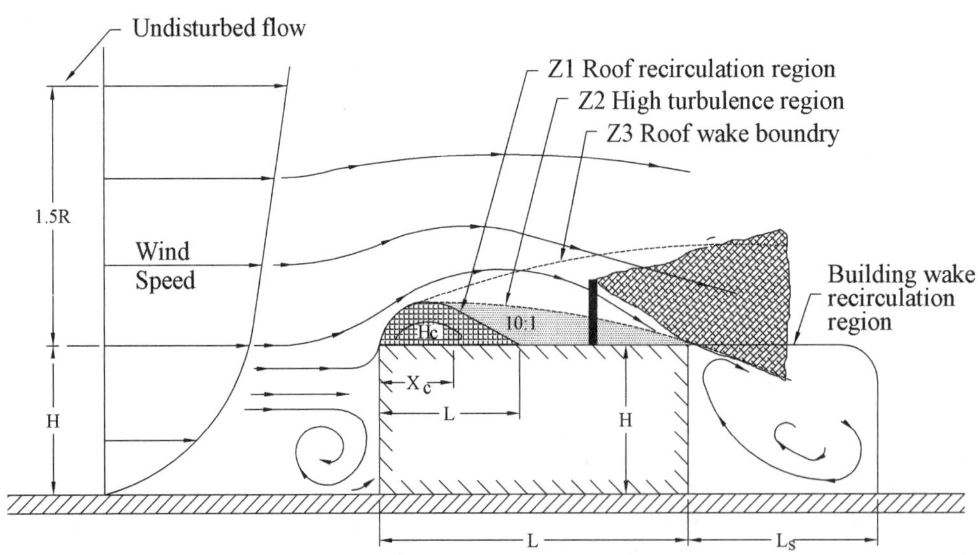

FIGURE 5-10. Effects of building on stack discharge

Air flows away from the stagnation zone resulting in a down draft near the ground. Vortices form by the wind action resulting in a recirculation zone along the front of the roof or roof obstructions, down flow along the downwind side, and forward flow along the upwind side of the building.

The USEPA uses computer modeling/simulations that utilize Gaussian distribution (such as PTMax) to predict resulting ground level concentrations of pollutants emitted from stacks. These predictive tools show 10 to 100 times the normal ground level concentrations when building wake effects are included (due to stacks being too short). More guidance in using these tools can be found at www.epa.gov/ttn/scram/, the site for SCRAM (Support Center for Regulatory Atmospheric Modeling).

A recirculation zone forms at the leading edge of the building. A recirculation zone is an area where a relatively fixed amount of air moves in a circular fashion with little air movement through the boundary. A stack discharging into the recirculation zone can contaminate the zone. Consequently, all stacks should penetrate the recirculation zone boundary.

The high turbulence region is one through which the air passes, however, the flow can be highly erratic with significant downward flow. A stack that discharges into this region will contaminate anything downwind of the stack. Consequently, all stacks should extend high enough that the resulting plume does not enter the high turbulence region upwind of an air intake.

Because of the complex flow patterns around simple buildings, it is almost impossible to locate a stack that is not influenced by vortices formed by the wind. Tall stacks are often used to reduce the influence of the turbulent flow, to release the exhaust air above the influence of the building and to prevent contamination of the air intakes. Selection of the proper location is made more difficult when the facility has several supply and exhaust systems, and when adjacent buildings or terrain cause turbulence around the facility itself.

When locating the stack and outdoor air inlets for the air handling systems, it is often desirable to locate the intakes upwind of the source. However, often there is no true upwind position. The wind direction in all locations is variable. Even when there is a natural prevailing wind, the direction and speed are constantly changing. If stack design and location rely on the direction of the wind, the system will clearly fail.

The effect of wind on stack height varies with speed:

1) At very low wind speeds, the exhaust jet from a vertical stack will rise above the roof level resulting in significant dilution at the air intakes.
2) Increasing wind speed can decrease plume rise and consequently decrease dilution.
3) Increasing wind speed can increase turbulence and consequently increase dilution.

The prediction of the location and the form of the recirculation cavity, high turbulence region and roof wake is difficult. However, for wind perpendicular to a rectangular building, the height (H) and the width (W) of the upwind building face determine the airflow patterns. The critical dimensions are shown in Figure 5-10. According to Wilson,[5.6] the critical dimensions depend on a scaling coefficient (R) and are given by:

$$R = B_S^{0.67} \times B_L^{0.33} \quad \textbf{[5.1]}$$

where B_S is the smaller and B_L is the larger of the dimensions H and W. When B_L is larger than 8 B_S, use $B_L = 8\ B_S$ to calculate the scaling coefficient. For a building with a flat roof, Wilson[5.9] estimated the maximum height (H_C), center (X_C), and lengths (L_C) of the recirculation region as follows:

$$H_C = 0.22\ R \quad \textbf{[5.2]}$$

$$X_C = 0.5\ R \quad \textbf{[5.3]}$$

$$L_C = 0.9\ R \quad \textbf{[5.4]}$$

In addition, Wilson estimated the length of the building wake recirculation region by:

$$L_R = 1.0\ R \quad \textbf{[5.5]}$$

The exhaust air from a stack often has not only an upward momentum due to the exit velocity of the exhaust air but buoyancy due to its density as well. The effective height is used for the evaluation of the stack height (Figure 5-11). The effective height is the sum of:

1) actual stack height (H_S),
2) the rise due to the vertical momentum of the air, and
3) any wake downwash effect that may exist.

Note that these equations are valid in IP units. Final translation to metric units would be done after completion in IP units.

A wake downwash occurs when air passing a stack forms a downwind vortex. The vortex will draw the plume down, reducing the effective stack height (Figure 5-12). This vortex effect is eliminated when the exit velocity is greater than 1.5 times the wind velocity. If the exit velocity exceeds 3000 fpm [15.24 m/s], the momentum of the exhaust air reduces the potential downwash effect.

The ideal design extends the stack high enough so that the expanding plume does not meet the wake region boundary. More realistically, the stack is extended so that the expanding plume does not intersect the high turbulence region or any recirculation cavity. According to Wilson,[5.7] the high turbulence region boundary (Z_2) follows a 1:10 downward slope from the top of the recirculation cavity.

To avoid entrainment of exhaust gas into the wake, stacks must terminate above the recirculation cavity.[5.8] The effective stack height to avoid excessive re-entry can be calculated by assuming that the exhaust plume spreads from the effective stack height with a slope of 1:5 (Figure 5-10). The first step is to raise the effective stack height until the lower edge of the 1:5 sloping plume avoids contact with all recirculation zone boundaries. The zones can be generated by roof top obstacles

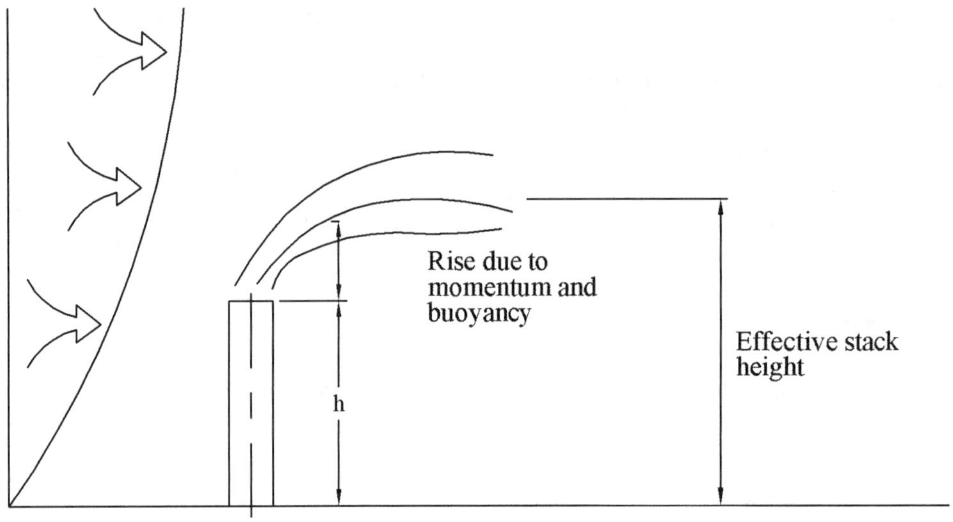

FIGURE 5-11. Effective stack height

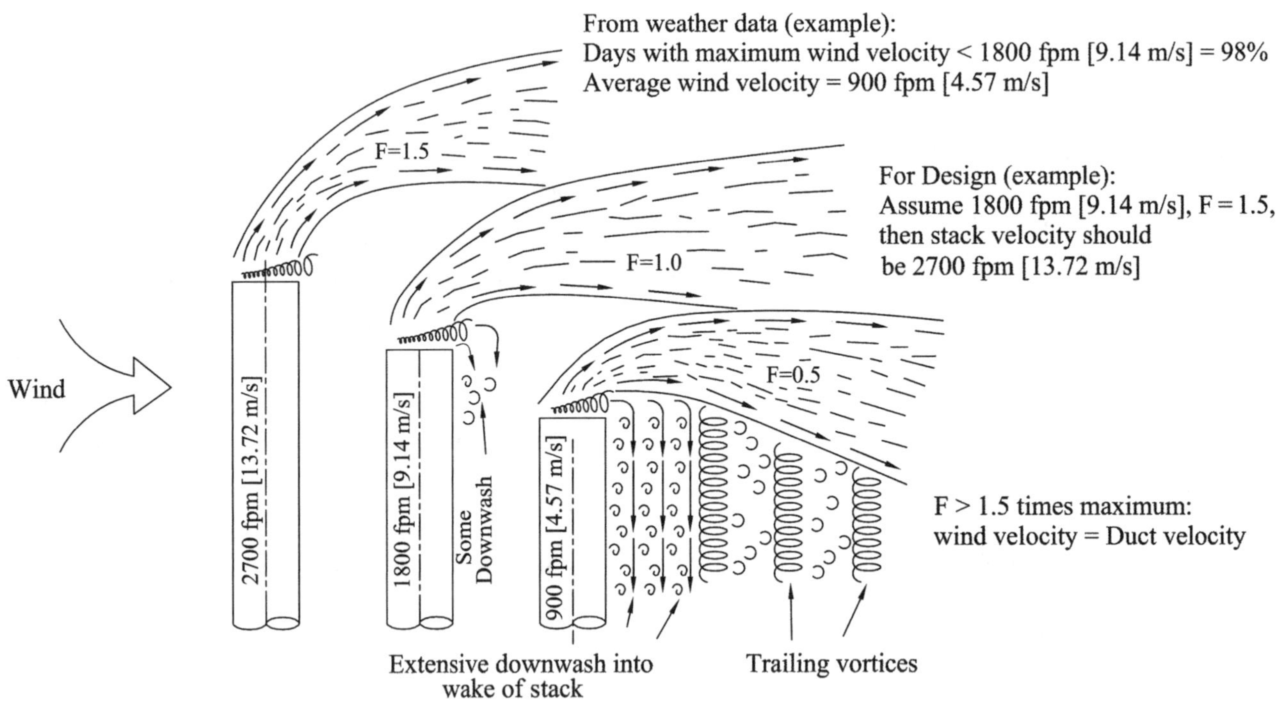

FIGURE 5-12. Wake downwash effects

such as air handling units, penthouses or architectural screens. The heights of the cavities are determined by Equations 5.2, 5.3 and 5.4 using the scaling coefficient for the obstacle. Equation 5.5 can be used to determine the length of the wake recirculation zone downwind of the obstacle.

If the air intakes, including windows and other openings, are located on the downwind wall, the lower edge of the plume with a downward slope of 1:5 should not intersect with the recirculation cavity downwind of the building. The length of the recirculation cavity (L_R) is given by Equation 5.5. If the air intakes are on the roof, the downward plume should not intersect the high turbulence region above the air intakes. When the intake is above the high turbulence boundary, extend a line from the top of the intake to the stack with a slope of 1:5. When the intake is below the high turbulence region boundary, extend a vertical line to the boundary, then extend back to the stack with a slope of 1:5. This allows the calculation of the necessary stack height. The minimum stack height can be determined for each air intake. The maximum of these heights would be the required stack height. In addition, the heights may need to be increased to ensure that plume does not intersect with the wake zone, as discussed above.

In large buildings with many air intakes, the above procedure will result in the specification of very tall stacks. An alternate approach is to estimate the amount of dilution that is afforded by stack height, distance between the stack and the air intake, and internal dilution that occurs within the system itself. This approach is presented in the "Airflow Around Buildings" chapter in the *Fundamentals* volume of the *ASHRAE Handbook*.(5.9)

In summary, the following should be considered for proper stack design:

1) Discharge velocity and gas temperature influence the effective stack height.
2) Wind can cause a downwash into the wake of the stack reducing the effective stack height. Stack velocity should be at least 1.5 times the wind velocity to prevent downwash.
3) A good stack velocity is 3000 fpm [15.24 m/s] because it prevents downwash for winds up to 2000 fpm [10.16 m/s] (22 mph). (Higher wind speeds have significant dilution effects.) It also increases effective stack height and allows selection of a smaller centrifugal exhaust fan. It can also provide transport velocity if there is any particulate in the exhaust or if there is a failure of the air-cleaning device.
4) High exit velocity is a poor substitute for stack height. For example, a stack located at roof elevation requires a velocity over 8000 fpm [40.64 m/s] to penetrate the recirculation cavity boundary.
5) The terminal velocity of rain is about 2000 fpm [10.16 m/s]. A stack velocity above 2600 fpm [13.20 m/s] should prevent rain from entering the stack when the fan is operating. (Rain can enter if fan is off.)
6) Locate stacks on the highest roof of the building when possible. If not possible, a much higher stack is required to extend beyond the wake of the high bay, penthouse or other obstacle.
7) The use of an architectural screen should be avoided. The screen becomes an obstacle and the stack must be raised to avoid the wake effect of the screen.
8) The best stack shape is a straight cylinder. If a drain is required, a vertical stack head is preferred (see Figure 5-13). In addition, the fan should be provided with a drain hole and the duct should be slightly sloped toward the fan.
9) Rain caps should not be used (Figure 5-14). The rain cap directs the air toward the roof, increases the possibility of re-entry, and causes potential exposures to maintenance personnel on the roof. Moreover, rain caps are not effective. A 12-inch [304.8 mm] diameter stack can pass as much as 16% of all rain and almost 45% during individual storms.

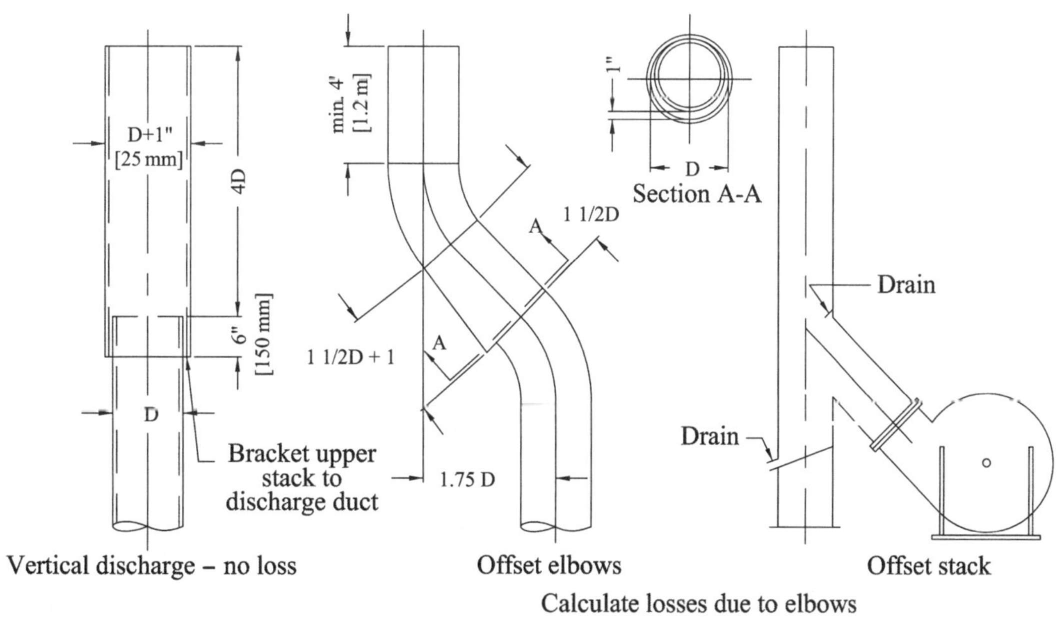

1. Rain protection characteristics of these caps are superior to a deflecting cap located 0.75D from top of a stack.
2. The length of upper stack is related to rain protection. Excessive additional distance may result in "blowout" of effluent at the gap between upper and lower sections.

FIGURE 5-13. Stackhead design

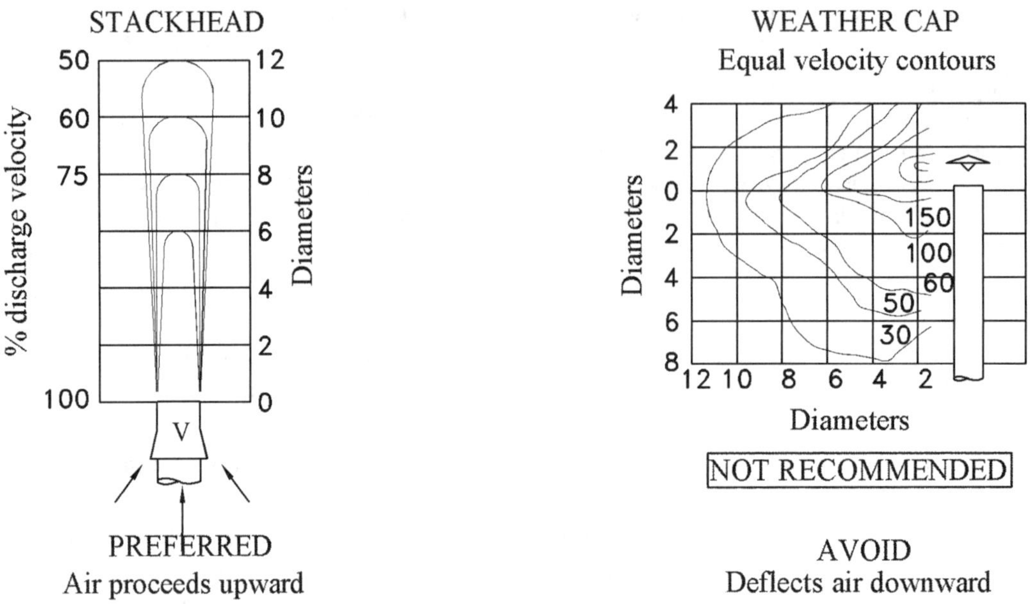

FIGURE 5-14. Rain caps

10) Separating the exhaust points from the air intakes can reduce the effect of re-entry by increasing dilution.

11) In some circumstances, several small exhaust systems can be placed in a single manifold to provide internal dilution thereby reducing re-entry.

12) A combined approach of vertical discharge, stack height, remote air intakes, proper air-cleaning device and internal dilution can be effective in reducing the consequences of re-entry.

A tall stack is not an adequate substitute for good emission control. The reduction achieved by properly designed air-cleaning devices can have a significant impact on the potential for re-entry. (This may not apply to scrubber exhaust because of moisture.)

5.13 DUCT CONSTRUCTION CONSIDERATIONS

The duct in an exhaust system will operate under the same conditions as the exhaust from hoods and enclosures. This can include conditions of extreme heat, erosion by the action of the dusts, and corrosion from solids, liquids, and vapors in the air stream. The designer must address the construction details including materials and methods of construction.

Ducts are specified most often for use in the low static pressure range (-20 "wg to +20 "wg) [-4982 Pa to +4982 Pa]; but higher static pressures are occasionally encountered. The duct can also convey air or gas at high temperatures and contaminated with abrasive particulate or corrosive aerosols and vapors. Whether conditions are mild or severe, correct design and competent installation of all system components are necessary for proper functioning of any local exhaust ventilation system.

Exhaust system components should be constructed with materials suitable for the conditions of service and installed in a permanent and workmanlike manner. To minimize friction loss and turbulence, the interior of all ducts should be smooth and free from obstructions — especially at connections between components.

5.13.1 Materials of Construction. Duct, hoods and other fabrications are to be constructed of black iron or welded galvanized sheet steel (flanged and proper gaskets included), unless the presence of corrosive gases, vapors and mists or other conditions make such material impractical. In those cases, stainless steel, PVC, special coatings or some other material compatible with the gas stream components will be used. Arc welding of black iron lighter than 18 gauge is not recommended. Galvanized construction is not recommended for temperatures exceeding 400 F [204 C]. It is recommended that a specialist be consulted for the selection of materials best suited for applications when corrosive atmospheres are anticipated. Table 5-3 provides a guide for selection of plastic materials for corrosive conditions.

There are four classifications for exhaust systems handling non-corrosive applications:

Class 1 (Light Duty): Includes nonabrasive applications, e.g., replacement air, general ventilation, gaseous emissions control with no oil mist or condensing vapors.

Class 2 (Medium Duty): Includes applications with moderately abrasive particulate in light concentrations, e.g., buffing

TABLE 5-3. Typical Physical and Chemical Properties of Fabricated Plastics and Other Materials

				Resistance To						
Chemical Type	**Trade Names**	**Max. Opr. Temp., F [C]**	**Flammability**	**Gasoline**	**Mineral Oil**	**Strong Alk.**	**Weak Alk.**	**Strong Acid**	**Weak Acid**	**Solvents**
Urea Formaldehyde	Beetle Plaskon Sylplast	170 [77]	Self Ext.	Good	Good	Unacc.	Fair	Poor	Poor	Good
Melamine Formaldehyde	Cymel Plaskon Resimene	210-300 [99-149]	Self Ext.	Good	Good	Poor	Good	Poor	Good	Good
Phenolic	Bakelite Durite Durez G.E. Resinox	250-450 [121-232]	Self Ext.	Fair	–	Poor	Fair	Poor	Fair	Fair
Alkyd	Plaskon	–	Self Ext.	Good	–	Unacc.	Poor	–	Good	Good
Silicone	Bakelite G.E.	550 [288]	–	Good	Good	–	–	Good	Good	Unacc.
Epoxy	Epiphem Araldite Maraset Renite Tool Plastik Epon Resin	50-200 [10-93]	Self Ext.	Good	–	Good	Good	Good	Good	Good
Cast Phenolic	Marblette	–	Self Ext.	–	–	Unacc.	Fair	Good	Good	Good to Fair
Allyl & Polyester	Laminac Bakelite Plaskon Glykon Paraplex	300-450 [149-232]	Self Ext.	–	–	Poor	Fair	Poor	Fair	Fair
Acrylic	Lucite Plexiglas Wascoline	140-200 [60-93]	0.5-2.0 in/min [0.5-9 mm/sec]	–	–	–	Good	–	Good	Good to Unacc. Unacc.
Polyethylene	Tenite Irrathene	140-200 [60-93]	Slow Burning	–	–	–	–	–	–	Unacc.
Tetrafluoroethylene	Teflon	500 [260]	Non-Fl.	Good	–	Good	Good	Good	Good	Good
Chlortrifluoroethylene	Kel F	–								
Polyvinyl Formal & Butyral	Vinylite Butyral Saflex Butvar Formuaré Butacite	– –	Slow Burning	Good	Good	Good	Good	Unacc.	Unacc.	Unacc.
Vinyl Chloride	Krene	130-175 [54-79]	Slow Burning	–	–	Good	Good	Good	Good	Unacc.
Polymer & Copolymer	Bakelite Vinyl Dow PVC Vygen	–								
Vinylidene Chloride	Saran	160-200 [71-73]	Self Ext.	Good	Good	Good	Good	Good	Good	Fair
Styrene	Bakelite Catalin Styron Dylene Luxtrex	150-165 [66-74]	0.5-2.0 in/min [0.5-9 mm/sec]	Unacc.	Fair	Good	Good	–	–	Fair
Polystyrene Reinforced with fibrous glass	–	–	–	Unacc.	Fair	Good	Good	–	–	Poor
Cellulose Acetate	Celanese Acetate Tenite Thermo Plastic	–	0.5-2.0 in/min [0.5-9 mm/sec]	Good	Good	Unacc.	Unacc.	Unacc.	Fair	Poor
Nylon	Plaskon Zytol Tynex	250 [121]	Self Ext.	Good	Good	Good	Good	–	Good	Good
Glass	Pyrex	450 [232]	Non-Fl.	Good	Good	Good	Good	Good	Good	Good

NOTE: Each situation must be thoroughly checked for compatibility of materials during the design phase if usage is changed.

and polishing, woodworking, grain dust.

Class 3 (Heavy Duty): Includes applications with highly abrasive particulate in low concentrations, e.g., abrasive cleaning operations, dryers and kilns, boiler breeching, foundry sand handling.

Class 4 (Extra Heavy Duty): Includes applications with highly abrasive particles in high concentrations, e.g., materials conveying high concentrations of particulate in all examples listed under Class 3 (usually used in heavy industrial plants such as steel mills, foundries, mining and smelting).

5.13.2 Duct Fabrication Methods. For most conditions, round duct is recommended for industrial ventilation, air pollution control and dust collecting systems. Compared to non-round duct, it provides for lower friction loss and its higher structural integrity allows lighter gauge materials and fewer reinforcing members. Round duct should be constructed in accordance with SMACNA Standards.(5.3) Metal thickness required for round industrial duct varies with classification, static pressure, reinforcement and span between supports. Metal thicknesses required for the four classes are based on design and use experience.

Rectangular ducts should only be used when space requirements preclude the use of round construction. Rectangular ducts should be as nearly square as possible to minimize resistance, and they should be constructed in accordance with SMACNA Standards.(5.4)

For many applications, spiral wound duct is adequate and less expensive than custom construction. However, spiral wound duct should not be used for Classes 3 and 4 because it does not withstand abrasion as well as smooth metal duct. It also should not be used for applications involving the carrying of oil mists or other vapors that may condense and appear through seams. Applications where materials may collect on the interior surfaces, such as paper trim and stringy materials, may also not be suitable for spiral duct. Elbows, branch entries, and similar fittings should be fabricated, if necessary, to achieve good design. Special considerations concerning use of spiral duct in local exhaust ventilation systems are as follows:

1) Unless flanges are used for joints, the duct should be supported close to each joint, usually within 2 inches [50.8 mm]. Additional supports may be needed. See Reference 5.4.
2) Joints should be sealed by methods shown to be adequate for the service.
3) Systems can be leak tested after installation at the maximum expected static pressure. The acceptable leakage criteria, often referred to as leakage class, should be carefully selected based on the hazards associated with the contaminant.
4) Fittings and elbows must be built with proper entry angles and throat radius to duplicate Round Duct Standards. This includes entry on the taper and not in round duct after or before the taper.

Where condensation may occur (moisture laden air or oil mist systems, etc.), the duct system should be liquid tight and provisions made for proper sloping and drainage. Spiral duct should not be used for these applications.

Ducts using clamp flanges may be used for small duct operations, particularly where hoods or machines are frequently moved, or if frequent removal for cleaning is required. This design incorporates a quick over-center levered clamp to join the rolled lips of all components. These duct systems can be fabricated in stainless steel or galvanized steel and generally are available only in small sizes (< 24" [600 mm] diameter). If this design is used, the rolled lips for connections must be mechanically formed on the end of the components by rolling the duct back on itself. Duct is to be longitudinally lock-seamed. Sleeves may be used for field adjustments, but sealing of the duct must meet the standards as required for standard SMACNA installations. There may be requirements for more hangers to provide the same structural integrity as traditional round duct standards. Metal thickness must be at least the same as standard round duct built to SMACNA standards.

5.13.3 Fabrication Standards for Materials Other Than Steel (IP Units). Equation 5.6 can be used for specifying ducts to be constructed of metals other than steel. For a duct of infinite length, the required thickness may be determined from:

$$\frac{t}{D} = \sqrt[3]{\frac{0.035714\, p(1-v^2)(52+D)}{E}} \qquad \textbf{[5.6] IP}$$

where:
- t = thickness of the duct in inches
- D = diameter of the duct in inches
- p = intensity of the negative pressure on the duct (psi)
- E = modulus of elasticity in psi
- v = Poisson's ratio (a dimensionless material constant)

The above equation (for Class 1 duct) incorporates a safety coefficient that varies linearly with the diameter (D), beginning at 4 for small ducts and increasing to 8 for duct diameters of 60 inches. This safety coefficient has been adopted by the sheet metal industry to provide for lack of roundness, excesses in negative pressure due to particle accumulation in the duct and other manufacturing or assembly imperfections unaccounted for by quality control, and tolerances provided by design specifications.

Additional metal thickness must be considered for Classes 2, 3 and 4. The designer is urged to consult the SMACNA standards for complete engineering design procedures.

Longitudinal joints or seams should be welded. All welding should conform to the standards established by the American Welding Society (AWS) structural code.(5.10) Double lock seams are limited to Class 1 applications.

5.13.4 Duct Component Considerations. Duct systems subject to wide temperature fluctuations should be provided with expansion joints. Flexible materials used in the construction of expansion joints should be selected with temperature and corrosion conditions considered.

Elbows and bends should be a minimum of two gauges heavier than straight lengths of equal diameter and have a centerline radius of at least two and preferably two and one-half times the duct diameter. Large centerline radius elbows are recommended where highly abrasive dusts are being conveyed (Figure 5-15).

Elbows of 90° should be five-piece construction for round duct up to six inches and seven-piece construction for larger diameters. Turns of less than 90° (known as "angles") should have a proportional number of pieces. Prefabricated angles and elbows of smooth construction may be used. Reinforced Flat Back Elbows can be used where high particulate loading is encountered (Figure 5-16).

Where the air contaminant includes particulate that may settle in the duct, clean-out doors should be provided in horizontal runs, near elbows, junctions and vertical runs (Figure 5-17). The spacing of clean-out doors should not exceed 12 feet [3.7 m] for ducts of 12 inches [305 mm] diameter and less, but may be greater for larger duct sizes. Removable caps should be installed at all terminal ends and the last branch connection should not be more than six inches from the capped end.

Transitions in mains and sub-mains should be tapered. The taper should be at least five units long for each one unit change in diameter or 45° maximum included angle (Figure 5-18).

All branches should enter the main at the center of the transition at an angle not to exceed 45° with 30° preferred in most cases (Figure 5-19). Smaller angles may be specified for abrasive materials. To minimize turbulence and possible particulate fall out, connections should be to the top or side of the main with no two branches entering at opposite sides.

A straight duct section of at least six equivalent duct diameters should be used when connecting to a fan (see Chapter 7 for discussion of System Effects). Elbows or other fittings at the fan inlet will seriously reduce the volume discharge (Figure 5-20). The diameter of the inlet duct should be approximately equal to the fan inlet diameter.

Hoods should be fabricated from the same materials as the duct and a minimum of two gauges heavier than straight sections of connecting branches. They should also be free of sharp edges or burrs, and reinforced to provide necessary stiffness. Ergonomic considerations for operator access and maintenance should be considered in all hood designs.

Discharge stacks should be vertical and terminate at a point where height or air velocity limits re-entry into supply air inlets or other plant openings (see Section 5.12).

Avoid use of flexible duct especially where the formation of severe bends is not restricted. Where required, use a non-collapsible type that is no longer than necessary to perform the required flexibility of the connection (< 2 feet [0.6 m]). Refer to the manufacturer's data for friction and bend losses.

Commercially available seamless tubing for small duct sizes (i.e., up to 8 inches) may be more economical on an installed cost basis than other types. Plastic pipe may be the best choice for some applications (e.g., corrosive conditions at low temperature) but could be a bad application for abrasive dusts.

Friction losses for duct not built to SMACNA standards can be different than standard construction. For specific information, consult manufacturer's data.

Where blast gates or dampers are used, locate them at least 5 diameters away from elbows or other interferences. Ensure that dampers cannot be adjusted after setting by locking in place (Figure 5-21).

5.13.5 Ancillary Equipment Design Considerations. Provide duct supports of sufficient capacity to carry the weight of the system plus the weight of the duct half filled with material and with no load placed on the connecting equipment at the hood.[5.3,5.4] Where quick clamp systems are used, more supports may be necessary.

Provide adequate clearance between ducts and ceilings, walls and floors for installation and maintenance. Install fire dampers, explosion vents, etc., in accordance with the National Fire Protection Association (NFPA) Codes and other applicable standards and manufacturers' instructions. Exhaust fans handling explosive or flammable atmospheres require special construction (see AMCA[5.5] for spark-resistant fan construction guidelines). Consult NFPA and other sources for correct specifications.

Minimize the use of blast gates or other dampers, if possible. However, if blast gates are used for system adjustment, place each in a vertical section midway between the hood and the next junction. To reduce tampering, provide a means of locking dampers in place after the adjustments have been made. Blast gates or orifice plates are mandatory if air balancing is required. Blast gates should be included in all ducts where adjustment is required.

Allow for vibration and expansion. If no other considerations make it inadvisable, provide a flexible connection between the duct and the fan. The fan housing and drive motor should be mounted on a common base of sufficient weight to dampen vibration, or on a properly designed vibration isolator.

Do not allow hoods and duct to be added to an existing exhaust system unless specifically provided for in the original design or unless the system design is modified. If changes are made to the duct system, use methods shown in Chapter 8 of the O&M Manual and Chapter 9 of this Manual. Locate fans and filtration equipment such that maintenance access is easy. Provide adequate lighting in penthouses and mechanical rooms.

Where federal, state, or local laws conflict with the preceding, the more stringent requirement should be followed. Deviation from existing regulations may require approval by local regulators.

5.14 TESTING AND BALANCING (TAB) OF LOCAL EXHAUST VENTILATION SYSTEMS

The exhaust system should be tested and balanced before operation (see Chapter 3 of the O&M Manual). Openings for sampling should also be provided in the discharge stack and/or duct network to test for compliance with air pollution codes or ordinances. Test ports should be located as required to verify flow and pressure of the fan and duct system.

REFERENCES

5.1 Hemeon, W.L.C.: Plant and Process Ventilation, 3rd Edition, pp. 215-218. Lewis Publishers (1999).

5.2 The Kirk and Blum Mfg. Co.: Woodworking Plants, pp. W-9. Cincinnati, OH (1964).

5.3 Sheet Metal and Air Conditioning Contractors' National Assoc., Inc.: Round Industrial Duct Construction Standards. Tysons Corner, Vienna, VA (1982).

5.4 Sheet Metal and Air Conditioning Contractors' National Assoc., Inc.: Rectangular Industrial Duct Construction Standards. Tysons Corner, Vienna, VA (1980).

5.5 Air Movement and Control Association, Inc.: AMCA Standard 210-74. Arlington Heights, IL (2005).

5.6 Wilson, D.J.: Flow Patterns Over Flat Roof Buildings and Application to Exhaust Stack Design. ASHRAE Transactions, 85:284-95 (1979).

5.7 Wilson, D.J.: Contamination of Air Intakes from Roof Exhaust Vents. ASHRAE Transactions, 82:1024-38 (1976).

5.8 Clark, J.: The Design and Location of Building Inlets and Outlets to Minimize Wind Effect and Building Reentry. Journal of the American Industrial Hygiene Society, 26:262 (1956).

5.9 American Society of Heating, Refrigerating and Air-Conditioning Engineers: 2001 Fundamentals Volume, Section 16.1. ASHRAE, Atlanta, GA (2001).

5.10 American Welding Society: (AWS D1.1-72) Miami, FL (2008).

FIGURE 5-15

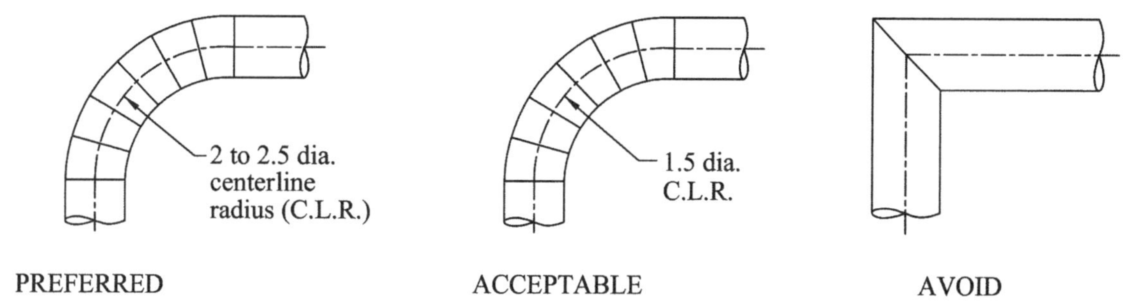

ELBOW RADIUS

Elbows should be 2 to 2.5 diameter centerline radius except where space does not permit. See Chapter 9, Fig. 9-e for loss factor.

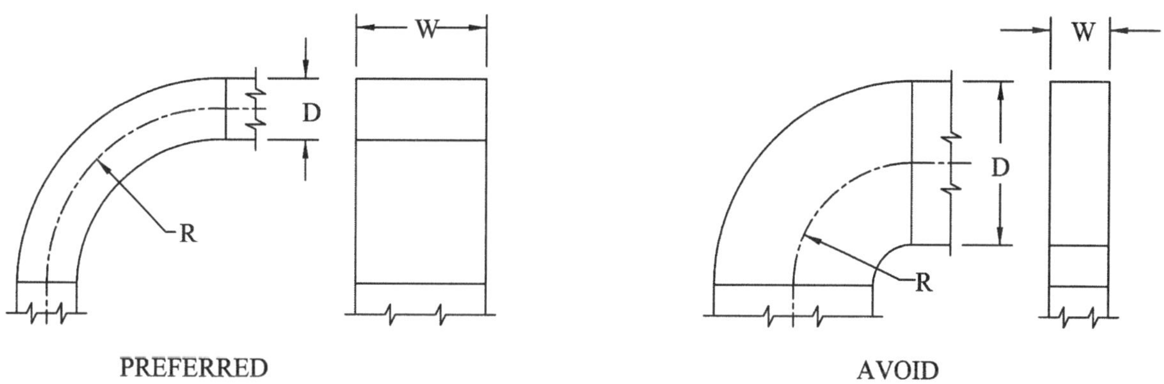

ASPECT RATIO $\left[\frac{W}{D}\right]$

Elbows should have $\left[\frac{W}{D}\right]$ and $\left[\frac{R}{D}\right]$ equal to or greater than (1).
See Chapter 9, Fig. 9-e for loss factor.

Note: Avoid mitered elbows. If necessary, use only with clean air and provide turning vanes. Consult mfg. for turning vane loss factor.

ACGIH®	TITLE: PRINCIPLES OF DUCT DESIGN ELBOWS	FIGURE: 5-15
		DATE: 1-12

CHECK CODES, REGULATIONS, AND LAWS (LOCAL, STATE, AND NATIONAL) TO ENSURE THAT DESIGN IS COMPLIANT.

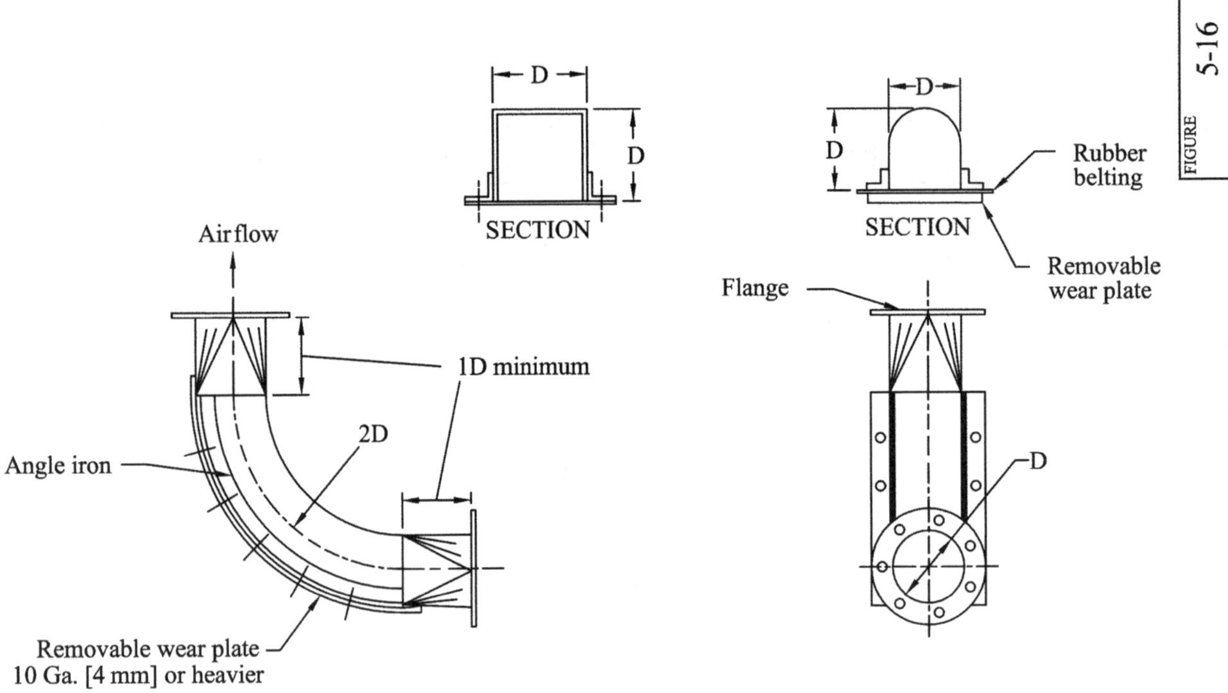

FLAT BACK ELBOW

CONCRETE REINFORCED ELBOW

Note: Provide solid mounting for concrete reinforced elbows

ACGIH®	TITLE: HEAVY DUTY ELBOWS	FIGURE 5-16 / DATE 1-12

CHECK CODES, REGULATIONS, AND LAWS (LOCAL, STATE, AND NATIONAL) TO ENSURE THAT DESIGN IS COMPLIANT.

FIGURE 5-17

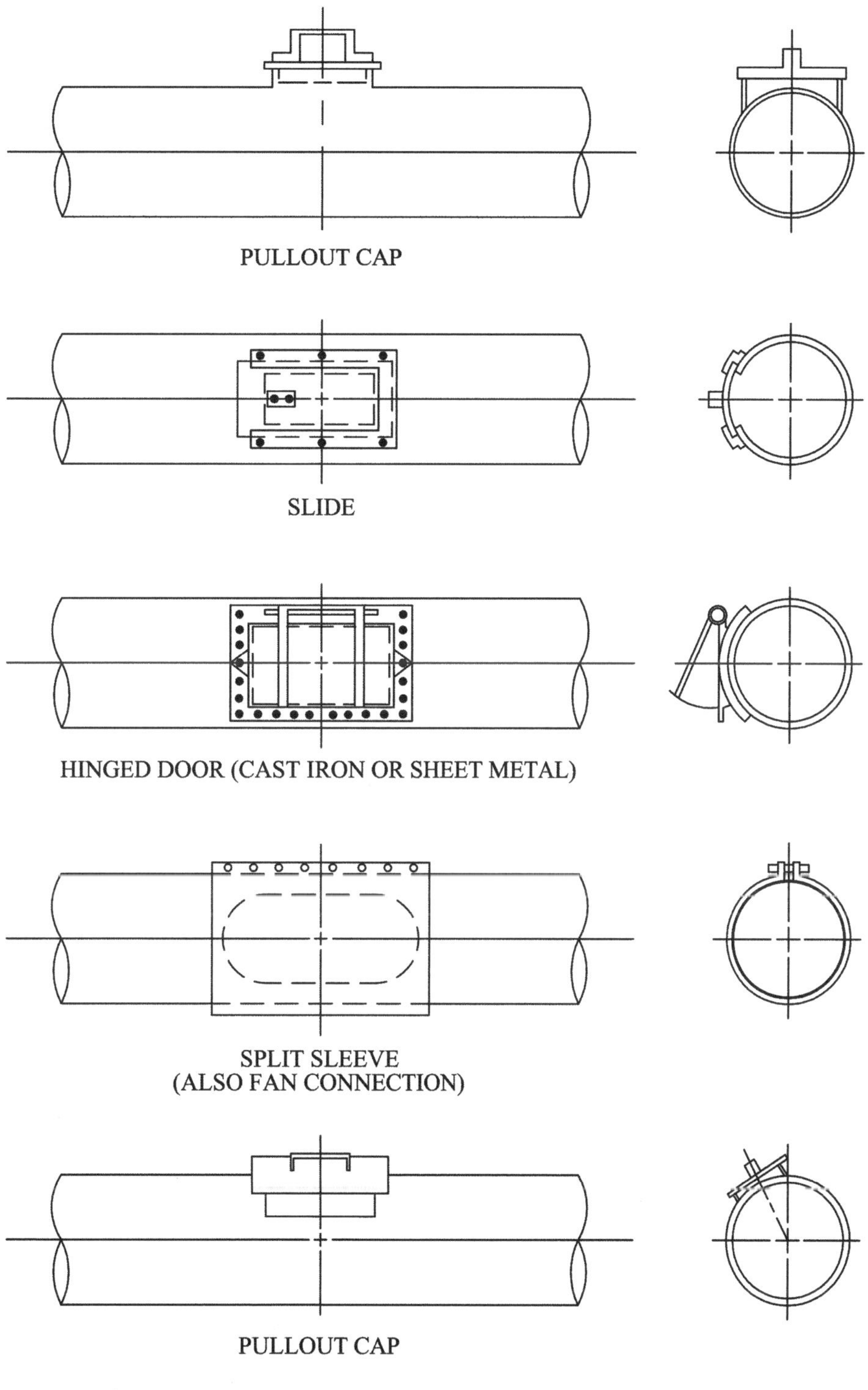

	TITLE	FIGURE 5-17
ACGIH®	CLEANOUT OPENINGS	DATE 1-12

CHECK CODES, REGULATIONS, AND LAWS (LOCAL, STATE, AND NATIONAL) TO ENSURE THAT DESIGN IS COMPLIANT.

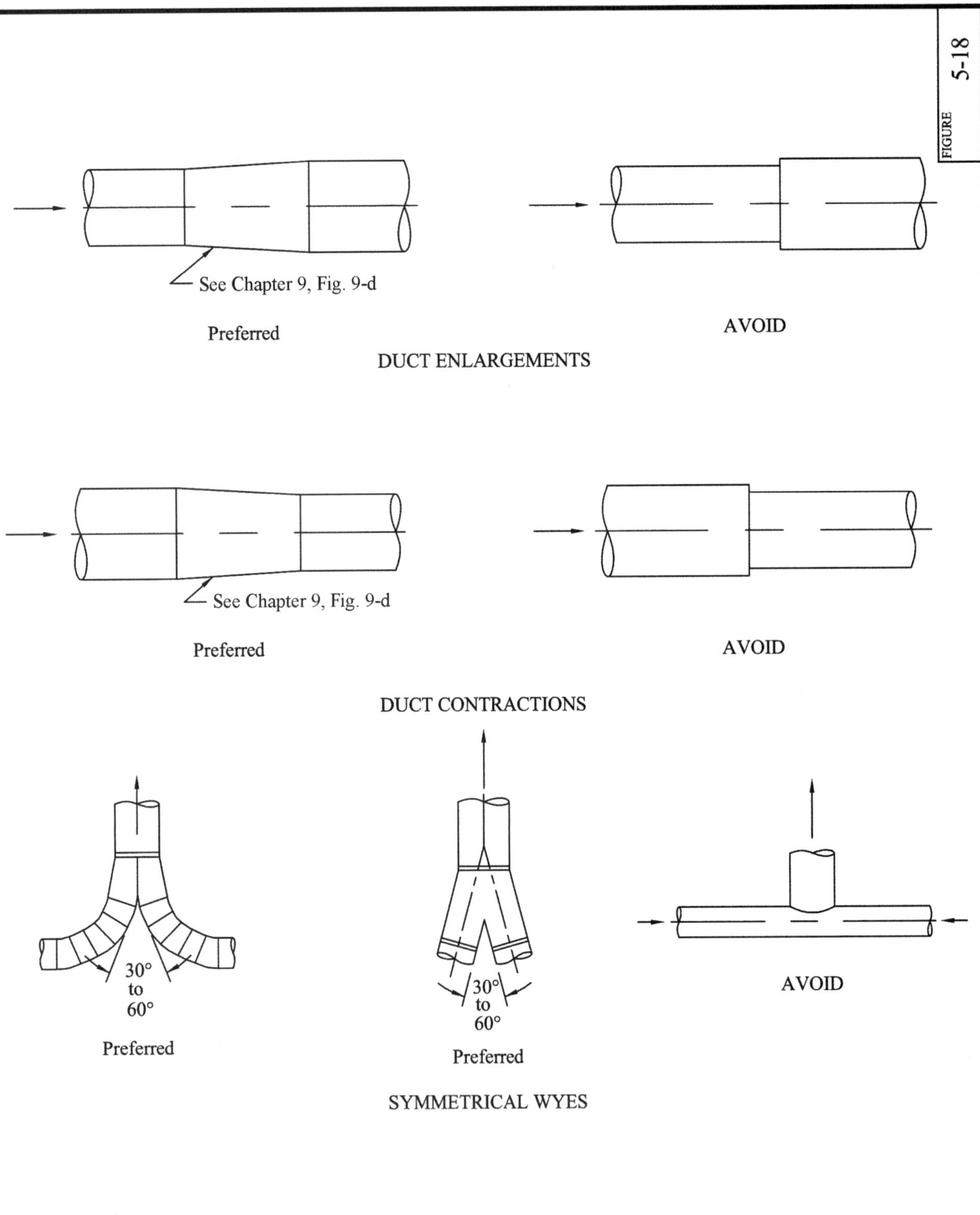

ACGIH®

TITLE: PRINCIPLES OF DUCT DESIGN

FIGURE 5-18

DATE 1-12

CHECK CODES, REGULATIONS, AND LAWS (LOCAL, STATE, AND NATIONAL) TO ENSURE THAT DESIGN IS COMPLIANT.

FIGURE 5-19

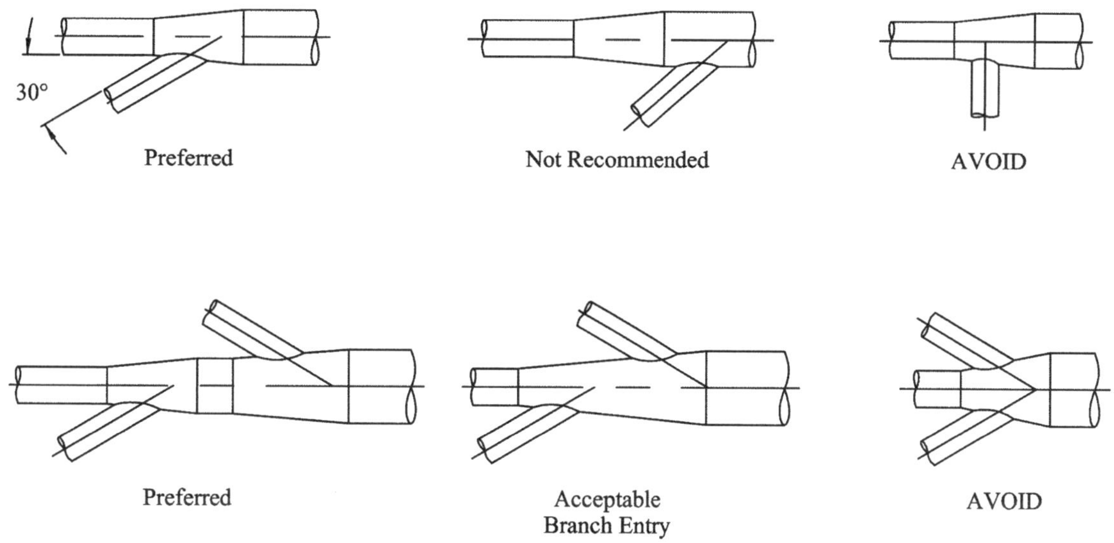

Branches should enter at gradual expansions and at an angle of 30° or less (preferred) to 45° if necessary. Expansion should be 15° maximum. See Chapter 9, Fig. 9-f for loss coefficients.

V_m = Minimum transport velocity
A = Cross section area

PROPER DUCT SIZE

Size the duct to maintain the proper transport velocity.

ACGIH®	TITLE PRINCIPLES OF DUCT DESIGN BRANCH ENTRY	FIGURE 5-19
		DATE 1-12

CHECK CODES, REGULATIONS, AND LAWS (LOCAL, STATE, AND NATIONAL) TO ENSURE THAT DESIGN IS COMPLIANT.

FIGURE 5-20

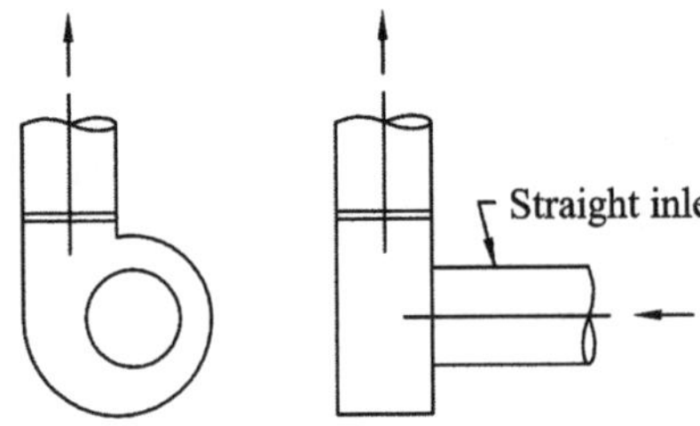

PREFERRED

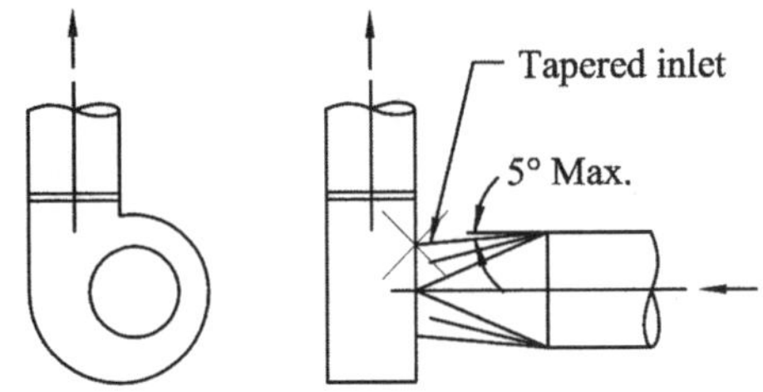

PREFERRED

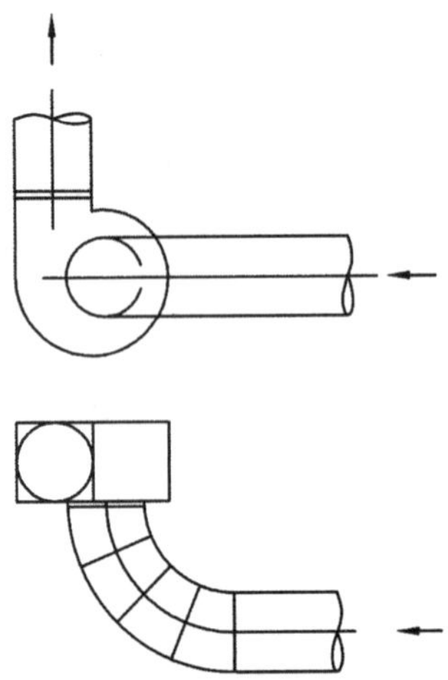

ACCEPTABLE

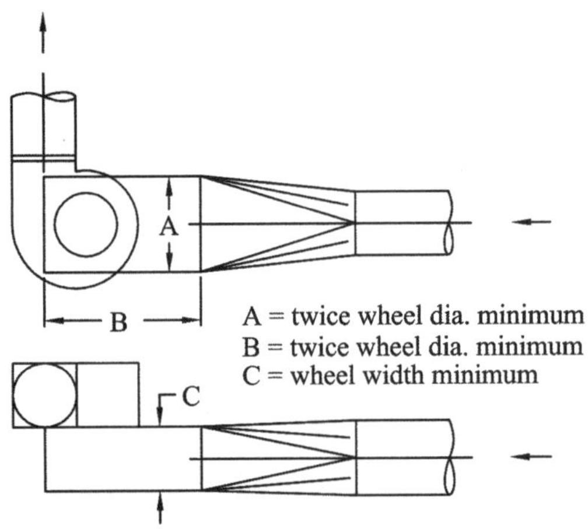

ACCEPTABLE

See Chapter 7 for system effect factors based on inlet and outlet duct arrangements

Consult fan manufacturer for actual effect on selected fan

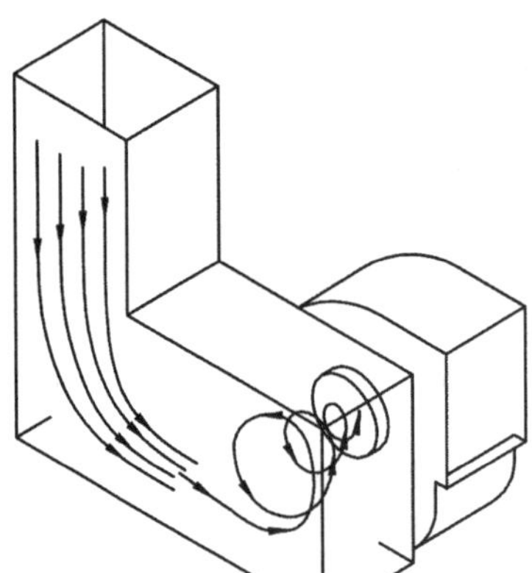

Use duct turn vanes to eliminate air spin or uneven loading of fan wheel

ACGIH®	TITLE PRINCIPLES OF DUCT DESIGN FAN INLETS	FIGURE 5-20 DATE 1-12

CHECK CODES, REGULATIONS, AND LAWS (LOCAL, STATE, AND NATIONAL) TO ENSURE THAT DESIGN IS COMPLIANT.

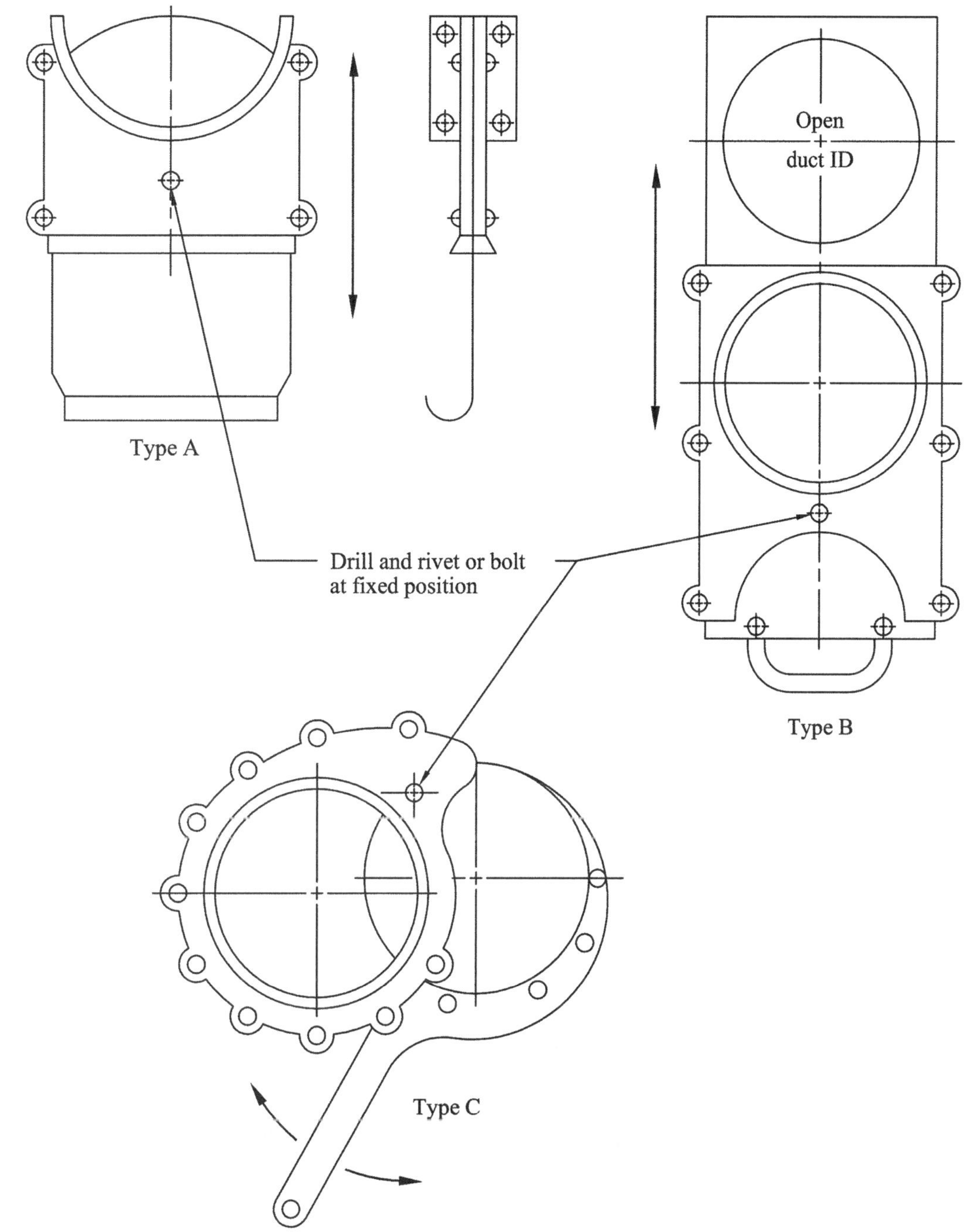

TITLE	FIGURE 5-21
BLAST GATES AND CUTOFFS	DATE 1-12

CHECK CODES, REGULATIONS, AND LAWS (LOCAL, STATE, AND NATIONAL) TO ENSURE THAT DESIGN IS COMPLIANT.

APPENDIX A5 COMPUTATIONAL FLUID DYNAMICS IN VENTILATION

A5.1 INTRODUCTION

Fluid dynamics is a discipline dealing with the complex behavior of airflows in occupational and industrial environments. In ventilation engineering, it may be necessary to make simplifying assumptions about airflows to make a practical solution possible. One method for understanding and predicting fluid behavior that is *relatively* free of simplifying assumptions is computational fluid dynamics (CFD). CFD uses numerical methods to solve the theoretical equations that describe fluid behavior.

The core concept of CFD is discretization: dividing the fluid (e.g., the air in a room or duct) into a grid of cells, with grid lines intersecting to form the cell corners called nodes. The partial differential equations that describe the conservation laws determining fluid behavior (Navier-Stokes equations) can be written approximately in a simple algebraic form in terms of a cell and its neighbor cells. This can be as basic as "what flows out of a cell must be the sum of what flows into its neighbors" or can be more sophisticated techniques that solve a function over a set of nodes.

While long established formulas and guidelines already exist, designers may encounter situations where accepted best practices are not possible to implement. For example, intuition and experience informs the designer that ventilation of a hot process such as foundry shakeout should involve overhead capture. If locating the hood directly above is not possible due to machinery such as a crane on overhead rails that move the work piece, CFD may provide a tool to quantitatively evaluate an alternative approach.

A5.1.1 Software and User Interface. CFD simulations are performed using commercial software, shareware, codes developed by research institutions, or codes written by individual users. User-friendly commercial CFD software may be the most efficient path. Any software that is employed must be capable of geometry set-up, grid formation, numerical solution, and reporting and graphical display of results, although different packages can be used for each task.

Whereas a large scientific computing cluster is still needed for the most computationally intense simulations, a desktop personal computer has become a practical, inexpensive platform that can handle many problems of interest.

A5.1.2 Geometry and Grid Formation. In this phase of a simulation, the user determines the required level of geometric realism to represent the relevant features of the flow field. The most important decision is dimensionality—is it necessary to model the space in three dimensions or is there a two-dimensional plane or planes cutting through the space that captures most of the important features? If the latter is sufficient, there is a very large savings in computation costs/time for a given level of accuracy.

The level of detail with which objects are resolved affects the ability to create a grid or mesh. For example, when a realistic geometry is created in a CAD package and then imported into a mesh generator, the CAD geometry must be "cleaned-up" before it is simple enough to allow meshing. In addition, finer details require smaller cells, and consequently more nodes. As more cells are needed to fill the computational domain, the computation time increases. If the mesh is not fine enough, the numerical solution can converge to an incorrect conclusion. The capability of the software to handle varying cell sizes and a variety of cell shapes helps somewhat with this problem and is an important feature for modeling complex geometries.

A5.1.3 Numerical Solutions. Most CFD codes use the Control Volume method.(A5.1) As an alternative, the Finite Element method offers some advantages.(A5.2) A technique known as the Discrete Vortex method has been used with great success in the special case of flow around a worker.(A5.3, A5.4)

In the Control Volume method, the computational element is a small volume or cell. While the Navier-Stokes equations (fundamental conservation equations that describe fluid motion) apply to a fluid continuum, the Control Volume method adapts these equations to a set of grid points. These discretized equations are written for each cell in terms of the neighboring cells. Using ϕ as a general variable to represent mass, momentum, or energy, the continuum form of the general conservation equation in Cartesian coordinates is:(A5.5)

$$\frac{\partial}{\partial t}(\rho\phi)+\frac{\partial}{\partial x}(\rho u\phi)+\frac{\partial}{\partial y}(\rho v\phi)+\frac{\partial}{\partial z}(\rho w\phi)= \frac{\partial}{\partial x}(\Gamma_\phi\frac{\partial\phi}{\partial x})+\frac{\partial}{\partial y}(\Gamma_\phi\frac{\partial\phi}{\partial y})+\frac{\partial}{\partial z}(\Gamma_\phi\frac{\partial\phi}{\partial z})+S_\phi \quad \text{[A5.1]}$$

When this partial differential equation is discretized in the Constant Volume method, it becomes

$$\phi_P\sum_i(A_i-S_P)=\sum_i(A_i\phi_i)+S_C \quad \text{[A5.2]}$$

where $i = N, S, E, W, F, B$ (or North, South, East, West, Front, Back) relative to the grid point P.(A5.6)

The A's are convective and diffusive flux coefficients between cells, and S_C and S_P are the components of the linearized source term, $S_\phi = S_C + S_P\phi$.

The starting point for the calculation is an explicit value (provided by the user) of the solution variables at a boundary of the area, for example, the air velocity and turbulence parameters over the face of a ceiling diffuser or the air contaminant mole fraction at the boundary of a source.

The solution proceeded through the following steps:

1) Solve Equation A5.2 for each velocity component, by

substituting that component for ϕ, using the current pressure field. These conservation of momentum equations update the velocity components.

2) Solve a "pressure correction" equation to adjust the pressure and velocity fields so that conservation of mass is achieved locally.
3) Solve Equation A5.2 for the kinetic energy of turbulence, 'k', and the eddy dissipation rate, 'ε', using the new velocity field.
4) Solve the contaminant conservation equation and update the concentration field.
5) Update the density and viscosity of the fluid based on the concentration.
6) Check the solution for convergence. If yes, stop. If no, repeat the process, starting at step 1.

This process is the SIMPLE algorithm of Patankar and Spalding.(A5.7)

If the problem is time-dependent, the strategy is to calculate the solution at a series of time steps that span the period of interest. The process mapped above will be followed for each discrete time step. In choosing the step length, it is important to choose the right timeframe to get good data.

A5.1.4 Physical and Chemical Modeling. Flows of concern to ventilation engineers are almost never laminar. Addressing turbulent fluctuations is a major concern and often a difficult issue. Simplifying assumptions are often required. Turbulence is modeled within CFD simulations empirically rather than calculated directly. The most widely used turbulence model is the k ε model, which was the assumed model in Step 3 of the SIMPLE procedure in the Numerical Methods section above.(A5.8) When turbulence is modeled, the grid spacing in the boundary layer near walls and other surfaces must be consistent with the assumptions of the wall function.

Air contaminants can be treated in several ways. Gas and vapor transport and concentration in room air can be modeled as a component mixture, with fluid properties such as density and viscosity determined by the volume fraction of each component present in a cell. The property of each component can be assumed constant, or a function such as the ideal or perfect gas law can be evoked.

A5.1.5 Boundary Conditions. The boundary conditions comprise all parts of the computational domain other than the cells. They contain an internal flow, such as a duct would; and they are contained by an external flow, as a manikin in a wind tunnel would be contained by the airflow. They hold the flow information that is input rather than computed in the course of the solution. Examples are walls, velocity inlets or outlets, pressure inlets or outlets, and outflows. All that is necessary is the known value of each flow variable of interest at that location in the domain. Closely related are the initial conditions, the value of the flow variables at the beginning of a time-dependent simulation. In addition, boundary conditions can vary in time, as might the emission rate of a contaminant source. Boundary conditions are important determinants of CFD accuracy.

A5.2 LIMITATIONS AND INTENDED USE

A perfect CFD simulation of the conditions used as inputs to the model will still fall short of capturing all variables that affect the flow under study, because the real scenario involves some degree of variability of conditions that eludes the modeler. However, it can be used to isolate the effect of the variables that are important to the investigator.

A5.2.1 Restrictions and Advantages of CFD for Industrial Ventilation Applications. The use of CFD for evaluation of systems and hood designs carries certain restrictions and advantages. These include:

A. Restrictions

Uncertain Input Quality: Like any predictive model, CFD is limited by the assumptions and input variables entered into the model. Erroneous inputs for initial or boundary conditions may still produce a result (the equations converge to an answer) but the veracity of the result is unknown. This concern may be of greater importance with "user friendly" computer codes developed with default values written into the code.

Assumptions for Handling Turbulence: Turbulence models require assumptions in order to predict the erratic behavior of turbulent fluids. The impact of these assumptions will depend on how the user applies the model. For instance, the popular "Two-equation Kinetic Energy (TKE)" model assumes that the entire flow field is fully turbulent. While this may be appropriate for flow over an aircraft fuselage, experience tells us that indoor flow is more likely to be weak-to-moderately turbulent with eddy formations varying widely in scale. Fully turbulent conditions in the indoor work environment are generally found in relatively few locations (supply/exhaust ducts, exhaust hoods, near obstacles to airflow). The TKE model's fully turbulent assumption will over-predict diffusion within the weak-moderately turbulent areas.

Time Consuming for Complex Geometries: When the objective relating to a complex geometry is limited, such as determining the pressure drop across some apparatus, it may be more appropriate to obtain the value experimentally.

Output is Only an Approximation: The closeness of the approximation depends on the accuracy of the model; however, the only way to get the "true" solution is through accurate experimental study.

B. Advantages

Good Results Under Non/Low Turbulent Conditions: The Navier-Stokes computational model (without using the turbulence model) has been shown to produce appropriate results when the fluid flow is within or close to laminar conditions (and the appropriate boundary conditions were employed). In this regard, CFD modeling could be a tool to studying indoor contaminant dispersion.

Potentially Cost Advantageous: Although the initial software packages are expensive, once acquired, the cost of modeling an individual scenario can be lower than the cost of conducting the actual experiment. This advantage becomes even more important when the situation under study is large and/or complex.

Speed: An experienced investigator can compare multiple scenarios or configurations using CFD in a fraction of the time it would take to build and conduct the corresponding experimental studies.

Detailed Information for the Entire Study Space: The output from an appropriately run CFD model can provide detailed information of all the relevant fluid variables (velocity, temperature, pressure, contaminant concentration, turbulence intensity) throughout the entire study space. It would be impossible to imitate this feature completely using experimental measurement methods and attempts to approximate this level of detail would be very time consuming.

Computer Is Not Affected By Hostile Environments: Hazardous environmental conditions such as high temperature and explosive or unsafe contaminant concentrations can be tested without human exposures.

A5.3 EFFECTIVE CFD APPLICATION TO PRACTICAL SYSTEM CONFIGURATIONS

It is important to remember that CFD model predictions are simply approximations limited by the accuracy of the user inputs, the appropriateness of the mathematical model and the limitations of the employing computer code. Combined with the known limitations concerning turbulent conditions, it would be inadvisable to use a CFD model prediction as the sole determinant in most applications. This is especially true for applications concerning safety and health. Despite this inadequacy as the definitive design tool, there are still several ways in which the CFD output can play a positive role in the evaluation and design of real-world systems. To ensure good practical application and review of CFD practices, the designer should:

Compare CFD Output With Prior Experience: After receiving the output of your CFD model, look closely at individual areas within the model and compare the CFD-predicted flow behavior with what you would expect to see. Laminar flow in areas of expected turbulence may be a good sign that something is wrong with the model. At the same time, if the flow behavior tends to agree with expectations, then there is added confidence that the model predictions are sufficiently close to accurate.

Compare CFD Output With Experimental Validation: If there are multiple scenarios of relatively similar variations to evaluate, compare the output from the CFD model with measured values obtained by physically conducting just one or two of the scenarios under study. Consistent results are evidence of a well-designed model.

Use Multiple Runs to Improve Accuracy: When complex geometries indicate uncertainties about the appropriate mesh density or boundary conditions, use multiple model runs while incrementally changing individual settings. Once changes consistently show a minimal effect upon model flow predictions, there is increased likelihood that the proper model parameters have been found.

Use CFD to Identify Designs With Highest Potential: After incorporating one of the previous steps to provide a confidence in the CFD model, use the model to compare among multiple variations of a prospective design or physical arrangement. Once top prospects have been identified, it is possible to build the physical model and test its performance.

REFERENCES

A5.1 Patankar, S.V.: Numerical Heat Transfer and Fluid Flow, pp. 30-40. Hemisphere: New York (1980).

A5.2 Baker, A.J.: Finite Element Computational Fluid Dynamics, pp. 11-13. Hemisphere: New York (1983).

A5.3 Kim, T.; Flynn, M.R.: Airflow Pattern Around a Worker in a Uniform Freestream, pp. 187-296. Amer. Ind. Hyg. Assoc. J. 52:7 (1991).

A5.4 George, D.K.; Flynn, M.R.; Goodman, R.: The Impact of Boundary Layer Separation on Local Exhaust Design and Worker Exposure. Appl. Occup. and Env. Hyg. 5:501-509 (1990).

A5.5 Awbi, H.B.: Ventilation of Buildings. London: E & FN Spon (1991).

A5.6 Fluent, Inc.: Fluent 4.4 User's Guide, vol. 3. Lebanon, NH: Fluent, Inc. (1997).

A5.7 Patankar, S.V.: Numerical Heat Transfer and Fluid Flow, p. 126. Hemisphere: New York (1980).

A5.8 Launder, B.E.; Spalding, D.B.: Lectures in Mathematical Models of Turbulence. London: Academic Press (1972).

Chapter 6

HOOD DESIGN: DESIGN OF INDUSTRIAL EXHAUST VENTILATION SYSTEMS

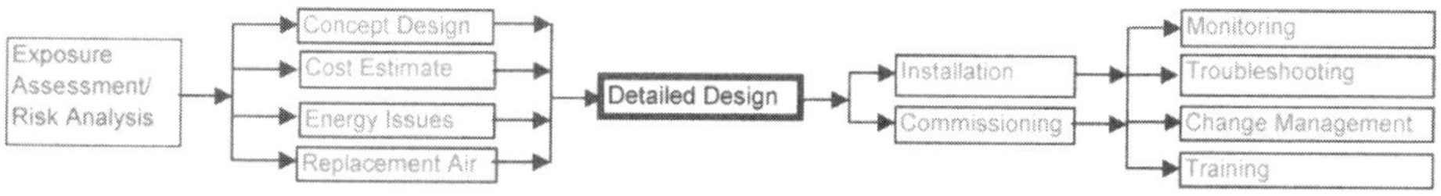

NOTE: Equations with notation followed by (IP) are designated for inch-pound system only; Equations followed by (SI) are designated for metric use only. If Equation bears neither, then it applies to both systems.

TABLE 6-1. Abbreviations Used in Chapter

A_f	= face area of hood opening	F_a	= acceleration (or Bernoulli) coefficient = 1
C_{cap}	= hood configuration flow factor	F_h	= hood entry to duct loss coefficient
h_{slot}	= slot height	F_s	= slot loss coefficient
L	= hood or table length	h_e	= overall hood entry loss = $h_h + h_s$
Q	= airflow requirement	h_h	= hood entry to duct loss = overall hood entry loss = $F_h\ VP_h$
W	= hood opening or table width	h_s	= slot or slot type opening loss = $F_s\ VP_s$
V_f	= hood face velocity	SP_h	= hood static pressure
V_x	= capture velocity necessary at distance X from the hood face	SP_f	= hood filter pressure loss
X	= greatest distance between contaminant and hood face	VP_d	= duct velocity pressure, "wg [Pa]
C_e	= hood flow coefficient	VP_s	= slot or opening velocity pressure, "wg [Pa]

6.1 INTRODUCTION

When air is exhausted from an area, the exhaust and supply air points set in motion a pattern of airflow that can be divided into three zones: the Supply Zone, the General Zone, and the "Local Exhaust" Zone.

The Supply Zone is created by the turbulent air motion of supply air as it enters the room from the supply point. It is marked by high velocity (> 200 fpm [1.00 m/s]) and turbulent mixing. Depending on supply fixture design, outlet velocity and orientation, supply air will continue in its initial direction from the supply point until its energy is lost in the general area at 20–30 feet [6–9 m] or more. Contaminated air released from a source located in the Supply Zone will be rapidly mixed (diluted) with the supply air. The Supply Zone ends where the air velocity is less than that of competing air currents induced by outside influences (traffic, thermal air currents, motion of material or equipment, etc.). This is called the General Zone. The first two zones are the province of "general ventilation" (see Chapter 4). The Local Exhaust Zone is the subject of this chapter.

6.1.1 Local Exhaust Hoods Compared to Dilution Ventilation. Local exhaust systems are ventilation systems that employ local exhaust hoods to control emissions from sources of airborne contaminants, not allowing most of the contaminants to mix with room air prior to collection by the hoods. Hoods control contaminant exposures by controlling airborne contaminants at their source and exhausting them from the area. This is very different from dilution ventilation (see Chapter 4), which allows contaminated air to mix with room air and then exhausts the mixture.

Local exhaust ventilation systems generally need only a fraction of the airflow required for dilution ventilation of the same sources. Furthermore, since dilution systems not only allow but promote mixing of the contaminated air with room air, dilution is generally adequate only when the contaminant is released at relatively low concentrations or can be well-mixed with clean air before reaching workers' breathing zones. Thus, a concentrated source very near a worker is likely to produce high exposures to that worker if only dilution ventilation is employed. By contrast, local exhaust hoods typically can adequately control emissions even when the source is within arm's reach of the worker as long as the contaminant cloud is relatively small or is projected away from the worker's breathing zone.

6.1.2 Local Exhaust System Effectiveness. The ability of a local exhaust ventilation system to reduce exposure to air contaminants is determined primarily by three factors:

1) The effectiveness of the hoods (if they have been provided sufficient airflow to contain and capture contaminants),
2) The ability of the fan/duct system to deliver sufficient airflow to each hood,
3) Whether workers use the hood when needed, which is strongly affected by the convenience of the hood during work.

This chapter addresses these issues, as well as issues important to the operation of hoods. Chapter 13 contains recommendations for hoods for specific processes and tasks based on the general principles in this chapter. The designs in Chapter 13 often can be adapted for different processes and tasks, especially if the operating conditions are similar and the degree of hazard is about the same. If there are important differences, any design should be adapted and airflows selected in conformance with the recommendations given in this chapter.

There are many hoods designed for specific applications and there are many more where the same basic design is used for different purposes but are given different names, leading to diverse and sometimes contradictory terminologies. For that reason, the terms used here are mostly descriptive and do not follow any specific conventions.

6.1.3 Design Goals. Exhaust airflow into a hood should reduce the worker-user's exposure while working at the hood to "acceptable" levels or lower. The owner of the ventilation system may choose to set acceptable levels of exposure to values required for compliance with governmental regulations (e.g., OSHA, EPA), conformance with recommended practices (e.g., ACGIH® TLVs®), or other levels. The latter most often occurs when there is no government regulation and no widely accepted recommended standard.

As in all other engineering design, the goal for hoods is an optimal tradeoff of effectiveness and overall costs while meeting the following goals:

1) The hood should not introduce a substantial new hazard to workers and should reduce safety hazards where possible.
2) The hood should not increase ergonomic stresses and should reduce them where possible.
3) The hood should use the minimum airflow as required to meet goals. Operating and installation costs are roughly proportional to system airflow.
4) Hoods should be designed to minimize the time and effort required for maintenance activities and other interferences with the process.
5) The hood and the materials handling system (entering and exiting the area) must be compatible.

6.1.4 Wake Zones. Understanding wake and separation zones are important to understand when designing or operating hoods that rely on plug flow to protect workers and, to a much lesser degree, capturing hoods. Air passing around any blunt obstruction, including the human body, creates a complex downstream counter-flow known as a wake zone that includes more or less stable recirculating airflow patterns called vortices as well as flow back towards the obstruction. Wake zones are crucial to exposure to contaminants and understanding them is crucial when designing hoods.

If the contaminant is released within the wake zone downstream of a human body it can circulate in that zone. Gradually it will dissipate due to dilution and sudden downstream movement of vortices. This is called shedding. Meanwhile, the backflow can also carry contaminants released several feet downstream of the obstruction back towards the body and up to the breathing zone.

If the flow is from the side or front of the body, the wake zone is on the other side or the back of the body, respectively. Since the mouth and nose generally face towards the front, they are not in the wake zone unless flow is from the back. Thus wake zones are generally of concern when the flow is from the back, (sometimes the case when standing in cross-drafts and nearly always the case when standing in front of enclosing hoods and especially when standing inside of a spray booth). Facing 90° to the cross-draft may provide the lowest exposures, depending on how the contaminant is released to the air. Facing upstream often will produce lower exposures than facing downstream if the contaminant cloud does not extend above waist height. That is likely only if the contaminant is released at low velocity exclusively below waist height, and immediately upstream of the body. Even under those conditions having the worker face upstream is discouraged because seemingly minor changes in work practices may cause at least some contaminant to be dispersed well above waist height.

Every blunt body in a flow pattern that is mostly in one direction will have a wake zone, including blunt bodies near the face of capturing hoods and most especially bodies just outside the face of enclosing hoods and within enclosing hoods that are designed to have plug flow. Plug flow will be used here to describe a flow without large scale eddies or swirling. Note that an "aerodynamically" rounded body will produce minimal wake zones, with the rounding on the downstream side being more important than the upstream side.

Although there may be larger blunt bodies with larger wake zones than the human body, the human body is the most important because the backflows in its wake zone can draw contaminants to the person's breathing zone.

Separation of flows from surfaces produces conditions with some similarities to the wake zones downstream of blunt bodies. Anytime airflow changes direction when flowing around a surface its momentum causes some degree of separation of the flow from that surface. The volume between the separation boundary and the surface ("separation zone" or "separation region") will be similar to those due to flow around rounded bodies. The greater the change in direction and the more abrupt it is, the greater the size of the separation region. The circulation velocity of the vortices within the wakes will increase with greater velocities. Contaminants released into a separation zone will only gradually dissipate with time.

For enclosing hoods there are separation zones associated with the momentum-induced separation of flows around the perimeter of the hood (see Figures 6-1 and 6-2). Ordinarily, contaminant that reaches those zones probably would not be a problem if the worker is centered on the hood or if the contaminant never reaches the perimeter. However, if a high-velocity cross-draft approaches the hood from 90°, the size of the separation zone on that side may be large enough to intersect the wake zone of the worker's body, allowing transfer of contaminants between the two zones.

6.1.5 Hood Types. Hoods may have a wide range of physical configurations but can be grouped into two main categories: enclosing and capturing (sometimes called 'external'). If a contaminant is released in front of the opening into which exhaust air flows, the exhaust opening is said to be a "capturing" hood since the movement of air induced to flow into the opening carries or "captures" some or all of the contaminant

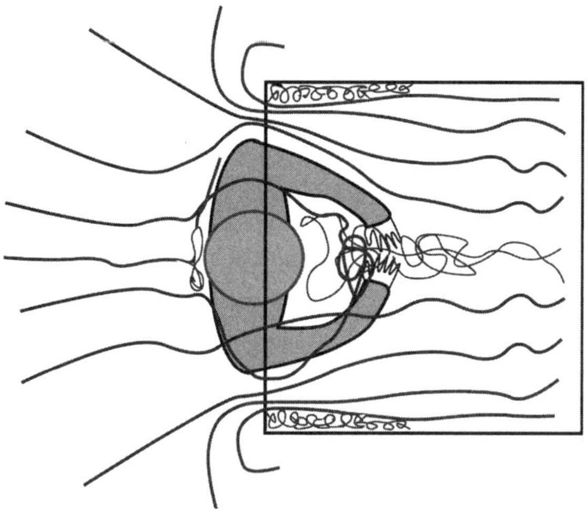

FIGURE 6-1. Flow with no crossdraft

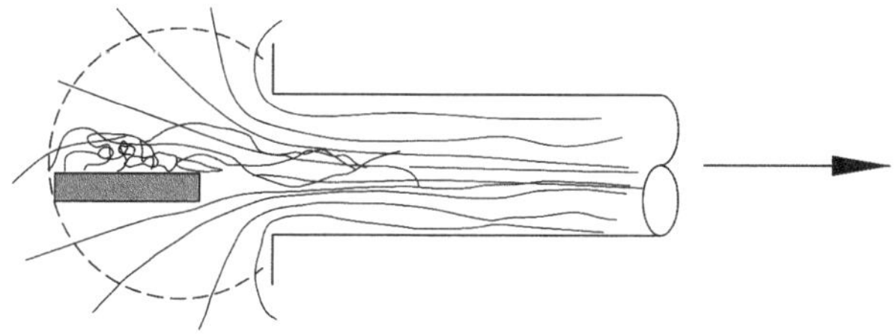

FIGURE 6-3a. Flow into a capturing hood

released in front (see Figure 6-3a). If the contaminant is pushed by moving air, thermal buoyancy, or the momentum from the contaminant release towards the capturing hood, the capture hood is called a "receiving" hood. A capturing hood can be large or small, depending mostly on the size of the source and its distance from the hood opening. Some capturing hoods protect workers working very near them (e.g., welding hoods) and others serve to reduce background concentrations (e.g., high canopies over furnaces).

If the contaminant is released within the confines of a ventilated structure that is exhaust ventilated, the structure can be called an "enclosing" hood (see Figure 6-3b). Enclosing hoods can be large or small, intended to allow frequent access to equipment inside or intended to completely minimize openings, allowing little or no access. With careful design, some enclosing hoods can be used with workers inside.

There are many specific terms to label different hoods within those two categories based on their specific purpose or aspects of their designs. Both types of hoods can be effective for cases where either the contaminant generation rate or the amount of dispersion or both are relatively low. If the contaminant generation rate is very high and highly dispersed, then only enclosing hoods are likely to be effective, and then only if their openings are minimized and the worker does not enter the hood without appropriate respiratory protection (also see Appendix A6).

6.2 ENCLOSING HOODS – INTRODUCTION

Enclosing hoods are ventilated boxes completely or partially enveloping one or more contaminant generation points. Enclosing hoods prevent the escape of contaminant by physically limiting the openings through which contaminated air can escape and by the movement of air through those openings to prevent its escape. Enclosing hoods can be large or small, depending on the needs of the process and materials handling. There are many hoods used for very specific applications to meet special requirements for materials handling or high toxicity.

In general, enclosing hoods are the most effective means of contaminant control but an important functional consideration that drives selection or design is the degree and quality of the necessary access by workers (see Section 6.10 – Ergonomics).

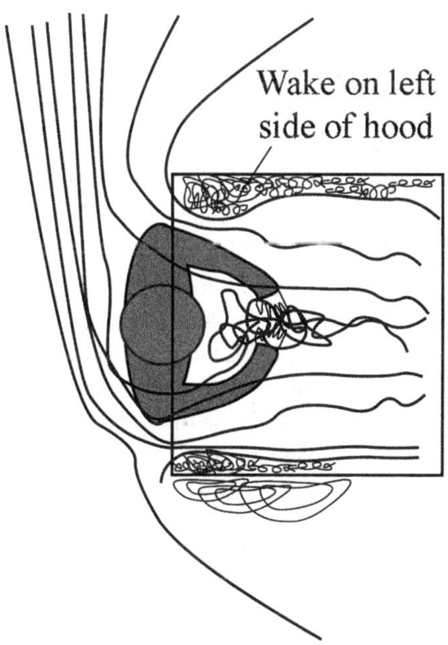

FIGURE 6-2. Flow with crossdraft

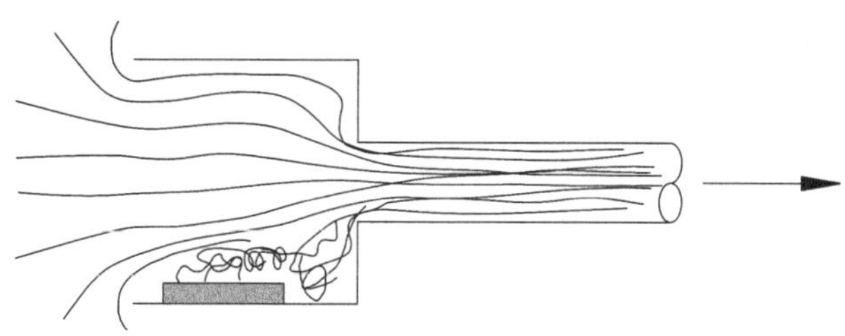

FIGURE 6-3b. Flow into an enclosing hood

Generally, the smaller the total area of permanent openings, the less airflow is required and the better the containment of the contaminants inside the enclosure. The trade off for the containment efficiency of such hoods is generally very high concentrations inside the hood, making them unsafe for worker entry without appropriate protection (see Section 6.3 – Total Enclosures).

If workers spend substantial durations reaching through permanent openings to manipulate objects inside the enclosure, then those openings must be large enough to access efficiently and conveniently. To be safe for this use, there are important issues in design and operation that must be addressed. Very often such hoods are mounted on stands, cabinets, or tables so that the opening extends from roughly waist height to above the head of the worker. For that reason, such hoods can be referred to as "bench top" hoods. Because of the importance of designing for work efficiency, they could also be called "bench top workstation hoods." A "laboratory" hood is a bench top workstation hood.

If workers must occupy a hood to work, then one side of the enclosure may be completely open for ease of access, material handling and to allow uniform flow of air into the hood. Protection of workers in such hoods depends very much on the uniformity of flows down the length of the hood since lateral and upstream flows will draw contaminants from downstream sources to the workers' positions. Because workers generally spend a great deal of time inside these hoods, they also should be considered workstations and designed accordingly. The most noteworthy application of this design is for spray painting large objects, so these hoods are sometimes called "spray booths" even when not used for spray painting. Since bench top workstation hoods also can be used for spray-painting, they will be henceforth referred to as "occupied enclosing" hoods. As with bench top enclosing hoods, critical design and operational issues must be addressed for workers to use these hoods safely (see Section 6.10 – Ergonomics).

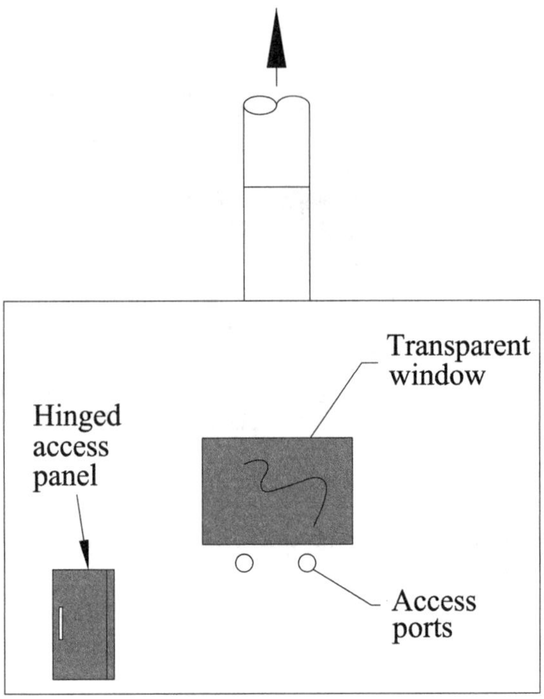

FIGURE 6-4. Near-total enclosure

6.3 ENCLOSING HOODS THAT RELY ON PLUG FLOW TO PROTECT USERS

In many cases work tasks require workers either to stand or sit and reach into the enclosure frequently or to work inside the hood for very large enclosures. In these applications, the hood can sufficiently protect the worker only if great care is taken in the design and operation of the hood. In particular, the contaminant cloud inside the hood must be largely prevented from reaching the breathing zone. This is best accomplished by preventing the contaminant from mixing with the wake zone of the worker as much as possible.

6.3.1 Importance of Plug Flow. If the worker is at the face of the hood reaching into it to work, then the inflowing air must push the contaminant towards the back of the hood and the contaminant should not recirculate to the face of the hood once it enters into the enclosure. The primary strategy is to provide relatively uniform velocities at the face of the hood and well into the enclosure. A flow that has a uniform velocity will show little swirl (spiraling flow), no large-scale eddy currents (thus no stagnation zones with rotating flow) and no flow back toward the face. The air is said to move as if it were a fixed volume or "plug." Obstructions and competing air movements tend to disrupt the uniformity of the airflow and thus reduce the protection provided by the hood.

The same issues apply to personnel who work inside a large enclosing hood. It is imperative that the movement of air separate their wake zones from the contaminant cloud. The separation is best achieved by distance, uniformity of velocities through the enclosure, and by keeping contaminant clouds downstream of or to the side of workers. Obstructions and competing air motions can disrupt the flow in ways that move the cloud toward the workers inside the enclosure.

To better accomplish plug flow, hoods generally have a completely open face that is the same cross-section as the enclosure. For occupied hoods, the face can be a wall of filters to remove room air dust, especially for spray-paint booths. While not "open," a cross-section of the wall of filters equal to enclosure size can provide relatively uniform flow.

If the hood face is partially blocked so that little or no flow passes through substantial portions of the face, the result will be large-scale eddy currents, along with accompanying stagnant zones as well as lateral and vertical movement of contaminated air. If workers are inside the hood such blockages will increase their exposures. If the worker is at the face of the hood, such partial blockages can draw contaminants toward the face of the hood, increasing exposures to the worker. In the case of a laboratory hood, the barrier is a sash that can be raised and lowered or moved laterally. It is intended to keep the user's face out of the hood. A vertical sash also serves to keep the worker's face well above the bottom of the sash. That is critical because the sash (and to a lesser degree, the "dome" inside the hood) produces a large vortex that rolls on a horizontal axis just behind the sash, bringing contaminants from throughout the hood to the bottom of the sash. Lab hoods (see Chapter 13, Section 13.35) have non-plug flow inside the enclosure, but the sash protects users by keeping their heads outside of the hood well above the bottom of the sash.

Every aspect of the design of such hoods is affected by the need to develop plug flow. If the contaminant is carried by air-currents back toward the user, the hood may provide very poor protection. This is likely to happen if: 1) the air flowing into the hood is highly disturbed by cross-drafts, 2) obstructions within the hood partially block the flow and channel it in undesirable locations, 3) the locations of the face(s) and exhaust are not capable of producing plug flow, and 4) the source is very large or produces copious or energetic dispersion of contaminants and is located adjacent to or upstream of the worker. Combinations of one or more of these factors can jeopardize the worker's safety.

6.3.2 Plug Flow Enclosing Hood Face Velocity. Air movement prevents the escape of contaminated air through the open face of the enclosure. Within limits, the higher the airflow (i.e., velocity) through the face, the less contaminant escapes. The design face velocity (V_f) should be based on the effectiveness (concentration outside/concentration inside) required to protect workers. In general, the minimum acceptable face velocity should be determined by the:

1) Toxicity of the contaminant. A higher hazard *generally* requires a higher velocity, but no studies have established how much more effectiveness is gained for additional increments of velocity. It is likely but not clearly demonstrated that the gain is very small when the velocity already exceeds 150 fpm. This is probably also true for all of the following determinants. USEPA Method 204 provides a requirement for 200 fpm [1.01 m/s] for containment of volatile organic compounds (VOC) and may be required for certain applications. This requirement has also been applied by some regulatory agencies for control of other materials.
2) Rate of generation of the contaminant. A higher generation rate generally requires velocity closer to the top of the recommended range.
3) Strength of competing air motions inside the enclosure (e.g., pneumatic spraying) and outside the hood (e.g., cross-drafts, personnel cooling fans, passing vehicles). Very strong competing air motions may warrant face velocities above the range typically recommended. It is also quite possible that for very poor conditions exposures simply cannot be controlled sufficiently to protect a worker who is very close to the source. Based on a study of laboratory hoods, one source[6.2] recommends taking steps to reduce cross drafts to no more than half of the hood face velocity. However, it is likely that to avoid having cross draft approach the hood from 90° is equally important.
4) Degree of enclosure employed. If the source is poorly enclosed, the face velocity must be higher to compensate for the lack of shielding from competing air currents. As with other determinants, it is likely that increasing velocities cannot completely compensate for poor enclosure.
5) Size of the hood used. A bench top hood is generally small enough that the user blocks a substantial fraction of the opening. It is likely that wake effects from air flowing over the back are worsened by that blockage, though probably to a lesser degree than for a person inside a booth.

For bench top hoods, hood effectiveness increases significantly with face velocity within the range of 75 to 130 ft/min [0.38 to 0.66 m/s] and sometimes higher. For cases where cross-drafts or competing air motions near or at the source are severe (e.g., pneumatic paint spraying), it is quite possible that face velocities above 150 ft/min [0.76 m/s] will be required unless the user stands well away from the region of contaminant dispersion.

For occupied hoods with relatively undisturbed flow from the face to the plenum (e.g., spray booths), a range of 100 to 150 ft/min [0.51 to 0.76 m/s] is usually adequate for typical applications and conditions. (See VS plates in Chapter 13 for other values on special hoods.)

If a hood's performance is not adequate and the face velocity is already above 120 ft/min [0.60 m/s], even substantial increases in face velocity can reduce exposures only modestly. Instead of continuing to increase the face velocity, obtaining substantial improvements in effectiveness is likely to require changing the design of the hood, improving the work practices of those using the hood, reducing crossdrafts or reducing the rate of generation of airborne contaminants within the hood.

There may be extra regulatory requirements for hood velocities, such as USEPA Method 204. Consult those references for hood designs that are required to meet those criteria. Also, environmental regulatory agencies require that for some regulated processes hoods should meet requirements to insure effectiveness in preventing the escape of contaminants into the general environment (i.e., "fugitive" gases and dusts). EPA Method 204 [Sections 5 and 6] describes the requirements of a permanent or temporary total enclosure from the EPA perspective. In general, a reduction in fugitive emissions from a hood will be associated with reduced ambient concentrations in the immediate area near the hood and consequently to reductions in exposures to workers standing near the hood. On the other hand, although a reduction in escape from a hood probably will be associated with a reduction in exposures to someone working at the face of the hood, there may be exceptions. Most importantly, if the worker must enter the enclosure, changes made to reduce escape from the hood may actually increase exposures to a worker while inside the hood unless done with great care and understanding of airflow.

6.3.3 Airflow Requirements for Enclosing Hoods. The system should deliver enough airflow to maintain the desired target airflow at the face of the hood (V_f) over the area of the hood face (A_f). Thus, airflow rate (Q) is computed from:

$$Q = V_f A_f \qquad [6.1]$$

where: Q = airflow rate, acfm

V_f = desired average velocity at the face, fpm [m/s]

A_f = total open area at the hood face, ft^2 [m^2]

For example, if the open face is 10' × 15' [3.0 m × 4.6 m] and the face velocity (V_f) is 100 fpm [0.51 m/s], then:

$$Q = (100\ ft/min)(10\ ft)(15\ ft) = 15{,}000\ acfm$$

$$[Q = (0.51\ m/s)(3.0\ m)(4.6\ m) = 7.01\ am^3/s]$$

To keep airflow rate (Q) to a minimum, the open area must be kept to a minimum consistent with the requirements of the process.

Note the airflow requirements for hoods are not affected at all by density, which is another way of saying that it is the velocity into the hood that determines the effectiveness of the hood, not the mass-rate. For example, the same hood used for the same purpose in both New Orleans (Sea Level) and Denver (5000 ft above Sea Level) [1524 m] should both have the same face velocity (e.g., 100 ft/min [0.51 m/s]). Note that the same is true for the minimum velocities required to keep contaminants moving in ducts. For that reason, standardized airflows (e.g., "scfm") should be used only for a few specialized applications. To emphasize this point, airflows are sometimes called "actual airflows" and given the units of "acfm" (actual cubic feet per minute) [am^3/s] instead of just "cfm" [m^3/s] to avoid confusion with standardized airflow.

Note that airflow rate on all figures in Chapters 3 and 13 are stated in acfm unless specified otherwise to make it clear that one should not consider the altitude and temperature of the air entering the hood when setting hood airflow requirements.

Finally, the airflow requirements must also maintain proper conditions inside the hood. This includes LEL or other exposure requirements even if face velocity values are maintained.

6.3.4 Achieving Uniform Face Velocities in Plug Flow Enclosing Hoods. The area of the face of most enclosing hoods designed for frequent worker access (i.e., plug flow hoods) is very large compared to the cross-sectional area of the connecting duct. Air passing through a hood face must converge to the much smaller area of the duct while accelerating to the higher velocity in that duct. Even without the effects of cross-drafts, the face velocity is not likely to be uniform across the face and the velocity could be very low at some points across the face. In those cases the contaminant might escape at weak points. To improve the uniformity of the flow velocities at the face and inside the hood:

1) Make the hood relatively deep by setting a minimum enclosure depth (d_{encl} in Figure 6-5) of at least 0.75 times the face height or face width – whichever is greater. Even if the velocity at the back of the hood (see the right side of Figure 6-5) is not uniform, an adequate depth will assist in providing a relatively uniform velocity near the face of the hood.

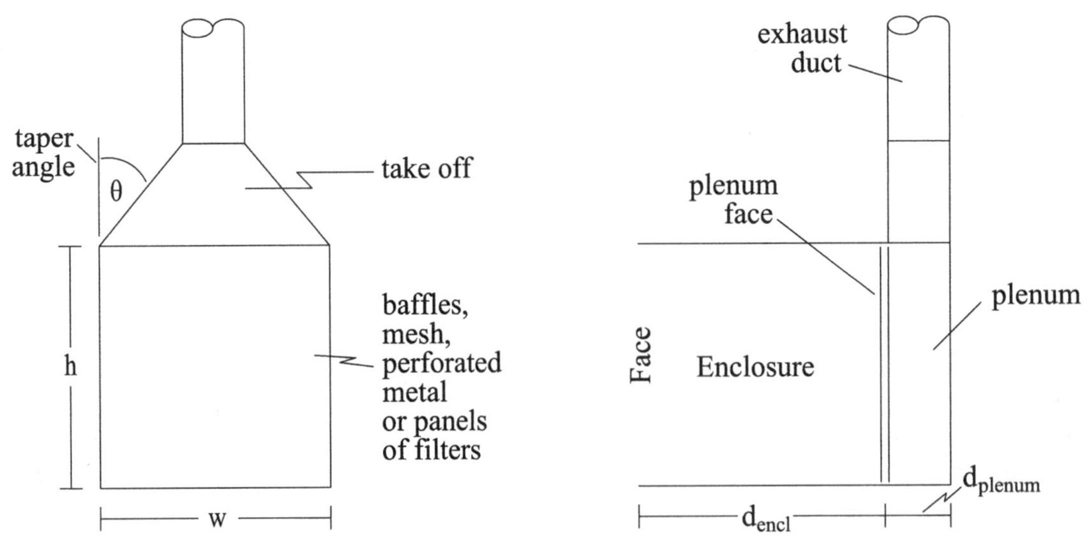

FIGURE 6-5. Parts of an enclosing hood

2) Install a plenum. The plenum is the section at the back of the hood formed by a wall of filters, baffles or slots (see Figure 6-5). Filters and baffles force the air to spread out at the back of the booth. If filters are used, the static pressure drop across the filters when clean should be > 0.10 in. w.g. [25 Pa].

   If baffles or slots are used instead of filters, the total cross-sectional area of the baffles should be 90–95% of the face area. If slots are used, there should be at least three, and they should be spaced evenly over the plenum face. Steel mesh, expanded metal, and perforated metal can be use instead of baffles, slots or filters.

3) To guide the air converging from the plenum section, design the transition to the duct or "takeoff" (see Figure 6-5) with an included angle of 90° (taper angle of 45°). If vertical space is not sufficient for a 90° included angle take-off consider multiple take-offs across the width of the plenum (see Figure 6-6).

4) Install a rounded or tapered entry at the hood face with a radius greater than 2" [50 mm] (see Figure 6-7) to reduce the separation zones that are inside the hood at the perimeter of the face. If the hood must be extremely effective and the contaminant may be released near the sides or top, consider installing airfoils to the perimeter of the hood face (installing a sash may be more effective).

5) The hood face should extend the full width and height of the enclosure to reduce wake zones, where possible.

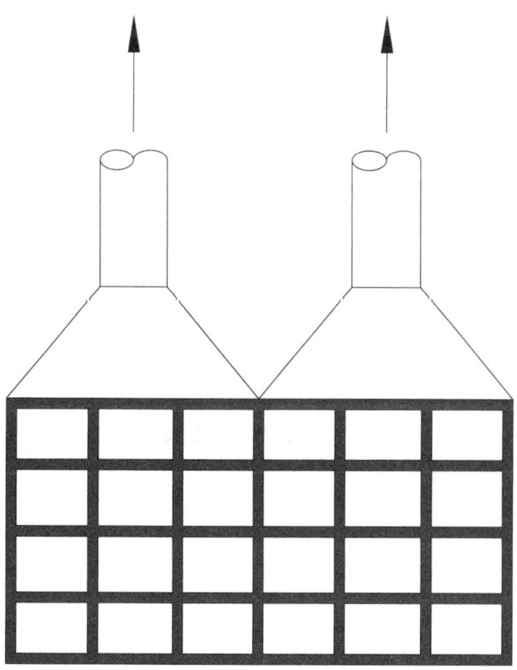

FIGURE 6-6. Multiple takeoffs for wide hoods

6.3.5 Effect of Supply Air on Uniformity of Flows at the Hood Face. If the path of the supply air to the hood is at some angle to the hood face, the airflow distribution at the face will be skewed (see Figure 6-8). The greater the velocity of the cross-draft and the closer its angle is to 90 degrees, the more disruptive the supply air will be. In general, approach velocities should be less than 30% of the hood face velocity.(6.2)

Even if its pathway is straight into the hood, supply air at high velocity near the hood can be a problem. Excess airflow can actually reverse course and exit through the face of the hood, carrying contaminants with it. There is anecdotal evidence that suggests that flow straight into a laboratory hood may be more disruptive than air approaching at 90°.

The supply air should be delivered to the room through a supply air duct system with its own fan and should be released with a low initial momentum in the direction of the exhaust hood but at a substantial distance from the hood.

6.3.6 Large "Spray Booth" Hood Airflow Patterns. For some operations (e.g., paint spraying) workers must occupy the hood to do their work. To prevent transport of contaminants towards the workers, hoods should be designed to insure that flow is relatively uniform and without back flow or large scale swirling. Hence, the flows should be aligned with the sidewalls all along the length of the hood. Although this plug flow is effective in carrying contaminated air away from the worker when the source is downstream, the tradeoff is that uniform flows produce more substantial wake zones downstream of blunt bodies, including the workers' own bodies. These wake zones are likely to be much more stagnant and larger than those seen in front of workers standing at the face of a bench top hood that has no sash. The reason is that much of the air entering a bench top hood comes from the perimeter and flows inward toward the center of the hood, partially filling the wake zone in front of the user. By contrast, a bench top hood with a sash partially blocking the face presents a much more complicated picture in which the wake of the body is perhaps less important than the complex separation zones and vortices induced by the flow under or around the sash.

Air flowing through an occupied hood should be parallel to the walls to avoid producing large eddy currents, especially if both the worker and the source are within the same eddy or wake. This is done by making the hood relatively deep and by making the flow at the back of the hood as uniform as reasonably possible by the use of panels, baffles or filters and by using 45° tapered takeoffs. Large objects in the hood can also produce a stagnation zone upstream of their leading, so their placement is also important.

6.3.7 Bench Top Enclosing Hood Airflow Patterns. For enclosing hoods small enough that the worker is stationed at the face of the hood (i.e., bench top enclosing hoods, including lab hoods), some of the air entering the hood must flow around the user's body to get into the face of the hood (see Figure 6-1). The air that flows around the operator's body cre-

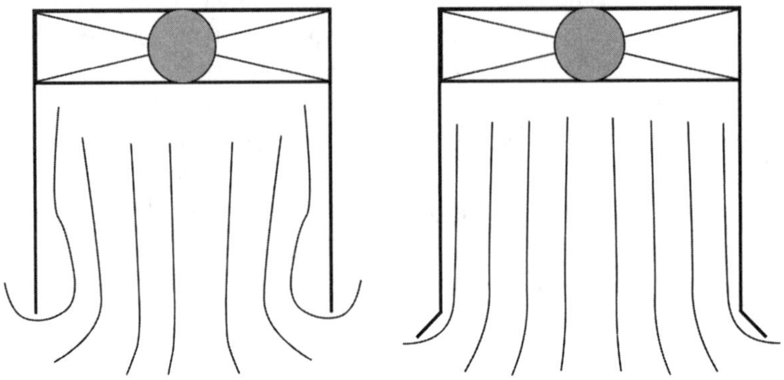

FIGURE 6-7. Tapered entry

ates a wake zone in front of the operator (see Section 6.1.4).

Performance is more vulnerable to conditions at the face of the hood than to conditions at the back of the hood.[(6.3)] Moving the source closer to the front of the hood generally will increase contaminant concentrations at the face. Extending the sides out past the operator's position could also be detrimental since the wake zone would be moved to the back of the user, giving it a greater chance to interact with him.

Objects that serve to guide air smoothly into the hood will reduce the size of the separation from the sides and top of a plain enclosing hood, possibly reducing exposures as a result. At high cross-draft velocities, it is likely that a flange will make little difference and that the only effective enhancement is a broad airfoil shape such as those found on laboratory hoods (see Chapter 13, Section 13.35). A deep airfoil or bevel at the bottom edge of the hood may reduce the vena cava at the floor of the hood, but it also may actually increase exposures because it pushes the worker away from the hood face, typically bringing the source closer to the face with them.

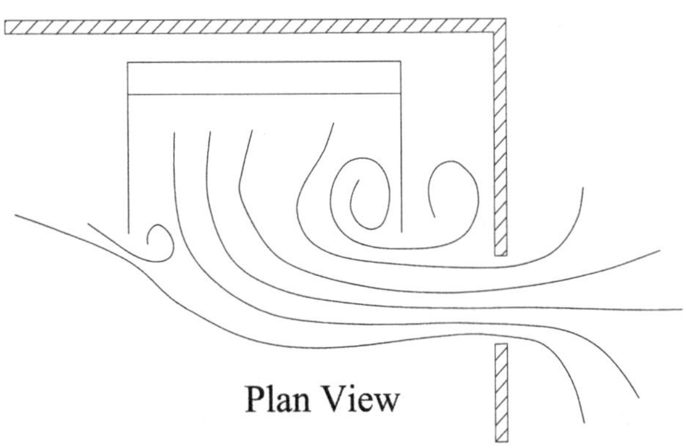

FIGURE 6-8. Skewed entry

Eddy currents produced by the body of the operator potentially can be reduced by directing 20–40% of the supply ("auxiliary flow") air in front of his/her body (see Figure 6-9 and Chapter 13, Section 13.35). The auxiliary flow from the top would increase the flow separation on the top of the inside of the hood. If the contaminant is mostly near the floor of the hood, the net result could be a reduction in exposure if the aux-

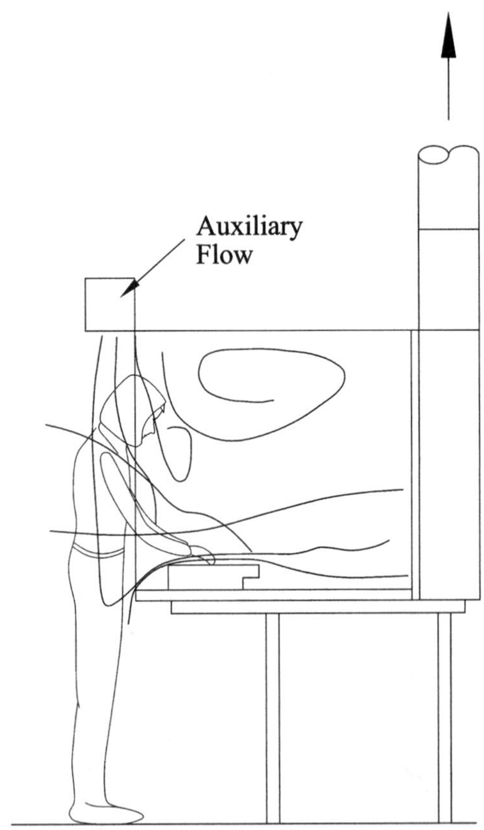

FIGURE 6-9. Auxiliary flow hood

iliary flow is not released in excessive amounts or with excessive velocity. On the other hand, if the contaminant is released with enough energy to reach the top of the hood, the enlarged flow separation could pull the contaminant to the user's face, potentially greatly increasing exposures.

In practice, auxiliary airflow is difficult to adjust properly, and when it is adjusted poorly it is likely to lead to higher exposures. Furthermore, the common practice of drawing unfiltered, unheated outside air to the auxiliary flow attachments is an extremely poor practice. In particular, workers are likely to block the flow if the auxiliary air is significantly colder or warmer than room air. In addition, if unfiltered, the air can contain surprisingly high levels of ambient dust. Finally, the additional costs of ductwork, an inlet into the building, the auxiliary attachments, and a separate fan would eliminate any significant savings due to the modestly reduced energy costs even for air-conditioned and heated spaces. For those reasons, flow from the top and sides is not recommended.

Channeling or blowing air from the leading edge of the floor of the hood is another approach. The upward flow normally increases dilution of the wake zone in front of the worker, potentially reducing exposures. However, it is important once again not to blow the air so high that it pushes the contaminant up to the worker's face, thereby increasing exposures.

6.3.8 Steps for Designing a Plug Flow Enclosing Hood. For hoods where workers must frequently reach into or work inside (see Figures 6-9a and 6-9b), the steps are:

1) Observe the operation through several cycles and question workers and maintenance personnel about access needs, work practices, materials handling, emergency conditions, and maintenance. Check the size of the enclosure by watching the process and its operators.
2) On the side where operators must frequently reach into the enclosure, install an opening (called a "face") to give operators the access they need. It is highly desirable to have only one side left open. Make sure the open face gives the operators sufficient room to perform tasks.
3) For maintenance and operator tasks that are done no more than a few times an hour, include additional openings that give access where needed, but cover with doors or panels that are easy to open and close.
4) At points where it is necessary only to see inside the enclosure, consider installing clear plastic or laminated safety glass windows or doors. Occasionally it is worthwhile to construct all sides of a hood of transparent material.
5) Provide light inside of the enclosure. Install the fixture on the outside of the hood so that its light shines through a plastic or laminated safety glass window. If a fixture must be inside the enclosure, consider whether explosion proof fixtures and wiring are required by Code. In locating the fixture, reduce glare as needed.
6) Make the enclosure convenient. Without blocking the airflow, install holders for hand tools, work rests, and for anything else that will be handy for the operator. To avoid blocking or disturbing the airflow, it may be necessary to have the shelves, holders, etc. flush with the inside side walls of the enclosure so that they look like built-in shelves or a medicine cabinet. (Avoid storing objects in the back of the hood since they may disturb the airflow through the hood.)
7) Force the air to flow evenly at the face of the hood so that the face velocity is reasonably uniform. For this reason, a 45° taper from the enclosure to the duct should be standard. Likewise, the plenum of the hood should be a wall of appropriate filters, if needed, or baffles, perforated panels, or other materials with 5–10% openings. Consider multiple take-offs to improve airflow patterns and better duct transition if overhead space is insufficient for a 45° tapered takeoff.
8) Particularly for laboratory hoods and large hoods, ensure that the supply air enters the hood at low velocity. If velocities are relatively high, it is possible that 90° entry is preferred.
9) Choose a target face velocity for the hood (see Section 6.3.2).
10) For extremely toxic materials, consider commercial laboratory hoods or glove boxes and follow the manufacturer's instructions.
11) To avoid product pickup, extend the length or height of the hood so that the duct opening is a sufficient distance away from the source.
12) To avoid exhausting materials that might plug the duct (rags, etc.), install expanded or perforated metal screens with small diameter openings at the back of the hood instead of baffles. Provide access for inspection and cleaning the screen.
13) For sections of duct that will very likely be coated by sticky material or are otherwise likely to plug, consider installing 5' [1.5 m] lengths of duct manufactured to be easily removed and re-installed (e.g., with built-in clamp connections).
14) After the hood is installed and periodically thereafter, evaluate its performance both for ventilation effectiveness and worker acceptance. If either is unacceptable, make revisions to meet all design and operational goals.

6.4 TOTALLY ENCLOSING HOODS

Total enclosures actually include a broad range of hoods with varying degrees of enclosure. True "total" enclosure (i.e., no openings whatsoever) would be a mistake in many cases

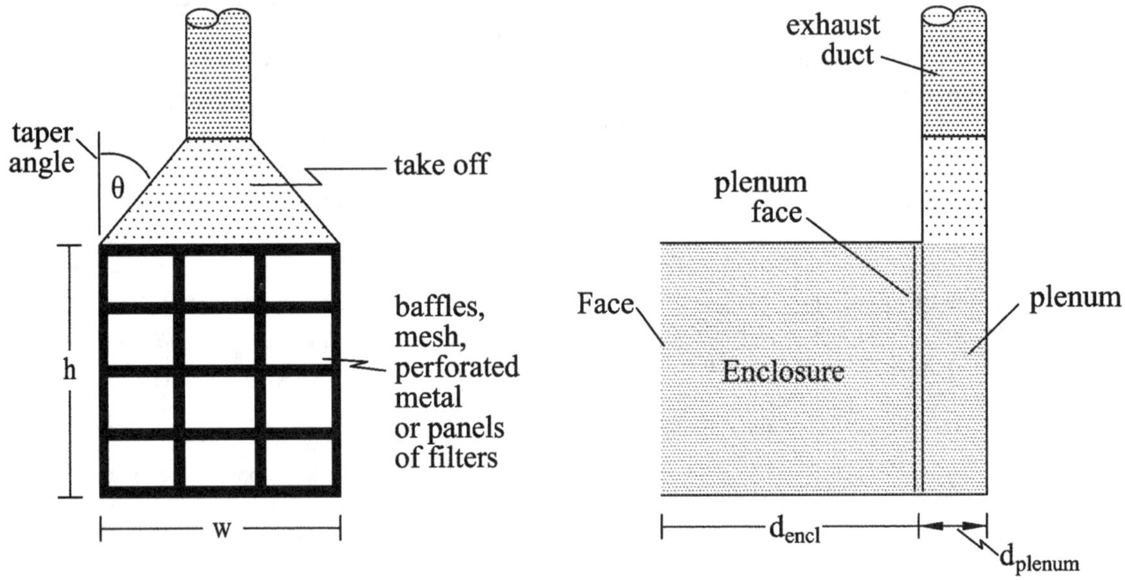

Face Velocity: V = 75 - 125 ft/min [0.38 - 0.63 m/s]
Low V values for hoods with good conditions outside the hood. Higher values for smaller hoods and for poor conditions outside the hood.

Height and width determination:
W = largest stock width plus 3' [0.9 m] on each side if needed for access or spraying.
Ergonomic minimum for "elbow room" is 6' [1.8 m].

H = largest stock height plus 3' [0.9 m] above if needed for access or spraying.
Ergonomic minimum for head clearance is 7' [2.1 m].

Very large pieces can distort airflow distribution. Since they can act as baffles, they should be located and centered as if they were baffles (i.e., allow d_{encl} in in front and d_{pl} behind).

Other dimensions:
$d_{encl} \geq 0.75$ W and $d_{encl} \geq 0.75$ H
total baffle area = 0.75–90 WH
$d_{pl} \geq 0.75$ duct diameter
"baffles" may be as pictured or any other configuration that forces air to flow uniformly in the back of the hood. Mesh, perforated metal, or panels of filters may also be used.

Taper angle: best choice is θ = 45°

Airflow determination:
Q = W H V

FIGURE 6-9a. User-occupied plug flow enclosing hood recommendations

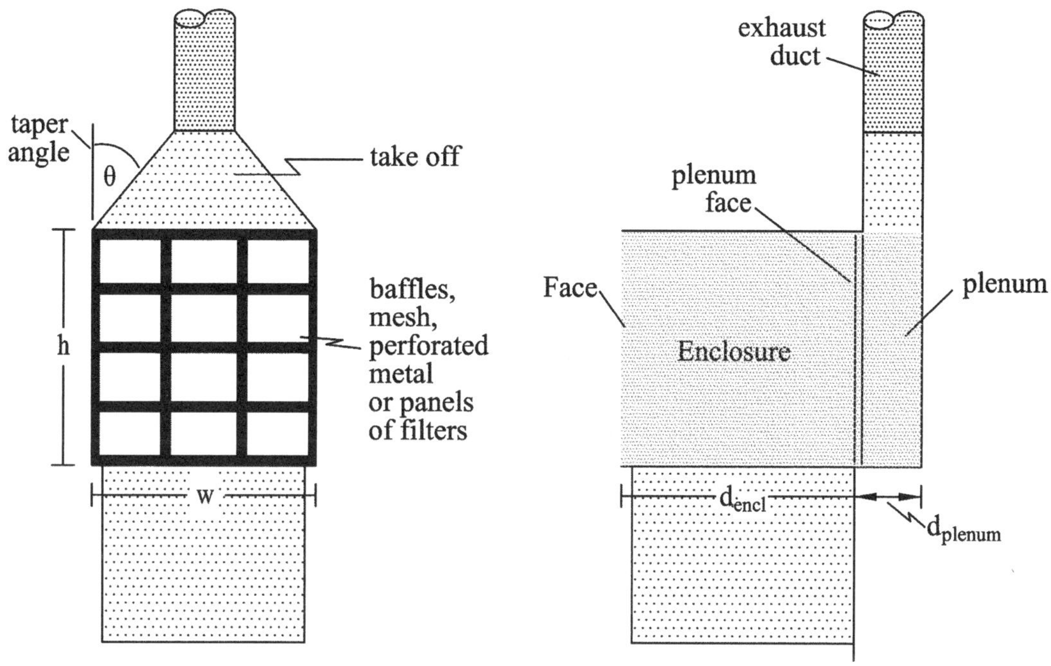

Face Velocity: V = 100 - 200 ft/min [0.5 - 1.0 m/s]
Low V values for hoods with good conditions outside the hood. Higher values for smaller hoods and for poor conditions outside the hood.

Height and width determination:
W = largest stock width plus 3' [0.9 m] on each side if needed for access or spraying.
Ergonomic minimum for "elbow room" is 3' [0.9 m].

H = largest stock height plus 3' [0.9 m] above if needed for access or spraying.
Ergonomic minimum for head clearance is 3' [0.9 m] above the table top.

Very large pieces can distort airflow distribution. Since they can act as baffles, they should be located and centered as if they were baffles (i.e., allow d_{encl} in in front and d_{pl} behind).

Other dimensions:
$d_{encl} \geq 0.75$ W and $d_{encl} \geq 0.75$ H
total baffle area = 0.75–90 WH
$d_{pl} \geq 0.75$ duct diameter
"baffles" may be as pictured or any other configuration that forces air to flow uniformly in the back of the hood. Mesh, perforated metal, or panels of filters may also be used.

Taper angle: best choice is θ = 45°

Airflow determination:
Q = W H V

FIGURE 6-9b. Benchtop plug flow enclosing hood recommendations since there could be no airflow and concentrations would build up during use. Whenever access was finally needed, the user would be exposed to a potentially very high concentration. If the materials were volatile (e.g., solvents), their evaporation could develop significant pressures in a tightly sealed enclosure, leading to an outward rush of air when the enclosure is opened for any reason.

Instead, total enclosures have varying degrees of completeness in physical or architectural enclosure and varying levels of stringency in attempting to prevent contaminant escape through whatever openings are in place (see Figure 6-4).

6.4.1 Issues in Common. A crucial commonality in totally enclosed hoods is the relatively high concentration of contaminant inside them when compared to hoods that are designed to have uniform velocities (i.e., plug flow) as the air flows through them. The concentrations are generally higher because their high containment efficiency leads to use of relatively low airflows compared to the generation rate of contaminants, and they are not designed for plug flow so there often are back flows and eddies. The stagnation zones due to eddies can produce large differences in concentrations within the structure and therefore "hot spots" of high concentrations. Hence, because of back flows and stagnation zones, concentrations at openings can be relatively high, making these hoods unsuitable for operations where workers must frequently reach into the hood through openings.

In almost all total enclosures, at least some material handling typically is done through openings that are kept blocked by panels and doors. When uncovered, these larger openings can dramatically compromise the effectiveness of the hood.

When larger openings are covered, well-designed total enclosures often are the only hoods capable of adequately controlling sources that are highly hazardous due to toxicity, rate of emissions, and energetic dispersion. The amount of air continuously exhausted from the hood must also exceed the amount of contaminant produced by evaporation and other mechanisms, including rapid displacement of the air in the hood due to rapid inflow of materials and thermal expansion. The recommended airflow could be specified (see Chapter 13) based on the expected effluent level for specific applications but typically is stated as a minimum velocity through ports and other openings (see USEPA Method 204 as used for the determination of hoods containing volatile compounds). However, the inlet velocity must be great enough to overcome the momentum of airflows that impact areas near the openings or the ports should be shielded from such impacts. Likewise, inward velocities should be higher for hot processes to overcome hydraulic forces due to the buoyancy of air and gasses.

The inlet air ports and the exhaust port should be located to ensure that stagnant regions do not develop especially if volatile materials are being contained. If the equipment inside the hood allows, it is desirable to create plug flow. For cases where high velocity air movements are created inside the enclosure by the process, it is important to either avoid placing ports where high velocity air can impact them or to shield the ports so that high velocity air cannot impact them directly. Normally sources in the enclosure should be at least 4 equivalent diameters away from any opening.

It is also recommended that it be large enough to allow highly energetic contaminant releases to dissipate momentum before striking the sides of the enclosure at locations with ports or other pathways to escape. Any hood can lose containment if it is too small considering the energy level and generation rate of the contaminants it seeks to control. At the same time, all enclosures must be evacuated at a flow rate that will provide operation at safe levels of LEL and other safety standards.

Assuming adequate levels of airflow and avoidance of stagnant regions, the range of containment efficiency with total enclosures of different types is strongly affected by the care taken in minimizing opportunities for contaminants to escape. For purposes of discussion, they are divided here into the functional groups: Extremely High Control, Very High Control, High Control, and Moderately High Control. The actual degree of control of each is determined not only by their initial design but how well they are installed and operated. The adequacy of the degree of control is strongly affected by the thermal and kinetic energy of the contaminant, as well as its generation rate and toxicity. A hood designed for one contaminant and set of conditions may fall short of requirements when another material is to be contained or the generation rate or other conditions are changed.

6.4.2 Extremely Effective Total Enclosures. Some processes are so hazardous that extreme care must be taken in minimizing any escape from the hood. Examples are the handling of radioactive dusts and gases, deadly bacteria and viruses. To achieve extremely high containment effectiveness, hoods must have a high degree of enclosure and extreme care must be taken to minimize escape through ports and openings.

The highest containment efficiencies within this group are obtained by using ventilated boxes and by restricting access when contaminants are inside the enclosure. Manual access may be provided by manipulators inside the enclosure controlled from outside the enclosure. The enclosure is opened only after a substantial purge period and thorough internal vacuuming of toxic dusts, viruses or bacteria. To assure constant dilution, the inlet ports should be very numerous and small. Ideally, outflow due to the momentum of air movements or pressure waves inside the enclosure are minimized by forcing circuitous paths to the ports and by resistance to flow through the port. The latter can be provided by filter media such as high efficiency particle arrestors. Filters at the ports would also provide secondary protection in case of fan failure. Special regulations and standards should be consulted for these design requirements.

6.4.3 Highly Effective Total Enclosures. Somewhat lower protection but still extremely high control is offered by "glove boxes" (see Chapter 13, VS-35-20), which are total enclosures with impregnable gloves securely attached to internal ports. The operator inserts his or her arms into the gloves and views the inside of the glove box through a plastic glass or laminated safety glass window. In most designs adding or removing materials or equipment to or from the glove box is done through an "air-lock" of two small doors in series. The user opens the outside door, places the object in the space between the two doors, closes the outside door, then opens the inside door and retrieves the object with the built in glove. Even this arrangement will allow some transfer of airborne contaminants to the room unless grilles are placed in the airlock doors to provide continuous dilution of the chamber between them. Sometimes the chambers have their own duct takeoffs, if it is desirable to minimize contamination. Settled or condensed material in the airlocks is likely unless extraordinary measures are made to clean the chamber, preferably using an internally mounted vacuum cleaner hose. Even without such cleaning, the amount transferred by handling should be very small.

The gloved port with window idea can be applied to almost any enclosure to good effect. For example, one can place a gloved port insert under the sash of a lab hood. Likewise, one could place glove ports on the wall outside a room, allowing manipulation of objects within reach of the gloves.

Glove boxes are not necessarily highly effective. For example, the level of control of lab-hood glove box inserts would be determined mostly by the quality of the seals for the insert and by the care taken to purge the hood before removing the inserts. Manufacturer standards and regulatory standards must be checked before usage and specification.

6.4.4 High Control Total Enclosures. If a hood has a high degree of enclosure but less care is taken to prevent contaminants from escaping through ports and other openings, it can still be capable of providing an effective control of containment. Assuming that the enclosure must be opened substantially at regular intervals, the most critical determinants of effectiveness are: 1) sufficient exhaust airflow, 2) prevention of outward flow through the inlet ports, 3) the quality of the seals of doors, panels, and windows, and 4) allowing sufficient purge time before opening.

An example is what could be called "rough glove boxes" used to handle hazardous (but not extremely hazardous) processes. For example, some sandblasting can be done in small rooms with the operator standing outside the room to manipulate the sand blast hose through gloved ports. An example application of the latter is "sandblast sheds" used in the manufacture of grave markers and other stone monuments. Because the seals in some cases may not be tight and because the operation depends on effective work practices (i.e., waiting for the enclosure to purge itself of dust before access), exposure levels can be exceeded.

Ventilated storage cabinets can be designed with an exhaust port and multiple grilles to allow entry of supply air into the cabinet. The exhaust port and grilles should be positioned at each end of the cabinet with the grilles placed to avoid stagnant zones within the cabinet. A door to access the stored chemicals or gas cylinders is a potential vulnerability for two reasons: 1) if it is not shut, the control offered by the cabinet will be poor, and 2) if a stored liquid spills or leaks from the storage vessel, the fluid can seep underneath the door unless the vessel stands in a bucket with sufficient volume to hold spilled liquids. Storage cabinets also can fail under other conditions. For example, if a gas cylinder stored in such a cabinet developed a massive leak, the resulting pressure could exceed the negative pressure in the cabinet, allowing toxic gases to flow through the grille. If the pressure were high enough, the escape could be at high speed.

6.4.5 Moderate Control Total Enclosures. If the total enclosure has a somewhat lower degree of enclosure and still less effective measures to prevent escape through ports and openings, it can be moderately effective compared to the preceding hoods, though potentially much more effective than plug flow hoods and capturing hoods. If the enclosure is relatively large and the velocity through openings is relatively high (e.g., 150–200 fpm) [approx. 0.75 – 1.00 m/s], it can provide a sufficiently high degree of reliable control. They also generally provide the most reliable control of very hot and large quantities of contaminated air. Examples of these hoods are shown in Section 13.27, Hot Processes.

There are three critical points concerning these hoods:

1) Even if large enough for operator entry, they are seldom designed or suitable for human occupancy. For a worker to enter one safely, the process may have to be shut down. In some cases, it may be possible to enter safely while wearing appropriate protection.
2) The location of the entry points and the exhaust point are important. Generally, they should be designed for flow from one end to another. If the inlet end is blocked, airflow can be drawn around the perimeter at the blocked end. Because they are often filled with large pieces of process equipment (e.g., melting furnaces, etc.), it is sometimes necessary to add additional inflow locations to ensure that air flows through otherwise blocked areas. In placing any opening, it is important that the opening not be in-line with a jet of contaminated air issued within the enclosure. A jet of high velocity air will blow through any opening and overpower lower velocity air drawn into that opening.
3) To operate with very high effectiveness, all openings with substantial areas must be opened only for short periods of time. Avoid times when the emissions are highly concentrated or energetic.

Note that the opening for supply air can be quite large yet still be effective for large, energetic sources if plug flow or near plug flow is established and the inlet is far from the workers' breathing zones and is not used for worker access.

6.5 DOWNDRAFT OCCUPIED HOODS ("ROOMS")

Downdraft occupied hoods are enclosures designed to have a plug flow that is vertical instead of horizontal. Downdraft hoods that rely on plug flow to protect the worker or to minimize unwanted dispersion of the contaminants generally should be designed to deliver airflow uniformly through the ceiling face and removed uniformly from the floor (see Figure 6-10). Downdraft designs have an advantage over horizontal flow in that wake zones from the worker and objects on the floor are mostly under the floor. The direction of flow is almost always downward but it is conceivable that there are situations where upwards flow could be better.

Non-uniform release of the supply air generally will not produce plug flow. Zero and low velocity regions will be marked by very large eddies in stagnation zones. High velocity releases of supply air can produce flows with sufficient momentum to "splash" from large items on the floor to the position of the worker. Large scale eddies also will transfer contaminants laterally, making it altogether difficult to separate the worker from contaminant clouds. For that reason, airflow should be released as uniformly as possible from the entire area of the ceiling.

It is still important to exhaust air from the room uniformly. Exhausting from a limited region in the floor will also produce stagnant zones in the non-exhausted area. Any contaminant reaching those zones will only slowly be diluted, potentially producing high exposures to workers standing in them.

6.6 HOT PROCESSES IN ENCLOSING HOODS

Enclosures with small amounts of heat added (soldering and welding) usually do not require special consideration for the effects of buoyancy on calculations and design. However, if a large area near the floor of the hood is heated to a high temperature (e.g., > 300 F [approx. 150 C]), the inward movement of air at an open vertical face of the hood may be insufficient to move the heated air towards the back of the hood. Instead, heated air may spill out of the opening near the top of the face (see Figure 6-11) since the upward velocity would be at least as great as the inward velocity of the air flowing through the face. The positive pressure exerted by the buoyant force of the hot air (a "hydrostatic pressure") can easily exceed the negative pressure in the upper sections of the hood, forcing hot air to leak from cracks and other openings near the top of the hood.

For that reason, it is important that openings in the vertical faces be as close to the bottom as possible and there be no permanent openings near the top of the enclosure. It is best to locate the takeoff at or near the top (see Figure 6-12), so that the exhaust direction is aligned with the buoyant air movement. Exhaust from front to rear (see Figure 6-11) is not recommended.

In addition, considerations must be made for the creation of hot gasses by the process inside the enclosure and the decrease in density (and therefore the increase of volume) as air is heated inside the enclosure. See Chapter 13, Section 13.27 for a comprehensive discussion of controls for heated processes.

6.7 CAPTURING HOODS

Capturing hoods do not enclose the source, but instead rely on a flow of air into the hood opening to carry the contaminated air into the hood. Note that the air converging on an exhaust point accelerates more rapidly as it approaches the hood face. As a result hood effectiveness in capturing the contaminated air improves rapidly with decreasing distances from the hood opening. The effectiveness falls off sharply at distances far enough from the hood face that the inward velocity is not significantly greater that the competing velocities induced by traffic, man-cooling fans, process machinery or other influences. The higher the velocity and the less the competition from outside air currents, the more contaminant will be collected and the more efficient the hood.

6.7.1 Shapes of Capturing Hoods. Capturing hoods can be shaped many different ways to fit specific geometric constraints and needs, but the main types are:

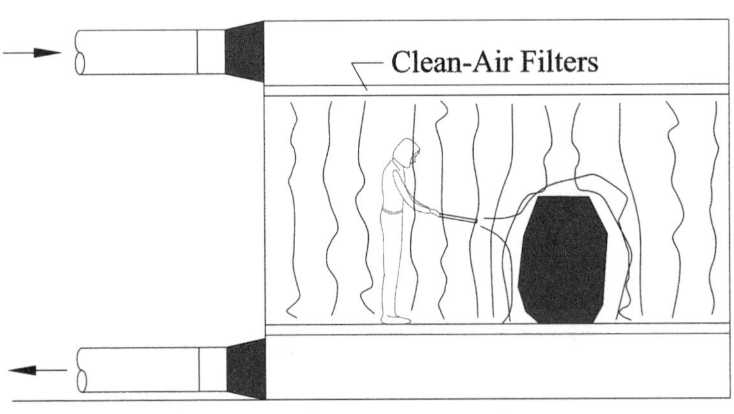

FIGURE 6-10. Downdraft room

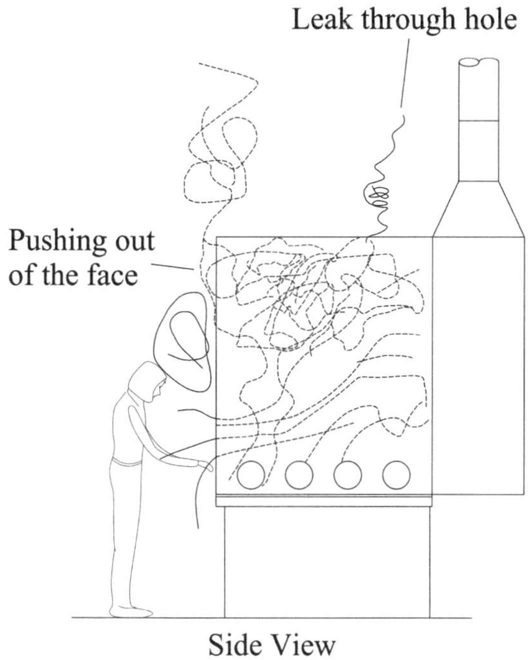

FIGURE 6-11. Ineffective hot process hood

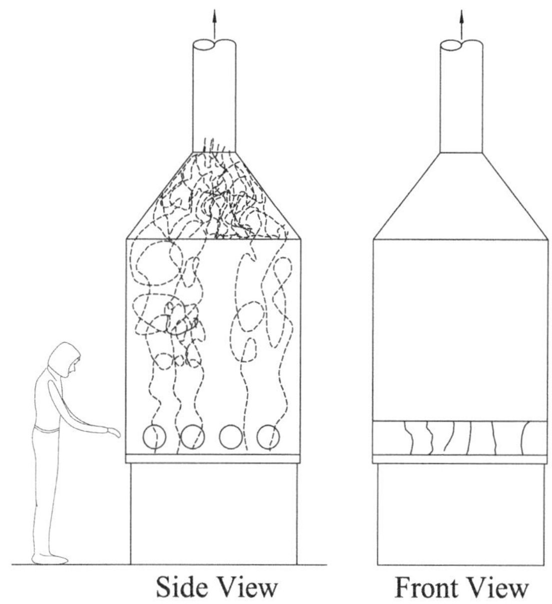

FIGURE 6-12. Enclosing hood designed for hot source

1) Plain opening (see Figure 6-13): Hood with a round opening or a rectangular opening with W/L (Width/Length) > 0.2. The open face can remain a fixed cross-sectional area for some distance or immediately converge to fit the duct.

2) Slot hood (see Figure 6-14): Hood with a relatively narrow slot width (W) compared to its length (L) followed by a straight or converging transition to the duct. An opening with W/L ≤ 0.2 is classified as a slot. However, it should be understood that the airflow behavior actually changes gradually with changes in aspect ratio and the demarcation at W/L = 0.2.

3) Slot hood with plenum (see Figure 6-15): Hood with one or more relatively narrow openings followed by a sudden expansion into a plenum. Airflow characteristics in front of the hood are similar to a flanged slot with no plenum.

Note that each of these hoods has a tapered transition from the hood face down to the duct size. The tapering has little or no effect on airflow but does affect static pressure requirements.

6.7.2 Capture Velocity. The minimum hood induced air velocity necessary to capture and convey the contaminant into the hood is referred to as "capture velocity" (see Figures 6-16, 6-17 and 6-18). In general, the effectiveness of capturing hoods increases with increasing airflow levels and therefore with increasing capture velocities (V_x). It is probable that an increased capture velocity can also offset the effects of competing air currents, buoyancy, and contaminant momentum, so for higher cross-draft velocities higher capture velocities should be used. On the other hand, to minimize the effects of cross-drafts, it may be still more effective to take other measures, such as reducing cross-drafts. For buoyant plumes, it may be more effective to place the hood above the level of the source.

Table 6-2[6.4,6.5,6.6] provides ranges of recommended velocities for each of several examples with increasing energies that serve to disperse the contaminated air. The ranges are quite broad for each example dispersion condition. The higher end of the ranges should be used for unfavorable conditions, such as:

1) High cross-draft velocities,
2) Strong competing air motions due to traffic, mechanical motions, etc.,
3) Hazardous contaminant generation due to the toxicity of material, its generation rate, and the duration of potential exposures.

The capture velocity should be at least 75 fpm [0.38 m/s] except under ideal conditions. A velocity of 100 fpm [0.51 m/s] may be a more realistic minimum for typical conditions (moderate toxicity, cross-drafts, etc.). It should be noted that a capture velocity can also be excessive for some conditions. In particular, very high capture velocities near dusty materials can cause "product pickup." The problem is most likely to occur when the airflow through the hood is relatively low and the hood must be kept very close to the source for the capture velocity to be high enough for the hood to be effective. Using

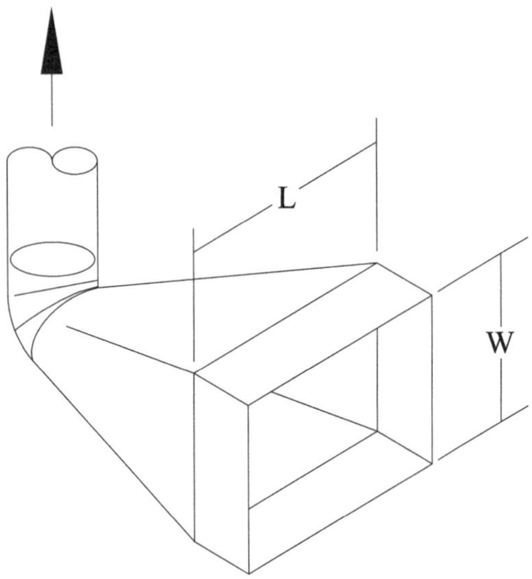

FIGURE 6-13. Plain opening

Figure 6-16, 1 sq ft hood [0.03 m^2] with a capture velocity of 100 ft/min [0.51 m/s] at 12" distance [0.3 m], the velocity at 6" [150 mm] would be roughly 310 ft/min [1.55 m/s]. That velocity could pick up powdery products such as flour or talcum powder. A better solution is to enclose the source and make the hood height high enough that the region of high velocities near the duct entry is far from the product. Not only would product pickup be eliminated but the required airflow for acceptable performance generally would be substantially lower than would have been required for the capturing hood.

EXAMPLE PROBLEM 6-1 (Capture Velocity)

Determine capture velocity welding on mild steel, moderate production, good conditions. The work table is 3' × 3'.

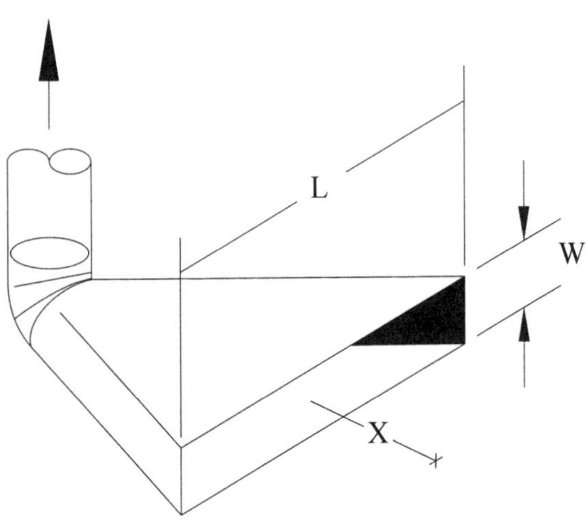

FIGURE 6-14. Slot hood

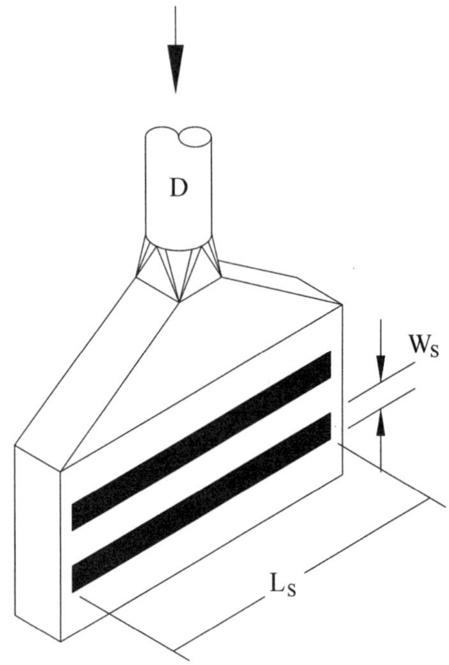

FIGURE 6-15. Slot-plenum hood

Solution (from Table 6-2) (Average Motion): V_X = 100 – 200 ft/min [0.51 to 1.02 m/s]. Based on the stated conditions, the low end of the range should be adequate, V_X = 100 ft/min [0.51 m/s].

6.7.3 Effective Zone of Capturing Hoods. The effective zone of a capturing hood is the region in front of the hood that is adequately controlled by the flow of air into the hood (see Figure 6-19). The boundary of the effective zone can coincide with the boundary where the induced velocity into the hood equals the recommended capture velocity (V_X). However, the two boundaries may be distinctly different if the contaminant is highly buoyant, has its own momentum, or if there are disturbing airflows due to cross-drafts, blowing air, mechanical movement, traffic, etc.

Since users cannot sense cross-drafts and seldom can see the contaminant, seldom are there visual indicators to let them know whether the source is within the effective zone. To further complicate the picture, the shape and extent of the effective zone is affected by the exhaust flow rate, the shape of the hood, nearby surfaces, cross-drafts, and potential convection from hot sources. In addition, if the contaminant is toxic or its generation rate is high, the hood efficiency must be increased. Thus, the zone considered to be effective would be smaller for the same airflow and the effectiveness of capturing hoods is affected by the following factors:

1) Distance from the source — capture velocity decreases

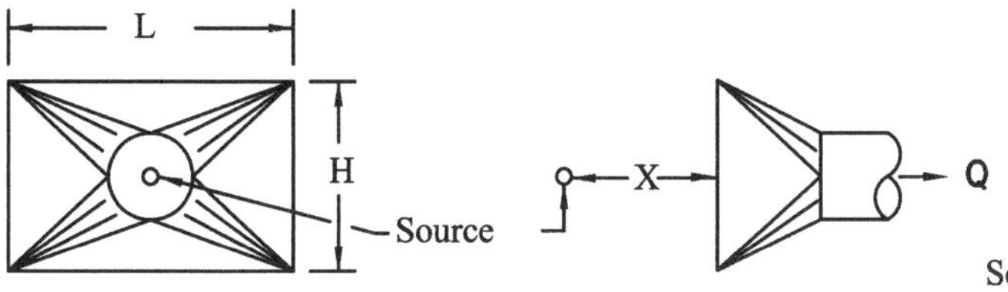

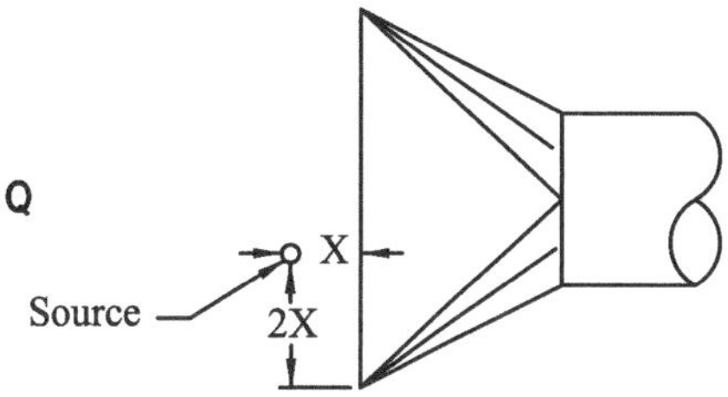

FREELY SUSPENDED HOOD

$$Q = V_x(10X^2 + A)$$

LARGE HOOD

Large hood, X small–measure X perpendicular to hood face, not less than 2X from hood edge.

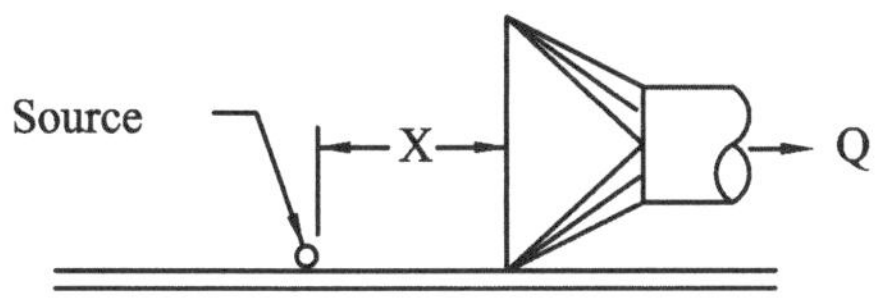

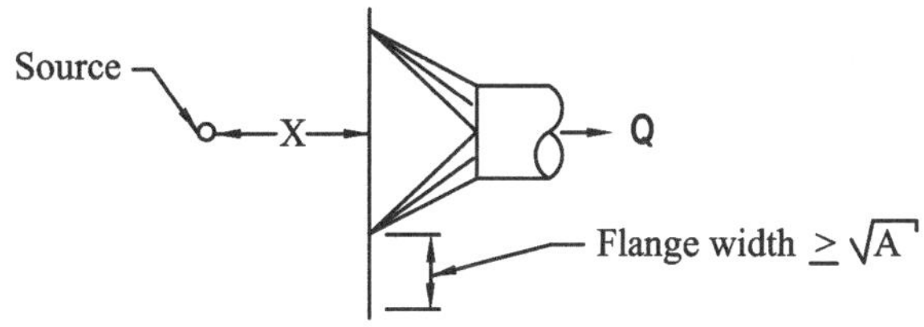

HOOD ON BENCH OR FLOOR

$$Q = V_x(5X^2 + A)$$

HOOD WITH WIDE FLANGE

$$Q = V_x\, 0.75(10X^2 + A)$$

SUSPENDED HOODS
(Small side-draft hoods)

Q = Required exhaust airflow, acfm [m^3/s]
X = Distance from hood face to farthest point of contaminant release, ft [m]
A = Hood face area, ft^2 [m^2]
V_x = Capture velocity, fpm [m/s], at distance X

Note: Airflow rate must increase as the square of distance of the source from the hood. Baffling by flanging or by placing on bench, floor, etc. has a beneficial effect.

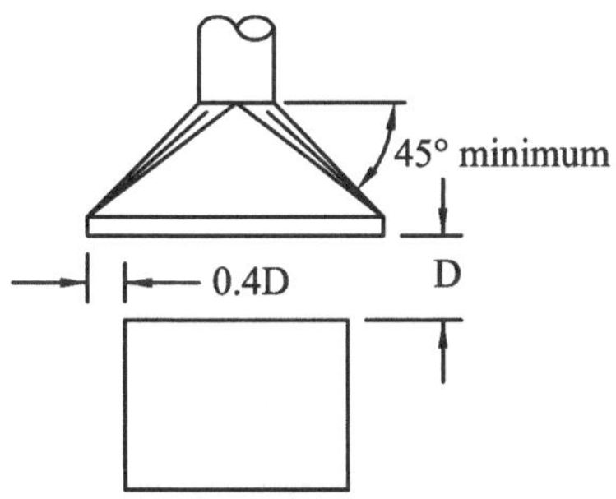

CANOPY HOOD

Q = 1.4 PDV (P = Perimeter of tank, ft [m])
Not recommended if workers must bend over source. V ranges from 50 to 500 fpm [0.25 to 2.50 m/s] depending on crossdrafts. Side curtains on two or three sides to create a semi-booth or booth are desirable.

ACGIH®	TITLE FLOW/CAPTURE VELOCITY	FIGURE 6-16 DATE 1-12

CHECK CODES, REGULATIONS, AND LAWS (LOCAL, STATE, AND NATIONAL) TO ENSURE THAT DESIGN IS COMPLIANT.

FIGURE 6-17

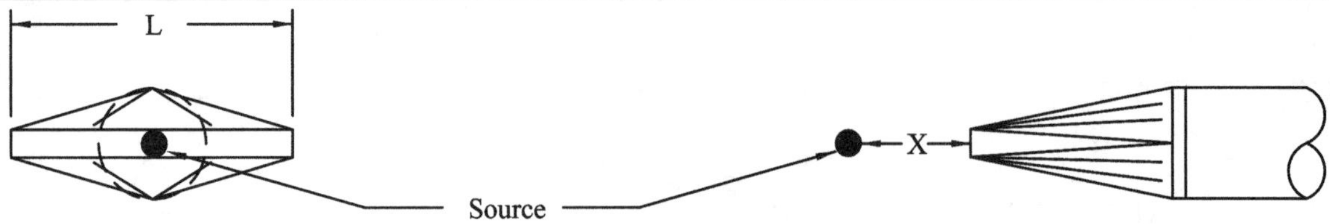

FREELY SUSPENDED SLOT
$Q = 3.7LV_xX$

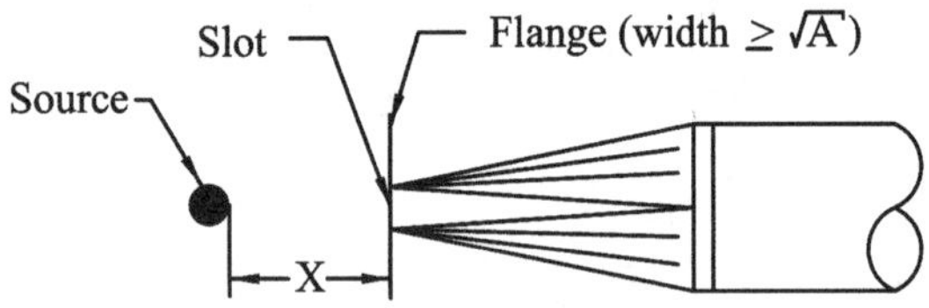

FLANGED SLOT
$Q = 2.6LV_xX$

Q = Required exhaust flow rate, acfm [m³/s]
X = Distance, hood face to farthest point of source (usually on centerline of hood), ft [m]
V_x = Capture velocity at distance X, fpm [m/s]
L = Length, of hood, slot, table, tank, etc., ft [m]
W = Width of table, tank, etc., ft [m]
A = Hood face area, ft^2 [m^2]

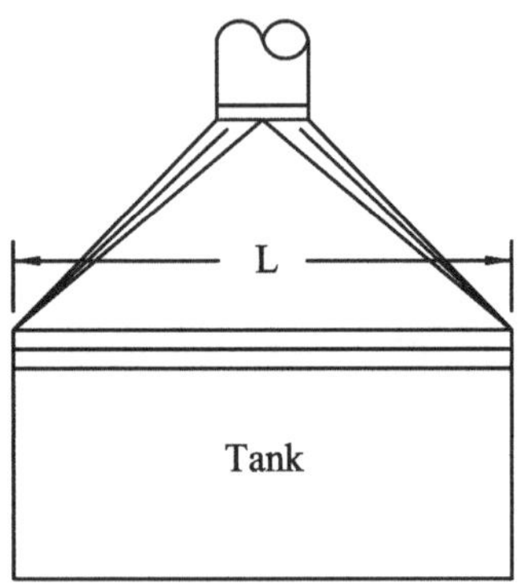

FLANGED SLOT
Q = CLW

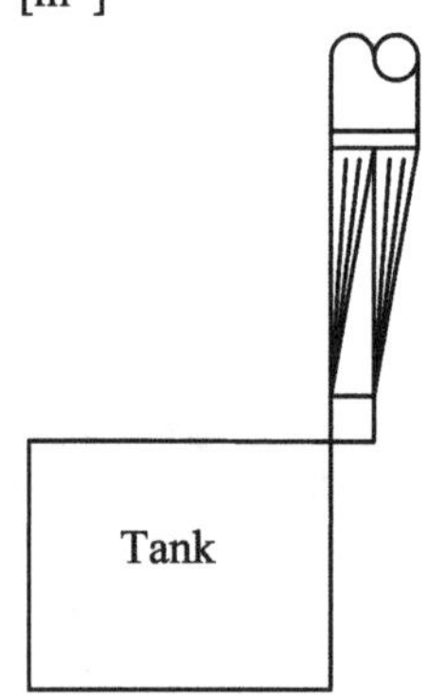

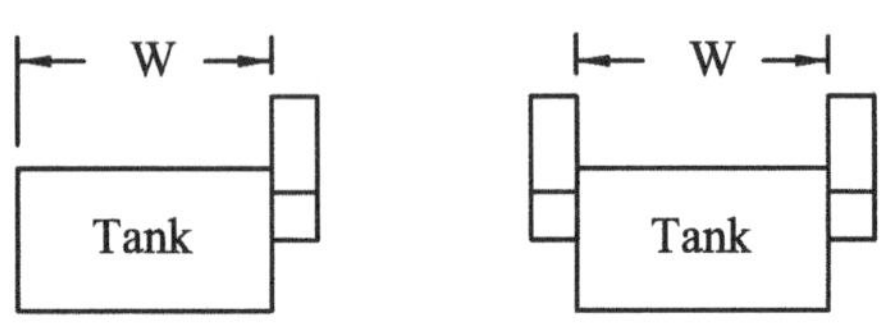

SLOT ON TANK
Q = CLW
One-half Q in each slot if slots on both sides

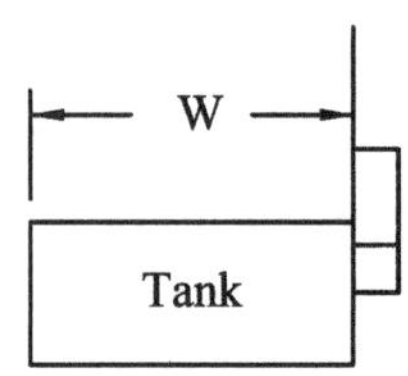

FLANGED SLOT
Q = CLW

L = Length of slot, ft [m]
W = Width of table or tank, ft [m]
C = Constant, varies from 50 to 500, usual choice is 150 to 250. Flanged slots require lowest exhaust. See Section 13.70.

ACGIH®	TITLE: FLOW/CAPTURE VELOCITY	FIGURE: 6-17 / DATE: 1-12

CHECK CODES, REGULATIONS, AND LAWS (LOCAL, STATE, AND NATIONAL) TO ENSURE THAT DESIGN IS COMPLIANT.

FIGURE 6-18

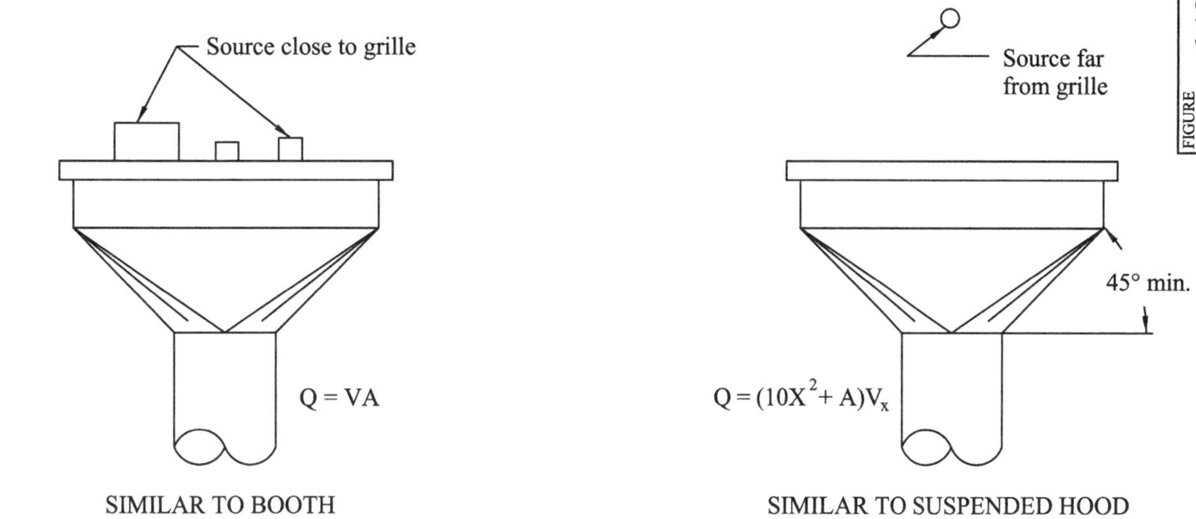

DOWNDRAFT HOODS

Not recommended for hot or heat-producing operations if downdraft area is large. See "Capture Velocity" in this section.

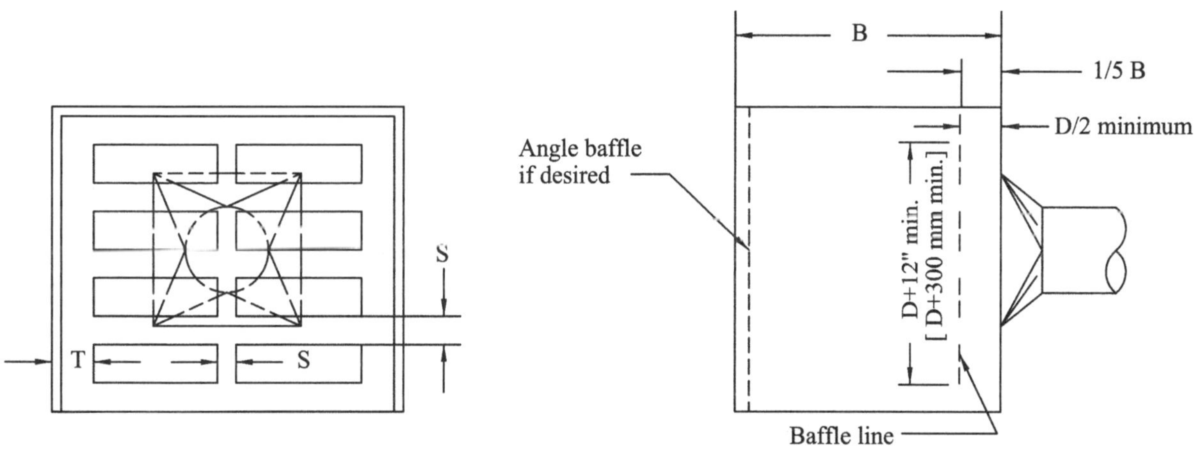

BOOTH-TYPE HOODS

Q = AV (A = Face area, ft^2 [m^2]; V = Face velocity, fpm [m/s])
Baffles are for air distribution. Baffles are not required if a water wall or other means of distribution is provided.
S varies from 4 inches to 8 inches [100 to 200 mm], depending on size of booth.
T varies from 6 inches to 12 inches [150 to 300 mm], depending on size of booth.
Increase the number of panels with size of booth.

ACGIH®	TITLE: FLOW/CAPTURE VELOCITY	FIGURE: 6-18 / DATE: 1-12

CHECK CODES, REGULATIONS, AND LAWS (LOCAL, STATE, AND NATIONAL) TO ENSURE THAT DESIGN IS COMPLIANT.

TABLE 6-2. Recommended Capture Velocities*

Energy of dispersion	Examples	V_x ft/min [m/s]
Little motion	Evaporation from tanks, degreasing	75–100 [0.38–0.51]
Average motion	Intermittent container filling; low speed conveyor transfers; welding; plating; pickling	100–200 [0.51–1.02]
High	Barrel filling; conveyor loading; crushers	200–500 [1.02–2.54]
Very high	Grinding; abrasive blasting; tumbling	500–2000 [2.54–10.2]

Factors affecting choices within ranges

Strength of cross-drafts due to makeup air, traffic, etc.

Need for effectiveness in collection:
toxicity of contaminants produced by the source
exposures from other sources, which reduce acceptable exposure from this source quantity of air contaminants generated – production rate, volatility, time generated

* see also ANSI Z9.2–1979

dramatically with increasing distance from the hood face (see Figures 6-20 and 6-21).

2) Location of the source — the source should be centered immediately in front of the hood.

3) Shape of the hood — for some distance in front of the hood, the velocity profile will differ depending on whether the hood face is a slot or a plain opening. Long slots produce velocity profiles that extend somewhat farther straight out from the opening than do plain openings. On the other hand, the effective zone of a plain hood will tend to be greater in the vertical plane.

   As the distance from the hood face becomes greater, all hoods begin to exhibit the performance profile of a plain hood. At near distances a narrow slot produces a cylindrical velocity contour for some distance in front of the slot. The slot opening is more effective at distances on the same level as the slot, but less effective for vertical distance above and below the level of the slot. Using two or more parallel horizontal slots, as in Figure 6-22, increases the effectiveness of the hood in vertical plane. However, unless the slots are relatively far apart (e.g., more than the desired effective zone in the horizontal direction), the two slots will behave more like a single large rectangular opening than two single slot openings. The relationship between capture velocity and airflow (Q) for several hood shapes is shown in Figures 6-16, 6-17 and 6-18 and Table 6-3. Note that the source is assumed to be DIRECTLY in front of the hood opening.

4) Presence of surfaces near the hood that do not block the flow — depending on their placement, such surfaces may channel more of the airflow over the source, reducing the required airflow. For example, a flange partially blocks the flow from behind the opening, increasing the velocities in front of the hood. Likewise, resting the hood on a tabletop can reduce the exhaust airflow requirement because the airflow is channeled into the hood and side baffles also can channel airflow to the hood face, reducing the exhaust airflow requirement.

   Baffles perpendicular to the hood opening are sometimes used to block cross-drafts (see Figure 6-23). They can channel air over the source and into the hood opening if cross-draft velocities are low. However, it is possible that if cross-draft velocities are high, the upstream baffle will create a strong wake zone that could reduce the effectiveness of the hood rather than enhance it.

5) Objects and surfaces that impede flow across the source and into the hood face – an object placed between the source and the hood can channel the airflow so that it misses the contaminant.

6) Competing air currents — a high velocity cross-draft (e.g., greater than 25% of the capture velocity) may

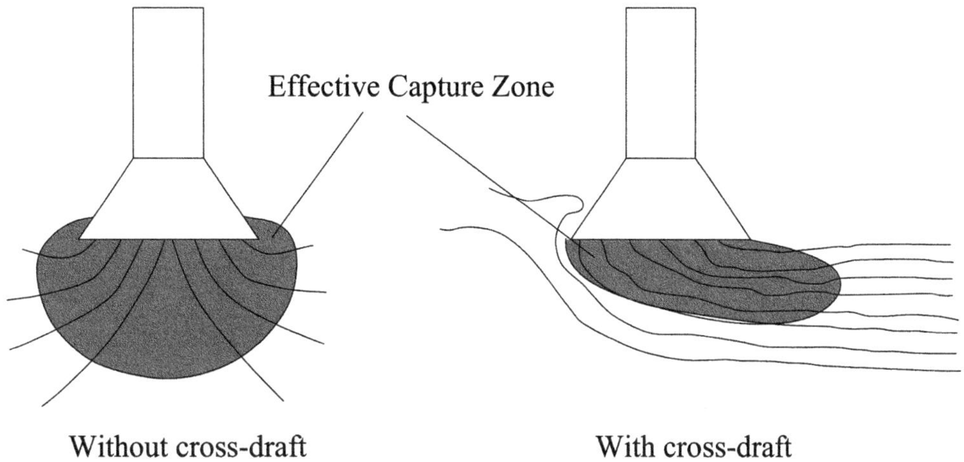

FIGURE 6-19. Effective capture zone

substantially distort the effective zone unless it is blocked by other surfaces or objects. Likewise, competing air currents near the hood due to blowing air, mechanical or operator movements, etc. also can distort and shrink the effective zone.

7) Motion of the contaminant — if the contaminant is released at high velocity, it may fly away from the hood despite the flow of air into the hood. The real problem is generally not the velocity imparted to the particle but that a competing air current has been simultaneously created.

8) Buoyancy of the contaminated air — if the contaminated air is rising rapidly because it is much warmer than room air, its path becomes a complex function of the velocity components in each direction induced by the air drawn into the hood face and the upward velocity of the buoyant air. If the hood is drawing air solely in the horizontal plane, the buoyant air may escape capture (see Figure 6-24). In those cases, the hood generally should be placed above the source with its face angled approximately 45 degrees with the vertical plane as is shown in Figure 6-25.

6.7.4 Capturing Hood Shape and Placement. The hood should be located so that the preponderance of the emissions

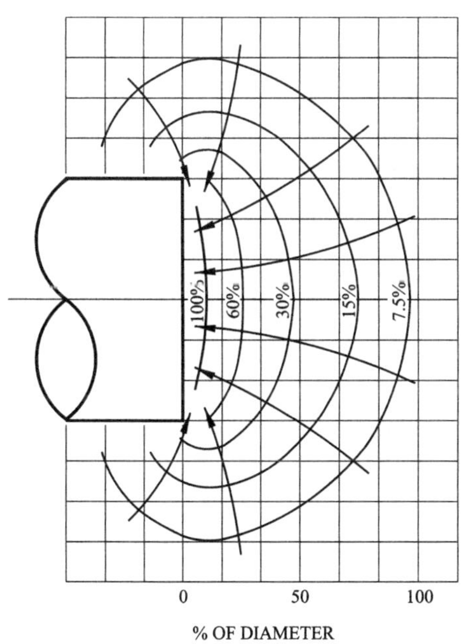

FIGURE 6-20. Velocity Contours – Plain Circular Opening – % of Opening Velocity

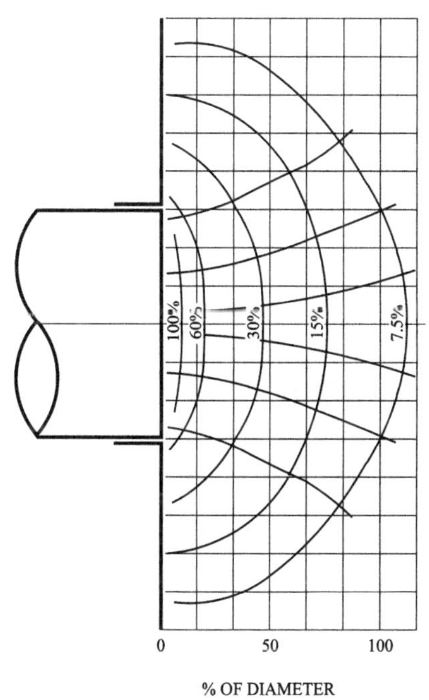

FIGURE 6-21. Velocity Contours – Flanged Circular Opening – % of Opening Velocity

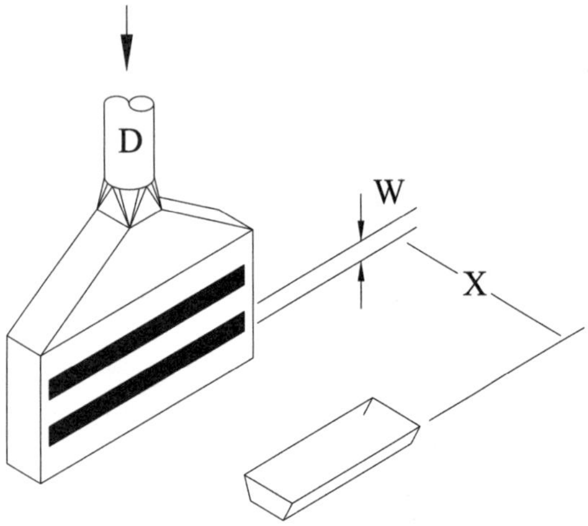

FIGURE 6-22. Multiple slot hood

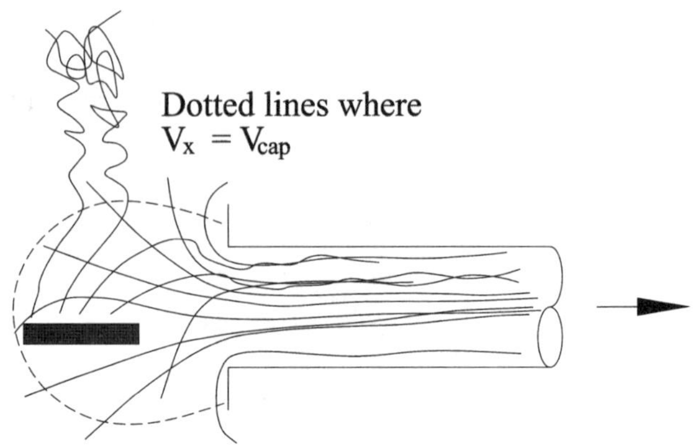

FIGURE 6-24. Buoyant source and horizontal flow

is in the effective zone of the hood. Considering the effects of cross-drafts and other disturbances on the effective zone, the hood should be placed so that contaminants are well within the effective zone.

In general, the capturing hood should be at least 50% wider than the anticipated diameter of the contaminant cloud. It also should be at least as wide as the distance "X" (indicating the greatest distance of contaminant from the hood face) (see Figure 6-22). If the source can be placed anywhere on a work bench, the width of the hood should be equal to the bench width if possible. For example, if the source is constrained to be within a 2 ft [0.6 m] width on the work bench and the cloud of released contaminants is less than 2 ft wide [0.6 m] and the value of "X" is less than 2 ft [0.6 m], then the width of the hood face should be 3 ft [0.9 m] (i.e., 50% wider).

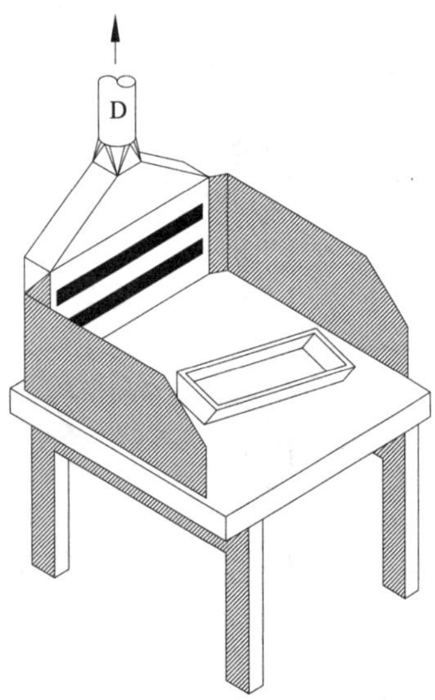

FIGURE 6-23. Slot hood with baffles

The height needed for a capturing hood depends on the type of hood, the vertical height of the bulk of the emissions, and the buoyancy or upward momentum of the contaminated air. For the same exhaust airflow, the height of the effective zone for a hood with horizontal slots will be smaller than for a plain hood. If the source is dispersed or rising over a significant vertical distance, more than one slot may be required. If a slot-plenum hood is used, the plenum must extend high enough to accommodate both openings. Note that if slots are located close together (e.g., distance between midlines of slots less than the distance X), their effective zones will merge and it will act like a plain hood.

For a plain hood, the effective zone vertically will be roughly proportional to the vertical size of the opening for a given exhaust volume. Greater exhaust volumes increase the effective size proportionally vertically and horizontally if the source is relatively close to the hood.

The hood should be centered on the contaminant cloud if the contaminated air is at room temperature and has no significant momentum. If the source rests on a table top or other work surface, the hood can be placed somewhat above the

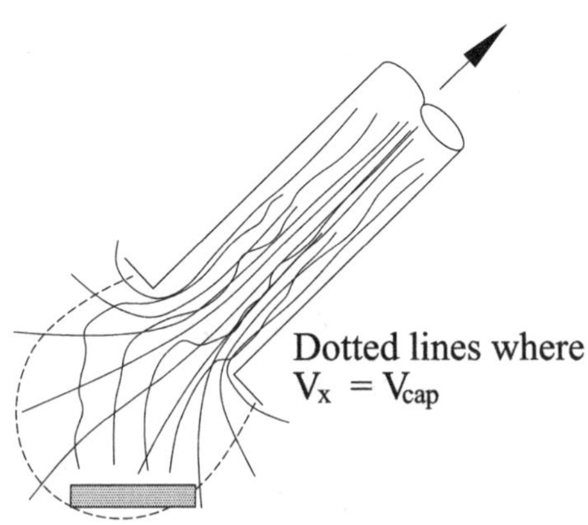

FIGURE 6-25. Incline and elevate capturing hoods for buoyant sources

emissions cloud, especially if the flange nearly touches the table. If the contaminated air is buoyant or has upward momentum, the hood should be placed above the source as close to it as possible without interfering with the work (see Figure 6-25).

6.7.5 Use of Slots in Slot Plenum Hoods. The length of the slot should be greater than the width of the source in front of it and its length also should increase with increasing distance of the source from the hood face. The primary reason to employ slots in a hood face is to force uniformity of flow along the length of the slot.

Some slot/plenum hoods have more than one slot, each parallel to the long side of the hood. Since the effective zone of a slot hood is limited above and below the plane of the slot, then to ventilate sources at two heights, a slot should be placed at each of the two heights (see Figure 6-18).

For a given plenum, the higher the velocity through the slot (V_s), the more uniform the velocities down the length of the slot and the more uniform the flow in front of the hood. Since the airflow requirement is determined based on other factors and the length is determined by geometry, the velocity through the slot(s) can be influenced only by setting the slot width (i.e., the smaller dimension).

The relatively low value of V_s = 1000 ft/min [approx. 5 m/s] can produce adequate uniformity if:

1) The plenum has a relatively low velocity (e.g., less than one-half of the slot velocity), which occurs if the depth of the plenum is at least twice the width of the sum of slot widths at that point and upstream (i.e., the plenum can be tapered).
2) The takeoff to the duct is centered on the slots and perpendicular to the slots (i.e., air makes a 90° turn after entering through the slots) or the plenum is very deep (e.g., depth = slot length).
3) The takeoff has a 45° or less taper angle.
4) The closest slot is at least 1/2 of the slot length distance from the taper.

If these conditions are not met, the slot velocity should be higher. Velocities above 2000 ft/min [approx. 10 m/s] are probably only marginally more effective than 2000 ft/min [10 m/s].

If it is necessary to use an undersized plenum or if the takeoff will be at one end of the slots rather than centered on the slots, it is likely that velocities down the length of the slot will be progressively higher as the takeoff is approached even if V_s = 2000 ft/min [10 m/s]. It is possible that extremely high slot velocities (e.g., V_s = 4000 ft/min [20 m/s]) could be effective in providing reasonably uniform velocities. Some practitioners instead use a slot width that decreases in size as it approaches the takeoff. Both seem reasonable, but there is no available empirical evidence that demonstrates the efficacy of either practice.

6.7.6 Airflow Requirements for Slot Hoods (Aspect Ratio < 0.2). For hoods having an aspect ratio (width divided by length (W/L) of 0.2 or less), only a small fraction of the air flows from the ends (see Figure 6-22) into the face, so the air behaves to a large degree as if it were flowing into a line sink. Therefore, at a distance "X" for a slot of length "L" the control volume would be a cylindrical shape with a surface area of $A = \pi X L$. Since $Q = V A$, the airflow (Q) required at a given distance would be $Q = \pi X L V$ and would fall linearly with distance from the hood (see Figure 6-17). The approximate airflow (Q) required to achieve a specific velocity (V_X) at a distance "X" for a slot of length "L" (upstream of the midpoint of a freely suspended slot with no flange and with no nearby obstructions) is:

$$Q = 3.7\ V_X\ L\ X \qquad \textbf{[6.2]}$$

For slot/plenum hoods, if the slot is in the center of a large flange, air is prevented from flowing from behind the hood, thus improving its effectiveness in front. This can reduce airflow requirements by as much as 20% for slots with aspect ratios equal to 4 and 35% for slots with aspect ratios equal to 16.[6.7] For a flange width (W_{fl}) greater than the square root of the hood face area (i.e., $W_{fl} \geq \sqrt{A_f}$), a reasonable approximation is a reduction of 25% from Equation 6.2:

$$Q = 2.8\ V_X\ L\ X \qquad \textbf{[6.3]}$$

If the slot is in a large wall (e.g., is cut into the plenum of a slot/plenum hood), the airflow requirement should be lower than predicted by Equation 6.3. The maximum possible reduction is 50% of the levels predicted in Equation 6.2. Other surfaces near the hood can also reduce the airflow requirement by channeling the air through the source and to the hood opening. The most important example is the surface of a table when a hood is on or very close to the surface.

Increasing slot velocity (by reducing slot height) while holding Q constant will not improve the "reach" of the capturing hood. The total airflow requirement for a slot/plenum hood with multiple slots is the sum of the requirements for the slots. If the slots are the same size and the plenum is of adequate size, the airflow through each slot will be the same. When the slots are less than 0.5X apart they will act as a plain opening. Note also that tapering from the hood face down to the duct has little or no effect on airflow requirements.

6.7.7 Airflow Requirements for Aspect Ratios Greater Than 0.2. The simplest possible hood would be a free standing exhaust point. If we neglect the duct, the hood acts as a point sink (see Figure 6-26). In the absence of disturbing air currents, the airflow would move toward the point sink uniformly from all directions. At any distance "X" from the exhaust point, the control volume would be a sphere with radius X and a surface area of $4\pi X^2$. The mean velocity through the surface of the imaginary sphere would be Q/A_{sphere}. Thus, to establish any given velocity "V_x" at a dis-

FIGURE 6-26. Plain opening acts as a point sink

tance of "X" the required airflow would be $Q = V_x (4\pi X^2)$. Since hoods have a finite size, the geometry is not so simple (see Figure 6-20). It can be shown that if the hood has an aspect ratio greater than 0.2 or is round, then a hood hanging in space with no nearby obstructions requires the airflow rate to be (with V_x at distance "X") estimated by:[(6.8)]

$$Q = V_x [10 X^2 + A_f] \quad [6.4]$$

where A_f = area of face opening

However, capturing hoods often have relatively large flanges that serve to block flow from the back of the hood, increasing the flow from the front (see Figure 6-16). For a flanged hood in unobstructed space with $W_{fl} \geq \sqrt{A_f}$, the required airflow is somewhat reduced:

$$Q = 0.75 V_x [10 X^2 + A_f] \quad [6.5]$$

Capturing hoods also are often resting on a surface such as a table top or are placed at some distance just above the surface (e.g., dip tanks, tabletops, etc.). If the hood rests on the table, the airflow requirement reduces to:

$$Q = V_x [5 X^2 + A_f] \quad [6.6]$$

And if the hood rests on the table and is flanged, the airflow requirement reduces further to:

$$Q = 0.75 V_x [5 X^2 + A_f] \quad [6.7]$$

Since the contaminant from even very small sources may be dispersed over a vertical height of several inches, it is usually not advisable to place the hood directly on the work surface unless it has a large flange resting on that surface. If the contaminant is buoyant, the hood should be elevated above the work surface (e.g., 12 to 24" [0.3 to 0.6 m]), with the height increasing to a point with increasing thermal rise velocity. If that is done, the airflow requirement should be somewhere between Equations 6.5 and 6.6 since the work surface still channels flows to some degree but not as much as when closer to the surface.

For both slot/plenum hoods and "rectangular" hoods, distance ("X") is crucial. For example, a 4" × 9" [101 mm × 229 mm] flanged hood that draws 206 acfm [0.097 am³/s] will induce a velocity of 100 fpm [0.508 m/s] at a distance of 6 inches [0.152 mm], but only 27 fpm [0.137 m/s] at a distance of 12 inches [305 mm]. Any measure that reduces the distance between hood face and the source is likely to vastly improve the performance of the hood.

The hood airflow equations are summarized in Table 6-3. Refer to Appendix A6 for alternative hood flow equations.

6.7.8 Caveats to Capturing Hood Airflow Equations. Equations 6.2 through 6.6 are based on the velocity perpendicular to the hood face at the midpoint of the face for ideal conditions. There is little research at this writing that can be used to determine if the current recommendations are optimal. It also should be noted that:

1) The equations model the velocity along the centerline of the hood face, not at other points in the expected control region. Real sources release contaminants that may be spread over a substantial lateral range. Practitioners should keep in mind that the capture velocity at the same distance from the hood face but not at the centerline will be increasingly lower than the midline velocity at increasing distances laterally from the midline, especially for square and round hood faces.
2) The equations do not consider the effects of cross-draft velocities. It is reasonable to assume that the value of capture velocity required to obtain the same effectiveness would increase substantially with higher cross-draft velocities.
3) The equations do not consider the effects of the worker's body or the effects of work items placed between the source and the hood.
4) The equations do not consider the effects of convection air currents due to hot surfaces or effluents (e.g., welding plume) nor the effects of competing air currents due to mechanical motions (e.g., spinning grinding wheel).
5) The equations for low aspect ratio hood openings probably apply much better to slot/plenum openings than to slots that are not the open face of a plenum.
6) Two slots that are relatively close together (e.g., distance between them less than 0.5 L) will behave more like a plain opening than a slot opening.
7) The equations may overestimate airflow requirements to some degree when the distance from the hood face exceeds 1.5 times the hydraulic diameter (i.e., 4 times the area of the hood face divided by its perimeter) of the hood face.

On the other hand, there is also no clear evidence that computing required airflows based on the current recommendations is leading to widespread failures to control contaminants.

TABLE 6-3. Summary of Hood Airflow Equations

HOOD TYPE	DESCRIPTION	ASPECT RATIO, W/L	AIRFLOW
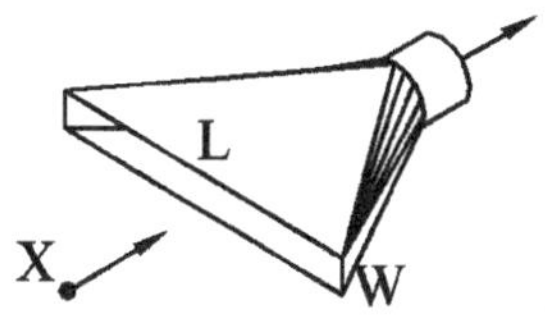	Slot	0.2 or less	$Q = 3.7\ LV_xX$
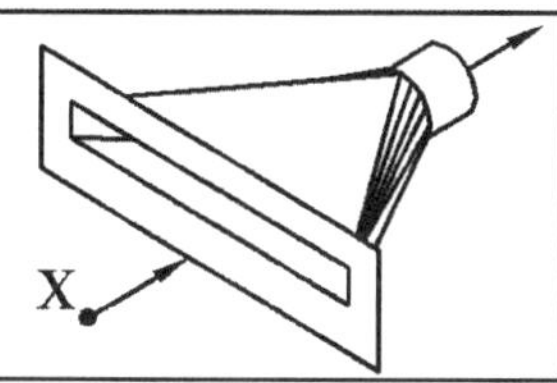	Flanged slot	0.2 or less	$Q = 2.6\ LV_xX$
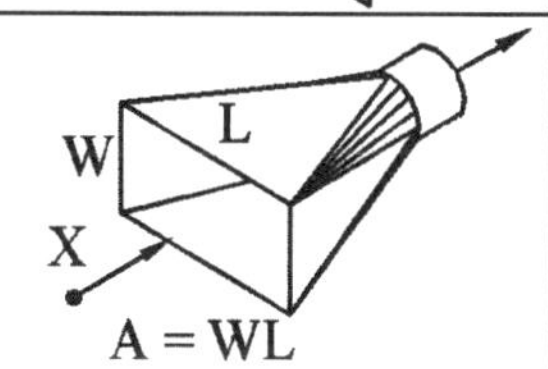	Plain opening	0.2 or greater and round	$Q = V_x(10X^2 + A)$
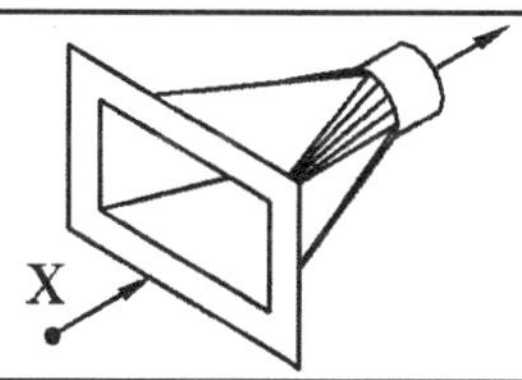	Flanged opening	0.2 or greater and round	$Q = 0.75V_x(10X^2 + A)$
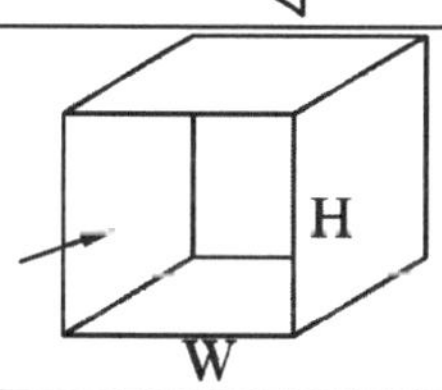	Booth	To suit work	$Q = VA = VWH$
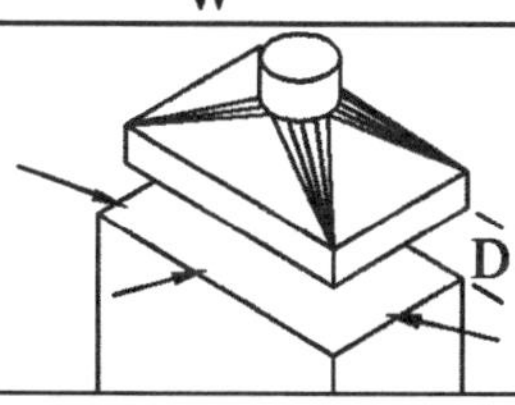	Canopy	To suit work	$Q = 1.4\ PVD$ P = Perimeter D = Height above work
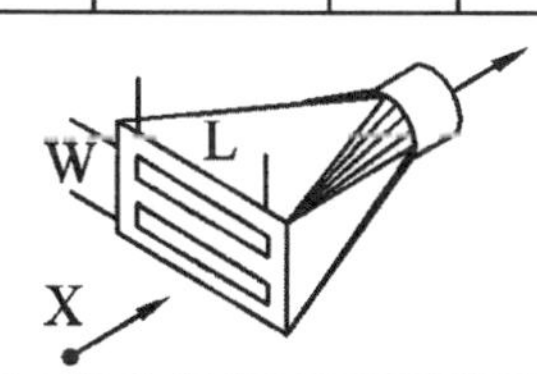	Plain multiple slot opening 2 or more slots	0.2 or greater	$Q = V_x(10X^2 + A)$
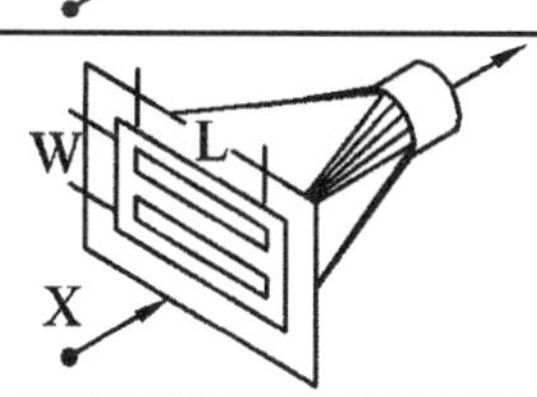	Flanged multiple slot opening 2 or more slots	0.2 or greater	$Q = 0.75V_x(10X^2 + A)$

When capturing hoods are ineffective, it is far more likely that failure to keep the source within the effective zone of the hood is the problem.

6.7.9 Example Airflow Calculations. In these three applications the same conditions are ventilated with three different types of hoods (Figure 6-27).

EXAMPLE PROBLEM 6-2a (Rectangular Capture Hood on a Table Top)

Find hood height (L), width (W), capture velocity (V_x), and Q for a high-aspect ratio capturing hood.

Information determined for the application:

1. For the table top (see Figure 6-27), W_{table} = 48" [1.2 m] and D_{table} = 36" [0.9 m]. The contaminant source is 24" [0.6 m] long by 6" [152 mm] wide by 6" [152 mm] high.
2. The contaminant is moderately toxic.
3. Cross-draft velocities are expected to be 20–40 ft/min [0.10–0.20 m/s].
4. The contaminant will be dispersed several inches vertically and horizontally by worker hand movements.
5. The contaminant source may be moved to any point on the entire area of the table. It would be inconvenient for the worker if they could not move it freely to accommodate different task requirements.
6. The hood may be flanged, if desired, and the hood can be suspended from any height desired.
7. The worker always works from just one side of the table.

Solution (see Figure 6-28)

1. From Table 6-2 for moderate toxicity and low initial dispersion rates for the contaminant, the range of capture velocities should be in the range of 75 to 100 ft/min [0.38 to 0.51 m/s]. Given the moderate cross-draft velocities, V_x = 100 ft/min [0.51 m/s] is selected.
2. The hood width (L = dimension parallel to the table) should be somewhat greater than the corresponding dimension of the source, which in this case is 24" [0.6 m]. Hence the minimum value of "W" should be at least 30" [0.7 m]. Since the source can be moved across the width of the table and workers normally do not move hoods to keep them in front of sources, a conservative choice for the hood length would be the entire width of the table, L = 48" [1.2 m].
3. Because the contaminant will be dispersed somewhat (e.g., 6" [approx. 150 mm] up and down) and begins from a height of 6 inches [approx. 150 mm], the hood should be centered at a height of perhaps 6–9" [approx. 150–230 mm] and have a hood vertical dimension "H" of approximately 12" [300 mm]. With a 6–9" [approx. 150–230 mm] flange, the hood face would be centered on the source if the flange rested on the table.
4. In this case the worker must be able to place the source anywhere on the table. If that is true, the hood should be affixed to the back of the table or it should be mobile to allow movement as needed. If it is placed on the back of the table, the value of X could be as much as 36" [0.9 m]. If it the hood is moveable, the worker could be instructed to keep it within some specified distance from the source. Again, because workers can be unreliable in moving hoods as needed, the cautious choice is to fix the hood in place as close as possible to the source.

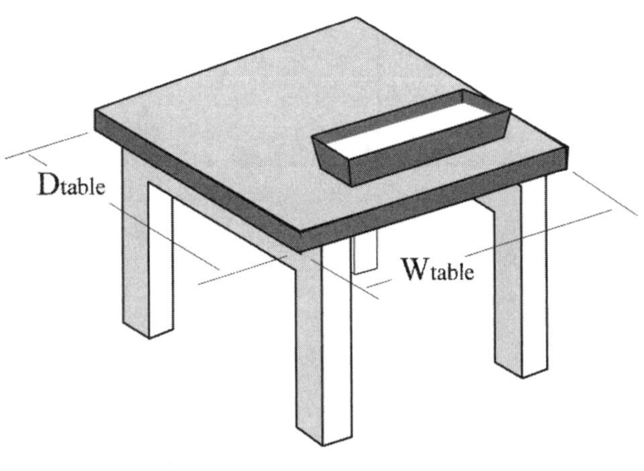

FIGURE 6-27. Work station for all three examples

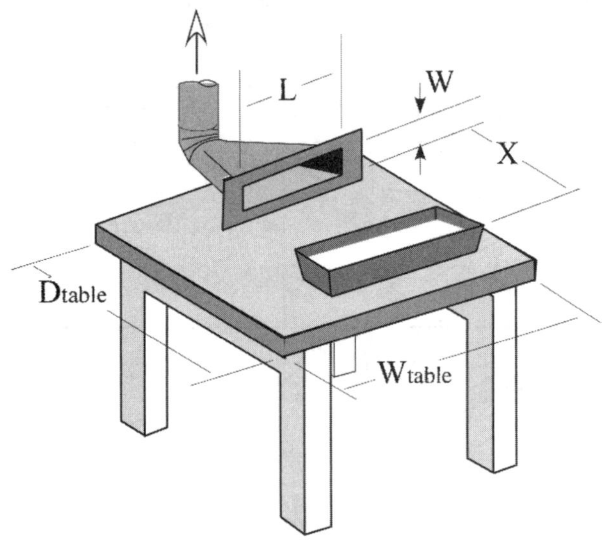

FIGURE 6-28. Rectangular capturing hood for Example Problem solution

In this case, an experienced practitioner would revisit the claim that the worker must be free to place the source anywhere on the table. Careful observation of the work and questioning of the worker would probably reveal that the worker actually would need far less latitude. After all, it is very unlikely they would want the source at the forward edge of the table or the back of the table or at the extreme left or right. The range of distances from the back of the hood is more likely to be 18–30" [0.5–0.8 m] from the back of the table and it is likely that the hood can be placed at least 6" [approx. 150 mm] from the back edge of the table, reducing the range of "X" to 12–24" [0.3–0.6 m]. In that case, 24" [0.6 m] would be a prudent choice. If the full table truly is required, then X = 36" [0.9 m].

5. Because the hood is flanged and almost, but not quite resting on the work surface, Equation 6.5 probably would overestimate Q and Equation 6.7 would underestimate Q. Equation 6.6 is probably a good compromise in this case.

6. Computations:

For $A_f = L * W = (12/12 \text{ ft})(48/12 \text{ ft}) = 4 \text{ ft}^2$, $V_x = 100$ ft/min, and $X = (24/12)$ ft = 2 ft

$Q = V_x [5X^2 + A_f] = (100 \text{ ft/min}) [5 (2 \text{ ft})^2 + 4 \text{ ft}^2] =$ 2400 acfm

For SI units the solution would be:

$A_f = L \times W = (0.3 \text{ m})(1.2 \text{ m})\ 0.36 \text{ m}^2$ $X = 0.6$ m

$V_x = 0.51$ m/s

$Q = V_x [5x^2 + A_f] = 0.51[5(0.36) + 0.36] = 1.10 \text{ am}^3/\text{s}$

A reasonable choice for the flange would be:

$W_{fl} = \sqrt{A_f} = (4 \text{ ft}^2)^{0.5} = 2 \text{ ft} = 24$ in. [approx. 600 mm]

If a smaller, mobile hood had been selected: $X = (12/12)$ ft [0.3 m],

$A_f = (12/12 \text{ ft})(30/12 \text{ ft}) = 2.5 \text{ ft}^2$ [0.23 m^2],

A reasonable choice for the flange would be:

$W_{fl} = \sqrt{A_f} = (2.5 \text{ ft}^2)^{0.5} \sim 20$ in. [approx. 500 mm]

$Q = V_x [5X^2 + A_f] = (100 \text{ ft/min}) [5 (1 \text{ ft})^2 + 2.5 \text{ ft}^2]$ = 750 acfm [0.35 am^3/s]

If the source could be anywhere on the full width and depth of the table and the hood were fixed at the back of the table:

$X = (36/12)$ ft = 3 ft,

$A_f = (12/12 \text{ ft})(48/12 \text{ ft}) = 4 \text{ ft}^2$ [0.372 m^2],

A reasonable choice for the flange would be:

$W_{fl} = \sqrt{A_f} = (4 \text{ ft}^2)^{0.5} = 24$ in. [approx. 600 mm]

$Q = V_x [5X^2 + A_f] = (100 \text{ ft/min}) [5 (3 \text{ ft})^2 + 4 \text{ ft}^2]$ = 4900 acfm [2.31 am^3/s]

If one assumes that the worker is unlikely to place the source closer than 6" to the front of the table but it otherwise could be anywhere on the full width and depth of the table and the hood were fixed at the back of the table:

$X = (30/12)$ ft = 2.5 ft [0.800 m],

$A_f = (12/12 \text{ ft})(48/12 \text{ ft}) = 4 \text{ ft}^2$ [0.372 m^2],

A reasonable choice for the flange would be:

$W_{fl} = \sqrt{A_f} = (4 \text{ ft}^2)^{0.5} = 24$ in. [approx. 600 mm]

$Q = V_x [5X^2 + A_f] = (100 \text{ ft/min}) [5 (2.5 \text{ ft})^2 + 4 \text{ ft}^2]$ = 3525 acfm [1.66 am^3/s]

7. Deciding which dimension values to use should depend on which assumptions more accurately represent reality and, for the mobile hood, whether one is confident that the worker will move the hood as needed.

EXAMPLE PROBLEM 6-2b (Slot/Plenum Hood on a Table Top)

For the same conditions and dimensions listed in Example Problem 6-2a, find hood width (W) and length (L) and Q for a slotted capturing hood.

Solution (see Figure 6-29):

1. The plenum should extend along all or nearly all of the back of the table.

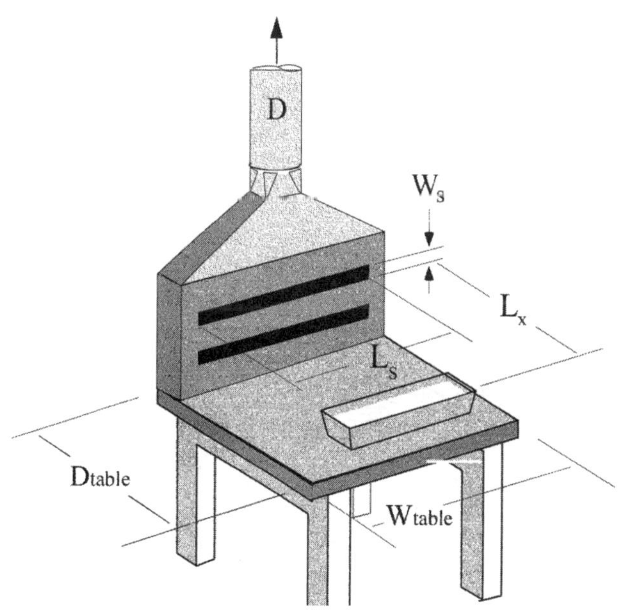

FIGURE 6-29. Slot/plenum hood solution

2. It may be desirable to install baffles on each side of the table to block cross-drafts (see Figure 6-23).
3. As with Example Problem 6-2a, V_x = 100 ft/min [0.51 m/s] and it is possible the source could be placed anywhere on the table.
4. The value of "X" could be less than 36" [0.9 m], depending on the actual placement of the source, but since the worker is unlikely to place it on the forward edge of the table, assume X = 30"/12 = 2.5 ft [0.800 m].
5. One of two slots should extend for all or nearly all of the width of the plenum,

   L_{slot} = 48"/12 = 4 ft [1.2 m].
6. Equation 6.3 would probably overestimate the airflow requirements since it does not account for channeling due to the table top and may underestimate the effect of the plenum in channeling flows. It also ignores the effects of the side baffles if they are used.

   If just one slot:

   $Q = 2.6\ V_x\ L\ X\ N_{slots} = 2.6\ (100\ \text{ft/min})(4\ \text{ft})(3\ \text{ft})(1) = 2600\ \text{acfm}\ [1.3\ \text{am}^3/\text{s}]$

   If two slots are required to ventilate two heights for contaminant release:

   $Q = 2.6\ V_x\ L\ X\ N_{slots} = 2.6\ (100\ \text{ft/min})(4\ \text{ft})(3\ \text{ft})(2) = 5200\ \text{acfm}\ [2.6\ \text{am}^3/\text{s}]$

   Based on the geometry of the channeling, it is likely that much less (e.g., perhaps as much as 30–50% less) than the airflow would be necessary in this case, especially for the calculation with two slots. However, there is no empirical basis for such an estimate.

EXAMPLE PROBLEM 6-2c (Enclosing Hood on a Table Top)

For the same conditions and dimensions listed in Example Problem 6-2b, find hood length (L) and width (W) and Q for a plug flow bench top enclosing hood.

Solution:

1. Make the hood enclose the entire table top with the plenum section extending beyond the back end of the table. This would allow placement of the source at any location on the table top. Placement within 6" [150 mm] of the front edge should be avoided and probably would be avoided by the user, anyway.
2. A plug flow bench top enclosing hood with a width of 4 ft [1.2 m], height of 3 ft [0.9 m] and face velocity of 100 ft/min would require Q = VA = (100 ft/min) (12 ft^2) = 1200 acfm [0.6 am^3/s]. Note that only the mobile hood in Example Problem 6-2a is lower (Q = 750 acfm [0.375 am^3/s]).

6.7.10 Push-Pull Hoods. Air emerging at high velocity from a duct or nozzle can travel 30 diameters before turbulence and expansion reduces its velocity to less than 10% of its initial value. On the other hand, air drawn into the face of a hood will have a velocity of less than 10% of the face velocity at a distance of as little as one duct diameter. "Push-pull" systems (see Figure 6-30) take advantage of this by containing and pushing contaminated air towards the capturing hood. Airflow reductions are possible with short push distances but can be quite substantial for large distances. See Chapter 13, Section 13.72 for detailed descriptions and formulae for Push-Pull hood systems. As is shown in Figure 6-30, a very large obstruction can reflect the push air away from the capturing hood, especially if it is very close to the push jets. Air generally will flow around a moderate size object, especially if relatively far from the jets (e.g., more than five times the smaller cross-sectional dimension of the obstruction).

6.7.11 Compensating Air Hood. Another type of hood blows clean air at low velocities at or near a capturing hood to improve its effectiveness. An example approach is shown in Figure 6-31. This strategy can be more effective than the cap-

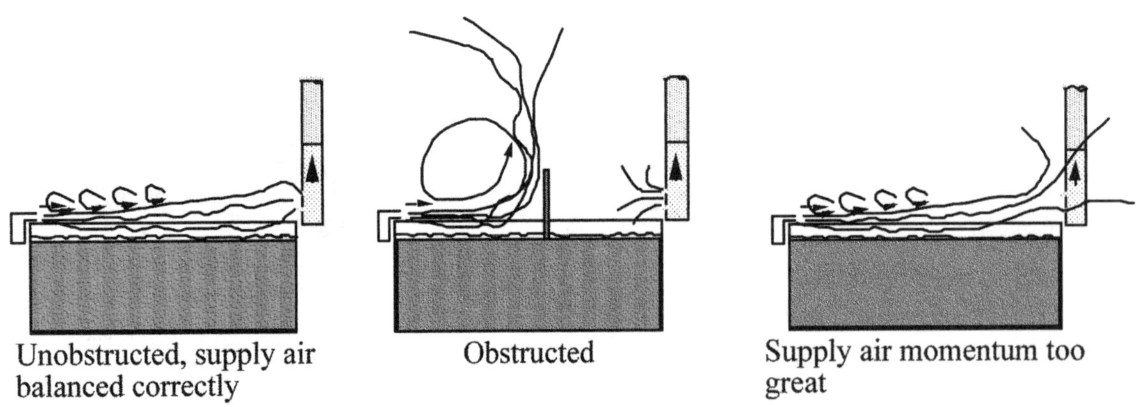

FIGURE 6-30. Push-pull ventilation for diptanks

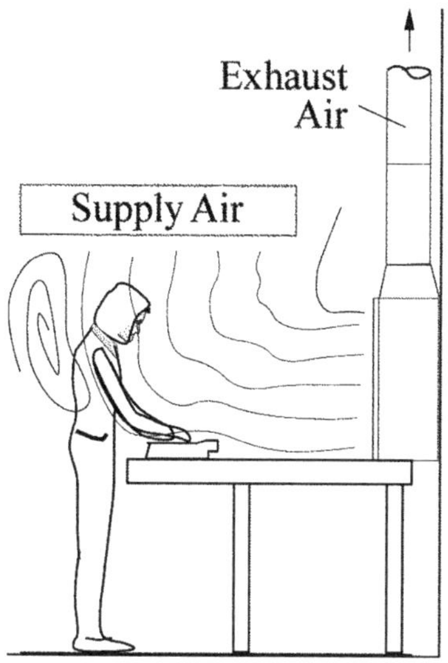

FIGURE 6-31. Compensating air hood

turing hood alone if done carefully. First, there should be only low velocity cross-drafts (e.g., < 35 ft/min [0.17 m/s]). Second, the supply airflow rate should be adjusted carefully to avoid blowing past the hood. If that happens, the exposure to the user may or may not increase, but the background concentration in the room almost certainly will. The exhaust airflow rate should be at least 30% larger than the supply airflow rate, and the release velocity of the supply air should be less than 50 ft/min [0.25 m/s]. These types of hoods have been used successfully in foundries on shakeout and pouring side-draft designs.

6.7.12 Downdraft Hoods. A downdraft hood is a type of capturing hood with the air flowing downwards through a horizontal face into the hood body (see Figure 6-32). The perceived advantage of a downdraft hood is that large particles will fall down through grille covering the face to be collected in cleanout drawers. It is also sometimes assumed that the required airflow will be very low since the distance to the source appears to be very low. As a capturing hood, the necessary airflow can be computed using Equation 6.6 where "X" is the maximum distance above the hood face where contaminant will be released.

It is sometimes assumed that if the work is done directly on top of the grille that the value of X will be zero and the airflow requirement will be $Q = V_x A_{grille}$. In reality, it is very unlikely that the maximum distance above the grille for contaminant release is ever zero. It is likely to be at least a foot [approx. 300 mm] above the table due to dispersion by tools and hand movements, not to mention the likelihood that work practices will include lifting contaminant-producing components.

It is important to note that "X" can be much higher than the maximum height of the source if the work disturbs the air. For example, a hand-held grinding wheel agitates the air directly

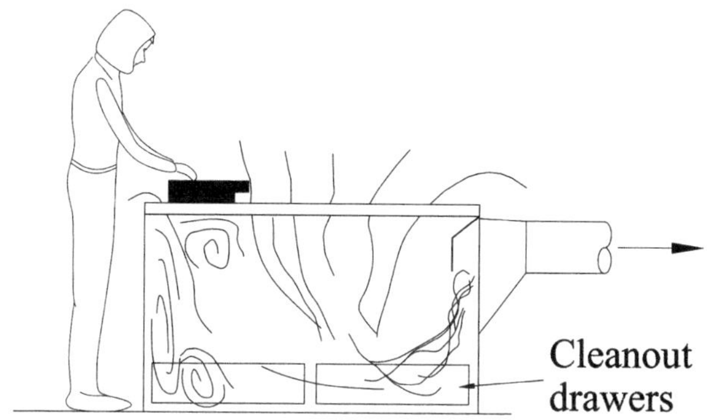

FIGURE 6-32. Downdraft hood

by the high rotation rate grinding surface.

Finally, it is important to recognize that operators may lay materials or tools over the grille, blocking the airflow where it is needed most, possibly rendering the hood useless.

6.7.13 Receiving Hoods. "Receiving hoods" are capture hoods positioned so that: 1) particulate contaminants are thrown into the hood opening from a distance, or 2) gas or vapor contaminants are lifted by convection towards the hood opening. Overhead canopy hoods (Figures 6-16 and 6-33) are typically used to receive contaminants mixed with heated air. Use of canopy hoods for very hot processes (e.g., as found in work with molten metal) is discussed in Chapter 13, Section 13.27.

Overhead canopy hoods are less effective for both warm and ambient temperature air because:

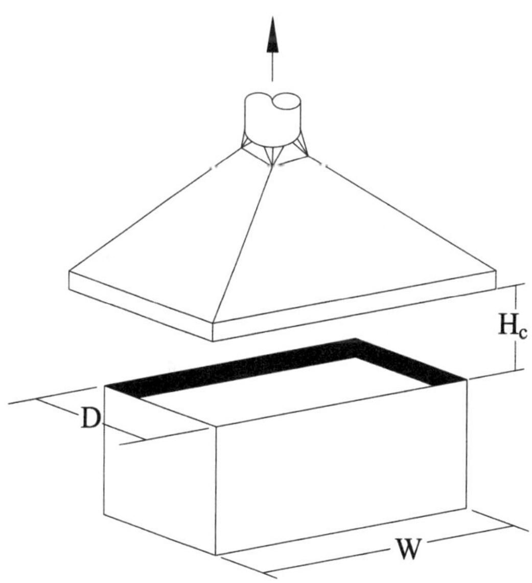

FIGURE 6-33. Overhead canopy hoods

1) Distribution of airflow is poor. There are no positive measures taken to spread air out. Air will flow preferentially near the top of the face, not near the source where it may be most needed.
2) The open faces of this hood are the planes formed by the perimeter of the source and the perimeter of the canopy. Airflow enters from all four sides, so the air volume requirements are correspondingly very large:

$$Q = (1.4)(\text{perimeter})(\text{height})(V)$$

Often the canopy is positioned five feet [approx. 1.5 m] or more above the source. The hood must "reach" several feet to control all avenues of escape of the contaminant. Its volume requirements can be extremely large.
3) If workers bend over the source to work, contaminated air may be directed into their breathing zone.
4) Since all sides are open, the hood is vulnerable to crossdrafts from all four sides.

The canopy hood can be vastly improved by adding three sides to it, but the distribution of velocities at the remaining face will not be good (see distribution for enclosures, Section 6.3.3).

6.7.14 Steps to Designing a Capture Hood. When designing a capturing hood and selecting the airflow for it, consider that crucial to its effectiveness is that the distance ("X") between the open face of the hood and the greatest distance to a point of contaminant generation be kept as low as possible. The steps to follow in designing a capturing hood are:

1) Observe the operation through several cycles and question workers and maintenance personnel about access needs, work practices, materials handling, emergency conditions, and maintenance.
2) Channel the airflow to it as much as possible by employing flanges and placing the work on a horizontal surface. Put a panel in the back and top if possible. Use side barriers only if the distance is great and the airflow is relatively low. The more you direct the airflow over the contaminant source and into the hood opening, the better the hood will be.
3) Since the source should be located directly in front of the hood opening, make the hood opening large enough that the operator doesn't have to keep moving the hood to keep it near a moving source of contamination.
4) Fix the hood and the source in place, if possible, so that the distance from it to the farthest point of contaminant generation is always within the hood's effective range.
5) To determine airflow (Q) requirements, first determine the capture velocity needed considering the crossdrafts, the toxicity of the contaminant, and the amount of the contaminant. Recommended capture velocities are shown in Table 6-2.
6) Cover the face of the hood with expanded metal or mesh to avoid picking up papers, caps, rags, etc.
7) When the hood is installed (and periodically thereafter), evaluate its performance both for ventilation effectiveness and worker acceptance.

6.8 CHOOSING BETWEEN CAPTURING AND ENCLOSING HOODS

If the contaminant is a combination of copious, energetic, and toxic, it is generally highly advisable to control it with a highly enclosed hood with carefully placed, small ports for access and inflow of supply air. Capturing hoods should not be used in such cases, not because they are incapable of a high degree of effectiveness, but because they are too likely to be moved to a distance at which they would fail.

If the contaminant source is considerably less hazardous and manual access is required, the choice is much less clear. Plug flow enclosing hoods require greater care in design and operation but are likely to be less vulnerable to poor work practices. If the worker must move the capturing hood frequently for it to reliably control the source, it is probably best to use a plug flow enclosing hood or a capturing hood so large and with so much airflow that it need not be moved. Unless a capture hood is small and very close to the source, it is likely that the airflow requirement for an enclosing hood would be less, sometimes considerably less.

The main disadvantages of enclosing hoods are that they are typically more expensive than capturing hoods, take up more floor space, and require much more imagination and effort to design well. Making enclosing hoods acceptably convenient to use sometimes can require extended design effort and sometimes can be accomplished only in conjunction with significant modifications to the material handling system. If the latter are not feasible, a capture hood may be the only suitable choice. Also, if the workers' tasks bring their faces very close to the source (i.e., work requiring close visual inspection), a plug flow enclosing hood may be much less protective than a capturing hood kept very close to the source.

The main advantages of capturing hoods when compared to enclosing hoods are that they: 1) require less airflow if they are small and close to the source, 2) typically can be used without modifying materials handling, 3) are less expensive to purchase or build, and 4) require much simpler selection, design, and installation.

Capturing hoods can be extremely effective if the contaminant is released: 1) with no velocity, 2) well within the hood's effective range, and 3) at locations with relatively low velocity competing air motions. The disadvantages of capturing hoods compared to enclosing hoods are that their performance typically can be strongly degraded more by: 1) seemingly small changes in positioning either the source or the hood, yet they often are used well outside their effective range; 2) crossdrafts and other competing air motions; and 3) significant reductions in exhaust airflow.

Because of their greater reliability, enclosures should be preferred over capturing hoods in situations where it is possible to install them. An enclosing hood also generally is more reliable in limiting escape (i.e., fugitive emissions), especially for high cross-draft velocity conditions. A capturing hood can be more effective in protecting workers if conditions are ideal for it and much worse if not.

6.9 ERGONOMIC DESIGN OF HOODS USED BY WORKERS

If workers must frequently reach into a hood or stand in it to work, considerations of ergonomics and human factors should be employed to make the hood as "user-friendly" as possible. A hood that is awkward or difficult to use is 1) likely to be modified by users or maintenance personnel, 2) may not be used when needed, and 3) could reduce work output and quality. Key issues are the dimensions of the enclosures and the integration of the design with materials handling.

For the dimensions of the hood and for work surfaces, flexibility in design is a key ergonomic consideration since different workers with varying physical characteristics may use the same workstation over time. Tables 6-4 and 6-5[6.1] present anthropometric data collected mostly from Caucasians. Other populations (e.g., South Asian) have somewhat different means and standard deviations, but it is clear that no one size can fit most of the workforce. Hence, it is highly desirable in many cases to make work heights and other critical dimensions adjustable.

Hoods, especially enclosing hoods, should allow clear sight lines and sufficient light (without glare) for the task.

Both reach-in and occupied hoods must be convenient and comfortable for the worker to use. The width and height of the hood should be large enough that the worker can conveniently handle materials or equipment inside it. Usually this will result in a minimum width of at least three feet [0.9 m]. If the worker must lean into the hood and lift relatively heavy objects, the hood should be wider. It would be better still to provide assists to any lifting required within the hood. Occupied hoods should be at least 6 ft [approx. 1.8 m] wide to reduce claustrophobic reactions and allow room for swinging the arms and bending the torso to the left and right.

Often the hood should be wider than the minimum needed for the operator. In particular, the width of the hood should be great enough to allow necessary access for manipulation of work objects inside the hood. If the objects are large, then the hood width must accommodate their widths as well. The additional width needed to manipulate objects depends on exactly what motions are necessary and can be determined by an understanding of the actual tasks that will be done. However, there are some general guidelines that can be helpful. If the worker will be standing to the side of the large object, they will require at least 3 ft [approx. 0.9 m] to comfortably reach forward, crouch, or bend over. If the worker will be reaching toward the object with a long probe or tool, the length of the tool will probably add to the width needed for the work.

Spraying the side of a large object (e.g., furniture) generally also requires at least 3 ft [0.9 m] of clearance on a side of that object if the worker is standing in front and reaching around to spray (unless the height of the object is less than roughly waist height). If the worker must spray while walking down the length of the enclosure, more clearance will probably be necessary to avoid having overspray invade his breathing zone.

Thus if the user must spray or access the side of large objects while within an enclosing hood, the hood width often should be greater than or equal to the width of the object plus 3 ft [0.9 m] on each side for which access or spraying is required.

If workers sit or stand and reach frequently into the enclosing hood to work, the height of the hood opening should be sufficient to prevent bumping their head on its top surface. A height of 7 ft [2.1 m] above the surface the worker is standing on is generally sufficient. The height of the hood opening is then simply the distance from the floor of the hood plus 7 ft [2.1 m].

If workers will enter a hood, the hood's height must be sufficient to allow headroom also. A height of 7 ft [2.1 m] will usually provide adequate headroom if the worker will not be doing anything that requires moving the arms, materials or a tool over the head. In the latter cases, clearance must be provided for the arms, materials or tools. If objects in the hood are taller than 7 ft [2.1 m], then the ceiling must be high enough to accommodate not only the objects but the clearances required for all tasks done on those objects. For example, if a worker must spray the top of a 7 ft [2.1 m] high machine chassis, he or she will probably have to stand on a platform to spray down the length of the top of the chassis. For very large objects, it is desirable to place them on legs to allow airflow under the object. That additional height must be accommodated, as must the height added by conveyors, turntables, etc.

Similarly, the heights of work tables, the floor of bench top hoods, and other work surfaces should be set to accommodate all workers and platforms used to augment the height of shorter workers should be provided. The necessary height of the work surface depends on whether the worker is sitting or standing. When sitting, a chair or stool with adjustable heights and footrests can be used to set operator height relative to the work surface. When the worker is standing, the optimal height of the work surface is a fraction of the worker's height, varying with the weight of the object being handled and the closeness required for adequate visual acuity. Heavy objects generally are manipulated at hip height; light objects with close work are held at just below sternum height. Note that these heights are specific to each worker. If at all possible, working heights should be adjustable by the worker using adjustable

TABLE 6-4. Anthropometric Data

Measurement, in. [mm]	Male				Female			
	Mean	Std	5th	95th	Mean	Std	5th	95th
Forward function reach (includes body depth at shoulder)	31.2 [792]	2.2 [55]	27.6 [701]	34.9 [886]	28.0 [711]	1.5 [381]	25.5 [647]	30.5 [774]
Waist height	41.3 [1041]	2.5 [63]	37.2 [994]	45.4 [1153]	38.8 [965]	2.2 [55]	35.2 [894]	42.4 [1074]
Elbow height	45.1 [1145]	2.5 [63]	41.0 [1041]	49.2 [1503]	42.2 [1071]	2.7 [68]	37.7 [957]	46.6 [1183]
Eye height	57.6 [1463]	3.1 [78]	52.6 [1320]	62.7 [1592]	53.3 [1351]	2.6 [66]	49.0 [1244]	57.6 [1463]
Stature	69.9 [1775]	2.6 [65]	65.5 [1666]	74.2 [1879]	64.8 [1625]	2.8 [71]	60.1 [1526]	69.4 [1762]
Seated-elbow height	9.5 [241]	1.3 [33]	7.4 [188]	11.6 [294]	9.1 [231]	1.2 [30]	7.1 [180]	11.0 [279]
Seated-eye height	31.0 [787]	1.4 [35]	28.6 [726]	33.3 [845]	29.0 [736]	1.2 [30]	27.0 [685]	31.0 [787]
Weight, lbs	183.4 [83.2]	33.3 [15.1]	128.5 [52.3]	238.4 [108.1]	146.4 [66.4]	30.6 [13.9]	95.8 [43.5]	196.9 [89.3]

Adapted from Ergonomic Design for People at Work by Eastman Kodak Company (Van Nostrand Reinhold, NY (1983).

standing platforms or adjustable work surface heights. For the latter, in many cases the table or hood floor height can be set at a low value for short workers with taller workers accommodated by adjustable height jigs.

Convenient visual access is also important. For example, the small enclosing hood shown in Figure 6-34 should allow necessary sight-lines even though it is relatively small. Like most other enclosing hoods, the inside of the hood should be well lit and without glare. In some cases, transparent sides (e.g., laminated safety glass or clear polycarbonate) may provide sufficient visual access.

Other considerations for hood design:

1) Provide a lean bar and foot rail where appropriate.
2) Have built in holders for tools and supplies (e.g., welding rods).
3) Suspend and counter balance heavy cables, tubing, etc. that the worker must move around (e.g., electrical lines for welders).
4) Counter balance movement arms for mobile hoods (e.g., "welding" hoods).
5) For enclosing hoods, use transparent plastic glass or laminated safety glass for sides to allow visual communication with nearby co-workers or to see items that must be kept under surveillance (e.g., indicators).
6) Move controls and indicators (e.g., hood static pressure display) so they are close and easy to see while not

TABLE 6-5. Recommended Values for Work Surfaces and Enclosure Dimensions

Parameter	Approach	Reasonable Range of Values (McCormick, 1993)
Work surface height, standing	Adjustable height surface or provide a moveable standing base for shorter workers. Work surface height higher if close inspection necessary, lower for coarse inspection and handling of heavy parts.	fine work: 37" – 49.5" [939 – 1257 mm] light assembly: 32" - 42" [812 – 1066 mm] heavy work: 29" to 30" [736 – 762 mm]
Work surface height, sitting	Adjust height of chairs and provide foot rests. Work surface height higher if close inspection necessary, lower for coarse inspection and handling of heavy parts.	fine work: 39" – 41.5" [990 – 1054 mm] light assembly: 32.5" - 37" [825 – 939 mm] medium coarse work: 26" to 28.5" [660 – 723 mm]
Work surface width	Fixed width based on size of the pieces and the tools the worker handles. The maximum value is based on workers' maximum reach while holding light objects. Effective grasp distance falls sharply with progressively heavier objects.	24" – 48" [609 – 1219 mm]
Work surface depth	Fixed depth based on size of the pieces and the tools the worker handles. Worker effective, comfortable reach distance for shorter workers is less than 16" inches [406 mm] for light objects and is progressively lower for heavier objects.	≥ 24" [609 mm]
<u>Enclosing hoods:</u>		
Hood width	Minimum based on ergonomics; add for access to materials.	≥ 36" [914 mm]
Hood depth	At least ¾ times height or width, whichever is larger.	≥ 36" [914 mm]
Hood height inside the hood	Should clear workers head by ≥ 3" [76 mm] for tabletop hoods. Minimum for walk-in hoods is 84" [2133 mm]. Greater heights required if tools or work pieces must be lifted overhead.	≥ 78" [1981 mm]

interfering with the work.

7) Make it easy to switch the layout to accommodate both left- and right-handed workers.

8) Avoid sharp or abrupt edges, especially at head and shin height.

9) If feasible, locate away from strong noise sources as well as sources of excessive vibration and temperature.

10) Where necessary, use safety controls such as "Hands Off" buttons, "dead man switches," etc.

11) Place outlets and controls for required utilities (compressed air, water, coolants, etc.) at convenient and safe locations.

12) Consider ease of required cleaning or decontamination tasks within and near the hood when selecting materials. If it will be cleaned with flowing water, provide a pathway to a sump pump or receptacle.

For large hoods, design considerations include:

1) Prevent heavy doors, sashes, work materials, etc., from falling by using safety cables and counter weights.

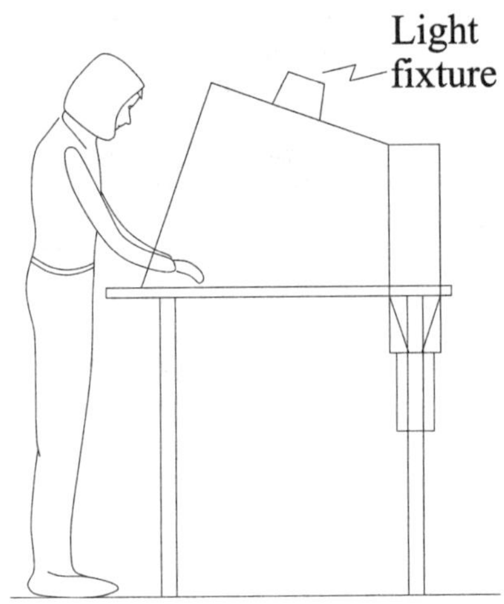

FIGURE 6-34. Small enclosing hood

2) Position doors and sashes for easy access to enclosures for both routine operations and maintenance.
3) Have observation windows in doors to prevent collisions and to allow visual inspection of the inside.
4) If the inside of the hood would be hazardous during operations, provide lockouts, interlocks, and warning lights as needed. If severely hazardous, the process and machinery should shut down if the doors are opened.

Note that enclosing hoods sometimes also act as machine guards. In those cases, safety personnel knowledgeable about machine guarding should be consulted.

6.10 WORK PRACTICES

Hoods used as or near work stations (i.e., bench top enclosures and occupied hoods) should be designed and operated with strong consideration of work practices. If they are not, the hoods often will fail to adequately protect the user. In some cases work practices should be modified to accommodate the hood. Any modification of work practices must be acceptable to workers and enforced by supervisors. Any modification that reduces productivity or makes the work more difficult will probably not be sustainable.

The most important work practice is to use the hood in its intended manner. This can be encouraged by making the hood a convenient and comfortable place to work. For example, tool-holders and rests should be attached to the top and sides of the inside of the hood to make the hood more convenient and productive for the worker.

6.11 MATERIAL HANDLING IN AND NEAR HOOD WORKSTATIONS

Moving products or materials into and out of the hood must be convenient and efficiently performed. Workers will avoid using hoods if they are awkward to use, increase work stress, or reduce productivity. All of those are strongly affected by the fit of the hood to the material handling and vice-versa. Ideally, hood design, material handling, and work practices should be considered as an integrated package. In many cases, the material handling should be improved regardless of ventilation issues, especially if there is poor work flow or excessive ergonomic stress on workers. Work practices are generally affected by material handling, and in many cases work practices that undermine ventilation effectiveness can be improved only by changing the material handling.

Note that changing material handling does not necessarily require elaborate or costly solutions. For example, lifting from below the knees to above the waist is highly stressful. Rotating the trunk of the body while holding a heavy or bulky object is very stressful. Both usually can be completely avoided by placing the object to be lifted at the same height as the receiving surface and by locating it so that the work does not have to rotate. Conveyors or slides should be used to avoid lifting and placing inside the hood where possible.

Where conveyance directly to the hood is not feasible, a hoist may be necessary for large or very heavy pieces. If a hoist is used with an enclosing hood, a V-shaped slot can be cut into the roof of the enclosure and reinforced with angle iron or steel pipes (see Figure 6-35). Alternatively, the bottom of the hood can roll out to receive the object (see Figure 6-36). For the latter solution, it is important that the roller wheels or bearings be used to minimize friction and that the hood be counterbalanced or bolted to the floor to prevent tipping. If the worker requires access to all sides of a work piece, consider a turntable (see Figure 6-37).

Some capturing hoods have articulated arms to allow repositioning the new face over an area. It is all too common that

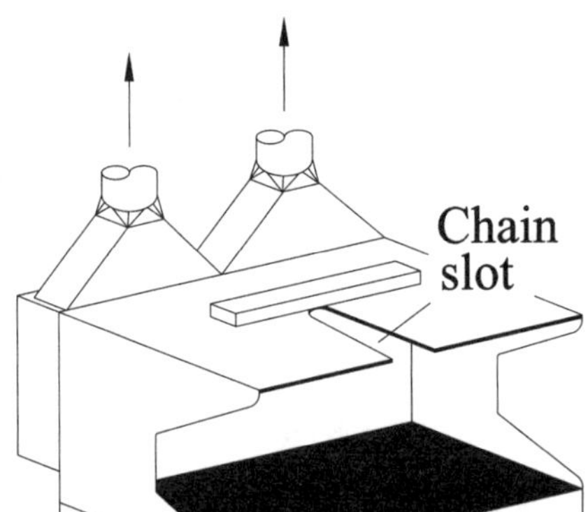

FIGURE 6-35. Chain slot

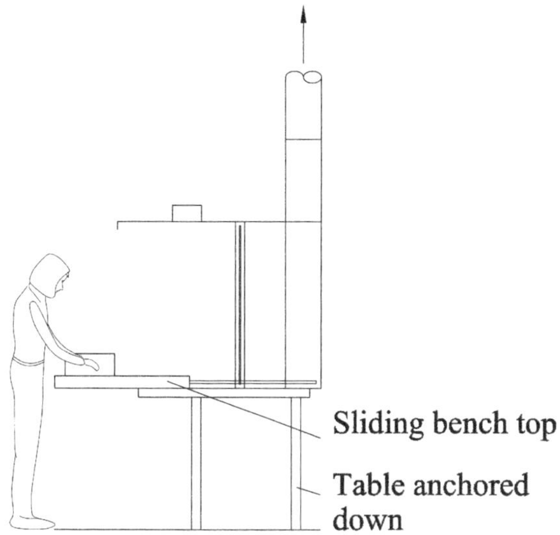

FIGURE 6-36. Roll out hood

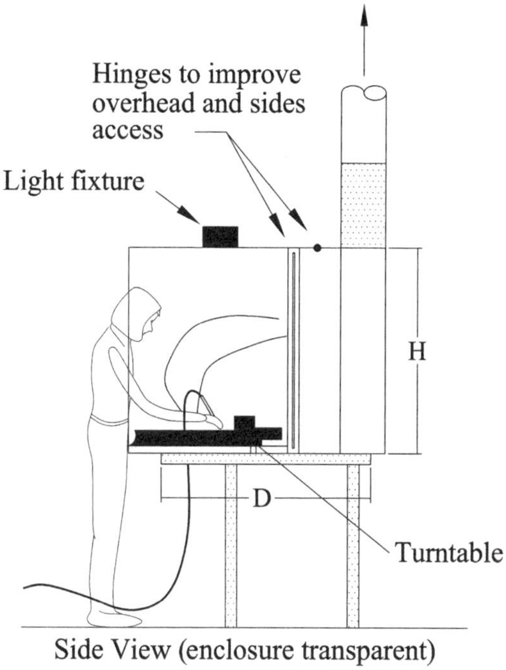

FIGURE 6-37. Turntable

the user fails to move the arm as often as needed or at all. To encourage needed movement, it is important to make the movement as easy and as smooth as possible. For that reason, the moveable capturing hood should be as lightweight as possible and no bulkier than necessary. The articulated arms also should be counter-weighted. Finally, it is important to maintain the articulations so that they bend or slide easily.

Note that large objects in the hood can act as dams, creating stagnant zones near them. If the contaminant is applied to the top of a large object that is above head height, the contaminated air may travel to the worker's breathing zone on its way down to the floor.

Downdraft flows can fail even if the supply air is uniformly released and the exhaust is designed to achieve uniform flow. They can fail because the contaminant is too copious and too widely dispersed to prevent invasion of the worker's breathing zone. A common example is abrasive blasting "rooms," that sometimes only prevent dusts from traveling to adjacent occupied areas.

6.12 MAINTENANCE AND CLEANING FOR ALL HOODS

Collection of settled material should be convenient whenever particulates (e.g., dusts or mist) may settle inside the enclosure or from the plenum of a slot plenum hood. For liquids, provide inclined pathways to a drain (see Figure 6-38). For dusts, relatively steep slopes should be used where possible to encourage settled material to slide to the bottom (see Figure 6-39). Access via panels or doors should be convenient. The panel or

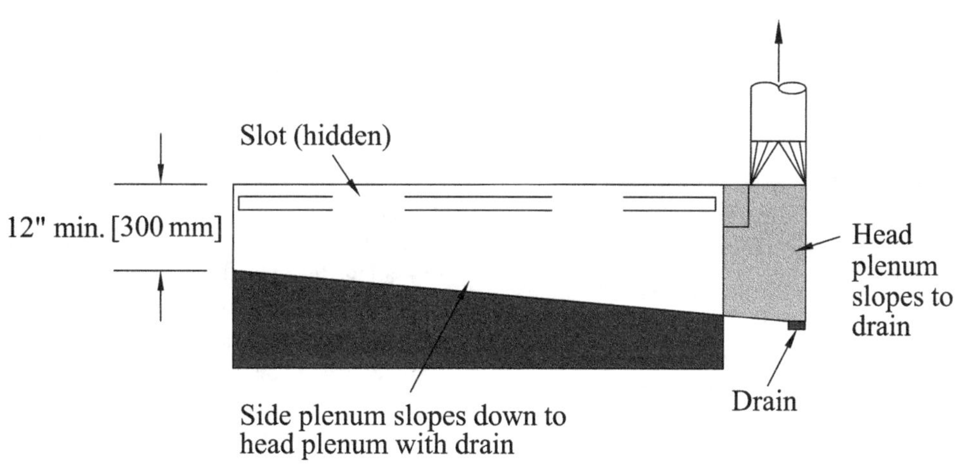

FIGURE 6-38. Diptank with draining for water that enters through ventilation slots on sides and front

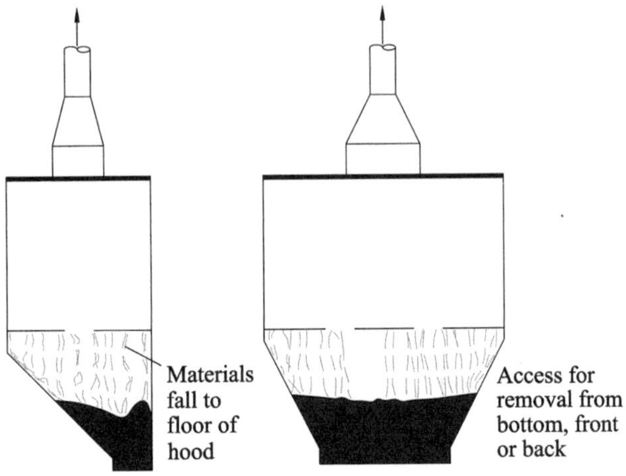

FIGURE 6-39. Hopper bottom to ease removal of settled materials

door should be hinged rather than hung from supports to assure that it will be returned to closed position during operation.

Access for maintenance of the hood and equipment within should be considered in the initial hood design. For example, easy access should be provided for maintenance of the lighting apparatus, preferably by installing on top of the hood and letting the light shine through a sealed plastic glass or laminated safety glass window.

For infrequent access for maintenance of items within arm's reach, provide a sliding or hinged panel. Avoid bolts or complex fasteners for securing serviceable parts. If particulates or liquids are released or sprayed in the hood, provide a means to remove them periodically or continuously.

Also, if it will be necessary to work on the top of an enclosure (e.g., to change the lights) or in any case where falls from an elevation are possible, ensure that proper fall protection access and clip on points are provided.

6.13 MAN-COOLING FANS

Although man-cooling fans are not "hoods" their use in the area near hoods is worth noting. Man-cooling fans move large quantities at very high velocities. That air movement will overwhelm enclosing hoods, blowing the contaminant out of it even if blown directly at the face. Even if the flow is perpendicular to the hood face, it is likely to radically reduce the effectiveness of the hood and prevent the escape of contaminant.

Although man-cooling fans are likely to reduce the effectiveness of both enclosing and capturing hoods, the hood user will not necessarily be overexposed to airborne contaminants as a result. The fan may simply blow the contaminant away and cause it to mix with the ambient air of the room. If the generation rate of the source is relatively small, the rise in room concentrations may be acceptably low, especially if there is an effective dilution ventilation system in operation. If the emission source is large the man-cooling fan may simply spread the contaminant around the worker, raising the area concentrations considerably.

A simple response to the effects of man-cooling fans is to ban them in areas ventilated by hoods. However, if the worker would experience heat stress without a man-cooling fan, it may be better to measure worker exposures and ambient concentrations with the fans both on and off. If the results are acceptably low, one could consider allowing use of the man-cooling fans despite their disruption of hood performance. However, it would also be worthwhile to investigate other means to reduce heat stress. In some cases, heated air and water vapor can be ventilated away from the space. Radiant heat from hot surfaces can be reduced by insulating the surface or by using aluminum shields to block infrared energy from them.

6.14 VENTILATION OF RADIOACTIVE AND HIGH TOXICITY PROCESSES

Ventilation of radioactive and high toxicity processes requires knowledge of the hazards, the use of extraordinarily effective control methods, and adequate maintenance that includes monitoring. Only the basic principles can be covered in this text and other resources should be reviewed, including published requirements of regulatory agencies.

Local exhaust hoods should be of the enclosing type with the maximum enclosure possible. Where complete or nearly complete enclosure is not possible, control velocities from 50% to 100% higher than the minimum recommended values in this Manual should be used. Supply air should be introduced at low velocity and in a direction that does not cause disruptive cross-drafts at the hood opening.

6.15 LABORATORY OPERATIONS

Glove boxes should be used for high activity alpha or beta emitters and highly toxic and biological materials. The air locks used with the glove box should be exhausted if they open directly to the room. For low activity radioactive laboratory work, a laboratory fume hood may be acceptable. For such hoods, a minimum average face velocity of 80–100 fpm [0.41–0.51 m/s] is recommended. See Chapter 13, Section 13.35, VS-35-01, VS-35-02, VS-35-04 and VS-35-20.

For new buildings, it is frequently necessary to estimate the air conditioning requirements early — before the detailed design and equipment specifications are available. For early estimating, the guidelines provided in Chapter 13, Section 13.35 for hood airflow and replacement airflow can be used and other regulatory standards should be consulted. These values may need to be revised as design conditions are firmed.

6.16 HOOD PRESSURE LOSSES

Air flowing through a hood will cause pressure changes that

must be considered when connecting the hood to the system duct. The sum of these changes is called "Hood Static Pressure" (SP_h) that occurs just downstream of the hood/duct connection. Simple hoods such as plain duct openings, flanged duct openings, canopies, and similar hoods have only a single loss point that is located at the duct inlet. Compound hoods (with slots) are hoods that have two or more points of energy loss that must be considered separately and added together to arrive at the total loss for the hood.

The hood entry loss (h_e) is expressed in terms of hood loss coefficients (F_s and F_h) that, when multiplied by the slot or duct velocity pressure (VP), will give the entry loss in inches of water ("H_2O or "w.g.) [Pa]. The hood entry loss may also contain a pressure loss due to the presence of a hood filter (SP_f). The Hood Static Pressure (SP_h) is equal to the hood entry loss (h_e) plus the energy required to transfer static to kinetic energy as the air moves from zero to duct velocity. This last term is defined as $(F_a)(VP_d)$. F_a is known as the acceleration factor and has a value of 1.0 (see Chapter 3, Section 3.5).

The duct velocity pressure (VP_d), utilized to determine hood losses in the following examples, is determined from the air velocity in the duct immediately downstream of the hood to duct connection. In equation form, these relationships are defined as:

$$SP_h = -(h_e + F_aVP_d)$$

$$SP_h = -(h_s + h_h + SP_f + F_aVP_d)$$

$$SP_h = -[(F_s)(VP_s) + (F_h)(VP_d) + SP_f + (F_a)(VP_d)] \quad [6.8]$$

where:

h_e = overall hood entry loss = $h_s + h_h$

SP_f = hood filter loss, "wg [Pa]

F_s = slot loss coefficient

h_s = slot or slot type opening loss = F_sVP_s

F_h = hood entry to duct loss coefficient

h_h = hood entry to duct loss = F_hVP_d

F_a = acceleration (or Bernoulli) coefficient = 1*

VP_s = slot or opening velocity pressure, "wg [Pa]

VP_d = duct velocity pressure, "wg [Pa]

*In a compound hood the acceleration coefficient (F_a) is applied only to the duct or other entry point (slot) with the higher velocity.

The "hood entry loss" (h_h) is the loss from air flowing from the hood into the duct (see Figures 6-40 and 6-9a) and is due to separation of the flow from the sides of the duct due to the momentum of the flow.[6.9,6.10] The more abruptly air must change direction to follow the transition to the duct, the greater the separation and the greater the value of F_h (see Table 6-6). For example, a 45° taper from a round entry to a smaller diameter duct allows a maximum turning angle of 45°, thus has a lower value at F_h = 0.18. A flanged duct opening allows air to flow from all directions up to a 90° angle to the duct. The momentum of the air approaching from 90° pushes air coming from the front towards the front and creates a much larger separation zone than a 45° tapered entry, as is evidenced by the value of F_h for a flanged entry: F_h = 0.5. Note that a flange that is 20% wider than the duct is almost as effective in affecting pressure requirements as one 10 times the duct size. Table 6-6 and Chapter 9, Figure 9-a give hood entry loss coefficients for several typical hood types.

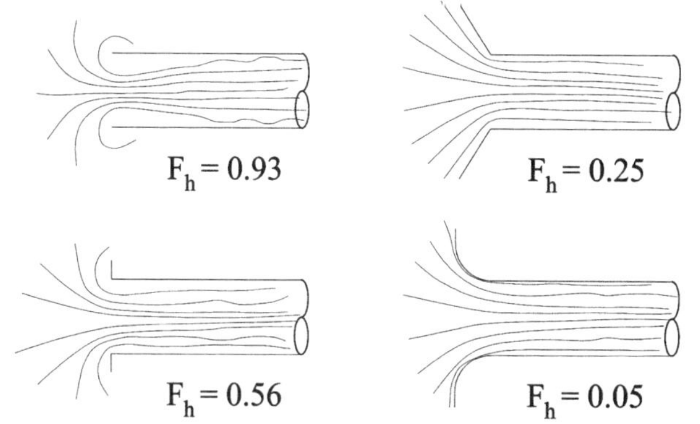

FIGURE 6-40. Separation of flows at the duct inlet and hood loss coefficients

The value of SP_{filter} for a filter in a hood (see Figures 6-41a and 6-41b) will vary from a minimum value when the hood is new or recently cleaned to a maximum value when it should be replaced or cleaned. When computing SP_h for purposes of

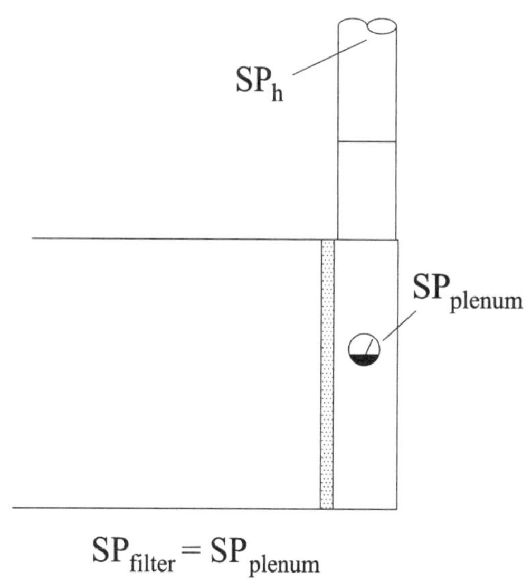

FIGURE 6-41a. Measurement location for SP_{filter} in typical enclosing hood

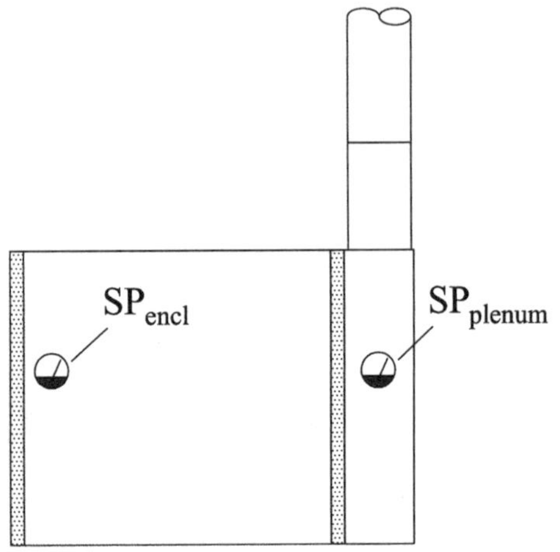

$$SP_{filter_1} = SP_{encl} \qquad SP_{filter_2} = SP_{plenum} - SP_{encl}$$

FIGURE 6-41b. Measurement locations for SP_{filter} with filter at entrance to hood and at the plenum face

sizing ducts when there are many hoods in the system and one or more has a filter, it is advisable to use the middle of the range of values. If it is a one-branch system or a multiple branch system for which all filters will be replaced or cleaned at once, then the maximum value of the range should be used for fan selection.

The "slot loss" (h_s) is the change in total pressure due to the sudden expansion after air passes through a restricted opening at the face of a hood (see Figure 6-42). If there is a continuous transition to the duct without a sudden transition, it does not occur. It is called a slot loss because the restriction typically is an opening with a low aspect ratio (i.e., a slot), but any opening with a significant velocity pressure (e.g., VP > 0.05 in. w.g.) [12 Pa] followed by a sudden expansion probably would behave much the same.

6.16.1 Pressure Loss in Simple Hoods. A simple hood is shown in Example 6-3. If the hood face velocity for a simple hood is less than 1000 fpm [5.08 m/s], loss at the face will be negligible and the loss will be dependent on h_h only. If the hood face velocity is greater than 1000 fpm [5.08 m/s], both the face loss and slot loss (h_s and h_h) should be considered.

EXAMPLE PROBLEM 6-3 (Simple Hood Loss)

Given: Simple hood, taper entry angle = 90°

No hood filter ($SP_f = 0$)

Face Velocity (V_f) = Q/A_f = 250 fpm [1.27 m/s]

Duct Velocity (V_d) = Q/A_d = 3000 fpm [15.24 m/s]

df = 1.0 (see Chapter 3)

$VP_d = df\,(V_d/4005)^2 = (1.0)(3000/4005)^2 = 0.56$ "wg

$[VP_d = (1.0)(15.24/1.29)^2 = 140$ Pa]

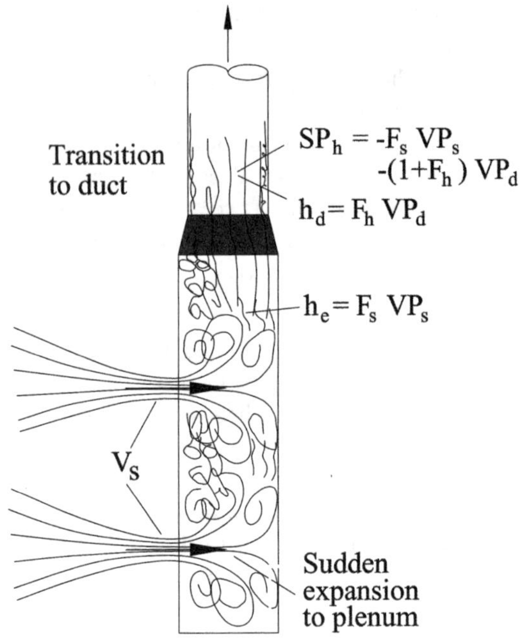

FIGURE 6-42. Compound losses in slot/plenum hood

$F_h = 0.25$ as shown in Figure 6-36

$SP_h = -[h_e + F_a VP_d]$

$h_e = h_h$ (slot or hood filter)

$SP_h = -(h_h + SP_f + F_a VP_d) = -[F_h VP_d + SP_f + F_a VP_d]$

$SP_h = -[(0.25)(0.56) + 0 + (1)(0.56)]$

$= -0.70$ "wg

$[SP_h = -(0.25)(140) + 0 + (1)(140) = -175$ Pa]

Note: Static pressure is negative on the inlet side of the fan and positive on the outlet side in a single fan system.

6.16.2 Pressure Loss in Compound Hoods. Example Problem 6-3 illustrates how air flows through a double entry loss (compound) hood. This is a single slot hood with a plenum and a transition from the plenum to the duct. The purpose of the plenum is to give uniform velocity across the slot opening. Air enters the slot, in this case a sharp edged orifice, and loses energy due to the vena contracta at this point. For this type of hood, losses occur at both the slot and the duct entry. Both h_s and h_h must be considered.

EXAMPLE PROBLEM 6-4 (Compound Hood Loss)

Given: Compound hood taper entry angle = 90°

No hood filter ($SP_f = 0$)

df = 1.0

Slot Velocity (V_s) = 2000 fpm [10.16 m/s]

Duct Velocity (V_d) = 3500 fpm [17.78 m/s] (V_d is greater than V_s; therefore, apply the 'acceleration or Bernoulli' coefficient (F_a) to the duct entry).

VP_s = df$(V_s/4005)^2$ = $(1.0)(2000/4005)^2$ = 0.25 "wg

[VP_s = df $(V_s/1.29)^2$ = $(1.0)(10.16/1.29)^2$ = 62 Pa]

F_s, for slot (sharp edged orifice)
= 1.78 (from Chapter 9, Figure 9a)

VP_d = df $(V_d/4005)^2$ = $(1.0)(3500/4005)^2$
= 0.76 "wg

[VP_d = df $(V_d/1.29)^2$ = $(1.0)(17.78/1.29)^2$ = 190 Pa]

F_h = 0.25 as shown in Chapter 5, Figure 5-15

$SP_h = -[h_e + VP_d] = -[h_s + h_h + SP_f + F_aVP_d]$

$SP_h = -[F_sVP_s + F_hVP_d + SP_f + F_aVP_d]$

$SP_h = -[(1.78)(0.25) + (0.25)(0.76) + 0 + (1)(0.76)]$

SP_h = -1.40 "wg

[SP_h = -(1.78)(62) + (0.25)(190) + (1)(190) = 348 Pa]

6.16.3 Hood Flow Coefficient. The hood flow coefficient is the ratio of the actual airflow through the hood to the theoretical flow if there were no hood losses and can provide the hood airflow of an operating system from the hood static pressure (see Chapter 3). The coefficient is a characteristic of the hood and can be calculated from the hood configuration. If the hood shape is altered or changes are made around the hood (cardboard on face, etc.), then the value of C_e is changed.

$$C_e = \sqrt{\frac{VP_d}{|SP_h|}} \quad [6.9]$$

$$C_e = \sqrt{\frac{VP_d}{(F_s)(VP_s)+(F_h)(VP_d)+(F_a)(VP_d)}} \quad [6.10]$$

The value for the hood efficiency would be identical if metric numbers were inserted in Equation 6.10. Hood flow coefficients are shown for a number of common hood types in Figure 6-43. These calculated values and those calculated as shown above should be considered as estimates. Hood construction variations and actual field conditions may alter the hood design and operating characteristics. Actual C_e values should be determined during system operating conditions by measuring actual conditions and using Equation 6.9.

6.16.4 Hood Flow Calculation. Once C_e is determined hood flow can be calculated from:

$$Q = 4005(C_e)\sqrt{\frac{SP_h}{df}}(A) \quad [6.11]\ IP$$

$$[Q = 1.29(C_e)\sqrt{\frac{SP_h}{df}}(A)] \quad [6.11]\ SI$$

It must be noted that "C_e" flow determination with hoods containing a hood filter is inappropriate as the filter static pressure will continually change with operation.

EXAMPLE PROBLEM 6-5 (Hood Flow Calculation [Use of C_e to calculate Q])

C_e = 0.76 (Calculated during system operation)

SP_h = -1.15 "wg [286 Pa]

6" diameter duct area = 0.1963 ft^2 [0.0182 m^2]

df = 1.0

$$Q = 4005(0.76)\sqrt{\frac{1.15}{1}}(0.1963) = 640 \text{ acfm}$$

$$[Q = (1.29)(0.76)\sqrt{\frac{286}{1}}(0.180) = 0.3 \text{ am}^3\text{/s}]$$

REFERENCES

6.1 Sanders, M.S.; McCormick, E.J.: Human Factors Engineering, 7th Edition. McGraw-Hill Book Company, New York (1993).

6.2 Caplan, K.J.; Knutson, G.W.: ASHRAE Trans.84(I), 511-521 (1978).

6.3 Guffey, S.E.; Barnea, N.: Effects of Face Velocity, Flanges, and Mannikin Position on the Effectiveness of a Benchtop Enclosing Hood in the Absence of Cross-Drafts. Am. Ind. Hyg. Assoc. J. 55(2):132-139 (1994).

6.4 Brandt, A.D.: Industrial Health Engineering. John Wiley and Sons, New York (1947).

6.5 Kane, J.M.: Design of Exhaust Systems. Health and Ventilating 42:68 (November 1946).

6.6 Djamgowz, O.T.; Ghoneim, S.A.A.: Determining Pickup Velocity of Mineral Dusts. Canadian Mining J. (July 1974).

6.7 Silverman, L.: Velocity Characteristics of Narrow Exhaust Slots. J. Ind. Hyg. Toxicology 24:276 (November 1942).

6.8 DallaValle, J.M.: Exhaust Hoods. Industrial Press, New York (1946).

6.9 Brandt, A.; Steffy, R.: Energy Losses at Suction Hoods. Heating, Piping & Air-Conditioning – Am. Soc. Heat. Vent. Eng. J. Section, Sept: 105-119 (1946).

6.10 McLoone, H.E.; Guffey, S.E.; Curran, J.C.: Effects of Shape, Size, and Air Velocity on Entry Loss Factors of Suction Hoods. Am. Ind. Hyg. Assoc. J. 54(3):87-94 (1993).

HOOD TYPE	DESCRIPTION	HOOD ENTRY LOSS (F_h) COEFFICIENT
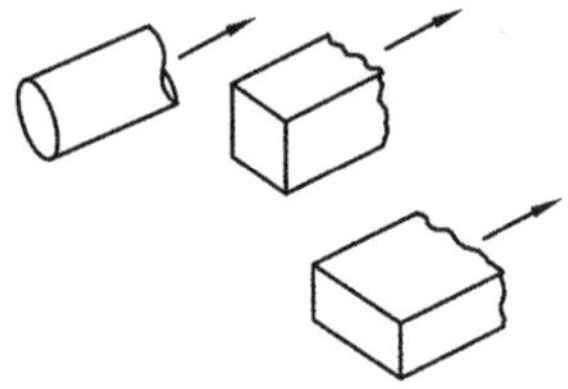	Plain opening	0.93
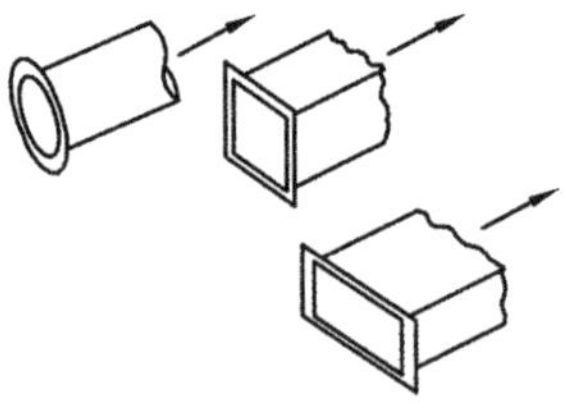	Flanged opening	0.49
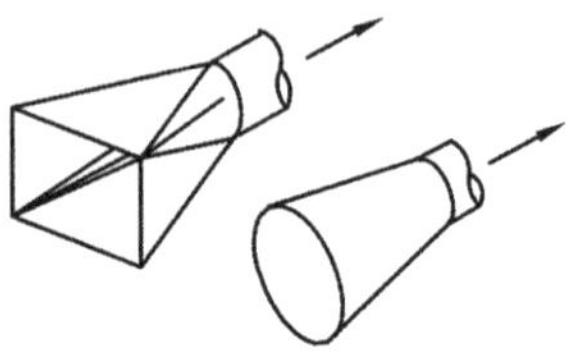	Taper or Cone hood	See Chapter 5
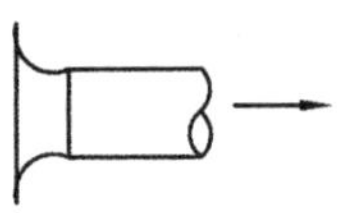	Bell mouth inlet	0.04
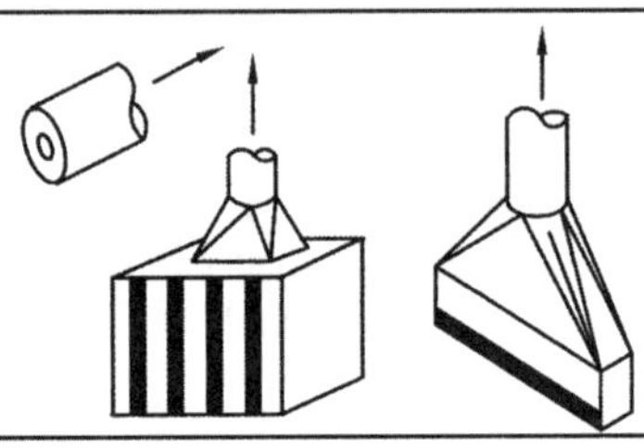	Orifice	See Chapter 10
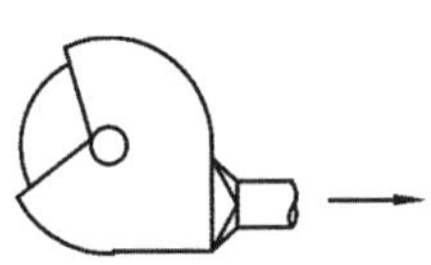	Typical grinding hood	(Straight takeoff) 0.65 (Tapered takeoff) 0.40

FIGURE 6-43

TITLE	FIGURE	DATE
HOOD LOSS COEFFICIENTS	6-43	1-12

CHECK CODES, REGULATIONS, AND LAWS (LOCAL, STATE, AND NATIONAL) TO ENSURE THAT DESIGN IS COMPLIANT.

FIGURE 6-43. Hood loss coefficients

APPENDIX A6 LOCAL EXHAUST HOOD CENTERLINE VELOCITY (IP UNITS)

A6.1 INTRODUCTION

Velocity characteristics of local exhaust hoods have been studied by many individuals during the past 60 years. These empirical studies provide an approximation of the actual situation. Review of a number of these studies is provided in Reference A6.1.

The works of DallaValle and Silverman have been the basis for the centerline velocity equations presented in this Manual. They are simple in format and have provided acceptable approximations for use in hood design and evaluation and are retained as a recommended method of determining exhaust hood centerline velocity. However, additional methods developed by Fletcher(A6.2) and Yousefi(A6.3) have found use by the European community.

A6.2 FLETCHER

The Fletcher centerline velocity equations for freely suspended non-flanged hoods are provided as follows:

$$\frac{V_x}{V_o} = \frac{1}{0.9318.58a^2}$$

$$Q = V_x A\left(0.93 + 8.58a^2\right)$$

$$a = \frac{x}{\sqrt{A}}\left(\frac{W}{L}\right)^{-B}$$

$$B = 0.2\left(\frac{x}{\sqrt{A}}\right)^{-1/3}$$

where: Q = hood flow rate, acfm

x = distance from hood along centerline, ft

V_o = average velocity at hood face, fpm

V_x = average velocity at centerline distance x (fpm)

A = hood face area, ft²

W = hood face width, ft

L = hood face length, ft

The effects of hood flanges on centerline velocities calculated from the Fletcher equations are shown in Figures A6-1, A6-2, and A6-3. The figures show the percent increase in centerline velocity in terms of hood dimensions.(A6.4)

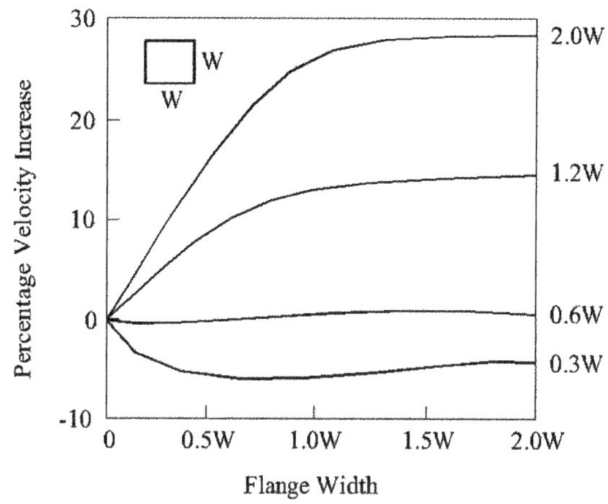

Distance from Hood expressed as Function of Hood width (W)

FIGURE A6-1. Effect of flange width on velocity in front of square hood

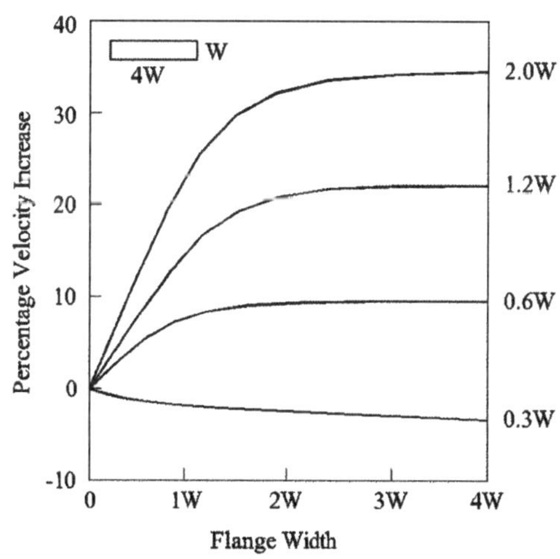

Distance from Hood expressed as Function of Hood width (W)

FIGURE A6-2. Effect of flange width on velocity in front of a 4:1 aspect ratio hood

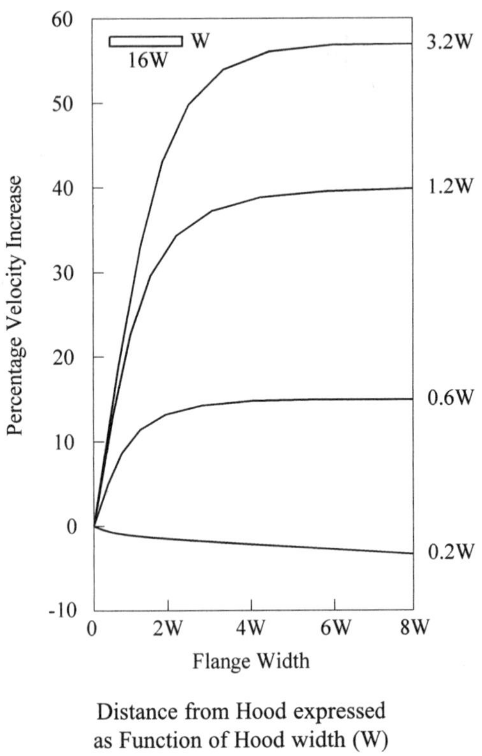

FIGURE A6-3. Effect of flange width on velocity in front of a 16:1 aspect ratio hood

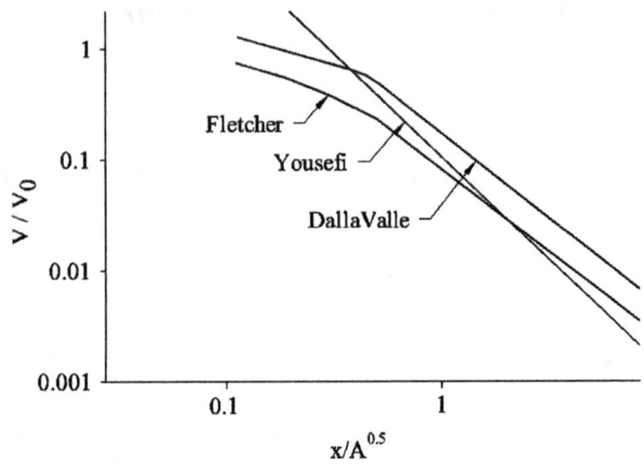

FIGURE A6-4. Effect of area and distance from hood face on velocity

Example A6-1 (*Fletcher*)

x = 1 ft

Vx = 100 fpm

W = 0.5 ft

L = 2 ft

Find Q for an unflanged rectangular hood.

$$\frac{V_x}{V_0} = \frac{1}{0.93 + 8.58a^2}$$

$$B = 0.2\left(\frac{x}{\sqrt{A}}\right)^{-1/3}$$

$$a = \frac{x}{\sqrt{A}}\left(\frac{W}{L}\right)^{-B} = 1.32$$

$$\frac{V_x}{V_0} = 0.063$$

V_o = 1587 fpm

Q = VA = 1587 acfm

A6.3 YOUSEFI

The Yousefi centerline equations for freely suspended non-flanged hoods are provided as follows:

A6.3.1 Rectangular Unflanged Hoods

$$\frac{V_x}{V_o} = \frac{1.8\left(\frac{x}{HR}\right)^{-2.04}}{1 + 0.16\left(\frac{x}{HR}\right)^{-2.04}}$$

A6.3.2 Circular Unflanged Hoods

$$\frac{V_x}{V_o} = \frac{1}{9.78 + 3.497\left(\frac{x^2}{A}\right)^{1.86}}$$

A6.3.3 All Shape Flanged Hoods

$$\frac{V_x}{V_o} = \frac{1.022\left(\frac{x}{HR}\right)^{-1.56}}{1 + 1.825\left(\frac{x}{HR}\right)^{-1.56}}$$

where:
- x = distance from hood along centerline, ft
- V_o = average velocity at hood face, fpm
- A = hood face area, ft²
- HR = hydraulic radius = (WL)/2(W+L)
- L = hood length
- V_x = average velocity at centerline distance at x, fpm

Example A6-2 (*Yousefi*)

$x = 1$ ft

$W = 0.5$ ft

$L = 2.0$ ft

$V_x = 100$ fpm

Find Q for an unflanged rectangular hood.

$HR = (WL)/2(W+L) = 0.2$

$$\frac{V_x}{V_o} = \frac{1.8\left(\frac{x}{HR}\right)^{-2.04}}{1+0.16\left(\frac{x}{HR}\right)^{-2.04}} = 0.0675$$

$V_o = 1481$ fpm

$Q = VA = 1481$ acfm

Example A6-3 (*DallaValle*)

$x = 1$ ft

$W = 0.5$ ft

$L = 2.0$ ft

$V_x = 100$ fpm

$Q = V_x(10X^2 + A) = 1100$ acfm

A comparison of the Fletcher, Yousefi and DallaValle velocity characteristics for the example hood is shown in Figure A6-4.

REFERENCES

A6.1 Branconnier, R.: Bibliographic Review of Velocity Fields in the Vicinity of Local Exhaust Hood Openings. Am. Ind. Hyg. Assoc. J. (49) (April, 1988).

A6.2 Fletcher, B.: Centerline Velocity Characteristics of Rectangular Unflanged Hoods and Slots Under Suction. Ann. Occu. Hyg. Vol. 20, pp. 141-146 (1977).

A6.3 Yousefi, V.; Annegarn, H.J.: Aerodynamic Aspects of Exhaust Ventilation. Ventilation '91, 3rd International Symposium on Ventilation for Contaminant Control; American Conference of Governmental Industrial Hygienists (ACGIH®), Cincinnati, OH (1991).

A6.4 Fletcher, B.: Effect of Flanges on the Velocity in Front of Exhaust Ventilation Hoods. Ann. Occup. Hyg. Vol. 1, pp. 265-269 (1978).

Chapter 7

FANS

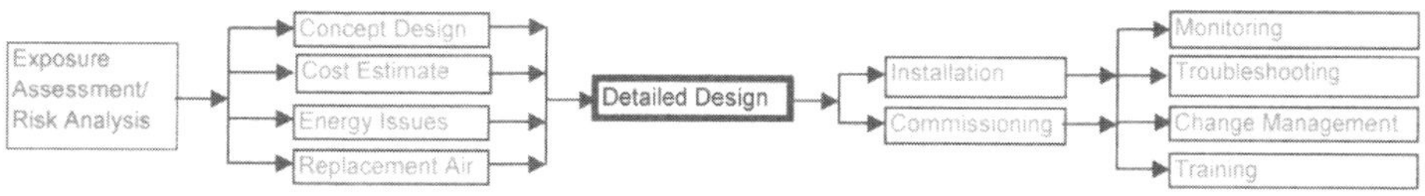

NOTE: Equations with notation followed by (IP) are designated for inch-pound system only; Equations followed by (SI) are designated for metric use only. If Equation bears neither, then it applies to both systems.

7.1 INTRODUCTION

To move air in a ventilation or exhaust system, energy is required to overcome the system losses. This energy can be in the form of natural convection or buoyancy. Most systems, however, require some powered air moving device such as a fan or an ejector.

This chapter will describe the various air moving devices that are used in industrial applications, provide guidelines for the selection of the air moving device for a given situation, and discuss the proper installation of the air moving device in the system to achieve desired performance.

Selection of an air moving device can be a complex task and the specifier is encouraged to take advantage of all available information from applicable trade associations as well as from individual manufacturers.

7.2 BASIC DEFINITIONS

Air moving devices can be divided into two basic classifications: ejectors and fans. Ejectors have low operating efficiencies and are used only for special material handling applications. Fans are the primary air moving devices used in industrial applications.

Fans can be divided into three basic groups: axial, centrifugal, and special types. As a general rule, axial fans are used for higher flow rates at lower resistances and centrifugal fans are used for lower flow rates at higher resistances.

7.2.1 Ejectors **(Figure 7-1).** These are sometimes used when it is not desirable to have contaminated air pass directly through the air moving device. Ejectors are utilized for air streams containing corrosive, flammable, explosive, hot, or sticky materials that might damage a fan, present a dangerous operating situation, or quickly degrade fan performance. Ejectors are also used in pneumatic conveying systems.

7.2.2 Axial Fans. There are three basic types of axial fans: propeller, tubeaxial, and vaneaxial **(Figures 7-2, 7-4a, 7-4b and 7-4c)**.

Propeller Fans are used for moving air against low static pressures and are used commonly for general ventilation. Two types of blades are available: disc blade types when there is no duct present; narrow or propeller blade types for moving air against low resistances (less than 1 "wg). Performance is very sensitive to added resistance and a small increase will cause a marked reduction in flow rate.

Tubeaxial Fans (Duct Fans) contain narrow or propeller type blades in a short, cylindrical housing normally without any type of straightening vanes. Tubeaxial fans will move air against moderate pressures (less than 2 "wg).

Vaneaxial Fans have propeller configurations with a hub and airfoil blades mounted in cylindrical housings that normally incorporate straightening vanes on the discharge side of the impeller. Compared to other axial flow fans, vaneaxial fans are more efficient and generally will develop higher pressures (up to 8 "wg). They are limited usually to clean air applications.

7.2.3 Centrifugal Fans **(Figures 7-2, 7-3, 7-4a, 7-4b and 7-4c).** These fans have three basic impeller designs: forward curved, radial, and backward inclined/backward curved.

Forward curved (commonly called "squirrel cage") impellers have blades that curve toward the direction of rotation. These fans have low space requirements, low tip speeds, and are quiet during operation. They are usually used against low to moderate static pressures such as those encountered in heating and air conditioning work and supply air systems. This type of fan is not recommended for dusts or particulate that could adhere to the short curved blades, cause imbalance, or reduce performance.

Radial Impellers have blades that are straight or radial from the hub. The housings are designed with their inlets and outlets sized to produce material conveying velocities. There are a variety of impeller types available ranging from "high efficiency minimum material" to "heavy impact resistance" designs. The radial blade shape will resist material buildup. This fan design is used for most exhaust system applications when particulate will pass through the fan. These fans usually have medium tip speeds and are used for a variety of exhaust systems that handle either clean or dirty air.

Backward Inclined/Backward Curved impeller blades are inclined opposite to the direction of fan rotation. This type usually has higher tip speeds and provides high fan efficiency and relatively low noise levels with "non-overloading" horsepower characteristics. In a non-overloading fan, the maximum horsepower occurs near the optimum operating point so any variation from that point due to a change in system resistance will result in a reduction in operating horsepower. The blade shape is conducive to material buildup so fan use in this group should be limited as follows:

- *Single Thickness Blade:* Solid blades allow the unit to handle light dust loading or moisture. It should not be used with particulate that would build up on the underside of the blade surfaces.
- *Airfoil Blade:* Airfoil blades offer higher efficiencies and lower noise characteristics. Hollow blades erode more quickly with material and can fill with liquid in high humidity applications. These should be limited to clean air service.

7.2.4 Special Type Fans **(Figures 7-2, 7-4a, 7-4b and 7-4c).** *Tubular Centrifugal fans* have backward inclined blades with special housings that permit a straight line duct installation. Pressure versus flow rate versus horsepower performance curves are similar to a scroll type centrifugal fan of the same blade type. Space requirements are similar to vaneaxial fans.

Power Exhausters, Power Roof Ventilators are packaged units that can be either axial flow or centrifugal type. The centrifugal type does not use a scroll housing but discharges around the periphery of the ventilator to the atmosphere. These

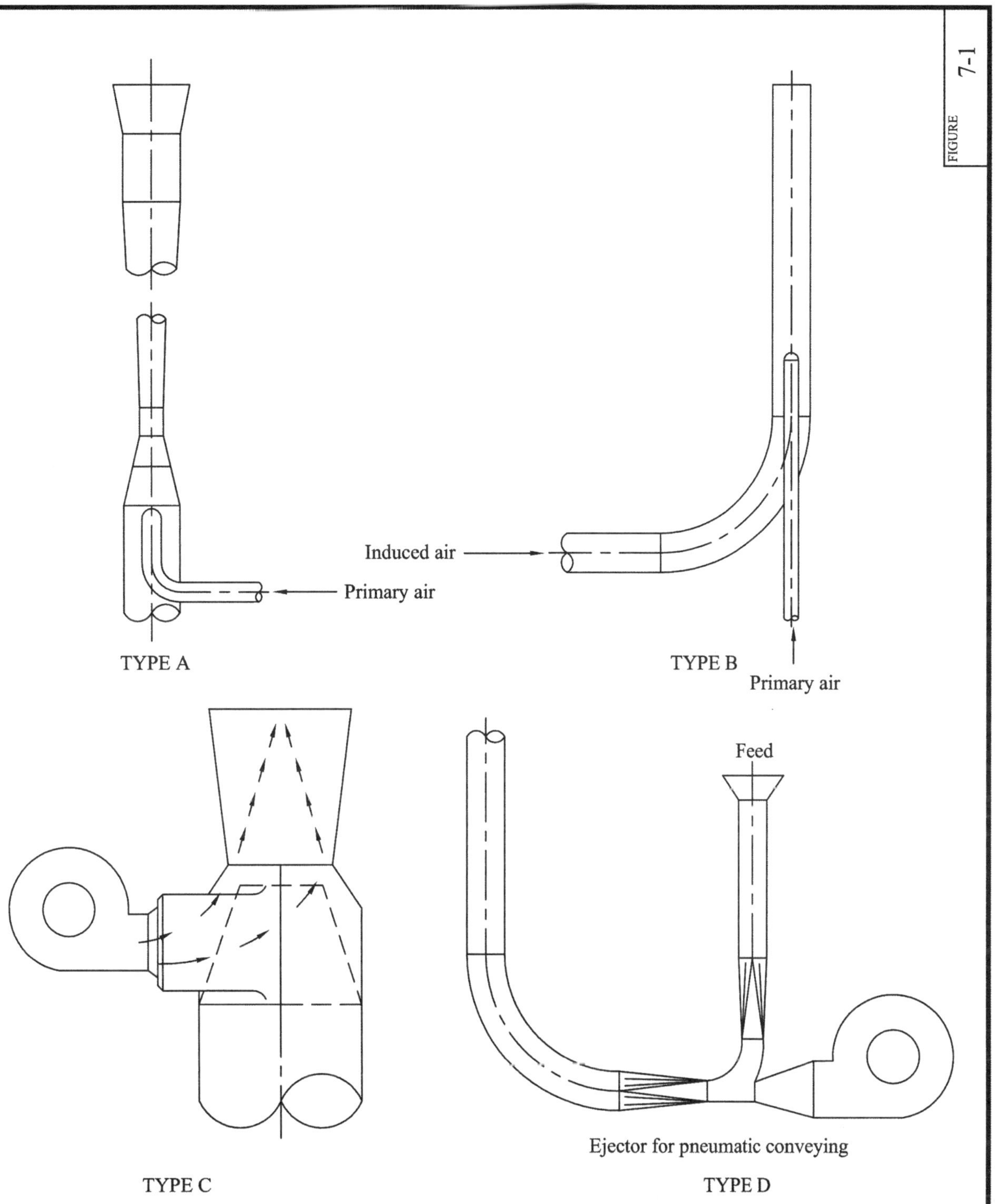

TITLE

AIR EJECTORS

FIGURE 7-1

DATE 1-12

CHECK CODES, REGULATIONS, AND LAWS (LOCAL, STATE, AND NATIONAL) TO ENSURE THAT DESIGN IS COMPLIANT.

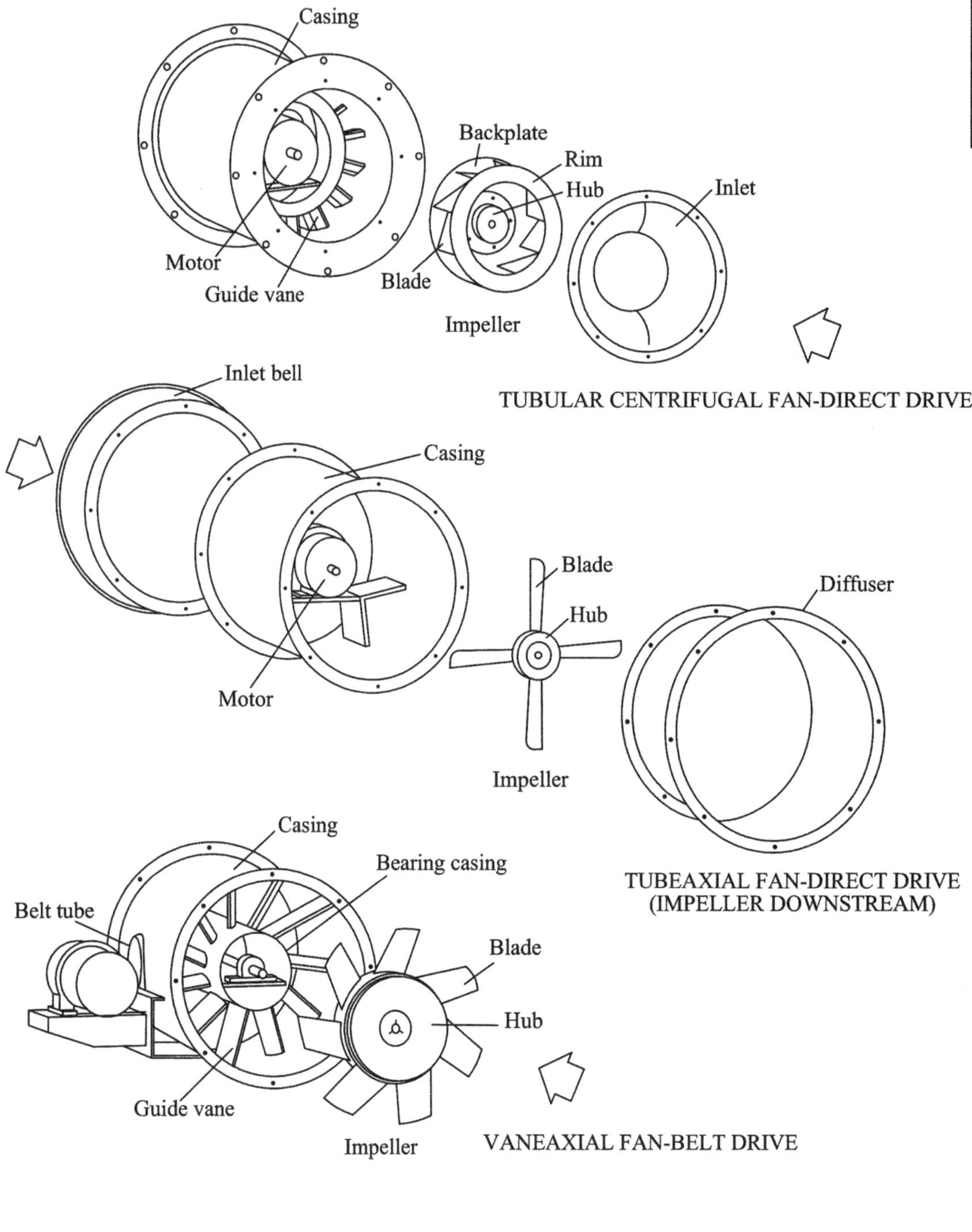

Reprinted from AMCA Publication 201-90, FANS AND SYSTEMS, by permission of the Air Movement and Control Association, Inc.(7.3)

ACGIH®	TITLE: TERMINOLOGY FOR FANS; AXIAL, TUBULAR, AND CENTRIFUGAL	FIGURE 7-2 / DATE 1-12

CHECK CODES, REGULATIONS, AND LAWS (LOCAL, STATE, AND NATIONAL) TO ENSURE THAT DESIGN IS COMPLIANT.

FIGURE 7-3

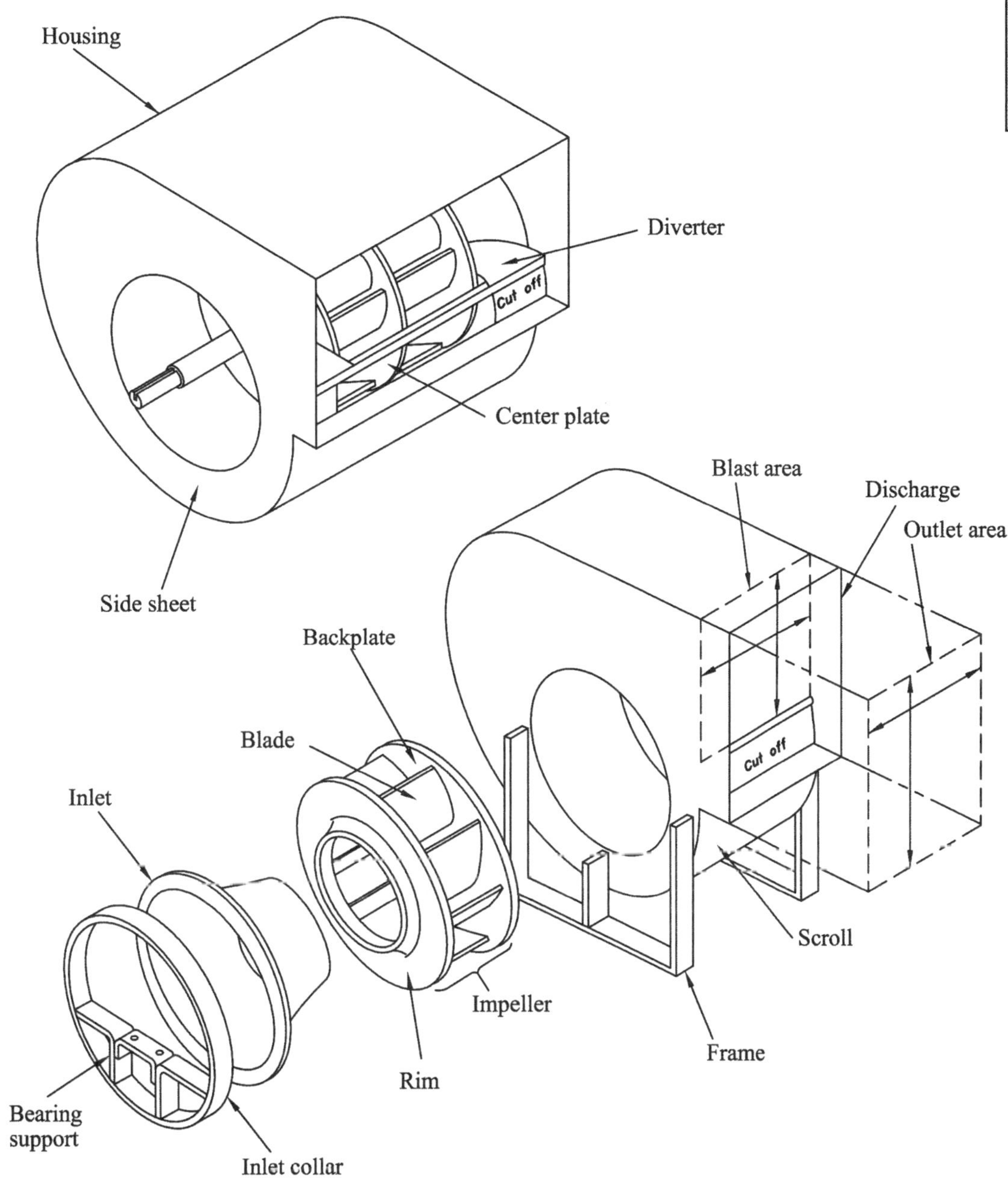

Reprinted from AMCA Publication 201-90, FANS AND SYSTEMS, by permission of the Air Movement and Control Association, Inc.(7.3)

ACGIH®	TITLE: TERMINOLOGY FOR CENTRIFUGAL FAN COMPONENTS	FIGURE: 7-3
		DATE: 1-12

CHECK CODES, REGULATIONS, AND LAWS (LOCAL, STATE, AND NATIONAL) TO ENSURE THAT DESIGN IS COMPLIANT.

units can be obtained with either downward deflecting or upblast discharges.

Fan and Dust Collector Combination: There are several designs in which fans and dust collectors are packaged in a unit. If use of such equipment is contemplated, the manufacturer should be consulted for proper application and performance characteristics.

7.3 FAN SELECTION

Fan selection involves not only finding a fan to match the required flow and pressure considerations but all aspects of an installation including the air stream characteristics, operating temperature, drive arrangement, and mounting. Section 7.2 discussed the various fan types and why they might be selected. This section offers guidelines to fan selection; however, the exact performance and operating limitations of a particular fan should be obtained from the original equipment manufacturer.

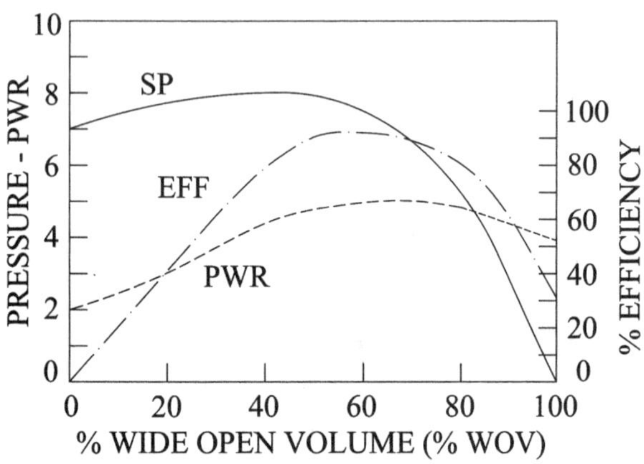

Figures 7-4b and 7-4c provide performance characteristics, curves and application information for a variety of fans. The performance curves show fan pressure, power and efficiency as a function of percent fan "wide open" volumetric flow.

Pressure (SP) represents the fan static pressure vs percent wide open fan volume at a given fan speed.

Power (PWR) represents the continuous power required to achieve the volumetric flow rate and pressure at the given fan speed.

Efficiency (EFF) (see Equation 7.4) represents the static efficiency achieved at the given fan speed.

Wide Open Volume (WOV) represents Fan Volumetric Flow Rate at free air (SP = 0).

% Wide Open Volume and % Efficiency are shown on a 0-100 scale and are read directly in percent. They are indicative for fans of the same type.

Pressure and Power are shown on a 0-10 scale and show the relationship of these values to the Volume Flow Rate of a specific fan type. For specific values, refer to the fan manufacturers' data.

FIGURE 7–4a. Key to Figures 7-4b and 7-4c[7.8]

7.3.1 Considerations for Fan Selection.

CAPACITY

Flow Rate (Q): Based on system requirements and expressed as actual cubic feet per minute (acfm) at the fan inlet.

Pressure Requirements: Based on system pressure requirements that normally are expressed as Fan Static Pressure (FSP) or Fan Total Pressure (FTP) in inches of water gauge at standard conditions (0.075 lbm/ft^3 [1.204 kg/m^3]). If the required pressure is known only at non-standard conditions, a density correction (see Section 7.3.8) must be made.

AIRSTREAM

Material Handled Through the Fan: When the exhaust air contains a small amount of smoke or dust, a backward inclined centrifugal or axial fan should be selected. With light dust, fume or moisture, a backward inclined or radial centrifugal fan would be the preferred selection. If the particulate loading is high, or when material is handled, the normal selection would be a radial centrifugal fan.

Explosive or Flammable Material: Use spark resistant construction (explosion proof motor if the motor is in the airstream). Conform to the standards of the National Board of Fire Underwriters, the National Fire Protection Association and governmental regulations (see Section 7.3.9 and Chapter 8, Section 8.11).

Corrosive Applications: May require a protective coating or special materials of construction (stainless, fiberglass, etc.).

Elevated Airstream Temperatures: Maximum operating temperature affects strength of materials and, therefore, must be known for selection of correct materials of construction, arrangement, and bearing types.

PHYSICAL LIMITATIONS

Fan size should be determined by performance requirements. Inlet size and location, fan weight, and ease of maintenance also must be considered. The most efficient fan size may not fit the physical space available.

DRIVE ARRANGEMENTS

All fans must have some type of power source — usually an electric motor. On packaged fans, the motor is furnished and mounted by the manufacturer. On larger units, the motor is mounted separately and coupled directly to the fan or indirectly by a belt drive. A number of standard drive arrangements are shown in Figures 7-5a, 7-5b and 7-5c.

Direct Drive offers a more compact assembly and assures constant fan speed. Fan speeds are limited to available motor speeds (except in the case of variable frequency controllers).

Capacity is set during construction by variations in impeller geometry and motor speed.

Belt Drive offers flexibility in that fan speed can be changed by altering the drive ratio. This may be important in some applications to provide for changes in system capacity or pressure requirements due to changes in process, hood design, equipment location, or air cleaning equipment. V-belt drives must be maintained and have some power losses which can be estimated from the chart in Figure 7-6. The following equation applies to Figure 7-6:

Motor Power Output, H_{mo} = Fan Power Output H + Drive Losses, H_L

NOISE

Fan noise is generated by turbulence within the fan housing and will vary by fan type, flow rate, pressure, and fan efficiency. Because each design is different, noise ratings must be obtained from the fan manufacturer. Most fans produce a "white" noise, which is a mixture of all frequencies. In addition to white noise, when 60 Hz power is encountered radial blade fans also produce a pure tone at a frequency equal to the blade passage frequency (BPF):

$$BPF = RPM \times N \times CF \qquad [7.1]$$

where: BPF = blade passage frequency, Hz

RPM = rotational rate, rpm

TYPE		IMPELLER DESIGN		HOUSING DESIGN	
CENTRIFUGAL FANS	AIRFOIL		Highest efficiency of all centrifugal fan designs. 9 to 16 blades of airfoil contour curved away from the direction of rotation. Air leaves the impeller at a velocity less than its tip speed and relatively deep blades provide for efficient expansion within the blade passages. For given duty, this will be the highest speed of the centrifugal fan designs.	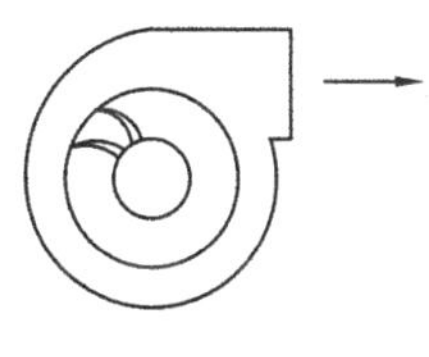	Scroll-type, usually designed to permit efficient conversion of velocity pressure to static pressure, thus permitting a high static efficiency; essential that clearance and alignment between wheel and inlet bell be very close in order to reach the maximum efficiency capability. Concentric housings can also be used as in power roof ventilators, since there is efficient pressure conversion in the wheel.
	BACKWARD-INCLINED BACKWARD-CURVED		Efficiency is only slightly less than that of airfoil fans. Backward-inclined or backward-curved blades are single thickness. 9 to 16 blades curved or inclined away from the direction of rotation. Efficient for the same reasons given for the airfoil fan above.	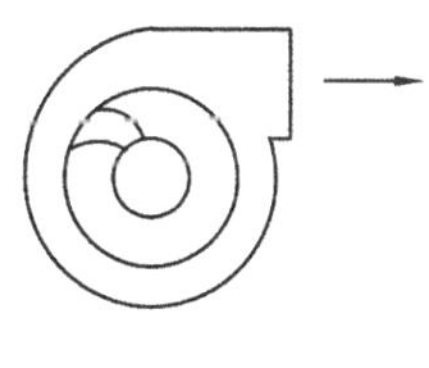	Utilizes the same housing configuration as the airfoil design.
	RADIAL	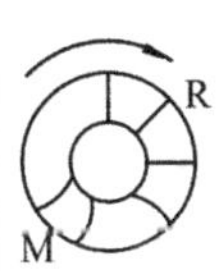	Simplest of all centrifugal fans and least efficient. Has high mechanical strength and the wheel is easily repaired. For a given point of rating, this fan requires medium speed. This classification includes radial blades (R) and modified radial blades (M), usually 6 to 10 in number.	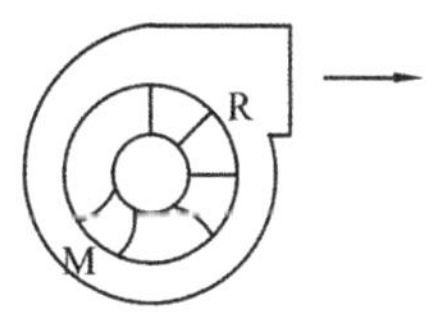	Scroll-type, usually the narrowest design of all centrifugal fan designs described here because of required high velocity discharge. Dimensional requirements of this housing are more critical than for airfoil and backward-inclined blades.
	FORWARD-CURVED		Efficiency is less than airfoil and backward-curved bladed fans. Usually fabricated of lightweight and low cost construction. Has 24 to 64 shallow blades with both the heel and tip curved forward. Air leaves wheel at velocity greater than wheel. Tip speed and primary energy transferred to the air is by use of high velocity in the wheel. For given duty, wheel is smallest of all centrifugal types and operates at lowest speed.	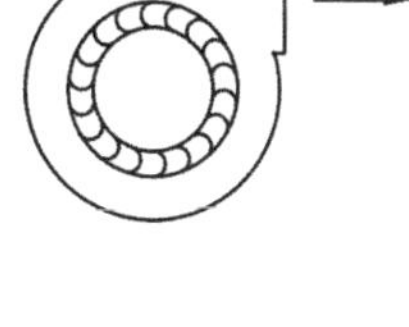	Scroll is similar to other centrifugal-fan designs. The fit between the wheel and inlet is not as critical as on airfoil and backward-inclined bladed fans. Uses large cut-off sheet in housing.

FIGURE 7–4b. Centrifugal fans: impeller and housing designs

N = number of blades

CF = conversion coefficient, 1/60

This tone can be very noticeable in some installations and should be considered in the system design.

Because of its higher efficiency, the backward inclined type of impeller design is generally the quietest. However, for all fan types, non-uniform airflow at the fan inlet or outlet can increase the fan noise level. This is another problem related to "system effect" (see Section 7.5.2).

Most fan manufacturers publish sound ratings for their products. There are a variety of ways to present the ratings. One popular way is to list sound power levels for eight ANSI standard octave bands. The sound power levels are typically in units called "decibels" (dB). The sound **power** level is a characteristic of a fan that varies with the fan speed and point of operation.

For an installed fan, the surrounding environment affects the sound level that is measured or heard. Walls, floors, and other equipment reflect and absorb sound to varying degrees. The sound that reaches the listener will be different than the fan's rated sound power level. Typical sound measuring devices detect sound with a microphone and display sound pressure level in decibels. This sound **pressure** is an environ-

PERFORMANCE CURVES	PERFORMANCE CHARACTERISTICS*	APPLICATIONS
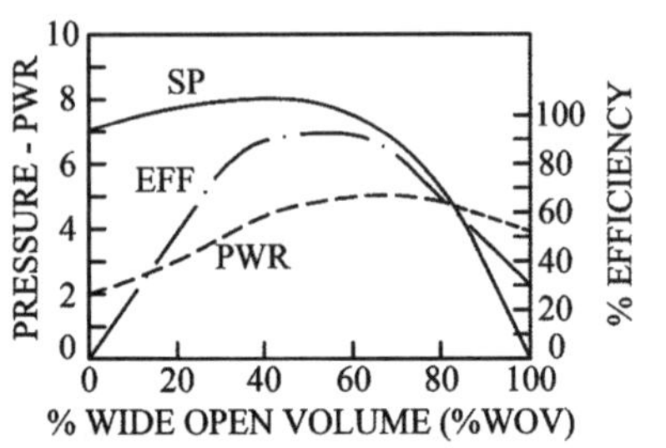	Highest efficiencies occur 50 to 60% of wide-open volume. This is also the area of good pressure characteristics; the horsepower curve reaches a maximum near the peak efficiency area and becomes lower toward free delivery, a self-limiting power characteristic as shown.	General heating, ventilating and air-conditioning systems. Used in large sizes for clean air industrial applications where power savings are significant.
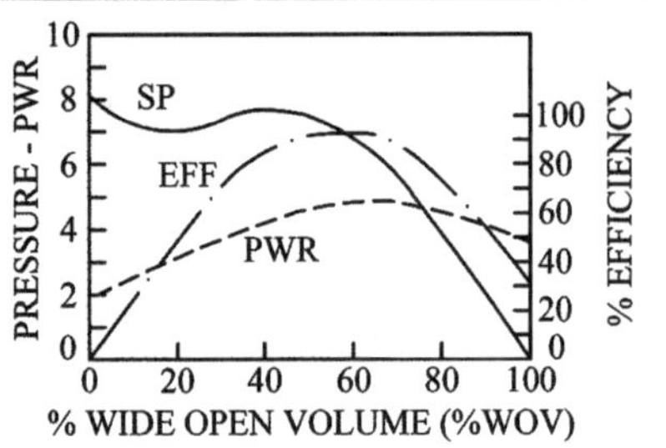	Operating characteristics of this fan are similar to the airfoil fan mentioned above. Peak efficiency for this fan is slightly lower than the airfoil fan. Normally unstable left of peak pressure.	Same heating, ventilating, and air-conditioning applications as the airfoil fan. Also used in some industrial applications where the airfoil blade is not acceptable because of corrosive and/or erosion environment.
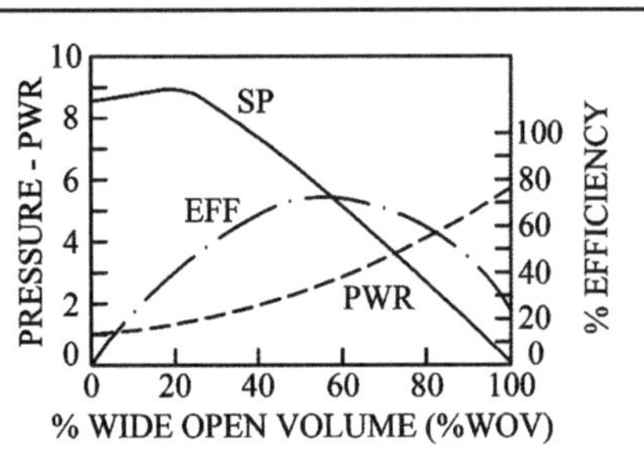	Higher pressure characteristics than the above mentioned fans. Power rises continually to free delivery.	Used primarily for material handling applications in industrial plants. Wheel can be of rugged construction and is simple to repair in the field. Wheel is sometimes coated with special material. This design also used for high-pressure industrial requirements. Not commonly found in HVAC applications.
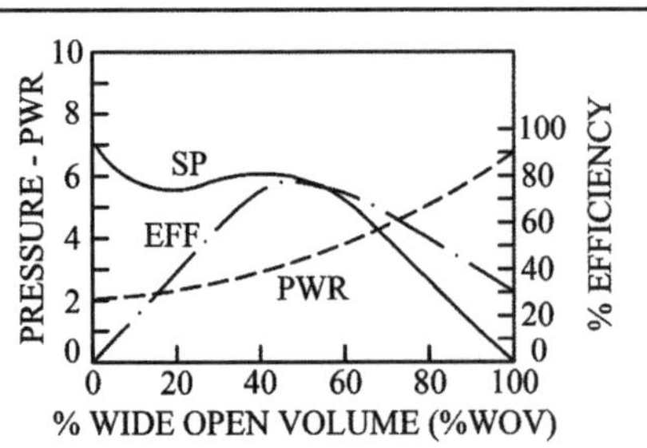	Pressure curve is less steep than that of backward-curved bladed fans. There is a dip in the pressure curve left of the peak pressure point and highest efficiency occurs to the right of peak pressure, 40 to 50% of wide-open volume. Fan should be rated to the right of peak pressure. Power curve rises continually toward free delivery and this must be taken into account when motor is selected.	Used primarily in low-pressure heating, ventilating and air-conditioning applications such as domestic furnaces central station units, and packaged air-conditioning equipment from room air-conditioning units to roof top units.

FIGURE 7–4b (Cont.). Centrifugal fans: impeller and housing designs. (*These performance curves reflect the general characteristics of various fans as commonly employed. They are not intended to provide complete selection criteria for application purpose, since other parameters, such as diameter and speed, are not defined. See Figure 7-4a.)

TYPE			IMPELLER DESIGN	HOUSING DESIGN
AXIAL FANS	PROPELLER		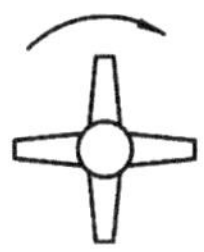 Efficiency is low. Impellers are usually of inexpensive construction and limited to low pressure applications. Impeller is of 2 or more blades, usually of single thickness attached to relatively small hub. Energy transfer is primarily in form of velocity pressure.	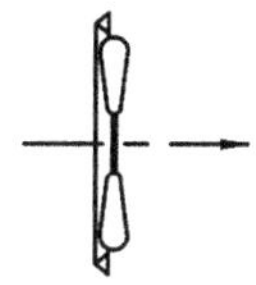 Simple circular ring, orifice plate, or venturi design. Design can substantially influence performance and optimum design is reasonably close to the blade tips and forms a smooth inlet flow contour to the wheel.
AXIAL FANS	TUBEAXIAL		 Somewhat more efficient than propeller fan design and is capable of developing a more useful static pressure range. Number of blades usually from 4 to 8 and hub is usually less than 50% of fan tip diameter. Blades can be of airfoil or single thickness cross section.	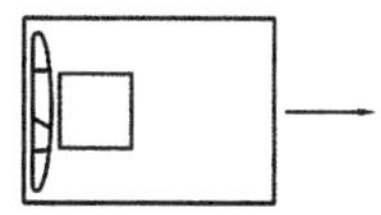 Cylindrical tube formed so that the running clearance between the wheel tip and tube is close. This results in significant improvement over propeller fans.
AXIAL FANS	VANEAXIAL		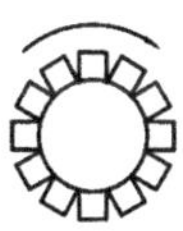 Good design of blades permits medium- to high-pressure capability at good efficiency. The most efficient fans of this type have airfoil blades. Blades are fixed or adjustable pitch types and hub is usually greater than 50% of fan tip diameter.	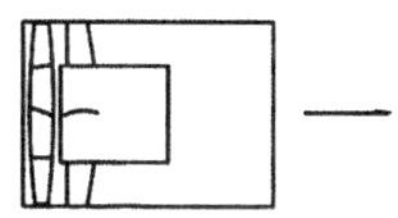 Cylindrical tube closely fitted to the outer diameter of blade tips and fitted with a set of guide vanes. Upstream or downstream from the impeller, guide vanes convert the rotary energy imparted to the air and increase pressure and efficiency of fan.
SPECIAL DESIGNS	TUBULAR	CENTRIFUGAL	 This fan usually has a wheel similar to the airfoil backward-inclined or backward-curved blade as described above. (However, this fan wheel type is of lower efficiency when used in fan of this design.) Mixed flow impellers are sometimes used.	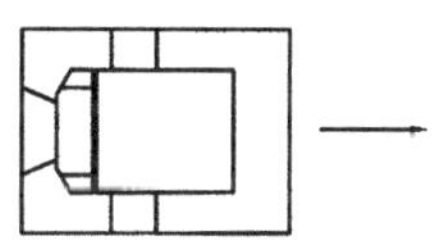 Cylindrical shell similar to vaneaxial fan housing, except the outer diameter of the wheel does not run close to the housing. Air is discharged radially from the wheel and must change direction by 90 degrees to flow through the guide vane section.
SPECIAL DESIGNS	POWER ROOF VENTILATORS	CENTRIFUGAL	 Many models use airfoil or backward-inclined impeller designs. These have been modified from those mentioned above to produce a low-pressure, high-volume flow rate characteristic. In addition, many special centrifugal impeller designs are used, including mixed-flow design.	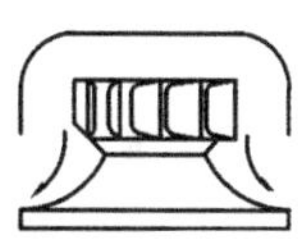 Does not utilize a housing in a normal sense since the air is discharged from the impeller in a 360 degree pattern and usually does not include a configuration to recover the velocity pressure component.
SPECIAL DESIGNS	POWER ROOF VENTILATORS	AXIAL	 A great variety of propeller designs are employed with the objective of high-volume flow rate at low pressure.	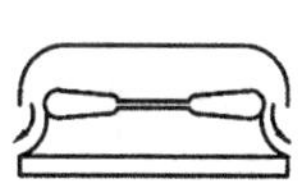 Essentially a propeller fan mounted in a supporting structure with a cover for weather protection and safety considerations. The air is discharged through the annular space around the bottom of the weather hood.

FIGURE 7–4c. Axial and special types of fan designs: impeller and housing designs

PERFORMANCE CURVES	PERFORMANCE CHARACTERISTICS*	APPLICATIONS
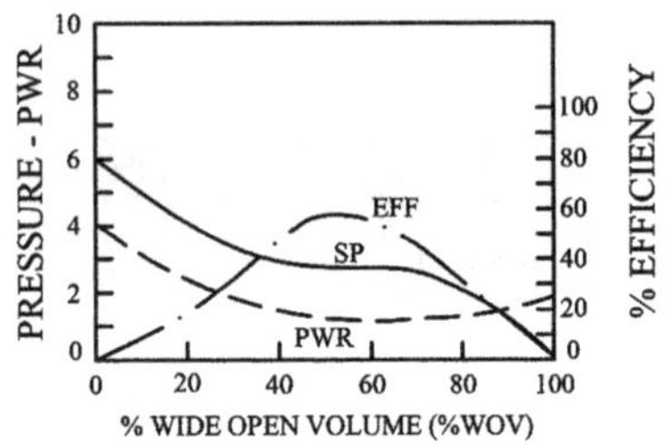	High flow rate but very low-pressure capabilities and maximum efficiency is reached near free delivery. The discharge pattern of the air is circular in shape and the air stream swirls because of the action of the blades and the lack of straightening facilities.	For low-pressure, high-volume air moving applications such as air circulation within a space or ventilation through a wall without attached duct work. Used for replacement air applications.
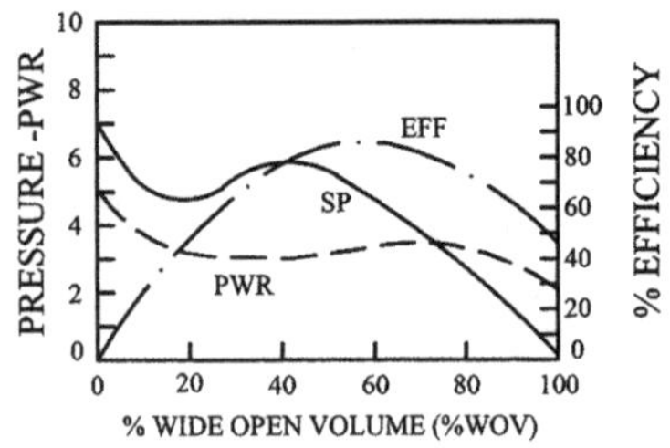	High flow-rate characteristics with medium-pressure capabilities. Performance curve includes a dip to the left of peak pressure, which should be avoided. The discharge air pattern is circular and is rotating or whirling because of the propeller rotation and lack of guide vanes.	Low- and medium-pressure ducted heading, ventilating, and air-conditioning applications where air distribution on the downstream side is not critical. Also used in some industrial applications such as drying ovens, paint spray booths, and fume exhaust systems.
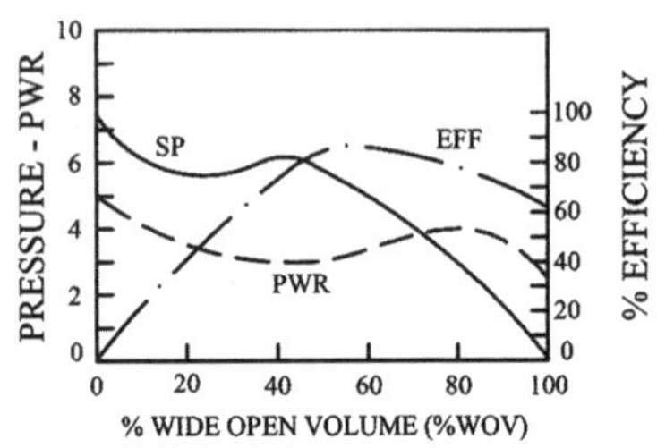	High-pressure characteristics with medium volume flow rate capabilities. Performance curve includes a dip caused by aerodynamic stall to the left of peak pressure, which should be avoided. Guide vanes correct the circular motion imparted to the air by the wheel and improve pressure characteristics and efficiency of the fan.	General heating, ventilating, and air-conditioning systems in low-, medium-, and high-pressure applications is of advantage where straight-through flow and compact installation are required; air distribution on downstream side is good. Also used in industrial application similar to the tubeaxial fan. Relatively more compact than comparable centrifugal-type fans for same duty.
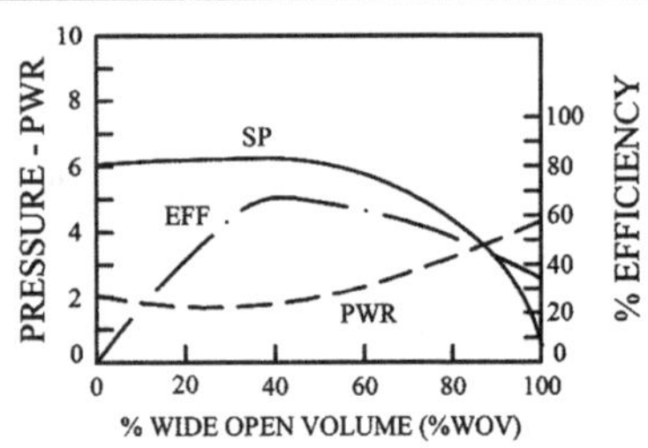	Performance is similar to backward-curved fan, except lower capacity and pressure because of the 90 degree change in direction of the airflow in the housing. The efficiency will be lower than the backward curved fan. Some designs may have a dip in the curve similar to the axial-flow fan.	Used primarily for low-pressure return air systems in heating, ventilating, and air-conditioning applications. Has straight through flow configuration.
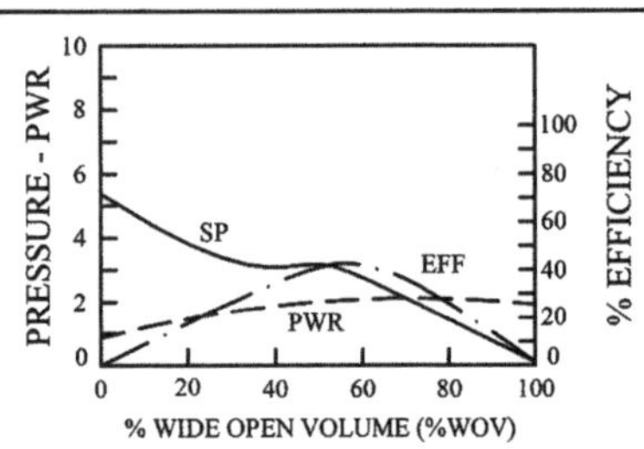	Usually intended to operate without attached duct and therefore to operate against a very low-pressure head. It is usually intended to have a rather high-volume flow rate characteristic. Only static pressure and static efficiency are shown for this type of product.	For low-pressure exhaust systems such as general factory, kitchen, warehouse, and commercial installations where the low-pressure rise limitation can be tolerated. Unit is low in first cost and low in operating cost and provides positive exhaust ventilation in the space, which is a decided advantage over gravity-type exhaust units. The centrifugal unit is somewhat quieter than the axial unit described below.
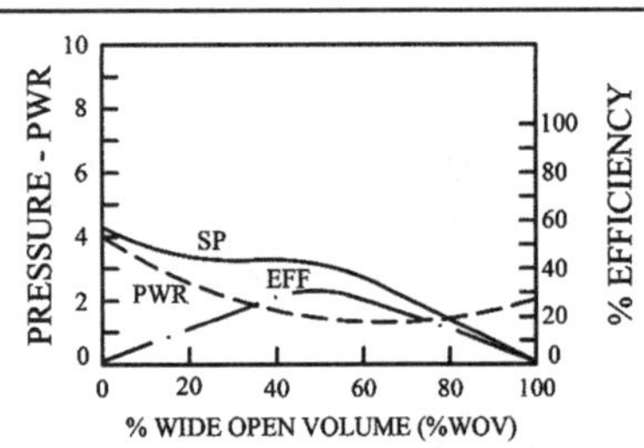	Usually intended to operate without attached duct and therefore to operate against very low-pressure head. It is usually intended to have a high-volume flow rate characteristic. Only static pressure and static efficiency are shown for this type of product.	For low-pressure exhaust systems such as general factory, kitchen, warehouse, and some commercial installations where the low-pressure rise limitations can be tolerated. Unit is low in first cost and low in operating cost and provides positive exhaust ventilation in the space, which is a decided advantage over gravity-type exhaust units.

FIGURE 7–4c (Cont.). Axial and special types of fan designs: impeller and housing designs (*These performance curves reflect the general characteristics of various fans as commonly employed. They are not intended to provide complete selection criteria for application purpose, since other parameters, such as diameter and speed, are not defined. See Figure 7-4a.)

FIGURE 7-5a

SW -Single Width DW -Double Width
SI -Single Inlet DI -Double Inlet

Arrangements 1,3,7 and 8 are also available with bearings mounted on pedestals or base set independent of the fan housing.

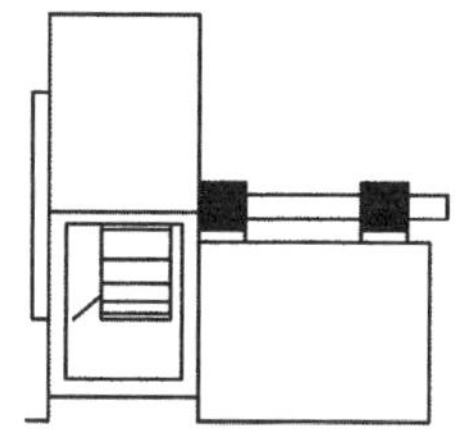

ARR. 1 SWSI For belt drive or direct connection. Impeller overhung. Two bearings on base.

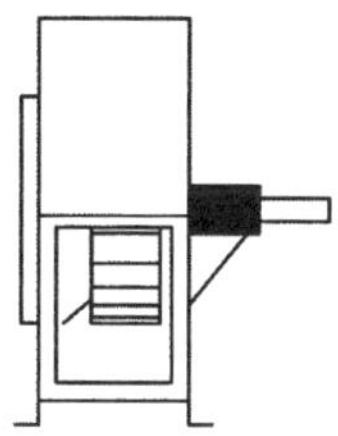

ARR. 2 SWSI For belt drive or direct connection. Impeller overhung. Bearing in bracket supported by fan housing.

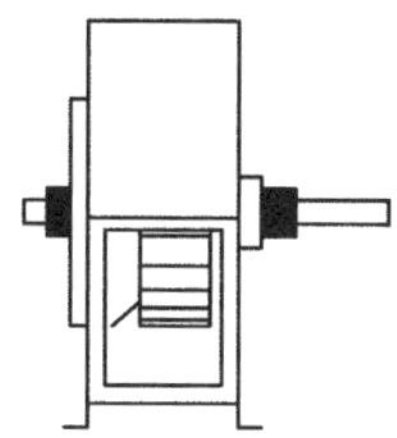

ARR. 3 SWSI For belt drive or direct connection. One bearing on each side and supported by fan housing.

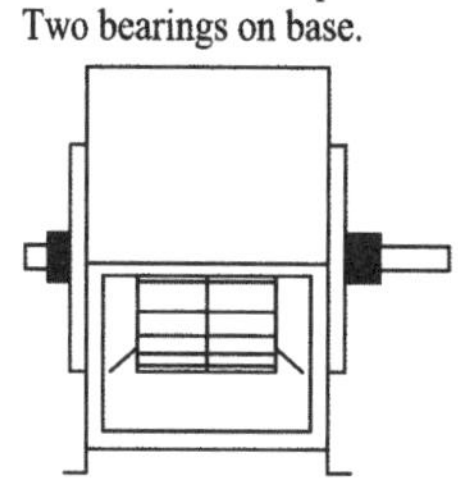

ARR. 3 DWDI For belt drive or direct connection. One bearing on each side and supported by fan housing.

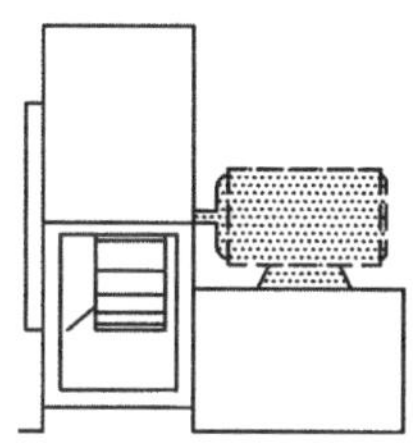

ARR. 4 SWSI Impeller overhung on prime mover shaft. No bearing on fan. Prime mover base mounted or integrally directly connected.

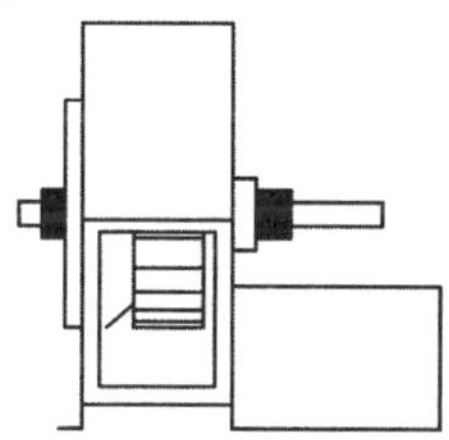

ARR. 7 SWSI For belt drive or direct connection. One bearing on each side and supported by fan housing.

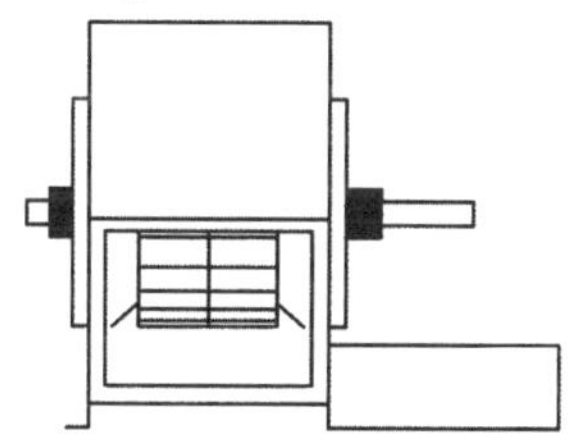

ARR. 7 DWDI For belt drive or direct connection. Arrangement 3 plus base for prime mover.

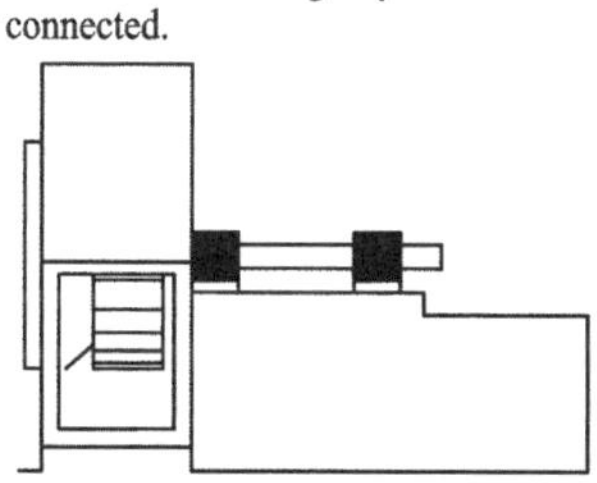

ARR. 8 SWSI For belt drive or direct connection. Arrangement 1 plus extended base for prime mover.

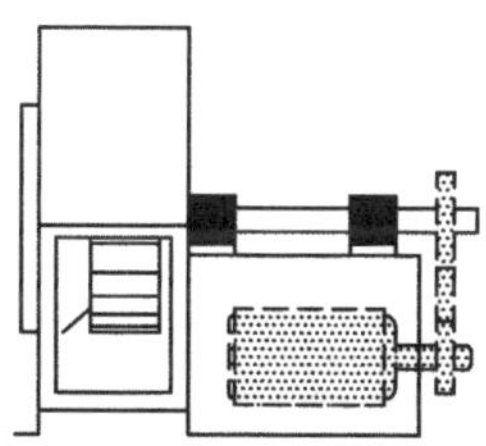

ARR. 9 SWSI For belt drive. Impeller overhung, two bearings, with prime mover outside base.

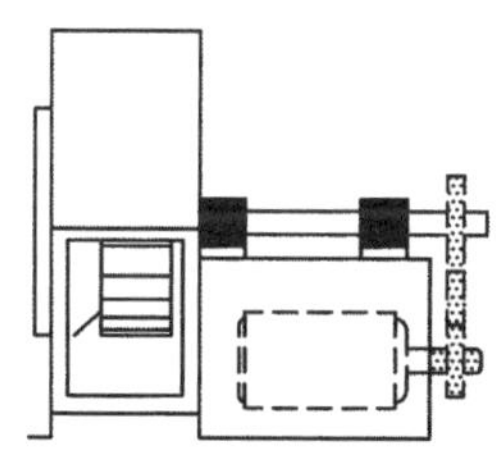

ARR. 10 SWSI For belt drive. Impeller overhung, two bearings, with prime mover inside base.

Reprinted from AMCA Publication 99-86, STANDARDS HANDBOOK, by permission of the Air Movement and Control Association Inc.[7.2]

ACGIH®	TITLE DRIVE ARRANGEMENTS FOR CENTRIFUGAL FANS	FIGURE 7-5a DATE 1-12

CHECK CODES, REGULATIONS, AND LAWS (LOCAL, STATE, AND NATIONAL) TO ENSURE THAT DESIGN IS COMPLIANT.

FIGURE 7-5b

SW -Single Width DW -Double Width
SI -Single Inlet DI -Double Inlet

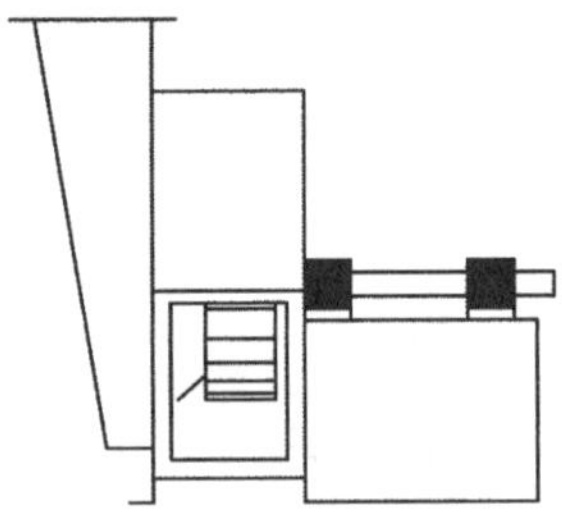

ARR. 1 SWSI WITH INLET BOX For belt drive or direct connection. Impeller overhung, two bearings on base. Inlet box may be self-supporting.

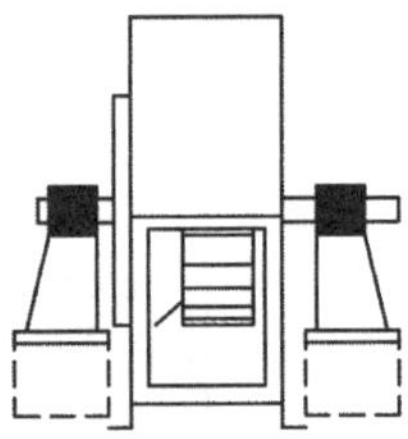

ARR. 3 SWSI WITH INDEPENDENT PEDESTAL For belt drive or direct connection fan. Housing is self-supporting. One bearing on each side supported by independent pedestals.

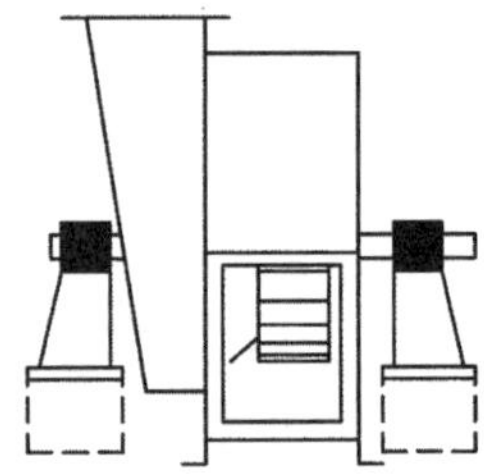

ARR. 3 SWSI WITH INLET BOX AND INDEPENDENT PEDESTALS For belt drive or direct connection fan. Housing is self-supporting. One bearing on each side supported by independent pedestals with shaft extending through inlet box.

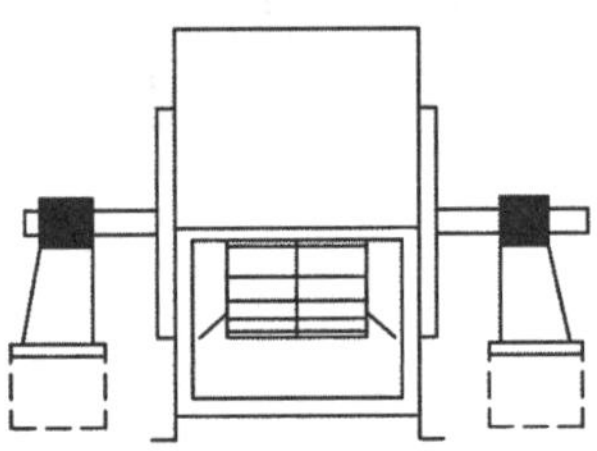

ARR. 3 DWDI WITH INDEPENDENT PEDESTAL For belt drive or direct connection fan. Housing is self-supporting. One bearing on each side supported by independent pedestals.

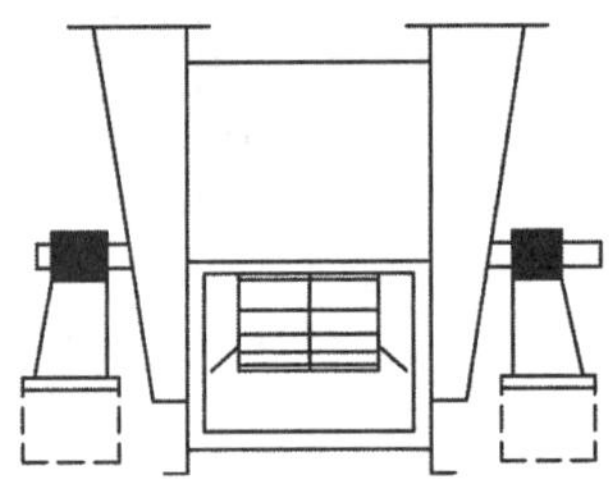

ARR. 3 DWDI WITH INLET BOX AND INDEPENDENT PEDESTALS For belt drive or direct connection fan. Housing is self-supporting. One bearing on each side supported by independent pedestals with shaft extending through inlet box.

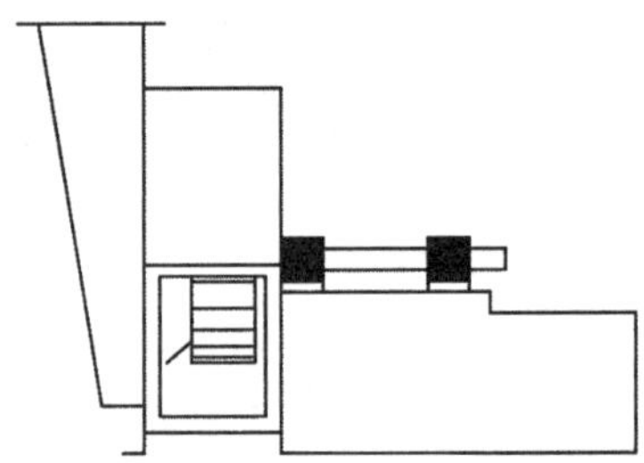

ARR. 8 SWSI WITH INLET BOX For belt drive or direct connection. Impeller overhung, two bearings on base plus extended base for prime mover. Inlet box may be self-supporting.

Reprinted from AMCA Publication 99-86, STANDARDS HANDBOOK, by permission of the Air Movement and Control Association Inc.[7.2]

ACGIH®

TITLE DRIVE ARRANGEMENTS FOR CENTRIFUGAL FANS

FIGURE 7-5b

DATE 1-12

CHECK CODES, REGULATIONS, AND LAWS (LOCAL, STATE, AND NATIONAL) TO ENSURE THAT DESIGN IS COMPLIANT.

FIGURE 7-5c

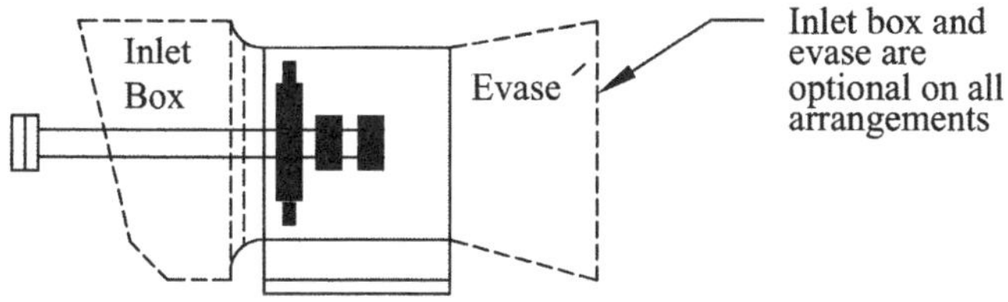

ARR. 1

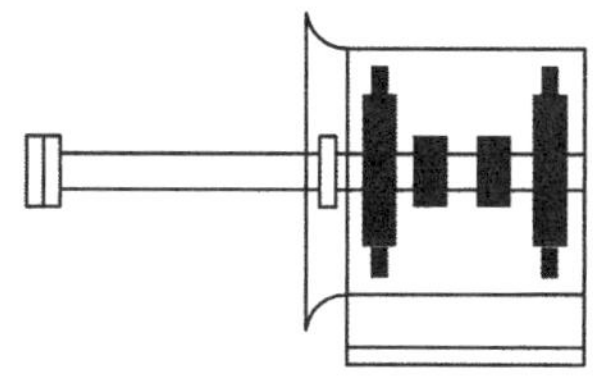

ARR. 1 TWO STAGE

For belt drive or direct connection. Impeller overhung. Two bearings located either upstream or downstream of impeller.

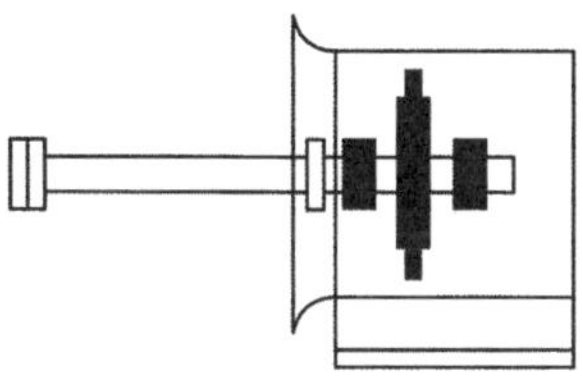

ARR. 3

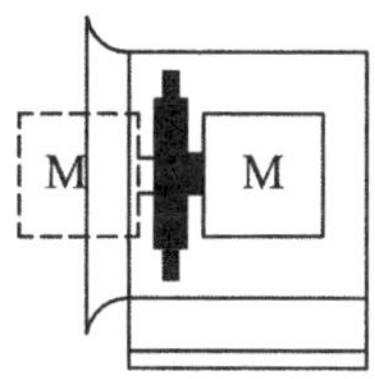

ARR. 4

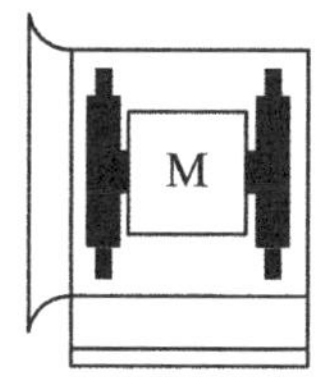

ARR. 4 TWO STAGE

For belt drive or direct connection. Impeller between bearings that are on internal supports. Drive through inlet.

For direct connection. Impeller overhung on motor shaft. No bearings on fan. Motor on internal supports.

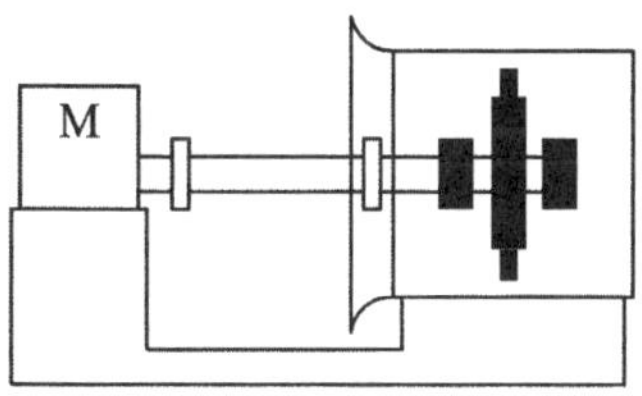

ARR. 7

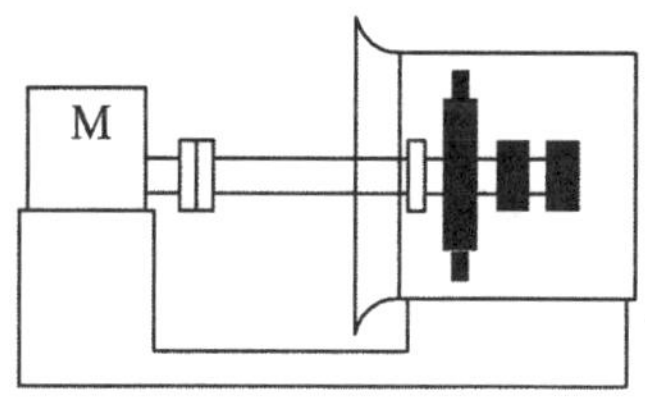

ARR. 8 (1 or 2 stage)

For belt drive or direct connection. Arr. 3 plus common base for prime mover.

For belt drive or direct connection. Arr. 1 plus common base for prime mover.

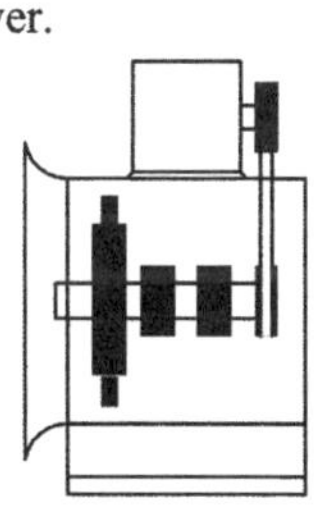

ARR. 9 Motor on Casing

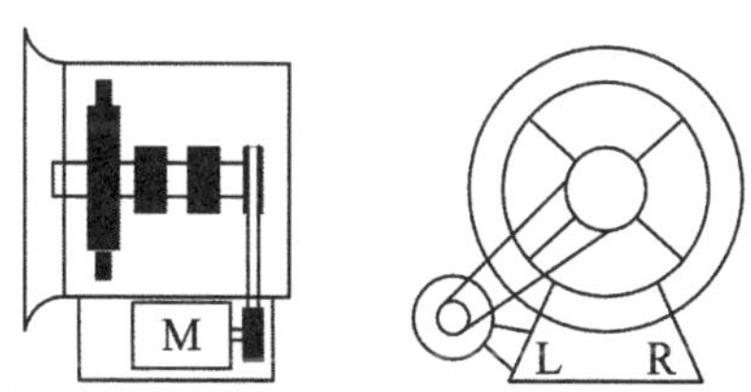

ARR. 9 Motor on Integral Base

For belt drive. Impeller overhung. Two bearings on internal supports. Motor on casing or on integral base. Drive through belt fairing.

NOTE: All fan orientations may be horizontal or vertical.

Reprinted from AMCA Publication 99-86, STANDARDS HANDBOOK, by permission of the Air Movement and Control Association Inc.(7.2)

ACGIH®	TITLE DRIVE ARRANGEMENTS FOR AXIAL FANS WITH OR WITH-OUT EVASE' AND INLET BOX	FIGURE 7-5c DATE 1-12

CHECK CODES, REGULATIONS, AND LAWS (LOCAL, STATE, AND NATIONAL) TO ENSURE THAT DESIGN IS COMPLIANT.

FIGURE 7-6 (IP)

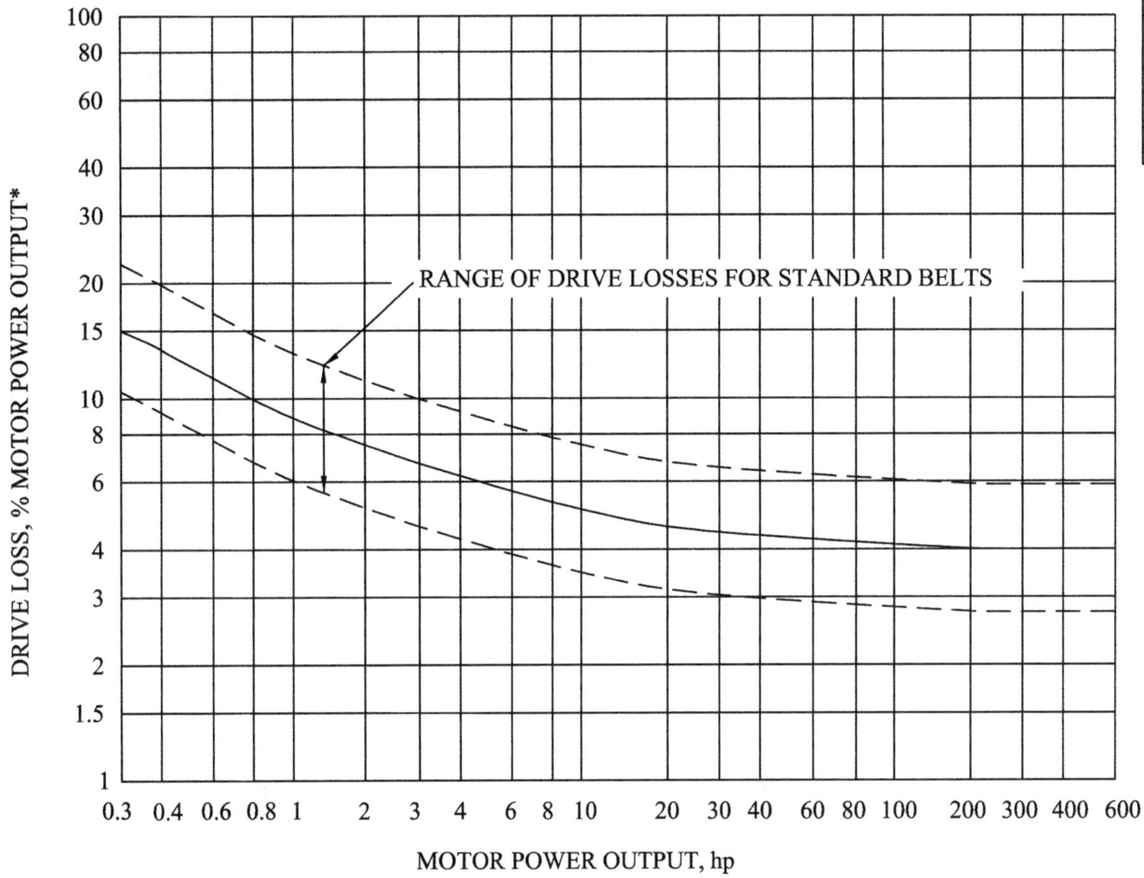

* Higher belt speeds usually have higher losses than lower belt speeds at the same horsepower

Drive losses are based on the conventional V-belt which has been the "work horse" of the drive industry for several decades.

EXAMPLE

- Motor power output, H_{MO}, is determined to be 13.3 hp
- The belts are the standard type and just warm to the touch immediately after shutdown
- From chart, drive loss = 5.1%
- Drive loss H_L = 0.051×13.3 = 0.7 hp
- Fan power input, H = $13.3 - 0.7$ = 12.6 hp

Reprinted from AMCA Publication 203-90, FIELD PERFORMANCE MEASUREMENT OF FAN SYSTEMS, by permission of the Air Movement and Control Association, Inc.(7.6)

ACGIH®	TITLE: ESTIMATED BELT DRIVE LOSS	FIGURE: 7-6 (IP)
		DATE: 1-12

CHECK CODES, REGULATIONS, AND LAWS (LOCAL, STATE, AND NATIONAL) TO ENSURE THAT DESIGN IS COMPLIANT.

FIGURE 7-6 (SI)

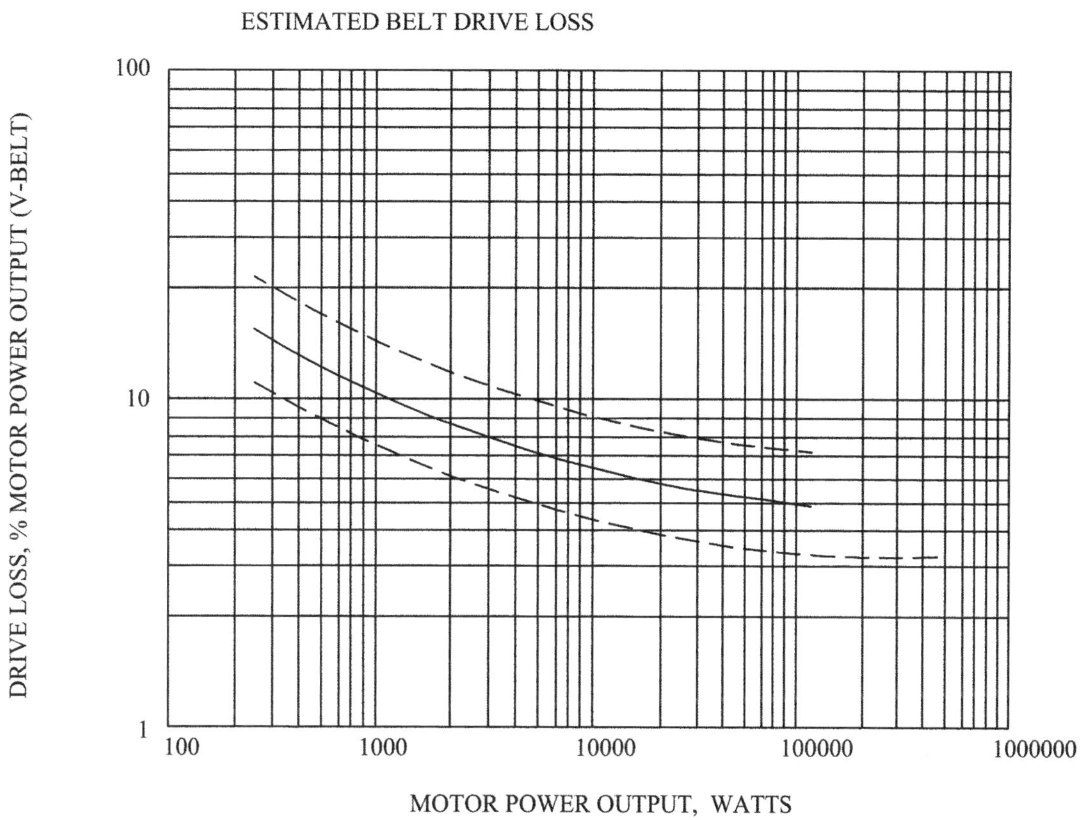

Higher belt speeds usually have higher losses than lower belt speeds at the same power

- Drive losses are based on the conventional V-belt which has been the "work horse" of the drive industry for several decades.

EXAMPLE

- Motor power output, H_{MO}, is determined to be 9.9 kw
- The belts are the standard type and just warm to the touch immediately after shutdown
- From chart, drive loss = 5.1%
- Drive loss H_L = 0.051×9.9 = 0.50 kw
- Fan power input, H = $9.9 - 0.5$ = 9.4 kw

Adopted for metric from AMCA Publication 203-90, FIELD PERFORMANCE MEASUREMENT OF FAN SYSTEMS, by permission of the Air Movement and Control Association, Inc.[7.6]

ACGIH®	TITLE ESTIMATED BELT DRIVE LOSS	FIGURE 7-6 (SI) DATE 1-12

CHECK CODES, REGULATIONS, AND LAWS (LOCAL, STATE, AND NATIONAL) TO ENSURE THAT DESIGN IS COMPLIANT.

ment-dependent measurement that changes with listener location and/or environment changes.

While the decibel unit is used for sound power and sound pressure, the two measures are not interchangeable. Seventy dB sound power is not seventy dB sound pressure. The decibel is not an absolute unit of measure. It is a ratio between a measured quantity and an agreed reference level. Both dB scales are logarithmic. The sound **power** is the log of the ratio of two power levels. The sound **pressure** is the log of the ratio of two pressure levels. The sound power scale uses a reference of 10^{-12} watts. The sound pressure scale uses a reference of 20×10^{-6} N/M^2.

For an installed fan, the sound pressure levels are usually measured in dB using the "A" weighting scale. Measurements obtained using the A-weighting scale provide a better estimation of the threat to human hearing than do other weighting scales. As a result, most criteria for a worker's exposure to noise are expressed in "A" weighted sound pressure levels. A sound level meter set on the "A" scale automatically integrates the noise of all frequencies to give a single dBA noise measurement. Expanded detail can be obtained by taking noise measurements with a meter capable of measuring the sound pressure level in each octave band. Such detail can help indicate the predominant source of a noise.

The topic of sound is quite broad and there are many reference texts available to cover it. For a concise introduction, the *ASHRAE Fundamentals Handbook*[7.1] is a good starting point.

SAFETY AND ACCESSORIES

Safety Guards are required. Consider all danger points such as inlet, outlet, shaft, drive, and cleanout doors. Construction should comply with applicable governmental safety requirements and attachment must be secure.

Accessories can help in the installation and in future maintenance requirements. Examples might include drains, cleanout doors, split housings, and shaft seals.

FLOW CONTROL

There are various accessories that can be used to change fan performance. Such changes may be required on systems that vary throughout the day or for reduction in flow rate in anticipation of some future requirement. Dampers, variable pitch blades, and speed control are three common accessories used with fans.

Variable pitch blades are available with some axial type fans. The fan impellers are designed to allow manual or automatic changes to the blade pitch. "Adjustable" impellers have a blade pitch that can be manually changed when the fan is not running. "Variable" impellers include devices to allow the blade pitch to be changed pneumatically or hydraulically while the fan is operating.

Dampers are installed directly on the fan inlet or outlet. Because they are in the air stream, dampers can build up with material and may not be acceptable on material handling fans. Two types of dampers are available:

Outlet Dampers mount on the fan outlet to add resistance to the system when partially closed. These are available with both parallel and opposed blades. Selection depends on the degree of control required (opposed blade dampers will control the flow more evenly throughout the entire range from wide open to closed).

Variable Inlet Vane and Inlet Box Dampers mount on the fan inlet or inlet box to pre-spin air into the impeller. This simultaneously reduces fan output and lowers operating horsepower. In fact, because these devices pre-spin the air in the direction of fan rotation, extra benefits are realized. Variable Inlet Vane (VIV) dampers actually change the shape of the fan curve, allowing the intersection of the fan pressure curve and allows the fan to continue to develop much of the static pressure ability. The effect is to allow the fan to operate at as much as a 5:1 turndown without causing the fan to surge due to operation within the "area of unpredictable performance" to the left of the fan static pressure peak at reduced air volumes (Figure 7-9a). This is especially important in backwardly-inclined fans, where surging can take place at as much as 50% of the fan's full volume. Inlet box dampers operate in a similar fashion to inlet vane dampers as both pre-spin the air in the direction of the fan rotation. Because of the power savings, inlet dampers should be considered when the fan will operate for long periods at reduced capacities.

A Variable Frequency Drive (VFD) may also be used to control flow. A VFD will control the fan speed, rather than varying the fan inlet flow conditions or the outlet area to change the fan's point of operation. This type of control varies both the flow rate and the fan static pressure. The fan static pressure curve shape will vary in a homologous fashion, that is, they will have a similar shape, but be larger or smaller as the fan speed ramps up or down. Because the shape is similar, but smaller, reduction of the fan rpm will mean that both flow and static pressure will see simultaneous reduction and static pressures available at a higher rpm may not be attainable. In a case where both high static pressures at lower air volumes and maximum energy savings at low volumes are possible, both variable inlet vane dampers and VFD controls may be necessary (Figure 7-9b). In recent years, VFD controls have become increasingly more affordable. Consequently, they can be used to make dynamic industrial ventilation controls that accommodate changes in fluctuating system resistance, i.e., filter resistance, without dramatically affecting the cost of the system. A significant monetary return is realized when filters are clean and lower in resistance. When used with control pressure sensors (transducers) that give continuous electrical output, one can use either hood static pressure, static pressure upstream of the filters (i.e., baghouse) or velocity pressure to give dependable continuous volume control for an otherwise dynamic system.

The VFD control unit is connected in-line between the electric power source and the fan motor. It is used to vary the voltage and frequency of the power input to the motor. The motor speed will vary linearly with the line frequency. Most VFD applications use a direct drive arrangement; however, belt drives are occasionally used.

For a typical system with fixed physical characteristics, the attainable points of operation will fall on the system curve. For example, Figure 7-11 shows points A1 and A2 on a system curve. These two points of operation can be attained with a VFD by adjusting it for speeds of RPM_1 or RPM_2. This will result in fan curve PQ_1 or PQ_2, respectively.

Variable Frequency Drives do have disadvantages. They may have a low speed limitation. Most AC motors are designed to operate at their nameplate speeds. If a VFD is used to run a motor well below its nominal speed, the motor's efficiency will be reduced, and losses will increase. This can increase motor heating and may cause damage.

The VFD can cause harmonic distortion in the electrical input lines from the power source. This may affect other electrical equipment on the same power system. Such distortion can be reduced with the addition of isolation transformers or line inductors.

To properly apply a VFD, the equipment supplier needs to know about its intended usage, about the building's power supply, and about other electrical equipment in use. In general, for applications where the minimum system airflow is 80% or more of the maximum system airflow, the VFD's losses and higher initial cost may make use of the inlet damper a better choice for flow control.

An advantage of the VFD or the Variable Pitch Blade over the dampers is often a dramatic power and noise reduction. However, these accessories usually require additional controlling equipment. An advantage of dampers is their relatively simple installation and use and their lower initial costs.

7.3.2 Rating Tables. Fan size, operating RPM and Power are usually obtained from a rating table based on required airflow and pressure. Tables are based on Fan Total Pressure or Fan Static Pressure:

$$FTP = (SP_{outlet} + VP_{outlet}) - (SP_{inlet} - VP_{inlet}) \quad [7.2]$$

$$FSP = SP_{outlet} - SP_{inlet} - VP_{inlet} \quad [7.3]$$

Fan Rating Tables are based on requirements for air at standard conditions (0.075 lbm/ft^3 [1.204 kg/m^3]). If other than standard conditions exist, the actual pressure must be converted to standard conditions. See Section 7.3.8, "Selection at Air Densities Other Than Standard."

The most common form of fan rating table is a "multi-rating table" (Table 7-1), which shows a range of capacities for a particular fan size. For a given pressure, the highest mechanical efficiency will usually be in the middle third of the "ACFM" column. Some manufacturers show the rating of maximum efficiency for each pressure by underscoring or similar indicator. In the absence of such a guide, the design engineer must calculate the efficiency from the total efficiency equation

$$\eta = \frac{Q \times FTP}{CF \times PWR} = \frac{Q \times (FSP + VP_{out})}{CF \times PWR} \quad [7.4]$$

where: η = Fan efficiency

Q = Volumetric flow rate, acfm [am^3/s]

FTP = Fan total pressure, "wg [Pa]

FSP = Fan Static Pressure, "wg [Pa]

PWR = Power requirement, hp [watts]

CF = Conversion Coefficient, 6362 [1]

Even with a multi-rating table, it is usually necessary to interpolate in order to select fan RPM and BHP for the exact conditions desired. In many cases, a double interpolation will be necessary. Straight line interpolations throughout the multi-rating table will introduce negligible errors.

Certain types of fans may be offered in various Air Movement and Control Association (AMCA)[7.2] performance classes identified as I through IV. A fan designated as meeting the requirements of a particular class must be physically capable of operating at any point within the performance limits for that class. Performance limits for each class are established in terms of outlet velocity and static pressure. Multi-rating tables will usually be shaded to indicate the selection zones for various classes or will state the maximum operating RPM. This can be useful in selecting equipment, but class definition is only based on performance and will not indicate quality of construction.

In fact, many high pressure fans utilize lightweight aluminum with riveted construction because of weight/strength considerations.

Capacity tables that attempt to show the ratings for a whole series of homologous fans on one sheet cannot be used accurately unless the desired rating happens to be listed on the chart. Interpolation is practically impossible since usually only one point of the fan curve for a given speed is defined in such a table.

Today, most fan manufacturers have "electronic catalogs" available. These catalogs are computer programs that can be used to calculate the correct fan speed and horsepower based on input data such as desired flow rate and fan static pressure or fan total pressure. Some electronic catalogs include estimates of the affects of various fan accessories such as dampers and inlet boxes.

7.3.3 Point of Operation. Fans are usually selected for operation at some fixed condition or single "Point of Operation." Both the fan and the system have variable performance characteristics that can be represented graphically as curves depicting an array of operating points. The actual

TABLE 7-1 (IP). Example of Multi-Rating Table

FAN DIMENSIONS: Inlet diameter: 13" O.D. Outlet area: .930 sq. ft. inside																	Wheel diameter: 22.63" Wheel circumference: 5.92 ft.						
ACFM	OV	2" SP		4" SP		6" SP		8" SP		10" SP		12" SP		14" SP		16" SP		18" SP		20" SP		22" SP	
		RPM	BHP	RPM	BHP	RPM	BHP	RPM	BHP	RPM	BHP	RPM	BHP	RPM	BHP	RPM	BHP	RPM	BHP	RPM	BHP	RPM	BHP
930	1000	843	0.57	1176	1.21	1434	1.93	1653	2.75	1846	3.64	2021	4.59	2184	5.62	2333	6.68	2475	7.81	2610	9.01	2738	10.2
1116	1200	853	0.67	1183	1.35	1439	2.12	1656	2.98	1848	3.90	2022	4.89	2182	5.95	2333	7.07	2473	8.23	2606	9.45	2733	10.7
1302	1400	866	0.77	1191	1.51	1445	2.33	1660	3.22	1852	4.20	2025	5.23	2183	6.31	2333	7.47	2474	8.68	2606	9.95	2731	11.2
1488	1600	882	0.89	1201	1.69	1453	2.56	1668	3.50	1857	4.51	2030	5.59	2188	6.72	2337	7.92	2474	9.13	2606	10.4	2734	11.8
1674	1800	899	1.01	1213	1.88	1463	2.81	1676	3.81	1863	4.86	2035	5.98	2194	7.16	2340	8.38	2479	9.67	2610	11.0	2735	12.4
1860	2000	917	1.14	1227	2.09	1474	3.09	1685	4.13	1872	5.24	2040	6.39	2199	7.62	2344	8.89	2484	10.2	2613	11.6	2735	13.0
2046	2200	937	1.29	1242	2.32	1484	3.37	1694	4.48	1879	5.63	2048	6.84	2206	8.13	2351	9.43	2487	10.8	2618	12.2	2741	13.6
2232	2400	961	1.45	1257	2.56	1497	3.68	1704	4.85	1889	6.07	2056	7.33	2212	8.64	2357	10.0	2493	11.4	2622	12.8	2745	14.3
2418	2600	984	1.62	1275	2.81	1513	4.02	1717	5.25	1900	6.53	2065	7.84	2222	9.22	2364	10.6	2501	12.1	2631	13.6	2750	15.1
2790	3000	1038	2.02	1313	3.36	1543	4.73	1744	6.11	1924	7.52	2088	8.96	2241	10.4	2383	12.0	2517	13.5	2644	15.1	2766	16.7
3162	3400	1099	2.50	1358	3.99	1580	5.52	1775	7.05	1952	8.60	2115	10.2	2265	11.8	2405	13.4	2538	15.1	2665	16.8	2783	18.5
3534	3800	1164	3.07	1407	4.69	1620	6.37	1812	8.09	1984	9.79	2144	11.5	2290	13.3	2428	15.0	2562	16.8	2684	18.6	2803	20.5
3906	4200	1232	3.75	1462	5.48	1665	7.31	1851	9.19	2018	11.0	2174	12.9	2320	14.8	2458	16.8	2587	18.7	2708	20.6	2825	22.5
4278	4600	1306	4.56	1520	6.39	1717	8.38	1894	10.4	2058	12.4	2209	14.5	2355	16.5	2489	18.6	2614	20.6	2736	22.7	2852	24.8
4650	5000	1380	5.49	1582	7.41	1770	9.53	1941	11.7	2100	13.9	2247	16.1	2390	18.3	2521	20.5	2645	22.7	2766	25.0	2883	27.3
5022	5400	1457	6.56	1647	8.57	1827	10.8	1990	13.1	2146	15.5	2291	17.8	2428	20.2	2558	22.6	2681	25.0	2798	27.3		
5394	5800	1535	7.79	1719	9.93	1885	12.2	2045	14.7	2194	17.2	2334	19.7	2469	22.2	2594	24.7	2717	27.3	2830	29.8		

TABLE 7-1 (SI). Example of Multi-Rating Table

FAN DIMENSIONS Inlet Diameter: 330 mm Outlet Area; 0.086 m²																Wheel diameter: 575 mm Wheel Circumference: 1.8 m					
		100 Pa		1000 Pa		1500 Pa		2000 Pa		2500 Pa		3000 Pa		4000 Pa		4500 Pa		5000 Pa		5500 Pa	
am³/s	OV	RPM	KW	RPM	KW	RPM	KW	RPM	KW	RPM	KW	RPM	KW	RPM	KW	RPM	KW	RPM	KW	RPM	KW
0.439	5.08	843	0.43	1176	0.90	1434	1.44	1653	2.10	1653	2.70	2021	3.40	2333	5.00	2475	5.82	2610	6.72	2738	7.61
0.527	6.1	853	0.50	1183	1.04	1430	1.58	1658	2.20	1658	2.90	2022	3.60	2333	5.30	2473	6.10	2606	7.02	2733	7.98
0.614	7.11	866	0.57	1191	1.13	1445	1.74	1660	2.40	1680	3.10	2025	3.90	2333	6.60	2474	6.50	2606	7.40	2731	8.40
0.702	8.13	882	0.66	1201	1.26	1453	1.91	1668	'2.60	1688	3.40	2030	4.20	2337	5.90	2474	6.80	2606	7.80	2734	8.80
0.79	9.14	899	0.75	1213	1.40	1463	2.10	1678	2.80	1678	3.60	2035	4.50	2340	6.20	2479	7.20	2610	8.20	2735	9.20
0.878	10.2	917	0.85	1227	1.56	1474	2.50	1685	3.10	1685	3.90	2040	4.80	2344	6.60	2484	7.60	2613	8.70	2735	9.70
0.968	11.2	937	0.96	1242	1.73	1484	2.51	1694	3.30	1694	4.20	2048	5.10	2351	7.00	2487	8.10	2618	9.10	2741	10.14
1.06	12.2	961	1.06	1257	1.91	1497	2.74	1707	3.60	1704	4.50	2058	5.50	2357	7.50	2493	8.50	2622	9.50	2745	10.66
1.14	13.2	984	1.21	1275	2.10	1513	3.00	1717	3.90	1717	4.90	2085	5.80	2364	7.90	2501	9.02	2631	10.14	2750	11.26
1.32	15.2	1038	1.51	1313	2.61	1543	3.53	1744	4.60	1744	5.80	2088	6.70	2383	8.90	2517	10.07	2644	11.26	2766	12.45
1.49	17.3	1099	1.86	1358	2.89	1580	4.72	1775	5.30	1775	6.40	2115	7.60	2405	9.99	2538	11.26	2665	12.53	2783	13.80
1.67	19.3	1164	2.29	1407	3.50	1620	4.75	1812	6.00	1812	7.30	2144	8.60	2428	11.19	2562	12.53	2684	13.87	2803	15.29
1.84	21.3	1232	2.80	1462	4.09	1665	5.45	1851	6.90	1851	8.20	2174	9.60	2458	12.53	2587	13.94	2708	15.36	2825	16.78
2.02	23.4	1306	3.40	1520	4.77	1717	6.25	1894	7.80	1894	9.20	2209	10.81	2489	13.87	2614	15.36	2736	16.93	2825	18.49
2.19	25.4	1380	4.09	1562	5.53	1770	7.11	1941	8.72	1941	10.37	2247	12.02	2521	14.29	2645	16.93	2766	18.64	2883	20.36
2.37	27.4	1457	4.89	1647	6.39	1827	8.05	1990	9.77	1990	11.96	2291	13.27	2558	16.85	2681	18.64	2798	20.36		
2.55	29.5	1535	5.81	1719	7.40	1885	8.10	2045	10.96	2045	12.83	2334	14.69	2594	18.42	2717	20.36	2830	22.22		

"point of operation" will be the one single point at the intersection of the fan curve and the system curve.

Fan Performance Curves: Certain fan performance variables are usually related to volumetric flow rate in graphic form to represent a fan performance curve. Figure 7-7 is a typical representation where Pressure (P) and power requirement (PWR) are plotted against flow rate (Q). Other variables may also be included and more detailed curves representing various fan designs are provided in Figures 7-4a, 7-4b and 7-4c. Pressure can be either fan static pressure (FSP) or fan total pressure (FTP). This depends on the manufacturer's method of rating.

It should be noted that a fan performance curve is always specific to a fan of given size operating at a single rotation rate

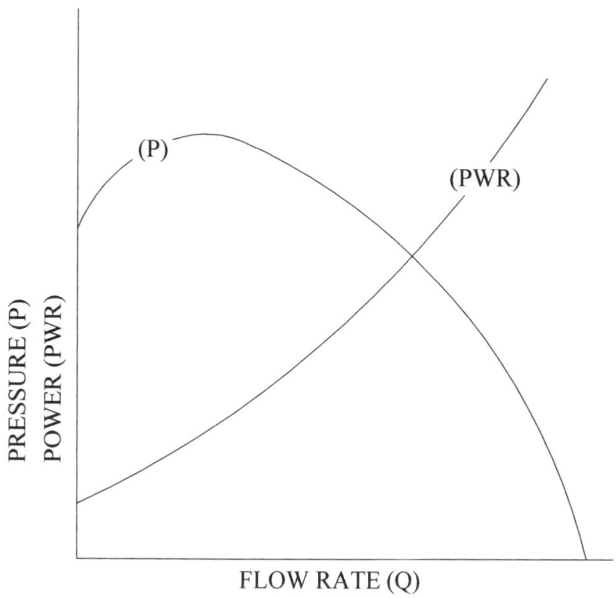

FIGURE 7-7. Typical fan performance curve

(RPM). Even with size and rotation rate fixed, it should be obvious that pressure and power requirements vary over a range of flow rates.

System Requirement Curves: The duct system pressure also varies with volumetric flow rate. Figure 7-8 illustrates the variation of pressure (P) with flow rate (Q) for three different situations. The turbulent flow condition is representative of duct losses and is most common. In this case, the pressure loss varies as the square of the flow rate. The laminar flow condition is representative of the flow through low velocity filter media. Some wet collector designs operate at or close to a constant loss situation.

The overall system curve results from the combined effects of the individual components.

Fan Selection at Standard Conditions: Utilizing the approach for calculating system static pressure found in Chapter 9 of this Manual, a designer will take the calculated volumetric flow rate at static pressure and this shall be designated the "System Static Pressure" (see Section 9.9.3). This System Static Pressure (SSP) point (described as Point A in Figure 7-9a) is the point derived directly from the calculation sheet. However, as a matter of practicality, the designer typically adds a safety factor in volume and/or pressure. It is also found in Figure 7-9a as "Point B" and will be found "up and to the right" of Point A. Finally, what we would expect to find in the field would be the intersection of the Pressure Curve selected for "Point B" and the real system curve — a third point that we will call "Point C." "Point C" will remain the operating point for the system except for degradation to the system due to a system component (hoods, ductwork, or fan) change, or dynamic portions of the system such as blast gates or plugged filters cause the system curve to fluctuate.

7.3.4 Matching Fan Performance and System Requirements. A desired point of operation results from the process of designing a duct system and selecting a fan. Considering the system requirement or fan performance curves individually, this desired point of operation has no special status relative to any other point of operation on the individual curve. Figure 7-10 depicts the four general con-

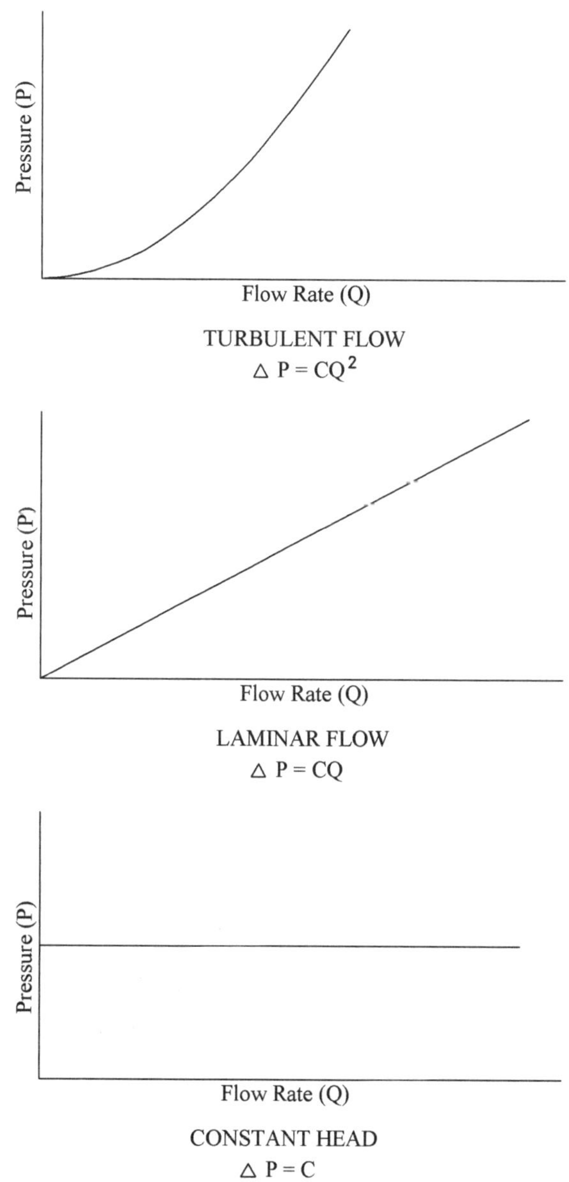

FIGURE 7-8. System requirement curves

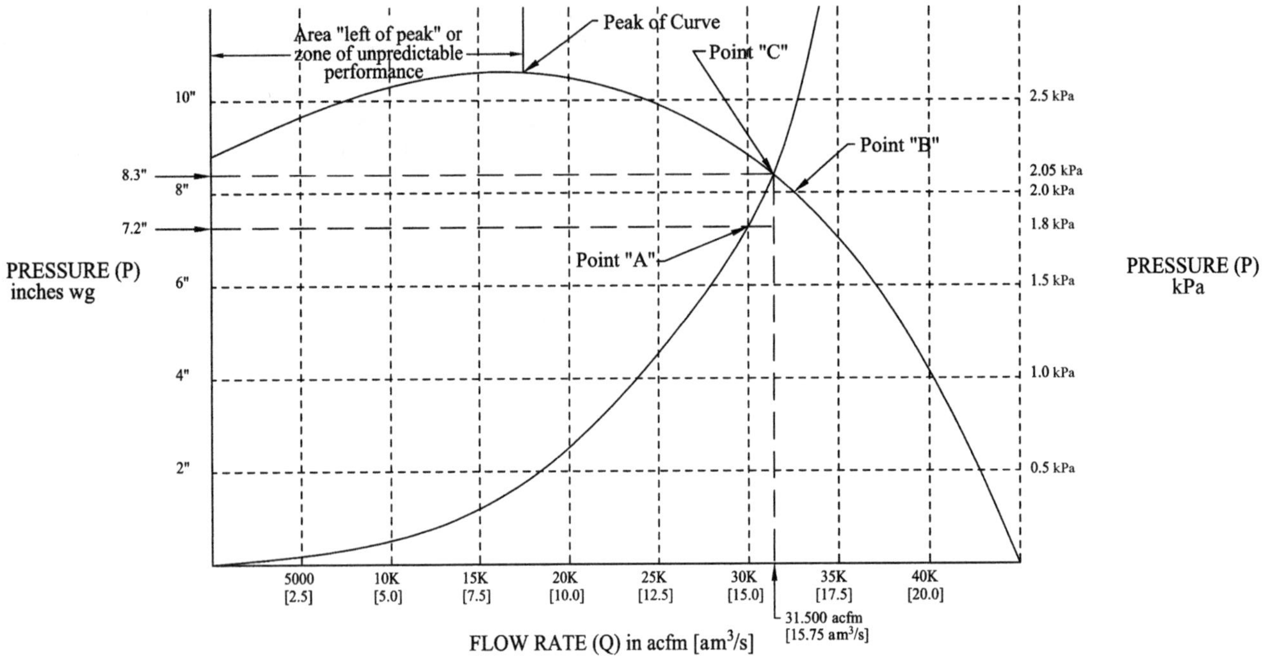

Assume:

Point "A" - (from a calculation sheet) System design is 30,000 acfm @ 7.2 "wg [15.0 am^3/s @ 1.8 kPa] SP. Point "A" and system curve go through that point.

Point "B" - Designer selects a fan rpm curve with a factor of safety (+10%) in published fan data at 32,500 acfm @ 8 "wg [16.5 am^3/s @ 2.0 kPa] SP (Point "B").

Point "C" - Actual operation will be at the intersection of the pressure curve and the system curve - Point "C" (31,500 @ 8.3 "wg [15.75 am^3/s @ 2.05 kPa])

FIGURE 7-9a. Fan selection at standard conditions

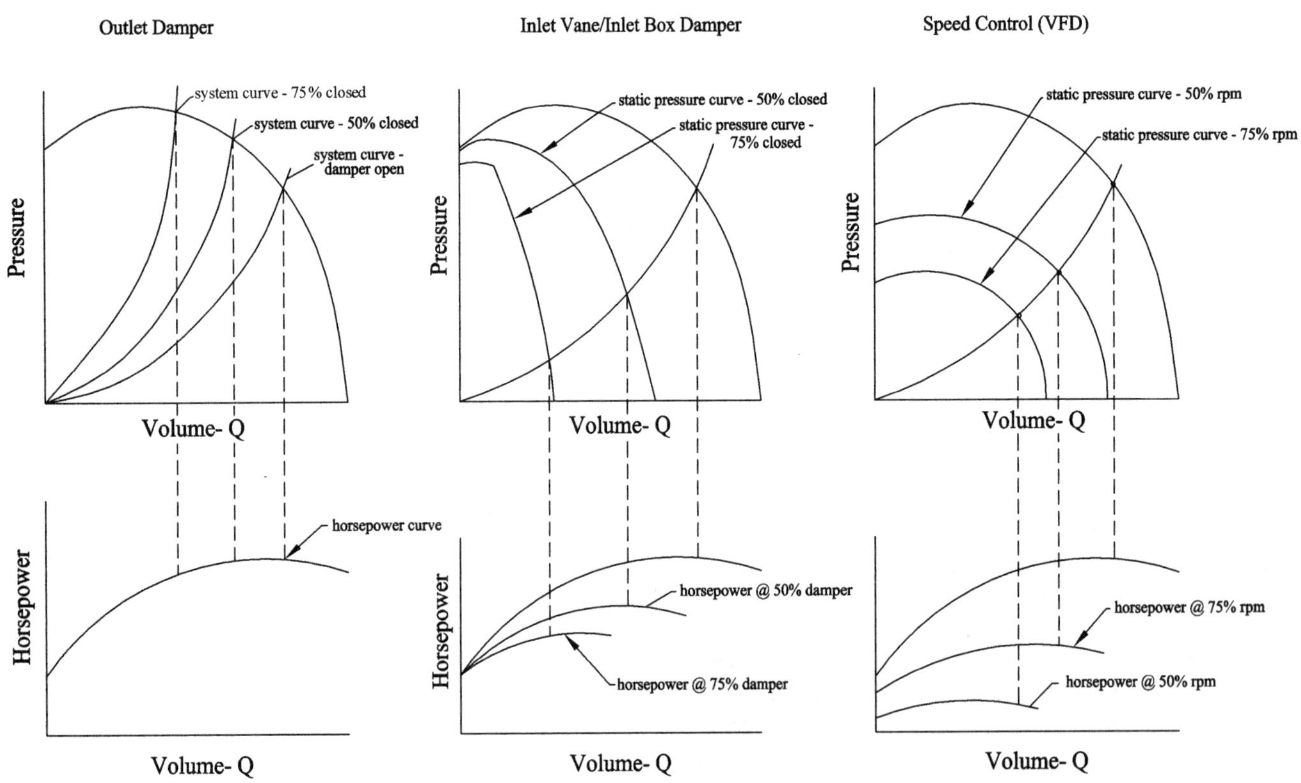

FIGURE 7-9b. Typical backwardly-inclined fan curves with volume controls

FIGURE 7-10

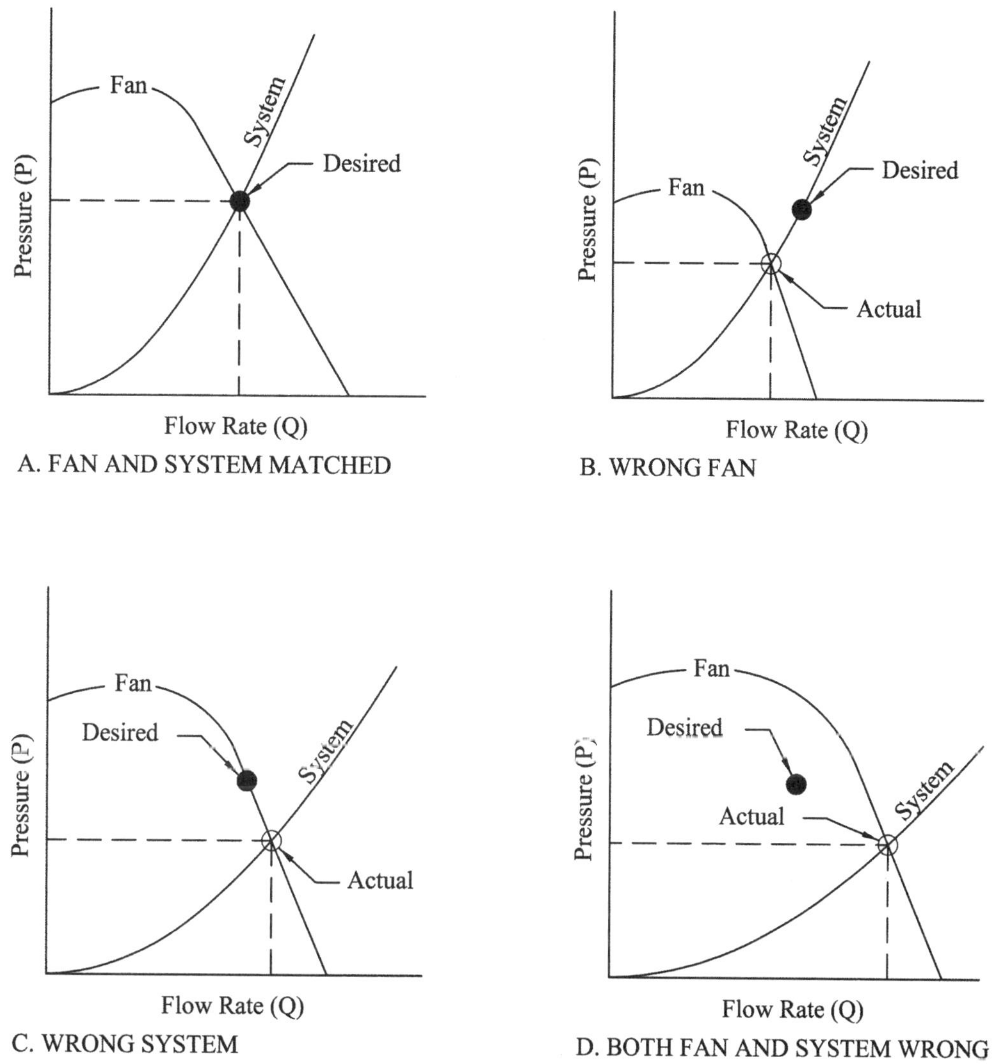

TITLE	FIGURE
ACTUAL VERSUS DESIRED POINT OF OPERATION	7-10
	DATE 1-12

CHECK CODES, REGULATIONS, AND LAWS (LOCAL, STATE, AND NATIONAL) TO ENSURE THAT DESIGN IS COMPLIANT.

ditions that can result from the system design fan selection process.

There are a number of reasons why the system design, fan selection, fabrication, and installation process can result in operation at some point other than design. When this occurs, it may become necessary to alter the system physically, which will change the system requirement curve and/or cause a change in the fan performance curve. Because the fan performance curve is not only peculiar to a given fan but specific to a given rotation rate (RPM), a change of rotation rate can be relatively simple if a belt drive arrangement has been used. The "Fan Laws" are useful when changes of fan performance are required.

7.3.5 Fan Laws. Fan laws relate the performance variables for any homologous series of fans. A homologous series represents a range of sizes where all dimensional variables between sizes are proportional. The performance variables involved are fan size (SIZE), rotation rate (RPM), gas density (ρ), flow rate (Q), pressure (P), power requirement (PWR), and efficiency (η). Pressure (P) may be represented by total pressure (TP), static pressure (SP), velocity pressure (VP), fan static pressure (FSP), or fan total pressure (FTP).

At the same relative point of operation on any two performance curves in this homologous series, the efficiencies will be equal. The fan laws are mathematical expressions of these facts and establish the inter-relationship of the other variables. They predict the effect of changing size, capacity, or gas density on speed, pressure, and power requirement as follows:

$$RPM_2 = RPM_1 \left(\frac{SIZE_2}{SIZE_1}\right)^3 \left(\frac{Q_2}{Q_1}\right) \qquad [7.5]$$

$$P_2 = P_1 \left(\frac{SIZE_2}{SIZE_1}\right)^2 \left(\frac{Q_2}{Q_1}\right)^2 \left(\frac{\rho_2}{\rho_1}\right) \qquad [7.6]$$

$$PWR_2 = PWR_1 \left(\frac{SIZE_2}{SIZE_1}\right)^5 \left(\frac{Q_2}{Q_1}\right)^3 \left(\frac{\rho_2}{\rho_1}\right) \qquad [7.7]$$

As these expressions involve ratios of the variables, any convenient units may be employed so long as they are consistent. Size may be represented by any linear dimension since all must be proportional in homologous series. However, impeller diameter is the most commonly used dimension.

7.3.6 The Effect of Changing Rotation Rate or Gas Density. In practice, these principles are normally applied to determine the effect of changing only one variable. Most often the fan laws are applied to a given fan size and may be expressed in the simplified versions that follow:

- For changes of flow rate:

  Rotation rates vary directly with flow rates; pressure varies as the square of the flow rate; and power varies as the cube of the flow rate:

$$RPM_2 = RPM_1 \left(\frac{Q_2}{Q_1}\right) \qquad [7.8]$$

$$P_2 = P_1 \left(\frac{Q_2}{Q_1}\right)^2 \qquad [7.9]$$

$$PWR_2 = PWR_1 \left(\frac{Q_2}{Q_1}\right)^3 \qquad [7.10]$$

- For changes of gas density:

  Flow is not affected by a change in density; pressure and power vary directly with density:

$$Q_2 = Q_1 \qquad [7.11]$$

$$P_2 = P_1 \left(\frac{\rho_2}{\rho_1}\right) \qquad [7.12]$$

$$PWR_2 = PWR_1 \left(\frac{\rho_2}{\rho_1}\right) \qquad [7.13]$$

7.3.7 Limitations on the Use of Fan Laws. These expressions are equations which rely on the fact that the performance curves are homologous and that the ratios are for the *same relative points of rating* on each curve. Care must be exercised to apply the laws between the same relative points of rating.

Figure 7-11 contains a typical representation of two homologous fan performance curves, PQ_1 and PQ_2. These could be the performances resulting from two different rotation rates, RPM_1 and RPM_2. Assuming a point of rating indicated as A_1 on PQ_1, there is only one location on PQ_2 with the same relative point of rating and that is at A_2. The A_1 and A_2 points of rating are related by the expression

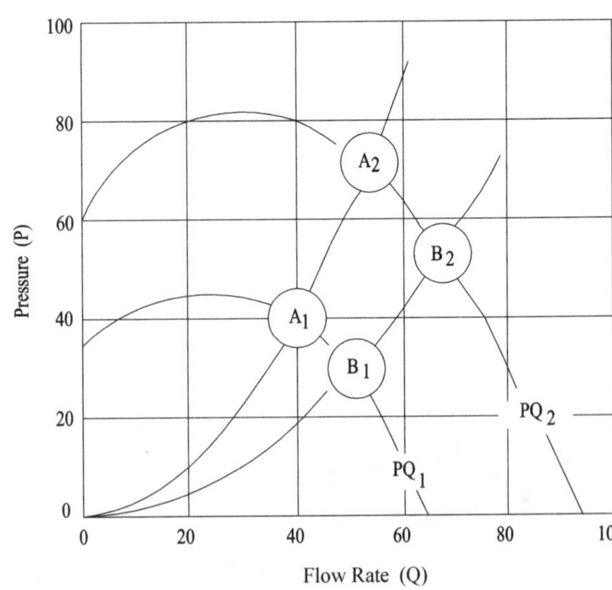

FIGURE 7-11. Homologous performance curves

$$P_{A_2} = P_{A_1}\left(\frac{Q_{A_2}}{Q_{A_1}}\right)^2 \quad [7.14]$$

The equation can be used to identify every other point that would have the same relative point of rating as A_1 and A_2. The dashed line passing through "A_2, A_1" and the origin locates all conditions with the same relative points of rating. These lines are more often called "system lines" or "system curves." As discussed in Section 7.3.3, there are a number of exceptions to the condition where system pressure varies as the square of flow rate. These lines representing the same relative points of rating are "system lines" or "system curves" for turbulent flow conditions only.

Where turbulent flow conditions apply, it must be understood that the system curves or lines of relative points of rating represent a system having fixed physical characteristics. For example, the "B_2-B_1" line defines another system that has lower resistance to flow than the "A_2-A_1" system.

Special care must be exercised when applying the fan laws in the following cases:

1) Where any component of the system does not follow the "pressure varies as the square of the flow rate" rule.
2) Where the system has been physically altered or for any other reason operates on a different system line.

7.3.8 Fan Selection at Air Density Other Than Standard. As discussed in Section 7.3.6, fan performance is affected by changes in gas density. Variations in density due to normal fluctuations of ambient pressure, temperature, and humidity are small and need not be considered. Where temperature, humidity, elevation, pressure, gas composition, or a combination of two or more cause density to vary by more than 5 percent from the standard 0.075 lbm/ft³ [1.204 kg/m³], corrections should be employed.

Rating tables and performance curves as published by fan manufacturers are based on standard air. Performance variables are always related to conditions at the fan inlet. Fan characteristics are such that volumetric flow rate (Q) is unaffected but pressure (P) and power (PWR) vary directly with changes in gas density. Therefore, the selection process requires that rating tables are entered with *actual* volumetric flow rate but with a corrected or equivalent pressure.

The equivalent pressure is that pressure corresponding to standard density and is determined from Equation 7.12 as follows:

$$P_e = P_a\left(\frac{0.075}{\rho_a}\right) \quad [7.15]\ \text{IP}$$

$$[P_e = P_a\left(\frac{1.204}{\rho_a}\right)] \quad [7.15]\ \text{SI}$$

where: P_e = Equivalent Pressure
P_a = Actual Pressure
ρ_a = Actual density, lbm/ft³ [kg/m³]

The pressures (P_e and P_a) can be either Fan Static Pressure or Fan Total Pressure in order to conform with the manufacturer's rating method.

The fan selected in this manner is to be operated at the rotation rate indicated in the rating table and actual volumetric flow rate is that indicated by the table. However, the pressure developed is not that indicated in the table but is the actual value. Likewise, the power requirement is not that of the table as it also varies directly with density. The actual power requirement can be determined from Equation 7.13 as follows:

$$PWR_a = PWR_t\left(\frac{\rho_a}{0.075}\right) \quad [7.16]\ \text{IP}$$

$$[PWR_a = PWR_t\left(\frac{\rho_a}{1.204}\right)] \quad [7.16]\ \text{SI}$$

where: PWR_a = Actual Power Requirement
PWR_t = Power Requirement in Rating Table
ρ_a = Actual Density, lbm/ft³ [kg/m³]

Fan selection at non-standard density requires knowledge of the actual volumetric flow rate at the fan inlet, the actual pressure requirement (either FSP or FTP, depending on the rating table used), and the density of the gas at the fan inlet. The determination of these variables requires that the system design procedure consider the effect of density as discussed in Chapter 9.

EXAMPLE PROBLEM 7-1 (IP Units)

Consider the system illustrated in Figure 7-12 where the heater causes a change in volumetric flow rate and density. For simplicity, assume the heater has no resistance to flow and that the sum of friction losses will equal FSP. Using the Multi-Rating Table, Table 7-1, select the rotation rate and determine power requirements for the optional fan locations ahead of or behind the heater.

Location 1: *Fan ahead of the heater* (side "A" to "B" in Figure 7-12).

Step 1. Determine actual FSP

FSP = 1 "wg + 3 "wg = 4 "wg at 0.075 lbm/ft³

$$Q = 1000\left(\frac{0.075}{0.075}\right) = 1000\ \text{acfm}$$

Step 2a. Density at fan inlet is standard. Therefore, enter rating table with actual volumetric flow rate at fan inlet, 1000 acfm, and FSP of 4 "wg.

2b. Interpolation from Table 7-1 (IP) results in:

RPM = 1182 rpm
PWR = 1.32 bhp

Step 3. The fan should be operated at 1182 rpm and actual power requirement will be 1.32 bhp.

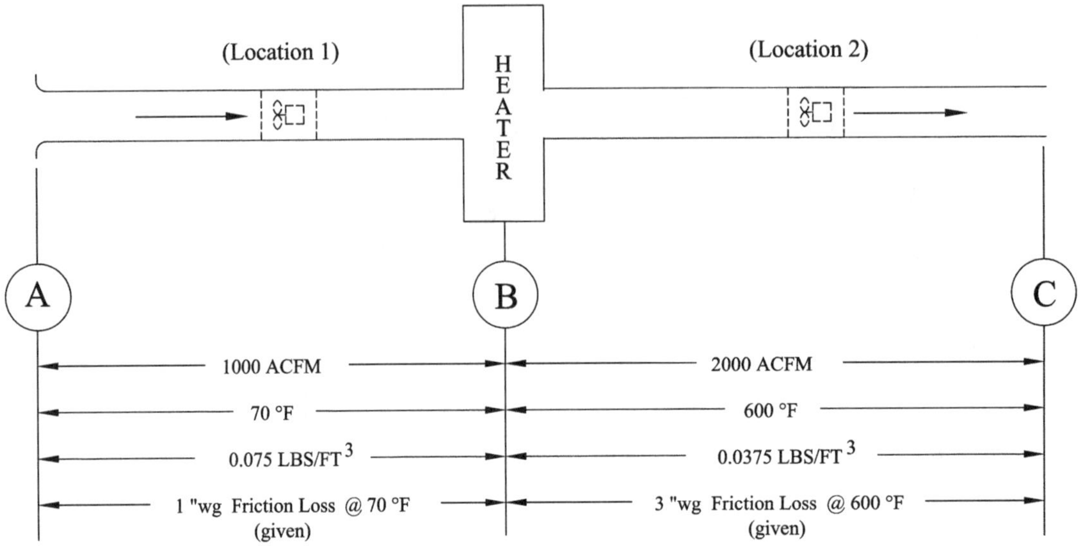

FIGURE 7–12 (IP). In-duct heater

Location 2: *Fan behind the heater.*

Step 1. Determine actual FSP

FSP = 1 "wg + 3 "wg (as in explanation) = 4 "wg at 0.0375 lbm/ft^3

Step 2a. Density at fan inlet is not standard and a pressure correction must be made (using Equation 7.12) to determine equivalent FSP.

$$FSP_e = FSP_a \left(\frac{0.075}{\rho a}\right) = 4 \text{ "wg} \left(\frac{0.075}{0.0375}\right) = 8 \text{ "wg}$$

In addition, system volume has increased at the ratio of the densities, therefore:

$$Q = 1000 \left(\frac{0.075}{0.0375}\right) = 2000 \text{ acfm}$$

Now, enter rating table with actual volumetric flow rate at fan inlet, 2000 acfm, and equivalent FSP, 8 "wg.

Step 2b. Interpolation from Table 7-1 (IP) results in:

RPM = 1692 rpm

PWR = 4.39 bhp

Step 3a. The fan should be operated at 1692 rpm, but actual power requirements will be affected by the density and can be determined by using Equation 7.13.

$$PWR_2 = PWR_t \left(\frac{\rho a}{0.075}\right) = 4.39 \left(\frac{0.0375}{0.075}\right) = 2.2 \text{ bhp}$$

Remember that this is the power required when the air is hot. If it is necessary to start the fan with the heater off, when the air is cold, the fan motor should be sized for the cold power calculated in Step 2b.

Step 3b. It should also be noted that a measurement of FSP will result in the value of 4 "wg (actual) and not the equivalent value of 8 "wg.

It will be noted that, regardless of location, the fan will handle the same mass flow rate. Also, the actual resistance to flow is not affected by fan location. It may appear then that an error is responsible for the differing power requirements of 1.32 bhp versus 2.2 bhp. In fact, the fan must work harder at the lower density to move the same mass flow rate. This additional work results in a higher temperature rise in the air from fan inlet to outlet. A fan located ahead of the heater will require less power and may be quieter due to the lower rotational speed.

EXAMPLE PROBLEM 7-1 (SI Units)

Consider the system illustrated in Figure 7-12 (SI) where the heater causes a change in volumetric flow rate and density. For simplicity, assume the heater has no resistance to flow and that the sum of friction losses will equal FSP. Using the Multi-Rating Table, Table 7-1 (SI), select the rotation rate and determine power requirements for the optional fan locations ahead of or behind the heater.

Location 1: *Fan ahead of the heater* (side “A” to “B” in Figure 7-12 (SI)).

Step 1. Determine actual FSP

FSP = 250 Pa + 750 Pa

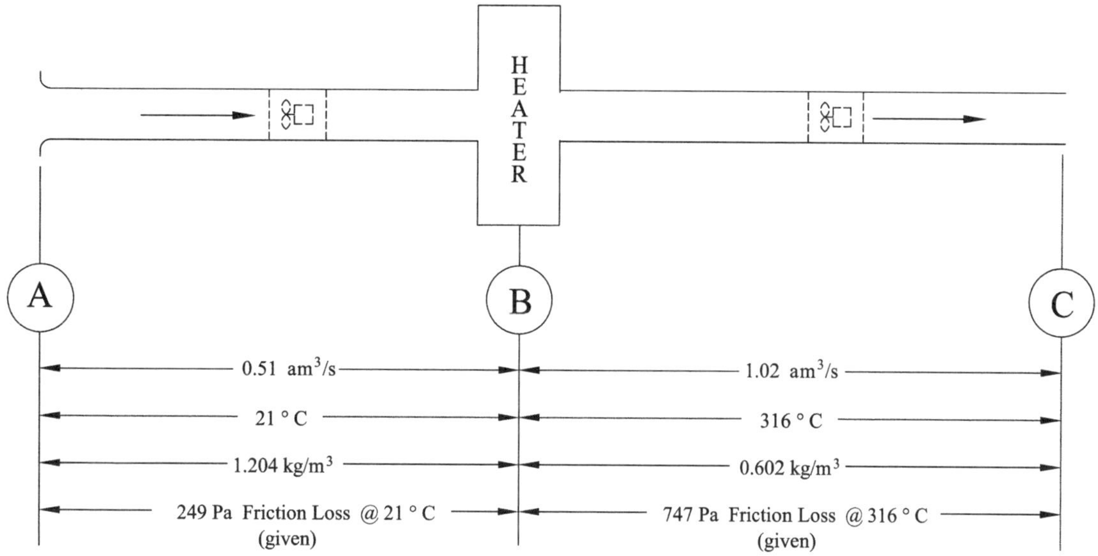

FIGURE 7–12 (SI). In-duct heater

= 1000 at 1.204 kg/m^3

Step 2a. Density at fan inlet is standard. Therefore, enter rating table with actual volumetric flow rate at fan inlet, 0.5 m^3/s, and FSP of 1000 Pa.

Step 2b. Interpolation from Table 7-1 (SI) results in:

RPM = 1181 rpm

PWR = 0.98 kW

Step 3. The fan should be operated at 1181 rpm and actual power requirement will be 0.98 kW.

Location 2: *Fan behind the heater* (side "B" to "C" in Figure 7-12 (SI).

Step 1. Determine actual FSP

FSP = 250 Pa + 750 Pa

= 1000 Pa at 1.204 kg/m^3

Step 2a. Density at fan inlet is not standard and a pressure correction must be made (using Equation 7.12) to determine equivalent FSP.

$$FSP_e = FSP_a\left(\frac{1.204}{a}\right) = 1000 \text{ Pa}\left(\frac{1.204}{0.602}\right) = 2000 \text{ Pa}$$

Now, enter rating table with actual volumetric flow rate at fan inlet, 1.0 m^3/s, and equivalent FSP, 2000 Pa.

Step 2b. Interpolation from Table 7-1 (SI) results in:

RPM = 1698 rpm

PWR = 3.45 kW

Step 3a. The fan should be operated at 1698 rpm, but actual power requirements will be affected by the density and can be determined by using Equation 7.13.

$$PWR_a = PWR_t\left(\frac{\rho_a}{1.204}\right) = 3.45\left(\frac{0.602}{1.204}\right)$$

= 1.73 kW

Remember that this is the power required when the air is hot. If it is necessary to start the fan with the heater off, when the air is cold, the fan motor should be sized for the cold power calculated in Step 2b.

Step 3b. It should also be noted that a measurement of FSP will result in the value of 1000 Pa (actual) and not the equivalent value of 2000 Pa.

It will be noted that, regardless of location, the fan will handle the same mass flow rate. Also, the actual resistance to flow is not affected by fan location. It may appear then that there is an error responsible for the differing power requirements of 0.98 kW versus 1.73 kW. In fact, the fan must work harder at the lower density to move the same mass flow rate. This additional work results in a higher temperature rise in the air from fan inlet and outlet. A fan located ahead of the heater will require less power and may be quieter due to the lower rotational speed.

Fan Cold Starts – Under some circumstances, fans designated for hot applications must be operated for a period of time without a hot air stream. If the air density is significantly lower during these colder operations, then the fan's motor may be in danger of operating beyond its capabilities and motor windings have burned under just such situations. Alternative controls such as VFD fan speed controls or mechanical controls such as the use of fan dampers for these occasions are suggested. In addition, atmospheric temperatures in very cold climates sometimes provide enough density change to affect motor workloads. In some cases, when fans are selected at standard temperature (70 F) [21 C], winter temperatures can increase the air density to as much as 0.1 lbs/cu. ft. [1.605 kg/m^3] (25% more than standard density at 30 degrees [-32 C] below zero). This can increase horsepower requirements 33% over standard conditions on the fan curve.

7.3.9 Explosive or Flammable Materials. When conveying explosive or flammable materials, it is important to recognize the potential for ignition of the gas stream. This may be from airborne material striking the impeller or by the physical movement of the impeller into the fan casing. AMCA(7.3) and other associations offer guidelines for both the manufacturer and the user on ways to minimize this danger. These involve more permanent attachment of the impeller to the shaft and bearings and the use of buffer plates or spark resistant alloy construction. Because no single type of construction fits all applications, it is imperative that both the manufacturer and the user are aware of the dangers involved and agree on the type of construction and degree of protection that is being proposed.

NOTE: For many years aluminum alloy impellers have been specified to minimize sparking if the impeller were to contact other steel parts. This is still accepted, but tests by the U. S. Bureau of Mines(7.4) and others have demonstrated that impact of aluminum with rusty steel creates a "Thermite" reaction and thus possible ignition hazards. Special care must be taken when aluminum alloys are used in the presence of steel.

7.3.10 Fans in Series or Parallel Operation. For fans operated in parallel or in series operation please refer to Figures 7-13 and 7-14, respectively. Note that identical fans for either operation will perform better than dissimilar ones.

7.4 FAN MOTORS

Most fans are driven by electric motors. There are many types of motors available on the market. Selecting the right motor for a given installation requires information on the electrical power available, the fan power requirement, the desired fan speed, how the fan is to be driven (belt or direct drive), and the environmental conditions where the fan and motor are to be located. Most motors conform to standards established by the National Electric Manufacturers Association (NEMA). These standards apply to motor design parameters such as dimensions, enclosures, power requirements, and insulation. In addition, the Energy Policy Act (EPACT) of 1999 mandates energy efficiency standards for most types of motors used in the United States.

7.4.1 Considerations for Motor Selection.

POWER SUPPLY

Current: By far the most common form of supply power has alternating current (AC). Traditionally, direct current (DC) has been used in special-purpose cases where variable speed was needed, but this has been changing due to the availability and cost of variable frequency AC drives.

Voltage: The supply voltage must be known to properly select motors and motor controls. In the United States, three phase power is generally either 230 or 460 volts, although some very high horsepower systems use 575 or 2300 volts. Single phase power is usually either 115 or 230 volts.

Phase: Power is supplied by either a three wire, three phase system or a two wire, single phase system. Three phase is commonly used on motors one horsepower and larger. It is economical because it requires smaller lead wires. Single phase is most commonly used for fractional horsepower motors.

Frequency: The standard frequency for AC current in the United States is 60 cycles per second (Hz). Some foreign countries use 50 cycles.

MOTOR CONSTRUCTION

Power Rating: The power capacity of the motor must be greater than the power requirement of the fan it drives. In the United States, motors are rated in horsepower (hp); in much of the rest of the world they are rated in kilowatts (kW).

Speed: Another important consideration is the motor speed, especially with direct driven fans. The speed of AC motors is a function of the frequency and the number of poles in the motor.

$$N_X = \frac{120 \times f}{P} \qquad [7.17]$$

where: N_x = synchronous speed (rpm)

f = frequency (Hz)

P = number of poles

Motors run at speeds slightly below the synchronous speed. For example, a four pole, 60 Hz motor has a synchronous speed of 1800 fpm. Most of these motors run between 1725 and 1780 rpm (U.S. values).

Frame: NEMA sets industry standards for motor dimensions, and designates them as frame sizes. Motors with common frame sizes have the same shaft diameter, centerline height, and feet mounting dimensions.

Enclosure: The type of enclosure indicates how much protection there is for the internal motor components from the surrounding environment, and the method of motor cooling.

Open Drip-Proof (ODP) motors allow a free exchange of air through the motor. Air is drawn into the motor and across the windings for cooling. ODP motors should be used for clean, indoor applications.

Totally Enclosed Fan Cooled (TEFC) motors do not have openings in the motor enclosure, but are not necessarily airtight. An integral fan blows air over the enclosure to cool the motor. TEFC motors are used in outdoor, damp, and dirty applications.

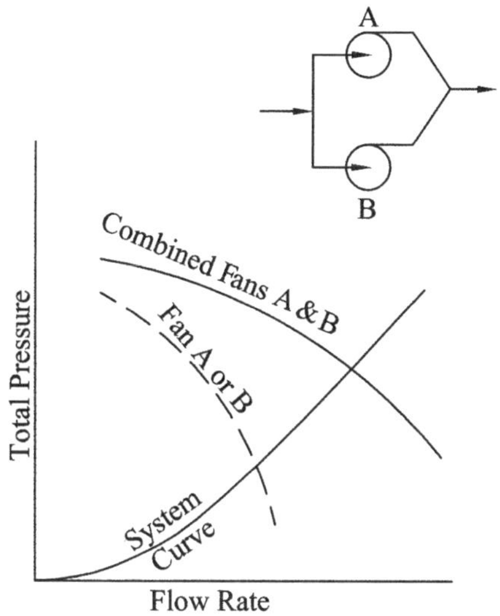

TWO IDENTICAL FANS
RECOMMENDED

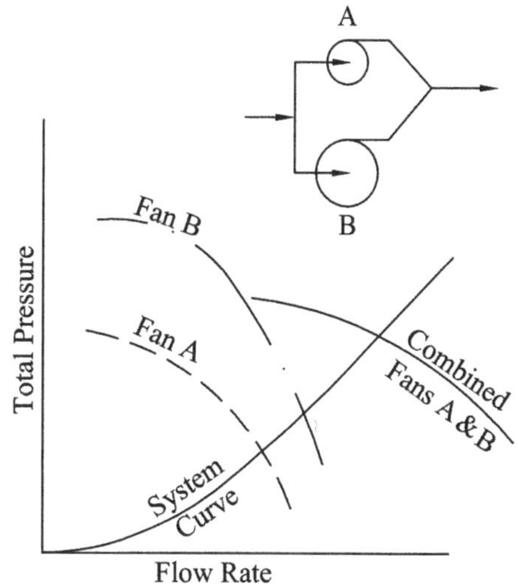

TWO DIFFERENT FANS
SATISFACTORY

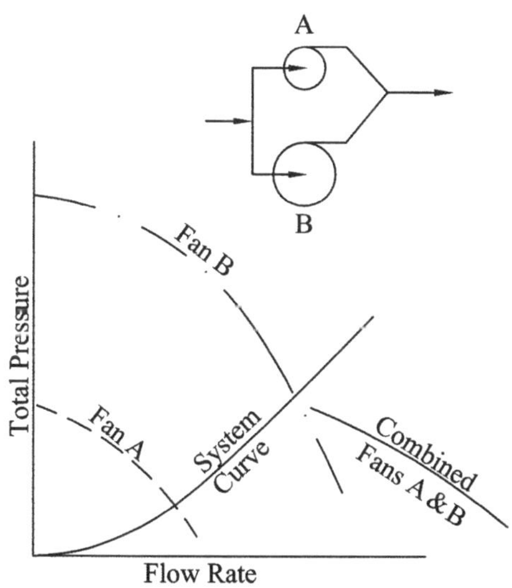

TWO DIFFERENT FANS
UNSATISFACTORY

When system curve does not cross combined fan curve, or crosses projected combined curve before Fan B, Fan B will handle more air than Fans A and B in parallel.

Notes:
1. To establish combined fan curve, the combined airflow rate, Q, is the sum of individual fan airflow rates at points of equal pressure.
2. To establish system curve, include losses in individual fan connections.
3. System curve must intersect combined fan curve or higher pressure fan may handle more air alone.
4. Consider system effects.

ACGIH®	TITLE: FANS PARALLEL OPERATION	FIGURE 7-13 DATE 1-12

CHECK CODES, REGULATIONS, AND LAWS (LOCAL, STATE, AND NATIONAL) TO ENSURE THAT DESIGN IS COMPLIANT.

FIGURE 7-14

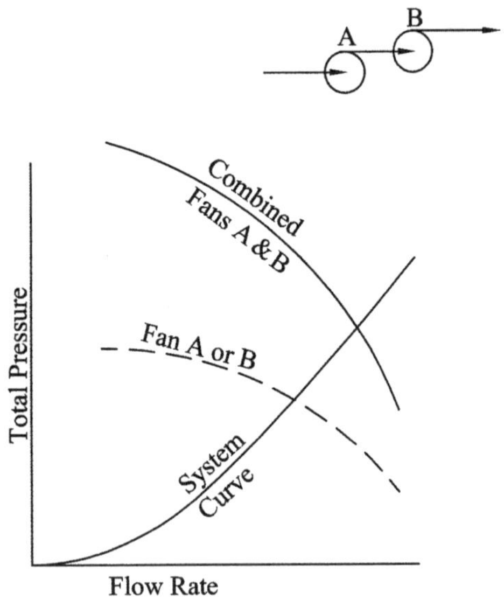

TWO IDENTICAL FANS
RECOMMENDED FOR BEST EFFICIENCY

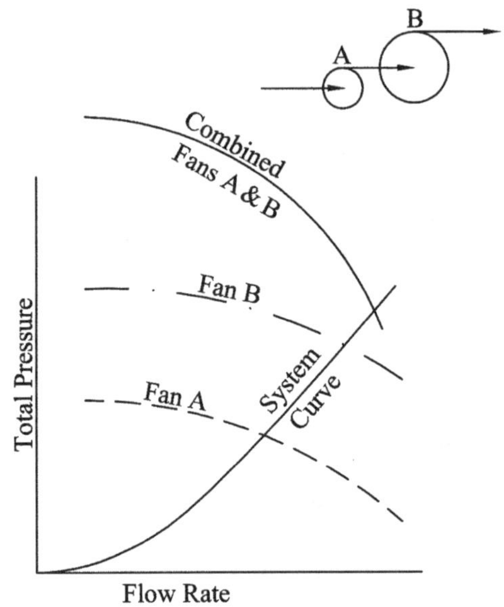

TWO DIFFERENT FANS
(SATISFACTORY)

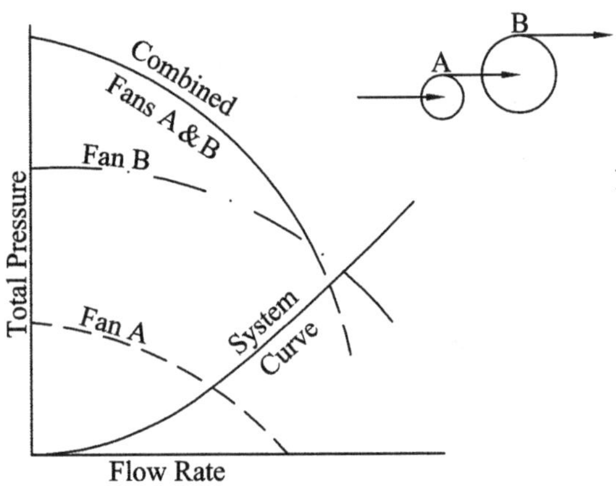

TWO DIFFERENT FANS
(UNSATISFACTORY)

Notes:

1. To establish combined fan curve, the combined total pressure is the sum of individual fan pressures at equal airflow rates, less the pressure loss in the fan connections.
2. Airflow rate through each fan will be the same, since air is considered incompressible.
3. System curve must intersect combined fan curve or large flow rate fan may handle more air alone.
4. Consider system effects.

When system curve does not intersect combined fan curve, or crosses projected comvined curve before Fan B curve, Fan B will move more air than Fans A and B in series.

ACGIH®	TITLE FANS SERIES OPERATION	FIGURE 7-14 DATE 1-12

CHECK CODES, REGULATIONS, AND LAWS (LOCAL, STATE, AND NATIONAL) TO ENSURE THAT DESIGN IS COMPLIANT.

Total Enclosed Air Over (TEAO) motors are similar to TEFC motors except that there is no integral cooling fan. These motors are frequently used on fans where the fan provides the cooling airflow over the motor.

Explosion Proof motors are special versions of TEFC motors, with design features to make them suitable for applications where explosive dust or gases are present. The enclosure is designed to withstand an explosion inside the motor, and contain the flame and sparks within the motor. There are different classifications of explosion proof construction, depending on the characteristics of the explosive gas or dust.

Severe Duty Motors are another variation of TEFC motors that have features that make them durable in hostile environments. They have better shaft seals, corrosion resistant paint, and some are available with stainless steel shafts.

Inertia Load Capacity: In some cases, it is not the horsepower requirements that determine the size of motor needed but the motor's ability to accelerate the fan to full speed. This is particularly true when using low horsepower motors on large, heavy fans. Motors must have an inertia load capability greater than the inertia of the fan corrected for the drive ratio, as shown in the equation below:

$$WR^2_{motor} \geq WR^2_{fan} \times \left(\frac{RPM_{fan}}{RPM_{motor}}\right)^2 \times 1.1 \qquad \textbf{[7.18]}$$

where: WR^2_{motor} = inertia load at motor shaft

WR^2_{fan} = inertia of the fan

RPM_{motor} = motor speed

RPM_{fan} = fan speed

The 1.1 factor is an allowance for belts and sheaves. If the motor does not have enough inertia load capacity, either it will not be able to start the fan, or it will take an excessive amount of time (20 seconds or more) to bring the fan up to speed.

7.4.2 Motor Installation. The National Electric Code calls out the special requirements of motor installation and wiring. The sizing of motor lead wires and overload protection must take into account the higher than normal amp draw that occurs when a motor is started and brought up to full speed. As a result, motor branch circuits are sized differently than other types of branch circuits. There are also requirements that specify how close to the motor disconnects should be located. These are very important since they provide protection for workers who must service the fan and motor. Some fans can be provided with integral motor disconnects.

If a fan is belt driven, the motor must be mounted on an adjustable base. This base allows the motor to move with respect to the fan and allows for the adjustment and replacement of the belts.

7.5 FAN INSTALLATION AND MAINTENANCE

Fan rating tests for flow rate, static pressure, and power requirements are conducted under ideal conditions that include uniform straight airflow at the fan inlet and outlet. However, if in practice, duct connections to the fan cause non-uniform airflow, fan performance and operating efficiency will be affected. Location and installation of the fan must consider the location of these duct components to minimize losses. If adverse connections must be used, appropriate compensation must be made in the system calculations. Once the system is installed and operating, routine inspection and maintenance will be required if the system is to continue to operate at original design levels.

7.5.1 Fan Installation. It is important to install a fan on a structure strong enough to support the loads produced by the fan. The support structure must be designed to carry not only the weight of the fan, but also the dynamic loads produced while the fan is operating. A well-designed support is rigid enough to keep vibration levels low. A wood stud wall may be adequate for a small lightweight wall propeller fan, but a large industrial exhaust fan requires more consideration. The ideal mounting for large fans is a concrete pad mounted on grade with a weight of at least three times the fan. Often this is not possible, and structural steel supports are used. To avoid problems with vibration, it is important that the dynamic loads are considered in the design of the support.

Consider maintenance when deciding how and where to mount the fan. Provide ample room around the fan to gain access to the motor, drives, and bearings. Include safety features such as guards, electrical disconnects, and safety railings where necessary.

7.5.2 System Effects. System effect is defined as the estimated loss in fan performance from this non-uniform airflow. Figure 7-15 illustrates deficient fan system performance. The system pressure losses have been determined accurately and a suitable fan selected for operation at Point 1. However, no allowance has been made for the effect of the system connections on fan performance. The point of intersection between the resulting fan performance curve and the actual system curve is Point 3. The resulting flow rate will, therefore, be deficient by the difference from 1 to 3. To compensate for this system effect, it will be necessary to add a "system effect loss" to the calculated system pressure. This will be equal to the pressure difference between Points 1 and 2 and will have to be added to the calculated system pressure losses. The fan will then be selected for this higher pressure (Point 2) but will operate at Point 1 due to loss in performance from system effects.

EXAMPLE PROBLEM 7-2 (IP Units Only)

One commonly neglected system effect is a duct elbow at the fan inlet. For example, consider the fan shown in Figure 7-17.

Inlet boxes: In an attempt to reduce system effects due to elbows at the inlets of centrifugal fans, fan manufacturers design and provide special appurtenances called inlet boxes. Most fan manufacturers recommend the addition of

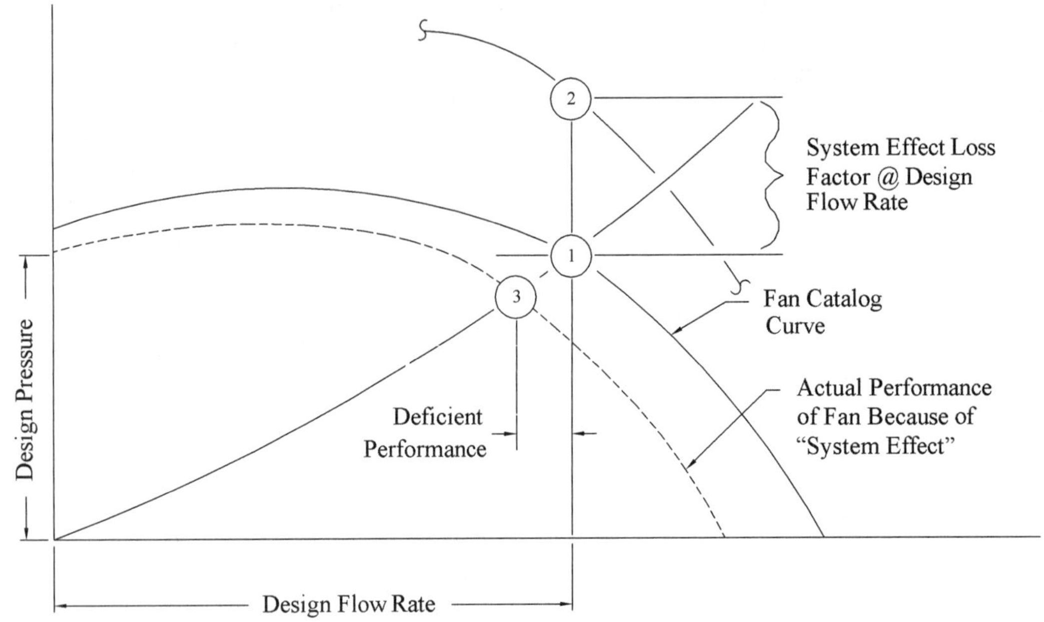

FIGURE 7–15. System effect factor

an additional 0.75 velocity pressure loss due to system effects (SEF) of even the best-designed inlet boxes and losses in excess of 1 VP are not atypical.

This fan has a four-piece 90° round duct elbow immediately in front of the inlet. There are no turning vanes inside the duct. The required flow rate is 5000 acfm and the system pressure losses are 8 "wg at standard conditions (0.075 lb/ft^3). Selecting a fan without the system effect, using Table 7-1, would result in a fan speed of 1987 rpm and power consumption of 13.02 hp.

With the elbow at the inlet, the airflow into the fan inlet will be degraded. Such a change in the airflow requires use of a system effect factor (SEF) to select a fan that overcomes the degradation in performance. The system effect factor is used to determine a correction value, in inches water gauge, to be added to the system pressure losses.

In this example, the duct diameter is 24" with a center line turning radius of 48". This is a radius-to-diameter (r/d) ratio of 2.0. From Figure 7-21, Item C, the system effect is 1.0 times the velocity pressure at the inlet of the fan. The duct area is 3.142 ft^2 and the velocity is 1592 fpm (5000 acfm ÷ 3.142 ft^2 = 1592 fpm) giving a velocity pressure of 0.16 "wg and a system effect correction of 0.16 (1.0 × 0.16). This 0.16" value is added to the fan static pressure when selecting the fan from the multi-rating table. Select the fan for a static pressure of 8.16 "wg. Interpolating in Table 7-1, we find a selection for 5000 acfm and 8.16 "wg at 1999 rpm and 13.22 hp. This selection for a fan with an elbow at the inlet will result in operation at 5000 acfm and 8 "wg drawing 13.22 hp.

NOTE: The system effect factor compensates for the affect on the fan of an irregular air stream. This system effect factor is taken in addition to the friction loss used to calculate the system loss.

Figure 7-16 illustrates typical discharge conditions and the losses that may be anticipated. The magnitude of the change in system performance caused by elbows and other obstructions placed too close to a fan inlet or outlet can be estimated for the conditions shown on Figures 7-18 through 7-25.

If the system effect factor is identified by a letter, use the corresponding loss coefficient found in Figure 7-26 or 7-27 to determine the additional static pressure. Follow the instructions provided in Figure 7-26 or 7-27.

A vortex or spin of the air stream entering the fan inlet may be created by non-uniform flow conditions as illustrated in Figure 7-24. These conditions may be caused by a poor inlet box, multiple elbows or entries near the inlet, or by other spin producing conditions. Since the variations resulting in inlet spin are many, no System Effect Factors are tabulated. Where a vortex or inlet spin cannot be avoided or is discovered at an existing fan inlet, the use of turning vanes, splitter sheets, or egg crate straighteners will reduce the effect.

7.5.3 Inspection and Maintenance. Material accumulation or abrasive wear on an impeller can cause a fan to "go out of balance." This imbalance will cause vibration of the fan. This

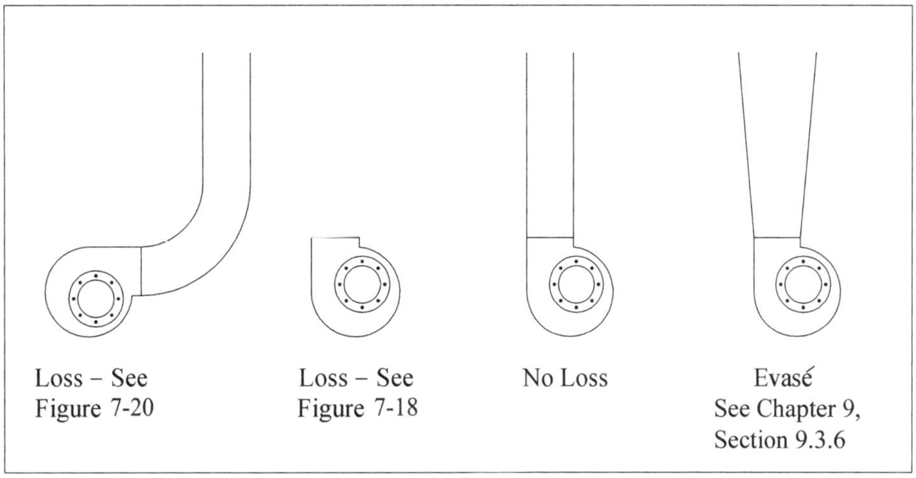

FIGURE 7–16. System effect factor

may result in damage to or failure of the fan impeller, housing, bearings, or pedestal. Periodic cleaning and rebalancing of fans operating in air streams handling abrasive, sticky, or wet materials is recommended.

Regular observation of fan vibration levels can detect problems before they increase in amplitude to the point where fan components become damaged. Different types of fans and fan installations can tolerate higher levels of vibration than others. Table 7-2 shows fan application categories for determining acceptable levels of vibration. Most fans used in industrial ventilation systems fall in category BV 3. Once the fan application category is determined, use Table 7-3 to determine acceptable levels of vibration. The levels shown are for filter-out readings, which take into account vibrations at all frequencies. The rigidly mounted column is for fans mounted directly to structural steel or concrete. The flexibly mounted column applies to fans mounted on spring or rubber-in-shear isolators. The start-up row gives acceptable levels of vibration for new or recently repaired fans. As the fan operates over time, parts wear, material builds up on the impeller, and vibration levels increase. When the vibration levels reach the level shown in the alarm row, corrective action should be taken at the next available shut down. If corrective action is not taken, and the vibration levels increase to the shutdown levels, the fan should be shut down immediately and the problem must be found and corrected. Failure to do so could lead to catastrophic failure of fan components. Refer to AMCA Standard 204[7.5] for more information on fan balancing and vibration levels.

Modern maintenance equipment permits the inspector to record vibration spectra. Review of changes in these spectra taken over time can indicate specific areas of developing problems with bearings, balance, belts, or motors. Electronic or computerized vibration monitors are available to mount on fans used in critical operations. These devices can be set up with automatic alarm functions and/or to provide continuous information about a unit's vibration level.

It is not uncommon, during fan installation or motor/starter maintenance, for the fan impeller rotation direction to be inadvertently reversed. Since fans do move a fraction of their rated capacity when running backward, incorrect rotation often goes unnoticed in spite of less effective performance of the exhaust system.

Scheduled inspection of fans is recommended. Items checked should include:

1) Bearings for proper operating temperature (lubricate them on the manufacturer's recommended schedule).
2) Excessive vibration of bearings or housing.
3) Belt drives for proper tension and minimum wear.

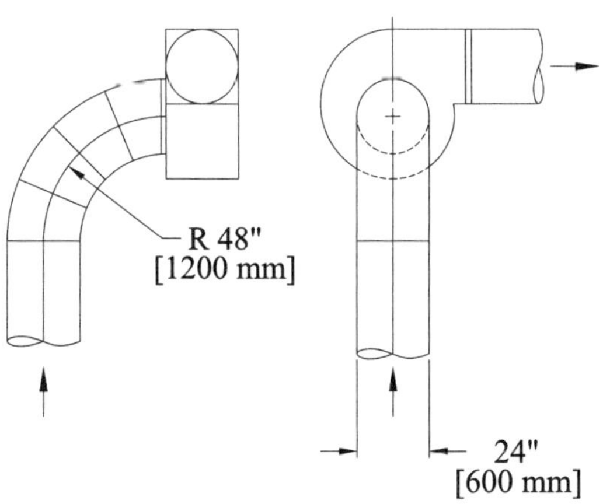

FIGURE 7–17. Inlet elbow

4) Correct coupling or belt alignment.
5) Fan impeller for proper alignment and rotation.
6) Impeller free from excess wear or material accumulation.
7) Tight fan hold-down bolts.
8) Tight fan impeller set screws or bushings.
9) Proper installation of safety guards.

Standard lockout/tagout procedures should be observed when servicing fan equipment or its associated duct. The electrical supply must be shut off and locked out at a disconnect near the fan. When opening access doors or reaching into the fan inlet or outlet, the fan must be mechanically locked out by blocking the impeller from rotating. A warning tag should be used when blocking a fan. Do not open an access door while the fan is operating or coasting down.

BE SURE to remove any inserted obstructions used to block impeller rotation when servicing is complete.

REFERENCES

7.1 American Society of Heating, Refrigerating, and Air-Conditioning Engineers, Inc.: Fundamentals Handbook 1993. Atlanta, GA.

7.2 Air Movement and Control Association, Inc.: Standards Handbook, Publication 99-86. Arlington Heights, IL.

7.3 Air Movement and Control Association, Inc.: AMCA Publication 201-90, Fans and Systems. Arlington Heights, IL. Publications (847) 394-0404.

7.4 Gibson, N.; Lloyd, F. C.; Perry, G. R.: Fire Hazards in Chemical Plants from Friction Sparks Involving the Thermite Reaction. Symposium Series No. 25. Insn. Chem. Engrs. London (1968).

7.5 Air Movement and Control Association, Inc.: ANSI/AMCA Standard 204-05, Balance Quality and Vibration Levels for Fans. Arlington Heights, IL. Publications (847) 253-0088.

7.6 Air Movement and Control Association, Inc.; AMCA Publication 203-90, Field Performance Measurement of Fan Systems. Arlington Heights, IL.

7.7 Fan Engineering: Buffalo Forge Company, Buffalo, New York (1961).

7.8 Wessel, W.: Private Communication to J.F. Hale (August 2011).

TABLE 7-2. Fan Balancing and Vibration Categories

Application	Examples	Driver Power Limits HP [kW]	Fan Application Category
Residential	Ceiling fans, attic fans, window air conditioners	0.2 [≤ 0.15] 0.2 [> 0.15]	BV-1 BV-2
HVAC & Agricultural	Building ventilation and air conditioning; commercial systems	5.0 [≤ 3.7] 5.0 [> 3.7]	BV-2 BV-3
Industrial Process & Power Generation, etc.	Baghouse scrubber, mine, conveying, boilers, combustion air, pollution control, wind tunnels	400 [≤ 300] 400 [> 300]	BV-3 BV-4
Transportation & Marine	Locomotives, trucks, automobiles	20 [≤ 15] 20 [> 15]	BV-3 BV-4
Transit/Tunnel	Subway emergency ventilation, tunnel fans, garage ventilation, tunnel jet fans	100 [≤ 75] 100 [> 75] ANY	BV-3 BV-4 BV-4
Petrochemical Process	Hazardous gases, process fans	50 [≤ 37] 50 [> 37]	BV-3 BV-4
Computer Chip Mfg.	Clean room fans	ANY	BV-5

TABLE 7-3. Vibration Limits for Tests Conducted In-Situ (Values shown are peak velocity, mm/s [inches/s], Filter-Out)

Condition	Fan Application Category	Rigidly Mounted Inches/s [mm/s]	Flexibly Mounted Inches/s [mm/s]
Start-Up	BV-1 BV-2 BV-3 BV-4 BV-5	0.55 [14.0] 0.30 [7.6] 0.25 [6.4] 0.16 [4.1] 0.10 [2.5]	0.60 [15.2] 0.50 [12.7] 0.35 [8.8] 0.25 [6.4] 0.16 [4.1]
Alarm	BV-1 BV-2 BV-3 BV-4 BV-5	0.60 [15.2] 0.50 [12.7] 0.40 [10.2] 0.25 [6.4] 0.20 [5.7]	0.75 [19.1] 0.75 [19.1] 0.65 [16.5] 0.40 [10.2] 0.30 [7.6]
Shut Down	BV-1 BV-2 BV-3 BV-4 BV-5	Note 1 Note 1 0.50 [12.7] 0.40 [10/2] 0.30 [7.6]	Note 1 Note 1 0.70 [17.8] 0.60 [15.2] 0.40 [10.2]

NOTE 1: Shutdown levels for Fan Applications categories BV-1 and BV-2 must be established based on historical data.

FIGURE 7-18

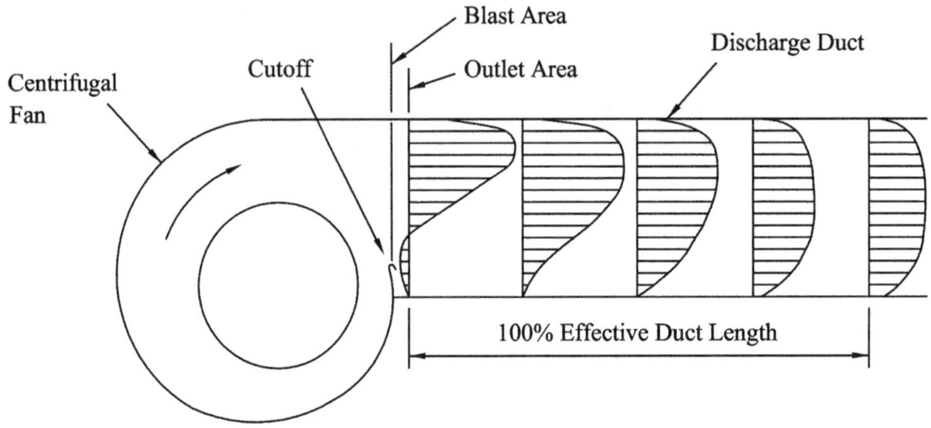

To calculate 100% effective duct length, assume a minimum of 2.5 duct diameters for 2500 fpm [12.5 m/s] or less, add 1 duct diameter for each additional 1000 fpm [5 m/s].

Example: 5000 fpm [25 m/s] = 5 equivalent duct diameters. If the duct is rectangular with side dimensions a and b, the equivalent duct diameter is equal to $(4ab/\pi)^{0.5}$

	No Duct	12% Effective Duct	25% Effective Duct	50% Effective Duct	100% Effective Duct
Pressure Recovery	0%	50%	80%	90%	100%
Blast Area / Outlet Area	System Effect Curve				
0.4	P	R-S	U	W	-
0.5	P	R-S	U	W	-
0.6	R-S	S-T	U-V	W-X	-
0.7	S	U	W-X	-	-
0.8	T-U	V-W	X	-	-
0.9	V-W	W-X	-	-	-
1.0	-	-	-	-	-

Determine the System Effect Factor by using Figure 7-26 or 7-27

Reprinted from AMCA Publication 201-90, FANS AND SYSTEMS, by permission of the Air Movement and Control Association, Inc.[7.3]

ACGIH®

TITLE
SYSTEM EFFECT FACTORS FOR OUTLET DUCTS— CENTRIFUGAL FANS

FIGURE 7-18

DATE 1-12

CHECK CODES, REGULATIONS, AND LAWS (LOCAL, STATE, AND NATIONAL) TO ENSURE THAT DESIGN IS COMPLIANT.

FIGURE 7-19

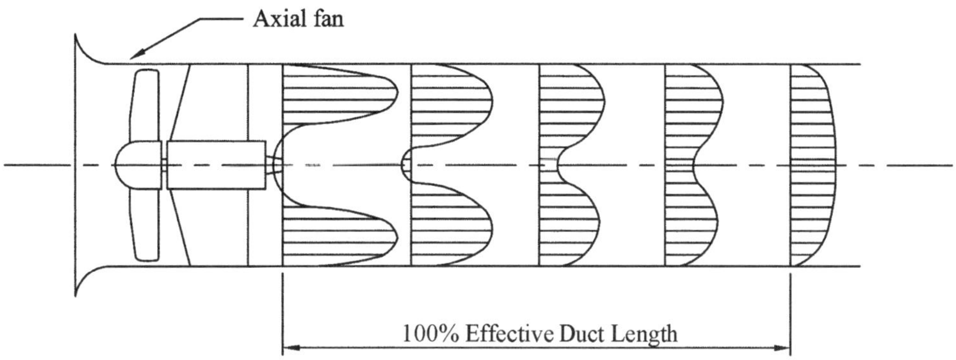

To calculate 100% effective duct length, assume a minimum of 2.5 duct diameters for 2500 fpm [12.5 m/s] or less. Add 1 duct diameter for each additional 1000 fpm [5 m/s].

EXAMPLE: 5000 FPM [25 M/S] = 5 EQUIVALENT DUCT DIAMETERS

	No Duct	12% Effective Duct	25% Effective Duct	50% Effective Duct	100% Effective Duct
Tubeaxial Fan	–	–	–	–	–
Tubeaxial Fan	U	V	W	–	–

Determine System Effect Factor by using Figure 7-26 or 7-27

Reprinted from AMCA Publication 201-90, FANS AND SYSTEMS, by permission of the Air Movement and Control Association, Inc.[7.3]

ACGIH®

TITLE

SYSTEM EFFECT FACTORS FOR OUTLET DUCTS– AXIAL FANS

FIGURE 7-19

DATE 1-12

CHECK CODES, REGULATIONS, AND LAWS (LOCAL, STATE, AND NATIONAL) TO ENSURE THAT DESIGN IS COMPLIANT.

FIGURE 7-20

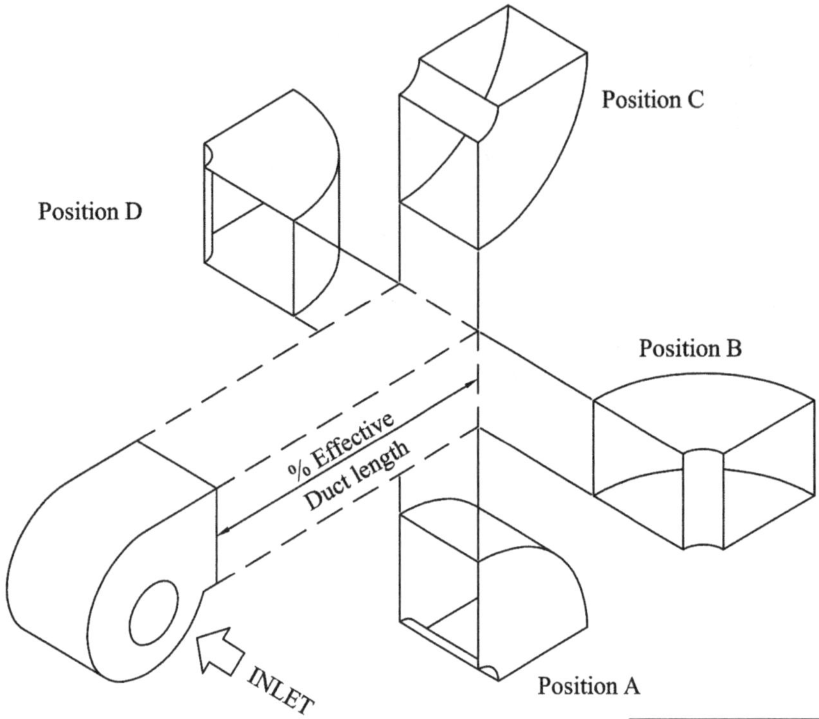

Blast Area / Outlet Area	Outlet Elbow Position	No Outlet Duct	12% Effective Duct	25% Effective Duct	50% Effective Duct	100% Effective Duct
0.4	A	N	O	P-Q	S	NO SYSTEM EFFECT FACTOR
	B	M-N	N	O-P	R-S	
	C	L-M	M	N	Q	
	D	L-M	M	N	Q	
0.5	A	O-P	P-Q	R	T	
	B	N-O	O-P	Q	S-T	
	C	M-N	N	O-P	R-S	
	D	M-N	N	O-P	R-S	
0.6	A	Q	Q-R	S	U	
	B	P	Q	R	T	
	C	N-O	O	Q	S	
	D	N-O	O	Q	S	
0.7	A		S	T	V	
	B	Q-R	R-S	S-T	U-V	
	C	P	Q	R-S	T	
	D	P	Q	R-S	T	
0.8	A	S	S-T	T-U	W	
	B	R-S	S	T	V	
	C	Q-R	R	S	U-V	
	D	Q-R	R	S	U-V	
0.9	A	T	T-U	U-V	W	
	B	S	S-T	T-U	W	
	C	R	S	S-T	V	
	D	R	S	S-T	V	
1.0	A	T	T-U	U-V	W	
	B	S-T	T	U	W	
	C	R-S	S	T	V	
	D	R-S	S	T	V	

Determine System Effect Factor by using Figures 7-26 and 7-27

For DWDI fans determine SEF using the curve for SWSI fans. Then apply the appropriate multiplier from the tabulation below

Multipliers for DWDI Fans

Elbow Position A = $\Delta P \times 1.00$
Elbow Position B = $\Delta P \times 1.25$
Elbow Position C = $\Delta P \times 1.00$
Elbow Position D = $\Delta P \times 0.85$

Obtain blast area (cut-off) from fan manufacturer

Reprinted from AMCA Publication 201-90, FANS AND SYSTEMS, by permission of the Air Movement and Control Association, Inc.[7.3]

ACGIH®

TITLE
SYSTEM EFFECT FACTORS FOR OUTLET ELBOWS ON CENTRIFUGAL FANS

FIGURE 7-20

DATE 1-12

CHECK CODES, REGULATIONS, AND LAWS (LOCAL, STATE, AND NATIONAL) TO ENSURE THAT DESIGN IS COMPLIANT.

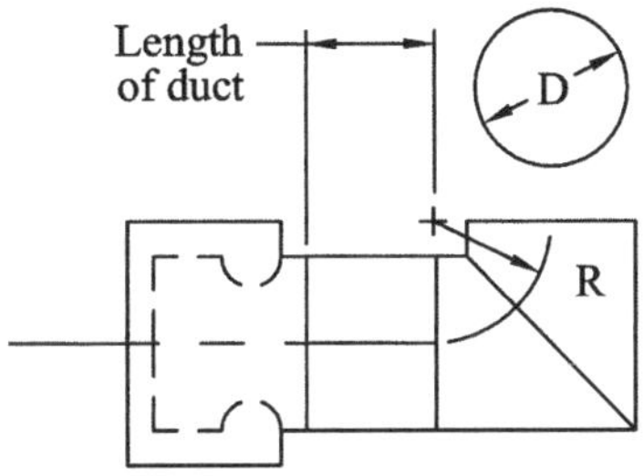

R/D	No Duct	2D Duct	5D Duct
-	N	P	R-S

System Effect Factor

A. TWO-PIECE MITERED 90° ROUND SECTION ELBOW – NOT VANED.

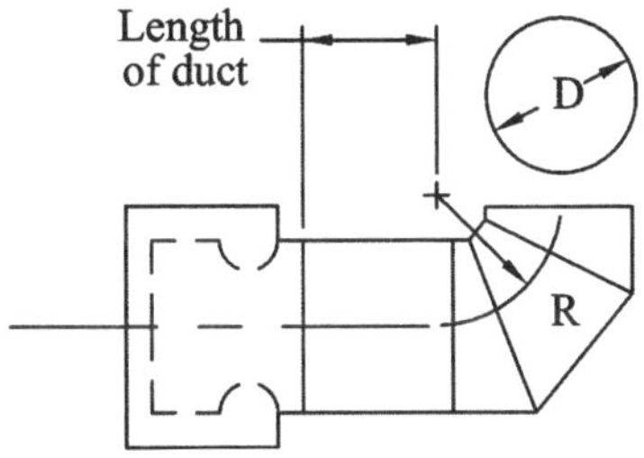

R/D	No Duct	2D Duct	5D Duct
0.5	O	Q	S
0.75	Q	R-S	T-U
1.0	R	S-T	U-V
2.0	R-S	T	U-V
3.0	S	T-U	V

System Effect Factor

B. THREE-PIECE MITERED 90° ROUND SECTION ELBOW – NOT VANED.

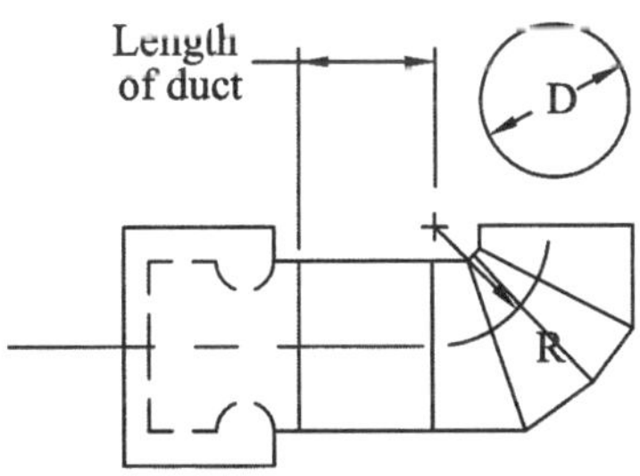

R/D	No Duct	2D Duct	5D Duct
0.5	P-Q	R-S	T
0.75	Q-R	S	U
1.0	R	S-T	U-V
2.0	R-S	T	U-V
3.0	S-T	U	V-W

System Effect Factor

C. FOUR OR MORE PIECE MITERED 90° ROUND SECTION ELBOW – NOT VANED.

D = Diameter of the inlet collar.
The inside area of the square duct (H × H) should be equal to the inside area of the fan inlet collar.
The maximum permissible angle of any converging element of the transition is 15°, and for a diverging element 7°.

Reprinted from AMCA Publication 201-90, FANS AND SYSTEMS, by permission of the Air Movement and Control Association, Inc.[7.3]

ACGIH®

TITLE
SYSTEM EFFECT FACTORS FOR ROUND ELBOWS AT FAN INLET

FIGURE 7-21

DATE 1-12

CHECK CODES, REGULATIONS, AND LAWS (LOCAL, STATE, AND NATIONAL) TO ENSURE THAT DESIGN IS COMPLIANT.

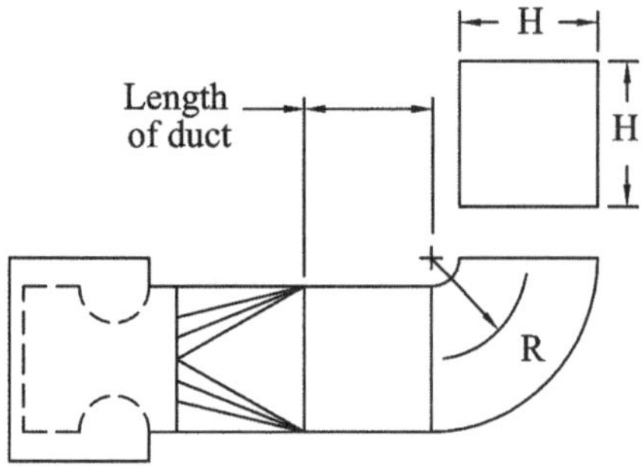

R/D	System Effect Factor No Duct	2D Duct	5D Duct
0.5	O	Q	S
0.75	P	R	S-T
1.0	R	S-T	U-V
2.0	S	T-U	V

A. SQUARE ELBOW WITH INLET TRANSITION – NO TURNING VANES.

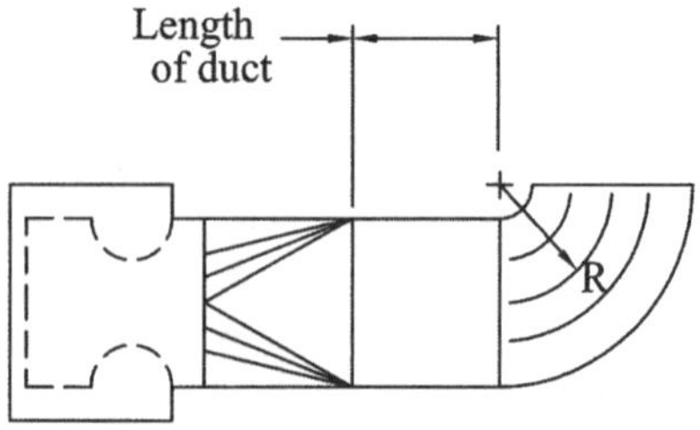

R/D	System Effect Factor No Duct	2D Duct	5D Duct
0.5	S	T-U	V
1.0	T	U-V	W
2.0	V	V-W	W-X

B. SQUARE ELBOW WITH INLET TRANSITION – 3 LONG TURNING VANES.

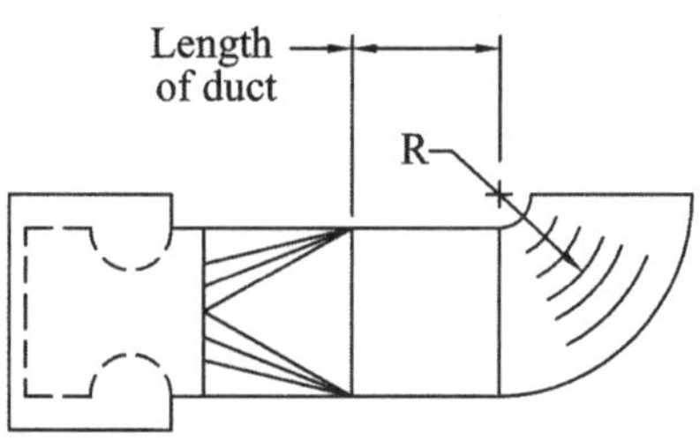

R/D	System Effect Factor No Duct	2D Duct	5D Duct
0.5	S	T-U	V
1.0	T	U-V	W
2.0	V	V-W	W-X

C. SQUARE ELBOW WITH INLET TRANSITION – SHORT TURNING VANES.

The inside area of the square duct (H × H) is equal to the inside area circumscribed by the fan inlet collar. The maximum permissible angle of any converging element of the transition is 15°, and for a diverging element 7.5°.

Reprinted from AMCA Publication 201-90, FANS AND SYSTEMS, by permission of the Air Movement and Control Association, Inc.[7.3]

ACGIH®

TITLE
SYSTEM EFFECT FACTORS FOR ELBOWS AND TRANSITIONS AT FAN INLET

FIGURE 7-22

DATE 1-12

CHECK CODES, REGULATIONS, AND LAWS (LOCAL, STATE, AND NATIONAL) TO ENSURE THAT DESIGN IS COMPLIANT.

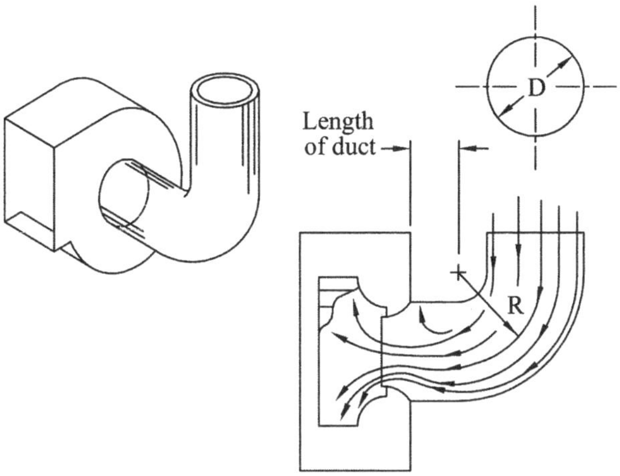

A. Round Inlet Duct
Non-uniform flow into fan inlet

System Effect Factors*

R/D	No DUCT	2D DUCT	5D DUCT
0.75	Q-R	S	U
1.0	R	S-T	U-V
2.0	R-S	T	U-V
3.0	S-T	U	V-W

*Values shown are in modification of the original chart.

*Determine the SEF by using Figure 7-26 or 7-27.

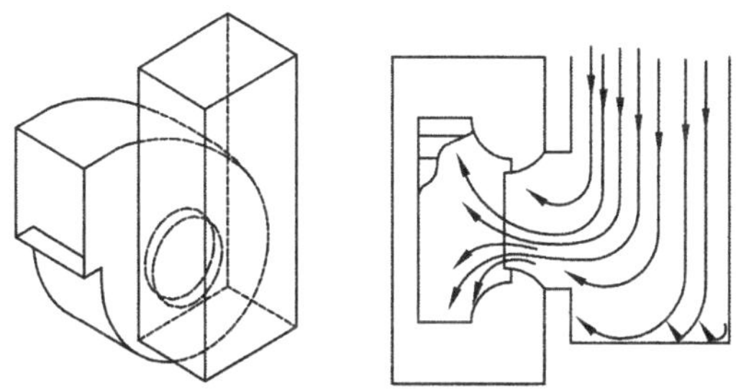

B. Rectangular Inlet Duct
Non-uniform flow into fan inlet

The reduction in flow rate and pressure for this type of inlet condition is impossible to tabulate. The many possible variations in width and depth of the duct influence the reduction in performance to varying degrees and therefore this inlet should be avoided. Flow rate losses as high as 45% have been observed. Existing installations can be improved with guide vanes or the conversion to square or mitered elbows with guide vanes.

Reprinted from AMCA Publication 201-90, FANS AND SYSTEMS, by permission of the Air Movement and Control Association, Inc.[7.3]

ACGIH®	TITLE: SYSTEM EFFECT FACTORS FOR NON-UNIFORM INLET FLOWS	FIGURE: 7-23 / DATE: 1-12

CHECK CODES, REGULATIONS, AND LAWS (LOCAL, STATE, AND NATIONAL) TO ENSURE THAT DESIGN IS COMPLIANT.

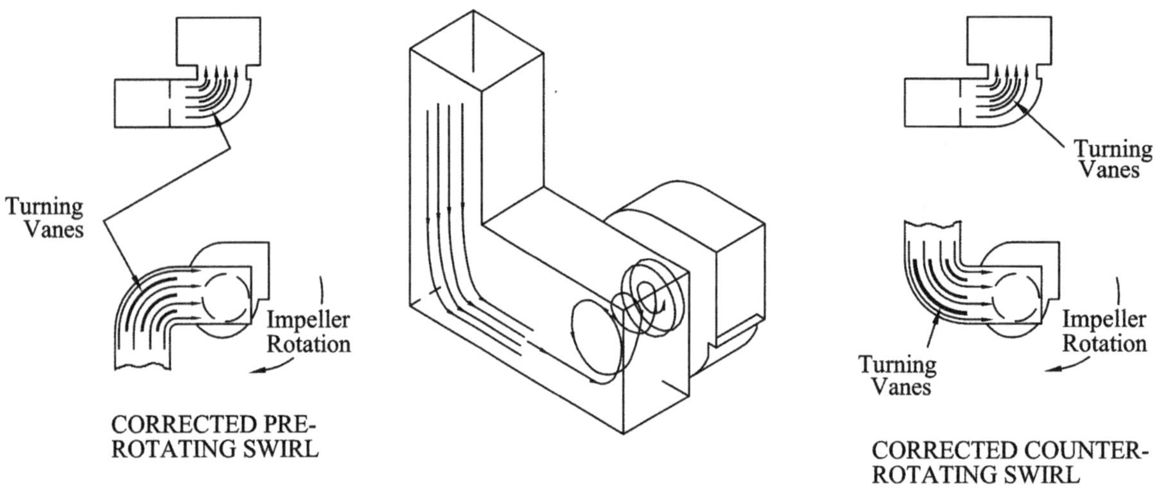

C. Induced Flow Design

Reprinted from AMCA Publication 201-90, FANS AND SYSTEMS, by permission of the Air Movement and Control Association, Inc.[7.3]

ACGIH®	TITLE: NON-UNIFORM FAN INLET CORRECTIONS	FIGURE 7-24
		DATE 1-12

CHECK CODES, REGULATIONS, AND LAWS (LOCAL, STATE, AND NATIONAL) TO ENSURE THAT DESIGN IS COMPLIANT.

FIGURE 7-25

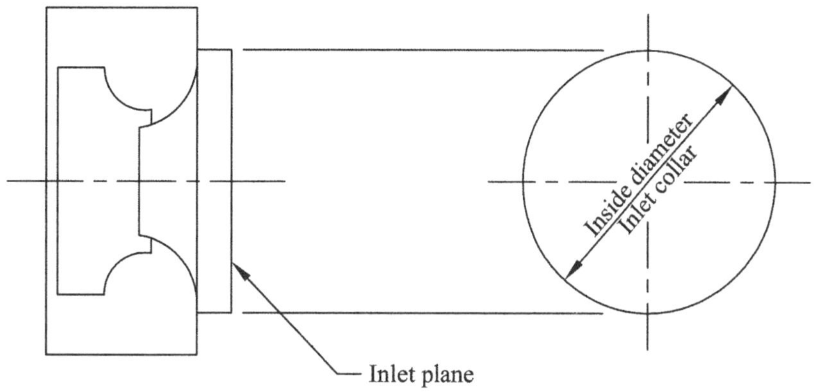

A. FREE INLET AREA PLANE – FAN WITH INLET COLLAR.

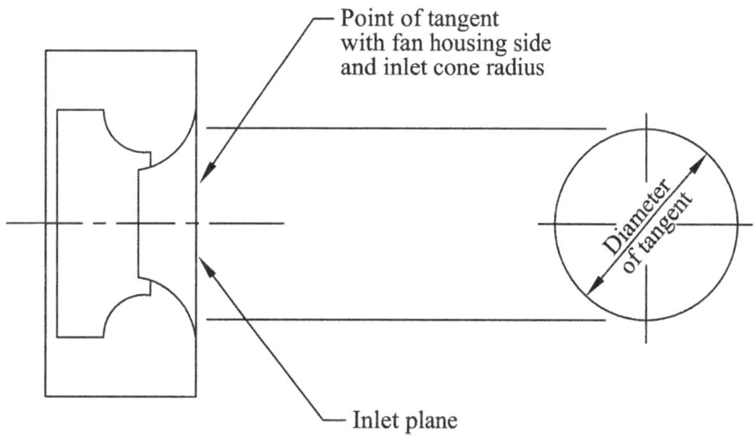

B. FREE INLET AREA PLANE – FAN WITHOUT INLET COLLAR.

PERCENTAGE OF UNOBSTRUCTED INLET AREA	SYSTEM EFFECT FACTORS
100	NO LOSS
95	V
90	U
85	T
75	S
50	Q
25	P

Determine SEF by using Figure 7-26 or 7-27

Reprinted from AMCA Publication 201-90, FANS AND SYSTEMS, by permission of the Air Movement and Control Association, Inc.[7.3]

ACGIH®	TITLE: SYSTEM EFFECT FACTORS FOR INLET OBSTRUCTIONS	FIGURE: 7-25
		DATE: 1-12

CHECK CODES, REGULATIONS, AND LAWS (LOCAL, STATE, AND NATIONAL) TO ENSURE THAT DESIGN IS COMPLIANT.

FIGURE 7-26

SYSTEM EFFECT FACTORS*

Curve	F_{sys}	Curve	F_{sys}
F	16.0	P	1.98
G	14.3	Q	1.60
H	12.8	R	1.20
I	11.3	S	0.80
J	9.62	T	0.53
K	8.02	U	0.40
L	6.42	V	0.26
M	4.63	W	0.18
N	3.20	X	0.10
O	2.51		

To use this table:

1) Obtain the curve letter from Figures 7-18 through 7-23 or Figure 7-25.

2) For inlet system effects, multiply the equivalent loss coefficient from the above table by the fan inlet velocity pressure.

3) For outlet system effects, multiply the equivalent loss coefficient from the above table by the fan outlet velocity pressure.

*F_{sys} values are in number of velocity pressures. For loss directly in "wg, refer to Figure 7-27 (IP).

For loss directly in Pa, refer to Figure 7-27 (SI).

ACGIH®

TITLE: SYSTEM EFFECT FACTORS

FIGURE: 7-26

DATE: 1-12

CHECK CODES, REGULATIONS, AND LAWS (LOCAL, STATE, AND NATIONAL) TO ENSURE THAT DESIGN IS COMPLIANT.

FIGURE 7-27 (IP)

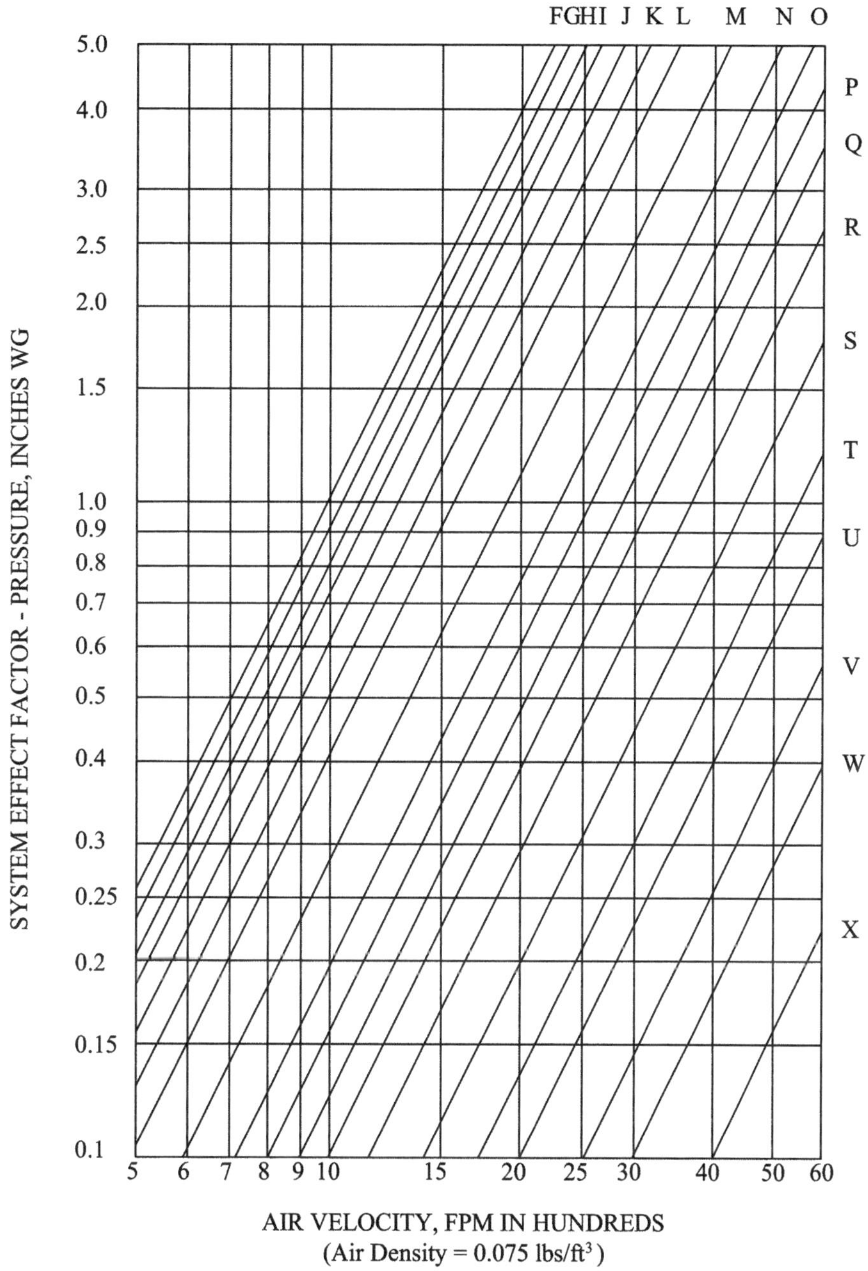

*Enter the chart at the appropriate air velocity (on the abcissa), read up to the application curve, then across from the curve (to the ordinate) to find the SEF at standard air density.

Reprinted from AMCA Publication 201-90, FANS AND SYSTEMS, by permission of the Air Movement and Control Association, Inc.[7.3]

ACGIH®	TITLE SYSTEM EFFECT FACTORS	FIGURE 7-27 (IP) DATE 1-12

CHECK CODES, REGULATIONS, AND LAWS (LOCAL, STATE, AND NATIONAL) TO ENSURE THAT DESIGN IS COMPLIANT.

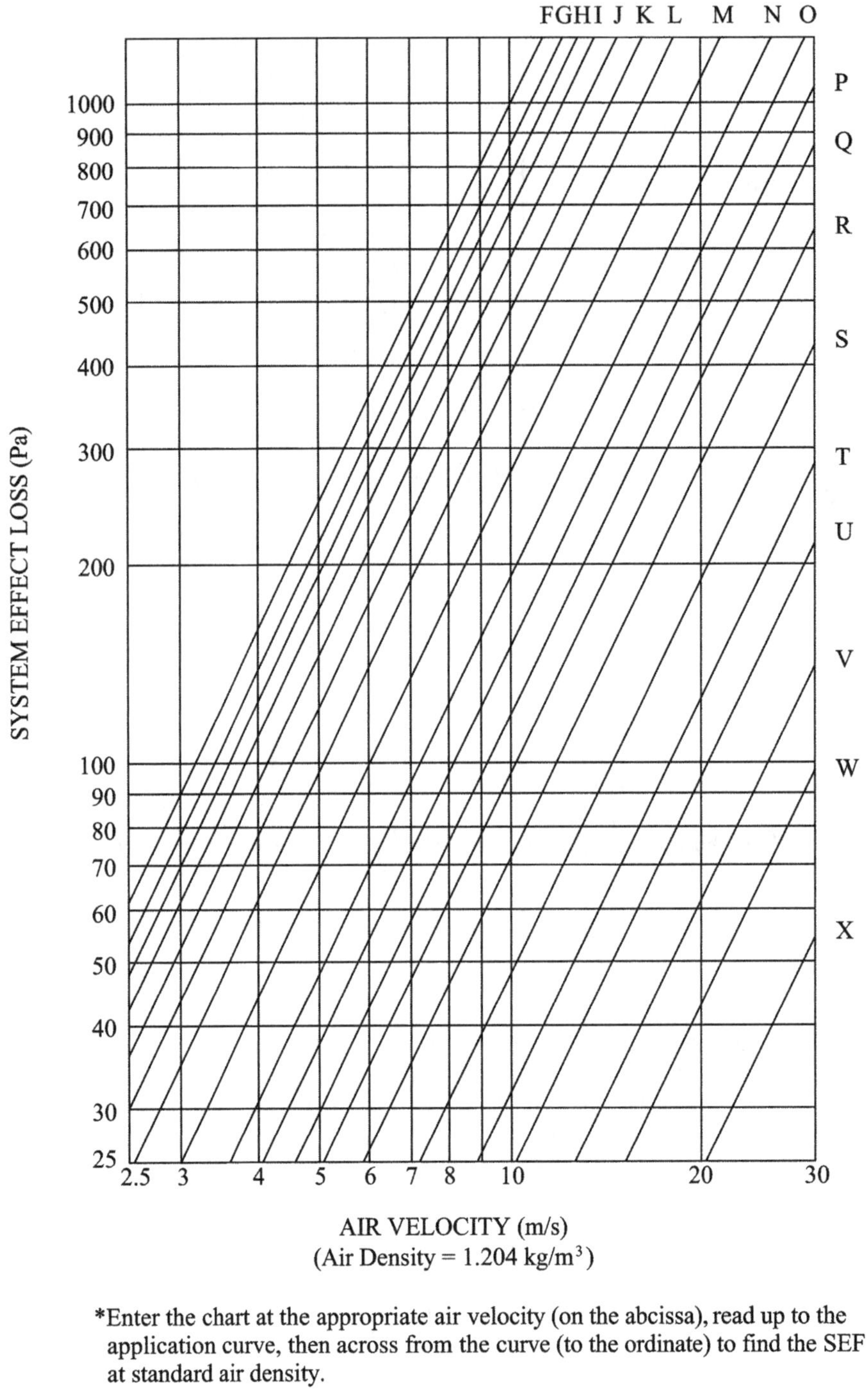

*Enter the chart at the appropriate air velocity (on the abcissa), read up to the application curve, then across from the curve (to the ordinate) to find the SEF at standard air density.

Adapted for metric from AMCA Publication 201-90, FANS AND SYSTEMS, by permission of the Air Movement and Control Association, Inc.[7.3]

FIGURE 7-27 (SI)

ACGIH®	TITLE SYSTEM EFFECT FACTORS	FIGURE 7-27 (SI) DATE 1-12

CHECK CODES, REGULATIONS, AND LAWS (LOCAL, STATE, AND NATIONAL) TO ENSURE THAT DESIGN IS COMPLIANT.

Chapter 8

AIR CLEANING DEVICES

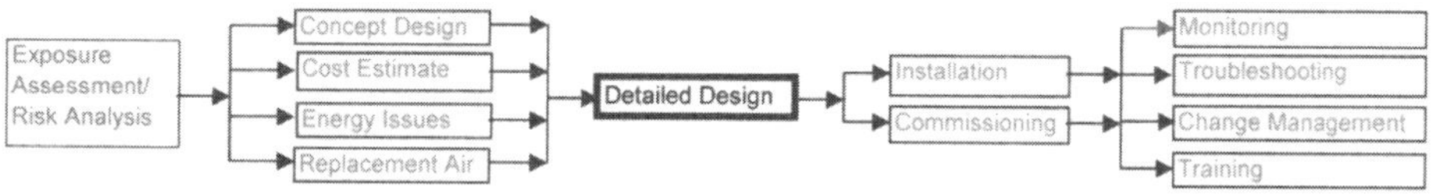

NOTE: Equations with notation followed by (IP) are designated for inch-pound system only; Equations followed by (SI) are designated for metric use only. If Equation bears neither, then it applies to both systems.

8.1 INTRODUCTION

Air cleaning devices remove contaminants from an air or gas stream. They are available in a wide range of designs to meet variations in air cleaning requirements. Degree of removal required, typically dictated by governmental standards, quantity and characteristics of the contaminant to be removed, and conditions of the air or gas stream will all have a bearing on the device selected for any given application. In addition, fire safety and explosion control must be considered in all selections. (See NFPA publications.)

This chapter will give an overview of major contaminant control devices, whether the contaminant is in solid, liquid (aerosol) or in a gaseous state. In order to choose the proper control device, it is of absolute importance to know the chemical constituents, particle or aerosol size distribution and relative concentration of those pollutants. The U.S. Environmental Protection Agency (U.S. EPA) has accepted methods of determining the constituents of different air streams. Testing done outside of these sanctioned test methods are likely not to be accepted as proof of compliance (see www.epa.gov).

For particulate contaminants, air cleaning devices are divided into two basic groups: AIR FILTERS and DUST COLLECTORS. Air filters are designed to remove low dust concentrations of the magnitude found in atmospheric air. They are typically used in ventilation, air-conditioning, and heating systems where dust concentrations seldom exceed 1.0 grains per thousand cubic feet of air, and are usually well below 0.1 grains per thousand cubic feet of air [0.28 gr/m^3]. (One pound equals 7000 grains. A typical atmospheric dust concentration in an urban area is 87 micrograms per cubic meter or 0.000038 grains per standard cubic feet of air.)

Dust collectors are usually designed for the much heavier loads from industrial processes where the air or gas to be cleaned originates in local exhaust systems or process stack gas effluents. Contaminant concentrations will vary from less than 0.1 to 100 grains or more for each cubic foot of air or gas. Therefore, dust collectors are, and must be, capable of handling concentrations 100 to 20,000 times greater than those for which air filters are designed.

Small, inexpensive versions of all categories of air cleaning devices are available. The principles of selection, application, and operation are the same as for larger equipment. However, due to the structure of the market that focuses on small, quickly available, and inexpensive equipment, much of the available equipment is of light duty design and construction. One of the major economies of unit collectors implies recirculation, for which such equipment may or may not be suitable. For adequate prevention of health hazards, fires, and explosions, application engineering is just as essential for unit collectors as it is for major systems.

8.2 SELECTION OF DUST COLLECTION EQUIPMENT

Dust collection equipment is available in numerous designs utilizing many different principles and featuring wide variations in effectiveness, first cost, operating and maintenance cost, space, arrangement, and materials of construction. Consultation with the equipment manufacturer is the recommended procedure in selecting a collector for any problem where extensive previous plant experience on the specific dust problem is not available.

8.2.1 Efficiency Required. Currently, there is no accepted standard for testing and/or expressing the "efficiency" of a dust collector. It is virtually impossible to accurately compare the performance of two collectors by comparing efficiency claims. The only true measure of performance is the actual mass emission rate, expressed in terms such as mg/m^3 or grains/ft^3. Evaluation will consider the need for high efficiency–high cost equipment requiring minimum energy high voltage electrostatic precipitators, high efficiency–moderate cost equipment such as fabric or wet collectors, or the lower cost primary units such as the dry centrifugal group. If either of the first two groups is selected, the combination with primary collectors should be considered.

When the cleaned air is to be discharged outdoors, the required degree of collection can depend on plant location; nature of contaminant (its salvage value and its potential as a health hazard, public nuisance, or ability to damage property); and the regulations of governmental agencies. In remote locations, damage to farms or contribution to air pollution problems of distant cities can influence the need for and importance of effective collection equipment. Many industries, originally located away from residential areas, failed to anticipate the construction of residential building which frequently develops around a plant. Such lack of foresight has required installation of air cleaning equipment at greater expense than initially would have been necessary. Today, the remotely located plant must comply, in most cases, with the same regulations as the plant located in an urban area. With present and future emphasis on public nuisance, public health, and preservation and improvement of community air quality, management can continue to expect criticism for excessive emissions of air contaminants whether located in a heavy industry section of a city or in an area closer to residential zones.

The mass rate of emission will also influence equipment selection. For a given concentration, the larger the exhaust volumetric flow rate, the greater the need for better equipment. Large central steam generating stations might select high efficiency electrostatic precipitators or fabric collectors for their pulverized coal boiler stacks while a smaller industrial pulverized fuel boiler might be able to use slightly less efficient collectors.

A safe recommendation in equipment selection is to select the collector that will allow the least possible amount of contaminant to escape and is reasonable in first cost and maintenance while meeting all prevailing air pollution regulations. For some applications, even the question of reasonable cost and maintenance must be sacrificed to meet established standards for air pollution control or to prevent damage to health or property. However, in areas designed above the established National Ambient Air Quality health limit (NAAQS) for a pollutant, for example, multiple control devices may be required in order to minimize emissions to the lowest achievable emission rate (LAER) as designated by the U.S. EPA.

It must be remembered that visibility of an effluent will be a function of the light reflecting surface area of the escaping material. Surface area per pound increases inversely as the square of particle size. This means that the removal of 80% or more of the dust on a weight basis may remove only the coarse particles without altering the stack appearance.

8.2.2 Gas Stream Characteristics. The characteristics of the carrier gas stream can have a marked bearing on equipment selection. Temperature of the gas stream may limit the material choices in fabric collectors. Condensation of water vapor will cause packing and plugging of air or dust passages in dry collectors. Corrosive chemicals can attack fabric or metal in dry collectors and when mixed with water in wet collectors can cause extreme damage.

8.2.3 Contaminant Characteristics. The contaminant characteristics will also affect equipment selection. Chemicals emitted may attack collector elements or corrode wet type collectors. Sticky materials, such as metallic buffing dust impregnated with buffing compounds, can adhere to collector elements, plugging collector passages. Linty materials will adhere to certain types of collector surfaces or elements. Abrasive materials in moderate to heavy concentrations will cause rapid wear on dry metal surfaces. Particle size, shape, and density will rule out certain designs. For example, the parachute shape of particles like the "bees wings" from grain will float through centrifugal collectors because their velocity of fall is less than the velocity of much smaller particles having the same specific gravity but a spherical shape. This difference is termed the "aerodynamic particle diameter" and drastically affects how some particles can be collected in the field. In addition, the combustible nature of many finely divided materials will require specific collector designs to assure safe operation.

Contaminants in exhaust systems cover an extreme range in concentration and particle size. Concentrations can range from less than 0.1 to much more than 100,000 grains of dust per cubic foot of air [2.29×10^5 g/m^3]. In low pressure conveying systems, the dust ranges from 0.5 to 100 or more microns in size. Deviation from mean size (the range over and under the mean) will also vary with the material.

8.2.4 Energy Considerations. The cost and availability of energy makes essential the careful consideration of the total energy requirement for each collector type that can achieve the desired performance. An electrostatic precipitator, for example, might be a better selection at a significant initial cost penalty because of the energy savings through its inherently lower pressure drop.

8.2.5 Dust Disposal. Methods of removal and disposal of collected materials will vary with the material, plant process, quantity involved, and collector design. Dry collectors can be unloaded continuously or in batches through dump gates, trickle valves, and rotary locks to conveyors or containers. Dry materials can create a secondary dust problem if careful thought is not given to dust-free material disposal or to collector dust bin locations suited to convenient material removal. See Figures 8-1, 8-2, and 8-3 for some typical discharge arrangements and valves. In addition, waste materials originating from air pollution control devices are hazardous waste as described by US regulators until they can be proven otherwise.

Wet collectors can be arranged for batch removal or continual ejection of dewatered material. Secondary dust problems are eliminated although disposal of wet sludge can be a material handling problem. Solids or dissolved toxins carry-over in waste water can create a sewer or stream pollution problem if waste water is not properly cleaned.

Material characteristics can influence disposal problems. Packing and bridging of dry materials in dust hoppers, and floating or slurry forming characteristics in wet collectors are examples of problems that can be encountered.

8.3 DUST COLLECTOR TYPES

The four major types of dust collectors for particulate contaminants are Electrostatic Precipitators, Fabric Collectors, Wet Collectors, and Dry Centrifugal Collectors.

8.3.1 Electrostatic Precipitators. In electrostatic precipitation, a high potential electric field is established between discharge and collecting electrodes of opposite electrical charge. The discharge electrode is of small cross-sectional area, such as a wire or a piece of flat stock, and the collection electrode is large in surface area such as a plate.

The gas to be cleaned passes through an electrical field that develops between the electrodes. At a critical voltage, the gas molecules are separated into positive and negative ions. This is called "ionization" and takes place at, or near, the surface of the discharge electrode. Ions having the same polarity as the discharge electrode attach themselves to neutral particles in the gas stream as they flow through the precipitator. These charged particles are then attracted to a collecting plate of opposite polarity. Upon contact with the collecting surface, dust particles lose their charge and then can be easily removed by washing, vibration, or gravity.

The electrostatic process consists of:

1) Ionizing the gas;
2) Charging the dust particles;
3) Transporting the particles to the collecting surface;
4) Neutralizing, or removing the charge from the dust particles; and
5) Removing the dust from the collecting surface.

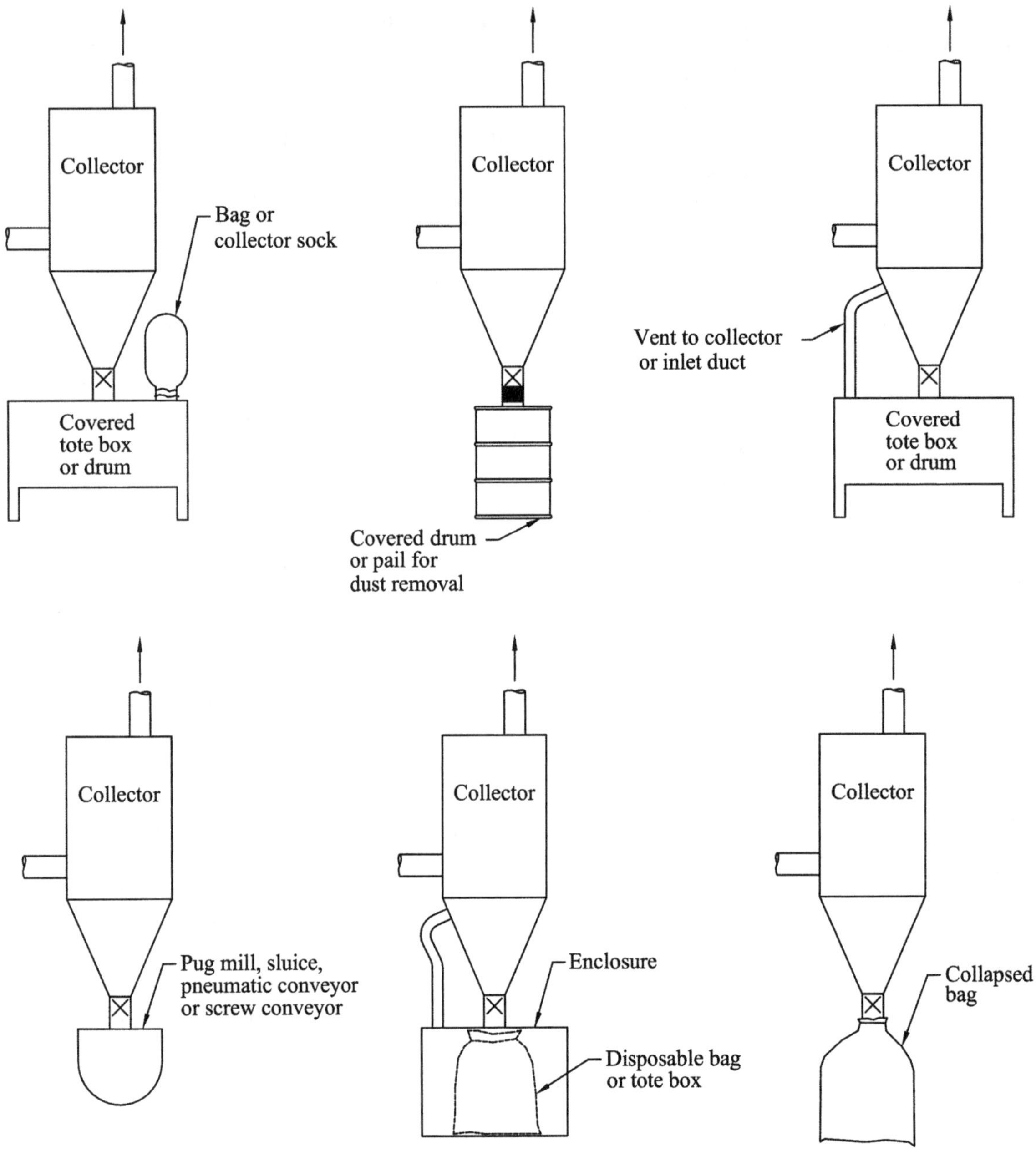

☒ — Rotary Valve

Do not store dust in collector hopper

ACGIH®	TITLE: DRY TYPE DUST COLLECTORS DUST DISPOSAL	FIGURE 8-1 / DATE 1-12

CHECK CODES, REGULATIONS, AND LAWS (LOCAL, STATE, AND NATIONAL) TO ENSURE THAT DESIGN IS COMPLIANT.

FIGURE 8-2

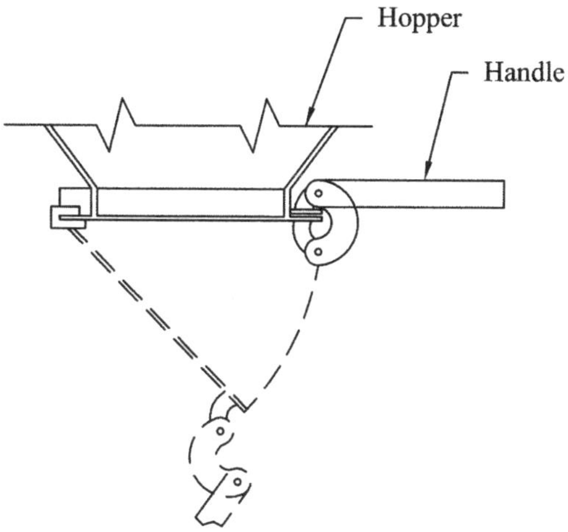

For intermittent manual dumping where dust loads are light.

DUST DOOR

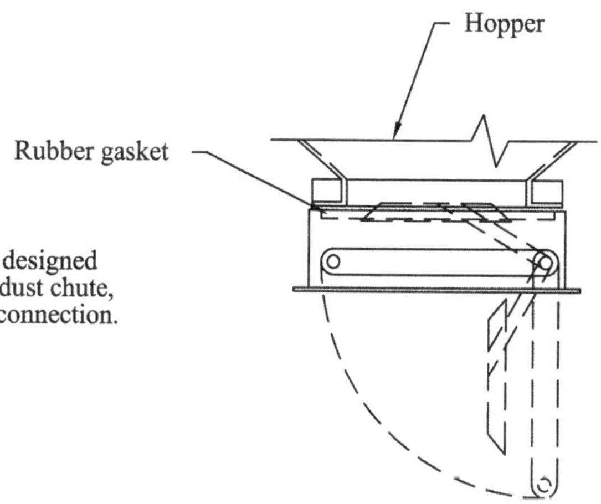

Similar to dust door but designed for direct attachment to dust chute, external pipe or canvas connection.

DUST GATE

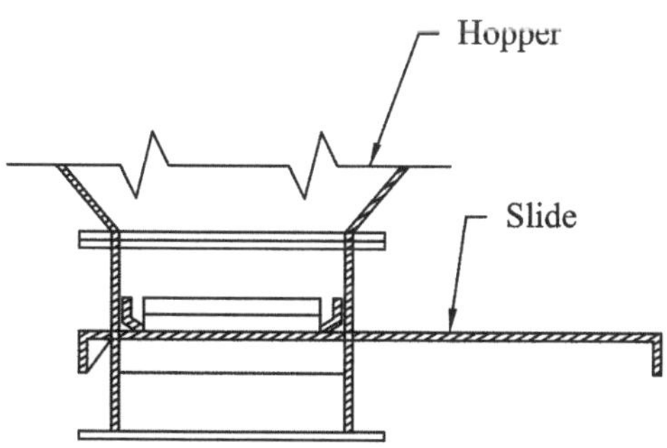

For intermittent, manual dumping where dust loads are light. Flange for connection to dust disposal chute.

SLIDE GATE

TITLE	FIGURE
DRY TYPE DUST COLLECTORS DISCHARGE VALVES	8-2
	DATE 1-12

CHECK CODES, REGULATIONS, AND LAWS (LOCAL, STATE, AND NATIONAL) TO ENSURE THAT DESIGN IS COMPLIANT.

FIGURE 8-3

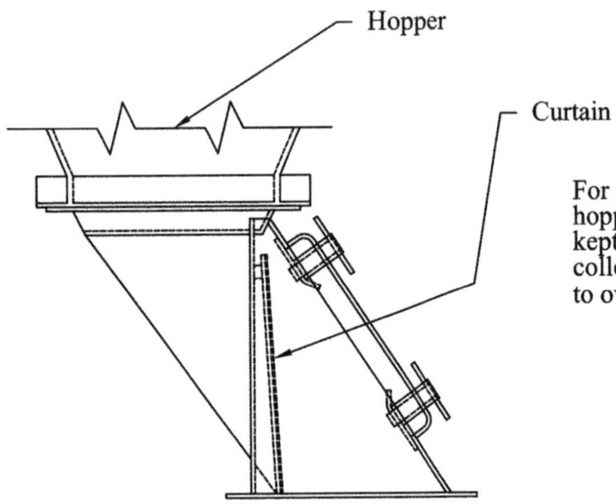

For continuous removal of collected dust where hopper is under negative pressure. Curtain is kept closed by pressure differential until collected material builds up sufficient height to overcome pressure.

TRICKLE VALVE

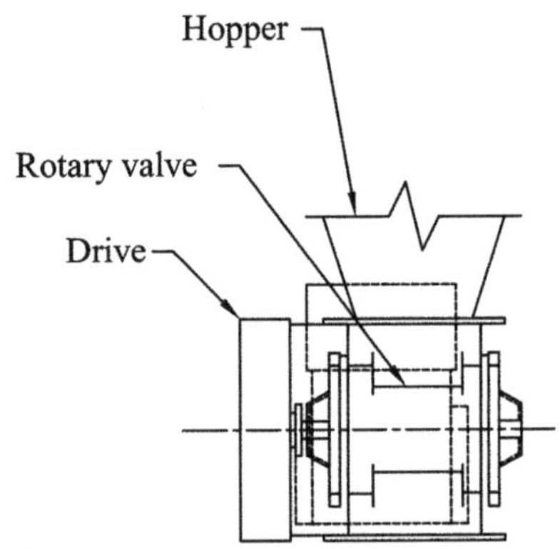

Motor driven multiple blade rotary valve provides an air seal while continuously dumping collected material. Can be used with hoppers under either positive or negative pressure. Flanged for connection to dust disposal chute.

ROTARY LOCK

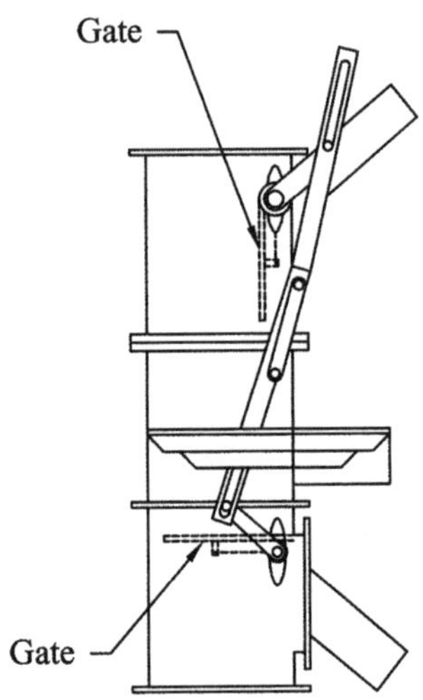

Motor driven, double gate valve for continuous removal of collected dust. Gates are sequenced so only one is open at a time in order to provide air seal. Flanged for connection to dust disposal chute.

DOUBLE DUMP VALVE

ACGIH®

TITLE	FIGURE	
DRY TYPE DUST COLLECTORS DISCHARGE VALVES	FIGURE 8-3	DATE 1-12

CHECK CODES, REGULATIONS, AND LAWS (LOCAL, STATE, AND NATIONAL) TO ENSURE THAT DESIGN IS COMPLIANT.

The two basic types of electrostatic precipitators are "Cottrell," or single-stage, and "Penny," or two-stage (Figures 8-4 and 8-5).

The "Cottrell," single-stage precipitator (Figure 8-4) combines ionization and collection in a single stage. Because it operates at ionization voltages from 40,000 to 75,000 volts DC, it may also be called a high voltage precipitator and is used extensively for heavy duty applications such as utility boilers, larger industrial boilers, and cement kilns. Some precipitator designs use sophisticated voltage control systems and rigid electrodes instead of wires to minimize maintenance problems.

The "Penny," or two-stage precipitator (Figure 8-5) uses DC voltages from 11,000 to 15,000 for ionization and is frequently referred to as a low voltage precipitator. Its use is limited to low concentrations, normally not exceeding 0.025 grains per cubic foot [0.057 gr/m^3]. It can be the most practical collection technique for the many hydrocarbon applications where an initially clear exhaust stack turns into a visible emission as vapor condenses. Some applications include plasticizer ovens, forge presses, die-casting machines, and various welding operations. Care must be taken to keep the precipitator inlet temperature low enough to insure that condensation has already occurred.

For proper results, the inlet gas stream should be evaluated and treated where necessary to provide proper conditions for ionization. For high-voltage units a cooling tower is sometimes necessary. Low voltage units may use wet scrubbers, evaporative coolers, heat exchangers, or other devices to condition the gas stream for best precipitator performance.

The pressure drop of an electrostatic precipitator is extremely low, usually less than 1 "wg [250 Pa]; therefore, the energy requirement is significantly less than for other techniques.

A modified style of Electrostatic Collector has come to the forefront on sticky submicron aerosol particulate that incorporates some properties of wet scrubbers and ESPs. It utilizes a continuous coating of the collection plates with water to cause particulate to collect on the water surface instead of sticking to the collection plates themselves. Wet electrostatic precipitation (WESP), once considered "experimental," has proven itself a very viable alternative on some very difficult particulate. As with scrubbers, water waste treatment is a significant issue; and wastewater treatability should be a part of every determination to use this (or any other) wet collection technology.

8.3.2 Fabric Collectors. Fabric collectors remove particulate by straining, impingement, interception, diffusion, and electrostatic charge. The "fabric" may be constructed of any fibrous material, either natural or man-made, and may be spun into a yarn and woven or felted by needling, impacting, or bonding. Woven fabrics are identified by thread count and weight of fabric per unit area. Non-woven (felts) are identified by thickness and weight per unit area. Regardless of construction, the fabric represents a porous mass through which the gas is passed unidirectionally such that dust particles are retained on the dirty side and the cleaned gas passes through.

The ability of the fabric to pass air is stated as "permeability" and is defined as the cubic feet of air that is passed through one square foot of fabric each minute at a pressure drop of 0.5 "wg [125 Pa]. Typical permeability values for commonly used fabrics range from 25 to 40 acfm [0.012 to 0.019 am^3/s].

A non-woven (felted) fabric is more efficient than a woven fabric of identical weight because the void areas or pores in the non-woven fabric are smaller. A specific type of fabric can be made more efficient by using smaller fiber diameters, a greater weight of fiber per unit area and by packing the fibers more tightly. For non-woven construction, the use of finer needles for felting also improves efficiency. While any fabric is made more efficient by these methods, the cleanability and permeability are reduced. A highly efficient fabric that cannot be cleaned represents an excessive resistance to airflow and is not an economical engineering solution. Final fabric selection is generally a compromise between efficiency and permeability.

Over the past 20 years, chemically inert membrane laminates of extended PTFE (Teflon) have shown value due to enhanced particulate release and ultra high efficiencies. Difficult particulate such as metal fumes or high temperatures are a good match for PTFE membrane technologies. However, condensable hydrocarbons and oils will foul the membranes (Table 8-1).

Choosing a fabric with better cleanability or greater permeability but lower inherent efficiency is not as detrimental as it may seem. The efficiency of the fabric as a filter is meaningful only when new fabric is first put into service. Once the fabric has been in service any length of time, collected particulate in contact with the fabric acts as a filter aid, defining the real collection efficiency. Therefore, compliance testing should never be attempted on new filters until they have been "seasoned" in service. Depending on the amount of particulate and the time interval between fabric reconditioning, it may well be that virtually all filtration is accomplished by the previously collected particulate — or dust cake — as opposed to the fabric itself. Even immediately after cleaning, a residual and/or redeposited dust cake provides additional filtration surface and higher collection efficiency than obtainable with new fabric. While the collection efficiency of new, clean fabric is easily determined by laboratory test and the information is often published, it is not representative of operating conditions and, therefore, is of little importance in selecting the proper collector.

Fabric collectors are not 100% efficient, but well-designed, adequately sized, and properly operated fabric collectors can be expected to operate at efficiencies in excess of 99%, and often as high as 99.9+% on a mass basis. The inefficiency, or penetration, that does occur is greatest during or immediately after reconditioning of the media. Fabric collector inefficiency is frequently a result of by-pass due to damaged fabric, faulty seals, or sheet metal leaks rather than penetration of the fabric. Where extremely high collection efficiency is essential, the fabric collector should be tested for mechanical leaks. In addition, when highly toxic dusts are involved, a designer should

FIGURE 8-4

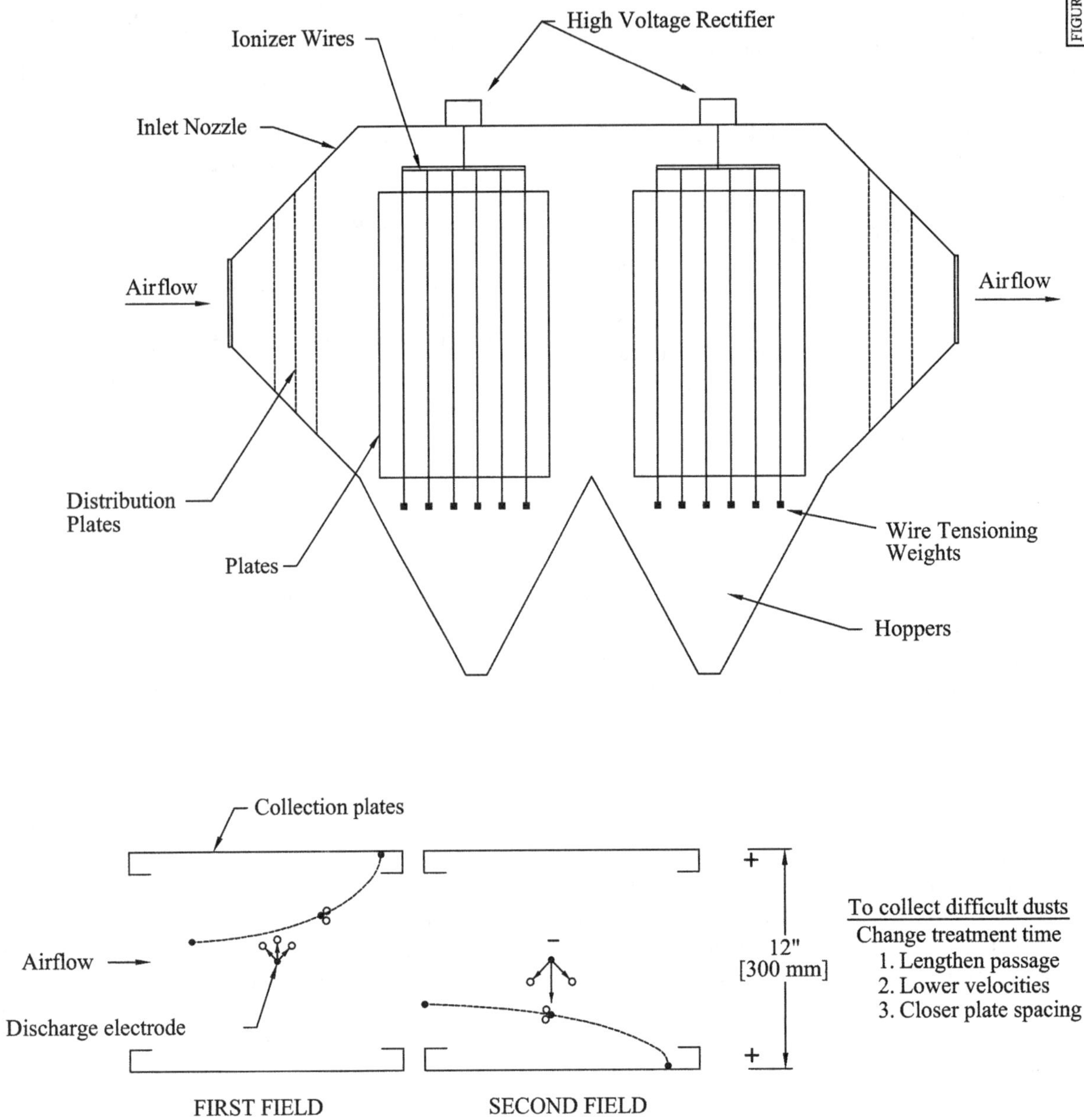

ACGIH®	TITLE ELECTROSTATIC PRECIPITATOR HIGH POWER DESIGN (40,000 TO 75,000 VOLTS)	FIGURE 8-4
		DATE 1-12

CHECK CODES, REGULATIONS, AND LAWS (LOCAL, STATE, AND NATIONAL) TO ENSURE THAT DESIGN IS COMPLIANT.

FIGURE 8-5

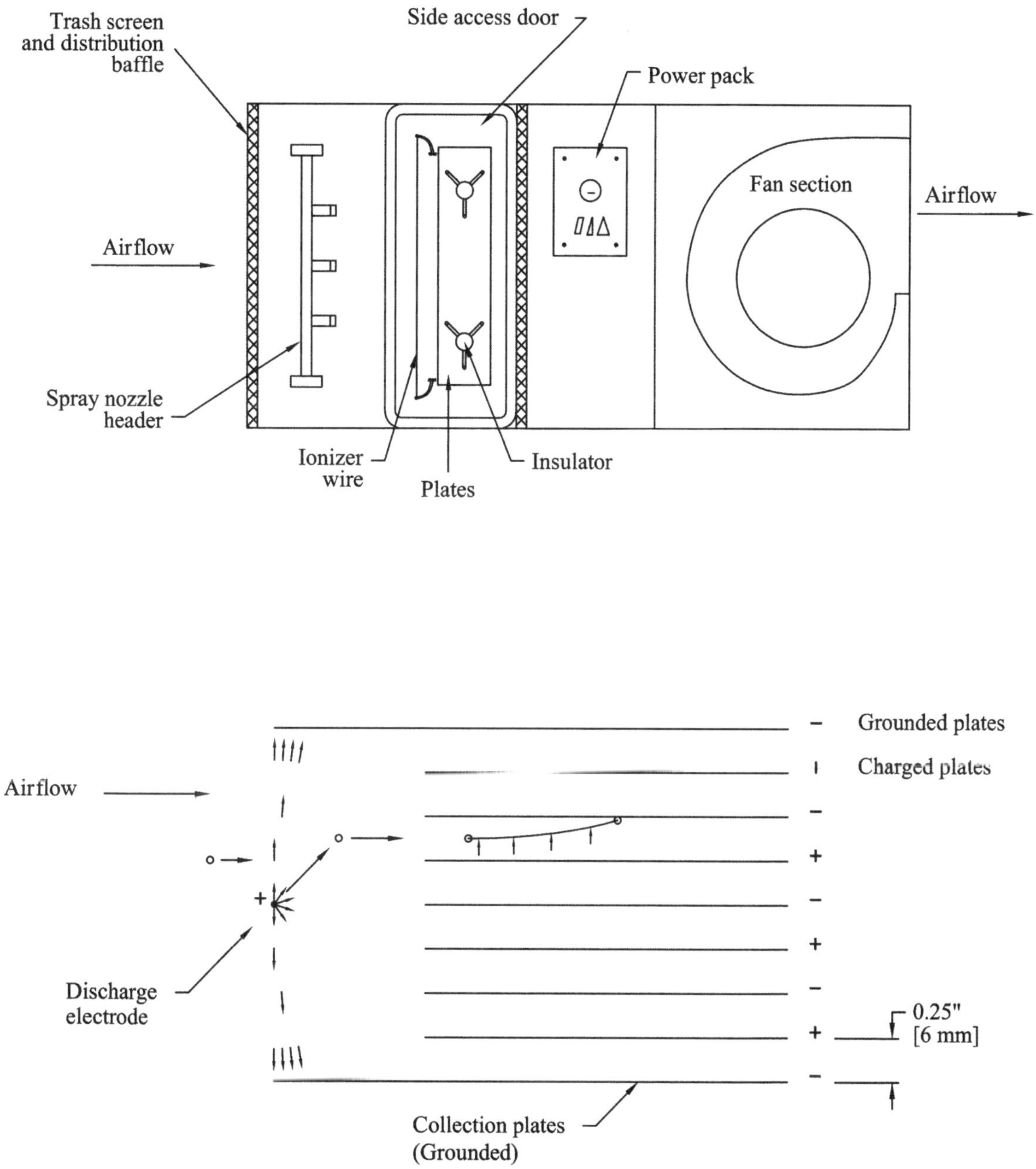

TITLE	FIGURE
ELECTROSTATIC PRECIPITATOR LOW POWER DESIGN (11,000 TO 15,000 VOLTS)	8-5
	DATE 1-12

CHECK CODES, REGULATIONS, AND LAWS (LOCAL, STATE, AND NATIONAL) TO ENSURE THAT DESIGN IS COMPLIANT.

TABLE 8-1. Characteristics of Filter Fabrics*

Generic Names	Example Trade Name Fabrics**	Max. Temp. F [C]		Resistance to Physical Action					Resistance to Chemicals				
		Continuous	Intermittent	Dry Heat	Moist Heat	Abrasion	Shaking	Flexing	Mineral Acid	Organic Acid	Alkalies	Oxidizing	Solvents
Cotton	Cotton	180 [86]	—	G	G	F	G	G	P	G	F	F	E
Polyester	Dacron[1] Fortrel[2] Vycron[3] Kodel[4] Enka Polyester[5]	275 [135]	—	G	F	G	E	E	G	G	F	G	E
Acrylic	Orlon[1] Acrilan[6] Creslan[7] Dralon T[8] Zefran[9]	275 [135]	285 [140]	G	G	G	G	E	G	G	F	G	E
Modacrylic	Dynel[10] Verel[4]	160 [71]	—	F	F	F	P-F	G	G	G	G	G	G
Nylon (Polyamide)	Nylon 6,6[1,2,6] Nylon 6[11,5,12]	225 [107]	—	G	G	E	E	E	P	F	G	F	E
	Nomex[11]	400 [204]	450 [232]	E	E	E	E	E	P-F	E	G	G	E
Polyimide	P-84[18]	500 [260]	580 [304]	E	P	G	G	E	P-F	G	F	G	E
Polypropylene	Herculon[13] Reevon[14] Vectra[15]	200 [93]	250 [121]	G	F	E	E	G	E	E	E	G	G
Teflon (Flurocarbon)	Teflon TFE[1]	500 [260]	550 [288]	E	E	P-F	G	G	E	E	E	E	E
	Teflon FEP[1]	450 [232]	—	E	E	P-F	G	G	E	E	E	E	E
Expanded PFTE	Rastex	500 [260]	550 [288]	E	E	P-F	G	G	E	E	E	E	E
Vinyon	Vinyon[16] Clevylt[17]	350 [177]	—	F	F	F	G	G	E	E	G	G	P
Glass	Glass	500 [260]	600 [316]	E	E	P	P	F	E	E	F	E	E
Fiberglass	Fiberglass[19]	550 [288]	550 [288]	E	E	P	P	G	G	G	G	E	G

*E = excellent; G = good; F = fair; P = poor

**Registered Trademarks

(1) Du Pont; (2) Celanese; (3) Beaunit; (4) Eastman; (5) American Enka; (6) Chemstrand; (7) American Cyanamid; (8) Farbenfabriken Bayer AG; (9) Dow Chemical; (10) Union Carbide; (11) Allied Chemical; (12) Firestone; (13) Hercules; (14) Alamo Polymer; (15) National Plastic; (16) FMC; (17) Societe Rhovyl; (18) Lenzing; (19) Huyglas

consider the use of secondary absolute filtration (safety monitoring filters) such as HEPA filters (or the like). Under some circumstances, even highly toxic particulate-laden air streams can be recirculated into the workplace (see Chapter 10, Section 10.8).

The combination of fabric and collected dust becomes increasingly efficient as the dust cake accumulates on the fabric surface. At the same time, the resistance to airflow increases. Unless the air moving device is adjusted to compensate for the increased resistance, the gas flow rate will be reduced. Figure 8-6 shows how efficiency, resistance to flow and flow rate change with time as dust accumulates on the fabric. Fabric collectors are suitable for service on relatively heavy dust concentrations. The amount of dust collected on a single square yard of fabric may exceed five pounds per hour. In virtually all applications, the amount of dust cake accumulated in just a few hours will represent sufficient resistance to flow to cause an unacceptable reduction in airflow.

In a well-designed fabric collector system, the fabric or filter mat is cleaned or reconditioned before the reduction in airflow is critical. The cleaning is accomplished by mechanical agitation or air motion, which frees the excess accumulation of dust from the fabric surface and leaves a residual or base cake. The residual dust cake does not have the same characteristics of efficiency or resistance to airflow as new fabric.

Commercially available fabric collectors employ fabric configured as bags or tubes, envelopes (flat bags), rigid elements, or pleated cartridges. Most of the available fabrics, whether woven or non-woven, are employed in either bag or envelope configuration. The pleated cartridge arrangement uses a paper-like fiber in either a cylindrical or panel configuration. It features extremely high efficiency on light concentrations. Earlier designs employed cellulose based media. Today, more conventional media, such as polypropylene or spun-bonded polyester, are frequently used.

The variable design features of the many fabric collectors available are:

1) Type of fabric (woven or non-woven)
2) Fabric configuration (bags or tubes, envelopes, cartridges)
3) Intermittent or continuous service
4) Type of reconditioning (shaker, pulse-jet/reverse-air)
5) Housing configuration (single compartment, multiple compartment)

At least two of these features will be interdependent. For example, non-woven fabrics are more difficult to recondition and, therefore, require high-pressure cleaning.

A fabric collector is selected for its mechanical, chemical, and thermal characteristics. Table 8-1 lists those characteristics for some common filter fabrics.

Fabric collectors are sized to provide a sufficient area of filter media to allow operation without excessive pressure drop. The amount of filter area required depends on many factors, including:

1) Release characteristics of dust
2) Porosity of dust cake
3) Concentration of dust in carrier gas stream
4) Type of fabric and surface finish, if any
5) Type of reconditioning
6) Reconditioning interval
7) Airflow pattern within the collector
8) Temperature and humidity of gas stream

Because of the many variables and their range of variation, fabric collector sizing is a judgment based on experience. The sizing is usually made by the equipment manufacturer, but at times may be specified by the user or a third party. Where no experience exists, a pilot installation is the only reliable way to determine proper size.

The sizing or rating of a fabric collector is expressed in terms of airflow rate versus fabric media area. The resultant ratio is called "air-to-cloth ratio" with units of acfm per square foot of fabric. This ratio represents the average velocity of the gas stream through the filter media. The expression "filtration velocity" is used synonymously with air-to-cloth ratio for rating fabric collectors. For example, an air-to-cloth ratio of 7:1 (7 acfm/sq ft) is equivalent to a filtration velocity of 7 fpm. In metric values the equivalent would be expressed in m/s. In that case 0.04 $m^3/s/m^2$ is a filtration velocity of 0.04 m/s.

Table 8-2 compares the various characteristics of fabric collectors. The different types will be described in detail later. Inspection of Table 8-2 now may make the subsequent discussion more meaningful. The first major classification of fabric collectors is intermittent or continuous duty. Intermittent duty fabric collectors cannot be reconditioned while in operation. By design, they require that the gas flow be interrupted while the fabric is agitated to free accumulated dust cake. Continuous duty collectors do not require shut down for reconditioning.

Shaker Fabric Collectors: Intermittent duty fabric collectors may use a tube, cartridge, or envelope configuration of woven fabric and will generally employ shaking or vibration for reconditioning. Figure 8-7 shows both tube and envelope shaker collector designs. For the tube type, dirty air enters the open bottom of the tube and dust is collected on the inside of the fabric. The bottoms of the tubes are attached to a tube sheet and the tops are connected to a shaker mechanism. Since the gas flow is from inside to outside, the tubes tend to inflate during operation and no other support of the fabric is required.

Gas flow for envelope type collectors is from outside to inside, therefore, the envelopes must be supported during operation to prevent collapsing. This is normally done by inserting wire mesh or fabricated wire cages into the envelopes. The opening of the envelope from which the

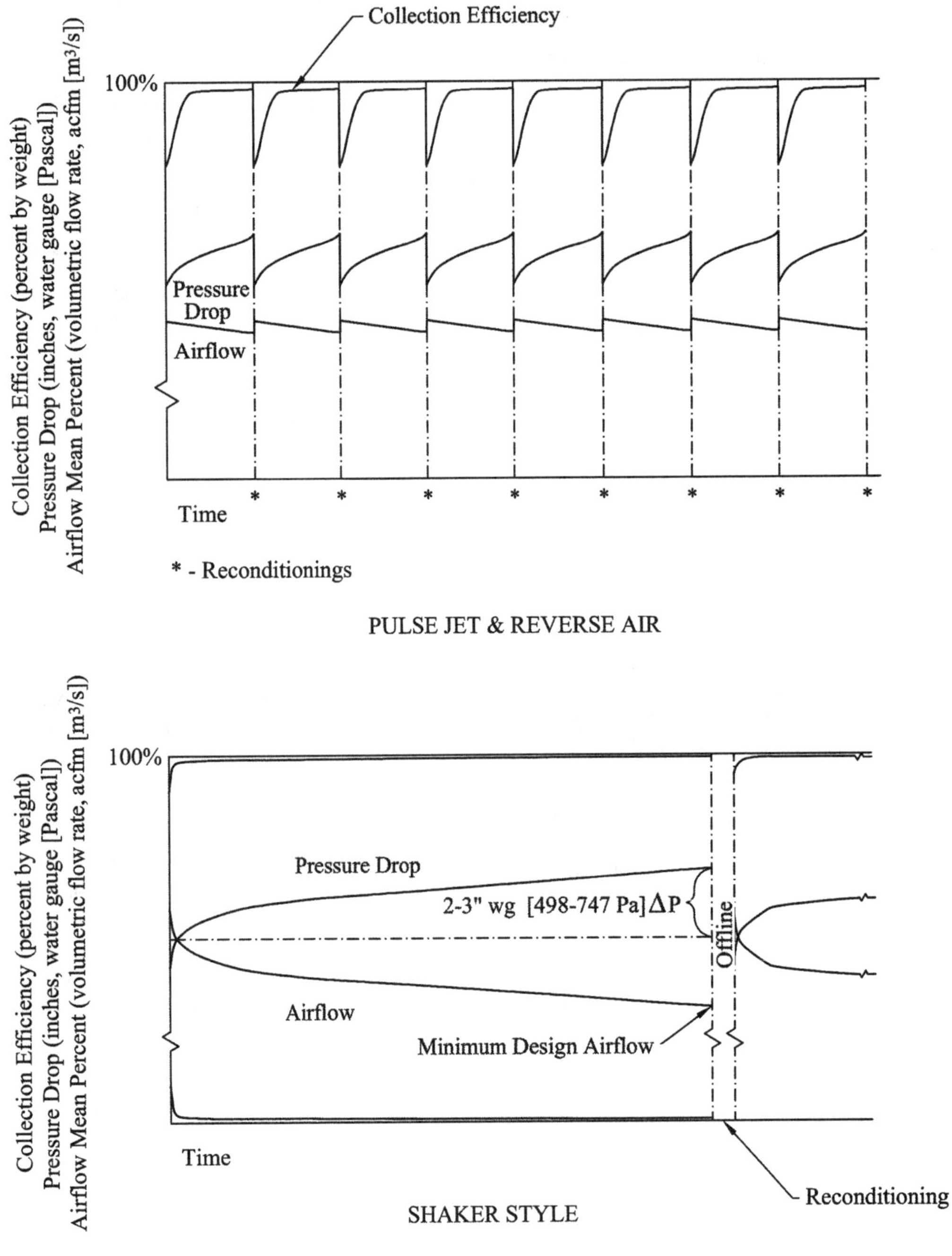

FIGURE 8-6

TITLE **PERFORMANCE vs. TIME BETWEEN RECONDITIONINGS FOR FABRIC COLLECTORS**

FIGURE 8-6

DATE 1-12

CHECK CODES, REGULATIONS, AND LAWS (LOCAL, STATE, AND NATIONAL) TO ENSURE THAT DESIGN IS COMPLIANT.

TABLE 8-2. Summary of Fabric Type Collectors and Their Characteristics

	INTERRUPTABLE OPERATION Light to Moderate Loading	INTERRUPTABLE OPERATION Heavy Loading		CONTINUOUS OPERATION Any Loading	
Fabric Reconditioning Requirement	Intermittent	Continuous			
Type of Reconditioning	Shaker	Shaker	Reverse Air (Low Pressure)	Reverse Pulse (High Pressure) Pulse Jet or Fan Pulse	
Collector Configuration	Single Compartment	Multiple Compartments with inlet or outlet dampers for each		Single Compartment	
Fabric Configuration	Tube, Cartridge or Envelope	Tube or Envelope	Tube	Tube or Envelope	Pleated Cartridge
Type of Fabric	Woven	Woven		Non-Woven (Felt)	Non-Woven
Airflow	Highly Variable	Slightly Variable		Virtually Constant	Virtually Constant
Normal Rating (filtration velocity, fpm) [m/s]	1 to 6 fpm [0.005 to 0.03 m/s]	1 to 3 fpm [0.005 to 0.015 m/s]	1 to 3 fpm [0.005 to 0.015 m/s]	5 to 12 fpm [0.027 to 0.061 m/s]	<1 to 7 fpm [0.005 to 0.036 m/s]

cleaned air exits is attached to a tube sheet and, depending on design, the other end may be attached to a support member or cantilevered without support. The shaker mechanism may be located in either the dirty air or cleaned air compartments.

Periodically (usually at 3- to 6-hour intervals) the airflow must be stopped to recondition the fabric. Figure 8-6 illustrates the system airflow characteristics of an intermittent-duty fabric collector. As dust accumulates on the fabric, resistance to flow increases and airflow decreases until the fan is turned off and the fabric reconditioned. Variations in airflow due to changing pressure losses is sometimes a disadvantage and, when coupled with the requirement to periodically stop the airflow, may preclude the use of intermittent collectors. Reconditioning seldom requires more than two minutes but must be done without airflow through the fabric. If reconditioning is attempted with air flowing it will be less effective and the flexing of the woven fabric will allow a substantial amount of dust to escape to the clean air side.

The filtration velocity for large intermittent duty fabric collectors seldom exceeds 6 fpm [0.03 m/s] and normal selections are in the 2 fpm [0.01 m/s] to 4 fpm [0.02 m/s] range. Lighter dust concentrations and the ability to recondition more often allow the use of higher filtration velocities. Ratings are usually selected so that the pressure drop across the fabric will be in the 2 to 5 "wg [500 to 1250 Pa] range between start and end of operating cycle.

With multiple-section, continuous-duty, automatic fabric collectors, the disadvantage of stopping the airflow to permit fabric reconditioning and the variations in airflow with dust cake build-up can be overcome. The use of sections or compartments, as indicated in Figure 8-7, allows continuous operation of the exhaust system because automatic dampers periodically remove one section from service for fabric reconditioning while the remaining compartments handle the total gas flow. The larger the number of compartments, the more constant the pressure loss and airflow. Either tubes or envelopes may be used and fabric reconditioning is usually accomplished by shaking or vibrating.

Figure 8-6 shows airflow versus time for a multiple-section collector. Each individual section or compartment has an airflow versus time characteristic like that of the intermittent collector, but the total variation is reduced because of the multiple compartments. Note the more constant airflow characteristic of the five-compartment unit as opposed to the three-compartment design. Since an individual section is out of service only a few minutes for reconditioning and remaining sections handle the total gas flow during that time, it is possible to clean the fabric more frequently than with the intermittent type. This permits the multiple-section unit to handle higher dust concentrations. Compartments are reconditioned in fixed sequence with the ability to adjust the time interval between cleaning of individual compartments.

One variation of this design is the low-pressure, reverse-air collector which does not use shaking for fabric reconditioning. Instead, a compartment is isolated for cleaning and the tubes collapsed by means of a secondary blower, which draws air from the compartment in a direction opposite to the primary airflow. This is a "gentle" method of fabric reconditioning and was developed primarily for the fragile glass cloth used for high temperature operation, but is now commonplace in the woodworking industry and other industries where clean, dry, compressed air is not readily available. The reversal of airflow and tube deflation is accomplished very gently to avoid damage to the glass fibers. The control sequence usually allows the deflation and re-inflation of tubes several times for complete removal of excess dust. Tubes

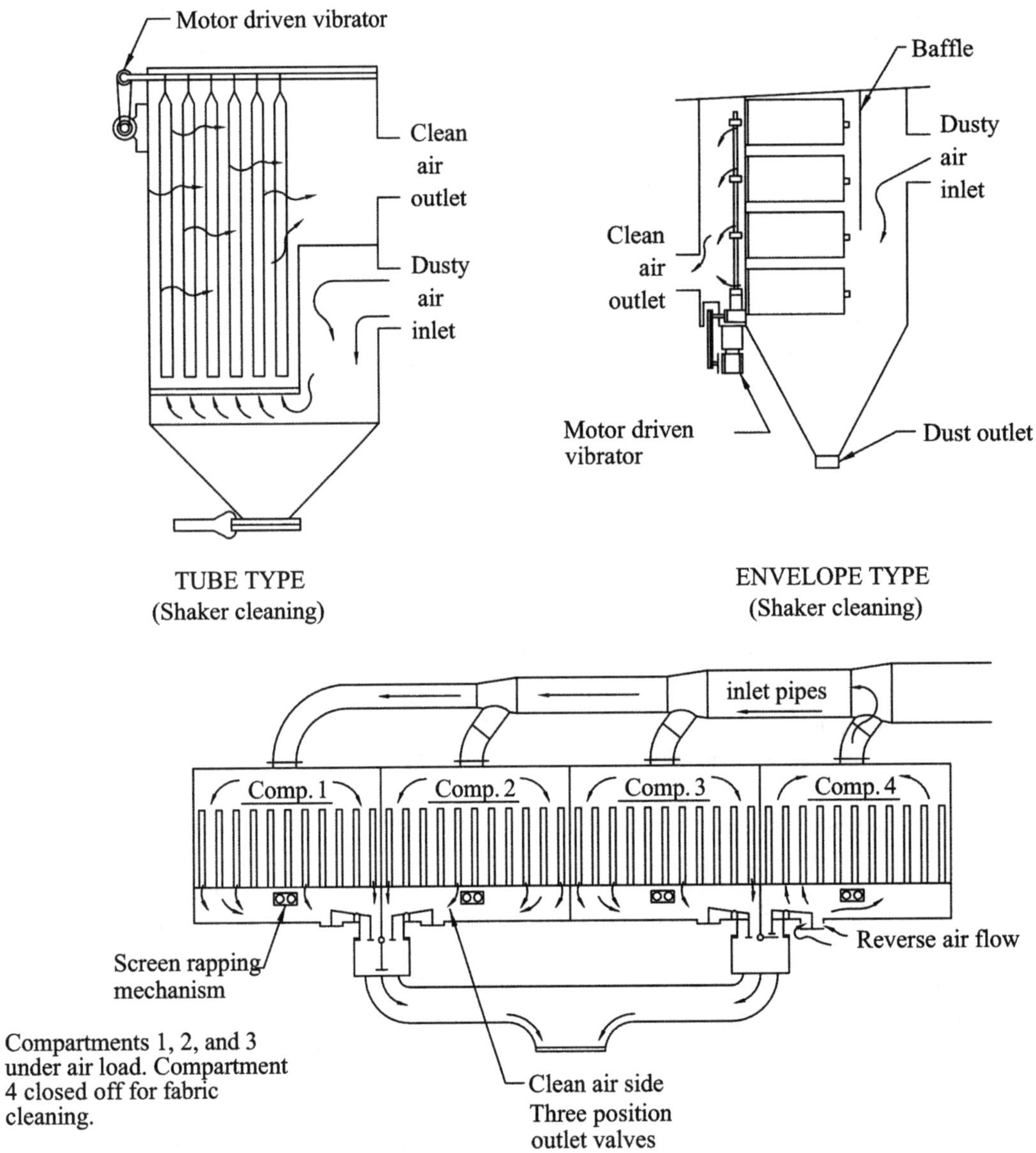

MULTIPLE SECTION CONTINUOUS AUTOMATIC
(Reverse air cleaning)

ACGIH®	TITLE: FABRIC COLLECTORS	FIGURE 8-7
		DATE 1-12

CHECK CODES, REGULATIONS, AND LAWS (LOCAL, STATE, AND NATIONAL) TO ENSURE THAT DESIGN IS COMPLIANT.

are 6 to 11 inches [152.4 to 279.4 mm] in diameter and can be as long as 30 feet [9.1 m]. For long tubes, stainless steel rings may be sewn on the inside to help break up the dust cake during deflation. A combination of shaking and reverse airflow has also been utilized.

When shaking is used for fabric reconditioning, the filtration velocity usually is in the 1 fpm to 4 fpm [0.005 to 0.02 m/s] range. Reverse air collapse type reconditioning generally necessitates lower filtration velocities since reconditioning is not as complete. They are seldom rated higher than 3 fpm [0.015 m/s]. The air to cloth ratio or filtration velocity is based on *net* cloth area available when a compartment is out of service for reconditioning.

Reverse Pulse Jet Fabric Collectors: Reverse-jet, continuous-duty, fabric collectors may use envelopes or tubes of non-woven (felted) fabric, pleated cartridges of non-woven mat (paper-like) in cylindrical or panel configuration, or rigid elements such as sintered polyethylene. They differ from the low pressure reverse air type in that they employ a brief burst of high pressure air to recondition the fabric. Woven fabric is not used because it allows excessive dust penetration during reconditioning. The most common designs use compressed air at 80 to 100 psig [470 to 590 kPa], while others use an integral pressure blower at a lower pressure but higher secondary flow rate. Those using compressed air are generally called pulse-jet collectors and those using pressure blowers are called fan-pulse collectors.

All designs collect dust on the outside and have airflow from outside to inside the fabric. All recondition the media by introducing the pulse of cleaning air into the opening where cleaned air exits from the tube, envelope, or cartridge. In many cases, a venturi shaped fitting is used at this opening to provide additional cleaning by inducing additional airflow. The venturi also directs or focuses the cleaning pulse for maximum efficiency.

Figure 8-8 shows typical pulse-jet collectors. Under normal operation (airflow from outside to inside) the fabric shape will tend to collapse, therefore, a support cage is required. The injection of a short pulse of high pressure air induces a secondary flow from the clean air compartment in a direction opposite to the normal airflow. Reconditioning is accomplished by the pulse of high pressure air which stops forward airflow, then rapidly pressurizes the media, breaking up the dust cake and freeing accumulated dust from the fabric. The secondary or induced air acts as a damper, preventing flow in the normal direction during reconditioning. The entire process, from injection of the high pressure pulse and initiation of secondary flow until the secondary flow ends, takes place in approximately one second. Solenoid valves which control the pulses of compressed air through the diaphragm valves may be open for a tenth of a second or less. An adequate flow rate of clean and dry compressed air of sufficient pressure must be supplied to ensure effective reconditioning.

Reverse-jet collectors normally clean no more than 10% of the fabric at any one time. Because such a small percentage is cleaned at any one time and because the induced secondary flow blocks normal flow during that time, reconditioning can take place while the collector is in service and without the need for compartmentalization and dampers. The cleaning intervals are adjustable and are considerably more frequent than the intervals for shaker or reverse-air collectors. An individual element may be pulsed and reconditioned as often as once a minute to every six minutes.

Due to this very short reconditioning cycle, higher filtration velocities are possible with reverse-jet collectors. However, with all reverse-jet collectors, accumulated dust that is freed from one fabric surface may become re-entrained and redeposited on an adjacent surface, or even on the original surface. This phenomenon of redeposition tends to limit filtration velocity to something less than might be anticipated with cleaning intervals of just a few minutes.

Laboratory tests[8.1] have shown that, for a given collector design, redeposition increases with filtration velocity. Other test work[8.2] indicates clearly that redeposition varies with collector design and especially with flow patterns in the dirty air compartment. EPA–sponsored research[8.3] has shown that superior performance results from downward flow of the dirty air stream. This downward airflow reduces redeposition since it aids gravity in moving dust particles toward the hopper.

Many reverse pulse jet collectors operate successfully with upward airflow. The upward velocity (known as “can,” “tank,” or “interstitial” velocity) must be kept below the velocity range based on equipment vendor experience with dust having similar aerodynamic particle diameters and particle size distribution. Both air-to-cloth ratio (actual face velocity at the filter surface-filtration velocity) and interstitial velocity must be evaluated when selecting a reverse jet collector. Although a collector with fewer, longer bags may look economically attractive, problems caused by excessive interstitial velocity and consequential bridging of dust between bags can quickly lead to operating costs that can negate the perceived capital savings. Filtration velocities of 4 to 12 fpm [0.020 × 0.06 m/s] are normal for reverse-jet collectors. The pleated cartridge type of reverse-jet collector is limited to filtration velocities in the 7 fpm [0.036 m/s] range and are most often used in the 1 fpm to 3 fpm [0.005 to 0.015 m/s] range. The pleat configuration may produce very high approach velocities and greater redeposition. There are many particulate parameters that cause a challenge to fabric filters and require a more conservative filtration and interstitial velocity. Some of the most important particulate characteristics to be most cautious about are:

- Hygroscopic – The affinity of a dust to absorb moisture and become tacky
- Abrasive – Cause premature filter or collector failure
- Aerodynamic particle diameter – Is the particle more like a feather or a solid sphere?
- Small in size – Typically finer particulate causes more

FIGURE 8-8

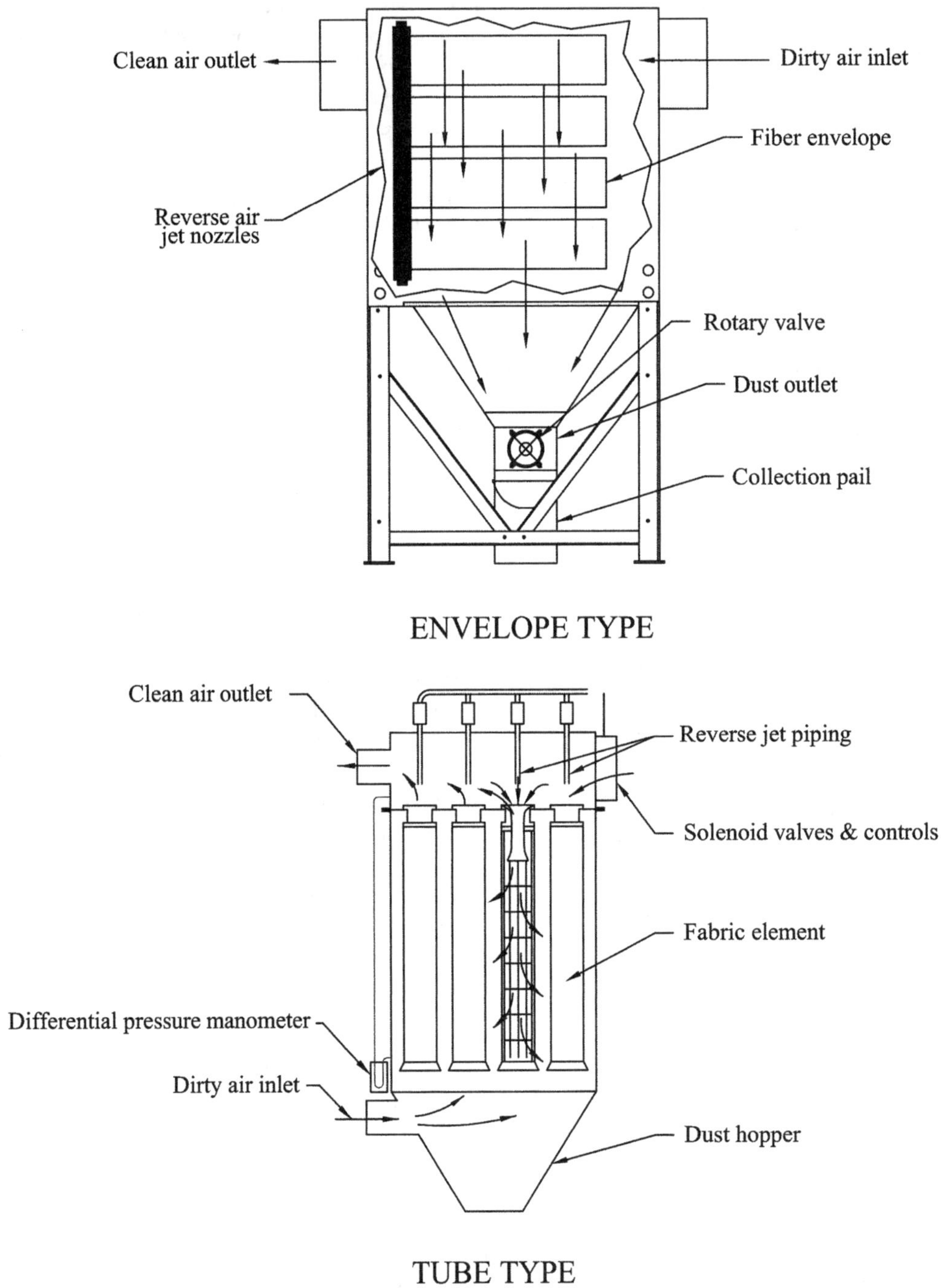

ACGIH®	TITLE FABRIC COLLECTORS PULSE JET TYPE	FIGURE 8-8 DATE 1-12

CHECK CODES, REGULATIONS, AND LAWS (LOCAL, STATE, AND NATIONAL) TO ENSURE THAT DESIGN IS COMPLIANT.

filter plugging and an inability to recover, especially particulate smaller than 3 microns in diameter

- Fibrous – Fibrous dust can have particularly low bulk densities and large aerodynamic particle diameters.

A newer type of dust pulse jet dust collector is now widely used with success and incorporates an enclosing hood built onto the dust collector itself. The hybrid could be termed a "dust collection booth" and is typically used on applications in which it is difficult to apply an exterior hood. One wall of a hopperless dust collector is open to the booth and the air is brought through the booth (and across the worker) at 100-150 fpm [0.50 to 0.76 m/s] (similar to a paint spray booth). Fans are typically incorporated, pulling through the media and recirculating it into the plant air space directly or through HEPA filters. This dust booth concept has been used with success on welding, sanding, and cutting materials and is coherent with the concept of enclosing hoods (Figure 8-9). Additionally, it does not require large energy considerations such as ducts, hood entry losses, elbows, etc. However, waste handling is significantly more difficult.

8.3.3 Wet Collectors. Wet collectors, or scrubbers, are commercially available in many different designs, with pressure drops from 1.5 "wg [374 Pa] to as much as 100 "wg [24.9 kPa]. There is a corresponding variation in collector performance. It is generally accepted that, for well-designed equipment, efficiency depends on the energy utilized in air to water contact and is independent of operating principle. Efficiency is a function of total energy input per cfm whether the energy is supplied to the air or to the water. This means that well-designed collectors by different manufacturers will provide similar efficiency if equivalent power is utilized.

Wet collectors have the ability to handle high-temperature and moisture-laden gases. The collection of dust in a wetted form minimizes a secondary dust problem in disposal of collected material. Some dusts represent explosion or fire hazards when dry. Wet collection minimizes the hazard; however, the use of water may introduce corrosive conditions within the collector and freeze protection may be necessary if collectors are located outdoors in cold climates. Space requirements are nominal. Pressure losses and collection efficiency vary widely for different designs.

Wet collectors, especially the high-energy types, are frequently the solution to air pollution problems. It should be realized that disposal of collected material in water without clarification or treatment may create water pollution problems and that dried sludges are considered hazardous waste until otherwise tested.

Wet collectors have one characteristic not found in other collectors — the inherent ability to humidify. Humidification, the process of adding water vapor to the air stream through

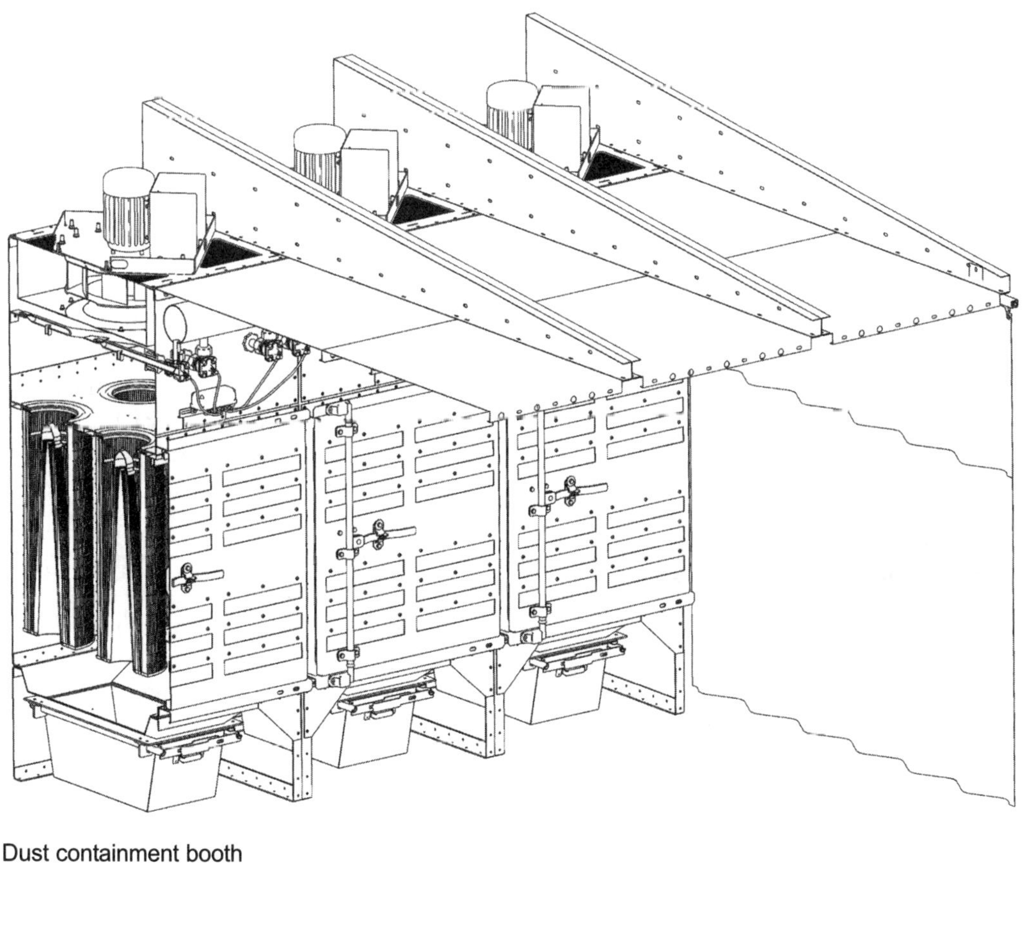

FIGURE 8-9. Dust containment booth

evaporation, may be either advantageous or disadvantageous depending on the situation. Where the initial air stream is at an elevated temperature and not saturated, the process of evaporation reduces the temperature and the volumetric flow rate of the gas stream leaving the collector. Assuming the fan is to be selected for operation on the clean air side of the collector, it may be smaller and will definitely require less power than if there had been no cooling through the collector. This is one of the obvious advantages of humidification; however, there are other applications where the addition of moisture to the gas stream is undesirable. For example, the exhaust of humid air to an air-conditioned space normally places an unacceptable load on the air conditioning system. High humidity can also result in corrosion of finished goods. Therefore, humidification effects should be considered before designs are finalized. While all wet collectors humidify, the amount of humidification varies for different designs. Most manufacturers publish the humidifying efficiency for their equipment and will assist in evaluating the results.

Chamber or Spray Tower: Chamber or spray tower collectors consist of a round or rectangular chamber into which water is introduced by spray nozzles. There are many variations of design, but the principal mechanism is impaction of dust particles on the liquid droplets created by the nozzles. These droplets are separated from the air stream by centrifugal force or impingement on water eliminators.

The pressure drop is relatively low (on the order of 0.5 to 1.5 "wg [125 to 375 Pa]), but water pressures range from 10 to 400 psig [60 to 2400 kPa]. The high pressure devices are the exception rather than the rule. In general, this type of collector utilizes low pressure supply water and operates in the lower efficiency range for wet collectors. Where water is supplied under high pressure, as with fog towers, collection efficiency can reach the upper range of wet collector performance.

For conventional equipment, water requirements are reasonable, with a maximum of about 5 gpm per thousand scfm of gas [40.1 l/m^3–gas]. Fogging types using high water pressure may require as much as 10 gpm per thousand scfm of gas [80.2 l/m^3–gas].

Packed Towers: Packed towers (Figure 8-10) are essentially contact beds through which gases and liquid pass concurrently, counter-currently, or in cross-flow. They are used primarily for applications involving gas, vapor, and mist removal. These collectors will capture solid particulate matter but they are not used for that purpose because dust plugs the packing and requires unreasonable maintenance.

Water rates of 5 to 10 gpm per thousand scfm [approx. 40 – 80 l/m^3–gas] are typical for packed towers. Water is distributed over V-notched ceramic or plastic weirs. High temperature deterioration is avoided by using brick linings, allowing gas temperatures as high as 1600 F [871 C] to be handled directly from furnace flues.

The airflow pressure loss for a four foot bed of packing, such as ceramic saddles, will range from 1.5 to 3.5 "wg [375 – 875 Pa]. The face velocity (velocity at which the gas enters the bed) will typically be 200 to 600 fpm [approx. 1.0 – 3.0 m/s].

Wet Centrifugal Collectors: Wet centrifugal collectors (Figure 8-11) comprise a large portion of the commercially available wet collector designs. This type utilizes centrifugal force to accelerate the dust particle and impinge it upon a wetted collector surface. Water rates are usually 2 to 5 gpm per thousand scfm of gas cleaned [approx. 16 to 40 l/m^3 – gas]. Water distribution can be from nozzles, gravity flow or induced water pickup. Pressure drop is in the 2 to 6 "wg [500 – 1500 Pa] range.

As a group, these collectors are more efficient than the chamber type. Some are available with a variable number of impingement sections. A reduction in the number of sections results in lower efficiency, lower cost, less pressure drop, and smaller space. Other designs contain multiple collecting tubes. For a given airflow rate, a decrease in the tube size provides higher efficiency because the centrifugal force is greater.

Wet Dynamic Precipitator: Sometimes called a "wet fan," the wet dynamic precipitator (Figure 8-12) is a combination fan and dust collector. Dust particles in the dirty air stream impinge upon rotating fan blades wetted with spray nozzles. The dust particles impinge into water droplets and are trapped along with the water by a metal cone while the cleaned air makes a turn of 180 degrees and escapes from the front of the specially shaped impeller blades. Dirty water from the water cone goes to the water and sludge outlet and the cleaned air goes to an outlet section containing a water elimination device.

Orifice Type Scrubber: In this group of wet collector designs (Figure 8-12) the airflow through the collector is brought in contact with a sheet of water in a restricted passage. Water flow may be induced by the velocity of the air stream or maintained by pumps and weirs. Pressure losses vary from 1 "wg [250 Pa] or less for a water wash paint booth to a range of 3 to 6 "wg [750 – 1500 Pa] for most of the industrial designs. Pressure drops as high as 20 "wg [5 kPa] are used with some designs intended to collect very small particles.

Venturi Scrubber: The venturi collector (Figure 8-11) uses a venturi-shaped constriction to establish throat velocities considerably higher than those used by the orifice type. Gas velocities through venturi throats may range from 12,000 to 24,000 fpm [60 – 120 m/s]. Water is supplied by piping or jets at or ahead of the throat at rates from 5 to 15 gpm per thousand scfm [40 – 120 l/m^3] of gas.

The collection mechanism of the venturi is impaction. As is true for all well-designed wet collectors, collection efficiency increases with higher pressure drops. Specific pressure drops are obtained by designing for selected velocities in the throat. Some venturi collectors are made with adjustable throats allowing operation over a range of pressure drops for a given flow rate or over a range of flow rates with a constant pressure drop. Systems are available with pressure drops as low as 5

FIGURE 8-10

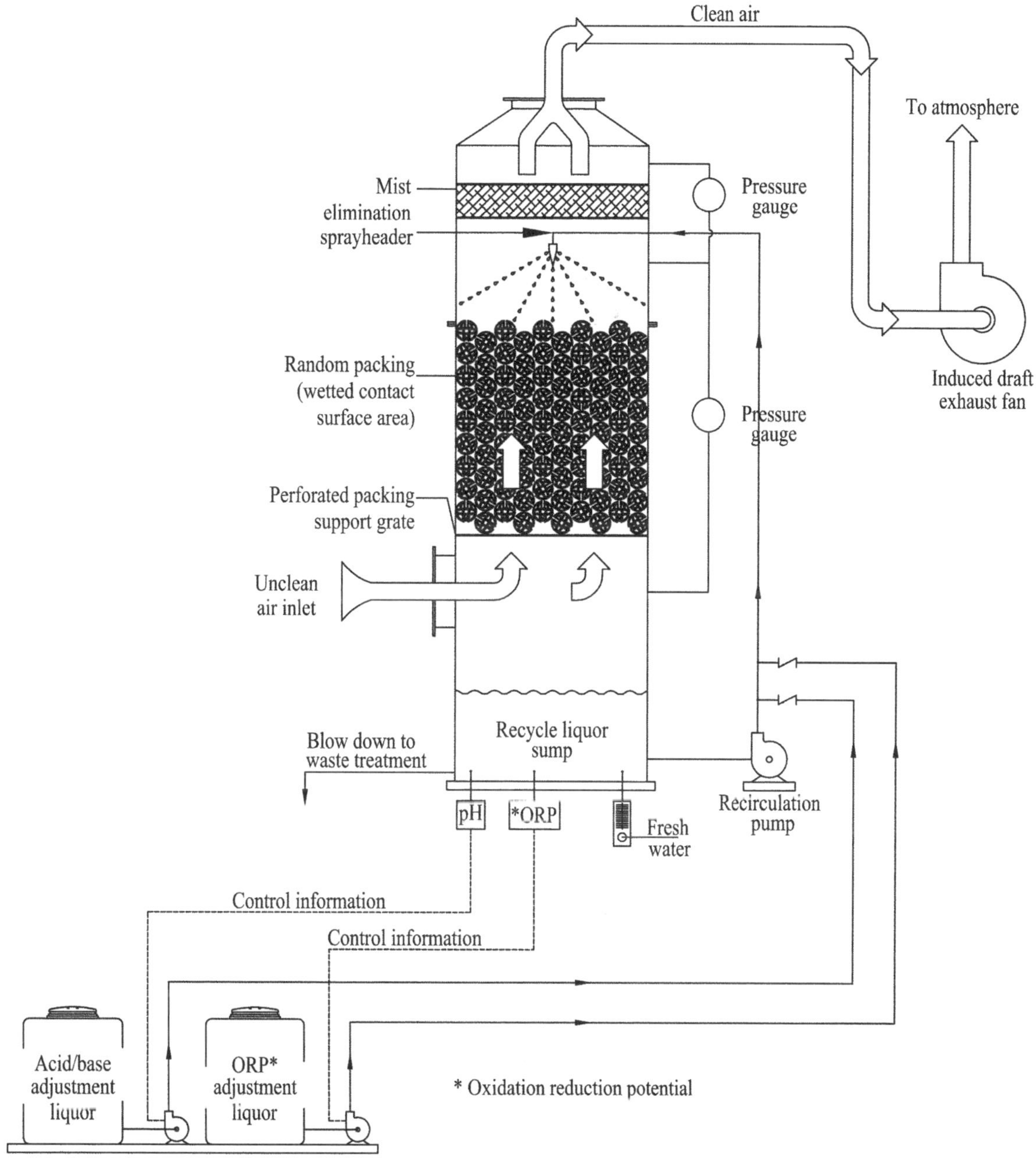

Single stage vertical tower scrubber system

ACGIH®	TITLE WET TYPE COLLECTOR (FOR GASEOUS CONTAMINANT)	FIGURE 8-10 DATE 1-12

CHECK CODES, REGULATIONS, AND LAWS (LOCAL, STATE, AND NATIONAL) TO ENSURE THAT DESIGN IS COMPLIANT.

FIGURE 8-11

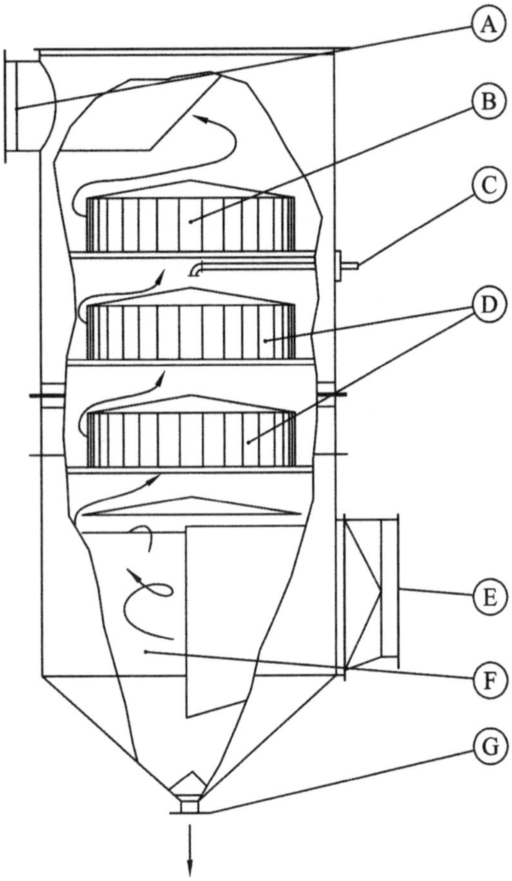

Wet centrifugal

Symbols	Parts
A	Clean air outlet
B	Entrainment separator
C	Water inlet
D	Impingement plates
E	Dirty air inlet
F	Wet cyclone for collecting heavy material
G	Water and sludge drain

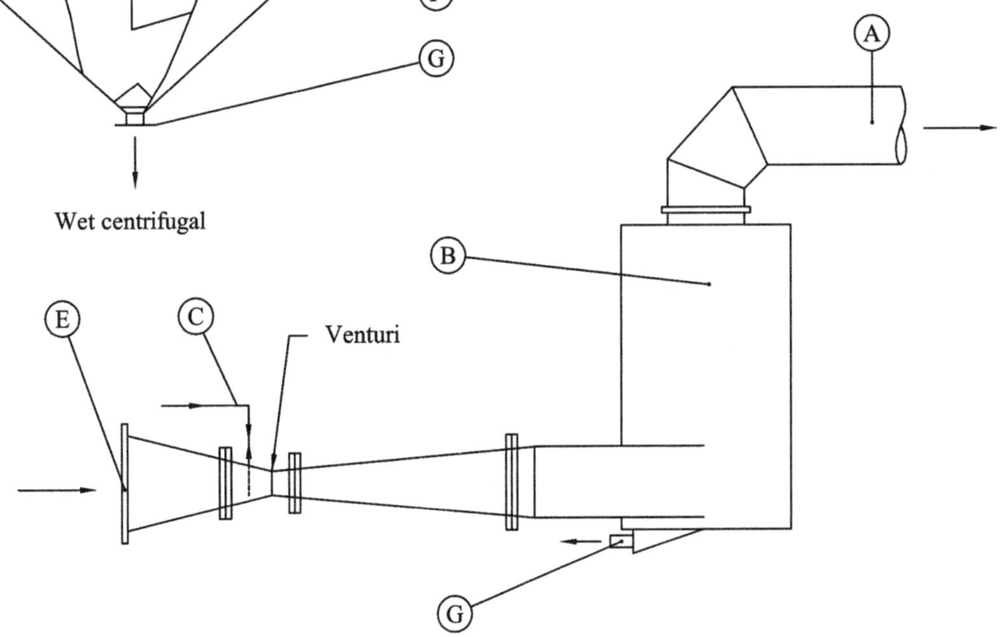

Venturi scrubber

ACGIH®

TITLE
WET TYPE DUST COLLECTORS (FOR PARTICULATE CONTAMINANTS)

FIGURE 8-11

DATE 1-12

CHECK CODES, REGULATIONS, AND LAWS (LOCAL, STATE, AND NATIONAL) TO ENSURE THAT DESIGN IS COMPLIANT.

FIGURE 8-12

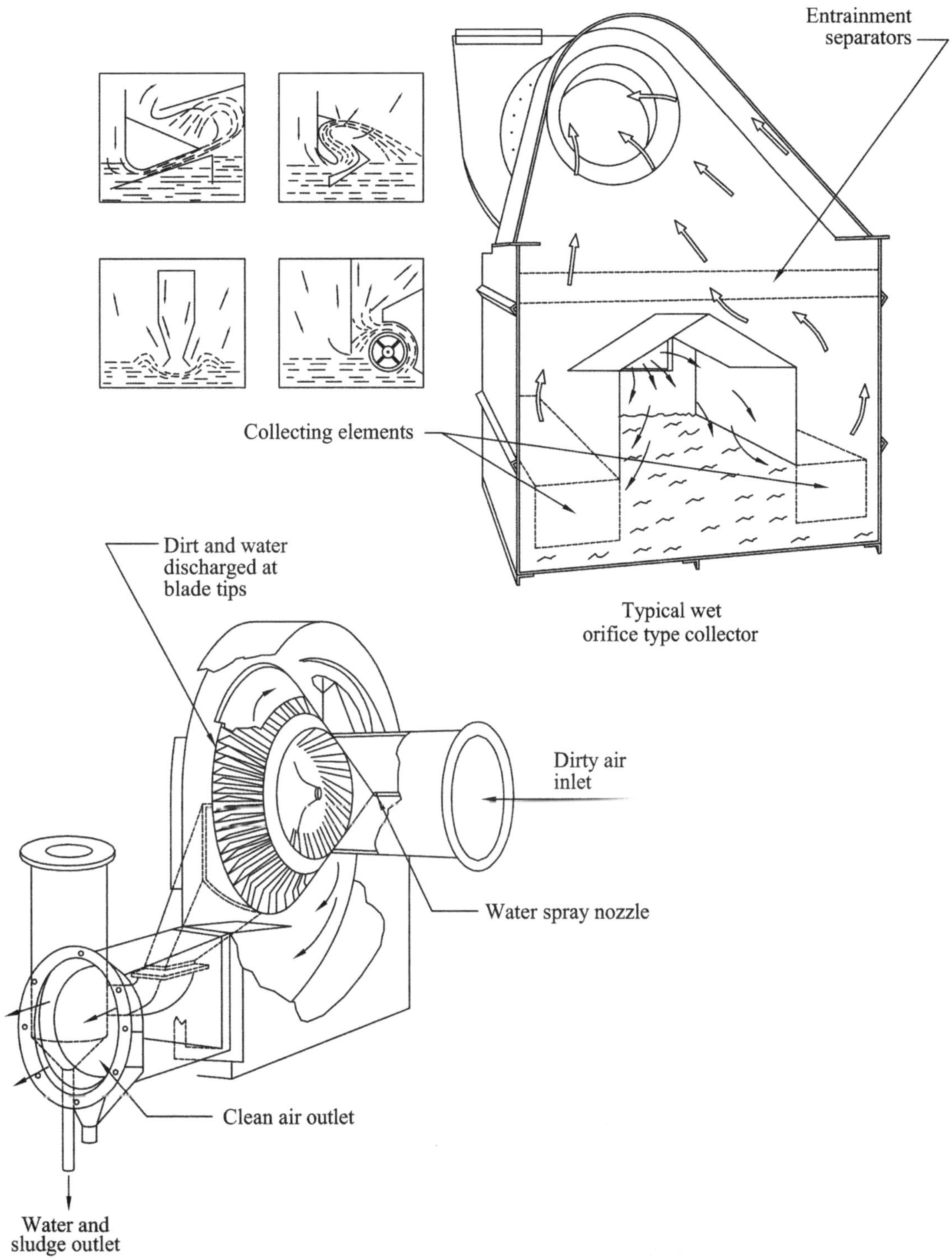

TITLE WET TYPE COLLECTORS (FOR PARTICULATE CONTAMINANTS)

FIGURE 8-12

DATE 1-12

CHECK CODES, REGULATIONS, AND LAWS (LOCAL, STATE, AND NATIONAL) TO ENSURE THAT DESIGN IS COMPLIANT.

"wg [1250 Pa] for moderate collection efficiency and as high as 100 "wg [25 kPa] for collection of extremely fine particles.

All scrubbers are gas conditioners causing intimate contact between the particulates in the gas and the multiple jet streams of scrubbing water. The resulting mixture of gases, fume-dust agglomerates and dirty water must be channeled through a separation section for the elimination of entrained droplets as shown in Figure 8-11.

8.3.4 Dry Centrifugal Collectors. Dry centrifugal collectors separate entrained particulate from an air stream by the use or combination of centrifugal, inertial, and gravitational force. Collection efficiency is influenced by:

1) Particle size, weight, and shape. Performance is improved as size and weight become larger and as the shape becomes more spherical.
2) Collector size and design. The collection of fine dust with a mechanical device requires equipment designed to best utilize mechanical forces and fit specific application needs.
3) Velocity. Pressure drop through a cyclone collector increases approximately as the square of the inlet velocity. There is, however, an optimum velocity that is a function of collector design, dust characteristics, gas temperature and density.
4) Dust Concentration. Generally, the performance of a mechanical collector increases as the concentration of dust becomes greater.

Gravity Separators: Gravity separators consist of a chamber or housing in which the velocity of the gas stream is made to drop rapidly so that dust particles settle out by gravity. Extreme space requirements and the usual presence of eddy currents nullify this method for removal of anything but extremely coarse particles.

Inertial Separators: Inertial separators depend on the inability of dust to make a sharp turn because its inertia is much higher than that of the carrier gas stream. Blades or louvers in a variety of shapes are used to require abrupt turns of 120 degrees or more. Well-designed inertial separators can separate particles in the 10 to 20 micron range with about 90% efficiency.

Cyclone Collector: The cyclone collector (Figure 8-13) is commonly used for the removal of coarse dust from an air stream, as a precleaner to more efficient dust collectors and/or as a product separator in air conveying systems. Principal advantages are low cost, low maintenance, and relatively low pressure drops (in the 0.75 to 1.5 "wg [approx. 190 – 370 Pa] range). It is not suitable for the collection of fine particles (Figure 8-14).

High Efficiency Centrifugal Collectors: High efficiency centrifugals (Figure 8-13) exert higher centrifugal forces on the dust particles in a gas stream. Because centrifugal force is a function of peripheral velocity and angular acceleration, improved dust separation efficiency has been obtained by:

1) Increasing the inlet velocity
2) Making the cyclone body and cone longer
3) Using a number of small diameter cyclones in parallel
4) Placing units in series.

While high efficiency centrifugal collectors are not as efficient on small particles as electrostatic, fabric, and wet collectors, their effective collection range is appreciably extended beyond that of other mechanical devices. Pressure losses of collectors in this group range from 3 to 8 "wg [750 – 2000 Pa].

8.4 ADDITIONAL AIDS IN DUST COLLECTOR SELECTION

The collection efficiencies of the five basic groups of air cleaning devices have been plotted against mass mean particle size (Figures 8-14 [IP] and 8-15). The graphs were found through laboratory and field testing and were not compiled mathematically. The number of lines for each group indicates the range that can be expected for the different collectors operating under the same principle. Variables, such as type of dust, velocity of air, water rate, etc., will also influence the range for a particular application.

Deviation lines shown in the upper right hand corner of the chart allow the estimation of mass mean material size in the effluent of a collector when the inlet mean size is known. Space does not permit a detailed explanation of how the slopes of these lines were determined, but the following example illustrates how they are used. The deviation lines should not be used for electrostatic precipitators but can be used for the other groups shown at the bottom of the figure.

EXAMPLE PROBLEM 1 (Use of Sylvan Chart)

A suitable collector is to be selected for a lime kiln. Referring to Figure 8-14, the concentration and mean particle size of the material leaving the kiln can vary between 3 and 10 grains per cubic foot [6.86 – 22.88 g/m^3], with 5 to 10 microns the range for mass mean particle size. Assume an inlet concentration of 7.5 grains per cubic foot [189.66 g/m^3] and a mean inlet size of 9 microns. Projection of this point vertically downwardly to the collection efficiency portion of the chart will indicate that a low resistance cyclone will be less than 50% efficient; a high efficiency centrifugal will be 60 to 80% efficient and a wet collector, fabric arrester and electrostatic precipitator will be 97+% efficient. A precleaner is usually feasible for dust concentrations over 5 grains per cubic foot [11.44 g/m^3] unless it is undesirable to have the collected dust separated by size. For this example, a high efficiency centrifugal will be selected as the precleaner. The average efficiency is 70% for this group, therefore the effluent from this collector will have a concentration of 7.5 (100 –

FIGURE 8-13

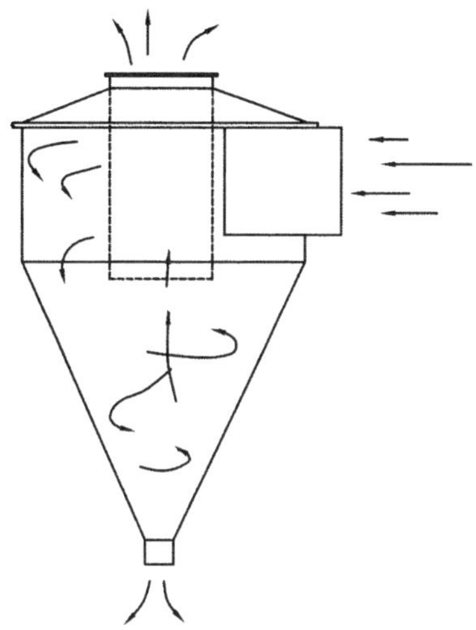

Low pressure cyclone

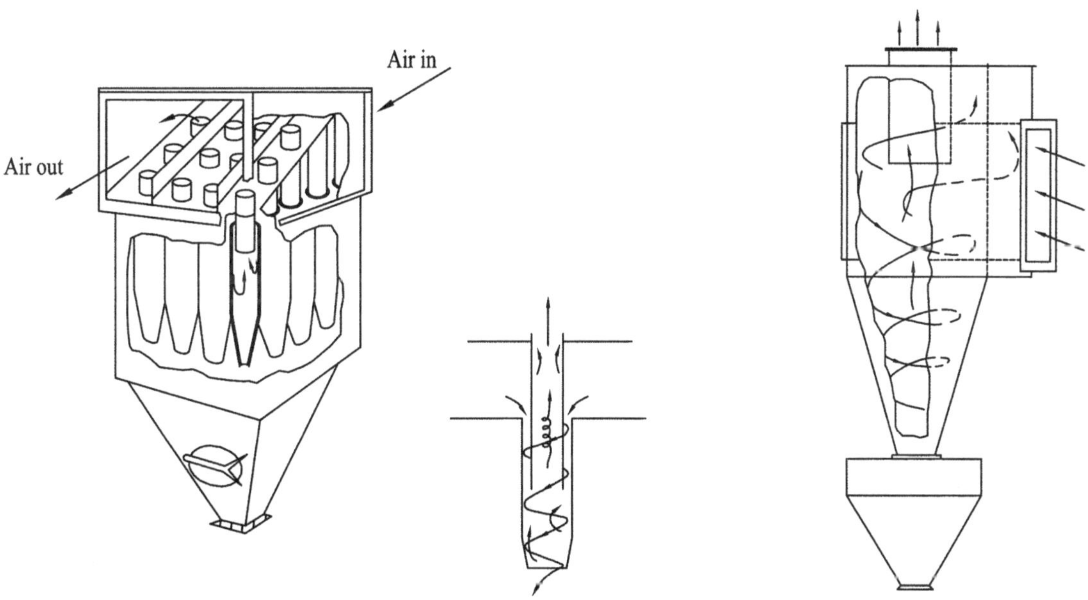

High efficiency cyclones and multi-clones

ACGIH®	TITLE DRY TYPE CENTRIFUGAL COLLECTORS	FIGURE 8-13
		DATE 1-12

CHECK CODES, REGULATIONS, AND LAWS (LOCAL, STATE, AND NATIONAL) TO ENSURE THAT DESIGN IS COMPLIANT.

FIGURE 8-14 (IP)

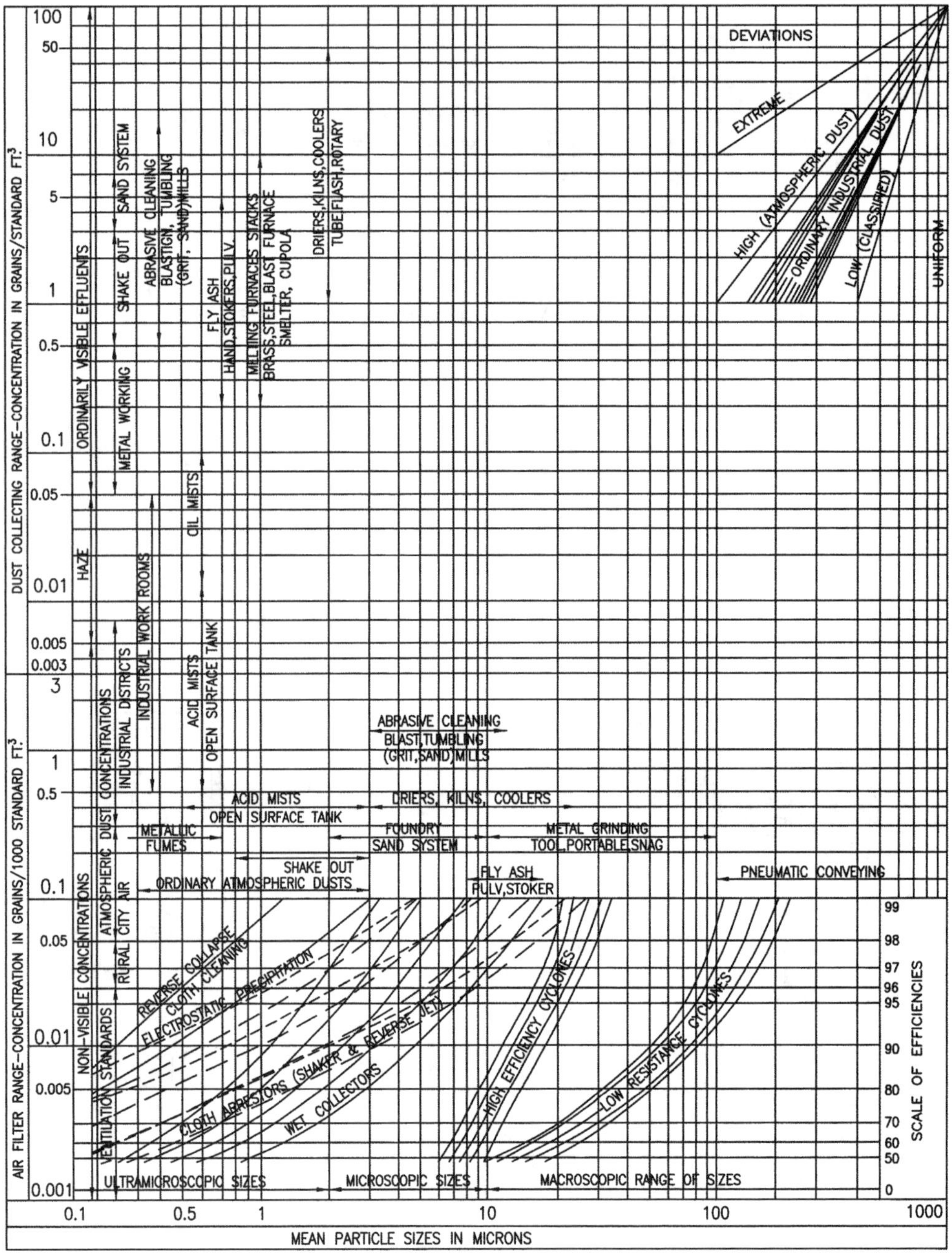

RANGE OF PARTICLE SIZES, CONCENTRATION, & COLLECTOR PERFORMANCE

ACKNOWLEDGEMENTS OF PARTIAL SOURCES OF DATA REPORTED:

1 FRANK W.G. -AMERICAN AIR FILTER - SIZE AND CHARACTERISTICS OF AIR BORNE SOLIDS - 1931
2 FIRST AND DRINKER - ARCHIVES OF INDUSTRIAL HYGIENE AND OCCUPATIONAL MEDICINE - APRIL 1952
3 TAFT INSTITUTE AND AAF LABORATORY TEST DATA - 1961 - '63
4 REVERSE COLLAPSE CLOTH CLEANING ADDED 1964

ACGIH®	TITLE SYLVAN CHART RANGE OF PARTICLE SIZE AND COLLECTOR EFFICIENCIES	FIGURE 8-14 (IP)
		DATE 1-12

CHECK CODES, REGULATIONS, AND LAWS (LOCAL, STATE, AND NATIONAL) TO ENSURE THAT DESIGN IS COMPLIANT.

FIGURE 8-14 (SI)

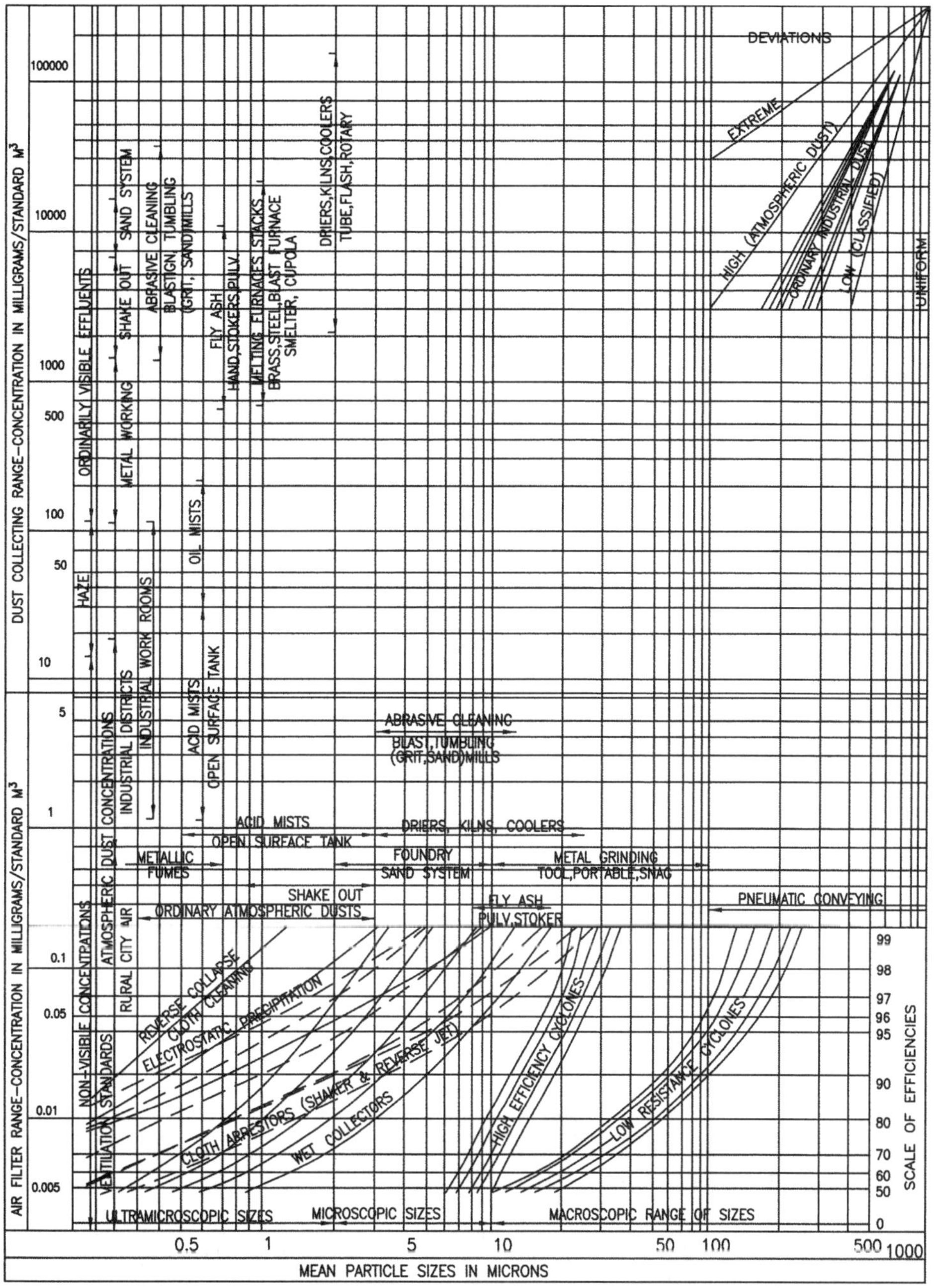

RANGE OF PARTICLE SIZES, CONCENTRATION, & COLLECTOR PERFORMANCE

COMPILED BY S. SYLVAN APRIL 1952 : COPYRIGHT 1952 AMERICAN AIR FILTER CO. INC.

ACKNOWLEDGEMENTS OF PARTIAL SOURCES OF DATA REPORTED:

1 FRANK W.G. -AMERICAN AIR FILTER - SIZE AND CHARACTERISTICS OF AIR BORNE SOLIDS - 1931
2 FIRST AND DRINKER - ARCHIVES OF INDUSTRIAL HYGIENE AND OCCUPATIONAL MEDICINE - APRIL 1952
3 TAFT INSTITUTE AND AAF LABORATORY TEST DATA - 1961 - '63
4 REVERSE COLLAPSE CLOTH CLEANING ADDED 1964

ACGIH®	TITLE SYLVAN CHART RANGE OF PARTICLE SIZE AND COLLECTOR EFFICIENCIES	FIGURE 8-14 (SI) DATE 1-12

CHECK CODES, REGULATIONS, AND LAWS (LOCAL, STATE, AND NATIONAL) TO ENSURE THAT DESIGN IS COMPLIANT.

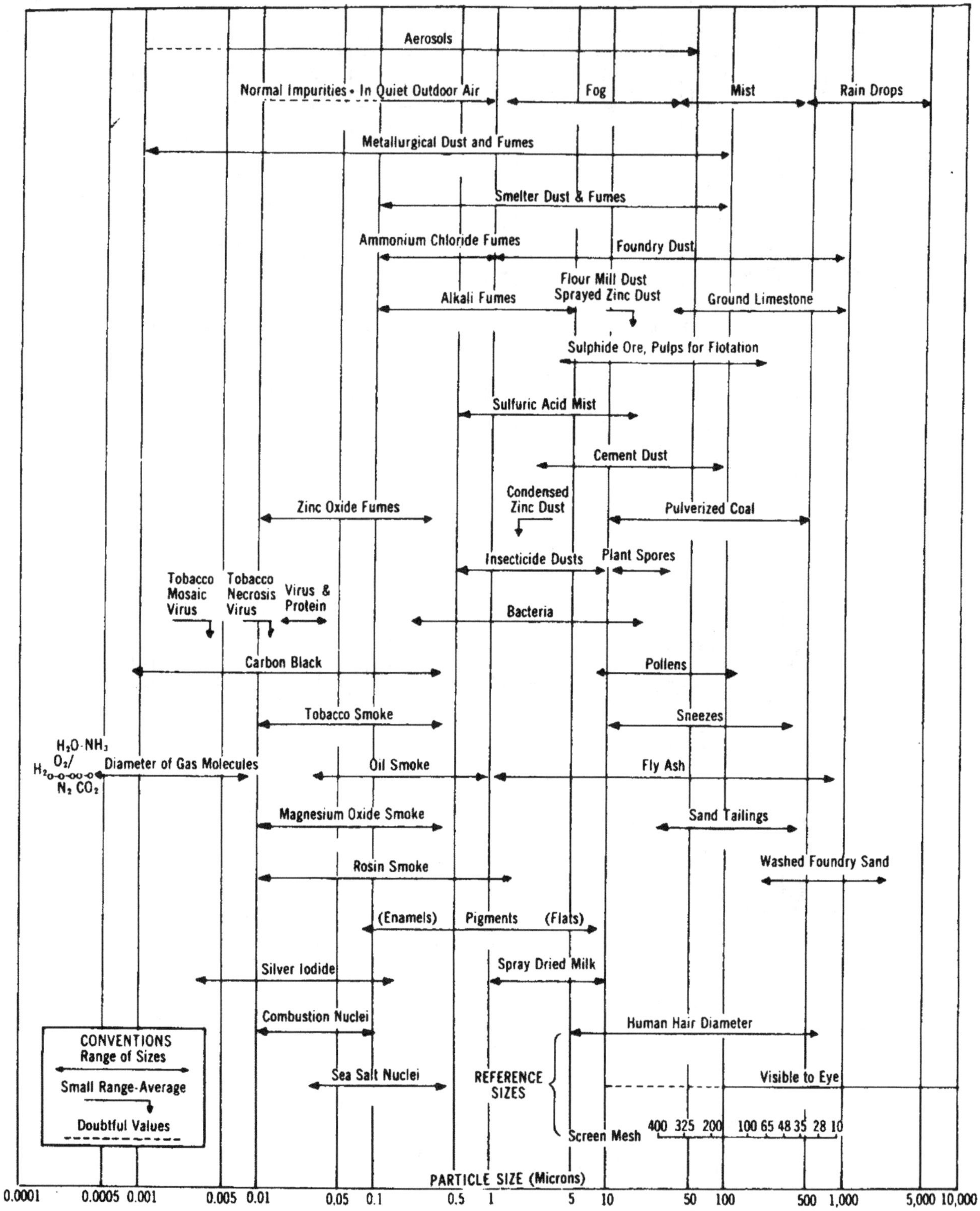

FIGURE 8–15. Characteristics of particles and particle dispersoids (Reprinted with permission from SRI International)

0.70) = 2.25 grains per cubic foot. Draw a line through the initial point with a slope parallel to the deviation lines marked "industrial dust." Figure 8-14 [IP] is in IP units and solution uses those values only. Where deviation is not known, the average of this group of lines normally will be sufficiently accurate to predict the mean particle size in the collector effluent. A vertical line from the point of intersection between the 2.25 grains per cubic foot horizontal and the deviation line to the base of the chart will indicate a mean effluent particle size of 6.0 microns.

A second high efficiency centrifugal collector in series would be less than 50% efficient on this effluent. A wet collector, fabric arrester, or electrostatic would have an efficiency of 94% or better. Assume that a good wet collector will be 98% efficient. The effluent would then be 2.25 (1.00 – 0.98) = 0.045 grains per cubic foot. Using the previous deviation line and its horizontal intersection of 0.045 grains per cubic foot yields a vertical line intersecting the mean particle size chart at 1.6 microns, the mean particle size of the wet collector effluent.

In Table 8-3, an effort has been made to report types of dust collectors used for a wide range of industrial processes. While many of the listings are purely arbitrary, they may serve as a guide in selecting the type of dust collector most frequently used.

8.5 CONTROL OF MIST, GAS AND VAPOR CONTAMINANTS

Previous discussion has centered on the collection of dust and fume or particulate existing in the solid state. Only the packed tower was singled out as being used primarily to collect mist, gas, or vapor. The character of a mist aerosol is very similar, aerodynamically, to that of a dust or fume aerosol, and the mist can be removed from an air stream by applying the principles that are used to remove solid particulate.

Standard wet collectors are used to collect many types of mists. Specially designed electrostatic precipitators are frequently employed to collect sulfuric acid or oil mist. Even fabric and centrifugal collectors, although not the types previously mentioned, are widely used to collect oil mist generated by high speed machining.

8.6 GASEOUS CONTAMINANT COLLECTORS

Industrial processes produce tremendous quantities of gaseous contaminants. In order to better understand the specific problems associated with the control of gaseous contaminants it would serve one well to look at the properties of gases and vapors. The terms "gas" and "vapor" are commonly incorrectly used interchangeably. Matter that takes both the shape and volume of its container is said to be in a gaseous state. Gas molecules contain enough energy to continue to move apart until they bounce off the sides of the container(s) holding them. The term gas describes those substances that exist in a gaseous state at room temperature. For example, air is a mixture of gases including oxygen and nitrogen. One characteristic property of a gas is its great compressibility.

The word "vapor" describes a substance that, although in the gaseous state, is generally a liquid or solid at room temperature. Steam, the gaseous form of water, is a vapor. Moist air contains water vapor. Partial pressure relationships described by Dalton's Law explain how water vapor and dry air coexist at room temperature and atmospheric pressure. (Refer to Chapter 9, Section 9.13 for further discussion of Psychrometric Principles.)

Numerous techniques have been developed to control gaseous contaminants. The more commonly used techniques include Absorption, Adsorption, Incineration/Oxidation, and more recently, Biofiltration. Lesser known control methods include Corona Reactors, Direct Electric Arcing, Plasma Treatment, Condensation, and Photochemical Oxidation.

8.6.1 Absorption. Absorption is a mass transfer process where transfer occurs through a phase boundary and the absorbed molecule is held within the absorbing medium. Absorbers remove soluble or chemically reactive gases from the gas stream through intimate contact with a suitable liquid so that one or more of the gas stream components will dissolve in the liquid. While all designs utilize intimate contact between the gaseous contaminant and the absorbent, they vary widely in configuration and performance. Removal may be by absorption if the gas solubility and vapor pressure promote absorption or chemical reaction. There are both dry and wet absorbers. In wet absorbers, water is the most frequently used absorbent, but additives are frequently required and occasionally other chemical solutions must be used. Typical wet absorber designs include packed scrubbers, staged devices, and high energy contactors (venturi scrubbers).

Packed Scrubbers: Variants of the packed scrubber are available in four configurations. They are the Horizontal Cocurrent Scrubber, the Vertical Cocurrent Scrubber, the Crossflow Scrubber, and the Countercurrent Scrubber. The horizontal cocurrent scrubber depends on the gas velocity to carry the liquid into the packed bed and operates as a wetted entrainment separator with limited gas and liquid contact time. A vertical cocurrent scrubber may be operated at pressure drops of 1 to 3 inches of water [250 to 750 Pa] per foot [0.3 m] of packing depth. Contact time is a function of packing depth in this configuration.[8.4]

Crossflow Scrubbers use a horizontal gas stream movement with the liquid scrubbing medium flowing down through the gas stream. Absorption efficiency for this design is generally somewhere between that of cocurrent and countercurrent flow scrubbers.

Countercurrent scrubbers have the gas flowing up through a downward liquid flow. The efficiency of countercurrent scrubbers is maximized because the exit gas is in contact with the fresh scrubbing liquor where the highest driving forces exist to aid the mass transfer process. Packed Towers are countercurrent scrubbers. The packed tower scrubber was previously dis-

TABLE 8-3. Dust Collector Selection Guide

Operation	Concentration Note 1	Particle Sizes Note 2	Collector Types Used in Industry: Dry Centrifugal Collector	Wet Collector	Fabric Collector	Low-Volt Electrostatic	Hi-Volt Electrostatic	See Remark No.
CERAMICS								
a. Raw product handling	light	fine	S	O	O	N	N	1
b. Fettling	light	fine-medium	S	S	O	N	N	2
c. Refractory sizing	heavy	coarse	N	S	O	N	N	3
d. Glaze & vitr. enamel spray	moderate	medium	N	O	O	N	N	
CHEMICALS								49
a. Material handling	light-moderate	fine-medium	S	O	O	N	N	4
b. Crushing, grinding	moderate-heavy	fine-coarse	O	S	O	N	N	5
c. Pneumatic conveying	very heavy	fine-coarse	O	S	O	N	N	6
d. Roasters, kilns, coolers	heavy	mid-coarse	O	O	O	N	N	7
COAL, MINING AND POWER PLANT								49
a. Material handling	moderate	medium	O	S	O	N	N	8
b. Bunker ventilation	light	fine	S	S	O	N	N	9
c. Dedusting, air cleaning	heavy	medium-coarse	S	O	O	N	N	10
d. Drying	moderate	fine	N	O	O	N	N	11
FLY ASH								
a. Coal burning—chain grate	light	fine	S	S	O	N	O	12
b. Coal burning—stoker fired	moderate	fine-coarse	S	S	O	N	O	
c. Coal burning—pulverized fuel	moderate	fine	S	S	O	N	O	13
d. Wood burning	varies	coarse	S	S	O	N	S	14
FOUNDRY								
a. Shakeout	light-moderate	fine	N	O	O	N	N	15
b. Sand handling	moderate	fine-medium	N	O	O	N	N	16
c. Tumbling mills	heavy	medium-coarse	N	S	O	N	N	17
d. Abrasive cleaning	moderate-heavy	fine-medium	N	S	O	N	N	18
GRAIN ELEVATOR, FLOUR AND FEED MILLS								49
a. Grain handling	light	medium	O	S	O	N	N	19
b. Grain dryers	light	coarse	S	S	O	N	N	20
c. Flour dust	moderate	medium	O	S	O	N	N	21
d. Feed mill	moderate	medium	O	S	O	N	N	22
METAL MELTING								49
a. Steel blast furnace	heavy	varied	N	O	S	N	S	23
b. Steel open hearth	moderate	fine-coarse	N	O	S	N	S	24
c. Steel electric furnace	light	fine	N	S	O	N	S	25
d. Ferrous cupola	moderate	varied	N	O	O	N		26
e. Non-ferrous reverberatory	varied	fine	N	S	O	N	N	27
f. Non-ferrous crucible	light	fine	N	S	O	N	N	28
METAL MINING AND ROCK PRODUCTS								
a. Material handling	moderate	fine-medium	N	O	O	N	N	29
b. Dryers, kilns	moderate	medium-coarse	O	O	O	N	O	30
c. Rock dryer	moderate	fine-medium	N	S	S	N	S	31
d. Cement kiln	heavy	fine-medium	N	N	O	N	S	32
e. Cement grinding	moderate	fine	N	N	O	N	N	33
f. Cement clinker cooler	moderate	coarse	O	N	O	N	N	34

TABLE 8-3 (Cont.). Dust Collector Selection Guide

Operation	Concen-tration Note 1	Particle Sizes Note 2	Collector Types Used in Industry					See Remark No.
			Dry Cen-trifugal Collector	Wet Collector	Fabric Collector	Low-Volt Electro-static	Hi-Volt Electro-static	
METAL WORKING								49
a. Production grinding, scratch brushing, abrasive cut off	light	coarse	O	O	O	N	N	35
b. Portable and swing frame	light	medium	S	O	O	N	N	
c. Buffing	light	varied	S	O	O	N	N	36
d. Tool room	light	fine	S	S	S	N	N	37
e. Cast iron machining	moderate	varied	O	O	O	S	N	38
PHARMACEUTICAL AND FOOD PRODUCTS								
a. Mixers, grinders, weighing, blending, bagging, packaging	light	medium	O	O	O	N	N	39
b. Coating pans	varied	fine-medium	N	O	O	N	N	40
PLASTICS								49
a. Raw material processing	(See comments under Chemicals)		O	S	O	N	N	41
b. Plastic finishing	light-moderate	varied	S	S	O	N	N	42
c. Extrusion	light	fine	N	S	N	O	N	
RUBBER PRODUCTS								49
a. Mixers	moderate	fine	S	O	S	N	N	43
b. Batchout rolls	light	fine	S	O	S	S	N	
c. Talc dusting and dedusting	moderate	medium	S	S	O	N	N	44
d. Grinding	moderate	coarse	O	O	O	N	N	45
WOODWORKING								49
a. Woodworking machines	moderate	varied	O	S	O	N	N	46
b. Sanding	moderate	fine	S	S	O	N	N	47
c. Waste conveying, hogs	heavy	varied	O	S	S	N	N	48

Note 1: Light: less than 2 gr/ft^3 [4.6 g/m^3]; Moderate: 2 to 5 gr/ft^3 [4.6 – 14.4 g/m^3]; Heavy: 5 gr/ft^3 [14.4 g/m^3] and up.
Note 2: Fine: 50% less than 5 microns; Medium: 50% 5 to 15 microns; Coarse: 50% 15 microns and larger.
Note 3: O = often; S = seldom; N = never.

Remarks Referred to in Table 8-3

1. Dust released from bin filling, conveying, weighing, mixing, pressing, forming. Refractory products, dry pan and screen operations more severe.
2. Operations found in vitreous enameling, wall and floor tile, pottery.
3. Grinding wheel or abrasive cut-off operation. Dust abrasive.
4. Operations include conveying, elevating, mixing, screening, weighing, packaging. Category covers so many different materials that recommendation will vary widely.
5. Cyclone and high efficiency centrifugals often act as primary collectors followed by fabric or wet type.
6. Cyclones used as product collector followed by fabric arrester for high overall collection efficiency.
7. Dust concentration determines need for dry centrifugal; plant location, product value determines need for final collectors. High temperatures are usual and corrosive gases not unusual.
8. Conveying, screening, crushing, unloading.
9. Remove from other dust producing points. Separate collector usually.
10. Heavy loading suggests final high efficiency collector for all except very remote locations.
11. Difficult problem but collectors will be used more frequently with air pollution emphasis.
12. Public nuisance from boiler blow-down indicates collectors are needed.
13. Large installations in residential areas require electrostatic in addition to dry centrifugal.
14. Cyclones used as spark arresters in front of fabric collectors.
15. Hot gases and steam usually involved.
16. Steam from hot sand, adhesive clay bond involved.
17. Concentration very heavy at start of cycle.
18. Heaviest load from airless blasting due to higher cleaning speed. Abrasive shattering greater with sand than with grit or shot. Amounts removed greater with sand castings, less with forging scale removal, least when welding scale is removed.
19. Operations such as car unloading, conveying, weighing, storing.
20. Collection equipment expensive but public nuisance complaints becoming more frequent.

Remarks Referred to in Table 8-3 (continued)

21. Operations include conveyors, cleaning rolls, sifters, purifiers, bins and packaging.
22. Operations include conveyors, bins, hammer mills, mixers, feeders and baggers.
23. Primary dry trap and wet scrubbing usual. Electrostatic is added where maximum cleaning required.
24. Use of this technique declining.
25. Air pollution standards will probably require increased usage of fabric arresters.
26. CAUTION! Recent design improvements such as coke-less, plasma-fired type, have altered emission characteristics.
27. Zinc oxide loading heavy during zinc additions. Stack temperatures high.
28. Zinc oxide plume can be troublesome in certain plant locations.
29. Crushing, screening, conveying involved. Wet ores often introduce water vapor in exhaust air.
30. Dry centrifugals used as primary collectors, followed by final cleaner.
31. Industry is aggressively seeking commercial uses for fines.
32. Collectors usually permit salvage of material and also reduce nuisance from settled dust in plant area.
33. Salvage value of collected material high. Same equipment used on raw grinding before calcining.
34. Coarse abrasive particles readily removed in primary collector types.
35. Roof discoloration, deposition on autos can occur with cyclones and less frequently with high efficiency dry centrifugal. Heavy duty air filters sometimes used as final cleaners.
36. Linty particles and sticky buffing compounds can cause pluggage and fire hazard in dry collectors.
37. Unit collectors extensively used, especially for isolated machine tools.
38. Dust ranges from chips to fine floats including graphitic carbon. Low voltage ESP applicable only when a coolant is used.
39. Materials vary widely. Collector selection depends on salvage value, toxicity, sanitation yardsticks.
40. Controlled temperature and humidity of supply air to coating pans makes recirculation deslrable.
41. Plastic manufacture allied to chemical industry and varies with operations involved.
42. Operations and collector selection similar to woodworking. See Item 13.
43. Concentration is heavy during feed operation. Carbon black and other fine additions make collection and dust-free disposal difficult.
44. Salvage of collected material often dictates type of high efficiency collector.
45. Fire hazard from some operations must be considered.
46. Bulking material. Collected material storage and bridging from splinters and chips can be a problem.
47. Dry centrifugals not effective on heavy concentration of fine particles from production sanding.
48. Dry centrifugal collectors required. Wet or fabric collectors may be used for final collectors.
49. See NFPA publications for fire hazards, e.g., zirconium, magnesium, aluminum, woodworking, plastics, etc.

---

cussed in Section 8.3.3. It consists of a cylindrical shell, a packed section held on a support plate, a liquid distributor, possibly a liquid redistributor, access manholes, gas inlet and outlet, and possibly a sump with recirculation pump and overflow. There are a wide variety of packing materials available. Packings providing more surface area per unit volume are generally regarded as superior. There are tradeoffs to consider when selecting a packing material which will impact the overall equipment height and pressure drop requirements to meet specific contaminant collection removal characteristics.[8.5]

Staged Scrubbers: Staged or "stagewise" equipment utilizes a group of horizontal metal plates arranged in a vertical series and generally placed in a cylindrical housing. Each horizontal plate is a stage. The plates can be sieves, bubble type or ballasts. Gas flow is countercurrent to the liquid flow in all cases. In each of these designs, the liquid is kept on the tray surface by a dam at the entrance to a downcomer or sealed conduit allowing overflow liquid to pass to the tray below.[8.6]

High Energy Scrubbers: High Energy Contactors (Venturi Scrubbers, Figure 8-12) were also described in Section 8.3.3. Although used predominately as particulate control devices they can simultaneously function as absorbers. Venturi scrubbers are cocurrent devices and their absorption characteristics are maximized when operating at low velocities with high liquid to gas ratios.

Dry Absorption: Dry Absorption systems include Dry Scrubbers, Spray Dryers and Fluid Bed Reactors. Dry Scrubbers involve injection of a dry solvent directly into a process gas stream. Spray Dryers inject a wet solvent into a hot gas stream where the liquid evaporates leaving a dry solvent in contact with the gas. Fluid Bed Reactors employ a bed of granulated solvent fluidized within a vessel and the process gas flows through the fluidized bed. All dry absorption systems must include an appropriate particulate removal device.

8.6.2 Adsorption. Adsorption is also a mass transfer process which removes contaminants by adhesion of molecules of one phase to the surface or interfaces of a solid second phase. Relatively weak adsorption, where the forces involved are intermolecular, is known as van der Waals Adsorption. Strong adsorption, where the forces involved are valence forces, is known as activated adsorption or chemisorption. No chemical reaction is involved as adsorption is a physical process that is normally thought of as reversible. Activated carbon, activated alumina, silica gel, Fuller's earth, and molecular sieves are popular adsorbents.

8.6.3 Incineration/Oxidation. These two terms, Incineration and Oxidation, are used interchangeably to describe the process of combustion. Combustion is a chemical process in which oxygen reacts with various elements or chemical compounds resulting in the release of light and heat. The combustion process readily converts volatile organic compounds (VOCs), organic aerosols, and most odorous materials to carbon dioxide and water vapor. It is a very effective means of eliminating VOCs. Typical applications for incineration devices include odor control, reduction in plume opacity, reduction in reactive hydrocarbon emissions, and reduction of explosion hazards. The equipment used for control of gaseous contaminants by combustion may be divided into three categories: Thermal Oxidizers, Direct Combustors, or Catalytic Oxidizers.

Thermal Oxidizers, or afterburners, may be used where the contaminant is combustible. The contaminated air stream is introduced to an open flame or heating device followed by a residence chamber where combustibles are oxidized producing carbon dioxide and water vapor. Most combustible contaminants can be oxidized at temperatures between 1,000 F [538 C] and 1,500 F [816 C]. The residence chamber must provide sufficient dwell time and turbulence to allow complete oxidation. Thermal oxidizers are often equipped with heat exchangers where combustion gas is used to preheat the incoming contaminated gas. If gasoline is the contaminant, heat exchanger efficiencies are limited to 25 to 35% and preheat temperatures are maintained below 530 F [277 C] to minimize the possibility of ignition occurring in the heat exchanger. Flame arrestors are always installed between the vapor source and the thermal oxidizer. Burner capacities in the combustion chamber range from 0.5 to 2 M BTU [527.9 to 2111.6 kJ] per hour. Operating temperatures range from 1,400 to 1,600 F [760 C to 871 C], and gas residence times are typically 1 second or less. This condition causes the molecular structure to break down into simple carbon dioxide and water vapor.

Regenerative Thermal Oxidation (RTO) units are distinguished from other thermal incinerators by their ability to recover heat at high efficiency. RTOs employ three, five, seven, or more chambers that store and recycle heat energy. RTO technology uses high temperature to convert VOCs into carbon dioxide and water vapor.

In RTO, contaminated process air enters a combustion chamber after being preheated through a ceramic bed, where the air is raised to a required temperature and held there for a specified period of time. The heat recovery chambers are outfitted with stoneware or ceramic beds that absorb most of the heat energy from the combustion chamber. The flow is then reversed, allowing the next contaminated batch of air to enter the combustion chamber through the stoneware bed that was heated from the last batch. The level of heat recovery varies, depending on the specific design of the system.

Using a Flameless Thermal Oxidation process, VOC-laden exhaust gas typically enters a single or multiple module thermal oxidation unit (or an RTO). The VOC gas stream is alternatively directed using valves to the top or bottom air plenum and is transported through a porous gravel heat exchange bed. In the gravel media, it is flamelessly oxidized and converted to carbon dioxide and water vapor. Reversal of the gas stream keeps the high temperature band centered in the gravel media. For start-up, natural gas/propane is injected into the heat transfer media to bring the temperature up to 1,800 F [982 C]. For low concentration streams of VOC exhaust, supplemental fuel is needed to maintain the proper oxidation temperature. For VOC streams above a concentration of 3.8%, the reaction is self-sustaining. The process attains greater than 98% VOC destruction and 95% heat recovery.

Direct Combustors (flares) differ from thermal oxidizers by introducing the contaminated gases and auxiliary air directly into the burner as fuel. Auxiliary fuel, usually natural gas or oil, is generally required for ignition. It may or may not be required to sustain burning and all of the waste gases react at the burner.

Catalytic Oxidation: Catalytic oxidation is a relatively new alternative for the treatment of VOCs in air streams resulting from remedial operations. It is very similar to thermal oxidation, except that with a catalyst present, the same reaction occurs at a lower temperature. Catalysts are substances that alter the rate of a chemical reaction without themselves being consumed in the reaction. VOCs are thermally destroyed at temperatures typically ranging from 600 to 1,000 F [316 C to 538 C] by using a solid catalyst. First, the contaminated air is directly preheated (electrically or, more frequently, using natural gas or propane) to reach a temperature necessary to initiate the catalytic oxidation of the VOCs. Then the preheated VOC-laden air is passed through a bed of solid catalysts where the VOCs are rapidly oxidized.

In most cases, the process can be enhanced to reduce auxiliary fuel costs by using an air-to-air heat exchanger to transfer heat from the exhaust gases to the incoming contaminated air. Typically, about 50% of the heat of the exhaust gases is recovered. Depending on VOC concentrations, the recovered heat may be sufficient to sustain oxidation without additional fuel. Catalyst systems used to oxidize VOCs typically use metal oxides such as nickel oxide, copper oxide, manganese dioxide, or chromium oxide. Noble Metals such as platinum and palladium may also be used. However, in a majority of remedial applications, non-precious metals (e.g., nickel, copper, or chromium) are used. Most commercially available catalysts are proprietary.

To use either thermal or catalytic oxidation, the combustible contaminant concentration must be below the lower explosive limit. Equipment specifically designed for control of gaseous or vapor contaminants should be applied with caution when the air stream also contains solid particles. Solid particulate can plug absorbers, adsorbers, and catalysts and, if noncombustible, will not be converted in thermal oxidizers and direct combustors.

8.6.4 Biofiltration.[8.7,8.8] Biofiltration process involves drawing contaminated air through a pretreatment unit to adjust its temperature and moisture content, and then through a filter in which the contaminants are transferred to microorganisms selected for their efficiency in treating those specific contaminants.

It is an emerging air pollution control technology suited for cleaning VOCs and other gases such as ammonia and hydrogen sulfide. These gases are considered responsible for odors associated with livestock and poultry production. Successful and common applications of biofilters in agricultural facilities, rendering plants, wastewater treatment plants, chemical, and food processing plants have been reported in Europe and Japan. In the United States, common applications are reported in water treatment plants. Some chemical manufacturing plants are also reported to be using biofilters. Few, if any, are currently being used in livestock and poultry facilities.

8.6.5 Other Gaseous Contaminant Controls. The most commonly used of the lesser known gaseous contaminant control methods referred to above is condensation. It has been widely used for recovery of and/or removal of gaseous specific constituents in a bulk gas flow. Specific examples would include the selective distillation of various hydrocarbons in refining processes and the drying of air. In order to remove a selected contaminant from a gas stream by this method the dew point of the pollutant must be significantly higher than that of the non-contaminant gases. This technique has been successfully applied as a control method for removal of some VOCs.

Application of the Corona Reactor, Photochemical Oxidation, Direct Electric Arcing, and Plasma Treatment techniques are largely experimental at this date. All of these techniques target VOCs and some inorganic gases such as hydrogen sulfide, mercaptans, trichloroethylene, and carbon tetrachloride.

Air streams containing both solid particles and gaseous contaminants may require appropriate control devices in series.

8.7 UNIT COLLECTORS

Unit collector is a term usually applied to small fabric collectors having capacities in the 200 to 2000 acfm [0.1 to 1.0 m^3/s] range. They have integral air movers, feature small space requirements and simplicity of installation. In most applications, cleaned air is recirculated, although discharge ducts may be used if the added resistance is within the capability of the air mover. One of the primary advantages of unit collectors is a reduction in the amount of duct required, as opposed to central systems. The addition of discharge ducts to unit collectors negates that advantage.

When cleaned air is to be recirculated, a number of precautions are required (see Chapter 10).

Unit collectors are used extensively to fill the need for dust collection from isolated, portable, intermittently used, or frequently relocated dust producing operations. Typically, a single collector serves a single dust source with the energy saving advantage that the collector must operate only when that particular dust producing machine is in operation.

Figure 8-16 shows a typical unit collector. Usually they are the intermittent duty, shaker-type in envelope configuration. Woven fabric is nearly always used. Automatic fabric cleaning is preferred. Manual methods without careful scheduling and supervision are unreliable.

8.8 DUST COLLECTING EQUIPMENT COST

The variations in equipment cost, especially on an installed basis, are difficult to estimate. Comparisons can be misleading if these factors are not carefully evaluated.

8.8.1 Price Versus Capacity. All dust collector prices per cfm of gas will vary with the gas flow rate. The smaller the flow rate, the higher the cost per cfm. The break point, where price per cfm cleaned tends to level off, will vary with the design. See the typical curves shown on Figure 8-17.

8.8.2 Accessories Included. Careful analysis of components of equipment included is very important. Some collector designs include exhaust fan, motor, drive, and starter. In other designs, these items and their supporting structure must be obtained by the purchaser from other sources. Likewise, while dust storage hoppers are integral parts of some dust collector designs, they are not provided in other types. Duct connections between elements may be included or omitted. Recirculating water pumps and/or settling tanks may be required but not included in the equipment price.

8.8.3 Installation Cost. The cost of installation can equal or exceed the cost of the collector. Actual cost will depend on the method of shipment (completely assembled, sub-assembled, or completely knocked down), the location (which may require expensive rigging), and the need for expensive supporting steel and access platforms. Factory installed media will reduce installation cost. The cost can also be measurably influenced by the need for water and drain connections, special or extensive electrical work, and expensive material handling equipment for collection material disposal. Items in the latter group will often also be variable, decreasing in cost per cfm as the flow rate of gas to be cleaned increases.

8.8.4 Special Construction. Prices shown in any tabulation must necessarily assume standard or basic construction. The increase in cost for corrosion resisting material, special high temperature fabrics, insulation, and/or weather protection for outdoor installations can introduce a multiplier of one to four times the standard cost.

A general idea of relative dust collector cost is provided in Figure 8-17. The additional notes and explanations included in

FIGURE 8-16

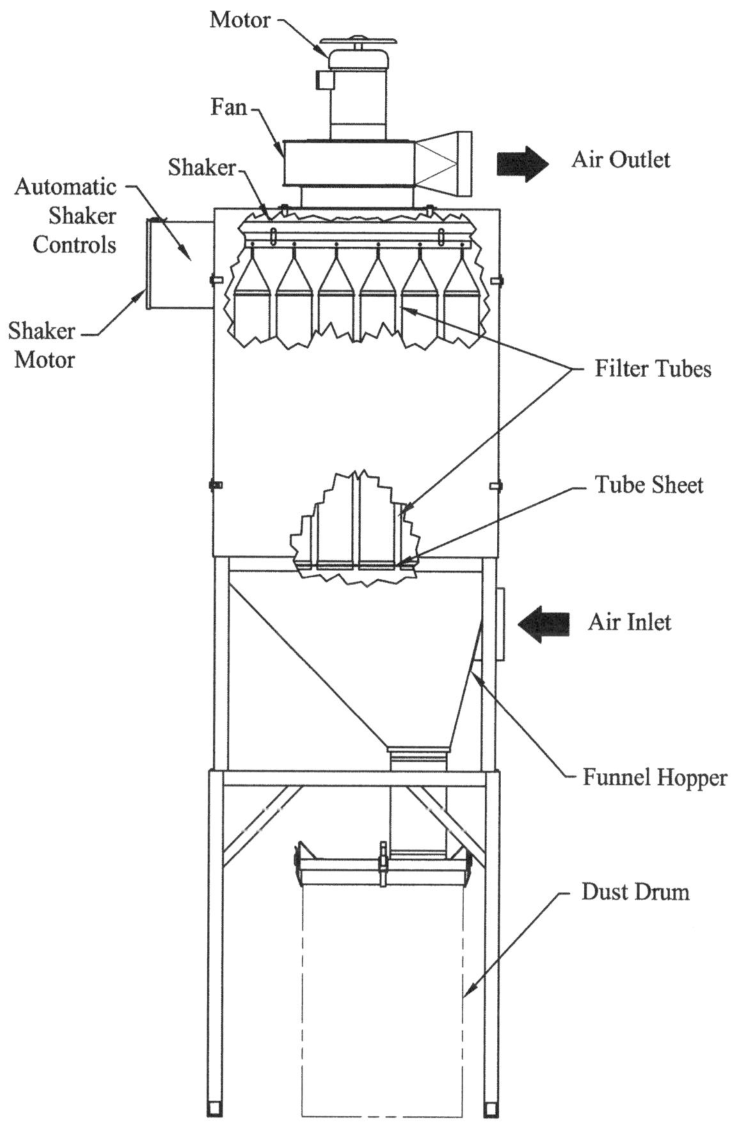

ACGIH®

TITLE

UNIT COLLECTOR
(SHAKER TYPE FABRIC)

FIGURE 8-16

DATE 1-12

CHECK CODES, REGULATIONS, AND LAWS (LOCAL, STATE, AND NATIONAL) TO ENSURE THAT DESIGN IS COMPLIANT.

FIGURE 8-17 (IP)

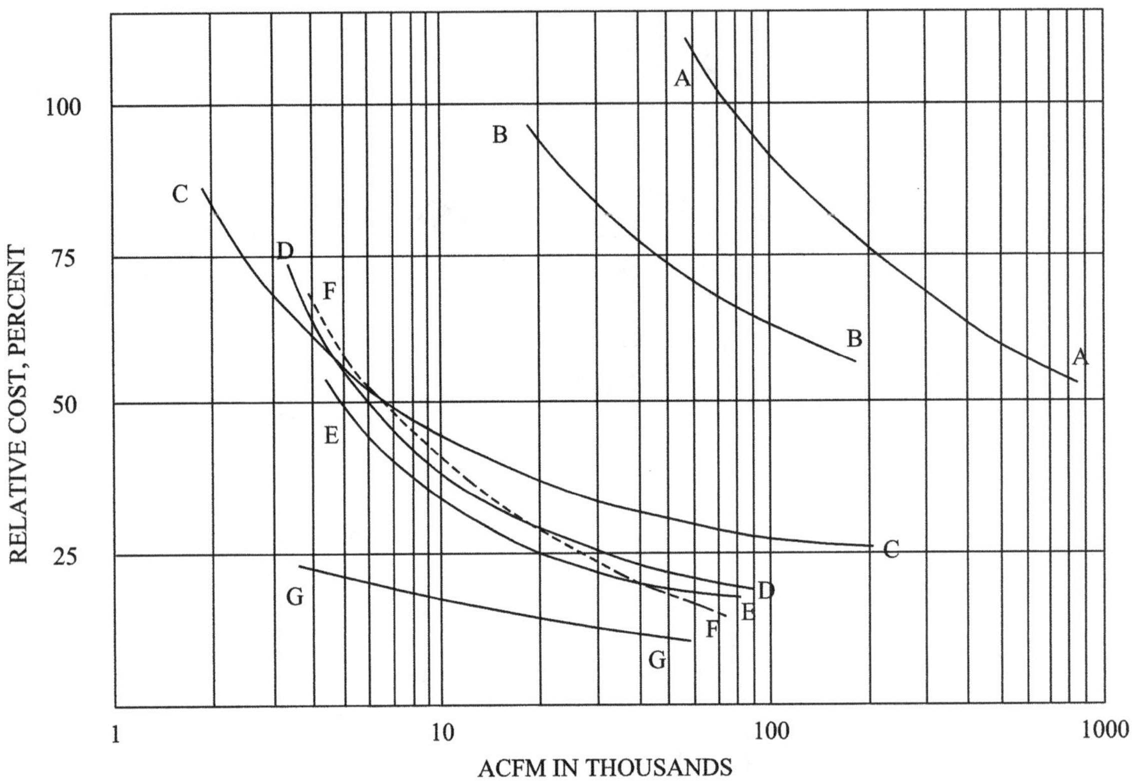

A. High voltage precipitator (minimum cost range)
B. Continuous duty high temperature fabric collector (air / cloth = 2.0:1)
C. Continuous duty reverse pulse (air / cloth = 8:1)
D. Wet collector
E. Intermittent duty fabric collector (air / cloth = 2.0:1)
F. Low voltage precipitator
G. Cyclone

Note 1: Cost based on collector section only. Does not include ducts, dust disposal devices, pumps, exhausters or other accessories not an integral part of the collector.

Note 2: Price of high voltage precipitator will vary substantially with applications and efficiency requirements. Costs shown are for fly ash applications where velocities of 200 to 300 fpm are normal.

ACGIH®	TITLE COST ESTIMATES OF DUST COLLECTING EQUIPMENT	FIGURE 8-17 (IP)
		DATE 1-12

CHECK CODES, REGULATIONS, AND LAWS (LOCAL, STATE, AND NATIONAL) TO ENSURE THAT DESIGN IS COMPLIANT.

FIGURE 8-17 (SI)

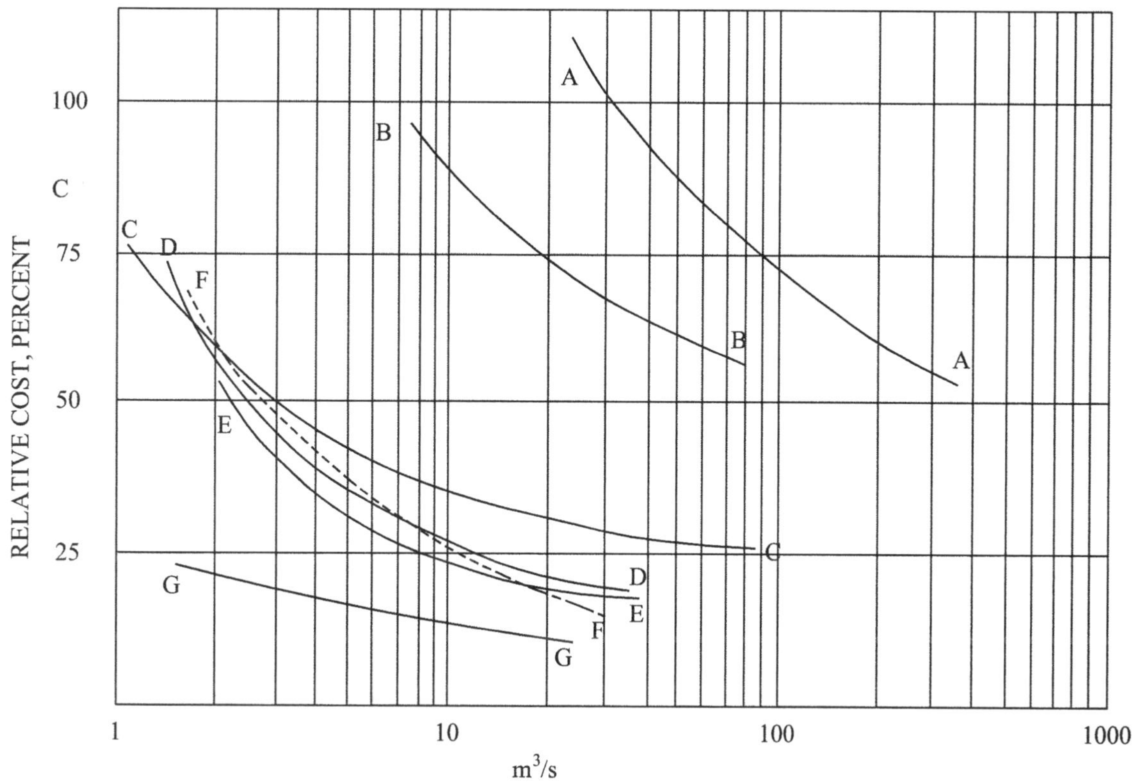

A. High voltage precipitator (minimum cost range)
B. Continuous duty high temperature fabric collector
C. Continuous duty reverse pulse
D. Wet collector
E. Intermittent duty fabric collector
F. Low voltage precipitator
G. Cyclone

Note 1: Cost based on collector section only. Does not include ducts, dust disposal devices, pumps, exhausters or other accessories not an integral part of the collector.

Note 2: Price of high voltage precipitator will vary substantially with applications and efficiency requirements. Costs shown are for fly ash applications where velocities of 1 m/s to 1.5 m/s are normal.

ACGIH®	TITLE COST ESTIMATES OF DUST COLLECTING EQUIPMENT	FIGURE 8-17 (SI) DATE 1-12

CHECK CODES, REGULATIONS, AND LAWS (LOCAL, STATE, AND NATIONAL) TO ENSURE THAT DESIGN IS COMPLIANT.

these data should be carefully examined before they are used for estimating the cost of specific installations. For more accurate data, the equipment manufacturer or installer should be asked to provide estimates or a past history record for similar control problems utilized. Table 8-4 lists other characteristics that must be evaluated along with equipment cost.

Price estimates included in Figure 8-17 are for equipment of standard construction in normal arrangement. Estimates for exhausters and dust storage hoppers have been included, as indicated in Notes 1 and 2, where they are normally furnished by others.

8.9 SELECTION OF AIR FILTRATION EQUIPMENT

Air filtration equipment is available in a wide variety of designs and capability. Performance ranges from a simple throwaway filter for the home furnace to the "clean room" in the electronics industry, where the air must be a thousand times as clean as in a hospital surgical suite. Selection is based on efficiency, dust holding capacity, and pressure drop. There are five basic methods of air filtration.

8.9.1 Straining. Straining occurs when a particle is larger than the opening between fibers and cannot pass through. It is a very ineffective method of filtration because the vast majority of particles are far smaller than the spaces between fibers. Straining will remove lint, hair, and other large particles.

8.9.2 Impingement. When air flows through a filter, it changes direction as it passes around each fiber. Larger dust particles, however, cannot follow the abrupt changes in direction because of their inertia. As a result, they do not follow the air stream and collide with a fiber. Filters using this method are often coated with an adhesive to help fibers retain the dust particles that impinge on them.

8.9.3 Interception. Interception is a special case of impingement where a particle is small enough to move with the air stream but, because its size is very small in relation to the fiber, makes contact with a fiber while following the tortuous airflow path of the filter. The contact is not dependent on inertia and the particle is retained on the fiber because of the inherent adhesive forces that exist between the particle and fiber. These forces, called van der Waals (J.D. van der Waals, 1837–1923) forces, enable a fiber to trap a particle without the use of inertia.

8.9.4 Diffusion. Diffusion takes place on particles so small that their direction and velocity are influenced by molecular collisions. These particles do not follow the air stream, but behave more like gases than particulate. They move across the direction of airflow in a random fashion. When a particle does strike a fiber, it is retained by the van der Waals forces existing between the particle and the fiber. Diffusion is the primary mechanism used by most extremely efficient filters.

8.9.5 Electrostatic. A charged dust particle will be attracted to a surface of opposite electrical polarity. Most dust particles are not electrically neutral, therefore, electrostatic attraction between dust particle and filter fiber aids the collection efficiency of all barrier type air filters. Electrostatic filters establish an ionization field to charge dust particles so that they can be collected on a surface that is grounded or of opposite polarity. This concept was previously discussed in Section 8.3.1.

Table 8-5 shows performance versus filter fiber size for several filters. Note that efficiency increases as fiber diameter decreases because more small fibers are used per unit volume. Note also that low velocities are used for high efficiency filtration by diffusion.

The wide range in performance of in-line media-style air filters made it necessary to agree on a new consolidated method of efficiency testing. The new adopted, industry-accepted method in the United States is the MERV (minimum efficiency reporting value) system developed by ASHRAE. This filter rating system ranges from 1 through 20, where a rating of 1 is a very coarse see-through style home HVAC filter and a rating of 20 exceeds even the ability of a HEPA (High Efficiency Particulate Air) filter. In a HEPA DOP Test, 0.3 micron particles of dioctylphthalate (DOP) are drawn through a HEPA filter. Efficiency is determined by comparing the downstream and upstream particle counts. To be designated as a HEPA filter, the filter must be at least 99.97% efficient, i.e., only three particles of 0.3 micron size can pass for every 10,000 particles fed to the filter.

MERV filters come in four typical filter types, as follows: *Flat or panel air filters* with a MERV of 1 to 4 are commonly used in residential furnaces and air conditioners. They are NOT typically used in industrial ventilation applications. Second, there are *pleated or extended surface filters*, with a MERV of 5 to 15 range from 1" [25 mm] deep pleated filters to true "box" and "envelope" filters. Third are *high efficiency "box and envelope filters,"* with a MERV of 14 to 16. Finally, there are *true HEPA filters* (MERV 17 to 20). Figure 8-18 shows the general relationship. Table 8-6 compares several important characteristics of commonly used air filters. Considerable life extension of an expensive final filter can be obtained by the use of one or more cheaper, less efficient, prefilters. For example, the life of a HEPA filter can be increased 25% with a throwaway prefilter. If the throwaway filter is followed by a 90% efficient extended surface filter, the life of the HEPA filter can be extended nearly 900%. This concept of "progressive filtration" allows the final filters in clean rooms to remain in place for 10 years or more.

8.10 RADIOACTIVE AND HIGH TOXICITY OPERATIONS

There are three major requirements for air cleaning equipment to be utilized for radioactive or high toxicity applications:

1) High efficiency
2) Low maintenance
3) Safe disposal

TABLE 8-4. Comparison of Some Important Dust Collector Characteristics

Type	Higher Efficiency Range on Particles Greater than Mean Size in Microns	Pressure Loss inches [Pa]	H_2O Gal. Per 1000 acfm [liters/am^3 – gas]	Space	Sensitivity to Q Change: Pressure	Sensitivity to Q Change: Efficiency	Humid Air Influence	Max. Temp. F [C] Standard Construction Note 4
Electrostatic:	0.25	0.5 [125]	—	Large	Negligible	Yes	Improves efficiency	500 [260]
Fabric:								
Intermittent—Shaker	0.25	3–6 [750–1500]	—	Large	∝ Q	Negligible	May make reconditioning difficult	See Table 8-1
Continuous—Shaker	0.25	3–6 [750–1500]	—	Large	∝ Q	Negligible		
Continuous—Reverse Air	0.25	3–6 [750–1500] Note 1	—	Large	∝ Q	Negligible		
Continuous—Reverse Pulse	0.25	3–6 [750–1500]	—	Moderate	∝ Q	Negligible		
Glass, Reverse Flow	0.25	3–8 [750–2000]	—	Large	∝ Q	Negligible		500 [260]
Wet:								
Packed Tower	1–5	1.5–3.5 [375–875]	5–10 [40-80]	Large	∝ Q	Yes		
Wet Centrifugal	1–5	2.5–6 [625–1500]	3–5 [25-40]	Moderate	∝ $(Q)^2$	Yes		
Wet Dynamic	1–2	Note 2	0.5–1 [4-8]	Small	Note 2	No	None	Unlimited
Orifice Types	1–5	2.5–6 [625–1500]	10–40 [80-320]	Small	∝ Q or less	Varies with Design		
Higher Efficiency:								
Fog Tower	0.5–5	2–4 [500–1000]	5–10 [40-80]	Moderate	∝ $(Q)^2$	Slightly	None	Note 3
Venturi	0.5–2	10–100 [2500–25000]	5–15 [40-120]	Moderate	∝ $(Q)^2$	Yes		Unlimited
Dry Centrifugal:								
Low Pressure Cyclone	20–40	0.75–1.5 [180–375]	—	Large	∝ $(Q)^2$	Yes	May cause condensation and plugging	400 [204]
High Eff. Centrifugal	10–30	3–6 [750–1500]	—	Moderate	∝ $(Q)^2$	Yes		400 [204]
Dry Dynamic	10–20	Note 2	—	Small	Note 2	No		

Note 1: Pressure loss is that for fabric and dust cake. Pressure losses associated with outlet connections to be added by system designer.
Note 2: A function of the mechanical efficiency of these combined exhausters and dust collectors.
Note 3: Precooling of high temperature gases will be necessary to prevent rapid evaporation of fine droplets.
Note 4: See NFPA requirements for fire hazards, e.g., zirconium, magnesium, aluminum, woodworking, etc.

∝ - Proportional to

TABLE 8-5. Media Velocity vs. Fiber Size

Filter Type	Filter Size (microns)	Velocity fpm [m/s]	Media Filtration Mechanism
Panel Filters	25–50	250–625 [1.25–3.20]	Impingement
Automatic Roll Filters	25–50	500 [2.50]	Impingement
Extended Surface Filters	0.75–2.5	20–25 [0.10–0.13]	Interception
HEPA Filters	0.5–6.3	5 [0.25]	Diffusion

High efficiency is essential because of extremely low tolerances for the quantity and concentration of stack effluent and the high cost of the materials handled. Not only must the efficiency be high, it must also be verifiable because of the legal requirement to account for all radioactive material.

The need for low maintenance is of special importance when exhausting any hazardous material. For many radioactive processes, the changing of bags in a conventional fabric collector may expend the daily radiation tolerances of 20 or more persons. Infrequent, simple, and rapid maintenance requirements are vital. Another important factor is the desirability of low residual build up of material in the collector since dose rates increase with the amount of material and reduce the allowable working time.

Disposal of radioactive or toxic materials is a serious and very difficult problem. For example, scalping filters loaded with radioactive dust are usually incinerated to reduce the quantity of material that must be disposed of in special burial grounds. The incinerator will require an air cleaning device, such as a wet collector of very special design, to avoid unacceptable pollution of air and water.

With these factors involved, it is necessary to select an air cleaning device that will meet efficiency requirements without causing too much difficulty in handling and disposal.

Filter units especially designed for high efficiency and low maintenance are available. These units feature quick change-out through a plastic barrier which is intended to encapsulate spent filters, thereby eliminating the exposure of personnel to radioactive or toxic material. A filtration efficiency of 99.97% by particle count on 0.3 micron particles is standard for this type of unit.

For further information on this subject, see Reference 8.10.

8.11 EXPLOSION VENTING/DEFLAGRATION VENTING

Two distinct types of explosions exist in nature. A detonation is an explosion that propagates at a velocity in excess of the speed of sound and cannot be controlled. In a deflagration, the combustion wave propagates more slowly (at less than the speed of sound) and can be controlled, if designed properly. Examples of detonations include dynamite, solid rocket fuel or other similar material. Examples of deflagrations include most organic dusts such as grain, wood, plastics, coal and many others. Metal dust deflagrations are especially dangerous and have there own NFPA designation (see Reference 8.11).

To begin taking precautions, sources of possible ignition must be identified and controlled to minimize the risk of a dust cloud explosion. Usual causes of explosions include static discharge, hot surfaces on machinery and sparks and flames from processes. After identifying possible sources of ignition, preventive measures should be taken. Static grounding of the equipment and spark traps are typical preventive measures. The addition of an inert gas to replace oxygen in a dust collector can prevent an explosion by ensuring the minimum oxygen content required for ignition is never reached. Inerting can be very effective in closed loop systems but is not economical in typical local exhaust systems because of the constant loss of expensive inerting gas. Should ignition occur, protective mea-

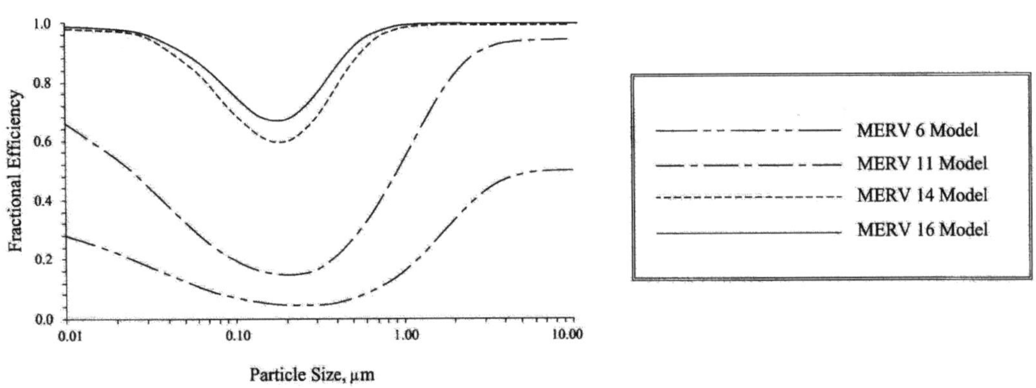

FIGURE 8–18. Comparison between various methods of measuring air cleaning capability

TABLE 8-6. Comparison of Some Important Air Filter Characteristics

Type	Pressure Drop "wg [Pa] (Notes 1 & 2) Initial	Final	ASHRAE Performance (Note 4) MERV (Note 5)	Arrestance	Efficiency	Face Velocity fpm [m/s]	Maintenance (Note 6) Labor	Material
Low/Medium Efficiency								
1. Glass Throwaway (2" deep)	0.1 [25]	0.5 [125]	2–3	77%	NA Note 7	300 [1.50]	High	High
2. High Velocity (permanent units) (2" deep)	0.1 [25]	0.5 [125]	2–3	73%	NA Note 7	500 [2.5]	High	Low
3. Automatic (viscous)	0.4 [100]	0.4 [100]	3	80%	NA Note 7	500 [2.5]	Low	Low
Medium/High Efficiency								
1. Extended Surface (dry)	0.15–0.60 [40–150]	0.5–1.25 [125–310]	8–12	90–99%	25–95%	300–625 [1.50–3.2]	Medium	Medium
2. Electrostatic								
a. Dry Agglomerator/ Roll Media	0.35 [90]	0.35 [90]	10–12	NA Note 8	90%	500 [2.50]	Medium	Low
b. Dry Agglomerator/ Extended Surface Media	0.55 [140]	1.25 [310]	13–16	NA Note 8	95%+	530 [2.70]	Medium	Medium
c. Automatic Wash Type	0.25 [60]	0.25 [60]	13–16	NA Note 8	95.5	400–600 [2.00–3.00]	Low	Low
Ultra High Efficiency								
1. HEPA	0.5–1.0 [125–250]	1.0–3.0 [250–750]	17–20	Note 3	Note 3	250–500 [1.25–2.50]	High	High

Note 1: Pressure drop values shown constitute a range or average, whichever is applicable.

Note 2: Final pressure drop indicates point at which filter or filter media is removed and the media is either cleaned or replaced. All others are cleaned in place, automatically, manually, or media renewed automatically. Therefore, pressure drop remains approximately constant.

Note 3: 95–99.97% by particle count, DOP test.

Note 4: ASHRAE Standard 52-76 defines (a) Arrestance as a measure of the ability to remove injected synthetic dust, calculated as a percentage on a weight basis and (b) Efficiency as a measure of the ability to remove atmospheric dust determined on a light-transmission (dust spot) basis.

Note 5: ASHRAE MERV (Minimum Efficiency Reporting Value) Efficiencies range from 1 (lowest) through 20 (highest).

Note 6: Compared to other types within efficiency category.

Note 7: Too low to be meaningful.

Note 8: Too high to be meaningful.

sures must be taken to limit the damage. Typical protective measures include: explosion suppression, explosion containment, and explosion venting.

Explosion suppression requires the early detection of an explosion, usually within the first 20 milliseconds. Once ignition is detected, an explosion suppression device injects a pressurized chemical suppressant into the collector to displace the oxygen and impede combustion. These are typically used in conjunction with fast acting isolation valves on the inlet and outlet ducts. These systems can be very useful when toxic dusts are being handled.

Explosion containment uses specialized dust collectors designed to withstand the maximum pressure generated and contain the explosion. Most pressure capabilities of commercially available dust collectors are not sufficient to contain an explosion in progress.

Explosion venting, the most common protection, is afforded by fitting pressure relief vents to the collector housing. As pressure increases quickly leading up to an explosion, a relief vent opens to allow the rapidly expanding gases to escape. This effectively limits the maximum pressure build up to less than the bursting pressure of the vessel. The necessary area for such a relief vent is a function of the vessel volume, vessel strength, the opening pressure of the relief vent and the rate of pressure rise characteristic of the dust in question. Most standard dust collectors will require reinforcing to withstand the reduced maximum pressure experienced during an explosion.

To choose the most reliable, economical, and effective means of explosion control, an evaluation of the specifics of the exhaust system and the degree of protection required is necessary.

The National Fire Protection Association (NFPA) Standards[8.11] are the most commonly recognized standards and should be studied and thoroughly familiar to anyone responsible for the design or evaluation of dust collectors applied to potentially explosive dusts. Always verify that the latest issue is being referenced for data and standards.

REFERENCES:

8.1 Leith, D.; First, M.K.W.; Feldman, H.: Performance of a Pulse-Jet at High Velocity Filtration II, Filter Cake Redeposition. J. Air Pollut. Control Assoc. 28:696 (July 1978).

8.2 Beake, E.: Optimizing Filtration Parameters. J. Air Pollut. Control Assoc. 24:1150 (1974).

8.3 Leith, D.; Gibson, D. D.; First, M. W.: Performance of Top and Bottom Inlet Pulse-Jet Fabric Filters. J. Air Pollut. Control Assoc. 24:1150 (1974).

8.4 American Society of Heating, Refrigerating and Air-Conditioning Engineers, Inc.: HVAC Systems and Equipment Handbook. Atlanta, GA (1996).

8.5 Lund, H.F.: Industrial Pollution Control Handbook. McGraw-Hill (1971).

8.6 Heumann, W.L.: Industrial Air Pollution Control Systems. McGraw-Hill (1997).

8.7 Gilliland, G.A.; Ramaswami, R.D.; Patel, D.N.: Removal of Volatile Organic Compounds (VOCs) Generated by Forest Product Industries Using Biofiltration Technology. In Proc. Emerging Technologies in Hazardous Waste Management VII, ACS Special Symposium: Atlanta, GA, September, 17-20, 1995. Tedder, D.W., Editor, Washington, DC (United States) American Chemical Society p. 921 (1352p) CONF-9509139.

8.8 Biofiltration. Air emissions from Wood and Wood-Based Products: Conducting Research and Sharing Information. 22 April 1998. USDA Forest Products Laboratory. 16 Dec 2000. http.fpl.fs.fed.us/voc/biofilt.html.

8.9 American Society of Heating, Refrigerating and Air-Conditioning Engineers: Method of Testing Cleaning Devices Used in General Ventilation for Removing Particulate Matter. ASHRAE Pub. No. 52-76. ASHRAE, Atlanta, GA (May 1976).

8.10 National Council on Radiation Protection and Measurement: NCRP Report No. 39, Basic Radiation Protection Criteria. NCRP Report No. 39. Publications, Bethesda, MD (January, 1971).

8.11 NFPA 654: Standard for the Prevention of Fire and Dust Explosions from the Manufacturing, Processing, and Handling of Combustible Particulate Solids (2006); NFPA 68: Guide for Venting of Deflagrations (2002); NFPA 69: Standard on Explosion Prevention Systems (2002); NFPA 91: Standard for Exhaust Systems for Air Conveying of Vapors, Gases, Mists, and Noncombustible Particulate Solids (2004); NFPA 484: Standard for Combustible Metals (2006); NFPA 497: Recommended Practice for the Classification of Flammable Liquids, Gases, or Vapors and of Hazardous (Classified) Locations for Electrical Installations in Chemical Process Areas (2004), National Fire Protection Association, Quincy, MA. NFPA 33: Standard for Spray Applications Using Flammable or Combustible Materials; NFPA 61: Standard for the Prevention of Fires and Dust Explosions in Agricultural and Food Processing Facilities

APPENDIX A8 CONVERSION OF POUNDS PER HOUR (EMISSIONS RATE) TO GRAINS PER DRY STANDARD CUBIC FOOT [g/nm³ (dry)] (EMISSION DENSITY OR "LOADING")

A collector is measured with a flow of 36,000 acfm of air at 120 F, 100% humidity with a particulate mass emissions rate of 1 pound per hour. What is the emissions rate in terms of grains per dry standard cubic foot (gr/dscf)?

120 F dB, 100% humidity → 0.0816 pounds H_2O/pound of dry air = 571 grains of water/pound dry air (psychrometric charts – Chapter 9)

Humid volume = 16.56 ft³ per pound of dry air

$$\frac{36{,}000\ \text{acfm}}{16.56\ \text{ft}^3/\text{pound}-\text{dry air}} = 2{,}174\ \text{pounds dry air}/\text{min}$$

$$\frac{2{,}174\ \text{pounds}-\text{dry air}}{0.075\ \text{pounds}-\text{air}/\text{dscf}} = 28{,}985\ \text{scfm}$$

$$\frac{1\ \text{pound}-\text{particulate}/\text{hr}}{60/\text{min}/\text{hr}} = 0.01667\ \text{pounds}/\text{min}$$

$$\frac{0.01667\ \text{pounds}/\text{min} \times 7{,}000\ \text{grs}/\text{pound}}{28{,}982\ \text{ft}^3/\text{min}}$$

Emission Density (Loading) = 0.004 grains/dscf

In SI units, the same problem would be set up as follows:

36,000 acfm ⇒ 16.9884 am³/s

120 F ⇒ 49 C

@ 100% humidity ⇒ 80 g/kg – dry air (Chapter 9, Figure 9-1 (SI))

Humid Volume ⇒ 1.03 m³/kg – dry air

$$\frac{16.9884\ \dfrac{m^3}{s}}{1.03\ \dfrac{m^3}{kg - \text{dry air}}} = 16.4936\ \text{kg} - \text{dry air/s}$$

$$16.4936\ \frac{\dfrac{\text{kg} - \text{dry air}}{s}}{\dfrac{1.204\ \text{kg}}{\text{nm}^3}} = 13.699\ \frac{\text{nm}^3}{s}$$

1 pound particulate ⇒ 453.6 grams

453 grams/hr × 1 hr/3600 s = 0.1258 grams/s

$$\frac{0.1258\ \dfrac{\text{grams}}{s}}{13.699\ \dfrac{\text{nm}^3}{s}} = 0.0092\ \frac{\text{grams}}{\text{nm}^3}(\text{dry})$$

Chapter 9

LOCAL EXHAUST VENTILATION SYSTEM DESIGN CALCULATION PROCEDURES

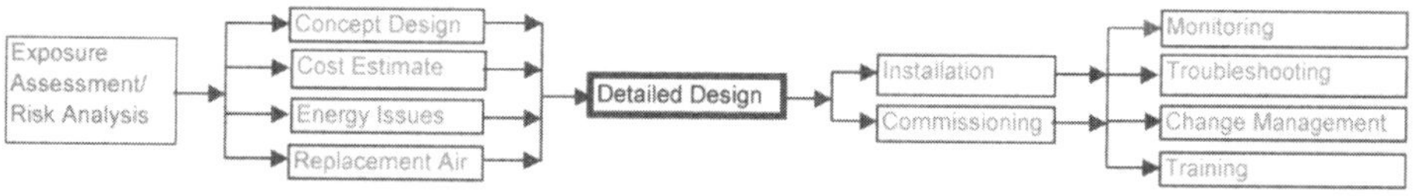

NOTE: Equations with notation followed by (IP) are designated for inch-pound system only; Equations followed by (SI) are designated for metric use only. If Equation bears neither, then it applies to both systems.

9.1 INTRODUCTION

The ventilation system that connects the hoods, duct, air cleaning device(s), and fan must be properly designed and balanced. This process is much more involved than merely connecting individual pieces of duct together. If the system is not carefully designed in a manner that inherently ensures that all design flow rates will be realized, contaminant control may not be achieved. In addition, minimum transport velocities must be maintained in all branches and main ducts at all times during operation if the system is handling particulate matter.

The procedures, criteria and organization for performing a detailed system design are included in Chapter 5. A thorough knowledge of the principles embodied in that chapter and in Chapter 3 is recommended before proceeding with the following calculation methods.

The results of the following design calculations will determine the duct sizes, the System Static Pressure (SSP), and the fan operating point (system flow rate and required pressure) required by the system. Chapter 7 describes how to select a fan based on these results.

Beginning with the 25th Edition, the reader will note that the revision includes the consideration of density changes in almost all calculations. Thus the user of the calculation methods will need to consider the changes in density of the air due to elevation, temperature, moisture and static pressure. "Standard Air" is defined as the condition of air at sea level, 70 F and with no moisture (Standard Temperature Pressure – STP).

Note also that in these cases "standard air" is not truly in units of "dry standard cubic feet per minute" (dscfm). This Manual uses dscfm and scfm interchangeably but if the designer uses the above ranges to define "standard" conditions, then be cautioned that results of calculations may be slightly changed.

If "Standard Air" is considered, a simple insertion of "1.0" for the value of the Density Factor (df) in the appropriate equations will yield the correct results. (See Chapter 3, Section 3.11 for a discussion of Density Factor.) Not all problems or calculations will require the consideration for density change, but it is strongly recommended that the designer investigate the relevant conditions or factors that might affect air density before beginning the design process. If the design involves gases other than air, the designer will need to consider the density of the gases involved. This could be the case when there are large concentrations of combustion products or other process gases. In these cases, the following calculation methods may not be appropriate.

When moisture is present in an air stream, different formulae and special techniques are required. This is because the water in the air can undergo phase changes (vapor to liquid and back to vapor, etc.). In those situations, all heat applied to the system must not only change the temperature of both the air and water (sensible heat), but also must supply energy to fuel the phase changes (i.e., the so called latent heat). When significant moisture from a process is introduced into the system (more than 0.02 pounds of water in a pound of dry air [kg – water per kg – dry air] – this ratio is defined as "ω"), the designer must use additional tools to predict the conditions and design the duct system. This includes the use of the Psychrometric Chart (see Examples in this chapter) and equations that consider the enthalpy of the gas/water mixture. A thorough understanding of these concepts should be accomplished before working with air streams that contain significant levels of water vapor.

Phase changes associated with dry-bulb temperatures close to dew points may cause condensation on duct walls and filter bags. The avoidance of this condition in a system is critical and required for the reliability and maintainability of any industrial ventilation system (IVS).

9.2 PRELIMINARY STEPS TO BEGIN CALCULATIONS

Chapter 5 details preliminary steps for the beginning of the detailed design process. These include:

- A layout of the operations, workroom, building (if necessary).
- The available location(s) for the air cleaning device and fan.
- A line sketch of the duct system layout, including plan and elevation dimensions, fan location, air cleaning device location, etc. Number, letter, or otherwise identify each branch and section of main duct on the line sketch for convenience. (The examples show hoods numbered and other points lettered.)
- A design or sketch of the desired hood for each operation with direction and elevation of outlet for duct connection.
- Information and specific details about required flow rate, minimum required duct velocity, entry losses and required capture velocities of all operations.
- Information about the elevation of the plant above sea level and also the temperature and moisture conditions from each process and duct branch.
- The method and location of the replacement air distribution devices because they affect the hood's performance. The type and location of these fixtures can dramatically lower contaminant control by creating undesirable turbulence at the hood (see Chapter 10).

9.3 DESIGN METHOD AND USE OF LOSS COEFFICIENTS

The basic goal of system design is to size and specify all the duct segments in the industrial ventilation system (IVS) by a series of calculations. The procedure used is known as the

"Velocity Pressure Method." Two primary factors or variables in design are airflow, "Q", (sometimes called "volume" or "airflow rate") and static pressure, "SP". Appropriate airflows are determined either by formulae for specific hood designs or by empirical data and experience with the process itself. Methods to calculate these hood airflows are provided in Chapter 6 and certain process airflows have been compiled in Chapter 13.

With the airflow selected, the air must also overcome the resistance of the duct and other parts of the system as it travels to the collection device. This total resistance is known as System Static Pressure (SSP) and is measured in inches-water gauge ("wg). This pressure could also be measured in other units (pounds per square inch (psi), atmospheres, Pascals (in metric units), etc.), but, in the IP units of measure, inches of water is an appropriate and easy-to-use unit of measurement. As a reference, 1 "wg equals about 0.0361 psi. One atmosphere (1 atm) equals 407.5 "wg [101.384 kPa]. Most IVSs operate in the range of 2 "wg to 30 "wg [500 to 7500 Pa].

The designer will make a sketch or drawing of the intended routing of duct and location of hoods, fan(s) and air control devices (see Chapter 8). All of these are connected in a network. The design procedure in this chapter will allow the designer to size each component and calculate the airflow and resistance in each segment and for the entire system. This will provide the specification of key pieces of equipment such as the fan and air control device.

9.3.1 System Component Loss Coefficients. The Velocity Pressure Method is based on the fact that all frictional and dynamic (turbulence) losses in ducts and hoods are functions of Velocity Pressure (VP) (see Chapter 3 and Figure 9-1). Values for Loss Coefficients (F) are shown in the Appendix to this chapter and included in Table 9-4 and Figures 9-a through 9-f. System losses (resistance) can then be calculated by the loss coefficients multiplied by VP. For example, the loss coefficient (F_{el}) for a 5-piece 90-degree elbow of R/D = 1.5 is shown to be 0.24 in Figure 9-e. When multiplied by the Velocity Pressure in that segment, the resulting value is the loss in "wg. Figure 9-2 shows the location and application of these coefficients in a simple hood and branch fitting.

Coefficients for different shaped hoods are shown on Figure 9-a and on the individual sheets of Chapter 13. Loss coefficients for straight duct, elbows, branch entries, contractions, and expansions are shown in Figures 9-b through 9-f. For convenience, loss coefficients for many components are also presented on the right edge of the ACGIH® calculation sheet (Figure 9-6).

9.3.2 Friction Loss Coefficients for Round Straight Duct. Duct friction coefficients for this method are presented in table, chart and equation form. They give the loss coefficients per foot of metal, plastic or flexible duct. In the past, there had been separate coefficients for various metal materials, but the system has been simplified to use one value for all metal and plastic ducts. In reality, a ventilation system will be coated with dust and other materials after some period of operation so a single value is appropriate.

The equations used to determine values for the chart and table are listed on the chart and also on the calculation sheet (Figure 9-5). For convenience, a visual representation has been developed and used to describe the friction tables. It has also been called the "three-eye chart" (Figures 9-b and 9-c). A Table presentation is also possible (Table 9-1) because, for a specific diameter, the friction loss coefficient changes only slightly with velocity. The Table lists the friction coefficient as a function of diameter for six different velocities. If desired, a linear interpolation between velocity values can be performed.

9.3.3 Friction Loss Coefficients for Non-Circular Straight Duct. Round ducts are strongly recommended for industrial exhaust systems because of a more uniform air velocity to resist settling of material and an ability to withstand higher static pressure. At times, however, the designer must use other duct shapes.

Rectangular duct friction can be calculated by using Table 9-5 in conjunction with Table 9-4 to obtain circular equivalents for rectangular ducts on the basis of equal friction loss. It should be noted that the area of the rectangular duct would be larger than the equivalent round duct; consequently, the actual air velocity in the duct will be reduced. Therefore, it is still necessary to use care to maintain minimum transport velocities. Even if the average velocity requirements are met, the flow characteristics in rectangular ducts could yield dead spots and potential locations for material to settle out in corners.

Occasionally the designer will find it necessary to estimate the air handling ability of odd-shaped ducts. The following procedure will be helpful in determining the frictional pressure losses for such ducts. The wetted perimeter in the following discussion is the inside perimeter of the odd-shaped duct corresponding to the cross-sectional area.

$$\text{Equivalent Diameter (inches) [mm]} = d = 1.3\frac{(a \times b)^{0.625}}{(a+b)^{0.25}}$$

where: a = duct width, in [mm]

b = duct height, in [mm]

See Table 9-5.

9.3.4 Friction Loss Coefficients for Straight Flexible Duct. The loss coefficient for flexible duct with the wires covered is shown to average:

$$F_{flex-duct} = 0.0311\left(\frac{V^{0.604}}{Q^{0.639}}\right)$$

where: V = velocity, ft/min

Q = airflow, acfm

In SI units the value is:

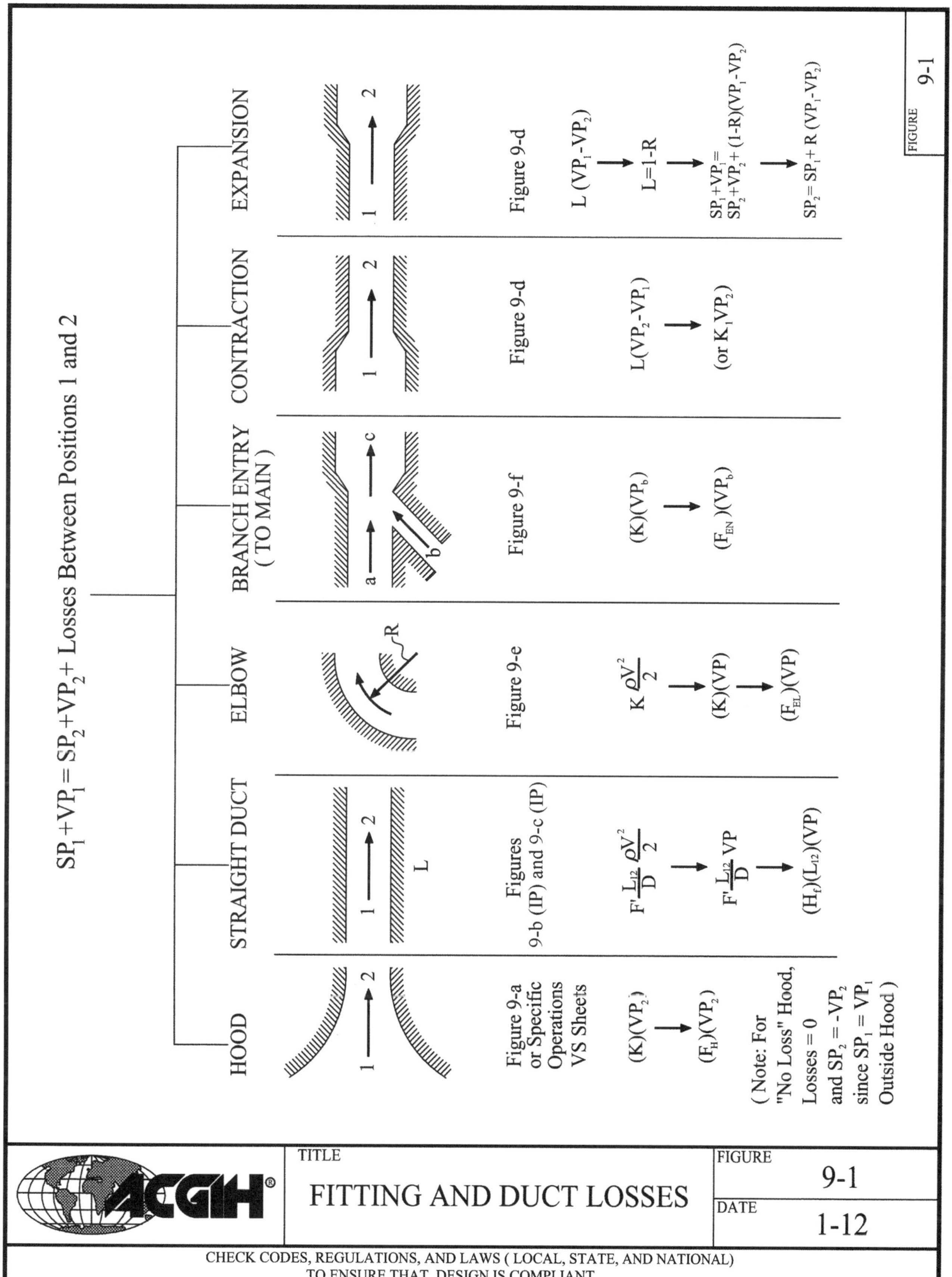
$SP_1 + VP_1 = SP_2 + VP_2 +$ Losses Between Positions 1 and 2
HOOD
STRAIGHT DUCT
ELBOW
BRANCH ENTRY (TO MAIN)
CONTRACTION
EXPANSION
1
2
L
R
a
b
c
Figure 9-a or Specific Operations VS Sheets
$(K)(VP_2)$ → $(F_H)(VP_2)$
(Note: For "No Loss" Hood, Losses = 0 and $SP_2 = -VP_2$ since $SP_1 = VP_1$ Outside Hood)
Figures 9-b (IP) and 9-c (IP)
$F' \frac{L_{12}}{D} \frac{\rho V^2}{2}$ → $F' \frac{L_{12}}{D} VP$ → $(H_f)(L_{12})(VP)$
Figure 9-e
$K \frac{\rho V^2}{2}$ → $(K)(VP)$ → $(F_{EL})(VP)$
Figure 9-f
$(K)(VP_b)$ → $(F_{EN})(VP_b)$
Figure 9-d
$L(VP_2 - VP_1)$ → $(\text{or } K_1 VP_2)$
Figure 9-d
$L(VP_1 - VP_2)$ → $L = 1 - R$ → $SP_1 + VP_1 = SP_2 + VP_2 + (1-R)(VP_1 - VP_2)$ → $SP_2 = SP_1 + R(VP_1 - VP_2)$
FIGURE 9-1
TITLE
FITTING AND DUCT LOSSES
FIGURE 9-1
DATE 1-12
CHECK CODES, REGULATIONS, AND LAWS (LOCAL, STATE, AND NATIONAL) TO ENSURE THAT DESIGN IS COMPLIANT.

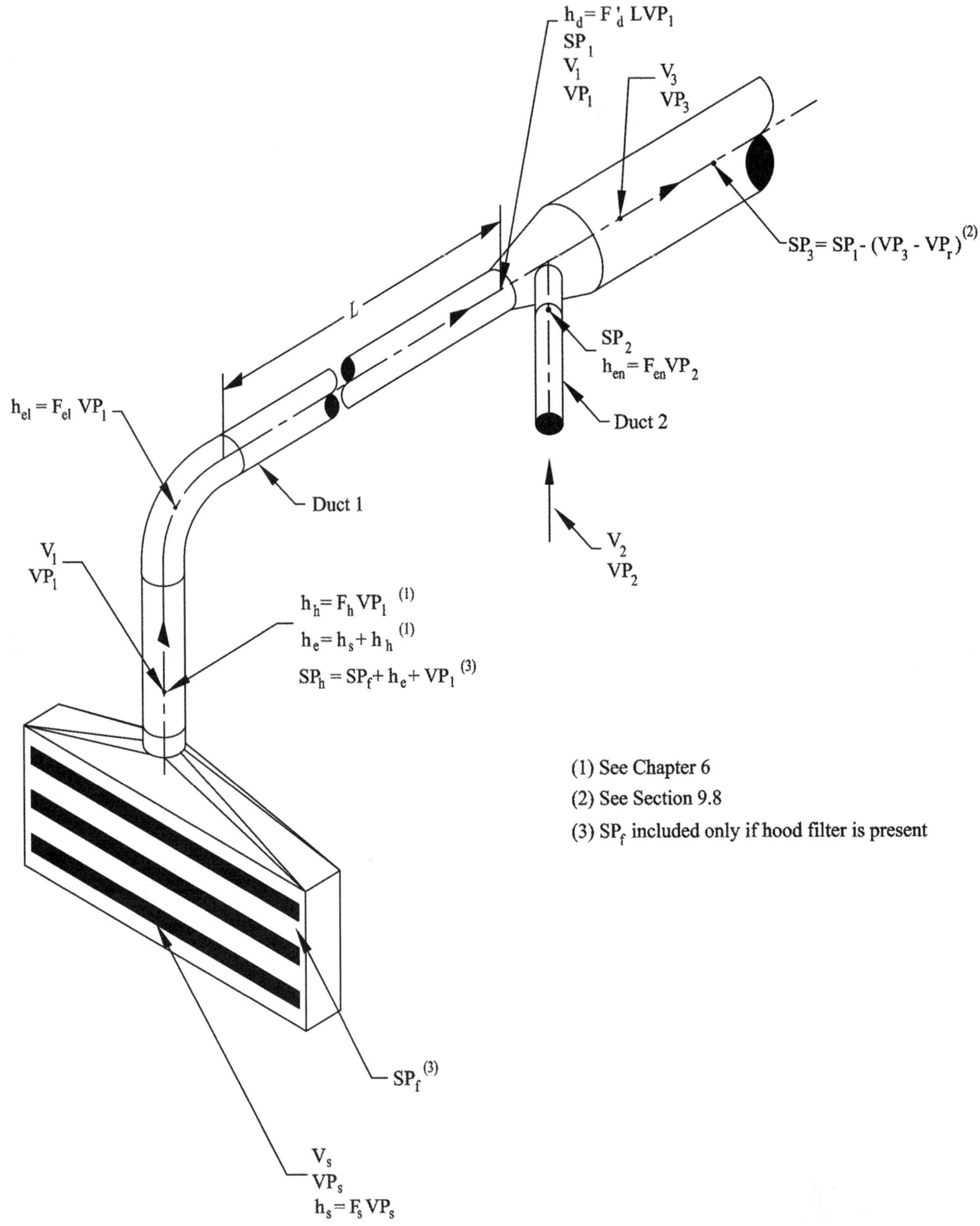

FIGURE 9-2. System duct calculation parameter location

$$\left[F_{flex-duct} = 0.0166\left(\frac{V^{0.533}}{Q^{0.612}}\right)\right]$$

Where: V = Velocity, m/s
Q = airflow, am³/s

This value does not reflect the wide varieties of material and wires and construction methods from manufacturer to manufacturer, so it must be reviewed closely if there is a significant amount of flexible duct in the design. In those cases, consult the manufacturer for actual loss data. Note that this loss coefficient is stated as straight duct length and flexible duct, by its very nature, is seldom straight. Typically, bends in flexible duct can produce extremely large losses that cannot be easily predicted. Be very careful to keep the flexible duct as straight and as short as possible. (Even straight sections of flexible duct have almost twice the losses of similarly sized metal duct and should be avoided except where necessary.)

9.3.5 Friction Loss Coefficients through Contractions and Expansions. Contractions are used when the size of the duct must be reduced to fit into tight places, to fit equipment, or to provide a high discharge velocity at the end of the stack. Expansions are used to fit a particular piece of equipment or to reduce the energy consumed in the system by reducing velocity and friction. Expansions are not usually desirable in particulate systems since the duct velocity may become less than the minimum transport velocity and material may settle in the ducts.

Regain or loss of pressure in a transition system is possible because static pressure and velocity pressure are mutually convertible. This conversion is accompanied by some energy loss or regain. The amount of this loss is a function of the geometry of the transition piece (the more abrupt the change in velocity, the greater the loss), and depends on whether air is accelerated or decelerated. Loss is expressed as a loss coefficient multiplied by the velocity pressure in the smaller area duct of the transition piece. One minus the loss coefficient is the efficiency of the energy conversion or regain.

A perfect (no loss) contraction or expansion would cause no change in the total pressure in the duct. There would be an increase or decrease in static pressure corresponding exactly to the decrease or increase in velocity pressure of the air. In practice, the contraction or expansion will not be perfect, and there will be a change in total pressure (Figure 9-3). In each example, total pressure and static pressure are plotted in order to show their relationship at various points in each system. See Figure 9-d for design data and determination of loss or regain coefficients for expansions and contractions. Note that in applications of these formulae there is a calculation for change in static pressure rather than the classic value of “F”. That is because there is more than one value of VP since there is a change in duct size.

9.3.6 Special Expansion Consideration – Evasé Discharge. An evasé discharge is a gradual enlargement at the outlet of the local exhaust system (Figure 9-d). The purpose of the evasé is to reduce the air discharge velocity efficiently; thus, the available velocity pressure can be regained and credited to the local exhaust system instead of being wasted. Practical considerations usually limit the construction of an evasé to approximately a 10° angle (5° side angle) and a discharge velocity of about 2,000 fpm [10 m/s] (0.25 "wg [63 Pa] Velocity Pressure) for normal local exhaust systems. Further streamlining or lengthening the evasé yields diminishing returns.

It should be noted, however, that for optimum vertical dispersion of contaminated air, many designers feel that the discharge velocity from the stack should not be less than 3,000 [15 m/s] to 3,500 fpm [18 m/s]. When these considerations prevail, the use of an evasé is questionable. In addition, the structural requirements for the support of an evasé may add more initial costs than can be realized in energy savings over the life of the project.

The following example indicates the application of the evasé fitting. It is not necessary to locate the evasé directly after the outlet of the fan. It should be noted that, depending on the evasé location, the static pressure at the fan discharge may be below atmospheric, i.e., negative (-), as shown in this example.

EXAMPLE PROBLEM 1 (Effects of Evasé) (IP Units)

Determine the effects of adding a 40"-long evasé to the discharge of a centrifugal fan with the following conditions:

Point		D	Q	V	VP	SP
1	Fan Inlet	20	8300	3800	0.90	7.27
2	Fan Discharge (16.5" × 19.5")		8300	3715	0.86	
3	Round Duct Connection (fan outlet)	20		3800	0.90	
4	Evasé Outlet	28		1940	0.23	0

To calculate the effect of the evasé, see Figure 9-d for expansion at the end of the duct where the Diameter Ratio, D4/D3 = 28/20 = 1.4 and Taper Length L/D = 40/20 = 2.0.

R = 0.52 × 70% (since the evasé is within 5 diameters of the fan outlet)

VP_3 = 0.9 "wg

SP_4 = 0" (since the end of the duct is at atmospheric pressure)

$SP_3 = SP_4 - R\,(VP_3)$

= 0.0" – (0.52)(0.70)(0.90")

= -0.33 "wg

FIGURE 9-3 (IP)

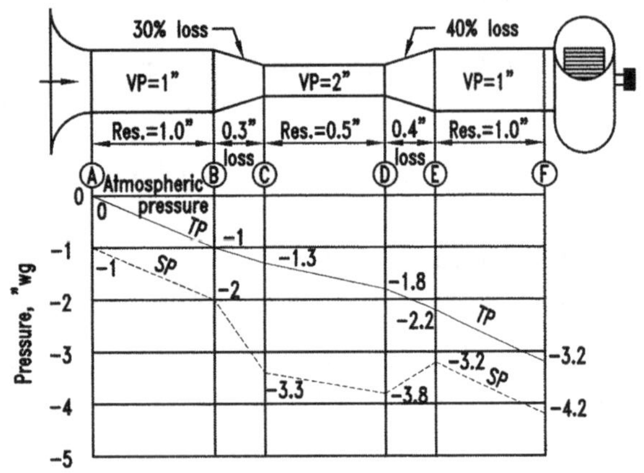

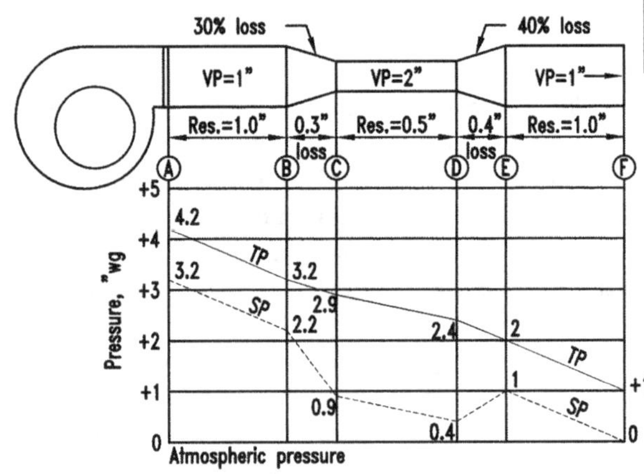

EXAMPLE 1-DUCT LOCATED ON SUCTION SIDE OF FAN
Velocity changes as indicated. Since all the duct is on the suction side of the fan, TP at the fan inlet (point F) is equal to VP at the fan inlet plus the total duct resistance up to that point. This equals -4.2" SP since static pressure on the suction side of the fan is always negative. The duct system is the same as was used in Example 2 and therefore has the same overall resistance of 3.2. If it is again assumed that the inlet and discharge of the fan are equal areas, the total pressure across the fan will be the same as in Example 2 and in each case, the fan will deliver the same air horsepower when handling equal volumes of air.

Static pressure conversion between B and C is obtained from the contraction formula (Chapter 5, Figure 9-d). There must be sufficient SP at B to furnish the additional VP required at C. In addition, the energy transfer between these two points is accompanied by a loss of 0.3. Since SP at B = -2.0", SP at C = -2.0 + (-1.0) + (-0.3) = -3.3".

Static pressure regain between D and E is obtained from the regain formula (Chapter 5, Figure 9-d). If there were no losses in the transition piece, the difference of 1" velocity pressure would be regained as the static pressure at E, and SP at that point would be -2.8". However, the transition is only 60% efficient (0.4 loss) so the SP at E = -2.8 + (-0.4) = -3.2".

EXAMPLE 2-DUCT LOCATED ON DISCHARGE SIDE OF FAN. Velocity changes as indicated. The duct is located on the discharge side of the fan. Total pressure at the fan discharge (point A) is equal to the velocity pressure at the discharge end of the duct (point F) plus the accumulated resistances. These add up to 1.0 + 1.0 + .4 + .5 + .3 + 1.0 = 4.2".

Static pressure regain between D and E is obtained from the regain formulae (Chapter 5, Figure 9-d). If there were no energy loss in the transition piece, static pressure at D would be 0 because the difference in VP of 1" would show up as static pressure regain. However, the transition is only 60% efficient which means a loss of 0.4", so SP at point D = 0 + 0.4 = 0.4".

Conversion of static pressure into velocity pressure between B and C is obtained from the contraction formulae (Chapter 5, Figure 9-d). There must be sufficient static pressure at B to furnish the additional velocity pressure required at C. In addition, transformation of energy between these two points is accompanied by a loss of 0.3". Since SP at C = 0.9", SP at B = 0.9 + 0.3 + 1.0 = 2.2". Since there is no duct on the suction side of the fan, total pressure against which the fan is operating is 4.2".

ACGIH®	TITLE: EXPANSIONS AND CONTRACTIONS	FIGURE: 9-3 (IP)
		DATE: 1-12

CHECK CODES, REGULATIONS, AND LAWS (LOCAL, STATE, AND NATIONAL) TO ENSURE THAT DESIGN IS COMPLIANT.

FIGURE 9-3 (SI)

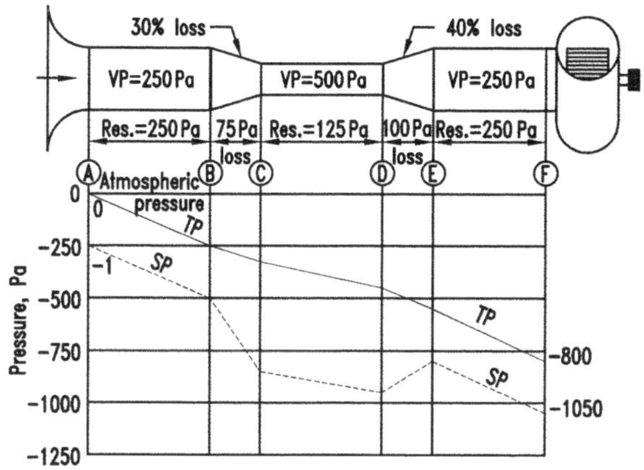

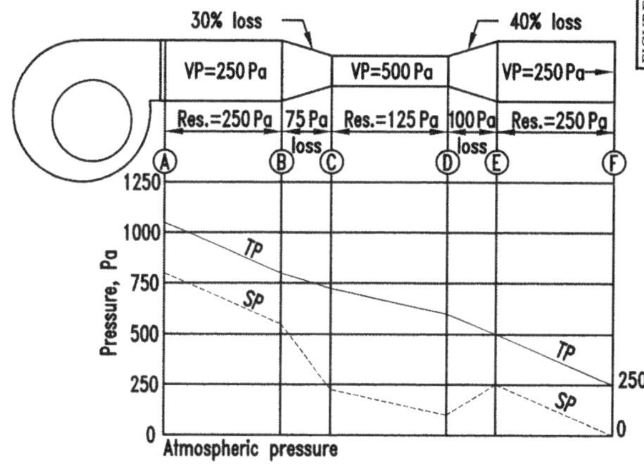

EXAMPLE 1-DUCT LOCATED ON SUCTION SIDE OF FAN Velocity changes as indicated. Since all the duct is on the suction side of the fan, TP at the fan inlet (point F) is equal to VP at the fan inlet plus the total duct resistance up to that point. This equals -1050 Pa since static pressure on the suction side of the fan is always negative. The duct system is the same as was used in Example 2 and therefore has the same overall resistance of 800 Pa. If it is again assumed that the inlet and discharge of the fan are equal areas, the total pressure across the fan will be the same as in Example 2 and in each case, the fan will deliver the same air horsepower when handling equal volumes of air.

Static pressure conversion between B and C is obtained from the contraction formula (Chapter 5, Figure 9-d). There must be sufficient SP at B to furnish the additional VP required at C. In addition, the energy transfer between these two points is accompanied by a loss of 75 Pa. Since SP at B = -500 Pa, SP at C = -500 + (-250) + (-75) = -825 Pa.

Static pressure regain between D and E is obtained from the regain formula (Chapter 5, Figure 9-d). If there were no losses in the transition piece, the difference of 250 Pa velocity pressure would be regained as the static pressure at E, and SP at that point would be -700 Pa. However, the transition is only 60% efficient (100 Pa loss) so the SP at E = -700 + (- 100) = -800 Pa.

EXAMPLE 2-DUCT LOCATED ON DISCHARGE SIDE OF FAN. Velocity changes as indicated. The duct is located on the discharge side of the fan. Total pressure at the fan discharge (point A) is equal to the velocity pressure at the discharge end of the duct (point F) plus the accumulated resistances. These add up to 250 + 250 + 100 + 125 + 75 + 250 = 1050 Pa.

Static pressure regain between D and E is obtained from the regain formulae (Chapter 5, Figure 9-d). If there were no energy loss in the transition piece, static pressure at D would be 0 because the difference in VP of 250 Pa would show up as static pressure regain. However, the transition is only 60% efficient which means a loss of 100 Pa, so SP at point D = 0 + 100 = 100 Pa.

Conversion of static pressure into velocity pressure between B and C is obtained from the contraction formulae (Chapter 5, Figure 9-d). There must be sufficient static pressure at B to furnish the additional velocity pressure required at C. In addition, transformation of energy between these two points is accompanied by a loss of 75 Pa. Since SP at C = 225 Pa, SP at B = 225 + 75 + 250 = 550 Pa. Since there is no duct on the suction side of the fan, total pressure against which the fan is operating is 1050 Pa.

ACGIH®	TITLE EXPANSIONS AND CONTRACTIONS	FIGURE 9-3 (SI) / DATE 1-12

CHECK CODES, REGULATIONS, AND LAWS (LOCAL, STATE, AND NATIONAL) TO ENSURE THAT DESIGN IS COMPLIANT.

$$FSP = (SP_{outlet} - SP_{inlet}) - VP_{inlet}$$
$$= -0.33" - (-7.27") - 0.9" = 6.04 \text{ "wg}$$

If only a 'no-loss' stack was added to the fan (see Chapter 5, Figures 5-13 and 5-14) and the effects of the evasé were not considered, then the Fan Static Pressure would have been (see Chapter 7 for discussion of Fan Static Pressure):

$$FSP = (SP_{outlet} - SP_{inlet}) - VP_{inlet}$$
$$= -0.0 - (-7.27) - 0.9 = 6.37 \text{ "wg}$$

or 5% higher than the fan with the evasé (and 5% higher operating horsepower over the life of the installation).

9.4 BASIC CALCULATIONS AND PROCEDURES REQUIRED FOR SYSTEM DESIGN

Before the calculation sheet is used there are basic determinations that must be made. Primary are the actual air conditions in the hood and duct systems. These can include effects of temperature, moisture, elevation and absolute pressure in the duct and the resulting density factor derived from these conditions. Density Factor (df) is a dimensionless term calculated by the following equation:

$$df = \frac{\rho_{act}}{\rho_{std}} \qquad [9.1]$$

where $\rho_{std} = 0.075 \frac{lbm}{ft^3}$ [1.204 kg/m³] and ρ_{act} is the density of the gas at its actual conditions.

Calculations and use of df are shown in Chapter 3, Section 3.11. The use and determination of df for the air and gas streams (in this Manual, air stream and gas stream are used interchangeably) is crucial for accurate calculations in this section.

Because there are many times when air is measured or specified in either actual or standard conditions, the Manual must set a basis for how calculations are done in the procedure to follow:

1) Air will be specified in acfm for a base value and entered in calculation sheet.
2) Density Factor (df) will be calculated for all appropriate conditions for that gas stream and inserted as one value on the calculation sheet.
3) Absolute pressure will not be considered except at the fan when calculating Density Factor for absolute pressure (df_p).
4) dscfm will be calculated after the base acfm is known and all conditions (i.e., Density Factor) are known for the air stream.
5) acfm will be used for determination of duct size (using appropriate transport velocities as discussed in Chapter 5).
6) acfm will be used at the Air Cleaning Device and Fan for the specification of equipment size.

It should be noted that sometimes this would seem like Step 1 requires a double calculation when air is originally specified in acfm. However, the return to basic standard conditions for all gas streams will allow for easier manipulation of mixtures and consistent methods.

9.4.1 Hood Airflow at Non-Standard Conditions. The control of dust, fumes and vapors requires an airflow that will achieve the velocity necessary to capture and carry the contaminant into the hood (or contain the contaminant inside an enclosure or enclosing hood) and then convey it through the hood and duct system. In particular, at high elevations, the air providing this containment is already at lower density. The methods defined in the Calculation Sheet in this chapter use the air "Actual" conditions and that airflow is inserted in the calculation sheet (Row 3). Dry standard cubic feet per minute (dscfm) is used primarily when mixing air streams of different densities. In most cases, it will not be necessary to calculate dscfm.

The values shown in Chapter 13 are in acfm under local conditions. For example, the volume shown in VS-15-02 is listed as 400 to 500 acfm [0.20 to 0.25 am³/s] for non-toxic dust.

When selecting the capture velocities based on the guidelines in Chapter 6 (Table 6-2), the designer should consider the upper end of the range when working with large dust particles at high temperatures or elevation (> 5000 feet [> 1500 m] above sea level).

Hood airflows can be determined by the formulae in Chapter 6 or the VS plates shown in Chapter 13 of this Manual. Many times these hoods are located at plants at higher elevations. In those cases, the airflow shown in these recommendations are at actual conditions in the plant (acfm) [am³/s] (not standard air – scfm [nm³/s]).

Volume (acfm) [am³/s] is required for calculating the size of duct, determining air/cloth ratio for fabric filters and providing the correct size of fans. Mass flow (pounds/per minute or dscfm) is required to determine air conditions (amount of moisture, enthalpy, etc.) from a single or mix of many air streams. So there are cases where either or both values may be required on the calculation sheet. Knowing the Density Factor (as a function of elevation, moisture, temperature, absolute pressure) and the moisture content will allow a calculation of actual conditions from standard conditions or vice versa (see Chapter 3, Sections 3.11 and 3.12, Equations 9.2 and 9.3).

EXAMPLE PROBLEM 2 (ACFM into Hood)

A hood designed as shown in VS-55-01 is located over a melting furnace. The hood has a required capture velocity at all openings of 200 fpm [1.0 m/s] per the VS plate and the opening sizes total 52 ft² [4.83 m²]. The hood is located in a plant that is 4,000 feet [1200 m] Above Sea Level (ASL) and the plant air temperature going into the hood is assumed to be

70 F [21 C] with no moisture. Calculate the required hood control airflow from the VS plate requirements.

Airflow into the hood = Q = AV = (52 ft^2)(200 ft/min)
= 10,400 acfm

[Q = AV = (4.83 m^2) (1.0 m/s) = 4.83 am^3/s]

The procedure requires the change of the acfm back to scfm for the beginning of the system calculation procedure. This allows for a base value to be manipulated by all density conditions before designing the duct and other equipment. After the airflow is selected from the hood requirements (using VS plates, Chapter 6 or process requirements), the value in acfm must be returned to its standard conditions for entrance into the calculation sheet. In effect, scfm is used primarily to calculate conditions after mixing two air streams of different density factors.

EXAMPLE PROBLEM 3 (DSCFM Calculation)

For the system in Example 2, determine the airflow into the hood in standard conditions (dscfm). Density Factor (df) for the air in the plant at 70 F [21 C] and 4,000 ft [1200 m] ASL is 0.86 (Table 9-6 (IP & SI) or Chapter 3, Equation 3.23 (IP & SI).

Flow in acfm is calculated using:

$$Q_{act} = \frac{Q_{std}(1+\omega)}{df} \qquad [9.2]$$

where:

Q_{std} = flow, dscfm [nm^3/s]

ω = moisture content, $\frac{lbm-water}{lbm-dry\ air}\left[\frac{kg-water}{kg-dry\ air}\right]$

df = Density Factor

Equation 9.2 can be used to solve for standard air:

$$Q_{std} = \frac{Q_{act}(df)}{(1+\omega)} \qquad [9.3]$$

In this example, there is no moisture and the only effect on conditions is the elevation since temperature is 70 F [21 C]. So Equation 9.3 can be solved:

$$Q_{std} = \frac{10{,}400\ acfm\ (0.86)}{1+(0.0)} = 8{,}944\ dscfm$$

In metric units:

$$\left[Q_{std} = \frac{(4.83\ am^3/s)\ (0.86)}{1+0} = 4.15\ nm^3/s\right]$$

In the field, there are cases where the mass flow rate is required, particularly in processes involving moisture. The transfer of volumetric flow rate (acfm) [am^3/s] to mass flow rate (pounds per minute) [kilograms/s] adds the concept of density. Note that the gas stream is first converted to dscfm [nm^3/s] to determine pounds [kilograms] of dry air (no moisture):

EXAMPLE PROBLEM 4 (Calculate Pounds Per Minute)

Determine the mass flow rate of the air stream in Example Problem 3 (pounds of dry air per minute).

$$\dot{m}_a = \rho_{std}(Q_{std}) \qquad [9.4]$$

$$\dot{m}_a = \left(0.075\frac{lbm}{ft^3}\right)\left(8{,}944\frac{ft^3}{min}\right) = 670.8\frac{lbm}{min}$$

[$\dot{m}_a$ = (1.204 kg/m^3) (4.15 nm^3/s) = 500 kg/s]

In the three examples it was determined that the correct flow into the furnace hood is 10,400 acfm [4.83 am^3/s] (from VS plate) which calculates to 8,944 scfm [4.15 nm^3/s] and 670.8 pounds per minute [5.00 kg/s] of air.

9.4.2 Addition of Materials inside the Hood. In some cases, an enclosed process may add gases or moisture to the calculated control airflow going into the face of the hood. These materials must be accounted for in the calculation of the connected duct system in order to properly size the duct and air handling equipment.

EXAMPLE PROBLEM 5 (Density Change Inside Hood)

The enclosure in Example Problem 2 contains a furnace with an induction heater that is generating 3,000 acfm [1.42 am^3/s] of gases at 1900 F [1038 C] with no moisture. The standard density of this process gas is the same as air (0.075 lbm/ft^3) [1.204 kg/m^3]. Determine the total pounds of material (air plus gases) exiting at the hood's duct connection.

It was determined in Example Problem 4 that the air coming into the hood from the plant totals 670.8 lbm/min [5.00 kg/s]. The gases being generated inside the hood must be added to this value. The density factor for the gas at 1900 F [1038 C] (see Chapter 3, Equation 3.22) is:

$$df_t = \frac{\rho_{act}}{\rho_{std}} = \left(\frac{T_{std}}{T_{act}}\right) = \left(\frac{70+460}{1900+460}\right) = 0.22$$

$$\left[df_t = \frac{\rho_{act}}{\rho_{std}} = \left(\frac{T_{std}}{T_{act}}\right) = \left(\frac{21+273}{1038+273}\right) = 0.22\right]$$

This value would be determined by the process requirements and, therefore, would be independent of the elevation of the plant where the furnace is located. So the df_t would be the only consideration. Solving Equation 9.3 for the standard conditions:

$$Q_{std} = \frac{Q_{act}(df)}{(1+\omega)} = \frac{3{,}000\ acfm\ (0.22)}{(1+0.0)} = 660\ dscfm$$

$$[Q_{normal} = \frac{Q_{act}(df)}{(1+\omega)} = \frac{(1.42\ am^3/s)\ (0.22)}{(1+0)} = 0.31\ nm^3/s]$$

From Equation 9.4:

$$\dot{m}_f = \left(0.075\frac{lbm}{ft^3}\right)\left(660\frac{ft^3}{min}\right) = 49.5\frac{lbm}{min}$$

$[\dot{m}_f = (1.204\ kg/m^3)\ (0.31\ nm^3/s) = 0.37\ kg/s]$

Determining total mass flow rate of the extra gases generated by the process itself and those entering through the hood face is now a simple addition of masses:

$$\dot{m}_a + \dot{m}_f = \dot{m}_{total} = 670.8 + 49.5 = 720.3\frac{lbm}{min}$$

$[\dot{m}_a + \dot{m}_f = \dot{m}_{total} = 5.00\ kg/s + 0.37\ kg/s = 5.37\ kg/s]$

9.4.3 Mixing Gases of Different Densities Due to Temperature. There will be conditions when input information will require calculations even before system design is determined. These can include the mixing of hot and cold gases and mixtures of dry and moist air streams. The results of the calculations would be used as input for the system calculations.

Example Problems 2 through 5 in Sections 9.4.1 and 9.4.2 show the effects of density and how to combine the mass flows of two streams of gases. The principle for this combination is the Law of the Conservation of Mass:

$$\dot{m}_a + \dot{m}_b = \dot{m}_c \qquad [9.5]$$

In addition to Conservation of Mass, there is a Conservation of Energy in this system (assuming no heat loss through the walls of the hood). In equation form this is:

$$\dot{m}_a(h_a) + \dot{m}_b(h_b) = \dot{m}_c(h_c) \qquad [9.6]$$

where $\dot{m}$ equals the mass flow rate and h is the total energy (enthalpy) of the gases. Definitions and descriptions of these items are included in Chapter 3. Subscript "a" is used for the conditions of one component gas and subscript "b" is used for another component. Subscript "c" indicates the conditions of the mixture of the two streams. Note that these can be two air streams meeting at a junction or two air streams mixed inside a hood or vessel.

For an Ideal Gas (see Chapter 3, Sections 3.10 and 3.11), Equation 9.6 is rewritten as 9.6a (note this is a gas that contains no moisture; if moisture is present then Equation 9.6 must be used):

$$\dot{m}_a(C_p)(T_a) + \dot{m}_b(C_p)(T_b) = \dot{m}_c(C_p)(T_c) \qquad [9.6a]$$

Since C_p will cancel out of the equation, this yields:

$$\dot{m}_a(T_a) + \dot{m}_b(T_b) = \dot{m}_c(T_c) \qquad [9.7]$$

NOTE: All temperatures should be in degrees Rankin (R) or degrees Kelvin (K) in SI system.

EXAMPLE PROBLEM 6 (Mixing of Airstreams) (IP Units)

Determine the exit temperature, density factor and airflow of the mixture of hot and cold gases coming from the enclosure defined in Example Problems 1 and 4.

Mass of 70 F [21 C] air coming through the face of the hood was determined to be 670.8 lbm/min [5.00 kg/s]. The mass flow of the furnace exhaust gases was determined to be 49.5 lbm/min [0.37 kg/s] at 1900 F [1038 C].

Solving from Equation 9.7 and changing F (Fahrenheit) to R (Rankin) [C (Celsius) to K (Kelvin)]:

$$\dot{m}_a(T_a) + \dot{m}_f(T_f) = \dot{m}_{mix}(T_{mix}) =$$
$$(670.8)(530) + (49.5)(2360) = (720.3)(T_{mix})$$

$[(5.0)\ (294) + (0.37)\ (1311) = (5.37)\ (T_{mix})]$

$T_{mix} = 656\ R = 196\ F$

$[T_{mix} = 364\ K = 91\ C]$

The conditions of the mixture leaving the hood:

720.3 pounds of gas per minute @ 196 F

[5.37 kg/s @ 91 C]

Solving for standard airflow (Q_{std}) in Equation 9.4:

$$Q_{std} = \frac{\dot{m}}{\rho_{std}} = \frac{720.3}{0.075} = 9{,}604\ dscfm$$

$[Q_{normal} = (5.37)\ /\ (1.204) = 4.46\ nm^3/s]$

NOTE: This is the sum of dscfm from both air streams.

Information for the calculation of df is shown in Chapter 3, Section 3.11. There are two items affecting density of the gases exiting this furnace in these examples. The gas is at an elevated temperature (196 F) [91 C] and the hood is located at 4000' ASL [1200 m]. The density factor for elevation was determined previously (0.86) in Example Problem 3. The Density Factor for temperature (196 F) [91 C] is calculated using Chapter 3, Equation 3.22:

$$df_T = \frac{T_{std}}{T_{act}} = \frac{460+70}{460+196} = 0.81$$

$[df_T = (273 + 21)\ /\ (273 + 91) = 0.81]$

The density factor of the mixture considering both temperature and elevation is:

$$df = (df_T)(df_e) = (0.86)(0.81) = 0.70$$

Actual airflow condition is determined by Equation 9.2:

$$Q_{act} = \frac{Q_{std}(1+\omega)}{df} = \frac{(9{,}604\ scfm)(1+0.0)}{0.70}$$

= 13,720 acfm @ 196 F @ 4,000' ASL

[= (4.46 nm^3/s) (1 + 0) / (0.7) = 6.37 am^3/s @ 91 C @ 1200 m ASL]

NOTE: Whenever airflow is specified in actual conditions it is important to list the conditions immediately following (196 F and 4,000' ASL [91C and 1200 m ASL]). This is not required when listing standard [or normal] conditions although it is good practice to provide a notation of the conditions whenever defining flow.

9.5 CALCULATION SHEET DESIGN PROCEDURE

A simple local exhaust system is comprised of a hood, duct segment and special fittings leading to and from an exhaust fan. A complex system is merely an arrangement of several simple local exhaust systems connected to a common duct.

The calculation procedure is a continuing/iterative process and does not end with the first system problem solution. It might be repeated several times including the original conceptual design and final drive speed specification from "as-built" drawings, as well as a tool for the air balance technician. In addition, the designer must not consider this only a simple tool to size ducts and fan. It should be used to identify ducts with very high velocities that could wear prematurely, and to analyze the branches with the highest pressure drop so system pressure could be reduced. For example, a small branch duct in a large system may represent the highest static pressure loss (determining leg or branch of the system). By increasing the flow at the hood, making the duct larger and reducing the friction losses in the duct, the overall system pressure may go down with very little increase in flow. The result would be a reduction in the required system horsepower.

Similarly, the system design usually considers only the conditions at initial start-up and installation. However, after the system is in use it will lose some effectiveness as dust covers the duct walls (changing friction losses) and fan impellers and dust collectors begin to wear. The designer must consider the conditions during the operating life of the system. For instance where airflows, face velocities or transport velocities are selected from a range of values, the upper end of the range should be considered if the system cannot be shut down for normal maintenance.

The system itself is dynamic and continuously changing. The system calculations define a single point of operation but the actions of the components yield a varying value of volume and pressure. Readings taken at start-up and commissioning may not be repeated again as the system ages. The readings themselves are open to interpretation (see Appendix C, Testing and Measurement of Ventilation Systems). The calculation method should be considered a tool to determine duct sizes and fan requirements rather than a prediction of exact operating conditions in all branches throughout the life of the system.

9.5.1 Using the Calculation Sheet. The procedure uses a basic calculation method to determine duct sizes and fan conditions. The 'VP' or Velocity Pressure Method inputs coefficients (F) for losses as expressed in numbers of VPs. (See Chapters 3 and 6 for the derivation and use of these coefficients and VP.) The values for these coefficients are acquired from laboratory and mathematical methods as well as experiences seen under field conditions. The coefficients are totaled in the calculation sheet and then the sum is multiplied by the Velocity Pressure in that segment to obtain the actual losses (in "wg [Pa]).

The calculation sheet is built as a series of columns (normally one column for each duct segment) and rows (data for a particular column). The cell location for inserted and calculated values is made using a matrix notation. The first value in the matrix would be the Duct Segment Identification (column), and the second value would be the row. For example, in Problem 1 (Figure 9-6), the value at cell A-B/25 would be -2.06 "wg. It is found in Column A-B and Row 25. Similarly, the value in cell E-F/10 would be 5" diameter.

For consistency, the following rules should be used when entering data in the calculation sheet:

- Area of duct (square feet): 3 decimal places, i.e., **0.196 ft^2**
- Static Pressure ("wg): 1 decimal place, i.e., **2.4 "wg**
- Velocity Pressure ("wg): 2 decimal places, i.e., **2.37 "wg**
- Velocity (feet per minute): whole values with no decimals, i.e., **3562 fpm**
- Temperature (F): no decimal places, i.e., **77 F**
- Volumetric Flow Rate (acfm): whole values with no decimals, i.e., **21,456 acfm**
- Factors (no dimensions): 2 decimal places, i.e., **1.78**
- System and Fan Static Pressure ("wg): 1 decimal place, i.e., **21.6 "wg**
- Length: no decimal places, i.e., **5 feet**

In metric (SI) units we use the following values:

- Area (m^2): 3 decimal places, i.e., **0.221 m^2**
- All Pressure (Pa): no decimal places, i.e., **247 Pa**
- Velocity (m/s): 2 decimal places, i.e., **24.05 m/s**
- Volumetric Flow Rate (m^3/s): 2 decimal places, i.e., **14.84 m^3/s**
- Temperature (C): no decimal places, i.e., **21 C**
- Factors (no dimensions): 2 decimal places, i.e., **1.78**
- Length (meters): 1 decimal place, i.e., **4.2 meters**

When using the Calculation Sheet, the normal procedure is to work from the top to the bottom of each column. The designer inputs known data from sketches, VS-Plates and other resources into the appropriate row at the top of the column. Note that certain Rows (1, 2, 3, 4, 5, 6, 14, 28, etc.) contain asterisks next to the Row number. This asterisk indicates data entry points needed for the design in certain cases. Other row values are normally calculated from these input points. The calculation sheet also includes shaded rows (5, 6, 7, 8, 14). These are required when non-standard air is encountered (when Density Factor (df) does not equal 1.0). If df equals 1.0 then these rows can be bypassed and acfm = dscfm [am^3/s = nm^3/s].

These data entry points can all be inserted before doing calculations for the column or can be placed as the calculations proceed down the column. In either case, a series of calculations are performed working down from the top of the column to obtain a static pressure (resistance) for that segment (Row 41).

Once a segment is complete, the designer then moves to the branch meeting that junction. Static pressures are compared at that junction (Row 41) versus the value from the joining branch and adjustments are made as required (balancing of the two branches so that there is only one static pressure). The pressure is noted and then airflows are added to proceed to the next segment. This procedure continues until the fan segment is reached where inlet Static Pressure to the fan is calculated. The same procedure starts beginning with the outlet of the fan. After inlet and outlet Static Pressures are determined for the system, a System Static Pressure (SSP) can be calculated and a Fan Static Pressure (FSP) specified.

9.5.2 Calculation Procedure and Input to the Calculation Sheet. Note the elevation for the plant location and input the value at the top of the calculation sheet ("z"). This will be used to calculate the Density Factor (df_e) for elevation in the plant (see Chapter 3). Input all other pertinent data for the system in the appropriate places in the title block.

Start with the duct segment that has the greatest number of duct segments between it and the fan. A duct segment is defined as the constant diameter round (or constant area rectangular) duct that separates points of interest such as hoods, entry points, fan inlet, etc. The calculation sheet includes asterisks next to certain lines. These describe the requirements for input information. For example, the asterisk at Row 8 indicates that the designer must input the airflow (acfm) [am^3/s] for that branch.

1) Select a duct segment identification number; this is usually specified by a single letter (A, B, etc.) for the hood and a number for the junction at the end of the segment). Input the number in Row 1.

2) If the column involves a hood design or other source of air (bleed-in, etc.), select an airflow based on the toxicity, physical and chemical characteristics of the material and the ergonomics of the process. Determine its design flow rate, minimum transport velocity and entry losses (see information in Chapters 6 and 13 for selection of hood airflows). Airflow is input in Row 3. Note that minimum velocity is only important for systems transporting particulate, condensing vapors or mist and to prevent explosive concentrations from building up in the duct (see Chapter 5, Section 5.3.5 for a discussion on economic velocities for non-particulate systems). Input the values for Minimum Transport Velocity into Row 4 (see Chapter 5, Table 5-1). Hood loss factor can be immediately input into Lines 16 (if Compound Hood) and/or Line 22. Account for the volume of contaminants generated inside the hood enclosure – defined in scfm @ 0.75 lbs/ft^3 [1.204 kg/m^3] plus any moisture added. Note that this may differ from the actual contaminants being generated and the designer will be required to re-state these contaminants in terms of scfm of air. The calculation sheet uses acfm as a start point (Row 3) because the face velocities and airflow going into the hood are at local conditions (acfm) [am^3/s]. This allows for one density factor in all of the eventual calculations for that branch (see Section 9.4.1).

3) Calculate the branch Density Factor (Row 7) considering the effects of elevation, temperature, and moisture for the air stream coming from the hood (see Chapter 3). Use these actual conditions (acfm) [am^3/s] for sizing duct. *NOTE: The Density Factor is affected by the absolute pressure inside the duct. However, for most calculations, the absolute pressure will only be considered at the fan inlet where the effects are usually the greatest and the information is needed to specify the fan. If more detailed system calculation is considered or if there are very low pressures throughout the system (< -20 "wg) [-5 kPa], then the designer may opt to consider these effects in all ducts.*

4) Calculate scfm [nm^3/s] if required for mixing of airstreams with two different temperatures (Row 8).

5) Determine the duct area by dividing the design flow rate (in acfm) [am^3/s] by the minimum duct velocity. Convert the resultant cross-sectional area into a tentative duct diameter. A commercially available duct size (Table 9-1) should be selected. If solid or liquid particulate or condensable vapor is being transported through the system, a minimum transport velocity is required (see Chapter 13 and Chapter 5, Table 5-1) and listed in Row 4. If the tentative duct diameter is not a standard size, select the next smaller size (from Table 9-1 [IP & SI]) to ensure that the actual duct velocity is equal to or greater than the minimum required. At this point, calculate velocity pressure for the selected velocity (remember to include effects of density).

6) Using the line sketch, determine the length for the duct segment and the number and type of fittings needed (elbows, entries and other special fittings). Design length is the centerline distance along the duct (the distance between the intersection of the centerlines of the straight duct components). Input to Rows 28, 30, and 33.

7) Calculate the pressure losses for the duct segments that merge at a common junction point (see Figures 9b-9f for information about these components).

8) Calculate the condition of the air at each branch by con-

sidering moisture, heat and mass flow in the mixture from the two branches and balancing mass (dscfm) [nm^3/s] or #/min [kg/s], moisture and heat. Review these conditions to ensure that the air is safely above the dew point if moisture is present from the process. Use the mixed air conditions for designing the next segment.

9) Directly at each junction point, there will be one and only one value for static pressure (SP), *regardless of the path taken to reach that point.* If not ensured by the design process, the system will "self-balance" by reducing the flow rate in the higher-resistance duct segment(s) and increasing the flow rate in the lower-resistance duct segment(s) until there is a single SP in the duct downstream of each junction point.

   SP balance at any junction point can be achieved by either one of two fundamental design methods (see Chapter 5, Section 5.4):

   a) Adjust the flow rate through the branch(es) until the static pressures at the junction point are the same ("Balance by Design"), or

   b) Increase the static pressure in the lower resistance duct segment(s) by means of some artificial device such as a blast gate, orifice plate or other obstruction in the segment or a reduction in duct size. Investigate whether system static pressure can be reduced by increasing flow at one or both hoods and increasing duct sizes. Consider the effect on total system horsepower and capital costs.

Select both the air cleaning device and fan based on final calculated system flow rate in acfm [am^3/s] (considering temperature, elevation, static pressure, moisture condition, contaminant loading, physical and chemical characteristics, and overall system resistance).

Check the duct sizes designed against the available space and resolve any interference problems (i.e., will the elbow or duct size desired actually fit into the available space). This may cause a redesign of a part of the system. Consider fan inlet and outlet conditions and the System Effects that will derate the fan (see Chapter 7, Section 7.5.2).

EXAMPLE PROBLEM 7 (Input to Calculation Sheet) (IP Units)

Input the data for the hood in Example Problems 2 through 5 into an ACGIH® calculation sheet (Figure 9-4). Note the method of entering data from top to bottom. First, the elevation of the system (4000' ASL) is added to the top of the sheet. This is the reference for the calculation of Density Factor due to elevation (df_e). Then a Duct Segment Identification (DSI) number is assigned by the designer. This usually includes a start and end number separated by a hyphen. In this case, A-1 indicates a hood (first segment designated by a letter) and the "1" is the end point of the duct connected to the next duct segment. This is placed at the top of the column. Sometimes other numbers, letters or combinations may be used for hoods, including machine names or numbers. The remaining data are entered vertically down the column into individual cells. These cells are identified by a matrix designation (see Section 9.5.1). The dry-bulb temperature of the mixture from the hood was calculated in Example Problem 6 and its value (196 F) inserted into cell A-1/2 (Column A-1 and Row 2). Similarly, the values for dscfm (9,604 – from Example Problem 6), pounds (720.3 from Example Problem 5), df (0.70 from Example Problem 6) and Duct Flow Rate (13,720 acfm from Example Problem 6) were added to their respective cells. Note that simple systems (no heat or moisture and elevation below 1000' ASL) may have values simply transferred from the VS plates (Chapter 13) or other calculated values and that more "complicated" values of acfm

1*		Duct Segment identification			A-1	
2*	T	Dry-Bulb Temperature		F	196	
3*	Q	Flow Rate (dry air)	Equation 3	acfm	13720	
4*	V_t	Minimum Transport Velocity	Chapter 13	fpm		
5*	$\dot{m}_{H_2O}$	Lbs Water per minute		$\#H_2O/min$		
6*	$\dot{m}_{da}$	Lbs Dry Air per Minute		#da/min	720.3	
7	df	Density Factor	Eqn 2		0.70	
8	Q_{duct}	Duct Flow Rate	Eqn 3	scfm	9600	
9	A_t	Targe Duct Area	(3/4)	ft^2		
10	d	Selected Diameter	Section 9.5.2 (#5)	inches		
11	A	Selected Duct Area	Table 9-1	ft^2		
12	V_t	Duct Velocity	(3/11)	fpm		
13	VP_d	Duct Velocity Pressure	Eqn 4	"wg		
14*	h	Total heat	Eqn 5	BTU/#da		

FIGURE 9-4. Data entry to calculation sheet (Example Problem 7)

(heat and moisture and elevation, etc.) may require calculation of acfm on a separate sheet. As mentioned previously, these simple systems do not require that information be placed into the shaded areas of the sheet. The equations referenced on the calculation sheet are shown on the right edge of the calculation sheet form (Figure 9-6).

9.6 SAMPLE SYSTEM DESIGN #1 (SINGLE BRANCH SYSTEM/"STANDARD" AIR CONDITIONS)

The following steps will establish the overall pressure loss of a duct segment that starts at a hood. Figures 9-5 (IP & SI) show a simple ventilation system with a single hood. **NOTE: The examples in metric (SI) and (IP) units are unique and do not correspond in size or description. However, methods are identical.** The use of the calculation sheet can be very beneficial when performing the calculations manually. Figures 9-6 (IP & SI) are calculation sheets that show the details of the calculations for each component of the system. In Figures 9-5 (IP & SI) there are also graphical representations through the system showing the magnitude and relationships of Total, Static and Velocity Pressures on both the "suction" and the "pressure" sides of the fan. It should be noted that Velocity Pressure (VP) is always positive. Total and Static Pressure may be either negative or positive with respect to atmospheric pressure. Total Pressure (TP) is always greater than Static Pressure (SP) (i.e., TP = SP + VP). Also note that VP can be affected by the air conditions (moisture, heat, temperature, elevation), but in this example "Standard Air" with df = 1.0 and $\omega = 0.0$ is considered.

The following steps refer to the ACGIH® Calculation Sheets shown in Figures 9-6 (IP & SI). Data are entered in rows with an asterisk included. The other rows require calculations to complete data. Not all rows need to be used based on the requirements of the system (i.e., if there are no elbows in the system then no data are required in Rows 30, 31 and 35).

Step 1. In the column for the first duct segment (from the hood at "A" to the inlet of the filter at "B"), name the duct segment (A-B) and place in Row 1. Since all air conditions in this problem are 'Standard', input 70 F [21 C] for the air temperature in cell A-B/2.

Step 2. Input the required flow (in acfm) [am^3/s] into cell A-B/3. This value comes from the information in the VS Plate (VS-80-11) for a grinding wheel hood in Chapter 13. From the same VS plate, input the minimum transport velocity (4,000 fpm) [20 m/s] into cell A-B/4.

Step 3. Determine duct size in a two-stage process. When 390 acfm [0.25 am^3/s] is carried exactly at 4000 fpm [20 m/s], the duct area required is 0.098 square feet [0.013 m^2] (A = Q/V) and shown in Cell A-B/9. Solving for duct diameter at that area yields a value of 4.228" [128 mm] across. Since this size duct is impractical for fabricators, a more standard size is considered. If the selection were for a larger duct, the velocity would not meet our minimum requirements for 4,000 fpm [20 m/s]. Thus, the next smaller commercially available duct is chosen – in this case 4.0" [120 mm] diameter – and that is entered in Cell A-B/10. This yields an area of 0.087 square feet [0.011 m^2] (Table 9-1) and a velocity of 4483 fpm [22.73 m/s]. These data are inserted in Rows A-B/11 and A-B/12, respectively. Then the velocity pressure in the duct is calculated from Table 9-3 (only if "Standard Air") or Equation 3.6a in Chapter 3 or Equation 5 on right edge of Calculation Sheet. This velocity pressure is placed in Cell A-B/13. This completes all of the basic system entry for the segment and now the static pressure losses can be calculated.

Step 4. The first component of the system loss in this segment is the hood static pressure (SP_h) Determine the hood static pressure from the equations in Chapter 6, or available information in Chapter 13 (VS-80-11 for this problem shows a value of 0.65). There are no slots (Rows 15-21), so the hood entry coefficient (F_h) is as entered into cell A-B/22.

Step 5. The hood SP (SP_h) is the sum of the hood loss (h_h) and the energy transfer as air moves from stillness outside the hood to the energy as it travels at the velocity in the duct ($F_a \times VP_d = 1VP_d$). This is derived in Example Problem 4 of Chapter 3. This value of 1VP represents the losses due to this energy transfer and must be added to the calculation. It is inserted in Cell A-B/23.

Step 6. The hood static pressure can then be determined by adding the values of Rows 22 and 23 (1+ 0.65 = 1.65) and inserting that value in Cell A-B/24. It is then multiplied by the VP for that segment (in A-B/13) to get the Hood Static Pressure (-2.1 "wg in Cell A-B/25) [512 Pa]. If there were other losses in the hood (i.e., a filter section or spray section that had resistance) they would be added in A-B/26. Since this system does not have those losses, the same value (-2.1 "wg) [-512 Pa] is placed in the Hood Static Pressure cell (A-B/27).

(The following steps then add any other cumulative losses as the evaluation proceeds to point "B". These include the losses for duct components (straight duct, elbows, contractions, expansions, etc.). In this Example Problem there is only straight duct.)

Step 7. From the drawing information in Figure 9-5 input 15' [4.5 m] into Cell A-B/28. Multiply this length by the duct friction factor (input into A-B/29) obtained from the tabulated data of Table 9-4 (IP or SI), or use Equation 9 included on the calculation sheet. The use of sheet metal duct is assumed throughout this chapter.

Step 8. Determine the number and type of fittings in the duct segment. For each fitting type (Figures 9-d, 9-e, 9-f), determine the loss coefficient and multiply by the number of fittings (as mentioned above there were none in this example). Input the data into Rows 31 through 33. All of the coefficients for the components for the segment are compiled in Cell A-B/36.

Step 9. Multiply the total in A-B/36 by the duct VP (in A-B/13). This is the actual loss in inches of water for the duct

FIGURE 9-5 (IP)

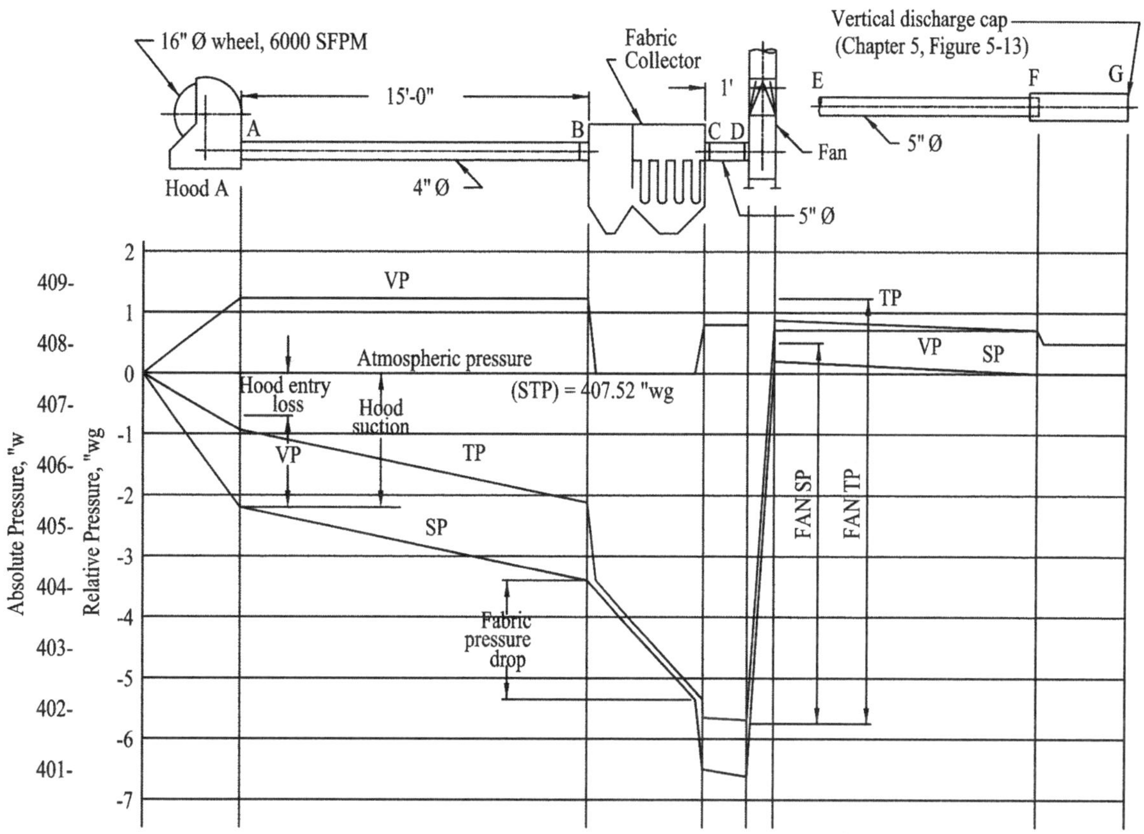

Details of Operation

NO.		Hood No.	VS-Print	Required Air-flow, ACFM
1	16" Diameter Grinding Wheel, 2" Wide	A	80-11	390

Dimensions

No. of Branch or Main	Straight Run, Ft	ACFM Required	Elbows	Entries
A-B	15	390	--	--
B-C	Collector	390	--	--
C-D	1	390	--	--
E-F	10	390	--	--
F-G	Stack Head	390	--	--

ACGIH®

TITLE DOUBLE LINE SKETCH EXAMPLE PROBLEM #1

FIGURE 9-5 (IP)

DATE 1-12

CHECK CODES, REGULATIONS, AND LAWS (LOCAL, STATE, AND NATIONAL) TO ENSURE THAT DESIGN IS COMPLIANT.

FIGURE 9-5 (SI)

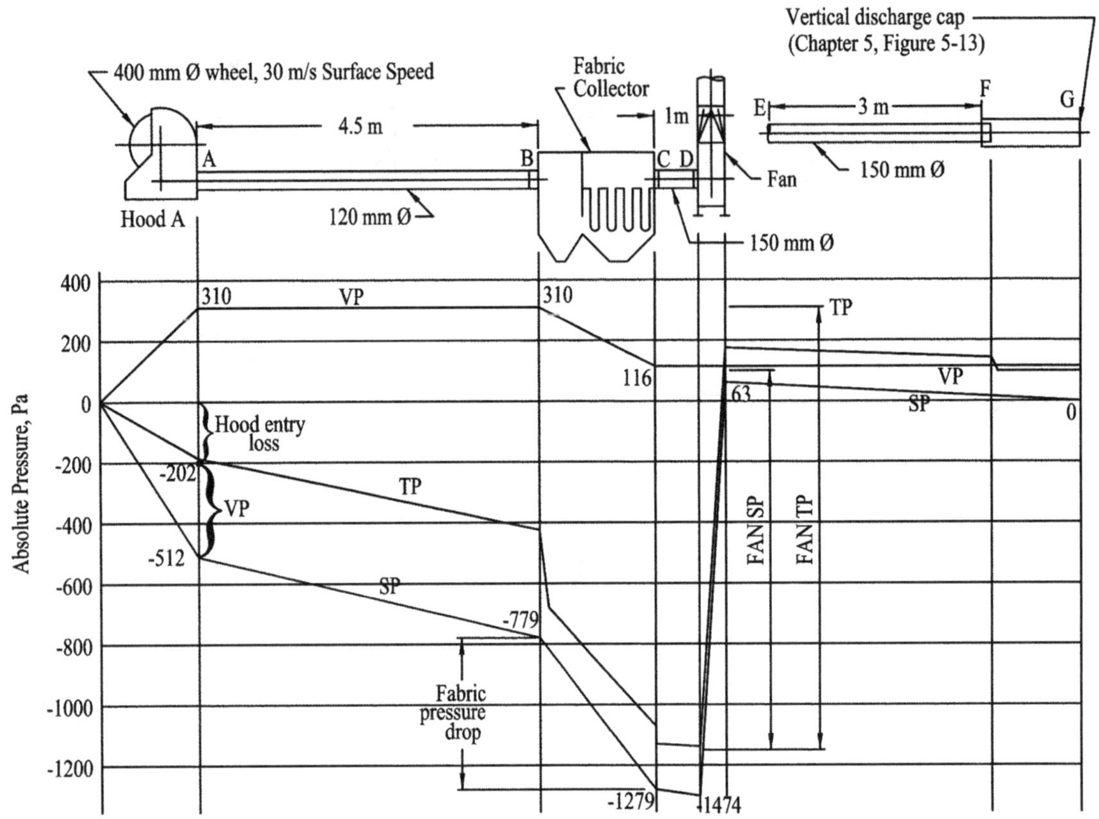

Details of Operation

NO.		Hood No.	VS-Print	Required Air-flow, am³/s
1	400 mm Diameter Grinding Wheel, 50 mm Wide	A	80-11	0.25

Dimensions

No. of Branch or Main	Straight Run, m	am³/s Required	Elbows	Entries
A-B	4.5	0.25	--	--
B-C	Collector	0.25	--	--
C-D	1	0.25	--	--
E-F	3.5	0.25	--	--
F-G	Stack Head	0.25	--	--

ACGIH®	TITLE: DOUBLE LINE SKETCH EXAMPLE PROBLEM #1	FIGURE: 9-5 (SI)
		DATE: 1-12

CHECK CODES, REGULATIONS, AND LAWS (LOCAL, STATE, AND NATIONAL) TO ENSURE THAT DESIGN IS COMPLIANT.

ACGIH® Velocity Pressure Method Calculation Sheet (IP)

Figure 9-6 (IP) Example Problem #1 — Page 1 of 1 — Date — Elevation (z, ft) = 0 ft — * Input Data

No.	Symbol	Group	Description	Reference	Units	A-B	B-C	C-D	E-F	Row
1*			Duct Segment Identification			A-B	B-C	C-D	E-F	1
2*	T		Dry-Bulb Temperature		F	70	70	70	70	2
3*	Q_{act}		Actual Duct Flow Rate	Eqn 3	acfm	390	390	390	390	3
4*	V_t		Minimum Transport Velocity	Chapter 13	fpm	4000		3000	3000	4
5*	$\dot{m}_{H2O}$		Pounds Water per Minute		lbs-H_2O/min					5
6*	$\dot{m}_{da}$		Pounds Dry Air per Minute		lbs-da/min					6
7	df		Density Factor	Eqn 2		1.00	1.00	1.00	1.00	7
8	Q_{std}		Standard Duct Flow Rate (Dry Air)	Eqn 3	dscfm	390	390	390	390	8
9	A_t		Target Duct Area	3/4	ft²	0.098		0.130	0.130	9
10*	D		Selected Duct Diameter	Section 9.5.2.5	in	4.0		5.0	5.0	10
11*	A		Selected Duct Area	Table 9.1	ft²	0.087		0.136	0.136	11
12	V_d		Duct Velocity	(3/11)	fpm	4483		2868	2868	12
13	VP_d		Duct Velocity Pressure	Eqn 4	"wg	1.25		0.51	0.51	13
14	h		Total Heat	Eqn 5 or Psych Chart	BTU/lbs-da					14
15	A_s	HOOD LOSSES – SLOT	Slot Area	Chapters 6 & 13	ft²					15
16*	F_s	HOOD LOSSES – SLOT	Slot Loss Coefficient	Figure 9a	VP/slot					16
17*	F_a	HOOD LOSSES – SLOT	Acceleration Coefficient	Chapter 3	0 or 1					17
18*	V_s	HOOD LOSSES – SLOT	Slot Velocity	Chapters 6 & 13	fpm					18
19	VP_s	HOOD LOSSES – SLOT	Slot Velocity Pressure	Eqn 4	"wg					19
20		HOOD LOSSES – SLOT	Slot Loss in VP	16+17	VPtotal					20
21	h_s	HOOD LOSSES – SLOT	Slot Loss	20x19	"wg					21
22*	F_h	HOOD LOSSES	Hood Entry Loss Coefficient	Fig 9a or Ch 13	F_h or 0	0.65		0.50		22
23*	F_a	HOOD LOSSES	Acceleration Coefficient	Chapters 3 & 6	1 or 0	1		1		23
24		HOOD LOSSES	Hood Entry Loss in VP	22+23	VPtotal	1.65		1.50		24
25	h_h	HOOD LOSSES	Hood Entry Loss	24x13	"wg	2.1		0.8		25
26*		HOOD LOSSES	Other Losses		"wg					26
27	SP_h	HOOD LOSSES	Hood Static Pressure	21+25+26	"wg	2.1		0.8		27
28*	L	DUCT LOSSES	Straight Duct Length		ft	15.0		1.0	10.0	28
29	F'_d	DUCT LOSSES	Duct Friction Factor	Eqn8	VP/ft	0.070		0.056	0.056	29
30*		DUCT LOSSES	No. of 90 Degree Elbows							30*
31*	F_{el}	DUCT LOSSES	Elbow Loss Coefficient	Table 6	F_{el} or 0					31
32*	F_{en}	DUCT LOSSES	Branch Entry Loss Coefficient	Table 7	VP/entry					32
33*		DUCT LOSSES	Special Fitting Loss Coefficient	i.e. Exp./Cont., etc.	VP/fitting					33
34	F_d	DUCT LOSSES	Duct Loss Coefficient	28X29	VP/duct	1.05		0.06	0.56	34
35		DUCT LOSSES	Elbow Loss in VP	30x31	VP					35
36		DUCT LOSSES	Total Duct Loss in VP	32+33+34+35	VP total	1.05		0.06	0.56	36
37		DUCT LOSSES	Total Duct Loss	36x13	"wg	1.3		0.0	0.3	37
38*		DUCT LOSSES	Other Losses		"wg		2.0			38
39*	VP_r	DUCT LOSSES	Weighted Average VP	Eqn 10	"wg					39
40*		DUCT LOSSES	Loss From Velocity Increase	13-39 (if > 0)	"wg					40
41		DUCT LOSSES	Segment Pressure Loss	27+37+38+40	"wg	-3.4	-2.0	-0.8	0.3	41
42*	SP_{gov}		Governing Static Pressure		"wg					42
43*	SP_{cum}		Cumulative Static Pressure		"wg	-3.4	-5.4	-6.2	0.3	43
44	Q_{corr}		Corrected Volumetric Flow	Eqn 9	acfm					44
45	V_{corr}		Corrected Velocity	44/11	fpm					45
46	VP_{corr}		Corrected Velocity Pressure	Eqn 4	"wg					46

Designer

1. $\omega = lbs - H_2O / lbs - DryAir$

2. $df = df_e * df_p * df_T * df_m$
   $df_e = [1 - (6.73 * 10^{-6})(z)]^{5.258}$
   $df_p = (407 + SP)/(407)$
   $df_T = 530/(T + 460)$
   $df_m = (1+\omega)/(1+1.607\omega)$

3. $Q_{std} = (Q_{act}/(1+\omega))df$
   $Q_{act} = (Q_{std})(1+\omega)/df$

4. $V = 4005\sqrt{VP/df}$
   $VP = (V/4005)^2 df$

5. $h = 0.24T + \omega(1061 + 0.444T)$

6. 90 Degree Elbow Loss Coefficient (5 Piece)

R/D	F_{el}
1.5	0.24
2.0	0.19
2.5	0.17

7. Branch Entry Loss Coefficients

Angle	F_{en}
15°	0.09
30°	0.18
45°	0.28
90°	1.00

8. $F_d\left(\frac{metal}{plastic}\right) = 0.0307\left(V^{0.533}/Q^{0.612}\right)$
   $F_d(flexible) = 0.0311\left(V^{0.604}/Q^{0.639}\right)$

9. $Q_{corrected} = Q_{design}\sqrt{SP_{gov}/SP_{lower}}$

10. $VP_r = (Q_1/Q_3)VP_1 + (Q_2/Q_3)VP_2$

11. $SP_{System} = SP_{out} - SP_{in} - VP_{in}$

SSP = +.3 - (-6.2) - 0.5 = 6.0"w.g.

FIGURE 9-6 (IP). Velocity Pressure Method Calculation Sheet

ACGIH® Velocity Pressure Method Calculation Sheet (SI)

Figure 9-6 (SI) Example Problem #1 | Page 1 of 1 | Date | Elevation (z), m= 0 m | * Input Data | Designer

No.	Symbol	Group	Description	Reference	Units	A-B	B-C	C-D	E-F	Row
1*			Duct Segment Identification			A-B	B-C	C-D	E-F	1
2*	T		Dry-Bulb Temperature		C	21	21	21	21	2
3*	Q_{act}		Actual Duct Flow Rate	Eqn 3	am³/s	0.25	0.25	0.25	0.25	3
4*	V_t		Minimum Transport Velocity	Chapter 13	m/s	20.00		15.00	15.00	4
5*	$\dot{m}_{H2O}$		kg Water per Second		kg-H_2O/s					5
6*	$\dot{m}_{da}$		kg Dry Air per Second		kg-da/s					6
7	df		Density Factor	Eqn 2		1.00	1.00	1.00	1.00	7
8	Q_{std}		Standard Duct Flow Rate (Dry Air)	Eqn 3	dnm³/s					8
9	A_t		Target Duct Area	3/4	m²	0.013		0.017	0.017	9
10*	D		Selected Duct Diameter	Section 9.5.2.5	mm	120		150	150	10
11*	A		Selected Duct Area	Table 9.1	m²	0.011		0.018	0.018	11
12	V_d		Duct Velocity	(3/11)	m/s	22.73		13.89	13.89	12
13	VP_d		Duct Velocity Pressure	Eqn 4	Pa	310		116	116	13
14	h		Total Heat	Eqn 5 or Psych Chart	kJ/kg-da					14
15	A_s	HOOD LOSSES – SLOT	Slot Area	Chapters 6 & 13	m²					15
16*	F_s	HOOD LOSSES – SLOT	Slot Loss Coefficient	Figure 9a						16
17*	F_a	HOOD LOSSES – SLOT	Acceleration Coefficient	Chapter 3	0 or 1					17
18*	V_s	HOOD LOSSES – SLOT	Slot Velocity	Chapters 6 & 13	m/s					18
19	VP_s	HOOD LOSSES – SLOT	Slot Velocity Pressure	Eqn 4	Pa					19
20		HOOD LOSSES – SLOT	Slot Loss in VP	16+17	VPtotal					20
21	h_s	HOOD LOSSES – SLOT	Slot Loss	20x19	Pa					21
22*	F_h	HOOD LOSSES	Hood Entry Loss Coefficient	Fig 9a & Ch 13	F_h	0.65		0.50		22
23*	F_a	HOOD LOSSES	Acceleration Coefficient	Chapters 3 & 6	1 or 0	1		1		23
24		HOOD LOSSES	Hood Entry Loss in VP	22+23	VPtotal	1.65		1.50		24
25	h_h	HOOD LOSSES	Hood Entry Loss	24x13	Pa	512		174		25
26*		HOOD LOSSES	Other Losses		Pa					26
27	SP_h	HOOD LOSSES	Hood Static Pressure	21+25+26	Pa	512		174		27
28*	L	DUCT LOSSES	Straight Duct Length		m	4.5		1.0	3.0	28
29	F'_d	DUCT LOSSES	Duct Friction Factor	Eqn 8	VP/m	0.190		0.180	0.180	29
30*		DUCT LOSSES	No. of 90 Degree Elbows							30*
31*	F_{el}	DUCT LOSSES	Elbow Loss Coefficient	Table 6	F_{el}					31
32*	F_{en}	DUCT LOSSES	Branch Entry Loss Coefficient	Table 7	VP/entry					32
33*		DUCT LOSSES	Special Fitting Loss Coefficient		VP/fitting					33
34	F_d	DUCT LOSSES	Duct Loss Coefficient	28X29	VP/duct	0.86		0.180	0.540	34
35		DUCT LOSSES	Elbow Loss in VP	30x31	VP					35
36		DUCT LOSSES	Total Duct Loss in VP	32+33+34+35	VP total	0.86		0.180	0.540	36
37		DUCT LOSSES	Total Duct Loss	36x13	Pa	267		21	63	37
38*		DUCT LOSSES	Other Losses	i.e. Exp./Cont.,etc.	Pa		500			38
39*	VP_r	DUCT LOSSES	Weighted Average VP	Eqn 10	Pa					39
40*		DUCT LOSSES	Loss From Velocity Increase	13-39 (if>0)	Pa					40
41		DUCT LOSSES	Segment Pressure Loss	27+37+38+40	Pa	-779	-500	-195	63	41
42*	SP_{gov}		Governing Static Pressure		Pa					42
43*	SP_{cum}		Cumulative Static Pressure		Pa	-779	-1279	-1474	63	43
44	Q_{corr}		Corrected Volumetric Flow	Eqn 9	am³/s					44
45	V_{corr}		Corrected Velocity	44/11	m/s					45
46	VP_{corr}		Corrected Velocity Pressure	Eqn 4	Pa					46

Designer equations:

1. $\omega = kg-H_2O/kg-DryAir$

2. $df = df_e * df_p * df_T * df_m$
$df_e = [1-(22.08*10^{-6})(z)]^{5.256}$
$df_p = (101384+SP)/(101384)$
$df_T = 294/(T+273)$
$df_m = (1+\omega)/(1+1.607\omega)$

3. $Q_{std} = (Q_{act}/(1+\omega))df$
$Q_{act} = (Q_{std})(1+\omega)/df$

4. $V = 1.29\sqrt{VP/df}$
$VP = (V/1.29)^2 df$

5. $h = 1.006T + \omega(2501+1.84T)$

6. 90 Degree Elbow Loss Coefficient (5 Piece)

R/D	F_{el}
1.5	0.24
2.0	0.19
2.5	0.17

7. Branch Entry Loss Coefficients

Angle	F_{en}
15°	0.09
30°	0.18
45°	0.28
90°	1.00

8. $F_d\left(\frac{metal}{plastic}\right) = 0.0155\left(V^{0.533}/Q^{0.612}\right)$
$F_d(flexible) = 0.0186\left(V^{0.604}/Q^{0.639}\right)$

9. $Q_{corrected} = Q_{design}\sqrt{SP_{gov}/SP_{lower}}$

10. $VP_r = (Q_1/Q_3)VP_1 + (Q_2/Q_3)VP_2$

11. $SP_{System} = SP_{out} - SP_{in} - VP_{in}$

SSP = +63 - (-1474) - 116 = 1421 Pa

FIGURE 9-6 (SI). Velocity Pressure Method Calculation Sheet

segment and should be placed in A-B/37.

Step 10. Add the result of Steps 6 and 9. This combines the hood and duct losses for the segment. If there are any additional losses (expressed in "wg [Pa]), such as an air cleaning device, include them also. This establishes the cumulative energy required, expressed as static pressure, to move the design airflow through the duct segment and is input into Cell A-B/41.

NOTE: The value in that cell is negative. This value of -3.4" [-512 Pa] would be used to begin the system calculation and is close to representing a value that would be seen if a measurement of pressure were taken at point "B". (See Appendix C for measurement methods.) The value represents the negative pressure required to pull 390 acfm [0.25 m³/s] through the duct and hood as designed.

Similar input is placed into the next columns. The second column is designated as "B-C" and covers the next segment, the fabric filter in this case. Because the only loss given in the example is across the filter media (sometimes called ΔP) of 2" [500 Pa], this information is placed in the cell designated for "Other Losses" (B-C/38). This 2.0 "wg [500 Pa] is added to the losses already accumulated from "A" to "B" to arrive at the -5.4 "wg [-1279 Pa] value inserted into B-C/43.

The process continues for all segments up to the inlet of the fan. Then the accumulation of negative static pressure is noted in Row 43 (in this problem that is shown at C-D/43) and is designated as System Static Pressure into the fan (SP_i). Second, a new set of pressures are calculated on the positive or discharge side of the fan. These are shown cumulatively in the same designated row (in this problem in E-F/43) and are designated as Static Pressure out of the fan (SP_o). SP_i and SP_o as well as the Velocity Pressure going into the fan (VP_i) (as shown in Cell C-D/13) are used to calculate the System Static Pressure. This is explained in Section 9.9.

9.7 DISTRIBUTION OF AIRFLOW IN A MULTI-BRANCH DUCT SYSTEM (IP Units Only)

System Sample Problem 1 (Figure 9-5) had no branch fittings (where two ducts are combined into a single duct). Most systems do have multiple branches and care must be taken to provide the correct balance of flows and pressures at each branch and hood. In a multiple hood system it is necessary to provide a means of distributing airflow between the branches either by balanced design or by the use of blast gates or orifice plates. Air will always take the path of least resistance. A natural balance at each junction will occur; that is, the exhaust flow rate will distribute itself automatically according to the pressure losses of the individual flow paths. The designer must provide distribution such that the required airflow at each hood will never fall below the minimums listed in Chapters 6 and/or 13 so the duct paths to these hoods must be properly sized.

To accomplish this, ensure that all flow paths (ducts) entering a junction will have equal calculated static pressure requirements for their required airflows. A junction can have only one Static Pressure available for the connected branches. For example, one duct may be entering from Hood A where all of the static pressure (SP) requirements for hood and duct is -3.5 "wg [875 Pa]. If the other joining branch and Hood B has an SP requirement of -5.0 "wg [1250 Pa] then a system fan must be able to deliver the higher (-5.0 "wg [1250 Pa]) SP or else Hood B will not have enough energy to provide all of the design flow. This higher value of pressure at the junction is called the "governing" pressure and the branch called the "governing branch." If the fan is selected to provide -5.0 "wg [1250 Pa] there is now excess pressure pulling at Hood A. The designer must provide 1.5 "wg (5.0" – 3.5") [1250 – 875 = 375 Pa] more resistance in the branch serving Hood A to balance the flow conditions.

There are two primary methods for accomplishing this balance and they are discussed in detail in Chapter 5, Section 5.4. The Balance by Design Method would create additional resistance by decreasing the duct size and creating more duct frictional resistance. Properly accomplished, there should be a duct size change that will result in the added SP of 1.5" [375 Pa]. The Blast Gate Method uses a partially closed gate to add the 1.5 "wg [375 Pa] of resistance. The closing of the gate creates turbulence (and resistance). Closed to the proper setting, it should be able to meet the requirements for balancing the branches.

The object of both methods is the same: to obtain the desired flow rate at each hood in the system while maintaining the desired velocity in each branch and main.

9.7.1 Balance by Design Method. This procedure provides for achievement of the desired airflow (a "balanced" system) without the use of blast gates or orifice plates. It is often called the "Static Pressure Balance Method." In this type of design, the calculation usually begins at the hood farthest from the fan in terms of number of duct segments and proceeds, segment by segment, to the fan. At each junction, the static pressure necessary to achieve desired flow in one stream must equal the static pressure in the joining air stream. The static pressures are balanced by suitable choice of duct sizes, elbow radii, etc.

The static pressure (SP) loss of each duct segment is calculated from a local exhaust hood to the junction with the next branch based on hood design data, fittings, and total duct length. At each junction, the SP for each parallel path of airflow must be the same. The governing SP is referred to as SP_g. The lower value of SP at the junction is called lower static pressure (SP_o).

When the **ratio** of the value of the governing SP to the original SP is:

a) **greater than 1.2:** redesign of the branch with the lower pressure loss should be considered. This may include a change of duct size, selection of different fittings and/or modifications to the hood design.

b) **unequal but less than 1.2:** balance can be obtained by increasing the airflow through the run with the lower resistance. This change in flow rate is calculated by noting that pressure losses vary with the velocity pressure and, therefore, as the square of the flow rate, so:

$$Q_{corrected} = Q_{original}\sqrt{\frac{SP_{governing}}{SP_{lower}}} \quad [9.8]$$

where the governing SP is the desired (higher) SP at the junction point and the "original" or lower SP is that calculated for the duct segment being designed.

NOTE: The value under the square root is always greater than 1.0.

9.7.2 Blast Gate/Orifice Plate Method. The design procedure depends on the use of blast gates and/or orifice plates located in branches or mains to provide the restrictions to balance static pressures. Blast gates (also called "cut-offs") (see Chapter 5, Figure 5-21) must be adjusted after installation in order to achieve the desired flow at each hood. At each junction, the flow rates of two joining ducts are achieved by blast gate adjustment that results in the desired static pressure balance. Similarly, orifice plate opening sizes may be changed to reflect actual requirements at start-up or when system revisions are made, but their design usually infers more permanent installation with less chance of operator adjustment.

NOTE: The corrosiveness or abrasiveness of the air stream should also be considered when using the blast gate/orifice plate method.

Data and calculations involved are the same as for the balanced design method except that the duct sizes, fittings and flow rates are not adjusted; the blast gates are set after installation to provide the design flow rates. This main advantage of blast gate design provides actual flows equal to design flows with a resultant savings in flow (and power requirement). It should be noted that a change in any of the blast gate settings could change the flow rates in all of the other branches. Readjusting the blast gates can also result in increases to the actual fan static pressure and increased fan power requirements.

NOTE: It is a common practice to design systems on the assumption that only a fraction of the total number of hoods will be used simultaneously and the flow to the branches not used will be shut off with dampers. For tapered system designs, where particulate is transported, this practice may lead to plugging in the main and branch ducts due to settled particulate. This procedure is not recommended unless minimum transport velocity can be assured in all ducts during any variation of opened and closed blast gates. It is better to design these systems with individual branch lines all converging very close to the fan inlet so that lengths of duct mains are minimized or use a Plenum System design (see Chapter 5, Section 5.5.2).

9.8 INCREASING VELOCITY THROUGH A JUNCTION (WEIGHTED AVERAGE VELOCITY PRESSURE)

Variations in duct velocity occur at many locations in local exhaust systems because of necessary limitations in available standard duct sizes (area) or due to duct selections based on balanced system design. As noted earlier, small accelerations and decelerations are usually compensated automatically in the system where good design practices and proper fittings are used. There are times, however, when special circumstances require the designer to have a knowledge of the energy losses and regains that occur, since these may work to his/her advantage or disadvantage in the final performance of the system.

Sometimes the final main duct velocity exceeds the weighted average of the two velocities in the branches entering the main. Air speed cannot be increased through the fitting without an expenditure of kinetic energy. If the difference between the weighted average of the branch velocities and the final velocity is greater than zero, additional static pressure is required to produce the increased velocity. This extra loss is shown in Row 40 of the Calculation Sheet. In previous editions, this calculation was called Resultant Velocity Pressure and is now more correctly designated Weighted-Average Velocity Pressure. It still maintains the symbol of VP_r.

Energy must be conserved at any junction point. The energy entering each of the two air streams would be Q(TP) = Q(SP +VP). The first law of thermodynamics states that the sum of these must equal the energy leaving, or

$$Q_1(VP_1 + SP_1) + Q_2(VP_2 + SP_2) = Q_3(VP_3 + SP_3) + \text{Losses}$$

Note that the overall losses would be:

$$\text{Losses} = F_1Q_1VP_1 + F_2Q_2VP_2$$

where the subscripts refer to the ducts shown in Figure 9-7.

In this Manual, F_1 is considered to be zero and F_2 is given on Figure 9-f. Assuming it is balanced and the junction losses are included such that $SP_1 = SP_2$ and $Q_3 = Q_1 + Q_2$ (Figure 9-7), there might be an additional change in static pressure due to the acceleration or deceleration of the gas stream. The following equation shows this effect:

$$SP_3 + VP_3 = SP_1 + \left(\frac{Q_1}{Q_3}\right) VP_1 + \left(\frac{Q_2}{Q_3}\right) VP_2$$

The last two terms on the right are defined as the weighted average velocity pressure, VP_r; this can be simplified to

$$VP_r = \left(\frac{Q_1}{Q_3}\right) VP_1 + \left(\frac{Q_2}{Q_3}\right) VP_2 \quad [9.9]$$

where: VP_r ≡ weighted average velocity pressure of the combined branches, "wg [Pa]

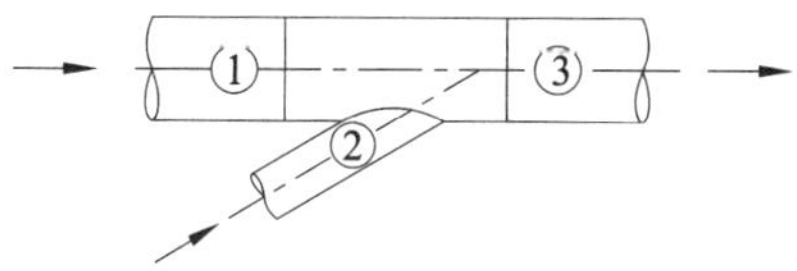

Duct No.	Dia.	Area	Q	V	VP	SP
(1)	10	0.545	1935	3550	0.79	-2.1
(2)	4	0.087	340	3890	0.94	-2.1
Main (3)	10	0.545	2275	4170	1.08	-

FIGURE 9-7 (IP). Branch entry velocity correction

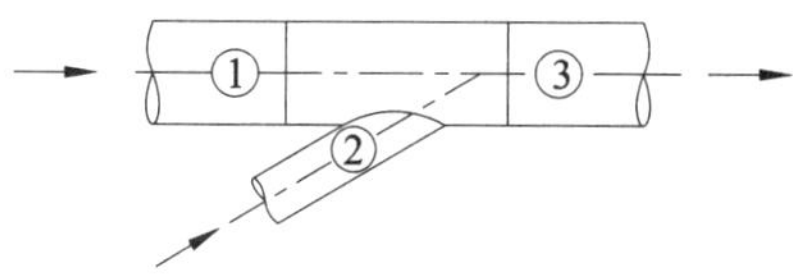

Duct No.	Dia.	Area	Q	V	VP	SP
(1)	250	0.049	0.97	19.80	235	-525
(2)	100	0.008	0.17	21.25	270	-525
Main (3)	250	0.049	1.14	23.27	325	-

FIGURE 9-7 (SI). Branch entry velocity correction

Q_1 ≡ flow rate in branch #1, acfm [am³/s]
Q_2 ≡ flow rate in branch #2, acfm [am³/s]
Q_3 = combined flow rate leaving the junction, acfm [am³/s]

Note that the above equation is valid for all conditions, including merging different density gas streams, as long as the velocity pressures include the density effects. Also note that, if the flow rate through one branch was changed to balance at the branch entry, the corrected velocity pressure and corrected flow rates should be used in Equation 9.9.

The Weighted Average Velocity Pressure (VP_r) (previously called "Resultant Velocity Pressure") is computed using Equation 9.9. Note that VP_r is not a measurable value in the system. It is a computed value only. When VP_3 is less than VP_r, a deceleration has occurred. No adjustment is made in the calculations in this case. However, if VP_3 is greater than VP_r, an acceleration of the air stream has occurred through the fitting. The difference between VP_3 and VP_r is the necessary loss in SP required to produce the increase in kinetic energy as air travels from the branches into the main duct. The correction is made as follows:

$$SP_3 = SP_1 - (VP_3 - VP_r) \qquad [9.10]$$

where: SP_3 = SP in main #3, "wg [Pa]
SP_1 = SP at branch #1 – SP at branch #2, "wg [Pa]
VP_3 = velocity pressure in main #3, "wg [Pa]
VP_r = weighted average velocity pressure, "wg [Pa]

In the Calculation Sheet, this is now shown in Row 40 of the downstream branch where the increase in velocity is considered. If this downstream value is higher than the VP_r into that branch, then the difference is added to the static pressure losses of that branch.

EXAMPLE PROBLEM 8 (Weighted Average VP)

With the data shown in Figure 9-7 (IP), determine the static pressure requirement at point 3.

$$VP_r = \frac{(1935)(0.79)}{2275} + \frac{(340)(0.94)}{2275} = 0.81 \text{ "wg}$$

$SP_3 = SP_1 - (VP_3 - VP_r) = -2.11 - (1.08 - 0.81) = -2.11 - 0.27 = -2.38$ "wg

The metric solution using data from Figure 9-7 (SI) is:

$$VP_r = \frac{(0.97)(235)}{1.14} + \frac{(0.17)(270)}{1.14} = 240 \text{ Pa}$$

$SP_3 = SP_1 - (VP_3 - VP_r) = -525 - (325\text{-}240) = -610$ Pa

Therefore, in this situation, an additional -0.27 "wg [325 – 240 = 85 Pa] should be added to the junction SP to account for losses in pressure due to acceleration of the air stream.

9.9 FAN AND SYSTEM PRESSURE CALCULATIONS

Local exhaust system calculations are based on static pressure; that is, all hood static pressures and balancing or governing pressures at the duct junctions are given as System Static Pressure (SSP). SSP can be measured directly in the field as described in Appendix C. Most fan rating tables are based on Fan Static Pressure (FSP). The SSP from the Calculation Sheet is the basis for the selection of the FSP. An additional calculation is required to determine the FSP before selecting the fan. This section describes the definition of FSP and FTP as provided by the Air Movement and Control Association (AMCA). (Chapter 7 details the pressure and other definitions required for fan selection.) FSP and FTP are then compared with the SSP and STP determined from the system calculation sheets to predict operating points.

9.9.1 Fan Total Pressure (FTP). FTP is the increase in total pressure through or across the fan and can be expressed by the equation:

$$FTP = TP_{outlet} - TP_{inlet} \qquad [9.11]$$

Discussions of Total Pressure (TP) are provided in Chapter

3, Section 3.4. Some fan manufacturers base catalog ratings on FTP. To select a fan on this basis the FTP is calculated noting that TP = SP + VP:

$$FTP = (SP_{outlet} + VP_{outlet}) - (SP_{inlet} + VP_{inlet}) \quad [9.12]$$

9.9.2 Fan Static Pressure (FSP). The AMCA Test Code defines the Fan Static Pressure (FSP) as follows: "the static pressure of the fan is the total pressure diminished by the fan velocity pressure. The fan Velocity Pressure is defined as the pressure corresponding to the air velocity at the fan outlet."(9.1) FSP can be expressed by the equation:

$$FSP = FTP - VP_{outlet} \quad [9.13]$$

or

$$FSP = SP_{outlet} - SP_{inlet} - VP_{inlet} \quad [9.14]$$

Fan Static Pressure is a term derived from the method of testing fans and is the value provided by most manufacturers in their fan selection tables (see Chapter 7). These are not from the system calculations but the laboratory or computer generated data for the fan.

NOTE: For the remainder of this chapter, the term fan pressure will apply to both FSP and FTP.

9.9.3 System Static Pressure (SSP). System Static Pressure (SSP) represents the pressure needed to overcome the losses in energy as a gas moves through the duct system and is determined from the data on the calculation sheet. It is the pressure used to specify the required fan pressure. To place SSP on the same graphic representation (fan/system curve), the units of measurement must be the same as FSP (Equation 9.14). This transposition provides SSP by also removing the effects of the VP_{in} (Equation 9.14). This yields the following equation for SSP:

$$SSP = SP_{outlet} - SP_{inlet} - VP_{inlet} \quad [9.15]$$

The values used for calculating SSP are taken from the calculation sheet whereas the values for calculating FSP are based on manufacturers' test data. Where these two data points intersect is the predicted operating point.

9.9.4 Use of System Static Pressure to Specify a Fan. The system pressure calculation is based on the same formula as used to determine fan pressure. Therefore, an estimate for the required FSP can be obtained by determining the system static pressure and then adding

1) factor of safety,
2) provisions for pressure variations (i.e., changing ΔP of baghouse during operation) will result in an estimate for required FSP.

In selecting a fan from catalog ratings, the rating tables should be examined to determine whether they are based on FSP or FTP. Most centrifugal fans used for industrial ventilation systems will be specified using FSP. Fan system effects (see Chapter 7) should also be considered when selecting a fan. Remember to give appropriate lengths of straight duct entering and leaving centrifugal fans, as they are especially sensitive to abrupt directional changes and will require more horsepower and tip speed if there are elbows or other interferences close to the fan's inlet or outlet. The proper pressure rating can then be calculated keeping in mind the proper algebraic signs; i.e., VP is always positive (+), SPinlet is usually negative (-), and SPoutlet is usually positive (+).

The final selection of the fan must also consider the air density. Most fan tables and curves are printed for standard conditions. The final SSP, calculated and then altered to meet the above FSP or FTP requirements, must then be adjusted for air density using the following formulas:

$$FSP_{specified} = FSP_{calculated}/df \quad [9.16a]$$

$$FTP_{specified} = FTP_{calculated}/df \quad [9.16b]$$

The values for df are those shown on the calculation sheet in the segment at the fan inlet and should include the factor for change in absolute pressure at the fan inlet – particularly if the fan inlet static pressure is below -20 "wg [-5 kPa].

Continuing Example Problem 1, the SSP and estimates for FSP can be made from values on the calculation sheet. Note that IP and SI units do not correspond. Example Problems have two different inlet conditions. At the outlet of the fan, the SP is +0.3 "wg [+63 Pa]. At the inlet to the fan, the SP is -6.2 "wg [-1474 Pa]. The VP at both locations is 0.51 "wg [116 Pa]. From Equation 9.15, the system pressure in terms of static pressure (SSP) = 0.3 – (-6.2) – 0.5 = 6.0 "wg [+63 – (-1474) – 116 = 1421 Pa]. This value is used to specify the required FSP for fan operation. In this example, the designer would use the SSP as 6.0 "wg [1421 Pa] and then may choose an FSP for specification of 6.0 "wg, 6.5 "wg or even 7.0 "wg [or similar metric equivalents] – based on factors of safety or other considerations. The specified FSP is the one selected from the fan tables after Equation 9.16 is completed and the fan is selected. (Assume df = 1.0 in this example.)

It must be understood that the actual Fan Pressure (Total or Static) is that value provided by the fan manufacturer from the designer's specification. In the example above, the SSP was determined to be 6.0 "wg [1421 Pa]. However, the actual fan selection (Fan Static Pressure) may be 6.5", 7.0", or even 20.0" (or similar metric equivalent) based on the decisions of the designer. The SSP is the actual calculation based on the system requirements. The FSP is a specified value based on the SSP and other process requirements (i.e., factors of safety, variances of pressure during operation, etc.). These descriptions are based on NOT using fan manufacturer's software since some calculations may be done internally to that particular software. Be alerted so that some factors are not considered twice.

9.10 SYSTEM CURVE/FAN CURVE RELATIONSHIP

The determination of the system pressure allows the development of the System Curve using the information in Chapter 3, Section 3.7. The fan curve depicts the relationship between Actual Volumetric Flow Rate (acfm) [am^3/s] and Pressure ("wg) [Pa] and is provided by the manufacturer of the fan (see Chapter 7; Section 7.3.2, and Figure 7-7). The system curve also states the relationship between these two factors. Since SSP and FSP are in the same units of measurement they can be plotted on the same graph.

The intersection of the System Curve and the manufacturer's provided Fan Curve will be the calculated (predicted) operating point (see Chapter 7). Note that the intersection of fan and system curves is an approximation. The Fan Curve is shown for the selected fan at a particular speed. It is an estimated point only because there could be a change in the SSP as the bag filter pressure or other values change during operation. Similarly, there may be multiple Fan Curves if a variable speed drive and/or fan dampers are utilized or if fan temperature is changing with the process. This could give varied operating points and these must be checked to ensure stable operation under all possible conditions.

NOTE: When accounting for System Effects (see Chapter 7), the fan curve is not altered from manufacturer's information. The impact of System Effects are considered in the system calculation sheet as additional system resistance, and this determines a new SSP curve and intersection point with the fan curve.

Where the fan and system controls are designed for constant airflow operation, some single point of stability may be accomplished, but most systems are dynamic with changing flows and pressures as the internal physical condition changes. As mentioned above, these may include items such as changes in damper settings (manual or automatic), changes in filter ΔP (differential pressure), changes in water flow in a scrubber, or changes in temperature or moisture from a process being ventilated.

In Chapter 7, Figure 7-11 is an example of two distinct fan and system curves that may be encountered. In this example, Fan Curves PQ_1 and PQ_2 could represent the same fan at two different speeds. Similarly, there could be multiple fan curves indicating different damper settings or temperatures. System Curves A_1-A_2 and B_1-B_2 could represent identical systems but with varying pressure value within the system. For example, B_1-B_2 could indicate the operation when the baghouse bags are relatively clean at startup. A_1-A_2 could indicate a more restrictive system as bags become laden with dust right before cleaning (higher ΔP). The system curve would then be a family of curves between the lines indicated by A_1-A_2 and B_1-B_2.

If the fan were selected at a constant speed (PQ_2 for example) with no damper controls, operating airflow and pressure in the system could vary between points B_2 at start-up and A_2 as the baghouse ΔP increases to a maximum. Please note in Chapter 7 that not all system components operate on the basis of Equation 7.14. In particular, the fluctuation in pressure with respect to changes in bag surface velocity (air/cloth ratio, fpm [m/s]) may be closer to a linear relationship (Chapter 7, Figure 7-8). So the overall System Curve may actually have a component that is not operating as the remainder of the dynamic losses in the system. If the filter bag losses are a significant value as part of the total system losses (more than 50%), the filter manufacturer may need to be consulted to assist in the expected values for changes in ΔP with respect to airflow change in the system. In those cases, the System Curve equation may need to have a square and linear component.

9.11 SAMPLE SYSTEM DESIGN # 2 (MULTI BRANCH SYSTEM/STANDARD AIR CONDITIONS)

A typical example using the local exhaust system shown in Figure 9-8 is helpful in discussing the calculations for a tapered duct method. Calculation sheets illustrate the orderly and concise arrangement of data and calculations (Figure 9-12) (IP & SI).

The procedure outlined in Section 9.5 was used to develop the design. Each column in the calculation sheet represents a constant diameter duct segment that starts at a hood, junction point, air-cleaning device, fan or transition point.

The problem considered is a bulk powder handling system. A minimum conveying velocity of 3,500 fpm [18.00 m/s] is used throughout the problem except after the discharge of the baghouse where clean air is handled (thus no requirement for transport velocities). The system has some hoods defined in Chapter 13 but Hood 1 required assumptions to be made for this special operation. This problem will consider the air at

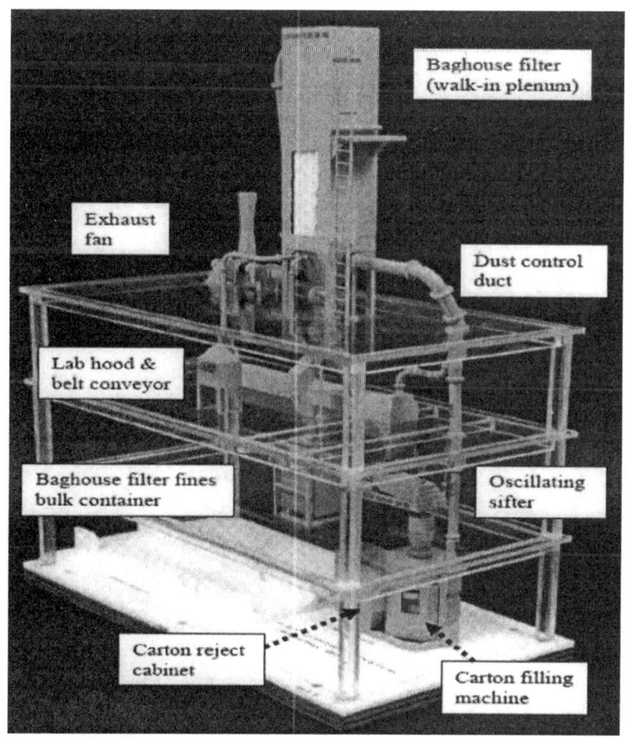

FIGURE 9-8. Sample system design – [Sample System Design # 2]

"Standard" conditions (70 F [21 C], no moisture and the system at sea level; df = 1). This seldom occurs under real conditions and most systems will require some adjustment for non-standard conditions. See Section 9.15 for an example of system design with non-standard air.

The first step to a normal design procedure is either to mark up a drawing of the system (Figure 9-10) or create a sketch or single line drawing of the system (Figure 9-9). This sketch will include the start and end numbers (Duct Segment Identification) for each segment eventually to be placed in Row 1 of the Calculation Sheet.

The operations, hood designations on the diagram, VS-print references, and required flow rates are then presented in table format either on a separate sheet or directly on the drawing or sketch. A sample from this problem is shown in Figure 9-11.

The following information is further clarification of key calculations on the ACGIH® Calculation Sheet as shown in Figure 9-12. The cell location for inserted and calculated values is made using a matrix notation. The first value in the matrix would be the Duct Segment Identification (column), and the second value would be the row. For example, in Figure 9-12 [IP or SI as appropriate], the value at cell 3-B/3 would be 500 acfm [0.25 am³/s]. It is found in Column 3-B and Row 3. Similarly, the value in cell 4-C/13 would be 0.74 "wg [211 Pa].

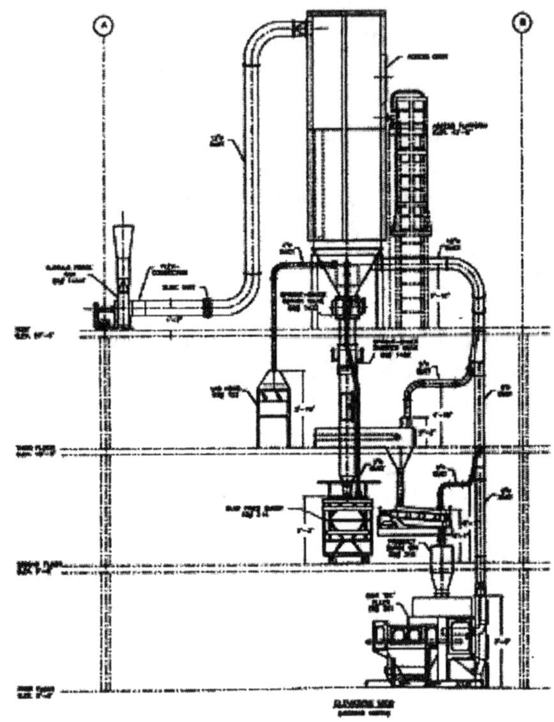

FIGURE 9-10. Elevation drawing – Sample Problem 2

With the information from the sketch and Figure 9-11, the data can be entered to the calculation sheet (Figure 9-12). The method would be to enter information from the top of each column. Normally, the designer will start with the hood farthest from the fan and/or with the most junctions between the hood and the fan. In this case, begin with Hood 1. From the sketch, Hood 1 duct combines with Hood 2 duct at junction "A" so the segment from Hood 1 would be designated "1-A" for the start and end number of the segment. This is placed in Row 1 of the first column of the Calculation Sheet.

The first 14 rows at the top of each column represent the basic information for that segment and include the flow, the duct size, the air conditions (temperature, moisture, etc.) and Velocity Pressure. Some of these data are taken from references, such as the VS plates, but others are calculated.

In Column 1-A and working down we could input the data as follows:

Row 2: Dry-Bulb Temperature for standard air is 70 F [21 C] by definition.

Row 3: Flow Rate (acfm) [am³/s] is taken from the data compiled in Figure 9-11 and includes effects of density due to elevation, temperature, moisture and absolute pressure.

Row 4: Minimum Transport Velocity also comes from the Table but originated in Chapters 5 and 13.

Row 5-8: These data are not required because Standard Air

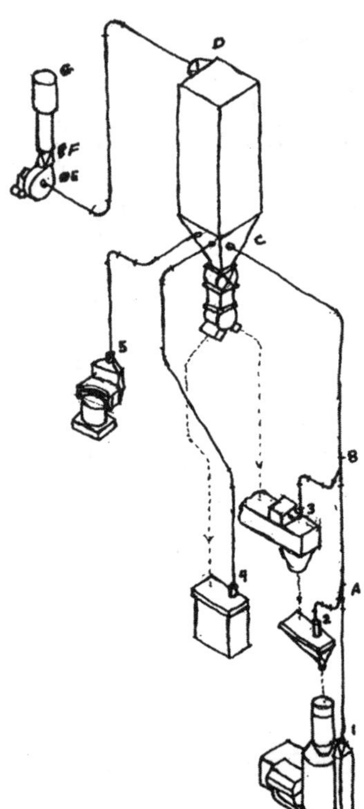

FIGURE 9-9. Single line sketch – [Sample System Design # 2]

Hood	Description	VS-Plate	h_e	acfm [am³/s]	Duct (ft.) [m]	Elbows	Fitting	V_t (fpm) [m/s]
1	Box Filler	*	0.25	1500 [0.70]	11 [3.4]			3500 [18.0]
2	Flat Deck Screen	VS-99-01	0.50	200 [0.10]	9 [2.7]	(1.67) 90	(1) 30	3500 [18.0]
3	Belt Conveyor	VS-50-21	0.40	500 [0.25]	20 [6.1]	(1.67) 90	(1) 30	3500 [18.0]
4	Fines Buggy	VS-15-01	0.25	300 [0.15]	15 [4.9]	(2) 90		3500 [18.0]
5	Fill Station	VS-15-02	0.25	300 [0.15]	21 [6.3]	(2) 90		3500 [18.0]

*Hood 1 had no similar design in the VS Plates so a straight 45 degree takeoff was assume.
Hood 4 actually had (1) 90 and (2) 45 degree elbows for a total of (2) 90 degree elbows.
Other duct lengths and components taken from sketch:

A-B	9 ft [2.7 m] of duct and no elbows
B-C	10 ft [3.0 m] of duct and no elbows
C-E	19 ft [5.8 m] of duct and (2) 90
F-G	10 ft [3.0 m]

FIGURE 9-11. Basic system information – Sample Problem 2

has no moisture or added heat and df = 1.0 (Sample System Design Problems 3 and 4 will consider these rows). Value for scfm is used only when balancing air streams.

Row 9: The target duct area is calculated using the formula Q = VA and solving for "A". Flow is taken from Row 3 and Minimum Velocity is taken from Row 4.

Row 10: "Selected Diameter" is determined by choosing the next smaller standard size after calculating Row 9. For example, in 1-A/9 the calculated duct size is 0.429 ft² [0.039 m²]. From Table 9-1 it can be seen that there is no regular duct size for that area. The designer would choose either an 8" [220 mm] diameter duct (area = 0.349 ft² [0.038 m²]) or a 9" [250mm] diameter duct (area = 0.442 ft² [0.049 m²]); because the larger duct will result in a velocity less than 3500 fpm [18 m/s] specified in Row 4, the smaller duct is chosen for this segment. (Note that IP and SI units will not match because of duct size selection differences.)

Row 11: After the 8" [220 mm] is selected the actual area is inserted from Table 9-1.

Row 12: Velocity is recalculated to match the flow rate in Row 3 and the actual duct area in Row 11.

Row 13: The duct Velocity Pressure (VP_d) is calculated from the velocity in Row 12 and Equation 5 from Calculation Sheet and/or Table 9-3. This VP becomes the base that is multiplied by loss coefficients for the remainder of the calculations. (1.15 "wg [204 Pa])

Row 14: Information required for non-standard air and not used in Example Problem 2.

The remainder of the column is then calculated based on the physical conditions of the system. The important data requested and input into the sheet include:

Rows 15-21: Data required for a slotted hood (see Chapter 6); this may include slot area, slot velocity and slot loss coefficient. (Not required for 1-A)

Rows 22-27: Required for all hoods with or without slots and includes the physical characteristics and shape factors for the hood. This information comes either from Chapter 6 or the VS Plates located in Chapter 13.

Rows 28-36: Data considering all of the physical aspects of the segment (length of duct, number of elbows, fitting losses, and any other special characteristics such as a filter). Note that all of the information in this section except for the length of duct and number of elbows are coefficients (dimensionless). These values are totaled in Row 36 and multiplied by the VP in Row 13. All losses in the segment that are a function of Velocity Pressure are accumulated and then multiplied by the VP in that segment to get the losses in "wg [Pa].

Row 37: This is the accumulated loss for the column and is stated in "wg [Pa].

Row 38: A cell where added losses can be placed (example the ΔP across a filter or spark arrester).

ACGIH® Velocity Pressure Method Calculation Sheet (IP)

Figure 9-12 (IP) Sample System Design #2 — Page 1 of 1 — Date — Elevation (z), ft = 0 ft — * Input Data — Designer

					1-A	2-A	1'-A	A-B	3-B	B-C	4-C	4'-C	5-C	C-E	F-G		
1*		Duct Segment Identification			1-A	2-A	1'-A	A-B	3-B	B-C	4-C	4'-C	5-C	C-E	F-G		1
2*	T	Dry-Bulb Temperature		F	70	70	70	70	70	70	70	70	70	70	70		2
3*	Q_{act}	Actual Duct Flow Rate	Eqn 3	acfm	1500	200	1500	1706	500	2298	300	300	300	2929	2929		3
4*	V_t	Minimum Transport Velocity	Chapter 13	fpm	3500	3500	3500	3500	3500	3500	3500	3500	3500	3000	3000		4
5*	$\dot{m}_{H2O}$	Pounds Water per Minute		lbs - H_2O/min													5
6*	$\dot{m}_{da}$	Pounds Dry Air per Minute		lbs - da/min													6
7	df	Density Factor	Eqn 2		1.00	1.00	1.00	1.00	1.00	1.00	1.00	1.00	1.00	1.00	1.00		7
8	Q_{std}	Standard Duct Flow Rate (Dry Air)	Eqn 3	dscfm	1500	200	1500	1706	500	2298	300	300	300	2929	2929		8
9	A_t	Target Duct Area	3/4	ft²	0.429	0.057	0.429	0.487	0.143	0.657	0.086	0.086	0.086	0.976	0.976		9
10*	D	Selected Duct Diameter	Section 9.5.2.5	in	8.0	3.0	7.0	9.0	5.0	10.0	4.0	3.5	4.0	13.0	13.0		10
11*	A	Selected Duct Area	Table 9.1	ft²	0.349	0.049	0.267	0.442	0.136	0.545	0.087	0.067	0.087	0.922	0.922		11
12	V_d	Duct Velocity	(3/11)	fpm	4298	4082	5618	3860	3676	4217	3448	4478	3448	3177	3177		12
13	VP_d	Duct Velocity Pressure	Eqn 4	"wg	1.15	1.04	1.97	0.93	0.84	1.11	0.74	1.25	0.74	0.63	0.63		13
14	h	Total Heat	Eqn 5	btu/#da													14
15	A_s	Hood Losses – Slot: Slot Area	Chapters 6 & 13	ft²									0.150				15
16*	F_s	Hood Losses – Slot: Slot Loss Coefficient	Figure 9a	VP/slot									1.78				16
17*	F_a	Hood Losses – Slot: Acceleration Coefficient	Chapter 3	0 or 1									0				17
18*	V_s	Hood Losses – Slot: Slot Velocity	Chapters 6 & 13	fpm									2000				18
19	VP_s	Hood Losses – Slot: Slot Velocity Pressure	Eqn 4	"wg									0.25				19
20		Hood Losses – Slot: Slot Loss in VP	16+17	VPtotal									1.78				20
21	h_s	Hood Losses – Slot: Slot Loss	20x19	"wg									0.4				21
22*	F_h	Hood Losses: Hood Entry Loss Coefficient	Fig 9a or Ch 13	F_h or 0	0.25	0.50	0.25		0.40		0.25	0.25	0.25	0.50			22
23*	F_a	Hood Losses: Acceleration Coefficient	Chapters 3 & 6	1 or 0	1	1	1		1		1	1	1	1			23
24		Hood Losses: Hood Entry Loss in VP	22+23	VPtotal	1.25	1.50	1.25		1.40		1.25	1.25	1.25	1.50			24
25	h_h	Hood Losses: Hood Entry Loss	24x13	"wg	1.4	1.6	2.5		1.2		0.9	1.6	0.9	0.9			25
26*		Hood Losses: Other Losses		"wg													26
27	SP_h	Hood Losses: Hood Static Pressure	21+25+26	"wg	1.4	1.6	2.5		1.2		0.9	1.6	1.3	0.9			27
28*	L	Duct Losses: Straight Duct Length		ft	11.0	9.0	11.0	9.0	20.0	10.0	15.0	15.0	38.0	19.0	10.0		28
29	F'_d	Duct Losses: Duct Friction Factor	Eqn8	VP/ft	0.030	0.101	0.035	0.026	0.054	0.023	0.072	0.083	0.072	0.017	0.017		29
30*		Duct Losses: No. of 90 Degree Elbows				1.67			1.67	1.00	2.00	2.00	2.00	2.00			30
31*	F_{el}	Duct Losses: Elbow Loss Coefficient	Table 6	F_{el} or 0		0.19			0.19	0.19	0.19	0.19	0.19	0.19			31
32*	F_{en}	Duct Losses: Branch Entry Loss Coefficient	Table 7	VP/entry		0.18			0.18								32
33*		Duct Losses: Special Fitting Loss Coefficient	i.e. Exp./Cont., etc.	VP/fitting													33
34	F_d	Duct Losses: Duct Loss Coefficient	28X29	VP/duct	0.33	0.91	0.39	0.23	1.08	0.23	1.08	1.25	2.74	0.32	0.17		34
35		Duct Losses: Elbow Loss in VP	30x31	VP		0.32			0.32	0.19	0.38	0.38	0.38	0.38			35
36		Duct Losses: Total Duct Loss in VP	32+33+34+35	VP total	0.33	1.41	0.39	0.23	1.58	0.42	1.46	1.63	3.12	0.70	0.17		36
37		Duct Losses: Total Duct Loss	36x13	"wg	0.4	1.5	0.8	0.2	1.3	0.5	1.1	2.0	2.3	0.4	0.1		37
38*		Duct Losses: Other Losses		"wg										6.0			38
39*	VP_r	Duct Losses: Weighted Average VP	Eqn 10	"wg				1.86		0.99							39
40*		Duct Losses: Loss From Velocity Increase	13-39 (if > 0)	"wg				0.0		0.1							40
41		Duct Losses: Segment Pressure Loss	27+37+38+40	"wg	-1.8	-3.0	-3.3	-0.2	-2.5	-0.6	-2.0	-3.6	-3.6	-7.3			41
42*	SP_{gov}	Governing Static Pressure		"wg	-3.1	-3.3		-3.5	-3.5		-4.1	-4.1	-4.1				42
43*	SP_{cum}	Cumulative Static Pressure		"wg	-1.8	-3.1	-3.3		-2.5	-4.1	-2.0			-11.4	0.1		43
44	Q_{corr}	Corrected Volumetric Flow	Eqn 9	acfm		206			592			320	320				44
45	V_{corr}	Corrected Velocity	44/11	fpm		4204			4353			4776	3678				45
46	VP_{corr}	Corrected Velocity Pressure	Eqn 4	"wg		1.10			1.18			1.42	0.84				46

Designer

1. $\omega = \text{lbs} - H_2O / \text{lbs} - \text{DryAir}$
2. $df = df_e * df_p * df_T * df_m$; $df_e = [1-(6.73*10^{-6})(z)]^{5.258}$; $df_p = (407+SP)/(407)$; $df_T = 530/(T+460)$; $df_m = (1+\omega)/(1+1.607\omega)$
3. $Q_{std} = (Q_{act}/(1+\omega))df$; $Q_{act} = (Q_{std})(1+\omega)/df$
4. $V = 4005\sqrt{VP/df}$; $VP = (V/4005)^2 df$
5. $h = 0.24T + \omega(1061 + 0.444T)$
6. 90 Degree Elbow Loss Coefficient (5 Piece)

R/D	F_{el}
1.5	0.24
2.0	0.19
2.5	0.17

7. Branch Entry Loss Coefficients

Angle	F_{en}
15°	0.09
30°	0.18
45°	0.28
90°	1.00

8. $F_d\left(\frac{\text{metal}}{\text{plastic}}\right) = 0.0307\left(V^{0.533}/Q^{0.612}\right)$; $F_d(\text{flexible}) = 0.0311\left(V^{0.604}/Q^{0.639}\right)$
9. $Q_{corrected} = Q_{design}\sqrt{SP_{gov}/SP_{lower}}$
10. $VP_r = (Q_1/Q_3)VP_1 + (Q_2/Q_3)VP_2$
11. $SP_{System} = SP_{out} - SP_{in} - VP_{in}$

SSP = +.1 - (-11.4) - 0.6 = 10.9"w.g.

FIGURE 9-12 (IP). Velocity Pressure Method Calculation Sheet

ACGIH® Velocity Pressure Method Calculation Sheet (SI)

Figure 9-12 (SI) Sample System Design #2 — Page *1* of *1* — Date — Elevation (z), m = *0 m* — * Input Data

1*		Duct Segment Identification			*1-A*	*2-A*	*1'-A*	*A-B*	*3-B*	*B-C*	*4-C*	*4'-C*	*5-C*	*C-E*	*F-G*	1
2*	T	Dry-Bulb Temperature		C	*21*	*21*	*21*	*21*	*21*	*21*	*21*	*21*	*21*	*21*	*21*	2
3*	Q_{act}	Actual Duct Flow Rate	Eqn 3	am^3/s	*0.70*	*0.10*	*0.70*	*0.34*	*0.25*	*1.10*	*0.15*	*0.15*	*0.15*	*1.46*	*1.46*	3
4*	V_t	Minimum Transport Velocity	Chapter 13	m/s	*18.00*	*18.00*	*18.00*	*18.00*	*18.00*	*18.00*	*18.00*	*18.00*	*18.00*	*15.00*	*15.00*	4
5*	$\dot{m}_{H2O}$	kg Water per Second		$kg\text{-}H_2O/s$												5
6*	$\dot{m}_{da}$	kg Dry Air per Second		kg-da/s												6
7	df	Density Factor	Eqn 2		*1.00*	*1.00*	*1.00*	*1.00*	*1.00*	*1.00*	*1.00*	*1.00*	*1.00*	*1.00*	*1.00*	7
8	Q_{std}	Standard Duct Flow Rate (Dry Air)	Eqn 3	dnm^3/s	*0.70*	*0.10*	*0.70*	*0.34*	*0.25*	*1.10*	*0.15*	*0.15*	*0.15*	*1.46*	*1.46*	8
9	A_t	Target Duct Area	3/4	m^2	*0.039*	*0.006*	*0.039*	*0.047*	*0.014*	*0.061*	*0.008*	*0.008*	*0.008*	*0.097*	*0.097*	9
10*	D	Selected Duct Diameter	Section 9.5.2.5	mm	*220*	*80*	*190*	*240*	*130*	*240*	*100*	*90*	*100*	*340*	*340*	10
11*	A	Selected Duct Area	Table 9.1	m^2	*0.038*	*0.005*	*0.028*	*0.045*	*0.013*	*0.045*	*0.008*	*0.006*	*0.008*	*0.091*	*0.091*	11
12	V_d	Duct Velocity	(3/11)	m/s	*18.42*	*20.00*	*25.00*	*18.67*	*19.23*	*24.44*	*18.75*	*25.00*	*18.75*	*16.04*	*16.04*	12
13	VP_d	Duct Velocity Pressure	Eqn 4	Pa	*204*	*240*	*376*	*209*	*222*	*359*	*211*	*376*	*211*	*155*	*155*	13
14	h	Total Heat	Eqn 5 or Psych Chart	kJ/kg-da												14
15	A_s	HOOD LOSSES – SLOT: Slot Area	Chapters 6 & 13	m^2									*0.020*			15
16*	F_s	Slot Loss Coefficient	Figure 9a										*1.78*			16
17*	F_a	Acceleration Coefficient	Chapter 3	0 or 1									*0*			17
18*	V_s	Slot Velocity	Chapters 6 & 13	m/s									*10.00*			18
19	VP_s	Slot Velocity Pressure	Eqn 4	Pa									*60*			19
20		Slot Loss in VP	16+17	VPtotal									*1.78*			20
21	h_s	Slot Loss	20x19	Pa									*107*			21
22*	F_h	HOOD LOSSES: Hood Entry Loss Coefficient	Fig 9a & Ch 13	F_h	*0.25*	*0.50*	*0.25*		*0.40*		*0.25*	*0.25*	*0.25*	*0.50*		22
23*	F_a	Acceleration Coefficient	Chapters 3 & 6	1 or 0	*1*	*1*	*1*		*1*		*1*	*1*	*1*	*1*		23
24		Hood Entry Loss in VP	22+23	VPtotal	*1.25*	*1.50*	*1.25*		*1.40*		*1.25*	*1.25*	*1.25*	*1.50*		24
25	h_h	Hood Entry Loss	24x13	Pa	*255*	*360*	*470*		*311*		*264*	*470*	*264*	*232*		25
26*		Other Losses		Pa												26
27	SP_h	Hood Static Pressure	21+25+26	Pa	*255*	*360*	*470*		*311*		*264*	*470*	*371*	*232*		27
28*	L	DUCT LOSSES: Straight Duct Length		m	*3.4*	*2.7*	*3.4*	*2.7*	*6.1*	*3.0*	*4.9*	*4.9*	*11.6*	*5.8*	*3.0*	28
29	F'_d	Duct Friction Factor	Eqn 8	VP/m	*0.091*	*0.312*	*0.107*	*0.083*	*0.179*	*0.080*	*0.237*	*0.275*	*0.236*	*0.054*	*0.054*	29
30*		No. of 90 Degree Elbows				*1.67*			*1.67*	*1.00*	*2.00*	*2.00*	*2.00*	*2.00*		30*
31*	F_{el}	Elbow Loss Coefficient	Table 6	F_{el}		*0.19*			*0.19*	*0.19*	*0.19*	*0.19*	*0.19*	*0.19*		31
32*	F_{en}	Branch Entry Loss Coefficient	Table 7	VP/entry		*0.18*			*0.18*							32
33*		Special Fitting Loss Coefficient	I.e Exp./Cont.,etc.	VP/fitting												33
34	F_d	Duct Loss Coefficient	28X29	VP/duct	*0.31*	*0.84*	*0.36*	*0.22*	*1.09*	*0.24*	*1.16*	*1.35*	*2.74*	*0.31*	*0.16*	34
35		Elbow Loss in VP	30x31	VP		*0.32*			*0.32*	*0.19*	*0.38*	*0.38*	*0.38*	*0.38*		35
36		Total Duct Loss in VP	32+33+34+35	VP total	*0.31*	*1.34*	*0.36*	*0.22*	*1.59*	*0.43*	*1.54*	*1.73*	*3.12*	*0.69*	*0.16*	36
37		Total Duct Loss	36x13	Pa	*63*	*322*	*135*	*46*	*353*	*154*	*325*	*650*	*659*	*107*	*25*	37
38*		Other Losses		Pa										*1500*		38
39*	VP_r	Weighted Average VP	Eqn 10	Pa				*399*		*216*						39
40*		Loss From Velocity Increase	13-39 (if>0)	Pa				*0.0*		*143*						40
41		Segment Pressure Loss	27+37+38+40	Pa	*-318*	*-682*	*-605*	*-46*	*-664*	*-296*	*-589*	*-1120*	*-1030*	*-1839*	*25*	41
42*	SP_{gov}	Governing Static Pressure		Pa	*-682*		*-682*		*-728*	*-1120*	*-1024*		*-1120*			42
43*	SP_{cum}	Cumulative Static Pressure		Pa	*-318*			*-728*	*-664*	*-1024*		*-1120*	*-1030*	*-2959*	*25*	43
44	Q_{corr}	Corrected Volumetric Flow	Eqn 9	am^3/s			*0.74*		*0.26*	*1.15*			*0.16*			44
45	V_{corr}	Corrected Velocity	44/11	m/s			*26.43*		*20.00*	*25.56*			*20.00*			45
46	VP_{corr}	Corrected Velocity Pressure	Eqn 4	Pa			*419*		*240*	*392*			*240*			46

Designer

1. $\omega = kg - H_2O/kg - DryAir$

2. $df = df_e * df_p * df_T * df_m$

   $df_e = [1-(22.08*10^{-6})(z)]^{5.258}$

   $df_p = (101384 + SP)/(101384)$

   $df_T = 294/(T+273)$

   $df_m = (1+\omega)/(1+1.607\omega)$

3. $Q_{std} = (Q_{act}/(1+\omega))df$

   $Q_{act} = (Q_{std})(1+\omega)/df$

4. $V = 1.29\sqrt{VP/df}$

   $VP = (V/1.29)^2 df$

5. $h = 1.006T + \omega(2501 + 1.84T)$

6. 90 Degree Elbow Loss Coefficient (5 Piece)

R/D	F_{el}
1.5	0.24
2.0	0.19
2.5	0.17

7. Branch Entry Loss Coefficients

Angle	F_{en}
15°	0.09
30°	0.18
45°	0.28
90°	1.00

8. $F_d\left(\frac{metal}{plastic}\right) = 0.0155\left(V^{0.533}/Q^{0.612}\right)$

   $F_d(flexible) = 0.0186\left(V^{0.604}/Q^{0.639}\right)$

9. $Q_{corrected} = Q_{design}\sqrt{SP_{gov}/SP_{lower}}$

10. $VP_r = (Q_1/Q_3)VP_1 + (Q_2/Q_3)VP_2$

11. $SP_{System} = SP_{out} - SP_{in} - VP_{in}$

SSP = +25 - (-2959) - 155 = 2829 Pa

FIGURE 9-12 (SI). Velocity Pressure Method Calculation Sheet

Row 39: The cell where the Weighted Average Velocity Pressure (VP_r) is calculated at a junction (see Section 9.8).

Row 40: If the velocity increases in a junction so that the downstream VP is higher than the value in Row 39, then the difference must be added in this cell; in effect the VP_r must be less than the VP of the upstream junction. It is good practice to calculate VP_r at every junction (see Section 9.8).

Row 41: This cell is the accumulation of all losses in that segment. In the case of Column 1-A in the example, it states that if -1.8 "wg [-318 Pa] of pressure is applied at junction A, then the duct and hood system from Hood 1 will exhaust 1,500 acfm [0.70 am³/s]. If more negative pressure is applied, then more air will flow, etc. The key to the proper design is to get the proper pressure at that point.

Additional notes:

Cells 1-A/3 and 1-A/8: Since the density factor is 1.0 (standard air) in this example, the acfm = scfm. The values in Rows 3 and 8 for all branches are equal. When designing systems with standard air only, the values in Row 8 can be left out of the calculation sheet and calculations are done with acfm.

Cells 1-A/41 and 2-A/41: *(NOTE: Do not consider the value of 2-A/42 at this time. See below.)* This is the classic example of a system balance issue (see Section 9.7). The calculation sheet states that -1.8 "wg [318 Pa] of pressure will deliver 1,500 scfm [0.70 am³/s] from Hood 1, but -3.1 "wg [682 Pa] of pressure is needed at the same junction to pull the 200 scfm [0.10 am³/s] from Hood 2. There can only be one value at SP at the junction and if the lower value (-1.8 "wg [318 Pa]) is selected then there will not be enough energy to pull the 200 scfm from Hood 2. At the same time, if the higher (governing) SP is selected we will pull more air than designed for Hood 1. First the ratio of the SP values for branch A are calculated using (value is always greater than 1.0 so higher value is in the numerator):

$$\frac{SP_{governing}}{SP_{lower}} = \frac{-3.1}{-1.8} = 1.72$$

$$[\frac{SP_{governing}}{SP_{lower}} = \frac{-682}{-318} = 2.14]$$

From Section 9.7.1, the branch with the lower SP should be redesigned since the ratio is higher than 1.2. This is accomplished in the third column of the calculation sheet and designated 1'-A. In this case, a smaller duct (decreased size from 8" to 7" diameter) [In SI units the size is reduced from 220 mm to 190 mm to increase velocity from hood 1] is selected and this increases the velocity in the duct segment from Hood 1. This increase in velocity increases the friction in the segment and when the new column is completed the required SP is now -3.3 "wg in the IP solution. Now 1'-A is the governing branch because its SP exceeds the -3.1" in 2-A. The ratio is again tested:

$$\frac{SP_{governing}}{SP_{lower}} = \frac{-3.2}{-3.1} = 1.03$$

$$[\frac{-696\ Pa}{-605\ Pa} = 1.15]$$

This falls below the 1.2 value required for segment redesign but we are also pulling more static pressure than Hood 2 [Hood 1] requires. Using Equation 9.8 in Section 9.7.1, we recalculate the actual flow from Hood 2 [Hood 1]:

$$Q_{corrected} = Q_{original}\sqrt{\frac{SP_{governing}}{SP_{original}}} = 200\sqrt{\frac{-3.3}{-3.1}} = 206 \text{ acfm}$$

(Note that this calculation is performed more than once in the SI Example to arrive at duct size and 2-A is governing branch.)

This value is entered in Row 44. Next, the velocity in the duct and VP are recalculated and placed in Rows 45 and 46, respectively.

Cell A-B/3: The new airflow of 1706 acfm [0.84 m³/s] required in this segment is the sum of the flow from 2-A (206 acfm) [0.10 m³/s] plus the value in 1'-A (1,500 acfm) [0.74 m³/s].

NOTE: This is one of the potential disadvantages listed in Chapter 5, Table 5-2 for 'Balance by Design' Method. In place of the design 1,700 acfm [0.80 m³/s] of (1500 + 200) [0.70 + 0.1] originally intended, this method now results in a recalculated design flow of 1706 acfm [0.84 m³/s] – a small increase. In the IP example, the pressure required for the junction is now -3.3 "wg [696 Pa] – an increase of about 3% over [6%] the original calculated need of -3.1 "wg at the junction. Even though small, the combined effects of increased volume and pressure will result in an increase of power consumption at the fan. If blast gates were used, the original -3.1 "wg [-605 Pa] would have remained at junction A and a blast gate added to branch 1-A to provide the deficit of pressure to balance (3.1" – 1.8") or 1.3 "wg. This would have resulted in a savings of airflow, pressure and horsepower.

Cells A-B/39 and B-C/39 (IP only): These show two possible situations when considering Weighted Average Velocity Pressure. The value for the Weighted Average Velocity Pressure of the branches entering Junction B (see Section 9.8 for explanation) is computed using the values in A-B/3 (Q_1), 3-B/44 (Q_2), B-C/3 (Q_3), A-B/13 (VP_1), and 3-B/46 (VP_2). From Equation 10 on the Calculation Sheet:

$$VP_r = \left(\frac{Q_1}{Q_3}\right)(VP_1) + \left(\frac{Q_2}{Q_3}\right)(VP_2) = \left(\frac{1706}{2298}\right)(0.93)$$

$$+\left(\frac{592}{2298}\right)(1.18) = 0.99 \text{ "wg}$$

This value is inserted in B-C/39 and compared to the VP in the next duct segment (B-C). Since VP_r is less than the VP calculated in the B-C segment, the difference between the two values (1.11 – 0.99 = 0.12 "wg) is added to the losses for the B-C segment and shown separately in B-C/40. Note: This calculation was not necessary in SI solution because velocity decreased through junction "C".

However, the Weighted Average Velocity Pressure at Junction A shows when the effects can be ignored. In that case, the VP_r for the branches 2-A and 1'-A is calculated with the same equation to be 1.86 "wg. The VP in the next segment after the combination (A-B) is 0.93 "wg (shown in A-B/13). Since this value is lower than the VP_r, air has basically slowed as it goes through the fitting so there is no added resistance/loss.

NOTE: This value is added only if there is an increase of velocity (and VP) while proceeding through the junction. If the VP in segment B-C had been less than VP_r (1.0 "wg) there would have been no insertion of a value in Row 40.

Cells 3-B/41 and 3-B/42 (IP only): Note that the governing Static Pressure at Junction "B" is -3.5 "wg. However, the SP requirement for 3-B is only -2.5 "wg. If a test is performed at that junction the ratio would be:

$$\frac{SP_{GOV}}{SP_{WON}} = \frac{-3.5}{-2.5} = 1.4$$

This would normally require a duct size change in branch 3-B, perhaps a reduction to 4.5". However, the designer ran the calculation and determined that this new pressure would now be governing and force even more volume from branch A-B. These types of decisions can be encountered during the system design. Rather than the smaller duct, the 1.4 ratio was applied to Equation 9.8 and a new volume (592 acfm) was calculated for branch 3-B.

Cell C-E/22: The baghouse in this case was specified with a maximum pressure drop across the filter media of 6.0 "wg [1500 Pa]. This is shown as "Other Losses" in Cell C-E/38. Since the baghouse loss is not 'flange to flange' there are other losses as the air is turbulent through the baghouse and adds resistance to the system. If the manufacturer does not provide the information for these losses, then a normal assumption is to use the losses for a tapered hood (Figure 9-a). That would allow for approximately 0.5VP of loss (added in Cell C-E/22) and another increase in velocity from the extremely low speed (usually 3.0 to 10.0 fpm [0.02 to 0.05 m/s]) through the filter bag and re-accelerated to 3177 fpm [16.04 m/s] in segment C-E. This increase in velocity requires the same consideration as the energy exchange in a hood, i.e., 1.0 VP. (See discussion of this coefficient in Chapters 3 and 6.) This is added in Cell C-E/23 and the baghouse losses are treated similarly to the total losses in a hood.

Cell F-G/33: Note that there is no added loss for the 'no-loss' stack as shown in the sketch (Figure 9-9). If a rain cap had been used (Figure 9-f) then more resistance would have been added in this section and more horsepower would be required.

Cells C-E/13, C-E/43, and F-G/43: These are the values used to determine the SSP (Equation 12 on the calculation sheet; see Section 9.3.3). When the value for SSP is determined, the designer can then select an FSP for specification of the fan. In this case, the value for SSP is (+0.1" – (-11.4") – 0.6") or 10.9 "wg [2829 Pa]. This could be rounded up to 11.0 "wg [2900 Pa] or even higher. Under normal conditions, a fan would not be purchased at this requirement. The selected FSP may include some factor of safety and rounding of values. In this case, it could be selected at 12.0" or some other value. After selection and review of the Fan and System Curves, the fan selection may be changed if it does not appear to be selected for a stable operating condition.

Cell C-E/3: This is the value for airflow used to select the fan. Since the air is at standard conditions, the fan would be specified:

2929 acfm @ 12" FSP @ standard conditions.

Note in the IP example that the FSP includes the small factor of safety increase from the SSP calculated at 10.9 "wg. [In the SI example, the fan would be specified at 1.41 am^3/s @ 2500 Pa @ STP]. The selection of 2500 Pa for the fan in the SI example includes a factor of safety above the 2444 Pa calculated. This value for FSP must reflect the maximum pressure drop to be encountered by the filter bags. When the system is first started there may not be 6" [1500 Pa] of resistance (ΔP) across the bags. In that case, the fan will operate at higher airflow than the design and can cause premature plugging of the filter media. A volume control damper or variable frequency drive should be considered to keep the system from operating at a volume in excess of design.

9.12 CALCULATION METHODS AND NON-STANDARD AIR DENSITY

The example shown in Sample System Design # 2 (Section 9.11) considers "standard" air density — something that rarely occurs in real system design. It simplifies the calculations by assuming that air is constantly at standard conditions (0.075 pounds/cubic feet and no moisture) [1.204 kg/m^3]. Even though the effects of moisture, elevation, pressure and temperature can be small when considered independently, they can have significant, additive effects when considered together.

Fan tables assume *standard air density* that corresponds to sea level elevation, no moisture, and 70 F [21 C]. Changes in air density can come from several factors, including elevation, temperature, internal duct pressure, changes in apparent molecular weight (moisture content, gas stream constituents, etc.), and amount of suspended particulate. The change in air density should be considered when calculating flow and pressure requirements.

Density Factors for different temperatures and elevations are listed in Table 9-6 (IP and SI). Internal duct pressures will also change air density and can have a significant effect, especially at the fan inlet. If there is excessive moisture in the air stream, the density will decrease. Suspended particulate is assumed to be only a trace impurity in industrial exhaust systems. If there are significant quantities of particulate in the duct system (> 20 grains/dscf [0.05 kg/m³]), this addition to the air stream density should be addressed. This field is called material conveying and is beyond the scope of this Manual. Note that at 20 grains/dscf [0.05 kg/m³] the particulate represents less than 0.4% of the air mass rate – significant amounts of air to move a small amount of particulate. In cases where there is more than 20 grains/dscf [0.05 kg/m³] of material in the air stream, a factor can be applied to the losses in this Manual for 'clean air'. This factor calculates as:

$$\text{Friction Loss of Mixture} =$$

$$\text{Friction Loss}_{\text{clean air}}\left(0.36\left(\frac{\text{wt.}-\text{conveyed solids}}{\text{wt.}-\text{conveying air}}\right)+1.0\right)$$

The density variation equations of Chapter 3 demonstrate that, for a constant mass flow rate, an increase in temperature or a reduction in absolute pressure will increase the actual flow. It is helpful to remember that a fan connected to a given system will exhaust the same volumetric flow rate regardless of air density. The mass of air moved, however, will be a function of the density.

9.12.1 Effects of Temperature and/or Altitude. Consider a local exhaust system at sea level where 5,000 scfm [25 nm³/s] of air at 70 F [21 C] is drawn into a hood. The air is then heated to 600 F [316 C] and the density of the air leaving the heater becomes 0.0375 lbm/ft³ [0.602 kg/m³]. The flow rate downstream of the heater would be 10,000 actual cubic feet per minute (acfm) [50 am³/s] at the new density of 0.0375 lbm/ft³ [0.602 kg/m³]. This is true because the 50% decrease in density must correspond to a twofold increase in the actual airflow since the mass flow rate has remained constant.

If this temperature effect is ignored and a fan selected for 5,000 acfm [25 am³/s] is placed in the system, the hood flow rate will be well below that required to maintain contaminant control. The exact operating point of such a system would have to be recalculated based upon the operating point of the incorrectly sized fan.

9.12.2 Effects of Elevated Moisture. When air temperature is below 100 F [38 C] correction for humidity is minimal and may be ignored for most industrial ventilation systems (if there are no other corrections for density changes). When air temperature exceeds 100 F [38 C] and moisture content is greater than 0.02 lbs H_2O per pound of dry air [0.02 kg–H_2O/kg–dry air] (Dew Point of 80 F [27 C]), correction is required to determine fan operating RPM and power. Correction factors may be read from the Psychrometric charts such as those illustrated in Figures 9-g through 9-j or from df_m in Equation Box 2 on the Calculation Sheet.

9.13 PSYCHROMETRIC PRINCIPLES

The properties of moist air are presented on the Psychrometric Chart at a single pressure. These parameters define the physical properties of an air/water vapor mixture. The actual gas flow rate and the density of the gas stream at the inlet of the fan must be known in order to select the fan. The Psychrometric Chart provides the information required to calculate changes in the flow rate and density of the gas as it passes through the various local exhaust system components. These properties are:

- Dry-Bulb Temperature (T or T_{db}) is the temperature observed with an ordinary thermometer. Expressed in degrees Fahrenheit (F) [C], it may be read directly on the chart and is indicated on the bottom horizontal scale.
- Wet-Bulb Temperature (T_{wb}) is the temperature at which liquid or solid water, by evaporating into air, can bring the air to saturation adiabatically at the same temperature. Also expressed in degrees Fahrenheit (F)[C], it is read directly at the intersection of the constant enthalpy line with the 100% saturation curve.
- Dew Point Temperature is that temperature at which the air in an air/vapor mixture becomes saturated with water vapor and any further reduction of dry-bulb temperature causes the water vapor to condense or deposit as drops of water. Expressed in degrees Fahrenheit, it is read directly at the intersection of the saturation curve with a horizontal line representing constant moisture content.
- Percent Saturation curves reflect the mass of moisture actually in the air as a percentage of the total amount possible at the various dry-bulb and moisture content combinations. Expressed in percent, it may be read directly from the curved lines on the chart.
- Density Factor (df) is a dimensionless quantity which expresses the ratio of the actual density of the mixture to the density of standard air (0.075 lbm/ft³) [1.204 kg/m³]. The lines representing density factor typically do not appear on low-temperature Psychrometric Charts when relative humidity or percent saturation curves are presented. A method of calculating the density of the gas defined by a point on the chart (when density factor curves are not presented) is discussed later in this section.

- Moisture Content, or weight of water vapor, is the amount of water which has been evaporated into the air. In ordinary air, it is very low-pressure steam and has been evaporated into the air at a temperature corresponding to the boiling point of water at that low pressure. Moisture content is expressed in grains of water vapor per pound of dry air (7,000 grains = one pound) or pounds of water vapor per pound of dry air [kg–H_2O/kg dry air] and is read directly from a vertical axis.
- Enthalpy (Total Heat) (h) as shown on the Psychrometric chart is the sum of the heat required to raise the temperature of a pound of air from 0 F to the dry-bulb temperature, plus the heat required to raise the temperature of the water contained in that pound of air from 32 F to the dew point temperature, plus the latent heat of vaporization, plus the heat required to superheat the vapor in a pound of air from the dew point temperature to the dry-bulb temperature. In the SI system, a Joule is defined as the amount of heat it takes to raise 4.186 grams of H_2O, 1°C (also equal to one calorie). Expressed in BTUs per pound of dry air [kJ/kg-DA], it is shown by following the diagonal wet-bulb temperature lines.
- Humid Volume (HV) is the volume occupied by the air/vapor mixture per pound [kg] of dry air and is expressed in cubic feet of mixture per pound of dry air [m³/kg – dry air]. It is most important to understand the dimensions of this parameter and realize that the reciprocal of humid volume is not density (first paragraph of this section). Humid volume is the parameter used most frequently in determining flow rate changes within a system as a result of mixing gases of different properties or when evaporative cooling occurs within the system.

Knowing dry-bulb temperature (T) and moisture content (ω), the value of enthalpy (h) can be calculated from the following equation:

$$h = 0.27 \times T + \omega \times (1061 + 0.444\,T) \qquad \textbf{[9.17] IP}$$

$$[h = 1.006\,T + \omega\,(2501 + 1.84\,T)] \qquad \textbf{[9.17] SI}$$

Use of this equation can eliminate errors sometimes occurring from difficulty in accurately reading a psychrometric chart.

When the quality of an air/vapor mixture is determined by a point on a Psychrometric chart having a family of density factor curves, all that must be done to determine the actual density of the gas at the pressure reference for which the chart is drawn is to multiply the density factor taken from the chart by the density of standard air (0.075 lbm/ft³) [1.204 kg/m³]. If relative humidity curves are presented on the chart in lieu of density factor curves, information available through dimensional analysis must be used to determine the actual density of the mixture. This can be done quite easily as follows: The summation of one pound of dry air plus the mass of the moisture contained within that pound of dry air divided by the humid volume will result in the actual density of the mixture.

$$\rho = \frac{1+\omega}{HV} \qquad \textbf{[9.18]}$$

where: ρ ≡ density of the mix (lbm/ft³) [kg/m³]
ω ≡ moisture content (lbm H_2O/lbm-dry air) [kg H_2O–kg dry air]
HV ≡ humid volume (ft³ mix/lbm-dry air) [m³/kg]

EXAMPLE PROBLEM 9 (Humid Volume)

The density of an air-water mixture is 0.061 pounds per ft³ [0.97 kg/m³]. The moisture content is 0.04 pounds-H_2O per pound of dry air [0.04 kg H_2O/kg DA]. Determine the Humid Volume.

$$HV = \frac{1+\omega}{\rho} = \frac{1+0.04}{0.061} = 17.05\ \text{ft}^3/\text{lbm} - \text{dry air}$$

$$[HV = \frac{1+0.04}{0.97} = 1.05\ \text{m}^3/\text{kg} - \text{dry air}]$$

EXAMPLE PROBLEM 10 (Moisture Level by Weight)

An air-water mixture is 15% moisture water by volume. Determine moisture level by weight (#-H_2O per #-dry air [kg – H_2O per kg – dry air]).

From Ideal Gas Laws (see Chapter 3, Equation 3.19):

$PV = nRT$

For air in mixture:

$PV_{air} = n_{air}RT$

For water in mixture:

$PV_{water} = n_{water}RT$

Partial volumes add to Total Volume and Temperature and Pressure are the same for both air and water so:

For mixture:

$PV_{mix} = (n_{air} + n_{water})RT$

and

$$\frac{V_{water}}{V_{air}} = \frac{n_{water}}{n_{air}} = \frac{0.15}{0.85}$$

By definition of a "mole":

MW = molecular weight

$m_{air} = n_{air}\,MW_{air}$ and $m_{water} = n_{water}\,MW_{water}$

MW_{air} = 28.8 lbs and MW_{water} = 18.0 lbs
and so:

$$\frac{m_{water}}{m_{air}} = \frac{n_{water}(MW_{water})}{n_{air}(MW_{air})} = \frac{(0.15)(18)}{(0.85)(28.8)} = 0.11\frac{\#\text{–water}}{\#\text{–air}}$$

The answer is independent of temperature or pressure of the mixture. The answer is identical in SI units (kg–H_2O/kg dry air).

9.14 MIXING GASES OF DIFFERENT CONDITIONS CONSIDERING TEMPERATURE AND MOISTURE

In cases where two air streams mix there can also be cases where moisture is added to an air stream. Section 9.4.3 considered mixing of air streams where little moisture was present. Industrial ventilation systems often combine a hot moist stream with a cooler dry mass. In some cases, the mixture can encourage condensation of the moisture from the hot stream and can be a problem for the design (condensed moisture mixing with dry dust can plug filters and coat the duct components). It is important to be able to predict moisture and heat conditions for these types of mixtures.

EXAMPLE PROBLEM 11 (Mixing of Air Streams at Different Conditions) (IP Units Only)

A hot gas stream of 19,000 acfm with a dry-bulb temperature of 400 F and containing moisture of 0.20 pounds of water per pound of dry air (ω) is mixed with 11,000 acfm of outside air entering the system in the winter at a temperature of -20 F. The outside air has virtually no moisture at those conditions. The plant is located at an elevation of 3,000' ASL. Determine the final conditions of the mix.

As mentioned in previous Example Problems in this chapter, there is conservation of both mass and energy. In the case of a moist air stream, the conservation of mass occurs with both air and water and must be considered individually.

The conditions of the hot, moist stream using Equation 9.3:

$$Q_{std} = \frac{Q_{act}(df)}{(1+\omega)} = \frac{19{,}000\text{ acfm (df)}}{1+0.2}$$

The density factor in this example is affected by (see Chapter 3, Section 3.12):

$$\text{Temperature: } df_T = \frac{T_{std}}{T_{act}} = \frac{460+70}{460+400} = 0.62$$

$$\text{Moisture: } df_m = \frac{1+\omega}{1+(1.607\omega)} = \frac{1+0.2}{1+(1.607)(0.2)} = 0.91$$

$$\text{Elevation: } df_e = [1-(6.73\times10^{-6})(z)]^{5.258}$$

$$= [1-(6.73\times10^{-6})(3000)]^{5.528} = 0.90$$

The density factor for the hot stream is the product of the three factors:

df = (0.62)(0.91)(0.90) = 0.51

NOTE: Sample System Design Problems 3 and 4 later in this chapter use the Psychrometric chart to determine many of the qualities of the gas stream. This chart is printed for conditions at sea level and must be altered for locations more than 1000' above sea level (ASL) by considering the density factor for elevation in addition to the df shown on the chart.

$$\text{then } Q_{std} = \frac{Q_{act}(df)}{(1+\omega)} = \frac{19{,}000\text{ acfm }(0.51)}{1+0.2} = 8{,}075\text{ scfm}$$

$$\dot{m}_{dry-air} = 0.075\frac{\text{lbm}}{\text{ft}^3}\left(8{,}075\frac{\text{ft}^3}{\text{min}}\right) = 605.6\frac{\text{lbm}}{\text{min}}$$

Water content (ω) equals 0.20 pounds of water per pound of dry air so water from hot gas stream is:

$$\dot{m}_{water} = \left(0.2\frac{\#\text{–water}}{\#\text{–dry air}}\right)\left(605.6\frac{\#\text{–dry air}}{\text{min}}\right)$$

$$= 121.1\frac{\text{lbm – water}}{\text{min}}$$

The conditions for the cold air stream are also considered using Equation 9.3:

$$Q_{std} = \frac{Q_{act}(df)}{(1+\omega)} = \frac{11{,}000\text{ acfm(df)}}{1+0.0}$$

The density factor in this example is affected by (see Chapter 3, Section 3.11):

$$\text{Temperature: } df_T = \frac{T_{std}}{T_{act}} = \frac{460+70}{460-20} = 1.20$$

$$\text{Elevation: } df_e = [1-(6.73\times10^{-6})(z)]^{5.258}$$

$$= [1-(6.73\times10^{-6})(3000)]^{5.528} = 0.90$$

The density factor for the cold stream is the product of the two factors:

df = (1.20) (0.90) = 1.08

$$Q_{std} = \frac{Q_{act}(df)}{(1+\omega)} = \frac{11{,}000\text{ acfm(1.08)}}{1+0.0} = 11{,}880\text{ scfm}$$

$$\dot{m}_{dry-air} = 0.075\frac{lbm}{ft^3}\left(11{,}880\frac{ft^3}{min}\right) = 891.0\frac{lbm}{min}$$

The mixture conditions (Equation 9.5) would be stated as:

Dry Air: $\dot{m}_a + \dot{m}_b = \dot{m}_c$ = (605.6) + (891.0)
= 1,496.6 lbm-dry air

Water: $\dot{m}_a + \dot{m}_b = \dot{m}_c$ = (121.1) + (0.0) = 121.1 lbm-water

So the new value for $\omega = \dfrac{121.1}{1{,}496.6} = 0.08\dfrac{lbm-water}{lbm-air}$

And standard air = $Q_{std} = \dfrac{1{,}496.6}{0.075} = 19{,}955$ scfm

Note that the dry air is only considered when calculating the scfm. Dry air is the base value from which other values are calculated.

The Conservation of Energy as stated in Equation 9.6 also uses only the dry air for the values of $\dot{m}$ so:

$$\dot{m}_a(h_a) + \dot{m}_b(h_b) = \dot{m}_c(h_c) = (605.6)(h_{hot})$$
$$+ (891)(h_{cold}) = (1{,}496.6)(h_{mix})$$

Now the Psychrometric Chart is required to determine the enthalpy. The charts in this Manual do not include values below 30 F. ASHRAE does print these values and enthalpy for -20 F is approximately -5.0 BTU per pound-dry air. The enthalpy for the 400 F hot air stream (from the Chart on Figure 9-i) is 340 BTU per lbm-dry air. This is determined by reading the value at the intersection of 400 F dry-bulb and moisture content of 0.2 pounds of water per pound of dry air.

Enthalpy of the mix can be calculated:

$$h_{mix} = \frac{(605.6)(340) + (891)(-5)}{1{,}496.6} = 134.6\frac{BTU}{lbm-air}$$

Knowing the enthalpy and ω for this mixture it is possible to find the remaining conditions. At the intersection of ω = 0.08 and h = 134.6 on Figure 9-i, the conditions of the mixture are:

Dry-Bulb Temperature ≈ 180 F

Dew Point ≈ 120 F

Wet-Bulb Temperature ≈ 124 F

df = 0.79

However, the density factor does not include the effects of elevation calculated previously (0.90) so corrected density factor equals:

df = (0.79)(0.90) = 0.71

Calculation of the actual flow of the mixture (from Equation 9.2):

$$Q_{act} = \frac{Q_{std}(1+\omega)}{df} = \frac{(19{,}955)(1.08)}{0.71} = 30{,}354 \text{ acfm}$$

9.15 SAMPLE SYSTEM DESIGN #3 (SINGLE-BRANCH SYSTEM/NON-STANDARD AIR CONDITIONS) (IP Units Only)

The example shown in Figure 9-13 illustrates the effect of elevation, moisture and temperature and a method of calculation for these systems. A calculation sheet showing the calculation is provided in Figure 9-14.

Given: The exit flow rate from a 60" × 24" dryer is 16,000 scfm plus removed moisture. The plant is located at 575 feet ASL. Exhaust air temperature is 500 F. The dryer delivers 60 tons/hr of dried material with capacity to remove 5% moisture. Required suction at the dryer hood is -2.0 "wg; minimum con-

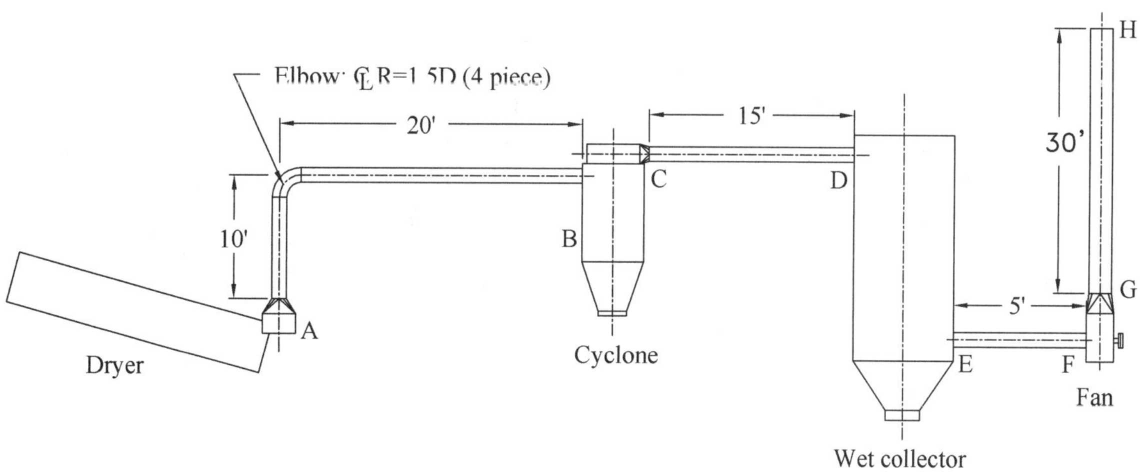

FIGURE 9-13. System layout

ACGIH® Velocity Pressure Method Calculation Sheet (IP)

Figure 9-14 (IP) Sample System Design #3 — Page *1* of *1* — Date — Elevation (z), ft = *575 ft* — * Input Data — Designer

No.	Symbol	Group	Item	Reference	Units	A-B	B-C	C-D	D-E	E-F	G-H	Line
1*			Duct Segment Identification			*A-B*	*B-C*	*C-D*	*D-E*	*E-F*	*G-H*	1
2*	T		Dry-Bulb Temperature		F	*500*	*500*	*500*	*180*	*180*	*180*	2
3*	Q_{act}		Actual Duct Flow Rate	Eqn 3	acfm	*33508*	*33508*	*33508*	*24600*	*26162*	*24600*	3
4*	V_t		Minimum Transport Velocity	Chapter 13	fpm	*4000*		*4000*		*4000*	*3500*	4
5*	$\dot{m}_{H_2O}$		Pounds Water per Minute		lbs - H_2O/min	*106.7*	*106.7*	*106.7*	*192.0*	*192.0*	*192.0*	5
6*	$\dot{m}_{da}$		Pounds Dry Air per Minute		lbs - da/min	*1200*	*1200*	*1200*	*1200*	*1200*	*1200*	6
7	df		Density Factor	Eqn 2		*0.52*	*0.52*	*0.52*	*0.74*	*0.70*		7
8	Q_{std}		Standard Duct Flow Rate (Dry Air)	Eqn 3	dscfm	*16000*	*16000*	*16000*	*16000*	*16000*	*16000*	8
9	A_t		Target Duct Area	3/4	ft^2	*8.377*		*8.377*			*7.028*	9
10*	D		Selected Duct Diameter	Section 9.5.2.5	in	*38.0*		*38.0*		*34.0*	*36.0*	10
11*	A		Selected Duct Area	Table 9.1	ft^2	*7.878*		*7.878*		*6.307*	*7.071*	11
12	V_d		Duct Velocity	(3/11)	fpm	*4253*		*4253*		*4148*	*3479*	12
13	VP_d		Duct Velocity Pressure	Eqn 4	"wg	*0.59*		*0.59*		*0.75*	*0.56*	13
14	h		Total Heat	Eqn 5 or Psych Chart	BTU/lbs-da	*234*	*234*	*234*	*234*	*234*	*234*	14
15	A_s	HOOD LOSSES – SLOT	Slot Area	Chapters 6 & 13	ft^2							15
16*	F_s	HOOD LOSSES – SLOT	Slot Loss Coefficient	Figure 9a	VP/slot							16
17*	F_a	HOOD LOSSES – SLOT	Acceleration Coefficient	Chapter 3	0 or 1							17
18*	V_s	HOOD LOSSES – SLOT	Slot Velocity	Chapters 6 & 13	fpm							18
19	VP_s	HOOD LOSSES – SLOT	Slot Velocity Pressure	Eqn 4	"wg							19
20		HOOD LOSSES – SLOT	Slot Loss in VP	16+17	VPtotal							20
21	h_s	HOOD LOSSES – SLOT	Slot Loss	20x19	"wg							21
22*	F_h	HOOD LOSSES	Hood Entry Loss Coefficient	Fig 9a or Ch 13	F_h or 0							22
23*	F_a	HOOD LOSSES	Acceleration Coefficient	Chapters 3 & 6	1 or 0							23
24		HOOD LOSSES	Hood Entry Loss in VP	22+23	VPtotal							24
25	h_h	HOOD LOSSES	Hood Entry Loss	24x13	"wg							25
26*		HOOD LOSSES	Other Losses		"wg	*2.0*						26
27	SP_h	HOOD LOSSES	Hood Static Pressure	21+25+26	"wg	*2.0*						27
28*	L	DUCT LOSSES	Straight Duct Length		ft	*30.0*		*15.0*		*5.0*	*30.0*	28
29	F'_d	DUCT LOSSES	Duct Friction Factor	Eqn8	VP/ft	*0.005*		*0.005*		*0.005*	*0.005*	29
30*		DUCT LOSSES	No. of 90 Degree Elbows			*1.00*						30*
31*	F_{el}	DUCT LOSSES	Elbow Loss Coefficient	Table 6	F_{el} or 0	*0.27*						31
32*	F_{en}	DUCT LOSSES	Branch Entry Loss Coefficient	Table 7	VP/entry							32
33*		DUCT LOSSES	Special Fitting Loss Coefficient	I.e Exp./Cont.,etc.	VP/fitting							33
34	F_d	DUCT LOSSES	Duct Loss Coefficient	28X29	VP/duct	*0.15*		*0.08*		*0.03*	*0.15*	34
35		DUCT LOSSES	Elbow Loss in VP	30x31	VP	*0.27*						35
36		DUCT LOSSES	Total Duct Loss in VP	32+33+34+35	VP total	*0.42*		*0.08*		*0.03*	*0.15*	36
37		DUCT LOSSES	Total Duct Loss	36x13	"wg	*0.2*		*0.0*		*0.0*	*0.1*	37
38*		DUCT LOSSES	Other Losses		"wg		*2.1*		*20.0*			38
39*	VP_r	DUCT LOSSES	Weighted Average VP	Eqn 10	"wg							39
40*		DUCT LOSSES	Loss From Velocity Increase	13-39 (if > 0)	"wg							40
41		DUCT LOSSES	Segment Pressure Loss	27+37+38+40	"wg	*-2.2*	*-2.1*	*0.0*	*-20.0*	*0.0*	*0.1*	41
42*	SP_{gov}		Governing Static Pressure		"wg							42
43*	SP_{cum}		Cumulative Static Pressure		"wg	*-2.2*	*-4.3*	*-4.3*	*-24.3*	*-24.3*	*0.1*	43
44	Q_{corr}		Corrected Volumetric Flow	Eqn 9	acfm							44
45	V_{corr}		Corrected Velocity	44/11	fpm							45
46	VP_{corr}		Corrected Velocity Pressure	Eqn 4	"wg							46

Designer

1. $\omega = lbs - H_2O / lbs - DryAir$

2. $df = df_e * df_p * df_T * df_m$

$df_e = [1 - (6.73 * 10^{-6})(z)]^{5.258}$

$df_p = (407 + SP)/(407)$

$df_T = 530/(T + 460)$

$df_m = (1 + \omega)/(1 + 1.607\omega)$

3. $Q_{std} = (Q_{act}/(1+\omega))df$

$Q_{act} = (Q_{std})(1+\omega)/df$

4. $V = 4005\sqrt{VP/df}$

$VP = (V/4005)^2 df$

5. $h = 0.24T + \omega(1061 + 0.444T)$

6. 90 Degree Elbow Loss Coefficient (5 Piece)

R/D	F_{el}
1.5	0.24
2.0	0.19
2.5	0.17

7. Branch Entry Loss Coefficients

Angle	F_{en}
15°	0.09
30°	0.18
45°	0.28
90°	1.00

8. $F_d\left(\frac{metal}{plastic}\right) = 0.0307\left(V^{0.533}/Q^{0.612}\right)$

$F_d(flexible) = 0.0311\left(V^{0.604}/Q^{0.639}\right)$

9. $Q_{corrected} = Q_{design}\sqrt{SP_{gov}/SP_{lower}}$

10. $VP_r = (Q_1/Q_3)VP_1 + (Q_2/Q_3)VP_2$

11. $SP_{System} = SP_{out} - SP_{in} - VP_{in}$

SSP = +0.1 - (-24.3) - 0.8 = 23.6"w.g.

FIGURE 9-14 (IP). Velocity Pressure Method Calculation Sheet

veying velocity is 4,000 fpm.

It has been determined that the air pollution control system should include a cyclone for dry product recovery and a high-energy wet collector. These devices have the following operating characteristics:

• *Cyclone:* Pressure loss is 4.5 "wg at rated flow rate of 35,000 scfm. (The pressure loss across most cyclones varies directly with any change in density and as the square of any change in flow rate from the rated conditions).

• *High-Energy Wet Scrubber:* The manufacturer has determined that a pressure loss of 20 "wg is required in order to meet existing air pollution regulations and has sized the collector accordingly. The humidifying efficiency of the wet collector is 90%.

NOTE: As a practical matter, a high energy scrubber as described in this example could have essentially 100% humidifying efficiency. The assumption of 90% humidifying efficiency along with a high pressure drop allows discussion of multiple design considerations in one example and was, therefore, adopted for instructional purposes.

- *Fan:* A size #34 "XYZ" fan with the performance shown in Figure 9-15 has been recommended.

REQUIRED:

Size the duct and select fan RPM and motor size.

SOLUTION:

Step 1: Find the actual gas flow rate that must be exhausted from the dryer. This flow rate must include both the air used for drying and the water, as vapor, which has been removed from the product. Since it is the actual flow rate, it must be corrected from standard air conditions to reflect the actual moisture, temperature and pressures that exist in the duct.

Step 1A: Find the amount (weight) of water vapor exhausted.

Dryer Discharge = 60 tons/hr of dried material (given)

Since the dryer has the capacity to remove 5% moisture, the dryer discharge is 95% × dryer feed rate.

60 tons/hr dried material = (0.95) × (dryer feed)

$$\text{dryer feed} = \frac{60 \text{ tons/hr}}{0.95} = 63.2 \text{ tons/hr}$$

Moisture removed = (feed rate) – (discharge rate)
= 63.2 tons/hr – 60 tons/hr
= 6,400 lbs/hr or 106.7 lbm/min

Step 1B: Find the amount (weight) of dry air exhausted.

Dry air exhausted = 16,000 scfm at 70 F and 29.92 "Hg (0.075 lbs/ft^3 density)

Exhaust rate, lbs/min = (16,000 scfm)(0.075 lbs/ft^3)
= 1,200 lbs/min dry air

FAN RATING TABLE

Fan Size #34 — Inlet Diameter = 34 — Max safe RPM = 1700

	20" SP		22" SP		24" SP		26" SP		28" SP		30" SP		32" SP		34" SP		36" SP		38" SP		40" SP	
ACFM	RPM	BHP	RPM	BHP	RPM	BHP	RPM	BHP	RPM	BHP	RPM	BHP	RPM	BHP	RPM	BHP	RPM	BHP	RPM	BHP	RPM	BHP
14688	1171	73.3	1225	81.4	1277	89.8	1326	98.3	1374	107	1421	116	1466	125	1510	134	1552	143	1594	153	1634	162
16524	1181	81.8	1234	90.2	1286	98.8	1335	107	1382	116	1428	126	1472	135	1516	145	1557	155	1600	165	1639	175
18360	1191	90.2	1244	99.5	1294	108	1344	118	1391	127	1437	137	1481	146	1524	157	1565	167	1606	178	1645	188
20196	1204	99.9	1256	109	1306	119	1354	129	1400	139	1446	149	1490	160	1532	170	1574	181	1615	191	1654	202
22032	1217	110	1268	120	1318	130	1366	141	1412	151	1456	162	1499	173	1542	184	1584	196	1624	207	1663	218
23868	1230	120	1282	131	1331	142	1378	154	1424	165	1468	176	1511	187	1553	199	1594	211	1633	223	1672	235
25704	1245	131	1296	143	1345	155	1391	167	1437	179	1481	191	1524	203	1565	215	1606	227	1645	239	1683	252
27540	1261	143	1311	156	1359	168	1406	181	1450	193	1494	206	1537	219	1578	232	1618	245	1658	258	1695	271
29376	1277	156	1327	169	1374	182	1421	196	1465	209	1508	222	1550	236	1591	249	1631	263	1670	277		
31212	1295	170	1344	184	1391	197	1436	211	1480	225	1523	239	1564	253	1605	268	1644	282	1683	297		
33048	1313	184	1361	198	1407	213	1453	228	1496	242	1538	257	1580	272	1620	287	1659	302	1697	317		
33884	1331	198	1379	214	1425	229	1469	245	1513	260	1555	276	1595	291	1635	307	1674	323				

FIGURE 9-15. Fan rating table

Step 1C: Knowing the water-to-dry air ratio and the temperature of the mixture, it is possible to determine other qualities of the air-to-water mixture. This can be accomplished by the use of the Psychrometric charts (Figures 9-g through 9-j) that are useful tools when working with humid air.

$\omega = 106.7/1{,}200 = 0.089$ lbs H_2O/lbm-dry air

Dry-Bulb temperature = 500 F (given)

The intersection of the 500 F Dry-Bulb temperature line and the 0.089 lbs H_2O/lb dry air line can be located on the Psychrometric chart (Figure 9-16). Point #1 defines the quality of the air and water mixture. Other data relative to this specific mixture can be read as follows:

Dew Point Temperature: 122 F

Wet-Bulb Temperature: 144 F

Humid Volume, ft^3 of mix/lbm-dry air: 27.5 ft^3/lb dry air

Enthalpy, BTU/lbm-dry air: 234 BTU/lbm-dry air

Density Factor, df: 0.53

The system is designed at an elevation of 575 feet ASL; this alters the df further to a value of 0.52. The density factor, Dry-Bulb temperature, mass of air and water, scfm and enthalpy are entered in the appropriate lines on the Calculation Sheet.

Step 2: Proceed with the system design using the calculation methods from previous Sample System Design Problems 1 and 2.

$$Q_{act} = \frac{Q_{std}(1+w)}{df} = \frac{(16{,}000)(1{,}089)}{0.52} = 33{,}508 \text{ acfm}$$

When considering the loss through the cyclone (B-C), the value is inserted in Row 38. The manufacturer provides the pressure loss (ΔP) of the cyclone. This is also called the pressure drop. In this example, the cyclone pressure loss is 4.5 "wg at a rated flow of 35,000 scfm. The pressure loss through a cyclone usually varies as the square of the change in flow rate and directly with the change in density.

Therefore, the actual loss through the cyclone would be:

$$(4.5)\left(\frac{33{,}508}{35{,}000}\right)^2(0.52) = 2.1\text{"wg}$$

and the static pressure at the cyclone outlet would be -4.3 "wg. There are no reacceleration losses.

The scrubber equipment manufacturer should provide the information for calculation of changes in flow rate and pressure drop across the wet collector, etc. An important characteristic of wet collectors is their ability to humidify a gas stream. The humidification process is generally assumed to be adiabatic (without gain or loss of heat to the surroundings). Water vapor is added to the mixture, but the enthalpy, expressed in BTU/lbm-dry air, remains unchanged. During the process of humidification, the point on the Psychrometric chart that

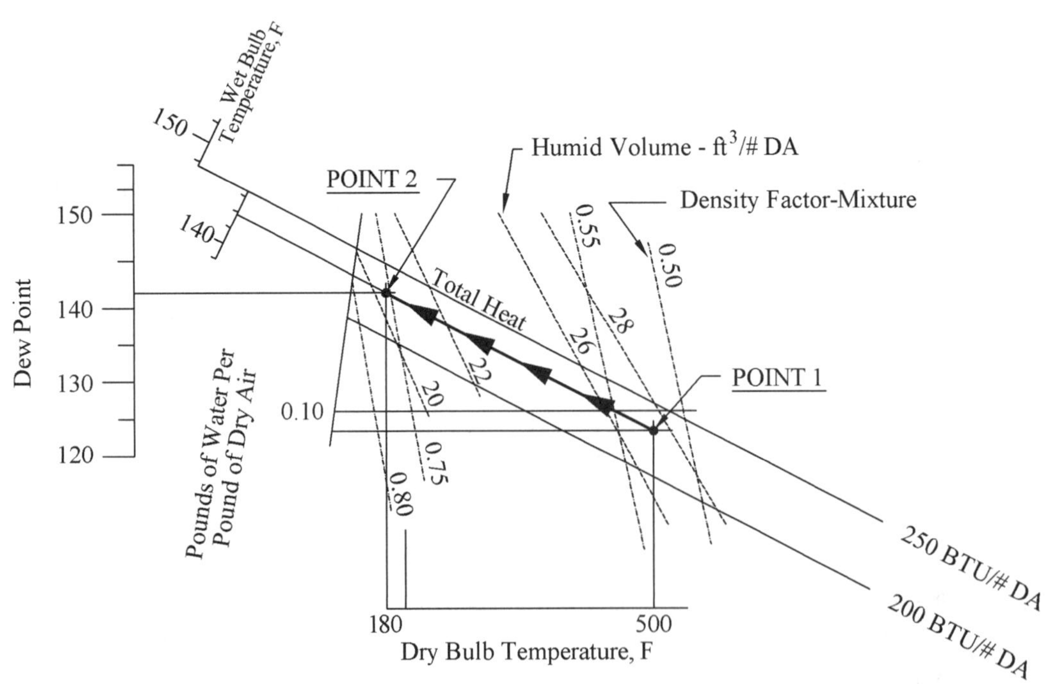

FIGURE 9-16. Psychrometric chart for humid air (see Figures 9-b (IP) through 9-j (IP))

defines the quality of the mixture moves to the left, along a line of constant enthalpy, toward saturation.

All wet collectors do not have the same ability to humidify. If a collector is capable of taking an air stream to complete adiabatic saturation, it is said to have a Humidifying Efficiency of 100%. The humidifying efficiency of a given device may be expressed by either of the following equations:

$$\eta_n = \frac{T_i - T_o}{T_i - T_s} \times 100 \quad [9.19]$$

where: η_n = Humidifying Efficiency, %
T_i = Dry-Bulb temperature at collector inlet, F
T_o = Dry-Bulb temperature at collector outlet, F
T_s = adiabatic saturation temperature, F

or

$$\eta_n = \frac{\omega_o - \omega_i}{\omega_s - \omega_i} \times 100$$

where: ω_i = moisture content in lb H_2O/lbm-dry air at inlet
ω_o = moisture content in lb H_2O/lbm-dry air at outlet
ω_s = moisture content in lb H_2O/lbm-dry air at adiabatic saturation conditions

Note: These formulae would apply in SI system also using temperature in C and ω in kg – H_2O/kg – dry air.

The designer must find the quality of the air to water mixture at Point 2, the collector outlet. Humidifying Efficiency = 90% (given). Dry-Bulb Temperature at Collector Inlet = 500 F (given). Adiabatic saturation temperature = 145 F from inspection of Psychrometric chart.

$$90\% = \frac{(500 - T_o)}{(500 - 144)} \times 100$$

thus: T_o = 180 F

Therefore, the air leaving the collector will have a Dry-Bulb temperature of 180 F and an enthalpy of 234 BTU/lbm of dry air as the humidifying process does not change the total heat or enthalpy.

The point of intersection of 180 F Dry-Bulb and 234 BTU/lbm-dry air on the Psychrometric chart (Figure 9-16) defines the quality of the air leaving the collector and allows other data to be read from the chart as follows:

Dew Point Temperature	143 F
Wet-Bulb Temperature	144 F
Humid Volume, ft³/lb dry air	20.5 ft³/lbm-dry air
Enthalpy, BTU/lb dry air	234 BTU/lbm-dry air
Density factor, df	0.76
ω	0.16

The density factor is recalculated at 0.74 to consider elevation. Required information is placed in the calculation sheet. (Formulas on calculation sheets can be used to obtain the density factor, knowing Dry-Bulb temperature, elevation, and moisture content.) With that information, the acfm can be calculated going into the scrubber.

Note: Water content in air is now (1200)(0.16)
= 192 #/min
Q = 20.5 × 1200
= 24,600 acfm

The scrubber loss was stated to be 20 "wg, so the static pressure at the wet collector outlet would be -24.3 "wg.

Step 3: Previously, in low-pressure local exhaust systems, (where the negative pressure at the fan inlet was less than -20 "wg), the effect of the negative pressure on air stream density was usually ignored (the effect was less than 5%). In practical system design, the other factors that affect density (temperature, moisture, elevation) can be additive so that the inlet pressure can be significant when specifying the fan. Systems designed at air temperatures less than 100 F and near sea level (df = 1) can still ignore fan inlet pressure if the values are between +10 and -10 "wg. However, as the pressures decrease (magnitude of negative pressures increase), it is understood that gases expand to occupy a larger volume. Unless this larger volume is anticipated and the fan is sized to handle the larger flow rate, it will have the effect of reducing the amount of air that is pulled into the hood at the front end of the system.

From the energy equation for flow in a duct without heat transfer (see Chapter 3 and Section 9.4.3):

$$\dot{m}_1(h_1) = \dot{m}_2(h_2) \text{ or}$$

$$\dot{m}_1(T_1) = \dot{m}_2(T_2)$$

$$\rho_1 Q_1 (T_1) = \rho_2 Q_2 (T_2)$$

Considering the Ideal Gas Equation, this would yield:

$$\frac{(w_1/Q_1)RT_1}{(w_2/Q_2)RT_2} = \frac{P_1}{P_2}$$

$$P_1 Q_1 = P_2 Q_2$$

or

$$\frac{P_1}{P_2} = \frac{Q_2}{Q_1}$$

Up to this point, the air has been considered to be at standard atmospheric pressure, which is 14.7 psi, 29.92 "Hg or 407 "wg. The pressure within the duct at Point F is -24.3 "wg and minus or negative only in relation to the pressure outside the duct which is 407 "wg. Therefore, the absolute pressure within the duct is 407 "wg - 24.3 "wg = 382.7 "wg.

$$\frac{407}{382.7} = \frac{Q_2}{24,600}$$

Q_2, the value at the fan inlet = 26,162 acfm

Note: If using Equations 9.3 and 9.6, values may vary slightly from psychrometric chart.

Step 4: Absolute pressure also affects the density of the air. From PQ = wRT, the relationship

$$\frac{(w_1/Q_1)RT_1}{(w_2/Q_2)RT_2} = \frac{P_1}{P_2}$$

can be derived. Assuming no heat transfer or change in temperature, the Density Factor is directly proportional to the density and the equation can be rewritten

$$\frac{P_1}{P_2} = \frac{df_2}{df_1}$$

If the pressure in the duct is compared to the absolute pressure at standard conditions (407 "wg), this can be calculated:

$$\frac{407}{382.7} = \frac{0.74}{df_2}$$

$df_2 = 0.70$.

This is now the 'real' value for Density Factor used in the fan specification. It considers temperature, moisture, elevation and now absolute pressure in the duct.

The duct from the wet collector to the fan can now be sized. The flow rate leaving the wet collector was 26,162 acfm. Since the fan selected has a 34-inch diameter inlet (area = 6.307 ft^2), it is logical to make the duct from the wet collector to the fan a 34-inch diameter.

After the system calculation has been completed, the fan can be selected.

$$\begin{aligned}\text{Actual SSP} &= SP_{out} - SP_{in} - VP_{in}\\ &= +0.1 - (-24.3) - 0.8\\ &= 23.6 \text{ "wg}\end{aligned}$$

Step 5: Specified fan static pressure is determined by dividing the actual fan static pressure by the density factor at the fan inlet (Equation 9.16a). This is necessary since all fan rating tables are based on standard air.

$$\text{Specified FSP} = \frac{23.6}{0.70} = 33.7 \text{ "wg}$$

Step 6: Interpolating the fan rating table (Figure 9-15) for 26,162 acfm at 33.7 "wg yields a fan speed of 1562 RPM at 217 BHP.

Since actual density is less than standard air density (and conveying air with less mass will require less work/energy), the actual required power is determined by multiplying by the density factor, or (217 BHP)(0.70) = 152 BHP. If a damper is installed in the duct to prevent overloading of the motor, at cold start the motor need only be a 200 HP (see Chapter 7, Section 7.3.8).

Additional Notes for Sample System Design Problem 3:

The following information is further clarification of key calculations in the ACGIH® Calculation Sheets. The cell location for inserted and calculated values is made using a matrix notation. The first value in the matrix would be the Duct Segment Identification (column), and the second value would be the row. For example, in Sample System Design Problem 2, the value at cell (2-A/3) would be 200 acfm. It is found in Column 2-A and Row 3. Similarly, the value in cell (A-B/13) would be 0.93 "wg.

Cell A-B/13: Note that the value for VP has already been corrected for density using Equation 5 on the calculation sheet. Because of this, it will represent a value close to the real check number when performing a balance. The calculation sheet is not the most accurate template for predicting actual field conditions. For one thing, the density would have to be exactly as calculated. If moisture levels or temperatures are different than calculated, this will affect these values. However, the 0.59 "wg is a good starting point to check airflows and conditions when commissioning the system and attempting to meet flow requirements.

Cell B-C/38: Care must always be taken when entering special losses or coefficients. In this case, the loss through the cyclone was calculated in "wg (see Step 2, Sample System Design Problem 2). Some equipment may be rated with losses in value of Velocity Pressures (i.e., 2.0 VP), as this is the most appropriate measurement unit. In those cases, the coefficient would be added in Row 33 (Special Fitting Coefficient) instead of Row 38.

9.16 SAMPLE SYSTEM DESIGN #4 (ADDING A BRANCH TO EXISTING SYSTEM/NON-STANDARD AIR CONDITIONS) (IP Units Only)

A second example is included where a new hood connection is added to the original duct system as an afterthought (Figure 9-17). This is not good practice under almost any circumstances. The original design is always compromised and there can be cases where material will settle in the duct, airflow will be reduced to other connections, and/or system changes in flow or pressure will cause the fan to operate in an unstable manner. If the addition of one or more ducts is made, the system calculation principles still apply. Losses can be calculated for the added flows required at the fan, and transport velocities must be considered for all ducts in the system. The following example should not in any way be considered an endorsement of this practice. It is included only to show that calculations and system adjustments can be made to get the system into balance (if suitable resources are available in the duct, fan, motor, collection device and electric power source).

In this case, a hood similar to the bagging hood shown in VS-15-02 is connected through a properly sized branch and tapped into the 38" diameter duct coming from the dryer. When the decision is made to proceed on this basis, many fac-

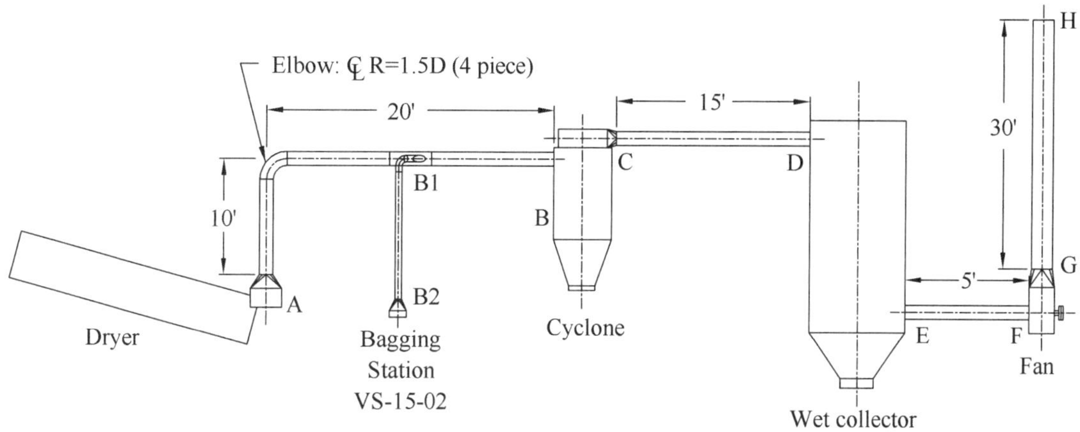

FIGURE 9-17. System layout (Sample Problem 4)

tors must be considered:

1) Mixing hot and moist air streams with cold air can cause condensation in duct or collectors. Under normal conditions, the dry-bulb (DB) temperature should be at least 35 F [20 C] above the dew point and preferably 50 F [28 C]. The system must also consider start-up and shut down when the system is especially susceptible to condensation.
2) Downstream velocities can be high enough to cause premature wear of duct and other parts.
3) Sufficient airflow and transport velocities must be maintained through all duct system components and at all hoods.

A new calculation sheet (Figure 9-18) shows the alterations that must be made. A new sketch inserting a new branch duct with a new numbering method is made. The bagging station is 60' away from the 38" main duct with (1) 90° elbow and requires 1500 acfm. The designer in this case has chosen to keep all duct the same size, i.e., no size increase in the main duct between the dryer and the cyclone. All calculations are done in the same manner as previous examples except a calculation must be done to allow for the mixing of the ambient air from the bagging station with the hot moist air from the dryer.

Knowing that mass and energy must be conserved, the conditions from downstream of the fitting can be calculated using Equation 9.6. The mass of dry air (Row 6) and enthalpy (Row 14) are known for the two branches and the mass is known for the downstream duct since it is simply the sum of the two branches.

$$(1200 \times 234)_1 + (110.3 \times 17)_2 = (1310.3 \times h)_3$$

$$h_3 = 216 \text{ BTU/lbm-dry air}$$

The mass of air and water downstream will be the summation of the two values from the new hood and the dryer (Rows 5 and 6 on the calculation sheet). Using this information, the conditions in duct B1-B can be determined from the Psychrometric chart as:

Dew Point Temperature	142 F
Dry-Bulb Temperature	470* F
Density factor, df	0.55
ω	0.081

*Calculated with equation 9.17 IP

$$Q_{act} = \frac{Q_{std}(1+20)}{df}$$

$$Q_{act} = \frac{(17{,}500)(1{,}081)}{0.52} = 36{,}380 \text{ acfm}$$

Density factor is again recalculated at 0.54 to consider elevation. Required information is placed in the calculation sheet.

Note that the static pressure requirement for the new branch is -3.9 "wg at junction B1 and the requirement at the same junction for the dryer is only -2.2 "wg. Normally there would be a change in duct design or selection of new airflows to increase the flow to the dryer and balance the pressure from each branch. In this case, the dryer is sensitive to the static pressure from the duct and cannot be altered. This is a good example of where dampers or orifice plates can be used to balance the system.

The remainder of the calculation process is identical to the first example and airflow at the fan inlet is now required to be 28,950 acfm. After the system calculation has been completed, the new system conditions can be determined:

$$\begin{aligned} SSP &= SP_{out} - SP_{in} - VP_{in} \\ &= +0.1 - (-26.5) - 0.9" \\ &= 25.7 \text{ "wg} \end{aligned}$$

$$FSP = \frac{25.7}{0.70} = 36.7 \text{ "wg}$$

ACGIH® Velocity Pressure Method Calculation Sheet (IP)

Figure 9-18 (IP) Sample System Design #4							Page *1* of *1*		Date			Elevation (z), ft = *575 ft*	* Input Data
1*		Duct Segment Identification			*A-B1*	*B2-B1*	*B1-B*	*B-C*	*C-D*	*D-E*	*E-F*	*G-H*	1
2*	T	Dry-Bulb Temperature		F	*500*	*70*	*470*	*470*	*470*	*470*	*175*	*175*	2
3*	Q_{act}	Actual Duct Flow Rate	Eqn 3	acfm	*33508*	*1500*	*34940*	*34940*	*34940*	*34940*	*28950*	*27020*	3
4*	V_t	Minimum Transport Velocity	Chapter 13	fpm	*4000*	*3500*	*4000*		*4000*				4
5*	$\dot{m}_{H2O}$	Pounds Water per Minute		lbs - H_2O/min	*106.7*	*0.0*	*106.7*	*106.7*	*106.7*	*106.7*	*209.6*	*209.6*	5
6*	$\dot{m}_{da}$	Pounds Dry Air per Minute		lbs - da/min	*1200*	*110.3*	*1310.3*	*1310.3*	*1310.3*	*1310.3*	*1310.3*	*1310.3*	6
7	df	Density Factor	Eqn 2		*0.52*	*0.98*	*0.54*	*0.54*	*0.54*	*0.54*	*0.70*	*0.75*	7
8	Q_{std}	Standard Duct Flow Rate (Dry Air)	Eqn 3	dscfm	*16000*	*1470*	*17470*	*17470*	*17470*	*17470*	*17470*	*17470*	8
9	A_t	Target Duct Area	3/4	ft²	*8.377*	*0.429*	*8.377*		*8.760*				9
10*	D	Selected Duct Diameter	Section 9.5.2.5	in	*38.0*	*8.0*	*38.0*		*38.0*		*34.0*	*36.0*	10
11*	A	Selected Duct Area	Table 9.1	ft²	*7.878*	*0.349*	*7.878*		*7.878*		*6.307*	*7.071*	11
12	V_d	Duct Velocity	(3/11)	fpm	*4253*	*4298*	*4435*		*4435*		*4590*	*3821*	12
13	VP_d	Duct Velocity Pressure	Eqn 4	"wg	*0.59*	*1.13*	*0.66*		*0.66*		*0.92*	*0.68*	13
14	h	Total Heat	Eqn 5 or Psych Chart	BTU/lbs-da	*234*	*17*	*216*	*216*	*216*	*216*	*216*	*216*	14
15	A_s	HOOD LOSSES – SLOT: Slot Area	Chapters 6 & 13	ft²									15
16*	F_s	Slot Loss Coefficient	Figure 9a	VP/slot									16
17*	F_a	Acceleration Coefficient	Chapter 3	0 or 1									17
18*	V_s	Slot Velocity	Chapters 6 & 13	fpm									18
19	VP_s	Slot Velocity Pressure	Eqn 4	"wg									19
20		Slot Loss in VP	16+17	VPtotal									20
21	h_s	Slot Loss	20x19	"wg									21
22*	F_h	HOOD LOSSES: Hood Entry Loss Coefficient	Fig 9a or Ch 13	F_h or 0		*0.25*							22
23*	F_a	Acceleration Coefficient	Chapters 3 & 6	1 or 0		*1*							23
24		Hood Entry Loss in VP	22+23	VPtotal		*1.25*							24
25	h_h	Hood Entry Loss	24x13	"wg		*1.4*							25
26*		Other Losses		"wg	*2.0*								26
27	SP_h	Hood Static Pressure	21+25+26	"wg	*2.0*	*1.4*							27
28*	L	DUCT LOSSES: Straight Duct Length		ft	*20.0*	*60.0*	*10.0*		*15.0*		*5.0*	*30.0*	28
29	F'_d	Duct Friction Factor	Eqn8	VP/ft	*0.005*	*0.030*	*0.005*		*0.005*		*0.005*	*0.005*	29
30*		No. of 90 Degree Elbows			*1.00*	*1.00*							30*
31*	F_{el}	Elbow Loss Coefficient	Table 6	F_{el} or 0	*0.27*	*0.24*							31
32*	F_{en}	Branch Entry Loss Coefficient	Table 7	VP/entry		*0.18*							32
33*		Special Fitting Loss Coefficient	i.e. Exp./Cont., etc.	VP/fitting									33
34	F_d	Duct Loss Coefficient	28X29	VP/duct	*0.10*	*1.80*	*0.05*		*0.08*		*0.03*	*0.15*	34
35		Elbow Loss in VP	30x31	VP	*0.27*	*0.24*							35
36		Total Duct Loss in VP	32+33+34+35	VP total	*0.37*	*2.22*	*0.05*		*0.08*		*0.03*	*0.15*	36
37		Total Duct Loss	36x13	"wg	*0.2*	*2.5*	*0.0*		*0.1*		*0.0*	*0.1*	37
38*		Other Losses		"wg				*2.4*					38
39*	VP_r	Weighted Average VP	Eqn 10	"wg			*0.61*			*20.0*			39
40*		Loss From Velocity Increase	13-39 (if > 0)	"wg			*0.1*						40
41		Segment Pressure Loss	27+37+38+40	"wg	*-2.2*	*-3.9*	*-0.1*	*-2.4*	*-0.1*	*-20.0*	*0.0*	*0.1*	41
42*	SP_{gov}	Governing Static Pressure		"wg	*-3.9*								42
43*	SP_{cum}	Cumulative Static Pressure		"wg	*-2.2*		*-4.0*	*-6.4*	*-6.5*	*-26.5*	*-26.5*	*0.1*	43
44	Q_{corr}	Corrected Volumetric Flow	Eqn 9	acfm	*add*								44
45	V_{corr}	Corrected Velocity	44/11	fpm	*blast*								45
46	VP_{corr}	Corrected Velocity Pressure	Eqn 4	"wg	*gate*								46

Designer

1. $\omega = \text{lbs} - H_2O / \text{lbs} - \text{DryAir}$

2. $df = df_e * df_p * df_T * df_m$
   $df_e = [1 - (6.73 * 10^{-6})(z)]^{5.258}$
   $df_p = (407 + SP)/(407)$
   $df_T = 530/(T + 460)$
   $df_m = (1+\omega)/(1+1.607\omega)$

3. $Q_{std} = (Q_{act}/(1+\omega))df$
   $Q_{act} = (Q_{std})(1+\omega)/df$

4. $V = 4005\sqrt{VP/df}$
   $VP = (V/4005)^2 df$

5. $h = 0.24T + \omega(1061 + 0.444T)$

6. 90 Degree Elbow Loss Coefficient (5 Piece)

R/D	F_{el}
1.5	0.24
2.0	0.19
2.5	0.17

7. Branch Entry Loss Coefficients

Angle	F_{en}
15°	0.09
30°	0.18
45°	0.28
90°	1.00

8. $F_d\left(\frac{\text{metal}}{\text{plastic}}\right) = 0.0307\left(V^{0.533}/Q^{0.612}\right)$
   $F_d(\text{flexible}) = 0.0311\left(V^{0.604}/Q^{0.639}\right)$

9. $Q_{corrected} = Q_{design}\sqrt{SP_{gov}/SP_{lower}}$

10. $VP_r = (Q_1/Q_3)VP_1 + (Q_2/Q_3)VP_2$

11. $SP_{System} = SP_{out} - SP_{in} - VP_{in}$

SSP = +0.1 - (-26.5) - 0.9 = 25.7"w.g.

FIGURE 9-18 (IP). Velocity Pressure Method Calculation Sheet

The fan will now be required to operate at increased airflow and pressure to meet the design requirements but with a significant increase in horsepower. The fan speed is recalculated at 1,642 RPM and the horsepower required under cold conditions is now 264.

The fan will now need to operate at increased airflow and pressure, but if a 250 HP motor was originally selected, it will not be large enough for a cold start-up (see Section 7.3.8). In that case, design and/or hardware changes will have to be made to damper the fan at start-up until sufficient heat is in the system to reduce the power requirements.

NOTE: For Cell A-B1/43: The static pressure required to deliver 33,508 acfm from segment A-B is -2.2 "wg. Since the system is not being balanced by design, the determining value at Junction B1 (-3.9 "wg from segment B2-B1) is not used to recalculate the conditions in A-B1. Instead, a blast gate, orifice, or other damper will be used to balance the system. The fan must be able to deliver -3.9" at this junction to pull all of the air designed for the Bagging Station. The difference between the junction determining pressure (-3.9 "wg) and the losses in the duct segment from A-B1 (-2.2 "wg) is the added amount of loss that will be required from the blast gate (1.7 "wg). An orifice with a loss of 1.7 "wg could also be calculated and inserted in segment A-B1 in place of the blast gate.

9.17 AIR BLEED DESIGN

Bleed-ins are used at the ends of branch ducts to provide additional airflow rates to transport heavy material loads or at the ends of a main duct to maintain minimum transport velocity. Some designs use bleed-ins to introduce additional air to a local exhaust system to reduce air temperature and/or to assist in balancing the system.

End cap bleed-in (Figure 9-19). Consider it to be an orifice or slot. From Figure 9-a, h_e = 1.78 VP.

1) Calculate SP for branch duct to junction (X).

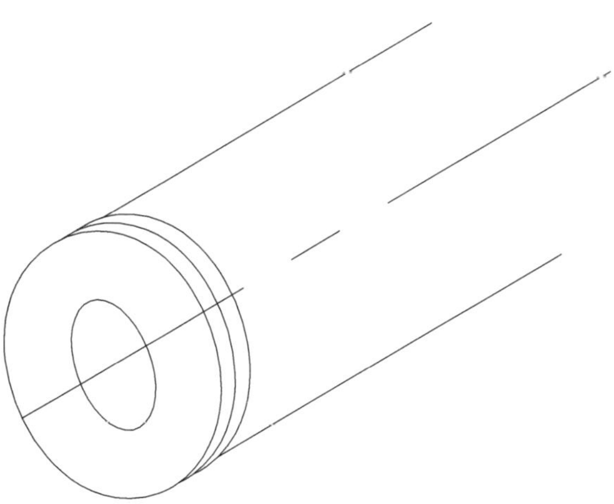

FIGURE 9-19. Air bleed opening

2) Determine flow rate in main duct according to design or future capacity or determine $Q_{bleed\text{-}in}$ directly from temperature or moisture considerations.
3) $Q_{bleed\text{-}in} = (Q_{main\ duct}) - (Q_{branch})$
4) SP bleed-in = SP branch as calculated = X = (h_e + 1 VP) = (1.78 + 1.0) VP
5) $\text{VP, bleed-in} = \frac{X}{(1.78 + 1.0)} = \frac{X}{2.78}$
6) Velocity, bleed-in from VP and Table 9-2
7) $\text{Area bleed-in} = \frac{Q_{bleed\text{-}in}}{V_{bleed\text{-}in}}$

EXAMPLE PROBLEM 12 (Air Bleed to Reduce Duct Temperature)

The furnace fume hood ventilated in Example Problem 6 has a temperature of 196 F [91 C]. An air bleed must be added to reduce the temperature to 125 F [52 C] for entry into the baghouse. Outside air temperature will be 70 F [21 C]. The air bleed will be placed in the duct system where accumulated losses from the furnace hood equal 3.2 "wg [800 Pa]. Calculate the size of the air bleed.

From Equation 9.7:

$$\dot{m}_a(T_a) + \dot{m}_f(T_f) = \dot{m}_{mix}(T_{mix}) =$$
$$(\dot{m}_a)(530) + (720.3)(656) = (720.3 + \dot{m}_a)(460 + 125)$$

$$[(\dot{m}_a)(294) + (5.37)(364) = (5.37 + \dot{m}_a)(273 + 52)]$$

Solving for $\dot{m}_a$ (the mass of bleed air) = 930 lbm per minute [6.77 kg/s].

$$\left(930\frac{\text{lbm}}{\text{min}}\right)\left(\frac{\text{ft}^3}{0.075\text{lbm}}\right) = 12{,}400 \text{ scfm @ 70 F}$$

[(6.77 kg/s)(m^3/1.204 kg) = 5.63 nm^3/s @ 21 C]

SP branch as calculated = X = 3.2 "wg = (h_e + 1 VP) = (1.78 + 1.0) VP = 800 Pa

$$\text{VP, bleed-in} = \frac{X}{(1.78 + 1.0)} = \frac{3.2}{2.78} = 1.15\text{ "wg}$$

$$[\text{VP}_{bleed\text{-}in} = \frac{800}{1 + 1.78} = 288 \text{ Pa}]$$

Velocity, bleed-in (from Table 9-2) = 4,295 fpm

[VP = 288 Pa V = 21.9 m/s]

$$\text{Area bleed-in} = \frac{Q_{bleed\text{-}in}}{V_{bleed\text{-}in}} = \frac{12{,}400 \text{ acfm}}{4{,}295 \text{ fpm}} = 2.89 \text{ ft}^2$$

[A_{bleed} = 5.63/21.9 = 0.26 m^2]

Unlike a duct size, this would be the actual size of the circular orifice opening, i.e., 23 3/16" diameter.

Note the new airflow required for specification of baghouse and fan is 24,287 acfm [11.07 am^3/s]. This example is for "dry" air only. If there is moisture present, then enthalpy will need to be used per Example Problem 3.

Please note that if calculations are done using computer (spreadsheet program, etc.), do not round numbers until final computation at fan and discharge point. If done manually, then significant numbers can be entered as referenced above for ease in calculation sheet but there could be different, though insignificant, values then achieved electronically.

REFERENCES

9.1 Air Movement and Control Association, Inc.: AMCA Standard 210-74. Arlington Heights, IL.

9.2 E. Ravert: Private Communication to G. Lanham (November, 2012).

APPENDIX A9 PRESSURE MEASUREMENT IN THE SI SYSTEM

In the research for this Manual, it was noted that within the SI system there are different units used worldwide for pressure. The classic measurement for pressure in the SI system (in ranges for ventilation systems) is the Pascal (Pa). A soft conversion (see Foreword and Definitions) from inches of water ("wg) to the Pascals is 249.1 Pa per "wg. In this Manual the hard conversion of 250 Pa per "wg was used, a value less than 0.0025 "wg from the soft (249.1) conversion. This can also be stated as 0.25 kilo-Pascals (kPa).

An alternative measurement method is millimeters of water (mmwg). This can also be referred to as millimeters water gauge. Using a U-Tube manometer yields a direct measurement of "wg. If a similar measurement is taken in the SI system, the values would be measured directly in mmwg. This becomes a conversion where (soft conversion) one inch = 25.4 mm.

Comparing the hard conversion of the two methods for 1 "wg (250 Pa and 25 mmwg) there is less than 1.5% from the soft conversion values. In this case 10 Pa is considered equal to 1 mmwg.

This Manual utilizes Pa for pressure measurement. Those using mmwg for pressure measurement need only to divide pressure in Pa to achieve a similar value in mmwg and calculate pressure density factor (df_p) from:

$$df_p = (10{,}338 + SP) / (10{,}338)$$

Those needing or desiring to do so can calculate velocity or velocity pressure mmwg from:

$$V = 4.043\sqrt{VP/df} = VP = \left(V/4.043\right)^2 df_p$$

The remaining SI values, airflow, velocity, dimensions, etc. are valid for both the Pa and mmwg methods.

TABLE 9-1. Area and Circumference of Circles

Dia	Area	Circ	Dia	Area	Circ	Dia	Area	Circ	Dia	Area	Circ
in	ft^2	in	mm	m^2	mm	in	ft^2	in	mm	m^2	mm
1	0.005	3.1				17	1.577	53.4			
			30	0.001					440	0.152	1382
1.5	0.012	4.7			94				450	0.159	1414
			40	0.001	126	18	1.768	56.5			
			50	0.002	157				460	0.166	1445
2	0.0.22	6.3							480	0.181	1508
			60	0.003	189	19	1.970	59.7			
			63	**0.003**	**198**				**500**	**0.196**	**1571**
2.5	0.034	7.9				20	2.182	62.8			
			70	0.004	220				550	0.238	1728
3	0.049	9.4				22	2.641	69.1	560	0.246	1759
			80	**0.005**	**251**				600	0.283	1885
3.5	0.067	11.0				24	3.143	75.4			
			90	0.006	283				**630**	**0.312**	**1979**
			100	0.008	314				650	0.332	2042
4	0.087	12.6				26	3.688	81.7			
			110	0.010	346				700	0.385	2199
4.5	0.110	14.1				28	4.277	88.0	710	0.396	2231
			120	0.011	377				750	0.442	2356
			125	**0.012**	**393**	30	4.910	94.2			
5	0.136	15.7							800	0.503	2513
			130	0.013	408	32	5.587	100.5			
5.5	0.165	17.3	140	0.015	440				850	0.567	2670
			150	0.018	471	34	6.307	106.8			
6	0.196	18.8							**900**	**0.636**	**2828**
			160	**0.020**	**503**	36	7.071	113.1			
			170	0.023	534				950	0.709	2985
7	0.267	22.0				38	7.878	119.4			
			180	0.025	566				1000	0.785	3142
			190	0.028	597	40	8.729	125.7			
			200	**0.031**	**628**				1050	0.866	3299
8	0.349	25.1				42	9.624	131.9			
			220	0.038	691				1100	0.950	3456
9	0.442	28.3							**1120**	**0.985**	**3519**
			240	0.045	754	44	10.563	138.2			
			250	**0.049**	**785**				1150	1.039	3613
10	0.546	31.4				46	11.545	144.5			
			260	0.053	817				1200	1.131	3770
11	0.660	34.6				48	12.570	150.8			
			280	0.062	880				1250	1.227	3927
			300	0.071	943	50	13.640	157.1			
12	0.786	37.7							1399	1.537	4395
			315	**0.078**	**990**	52	14.753	163.4			
			320	0.080	1005				1350	1.431	4241
13	0.922	40.8				54	15.909	169.6			
			340	0.091	1068				**1400**	**1.539**	**4398**
			355	**0.099**	**1115**	56	17.110	175.9			
14	1.069	44.0							1450	1.651	4555
			360	0.102	1131	58	18.354	182.2			
15	1.228	47.1	380	0.113	1194				1500	1.767	4713
			400	**0.126**	**1257**	60	19.641	188.5			
16	1.397	50.3							1550	1.887	4870
			420	0.139	1320	62					
									1600	**2.011**	**5027**

EU CEN 1506 (European Duct Standard Sizes)

TABLE 9-2 (IP). Velocity Pressure to Velocity Conversion — Standard Air

From $V = 4005\sqrt{VP/df}$

V = Velocity fpm df = 1
VP = Velocity Pressure "wg

VP	V	VP	V	VP	V	VP	V	VP	V	VP	V
0.01	401	0.51	2860	1.01	4025	1.51	4921	2.01	5678	2.6	6458
0.02	566	0.52	2888	1.02	4045	1.52	4938	2.02	5692	2.7	6581
0.03	694	0.53	2916	1.03	4065	1.53	4954	2.03	5706	2.8	6702
0.04	801	0.54	2943	1.04	4084	1.54	4970	2.04	5720	2.9	6820
0.05	896	0.55	2970	1.05	4104	1.55	4986	2.05	5734	3	6937
0.06	981	0.56	2997	1.06	4123	1.56	5002	2.06	5748	3.1	7052
0.07	1060	0.57	3024	1.07	4143	1.57	5018	2.07	5762	3.2	7164
0.08	1133	0.58	3050	1.08	4162	1.58	5034	2.08	5776	3.3	7275
0.09	1202	0.59	3076	1.09	4181	1.59	5050	2.09	5790	3.4	7385
0.1	1266	0.6	3102	1.1	4200	1.6	5066	2.1	5804	3.5	7493
0.11	1328	0.61	3128	1.11	4220	1.61	5082	2.11	5818	3.6	7599
0.12	1387	0.62	3154	1.12	4238	1.62	5098	2.12	5831	3.7	7704
0.13	1444	0.63	3179	1.13	4257	1.63	5113	2.13	5845	3.8	7807
0.14	1499	0.64	3204	1.14	4276	1.64	5129	2.14	5859	3.9	7909
0.15	1551	0.65	3229	1.15	4295	1.65	5145	2.15	5872	4	8010
0.16	1602	0.66	3254	1.16	4314	1.66	5160	2.16	5886	4.1	8110
0.17	1651	0.67	3278	1.17	4332	1.67	5176	2.17	5900	4.2	8208
0.18	1699	0.68	3303	1.18	4351	1.68	5191	2.18	5913	4.3	8305
0.19	1746	0.69	3327	1.19	4369	1.69	5207	2.19	5927	4.4	8401
0.2	1791	0.7	3351	1.2	4387	1.7	5222	2.2	5940	4.5	8496
0.21	1835	0.71	3375	1.21	4406	1.71	5237	2.21	5954	4.6	8590
0.22	1879	0.72	3398	1.22	4424	1.72	5253	2.22	5967	4.7	8683
0.23	1921	0.73	3422	1.23	4442	1.73	5268	2.23	5981	4.8	8775
0.24	1962	0.74	3445	1.24	4460	1.74	5283	2.24	5994	4.9	8865
0.25	2003	0.75	3468	1.25	4478	1.75	5298	2.25	6007	5	8955
0.26	2042	0.76	3491	1.26	4496	1.76	5313	2.26	6021	5.5	9393
0.27	2081	0.77	3514	1.27	4513	1.77	5328	2.27	6034	6	9810
0.28	2119	0.78	3537	1.28	4531	1.78	5343	2.28	6047	6.5	10211
0.29	2157	0.79	3560	1.29	4549	1.79	5358	2.29	6061	7	10596
0.3	2194	0.8	3582	1.3	4566	1.8	5373	2.3	6074	7.5	10968
0.31	2230	0.81	3605	1.31	4584	1.81	5388	2.31	6087	8	11328
0.32	2266	0.82	3627	1.32	4601	1.82	5403	2.32	6100	8.5	11676
0.33	2301	0.83	3649	1.33	4619	1.83	5418	2.33	6113	9	12015
0.34	2335	0.84	3671	1.34	4636	1.84	5433	2.34	6126	9.5	12344
0.35	2369	0.85	3692	1.35	4653	1.85	5447	2.35	6140	10	12665
0.36	2403	0.86	3714	1.36	4671	1.86	5462	2.36	6153	10.5	12978
0.37	2436	0.87	3736	1.37	4688	1.87	5477	2.37	6166	11	13283
0.38	2469	0.88	3757	1.38	4705	1.88	5491	2.38	6179	11.5	13582
0.39	2501	0.89	3778	1.39	4722	1.89	5506	2.39	6192	12	13874
0.4	2533	0.9	3799	1.4	4739	1.9	5521	2.4	6205	12.5	14160
0.41	2564	0.91	3821	1.41	4756	1.91	5535	2.41	6217	13	14440
0.42	2596	0.92	3841	1.42	4773	1.92	5549	2.42	6230	13.5	14715
0.43	2626	0.93	3862	1.43	4789	1.93	5564	2.43	6243	14	14985
0.44	2657	0.94	3883	1.44	4806	1.94	5578	2.44	6256	14.5	15251
0.45	2687	0.95	3904	1.45	4823	1.95	5593	2.45	6269	15	15511
0.46	2716	0.96	3924	1.46	4839	1.96	5607	2.46	6282	15.5	15768
0.47	2746	0.97	3944	1.47	4856	1.97	5621	2.47	6294	16	16020
0.48	2775	0.98	3965	1.48	4872	1.98	5636	2.48	6307	16.5	16268
0.49	2804	0.99	3985	1.49	4889	1.99	5650	2.49	6320	17	16513
0.5	2832	1	4005	1.5	4905	2	5664	2.5	6332	17.5	16754

TABLE 9-2 (SI). Velocity Pressure to Velocity Conversion — Standard Air

$V = 1.29\sqrt{VP / df}$ V = Velocity m^3/s VP = Velocity Pressure Pa df = 1

VP	V	VP	V	VP	V	VP	V	VP	V	VP	V
2	1.82	102	13.03	202	18.33	302	22.42	402	25.86	504	28.96
4	2.58	104	13.16	204	18.42	304	22.49	404	25.93	508	29.08
6	3.16	106	13.28	206	18.51	306	22.57	406	25.99	512	29.19
8	3.65	108	13.41	208	18.60	308	22.64	408	26.06	516	29.30
10	4.08	110	13.53	210	18.69	310	22.71	410	26.12	520	29.42
12	4.47	112	13.65	212	18.78	312	22.79	412	26.18	524	29.53
14	4.83	114	13.77	214	18.87	314	22.86	414	26.25	528	29.64
16	5.16	116	13.89	216	18.96	316	22.93	416	26.31	532	29.75
18	5.47	118	14.01	218	19.05	318	23.00	418	26.37	536	29.87
20	5.77	120	14.13	220	19.13	320	23.08	420	26.44	540	29.98
22	6.05	122	14.25	222	19.22	322	23.15	422	26.50	544	30.09
24	6.32	124	14.36	224	19.31	324	23.22	424	26.56	548	30.20
26	6.58	126	14.48	226	19.39	326	23.29	426	26.63	552	30.31
28	6.83	128	14.59	228	19.48	328	23.36	428	26.69	556	30.42
30	7.07	130	14.71	230	19.56	330	23.43	430	26.75	560	30.53
32	7.30	132	14.82	232	19.65	332	23.50	432	26.81	564	30.64
34	7.52	134	14.93	234	19.73	334	23.58	434	26.87	568	30.74
36	7.74	136	15.04	236	19.82	336	23.65	436	26.94	572	30.85
38	7.95	138	15.15	238	19.90	338	23.72	438	27.00	576	30.96
40	8.16	140	15.26	240	19.98	340	23.79	440	27.06	580	31.07
42	8.36	142	15.37	242	20.07	342	23.86	442	27.12	584	31.17
44	8.56	144	15.48	244	20.15	344	23.93	444	27.18	588	31.28
46	8.75	146	15.59	246	20.23	346	24.00	446	27.24	592	31.39
48	8.94	148	15.69	248	20.31	348	24.06	448	27.30	596	31.49
50	9.12	150	15.80	250	20.40	350	24.13	450	27.37	600	31.60
52	9.30	152	15.90	252	20.48	352	24.20	452	27.43	604	31.70
54	9.48	154	16.01	254	20.56	354	24.27	454	27.49	608	31.81
56	9.65	156	16.11	256	20.64	356	24.34	456	27.55	612	31.91
58	9.82	158	16.22	258	20.72	358	24.41	458	27.61	616	32.02
60	9.99	160	16.32	260	20.80	360	24.48	460	27.67	620	32.12
62	10.16	162	16.42	262	20.88	362	24.54	462	27.73	624	32.22
64	10.32	164	16.52	264	20.96	364	24.61	464	27.79	628	32.33
66	10.48	166	16.62	266	21.04	366	24.68	466	27.85	632	32.43
68	10.64	168	16.72	268	21.12	368	24.75	468	27.91	636	32.53
70	10.79	170	16.82	270	21.20	370	24.81	470	27.97	640	32.63
72	10.95	172	16.92	272	21.28	372	24.88	472	28.03	644	32.74
74	11.10	174	17.02	274	21.35	374	24.95	474	28.09	648	32.84
76	11.25	176	17.11	276	21.43	376	25.01	476	28.14	652	32.94
78	11.39	178	17.21	278	21.51	378	25.08	478	28.20	656	33.04
80	11.54	180	17.31	280	21.59	380	25.15	480	28.26	660	33.14
82	11.68	182	17.40	282	21.66	382	25.21	482	28.32	664	33.24
84	11.82	184	17.50	284	21.74	384	25.28	484	28.38	668	33.34
86	11.96	186	17.59	286	21.82	386	25.34	486	28.44	672	33.44
88	12.10	188	17.69	288	21.89	388	25.41	488	28.50	676	33.54
90	12.24	190	17.78	290	21.97	390	25.48	490	28.56	680	33.64
92	12.37	192	17.87	292	22.04	392	25.54	492	28.61	684	33.74
94	12.51	194	17.97	294	22.12	394	25.61	494	28.67	688	33.84
96	12.64	196	18.06	296	22.19	396	25.67	496	28.73	692	33.93
98	12.77	198	18.15	298	22.27	398	25.74	498	28.79	696	34.03
100	12.90	200	18.24	300	22.34	400	25.80	500	28.85	700	34.13

TABLE 9-3 (IP). Velocity to Velocity Pressure Conversion — Standard Air

$VP = (V / 4005)^2 df$

V = Velocity fpm   df = 1

VP = Velocity Pressure "wg

V	VP	V	VP	V	VP	V	VP	V	VP	V	VP
400	0.01	2600	0.42	3850	0.92	4880	1.48	5690	2.02	6190	2.39
500	0.02	2625	0.43	3875	0.94	4900	1.50	5700	2.03	6200	2.40
600	0.02	2650	0.44	3900	0.95	4920	1.51	5710	2.03	6210	2.40
700	0.03	2675	0.45	3925	0.96	4940	1.52	5720	2.04	6220	2.41
800	0.04	2700	0.45	3950	0.97	4960	1.53	5730	2.05	6230	2.42
900	0.05	2725	0.46	3975	0.99	4980	1.55	5740	2.05	6240	2.43
1000	0.06	2750	0.47	4000	1.00	5000	1.56	5750	2.06	6250	2.44
1100	0.08	2775	0.48	4020	1.01	5020	1.57	5760	2.07	6260	2.44
1200	0.09	2800	0.49	4040	1.02	5040	1.58	5770	2.08	6270	2.45
1300	0.11	2825	0.50	4060	1.03	5060	1.60	5780	2.08	6280	2.46
1400	0.12	2850	0.51	4080	1.04	5080	1.61	5790	2.09	6290	2.47
1450	0.13	2875	0.52	4100	1.05	5100	1.62	5800	2.10	6300	2.47
1500	0.14	2900	0.52	4120	1.06	5120	1.63	5810	2.10	6310	2.48
1550	0.15	2925	0.53	4140	1.07	5140	1.65	5820	2.11	6320	2.49
1600	0.16	2950	0.54	4160	1.08	5160	1.66	5830	2.12	6330	2.50
1650	0.17	2975	0.55	4180	1.09	5180	1.67	5840	2.13	6340	2.51
1700	0.18	3000	0.56	4200	1.10	5200	1.69	5850	2.13	6350	2.51
1750	0.19	3025	0.57	4220	1.11	5220	1.70	5860	2.14	6360	2.52
1800	0.20	3050	0.58	4240	1.12	5240	1.71	5870	2.15	6370	2.53
1825	0.21	3075	0.59	4260	1.13	5260	1.72	5880	2.16	6380	2.54
1850	0.21	3100	0.60	4280	1.14	5280	1.74	5890	2.16	6390	2.55
1875	0.22	3125	0.61	4300	1.15	5300	1.75	5900	2.17	6400	2.55
1900	0.23	3150	0.62	4320	1.16	5320	1.76	5910	2.18	6410	2.56
1925	0.23	3175	0.63	4340	1.17	5340	1.78	5920	2.18	6420	2.57
1950	0.24	3200	0.64	4360	1.19	5360	1.79	5930	2.19	6430	2.58
1975	0.24	3225	0.65	4380	1.20	5380	1.80	5940	2.20	6440	2.59
2000	0.25	3250	0.66	4400	1.21	5400	1.82	5950	2.21	6450	2.59
2025	0.26	3275	0.67	4420	1.22	5420	1.83	5960	2.21	6460	2.60
2050	0.26	3300	0.68	4440	1.23	5440	1.84	5970	2.22	6470	2.61
2075	0.27	3325	0.69	4460	1.24	5460	1.86	5980	2.23	6480	2.62
2100	0.27	3350	0.70	4480	1.25	5480	1.87	5990	2.24	6490	2.63
2125	0.28	3375	0.71	4500	1.26	5500	1.89	6000	2.24	6500	2.63
2150	0.29	3400	0.72	4520	1.27	5510	1.89	6010	2.25	6550	2.67
2175	0.29	3425	0.73	4540	1.29	5520	1.90	6020	2.26	6600	2.72
2200	0.30	3450	0.74	4560	1.30	5530	1.91	6030	2.27	6650	2.76
2225	0.31	3475	0.75	4580	1.31	5540	1.91	6040	2.27	6700	2.80
2250	0.32	3500	0.76	4600	1.32	5550	1.92	6050	2.28	6750	2.84
2275	0.32	3525	0.77	4620	1.33	5560	1.93	6060	2.29	6800	2.88
2300	0.33	3550	0.79	4640	1.34	5570	1.93	6070	2.30	6900	2.97
2325	0.34	3575	0.80	4660	1.35	5580	1.94	6080	2.30	7000	3.05
2350	0.34	3600	0.81	4680	1.37	5590	1.95	6090	2.31	7100	3.14
2375	0.35	3625	0.82	4700	1.38	5600	1.96	6100	2.32	7200	3.23
2400	0.36	3650	0.83	4720	1.39	5610	1.96	6110	2.33	7300	3.32
2425	0.37	3675	0.84	4740	1.40	5620	1.97	6120	2.34	7400	3.41
2450	0.37	3700	0.85	4760	1.41	5630	1.98	6130	2.34	7500	3.51
2475	0.38	3725	0.87	4780	1.42	5640	1.98	6140	2.35	7600	3.60
2500	0.39	3750	0.88	4800	1.44	5650	1.99	6150	2.36	7700	3.70
2525	0.40	3775	0.89	4820	1.45	5660	2.00	6160	2.37	7800	3.79
2550	0.41	3800	0.90	4840	1.46	5670	2.00	6170	2.37	7900	3.89
2575	0.41	3825	0.91	4860	1.47	5680	2.01	6180	2.38	8000	3.99

TABLE 9-3 (SI). Velocity to Velocity Pressure Conversion — Standard Air

$VP = (V / 1.29)^2 df$

V = Velocity m s df = 1
VP = Velocity Pressure Pa

V	VP	V	VP	V	VP	V	VP	V	VP	V	VP
2.00	2	13.50	110	18.90	215	23.90	343	28.90	502	33.90	691
2.25	3	13.70	113	19.00	217	24.00	346	29.00	505	34.00	695
2.50	4	13.90	116	19.10	219	24.10	349	29.10	509	34.10	699
2.75	5	14.10	119	19.20	222	24.20	352	29.20	512	34.20	703
3.00	5	14.30	123	19.30	224	24.30	355	29.30	516	34.30	707
3.25	6	14.40	125	19.40	226	24.40	358	29.40	519	34.40	711
3.50	7	14.50	126	19.50	229	24.50	361	29.50	523	34.50	715
3.75	8	14.60	128	19.60	231	24.60	364	29.60	527	34.60	719
4.00	10	14.70	130	19.70	233	24.70	367	29.70	530	34.70	724
4.25	11	14.80	132	19.80	236	24.80	370	29.80	534	34.80	728
4.50	12	14.90	133	19.90	238	24.90	373	29.90	537	34.90	732
4.75	14	15.00	135	20.00	240	25.00	376	30.00	541	35.00	736
5.00	15	15.10	137	20.10	243	25.10	379	30.10	544	35.10	740
5.25	17	15.20	139	20.20	245	25.20	382	30.20	548	35.20	745
5.50	18	15.30	141	20.30	248	25.30	385	30.30	552	35.30	749
5.75	20	15.40	143	20.40	250	25.40	388	30.40	555	35.40	753
6.00	22	15.50	144	20.50	253	25.50	391	30.50	559	35.50	757
6.25	23	15.60	146	20.60	255	25.60	394	30.60	563	35.60	762
6.50	25	15.70	148	20.70	257	25.70	397	30.70	566	35.70	766
6.75	27	15.80	150	20.80	260	25.80	400	30.80	570	35.80	770
7.00	29	15.90	152	20.90	262	25.90	403	30.90	574	35.90	774
7.25	32	16.00	154	21.00	265	26.00	406	31.00	577	36.00	779
7.50	34	16.10	156	21.10	268	26.10	409	31.10	581	36.10	783
7.75	36	16.20	158	21.20	270	26.20	412	31.20	585	36.20	787
8.00	38	16.30	160	21.30	273	26.30	416	31.30	589	36.30	792
8.25	41	16.40	162	21.40	275	26.40	419	31.40	592	36.40	796
8.50	43	16.50	164	21.50	278	26.50	422	31.50	596	36.50	801
8.75	46	16.60	166	21.60	280	26.60	425	31.60	600	36.60	805
9.00	49	16.70	168	21.70	283	26.70	428	31.70	604	36.70	809
9.25	51	16.80	170	21.80	286	26.80	432	31.80	608	36.80	814
9.50	54	16.90	172	21.90	288	26.90	435	31.90	612	36.90	818
9.75	57	17.00	174	22.00	291	27.00	438	32.00	615	37.00	823
10.00	60	17.10	176	22.10	293	27.10	441	32.10	619	37.10	827
10.10	61	17.20	178	22.20	296	27.20	445	32.20	623	37.20	832
10.30	64	17.30	180	22.30	299	27.30	448	32.30	627	37.30	836
10.50	66	17.40	182	22.40	302	27.40	451	32.40	631	37.40	841
10.70	69	17.50	184	22.50	304	27.50	454	32.50	635	37.50	845
10.90	71	17.60	186	22.60	307	27.60	458	32.60	639	37.60	850
11.10	74	17.70	188	22.70	310	27.70	461	32.70	643	37.70	854
11.30	77	17.80	190	22.80	312	27.80	464	32.80	646	37.80	859
11.50	79	17.90	193	22.90	315	27.90	468	32.90	650	37.90	863
11.70	82	18.00	195	23.00	318	28.00	471	33.00	654	38.00	868
11.90	85	18.10	197	23.10	321	28.10	474	33.10	658	38.10	872
12.10	88	18.20	199	23.20	323	28.20	478	33.20	662	38.20	877
12.30	91	18.30	201	23.30	326	28.30	481	33.30	666	38.30	881
12.50	94	18.40	203	23.40	329	28.40	485	33.40	670	38.40	886
12.70	97	18.50	206	23.50	332	28.50	488	33.50	674	38.50	891
12.90	100	18.60	208	23.60	335	28.60	492	33.60	678	38.60	895
13.10	103	18.70	210	23.70	338	28.70	495	33.70	682	38.70	900
13.30	106	18.80	212	23.80	340	28.80	498	33.80	687	38.80	905

TABLE 9-4 (IP). Duct Friction Loss Factors, F'_d

<table>
<tr><th colspan="7">Sheet Metal and Plastic Duct</th></tr>
<tr><th>Diameter</th><th colspan="6">Friction Loss, No VP per foot</th></tr>
<tr><th>inches</th><th>1000 fpm</th><th>2000 fpm</th><th>3000 fpm</th><th>4000 fpm</th><th>5000 fpm</th><th>6000 fpm</th></tr>
<tr><td>0.5</td><td>1.008</td><td>0.955</td><td>0.925</td><td>0.904</td><td>0.888</td><td>0.875</td></tr>
<tr><td>1</td><td>0.432</td><td>0.409</td><td>0.396</td><td>0.387</td><td>0.380</td><td>0.375</td></tr>
<tr><td>1.5</td><td>0.263</td><td>0.249</td><td>0.241</td><td>0.236</td><td>0.231</td><td>0.228</td></tr>
<tr><td>2</td><td>0.185</td><td>0.175</td><td>0.169</td><td>0.166</td><td>0.163</td><td>0.160</td></tr>
<tr><td>2.5</td><td>0.141</td><td>0.133</td><td>0.129</td><td>0.126</td><td>0.124</td><td>0.122</td></tr>
<tr><td>3</td><td>0.113</td><td>0.107</td><td>0.103</td><td>0.101</td><td>0.099</td><td>0.098</td></tr>
<tr><td>3.5</td><td>0.093</td><td>0.088</td><td>0.085</td><td>0.083</td><td>0.082</td><td>0.081</td></tr>
<tr><td>4</td><td>0.079</td><td>0.075</td><td>0.073</td><td>0.071</td><td>0.070</td><td>0.069</td></tr>
<tr><td>4.5</td><td>0.068</td><td>0.065</td><td>0.063</td><td>0.061</td><td>0.060</td><td>0.059</td></tr>
<tr><td>5</td><td>0.060</td><td>0.057</td><td>0.055</td><td>0.054</td><td>0.053</td><td>0.052</td></tr>
<tr><td>5.5</td><td>0.054</td><td>0.051</td><td>0.049</td><td>0.048</td><td>0.047</td><td>0.047</td></tr>
<tr><td>6</td><td>0.048</td><td>0.046</td><td>0.044</td><td>0.043</td><td>0.042</td><td>0.042</td></tr>
<tr><td>7</td><td>0.040</td><td>0.038</td><td>0.037</td><td>0.036</td><td>0.035</td><td>0.035</td></tr>
<tr><td>8</td><td>0.034</td><td>0.032</td><td>0.031</td><td>0.030</td><td>0.030</td><td>0.029</td></tr>
<tr><td>9</td><td>0.029</td><td>0.028</td><td>0.027</td><td>0.026</td><td>0.026</td><td>0.025</td></tr>
<tr><td>10</td><td>0.026</td><td>0.024</td><td>0.024</td><td>0.023</td><td>0.023</td><td>0.022</td></tr>
<tr><td>11</td><td>0.023</td><td>0.022</td><td>0.021</td><td>0.021</td><td>0.020</td><td>0.020</td></tr>
<tr><td>12</td><td>0.021</td><td>0.020</td><td>0.019</td><td>0.018</td><td>0.018</td><td>0.018</td></tr>
<tr><td>13</td><td>0.019</td><td>0.018</td><td>0.017</td><td>0.017</td><td>0.016</td><td>0.016</td></tr>
<tr><td>14</td><td>0.017</td><td>0.016</td><td>0.016</td><td>0.015</td><td>0.015</td><td>0.015</td></tr>
<tr><td>15</td><td>0.016</td><td>0.015</td><td>0.014</td><td>0.014</td><td>0.014</td><td>0.014</td></tr>
<tr><td>16</td><td>0.014</td><td>0.014</td><td>0.013</td><td>0.013</td><td>0.013</td><td>0.013</td></tr>
<tr><td>17</td><td>0.013</td><td>0.013</td><td>0.012</td><td>0.012</td><td>0.012</td><td>0.012</td></tr>
<tr><td>18</td><td>0.013</td><td>0.012</td><td>0.012</td><td>0.011</td><td>0.011</td><td>0.011</td></tr>
<tr><td>19</td><td>0.012</td><td>0.011</td><td>0.011</td><td rowspan="2">0.010</td><td rowspan="2">0.010</td><td rowspan="2">0.010</td></tr>
<tr><td>20</td><td>0.011</td><td rowspan="2">0.010</td><td rowspan="2">0.010</td></tr>
<tr><td>21</td><td rowspan="2">0.010</td><td rowspan="2">0.009</td><td rowspan="2">0.009</td><td rowspan="2">0.009</td></tr>
<tr><td>22</td><td rowspan="2">0.009</td><td rowspan="2">0.009</td></tr>
<tr><td>23</td><td rowspan="2">0.009</td><td rowspan="3">0.008</td><td rowspan="2">0.008</td><td rowspan="2">0.008</td></tr>
<tr><td>24</td><td rowspan="3">0.008</td><td rowspan="2">0.008</td></tr>
<tr><td>25</td><td rowspan="3">0.008</td><td rowspan="3">0.007</td><td rowspan="3">0.007</td></tr>
<tr><td>26</td><td rowspan="3">0.007</td><td rowspan="3">0.007</td></tr>
<tr><td>27</td><td rowspan="3">0.007</td></tr>
<tr><td>28</td><td rowspan="3">0.007</td><td rowspan="5">0.006</td><td rowspan="5">0.006</td></tr>
<tr><td>29</td><td rowspan="4">0.006</td><td rowspan="4">0.006</td></tr>
<tr><td>30</td><td rowspan="3">0.006</td></tr>
<tr><td>31</td><td rowspan="2">0.006</td></tr>
<tr><td>32</td></tr>
</table>

$$F'_d = 0.0307 \frac{V^{0.533}}{Q^{0.612}}$$

TABLE 9-4 (IP) (Cont.). Duct Friction Loss Factors, F'_d

<table>
<tr><th colspan="7">Sheet Metal and Plastic Duct</th></tr>
<tr><th>Diameter</th><th colspan="6">Friction Loss, No VP per foot</th></tr>
<tr><th>inches</th><th>1000 fpm</th><th>2000 fpm</th><th>3000 fpm</th><th>4000 fpm</th><th>5000 fpm</th><th>6000 fpm</th></tr>
<tr><td>33</td><td rowspan="3">0.006</td><td>0.006</td><td rowspan="6">0.005</td><td rowspan="6">0.005</td><td rowspan="5">0.005</td><td rowspan="5">0.005</td></tr>
<tr><td>34</td><td rowspan="6">0.005</td></tr>
<tr><td>35</td></tr>
<tr><td>36</td><td rowspan="6">0.005</td></tr>
<tr><td>37</td></tr>
<tr><td>38</td><td rowspan="9">0.004</td><td rowspan="8">0.004</td></tr>
<tr><td>39</td><td rowspan="9">0.004</td><td rowspan="8">0.004</td></tr>
<tr><td>40</td><td rowspan="9">0.004</td></tr>
<tr><td>41</td></tr>
<tr><td>42</td><td rowspan="9">0.004</td></tr>
<tr><td>43</td></tr>
<tr><td>44</td></tr>
<tr><td>45</td></tr>
<tr><td>46</td><td rowspan="9">0.003</td></tr>
<tr><td>47</td><td rowspan="9">0.003</td><td rowspan="9">0.003</td></tr>
<tr><td>48</td><td rowspan="9">0.003</td></tr>
<tr><td>49</td><td rowspan="9">0.003</td></tr>
<tr><td>50</td></tr>
<tr><td>52</td><td rowspan="8">0.003</td></tr>
<tr><td>54</td></tr>
<tr><td>56</td></tr>
<tr><td>58</td></tr>
<tr><td>60</td><td rowspan="16">0.002</td></tr>
<tr><td>62</td><td rowspan="15">0.002</td><td rowspan="15">0.002</td></tr>
<tr><td>64</td><td rowspan="14">0.002</td></tr>
<tr><td>66</td><td rowspan="13">0.002</td></tr>
<tr><td>68</td><td rowspan="12">0.002</td></tr>
<tr><td>70</td></tr>
<tr><td>72</td></tr>
<tr><td>74</td></tr>
<tr><td>76</td></tr>
<tr><td>78</td></tr>
<tr><td>80</td></tr>
<tr><td>82</td></tr>
<tr><td>84</td></tr>
<tr><td>86</td></tr>
<tr><td>88</td></tr>
<tr><td>90</td></tr>
</table>

$$F'_d = 0.0307 \frac{V^{0.533}}{Q^{0.612}}$$

TABLE 9-4 (SI). Duct Friction Loss Factors, F'_d

<table>
<tr><th colspan="7">Sheet Metal and Plastic Duct</th></tr>
<tr><th>Diameter</th><th colspan="6">Friction Loss, No Pa per meter</th></tr>
<tr><th>mm</th><th>5 m/s</th><th>10 m/s</th><th>15 m/s</th><th>20 m/s</th><th>25 m/s</th><th>30 m/s</th></tr>
<tr><td>20</td><td>1.900</td><td>1.799</td><td>1.742</td><td>1.703</td><td>1.674</td><td>1.650</td></tr>
<tr><td>40</td><td>0.814</td><td>0.770</td><td>0.746</td><td>0.729</td><td>0.716</td><td>0.706</td></tr>
<tr><td>60</td><td>0.495</td><td>0.469</td><td>0.454</td><td>0.444</td><td>0.436</td><td>0.430</td></tr>
<tr><td>80</td><td>0.348</td><td>0.330</td><td>0.319</td><td>0.312</td><td>0.307</td><td>0.302</td></tr>
<tr><td>100</td><td>0.265</td><td>0.251</td><td>0.243</td><td>0.238</td><td>0.233</td><td>0.230</td></tr>
<tr><td>120</td><td>0.212</td><td>0.201</td><td>0.194</td><td>0.190</td><td>0.187</td><td>0.184</td></tr>
<tr><td>140</td><td>0.176</td><td>0.166</td><td>0.161</td><td>0.157</td><td>0.155</td><td>0.152</td></tr>
<tr><td>160</td><td>0.149</td><td>0.141</td><td>0.137</td><td>0.134</td><td>0.131</td><td>0.129</td></tr>
<tr><td>180</td><td>0.129</td><td>0.122</td><td>0.118</td><td>0.116</td><td>0.114</td><td>0.112</td></tr>
<tr><td>200</td><td>0.113</td><td>0.107</td><td>0.104</td><td>0.102</td><td>0.100</td><td>0.098</td></tr>
<tr><td>220</td><td>0.101</td><td>0.096</td><td>0.093</td><td>0.090</td><td>0.089</td><td>0.088</td></tr>
<tr><td>240</td><td>0.091</td><td>0.086</td><td>0.083</td><td>0.081</td><td>0.080</td><td>0.079</td></tr>
<tr><td>260</td><td>0.082</td><td>0.078</td><td>0.075</td><td>0.074</td><td>0.072</td><td>0.071</td></tr>
<tr><td>280</td><td>0.075</td><td>0.071</td><td>0.069</td><td>0.067</td><td>0.066</td><td>0.065</td></tr>
<tr><td>300</td><td>0.069</td><td>0.065</td><td>0.063</td><td>0.062</td><td>0.061</td><td>0.060</td></tr>
<tr><td>325</td><td>0.063</td><td>0.059</td><td>0.057</td><td>0.056</td><td>0.055</td><td>0.054</td></tr>
<tr><td>350</td><td>0.057</td><td>0.054</td><td>0.052</td><td>0.051</td><td>0.050</td><td>0.050</td></tr>
<tr><td>375</td><td>0.053</td><td>0.050</td><td>0.048</td><td>0.047</td><td>0.046</td><td>0.046</td></tr>
<tr><td>400</td><td>0.049</td><td>0.046</td><td>0.045</td><td>0.044</td><td>0.043</td><td>0.042</td></tr>
<tr><td>425</td><td>0.045</td><td>0.043</td><td>0.041</td><td>0.040</td><td>0.040</td><td>0.039</td></tr>
<tr><td>450</td><td>0.042</td><td>0.040</td><td>0.039</td><td>0.038</td><td>0.037</td><td>0.037</td></tr>
<tr><td>475</td><td>0.039</td><td>0.037</td><td>0.036</td><td>0.035</td><td>0.035</td><td>0.034</td></tr>
<tr><td>500</td><td>0.037</td><td>0.035</td><td>0.034</td><td>0.033</td><td>0.033</td><td>0.032</td></tr>
<tr><td>550</td><td>0.033</td><td>0.031</td><td>0.030</td><td>0.029</td><td>0.029</td><td>0.029</td></tr>
<tr><td>600</td><td>0.030</td><td>0.028</td><td>0.027</td><td>0.027</td><td>0.026</td><td>0.026</td></tr>
<tr><td>650</td><td>0.027</td><td>0.025</td><td>0.025</td><td>0.024</td><td>0.024</td><td>0.023</td></tr>
<tr><td>700</td><td>0.024</td><td>0.023</td><td>0.022</td><td>0.022</td><td>0.022</td><td>0.021</td></tr>
<tr><td>750</td><td>0.023</td><td>0.021</td><td>0.021</td><td>0.020</td><td>0.020</td><td>0.020</td></tr>
<tr><td>800</td><td>0.021</td><td>0.020</td><td>0.019</td><td>0.019</td><td>0.018</td><td>0.018</td></tr>
<tr><td>850</td><td>0.019</td><td>0.018</td><td>0.018</td><td>0.017</td><td>0.017</td><td>0.017</td></tr>
<tr><td>900</td><td>0.018</td><td>0.017</td><td>0.017</td><td>0.016</td><td>0.016</td><td>0.016</td></tr>
<tr><td>950</td><td>0.017</td><td>0.016</td><td>0.015</td><td>0.015</td><td>0.015</td><td>0.015</td></tr>
<tr><td>1000</td><td>0.016</td><td>0.015</td><td>0.015</td><td>0.014</td><td>0.014</td><td>0.014</td></tr>
<tr><td>1050</td><td>0.015</td><td>0.014</td><td>0.014</td><td rowspan="3">0.013</td><td>0.013</td><td>0.013</td></tr>
<tr><td>1100</td><td>0.014</td><td>0.013</td><td>0.013</td><td rowspan="2">0.012</td><td rowspan="2">0.012</td></tr>
<tr><td>1150</td><td>0.013</td><td>0.013</td><td rowspan="2">0.012</td></tr>
<tr><td>1200</td><td>0.013</td><td>0.012</td><td rowspan="3">0.011</td><td rowspan="3">0.011</td><td rowspan="3">0.011</td></tr>
<tr><td>1250</td><td>0.012</td><td rowspan="2">0.011</td><td rowspan="2">0.011</td></tr>
<tr><td>1300</td><td>0.011</td></tr>
</table>

TABLE 9-4 (SI) (Cont.). Duct Friction Loss Factors, F'_d

<table>
<tr><th></th><th colspan="6">Sheet Metal and Plastic Duct</th></tr>
<tr><th>Diameter</th><th colspan="6">Friction Loss, No Pa per meter</th></tr>
<tr><th>mm</th><th>5 m/s</th><th>10 m/s</th><th>15 m/s</th><th>20 m/s</th><th>25 m/s</th><th>30 m/s</th></tr>
<tr><td>1350</td><td>0.011</td><td rowspan="3">0.010</td><td rowspan="2">0.010</td><td>0.010</td><td>0.010</td><td>0.010</td></tr>
<tr><td>1400</td><td rowspan="3">0.010</td><td rowspan="3">0.009</td><td rowspan="2">0.009</td><td rowspan="2">0.009</td></tr>
<tr><td>1450</td><td rowspan="2">0.009</td></tr>
<tr><td>1500</td><td rowspan="2">0.009</td><td rowspan="4">0.008</td><td rowspan="3">0.008</td></tr>
<tr><td>1550</td><td rowspan="3">0.009</td><td rowspan="4">0.008</td><td rowspan="3">0.008</td></tr>
<tr><td>1600</td><td rowspan="4">0.008</td></tr>
<tr><td>1650</td><td>0.007</td></tr>
<tr><td>1700</td><td rowspan="3">0.008</td><td rowspan="4">0.007</td><td rowspan="4">0.007</td><td rowspan="3">0.007</td></tr>
<tr><td>1750</td><td rowspan="4">0.007</td></tr>
<tr><td>1800</td></tr>
<tr><td>1850</td><td rowspan="5">0.007</td><td rowspan="6">0.006</td></tr>
<tr><td>1900</td><td rowspan="6">0.006</td><td rowspan="5">0.006</td></tr>
<tr><td>1950</td><td rowspan="6">0.006</td></tr>
<tr><td>2000</td><td rowspan="6">0.006</td></tr>
<tr><td>2050</td></tr>
<tr><td>2100</td><td rowspan="6">0.006</td></tr>
<tr><td>2150</td><td rowspan="8">0.005</td><td rowspan="7">0.005</td></tr>
<tr><td>2200</td><td rowspan="8">0.005</td></tr>
<tr><td>2250</td><td rowspan="8">0.005</td></tr>
<tr><td>2300</td><td rowspan="8">0.005</td></tr>
<tr><td>2350</td></tr>
<tr><td>2400</td><td rowspan="8">0.005</td></tr>
<tr><td>2450</td></tr>
<tr><td>2500</td><td rowspan="12">0.004</td></tr>
<tr><td>2550</td><td rowspan="11">0.004</td></tr>
<tr><td>2600</td><td rowspan="11">0.004</td></tr>
<tr><td>2650</td><td rowspan="11">0.004</td></tr>
<tr><td>2700</td><td rowspan="12">0.004</td></tr>
<tr><td>2750</td></tr>
<tr><td>2800</td><td rowspan="10">0.004</td></tr>
<tr><td>2850</td></tr>
<tr><td>2900</td></tr>
<tr><td>2950</td></tr>
<tr><td>3000</td></tr>
<tr><td>3050</td></tr>
<tr><td>3100</td><td rowspan="4">0.003</td><td rowspan="4">0.003</td></tr>
<tr><td>3150</td><td rowspan="3">0.003</td></tr>
<tr><td>3200</td><td rowspan="2">0.003</td></tr>
<tr><td>3250</td></tr>
</table>

$$F'_d = 0.0155 \frac{V^{0.533}}{Q^{0.612}}$$

TABLE 9-5 (IP). Circular Equivalents of Rectangular Ducts (in)

a \ b	4.0	4.5	5.0	5.5	6.0	6.5	7.0	7.5	8.0	8.5	9.0	9.5	10.0	10.5	11.0	11.5	12.0	12.5	13.0	13.5	14.0	14.5	15.0	15.5	16.0	16.5	17.0	17.5	18.0	18.5
3.0	3.8	4.0	4.2	4.4	4.6	4.7	4.9	5.1	5.2	5.3	5.5	5.6	5.7	5.9	6.0	6.1	6.2	6.3	6.4	6.5	6.6	6.7	6.8	6.9	7.0	7.1	7.2	7.3	7.3	7.4
3.5	4.1	4.3	4.6	4.8	5.0	5.2	5.3	5.5	5.7	5.8	6.0	6.1	6.3	6.4	6.5	6.7	6.8	6.9	7.0	7.1	7.2	7.3	7.5	7.6	7.7	7.8	7.9	7.9	8.0	8.1
4.0	4.4	4.6	4.9	5.1	5.3	5.5	5.7	5.9	6.1	6.3	6.4	6.6	6.7	6.9	7.0	7.2	7.3	7.4	7.6	7.7	7.8	7.9	8.0	8.2	8.3	8.4	8.5	8.6	8.7	8.8
4.5	4.6	4.9	5.2	5.4	5.7	5.9	6.1	6.3	6.5	6.7	6.9	7.0	7.2	7.4	7.5	7.7	7.8	7.9	8.1	8.2	8.4	8.5	8.6	8.7	8.8	9.0	9.1	9.2	9.3	9.4
5.0	4.9	5.2	5.5	5.7	6.0	6.2	6.4	6.7	6.9	7.1	7.3	7.4	7.6	7.8	8.0	8.1	8.3	8.4	8.6	8.7	8.9	9.0	9.1	9.3	9.4	9.5	9.6	9.8	9.9	10.0
5.5	5.1	5.4	5.7	6.0	6.3	6.5	6.8	7.0	7.2	7.4	7.6	7.8	8.0	8.2	8.4	8.6	8.7	8.9	9.0	9.2	9.3	9.5	9.6	9.8	9.9	10.0	10.2	10.3	10.4	10.6

a \ b	6.0	7.0	8.0	9.0	10.0	11.0	12.0	13.0	14.0	15.0	16.0	17.0	18.0	19.0	20.0	22.0	24.0	26.0	28.0	30.0	32.0	34.0	36.0	38.0	40.0	42.0	44.0	46.0	48.0	50.0
6.0	6.6																													
7.0	7.1	7.7																												
8.0	7.6	8.2	8.7																											
9.0	8.0	8.7	9.3	9.8																										
10.0	8.4	9.1	9.8	10.4	10.9																									
11.0	8.8	9.5	10.2	10.9	11.5	12.0																								
12.0	9.1	9.9	10.7	11.3	12.0	12.6	13.1																							
13.0	9.5	10.3	11.1	11.8	12.4	13.1	13.7	14.2																						
14.0	9.8	10.7	11.5	12.2	12.9	13.5	14.2	14.7	15.3																					
15.0	10.1	11.0	11.8	12.6	13.3	14.0	14.6	15.3	15.8	16.4																				
16.0	10.4	11.3	12.2	13.0	13.7	14.4	15.1	15.7	16.4	16.9	17.5																			
17.0	10.7	11.6	12.5	13.4	14.1	14.9	15.6	16.2	16.8	17.4	18.0	18.6																		
18.0	11.0	11.9	12.9	13.7	14.5	15.3	16.0	16.7	17.3	17.9	18.5	19.1	19.5																	
19.0	11.2	12.2	13.2	14.1	14.9	15.7	16.4	17.1	17.8	18.4	19.0	19.6	20.1	20.8																
20.0	11.5	12.5	13.5	14.4	15.2	16.0	16.8	17.5	18.2	18.9	19.5	20.1	20.6	21.3	21.9															
22.0	12.0	13.0	14.1	15.0	15.9	16.8	17.6	18.3	19.1	19.8	20.4	21.1	21.6	22.3	22.9	24.0														
24.0	12.4	13.5	14.6	15.6	16.5	17.4	18.3	19.1	19.9	20.6	21.3	22.0	22.5	23.3	23.9	25.1	26.2													
26.0	12.8	14.0	15.1	16.2	17.1	18.1	19.0	19.8	20.6	21.4	22.1	22.9	23.4	24.2	24.9	26.1	27.3	28.4												
28.0	13.2	14.5	15.6	16.7	17.7	18.7	19.6	20.5	21.3	22.1	22.9	23.7	24.3	25.1	25.8	27.1	28.3	29.5	30.6											
30.0	13.6	14.9	16.1	17.2	18.3	19.3	20.2	21.1	22.0	22.9	23.7	24.4	25.1	25.9	26.6	28.0	29.3	30.5	31.7	32.8										
32.0	14.0	15.3	16.5	17.7	18.8	19.8	20.8	21.8	22.7	23.5	24.4	25.2	25.8	26.7	27.5	28.9	30.2	31.5	32.7	33.9	35.0									
34.0	14.4	15.7	17.0	18.2	19.3	20.4	21.4	22.4	23.3	24.2	25.1	25.9	26.6	27.5	28.3	29.7	31.1	32.4	33.7	34.9	36.1	37.2								
36.0	14.7	16.1	17.4	18.6	19.8	20.9	21.9	22.9	23.9	24.8	25.7	26.6	27.3	28.2	29.0	30.5	32.0	33.3	34.6	35.9	37.1	38.2	39.4							
38.0	15.0	16.5	17.8	19.0	20.2	21.4	22.4	23.5	24.5	25.4	26.4	27.2	28.0	28.9	29.8	31.3	32.8	34.2	35.6	36.8	38.1	39.3	40.4	41.5						
40.0	15.3	16.8	18.2	19.5	20.7	21.8	22.9	24.0	25.0	26.0	27.0	27.9	28.6	29.6	30.5	32.1	33.6	35.1	36.4	37.8	39.0	40.3	41.5	42.6	43.7					
42.0	15.6	17.1	18.5	19.9	21.1	22.3	23.4	24.5	25.6	26.6	27.6	28.5	29.3	30.3	31.2	32.8	34.4	35.9	37.3	38.7	40.0	41.3	42.5	43.7	44.8	45.9				
44.0	15.9	17.5	18.9	20.3	21.5	22.7	23.9	25.0	26.1	27.1	28.1	29.1	29.9	30.9	31.8	33.5	35.1	36.7	38.1	39.5	40.9	42.2	43.5	44.7	45.8	47.0	48.1			
46.0	16.2	17.8	19.3	20.6	21.9	23.2	24.4	25.5	26.6	27.7	28.7	29.7	30.5	31.6	32.5	34.2	35.9	37.4	38.9	40.4	41.8	43.1	44.4	45.7	46.9	48.0	49.2	50.3		
48.0	16.5	18.1	19.6	21.0	22.3	23.6	24.8	26.0	27.1	28.2	29.2	30.2	31.1	32.2	33.1	34.9	36.6	38.2	39.7	41.2	42.6	44.0	45.3	46.6	47.9	49.1	50.2	51.4	52.5	
50.0	16.8	18.4	19.9	21.4	22.7	24.0	25.2	26.4	27.6	28.7	29.8	30.8	31.7	32.8	33.7	35.5	37.2	38.9	40.5	42.0	43.5	44.9	46.2	47.5	48.8	50.0	51.2	52.4	53.6	54.7
54.0	17.3	19.0	20.6	22.0	23.5	24.8	26.1	27.3	28.5	29.7	30.8	31.8	32.8	33.9	34.9	36.8	38.6	40.3	41.9	43.5	45.1	46.5	48.0	49.3	50.7	52.0	53.2	54.4	55.6	56.8
58.0	17.8	19.5	21.2	22.7	24.2	25.5	26.9	28.2	29.4	30.6	31.7	32.8	33.8	35.0	36.0	38.0	39.8	41.6	43.3	45.0	46.6	48.1	49.6	51.0	52.4	53.8	55.1	56.4	57.6	58.8
62.0	18.3	20.1	21.7	23.3	24.8	26.3	27.6	28.9	30.2	31.5	32.6	33.8	34.8	36.0	37.1	39.1	41.0	42.9	44.7	46.4	48.0	49.6	51.2	52.7	54.1	55.5	56.9	58.2	59.5	60.8
66.0	18.8	20.6	22.3	23.9	25.5	26.9	28.4	29.7	31.0	32.3	33.5	34.7	35.8	37.0	38.1	40.2	42.2	44.1	46.0	47.7	49.4	51.1	52.7	54.2	55.7	57.2	58.6	60.0	61.3	62.6
70.0	19.2	21.1	22.8	24.5	26.1	27.6	29.1	30.4	31.8	33.1	34.4	35.6	36.7	37.9	39.1	41.2	43.3	45.3	47.2	49.0	50.8	52.5	54.1	55.7	57.3	58.8	60.3	61.7	63.1	64.4
74.0	19.6	21.5	23.3	25.1	26.7	28.2	29.7	31.2	32.5	33.9	35.2	36.4	37.6	38.8	40.0	42.2	44.4	46.4	48.4	50.3	52.1	53.8	55.5	57.2	58.8	60.3	61.9	63.3	64.8	66.2
78.0	20.0	22.0	23.8	25.6	27.3	28.8	30.4	31.8	33.3	34.6	36.0	37.2	38.4	39.7	40.9	43.2	45.4	47.5	49.5	51.4	53.3	55.1	56.9	58.6	60.2	61.8	63.4	64.9	66.4	67.9
82.0	20.4	22.4	24.3	26.1	27.8	29.4	31.0	32.5	33.9	35.4	36.7	38.0	39.2	40.6	41.8	44.1	46.4	48.5	50.6	52.6	54.5	56.4	58.2	59.9	61.6	63.3	64.9	66.5	68.0	69.5
86.0	20.8	22.9	24.8	26.6	28.3	30.0	31.6	33.1	34.6	36.1	37.4	38.8	40.0	41.4	42.6	45.0	47.3	49.6	51.7	53.7	55.7	57.6	59.4	61.2	63.0	64.7	66.3	67.9	69.5	71.0
90.0	21.2	23.3	25.2	27.1	28.9	30.6	32.2	33.8	35.3	36.7	38.2	39.5	40.8	42.2	43.5	45.9	48.3	50.5	52.7	54.8	56.8	58.8	60.7	62.5	64.3	66.0	67.7	69.4	71.0	72.6

$$D_{equiv} = 1.3\frac{a \times b^{0.625}}{a + b^{0.25}}$$

D_{equiv} = equivalent round size of rectangular duct, in
a = one side of rectangular duct, in
b = adjacent side of rectangular duct, in

TABLE 9-5 (SI). Circular Equivalents of Rectangular Ducts (mm)

a \ b	100	120	140	160	180	200	220	240	260	300	320	340	360	380	400	420	440	460	480	500	520	540	560	580
80	98	107	115	122	129	135	141	146	151	161	165	170	174	178	182	185	189	193	196	199	202	206	209	212
100	109	120	129	137	145	152	159	165	171	183	188	193	198	202	207	211	215	219	223	227	231	235	238	242
120	120	131	142	151	160	168	176	183	190	202	208	214	219	224	229	234	239	244	248	252	257	261	265	269
140	129	142	153	164	173	182	191	199	206	220	227	233	239	245	250	256	261	266	271	276	280	285	289	294
160	137	151	164	175	185	195	204	213	221	237	244	251	257	264	270	276	281	287	292	298	303	308	312	317
180	145	160	173	185	197	207	217	227	235	252	260	267	274	281	288	294	300	306	312	318	323	329	334	339
200	152	168	182	195	207	219	229	239	249	266	275	283	290	298	305	312	318	325	331	337	343	349	354	360
220	159	176	191	204	217	229	240	251	261	280	289	297	305	313	321	328	335	342	349	355	362	368	374	380

a \ b	100	200	300	400	500	600	700	800	900	1000	1100	1200	1300	1400	1500	1600	1700	1800	1900	2000	2100	2200	2300	2400
100	109																							
200	166	257																						
300	183	266	328																					
400	207	305	378	437																				
500	227	337	420	488	547																			
600	245	365	457	533	598	656																		
700	261	391	490	573	644	708	765																	
800	275	414	520	609	687	755	818	875	984															
900	289	435	548	643	726	799	866	927	1037															
1000	301	454	574	674	762	840	911	976	1086	1093														
1100	313	473	598	703	795	878	953	1022	1133	1146	1202													
1200	324	490	620	731	827	914	993	1066	1177	1196	1256	1312												
1300	334	506	642	757	857	948	1031	1107	1220	1244	1306	1365	1421											
1400	344	522	662	781	886	980	1066	1146	1260	1289	1354	1416	1475	1530										
1500	353	536	681	805	913	1011	1100	1183	1298	1332	1400	1464	1526	1584	1640									
1600	362	551	700	827	939	1041	1133	1219	1335	1373	1444	1511	1574	1635	1693	1749								
1700	371	564	718	849	964	1069	1164	1253	1371	1413	1486	1555	1621	1684	1745	1803	1858							
1800	379	577	735	869	988	1096	1195	1286	1405	1451	1527	1598	1667	1732	1794	1854	1912	1968						
1900	387	590	751	889	1012	1122	1224	1318	1438	1488	1566	1640	1710	1778	1842	1904	1964	2021	2077					
2000	395	602	767	908	1034	1147	1252	1348	1605	1523	1604	1680	1753	1822	1889	1952	2014	2073	2131	2186				
2100	407	627	808	967	1112	1246	1372	1492	1501	1715	1820	1922	2020	2116	2209	2300	2389	2476	2561	2645	2726			
2200	410	625	797	945	1076	1195	1305	1406	1532	1591	1676	1756	1833	1906	1977	2044	2110	2173	2233	2292	2350	2405		
2300	417	636	812	963	1097	1218	1330	1434	1561	1623	1710	1793	1871	1947	2019	2088	2155	2220	2283	2343	2402	2459	2514	
2400	424	647	826	980	1116	1241	1355	1461	1561	1655	1744	1828	1909	1986	2060	2131	2200	2266	2330	2393	2453	2511	2568	2624

$$D_{equiv} = 1.3 \frac{a \times b^{\ 0.625}}{a + b^{\ 0.25}}$$

D_{equiv} = equivalent round size of rectangular duct, mm

a = one side of rectangular duct, mm

b = adjacent side of rectangular duct, mm

TABLE 9-6 (IP). Air Density Correction Factor (Temperature and Elevation Only), $df_T \times df_E$

	ALTITUDE RELATIVE TO SEA LEVEL, ft															
	-5000	-4000	-3000	-2000	-1000	0	1000	2000	3000	4000	5000	6000	7000	8000	9000	10000
	Barometric Pressure															
in hg	35,74	34.5	33.3	32.2	31.0	29.9	26.9	27.8	26.8	25.8	24.9	23.9	23.1	22.2	21.4	20.6
in wg	486.7	469.7	453.3	437.6	422.2	407.2	365.6	378.6	365.0	351.7	338.8	325.1	314.3	302.4	291.1	280.0
Temp F	DENSITY FACTOR df															
-40	1.50	1.45	1.40	1.35	1.31	1.26	1.22	1.18	1.13	1.09	1.05	1.02	0.98	0.94	0.91	0.87
0	1.37	1.32	1.28	1.24	1.14	1.15	1.11	1.07	1.04	1.00	0.96	0.93	0.89	0.86	0.83	0.80
40	1.26	1.22	1.18	1.14	1.10	1.06	1.02	0.99	0.95	0.92	0.89	0.85	0.82	0.79	0.76	0.73
70	1.19	1.15	1.11	1.07	1.04	1.00	0.97	0.93	0.90	0.87	0.84	0.81	0.78	0.75	0.72	0.69
100	1.13	1.09	1.05	1.02	0.98	0.95	0.91	0.88	0.85	0.82	0.79	0.76	0.73	0.71	0.68	0.66
150	1.03	1.00	0.97	0.93	0.90	0.87	0.84	0.81	0.78	0.75	0.73	0.70	0.67	0.65	0.63	0.60
200	0.96	0.92	0.89	0.86	0.83	0.80	0.78	0.75	0.72	0.70	0.67	0.65	0.62	0.60	0.58	0.56
250	0.89	0.86	0.83	0.80	0.77	0.75	0.72	0.70	0.67	0.65	0.62	0.60	0.58	0.56	0.54	0.52
300	0.83	0.80	0.77	0.75	0.72	0.70	0.67	0.65	0.63	0.60	0.58	0.56	0.54	0.52	0.50	0.48
350	0.78	0.75	0.73	0.70	0.68	0.65	0.63	0.61	0.59	0.57	0.55	0.53	0.51	0.49	0.47	0.45
400	0.73	0.71	0.68	0.66	0.64	0.62	0.59	0.57	0.55	0.53	0.51	0.50	0.48	0.46	0.44	0.43
450	0.69	0.67	0.65	0.62	0.60	0.58	0.56	0.54	0.52	0.50	0.49	0.47	0.45	0.44	0.42	0.40
500	0.66	0.63	0.61	0.59	0.57	0.55	0.53	0.51	0.50	0.48	0.46	0.44	0.43	0.41	0.40	0.38
550	0.62	0.60	0.58	0.56	0.54	0.52	0.51	0.49	0.47	0.45	0.44	0.42	0.41	0.39	0.38	0.36
600	0.60	0.57	0.56	0.54	0.52	0.50	0.48	0.47	0.45	0.43	0.42	0.40	0.39	0.37	0.36	0.35
700	0.54	0.53	0.51	0.49	0.47	0.46	0.44	0.43	0.41	0.40	0.38	0.37	0.35	0.34	0.33	0.32
800	0.50	0.48	0.47	0.45	0.44	0.42	0.41	0.39	0.38	0.36	0.35	0.34	0.33	0.31	0.30	0.29
900	0.46	0.45	0.43	0.42	0.40	0.39	0.38	0.36	0.35	0.34	0.33	0.31	0.30	0.29	0.28	0.27
1000	0.43	0.42	0.40	0.39	0.38	0.36	0.35	0.34	0.33	0.31	0.30	0.29	0.28	0.27	0.26	0.25

TABLE 9-6 (SI). Air Density Correction Factor (Temperature and Elevation Only), $df_T \times df_E$

	ALTITUDE RELATIVE TO SEA LEVEL, m															
	-2500	-2000	-1500	-1000	-500	0	500	1000	1500	2000	2500	3000	3500	4000	4500	5000
	Barometric Pressure															
mm hg	831	816	802	788	774	760	746	733	720	707	694	681	689	656	644	633
kPa	111	109	107	105	103	101	99.4	97.6	95.9	94.1	92.5	90.8	89.2	87.5	85.9	84.3
Temp C	DENSITY FACTOR df															
-10	1.47	1.39	1.32	1.25	1.18	1.11	1.05	0.99	0.94	0.88	0.83	0.78	0.74	0.69	0.65	0.61
0	1.10	1.34	1.27	1.20	1.14	1.07	1.01	0.96	0.90	0.85	0.80	0.75	0.71	0.67	0.63	0.59
10	1.06	1.29	1.22	1.16	1.10	1.04	0.98	0.92	0.87	0.82	0.77	0.73	0.68	0.64	0.60	0.57
20	1.03	1.25	1.18	1.12	1.06	1.00	0.94	0.89	0.84	0.79	0.75	0.70	0.66	0.62	0.58	0.55
30	0.99	1.21	1.14	1.08	1.02	0.97	0.91	0.86	0.81	0.77	0.72	0.68	0.64	0.60	0.56	0.53
40	0.96	1.17	1.11	1.05	0.99	0.94	0.88	0.83	0.79	0.74	0.70	0.66	0.62	0.58	0.55	0.51
50	0.93	1.13	1.07	1.01	0.96	0.91	0.86	0.81	0.76	0.72	0.68	0.64	0.60	0.56	0.53	0.50
60	0.90	1.10	1.04	0.98	0.93	0.88	0.83	0.78	0.74	0.70	0.66	0.62	0.58	0.55	0.51	0.48
70	0.88	1.07	1.01	0.96	0.90	0.85	0.81	0.76	0.72	0.68	0.64	0.60	0.57	0.53	0.50	0.47
80	0.85	1.04	0.98	0.93	0.88	0.83	0.78	0.74	0.70	0.66	0.62	0.58	0.55	0.52	0.48	0.46
90	0.83	1.01	0.95	0.90	0.85	0.81	0.76	0.72	0.68	0.64	0.60	0.57	0.53	0.50	0.47	0.44
100	0.81	0.98	0.93	0.88	0.83	0.79	0.74	0.70	0.66	0.62	0.59	0.55	0.52	0.49	0.46	0.43
120	0.77	0.93	0.88	0.83	0.79	0.75	0.70	0.66	0.63	0.59	0.56	0.52	0.49	0.46	0.44	0.41
140	0.73	0.89	0.84	0.79	0.75	0.71	0.67	0.63	0.60	0.56	0.53	0.50	0.47	0.44	0.41	0.39
160	0.70	0.85	0.80	0.76	0.72	0.68	0.64	0.60	0.57	0.54	0.51	0.48	0.45	0.42	0.40	0.37
180	0.66	0.81	0.76	0.72	0.68	0.65	0.61	0.58	0.54	0.51	0.48	0.45	0.43	0.40	0.38	0.35
200	0.64	0.77	0.73	0.69	0.66	0.62	0.59	0.55	0.52	0.49	0.46	0.44	0.41	0.39	0.36	0.34
250	0.58	0.70	0.66	0.63	0.59	0.56	0.53	0.50	0.47	0.44	0.42	0.39	0.37	0.35	0.33	0.31
300	0.53	0.64	0.60	0.57	0.54	0.51	0.48	0.46	0.43	0.41	0.38	0.36	0.34	0.32	0.30	0.28
350	0.48	0.59	0.56	0.53	0.50	0.47	0.44	0.42	0.40	0.37	0.35	0.33	0.31	0.29	0.27	0.26
400	0.45	0.54	0.51	0.49	0.46	0.44	0.41	0.39	0.37	0.35	0.33	0.31	0.29	0.27	0.25	0.24
450	0.42	0.51	0.48	0.45	0.43	0.41	0.38	0.36	0.34	0.32	0.30	0.28	0.27	0.25	0.24	0.22
500	0.39	0.47	0.45	0.42	0.40	0.38	0.36	0.38	0.32	0.30	0.28	0.27	0.25	0.24	0.22	0.21

FIGURE 9-a

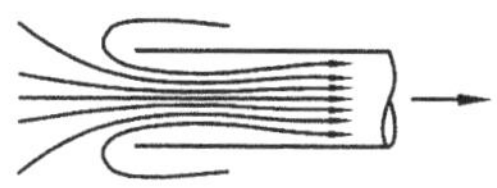

$h_e = 0.93\ VP_d$

$C_e = 0.72$

PLAIN DUCT END

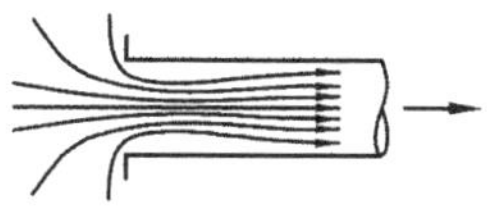

CONICAL	RECTANGULAR
$h_h = 0.56\ VP_d$	$h_h = 0.64\ VP_d$
$C_e = 0.80$	$C_e = 0.78$

FLANGED DUCT END

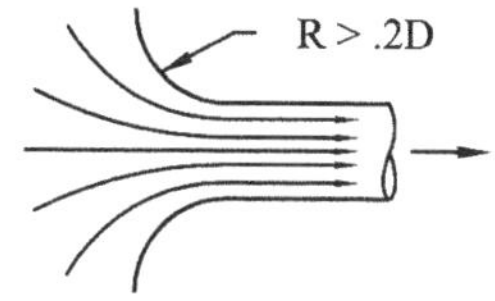

$h_h = 0.04\ VP_d$

$C_e = 0.96$

BELLMOUTH ENTRY

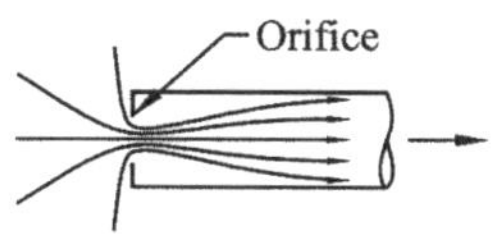

$h_h = 1.78\ VP_{Orifice}$

$C_e = 0.35$

SHARP-EDGED ORIFICE

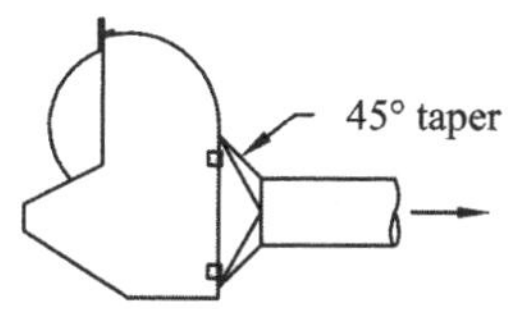

$h_h = 0.4\ VP_d$ (tapered take-off)

$C_e = 0.85$

$h_h = 0.65\ VP_d$ (no taper)

$C_e = 0.78$

STANDARD GRINDER HOOD

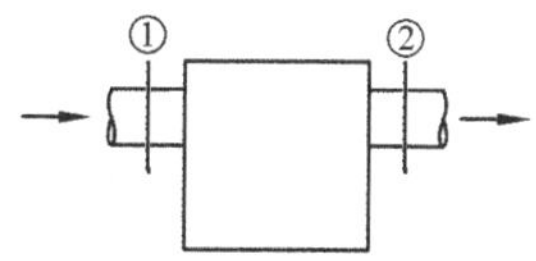

TRAP OR SETTLING CHAMBER

$SP_2 - SP_1 = 1.5\ VP_2$

$C_e = 0.40$

$$C_e = \sqrt{\frac{1}{F_h + 1}}$$

TAPERED HOODS

Flanged or unflanged; conical, square or rectangular. α is the major angle on rectangular hoods.

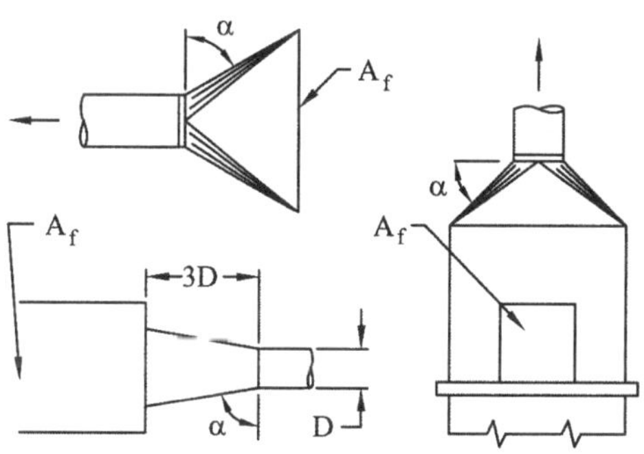

Face area (A_f) at least 2 times the duct area.

h_h Entry Loss Coefficient

α	Conical	Rectangular
90	0.93	0.93
80	0.11	0.19
70	0.06	0.13
60	0.08	0.16
50	0.12	0.21
45	0.15	0.25
40	0.18	0.28
30	0.25	0.35
20	0.34	0.43
10	0.44	0.52
0	0.56	0.64

Note: 0° values represent round ducts butted into back of booth or hood without a rectangular to round transition.

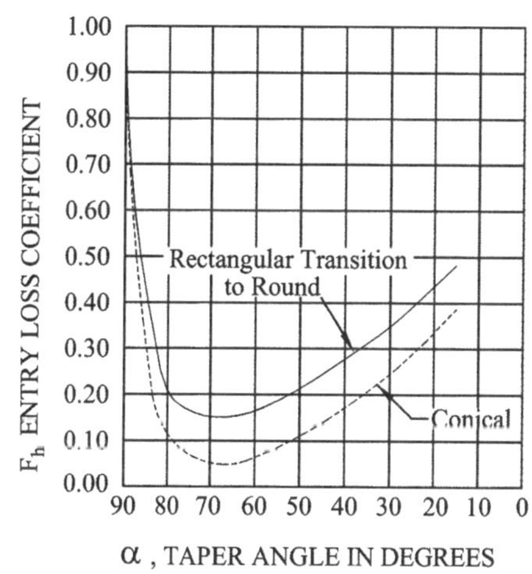

COMPOUND HOODS

A compound hood, such as the slot/plenum shown to the right, would have 2 losses, one through the slot and the other through the transition into the duct.

The slot entry loss coefficient, F_s, would have a value typically in the range of 1.00 to 1.78 (see Chapters 6 and 13).

The duct entry loss coefficient is given by the above data for tapered hoods.

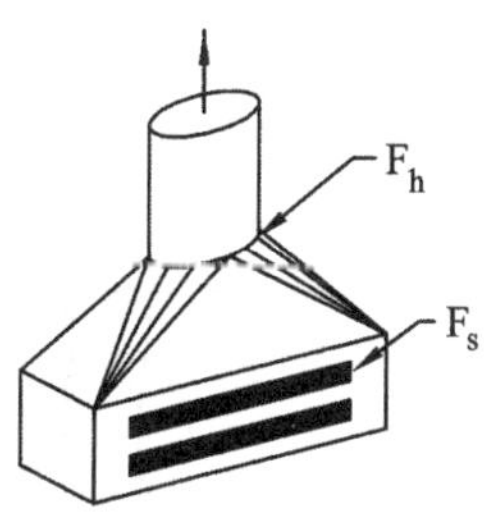

$h_e = F_s VP_s + F_h VP_d$

(See Chapter 6)

MISCELLANEOUS VALUES

HOOD	ENTRY LOSS COEFFICIENT F_h	HOOD FLOW COEFFICIENT C_e
Abrasive blast chamber	1.00	0.50
Abrasive blast elevator	2.30	0.31
Abrasive separator	2.30	0.31
Elevators (enclosures)	0.69	0.59
Flanged pipe plus close elbow	0.80	0.56
Plain pipe plus close elbow	1.60	0.38

ACGIH®

TITLE

HOOD ENTRY LOSS COEFFICIENTS

FIGURE 9-a

DATE 1-12

CHECK CODES, REGULATIONS, AND LAWS (LOCAL, STATE, AND NATIONAL) TO ENSURE THAT DESIGN IS COMPLIANT.

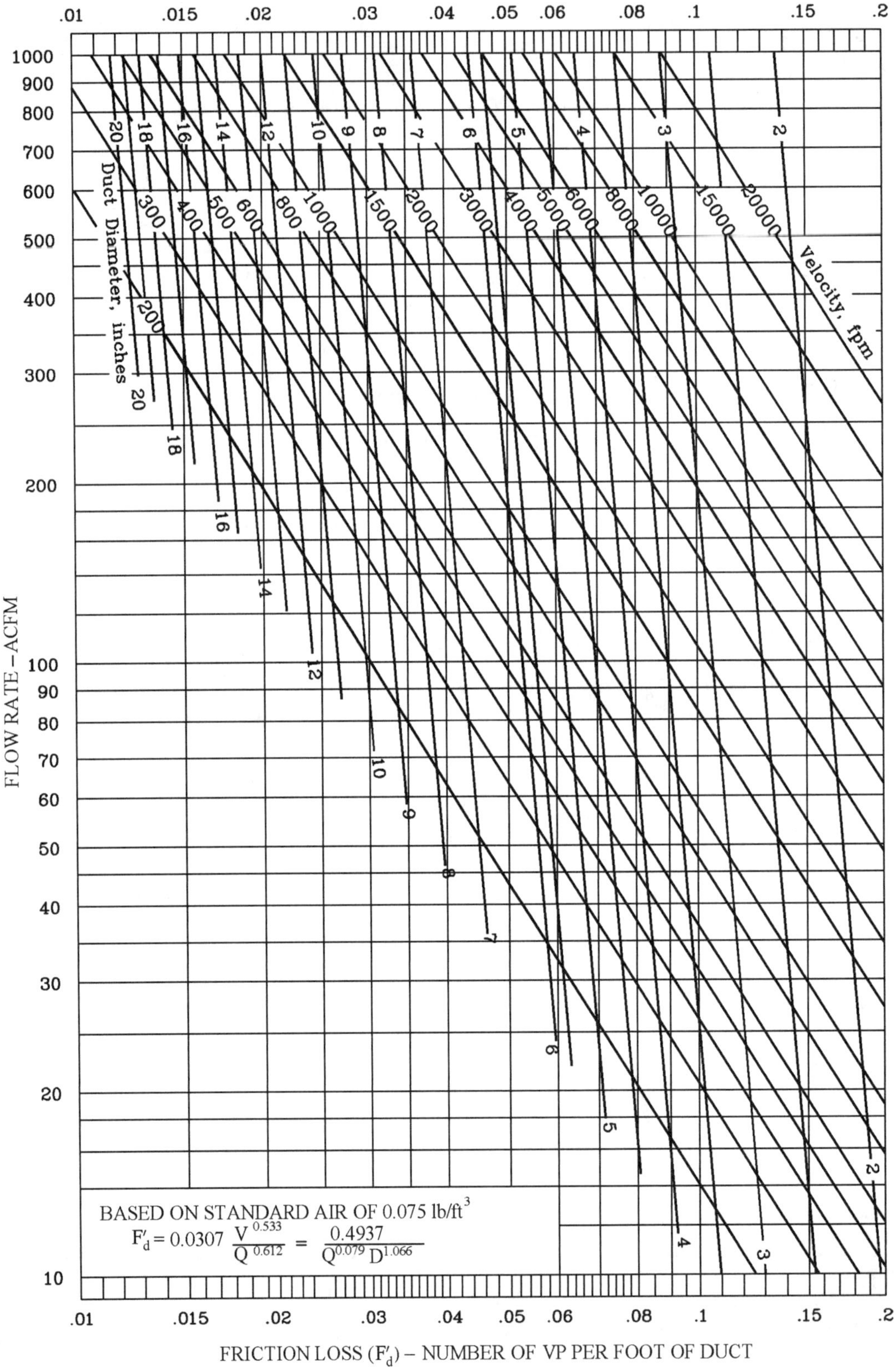

FIGURE 9-b (IP). Friction chart for sheet metal & plastic ducts (equivalent sand grain roughness height = 0.00015 feet)

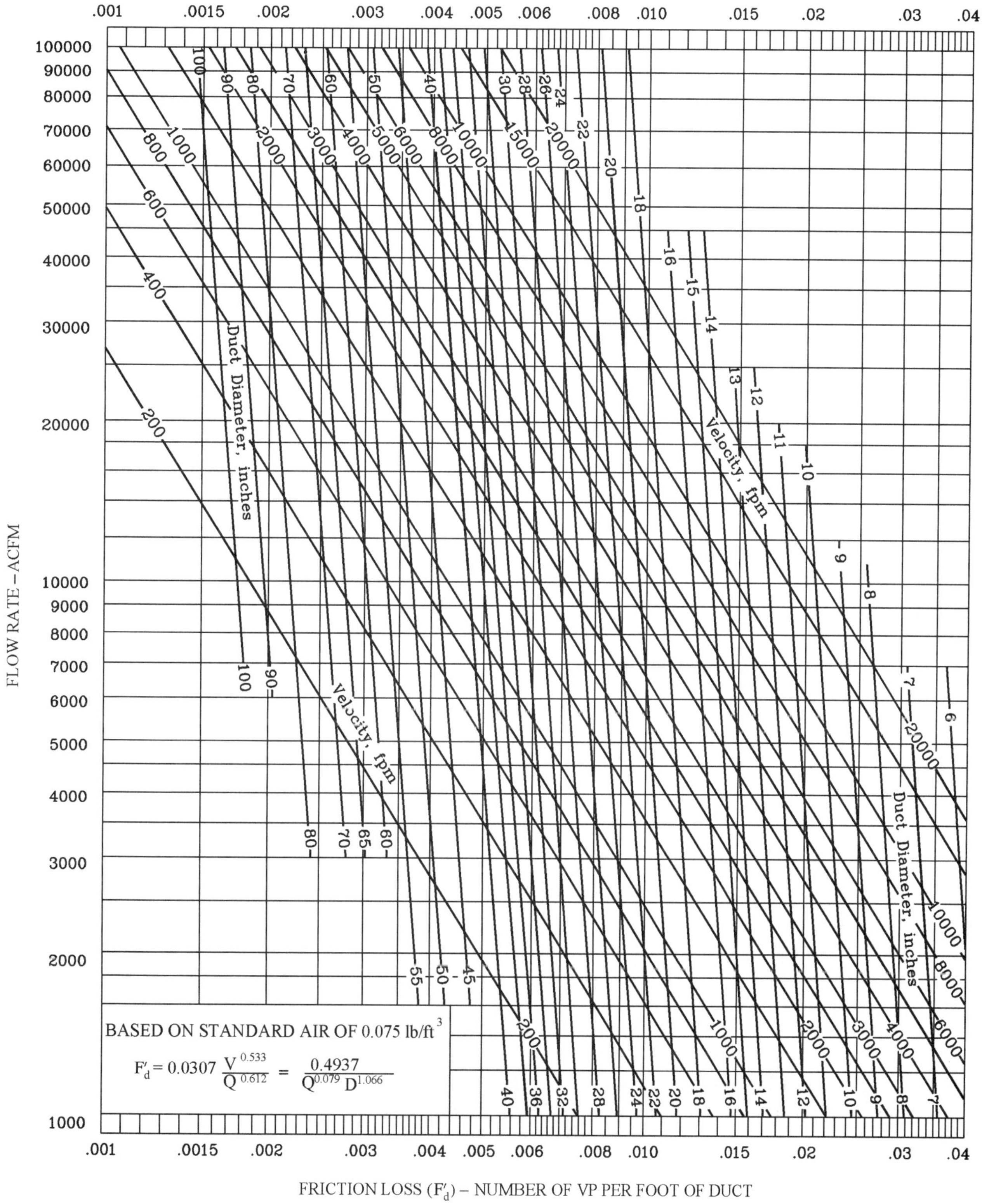

FIGURE 9-c (IP). Friction chart for sheet metal & plastic ducts (equivalent sand grain roughness height = 0.00015 feet)

STATIC PRESSURE REGAINS FOR EXPANSIONS

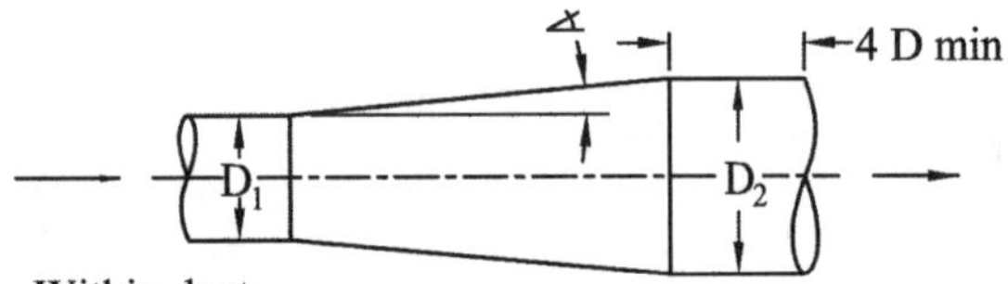

Within duct

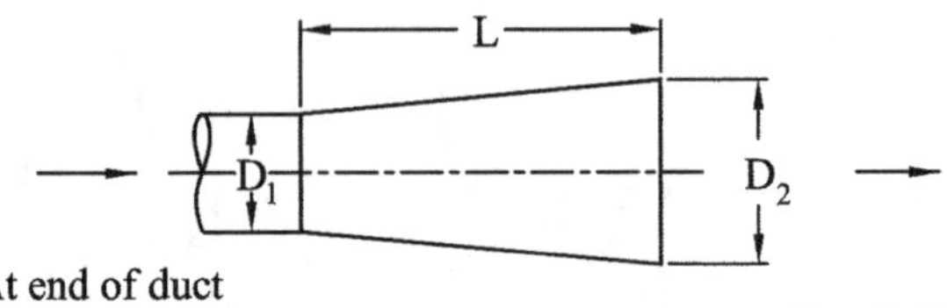

At end of duct

Regain (R), fraction of VP difference

Taper angle degrees	Diameter ratios D_2/D_1 1.25:1	1.5:1	1.75:1	2:1	2.5:1
3 1/2	0.92	0.88	0.84	0.81	0.75
5	0.88	0.84	0.80	0.76	0.68
10	0.85	0.76	0.70	0.63	0.53
15	0.83	0.70	0.62	0.55	0.43
20	0.81	0.67	0.57	0.48	0.43
25	0.80	0.65	0.53	0.44	0.28
30	0.79	0.63	0.51	0.41	0.25
Abrupt 90	0.77	0.62	0.50	0.40	0.25

Where: $SP_2 = SP_1 + R(VP_1 - VP_2)$

Regain (R), fraction of inlet VP

Taper length to inlet diam L / D	Diameter ratios D_2/D_1 1.2:1	1.3:1	1.4:1	1.5:1	1.6:1	1.7:1
1.0:1	0.37	0.39	0.38	0.35	0.31	0.27
1.5:1	0.39	0.46	0.47	0.46	0.44	0.41
2.0:1	0.42	0.49	0.52	0.52	0.51	0.49
3.0:1	0.44	0.52	0.57	0.59	0.60	0.59
4.0:1	0.45	0.55	0.60	0.63	0.63	0.64
5.0:1	0.47	0.56	0.62	0.65	0.66	0.68
7.5:1	0.48	0.58	0.64	0.68	0.70	0.72

Where: $SP_1 = SP_2 - R(VP_1)^*$

* When $SP_2 = 0$ (atmosphere) SP_1 will be (-)

The regain (R) will only be 70% of value shown above when expansion follows a disturbance or elbow (including a fan) by less than 5 duct diameters.

STATIC PRESSURE LOSSES FOR CONTRACTIONS

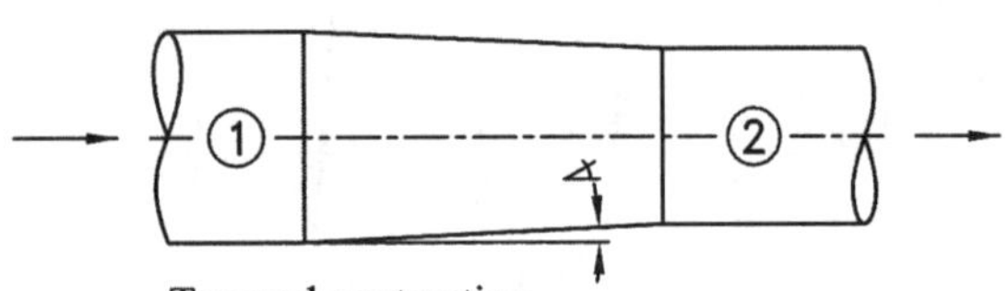

Tapered contraction

$SP_2 = SP_1 - (VP_2 - VP_1) - L(VP_2 - VP_1)$

Taper angle degrees	L(loss)
5	0.05
10	0.06
15	0.08
20	0.10
25	0.11
30	0.13
45	0.20
60	0.30
over 60	Abrupt contraction

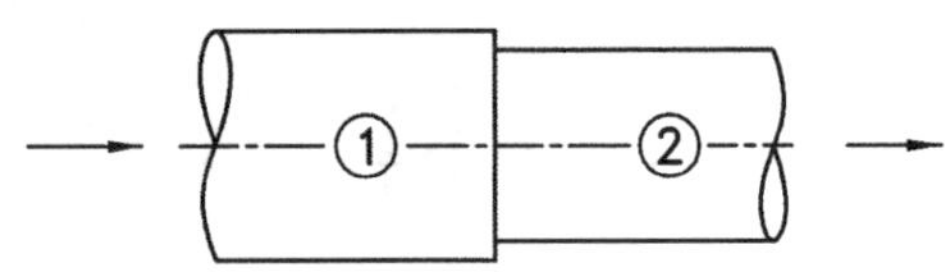

Abrupt contraction

$SP_2 = SP_1 - (VP_2 - VP_1) - K(VP_2)$

Ratio A_2/A_1	K
0.1	0.48
0.2	0.46
0.3	0.42
0.4	0.37
0.4	0.32
0.6	0.26
0.7	0.20

Note:

In calculating SP for expansion or contraction use algebraic signs: VP is (+), and usually SP is (+) in discharge duct from fan, and SP is (-) in inlet duct to fan.

ACGIH®

TITLE: EXPANSIONS AND CONTRACTIONS

FIGURE 9-d

DATE 1-12

CHECK CODES, REGULATIONS, AND LAWS (LOCAL, STATE, AND NATIONAL) TO ENSURE THAT DESIGN IS COMPLIANT.

FIGURE 9-e

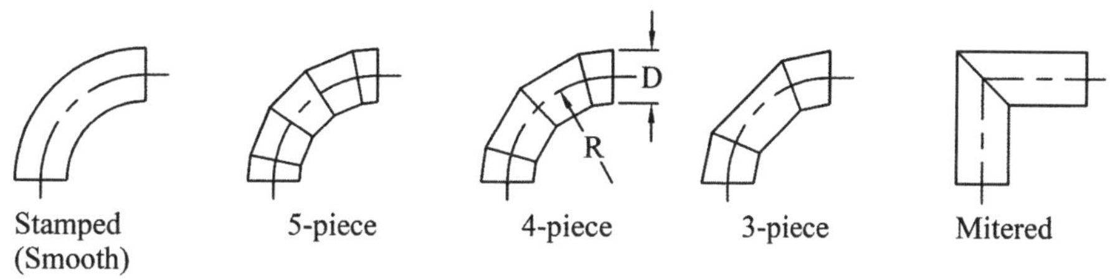

	R / D				
	0.75	1.00	1.50	2.00	2.50
Stamped	0.33	0.22	0.15	0.13	0.12
5-piece	0.46	0.33	0.24	0.19	0.17*
4-piece	0.50	0.37	0.27	0.24	0.23*
3-piece	0.54	0.42	0.34	0.33	0.33*

* extrapolated from published data

OTHER ELBOW LOSS COEFFICIENTS

Mitered, no vanes	1.2
Mitered, turning vanes	0.6
Flatback (R/D = 2.5)	0.05 (see Chapter 5, Figure 5-16)

NOTE: Loss factors are assumed to be for elbows of "zero length." Friction losses should be included to the intersection of centerlines.

ROUND ELBOW LOSS COEFFICIENTS

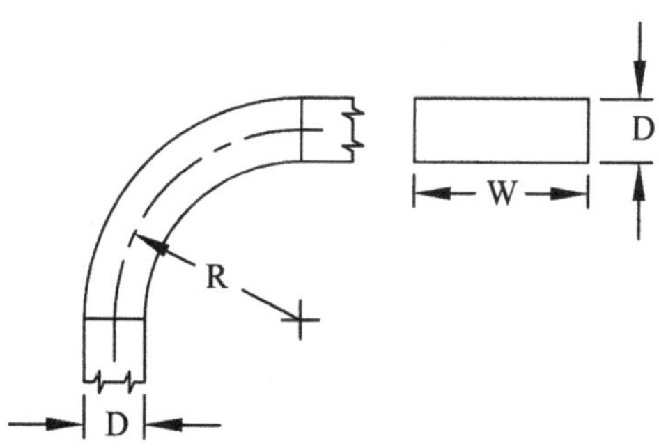

R/D	Aspect Ratio, W/D					
	0.25	0.5	1.0	2.0	3.0	4.0
0.0 (Mitered)	1.50	1.32	1.15	1.04	0.92	0.86
0.5	1.36	1.21	1.05	0.95	0.84	0.79
1.0	0.45	0.28	0.21	0.21	0.20	0.19
1.5	0.28	0.18	0.13	0.13	0.12	0.12
2.0	0.24	0.15	0.11	0.11	0.10	0.10
3.0	0.24	0.15	0.11	0.11	0.10	0.10

SQUARE & RECTANGULAR ELBOW LOSS COEFFICIENTS

ACGIH®	TITLE DUCT DESIGN DATA ELBOW LOSSES	FIGURE 9-e DATE 1-12

CHECK CODES, REGULATIONS, AND LAWS (LOCAL, STATE, AND NATIONAL) TO ENSURE THAT DESIGN IS COMPLIANT.

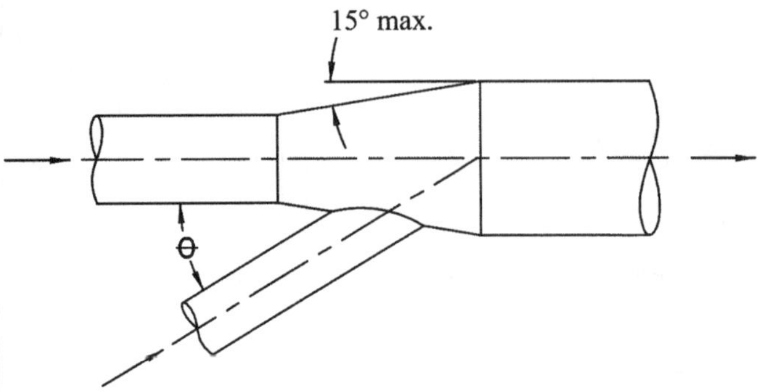

Angle Θ Degrees	Loss Fraction of VP in Branch
10	0.06
15	0.09
20	0.12
25	0.15
30	0.18
35	0.21
40	0.25
45	0.28
50	0.32
60	0.44
90	1.00

Note: Branch entry loss assumed to occur in branch and is so calculated.

Do not include a regain calculation for branch entry enlargements.

BRANCH ENTRY LOSSES

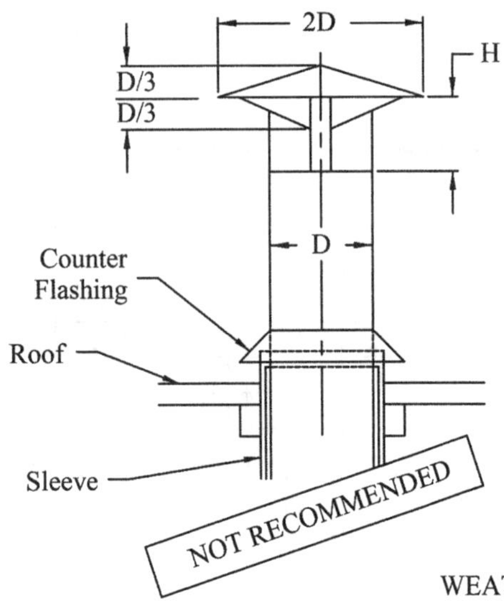

H, No. of Diameters	Loss Fraction of VP
1.0 D	0.10
0.75 D	0.18
0.70 D	0.22
0.65 D	0.30
0.60 D	0.41
0.55 D	0.56
0.50 D	0.73
0.45 D	1.0

WEATHER CAP LOSSES

See Chapter 5, Figure 5-14

TITLE	FIGURE	DATE
BRANCH ENTRY AND WEATHER CAP LOSSES	9-f	1-12

CHECK CODES, REGULATIONS, AND LAWS (LOCAL, STATE, AND NATIONAL) TO ENSURE THAT DESIGN IS COMPLIANT.

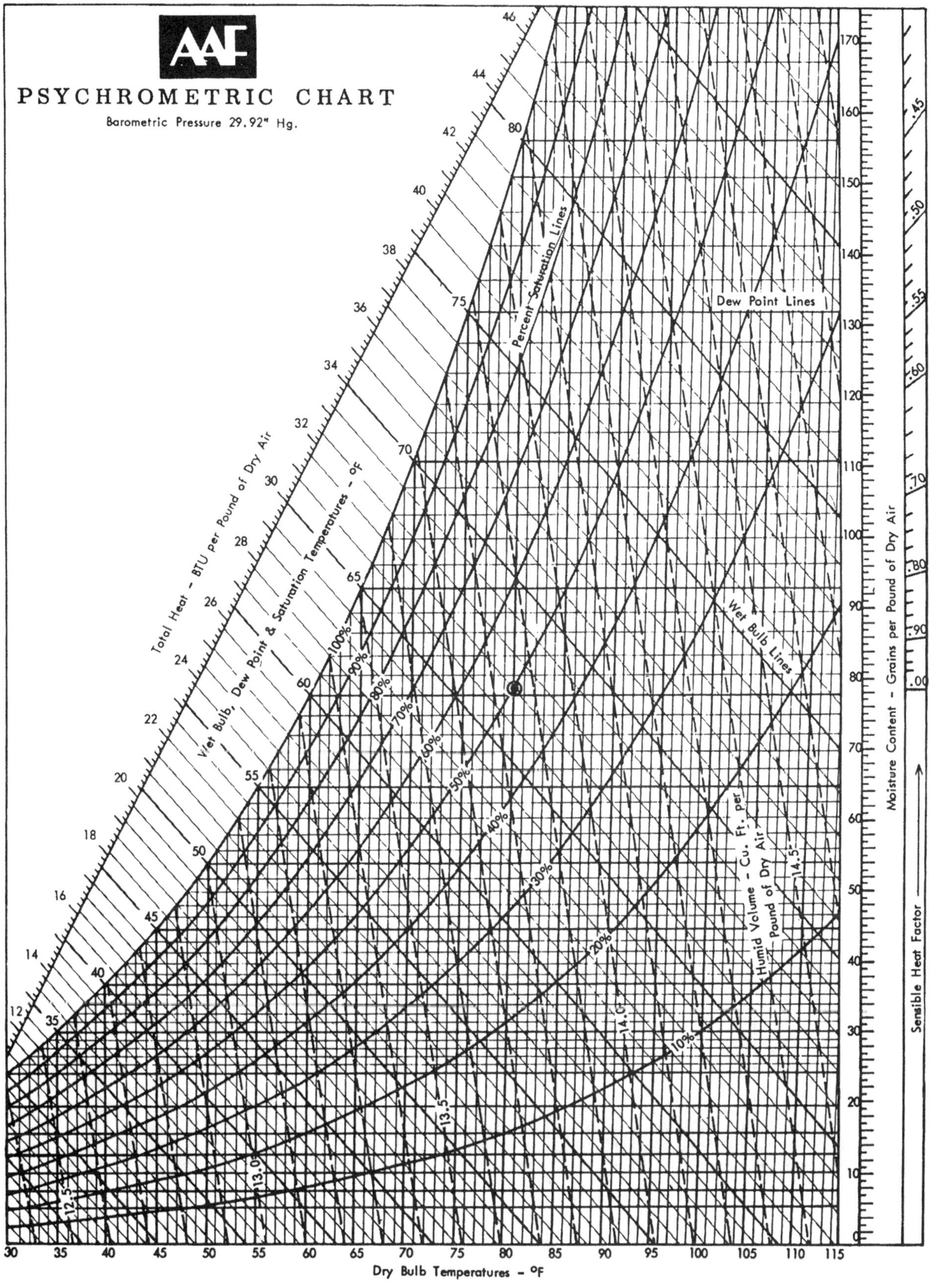

FIGURE 9-g (IP). Psychrometric chart — 30 F to 115 F

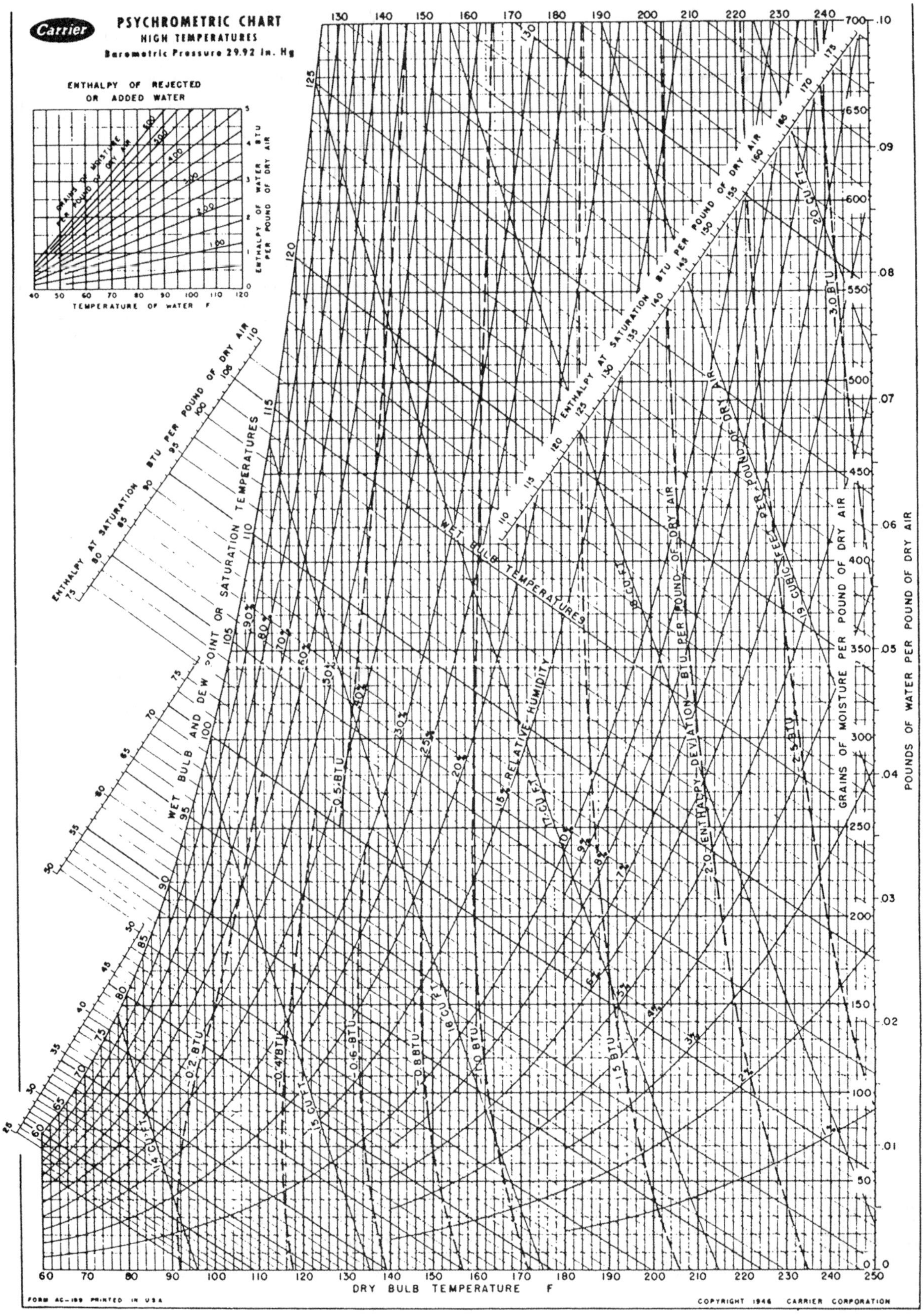

FIGURE 9-h (IP). Psychrometric chart — 60 F to 250 F

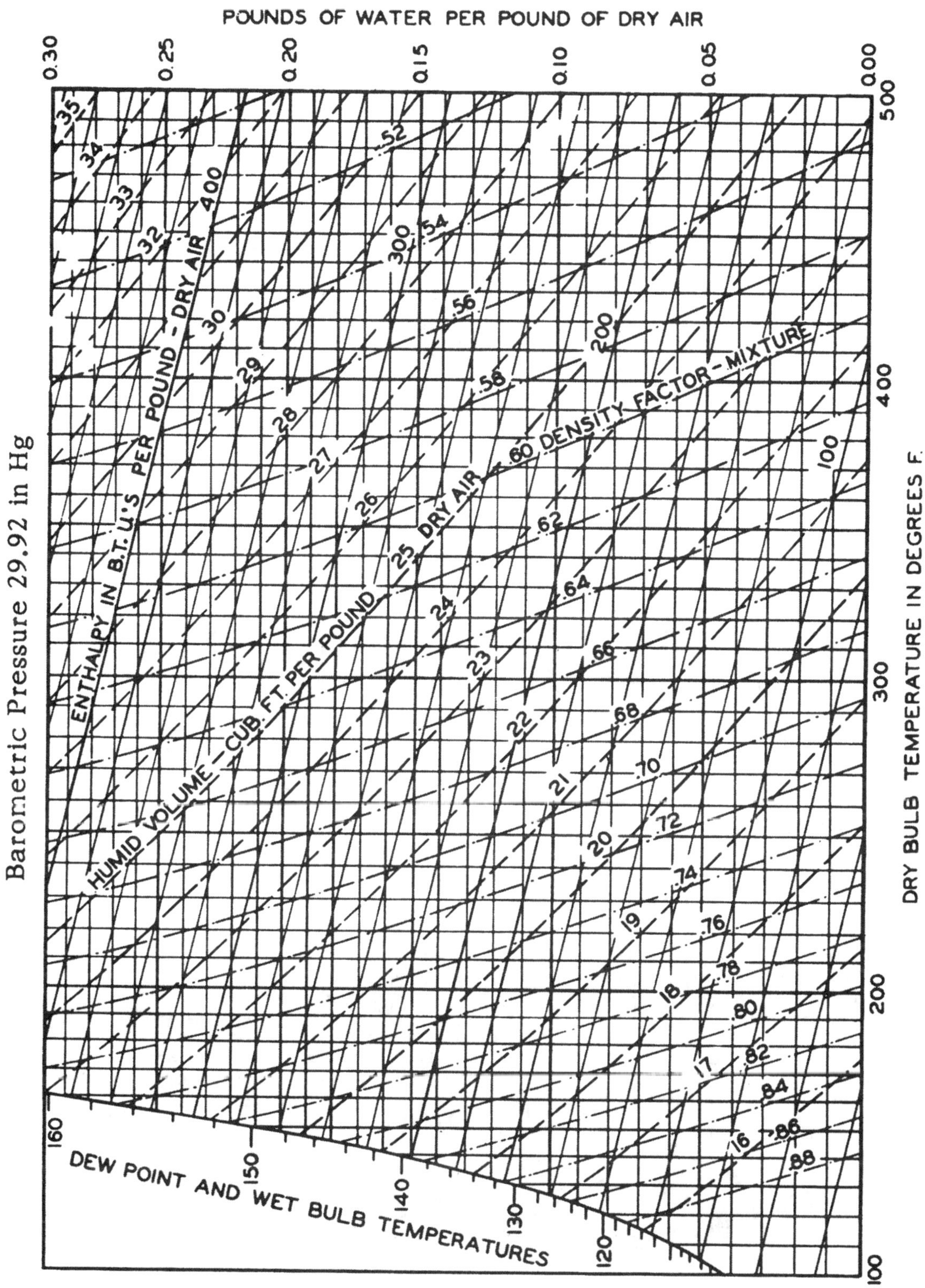

FIGURE 9-i (IP). Psychrometric chart — 100 F to 500 F

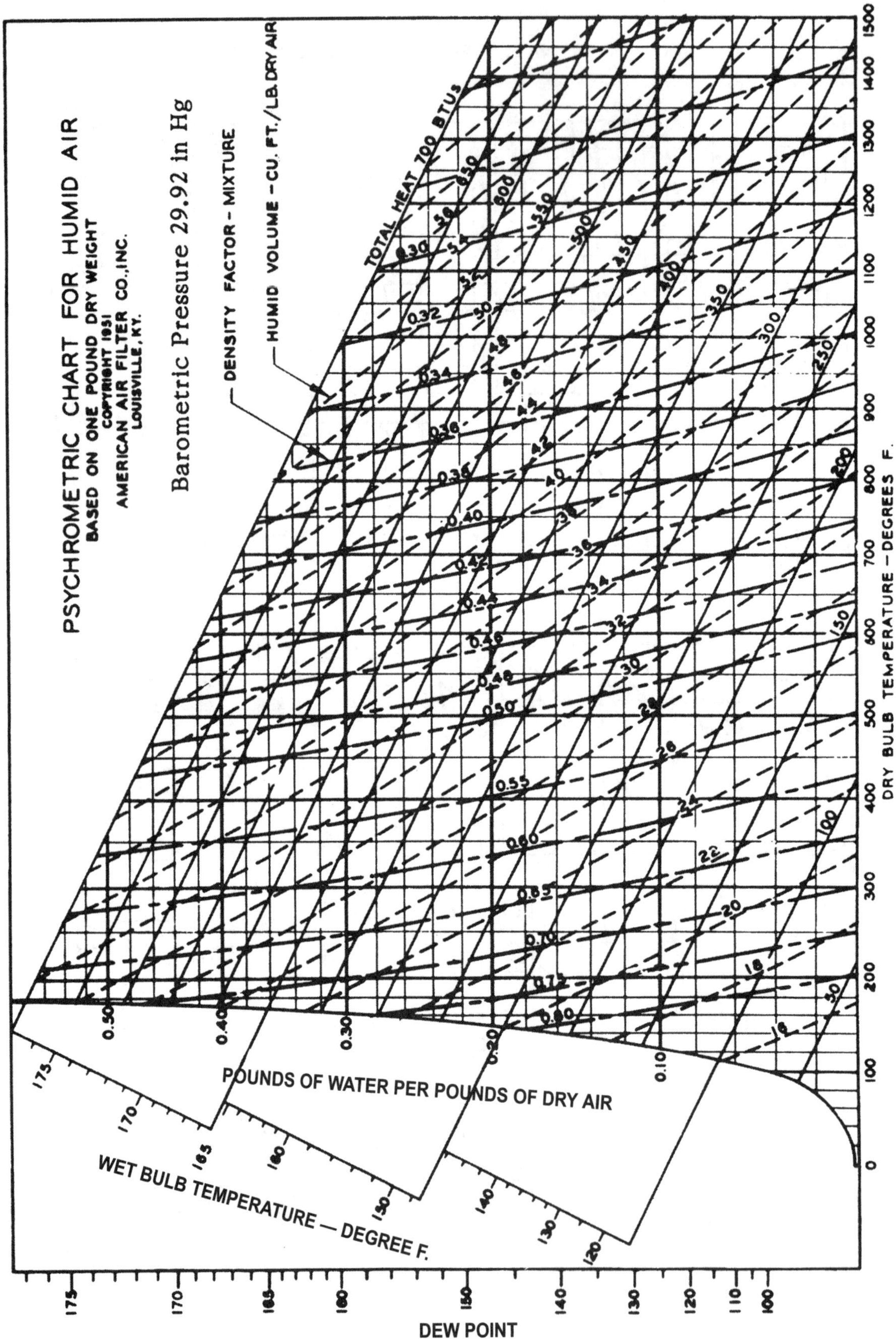

FIGURE 9-j (IP). Psychrometric chart — Up to 1500 F Temperatures

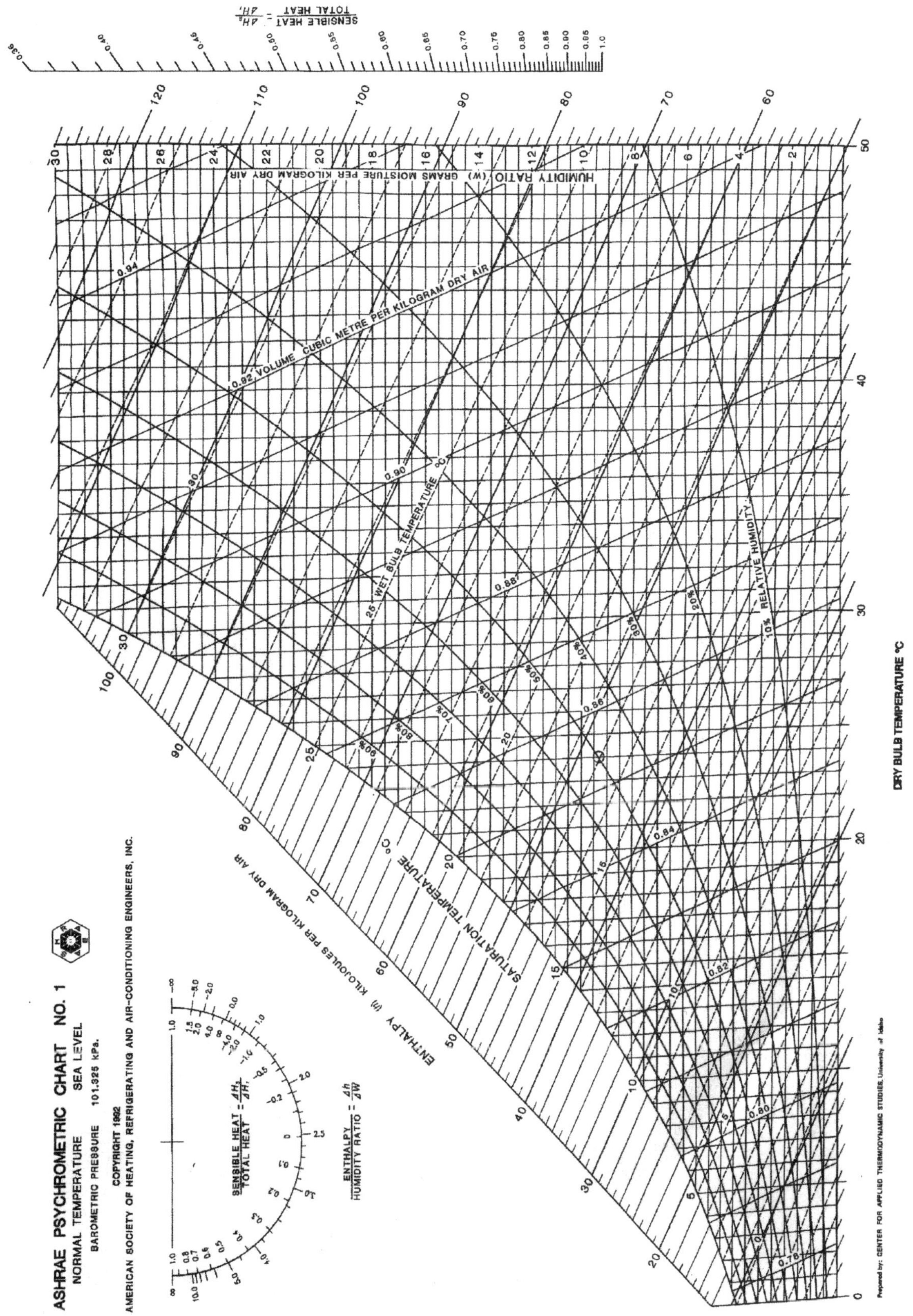

FIGURE 9-k (SI). Psychrometric chart – 0 C to 50 C

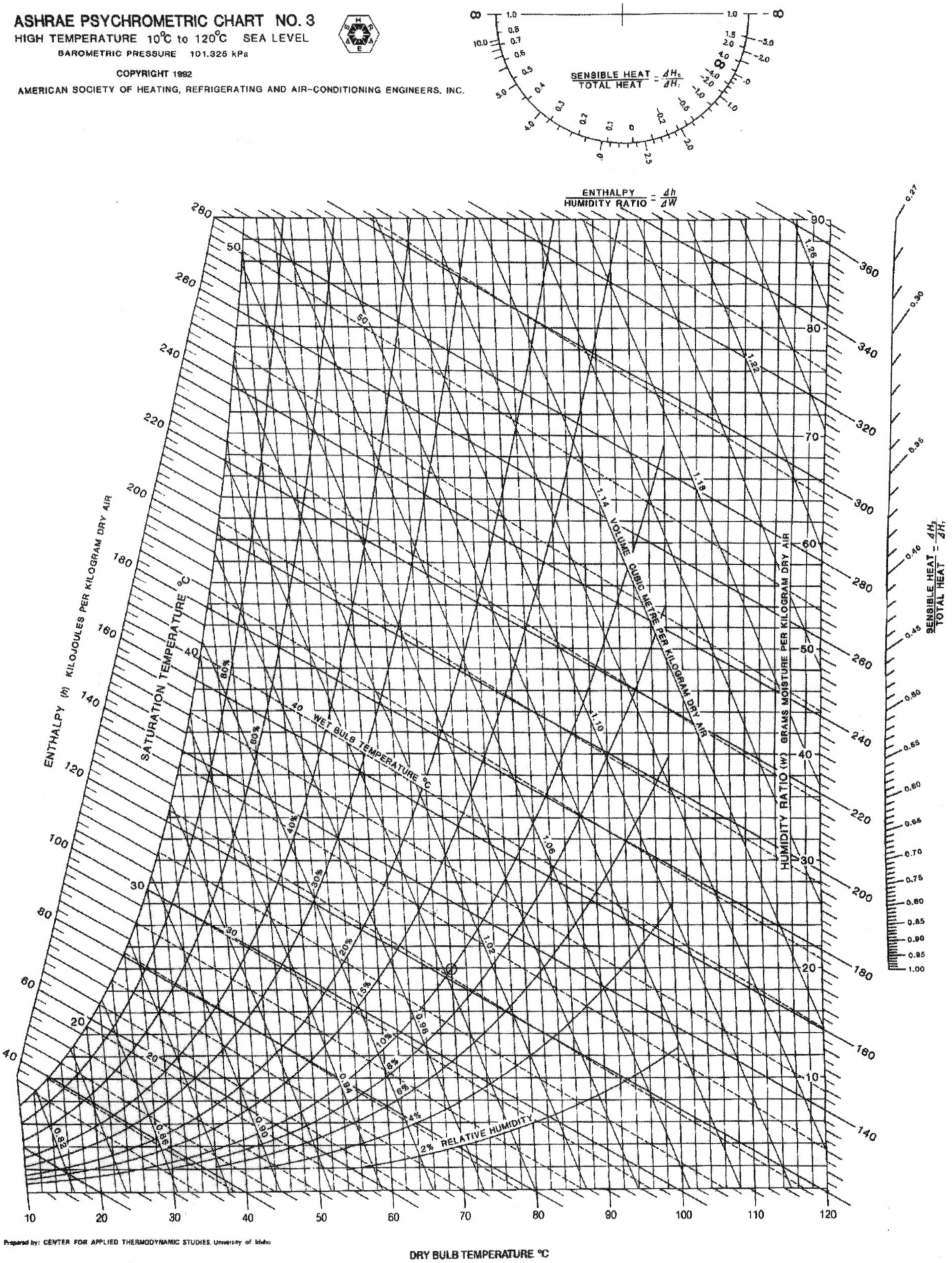

FIGURE 9-I (SI). Psychrometric chart – 10 C to 120 C

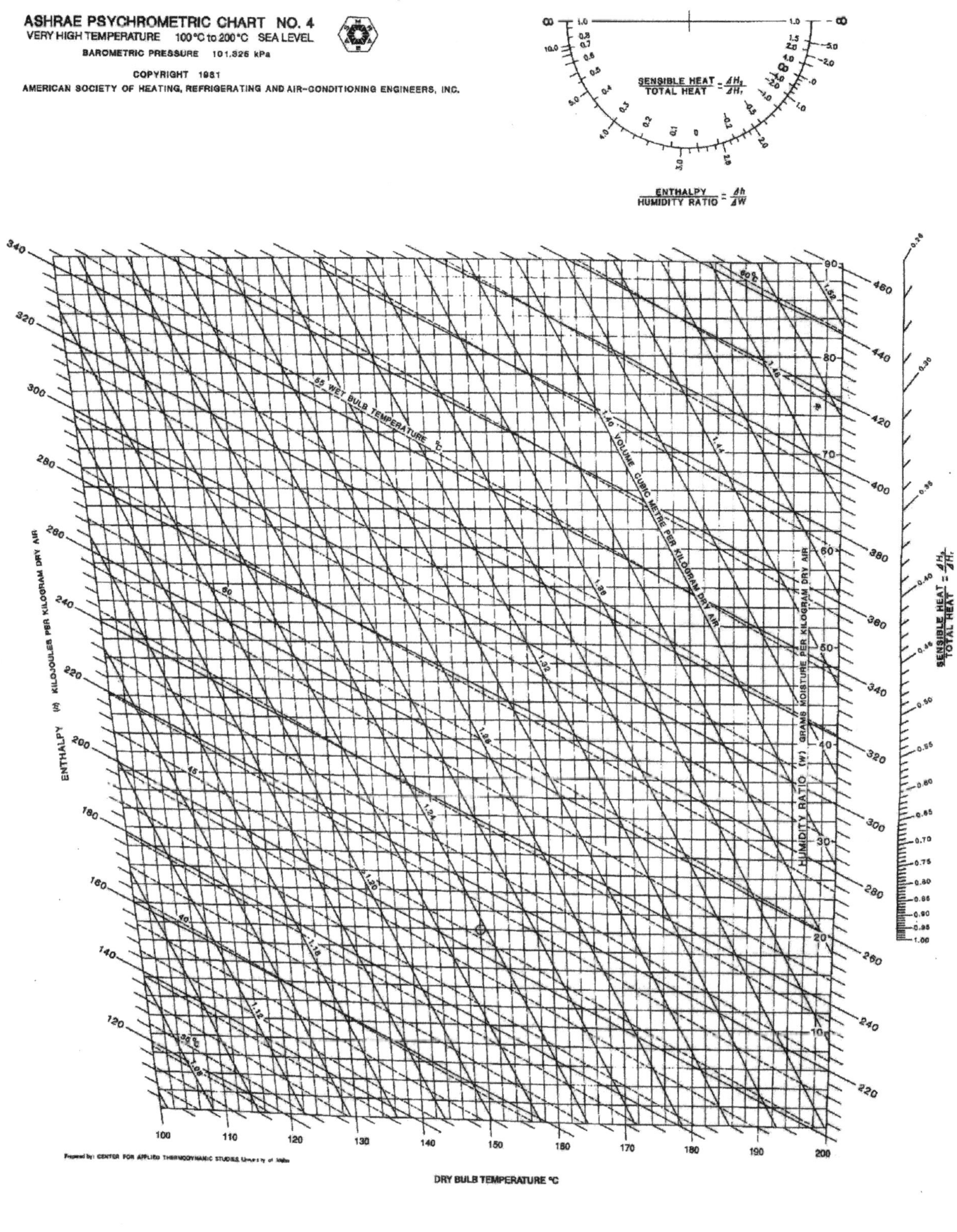

FIGURE 9-m (SI). Psychrometric chart – 100 C to 200 C

Chapter 10

SUPPLY AIR SYSTEMS

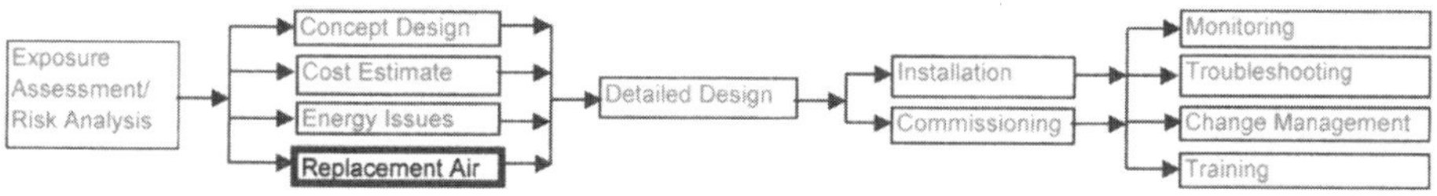

NOTE: Equations with notation followed by (IP) are designated for inch-pound system only; Equations followed by (SI) are designated for metric use only. If Equation bears neither, then it applies to both systems.

10.1 INTRODUCTION

Industrial buildings operating in the early 1900s had simple building mechanical systems. Ventilation was accomplished by opening a wall/roof section and letting the outside air naturally flow through the building. Heating systems consisted of radiators and unit heaters. As more automation was incorporated into the industrial process, buildings had to deal with increasing amounts of energy being consumed inside. Some process operations created potentially hazardous emissions in the worker's environment. This caused the need to install exhaust air systems to control these airborne emissions. With the use of powered exhaust systems, many buildings began to operate with a negative pressure. Supply air equipment was soon found to be critical to the success of industrial ventilation systems. They provide the air that allows exhaust systems to perform properly. In some situations, they also provide dilution of contaminants that escape into the general workspace. Over the years, heating and ventilating units advanced to provide a more comfortable building temperature at a lower energy use when compared to a system that uses only unit heaters.

Manufacturing facilities evolved to the point where there is now widespread use of automation/computers. Production of parts requiring tight tolerances is often required. These facilities require temperature control to perform at effective levels. Workers need to be cooled to relieve body heat caused by their activity. This heat exchange is easily accomplished with cool air. With warmer temperatures that occur in the summer or near hot industrial operations, maintaining a suitable rate of cooling becomes more difficult. Increasing air movement is a technique that will increase a person's rate of cooling. Velocities of 200 feet per minute [approx. 1.0 m/s] are common for workstations that use air movement to aid personal cooling. When using high velocity air for spot cooling, do not disturb the operation of each exhaust system.

Air cleanliness requirements have become more stringent to improve worker health, reduce housekeeping and maintenance costs, and increase product quality. The supply air ventilation system plays a significant role in balancing these collective needs and maintaining the proper work space environment.

In some industrial plants, ventilation systems are key elements of a process. A few are critical to the success of that process; this is the case in automotive painting. A number of years ago, automobiles were painted in an open booth by people who sprayed paint onto the vehicle body. Air was exhausted to remove solvent vapors so the workers would not be exposed to hazardous concentrations. The replacement make-up air entering the booth had a minimum degree of filtration and no significant temperature or humidity control. Supply air was distributed to provide good air exchange throughout the booth so the concentration of paint solvent vapors would be low. Over time, the quality of the paint coating became more important and the performance of ventilation systems began to improve. Currently supply air humidity and temperature are controlled to improve paint curing time. The air is well filtered to eliminate defects in the painted surface. Painting operations are conducted in a clean-room type space that is pressurized to maintain high levels of cleanliness.

In other plants, the distribution of supply air may not be as critical for product quality but will always be important for the proper operation of exhaust systems and plant comfort control. Poorly distributed supply air sometimes overwhelms a well-designed exhaust hood and destroys the hood's ability to capture contaminants. Therefore, the designer should pay equal attention to both the quantity and distribution of the supply air system.

10.2 PURPOSE OF SUPPLY AIR SYSTEMS

A proper supply air ventilation system can serve several purposes in an industrial facility: 1) exhaust air replacement, 2) plant ventilation, 3) building pressurization, 4) building heating, cooling, and humidification, and 5) space air cleanliness. The purposes of the supply air system are discussed in the following paragraphs. The total amount of supply air should be the amount that satisfies all the requirements of the supply air system. For example, a small amount of air may be required for replacing the exhaust air, but a much larger amount may be required to deliver enough tempered air for heating or cooling.

10.2.1 Exhaust Air Replacement. Air will enter a building in an amount equal to the flow rate of exhaust air whether or not provision is made for replacement. However, the actual exhaust flow rate will be less than the design value if the plant is under negative pressure. If the building perimeter is tightly sealed, thus blocking effective infiltration of outdoor air, a severe decrease of the exhaust flow rate will result. If, on the other hand, the building is relatively old with large sash areas, air infiltration may be quite pronounced and the exhaust system performance will decrease only slightly. However, other problems may occur as identified in Table 10-1. When the building is relatively open, the resulting in-plant environmental condition is often undesirable since the influx of cold outdoor air in the northern climates chills the perimeter of the building. Exposed workers are subjected to drafts, space temperatures are not uniform, and the building heating system is usually overtaxed (Figure 10-1). Under negative pressure conditions, workers in the cold zones turn up thermostats in an attempt to get heat. Because this will do nothing to stop leakage of cold air, they remain cold while the center of the plant is overheated. Although the air may eventually be tempered to acceptable conditions by mixing as it moves to the building interior, this is an ineffective way of transferring heat to the air and usually results in fuel waste. For an estimated value of the amount of air that enters a building through cracks that occur around doors or windows or other small openings in a building exterior, refer to Table 10-2. Figure 10-2 presents the force necessary to open a door against a building's negative pressure. The performance of a fan operation can also suffer as

TABLE 10-1. Negative Pressures That May Cause Unsatisfactory Conditions Within Buildings

Negative Pressure, "wg [Pa]	Adverse Conditions
0.01 to 0.02 [3–5]	Worker Draft Complaints—High velocity drafts through doors and windows.
0.01 to 0.05 [3–13]	Natural Draft Stacks Ineffective—Ventilation through roof exhaust ventilators, flow through stacks with natural draft greatly reduced.
0.02 to 0.05 [5–13]	Carbon Monoxide Hazard—Back drafting will take place in hot water heaters, unit heaters, furnaces, and other combustion equipment not provided with induced draft fan.
0.03 to 0.10 [8–25]	General Mechanical Ventilation Reduced—Airflows reduced in propeller fans and low pressure supply and exhaust systems.
0.05 to 0.10 [13–25]	Doors Difficult to Open—Serious injury may result from non-checked, slamming doors.
0.10 to 0.25 [25–63]	Local Exhaust Ventilation Impaired—Centrifugal fan exhaust airflow reduced.

shown in Figure 10-3.

For general plant ventilation, replacement airflow rate should be slightly more than the total airflow rate removed from the building by exhaust ventilation systems, process systems, and combustion processes. Determination of the actual flow rate of air removed usually requires an inventory of exhaust locations with airflow testing of these sources. When conducting the exhaust inventory, it is necessary not only to determine the quantity of air removed, but also to identify the need to upgrade any part of the ventilation system. At the same time, reasonable projections should be made of the total plant exhaust requirements for the next few years, particularly if process changes or plant expansions are contemplated. In such cases it can be practical to purchase a replacement air unit slightly larger than immediately necessary with the knowledge that the increased capacity will be required within a short time. The additional cost of a larger unit is relatively small and, in most cases, the fan drive can be adjusted to supply the desired quantity of air at the time of installation.

Having established the minimum air supply quantity necessary for replacement air purposes, many plants have found that

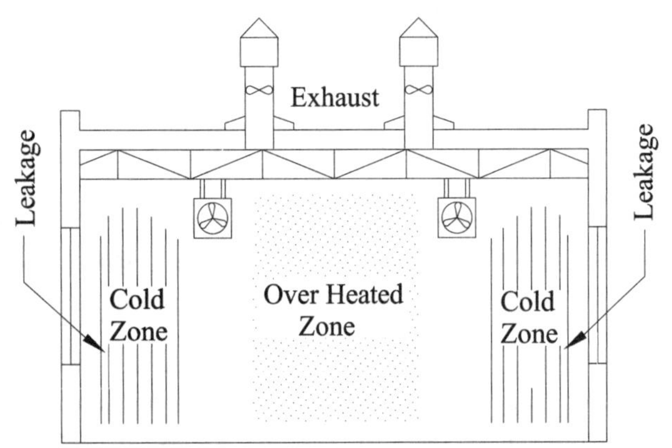

FIGURE 10–1. Cold zones vs. overheated zones (poor ventilation design)

it is wise to provide additional supply airflow to overcome natural ventilation leakage and further minimize drafts at the perimeter of the building. Conversely, some facilities deliberately design for a higher exhaust flow rate to prevent fugitive emissions from migrating into "clean" areas of the building or to the outdoors. In these situations, the control of the building

TABLE 10-2. Negative Pressures and Corresponding Velocities Through Crack Openings (Calculated with air at room temperature, standard atmospheric pressure, C_e = 0.6.)

Negative Pressure, "wg [Pa]	Velocity, fpm [m/s]
0.004 [1]	150 [0.75]
0.008 [2]	215 [1.08]
0.010 [3]	240 [1.20]
0.016 [4]	300 [1.50]
0.020 [5]	340 [1.70]
0.025 [6]	380 [1.90]
0.030 [8]	415 [2.08]
0.040 [10]	480 [2.40]
0.050 [13]	540 [2.70]
0.060 [15]	590 [2.95]
0.080 [20]	680 [3.40]
0.100 [25]	760 [3.80]
0.150 [38]	930 [4.65]
0.200 [50]	1080 [5.40]
0.250 [63]	1200 [6.00]
0.300 [75]	1310 [6.55]
0.400 [100]	1520 [7.60]
0.500 [125]	1700 [8.50]
0.600 [150]	1860 [9.30]

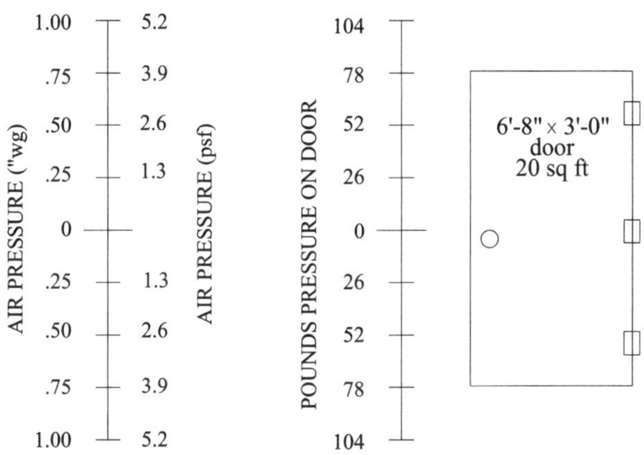

FIGURE 10–2 (IP). Relationship between air pressure and amount of force needed to open or close an average-sized door.

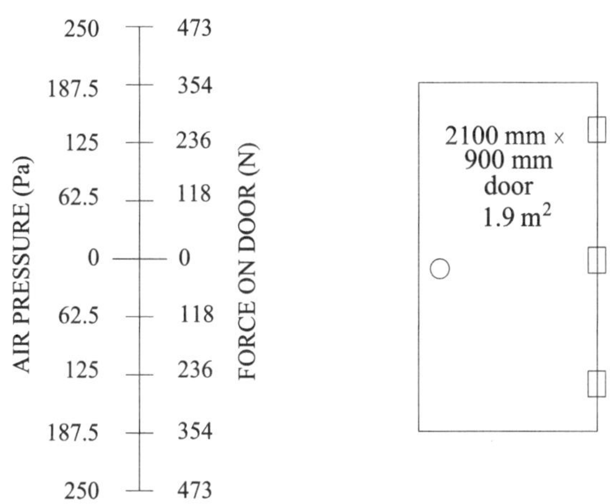

FIGURE 10–2 (SI). Relationship between air pressure and amount of force needed to open or close an average-sized door.

pressure is quite important.

10.2.2 Plant Ventilation. Outside air brought into an industrial plant is utilized to replace air exhausted, and may help dilute airborne contaminants present in the workspace. As discussed in Chapters 3, 5, 6, 8 and 9, exhaust air systems are used to remove unwanted airborne contaminants, heat, odors, and gases by placement as close to the source of generation as possible. The supply air system can aid in contaminant control by diluting remaining contaminants with outdoor air. Chapter 4 discusses the design approach for sizing the supply air rate for this purpose. Outdoor air can also be used to reduce the temperature by blending the warmer plant air with cooler outside air. The air can be blown across a person to achieve a greater cooling effect than still air. Chapter 4 also discusses heat relief and measurements of the air relating to the ability to cool a person.

Ventilation air is also needed to deliver oxygen for breathing. This is a concern with weather tight buildings, but most industrial plants have porous building shells, and outside air infiltration is normally more than adequate to provide fresh air for breathing. Air can easily flow through cracks around doors, operable windows, utility entrances, conveyor openings, and through roof mounted equipment components. Infiltration of air in this manner may cause drafts or cold/hot spots within the plant and should be avoided and may not concur with ASHRAE 62.1, Ventilation for Indoor Air Quality.

10.2.3 Building Pressure. While negative pressure can cause adverse conditions, there are situations where negative pressures are desired. An example is a room or area where a contaminant must be prevented from escaping into the surrounding area. It may also be desirable to maintain a room or area under positive pressure to maintain a clean environment. Either of these conditions can be achieved by setting and maintaining the proper exhaust/supply flow differential. Negative pressure can be achieved by setting the exhaust volumetric flow rate (Q) from the area to a level higher than the supply rate. A good performance standard for industrial processes is to set a negative pressure differential of 0.04 ± 0.02 "wg [10 ± 5 Pa]. Conversely, positive pressure is achieved by setting the supply airflow rate higher than the exhaust rate. The proper flow differential will depend on the physical conditions of the area, but a general guide is to set a 5% flow difference but no less than 50 acfm [0.03 m^3/s]. If the volume flows vary during either a negatively or positively pressurized process, it is easier to maintain the desired room pressure by adjusting the supply air.

10.2.4 Building or Process Temperature Control, Heating, and Cooling. In addition to contaminants, which are most effectively controlled by hoods, industrial processes may create an undesirable heat load in the workspace. Modern automated machining, conveying, and transferring equipment requires considerable horsepower. It is not uncommon for the process to have an electrical use of 10 to 20 watts per square foot [0.9–1.8 watts/m^2] of floor space. This equals a heat input of 34 to 68 BTU per hour in IP units. Precision manufacturing and assembling demand increasingly higher light levels in the plant with correspondingly greater heat release. The resulting in-plant heat burden raises indoor temperatures, often beyond the limits of efficient and healthful working conditions and, in some cases, beyond the tolerance limits for the product.

Environmental control of these factors can be accommodated through the careful planning and use of the supply air system. Industrial air conditioning may be required to maintain process specifications and reduce hot working conditions.

For a large industrial plant whose size is several hundred thousand square feet, the internal process heat may more than equal the heat loss through the building's walls and roof on the coldest of days. Therefore, this plant needs to be cooled

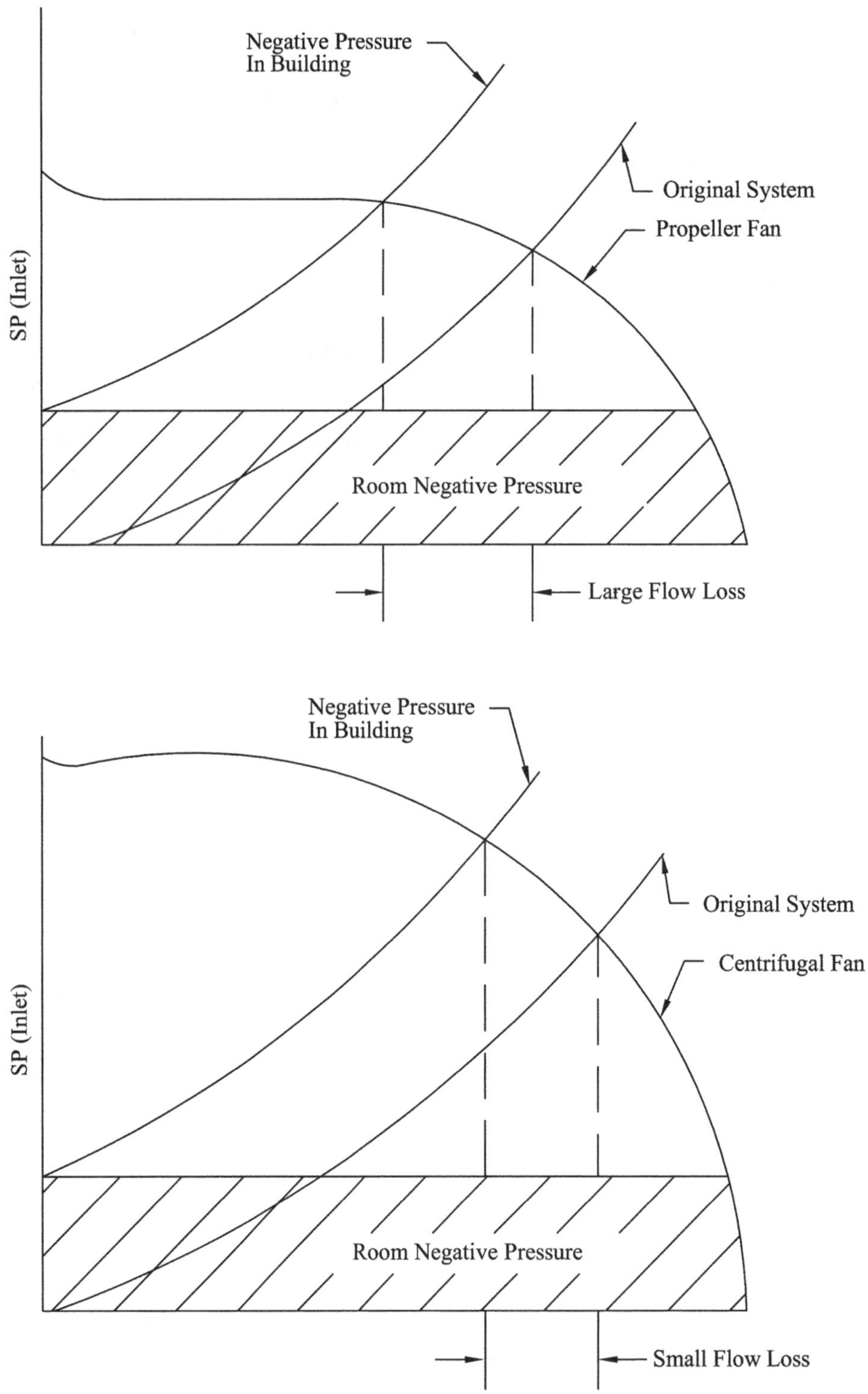

FIGURE 10-3

ACGIH®

TITLE HOW FAN PERFORMANCE DECREASES WITH NEGATIVE PRESSURE

FIGURE 10-3

DATE 1-12

CHECK CODES, REGULATIONS, AND LAWS (LOCAL, STATE, AND NATIONAL) TO ENSURE THAT DESIGN IS COMPLIANT.

throughout the year. The supply air must be heated to the degree that cold drafts are avoided. Heated air should also be utilized at door openings to reduce the cold drafts occurring with an open door. With these large facilities, the issue is how best to accomplish plant cooling.

The engineer in charge of providing suitable in-plant temperatures must understand and consider the building occupant needs as well as those of the building. Humans are warm blooded animals and must lose heat to survive and lose it at a controlled rate to be comfortable. Therefore, the design engineer who is trying to achieve human comfort sometimes has a heating concern, but always has a cooling problem.

Cooling the workspace in the summer is often more difficult than heating this space. In the heating season, the outdoor air temperatures are cool and it is relatively easy to obtain a 60 F to 70 F [16 C to 21 C] supply air temperature with normal process heat release to the space. In the summer when the outside temperature is in the 80s and 90s [approx. 27 C to 30 C], reasonable space temperatures can be obtained by bringing in additional outside air, increasing the air velocity over the person, or using evaporative coolers/refrigeration equipment to cool the supply air. Climate change studies indicate that there may be a 10 to 15 degree F [6 C to 9 C] increase in extreme weather events although the globally average temperature may only increase a degree or two.

When applying a cooling system to industrial operations, a common objective is to obtain a plant temperature of approximately 80 F [27 C]. The intent is not to try to provide a high level of comfort or control humidity; it is only to control heat. ASHRAE[10.1] gives basic criteria for industrial air conditioning in HVAC applications. Sensible and latent heat released by people and processes can be controlled to desired limits by proper use of air conditioning equipment. Radiant heat cannot be controlled by cooler air or increased ventilation, thus methods such as shielding, described in Chapter 4, are required.

To obtain the most cost-effective cooling system, a comparison should be performed between the use of extra air and cooling the air. The extra air approach often uses twice the wintertime airflow of outside air to dilute the summertime increase in workspace temperature due to process heat. This results in the operation of an oversized fan and air distribution year round. Compare this to a system with cooling capability that is used only when needed. Air distribution is not as important in this system, since either untreated or cooled outside air must be delivered into the occupied space if it is to be effective in providing acceptable conditions. The most satisfactory results are obtained when the air is delivered at the 10-foot level above the floor through industrial registers and where workers can adjust the quantity and direction.

For industrial plants larger than 400,000 square feet [36,000 m^2] in size, the supply inlet air temperature in a ventilation system is typically 5 to 10 degrees F [3 C to 6 C] warmer in the summer than the actual outside temperature. The combination of the building process heat and solar radiation heat on the roof results in the situation that the air taken into a rooftop supply unit intake is warmer than the surrounding ambient air. To minimize this temperature increase, the unit's outside air intake opening should be a distance above the roof that is equal to the sum of at least two feet plus the effective diameter of the intake opening. The fan and motor also increase the air temperature by approximately three degrees. If the motor is located outside the air stream, the temperature rise can be reduced by two degrees F [1 C].

10.2.5 Product Protection and Space Air Cleanliness. If a space requires a higher level of cleanliness than adjacent spaces, there should be an excess flow of clean air into the clean space, resulting in space pressurization and an outward airflow from the clean space to the less clean spaces. The clean air displaces the air in the space and the amount of airborne contaminants is reduced. To achieve a high degree of air cleanliness, special filters provide the final filtration. Refer to Chapter 8 for air cleaning characteristics of HEPA and other filter systems. The air exchange rate of cleanrooms must increase to achieve higher degrees of special air cleanliness depending upon the process and work practices involved. Balance the need for product cleanliness and worker protection. In most situations, the supply air enters the room from ceiling panels or diffusers and is exhausted near the floor. Room velocities in the range of 50 to 100 feet per minute are typically used in the cleanest spaces. Filters for typical commercial buildings have a minimum efficiency reporting value (MERV) around 7 or 8 based on the ASHRAE Standard 52.2, *Method of Testing General Ventilation Air Cleaning Devices for Removal Efficiency by Particle Size.*[10.4] The MERV indicates a filter's initial efficiency as a function of particle size, and gives a numeric value that allows a user or engineer to specify the appropriate rating. When a process requires a high level of cleanliness, such as food processing, painting, or assembly of parts where a fine dust is a detriment, a more efficient filtration system is required. Facilities such as hospitals and other clean operations may need filters with a MERV rating as high as 13 to 16. Refer to Chapter 8 for more discussion regarding air cleaning equipment.

10.3 SUPPLY AIR SYSTEM DESIGN FOR INDUSTRIAL SPACES

The design of the supply air system must satisfy several requirements for success. The air must enter the space without disturbing the performance of local exhaust systems or process equipment operation and without causing undesired drafts or excessive noise. High velocity airflows created by large volumes of supply air directed out of supply air registers can ruin the effectiveness of a local exhaust system. Processes involving powders, extrusions of thin membranes or the handling of objects easily dislodged by air movement are not tolerant of high velocity air streams. Employees who are reasonably comfortable often dislike high air velocities that result in unwanted drafts. The movement of high velocity air through the supply air system can also result in objectionable noise.

There are several types of spaces that occur in an industrial facility that require care in the design of supply air systems. They are discussed in the following paragraphs.

10.3.1 General Manufacturing Areas. This space is an open area with the process equipment and people spread throughout. The processes may or may not have local exhaust systems associated with them. The purpose of the supply air system is to provide exhaust replacement air, general ventilation, and temperature control. Several approaches to the supply air system design are shown in Figure 10-4. The ventilation system choices shown include the use of unit heaters that provide no ventilation and are poor at controlling the space temperature. Make up air enters the building by the infiltration of unconditioned outside air through doors, windows, and other openings. The only means of adjusting the rate of airflow is by opening or closing windows or some other building elements open to the outside. This type of system can produce cold drafts that cause employee complaints. It also has high energy usage since the cold drafts make the heating system work harder. This type of system often provides an uncomfortable plant interior since the hard working heating system raises the temperature in this area to excessive temperatures.

Another type of system (high level ventilation) is one that can bring in outside air, heat that air, and deliver it into the building using minimal air distribution ducts. This system has the ability to provide reasonably good thermal conditions during the heating season, but when the supply air is warmer than the air at the floor level, very little of the ventilation air is able to enter the worker zone. This system can be a poor replacement air system for local exhaust. The large mass of air released into the space causes high air velocities that can adversely affect the performance of exhaust hoods. Also, many processes release heat as they operate, causing the surrounding air to become warmer and forcing it to rise. As this air moves upward, it carries contaminants expelled by the process. By introducing the supply air in the truss space, the fresh air mixes with the process contaminants often resulting in the contaminants being pushed back down to the workers.

A third system is similar to the second system except that there is a ducted air distribution system. The results are similar to the second system since the air is still discharged in the upper level of the plant; and, like the second system, when the supply air is warmer than the space air, it will stay above the occupied zone unless it is forced down with high velocity outlets. Registers used as air outlets will entrain room air into the supply air stream.

A fourth design (low level distribution) is the same as the third, except the supply air outlets are dropped to the worker level. The purpose of this design is to place the air discharge low enough to provide a cooler air temperature and not disturb the warmer air located in the truss space. Good design for maintaining cool summer temperatures is to have a system that has the air discharged at approximately 8 to 10 feet [2.4 to 3.0 m] off of the floor.

When an older plant is being renovated for improved ventilation, a system for the entire plant should be considered. The system does not need to be installed at one time and can be constructed in phases. Since the ventilation air mixes readily with the plant air, the need to treat areas or spots in a manufacturing plant is normally not needed. This allows the installation of a repetitive system design without a significant reduction in performance. Using the repetitive system approach provides a lower cost system since many of the components can be duplicated. Each duct system register box and registers should be the same. The use of common devices simplifies the maintenance of the system and provides a more flexible system to operate.

10.3.2 Shipping and Receiving Areas. Plants looking to upgrade their ventilation system should first consider providing adequate door heaters or air curtains at their primary outside truck doors. For plants that have a negative pressure condition, the doors will be a source of cold drafts and lost building heat. The design of a door heater can take one of three approaches as shown in Figure 10-5. The first type of door heater uses air that is discharged at the top of the door blowing down over the opening. This type works reasonably well if the door is not too high (12 feet [3.6 m] or less) and the plant's building pressure is neutral or positive. The second type has a duct system that directs the heated air horizontally across the door opening from each side. This approach works better for taller doors since the throw of air is shorter and much of the heated air is provided close to the floor. The delivered air can readily mix with the incoming cold air that is dense and wants to flow along the floor. The third option is significantly more costly than the first two and should be reserved for very large doors. This door heater delivers the air through an opening that is in the floor running the width of the door. Since truck traffic using the door will ride over the opening, it must be covered with steel grating that can support the vehicle and allow the heated air to be blown through it. This type of door heater provides the best results since the warm air is blown upward at the door opening where it mixes with the incoming cold air warming the cold draft. With the air being warmed at the lowest elevation, people in the occupied zone get full benefit of the heat provided.

The required airflow for the proper performance of these door heaters is dependent on the amount of negative pressure in the building, the wind force commonly present and the outside temperature. Typically, a value of 100 acfm per square foot [0.5 $m^3/s/m^2$] of door opening is utilized for door heater sizing in a building with a neutral or positive pressure. The discharge air velocity at the outlet should be approximately 3,000 feet per minute [15.0 m/s] for those heaters that discharge air down or from the sides. The door heater type that discharges air up requires a lower air velocity, generally less than 1,000 feet per minute [50 m/s].

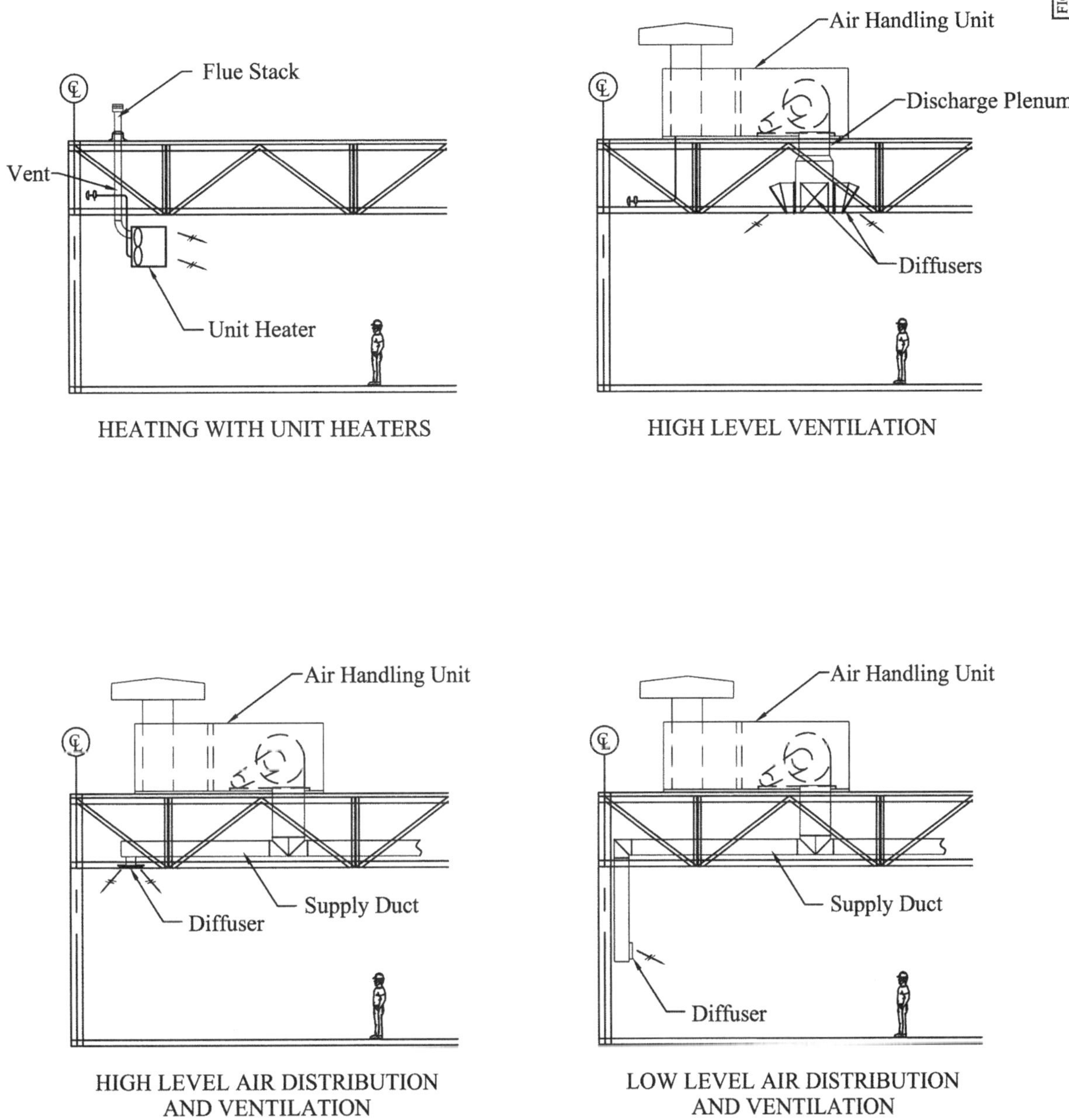

FIGURE 10-4

ACGIH®

TITLE

TYPES OF SUPPLY AIR SYSTEM DESIGNS

FIGURE 10-4

DATE 1-12

CHECK CODES, REGULATIONS, AND LAWS (LOCAL, STATE, AND NATIONAL) TO ENSURE THAT DESIGN IS COMPLIANT.

FIGURE 10-5

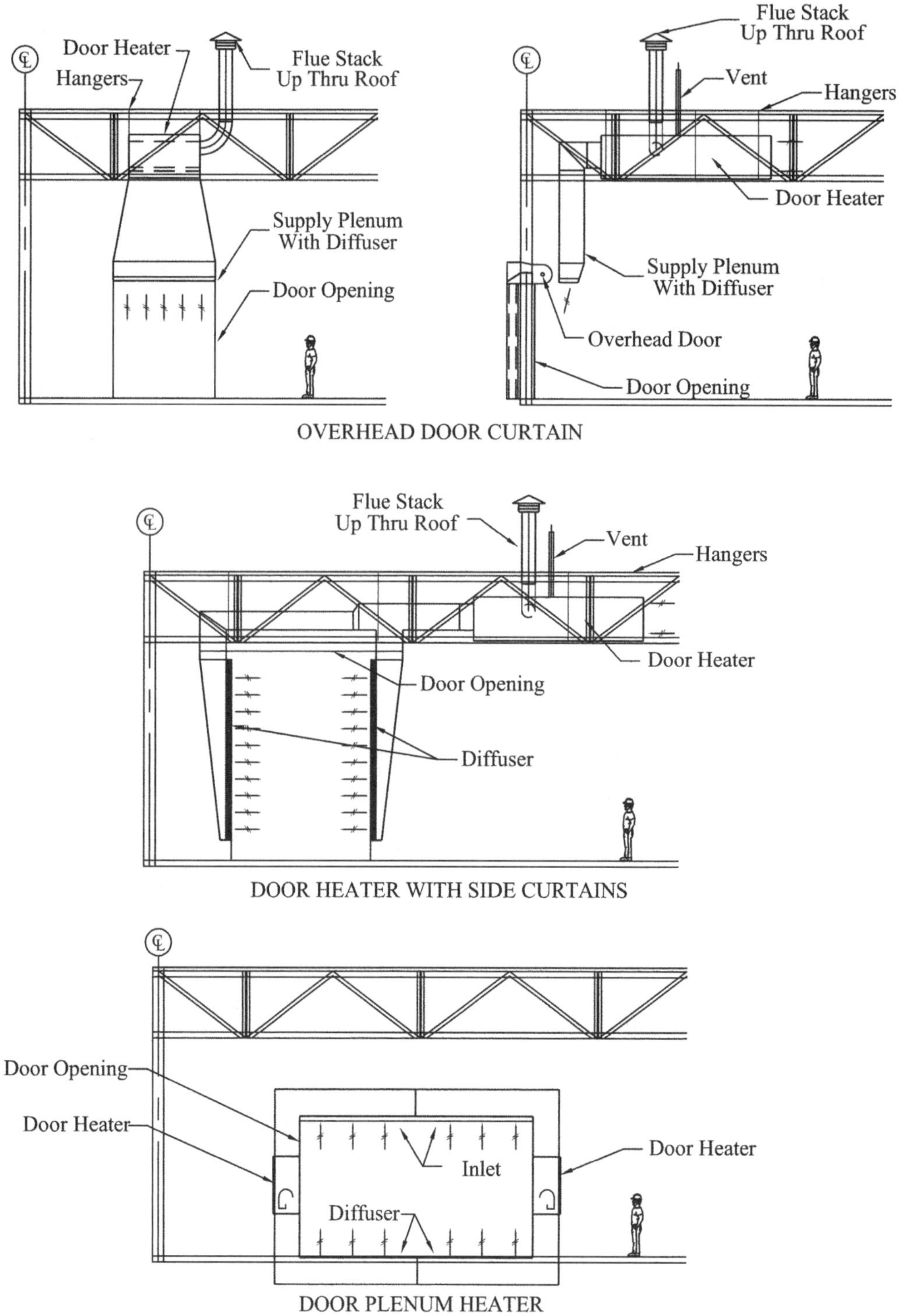

ACGIH®	TITLE TYPES OF DOOR HEATER DESIGNS	FIGURE 10-5 DATE 1-12

CHECK CODES, REGULATIONS, AND LAWS (LOCAL, STATE, AND NATIONAL)
TO ENSURE THAT DESIGN IS COMPLIANT.

10.3.3 Spaces with High Exhaust Volumes. Some spaces require large quantities of make-up air to satisfy the exhaust airflow requirement. A major issue is how to introduce the air into the room without adversely affecting the performance of those exhaust systems. Significant air velocities across the face of a hood can greatly affect its performance to capture the contaminants it was installed to control. In this type of space, supply air should be released at low velocities. One option is the use of a perforated duct that has a number of openings through which air is released into the room at low velocities. An alternate approach would be to use a plenum with perforated sides or bottom to release the air with little velocity. Both of these approaches work well in small spaces that have high levels of exhaust. The plenum or perforated duct should be placed behind or above the workers so that clean air movement is over or behind them on the way to the exhaust in enclosures such as paint booths, small laboratories, fiberglass lay up and spray up rooms, etc.

In large spaces, the supply air can be released in any manner that does not cause excessive air movement near the exhaust hoods. Care should be taken to assure that the discharge air is not hitting vertical surfaces and creating unwanted high velocities in the occupied level.

10.4 SUPPLY AIR EQUIPMENT

A supply ventilation system consists of the supply air handling unit, the air distribution duct, and the supply air outlet. The supply air unit has components to temper, clean, and move the air. A microprocessor normally controls these devices through sensors and actuators. Since there can be significant internal heat generation in many industrial plants, space cooling is the objective for most of the year.

There are several grades of air handling units: heavy industrial, light industrial, and commercial. The heavy industrial units are normally a custom or modular type and can provide many years of continuous service. If well maintained, they can easily operate for 20 years or more. The components are stronger and there is significantly more space for access to fans, filters, coils, and dampers. This facilitates the ability to maintain and repair the equipment. A light industrial grade unit typically provides less space for components maintenance making it difficult to change belts, motors, etc. Parts are less suitable for rugged industrial use. They are often mass-produced with some flexibility to make modifications. They offer the same wide choice of heating and cooling media as the heavy industrial unit. In contrast, the commercial unit has less of a choice of heating and cooling equipment, is mass produced, has a minimum amount of space for maintenance, and is structurally designed for non-industrial buildings.

Unit heaters and fan coils (Figure 10-6) are also utilized in an industrial space. The unit heater is a low cost heating only unit. It uses a propeller fan to push room air through a heating coil or fuel fired furnace. It is used for spot heating since each unit has a limited capacity. Typically, unit heaters are hung from the building structure and located to blow into a specific area. Fan coil units are similar in function except they are normally placed against a wall at floor level. They are most often found at building entrances, administrative areas, and similar spaces.

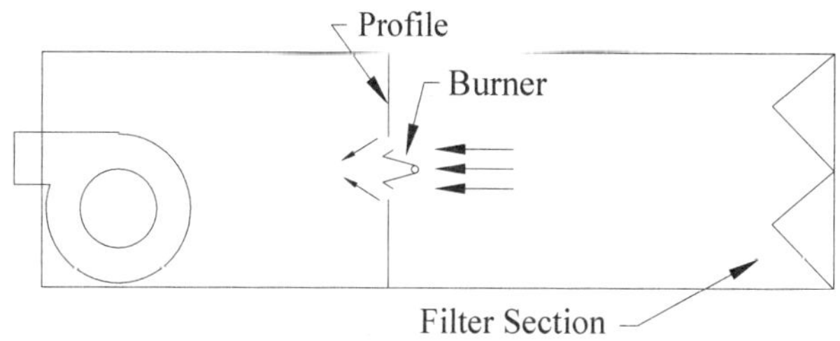

FIGURE 10–6. Direct-fired unit

Most air handling units are manufactured in a factory and shipped to the industrial site. Years ago, it was common to have units that were erected in the field. The fan, coils, filters and other components were delivered to the site and installed in a sheet metal enclosure that formed the air handling unit. Penthouses would be erected on the roof to house these air handling units. Today, field erected units are too expensive when compared to factory built units and are used only for special applications. Large factory units are designed to be split into a number of sections sized for ease of handling and shipping to the site. When installing these units at a large manufacturing plant, it is common to lift the air handling sections into place by helicopters. Most helicopters used for this purpose have a maximum lifting capability of approximately 8,000 to 10,000 pounds [3600 to 4500 kg]. Another consideration of air handling unit size is the dimension restrictions regarding over road travel. The maximum trailer width is typically 12 feet [3.6 m] and the normal height limit is 13.5 feet [4.05 m] off the road with the unit sitting on the trailer.

Factory manufactured air handling units are completely assembled and tested before they leave the plant. After they have passed the necessary tests and approvals, they are disassembled and made ready for shipment. These units are constructed so that sections can be unbolted from each other. The piping has joints to allow quick disassembly. The electrical wiring has junction boxes near the joints or wiring is pulled back from one of the connection points. Units as large as 100,000 to 150,000 acfm [50 to 75 am^3/s] capacity are constructed in this manner. When they are received at the construction site, they are lifted into place and all sections are reattached to form the air handling unit. The electrical wiring is reinstalled along with the necessary piping connections. Having tested the unit in the factory, equipment startup usually goes smoothly.

Units that are commercial grade are normally shipped in

one section and this limits their size. Their size is also limited by the sales demand. Since they are mass produced on an assembly line, significant demand is required to warrant production of a particular size unit. As a result, the highest volume units are in the size ranges below 40,000 acfm [20 m^3/s].

10.4.1 Fans. The heart of the air handling unit is the fan. It is the device that causes air to flow through the supply air system. To size the fan properly, the quantity of airflow must be identified as well as the static pressure loss due to the elements in the system that resist flow. The air quantity is determined by the purpose of the system. If the unit provides makeup air for an exhaust system, the airflow quantity depends on the system use as discussed in Sections 10.1, 10.2 and 10.3. If the system is to provide heating and/or cooling for a building, then the airflow becomes the quantity required to satisfy the heating/cooling load. Once the total airflow of the building space is identified and the number of units is chosen, the airflow required for each unit can be determined. The static pressure required for system flow is determined similar to the way exhaust system static pressure is calculated.

The fan selection for a supply fan is the same as that used to select an exhaust fan. The types of fans used are different since the static pressure is normally lower than that of an exhaust system and the air is cleaner. Common fans associated with ducted supply air systems are forward curved or backward inclined blade centrifugal fans. These fans have the capability to generate several inches of static pressure needed to move the air through the air handling unit and duct distribution system.

When selecting a fan, choose one that can be upgraded to meet more demanding operating conditions. This will give the system the flexibility to meet future needs. Fans are built to achieve different levels of service (Class I to Class IV). The Class IV fan is designed to be strong enough to handle the stresses of the highest fan outlet velocity and pressure. When selecting a fan for a range of service, one must consider the fan laws to understand the limitations for varied flow. It is normally the custom to select a fan that will operate at no more than 80% of its full rated speed. The motor selected should be able to provide the horsepower required to achieve that full speed. The electrical service for the fan should be designed to handle the horsepower [watts] required for the speed increase of 20%. The motor horsepower goes up as the cube of the increase in speed. Be sure to have the power required for a cold start of the fan, even if it is to operate continuously. All fans need to be shut down for maintenance. Refer to Chapter 7 in this Manual and the ASHRAE Handbooks for more information regarding fan selection.

10.4.2 Heating Systems. With the availability of piped natural gas, many new heating systems are of the direct gas-fired type instead of heated water flowing through a coil.

Air handling units (AHUs) are usually categorized according to the source of heat: steam, hot water, indirect gas and oil-fired units, and direct gas-fired units. Table 10-3 summarizes the basic differences of typical industrial hot processes. Each type of air heater has specific advantages and limitations that must be understood by the designer when making a selection. Each type must be capable of constant operation. Variations occur within each type in their capability of delivering a wide range of air temperatures, but they should be able to control the discharge air temperature within a range of 5 F [3 C]. Hot water and steam coil types are better able to achieve a narrow temperature range of desired room conditions due to superior modulation ability and low heat control.

10.4.3 Steam Coil Heating. Steam heating was used in the earliest air heaters applied to general industry as well as commercial and institutional buildings (Figure 10-7). When properly designed, selected, and installed, they are reliable and safe. They require a reliable source of clean steam at a dependable pressure. For this reason, they are applied most widely in large installations since smaller industrial plants often do not have sufficient boiler or steam capacity. Principal disadvantages of steam units are potential damage from freezing or water hammer in the coils, the complexity of controls when close temperature limits must be maintained, higher installed cost, and excessive piping.

Freezing and water hammer are the result of poor equipment selection and installation. Both can be minimized through careful design. Size the coil to provide the desired heat output at the available steam pressure and flow. Consider using a steam distributing coil with vertical tubes. Size the traps and return piping for the maximum condensate flow at minimum steam pressure plus a safety factor. Provide atmospheric vents to minimize the danger of a vacuum in the coil that would keep condensate from draining. Finally, never permit the condensate to be lifted by steam pressure. The majority of freeze-up and water hammer problems relate to the steam modulating type of unit that relies on throttling of the steam supply to achieve temperature control. When throttling occurs, a vacuum will be created in the coil; unless adequate venting is provided, condensate will not drain and can freeze rapidly under the influence of cold outdoor air. Most freeze-ups occur when outdoor air is in the range of 20 to 30 F [-7 C to -1 C] and the steam control valve is partially closed, rather than when the outdoor air is a minimum temperature and full steam supply is occurring (Figure 10-8).

Safety controls are often used to detect imminent danger from freeze-up. A thermostat in the condensate line or an extended bulb thermostat on the downstream side of the coil can be connected into the control circuit to shut the unit down when the temperature falls below a safe condition. An obvious disadvantage is that the plant air supply is reduced; if the building should be subjected to an appreciable negative pressure, unit freeze-up may still occur due to cold air leakage through the fresh air dampers.

Temperature control with steam coils is accomplished by

TABLE 10-3. Summary of Advantages and Limitations of Typical Industrial Heating Sources

Heat Sources	Advantages	Disadvantages
Steam Heating	• Reliable • Safe • Improves with multiple coils and face & bypass systems • Suitable for large industrial operations • Modulation schemes	• Requires clean regulated steam source • Requires piping to move steam through plant • Susceptible to freeze and water hammer damage • High installation costs • Complex controls needed for tight temp. requirements
Hot Water Heating	• Less susceptible to freezing than steam systems • Excellent temp. control in narrow ranges • Coil control is less complex than steam • Can modify with hybrid heat exchange fluids e.g. ethylene glycol	• Must have a dependable source of water • Must be smaller systems
Indirect-fire Gas or Oil Heating	• Easily applied to small industry of commercial applications • Economical for systems up to 10,000 acfm [5 am^3/s] • When appropriate room air recirculation is feasible • May be useful for areas with combustible and flammable materials • Good temperature control	• Venting required to prevent corrosion in heat exchanger from condensation • Requires constant energy use for minimum temp. in exchanger and flue
Direct-fire Gas Heating	• Economical since all fuel heat value is used • Good Temp Control • Best used above 10,000 acfm [5 am^3/s] • Bypass units only heat outside air, *then* mixed into air stream	• Below 10,000 acfm [5 am^3/s], heavy toll for safety and combustion control • Codes frequently prohibit recirculation • Flame controls have pressure sensors and valves • Not for areas with combustible or flammable operations

operating a valve that allows steam to flow into the coil. The steam condenses and the water drains away through a steam trap. Control is often an on/off modulation of the steam coil, which does not provide good close temperature control. To improve temperature control, use two control valves instead of one. One valve is usually sized for about two-thirds of the capacity and the other valve is sized for one-third of the capacity. Through suitable control arrangements, both valves will provide 100% steam flow when fully opened and various combinations will provide a wide range of temperature control. Controls are complex in this type of unit, and care must be taken to insure that pressure drop through the two valve circuits is essentially equal.

Multiple coil steam units (Figure 10-9) and bypass designs (Figure 10-10) are available to improve the temperature control range and help minimize freeze-up. With multiple coil units, the first coil (preheat) is usually sized to raise the air temperature from the design outdoor temperature to at least 40 F [4 C]. The coil is operated with an on-off valve that will be fully open whenever the outdoor temperature is below 40 F [4 C]. The second (reheat) coil is designed to raise the air temperature from 40 F [4 C] to the desired discharge condition. Refined temperature control can be accomplished by using a second preheat coil to split the preheat load. When less heat is required, it is best to

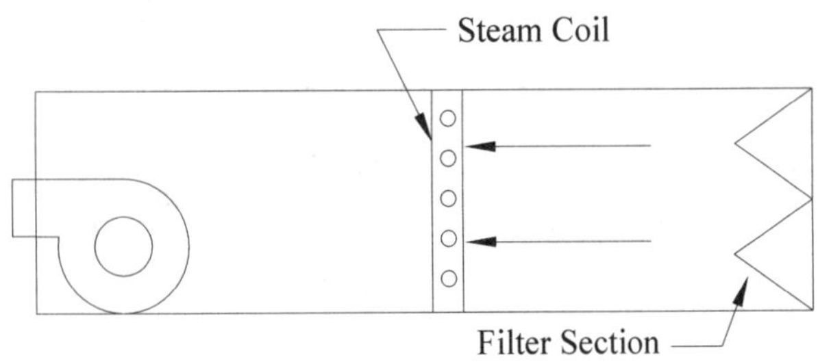

FIGURE 10–7. Single steam coil unit

reduce steam flow to the second or reheat coil by a modulating steam valve. When, and only when, this valve is closed, the modulating steam valve on the pre-heat or first coil begins to close. It is never allowed to close to the point where the air temperature leaving the coil, measured by the long (several feet) capillary tube located at the discharge side of the coil, is below a setting that will prevent freezing, usually 40 F [4 C].

Bypass units incorporate dampers to direct the airflow. When maximum temperature rise is required, all air is directed through the coil. As the outdoor temperature rises, more and more air is diverted through the bypass section until finally all air is bypassed. The principal disadvantage of this type of unit is the bypass is not always sized for full airflow at the same pressure drop as through the coil, thus (depending on the damper position) the unit may deliver differing airflow rates. Damper airflow characteristics are also a factor. An additional concern is that in some units, the air coming through the bypass and entering the fan compartment may have a nonuniform temperature characteristic that might affect the ability to deliver air within a close temperature range.

Another type of bypass design, called integral face and bypass (Figure 10-11), features alternating sections of coil and bypass. This design promotes more uniform mixing of the air stream, minimizes any nonuniform flow effect, and, through carefully engineered damper design, permits minimum temperature pickup of about 3 F [2 C], even at full steam flow and full bypass.

The same basic control system that has proven satisfactory for a two-coil system can be used for a face and by-pass system. The by-pass dampers are modulated closed when less heat is desired. Then, and only then, is the steam flow reduced to the coil by the steam modulating valve.

10.4.4 Hot Water Coil Heating. Hot water is an excellent heating medium for air heaters. As with steam, there must be a dependable source of water at predetermined temperatures for accurate coil sizing. Hot water units are less susceptible to freezing than steam because the pumped water flow ensures that the cooler water can be positively removed from the coil. Practical difficulties and pumping requirements thus far have limited the application of hot water to relatively small systems. For a 100 F [38 C] air temperature rise and an allowable 100 F [38 C] water temperature drop, 1 gpm [0.06 liters/s] of water will provide heat for only 450 acfm [0.23 am^3/s] of air. This range can be extended with high temperature hot water systems.

Temperature control for all applications is excellent with hot water coils. Temperatures are easily maintained in a narrow range since the temperature of the hot water can be varied. The operation of the coil control valve to reduce or increase flow for temperature changes does not need to be as precise as with a steam coil.

Hybrid systems using an intermediate heat exchange fluid, such as ethylene glycol water mixtures, also have been installed by industries with critical air supply problems and a desire to eliminate all freeze-up dangers. A primary steam system provides the necessary heat to a converter that supplies a secondary closed loop of the selected heat exchange fluid. The added equipment cost is at least partially offset by the less complex control system.

10.4.5 Indirect Gas/Oil-fired Heating. Indirect gas/oil-fired units (Figure 10-12) are widely applied in small industrial and commercial applications. Economics appear to favor their use up to approximately 10,000 acfm [5.00 am^3/s]; above this size the capital cost of direct-fired air heaters is normally lower. Indirect-fired heaters incorporate a heat exchanger, commonly stainless steel, which effectively separates the incoming air stream from the products of combustion. Positive venting of combustion products is usually accomplished with induced draft fans. Venting is required to minimize interior corrosion damage from condensation in the heat exchanger due to the chilling effect of the incoming cold air stream. The indirect-fired air heater permits the use of room air recirculation since the air stream is separated from the products of combustion. This separation also allows oil to be used as a heat source. Since the supply air is not exposed to an open flame, this type of heater is well suited to ventilate areas such as paint mix rooms and storage areas that have potentially explosive fumes released in the workspace.

Temperature control, "turn-down ratio," is limited to about 3:1 or 5:1 due to burner design limitations and the necessity to maintain minimum temperatures in the heat exchanger and flues. Turn-down ratio is a function of the heater's ability to modulate gas delivery from full gas delivery to zero (idle). If the burner design and other features permit a 50% reduction of gas delivery to the heater, the turn-down ratio is 2:1. If gas delivery can be reduced to 25% of the maximum and the burner still operates satisfactorily, the turn-down ratio is 4:1. Temperature control can be extended through the use of a bypass system similar to that described for single coil steam air heaters. Bypass units of this design offer the same advantages and disadvantages as the steam bypass units.

Another type of indirect-fired unit incorporates a rotating heat exchanger. Temperature control turn-down with these units can be as high as 20:1.

FIGURE 10-8

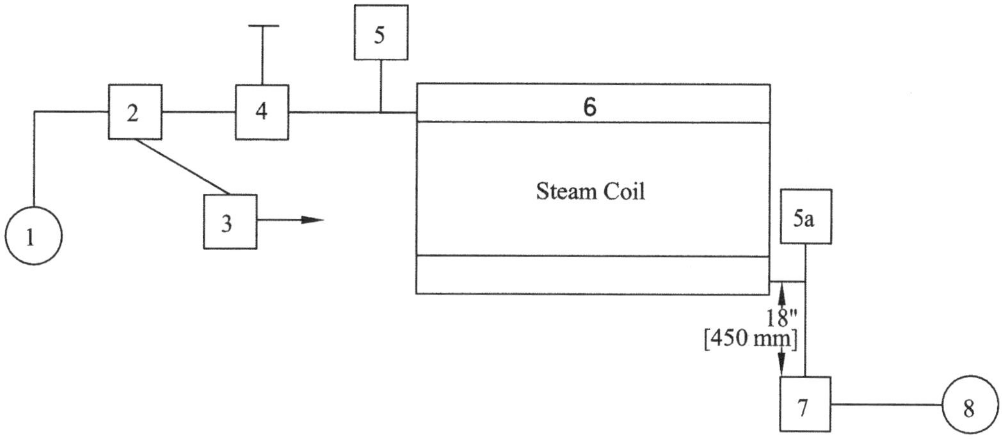

1. Steam supply
   - A. Provide steam from a clean source
   - B. Maintain constant pressure with reducing valves if required
   - C. Provide trapped drips from supply lines
   - D. Size supply piping for full load at available pressure
2. Strainer
   - A. 1/32" [1 mm] Diameter minimum perforations
3. Drip trap
   - A. Inverted bucket trap preferred
4. Control valve
   - A. Size for maximum steam flow
   - B. Maximum pressure drop equal to 50% inlet steam pressure
5. Vacuum breaker
   - A. 1/2" [13 mm] check valve to atmosphere

5a. Alternate vacuum breaker

6. Steam coil
   - A. Size for design capacity at inlet steam pressure (supply-valve drop)
   - B. Vertical coils preferred
   - C. Horizontal coils must be pitched 1/4" [6 mm] per foot toward drain. 6' [1.8 m] maximum length recommended
7. Condensate trap
   - A. Inverted bucket preferred
   - B. Size trap for three times maximum condensate load at pressure drop equal to 50% inlet pressure
   - C. Individual trap for each coil
8. Condensate return
   - A. Atmospheric drain only

ACGIH®

TITLE	FIGURE	DATE
STEAM COIL PIPING	10-8	1-12

CHECK CODES, REGULATIONS, AND LAWS (LOCAL, STATE, AND NATIONAL) TO ENSURE THAT DESIGN IS COMPLIANT.

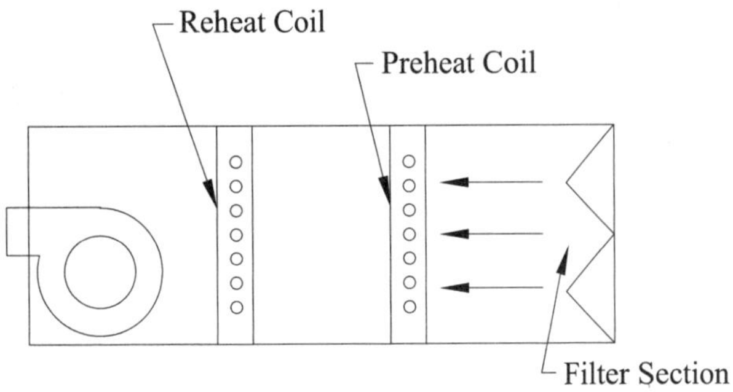

FIGURE 10–9. Multiple coil steam unit

10.4.6 Direct Gas-fired Heaters. Direct-fired heaters, where natural or liquid petroleum gas (LPG) gas is burned directly in the air stream and the products of combustion are released in the air supply, have been commercially available for some years (Figure 10-6). These units are economical to operate since all of the heating value of the fuel is available to raise the temperature of the air. This results in a net heating efficiency over 90+%. Commercially available burner designs provide turn-down ratios from approximately 25:1 to as high as 45:1 permitting good temperature control.

In sizes above 10,000 acfm [5.00 am^3/s], the units are relatively inexpensive on a cost per acfm basis; below this capacity, the costs of the additional combustion and safety controls weigh heavily against this design. A further disadvantage is that governmental codes often prohibit the recirculation of room air across the burner. Controls and sensors in these units are designed to provide 1) a positive proof of airflow before the burner can ignite, 2) a timed pre-ignition purge to insure that any leakage gases will be removed from the housing, and 3) a constantly supervised flame operation that includes both flame controls and high temperature limits. For safety purposes, the flame controls have a number of pressure sensors and valves in the gas piping to stop flow if significant changes in gas pressure are experienced.

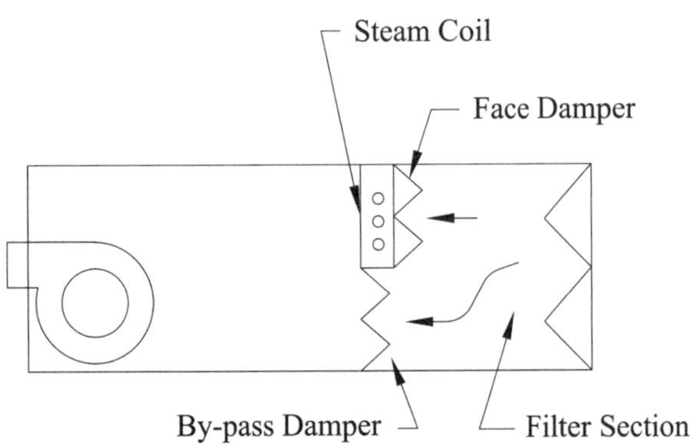

FIGURE 10–10. By-pass steam system

Concerns are often expressed with respect to potentially toxic concentrations of carbon monoxide, oxides of nitrogen, aldehydes, and other contaminants produced by combustion and the resulting gases released into the supply air stream. Practical field evaluations and detailed studies show that with a properly operated, adequately maintained unit, carbon monoxide concentrations should not exceed 5 ppm, and oxides of nitrogen and aldehydes should be well within acceptable limits.[10.2] Before specifying direct-fired equipment, evaluate all the expected contaminants to determine if direct-fired heating is appropriate in the space. For example, direct-fired heating should not be used in heating/ventilating paint mix rooms or fiberglass lay-up operations.

A variation of this unit, known as a bypass design, has gained acceptance in larger plants where there is a desire to circulate large airflows at all times (Figure 10-13). The large airflow is needed for summer ventilation with outdoor air to reduce hot plant temperatures. In the heating season, the outdoor air amount is reduced by recirculating plant air in the air-handling unit. In the bypass design, controls are arranged to reduce the flow of outdoor air with a certain percentage flowing across the burner and the balance of the airflow provided by the permit entry of room air into the fan compartment. In this way the fan airflow rate remains constant and circulation in the space is maintained. It is important to note that the bypass air does not cross the burner; only 100% outdoor air is allowed to pass through the combustion zone. Controls are arranged to regulate outdoor airflow to insure that burner profile velocity (the rate of airflow through the burner plates) remains within the limits specified by the burner manufacturer — usually in the range of 2,000 to 3,000 fpm [10 to 15 m/s]. This is accomplished by providing a variable profile that changes area as the damper position changes. A similar type unit has a fixed amount of outside air passing over the burner. This is mixed with return or unheated outside air. The total amount of outside air is varied to provide adequate replacement air and to achieve a building positive pressure. The air passing over the burner is heated to higher temperatures for mixing with the unheated air. A minimum of 20 percent of the total air must pass over the burner to maintain suitable carbon dioxide levels. Direct-fired heaters are not well suited for heating areas at outside doors unless they operate continuously since it takes two to three minutes before it can deliver warm air. This time period is required to purge the unit, have the safety devices in the natural gas line check themselves, and open the gas valve.

The related disadvantage of the direct gas-fired system is the requirement to use outside air. Since outside air is brought into the building, it must also be exhausted. In the situation where there is enough process exhaust to remove the outside air, which is heated by the burner, no energy loss occurs. If there is an excessive amount of supply air over the process exhaust, the

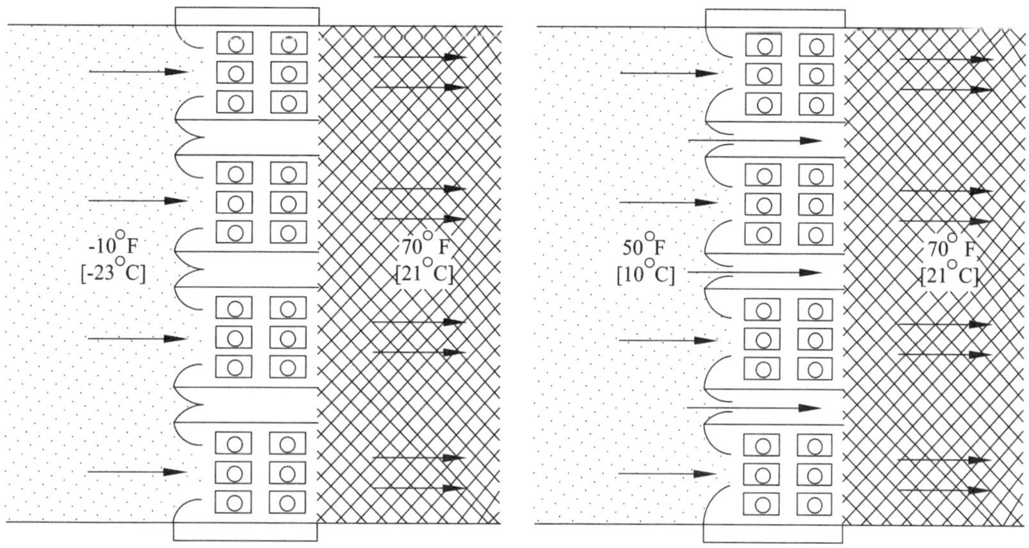

FIGURE 10–11. Integral face and by-pass coil[(10.3)]

excess air must be heated and then exhausted. This represents an energy loss.

Inasmuch as there are advantages and disadvantages to both direct-fired and indirect-fired replacement air heaters,[(10.2)] a careful consideration of characteristics of each heater should be made. A comparison of the heaters is given in Table 10-4.

10.4.7 Air Cooling Equipment. Since most industrial facilities have a process heat release, the supply air system is required to reduce the effect of this heat for temperature control in summer or, in some locations, all seasons. The ability to use untempered outside air to obtain space cooling depends upon the amount of heat release from equipment in the space and the outside air temperature. If the supply air temperature needs to be lowered, air-cooling is accomplished by means of a cooling coil (mechanical cooling) or an evaporative cooling unit. A detailed discussion regarding air-cooling can be found in Chapters 19 and 21 of the ASHRAE Handbook of HVAC Systems and Equipment.[(10.3)] Cooling is utilized for process requirements and to provide summer heat relief.

To provide summer relief of hot space temperatures, a greater amount of outside supply air may be needed than that required for replacement air purposes. In this situation, the use of cooling may be justified since a lower airflow is required compared to using untempered outdoor air ventilation to achieve reasonable space temperatures. The use of outside air for cooling is calculated on a temperature rise of 20 F [11 C] before being exhausted. If a cooling unit is used, the entering temperature is lower, allowing a supply air temperature rise of 30 F to 40 F [16 C to 22 C]. Thus, with cooling, less airflow is needed.

10.4.8 Mechanical Cooling. With mechanical cooling, the

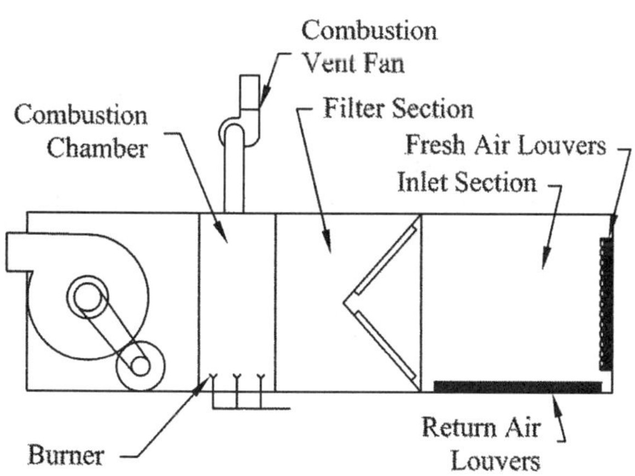

FIGURE 10–12. Indirect-fired unit

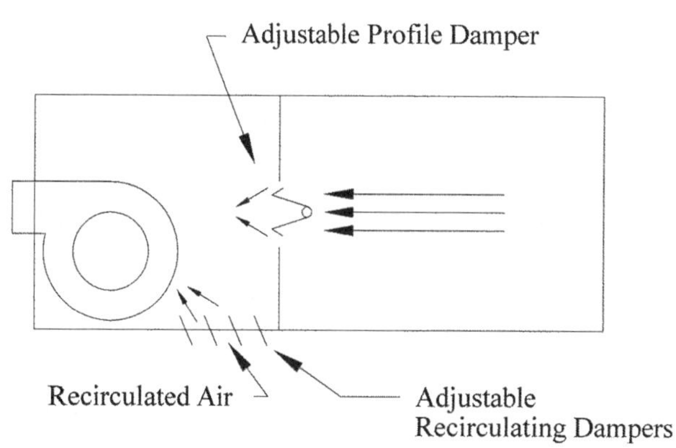

FIGURE 10–13. Direct-fired by-pass unit

TABLE 10-4. Comparison of Heater Advantages and Disadvantages

Advantages	Disadvantages
Direct-fired Unvented:	
1. Good turn-down ratio—8:1 in small sizes; 25:1 in large sizes. Better control; lower operating costs. 2. No vent stack, flue or chimney necessary. Can be located in sidewalls of the building. 3. Higher efficiency (90+%). Lower operating costs. (Efficiency based on available sensible heat.) 4. Can heat air over a wide temperature range. 5. First cost lower in large size units.	1. Products of combustion in heater air stream (some CO_2, CO, oxides of nitrogen, and water vapor present). 2. First cost higher in small size units. 3. May be limited in application by governmental regulations. Consult local ordinances. 4. Extreme care must be exercised to prevent minute quantities of chlorinated or other hydrocarbons from entering air intake or toxic products may be produced in heated air. 5. Can be used only with natural gas or LPG. 6. Burner must be tested to assure low CO and oxides of nitrogen content in air stream. 7. Outside air brought into building may be significantly more than process exhaust causing an excessive amount of heating energy use.
Indirect Exchanger:	
1. No products of combustion are discharged into building. 2. Allowable in all types of applications and buildings if provided with proper safety controls. 3. Small quantities of chlorinated hydrocarbons will not normally break down on exchanger to form toxic products in heated air. 4. Can be used with oil, LPG, and natural gas as fuel. 5. First cost lower in small size units. 6. Can be used in air recirculation mode as well as for makeup air.	1. First cost higher in large size units. 2. Turn-down ratio is limited — 3:1 usual, maximum 5:1. 3. Flue or chimney required. Can be located only where flue or chimney is available. 4. Low efficiency (80%). Higher operating cost. 5. Can heat air over a limited range of temperatures. 6. Heat exchanger may be subject to severe corrosion condition. Needs to be checked periodically for leaks after a period of use. 7. Difficult to provide combustion air from outdoors unless roof or outdoor mounted.

cooling coil has a chilled fluid flowing through it to remove the heat from the air stream. This heat exchange reduces the temperature of the air stream and warms the chilled fluid. The fluid is typically a refrigerant or water. Air handling units that use a refrigerant have a compressor and condenser nearby to change the refrigerant gas back into a liquid and reduce its temperature. The act of quickly reducing the pressure on the liquid allows it to change into a gas and become cold, thus chilling the coil. In a chilled water unit, water of approximately 45 F [7 C] flows through the cooling coil. The water is chilled by a central chiller and pumped through a pipe distribution system to each air-handling unit. Commercial and light industrial type AHUs most often use the refrigerant type system commonly called direct expansion (DX) cooling equipment. The first cost of the chilled water system is higher than the DX system, but it offers longer component life, reduced maintenance, lower energy costs, and is more suitable for larger installations.

The use of a cooling coil can often reduce both air temperature and humidity. The humidity reduction is caused by dropping the air temperature below its dew point. The objective is to get the air temperature cold enough so that the concentration of water vapor in the air can no longer be maintained. The air begins to fog and water droplets called condensate begin to form on the cooling coil. The condensing of the water vapor to reduces humidity requires additional cooling over and above that for reducing the air temperature.

10.4.9 Evaporative Cooling. Evaporative cooling systems rely upon the evaporation of water vapor to lower the air temperature. The air also becomes more humid since the water vapor evaporates into the air stream. In the evaporative cooling unit, air absorbs water vapor as it passes through a wetted pad or through a water spray zone. Energy is given up by the air to evaporate the water and the air temperature is reduced. Since evaporative coolers raise the relative humidity in the space, the impact on the industrial processes should be evaluated. Some evaporative cooling systems have their own pumps and water circulating systems. Others rely on the pressure in the water line to generate a water spray. Evaporative coolers are commonly used in dry areas of the world, but can be applied to almost all areas of the United States. They are also used in industrial applications that have high replacement airflow or large internal heat releases. For an evaporative cooling unit to operate at peak efficiency, the pads must be well wetted and reasonably clean. Spray nozzles must be kept free of clogging deposits. The following formulae can be used to identify the temperature leaving an evaporative cooler:

$$T_{exit} = T_{enter} - E(T_{enter} - T_w) \quad \textbf{[10.1]}$$

Where

T_{enter} = Dry-Bulb temperature entering, F [C]
T_{exit} = Dry-Bulb temperature leaving, F [C]
T_w = Wet-Bulb temperature entering, F [C]
E = Efficiency factor

The Wet-Bulb temperature is the value measured using a psychrometer as discussed in Chapter 4. The efficiency is normally 80%.

10.4.10 Air Filtration. Supply air filtration for workspaces is not a major concern for most industrial processes; however, seasonal factors such as insects, pollen, organic debris, etc., may require removal before the air is supplied. The filters are typically selected on the basis of keeping the supply air unit clean. However, in some cases, filters are selected for employee health considerations or process concerns. When outside air sources are contaminated, air cleaning is required to remove those contaminants. Filters for normal service typically have a minimum efficiency reporting value (MERV) of 6 to 8 as defined in ASHRAE Standard 52.2, "Method of Testing General Ventilation Air-Cleaning Devices for Removal Efficiency by Particle Size."(10.4) When a process requires a high level of cleanliness, such as food processing, painting, or assembly of parts where a fine dust is a detriment, a more efficient filtration system is required. Refer to Chapter 8 for more discussion regarding air cleaning equipment.

10.4.11 System Temperature Control. Some processes require a space that has close control of temperature and humidity. This often requires both heating and cooling of the supply air to achieve the desired thermal conditions. Often these spaces also require humidification if the air is too dry; a condition that most likely occurs in the winter. An example could be a powder painting operation that requires air entering the paint spray booth to be 70 F [21 C] and 50% relative humidity (RH). This temperature condition is necessary to achieve proper drying of the paint and to prevent arcing and sparks inside the booth (a fire prevention concern).

In summer, the supply air would need to be cooled below 51 F [11 C] to condense enough water vapor from the air to achieve the 50% RH. This air would then need to be reheated to raise the temperature to the 70 F [21 C] goal. The layout of the air handling unit would have the cooling coil followed by the reheat coil.

In winter, the air must be heated and water vapor added to the air to achieve the desired 70 F [21 C], 50% RH. A heating coil or gas-fired device can be utilized. Either a humidifier or an evaporative cooler is used to add humidity. If a humidifier is used, the heat in the vapor must be identified and the energy of the heater reduced accordingly. Refer to Figure 10-14 for a representation of the performance of this equipment during the cooling and heating seasons. As can be seen, on a cold day the air must be heated to a temperature of 99 F [37 C] to achieve a condition of 70 F [21 C] and 50% RH. The proper air handling arrangement would have the humidifier behind the heating device.

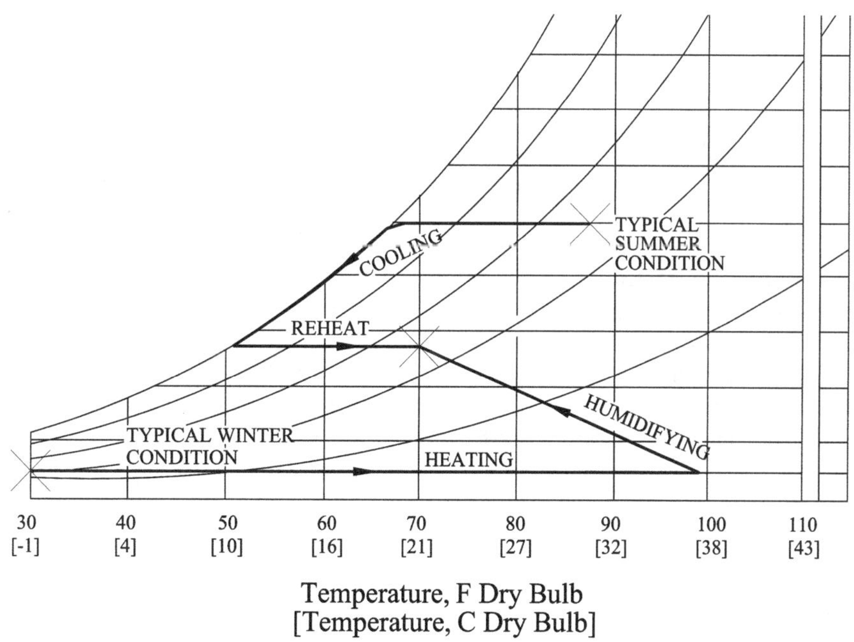

FIGURE 10–14. Air heating and cooling requirements

The closeness of control desired will dictate the component type to be utilized in the system. Heating and cooling water coils provide the best control with gas-fired equipment providing the least level of control.

10.4.12 Unit Location. Air supply units are normally located in the upper level of the plant or on the roof. In some recent designs, these units have been placed just below the roof (in the truss space) and have a catwalk system for ease of access. Rooftop units create the need for people to walk on the roofs. It is good practice to provide a walkway to minimize excessive wear on the single-ply roofs in common use today. Some systems have the unit placed along an outer wall inside the building. Outside air is mixed with room air to satisfy general building heating and replacement air requirements. This type of system has little distribution duct and its ventilation effectiveness is low except in small buildings.

10.4.13 Size and Cost Considerations. There are several cost considerations to a supply air system installed in an industrial facility. First, the relative cost for the supply air unit decreases as the size increases. Some cost elements of the unit increase with unit size: the unit housing, fan, filters, and coils. The unit's control cost depends on the control functions being performed and is approximately the same for all size units. Another major cost element is the air distribution system; i.e., the duct and registers. The duct and register costs increase as the system gets larger. The final cost consideration is installation, which includes lifting the unit; structural steel supports, electrical, natural gas, and other piping system hook-ups; unit start-up; and warranty. Installation cost is somewhat independent of unit size and increases at a rate slower than the unit size. For more information regarding system costs, see Chapter 12, Cost Estimating.

10.5 SUPPLY AIR DISTRIBUTION

In an industrial facility, the supply air distribution plays an important role in the success of controlling airborne contaminants. If contaminants are controlled by local exhaust ventilation, the supply/replacement air should be introduced into the space in a way that does not interfere with the capture effectiveness of the exhaust hoods. Interference is created when supply/replacement air is introduced at an excessive velocity into the vicinity of an exhaust hood, thus interrupting the protective flow path of the hood's exhaust air volume. If the supply/replacement air diffuser is blocked by equipment, replacement air will not reach the hood as designed. When the supply/replacement air diffuser is located too close to the exhaust outlet, the clean air may be "short-circuited" and not reach the workspace at all.

There are additional supply air design considerations when dilution ventilation is used rather than local exhaust ventilation to control contaminants. These include the location of the supply air outlets, the rate of airflow, and the placement of the exhaust air intakes. Refer to Chapter 4 for more discussion and system sizing considerations. The choice of dilution ventilation versus local exhaust ventilation depends on the nature and quantity of the contaminants and the workspace. Several supply air design approaches are discussed in the following sections. Regardless of the selected supply air design, emphasis should be given to avoid creation of a working environment undesirable for space inhabitants.

10.5.1 Unidirectional or Plug Airflow. The use of non-turbulent or laminar supply airflow is required in situations where high cleanliness or extreme contaminant control is desired. This approach has clean supply air moving across the space in a uniform direction and the air is removed from the space at a location opposite the supply air entry point. This design scheme is often referred to as unidirectional, laminar, or plug airflow. It is normally employed to protect workers and critical processes. In addition to careful consideration of the supply air distribution design, physical obstructions such as partitions or furniture should be minimized to avoid any turbulent airflow. Examples of this type of supply air design can be found in industries or activities associated with firing ranges, pharmaceutical manufacturing, semiconductor manufacturing, healthcare treatment, aerospace, and painting operations.

For areas that require non-turbulent air for proper exhaust system operation, one approach is to pass air through a supply air plenum built as part of a perforated ceiling and/or through perforated duct. The ceiling plenum or duct runs should cover as large an area as possible to diffuse the airflow. A plenum wall providing cross-flow ventilation should be used when the workers are positioned between the supply air system and the contaminant source or exhaust hood. This approach should not be used for design velocities at the worker over 100 fpm [0.5 m/s] since a low pressure zone can be created causing contaminants to be carried into the worker's breathing zone. See Chapter 6, Section 6.4.8 for more information on Worker Position Effects.

Perforated drop-type ceilings work best in spaces with ceiling heights of less than 15 feet [4.5 m]. Hoist tracks, lighting, and fire protection systems can be built into the ceiling. In some cases, fire protection will be required above and below the ceiling. Use the perforated duct approach when ceiling heights are over 15 feet [4.5 m]. Perforated duct manufacturers typically have computer programs to assist designers in determining duct sizes, shapes, and types as well as the location of pressure adjusting devices such as orifice plates and reducers. Airflow delivery in large bays may require supplemental air delivered at work stations to provide comfortable conditions for workers.

How the supply air is fed into a plenum is critical to its performance. High velocity flow into the plenum can cause turbulence problems similar to large diffusers. Air that is introduced into a plenum at an excessive velocity will bounce off the floor or an opposite wall causing turbulence inside the plenum. This can cause re-entrainment of contaminants from the room into the clean replacement air via a low-pressure area created near the introduction point. The low-pressure phenomenon also creates uneven replacement air distribution in the room. Providing a wide replacement air plenum and slowly introducing supply air into the plenum will reduce the problem. However, space for a wide plenum is frequently unavailable. One solution is to feed the plenum with a perforated duct to diffuse the air inside the plenum. Ensure that the proper pressure adjusting devices (e.g., orifice plates) are installed per the manufacturer's recommendations. Another approach to distribute air from either a ceiling or wall-mounted plenum is to design the plenum face with two overlapping perforated plates, one fixed and one adjustable, at the time of airflow balancing, located 2 to 6 inches [50 to 150 mm] apart. Air flowing through slightly offset holes will encounter more resistance; thus, air quantities passing through the low-flow areas will increase. The holes must be small enough to fine-tune the airflow from the plenum. Openings of 3/8" [9 mm] diameter in the adjustable plates with sufficient numbers to provide a velocity of 2000 fpm [10 m/s] seem to work well.

This approach is used in clean room and paint booth designs to achieve a high control on air cleanliness. For these applications, clean supply air flows through a grid of filters in the ceiling and is exhausted at floor level. Flow velocities in the range of 50 to 100 fpm [0.25 to 0.50 m/s] are common.

10.5.2 Mixing Ventilation Systems. The mixing approach to the supply air ventilation system relies on high-velocity air streams leaving supply registers as the means of delivering air to the workspace. These jets of supply air quickly entrain and mix with the space air. As shown in Figure 10-15, the average temperature of this air stream begins to approach the space temperature as the velocity of the jets slows. This example has air leaving a register at a velocity of 2000 fpm [10 m/s] and a temperature of 20 F [11 C] below room temperature. At a work station 27 feet [8 m] from the register, the average speed of the jet has dropped to 200 fpm [1.0 m/s] and the air temperature will approach the room temperature of 86 F [30 C]. The actual conditions depend on the register selected, but velocities of 100 to 200 fpm [0.5 to 1.0 m/s] and temperatures of one to two degrees below the room temperature are likely.

Mixing systems dilute airborne contaminants the same way that the air jets dilute temperature. Care should be taken to direct the supply air jets so as not to disturb the performance of local exhaust systems. Otherwise the resulting air currents can sweep contaminants away from exhaust hoods rendering the hoods less effective. The local exhaust hoods may then require additional airflow to control the contaminants. Increasing exhaust airflow also increases energy costs due to the need for larger fans and motors. In extreme cases of high room air motion, the hoods remain ineffective even with substantial increases in airflow. Hence, workers could still be overexposed even with a local exhaust system in place. Therefore, locate air supply discharges away from local exhaust hoods.

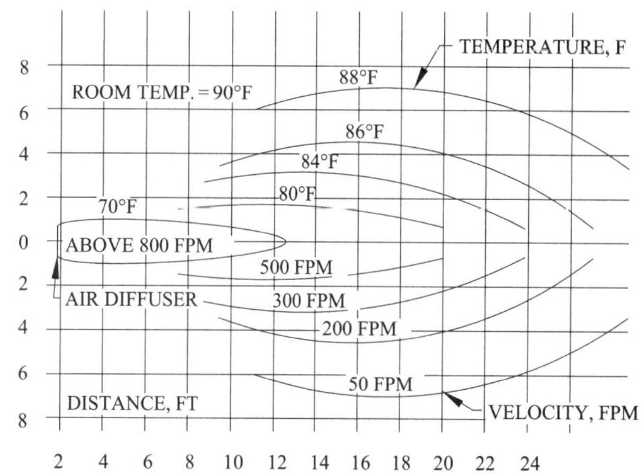

FIGURE 10–15. Example – Air jet temperature and velocity profile (IP units)

If the supply air system does not sufficiently cool the employees, pedestal fans are often used for greater air movement. Care must be used in their placement since pedestal fans can degrade contaminant control by causing turbulence near local exhaust hoods.

Mixing systems can have air outlets in the truss space (20 feet or higher) blowing downward or placed at lower levels. For those systems where the air is discharged below the truss, duct routing must be coordinated with the process layout and the needs of the process equipment. Quite often the use of cranes, gantries, conveyors, and other material handling equipment greatly reduces the access to space for routing duct below the truss. A common low-level discharge height is 10 feet [3 m] above the floor with the air directed horizontally with a downward deflection in the summer. During the winter, the air is directed upward approximately 5 degrees above horizontal. The lower-height-discharge approach provides a cooler workspace and should be considered to obtain a lower space temperature.

Supply air not removed by process exhaust systems is normally removed from the building through the use of roof-mounted exhaust fans. The discharge of supply air at the 10-foot [3 m] level is an approach often used for spot cooling. Spot cooling is the directing of a mass of supply air to a workstation with the purpose of keeping it as cool as possible. It is often used in operations that have high radiant heat exposures such as is found in metal casting, forging and steel making operations. The approach of high velocity discharge spot cooling is normally not very effective if the air discharge grille is located a distance from the workstation. More effective methods place the supply air outlet at the same height and adjacent to the workstation. Even better is the direction of the supply air up through the workstation from a grate on the floor. Both of these approaches have the air discharge much closer to the worker so little entrainment of room air takes place.

Care must be taken with spot cooling systems. Air delivery at high velocities from behind the operator will create a low pressure zone on the other side of the body (the person's breathing zone). Contaminants can be induced into this zone and inhaled by the worker. Blowing air up from below provides no opportunity to mix the supply air with room air, and thus provides the coolest thermal condition. Care must be taken with this system not to blow contaminants into the employee's eyes, so it should not be used in conjunction with processes that have airborne particles escaping. Spot cooling systems for these applications often have airflows in the range of 3,000 to 4,000 acfm [1.5 to 2.0 am^3/s] per workstation and velocities at 1500 to 2000 feet per minute [7.5 to 10.0 m/s] leaving the supply air register.

10.5.3 Air Displacement Ventilation Systems. Areas that require year-round cooling due to process heat can utilize a nonturbulent approach to adding air into the workspace called air displacement. Air displacement ventilation systems were first applied in the welding industry in 1978, and now are widely used in Scandinavian countries. This type of supply air system relies upon the natural effect of warm air rising. Provision is made to remove the warm air at the top of the space. The supply air is introduced into the space through low-velocity diffusers placed near the floor. The objective of the air displacement system is to achieve air quality conditions in the occupied zone that are similar to those of the supply air.

As illustrated in Figure 10-16, there are two air distribution zones in an air displacement system, the upper and lower strata. The upper zone is formed at the elevation where the supply air quantity equals the total air moving upward in the thermal plumes caused by the process heat. As this warm air rises, it entrains adjacent air and the total volume of moving air increases. When this total air volume equals the supply air, there is no more incoming air to feed the plume and recirculation of space air begins. The elevation where the recirculation starts is called the stratification level. Properly designed air displacement systems have the stratification level well above the occupied lower zone. The height of this lower zone is dependent on the amount of supply air, the nature of the heat sources, and the air distribution across the floor.

When designing a displacement ventilation system, the following parameters need to be considered: 1) supply airflow rate and temperature; 2) air temperature at floor level; 3) vertical temperature gradient; 4) maximum air velocity at floor level; and 5) first cost, operating cost, and energy consumption.[10.5]

The supply air temperature can be 4 F to 6 F [2 C to 3 C] warmer than that used in a mixing type system to achieve the same occupied space temperature.[10.6] The vertical temperature gradient or the temperature rise of the supply air compared to the exhaust is greater in the displacement type system. Typical temperature differences compared with increases in building height are:[10.7]

Building Height, ft [m]	Temperature Rise, F [C]
Less than 10 [3]	11–13 [6–7]
10–20 [3–6]	15–18 [8–10]
over 30 [> 9]	18–22 [10–12]

This increase in temperature difference will reduce the amount of exhaust air required.

The advantage of not significantly mixing the space air with the supply air is a workspace that is cooler and has less airborne contaminants. The process heat and many of its associated contaminants are carried away as the warm air rises. Special provisions must be made for supply air outlets. Since they are on the floor, they must be coordinated with the process equipment layout to allow access to operate, service and maintain the equipment. Air outlets need to be placed a reasonable distance from each other to avoid drafts caused by the high quantity of supply air leaving the diffusers.

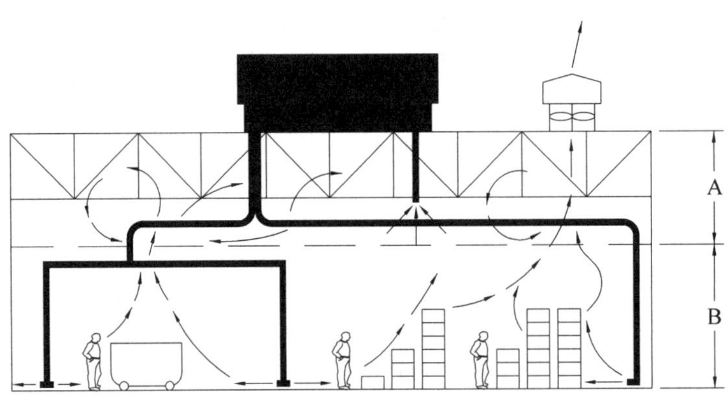

A - Stratification Level

B - Lower Zone or Displacement Level

FIGURE 10–16. Airflow in displacement ventilation system

10.5.4 Duct Materials. Supply duct materials are generally Sheet Metal and Air Conditioning Contractors National Association (SMACNA) Class I or II medium gauge sheet metal, but other materials such as specially coated cloth, may be used. The material does not need to be as strong as exhaust duct for several reasons:

1) It is not exposed to the transport of abrasive process contaminants.
2) The system operates at a relatively low pressure.
3) Much of the duct is on the downstream side of the fan and is under a positive pressure.
4) Duct leaks do not pose a health hazard and have little affect on system performance.

The duct needs to be strong enough to last in its environment. Often the abuse of plant operations requires a heavier duct system than one that is hidden above a ceiling. It needs to be able to withstand the rigors of pedestrian traffic and being hit by forklift trucks.

If too light a sheet metal gauge is selected, fan noise may be more pronounced due to vibration of the duct. Metal stiffeners attached to the outside of the duct help prevent this type of noise. SMACNA standards provide detailed information on duct construction.(10.8, 10.9)

10.5.5 Sheet Metal. Sheet metal materials typically include galvanized steel, uncoated steel, stainless steel and aluminum. In addition to round ducts, oval or rectangular ducts are often used in order to adjust the cross-sectional area to avoid obstructions in the building space. SMACNA standards use pressure ranges to classify duct thickness and construction methods. The allowable duct leakage and acoustical considerations also dictate the construction methods.

10.5.6 Plastic. Thermosetting (glass-fiber reinforced polyester) and thermoplastic (polyvinyl chloride, polyethylene, polypropylene, acrylonitrile butadiene styrene) construction standards are provided by SMACNA's Thermoplastic Duct Construction Manual.(10.10)

10.5.7 Fiberglass. For acoustical reasons, supply duct may be lined for better sound adsorption. The lining used is a fiberglass-based material treated to minimize moisture absorption. The coating is also mold and fungus resistant.

10.5.8 Textile. For certain applications, especially those with high open bays and/or areas requiring frequent cleaning, porous textile duct is a relatively new option. The supply air inflates the ductwork. The textile duct can be washed and reused and ultimately recycled. Hanging systems rely on engineered orifices, varying fabric porosity and linear vents to distribute airflow.

10.5.9 Supply Air System Design Considerations. A properly designed ventilation system must adhere to building codes, requirements established by the National Fire Protection Association (NFPA), and standards developed by SMACNA. There are also standards published by ASHRAE, ANSI, and AMCA as well as those developed by specific industrial corporations.

There are several design tools currently available to aid in the design of industrial ventilation systems. Many involve rules of thumb, spreadsheet evaluations, or simple one- or two-dimensional models. These design tools are used to evaluate building air balance, heating and cooling loads, special pressures, smoke/contaminant migration, and migration and dilution of gases and/or fumes.

With the recent availability of low-cost, powerful computers, designers have been applying Computational Fluid Dynamics (CFD) to the analysis of ventilation system performance. With CFD, the geometry of a space can be configured and airflow rates, temperatures, contaminant migration, and exhaust capture efficiency can be evaluated. The CFD model divides time and distance into discrete intervals. The modeled space is divided into many smaller volumes that interact with each other based on fundamental conservation equations. An iterative process is used to calculate the results and computation times can take a number of hours.(10.11)

There are numerous computer programs, nomagraphs, and other catalog-type data available from manufacturers for selecting and sizing system components. Air handling unit fans, coils, filters, and humidifiers are selected with this information. Elements in the air distribution system (duct, registers, etc.) are also selected with these aids. It should be noted that supply air duct pressure loss calculations are much easier to accomplish than those for exhaust air systems. The air is clean and dampers are used to adjust airflow through the duct system. Duct velocity is typically limited to 2500 to 3000 fpm [12.5 to 15.0 m/s] to avoid excessive noise and minimize horsepower requirements of the fan motor. Registers are selected to obtain the desired air throw or air distribution in the space.

Industrial rated duct components should be used in air distribution systems since the higher velocities being utilized will stress the materials found in lesser duty components. Balancing dampers may not be required, but if used in a system that has air velocities greater than 2000 feet per minute [10.0 m/s], industrial rated dampers are needed. For systems that serve large industrial spaces, often an air balance is not critical to the facilities operation, and in these situations, dampers provide little value to system performance. Dampers should not be installed behind grilles to minimize noise and stress on the blades found in a grille or diffuser.

10.6 AIRFLOW RATE

The design supply airflow rate depends on several factors, including health and comfort requirements. Sensible heat can be removed through simple air dilution (see Chapter 4, Section 4.15). "Nuisance" or undesirable contaminants can also be reduced by dilution with outdoor air as described in Chapter 4. For many industrial facilities, experience shows that when the air supply is properly distributed to the working level (i.e., in the lower 8 to 10 ft of the space [2.4 to 3.0 m]), outdoor air supply of 1 to 2 acfm/ft^2 [0.005 to 0.01 am^3/s/m^2] of floor space will give good results. This flow rate will normally satisfy the process exhaust quantity as well as circulate adequate air for building heating requirements and general ventilation. Specific quantities of minimum outdoor air supply may be obtained from building and health codes or from criteria developed by groups such as ASHRAE.

10.6.1 Air Changes. "Number of air changes per minute or per hour" is the ratio of the airflow ventilation rate (per minute or per hour) to the room volume. "Air changes per hour" or "air changes per minute" is a poor basis for ventilation criteria. The required ventilation depends on the generation rate and toxicity of the contaminant, not on the size of the room in which it occurs. For example, let us assume a situation where an airflow of 11,650 acfm [5.497 m^3/s] would be required to control solvent vapors by dilution. The operation may be conducted in either of two rooms, but in either case, 11,650 acfm [5.497 m^3/s] is the required ventilation. The "air changes," however, would be quite different for the two rooms. As can be seen in Table 10-5, for the same "air change" rate, a high ceiling space will require more ventilation than a low ceiling space of the same floor area. Thus, there is little relationship between "air changes" and the required contaminant control. Also, there are often great differences between the calculated air change per hour and the effective air changes, which is determined by location of air supply outlets, exhaust systems, cross-drafts and air temperature. Room air change rate is the same as dilution ventilation that is discussed in Chapter 4. The difference between the actual and effective air change rate is the same as applying mixing factor (m_i) values to increase the amount of airflow because of less than ideal ventilation conditions.

The "air change" basis for ventilation does have some applicability for relatively standard situations such as office buildings and school rooms where a standard ventilation rate is reasonable. Building codes for the design of certain types of buildings often use a minimum air change per hour for specific spaces. For example, a flammable storage room requires six air changes per hour using OSHA requirements. This approach is easily understood and reduces the engineering effort required to establish a design criteria for ventilation. It is this ease of application, in fact, which often leads to lack of investigation of the real engineering parameters involved with correspondingly poor results.

TABLE 10-5 (IP). Air Exchanges vs. Room Size

Room Size	Room ft^3	Air changes/ minute	Air changes/ hour
40 × 40 × 12 high	19,200	11,650/19,200 = 0.61	36
40 × 40 × 20 high	32,000	11,650/32,000 = 0.364	22

TABLE 10-5 (SI). Air Exchanges vs. Room Size

Room Size (m)	Room (m^3)	Air changes/s
12.192 × 12.192 × 3.657	543.5943	5.479/543.5943 = 0.0101
12.192 × 12.192 × 6.096	906.1391	5.497/906.1391 = 0.0061

10.7 HEATING, COOLING AND OTHER OPERATING COSTS

Operating the supply air can be a major expense for a manufacturing plant. In addition to heating the air during the cold weather and possibly cooling in the hotter months, there are other operational concerns. Perhaps the highest energy user is the electric motor that creates the air movement. A rough estimate value for motor size is one horsepower for every 1,000 acfm of airflow [approx. 1500 watts/m^3/s]. For a complete discussion of ventilation system energy, see Chapter 11. Other operational costs are air cleaning component replacement, control calibration, and component maintenance.

10.7.1 Estimating Heating Energy Use. Supply air temperature is controlled by the demand for heating and cooling. Factors to consider in maintaining a comfortable work environment for occupants are: setpoint temperature, humidity control, air distribution, and airflow rate. Where high internal heat loads are to be controlled, a low air supply temperature can be obtained by reducing the amount of heat supplied to the air during the winter months and by deliberately cooling the air in the summer. When a large airflow rate is delivered at approximately space temperatures or below, the distribution of the air becomes vitally important in order to maintain satisfactory environmental conditions for individuals in the space.

Maximum utilization of the supply air is achieved when the air is distributed in the "living zone" of the space — below the 10 foot [3 m] level (Figure 10-17).[10.3] When delivered in this manner, where the majority of the people and processes are located, maximum ventilation results with minimum air handling. During the warm months of the year, large airflow in the working space at relatively high velocities is welcomed by the workers. During the winter months, however, care must be taken to insure that air velocities over the person are kept within acceptable values (except when extremely high heat loads are involved) (see Chapter 4). To accomplish this, the air can be distributed uniformly in the space or where required for worker comfort. Heavy-duty, adjustable, directional grilles and louvers have proven to be very successful in allowing individual workers to direct the air as needed.[10.1] Light gauge, stamped grilles intended for commercial use are not satisfactory. Suitable control must be provided to accommodate seasonal and even daily requirements with a minimum of attention by supervision or maintenance forces. *Warning: When the work process generates a hazardous contaminant, worker adjustable supply grilles may be incompatible with the contaminant control scheme.*

10.7.2 Air Supply vs. Plant Heating Costs. Even if the supply air was drawn into the building only by the action of the

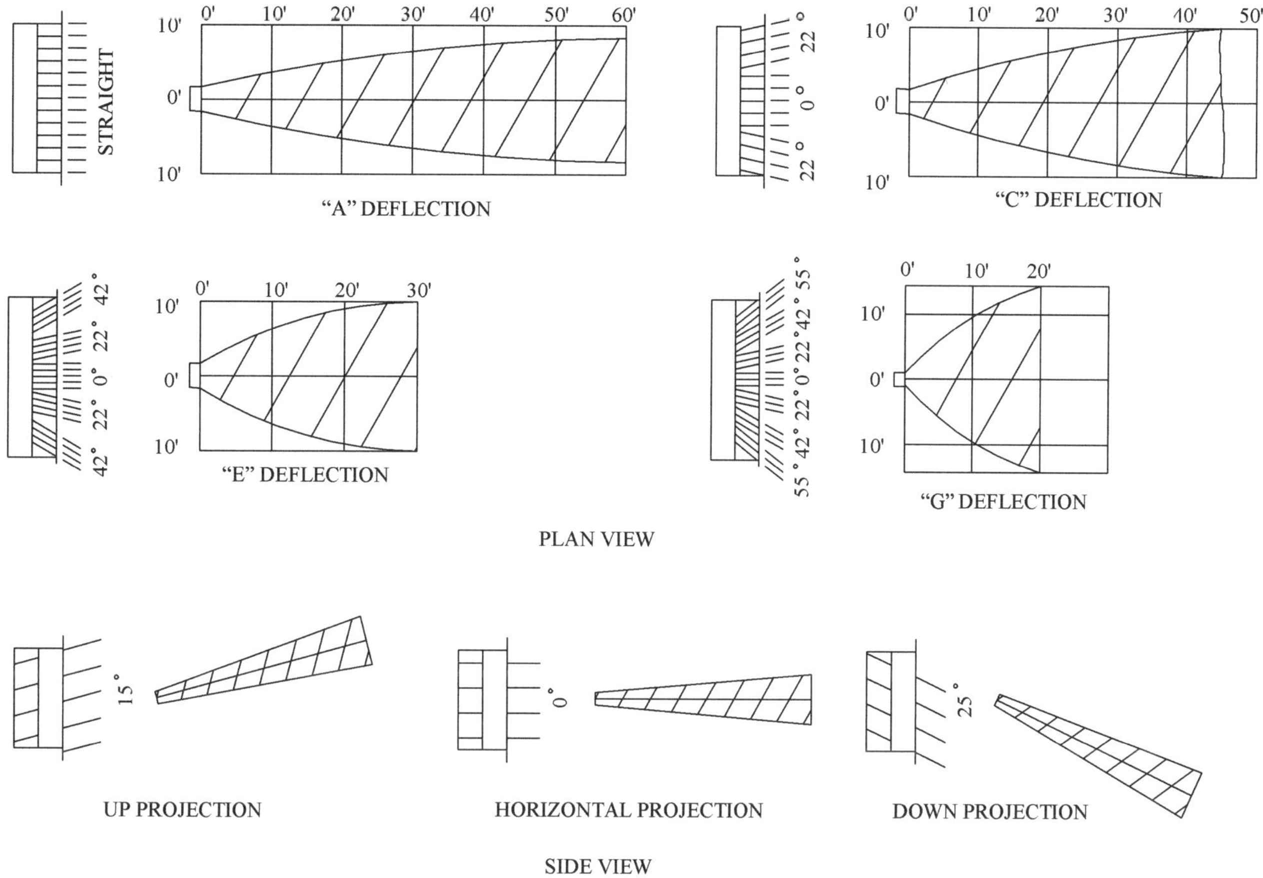

FIGURE 10–17. Register airflow patterns[(10.12)]

exhaust fans, there would be an added burden on the plant heating system during the winter months and fuel costs would rise. Normally such a system would use unit heaters for building temperature control and numerous openings in the building for outside air supply. Compared to a situation where the same flow rate of outdoor air was introduced through properly designed supply replacement air heaters, experience has shown that the overall fuel cost would be less than or equal to that consumed by the unit heater system. A partial explanation of this savings is more efficient heat transfer. The most important factor, however, is that a well-designed air supply system is not dependent on the plant space heating system; rather, the two systems operate in an independent fashion. The air supply system and the plant heating system can be understood best by considering the building as a whole. In order for a temperature equilibrium to be established, the heat outflow from the building must balance the heat inflow. To obtain additional energy saving during downtime, one should design the supply system to provide sufficient heating to counter air entering the building through infiltration and to prevent freezing.

10.7.3 Cost of Heating Supply Air. The cost of heating supply air is a significant portion of the annual operating cost of a ventilation system. Processes requiring cooling also need to be evaluated for their energy and operating costs. Occupant comfort is more important than saving a few dollars in energy costs. Recent indoor air quality studies quantify diminished productivity when workers are uncomfortable. A compilation of nine productivity and temperature reports indicates that people are most comfortable between 72 F and 77 F [22 C and 25 C].[(10.11a)] Acclimated industrial personnel may tolerate slightly higher temperatures. In addition to the equipment first cost, local building codes, and environmental regulations, designer experience in utility incentives and operating costs are involved in purchasing decisions.

10.7.4 Cooling Energy Considerations. The energy for heating air considers only raising the air temperature that is sensible heat energy. If the quantity of water vapor in the air is changed, a latent energy transfer takes place. In most cooling system operations, a latent energy transfer takes place to lower the air humidity level. This latent energy transfer occurs when the air is cooled to a temperature below the dew point of the air. When this occurs, water is condensed from the air stream onto the cooling coil.

In plant operations, the humid air can come from the outside or be created by a process that releases water vapor into the general air.

The cooling system is sized to overcome several energy sources: the heat released to the space by the process, the heat absorbed by the building structure (since the outdoor air is warmer than the inside space temperature), the heat absorbed by the building structure by solar radiation, and the heat caused by the outside supply air being warmer than the desired space temperature. The internal heat released by the process varies with the process rate of production.

10.7.5 Filter Replacement. As the filter works to remove airborne material from the air stream, it becomes saturated with that material and the pressure drop for air to pass through the filter increases. As the pressure drop rises, the airflow decreases unless the fan speed is increased or other adjustments are made. Unless a replacement filter is installed, the original filter will eventually become blinded allowing little or no air to pass. Refer to Chapter 8 for more information regarding supply air filter efficiency and application by type.

10.7.6 System Maintenance. Maintenance of the supply air system is mainly associated with the supply air unit and its operating controls. In addition to filter replacement, there are fan bearings to grease and belts to replace. Coils and drain pans must be cleaned and dampers lubricated. If the system includes humidifying equipment, there are spray nozzles to clean/replace and there may be a pump to service as well as controls to check and adjust. A major maintenance issue is the recalibration of the systems controls. If the controls are not routinely calibrated, operating set points may drift, resulting in energy waste and poor system performance.

10.7.7 Untempered Air Supply. In many industries utilizing hot processes, cold outdoor air is supplied untempered or moderately tempered to dissipate sensible heat loads on the workers and to provide temperature relief for workers. The air required for large compressors, as well as for cooling tunnels in foundries, can also come directly from outside the plant and thus eliminate a load of the tempered replaced air.

10.7.8 Energy Recovery. Energy recovery from exhaust air is accomplished through the use of heat exchange equipment to extract heat from the air stream before it is exhausted to the outside. The application of return or recirculated cleaned air from industrial exhaust systems is another method of recovering process heat for use in the building. The application of heat exchangers to industrial exhaust systems is discussed in Chapter 11.

10.8 INDUSTRIAL EXHAUST RECIRCULATION

Where large amounts of air are exhausted from a room or building in order to remove particulate, gases, fumes, or vapors, an equivalent amount of fresh tempered replacement air must be supplied to the room. If the amount of replacement air is large, the cost of energy to condition the air can be very high. Recirculation of the exhaust air after thorough cleaning is one method of reducing the amount of energy consumed. Acceptance of such recirculating systems will depend on the degree of health hazard associated with the particular contaminant being exhausted as well as other safety, technical, and economic factors. A logic diagram listing the factors that must be evaluated is provided in Figure 10-18.[10.13]

Essentially this diagram states that recirculation may be permitted if the following conditions are met:

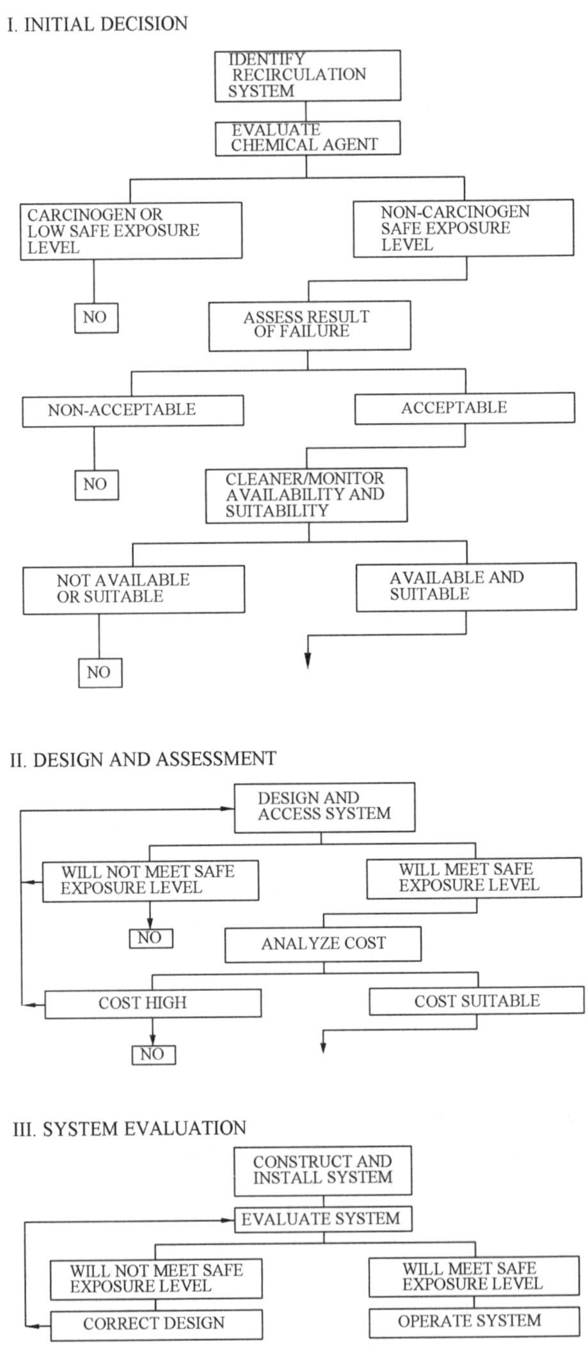

FIGURE 10–18. Recirculation decision logic[10.13]

1) The chemical, physical, and toxicological characteristics of the chemical agents in the air stream to be recirculated must be identified and evaluated. Exhaust air containing chemical agents whose toxicity is unknown or for which there is no established safe exposure level should not be recirculated. Exhaust air from processes using or generating explosive agents should not be recirculated if another process or system may add additional amounts of that agent to the air stream.
2) All governmental regulations regarding recirculation must be reviewed to determine whether recirculation is restricted or prohibited for the system under review.
3) The effect of a recirculation system malfunction must be considered. Recirculation should not be attempted if a malfunction could result in exposure levels that would cause worker health problems. Substances that can cause permanent damage or significant physiological harm from a short overexposure should not be recirculated.
4) The availability of a suitable air cleaner must be determined. An air cleaning device capable of providing an effluent air stream contaminant concentration sufficiently low to achieve acceptable workplace concentrations must be available. For example, a scrubber or mist collector may provide suitable cleaning efficiency but adds significant humidity to the air.
5) The effects of minor contaminants should be reviewed. For example, welding fumes can be effectively removed from an air stream with a fabric filter; however, if the welding process produces oxides of nitrogen, recirculation could cause a concentration of these gases to reach an unacceptable level.
6) Recirculation systems must incorporate a monitoring system that provides an accurate warning or signal capable of initiating corrective action or process shutdown before harmful concentrations of the recirculated agents build up in the workplace. Monitoring may be accomplished by a number of methods and must be determined by the type and hazard of the substance. Examples include area monitoring for nuisance type substances and secondary high efficiency filter pressure drop indicators or on-line monitors for more hazardous materials. Explosive agent concentration must be monitored to assure the concentration stays well below the Lower Explosive Limit (vapors) or the minimum explosive concentration (particulates).

10.8.1 Evaluation of Employee Exposure Levels. Under equilibrium conditions, the following equations may be used to determine the concentration of a contaminant permitted to be recirculated in the return air stream:

$$C_R = [(1 - \eta)(C_E - K_R C_M)] / [1 - (K_R)(1 - \eta)] \quad [10.2]$$

where: C_R = air cleaner discharge concentration after recirculation

η = fractional air cleaner efficiency

C_E = local exhaust duct concentration before recirculation

K_R = coefficient which represents a fraction of the recirculated exhaust stream that is composed of the recirculated air returning from the air cleaner (range 0–1.0)

C_M = replacement air concentration

NOTE: Units for C_R, C_E and C_M are in parts per million (ppm) or milligrams per cubic meter (mg/m^3) and all must be in the same unit system.

$$C_B = [(Q_B/Q_A)(C_G - C_M)(1 - f)] + [(C_O - C_M) f] + [K_B C_R + (1 - K_B)(C_M)] \quad [10.3]$$

where: C_B = 8-hr TWA worker breathing zone concentration after recirculation

Q_B = total ventilation airflow before recirculation (acfm) [am^3/s]

Q_A = total ventilation airflow after recirculation (acfm) [am^3/s]

C_G = general room concentration before recirculation

f = coefficient which represents the fraction of time the worker spends at the workstation

C_O = 8-hr TWA breathing zone concentration at workstation before recirculation

K_B = fraction of worker's breathing zone air that is composed of recirculated air returning from the air cleaner (range 0 to 1.0)

NOTE: Units for C_B, C_G and C_O are in parts per million (ppm) or milligrams per cubic meter [mg/m^3] and all the values must be in the same unit system.

The coefficients K_R, K_B, and f are dependent on the workstation and the worker's position in relation to the source of the recirculated air returning from the air cleaner and in relation to the exhaust hood. The value of K_R can range from 0 to 1.0 where 0 indicates no recirculated air entering the hood and 1.0 indicates 100% of recirculated air entering the hood. Similarly, the value of K_B can range from 0 to 1.0 where 0 indicates there is no recirculated air in the breathing zone and 1.0 indicates that the breathing zone air is 100% recirculated air from the air cleaner. The coefficient "f" varies from 0 where the worker does not spend any time at the workstation where the air is being recirculated to 1.0 where the worker spends 100% time at the workstation. In many cases, it will be difficult to attempt quantification of the values required for solution of these equations for an operation not yet in existence. Estimates based on various published and other available data for the same or similar operations may be useful. Furthermore, it may be difficult to determine accurate K_R and K_B values for existing systems unless contaminant is easily detected with direct-reading instruments or surrogate tracer gas testing is used. Tracer gas testing will require highly trained technicians. The final system must be tested to demonstrate that it meets design specifications.

EXAMPLE PROBLEM 1

An example of use of Equations 10.2 and 10.3 and the effect of the various parameters is as follows. Consider a system consisting of 5,000 acfm [2.5 m^3/s] of general exhaust and 5,000 acfm [2.5 m^3/s] of local exhaust. If the local exhaust is recirculated, the ventilation system of Q_A = 10,000 acfm [5.0 m^3/s] changes to 5,000 acfm [2.5 m^3/s] recirculated plus Q_B = 5,000 acfm [2.5 m^3/s] fresh airflow. Assume poor placement of the supply air register(s) delivering the cleaned recirculated air back from the air cleaner (K_R and K_B = 1) and that the worker spends all his time at the workstation (f = 1); the air cleaner efficiency $\eta = 0.90$; exhaust duct concentration (C_E) = 500 ppm; general room concentration (C_G) = 20 ppm; replacement air concentration (C_M) = 2 ppm; workstation (breathing zone) concentration before recirculation (C_O) = 14 ppm; and a contaminant TLV® of 25 ppm.

Equation 10.2 gives recirculation air return concentration:

Initially:

Q_B= 10,000 [5.0], Q_A = 5000 [2.5], C_E = 500, C_G = 20, C_M = 2, f = 1, C_o = 14, $K_B = K_R = 1$

$$C_R = [(1 - .9)(500 - (1)(2)] / [1 -(1)(1 - .9)]$$

$$C_R = [(.1)(498)] / [1 - .1] = 55.33 \text{ ppm} = 55 \text{ ppm}$$

Equation 10.3 gives the worker breathing zone concentration:

$$C_B = (10{,}000/5{,}000)(20 - 2)(1 - 1) + (14 - 2)(1) + (1)(55) + (1 - 1)(2)$$

$$C_B = (0) + 12 + 55 + 0 = 67 \text{ ppm}$$

The concentration of 67 ppm is over the TLV® of 25 ppm and, therefore, is not an acceptable value for an engineered ventilation solution to this situation. Note that value is same in IP and SI units.

In order to achieve lower concentrations (CB), two modifications in the design are made. First, the efficiency of the air cleaning device is improved to $\eta = 97\%$. Second, the system configuration is redesigned so that only 30% of the recirculation return air reaches the workstation. Thus, K_R and K_B are reduced to 0.3. Substituting these new data in Equations 10.2 and 10.3, the concentration in the air stream leaving the air cleaner drops to 15.0 ppm and the breathing zone concentration calculates as 18 ppm. This is less than the TLV® of 25 ppm and, therefore, the design would normally be acceptable. While this performance is technically acceptable, the owner may choose to further redesign the system to an even lower percentage of the TLV®. Ideally, an engineered design should reduce the exposure to below 50% of the TLV® to avoid administrative and medically related regulatory requirements.

After changing parameters: η $K_B = K_R = 0.3$

$$C_R = [(1 - .97)(500 - .3(2))] / [1 - [(.3)(1 - .97)]]$$

$$C_R = [(.03)(499.4)] / [1 - [(.3)(.03)]]$$

$$C_R = [14.98] / [.991] = 15.12 \text{ ppm} = 15 \text{ ppm}$$

$$C_B = (10{,}000/5000)(20 - 2)(1 - 1) + (14 - 2)1 + (.3)(15) + (1 - .3)(2)$$

$$C^B = 0 + 12 + (4.5 + 1.4) = 17.9 \text{ ppm} = 18 \text{ ppm}$$

10.8.2 Design Considerations for Air Recirculation. More requirements associated with the recirculation of exhaust can be found in published standards. The American National Standards Institute (ANSI) issued ANSI 9.7, *Recirculation of Air from Industrial Process Exhaust Systems.*(10.14)

Care must be taken in the design of recirculated air systems. Considerations for good system performance are:

1) Recirculating systems should, whenever practicable, be designed to bypass to the outdoors, rather than recirculate, when weather conditions permit. If a system is intended to conserve heat in winter months and if adequate window and door openings permit sufficient replacement air when open, the system can discharge to the outdoors in warm weather. In other situations where the workspace is conditioned or where mechanically delivered supply air is required at all times, continuous bypass operation may not be attractive.
2) Wet collectors also act as humidifiers. Recirculation of humid air from such equipment can cause uncomfortably high humidity and require auxiliary ventilation or some method to prevent excess humidity. Avoid the return of overly humidified air into air conditioned spaces.
3) The exit concentration of typical collectors can vary with time. Design data and testing programs should consider all operational time periods.
4) The layout, design and delivery of the recirculated air returning from the air cleaner should provide adequate mixing with other supply air and avoid uncomfortable drafts on workers or disruptive air currents which could

interfere with the capture velocity of local exhaust hoods.

5) Odors or nuisance values of contaminants should be considered as well as official exposure limit values. In some areas, adequately cleaned recirculated air, provided by a system with safeguards, may be of better quality than the ambient outdoor air available for replacement air supply.
6) Routine testing, maintenance procedures, and records should be developed for recirculating systems.
7) Periodic testing of the workroom air should be provided. Air cleaners should not be installed without proven (certifiable performance) expectations, data-driven knowledge regarding change-out and breakthrough and capability to detect when breakthrough has occurred.
8) An appropriate sign should be displayed in a prominent place, which should read as follows:

CAUTION

AIR CONTAINING HAZARDOUS SUBSTANCES IS BEING CLEANED TO A SAFE LEVEL IN THIS EQUIPMENT AND RETURNED TO THE BUILDING. SIGNALS OR ALARMS INDICATE MALFUNCTIONS AND MUST RECEIVE IMMEDIATE ATTENTION: STOP RECIRCULATION, DISCHARGE THE AIR OUTSIDE, OR STOP THE PROCESS IMMEDIATELY.

10.8.3 Recirculation Air Monitor Selection. While all system components are important, special consideration should be given to the monitor on any system recirculating a potentially hazardous material. The prime requisites are that the monitor be capable of sensing a system malfunction or failure, and of providing a signal that will initiate an appropriate sequence of actions to assure that overexposure does not occur. The sophistication of the monitoring system can vary widely. The type of monitor selected will depend on various parameters (i.e., location, nature of contaminant — including shape and size — and degree of automation). The safe operation of a recirculating system depends on the selection of the best monitor for a given system. Reference 10.14 describes four basic components of a complete monitoring system, which include signal transfer, detector/transducer, signal conditioner, and information processor. Figure 10-19 shows a schematic diagram of the system incorporating these four components. It is quite likely that commercially available monitors may not contain all of the above four components and may have to be custom engineered to the need.

In addition to the four monitoring system components, the contaminant samples must be collected from the air stream either as an extracted sample or in total. If a sample is taken, it must be representative of the average conditions of the air stream at that point in time. At normal duct velocities, turbulence assures good mixing so gas and vapor samples should be representative. For aerosols, however, the particle size discrimination produced by the probe may bias the estimated concentration unless isokinetic conditions are achieved.

The choice of detection methods depends on the measurable chemical and physical properties of the contaminants in the air stream. Quantifying the collected contaminants is generally much easier for particulate than for gases, vapors, or liquid aerosols.

Particulates. Where the hazardous contaminant constitutes a large fraction of the total dust weights, filter samples may allow adequate estimation of concentration in the recirculated air. If the primary collector (e.g., bag filters, cartridge filters) allows very low penetration rates, it may be more economical to use high efficiency filters as secondary filters. If the primary filter fails, the secondary filter not only will experience an easily measured increase in pressure drop, but will filter the penetrating dust as well — earning this design the sobriquet, "safety monitor" system (Figure 10-20).

Non-particulates. Continuously detecting and quantifying vapor and gas samples reliably and accurately is a complex subject beyond the scope of this Manual.

Air Sampling Instruments for Evaluation of Atmospheric Contaminants, published by ACGIH®,[10.15] describes and evaluates different air monitoring devices. The monitor in a recirculating system must be capable of reliably monitoring continuously and unattended for an extended period of time. It must also be able to quickly and accurately sense a change in system performance and provide an appropriate warning if a preselected safety level is reached. In order to function properly, monitors must be extremely reliable, accurate, and properly maintained. Monitors should be designed so that potential malfunctions are limited in number and can be detected easily by following recommended procedures. Required maintenance/recalibration should be simple, infrequent, and of short duration. Employees in the area should be able to determine if the monitor is functioning and have instructions to follow when there is a problem. Otherwise, the monitor should have a fail safe feature.

10.9 SYSTEM CONTROL

10.9.1 Building Air Balance. Proper building air balance is needed to control space cleanliness, contaminants, and temperatures. Air that is supplied to a building is also removed by a combination of methods. When there is adequate replacement air, any additional clean plant air can be returned to the supply air unit. Obviously, the contaminant-laden air should be captured by means of a local exhaust hood or ventilation system and removed from the building. Sensors are available to monitor the building pressure to maintain a building pressure. Another approach is to use a central control system that tracks which exhaust systems are operating and their associated airflow. The control system also operates the supply system to assure the necessary amount

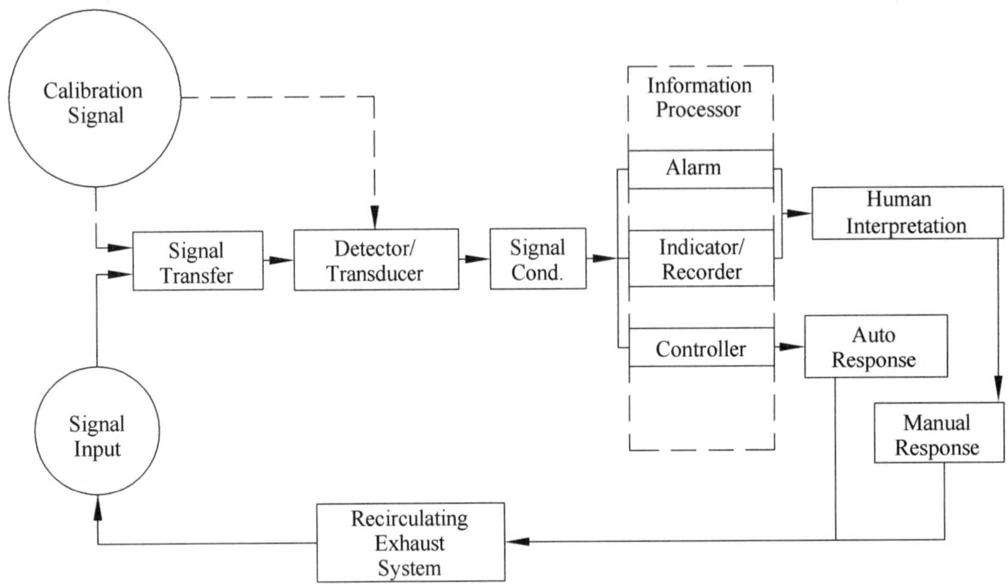

FIGURE 10–19. Schematic diagram of recirculation monitoring system

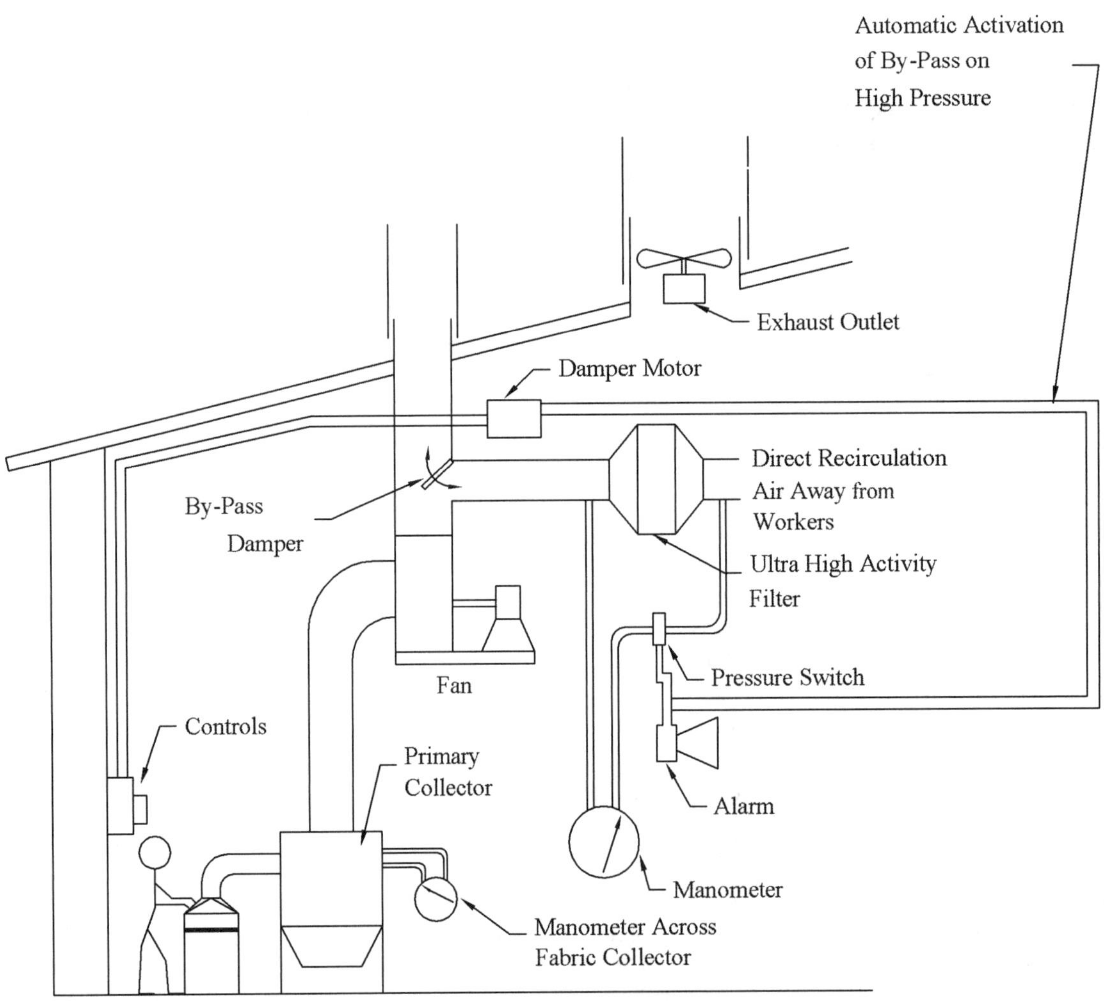

FIGURE 10–20. Schematic of recirculation from air cleaning devices (particulates)

of outside air is brought into the plant to obtain the desired building pressure. The remainder of the airflow is obtained from returning plant air to the air handler units.

In industrial plants where there is no concern for maintaining close temperature control or high air cleanliness levels, the maintenance of the building air balance is only required during the heating season. During other times of the year and when environmental permits allow it, the building temperature is not controlled and doors are left open to allow as much outside air to enter as possible.

10.9.2 Temperature. Return air temperature and humidity are other characteristics that will cause the discharge of clean return air outdoors. These air qualities can be monitored and used to vary the return airflow to the supply air unit. The HVAC industry uses automated building control and direct digital control (DDC) in many facilities. The technology can be applied to industrial ventilation with careful planning. DDC uses computers and microprocessors tied to sensors and actuators to form a feedback and control system. DDC can be useful in industrial ventilation systems to control temperature, humidity, and relative room pressures. DDC systems can also track the system performance at hoods, fans, heating and cooling, and air pollution control equipment. DDC is especially useful in preventive maintenance. However, DDC systems for industrial ventilation systems are complicated. Many are "one-of-a-kind" systems designed by a controls manufacturer and require trained personnel to install, calibrate, and operate.

10.9.3 Indoor Air Quality. Many industrial processes release minor amounts of "nuisance" contaminants which, at low concentrations, have no known health effects but are unpleasant or disagreeable to the workers or harmful to the product. The desire to provide a clean working environment for both the people and the product often dictates controlled airflow between rooms or entire departments. The air streams recirculated into the facility must be evaluated to determine if the air pollution control devices (e.g., filters, cyclones) are providing sufficient cleaning to prevent employee exposure to "nuisance" contaminants. In addition, systems with known contaminants require the controls listed in Section 10.8 under "Industrial Exhaust Recirculation." The facility must employ trained mechanics and support a preventive maintenance program to sufficiently protect the workers.

10.10 SYSTEM NOISE

Supply air systems have several sources for objectionable noise. Care must be taken in selecting the fan and avoiding high air velocities in the duct system. The fan may develop unwanted noise if it operates at high speeds or if it is not properly isolated from the rest of the system. Using a larger fan at a lower speed will usually avoid high frequency noise. Duct silencers and other acoustical treatment can be applied to noisy systems to achieve sound reduction. Isolation by springs or other vibration absorbing devices is needed to limit the transmission of fan noise into the building structure. Additional information regarding sound and vibration control can be found in the ASHRAE *Applications*.[10.1]

There are several places in a duct system where noise can occur. Diffusers/grilles that are providing long throws can create excessive noise due to the high air velocities required. Sound data are provided in the manufacturer's catalog to aid in the selection of the most appropriate diffuser. Specify devices that can operate under the higher static pressures found in industrial supply air systems. Commercial grade devices may become stressed or their components may loosen causing unwanted vibration and noise.[10.16]

REFERENCES

10.1 American Society of Heating, Refrigerating and Air-Conditioning Engineers: HVAC Application. ASHRAE, Atlanta, GA (2011).

10.2 Hama, G.: How Safe Are Direct-Fired Makeup Units? Air Engineering, p. 22 (September 1962).

10.3 American Society of Heating, Refrigerating and AirConditioning Engineers: HVAC Systems and Equipment. ASHRAE, Atlanta, GA (2008).

10.4 American Society of Heating, Refrigerating, and AirConditioning Engineers: Method of Testing General Ventilation Air Cleaning Devices for Removal Efficiency by Particle Size. ASHRAE Publication No. 52.2-2007. ASHRAE, Atlanta, GA (2007).

10.5 Yuan, X.; Chen, Q.; Glicksman, L.R.: A Critical Review of Displacement Ventilation. ASHRAE Transactions (January 1998).

10.6 Skistad, H.: Displacement Ventilation. Taunton, Somerset, England: Research Studies. Press, Ltd. (1994).

10.7 Kristensson, J.A.; Lindqvist, O.A.: Displacement Ventilation Systems in Industrial Buildings. ASHRAE Transactions, 99/1 (1993).

10.8 Sheet Metal and Air Conditioning Contractors National Association, Inc.: Rectangular Industrial Duct Construction Standards. SMACNA, Vienna, VA (2004).

10.9 Sheet Metal and Air Conditioning Contractors National Association, Inc.: Round Industrial Duct Construction Standards. SMACNA, Vienna, VA (1999).

10.10 Sheet Metal and Air Conditioning Contractors National Association, Inc.: Thermoplastic Duct (PVC) Construction Manual. SMACNA, Vienna, VA (1995).

10.11 Thomson, M.; Goodfellow, H.: Computational Fluid Dynamics as a Design Tool for Industrial Ventilation. Ventilation (1997).

10.11a Tom, S.: Managing Energy and Comfort. ASHRAE Journal (July 2008).

10.12 Hart and Cooley Manufacturing Co.: Bulletin E-6. Holland, MI.

10.13 Hughes, R.T.; Amendola, A.A.: Recirculating Exhaust Air: Guides, Design Parameters and Mathematical Modeling. Plant Engineering (March 18, 1982).

10.14 ANSI/AIHA: Recirculation of Air from Industrial Process Exhaust Systems. ANSI/AIHA Z9.7 – 2007. American National Standards Institute (2007).

10.15 American Conference of Governmental Industrial Hygienists: Air Sampling Instruments for Evaluation of Atmospheric Contaminants, 9th Edition. ACGIH®, Cincinnati, OH (2001).

10.16 Shaffer, M.: A Practical Guide to Noise and Vibration Control for HVAC Systems. ASHRAE Publications (2005).

Chapter 11

ENERGY CONSIDERATIONS

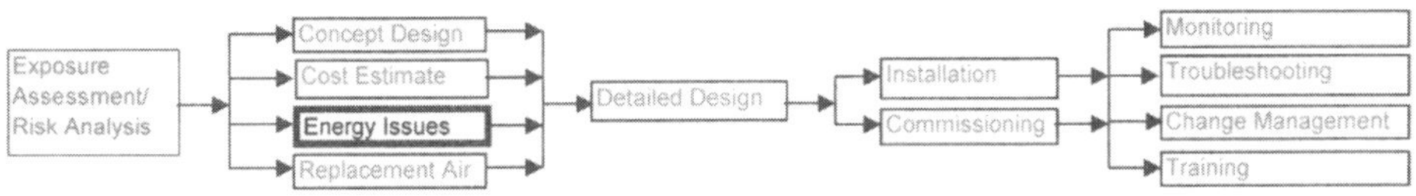

NOTE: Equations with notation followed by (IP) are designated for inch-pound system only; Equations followed by (SI) are designated for metric use only. If Equation bears neither, then it applies to both systems.

NOTE: Chapters 11 and 12 use mostly IP units because of the variety of cost values that could be used worldwide. Where appropriate, metric values are inserted but Example Problem solutions are in IP units only.

11.1 INTRODUCTION

Industrial exhaust systems are energy intensive in their operation. These systems require electricity to power the fan motor that removes the air from the building and this power is dependent on the amount of airflow and System Static Pressure (SSP). The exhaust air often requires cleaning before it can be discharged to the outside. This adds additional resistance to airflow and more power requirements. Supply air must be furnished to replace the exhaust air and also needs power to be delivered. These systems typically operate most of the time and the energy use can become significant.

This chapter will discuss the energy use of the ventilation system components and provide methods for calculating the actual amount. Approaches to reduce energy use of the exhaust system will also be presented. Safeguards are also listed for systems where air can be recirculated back to the plant.

11.2 EXHAUST SYSTEM ENERGY USE

Significant amounts of energy are used by both supply and exhaust systems. Exhaust systems use energy to move and clean the air. Wherever any exhaust system moves air from an enclosed space, replacement air will be drawn in. This replacement air may require energy for heating or cooling in order to achieve the proper plant conditions. This section contains the theoretical formulas used to determine energy requirements for industrial ventilation systems.

11.2.1 Energy Used to Move Air. The exhaust air system has a determined amount of air flowing against a resistance caused by hoods, a duct configuration and air cleaning devices (see Chapters 5 and 6). As the air flows through a duct system there is a resistance caused by friction in the elbows, straight runs of duct, air cleaning devices, etc. As the resistance to flow increases, the required static pressure the fan must generate also increases. As a result, a higher horsepower motor is required to operate the fan. The formulae for determining the horsepower for a fan is:

$$\text{Power} = \left(\frac{(Q)(FTP)(df)}{(CF)\eta}\right)$$

$$= \left(\frac{(Q)(FSP + VP_i)(df)}{(CF)\eta}\right) \quad [11.1]$$

where:

Power = Brake horsepower (BHP) [watts]
Q = Volumetric flow rate, acfm [am^3/s]
FTP = Fan Total Pressure ("wg) [Pa]
FSP = Fan Static Pressure ("wg) [Pa]
VP_i = Velocity Pressure @ fan inlet ("wg) [Pa]
df = Density Factor
CF = Conversion coefficient, 6362 [1.0]
η = Mechanical efficiency

The fan power in horsepower must be increased by the belt drive loss to reach the required motor horsepower to operate the fan. The belt drive loss can range from 3 to 8% for motors over 10 horsepower. It is very common for exhaust air systems to require a relatively high static pressure to operate and thus have a motor horsepower in the range of two to three horsepower for each 1,000-acfm of airflow.

EXAMPLE PROBLEM 1 (Fan Power Calculation) (IP Units)

The design of an exhaust system with multiple hoods has a total airflow of 18,500 acfm with a calculated Fan Static Pressure (FSP) of 9.5 "wg and Density Factor equal to 0.87. The inlet Velocity Pressure at the fan is 1.12 "wg The mechanical efficiency of the chosen fan is 70%. Determine the expected motor horsepower required for this system.

$$\text{Power} = \left(\frac{(Q)(FSP + VP_i)(df)}{(CF)\eta}\right)$$

$$= \left(\frac{(18{,}500)(9.5 + 1.12)(0.87)}{(6362)(0.70)}\right)$$

$$= 38.4 \text{ Horsepower}$$

A belt drive motor using standard type belts is chosen to power the fan. The estimated belt loss is 5% (see Chapter 7).

Motor Output (hp) = Fan Power + (belt loss %) (Fan Power)
= 38.4 + 0.05 (38.4) = 40.32

Select a 50 horsepower motor

11.2.2 Energy Used for Cleaning Exhaust Air. Depending on the type of air cleaning device being utilized, different types of energy will be required. The type of contaminant and the degree of cleanliness are criteria used to select the air cleaning type of equipment. The first four methods reduce particulate loading in the exhaust and the remaining control gaseous emissions. Details of operation of these devices are included in Chapter 8.

- **Air filters** are used to remove all sizes of airborne particulate from the air stream. The filter collects the removed particulate and must be cleaned or replaced at some level of dust loading. The greater the dust loading the more resistant to flow and the pressure drop increases. This is reflected in the FTP or FSP values in Equation 11.1. For an increase of 1.0 "wg [250 Pa] static pressure, the horsepower increase to achieve a constant airflow is approximately one horsepower per 5,000 acfm [300 watts/m^3/s].
- **Dry centrifugal collectors** are used to remove large and heavy dust particles from the air stream. This collector works on the premise that the air is slowed down and it

must go through sharp turns that cause the large particles to drop out into a collection bin. There is no component to change or clean other than emptying the collection bin. These collectors are not very efficient on small particles and normally have a pressure drop of 2 "wg to 6 "wg [500 to 1500 Pa]. The static pressure drop typically remains constant for a specific airflow rate and Density Factor.

- **Electrostatic Precipitators** are used to collect dust particulate of all sizes. They work on the principle that statically charged particles can be attracted to plates having the opposite charge. The dust particles initially pass through a high potential electrical field of discharge electrodes made of wire. As the dust-laden air moves through the collector it enters a zone where collection electrodes have the opposite electrical charge. Here the dust particles are collected on these plates and they drop off when their charge is neutralized. The resistance to airflow remains essentially the same throughout the life of this collector and for most systems is 0.5 "wg to 6 "wg [125 to 1500 Pa]. This collector also requires electrical energy to create the electrically charged fields.
- **Wet collectors** (Scrubbers) are used to collect all sizes of dust particulate. They use impingement of the dust particle in the wet solution as the means to remove the dust from the air stream. Generally, water is used as the solution unless there is a gaseous contaminant involved. The low efficiency collectors have a small pressure drop (< 6.0 "wg [1500 Pa]) and the air is only required to pass through a water bath. Higher efficiency units require a large expenditure of energy to remove the dust from the air stream (as high as 100 "wg [25 kPa]). Wet collectors also drop the temperature of the air stream and the leaving air has a high moisture level (affecting the Density Factor in Equation 11.1). The energy use of this equipment remains fairly constant, but the collected dust in the water poses a water pollution issue. With this equipment there is also a small energy use to power pumps for moving the water into and out of the collector.
- **Absorption** is the process of holding the emission molecule within a collecting medium. Soluble or chemically reactive gases can be removed from the air stream by either a wet or dry absorber. A wet collector called a scrubber can be used in some cases. Another type of system sprays a wet solvent, which dries in the air stream leaving the dry solvent in contact with the gas. The energy use is similar to that of the lower energy wet collectors and the efficiency is dependent on contact time the gas has with the solvent or water.
- **Adsorption** is the process of adhesion of gas molecules to the surface or interfaces of a solid such as activated carbon or silica gel. The efficiency of the process is dependent on the contact time so a deeper bed of activated carbon will provide a longer contact time period. With deeper beds there is more resistance to flow, which increases the energy use.

  The adsorption media can sometimes be recharged by heat as in the case of charcoal. By raising the temperature of the charcoal to approximately 250 F [121 C], most low boiling point organic compounds will be removed from the charcoal. A temperature of approximately 400 F [204 C] is needed for the release of higher boiling point organic compounds and compounds with a high molecular weight reactivation require temperatures as high as 1350 F [732 C].
- **Incineration/Oxidation** is normally an energy intensive air cleaning process. Burning destroys gaseous organic pollutants. The combustion process is used to turn these organic compounds into carbon dioxide and water. Typical temperatures required to achieve combustion are in the range of 1000 to 1500 F [537 to 816 C]. To obtain these temperatures as efficiently as possible different techniques of heat recovery may be used. Air to air heat exchangers or preheated ceramic beds are used to minimize the energy required. A catalyst may also be used in the oxidizer to reduce the temperature needed to achieve the chemical reaction. Normally, the pressure drop across the units is not high so fan energy impact may be less than other methods, but large amounts of energy may be required for raising the air stream temperature. Fortunately, with some gaseous contaminants, the fuel value of the organic pollutant is adequate to support the combustion process. Thus, little extra fuel is required for the oxidation process. One approach is to use an adsorption unit to concentrate the gaseous emission and then fire the oxidizer with this fuel rich air stream.

  For thermal oxidizers most organic gases oxidize with 95% conversion efficiency or better at a temperature in the range of 1400 to 1500 degrees F [760 to 816 C] with a residence time of ½ to one second. Catalytic oxidizers operate at a lower temperature since they utilize a catalyst to achieve oxidization. The required temperature depends on the catalyst and configuration of the equipment.
- **Biofiltration** is a relatively new process where the exhaust air is pretreated to achieve proper temperature and moisture content for entry into a filter that has microorganisms that destroy specific contaminants. The energy use of this type of air cleaner is similar to that of an adsorption unit.

EXAMPLE PROBLEM 2 (Oxidizer Fuel Cost Calculation) (IP Units)

A thermal oxidizer is used to control Volatile Organic Compounds (VOCs) from a solvent cleaning line. The oxidizer requires 1,500,000 BTU/hour (BTUH) to maintain the proper operating temperature. The solvent fumes contribute an average of 20% of this total. The oxidizer will operate 2000 hours per year. What are the annual energy requirements if natural gas is used as the supplemental fuel? What is the cost if natural gas is $5.20 per mcf (1000 ft^3)?

BTU requirement for the supplemental fuel =

$$\left(1{,}500{,}000\frac{\text{BTU}}{\text{hour}}\right)(1-0.2)\left(2000\frac{\text{hours}}{\text{year}}\right) = 2.4\times10^{9}\frac{\text{BTU}}{\text{year}}$$

Quantity of gas needed =

$$\frac{2.4\times10^{9}\dfrac{\text{BTU}}{\text{year}}}{1000\dfrac{\text{BTU}}{\text{ft}^3}} = 2.4\times10^{6}\text{ cubic feet per year}$$

Cost of the gas =

$$\left(2.4\times10^{6}\frac{\text{ft}^3}{\text{year}}\right)\left(\frac{\$5.20}{\text{mcf}}\right)\left(\frac{1\text{ mcf}}{1000\text{ ft}^3}\right) = \$12{,}480\text{ per year}$$

11.2.3 Energy Used for Heating Supply and Replacement Air. Supply air includes all air delivered into the building to provide ventilation, heat and cooling. Some of this air may be replacement air which is supplemental supply air needed to make up for the air exhausted. Details of the application and calculation of supply air systems are included in Chapter 10.

Supply air energy use consists of the electrical power to operate the fan and the heating/cooling energy to maintain the desired space temperature. The fan electrical use is determined similar to exhaust units (see Equation 11.1). The heating/cooling energy use is dependent on the amount of energy required to achieve a suitable temperature for discharge into the space. In a space such as a paint booth, process temperature and humidity requirements may define the conditions. Other spaces may be designed strictly for worker comfort.

11.2.3.1 Energy Calculation Based on Degree Days and Temperature Difference. The American Society of Heating, Ventilating and Air Conditioning Engineers (ASHRAE), the U.S. Department of Energy (DOE), and others have developed equations and computer programs to determine the life-cycle costs of various types of air-handling equipment. The equations and programs are not intended to determine the annual utility bill. Instead, they help compare the costs of various options in providing heating and cooling to a variety of industrial and commercial operations. Air that is delivered into the building via air hand-ling units requires fan horsepower as well as heating energy to be consumed. The rate of electrical use for fan energy is approximately one horsepower per 1000 cfm. Since this air is from the outside, heat is added in the winter and in temperature controlled spaces, and cooling is needed in the summer. Equations 11.2 and 11.3 may be used to estimate replacement air heating costs on an hourly and yearly basis. They are based on average usage schedules and typical weather conditions rather than worst case conditions and maximum usage.

$$\text{Energy Cost per hour} = C_h = \left(\frac{(1.08)(Q)(T_s - T_o)(c)}{q}\right) \quad \textbf{[11.2] IP}$$

where:

Q = airflow (acfm)
T_s = Supply air temperature, F
T_o = Outside temperature, F*
c = cost of fuel ($/unit)
q = available heat per unit of fuel (BTU) (Table 11-1)

*To determine the maximum heating cost per hour, the coldest temperature that would be experienced should be used. There are design heating outdoor temperatures for various locations available in the ASHRAE *Handbook of Fundamentals*.(11.1)

$$\text{Energy Cost per year} = C_a = \left(\frac{(0.154)(Q)(dg)(T)(c)}{q}\right) \quad \textbf{[11.3] IP}$$

where:

Q = airflow (acfm)
dg = annual degree days (Table 11-2)
T = operating time (hours per week)
c = cost of fuel ($/unit)
q = available heat per unit of fuel (BTU) (Table 11-1)

TABLE 11-1. Available Heat per Unit of Fuel (IP Units)

Fuel	BTU/Unit	Efficiency, %	Available BTU/Unit	Approximate Cost/Unit
Coal	12,000 BTU/lb	50	6,000	$50/ton
Oil	142,000 BTU/gal	75	106,500	$3.50/gallon
Natural Gas	1,000 BTU/cubic foot (cu ft)	Indirect fired = 80 Direct fired = 90	800 900	$5.00/1000 cu ft $5.00/1000 cu ft
Electricity	3,413 BTU/kWh	100	3,413	$0.08/kWh

TABLE 11-2. Heating Degree Day Normal and Average Winter Temperature

City	Albany	Boston	Chicago	Cleveland	Detroit	Minneapolis	NY	Philadelphia	Pittsburgh	St. Louis	Wash., DC
Avg Temp (F) Dec-Feb	24	22.4	25	28	25.9	16	33.2	33.3	29	32.2	33.4
Discharge Air Temp (F)	Heating Degree Days										
80	11782	10409	10613	11343	10959	13176	9284	9652	10797	8943	8422
79	11425	10049	10277	10982	10605	12826	8937	9300	10436	8624	8089
78	11062	9690	9940	10621	10256	12478	8596	8954	10076	8310	7764
77	10709	9242	9610	10265	9914	12135	8265	8619	9726	8003	7446
76	10356	8994	9283	9915	9581	11797	7938	8285	9379	7702	7139
75	10009	8652	8972	9570	9247	11475	7620	7959	9036	7413	6835
74	9669	8317	8656	9229	8920	11142	7308	7641	8702	7121	6538
73	9333	7790	8349	8898	8599	10816	7004	7328	8372	6839	6250
72	9007	7668	8046	8567	8291	10496	6706	7028	8050	6560	5974
71	8682	7354	7750	8248	7981	10180	6421	6728	7740	6289	5703
70	8364	7046	7468	7928	7678	9870	6146	6438	7429	6023	5438
69	8256	6749	7183	7617	7383	9567	5871	6158	7127	5767	5179
68	7750	6458	6905	7313	7100	9269	5606	5886	6833	5523	4929
67	7452	6175	6635	7016	6816	8975	5349	5618	6546	5277	4690
66	7162	5903	6373	6722	6543	8687	5101	5360	6272	5053	4455
65	6881	5633	6122	6445	6278	8410	4858	5109	5997	4822	4229
64	6607	5370	5875	6165	6020	8131	4621	4864	5734	4595	4014
63	6340	5118	5638	5897	5772	7858	4394	4628	5483	4379	3798
62	6081	4873	5399	5636	5533	7590	4176	4397	5234	4168	3588
61	5829	4643	5164	5381	5290	7339	3957	4172	5006	3963	3383
60	5586	4399	4936	5140	5054	7086	3747	3952	4769	3761	3182

Since there is an allowance for the efficiency of delivering heat to the supply air, these equations will tend to give an incorrect low result if air is allowed to enter by infiltration only. The annual cost value is based on normal indoor temperatures (65 F to 70 F), moisture, moisture ratios, and standard atmospheric pressure of 14.7 psi. The moisture content, sometimes called the humidity ratio (ω), is assumed to be 0.01 pounds of water per pound of dry air.

Fuel cost can vary and is often subject to market pressures that make it difficult to predict the future cost. If the energy cost analysis is at a location of an existing plant, review the current utility contracts and bills to identify the current energy costs. Information regarding future trends can be obtained from the National Institute of Standards and Technology (NIST) Annual Supplement to NIST Handbook 135 and NBS Special Publication 709.[11.2] This publication can be downloaded from a DOE website at: http://www.eere.doe.gov/femp/. Use the link "Information Resources," then "Access Tools" and finally "BLCC Programs" to access the above document. It provides information that is tailored for life cycle cost analysis. Energy cost information can also be obtained from the National Energy Information Center,[11.3] which is maintained by Energy Information Administration, DOE (www.eia.doe.gov/mer). Monthly use and cost data for a wide range of energy sources are available in this database.

EXAMPLE PROBLEM 3 (Replacement Air Fuel Cost)

Determine the hourly and yearly cost of tempering 10,000 acfm of supply air to 70 F in St. Louis, Missouri, using oil at $1.85 per gallon. Average winter temperature is 42 F. System operates 120 hours per week.

$$\text{Energy Cost per hour} = C_h = \left(\frac{(1.08)(10{,}000)(70-42)(1.85)}{(106{,}500)}\right) =$$

$5.25 per hour

Energy Cost per year = C_a =

$$\left(\frac{(0.154)(Q)(dg)(T)(c)}{q}\right) = \left(\frac{(0.154)(10{,}000)(6023)(120)(1.85)}{106{,}500}\right)$$

= $19,335 per year

The yearly cost is more representative because both the length and severity of the heating season are taken into account.

11.2.3.2 Energy Calculation Based on Air Volume and Discharge Temperature. Estimates of annual heating energy use for spaces controlled at any temperature or that have specific discharge air temperatures can be obtained by using Equation 11.4. A time period of 210 days or 30 weeks is used as the heating season in determining the degree day value. Degree day values for a few cities in the United States are provided in Table 11-2. Similar values for over 4,000 U. S. locations can be obtained online at no cost at http://www5.ncdc.noaa.gov. Click on "Free Data" and then on

"CLIM 81" for the degree day information.

Annual Heat Use (BTU) [kJ] =

$$\left(\frac{(CF)(Q)(T_s - T_w)\left(\frac{168\,\text{hours}}{\text{week}}\right)}{\eta} \right) \qquad [11.4]$$

where:

H = annual heating use, BTU [kJ]
Q = airflow, acfm [am^3/s]
T_s = Supply air temperature, F [C]
T_W = Average winter temperature, F [C]
η = Heating system efficiency
CF = 30 [1597]

If average winter temperature is not known, it can be calculated by heating degree-day data available (IP only):

$$T_w = 65 - \frac{(dg)}{210} \qquad [11.5]\ \text{IP}$$

where:

dg = annual degree days @ 65 F (Table 11-2)

The annual heating cost can be obtained by multiplying the annual heating energy use (therms, cubic feet, gallons, etc.) by the number of BTUs found in the type of heating energy source. That will provide the quantity of the heating energy source consumed in a heating season in BTUs. That value would need to be adjusted by the energy conversion efficiency factor provided in Table 11-1, which considers the type of heat generation equipment used. Common cost for various types of fuel is provided in Table 11-1 but these values can change frequently. More accurate values can be obtained from local energy suppliers or by looking at the sources identified above. The annual heating energy use in BTU per year multiplied by the cost per BTU will provide the annual cost for heating.

EXAMPLE PROBLEM 4 (Annual System Operating Costs) (IP Units)

The installation of a new 24,000 acfm welding operation exhaust system requires a replacement air system to be installed. The size of this new supply air system is 25,000 acfm. The location of this plant is in Chicago, Illinois. To avoid drafts the required temperature of supply air is 85 F. The selected unit is an indirect gas fired type and will have a 25 hp motor to power the fan. The fan will operate at 22.6 horsepower including drive losses. The exhaust system operation will be 20 hours per day and six days per week (6,000 hours per year). Determine the expected energy use of operating this system. The welding exhaust fan is powered by a 40 horsepower motor operating at full load. Electrical costs are \$0.06 per kWh (kilowatt-hour). Natural gas cost is \$6.00 per 1,000 cubic feet.

Electrical Use = (Exhaust fan HP + supply fan HP) × .746 kW/hp × 120 hrs/week × 50 weeks/yr
= (40 hp + 22.6 hp) × .746 kW/hp × 6000 hrs/yr
= 280,198 kWh/yr

$$\text{Annual Electric Costs} = \left(\frac{280{,}198\ \text{kWh}}{\text{year}}\right)\left(\frac{\$0.06}{\text{kWh}}\right)$$

$$= \$16{,}812 \text{ per year}$$

$$T_w = 65 - \frac{(dg)}{210} = 65 - \frac{6122}{210} = 35.8\ \text{F}$$

$$\text{Heating Energy Use} = \left(\frac{(30)(Q)(T_S - T_W)\left(\frac{120\,\text{hours}}{\text{week}}\right)}{\eta} \right)$$

$$= \left(\frac{(30)(25{,}000)(85 - 35.8)(120)}{0.8} \right)$$

$$= 5{,}535{,}000{,}000\ \text{BTU/hr (BTUH)}$$

Since there are approximately 1,000 BTU per cubic foot of natural gas, the annual natural gas use would be 5,535,000 cubic feet per year.

$$\text{Heating cost} = \left(\frac{5{,}535{,}000\ \text{ft}^3}{\text{year}}\right)\left(\frac{\$6.00}{1000\ \text{ft}^3}\right) = \$33{,}210 \text{ per year}$$

Total energy cost = \$16,812 + \$33,210 = \$50,022
(almost \$2 per acfm)

11.2.4 Energy Used for Cooling Supply Air. Cooling energy use is more difficult to estimate since the amount of moisture present in the air affects the cooling energy requirements to lower the air temperature. If the moisture level is high enough, the dew point of that moisture will be reached as the air temperature is lowered. Then cooling energy is required to condense some of the water vapor in the air as the air temperature is reduced below the dew point. Having good weather data that reflect the outside air temperature and corresponding humidity levels in a summary form is normally difficult to obtain. There is joint frequency or weather "bin data" available from ASHRAE and from an update of Air Force Manual 88-29 called the Engineering Weather Data (NCDC 1999).(11.4) The "bin data" provide corresponding wet-bulb temperatures for dry bulb temperature by month. Once the cooling demand is established then a logical system can be selected to provide that cooling. The energy requirements for the equipment providing the cooling can be estimated from information provided by the equipment manufacturer. Values are obtained at different rates of cooling load. An annual cooling energy use can be calculated by using the weather data to identify the amount of time spent at different levels of equipment load. Knowing the cost of that energy source, the annual cooling energy cost is obtained. A good measure of the monthly cooling energy requirement of an operating system is to observe the variance in the electrical use during the cooling season compared to time periods or months when there is little or no air tempering.

11.3 RECIRCULATION OF EXHAUST AIR

Some plants obtain their replacement air by cleaning the exhaust air and discharging or recirculating the air back into the plant. There are a number of concerns with this approach. Since replacement air is being directly discharged into occupied spaces its requirements for cleaning are much higher than needed for exhaust air outside. The additional level of cleanliness normally requires the use of HEPA type filters. This imposes additional resistance to flow in the exhaust system and extra cost for filter changes. There are specific requirements for recirculating air systems associated with monitoring, system design, etc. that are discussed in Chapter 10 and American National Standards Institute (ANSI) 9.7, Recirculation of Air from Industrial Process Exhaust Systems.[11.5]

The recirculation of exhaust air appears on the surface to be a simple, straightforward approach to avoid the need to add supply air when an exhaust system is being installed. It should be done with caution since it has a number of potential problems. By recirculating the exhaust air, only the cost of tempering the air is saved. Air is exhausted for a reason; usually there is some property of the air that is not desired. It could be too warm, too dirty, have odors or be harmful to the health of those exposed. Since recirculating the exhaust air does not physically remove the contaminants in the air, special care needs to be taken in the design of an air recirculating system. If the exhaust air contains a chemical agent or gas that is a known carcinogen, recirculation must be avoided. The same applies to explosive chemical agents or a substance whose toxicity is unknown. If the exhaust air has chemical agents or gases that are a low health risk, then recirculation can be considered. The following are the required components for a recirculation system:

- Process exhaust hood/enclosure,
- Exhaust duct,
- Air-cleaning device,
- Exhaust fan,
- Recirculation air cleaner,
- Clean air monitor,
- Recirculation air outlet/grille,
- Air by-pass to outside, and
- Outside exhaust stack.

This system is the same as an exhaust system except for the recirculation air cleaner and the clean air monitor. The recirculation system must be able to clean and monitor the air to a level where there is no buildup of any contaminant concentration above unacceptable levels. A complete discussion of how to apply recirculation systems, the required efficiency of the air-cleaning device, and other design considerations is provided in Chapters 8 and 10.

11.4 ENERGY CONSERVATION OPPORTUNITIES

The largest cost associated with a ventilation system is the cost of operation. In most cases, the expense of energy and maintenance is much greater than the system's original installation cost. The best time to implement actions that save these energy expenses is during the system design phase before any components are installed. These actions often reduce the system's first cost in addition to lowering the annual operating costs. If the exhaust airflow can be reduced through better hood design, the required replacement air for this exhaust can also be reduced. As a result there is an installation savings on the supply air side that can be used to offset any additional initial cost the exhaust system may experience. With less airflow, fan horsepower requirements are lower and heating and/or cooling energy use is reduced. When analyzing a ventilation system for energy conservation, both the exhaust and supply/replacement air systems must be assessed. Energy use of exhaust air systems can be reduced by the following methods:

1) Reduce Airflow/Static Pressure. The system's airflow or static pressure can be reduced by improvements in the design or adjusting the system to operate at current requirements. Improvements in the design could include:

   - Moving the hood closer to the source,
   - Changing the hood to a more enclosing type,
   - Adding improvements to the hood design such as flanges, curtains or barriers to reduce cross air currents,
   - Changing duct design to reduce elbow or entry losses,
   - Replacing the air cleaning unit with one that has a lower pressure drop,
   - Removing air balancing or flow control dampers,
   - Eliminating fan system effects.

   If the airflow can be reduced (do not reduce the velocity below the required minimum transport velocity) then the fan laws can be used to determine the new energy use of the fan motor. Otherwise a recalculation of the system would be required. Methods to use for improvement of the exhaust system effectiveness can be obtained in Chapters 5 and 9.

   Numerous hood designs presented in Chapter 13 are

intended to provide for adequate contaminant capture without unnecessarily high airflow rates. For instance, the use of a horizontal sliding sash in the laboratory hood can provide a 30% saving in exhaust airflow without impairing capture velocity. The saving is accomplished by having at least one-third of the face area covered by sash (see VS-35-01). The use of a tailored hood design, such as the evaporation hood shown in VS-35-40 provides good contaminant capture with far lower exhaust flow rates than would be required for a typical laboratory bench hood. Low Volume-High Velocity hoods and systems such as those illustrated in VS-40-01, -02, -03, -04, -05, -06, and -20 are used for many portable hand tool and fixed machining operations and can provide contaminant capture at far lower air handling requirements.

2) Process Change. Changing the process so an exhaust system is no longer required can also save energy. This type of approach normally results from several years' effort and is usually done with a significant upgrade to the process tooling. An example of such a change is the use of colored rubber coated bumpers instead of chrome units thus eliminating plating operations. This product change eliminated the requirement for acid mist exhaust. The first step in pollution prevention is the management of the process.

The supply/replacement air system also offers energy conservation opportunities. Depending upon the climate of the subject location, the cost of heating and/or cooling the supply air can be significantly more than the electrical cost of operating the fans. The costs associated with heating can be sizable in northern climates. There are four methods by which the cost of heating and cooling a large flow of outdoor air can be reduced: 1) reduction in the total flow of air handled, 2) delivery of non-tempered outdoor air to the space, 3) recovery of energy from the exhaust air, and 4) recirculation of non-contaminated air from processes back into the plant.

The successful application of these engineering methods without reduction in health hazard control and without impairing the in-plant environment requires careful consideration. Special care must be taken if mixing the air from several spaces where the combination of contaminants may result in a more hazardous condition.

11.4.1 Reducing the Flow Rate to Save Energy. A reduction of total airflow should be accomplished only after conducting a careful inventory of all exhaust and supply systems in the plant. Determine which processes and hoods are necessary, which hoods or systems can be replaced with more efficient designs, and which systems may have been rendered obsolete by changes. For example, window exhaust fans and power roof exhausters are often used to remove heat or nuisance contaminants that could be captured more readily at the source with lower airflow rates. Many roof exhausters were installed initially to combat problems caused by a lack of replacement air. When air supply and balanced ventilation conditions are established, their use may no longer be necessary.

After the inventory of the ventilation equipment is finished and the obsolete equipment removed, determine an air balance of the remaining supply and exhaust systems serving the plant. Also consider any planned exhaust system airflow reductions in this analysis and consider the desired airflow from one space to another to insure cleanliness and contaminant control. The final result will be a document that identifies the required exhaust of the various spaces in the plant and the required supply air to provide replacement air to those exhaust systems plus the extra air that moves from one space to another. Most buildings are designed to have a slight positive pressure compared to the outside for the reduction of cold drafts. The extra air to accomplish the positive pressure is added to the exhaust air quantity to become the total supply air quantity. Other situations may require pressurization of one room or department as compared to adjacent areas.

The obsolete equipment identified and the extra supply air not required can be turned off for an energy saving. Equipment can also be turned off depending upon the schedule of process operations, the use of different process materials and other process modifications as well as climate changes to achieve energy savings. A plan that considers the changing requirements of the manufacturing process compared to the schedule of the needed ventilation equipment will achieve the greatest energy savings.

11.4.2 Non-tempered Air. In operations where people are not present, non-tempered air can be utilized as replacement air for process exhaust systems. This air that is not heated and/or cooled must be acceptable to the process operation that is being exhausted. Sometimes air that is slightly heated and/or cooled is utilized to satisfy the process requirements, but if workers are exposed directly to the air or the drafts created by its movement in the plant it could become a comfort issue. Non-tempered air is used effectively for hot processes and laboratory hoods as well as kitchen ventilation systems.

11.4.3 Energy Recovery. There are many methods available for the capture of waste energy in exhaust gases. For recovered energy to have value an effective use must be found for it. Typical uses are process heating, tempering replacement air or heating/cooling a fluid that is used at the facility. Processes that have hot exhaust gas streams and, therefore, potential for heat recovery include:

- Heat treat furnaces
- Paint ovens
- Parts washers and dryers

To minimize the cost of the heat recovery system, the exhaust and supply stream should be concentrated so that the minimum number of heat recovery units are needed.

There are a number of heat recovery devices that can be utilized to recover the waste heat; the common ones are discussed in more detail in the following sections. Since most industrial exhaust contains an airborne contaminant, care must be exercised to ensure that the contaminant will not interfere with the desired heat transfer. Most heat transfer devices rely on a clean surface for maximum heat transfer. Heat transfer coils, tube sets and plate types use metal to separate the hot exhaust air from the incoming cooler supply air. If contaminants collect on the heat transfer surfaces, the rate of heat transfer will decrease. Air-cleaning devices may be required upstream of the heat recovery systems to keep them free of contamination and operating at high efficiency. The heat recovery unit is normally placed upstream of the heating and cooling equipment in an air handling unit as shown in Figure 11-1.

The heat or temperature transfer efficiency of these heat recovery units is called Effectiveness and is calculated using the following formula:[(11.6)]

$$\text{Effectiveness} = \frac{Q_s\rho_o(T_s - T_o)}{Q_e\rho_e(T_e - T_o)} = \frac{\dot{m}_s(T_s - T_o)}{\dot{m}_e(T_e - T_o)} \qquad [11.6]$$

Q_s = supply airflow, scfm [nm³/s]
Q_e = exhaust airflow, scfm [nm³/s]
ρ_o = density of supply air, lb/cubic foot [kg/m³]
ρ_e = density of exhaust air, lb/cubic foot [kg/m³]
T_s = temperature of the supply air, F [C]
T_o = temperature of the outside air, F [C]
T_e = temperature of the exhaust air, F [C]

The Effectiveness for moisture transfer or total heat exchange uses the same ratio of values.

EXAMPLE PROBLEM 5 (Simple Heat Exchange Calculation) (IP Units Only)

A flat plate heat exchanger has been selected to recover heat from a drying oven that has a 400 F exhaust stream (10,000 scfm). The Effectiveness of the heat exchanger is 70% and inlet flow is also 10,000 scfm. Determine the supply air

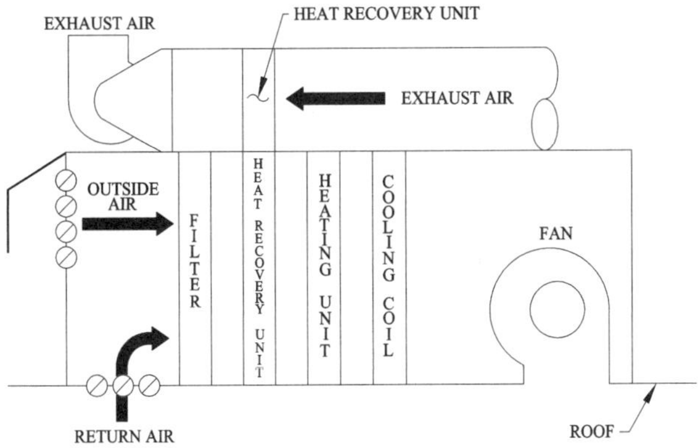

FIGURE 11–1. Heat recovery air handling unit

temperature when the outside temperature is 0 F. The density of 400 F air is 0.045 lb/cubic foot and the density of 0 F air is 0.086 lb/cubic foot.

From Equation 11.6:

$$(T_s - T_o) = \frac{(\text{Effectiveness})(Q_e\rho_e)(T_e - T_o)}{Q_s\rho_o}$$

$T_o = 0$ F

$$T_s = \frac{(0.7)(10{,}000)(0.045)(400 - 0)}{(10{,}000)(0.086)} = 147 \text{ F}$$

When the exhaust air stream is 10,000 scfm and the supply airflow is 25,000 acfm determine the resulting supply air temperature. The outside temperature is 0 F.

$$T_s = \frac{(0.7)(10{,}000)(0.045)(400 - 0)}{(25{,}000)(0.086)} = 59 \text{ F}$$

11.4.4 Heat Recovery Wheels. With heat or energy recovery wheels, the exhaust air streams need to be combined and discharged in the same location as the supply or replacement air streams. The temperature of the exhaust gases must be high enough to make recovery worthwhile and the exhaust gases need to be relatively free of airborne contaminants that could adhere to the wheel during the purge cycle. The heat recovery wheel is heated to a temperature that approaches the exhaust air as it rotates through the exhaust duct. As that section of the wheel continues to rotate, it enters the supply air duct where the media in the wheel begins to heat the outdoor air as it passes through the media as shown in Figure 11-2.

To help keep the wheel media clean there is a small purge zone placed after the exhaust section and before the supply air tempering zone. Cleaning is accomplished by the outdoor air passing through the wheel material and immediately being diverted back to join the exhaust air stream. This air movement sweeps through the wheel and blows off any material collected on the wheel, thereby reducing the chance of the exhaust air stream contaminating the supply air. Figure 11-3 shows the purge zone, which occupies approximately 12% of the wheel's circumference.

The heat recovery wheel can be made of a range of materials to handle different applications. Common wheel materials and their applications are as follows:

- Aluminum alloys – for sensible heat recovery only and applied to heating and ventilation systems or cooling industrial drying operations.
- Aluminum alloys with hygroscopic surface and desiccant – both sensible and latent heat recovery found in heating,ventilating and air conditioning (HVAC) systems having a high outside air requirement.
- Aluminum alloy with baked-on epoxy surface – for

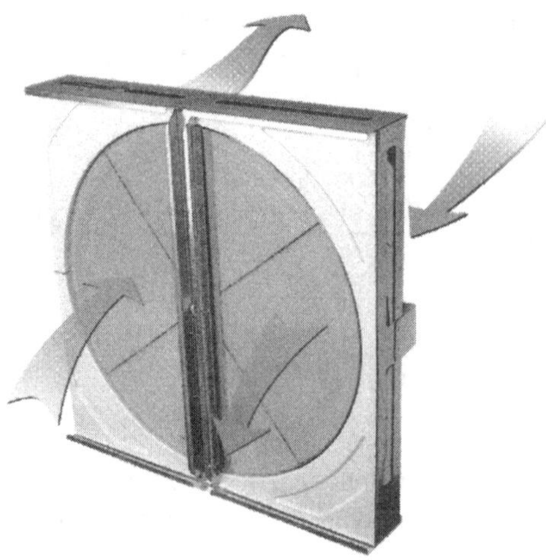

FIGURE 11–2. Heat wheel configuration

sensible heat recovery only with exhaust systems that have mildly corrosive air streams, commonly found in industrial curing systems and indoor pools.

- Aluminum alloys with hygroscopic surface – mainly sensible heat recovery with some latent capabilities when used on spray paint systems.
- Aluminum alloy or stainless steel – for sensible heat recovery only and used on industrial drying operations with leaving temperatures as high as 1200 F [649 C].

The Effectiveness of energy transfer is generally in the range of 50 to 85% depending on the wheel diameter (area), wheel depth and the amount of air that is used in the purge section. A small horsepower motor is used to rotate the wheel using a large belt at a speed ranging between 12 to 20 rpm. Systems used in special or high temperature environments use chain drives in place of a belt drive.

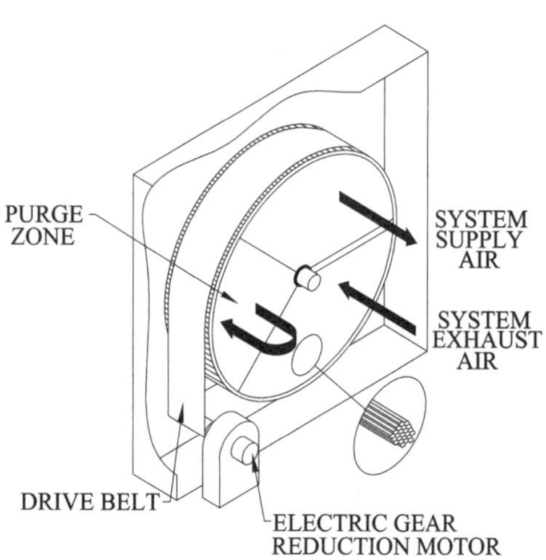

FIGURE 11–3. Heat wheel purge zone

The use of a heat wheel should be avoided where there are hazardous substances in the exhaust air stream since there is leakage between the exhaust and supply air zones. To isolate these air streams wipers are employed to seal the spaces on the upstream and downstream sides of the wheel. During the operating life of the wheel, seals must be inspected for adjustment and replacement. The casings in the energy wheel lose their Effectiveness and need to be replaced after several years. The drive motor and v-belt and chain also require inspection and maintenance.

Care also must be exercised when the exhaust air stream has a high moisture content and the incoming air stream could be lower than 32 F [0 C]. When it is below freezing outside, the cold incoming air could drop the wheel temperature below the dew point of the exhaust air causing water drops to form. These water drops could then freeze on the wheel causing deterioration of the wheel materials.

Energy used by the heat wheel is basically limited to the horsepower of the small drive motor and the system fan motor horsepower increase (due to the increase in system static pressure). The normal air velocity through a heat wheel is 500 fpm to 900 fpm [2.5 to 4.5 m/s] so this static pressure drop is relatively small.

Maintenance and operational considerations are straightforward with the following system checks recommended:

- Free movement of the rotor or wheel,
- Proper fitting of the seals,
- Correct position of the drive gear motor,
- Correct tension on the drive belt or chain,
- Correct rotational direction of the rotor,
- Lubrication of the two pillow block bearings that support the rotor.

The rotor should turn so that the portion of the wheel moving out of the exhaust air stream enters the purge section before entering the supply air section. If the particles in the exhaust air stream are dry, cleaning of the rotor should be minimal with only lint and other debris becoming caught. Normal designs will allow particles up to 800 microns in diameter to pass through the rotor's media. Moist/sticky particles can adhere to the media and their presence would require a periodic cleaning to reduce static pressure loss and increase heat transfer. Follow the manufacturer's instructions regarding rotor cleaning and the timing of other maintenance actions.

11.4.5 Fixed Plate Air/Air Heat Exchangers. Fixed plate heat exchangers have intertwined tunnels of the exhaust and supply air streams as illustrated in Figure 11-4. These tunnels are separated by plates with fins to keep one stream isolated from the other (see Figure 11-5). Often the arrangement of the heat recovery unit has the exhaust air stream flowing at an angle of 90 degrees to the supply. Like the heat wheel, the

FIGURE 11–4. Fixed plate air/air heat exchanger

FIGURE 11–6. Cross-flow heat exchanger arrangement

exhaust air stream must be brought close to the supply air stream. The 90-degree cross flow usually makes it easier to arrange the flow of the two air streams through the heat exchanger. Such an arrangement is shown in Figure 11-6. Exhaust air enters the unit's lower section and exits with the upper section's fan. The supply air enters the higher section and is delivered into the building by the lower fan.

Most of the plate type heat exchangers transfer sensible heat at an Effectiveness of up to 80% (see Equation 11.6). The temperature of one air stream is used to heat or cool the other. In the heating season, hot exhaust air will warm the adjacent metal sides of its tunnel. These metal sides also form the walls of the incoming supply air tunnel. Thus, the warm sides heat up the cooler outdoor air stream for the supply air unit.

These heat exchangers are typically made of aluminum or stainless steel plate, which have an upper thermal operating limit of approximately 450 F [232 C]. The choice of material

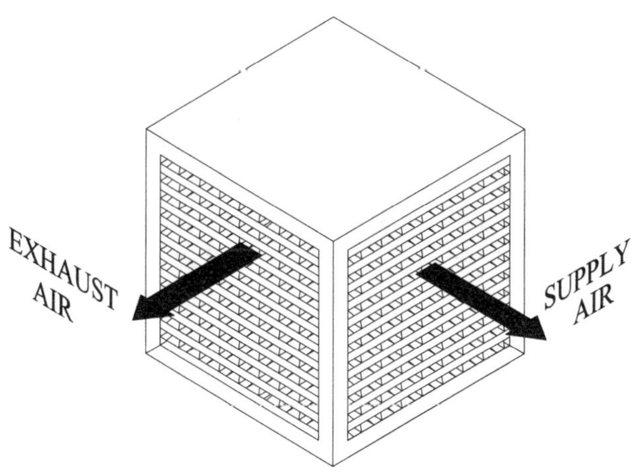

FIGURE 11–5. Fixed plate heat exchanger

would depend on the chemicals present in the exhaust air. For operations that have higher temperatures, welded plate and welded tubular types of cross flow heat exchangers are available. These units may be rated for operation at temperatures approaching 1800 F [980 C] depending upon the manufacturer. Some of the plate type heat exchangers can transfer temperature and humidity. These units have the same cross flow design, but they are constructed using a corrugated cellulose media impregnated with a polymeric desiccant in a steel or aluminum frame. This heat exchanger has an Effectiveness of approximately 50% and an upper temperature limit of 180 F [82 C].

The energy use of these heat exchangers is small. They will cause only a slight increase in static pressure under normal operations to both the supply and exhaust systems.

Flat plate heat exchangers have no moving parts and are easy to maintain. The major concern is keeping the metal plate surfaces free of dirt or any other materials that may coat the surfaces. This can affect the heat transfer capability and will eventually begin to add resistance to airflow causing additional static pressure across the unit. "Fouling" is the accumulation of dirt on the collector surfaces. Another issue of possible concern is frosting. If the air being exhausted is humid, the cold entering air could cause condensation of water in the exhaust tunnels of the heat exchanger. This water could freeze and begin to block airflow when the entering outdoor air is below freezing temperatures.

Cleaning of these units can be accomplished by careful use of compressed air or water flush per manufacturer's instructions.

11.4.6 Heat Pipe Heat Exchanger. The heat pipe heat exchanger will only transfer sensible heat. It is similar to the two previous heat exchanger designs in that the exhaust air and supply air streams must be brought to the same location to pass through this unit. The heat pipe system has no moving parts and there is a physical separation between the exhaust and supply air streams so no cross contamination can take place.

The unit consists of a bundle of heat pipes placed in a frame similar to a coil (see Figure 11-7). The heat pipes are finned tubes sealed at each end and placed in the coil frame horizontally to resemble a heating or cooling coil. Halfway across the coil shaped unit there is a physical barrier to separate the exhaust air from the supply air. Each heat pipe has the supply air flowing over one end and the exhaust flowing over the other end. Figure 11-8 shows the inside of the sealed heat pipe where there is a refrigerant that will condense in the colder air and evaporate in the warmer air. As the refrigerant condenses in the cold end it releases heat that warms the cool air. The refrigerant fluid then migrates to the warm air exhaust where it adsorbs heat as it evaporates to a vapor. Thus heat is transferred from the warm exhaust to the cooler supply air.

Heat pipes are normally made of the same materials as HVAC coils – aluminum and copper. For applications with the exhaust temperature above 125 F [52 C] the tubes and fins are made of the same material to avoid problems associated with different thermal expansions. The materials of construction change to steel tubes and fins when the operating temperature is over 425 F [218 C]. The unit can be covered with a protective coating to allow the use of aluminum in a corrosive atmosphere. The operating refrigerant must be carefully selected for the temperatures of operation. A wrong selection can cause the decomposition of the thermal fluids. This can cause the formation of non-condensable gases that will affect unit performance.

The normal Effectiveness range of the heat pipe unit is 45% to 65%. These values depend on the face velocity and the number of rows of heat pipes. By doubling the number of rows, Effectiveness of a unit can be increased from 50% to as much as 65%. The capacity of the heat pipe increases approximately with the square of the inside diameter of the pipe. Unless the tube is quite short, the heat transfer capacity is almost independent of the length. As a result, the heat transfer increases with a unit that is taller having more tubes than one which is wider with fewer tubes.

FIGURE 11–7. Heat pipe heat exchanger

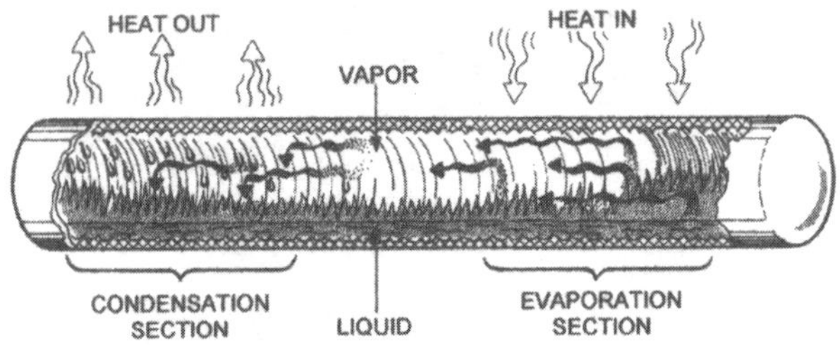

FIGURE 11–8. Heat pipe configuration

The selection of fin spacing should be based on the cleanliness of the two air streams. Industrial applications would generally have 8 to 10 fins per inch compared to an HVAC application using 11 to 14 fins per inch. The fin spacing can be different for the exhaust side compared to the supply side allowing design flexibility for pressure drop and maintenance concerns.

Cleaning of the heat pipe heat exchanger can be done using the same methods used to clean HVAC system coils. They can be flushed with compressed air or washed off with water or water and cleaning agent solution. These units do not usually have special coatings or materials that are harmed by wetting. Cleaning is required when the buildup on the surface of the heat exchanger creates excessive pressure drop or reduces the heat transfer to unacceptable levels. Since the exhaust air stream is more likely to have dirty air this side of the heat exchanger will likely require more frequent cleaning.

11.4.7 Coil Runaround Loops. The coil runaround loop has an HVAC coil placed in both the exhaust and supply air streams. A pumped piping system keeps the heat transfer fluid (usually water or a water/glycol mixture) flowing between the two coils. This is a very flexible system and the two coils can be located some distance apart. There is no need to route the exhaust air and supply air streams close together as required with the previous types of heat recovery systems.

There is only sensible heat transfer with this design. Each coil uses extended surface finned tube coils. The fluid enters the coil in a counter flow design thus allowing the coldest air to interact with the warmest fluid and vice-versa. The Effectiveness of this type of heat recovery system is 45% to 65%.

One concern with this system is freeze protection in the

warm air stream. The temperature of the cold air stream could be such that the leaving fluid temperature could be significantly below 32 F [0 C]. This could cause freezing in the warm air stream if it has high enough moisture content. The cold fluid could condense the moisture from the warm air and cause it to freeze on the coil surface. A way to avoid this is to monitor the cold fluid temperature and use a three-way valve to bypass some of the fluid around the cold coil keeping the fluid temperature above 32 F [0 C].

The coil materials of construction are the same as the materials used to build heat pipes. Temperature and corrosiveness are the major concerns. Excluding the fluid pumping system, the required maintenance on these units is similar to that of heat pipes.

11.4.8 Enthalpy Recovery Loop. The enthalpy recovery loop transfers moisture and temperature between two remote air streams. It utilizes a structure similar to a cooling tower. The air passes through a flow of a sorbent solution. This fluid is typically a special salt solution that can absorb or release water vapor to the air depending on the humidity of the air (Figure 11-9). Thus, a warm humid exhaust will heat the circulating solution and raise its moisture content. When this solution comes in contact with a cooler and dryer supply air stream it warms the air and adds water vapor. Generally, this equipment operates in a temperature range of -40 F to +115 F [-40 to +46 C].

The tower structures are quite large and heavy and need to be specially fitted into an exhaust and supply air system. Mist elimination pads are used to remove droplets of the sorbent solution entrained in the air stream leaving a tower. Lint or other solids need to be filtered from the air stream before the tower and filters are located in the sorbent circulation piping to remove contaminants. Some of the sorbent solutions such as lithium chloride can remove up to 94% of the bacteria in the air stream, making this heat recovery system effective for health care applications.

In cold climates, moisture losses from the exhaust air stream can dilute the sorbent mixture and heating the solution before

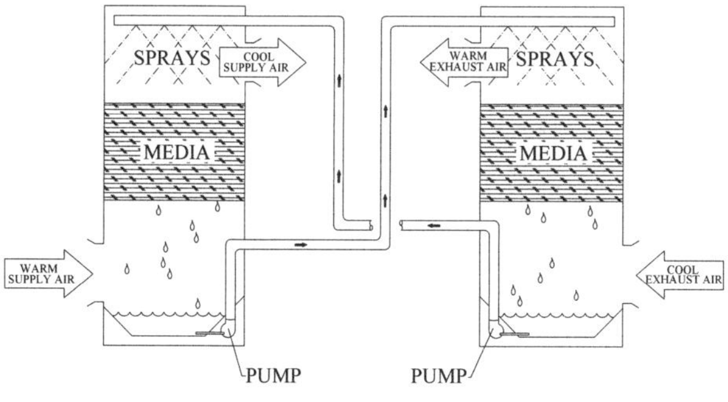

FIGURE 11–9. Enthalpy recovery loop

entering the supply air stream may be required. If either air stream has gaseous contaminants, their effects on the sorbent solution should be investigated.

11.4.9 Considerations with Contaminated Gas Streams. There are several concerns that arise when handling contaminated gas streams in heat recovery systems. These include:

1) The possibility of the contaminated gases entering the plant by leaking through the heat transfer walls into the supply air stream.
2) The possibility for a short-circuiting of exhaust gases directly into the supply air intake. This is because the exhaust and supply air streams are in close proximity in most designs.
3) The possibility of the exhaust gases having a harmful effect on the heat recovery unit. The gas stream could have chemicals that would attack the materials of the heat recovery unit and render it useless and possibly unsafe.
4) The possibility of the heat exchanger becoming fouled by the gas stream; material such as lint, sticky particulate, etc. could collect on the exchanger surfaces and begin to inhibit flow as well as limit the heat transfer.

To avoid these concerns, complete knowledge of the gas stream being exhausted is required so that a proper selection of the heat recovery unit can be made. The temperature range of the gas stream needs to be identified and the type, size and amount of particulate or corrosive materials in the gas stream needs to be determined. With this information, a decision can be made as to the ability to recover heat from the gas stream and the type of heat recovery system to be selected. Saving energy is an important issue, however, a poor assessment at the beginning can cause much more trouble than the savings realized.

To avoid the problem of plugging the heat exchanger, an air-cleaning device could be installed before the heat recovery system, but this adds more maintenance and pressure drop. The cost of the cleaning device increases the cost to recover the heat, can reduce some of the heat being recovered, and can affect the project's economics. Another approach can be to provide periodic cleaning of the heat recovery unit. Some systems have wash systems built in to spray off the heat exchanger so fouling is controlled within reasonable limits.

The choice of heat recovery unit is important. Some are less susceptible to fouling and easier to clean than others are. Some are made of materials that cannot withstand rigorous cleaning. It is common for the units with higher Effectiveness ratings also to be more susceptible to fouling and becoming plugged. Several factors are important in the selection of the appropriate heat exchanger:

1) The nature of the exhaust stream. A corrosive or dust laden stream may need to be pre-cleaned and special construction materials may be required.
2) The need to isolate the contaminated exhaust stream

from the clean supply replacement air stream.

3) The temperature of the exhaust stream. Unless the hot air stream is well above the desired delivery temperature of the supply replacement air stream and the exhaust air stream is at elevated temperatures whenever heat is demanded by the supply replacement air stream, additional heating capacity will be required.

4) Space requirements for some heat exchangers can be very extensive.

5) During failure mode or summer conditions a by-pass will be required.

6) Need good distribution of air to prevent ice, etc., in potential dead air spots in the building.

REFERENCES

11.1 American Society of Heating, Ventilating and Air Conditioning Engineers: ASHRAE Handbook Fundamentals. ASHRAE, Atlanta, GA (2005).

11.2 National Institute of Standards and Technology: Energy Price Indices and Discount Factors for Life-Cycle Cost Analysis – April 2005, Annual Supplement to NIST Handbook 135 and NBS Special Publication 709, NISTIR 85-3273-17 (Rev. 4/05). Washington, DC (2005).

11.3 Monthly Energy Review: National Energy Information Center, Energy Information Administration, United States Department of Energy, Washington, DC.

11.4 National Climate Data Center: NCDC, Engineering Weather Data (1999).

11.5 ANSI/AIHA: Recirculation of Air from Industrial Process Exhaust Systems. ANSI/AIHA Z9.7 – 1988 American National Standards Institute (April 1988).

11.6 American Society of Heating, Ventilating and Air Conditioning Engineers: ASHRAE Handbook HVAC Systems and Equipment. ASHRAE, Atlanta, GA (2004).

Chapter 12

COST ESTIMATING

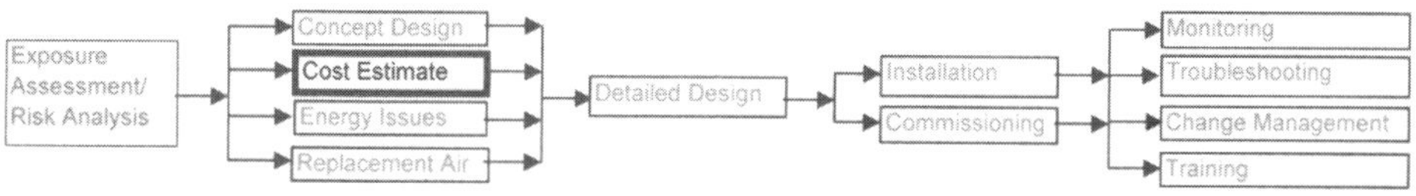

NOTE: Equations with notation followed by (IP) are designated for inch-pound system only; Equations followed by (SI) are designated for metric use only. If Equation bears neither, then it applies to both systems.

NOTE: Chapters 11 and 12 use mostly IP units because of the variety of cost values that could be used worldwide. Where appropriate, metric values are inserted but Example Problem solutions are in IP units only.

12.1 INTRODUCTION

An Industrial Ventilation System will have many costs associated with construction and operation. If cost estimates are not accurate, there could be dire financial results for the owner or installer. Unfortunately, the price tag for most installations is not easily determined prior to construction. Cost estimating methods and guidelines can be used to predict these costs. In addition, the accuracy and presentation of this data must be suitable for several different audiences and at different stages in the construction process.

The same project can have a number of different cost requirements depending on what type of information is needed. Plant accounting may want the total capital outlay for a project (before the project starts and during the installation). Maintenance will need to know the cost in labor-hours for start up, training and operation. Plant planning personnel will need information on annual operating expenses, utilities and manpower needs. The engineering manager will need to allocate labor-hours and cash for design services, construction supervision and technical services. The plant manager may need to determine the cost of downtime for production during construction. The environmental regulators may ask for the total cost of a project including both external costs and internal costs.

Some costs may be simple to estimate. A catalogue item can have a fixed and predictable price but that amount can change due to competition, delivery charges and accessories added to the item. Other costs will be more difficult to estimate with accuracy. These include the operating and maintenance requirements for the system. Some items can have costs that change in response to other factors. The price of equipment can vary significantly depending on the level of competition between vendors. Installation costs will vary significantly depending on the likelihood of unforeseen project modifications as well as local economic conditions and unknown site conditions. As local construction resources become scarce, the cost of installation can rise dramatically, sometimes over the time span of the project.

The accuracy of the cost estimate will also depend on the knowledge level associated with the system. During the initial phase of a project, only the emission sources will be known. Since the design details of the industrial ventilation system are not known, 'rule of thumb' system cost estimates are commonly applied. The United States Environmental Protection Agency (USEPA) and other sources have developed methods for making rough estimates on control costs based on the process type.[12.1] The level of accuracy can be very low, but it can be enough to begin to determine basic cost information and develop a preliminary budget.

During the Preliminary Design Phase (see Chapter 2), the design team will determine control options, potential emission sources, control equipment locations, and estimates of air volumes. This information can be used to refine the cost estimate to within a factor of ± 30%.

Once detailed engineering has been completed, a detailed cost analysis can be made. Depending on the method and accuracy of the estimator, this estimate can be within ± 5% of the project cost.

In this chapter, the basics of developing a project budget, determining economic advantages of multiple control options, estimating annual operating and maintenance costs, and reporting of costs to environmental agencies will be covered. Detailed cost estimating is a highly specialized task that is beyond the scope of this Manual.

12.2 CAPITAL COSTS

Determination of capital cost is one of the first requirements in determining the feasibility of a project. For some projects, the industrial ventilation system can contribute a significant percentage of the total cost. There can be several estimates made during the course of the project, usually with increasing accuracy. The first (and least accurate) is an "Order of Magnitude" estimate made very early in the project.

12.2.1 Order of Magnitude Pricing Methods. There are several methods for determining an order of magnitude cost estimate. None of these methods has a reliable or repeatable accuracy of less than ± 30%. While inaccurate, these methods can be an invaluable first step in determining the feasibility of the project. These methods are also sometimes called 'first cut' or 'rough cut' pricing. There are two main types of "Order of Magnitude" cost estimating.

"Rule of Thumb" Pricing. This method is based on historical price data on similar applications. If, for several past projects, a baseline unit and a ventilation system cost are known, a rule of thumb can be developed. The baseline can be one of several known performance indicators of a process. Commonly used baselines include: tons of processed material per day, heat output of a process, volumetric flow of exhaust system, or units per hour of production. As an example, consider a ventilation system with an estimated volumetric flow. Using several projects, an average cost can be developed using units such as control dollars per acfm ($/acfm). Then the size of the proposed system can be multiplied by the projected system size to determine the rule of thumb price.

EXAMPLE PROBLEM 1 (Rule of Thumb Pricing)

In the evaluation of five dust control projects at a gray iron foundry in the cleaning department, the average cost for the exhaust was found to be $15/acfm. A shot blast operation is being added for a new casting that will require 20,000 acfm of exhaust. Determine the rule of thumb cost estimate.

$$20{,}000\ \text{acfm}\left(\frac{\$15}{\text{acfm}}\right) = \$300{,}000\ \text{Estimated System Cost}$$

When applying a rule of thumb estimate the impact of expensive auxiliary items should be identified. If the item is a variable to the system it should be removed from the projected cost. Examples of such items are electrical substations, steel reinforcing, significant steel supports or relocation of significant utilities.

Scaled Factor Method. If the proposed system is similar to an existing system where the project cost and system size are known, the scaled method can be used. The only variable required to calculate in this estimate is a volume estimate for the proposed system. The cost can be determined using the "power rule." If a baseline cost and the corresponding system size are known, the cost for the proposed system can be estimated using the equation:

$$C_a = C_b\left(\frac{S_a}{S_b}\right)^n \qquad [12.1]$$

where:

C_b = baseline cost for previous project

C_a = estimated cost for new project

S_b = baseline system size for previous project

S_a = estimated system size for new project

The system size can be expressed in volumetric flow, production output, or any other scalable function associated with the controlled operation. The exponent "n" has been shown to average 0.6. This is commonly called the "sixth tenths factor." Estimates found using this method are only recommended if a single project cost for a similar application is known.

EXAMPLE PROBLEM 2 (Scaled Factor Example) (IP Units)

A system had been installed recently with an air volume of 24,000 acfm and an installed cost of $420,000. Included in the cost of the project was a new substation worth $100,000. A new similar system is being considered but 32,000 acfm is required.

Base system cost = C_b = $420,000 – $100,000 = $320,000.

$$C_a = (\$320{,}000)\left(\frac{32{,}000\text{ acfm}}{24{,}000\text{ acfm}}\right)^{0.6} = \$380{,}000$$

12.2.2 Equipment Factor Estimate Method. Determining a cost estimate for the system can be more accurate if the air control device and system volume are known. It is common for an industry to use one predominate type of control equipment for a given application. An example of this is the use of catalytic oxidizers for the control of Volatile Organic Compounds (VOCs) from printing operations or baghouses (fabric filters) for foundry surface treatment operations. For a given control method, the cost of equipment is multiplied by factors to estimate related costs. This method is documented in the publication "The EPA Air Pollution Control Cost Manual"[12.1] and is provided at no cost from the EPA website (www.epa.gov/ttn/catc/dir1/c_allchs.pdf). Estimates using this method can be as accurate as ± 20% depending on the project.

An example of this method is shown in Figure 12-1, which was taken from the referenced USEPA Cost Manual. The calculation shown is designed to determine the total project cost for the installation of a fabric filter. The only input to the sheet is the purchase price of the collection device. This is available from vendors if a detailed specification can be provided. Based on that figure, all of the other associated costs can be estimated as shown. The USEPA has prepared estimating spreadsheets for most commonly available types of control equipment, duct systems and exhaust stacks.

EXAMPLE PROBLEM 3 (EPA Cost Manual) (IP Units)

A process requires the use of a fabric filter for control of emissions from a process. The estimated equipment cost is $225,000. No building or special site preparation is required. Determine the estimated total capital investment for the installation and the estimated engineering cost using the values provided in Figure 12-1.

A = Purchased Price Estimate = $225,000

B = Purchased Equipment Cost = (1.18)(A) = (1.18)($225,000) = $265,500

Total Capital Investment = (Purchase Equipment Cost) (Installation Factor)

= (B) (2.19) = $581,400

Capital Cost Factors for Fabric Filters

Cost Item		Factor
Direct costs		
Purchased equipment costs		
Fabric filter (EC) + bags + auxiliary equipment as estimated		A
Instrumentation		0.10 A
Sales taxes		0.03 A
Freight		0.05 A
Purchased Equipment Cost (PEC)	B =	1.18 A
Direct installation costs		
Foundations & supports		0.04 B
Handling & erection		0.50 B
Electrical		0.08 B
Piping		0.01 B
Insulation for duct		0.07 B
Painting		0.04 B
Direct installation cost		0.74 B
Site preparation as required		SP
Buildings as required		Bldg.
Total Direct Cost	DC=	**1.74 B + SP + Bldg.**
Indirect Costs (installation)		
Engineering		0.10 B
Construction and field expense		0.20 B
Contractor fees		0.10 B
Start-up		0.01 B
Performance test		0.01 B
Contingencies		0.03 B
Total Indirect Cost	IC=	**0.45 B**
Total Capital Investment = DC + IC		**2.19 B + SP + Bldg.**

FIGURE 12-1. Equipment factored estimates

Engineering cost = 0.10(B = Purchased Equipment Cost)
= 0.10 ($265,500) = $26,550.

NOTE: Many of the cost factors can change due to special considerations; for example, sales tax varies by region and state.

12.2.3 Price Book Estimating Method. The methods in this section are more accurate (± 10%) but require considerably more time and effort to complete. The design phase of the project (see Chapter 2) must be complete before this method can be used. A price book is a listing of equipment, fabricated pieces and related installation costs in an indexed format. There are price books available commercially and many contractors and fabricators have prepared price books for their internal use.

The project must be broken down into the individual pieces and each separate piece must have an individual cost estimate prepared. The first step to preparing this type of estimate is called the "take off." Using the design drawings for the system, each individual component in the system is listed in a spreadsheet format. Information such as component size, length of component required, number of components needed are compiled on the spreadsheet. Using the price book, the cost of purchasing the item and the installation labor requirements can be found. If an item is not included in the price book, a determination of the purchase price and installation labor must be made. Since some special industrial ventilation components are not commonly purchased items (special hoods and fittings, etc.) they must be independently estimated. After the purchase price for equipment and labor requirements are calculated, overhead expenses and the desired profit are determined and added to complete the cost estimate.

Because of the time and expense involved, price book methods are rarely used by owners or designers to determine a cost estimate. These methods are primarily used by contractors and installers for determining their project bid.

There are a number of sources that provide cost information needed to develop a project estimate. A listing of some of those that publish this information is provided in Table 12-1. Be sure to use the most current information when preparing an estimate. Some of the listed sources provide annual updates of the costs provided. Also consider the regional influence on labor costs. Again, some of the pricing books have adjustment factors to modify labor cost according to the location of the project. Also be sure to include cost mark-ups for overhead and profit of the subcontractors and prime contractor. If the project is to be designed by an engineering firm those fees must be added. The same applies for the use of a construction manager.

12.3 TOTAL ANNUAL COSTS AND OPERATING COST METHODS

Initial Capital Costs are only one component of the total system cost impact. In some cases, annual operating costs can be higher than the original capital cost of the system. Each industrial ventilation system has specific ongoing costs associated with the installation. These include electrical power, natural gas, compressed air, waste disposal, maintenance and other items.

These estimates are not only made to predict the economic impact of the system but also can be used for comparison of vendors' proposals as well as comparisons between control strategies (baghouse versus scrubber, etc.).

12.3.1 Annual Operating Cost Components. Annual operating costs represent the cash outlay requirements for operating and maintaining the system. These costs are variable depending on the maintenance needs for the system and the hours of operation. Costs in this category include: electricity, compressed air, water, natural gas, waste disposal, operating labor, maintenance labor, supervision, replacement and wear parts, insurance, taxes and overhead.

Very soon after the air control device is selected (during preliminary or detailed engineering phase) an energy audit should

TABLE 12-1. Sources of Capital Cost Information

Equipment Cost and "Rule of Thumb" Guides
Innovative and Alternative Assessment Manual, USEPA, 430/9-78-009
Capital and Operating Costs of Selected Air Pollution Control Systems, USEPA, 450/3-76-014
The EPA Air Pollution Control Cost Manual, USEPA, 452/B-02-001
Equipment Factors
The EPA Air Pollution Control Cost Manual, USEPA, 452/B-02-001
Engineering Guide #46, Ohio EPA 1983
Price Book Guides
Craftsman Estimating Guides, at http://craftsman-book.com
R.S. Means Building Construction Data, at www.rsmeans.com
Manuals
Estimating Costs of Air Pollution Control, Vatavuk, 1990
The EPA Air Pollution Control Cost Manual, USEPA, 452/B-02-001

be conducted. This audit will include the determination of brake horsepower of motors, compressed air usage, and other utility requirements. For a thermal or catalytic oxidizer, the fuel usage requirements should also be determined.

Energy Costs. Energy costs for the system can be calculated by multiplying the projected energy usage by the cost of acquiring the energy. When determining electrical usage, the primary expense is the operation of the motors for the system fans.

Other electrical devices included on typical systems are controls, lighting, rotary locks, heaters, pumps, air compressors, shaker motors, and rectifiers (for electrostatic precipitator systems). Motor horsepower should be converted to kilowatts. Other items will typically specify the wattage requirements. Electricity is sold in kilowatt-hours. Since the equipment electrical usage has been calculated in kilowatts, the number of kilowatt hours is simply the hours of operation for the system times the total power requirement.

Operating Labor and Maintenance Costs. All industrial ventilation systems will require some operating and maintenance labor. The amount of labor can vary significantly depending on the type of control equipment selected, the size and complexity of the system and the type and quantity of material that is collected. Typical maintenance requirements for industrial ventilation systems that utilize different types of control equipment are provided in Table 12-2.[12.2]

Replacement Parts Costs. Some components of the industrial ventilation system will require replacement of worn parts on a regular basis. These items can have relatively low cost, such as fan belts, or have major cost impact in the case of replacement filter media (bags or cartridges). Determining these costs on an annual basis is done by dividing the cost of replacement parts by the estimated life in years. Note that labor required to install these replacements must also be considered.

EXAMPLE PROBLEM 4 (Yearly Cost of Media)

A fabric filter collector has polyester media with a useful life of 2.5 years. The cost (bags and labor) to replace the filter media is $10,000. The resulting annual replacement cost estimate for the filter media is $10,000 / 2.5 years or $4000 per year.

TABLE 12-2. Operating and Maintenance Requirements

Control Device	Operating Labor (hours per shift)	Maintenance Labor (hours per shift)
Fabric filter	2–4	1–2
Electrostatic Precipitator	0.5–2	0.5–1
Venturi scrubber	2–8	1–2
Oxidizer	0.5	0.5
Adsorption or absorption systems	0.5	0.5

Waste Disposal Costs. This cost can vary widely depending on the nature of the material collected. Hazardous waste disposal costs have increased dramatically and this trend is expected to continue. Estimating waste removal costs is dependent on the quantity of material collected and is typically charged by the ton of material disposed.

Waste requirements can be estimated in advance using the emission factors for the process under control. USEPA document AP-42[12.3] lists typical waste concentrations in the exhaust air for many industry classifications. This document can be found on the internet at www.epa.gov/ttn/chief/ap42/index.html. Using these emission factors and the volume of air collected, a rough estimate of the amount of collected waste can be determined.

EXAMPLE PROBLEM 5 (Dust Catch Disposal) (IP Units)

A proposed process has an emission factor of 2 pounds per ton of production. A fabric filter collector is used to remove the dust from the air. The efficiency is assumed to be almost 100%. If the production is 4 tons per hour and the period of operation is 6000 hours per year, determine the annual waste quantity.

Waste =

$$\left(\frac{2\text{ pounds-waste}}{\text{Ton-production}}\right)\left(\frac{4\text{ Tons-production}}{\text{hour}}\right)\left(\frac{6000\text{ hours}}{\text{year}}\right)$$

$$= 48{,}000\,\frac{\text{pounds-waste}}{\text{year}} = 24\,\frac{\text{Tons-waste}}{\text{year}}$$

12.3.2 Estimation of Total Annual Cost. The total annual cost of the system includes both the direct costs (components in Section 12.3.1) and the indirect costs of depreciation, taxes, insurance and administration.

There are two types of overhead: labor and facility. Labor overhead includes fringe benefits for the operation and maintenance staff such as workmen's compensation, Social Security and pension fund contributions, vacations and health insurance. Some of these are fixed costs. Payroll overhead is traditionally computed as a percentage of the total annual labor cost.

Facility overhead accounts for expenses not directly related to the operation and maintenance of the control system, including: plant security, maintenance of restrooms and break areas, lighting, and parking. The USEPA uses an estimating factor of 50%–70% of the wage expenses to calculate the total of labor and facility overhead.

Depreciation, taxes, insurance and administration are usually combined into a single factor for cost estimates. The USEPA uses 4% of the total capital investment. When the equipment has a short life, the depreciation component may require adjustment.

12.3.3 Factored Methods for Determining Annual Costs. The USEPA has created a series of spreadsheets for calculating direct and indirect annual costs for emission control systems.[12.1] This information is utilized in Figure 12-2, which provides an example calculation for a thermal oxidization system. The EPA publication includes forms for several other types of control equipment including fabric filters and wet scrubbers.

12.3.4 Annualizing the Capital Cost of a Project. In addition to the direct and indirect annual costs (electrical power, compressed air, overhead, etc.) listed in previous sections, there is also a method to annualize the initial capital investment of the project. This is not depreciation but a method to show an added annual effect from the actual purchase of the system. This effect will consider three main components:

- The capital invested for the purchase of the system
- The cost of this capital (interest charged to borrow the money for the purchase); this value is used even if the system is purchased without borrowing funds – it is the loss of use of this capital
- The expected life of the project (will the system last 15 or 20 years before requiring replacement?, etc.)

This information can be used to compare the annual direct costs to a value that considers the original purchase. By totaling all of these values the design team can determine the real cost impact of the installation. This type of analysis is very important when trying to choose the best of different control strategies. This method is also used for BACT (Best Available Control Technology) studies when assessing whether a control strategy is appropriate or economically feasible.

Direct Annual Costs, DC		
Operating Labor		
Operator	a	0.5 hr/shift
Supervisor	b	15% of operator
Operating Materials		
Maintenance		
Labor	c	0.5 hr/shift
Materials	d	100% of maintenance labor
Catalyst replacement	e	100% of catalyst replaced
Utilities		
Natural Gas	f	Calculated usage
Electricity	g	Calculated usage
Total DC		**= a+b+c+d+e+f+g**
Indirect Annual Cost, IC		
Overhead (Supervisor, maintenance labor & maintenance materials)	h	60% of DC
Administrative Charges	j	2% Total Capital Investment (TCI)
Property Taxes	k	1% TCI
Insurance	l	1% TCI
Total Indirect Cost		**= h+j+k+l**
Total Annual Cost		**= DC+IC**

FIGURE 12-2. Suggested factor unit cost: thermal fluid-bed catalyst unit

This basis for the annualized cost is the calculation of a Capital Recovery Cost Factor (CRCF):

$$CRCF = \frac{i(1+i)^n}{(1+i)^n - 1} \quad [12.2]$$

where:

i = annual interest rate

n = capital recovery period (depreciable life of the system)

This factor is multiplied by the capital invested to purchase/install the system to obtain the annual cost impact. Note that all of these costs are in present value dollars and do not consider inflation or time value of money. This simplified method can then be used for cost comparisons without adding these effects. *NOTE: This analysis does not include other annualized costs such as electric power and maintenance, but only covers the cost impact of the original project price. In effect, it is the mortgage price.*

EXAMPLE PROBLEM 6 (CRCF)

A scrubber system has an installed cost of \$250,000. The system is expected to last 15 years before it is replaced and the annual interest rate at the time of the project was 6.5%. Determine the CRCF and the annualized cost of the project.

$$CRCF = \frac{(0.065)(1+0.065)^{15}}{(1+0.065)^{15} - 1} = 0.106$$

Annualized Cost = CRCF (\$250,000) = \$26,500 per year for 15 years

12.4 COST COMPARISON METHODS

Cost comparison methods are required to help decide which option of competing systems, components, programs, etc. is the best to select. Obviously, the first or installed cost is of primary importance, but it could be unwise to choose a component that has a significantly shorter life, a much higher energy use or a greater operating labor requirement without having a corresponding lower first cost. It is possible to add all of these costs over the life of each option and make comparisons.

The most common way in the accounting field to compare competing options is to consider all costs over the equipment life and bring these costs back to an equivalent cost in today's dollars. This is called the present value analysis of costs. There are factors that are used to multiply future costs to obtain the present value of those costs. The single present value factor is used to take a future cost of X dollars and provide the equivalent current cost of that expenditure. The uniform present value factor takes an annual recurring cost that does not change and computes the equivalent cost of that expenditure over the years representing the system's life in today's dollars. The uniform present value factor is used on costs that are routine and will not change significantly from year to year.

These accounting methods are used to determine the "Life Cycle Cost" of a project. Life cycle cost analysis compares the cost of various alternative actions and identifies the least costly option by predicting the future costs as well as the current costs of each choice.[12.4] Considering the impact of inflation, all costs are expressed in today's dollars to achieve an understanding of the true relationship between alternative costs or the payback of an investment. The life cycle cost evaluations require a thorough understanding of the changing value of currency, and also require an estimation of the inflation influence on all components in the analysis.

Five types of economic analyses are commonly performed:

- Simple Payback Period — the total investment savings in the first year divided by the investment cost. This analysis is used to evaluate options for changing a current system.
- Abbreviated Life Cycle Cost Method — comparison of the annualized cost of competing systems.
- Life Cycle Payback — the time necessary for the alternative to payback the investment considering increases in operating and maintenance costs as well as the cost of capital.
- Internal Rate of Return — the interest being returned on the investment over the length of the study.
- Savings to Investment Ratio — the present worth of the investment's savings divided by the present worth of the investment's first cost.

All but the first two economic analyses utilize the present worth of an investment's savings and costs over a time period. These analyses are used to compare alternative investments or the value of making an investment compared to doing nothing. In doing some of these analyses, one must assume an inflation rate for the cost of money (called the discount rate).

Since most industrial ventilation systems are not an investment in the monetary sense (unless collected materials can be sold or returned to the process), the classic Life Cycle calculations are not easily applied. In most cases, the installation of the system would be selected by lowest cost alternative. The methods in this section emphasize that total annualized costs should still be considered for alternate vendor and control method analysis rather than just initial capital costs.[12.5]

12.4.1 Simple Payback Period Method. This is the easiest evaluation method for comparing changes to an existing system or systems that would replace an existing system. To perform these analyses add the annual cost and the cost savings that the alternative system would provide. This total annual savings would then be divided into the cost to make the change or install the alternative system. The result is the number of years required to pay for implementation of the change. The simple payback period analysis may not correctly lead to the best solution since the effect of inflation and cost of money are ignored.

EXAMPLE PROBLEM 7 (Simple Payback) (IP Units)

A heat recovery unit can be installed in an exhaust system at a cost of $200,000. This unit will provide an annual heating energy cost savings of $ 55,000. There also would be an associated increased maintenance cost of $ 5,000 per year. What is the simple payback of this system modification?

Annual Savings = $55,000 – $5,000 = $50,000 per year

Simple Payback Period = First Cost/Annual Savings
= $200,000/$50,000 per year = 4 years

The simple payback of this project can be compared with the payback of other projects that compete for corporate funding and decisions can be made as to which ones to implement.

12.4.2 Abbreviated Life Cycle Cost Method. Rather than considering the time value of money, a simpler method would be to keep all factors in present value dollars for comparison. This does not consider the changing of value of currency but assumes that the effects of inflation are constant for all components of the analysis. So even though energy costs may be increasing in value at a faster rate than the cost of maintenance labor, this method does not make that differentiation.

EXAMPLE PROBLEM 8 (Abbreviated Life Cycle Cost)

The scrubber system listed in Example Problem 6 is being compared to an alternate control strategy (fabric filter baghouse). The initial cost of the baghouse system is $320,000 so the scrubber system has a potential savings to the project of $70,000. The baghouse system will have a useful life of 20 years before needing replacement. The baghouse system will have an annual operating cost of $92,000 for electrical power, maintenance and all other costs. The scrubber system will require more horsepower because of higher pressure drop and annual operating cost was estimated at $105,000. Determine the best control strategy considering annualized costs.

Scrubber System

Annual Capital Cost = $26,500
(from Example Problem 6)
Annual Operating Cost = $105,000
Total Annual Cost = $131,500

Baghouse System

$$CRCF = \frac{(0.065)(1 + 0.065)^{20}}{(1 + 0.065)^{20} \quad 1} = 0.0908$$

Annualized Capital Cost = CRCF ($320,000) = $29,000 per year for 20 years

Annual Capital Cost =	$29,000
Annual Operating Cost =	$92,000
Total Annual Cost	$121,000

From this calculation, the baghouse system would have a total annual cost of about $10,500 less than the scrubber system over the life of the project. Note that this will not always favor one type of collection device over another, but does show that annual operating costs as well as other factors such as useful life of the equipment can have more impact than just initial project costs.

12.4.3 Life Cycle Payback Method. The life cycle payback method determines the time necessary to recover the investment required to install the selected system. The present value of future costs and savings are identified for each year in the future. These values are then added to the initial investment cost. The year in which savings repays the final part of the investment is noted as the time period required to achieve the payback.

To determine the future cost of an item that increases by a fixed percent, i_c, each year, multiply the cost for the previous year by $(1 + i_c)$. Then:

$$C(k) = C(k\text{-}1)\,(1+i_c) \qquad \textbf{[12.3]}$$

Where:

$C(k)$ = the cost in the year k
$C(0)$ = the initial cost
i_c = yearly increase in cost

In conducting the analysis, the cost of energy, E(k), and rates of change for energy, i_e, the cost of maintenance, M(k), and rate of change for maintenance, i_m, the cost of operating labor, L(k), and rate of change for labor, i_L, and the cost of replacement parts, R(k), and the rate of change in replacement parts, i_R, or any other recurring cost can be factored into the calculations. The discount rate or estimated inflation rate can also be input.

The present value for the cost in year k item is given by:

$$PVC(k) = C(k)/(1 + i)^k \qquad \textbf{[12.4]}$$

Where:

PVC(k) = Present value cost in year k
PVC(0) = Cost if purchased in year 0
i = Discount rate (cost of capital)

EXAMPLE PROBLEM 9 (Present Value Maintenance Costs)

If the annual maintenance for the equipment in Example Problem 7 is $ 5,000 in the first year, the expected increase in labor cost is 3% per year, and the discount rate is 3.9%, what are the maintenance costs and the present value costs for the first five years.

In year 1, maintenance costs would be $ 5,000 (1 + 0.03) = $5,150. The present value would be $5,150/(1 + 0.039) = $4,947. Table 12-3 tabulates the present values for the first five years.

The same approach can be used with each component of the total costs and variable changes in the rate of increase.

EXAMPLE PROBLEM 10 (Present Value Total Costs)

The heat recovery unit of Example Problem 7 can be installed in an exhaust system at a cost of $200,000. This unit will provide an annual heating energy cost savings of $ 55,000. There also would be an associated increased maintenance cost of $ 5,000 per year. In this analysis, no tax impact is considered. The discount rate (cost of capital) equals 3.9%. The annual increase in maintenance costs is 3% per year. The energy cost increases, which were obtained from Department of Energy (DOE) Tables,[12.4] are 2.62%, 2.88%, 2.24%, 2.37%, 1.73%, 1.92%, and 2.24%, respectively for the next seven years. The cash flow (present value cumulative) is the sum of the initial investment and the present values of the energy and maintenance savings for the preceding years cash flow.

Table 12-4 shows the calculations for the project. Since the cumulative cash flow changes from negative (loss) to positive (gain) between the fourth and fifth years, the payback period for the $200,000 investment was a little over four years.

12.4.4 Life Cycle Costing Considering Taxes. The previous analysis did not take into account the implications of taxes on the cash flow obtained from the implemented system. Taxes are paid on profits that a company makes when its income exceeds its costs. Therefore, when operating costs are reduced the savings are normally taxable income. In addition, the investment cost of the installed system can be depreciated over the system life. This depreciated amount is used to offset income that is taxable. There are different depreciation tables for installed equipment. The financial department for the specific plant should provide that information as well as the rate of taxes they pay.

TABLE 12-3. Example Problem 9, Yearly Costs and Present Value of the Costs

Year	Yearly Maintenance Costs		Present Value Maintenance Costs	
	Yearly Increase	Maintenance Costs	Discount Rate	Present Value Cost
k	i_m	$MC(k) = MC(k-1)*(1 + i_m)$	i_d	$PVC(k) = MC(k)/(1 + i_d)^k$
0		$5,000		$5,000
1	3.0%	$5,150	3.9%	$4,957
2	3.0%	$5,305	3.9%	$4,914
3	3.0%	$5,464	3.9%	$4,871
4	3.0%	$5,628	3.9%	$4,829
5	3.0%	$5,796	3.9%	$4,787

TABLE 12-4. Life Cycle Payback Analysis

Year	Discount Rate	Energy			Maintenance			Net	Cash Flow
		Yearly Increase	Yearly Costs	Present Value	Yearly Increase	Yearly Costs	Present Value	Present Value	Present Value
0			$55,000			$(5,000)			$(200,000)
1	3.9%	2.6%	$56,441	$54,322	3.0%	$(5,150)	$(4,957)	$49,366	$(150,634)
2	3.9%	2.9%	$58,067	$53,789	3.0%	$(5,305)	$(4,914)	$48,875	$(101,759)
3	3.9%	2.2%	$59,367	$52,930	3.0%	$(5,464)	$(4,871)	$48,059	$ (53,700)
4	3.9%	2.4%	$60,774	$52,150	3.0%	$(5,628)	$(4,829)	$47,321	$ (6,379)
5	3.9%	1.7%	$61,826	$51,361	3.0%	$(5,796)	$(4,787)	$46,274	$ 39,895
6	3.9%	1.9%	$63,013	$50,088	3.0%	$(5,970)	$(4,746)	$45,342	$ 85,237
7	3.9%	2.2%	$64,424	$49,288	3.0%	$(6,149)	$(4,705)	$44,583	$ 129,821

To help understand how taxes and depreciation affect the life cycle cost the heat recovery system will be evaluated again.

EXAMPLE PROBLEM 11 (Present Value with Taxes)

The heat recovery system of Example Problem 7 provides a \$55,000 energy savings in the first year of operation and there is an annual maintenance cost of \$5,000. The system has an installed cost of \$200,000, which has a \$10,000 per year depreciation if a 20 year straight-line depreciation is used. The resulting savings are taxed at a rate of 40%, which means 60% of the savings is used in the cash flow summation. What are the life cycle costs and how long is the payback?

Table 12-5 shows the analysis. From the information provided in Table 12-5, the system pays for itself during the 7th year. In this example, the discount factor is 3.9% and the energy and maintenance costs escalate at the same rate as with the other examples. The cash flow changes from negative to positive between the fifth and sixth years. The payback period is about 5.5 years.

REFERENCES

12.1 United States Environmental Protection Agency: The EPA Air Pollution Control Cost Manual, 6th edition, Publication No. EPA/452/B-02-001, USEPA, Washington, DC (January 2002).

12.2 Vatavuk, W.: Estimating Costs of Air Pollution Control (1990).

12.3 United States Environmental Protection Agency: Compilation of Air Pollution Emission Factors: AP-42, 5th Edition, Volume 1: Stationary Point and Area Sources, USEPA, Washington, DC (January 1995).

12.4 Department of Energy, Software Program, Building Life-Cycle Costs, BLCC 5.3-09, downloaded from http://www/eere.energy.gov/temp/information/cfm/register_blcc.cfm.

12.5 Mullen, M.E.: Moving Beyond Simple Payback, Addressing Clients' Financial Needs. ASHRAE Journal (June 2005).

TABLE 12-5. Example Problem 11, Life Cycle Payback Analysis with Tax Implications

		Energy			Maintenance			Depreciation	Cash Flow	Taxes	Savings	Cash Flow
Year	Discount Rate	Yearly Increase	Yearly Costs	Present Value	Yearly Increase	Yearly Costs	Present Value	Present Value	Total Cash		After Taxes	Present Value
Initial			$ 55,000			$ (5,000)						$(200,000)
1	3.9%	2.6%	$ 56,441	$51,484	3.0%	$ (5,150)	$(4,957)	$10,000	$56,528	$22,611	$33,917	$(166,083)
2	3.9%	2.9%	$ 58,067	$53,153	3.0%	$ (5,305)	$(4,914)	$10,000	$58,239	$23,296	$34,943	$(131,140)
3	3.9%	2.2%	$ 59,367	$54,496	3.0%	$ (5,464)	$(4,871)	$10,000	$59,625	$23,850	$35,775	$ (95,365)
4	3.9%	2.4%	$ 60,774	$55,945	3.0%	$ (5,628)	$(4,829)	$10,000	$61,116	$24,446	$36,670	$ (58,695)
5	3.9%	1.7%	$ 61,826	$57,038	3.0%	$ (5,796)	$(4,787)	$10,000	$62,251	$24,901	$37,351	$ (21,345)
6	3.9%	1.9%	$ 63,013	$58,267	3.0%	$ (5,970)	$(4,746)	$10,000	$63,521	$25,408	$38,113	$ 16,768
7	3.9%	2.2%	$ 64,424	$59,720	3.0%	$ (6,149)	$(4,705)	$10,000	$65,015	$26,006	$39,009	$ 55,777
8	3.9%	2.6%	$ 66,099	$61,435	3.0%	$ (6,334)	$(4,664)	$10,000	$66,771	$26,709	$40,063	$ 95,840

Chapter 13

SPECIFIC OPERATIONS

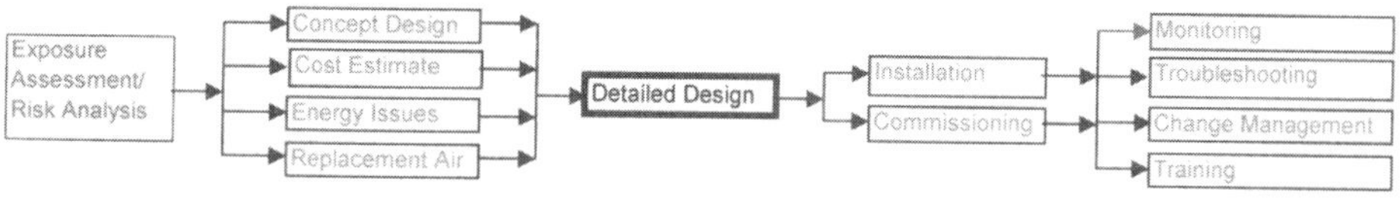

NOTE: Equations with notation followed by (IP) are designated for inch-pound system only; Equations followed by (SI) are designated for metric use only. If Equation bears neither, then it applies to both systems.

The following illustrations of hoods for specific operations are intended as guides for design purposes and apply to usual or typical operations. In most cases, they are taken from designs used in actual installations of successful local exhaust ventilation systems. Technology and shop equipment change over the years. Some drawings in Chapter 13 show ventilation solutions for older types of shop equipment. Manufacturers, designers, and end-users are encouraged to submit improved designs to the Industrial Ventilation Committee for inclusion in future editions of the Manual. All conditions of operation cannot be categorized, and because of special conditions (i.e., cross-drafts, motion, differences in temperature, or use of other means of contaminant suppression), modifications may be in order. Using principles discussed in earlier chapters, Manual users are encouraged to cautiously adapt the existing drawings to their specific needs. For instance, there are presently no drawings for fiberglass lay-up and spray-up operations. To collect the solvent vapors, the Manual user could adapt a paint booth for these operations. The designer should also recommend Low Volume-High Velocity (vacuum) tools and a separate collection system if workers also perform grinding in the same shop.

The flow rates specified in the various VS-prints are to be considered actual flow rates (acfm) at the local conditions existing at the places used.

Unless it is specifically stated, the design data are not to be applied indiscriminately to materials of high toxicity, e.g., beryllium and radioactive materials. Thus the designer may require higher or lower air flow rates or other modifications because of the peculiarities or the process in order to adequately control the air contaminant.

13.05 BATTERY CHARGING

Operations using rechargeable batteries are increasing in industry, commercially, and even in the home, as regulatory agencies encourage the move away from petroleum fuels. Traditional battery operated vehicles, such as warehouse forklifts and golf carts, are now joined by battery charging facilities in automotive, bus, and light vehicle fleet garages. The telecommunications industry, ships, and submarines also rely on rechargeable batteries. Operations requiring an uninterrupted power source (UPS) sometimes use rechargeable batteries as a back-up power source. For further information on recovery and recycling, see Reference 13.05.1.

13.05.1 End-user Operations. End-user battery maintenance operations are usually limited to inspections, electrolyte level adjustment, adding water, recharging, and battery replacement. The most commonly found rechargeable batteries are lead-acid and nickel-cadmium. The primary hazard is the build-up of hydrogen that could lead to an explosion if there is sufficient quantity and an ignition source. Charging lead-acid batteries also can produce sulfuric acid mists. Charging nickel-cadmium batteries can produce alkaline mists. Arsine and stilbine have been reported in poorly ventilated battery charging rooms. If the end-user makes frequent battery repairs (e.g., lead post maintenance and lead plate work), consider modifying a welding hood shown in VS-90-01 or VS-90-02.

Battery out-gassing occurs by inducing charging currents in excess of those needed to charge the cells, thereby converting the water to oxygen and hydrogen which are released into the room. When charging a large quantity of batteries in the same room, hydrogen gas can build up to the LEL. The LEL of hydrogen is 4.1%. In some cases, regulatory requirements, local practice or consensus standards require a lower safety factor. Typical safety factors are 25, 10, or 1% of the LEL.

Many modern rechargeable batteries are valve-regulated or sealed and they do not release gasses. However, they have a pressure release valve and the batteries can leak hydrogen if they are overcharged or charged in a warm (> 80 F [27 C]) environment. Many newer batteries are able to absorb the hydrogen generated internally. When charging older-style batteries, more hydrogen is generated as batteries age. Contact the manufacturer to determine the proper ventilation rates for older batteries.

13.05.2 Battery Charging Ventilation. Some code organizations require mechanically supplied replacement air. The user should move the battery to a ventilated table when performing operations such as electrolyte addition and post welding.

Dilution Ventilation: Typical electric and fire codes permit the use of the building's ventilation without local exhaust ventilation for the following:

- The engineered ventilation system (HVAC) approved by local fire officials that is intended to maintain a gas-air mixture of less than 25% (or 10% in certain cases) of the lower flammable limit.
- The charger system is approved for indoor charging of batteries by a testing or consensus organization such as Factory Mutual or the National Electric Code. These are normally a charger/battery combination that does not emit hydrogen (e.g., pagers, cell phones, etc.)

TABLE 13-05-1. Minimum Dilution Ventilation Requirements per Electric Vehicle [(13.05.2)]

Branch Circuit Ampere Rating	Minimum Ventilation Required in Actual Cubic Feet per Minute (acfm) [am³/s] for Each of the Total Number of Electric Vehicles that Can Be Charged at One Time — Branch Circuit Voltage						
	Single Phase			Three Phase			
	120V	208V	240V or 120/240V	208V or 208Y/120V	240V	480V or 480Y/277V	600V or 600Y/347V
15	37 [0.02]	64 [0.03]	74 [0.04]				
20	49 [0.03]	85 [0.04]	99 [0.05]	148 [0.05]	117 [0.06]	342 [0.17]	427 [0.21]
30	74 [0.04]	128 [0.06]	148 [0.07]	222 [0.11]	256 [0.13]	512 [0.27]	641 [0.32]
40	99 [0.05]	171 [0.09]	197 [0.10]	296 [0.15]	342 [0.17]	683 [0.34]	854 [0.43]
50	123 [0.06]	214 [0.11]	246 [0.12]	370 [0.19]	427 [0.21]	854 [0.43]	1066 [0.53]
60	148 [0.07]	256 [0.13]	296 [0.15]	444 [0.22]	512 [0.26]	1025 [0.51]	1281 [0.64]
100	246 [0.12]	427 [0.21]	493 [0.25]	740 [0.37]	854 [0.43]	1708 [0.85]	2165 [1.08]
150				1110 [0.56]	1281 [0.64]	2562 [1.28]	3203 [1.60]
200				1480 [0.74]	1708 [0.85]	3416 [1.71]	4270 [2.36]
250				1850 [0.93]	2135 [1.07]	4270 [2.36]	5338 [2.67]
300				2221 [1.11]	2562 [1.28]	5125 [2.56]	6406 [3.20]
350				2591 [1.30]	2989 [1.49]	5979 [2.99]	7473 [3.74]
400				2961 [1.48]	3416 [1.71]	6832 [3.42]	8541 [4.27]

- Open garages, carports, and other structures with two or more sides open.

Small Battery Charging Room or Small Battery Charging Areas in Large Rooms: To determine the amount of ventilation required for compliance with the health and safety requirements, follow the battery manufacturer's recommendation and the local building codes. The National Electric Code also cites the criteria for minimum mechanical ventilation in cubic feet per minute required for each parking space that is equipped to charge an electric vehicle.(13.05.3)

For voltages and currents not shown on Table 13-05-1, use the following formula:

$$Q_{\text{single phase}} = \frac{\text{(volts)(amperes)}}{48.7} \text{ acfm} \qquad \textbf{[13.05.1] IP}$$

$$[Q_{\text{single phase}} = \frac{\text{(volts)(amperes)}}{97400} \text{ am}^3\text{/s]} \qquad \textbf{[13.05.1] SI}$$

$$Q_{\text{three phase}} = \frac{1.732\text{(volts)(amperes)}}{48.7} \text{ acfm} \qquad \textbf{[13.05.2] IP}$$

$$[Q_{\text{three phase}} = \frac{(1.732)\text{(volts)(amperes)}}{97400} \text{ am}^3\text{/s]} \qquad \textbf{[13.05.2] SI}$$

The above calculated ventilation rate ($Q_{calculated}$) assumes efficient mixing of the air inside the battery charging facility. In most cases, this does not occur and a safety factor (m_i), to allow for incomplete mixing, must be used to determine the actual ventilation required. Values for m_i can be found in Chapter 4.

A properly designed industrial ventilation system will assist in completely mixing the air inside the battery charging facility, lowering the m_i value, and lowering the actual ventilation rate. Industrial ventilation design considerations to include in the battery charging facility are:

- High level and low level dilution exhaust. High level exhaust should ventilate all roof pockets. Low level exhaust should be a maximum of 12 inches [0.3 m] above the floor.
- Exhaust all air directly outdoors. Consider the required air pollution permits.
- The supply air rate should be approximately 95% of the exhaust ventilation rate to maintain a slightly negative room static pressure (relative to outside) to prevent fumes and gases from migrating outside the battery charging facility.

Passive Exhaust Systems: In temperate climates, some facilities employ passive ventilation systems where the stack effect of varying densities and rising heat carries the hydrogen to the uppermost point in the facility. Providing a supply air opening low in the building and a covered opening in the roof peak/opening usually provides sufficient natural ventilation. Since these systems depend on natural ventilation, it is impossible to predict a constant amount of airflow through the building. Passive systems generally are not recommended and some local code organizations prohibit them.

13.05.3 End-user Electrolyte Maintenance Hood. Electrolyte changing is usually a short term operation. Industrial ventilation is not generally required for this operation. However, personal protective equipment (i.e., apron, gloves, safety glasses, and shield) is imperative. A small laboratory hood (VS-35-01 or VS-35-02) can be adapted when the operator mixes and dispenses large quantities of electrolyte.

13.05.4 Supply Air Systems.

Small Dilution Ventilation Type Systems: The designer should evaluate the local supply air system balance in small battery recharging operations, such as forklift and electric cart recharging areas, to ensure that the system provides a sufficient air volume. Check the ventilation control cycle to ensure that at least a small amount of air is available even when the plant ventilation system is off such as night and weekend setback conditions. When only a few batteries are charged simultaneously, a dedicated supply air system may not be required.

Larger Operations with Dedicated Exhaust Hoods: Depending on the operation, supply air can be provided by a perforated plenum covering the whole room or through a low hanging diffuser providing laminar-like flow. The key point in supplying the replacement air is that air/contaminant mixing does not occur in the worker's breathing zone or the contaminant does not bounce back into the worker's breathing zone.

Do not recirculate air from a ventilation system installed to remove hydrogen.

REFERENCES

13.05.1 National Institute for Occupational Safety and Health: NIOSH Health Hazard Evaluation Report HETA 95-0097-2661 (October, 1997).

13.05.2 California Building Code, Section 1202.2.2.2.1; California Building Officials, 2215 21st Street, Sacramento, CA 95818.

13.05.3 National Fire Protection Association, National Electric Code Handbook, Section 625 (2005).

13.10 CLEANROOMS

The early cleanrooms were mostly built for government use and, thus, cleanliness testing methods were established by the government standard U.S. Federal Standard 209, which became widely accepted. In 2001, it was withdrawn and now International Organization for Standardization (ISO) standards are used. ISO 14644-1[13.10.1] is the new standard used to establish the class of air cleanliness for airborne particulates levels in cleanrooms and clean spaces. ISO 14644-1 has been adopted by the Institute of Environmental Sciences and Technology (IEST) and is used in their testing procedures. Refer to Table 13-10-1 for a comparison of cleanliness levels between U. S. Federal Standard 209 and ISO 14644-1. Table 13-10-1 illustrates various classes of cleanrooms that are categorized by the allowable number of particles equal to or greater than 0.5 micrometers in size per volume of air.

ISO 14644-1 does not address the physical, chemical, radiological, or viable nature of the airborne contaminants. It also does not address the occupational health concerns of employees working in cleanroom environments.

In order to meet the class limits, a High Efficiency Particulate Air (HEPA) or Ultra Low Penetration Air (ULPA) filter is required. A HEPA filter is a disposable, extended-media, dry-type filter in a rigid frame with a minimum particle collecting efficiency of 99.97% for 0.3 micrometer, thermally generated dioctyl phthalate (DOP), or specified alternate, aerosol particles at a resistance of 0.5 to 2.0 "wg [125 to 500 Pa] with a velocity of 250 fpm [1.25 m/s]. An ULPA filter is a disposable, extended-media, dry-type filter in a rigid frame with a minimum particle collection efficiency of 99.999% for particulate diameters between 0.1 and 0.2 micrometers in size.

Military specifications[13.10.2] and publications[13.10.3, 13.10.4] by the Institute of Environmental Sciences and Technology (IEST) define HEPA and ULPA filter construction and methods of testing. Filters with an efficiency even higher than an ULPA filter are available from some companies specializing in cleanrooms and air filtration.

The primary design considerations for cleanrooms are the supply airflow rate, the airflow patterns within the cleanroom, the method for recirculating the air from the cleanroom, and the filter efficiency.

Air is supplied to the cleanroom by an air handling system containing the components needed for heating, cooling, and humidity control. Noise is readily transmitted to the cleanroom so very low fan speeds, vibration isolation, and noise control devices are important design considerations. The air circulation system will also contain two or three stages of pre-filtration. This allows the final filters in the cleanroom ceiling to remain in place for very long periods of time. A final filter life of 10 years or more is typical for Class 100 and better cleanrooms.

Air from the supply system enters the cleanroom through either a ducted module or a pressurized plenum. VS-10-01 shows a ducted module arrangement. Ducted modules containing HEPA or ULPA filters are connected to the main air supply duct by flexible branch ducts. The modules usually contain an internal baffle for balancing the air exhaust that must be at a uniform velocity across the face of the filter. The ducted modules are mounted in a T-bar grid and sealed with either solid gaskets or liquid gel sealant. The ducted modules are usually considered to be throwaway items because of long filter life; however, some arrangements do permit the replacement of filters from within the cleanroom. A ducted module system offers maximum flexibility for cleanroom modification.

VS-10-02 shows a pressurized plenum arrangement. A heavy duty grid system is suspended from the ceiling with suspension rods and the HEPA or ULPA filters sealed in the grid with liquid gel or solid gaskets. The entire plenum is pressurized by the air supply system to allow a uniform flow of air through the filters to the cleanroom below. A pressurized plenum system will usually cost less than a ducted module system for large cleanrooms.

VS-10-03 shows raised floor and low sidewall arrangements. Air is returned through a utility chase to the cleanroom supply air system. To provide better particulate control, the raised floor arrangement is preferred. The low sidewall return should not be used for vertical downflow cleanrooms more than 14 feet [4.2 m] wide in order not to disrupt laminar flow at the work area.

TABLE 13-10-1. Maximum Number of Particles Allowed in Subject Volume by Particle Size for Class of Cleanroom

		0.1 um		0.2 um		0.3 um		0.5 um		1.0 um		5.0 um	
FS 209 Cleanroom Class	ISO 14644 Cleanroom Class	FS 209 Subject Volume, Cu. Ft.	ISO 14644 Subject Volume, Cu. Meter	FS 209 Subject Volume, Cu. Ft.	ISO 14644 Subject Volume, Cu. Meter	FS 209 Subject Volume, Cu. Ft.	ISO 14644 Subject Volume, Cu. Meter	FS 209 Subject Volume, Cu. Ft.	ISO 14644 Subject Volume, Cu. Meter	FS 209 Subject Volume, Cu. Ft.	ISO 14644 Subject Volume, Cu. Meter	FS 209 Subject Volume, Cu. Ft.	ISO 14644 Subject Volume, Cu. Meter
	1		10										
	2		100		24		10		4				
1	3	35	1,000	7.5	237	3	102	1	35		8		
10	4	350	10,000	75	2,370	30	1,020	10	352		83		
100	5		100,000		23,700		10,200		3,520		832		29
1,000	6		1,000,000		237,000		102,000		35,200		8320	7	293
10,000	7								352,000		83,200	65	2,930
100,000	8								3,520,000		832,000	700	29,300
	9								35,200,000		8,320,000		293,000

Values are number of particles equal to or larger than the size identified.

REFERENCES:

13.10.1 ISO. 1999, Cleanrooms and associated controlled environments, part 1: Classification of air cleanliness. Standard 14644-1. International Organization for Standardization, Geneva, Switzerland.

13.10.2 MIL-F-51068(EA), Specification Filters, Particulate, High-Efficiency, Fire Resistant, Biological Use, General Specifications For, Commander, U.S. Army Armament Research and Development Command, ATTN: DRDAR-TSC-S, Aberdeen Proving Ground, MD (October 4, 1982).

13.10.3 IES RR-CC 001.3 HEPA and ULPA Filters, Institute of Environmental Sciences and Technology, Mount Prospect, IL.

13.10.4 IES RR-CC 00.7.1 Testing ULPA Filters, Institute of Environmental Sciences and Technology, Mount Prospect, IL.

FIGURE VS-10-01

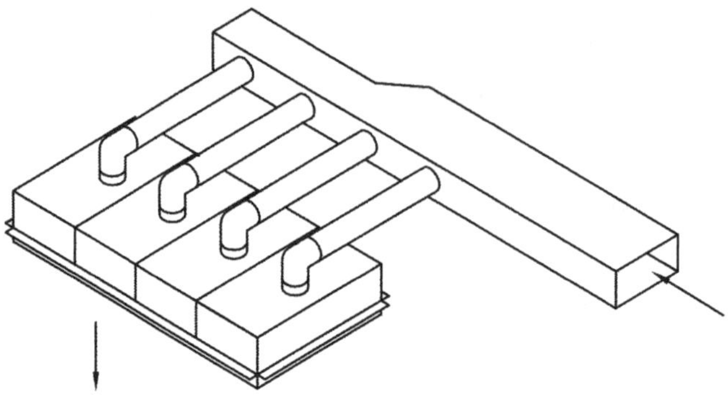

Air is supplied to the ducted modules from the air distribution duct through branch ducts, which are secured at the bottom ends by clamps. A damper (not shown) on the inside of the collar allows balancing of the air flowing from the module to the cleanroom.

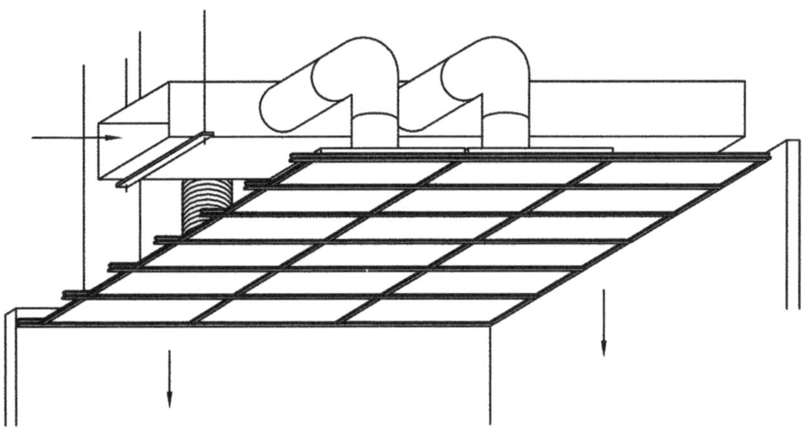

Ducted modules are usually mounted in grids and sealed with gaskets or a liquid gel sealant. Tear drop or recessed lighting provides illumination.

ACGIH®	TITLE: CLEANROOM DUCTED MODULE	FIGURE: VS-10-01 DATE: 1-12

CHECK CODES, REGULATIONS, AND LAWS (LOCAL, STATE, AND NATIONAL) TO ENSURE THAT DESIGN IS COMPLIANT.

FIGURE VS-10-02

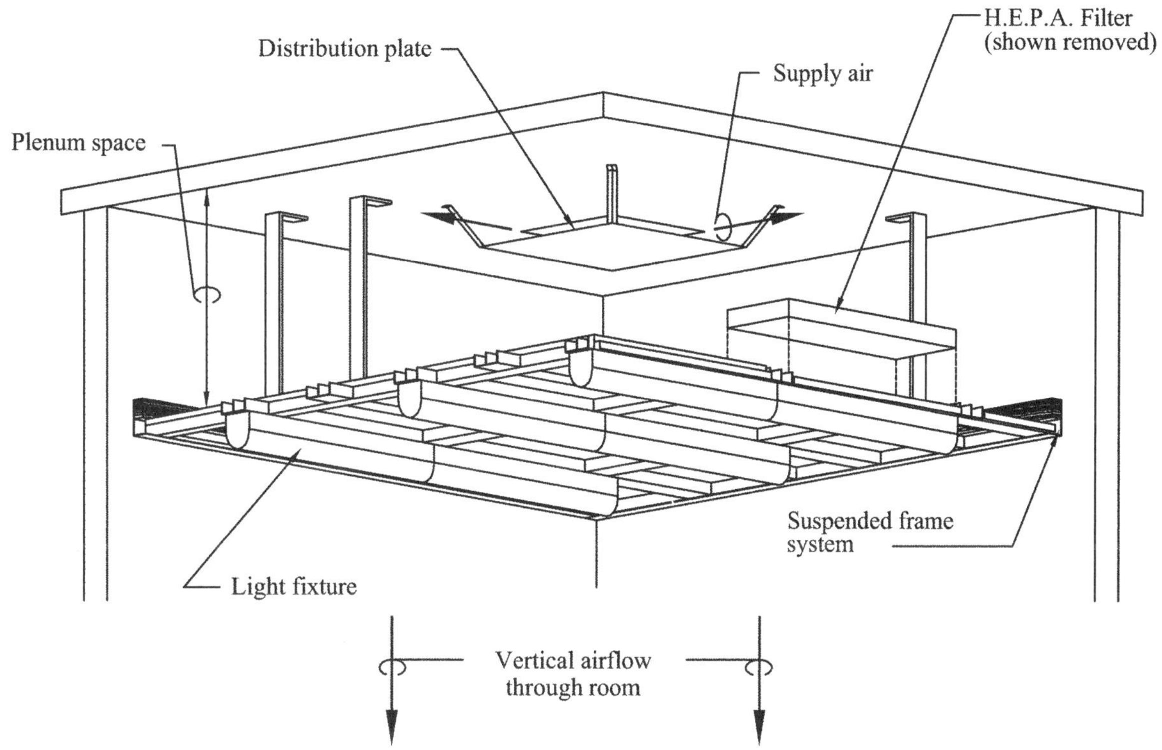

Supply air enters a pressurized plenum and strikes a distribution plate. Each opening in the support structure contains a HEPA filter. Filters are sealed around the perimeter with gaskets or a fluidic sealant. The framing structure is supported from the plenum ceiling by suspension rods.

ACGIH®

TITLE: CLEANROOM PRESSURIZED PLENUM

FIGURE: VS-10-02

DATE: 1-12

CHECK CODES, REGULATIONS, AND LAWS (LOCAL, STATE, AND NATIONAL) TO ENSURE THAT DESIGN IS COMPLIANT.

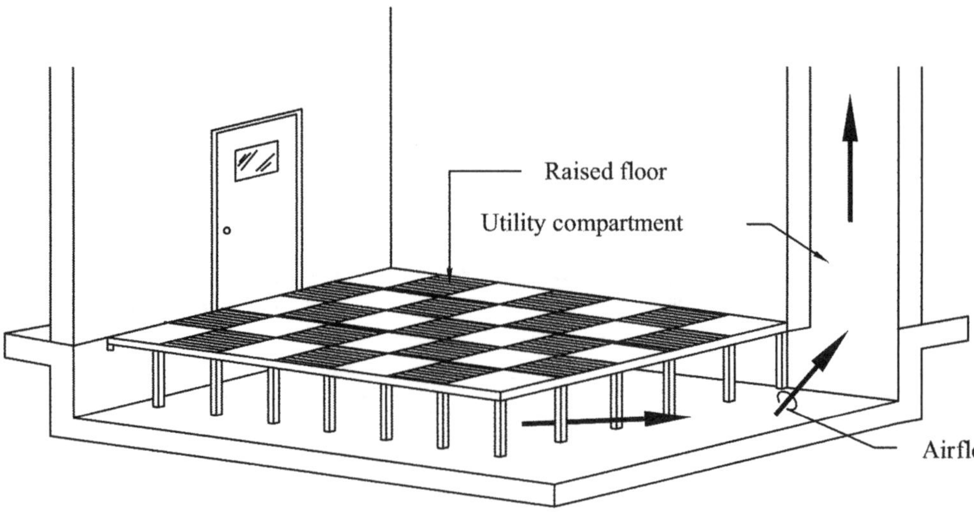

Raised floor with depressed slab. Air is returned through a utility compartment to the air supply system.

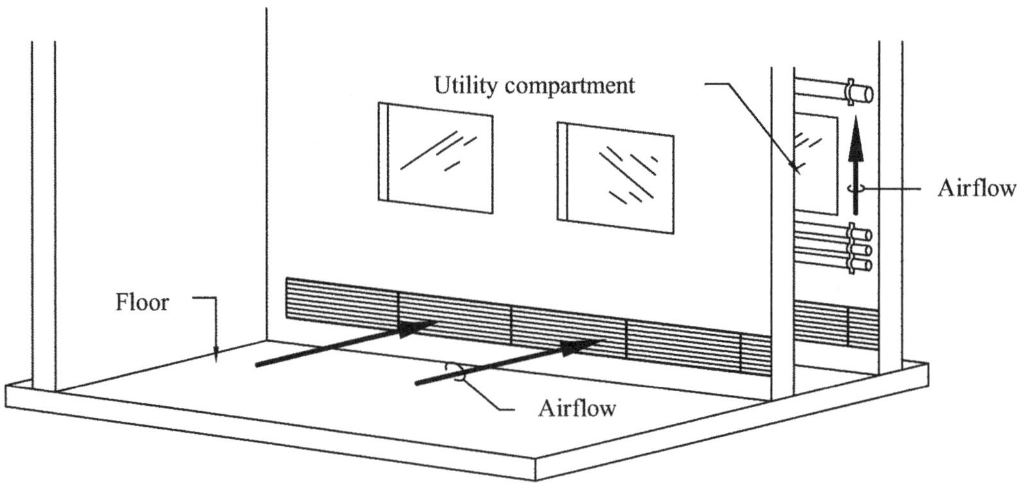

Low sidewall grille return through a utility compartment to the air supply system. Room width is limited to 14 feet [4.2 meters] if laminar airflow is to be achieved. The distance from the top of the grille to the floor should not exceed 18 inches [450 millimeters].

FIGURE VS-10-03

ACGIH®	TITLE CLEANROOM RETURN AIR ARRANGEMENTS	FIGURE VS-10-03
		DATE 1-12

CHECK CODES, REGULATIONS, AND LAWS (LOCAL, STATE, AND NATIONAL) TO ENSURE THAT DESIGN IS COMPLIANT.

13.15 FILLING OPERATIONS

Filling operations have special considerations that should be addressed when designing hoods. An enclosed space is not empty but rather is filled with air. When material enters the space, it forces the air out which in turn can carry some of the material with it. Also, additional air can be entrained by the material stream entering the enclosed space. This effect is a function of the size of the particles and the distance the material must fall. These two effects must be considered when designing hoods for material handling situations.

If there are any openings in the walls of the container which is being filled, some "splashing" of the material can occur. This can lead to loss of material through cracks and openings in the receiving vessel. The design of the ventilation system should take this effect into account.

The proper choice of exhaust flow rate is critical. If too little air is exhausted, the air displaced by the falling material may exceed the exhaust rate and the contaminant may not be adequately controlled. If too much air is exhausted, excess material could be entrained into the exhaust air stream. As this material often is the product, excess product loss could occur.

VS-15-01 illustrates four different ways of controlling barrel or drum filling operations. VS-15-02 illustrates bag filling and weighing. VS-15-03 depicts a bag tube packer. VS-15-10 and VS-15-11 depict a weighing hood where dry materials are removed from a bulk pack and weighed into smaller bags.[13.15.3]

These smaller bags are then packed into a container. Bags containing toxic materials can be opened within an enclosing hood such as shown on VS-15-20.[13.15.4]

VS-15-30 shows how to extract a toxic liquid from a process line or vessel for analysis and VS-15-21 shows a design for control of leaks around rotating shafts that enter containers.[13.15.5]

REFERENCES

13.15.1 Hama, G.M.: Ventilation Control of Dust from Bagging Operations, Heating and Ventilating, p. 91 (April, 1948).

13.15.2 Cooper, T.C.: Control Technology for a Dry Chemical Bagging and Filling Operation. Monsanto Agricultural Products Co., Cincinnati, OH (1983).

13.15.3 Gressel, M.G.; Fischback, T.J.: Workstation Design Improvements for the Reduction of Dust Exposures During Weighing of Chemical Powders. Applied Industrial Hygiene, 4:227-233 (1989).

13.15.4 Goldfield, J.; Brandt, F.E.: Dust Control Techniques in the Asbestos Industry. A paper presented at the American Industrial Hygiene Conference, Miami Beach, FL (May 12-17, 1974).

13.15.5 Langner, R.R.: How to Control Carcinogens in Chemical Production. Occupational Health and Safety (March-April 1977).

FIGURE VS-15-01

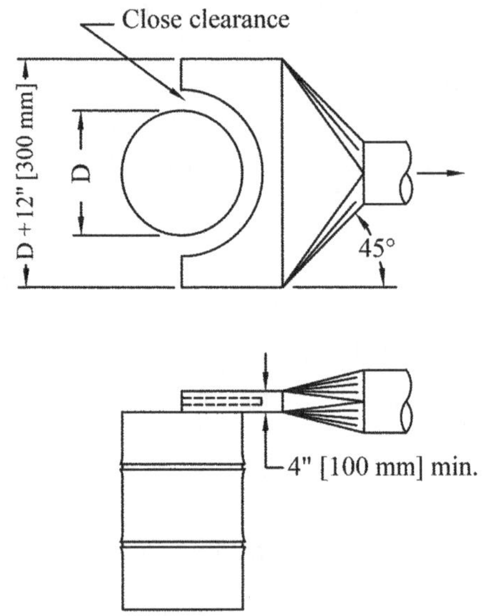

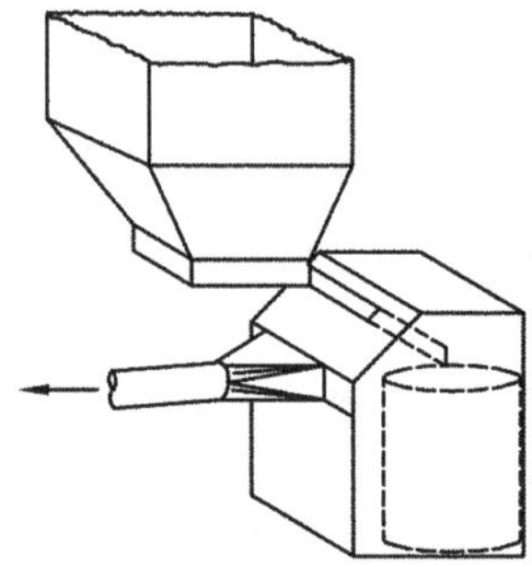

Q = 100 acfm/ft^2 [0.50 am^3/s/m^2] barrel top (minimum)
Minimum duct velocity = 3500 fpm [17.50 m/s]
h_e = 1.78 VP_s + 0.25 VP_d

Q = 150 acfm/ft^2 [0.75 am^3/s/m^2] of open face area
Minimum duct velocity = 3500 fpm [17.50 m/s]
h_e = 0.25 VP_d (45° taper)

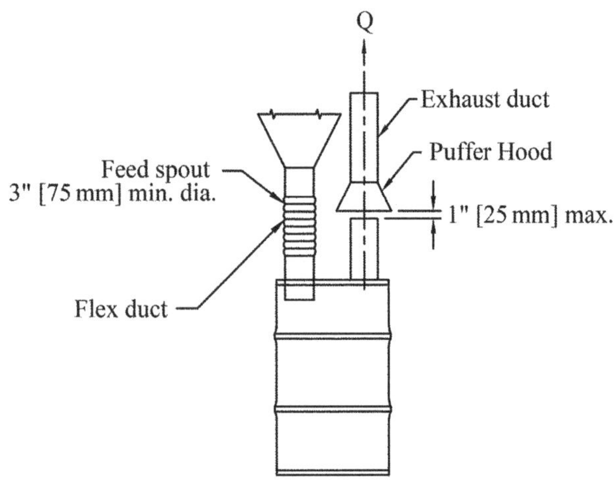

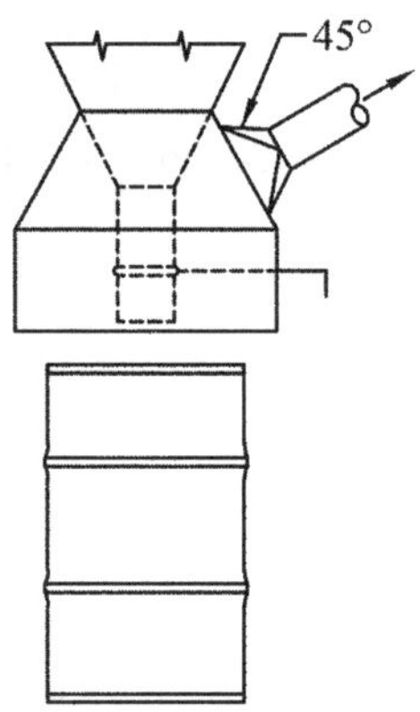

Q = 50 acfm [0.03 m^3/s] × drum diam. (ft) [m]
Minimum duct velocity = 3500 fpm [17.50 m/s]
h_e = 0.25 VP_d

Q = 300–400 acfm [0.15–0.20 m^3/s]
Minimum duct velocity = 3500 fpm [17.50 m/s]
h_e = 0.25 VP_d

Note 1: Air displaced by material feed rate may require higher exhaust flow rates.

Note 2: Excessive airflow can cause loss of product.

Note 3: When transferring flammable or combustible liquids, bonding and grounding requirements of NFPA Code 77 should be followed.

Reference: 13.15.4

ACGIH®	TITLE BARREL FILLING	FIGURE VS-15-01 DATE 1-12

CHECK CODES, REGULATIONS, AND LAWS (LOCAL, STATE, AND NATIONAL) TO ENSURE THAT DESIGN IS COMPLIANT.

FIGURE VS-15-02

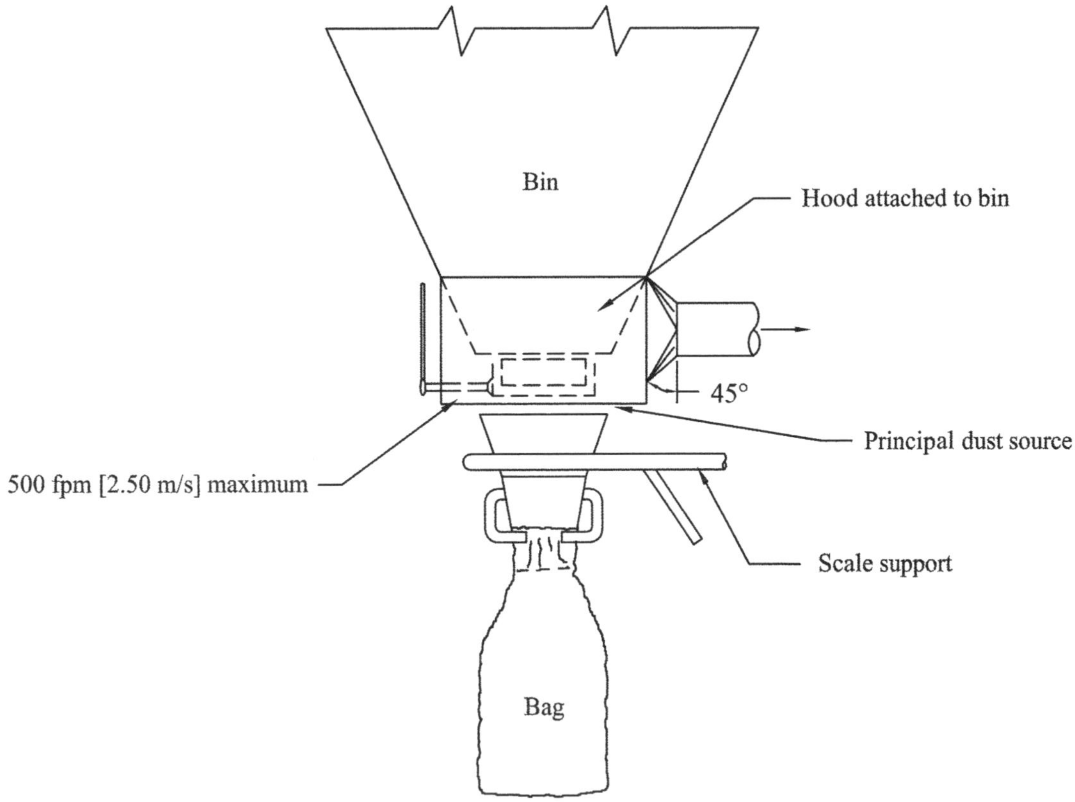

Q = 400–500 acfm [0.20–0.25 am^3/s] – non-toxic dust
1000–1500 acfm [0.50–0.75 am^3/s]–toxic dust

Minimum duct velocity = 3500 fpm [17.50 m/s]
$h_e = 0.25\ VP_d$

Note: Care must be taken so that too much air is not used, as valuable product will be pulled into the exhaust system.

Reference: 13.15.2

ACGIH®	TITLE BAG FILLING	FIGURE VS-15-02 DATE 1-12

CHECK CODES, REGULATIONS, AND LAWS (LOCAL, STATE, AND NATIONAL) TO ENSURE THAT DESIGN IS COMPLIANT.

FIGURE VS-15-03

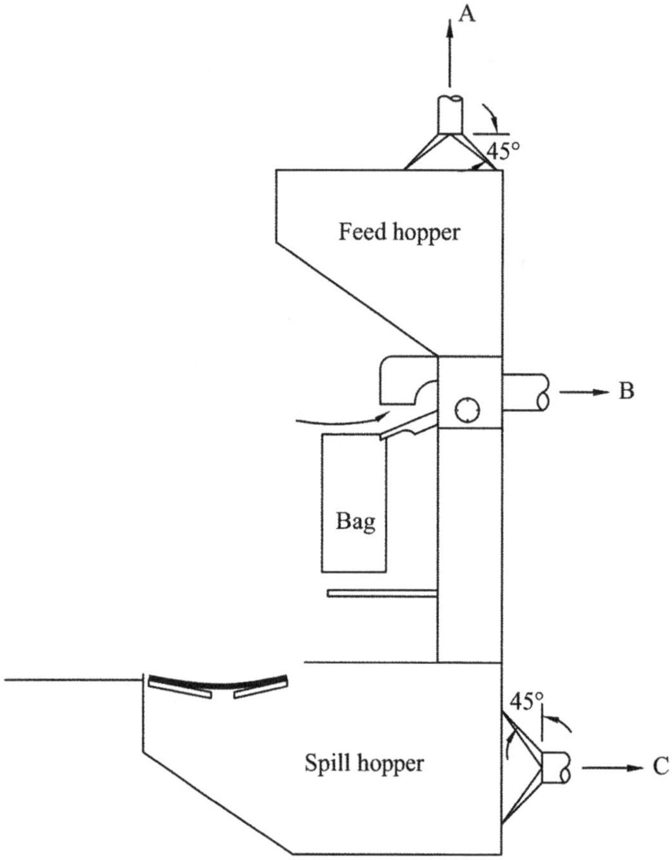

Q = 500 acfm [0.25 am^3/s] per Feed hopper (A)
= 500 acfm [0.25 am^3/s] at filling tube (B)
= 950 acfm [0.48 am^3/s] at Spill hopper (C)

Minimum duct velocity = 3500 fpm [17.50 m/s]

$h_e = 0.25\ VP_d$ for take-off at A and C
$1.0\ VP_d$ for take-off at B

Reference: 13.15.5

ACGIH®	TITLE BAG TUBE PACKER	FIGURE VS-15-03 DATE 1-12

CHECK CODES, REGULATIONS, AND LAWS (LOCAL, STATE, AND NATIONAL)
TO ENSURE THAT DESIGN IS COMPLIANT.

FIGURE VS-15-10

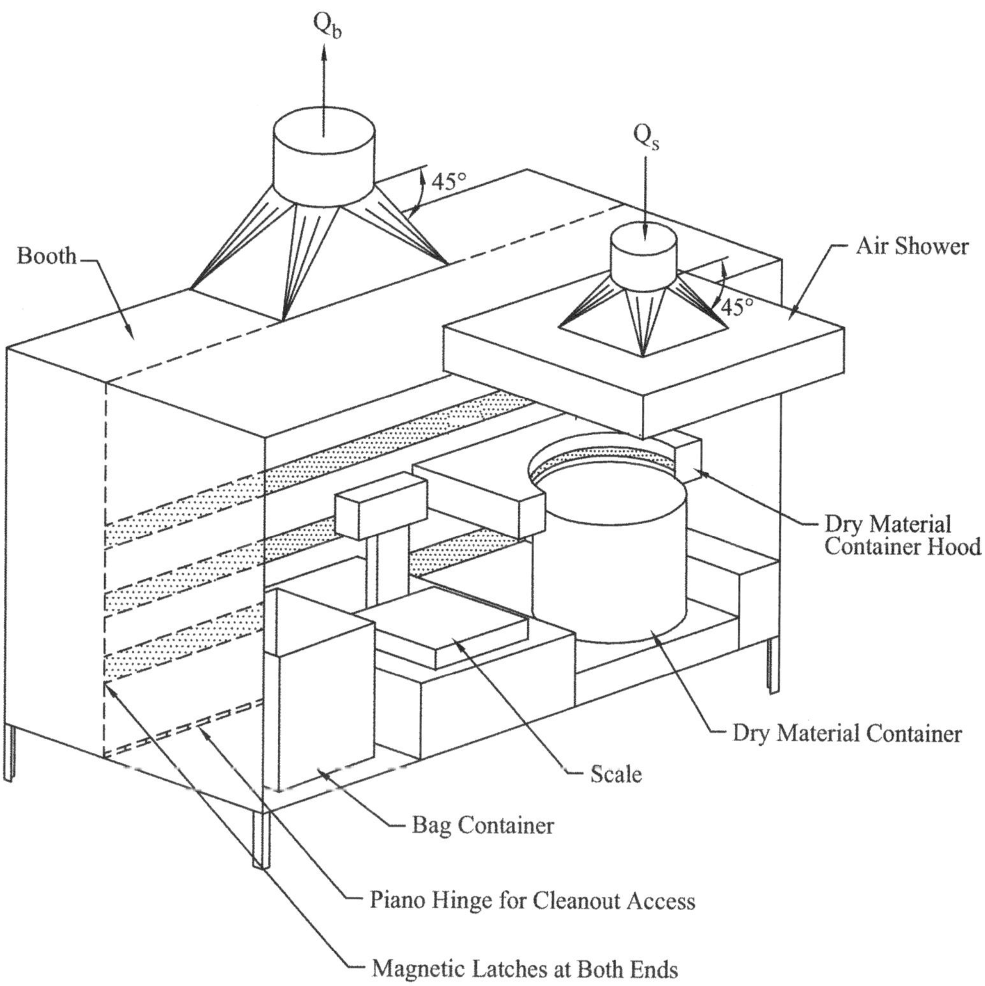

NOTE: See VS-15-11 for design details
See Chapter 6 for further details on hood plenum cleaning

Reference 13.15.1

ACGIH®

TITLE
WEIGH HOOD ASSEMBLY
DRY MATERIAL

FIGURE VS-15-10

DATE 1-12

CHECK CODES, REGULATIONS, AND LAWS (LOCAL, STATE, AND NATIONAL) TO ENSURE THAT DESIGN IS COMPLIANT.

FIGURE VS-15-11

BOOTH

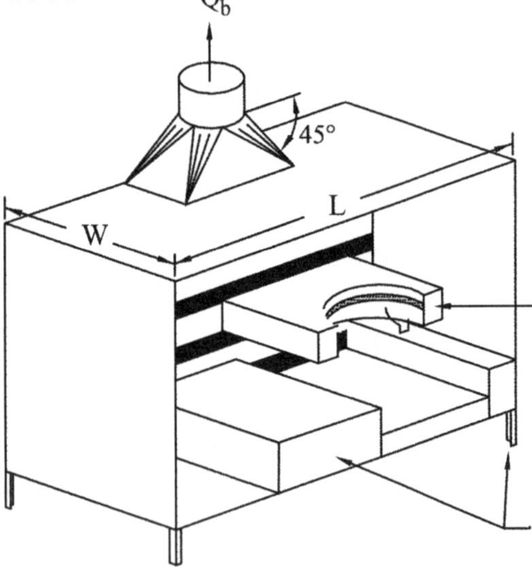

Q_b = 50 acfm/ft^2 [0.25 am^3/s/m^2] of face open area.

L and W to fit operation

Minimum duct velocity = 3500 fpm [17.50 m/s]

$h_e = 1.78\ VP_s + 0.25\ VP_d$

Dry material container hood is extension of booth slot

Configure to fit equipment

AIR SHOWER

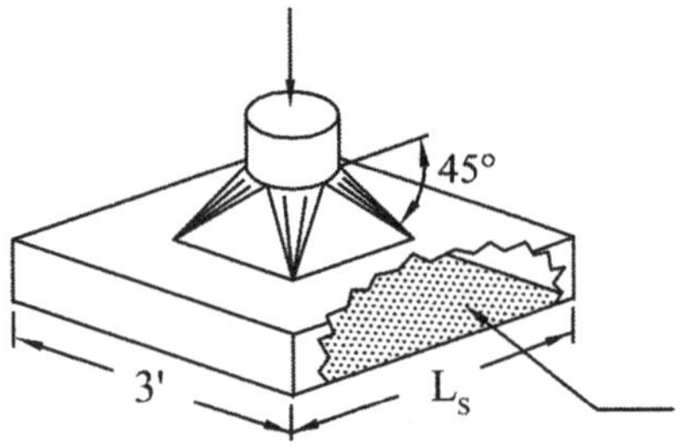

$Q_s = 100\ L_s$ acfm [$Q_s = 0.05\ L_s$ m^3/s]

L_s = 3 feet [0.9 m]. (Can be longer if required to fit workstation but do not exceed 1/2 booth length, L)

0.25" [6 mm] pegboard or equivalent, 20 percent maximum open area.

DRY MATERIAL CONTAINER HOOD

Hood is extension of booth slot. An additional takeoff(s) may be used if required for hood airflow distribution.

Airflow and hood slot design per VS-15-01

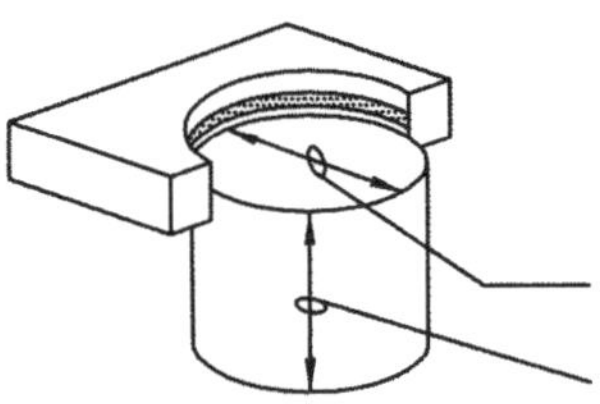

12" to 24" [300 mm to 600 mm] Diameter

24" [600 mm] Maximum

Reference 13.15.1

ACGIH®	TITLE WEIGH HOOD DETAILS DRY MATERIAL	FIGURE VS-15-11 DATE 1-12

CHECK CODES, REGULATIONS, AND LAWS (LOCAL, STATE, AND NATIONAL) TO ENSURE THAT DESIGN IS COMPLIANT.

FIGURE VS-15-20

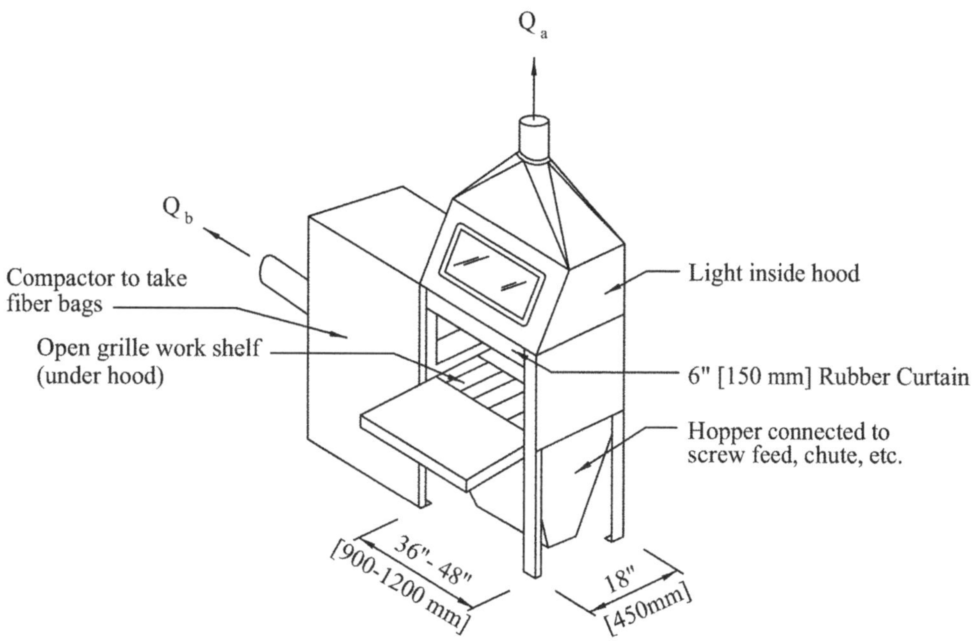

Q_a = minimum 250 acfm/ft^2 [1.25 am^3/s/m^2] of hood face open area
Minimum duct velocity = 3500 fpm [17.50 m/s]
$h_e = 0.25\ VP_d$

Q_b = minimum 250 acfm/ft [1.25 am^3/s/m^2] of compactor open area
Minimum duct velocity = 3500 fpm [17.50 m/s]
$h_e = 0.25\ VP_d$ (tapered takeoff, $\alpha = 45^\circ$)

Reference: 13.15.2

ACGIH®	TITLE TOXIC MATERIAL BAG OPENING	FIGURE VS-15-20
		DATE 1-12

CHECK CODES, REGULATIONS, AND LAWS (LOCAL, STATE, AND NATIONAL) TO ENSURE THAT DESIGN IS COMPLIANT.

FIGURE VS-15-21

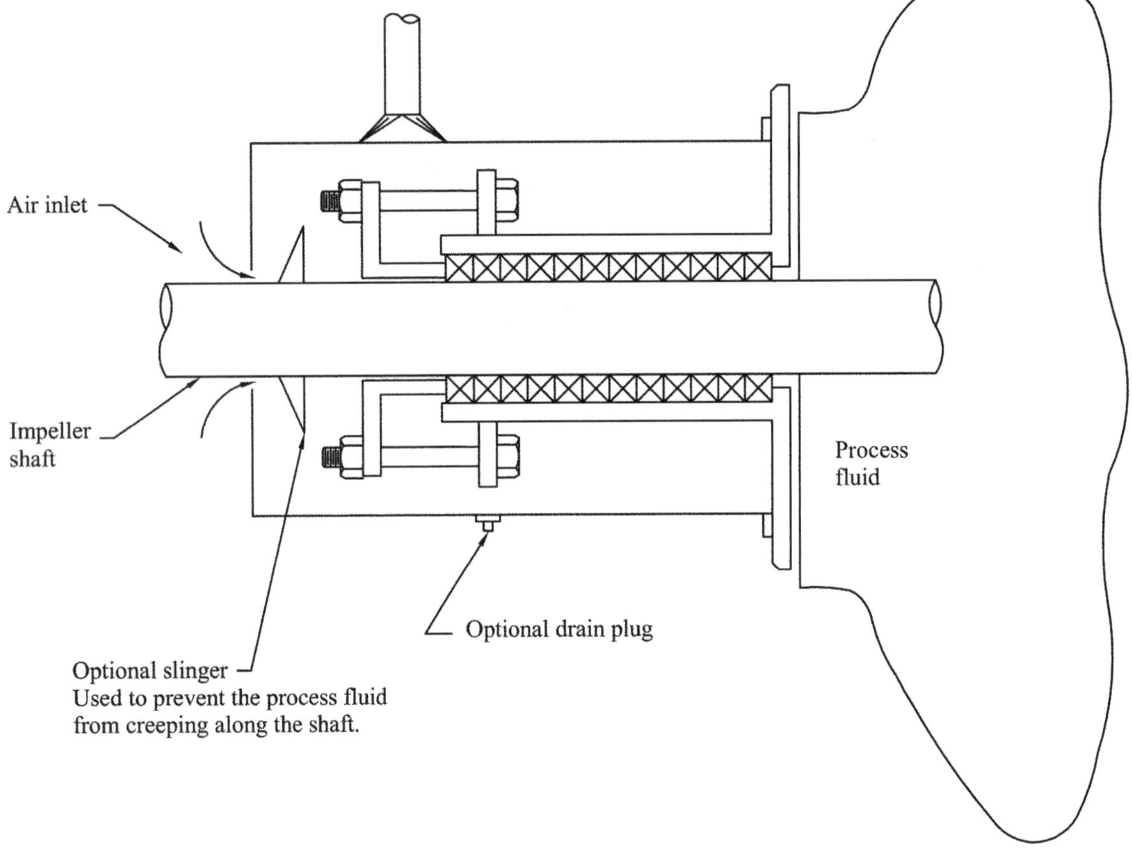

Q = 500 acfm/ft^2 [2.50 am^3/s/m^2] of open area
(typically 10–40 acfm [0.01–0.02 am^3/s])

Note: Sufficient air must be provided to dilute flammable gases and/or vapors to below 25% of LEL. See Chapter 4.

Duct velocity = 2000 fpm [10 m/s]

$h_e = 1.78VP_s + 0.25VP_d$

Note: Similar hood is appropiate for union fittings.

Reference: 13.15.3

ACGIH®	TITLE SHAFT SEAL ENCLOSURE	FIGURE VS-15-21 DATE 1-12

CHECK CODES, REGULATIONS, AND LAWS (LOCAL, STATE, AND NATIONAL) TO ENSURE THAT DESIGN IS COMPLIANT.

FIGURE VS-15-30

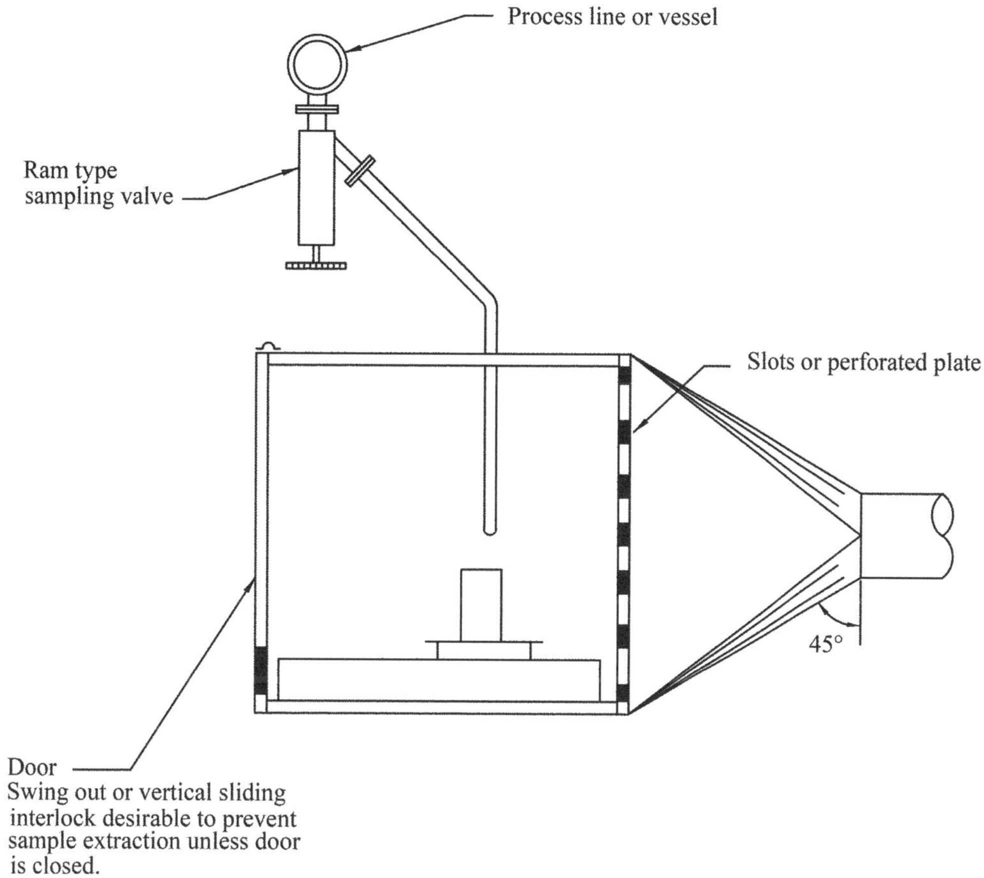

$Q = 125$ acfm/ft^2 [0.63 am^3/s/m^2] of open area (door area) minimum
Duct velocity = 2000 fpm [10 m/s]
$h_e = 1.78\ VP_s + 0.56\ VP_d$

NOTE: Sufficient air must be provided when door closed to dilute flammable gases and/or vapors to 25% of LEL and meet all NFPA requirements. See Chapter 4.

Reference: 13.15.3

ACGIH®

TITLE: SAMPLING BOX

FIGURE: VS-15-30

DATE: 1-12

CHECK CODES, REGULATIONS, AND LAWS (LOCAL, STATE, AND NATIONAL) TO ENSURE THAT DESIGN IS COMPLIANT.

13.20 FOUNDRY OPERATIONS

Foundry operations include many operations common to other industries. Some of these operations are covered in the following subsections of this chapter:

13.45 Machining
13.50 Material Transport
13.55 Metal Melting
13.60 Mixing
13.80 Surface Cleaning
13.90 Welding and Cutting

This subsection addresses operations that are more unique to the foundry industry: casting shakeout and core making.

13.20.1 Casting Shakeout. Foundry shakeout ventilation rates depend on the type of enclosure and the temperature of the sand and castings. The enclosing shakeout hood (VS-20-01) requires the smallest airflow rate. The side draft shakeout hood (VS-20-02) requires additional airflow rate but provides improved access for casting and sand delivery and for casting removal. The downdraft shakeout (VS-20-03) is the least effective in controlling contaminant and requires the highest ventilation rate. It is not recommended for hot castings. The shakeout hopper below the shakeout table requires additional exhaust ventilation equivalent to 10 percent of the shakeout hood exhaust rate.

Moisture released during shakeout operations can condense on duct and hood surfaces. Systems may require the addition of heat to keep moisture in vapor form, especially if duct is connected to dry filter (baghouse).

Particular attention should be paid to the conveyor removing sand from the shakeout. This conveyor requires hoods and ventilation as described in Section 13.55.

Rotary tumble mills used for shakeout should be treated as an enclosing hood with a minimum inward velocity of 150 fpm through any opening.

13.20.2 Core Making. Core making machines require ventilation to control reactive vapors and gases such as amines and isocyanates that are used in the core making process. A minimum capture velocity of 75 fpm [0.36 m/s] is required. However, a ventilation rate as high as 250 acfm/ft^2 [1.25 am^3/m^2] of opening may be necessary for adequate control of contaminant emissions. When cores are cured in ovens, adequate ventilation control of the oven is required.

REFERENCE

13.20.1 American Foundrymen's Society, Inc.: Managing the Foundry Inplant Air Environment. Des Plaines, IL (2004).

FIGURE VS-20-01

ENCLOSING HOOD

Provides best control with least flow rate
Minimum duct velocity = 4000 fpm [20 m/s]
$h_e = 0.25\ VP_d$

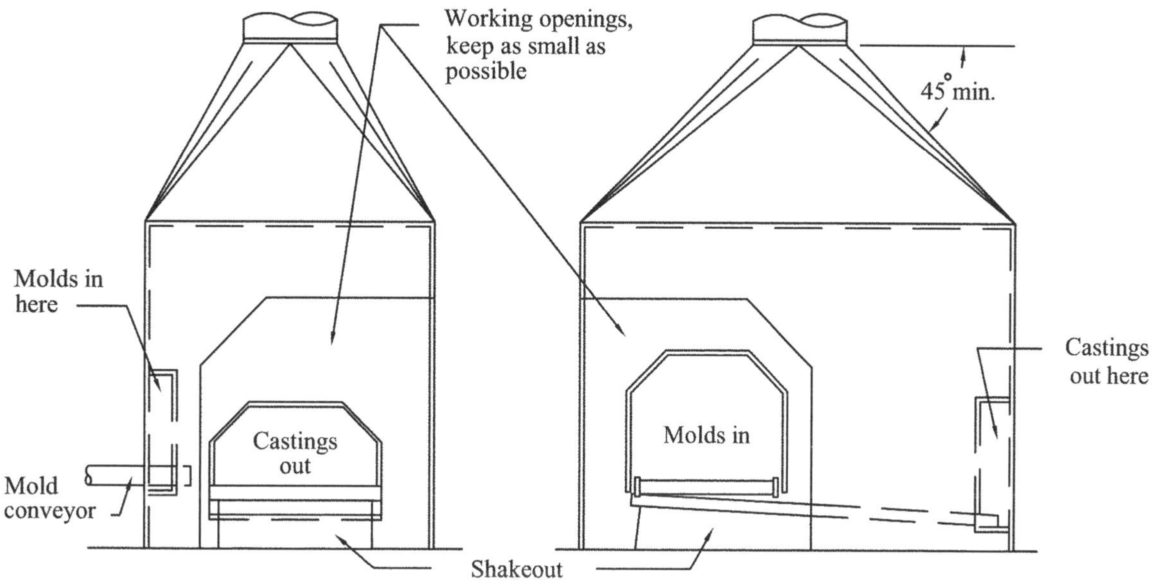

Shakeout exhaust, minimum *

Type of hood	Hot castings	Cool castings
Enclosing ** VS-20-01	200 acfm/ft^2 [1.0 am^3/s/m^2] opening At least 200 acfm/ft^2 [1.0 am^3/s/m^2] grate area	200 acfm/ft^2 [1.0 am^3/s/m^2] opening At least 150 acfm/ft^2 [0.75 am^3/s/m^2] grate area
Two sides and 1/3 top area enclosed ** VS-20-02	300 acfm/ft^2 [1.50 am^3/s/m^2] grate area	275 acfm/ft^2 [1.38 am^3/s/m^2] grate area
Side hood (as shown or equivalent) ** VS-20-02	400–500 acfm/ft^2 [2.0–2.5 am^3/s/m^2] grate area	350–400 acfm/ft^2 [1.75–2.00 am^3/s/m^2] grate area
Double side hood ** VS-20-02	400 acfm/ft^2 [2.0 am^3/s/m^2] grate area	300 acfm/ft^2 [1.50 am^3/s/m^2] grate area

* Choose higher values when
(1) Castings are hot
(2) Sand to metal ratio is low
(3) Cross-drafts are high

** Shakeout hoppers require an additional 10% exhaust.

ACGIH®

TITLE
FOUNDRY SHAKEOUT ENCLOSURE

FIGURE VS-20-01

DATE 1-12

CHECK CODES, REGULATIONS, AND LAWS (LOCAL, STATE, AND NATIONAL) TO ENSURE THAT DESIGN IS COMPLIANT.

FIGURE VS-20-02

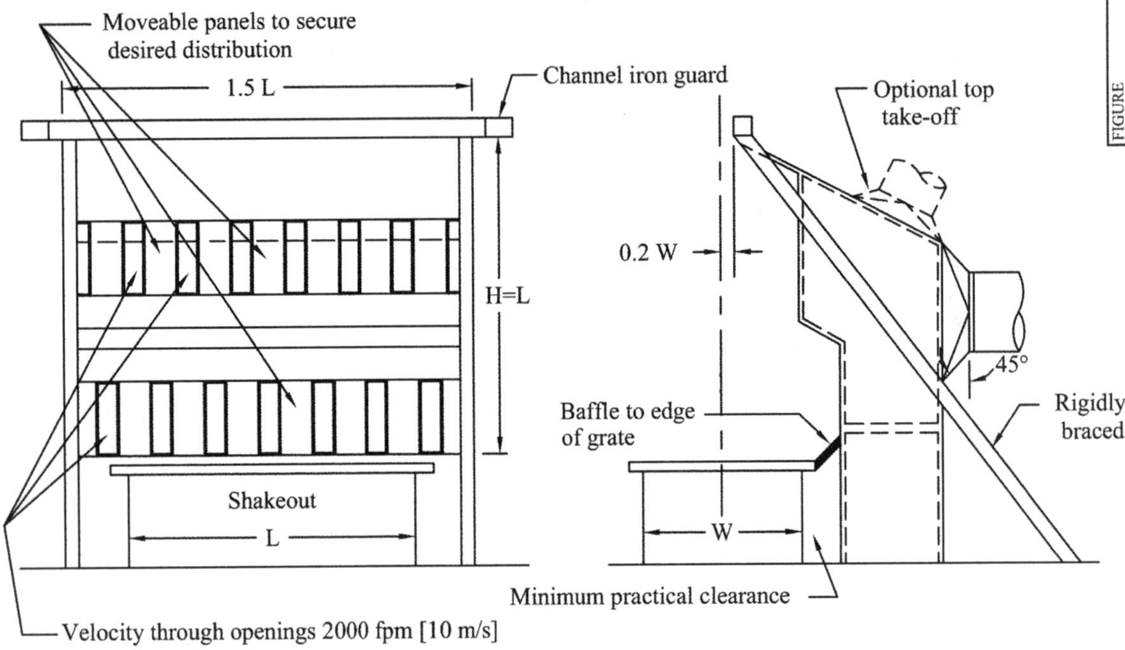

SIDE-DRAFT HOOD

Minimum duct velocity = 4000 fpm [20 m/s]
$h_e = 1.78\ VP_s + 0.25\ VP_d$

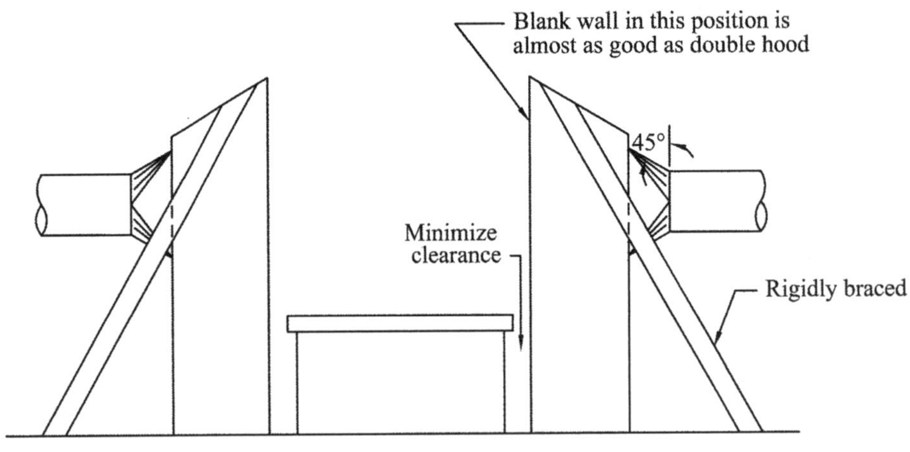

DOUBLE SIDE-DRAFT

Proportions same as single side-draft hood except for overhang

Minimum duct velocity = 4000 fpm [20 m/s]
Slots sized for 2000 fpm [10 m/s]
$h_e = 1.78\ VP_s + 0.25\ VP_d$

See VS-20-01 for exhaust rates

ACGIH®	TITLE FOUNDRY SHAKEOUT SIDE DRAFT HOOD	FIGURE VS-20-02 / DATE 1-12

CHECK CODES, REGULATIONS, AND LAWS (LOCAL, STATE, AND NATIONAL) TO ENSURE THAT DESIGN IS COMPLIANT.

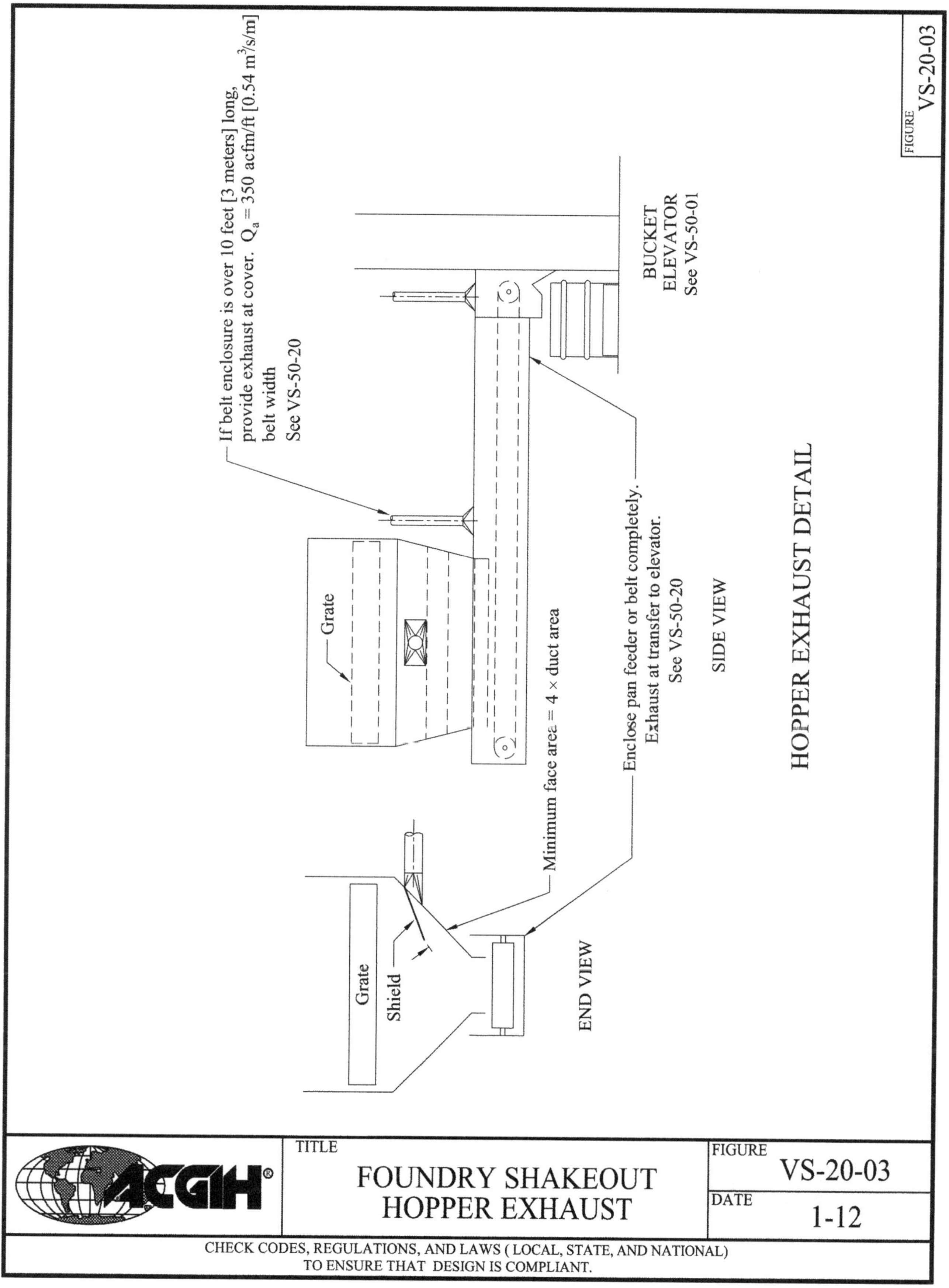

ACGIH®	TITLE FOUNDRY SHAKEOUT HOPPER EXHAUST	FIGURE VS-20-03 DATE 1-12

CHECK CODES, REGULATIONS, AND LAWS (LOCAL, STATE, AND NATIONAL)
TO ENSURE THAT DESIGN IS COMPLIANT.

FIGURE VS-20-10

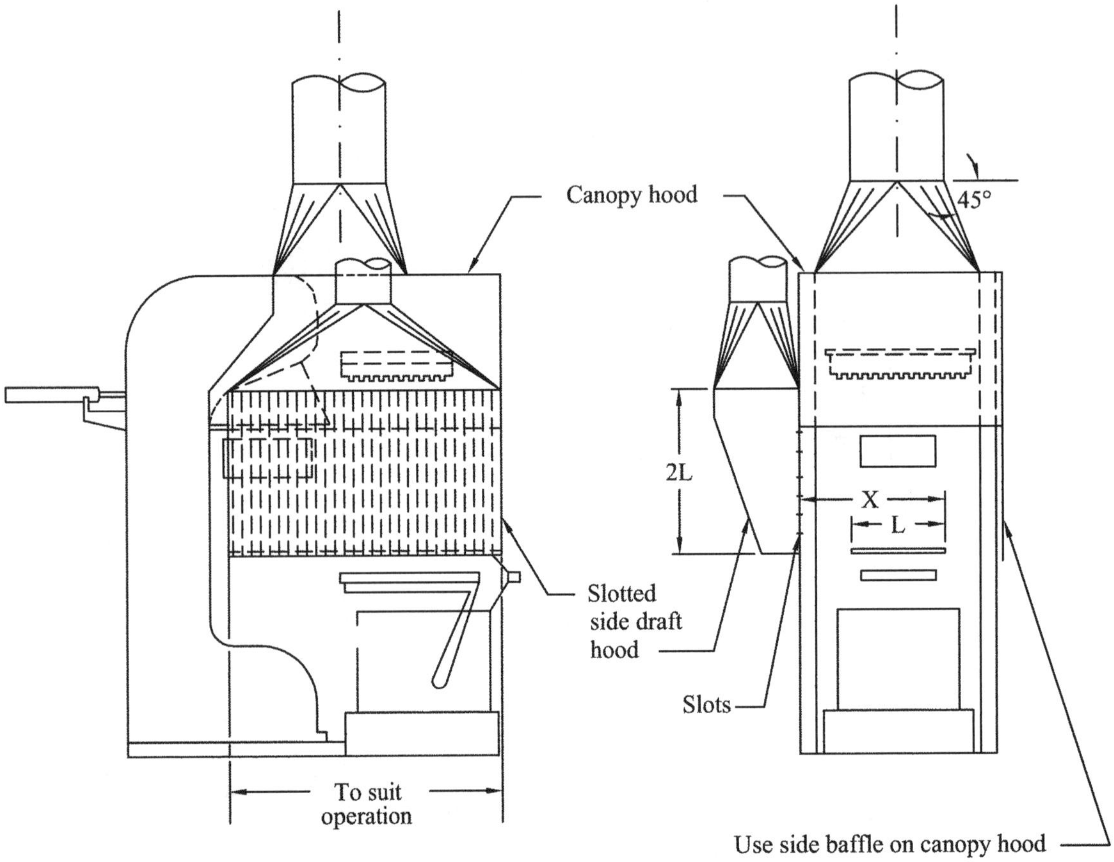

Canopy hood: $Q = 250$ acfm/ft^2 [1.25 am^3/s/m^2] canopy – single unit
150 acfm/ft^2 [0.75 am^3/s/m^2] canopy – double unit
$h_e = 0.25\ VP_d$

Note: Slotted side draft hoods required to remove smoke as hot cores emerge from machine.
Minimum capture velocity = 150 fpm [0.75 m/s]

Side draft hood: $Q = 150(10X^2 + A)$ acfm where A equals hood area (ft^2)
[$Q = 0.75(10X^2 + A)$ m^3/s where A equals hood area (m^2)]
$h_e = 1.78\ VP_s + 0.25\ VP_d$

Note: Conveyor or cooling area require ventilation for large cores. Scrap conveyor or tote boxes may also require additional ventilation.
Minimum duct velocity = 3500 fpm [17.50 m/s]

ACGIH®	TITLE: SHELL CORE MAKING	FIGURE: VS-20-10 / DATE: 1-12

CHECK CODES, REGULATIONS, AND LAWS (LOCAL, STATE, AND NATIONAL) TO ENSURE THAT DESIGN IS COMPLIANT.

FIGURE VS-20-11

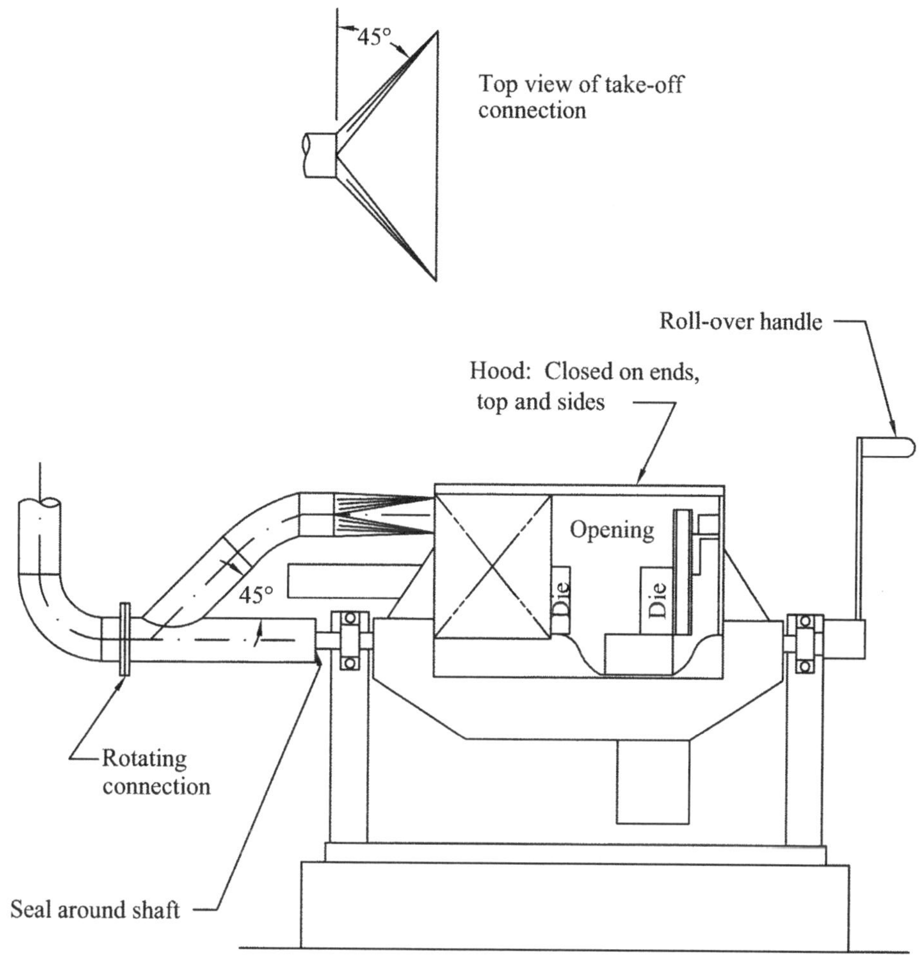

SIDE VIEW

Q = 200 acfm/ft^2 [1.00 am^3/s/m^2] of open face area
Minimum duct velocity = 3500 fpm [17.50 m/s]
$h_e = 0.25\ VP_d$

Note: Elbow and rotating connection losses not included

ACGIH®	TITLE CORE MAKING MACHINE SMALL ROLL-OVER TYPE	FIGURE VS-20-11 DATE 1-12

CHECK CODES, REGULATIONS, AND LAWS (LOCAL, STATE, AND NATIONAL) TO ENSURE THAT DESIGN IS COMPLIANT.

13.25 GAS TREATMENT

The handling of gas cylinders for industrial operations requires special attention. In addition to the potential safety problems associated with transportation and use of compressed gas cylinders, the gas inside the cylinders can escape through leaking valves and fittings. During connection and disconnection of the gas lines, due to the operating pressures, gas can be released.

This section of VS-prints illustrates uses of toxic gases during fumigation (VS-25-01 and -02) and during ethylene oxide sterilization (VS-25-10, -11, -12, and -13).

13.25.01 Ethylene Oxide Sterilizer – In hospitals the sterilizer rooms should be positively pressurized with respect to adjoining rooms to prevent contaminants from entering the sterile environment and the direction of contaminated air should go from clean to contaminates equipment rooms.[13.25.1, 13.25.2] A typical hospital system is shown in Figure VS-25-10.

In industrial type operations, the large sterilizer room is purged with nitrogen to bring the oxygen level below 3%. EtO becomes explosive below 3% oxygen. One method uses pressure control to create a vacuum in the system by flushing the nitrogen before the EtO is introduced. After sterilization the EtO is evacuated from the chamber and recycled using a filtration system via a back vent system. The material is moved to degassing chambers where ventilation is also required. Following degassing, material is inspected in a ventilated hood to prevent exposure to residual gasses.

Typical emission controls on industrial EtO systems are oxidizing emission control devices (OECD). Older systems may have an acidified wet scrubber that converts the EtO to ethylene glycol. The OECD may be used alone or in series with a wet scrubber. While the OECDs reduce EtO emissions, they increase the possibility of explosions. The EtO concentration in the system must remain at 7,500 ppm of EtO or below 25 of the lower flammable limit (LFL) of 30,000 ppm EtO, since the thermal or catalytic OECD is itself an ignition source. Some manufacturers recommend a lower percentage of the LFL. Overfeeding the OECD is the primary source of most explosions. Overfeeding occurs when:

- A back vent is open while a high concentration of EtO is in the sterilizer;
- No valve is used to control the EtO flow rate to the OECD; or
- An EtO-rich stream reaches the OECD

Safety features to consider are:[13.25.6]

- Interlock sterilizer door with the gas valve.
- Provide mechanisms to monitor minimum and maximum temperature, pressure and EtO in each step of the process.
- Use interlocks to prevent opening a sterilizer door before a cycle is complete.
- Vent the products in the sterilizer or aeration room to prevent EtO buildup.
- Use redundant control valves (double block with leak check) on the EtO line.
- Install valve position sensors on critical valves.
- Install flow-limiting device in the vacuum pump inlet to prevent overfeeding the OECD.
- Vent confined spaces to the outside after a power loss.
- Provide limited access to the override controls.
- Provide an abort cycle with diluents in the sterilizer control system using evacuation of the EtO and a dilution.
- Eliminate back vents, if possible, through equipment and cycle design. Most explosions have occurred when an automatic system opened a sterilizer door that triggered back vent operations,[13.25.1] a manual switch triggered back vent operations[13.25.2] or a sterilizer control sequence triggered an opening earlier than safely appropriate.[13.25.3]
- A sound preventative maintenance system is required. This is especially important to detect leaks.

REFERENCES

13.25.1 American Institute for Architect Press: Guidelines for the Design and Construction of Hospitals and Health Care Facilities. Washington, DC (2001).

13.25.2 Infection Control and Epidemiology, Association for Professionals in Infection Control and Epidemiology, Principles and Practice. Olmstead, R., Ed., Mosby Year Book Publications, Asepsis 20:1-10.4 St. Louis, MO.

13.25.3 National Institute for Occupational Safety and Health: Preventing Worker Injuries and Deaths from Explosions in Industrial Ethylene Oxide Sterilization Facilities. Environmental Protection Agency and Ethylene Oxide Sterilization Association (EOSA) ALERT; DHHS (NIOSH) 2000-119, EPA Publication 55-9-018 (April 2000).

13.25.4 Mortimer, V.D.; Kercher, S.L.; O'Brien, D.M.: Effective Controls for Ethylene Oxide: A Case Study. Applied Industrial Hygiene, 1(1):15-20 (1986).

13.25.5 Hama, G.M.: Ventilation for Fumigation Booths. Air Engineering (December 1964).

13.25.6 National Institute for Occupational Safety and Health: Current Intelligence Bulletin 52, Ethylene Oxide Sterilizers in Health Care Facilities. Engineering and Work Practices. NIOSH (July 13, 1989).

FIGURE VS-25-01

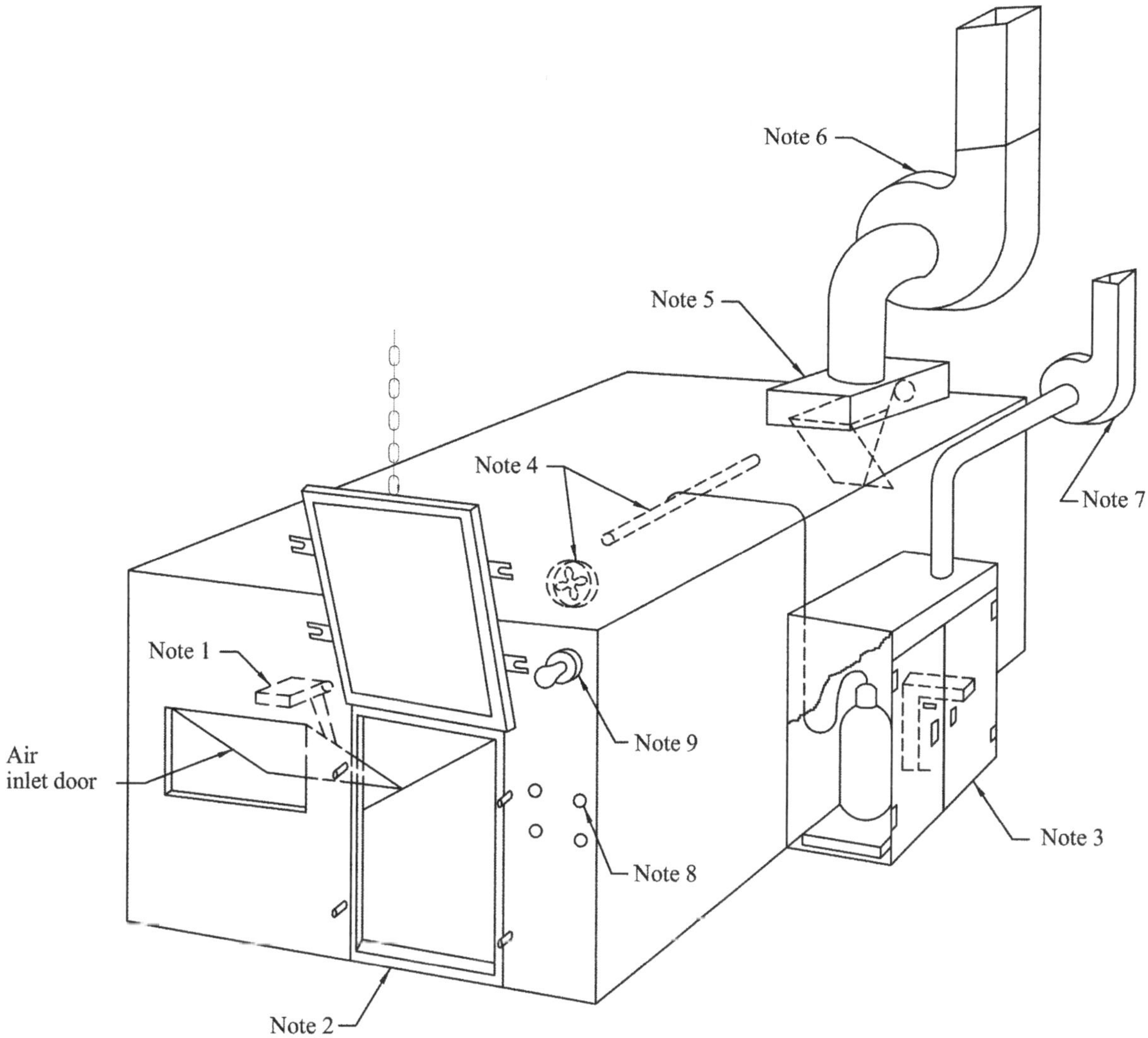

Ventilation Rates

Allow 60 minute purge time
Ventilation rate must be 20 air changes per hour or greater.
Design must provide:

(a) 500 fpm [2.50 m/s] velocity or greater through air inlet door when large access door is closed

and

(b) at least 100 fpm [0.50 m/s] through all openings when large access door is open.

Reference: 13.25.5

See VS-25-02 for explanation of notes.

	TITLE	FIGURE
ACGIH®	FUMIGATION BOOTH	VS-25-01
		DATE 1-12

CHECK CODES, REGULATIONS, AND LAWS (LOCAL, STATE, AND NATIONAL) TO ENSURE THAT DESIGN IS COMPLIANT.

FIGURE VS-25-02

NOTES:

1. Provide an air inlet with automatic damper closure; damper must be interlocked with fan circuit to open only when fan is turned on. Size opening for a minimum velocity of 500 fpm [2.50 m/s]. Air inlet must be located so purge air sweeps entire booth.

2. Loading door must be opened only when booth has been completely purged. Provide gaskets, screw clamps, and brackets for applying uniform pressure for a gas-tight fit.

3. Provide ventilated cabinet for gas cylinders in use and being stored. Fan must be on continuously and exhaust approximately 500 acfm [0.25 am^3/s] to produce a negative pressure in the cabinet when the doors are closed.

4. Provide nozzle openings for introducing fumigant gas. A circulating cabinet fan should also be provided for obtaining good mixture of fumigant gas.

5. Mechanical fan damper must be provided that closes tightly when fan is shut off during fumigation and opens when fan is turned on. Damper controls should be interlocked with fan controls.

6. Fan for ventilating fumigation booth must be sized to dilute air to safe limit in required time. Use vertical, outside, discharge stack away from windows, doors, and air intakes.

7. Fumigant gas cylinder cabinet fan must run continuously.

8. Control switches for fan and lights and an airflow switch-actuated pilot light are recommended.

9. Red warning light to indicate booth is under fumigation as a protection against careless entry is recommended.

10. To facilitate penetration of fumigant gas and subsequent airing out, items should be loaded with separators to allow free air space around each item.

11. Fumigants with no odor-warning properties should be used together with an odor-indicating chemical.

12. Where toxic fumigants are used, a leak test should be made on the booth. The booth first should be tested by lighting several large smoke candles in it with doors and dampers closed. Leaks can be noted by the presence of smoke at the point of escape. Where highly diffusible toxic gases are used, an additional test should be made with the booth under charge, at doors and dampers, with a sensitive detecting meter or sampling device.

ACGIH®	TITLE FUMIGATION BOOTH NOTES	FIGURE VS-25-02 DATE 1-12

CHECK CODES, REGULATIONS, AND LAWS (LOCAL, STATE, AND NATIONAL) TO ENSURE THAT DESIGN IS COMPLIANT.

FIGURE VS-25-10

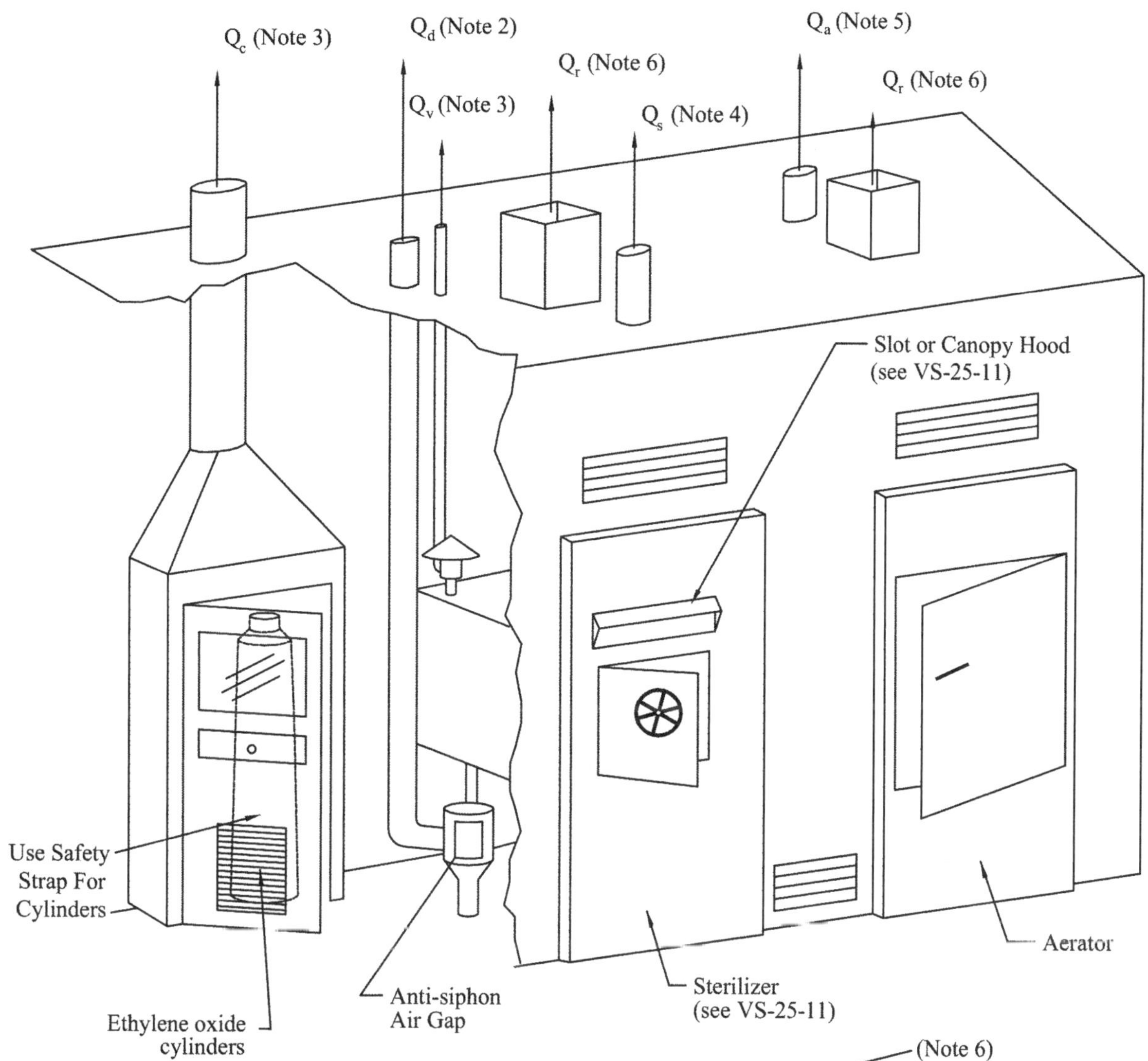

TYPE OF EQUIPMENT	EXHAUST FLOW RATE (Q_r) ACFM	[am^3/s]
Electrically heated gas sterilizers (< 10 ft^3[0.3 m^3])	200	[0.10]
Steam heated gas or steam sterilizers	500	[0.25]
Aerator and instrument washer units	300	[0.15]

See VS-25-11, 25-13 for notes.

See VS-25-12 for details.

See VS-25-12 for Cylinder Hood.

Reference: 13.25.4

TITLE	FIGURE	DATE
ETHYLENE OXIDE STERILIZERS	VS-25-10	1-12

CHECK CODES, REGULATIONS, AND LAWS (LOCAL, STATE, AND NATIONAL)
TO ENSURE THAT DESIGN IS COMPLIANT.

FIGURE VS-25-11

NOTES:

1. The ethylene oxide (EtO) supply cylinders should be placed in a ventilated cabinet or a partially enclosed hood with an exhaust rate, Q_C, of at least 100 acfm/ft^2 [0.50 am^3/s/m^2] of open area. Install a hinged plexiglass shield to protect maintenance personnel during cylinder change out. (See VS-25-13)

2. The anti-siphon air gap in the sterilizer evacuation drain line should be enclosed and ventilated. The enclosure should have one or two openings to allow air, Q_D, to enter and to prevent liquid, which might back up from the drain, from reaching the sterilizer evacuation line. In lieu of a greater value specified by the sterilizer/vacuum pump manufacturer, Q_D, should be approximately 50 acfm [0.03 m^3/s] and the openings sized to maintain approximately a 600 fpm [3 m/s] face velocity.

3. The overpressure relief valve should be vented to carry EtO out of the building if it should ever open. With a sealed line connecting the valve with the ventilation duct, there will be no ventilation volume, Q_V, except when the valve opens. Consult the sterilizer manufacturer for the proper size of this line; too much resistance could interfere with the proper venting of the chamber.

4. A hood should be placed above the sterilizer door to remove EtO rising from the chamber when the sterilizer door is "cracked" open a few inches approximately 15 minutes before the sterilized items are removed from the chamber. See VS-25-12 for a discussion of the exhaust volume, Q_S, requirements.

5. If an aerator is installed, its door should be hinged, and it should be placed beside the sterilizer so that the doors of the gas sterilizer and aerator opens away from each other to facilitate transferring the sterilized items. Consult the manufacturer for the required airflow, Q_A.

6. The room behind the wall enclosing the sterilizer(s) and other equipment should be exhausted adequately to handle the air driven to the ceiling by the thermal gradients caused by the heated equipment. The ideal arrangement would be to have a properly sized vent above each piece of heated equipment. The total Q_R should be the sum of the values for each piece of heated equipment (see VS-25-10) plus 100 acfm/ft^2 [0.50 am^3/s/m^2] of open area for transfer vents placed in the upper portion of the room. However, federal hospital standards specify that, for a recess room containing a gas sterilizer, the volume exhausted in one hour should be at least ten times the room volume. Transfer vents placed in the lower portion of the room will help the influx of air to supply the thermal air currents and would not add to the total exhaust requirement.

7. All air that could contain EtO should be exhausted through a ventilation system which does not have vents in any other rooms. The discharge of the fan on the roof should be located so that the exhausted air will not re-enter the building or expose people outside the building. This ventilation system should have a flow sensor/alarm to warn if it is not functioning properly. If there is the possibility of lint in the exhausted air, use a different pressure sensor or some other type that will not be clogged or stuck open by the accumulation of lint.

Reference: VS-25-10

ACGIH®	TITLE: ETHYLENE OXIDE STERILIZER NOTES	FIGURE: VS-25-11 / DATE: 1-12

CHECK CODES, REGULATIONS, AND LAWS (LOCAL, STATE, AND NATIONAL) TO ENSURE THAT DESIGN IS COMPLIANT.

FIGURE VS-25-12

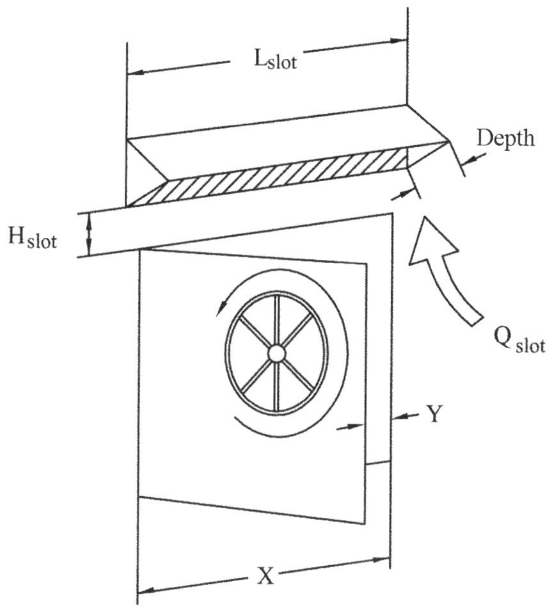

SLOT HOOD

$H_{slot} \leq$ 3" [75 mm]

$L_{slot} = X + 0.66H_{slot}$

Depth = 1" [25 mm]
slot face to be at 45°–90° angle with plane of enclosure

Q_{slot} = 75 acfm/ft [0.12 am^3/s/m] slot length

Y = 2" [50 mm] for airing out sterilizer chamber

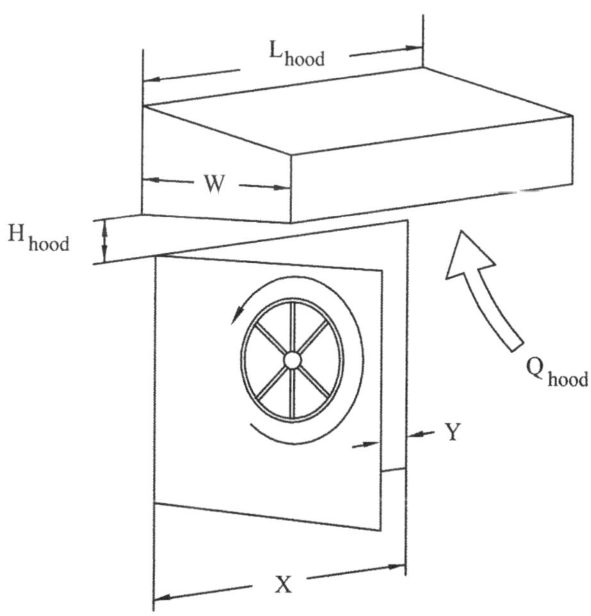

CANOPY HOOD

$Q_{hood} = 100(L_{hood}W)$ acfm

[$Q_{hood} = 0.5(L_{hood}W)$ am^3/s]

H_{hood} = 12" to 24" [300 to 600 mm]

$$W = \frac{H_{hood}}{3}$$

$L_{hood} = X + 0.66H_{hood}$

Y = 2" [50 mm] for airing out sterilizer chamber

ACGIH®	TITLE: ETHYLENE OXIDE STERILIZER HOOD DETAILS	FIGURE: VS-25-12 — DATE: 1-12

CHECK CODES, REGULATIONS, AND LAWS (LOCAL, STATE, AND NATIONAL) TO ENSURE THAT DESIGN IS COMPLIANT.

FIGURE VS-25-13

To dedicated exhaust

1 Cylinder: 175–200 fpm [0.88–1.0 m/s]
2 Cylinders: 250–400 fpm [1.25–2.0 m/s]
3 Cylinders not recommended

Provide a self closing door of sufficient size and appropriate dimensions to permit access to cylinders and gas manifold.

Manufacture cabinet and accessories of a noncombustible material.

Provide fixtures within the cabinet to secure cylinders.

Equip cabinets with a window to permit observation of valves and connections.

Provide a door to access valves and fittings without having to open the full door.

Face velocity = 200 fpm [1.0 m/s] at access door when open.

Review current OSHA, NFPA and CGA standards to ensure design compliance.

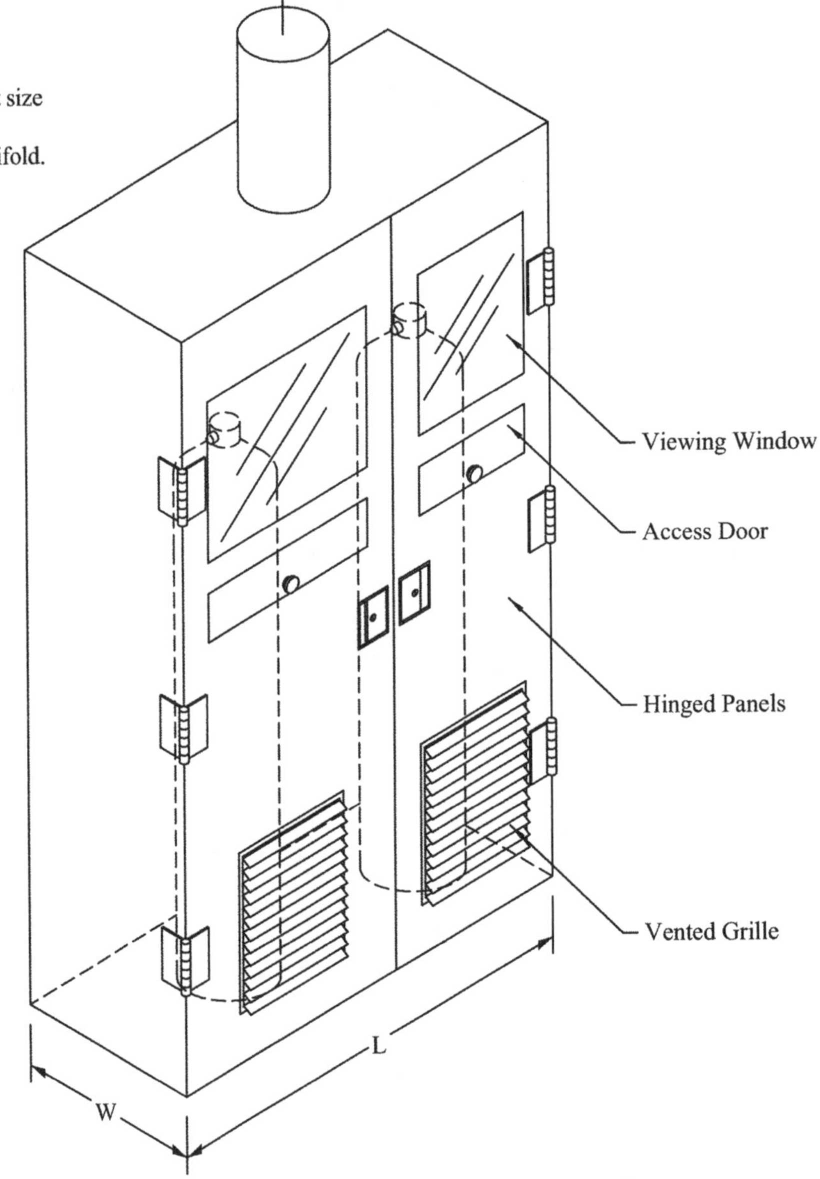

ACGIH®	TITLE: ETHYLENE OXIDE STERILIZER HOOD DESIGN	FIGURE: VS-25-13 / DATE: 1-12

CHECK CODES, REGULATIONS, AND LAWS (LOCAL, STATE, AND NATIONAL) TO ENSURE THAT DESIGN IS COMPLIANT.

13.27 HOT PROCESS VENTILATION

Designing ventilation controls for exothermic or heated (i.e., hot) processes must meet two sometimes competing goals. Controls should minimize heat transfer rates to ensure retaining a majority of the heat in the process, thereby reducing operating costs, while at the same time maintain a safe and healthy work environment for workers. If the ventilation system induces excessive flow, it negatively impacts operations in two ways; 1) it decreases the operating efficiency of the process by increasing heat transfer rates; and 2) wastes conditioned indoor air. In both cases this leads to increased energy consumption to maintain process and work environment temperatures. At the same time, the ventilation rate must be sufficient to capture airborne contaminants and reduce worker exposures. Therefore, the most accurate estimate of the volume of contaminated air generated by a process is desired to optimize hood efficiency and energy usage.

13.27.1 Hierarchy of Controls. Controlling the contaminated air generated by hot processes can be accomplished in a number of ways. The hierarchy of controls dictates a strategy of first substituting or eliminating the hazard (see Chapter 1, Section 1.8). In most instances where hot processes are used, this is not feasible. However, there may be exceptions, such as when a hot process is used to bind objects together (i.e., binding two pieces of plastic by heat). Substituting an adhesive may be a feasible solution for this example. It is important to weigh substitution options before proceeding with implementing ventilation control strategies, since ventilation involves a long-term commitment of capital resources. The secondary step in the hierarchy of controls is modifying the process. An example of this might be attempting to conduct the hot process at decreased temperatures to reduce the amount of air contaminants produced. The tertiary step is enclosing or isolating the hot process to limit worker interaction with the source of contamination (see Chapter 6 and VS-55-05). This may be feasible for processes that require little worker interaction. Retrofitting existing processes with mechanical loading and unloading features could be successful in eliminating exposures where workers are currently loading or unloading material from the process manually. Only when it is determined that all of the proceeding steps are infeasible for the hot process should other local ventilation techniques be considered. The remaining control techniques indicated in Chapter 1, administrative controls and personal protective equipment are not generally recommended as they do not attempt to reduce process contaminant emissions.

13.27.2 Ventilation Control Recommendations. Ventilation controls for hot processes may be accomplished through the use of ventilated enclosures (VS-55-05), slot hoods (VS-70-02), hoods with sidewalls (see Chapter 6, VS-55-20, and VS-55-21) or canopy receiving hoods. Canopy hoods can be effective when contaminant is released over a well defined area, and the contaminant is completely entrained within a rising, heated plume of air. Room cross-drafts can substantially deflect the rising air when the plume is created by a low-temperature process, or when cross-drafts are substantial (> 50 fpm or > 0.25 m/s) between the emission point and the canopy. Careful consideration should be given to such process information as the required worker access and the hazard potential of the process contaminants when determining the proper hood (see Chapter 6). When feasible, ventilated enclosures are recommended for hot processes. This has the two-fold advantage of eliminating contaminants from the workplace, and providing a physical barrier between the hot process and the worker (i.e., additional safety). Less desirable controls include capture hoods (with or without slots); however, these are advantageous in some cases because they can be incorporated at or near the surface of the process, while not interfering with the process or worker. These hoods provide good control at or near the slot face. However, capture velocity diminishes inversely with the square of the distance from the slot. This practically limits slot hood effectiveness to less than 3 to 4 feet (~ 1 m) from the hood face. Furthermore, slot hoods are diminished if the velocity of the heated plume of air rising from the process is moving rapidly (i.e., smelting processes) since the plume could be moving upward at > 20% of the slot face velocity. Although control may sometimes be achieved by implementing a side-draft multiple slot hood, it is not recommended unless the application is similar to specific heated operations documented in this chapter.

13.27.3 Canopy Hoods. Canopy hoods are the remaining choice if the process to be controlled does not fit applications for enclosures or capture hoods. There are four combinations of canopy hoods that will be discussed; low and high canopy hoods, with and without side walls. Low or high canopy hoods with side walls on all sides are preferred, in that they provide similar benefits as ventilated enclosures. The side walls should extend slightly below the top surface of the hot process to ensure capturing contaminants. If temporary access to the process is required, side walls can include hinged or sliding doors for access. Another variant is to partially enclose the process by installing side walls on one, two or three sides of the process. Although not ideal, partial enclosures are an improvement over canopy designs with no side walls.

13.27.4 Canopy Hoods Without Sidewalls. Low and high canopy hoods without side walls are the least effective and least efficient method of controlling hot processes. Although canopies are included in Chapter 6 as a recommended design, there are restrictions to their application. For low canopy hoods, the distance from the hood face to the surface of the hot process should be no greater than the diameter of the source or 3 feet, [0.9 m] whichever is smaller.[13.27.1, 13.27.2] It is also recommended that the size of low hoods and hoods with side walls should be a foot [0.3 m] larger on all sides and have the diameter of the hot process at its widest cross-section (Figure 13-27-1).[13.27.1, 13.27.2, 13.27.3] The limitation of any hood design with distance between the hood face and the surface of the

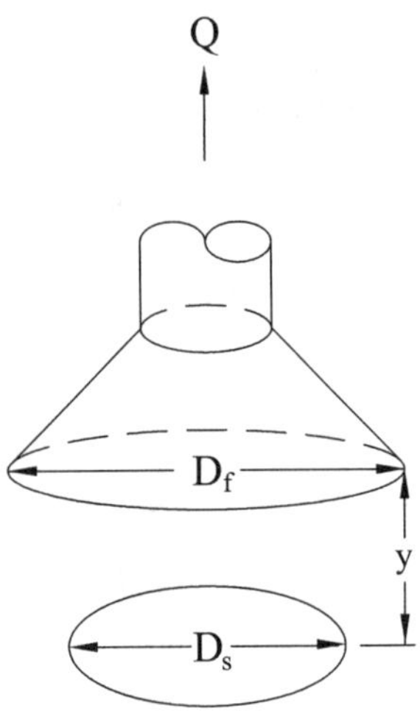

$D_f = D_s + 2$ ft [0.6 m]
$y < 3$ ft [0.9 m]

FIGURE 13-27-1. Geometry and design of low canopy hoods

source is the ability of cross-drafts to interfere with the capture of the contaminants rising from the hot process. Where substantial cross-drafts (> 50 fpm) are present, hood designs should include side walls. At a minimum, it is recommended that at least one side wall be included on the side of the process where the cross-draft originates (upstream side).

13.27.5 Canopy Hoods With Sidewalls. When side walls are included, or when the heated source is close to a structural wall, the thermal plume may attach to the wall. If this occurs, the entrainment will be reduced compared to the entrainment in an unbounded plume, and the flow in the attached plume can be calculated as half of the flow of an unbounded plume. If there are two walls attached at a right angle, the flow decreases to ¼ of the flow of an unbounded plume.(13.27.4)

13.27.6 High Canopy Hood Use as a Redundant Control Measure. The least favorable canopy design is the high canopy without side walls. This design can be successfully employed as a redundancy for controlling hot processes that generate contaminants in large volume, such as an electric arc furnace where the hood is used primarily for charging of materials (VS-55-03). It is not recommended as a primary control measure for hot processes, due to the large volume of air that must be displaced to remove contaminants from the workplace air. In the case of an arc furnace, canopy hood use can be limited to the times when the furnace is being charged to reduce the volume of replacement air required.

Ideally, when side walls are not used, high hood faces should be round because rising air from point sources and compact shapes (i.e., not line sources) become circular in cross-section as they rise.(13.27.5) This occurs, regardless of the source shape, because turbulence tends to sweep the plume edges inward to a minimal volume.(13.27.5) However, it is easier to manufacture and install square or rectangular hoods. Baffles are recommended at the face of the rectangular canopy to approximate the calculated area of an equivalent round hood face.

13.27.7 Ventilation Controls for Large-Scale Hot Processes. Existing equations to approximate the velocity, area, and volumetric flow of the rising air above a large-scale cylindrical heated process with excess air temperatures $\Delta T < 230$ F [110 C] have been in use for decades. These equations are the result of the compilation of empirical research conducted from 1950 to the 1970s.(13.27.1, 13.27.2, 13.27.6, 13.27.7, 13.27.8)

13.27.7.1 Circular High Canopy Hoods. As the heated air rises, it mixes turbulently with the surrounding air. This results in an increasing air column diameter and volumetric flow rate. The diameter of the column (Figure 13-27-2) can be approximated by:

$$D_c = 0.5X_c^{0.88} \qquad \text{[13.27.1] IP}$$

$$[D_c = 0.43X_c^{0.88}] \qquad \text{[13.27.1] SI}$$

where:
D_c = column diameter at hood face
X_c = y + z = the distance from the hypothetical point source to the hood face, ft [m]
y = distance from the process surface to the hood face, ft [m]
z = distance from the process surface to the hypothetical point source, ft [m]

"z" can be calculated from:

$$z = (2\ D_s)^{1.138} \qquad \text{[13.27.2] IP}$$

$$[z = 2.6\ D_s^{1.138}] \qquad \text{[13.27.2] SI}$$

where: D_s = diameter of hot source, ft [m]

The velocity of the rising hot air column can be calculated from:

$$V_f = 8(A_s)^{0.33}\frac{(\Delta T)^{0.42}}{X_c^{0.25}} \qquad \text{[13.27.3] IP}$$

$$[V_f = 0.084(A_s)^{0.33}\frac{(\Delta T)^{0.42}}{X_c^{0.25}}] \qquad \text{[13.27.3] SI}$$

where:
V_f = velocity of hot air column at the hood face, fpm [m/s]
A_s = area of the hot source, ft² [m²]
Δ_T = the temperature difference between the hot source and the ambient air, F [C]
X_c = y + z = the distance from the hypothetical point source to the hood face, ft [m]

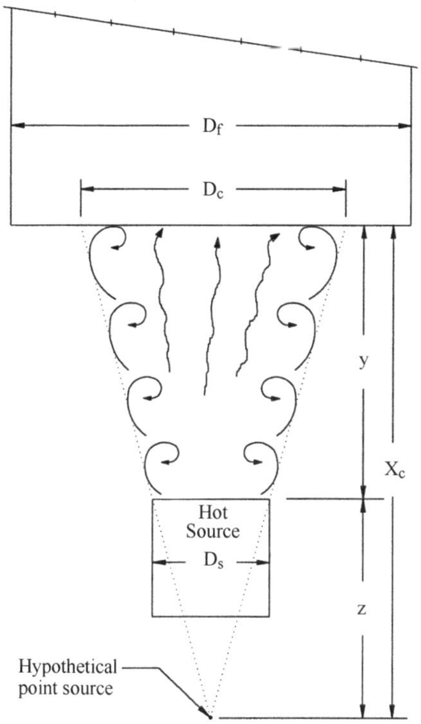

FIGURE 13-27-2. Dimensions used to design high-canopy hoods for hot sources

The diameter of the hood face must be larger than the diameter of the rising hot air column to assure complete capture. The hood diameter is calculated from:

$$D_f = D_c + 0.8y \qquad [13.27.4]$$

where: D_f = diameter of the hood face, ft [m]

Total hood airflow rate is:

$$Q_t = V_f A_c + V_r (A_f - A_c) \qquad [13.27.5]$$

where:
- Q_t = total volume entering hood, acfm [am³/s]
- V_f = velocity of hot air column at the hood face, fpm [m/s]
- A_c = area of the hot air column at the hood face, ft² [m²]
- V_r = the required air velocity through the remaining hood area, fpm [m/s]
- A_f = total area of hood face, ft² [m²]

EXAMPLE PROBLEM 1 (IP)

Note: Equivalent metric (SI) equations shown in brackets; solutions in IP units only.

Given: 4.0 ft diameter melting pot (D_s)
1000 F metal temperature
100 F ambient temperature
Circular Canopy hood located 10 ft above pot (y)
V_r = 100 fpm

Calculate X_c:

$$X_c = y + z = y + (2D_s)^{1.138}$$

$$[X_c = y + z = y + (2.6\,D_s)^{1.135}]$$

$$X_c = 10 + (2 \cdot 4)^{1.138}$$

$$X_c = 20.7 \text{ ft}$$

Calculate the diameter of the hot air column at the hood face:

$$D_c = 0.5\,X_c^{0.88}$$

$$[D_c = 0.43\,X_c^{0.88}]$$

$$D_c = 0.5(20.7)^{0.88}$$

$$D_c = 7.2 \text{ ft}$$

Calculate the velocity of the hot air column at the hood face:

$$V_f = 8\,(A_s)^{0.33}\,\frac{(\Delta T)^{0.42}}{(X_c)^{0.25}}$$

$$[V_f = 0.084\,(A_s)^{0.33}\,\frac{(\Delta T)^{0.42}}{(X_c)^{0.25}}]$$

$$A_s = 0.25\pi D_s^2$$

$$A_s = 0.25\pi\,(4.0)^2$$

$$A_s = 12.6 \text{ ft}^2$$

$$\Delta T = 1000 - 100 = 900 \text{ F}$$

$$V_f = 8(12.6)^{0.33}\,\frac{(900)^{0.42}}{(20.7)^{0.25}}$$

$$V_f = 8(2.3)\,\frac{(17.4)}{(2.13)}$$

$$V_f = 150 \text{ fpm}$$

Calculate diameter of hood face:

$$D_f = D_c + 0.8y$$

$D_f = 7.2 + (0.8 \times 10)$

$D_f = 15.2$ ft

Calculate total hood airflow rate:

$Q_t = V_f A_c + V_r (A_f - A_c)$

$A_c = 0.25\pi D_c^2$

$A_c = 0.25\pi (7.2)^2$

$A_c = 41$ ft²

$A_f = 0.25\pi D_f^2$

$A_f = 0.25\pi (15.2)^2$

$A_f = 181$ ft²

$Q_t = 150(41) + 100(181 - 41)$

$Q_t = 20{,}150$ acfm

*13.27.7.2 **Rectangular High Canopy Hoods.*** Hot air column from sources that are not circular may be better controlled by a rectangular canopy hood. Hood airflow calculations are performed in the same manner as for circular hoods except the dimensions of the hot air column at the hood (and the hood dimensions) are determined by considering both the length and width of the source. The remaining values are calculated in the same manner as for the circular hood.

EXAMPLE PROBLEM 2 (IP)

Note: Equivalent metric (SI) equations shown in brackets; solutions in IP units only.

Given: 2.5' × 4' rectangular melting furnace

700 F metal temperature

80 F ambient temperature

Rectangular canopy hood located 8' above furnace (y)

$V_r = 100$ fpm

Calculate X_c for each furnace dimension:

$X_{c2.5} = y + z_{2.5} = y + (2D_{s2.5})^{1.138}$

$[X_{c2.5} = y + z_{2.5} = y + (2.6D_{s2.5})^{1.138}]$

$= 8 + (2 \times 2.5)^{1.138}$

$= 14.2$ ft

$X_{c4} = 8 + (2 \times 4)^{1.138}$

$= 18.7$ ft

Calculate the width of the hot air column at the hood face:

$D_{c2.5} = 0.5\, X_{c2.5}^{0.88}$

$[D_{c2.5} = 0.43\, X_{c2.5}^{0.88}]$

$= 0.5(14.2)^{0.88}$

$= 5.2$ ft

$D_{c4.0} = 0.5(18.7)^{0.88}$

$= 6.6$ ft

Calculate the velocity of the hot air column at the hood face:

$$V_f = 8(A_s)^{0.33} \frac{(\Delta T)^{0.42}}{X_c^{0.25}}$$

$$\left[V_f = 0.084(A_s)^{0.33} \frac{(\Delta T)^{0.42}}{X_c^{0.25}}\right]$$

$A_s = 2.5 \times 4 = 10$ ft²

$\Delta t = 700 - 80 = 620$ F

$X_c = X_{c2.5} = 14.2$ ft

NOTE: $X_{c2.5}$ is used rather than $X_{c4.0}$ as it is smaller and will yield a slightly larger V_f, that results in a margin of safety.

$$V_f = 8(10)^{0.33} \frac{(620)^{0.42}}{(14.2)^{0.25}}$$

$$V_f = 8(2.1) \frac{(14.9)}{1.9}$$

$= 132$ fpm

Calculate hood face dimensions:

Hood width $= D_{c2.5} + 0.8y$

$= (5.2) + 0.8(8)$

$= 11.6$ ft

Hood length $= D_{c4.0} + 0.8y$

$= 6.6 + 0.8(8)$

$= 13.0$ ft

Calculate the total hood airflow rate:

$Q_t = V_f A_c + V_r (A_f - A_c)$

$A_c = (D_{c2.5})(D_{c4.0})$

$= (5.2)(6.6)$

$= 34$ ft²

A_f = (hood length)(hood width)

$= (11.6)(13.0)$

$= 151$ ft²

$Q_t = (132)(34) + 100(151 - 34)$

$= 4{,}488 + 11{,}700$

$= 16{,}188$ acfm

13.27.7.3 Low Canopy Hoods. If the distance between the hood and the hot source does not approximately exceed the diameter of the source or 3 ft [0.9 m], whichever is smaller, the hood may be considered a low canopy hood. Under such conditions, the diameter or cross-section of the hot air column will be approximately the same as the source. The diameter or side dimensions of the hood, therefore, need only be 1 foot [0.3 m] larger than the source.

The total flow rate for a circular low canopy hood is

$$Q_t = 4.7\ (D_f)^{2.33}\ (\Delta T)^{0.42} \quad \textbf{[13.27.6] IP}$$

$$[Q_t = 0.045\ (D_f)^{2.33}\ (\Delta T)^{0.42}] \quad \textbf{[13.27.6] SI}$$

where: Q_t = total hood airflow, acfm [am³/s]
D_f = diameter of hood, ft [m]
ΔT = difference between temperature of the hot source, and the ambient, F [C]

The total flow rate for a rectangular low hood is:

$$Q_t = 6.2\ (ab)^{1.33}\ \Delta T^{0.42} \quad \textbf{[13.27.7] IP}$$

$$[Q_t = 0.06\ (ab)^{1.33}\ \Delta T^{0.42}] \quad \textbf{[13.27.7] SI}$$

where: Q_t = total hood airflow, acfm [am³/s]
a = length of the rectangular hood, ft [m]
b = width of the rectangular hood, ft [m]
ΔT = difference between temperature of the hot source and the ambient, F [C]

13.27.8 Ventilation Controls for Small-Scale Hot Processes. New equations to approximate the velocity, area, and volumetric flow of the rising air above a small-scale cylindrical heated process have been developed, and validated within a range of excess air temperatures (2 F < ΔT < 54 F) [10 C < ΔT < 129 C].(13.27.3, 13.27.9, 13.27.10) The new equations build on modern research and principles applicable to designing engineering controls for heated processes, as well as historic work by Hemeon and others.(13.27.1, 13.27.2, 13.27.6, 13.27.8) The new, validated equations are particularly useful for determining the volumetric flow from discrete sources operating at lower temperatures. The historic equations provided by Hemeon and others continue to have utility for the traditional large-scale, high temperature processes (i.e., arc furnaces, tapping operations, etc.).

The diameter and radius of the heated air rising from the hot process can be approximated by:(13.27.3, 13.27.9)

$$D_c = 2R_c,\ \text{ft [m]} \quad \textbf{[13.27.8]}$$

$$R_c = 0.45 \bullet X_c^{0.86},\ \text{ft} \quad \textbf{[13.27.9] IP}$$

$$[R_c = 0.38 \bullet X_c^{0.86},\ \text{m}] \quad \textbf{[13.27.9] SI}$$

where: R_c = column radius at hood face, ft
X_c = z + y, ft
z = 2.51 $R_s^{1.16}$ = distance between the hypothetical point source and the process surface, ft
R_s = $D_s/2$; radius of the hot source, ft
y = distance between process surface and hood face, ft

Therefore, the area of the hot air column at the hood face is:

$$A_c = \pi \bullet R_c^2 = \pi(0.45X_c^{0.86})^2 = 0.63 \bullet X_c^{1.72}\ \text{ft}^2 \quad \textbf{[13.27.10] IP}$$

$$[A_c = \pi \bullet R_c^2 = \pi(0.38X_c^{0.86})^2 = 0.45 \bullet X_c^{1.72}\ \text{m}^2] \quad \textbf{[13.27.10] SI}$$

In recommending hood designs, if the hood is located above the source, the area of the hood face must be at least as large as the area of the hot source (A_s) to assure complete capture. Therefore, the hood should have an area equivalent to A_c or A_s, whichever is greater. Note that the new equations provide solutions for $A_c > A_s$ for values of $X_c > 0$ ft. Results from the new area equation are based on calculating the 99% area of the thermal plume, and not the 95% area as in previous equations.(13.27.3, 13.27.9)

The average velocity of the air rising from the hot process can be approximated by:

$$V_f = \frac{83.2}{X_c^{0.29}}\left(\frac{A_sP}{T_\infty}\right)^{0.33} \text{fpm} \quad \textbf{[13.27.11] IP}$$

$$\left[V_f = \frac{0.37}{X_c^{0.29}}\left(\frac{A_oP}{T_\infty}\right)^{0.33}\right] \text{m/s} \quad \textbf{[13.27.11] SI}$$

where:
A_s = $\pi(R_s^2)$ = area of the hot source, ft² [m²]
P = power per unit area or total heat flux from the source = $(2 \times 10^{-9}(T_s^4 - T_\infty^4)) + (0.22(\Delta T)^{1.33})$, BTU hr⁻¹ ft⁻²
[P = $(5.4 \times 10^{-8}(T_s^4 - T_\infty^4)) + (1.52(T_s - T_\infty)^{1.33})$], wm⁻²
T_s = source temperature, R [K]
T_∞ = ambient temperature, R [K]

The volumetric flow (*Q*) of the hot air at the hood face is approximated by:(13.27.3)

$$Q = V_fA_c = 52.3X_c^{1.43}\left(\frac{A_sP}{T_\infty}\right)^{0.33} \text{acfm} \quad \textbf{[13.27.12] IP}$$

$$\left[Q = 0.17X_c^{1.43}\left(\frac{A_sP}{T_\infty}\right)^{0.33}\right] \text{m}^3\text{/s} \quad \textbf{[13.27.12] SI}$$

No adjustments for additional flow are required in this flow calculation unless there are substantial cross-drafts (> V_f or 50 fpm, whichever is greater) in the work area. If cross-drafts are outside of this range, then the hood should be increased in size, and the flow increased as indicated in Section 13.27.7.

EXAMPLE PROBLEM 3 (IP)

Given: 4 ft melting pot diameter (D_s); R_s = 2 ft

1,000 F metal temperature (T_s) = 1,460 R

100 F ambient temperature (T_∞) = 560 R

Circular canopy hood located 10 ft above pot (y)

Calculate X_c:

$X_c = y + z = y + 2.51(R_s)^{1.16} = 15.6$ ft

Calculate cross-sectional area of the rising air at the hood face:

$A_c = \pi \cdot R_c^2 = 0.63 \cdot X_c^{1.72} = 0.63 \cdot (15.6)^{1.72} = 71.7$ ft^2

$R_c = 0.45 \cdot X_c^{0.86} = 0.45 \cdot (15.6)^{0.86} = 4.78$ ft

Calculate the average velocity of the rising air at the hood face:

$$V_f = \frac{83.2}{X_c^{0.29}}\left(\frac{A_sP}{T_\infty}\right)^{0.33}$$

$A_s = \pi \cdot R_s^2 = \pi \cdot 2^2 = 12.6$ ft^2

$P = [2 \times 10^{-9}\,(T_s^4 - T_\infty^4)] + [0.22(\Delta T)^{1.33}]$ (BTU hr^{-1})ft^{-2}

$P = [2 \times 10^{-9}\,(1{,}460^4 - 560^4)] + [0.22(1{,}460 - 560)^{1.33}]$

$= 10{,}800$ (BTU hr^{-1}) ft^{-2}

$$V_f = \frac{83.2}{15.6^{0.29}}\left(\frac{12.6 \cdot 10{,}800}{560}\right)^{0.33} = 230 \text{ fpm}$$

Calculate the volumetric flow of the rising air at the hood face:

$$Q = V_fA_c = 52.3X_c^{1.43}\left(\frac{A_sP}{T_\infty}\right)^{0.33}$$

$$= 52.3(15.6)^{1.43}\left(\frac{12.6 \cdot 10{,}800}{560}\right)^{0.33}$$

Q = 16,300 acfm

Calculate the diameter of the hood face:

$D_f = D_c + 0.8y$

$D_c = 2R_c = 2(0.45\ X_c^{0.86}) = 2(0.45 \cdot (15.6)^{0.86}) = 9.56$ ft

$D_f = 9.56 + (0.8 \cdot 10) = 17.6$ ft

Calculate the total hood airflow rate (crossdraft > V_f or 50 fpm):

$Q_t = V_f\,A_c + V_r\,(A_f - A_c)$ acfm

V_r = 100 fpm (recommended)

$A_f = \pi \cdot 0.25D_f^2 = 243$ ft^2

$Q_t = (230 \cdot 71.7) + (100 \cdot (243 - 71.7)) = 33{,}600$ acfm

EXAMPLE PROBLEM 3 (SI)

Given: 1.2 m melting pot diameter (D_s); R_s = 0.6 m

T_s = 811 K metal temperature

T_∞ = 311 K ambient temperature

Circular canopy hood located 3 m above pot (y)

Calculate X_c:

$X_c = y + z = 3 + 3.03(R_s)^{1.16} = 3 + 3.03(0.6)^{1.16} = 4.7$ m

Calculate cross-sectional area of the rising air at hood face:

$A_c = 0.45X_c^{1.72} = 0.45(4.7)^{1.72} = 6.45$ m^2

$R_c = 0.38X_c^{0.86} = 0.38(4.7)^{0.86} = 1.4$ m

Calculate the average velocity of the rising air at hood face:

$$V_f = \frac{0.37}{X_c^{0.29}}\left(\frac{A_sP}{T_\infty}\right)^{0.33}$$

$A_s = \pi R_s^2 = \pi(0.6)^2 = 1.17$ m^2

$P = (5.4 \times 10^{-8}(T_s^4 - T_\infty^4)) + (1.52\Delta T^{1.33})$

$P = (5.4 \times 10^{-8}(811^4 - 311^4)) + (1.52(500)^{1.33})$

$= 28800$ wm^{-2}

$$V_f = \frac{0.37}{4.7^{0.29}}\left(\frac{\pi \cdot (0.6)^2 \cdot 28800}{311}\right)^{0.33} = 0.23(4.64)$$

$= 1.4$ m/s

Calculate the volumetric flow of the rising air at hood face:

$$V_f = 0.17(X_c)^{1.43}\left(\frac{A_sP}{T_\infty}\right)^{0.33}$$

$$= 0.17(4.7)^{1.43}\left(\frac{1.17 \cdot 28800}{311}\right)^{0.33} = 7.29 \text{ am}^3\text{/s}$$

Calculate the diameter of the hood face:

$D_f = D_c + 0.8y$

$D_c = 2R_c = 2(0.38\ X_c^{0.86}) = 2(0.38 \cdot (4.7)^{0.86}) = 2.88$ m

$D_f = 2.88 + (0.8 \cdot 3.05) = 5.32$ m

Calculate the total hood airflow rate (crossdraft > 0.25 m/s):

$Q_t = V_f A_f + V_r (A_f - A_c)$

$V_r = 0.5$ m/s (recommended)

$A_f = \pi \cdot 0.25 D_r^2 = \pi \cdot 0.25 \cdot 5.32^2 = 22.23$ m^2

$Q_t = (1.06 \cdot 6.45) + 0.5(22.23 - 6.45) = 15.82$ am^3/s

REFERENCES

13.27.1 Hemeon, W.C.L.: Chapter 8: Exhaust for Hot Processes. In: Hemeon's Plant and Process Ventilation, 3rd ed., D.J. Burton, Ed. New York: Lewis Publishers, 117-47 (1999).

13.27.2 Hemeon, W.C.L.: Chapter 8: Exhaust for Hot Processes. In: Hemeon's Plant and Process Ventilation, 2nd ed., D.J. Burton, Ed. New York: Lewis Publishers, 160-96 (1963).

13.27.3 McKernan, J.L.; Ellenbecker, M.J.: Ventilation equations for improved exothermic process control. Ann Occup Hyg 51:269-79 (2007).

13.27.4 Nielsen, P.V.: Displacement ventilation: Theory and design. Aalborg: Aalborg University (1993).

13.27.5 Bill, R.G.; Gebhart, B.: The transition of plane plumes. International Journal of Heat and Mass Transfer 18:513-26 (1975).

13.27.6 U.S. Public Health Service: Air Pollution Engineering Manual. Publication No. 999-AP-40 (1973).

13.27.7 ACGIH®: Chapter 6: Design Issues – Hoods. In: Industrial Ventilation: A Manual of Recommended Practice for Design, 26th ed. Cincinnati, OH: American Conference of Governmental Industrial Hygienists 6-20–6-24 (2007).

13.27.8 Goodfellow, H.: Design of Ventilation Systems for Fume Control. In: Advanced Design of Ventilation Systems for Contaminant Control. New York: Elsevier 359-438 (1985).

13.27.9 McKernan, J.L.; Ellenbecker, M.J.; Holcroft, C.A.; Petersen, M.R.: Evaluation of a proposed area equation for improved exothermic process control. Ann Occup Hyg 51:725-38 (2007).

13.27.10 McKernan, J.L.; Ellenbecker, M.J.; Holcroft, C.A.; Petersen, M.R.: Evaluation of a proposed velocity equation for improved exothermic process control. Ann Occup Hyg 51:357-69 (2007).

13.30 KITCHEN EQUIPMENT

The purpose of an exhaust system for kitchen equipment is to control heat, smoke, odor, humidity, and grease vapor released into the space by cooking or dishwashing equipment. A secondary consideration is the control of combustion products associated with the heat source which may be vented separately or through the hood itself.

Kitchen hoods are generally designed to satisfy NFPA Standard 96 and model building codes or local codes where they apply. There are two types of kitchen hoods. Type 1 hoods are designed for the removal of grease and smoke. They have some type of grease removal system and fire suppression equipment. Type 2 hoods are for the removal of vapors, heat and odors. For Type 2 hoods, a means of grease removal and fire suppression is not required; thus, they are used over dishwashers and steam tables. Only a Type 1 hood can be used over fryers, griddles, broilers and other cooking equipment where smoke or grease vapors may be present. Type 1 hoods can be used where Type 2 hoods are round, but the reverse is not true.

Type 1 hoods fall into two categories: Conventional and Listed. Conventional hoods are built to satisfy national and local codes. Listed hoods are those listed in Underwriters Laboratories (UL) Standard 710 and may not satisfy the design as specified in the model codes. These hoods have been tested and have shown that they provide an equal or better safety performance than found in the model code requirements.

Type 1 hoods are designed to have an overhang of 12 inches [300 mm] in the front and the back if there is an open space in the rear. The overhang is normally 12 inches [300 mm] for a single island canopy hood, six inches for a double island canopy and wall mounted canopy. No overhang is required for eyebrow and back shelf hoods.

Hood exhaust flow rates vary on the hood style, the amount of overhang, the presence of side panels, the type of cooking, the distance from the cooking surface and the food being cooked. Hot cooking surfaces create thermal plumes that have an upward velocity which can approach 150 fpm [0.75 m/s]. The hood airflow is determined by the plume velocity with safety factors added for the style of hood, cross air currents and flare-ups. The plume velocities are categorized by a duty of cooking appliance:

- Extra heavy duty – solid burning equipment up to 700 F [371 C].
- Heavy duty – upright broilers, charbroilers and woks up to 600 F [316 C].
- Medium duty – large kettles, ranges, griddles and fryers up to 400 F [204 C].
- Light duty – ovens, steamers and small kettles up to 400 F [204 C].

The exhaust rate of a hood is based on the equipment under the hood. The highest duty appliance determines the duty for the hood. The VS prints on kitchen range hoods (VS-30-10 and -11) realize these duty categories to identify the exhaust airflow rate.

National Fire Protection Association (NFPA) Standard 96[13.30.1] describes grease filter construction as well as hood construction necessary to maintain hood integrity in the event of a fire. Welded seam construction is preferred and sometimes required by public health authorities to assure cleanability and ease of maintenance. The National Sanitation Foundation Standard No. 4[13.30.2] also lists hood construction requirements for cleanability and integrity in the cooking and food zones within the hood. In all cases, the local health authorities having jurisdiction should be consulted for construction requirements prior to hood fabrication.

Fire is a primary concern with all cooking equipment. Each hood will require some type of fire suppression consistent with local fire code requirements. The system selected must not compromise sanitation or endanger workers due to location or system activation. Hood or duct penetrations by fire suppression piping, etc., must be sealed to prevent short circuiting of air or loss of fire arrestance.

For high temperature situations such as exposed flames or charcoal, the grease filters must be sufficiently removed from the heat source to prevent ignition. Fan selection may require use of high temperature fan components and consideration of the effect of change in air density.

REFERENCES

13.30.1 National Fire Protection Association Standard: Ventilation Control and Fire Protection of Commercial Cooking Operations. Standard 96-94.

13.30.2 National Sanitation Foundation: Commercial Cooking and Hot Food Storage Equipment. Ann Arbor, MI (1986).

13.30.3 American Society of Heating, Refrigerating and Air Conditioning Engineers: 2003 HVAC Applications Handbook, Chapter 21 "Kitchen Ventilation." ASHRAE, Atlanta, GA (2003).

13.30.4 UL: Exhaust Hoods for Commercial Cooking Equipment, 5th Edition. Standard 710-95 (1995).

13.30.5 UL: Grease Ducts, 1st Edition. Standard 1978-95 (1995).

13.30.6 International Code Council: 2003 International Mechanical Code (2003).

FIGURE VS-30-01

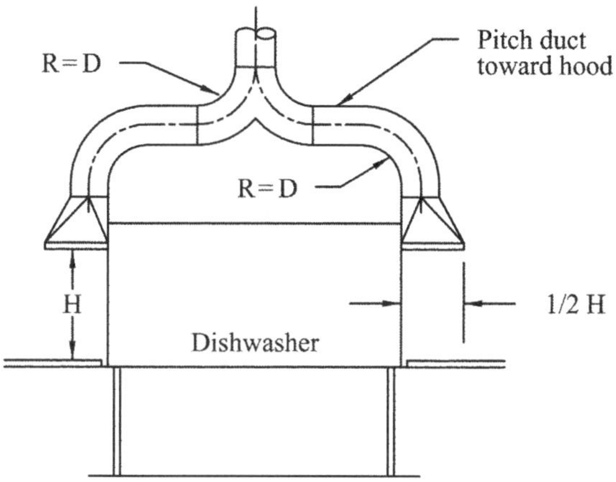

CANOPY HOODS

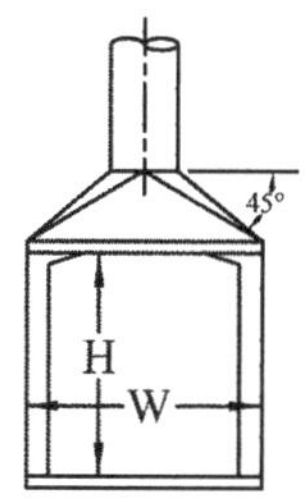

Q = 250 acfm/ft^2 [1.25 am^3/s/m^2] of door area – each end
Duct Velocity = 3000 fpm [15 m/s]
h_e = 0.25 VP_d

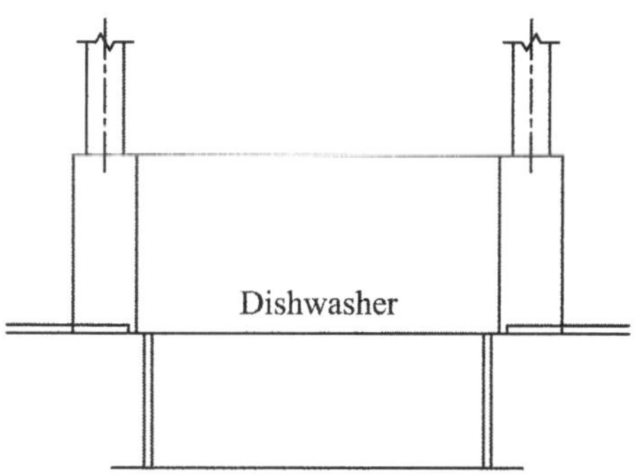

EXHAUSTED VESTIBULES

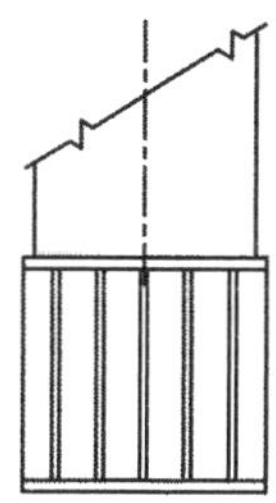

Q = 150 acfm/ft^2 [0.75 am^3/s/m^2] of entrance and exit area
Duct Velocity = 3000 fpm [15 m/s]
h_e = 0.50 VP_d

Note: If direct exhaust connections are provided from dishwasher body, cap these connections and use external hoods.

ACGIH®	TITLE DISHWASHER VENTILATION	FIGURE VS-30-01 DATE 1-12

CHECK CODES, REGULATIONS, AND LAWS (LOCAL, STATE, AND NATIONAL) TO ENSURE THAT DESIGN IS COMPLIANT.

FIGURE VS-30-05

Type of Kitchen Hood, acfm/Linear ft [am³/s/Linear meter] of Hood Length				
Kitchen Appliance Duty	Wall Mounted Canopy	Island	Canopy	Back Shelf
		Single	Double Per Side	
Type 1				
Extra Heavy Duty	550 [0.85]	700 [1.09]	550 [0.85]	NA NA
Heavy Duty	400 [0.62]	600 [0.93]	400 [0.62]	400 [0.62]
Medium Duty	300 [0.47]	500 [0.78]	300 [0.47]	300 [0.47]
Light Duty	200 [0.31]	400 [0.62]	250 [0.39]	250 [0.39]
Type 2	200 [0.31]	400 [0.62]	250 [0.39]	250 [0.39]

Hood exhaust flow rates vary on account of the hood style, the amount of overhang, the presence of side panels, the type of cooking, the distance from the cooking surface and the food that is being cooked. Hot cooking surfaces create thermal plumes that have an upward velocity that can approach 150 fpm [0.75 m/s]. The hood airflow is determined by the plume velocity, with safety factors added for the style of hood, cross air currents, and flare-ups. The plume velocities are categorized by a duty of cooking appliances:

- Extra heavy duty – solid burning equipment up to 700 F [371 C].
- Heavy duty – upright broilers, char broilers and woks up to 600 F [316 C].
- Medium duty – large kettles, ranges, griddles and fryers up to 400 F [204 C].
- Light duty – ovens, steamers and small kettles up to 400 F [204 C].

The exhaust rate of a hood is based on the equipment under the hood. The highest duty appliance determines the duty for the hood. The kitchen range hoods shown in VS-30-10 and VS-30-11 utilize these duty categories to identify the exhaust airflow rate.

ACGIH®

TITLE
KITCHEN HOOD EXHAUST FLOW RATES

FIGURE VS-30-05

DATE 1-12

CHECK CODES, REGULATIONS, AND LAWS (LOCAL, STATE, AND NATIONAL) TO ENSURE THAT DESIGN IS COMPLIANT.

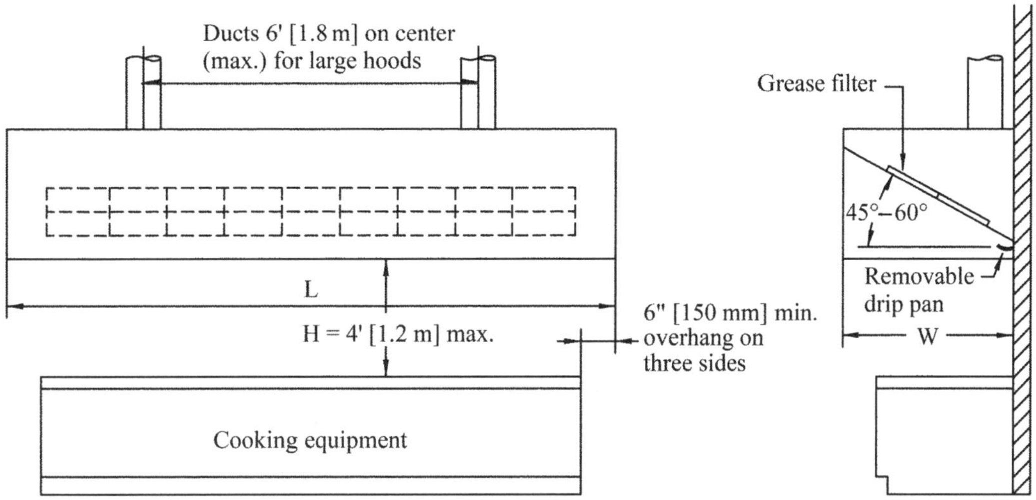

WALL MOUNTED CANOPY

See VS-30-05 for Hood Exhaust Flow

P = perimeter of hood = 2W + L

Duct velocity = 1000 – 4000 fpm [5–20 m/s], to suit conditions

h_e = (filter resistance + 0.1" [25 Pa]) + 0.56 VP_d (straight take off)

h_e = (filter resistance + 0.1" [25 Pa]) + 0.25 VP_d (tapered take off)

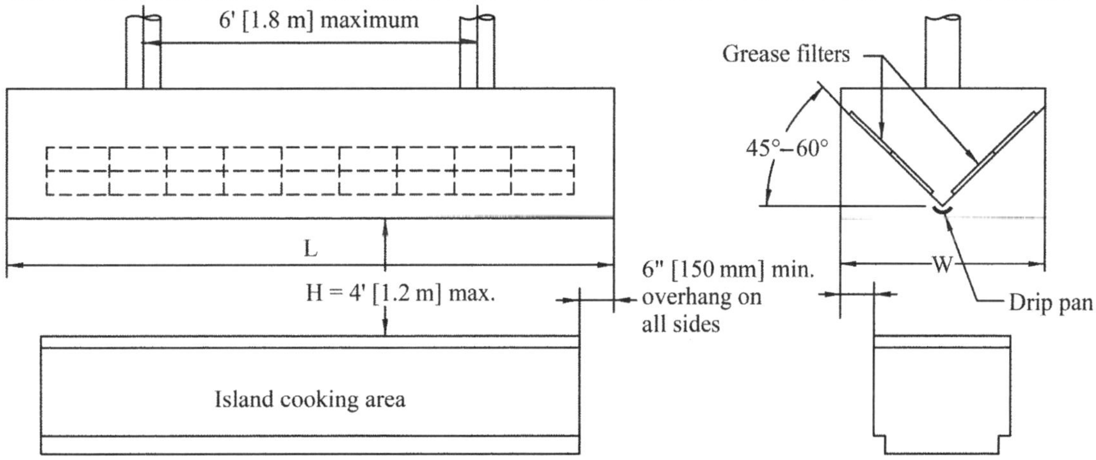

ISLAND TYPE HOOD

P = perimeter of hood
= 2W + 2L

See VS-30-05 for Hood Exhaust Flow

Duct velocity = 1000 – 4000 fpm [5–20 m/s], to suit conditions

h_e = (filter resistance + 0.1" [25Pa]) + 0.56 VP_d (straight take off)

h_e = (filter resistance + 0.1" [25 Pa]) + 0.25 VP_d (tapered take off)

Note: See VS-30-11 for information about filters and fans for range hoods.

ACGIH®	TITLE KITCHEN RANGE HOODS	FIGURE VS-30-10 DATE 1-12

CHECK CODES, REGULATIONS, AND LAWS (LOCAL, STATE, AND NATIONAL)
TO ENSURE THAT DESIGN IS COMPLIANT.

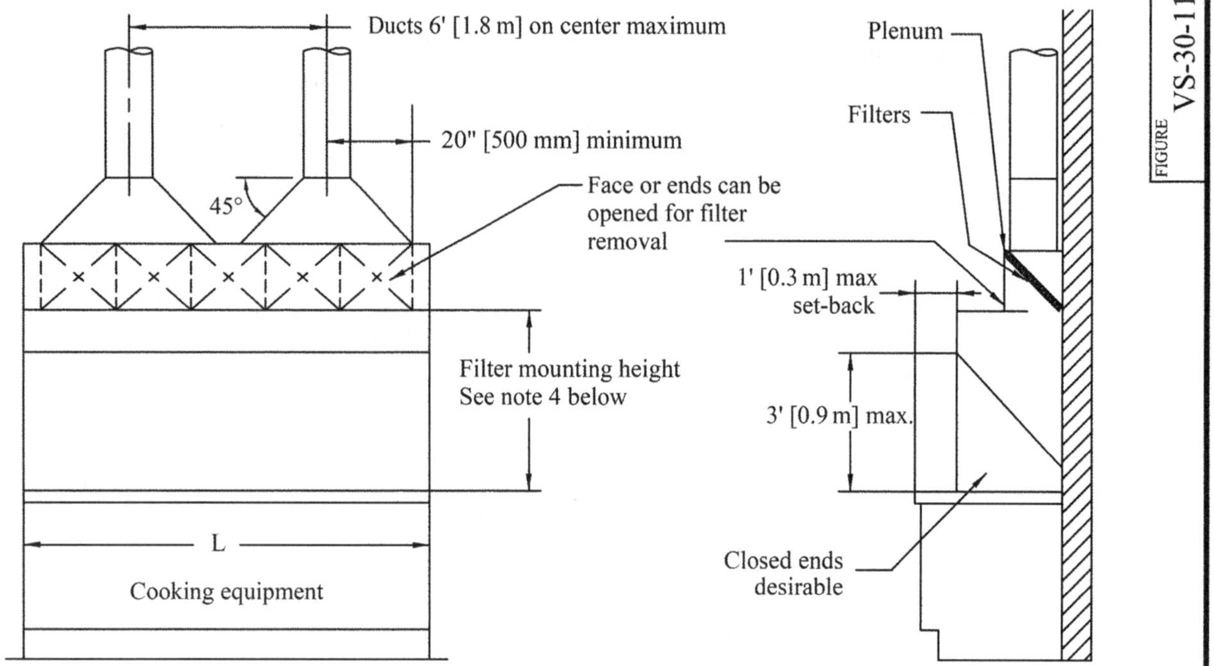

BACK SHELF HOOD

See VS-30-05 for Hood Exhaust Flow

Q = 200 acfm/lineal ft [0.31 am^3/s/lineal meter of cooking surface]

Minimum duct velocity = 1000 – 4000 fpm [5.00 – 20.00 m/s], to suit conditions

h_e = (filter resistance + 0.1" [25 Pa]) + .056 VP_d (straight take off)

h_e = (filter resistance + 0.1" [25 Pa]) +.0.25 VP_d (tapered take off)

NOTES FOR KITCHEN HOODS

Filters:

1. Select practical filter size.
2. Determine number of filters required from manufacturer's data. (Usually: 2 acfm maximum exhaust for each square inch [1.5 am^3/s for each square meter] of filter area.)
3. Install at 45°– 60° to horizontal. Never horizontal.
4. Filter mounting height (Reference 13.30.1)
   a. No exposed cooking flame — 18" [450 mm] minimum to lowest edge of filter.
   b. Charcoal and similar fires — 4' [1.2 m] minimum to lowest edge of filter.
5. Shield filters from direct radiant heat.
6. Provide removable grease drip pan.
7. Clean pan and filters regularly.

Fan:

1. Use upblast discharge fan. Downblast is not recommended.
2. Select fan for design Q and SP resistance of filters and duct.
3. Adjust fan specification for expected exhaust air temperature.

ACGIH®	TITLE KITCHEN RANGE HOOD	FIGURE VS-30-11 DATE 1-12

CHECK CODES, REGULATIONS, AND LAWS (LOCAL, STATE, AND NATIONAL) TO ENSURE THAT DESIGN IS COMPLIANT.

FIGURE VS-30-12

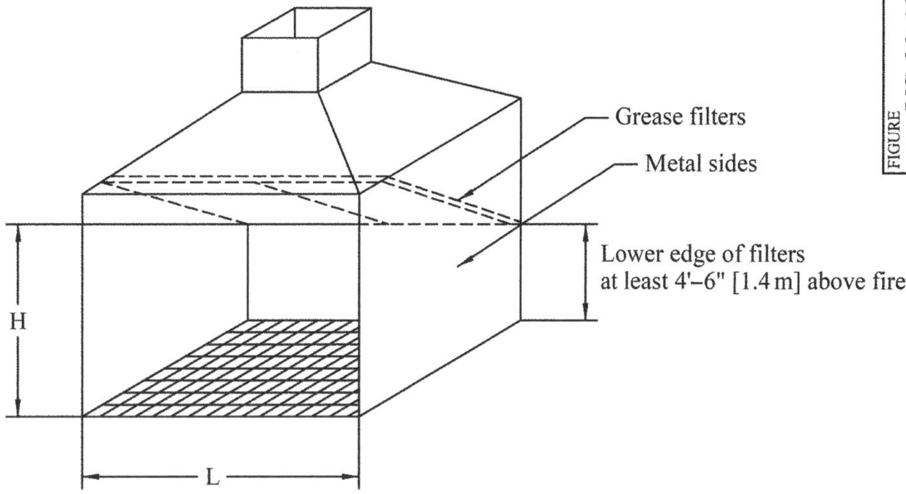

CHARCOAL BROILER

Q = 100 LH acfm [Q = 0.5 LH am^3/s]

Minimum duct velocity = 1000–3000 fpm [5.00–15.00 m/s]

h_e = (filter resistance + 0.1" [25 Pa]) + 0.56 VP_d (straight take off)

h_e = (filter resistance + 0.1" [25 Pa]) + 0.25 VP_d (tapered take off)

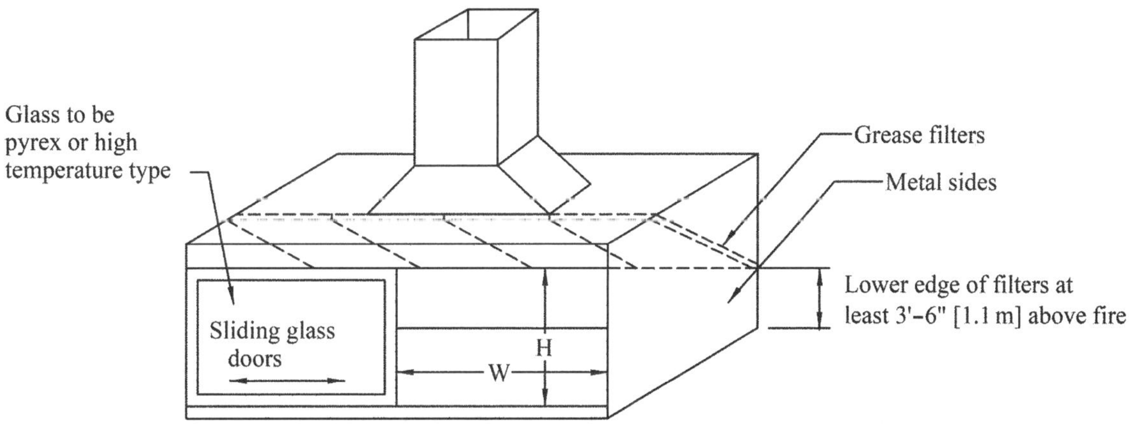

BARBECUE PITS

Notes: 1. If hood is more than 12 feet [3.6 m] long use multiple takeoffs 6' [1.8 m] on center.

2. See VS-30-11 for information about filters and fans for range hoods.

Q = 100 WH acfm [Q = 0.5 WH] (maximum open door area, ft^2 [m^2])

Minimum duct velocity = 1000– 3000 fpm [5.00–15.00 m/s]

h_e = (filter resistance + 0.1" [25 Pa]) + 0.56 VP_d (straight take off)

h_e = (filter resistance + 0.1" [25 Pa]) + 0.25 VP_d (tapered take off)

ACGIH®	TITLE CHARCOAL BROILER AND BARBECUE PIT VENTILATION	FIGURE VS-30-12
		DATE 1-12

CHECK CODES, REGULATIONS, AND LAWS (LOCAL, STATE, AND NATIONAL) TO ENSURE THAT DESIGN IS COMPLIANT.

13.35 LABORATORY VENTILATION

The primary method of contaminant control within the laboratory is exhaust ventilation and, in particular, laboratory hoods. This section presents information on laboratory hoods but expands to other types of ventilation control such as biological safety cabinets, clean benches, and other local exhaust systems found in the laboratory.

13.35.1 Laboratory Hoods. In most cases, laboratory hoods will be purchased from manufacturers specializing in the design and construction of laboratory hoods. VS-35-01 shows a typical laboratory hood design. VS-35-02 describes general use laboratory hoods and VS-35-03 describes perchloric acid hoods. VS-35-04 describes work practices for laboratory hoods.

Several features are essential to the proper performance of the hood. The most important aspect of the hood is the aerodynamic entry characteristics. For the hood to adequately control contaminants, the entry must be smooth. This is usually achieved with an airfoil sill at the leading edge of the work bench. Often, beveled jambs at the side wall entry will improve the airflow.

In many cases, good performance correlates with uniform face velocity. To achieve a uniform face velocity, many hood manufacturers provide adjustable slots in the plenum at the back of the hood. Although the adjustment will allow for unusual conditions such as large hot plates for sample digestions, inappropriate adjustment of the slots can have a detrimental effect on hood performance.[13.35.1]

Supply Air Distribution: For typical operation of a laboratory hood, the worker stands at the face of the hood and manipulates the apparatus in the hood. The indraft at the hood face creates eddy currents around the worker's body that can drag contaminants in the hood along the worker's body and up to the breathing zone. The higher the face velocity, the greater the eddy currents. For this reason, higher face velocities do not result in greater protection as might be supposed.

Room air currents have a large effect on the performance of the hood. Thus, the design of the room air supply distribution system is as important in securing good hood performance as is the face velocity of the hood. American Society of Heating, Refrigerating and Air Conditioning (ASHRAE) research project RP-70 results, reported by Caplan and Knutson,[13.35.2] concludes in part:

1) Lower breathing zone concentrations can be attained at 50 acfm/ft^2 [0.25 am^3/s/m^2] face velocities with good air supply distribution than at 150 acfm/ft^2 [0.75 am^3/s/m^2] with poor air distribution. With a good air supply system and tracer gas released at 8 liters per minute inside the hood, breathing zone concentrations can be kept below 0.1 ppm and usually below 0.05 ppm.
2) The terminal throw velocity of supply air jets should be no more than one-half the hood face velocity; such terminal throw velocities are far less than conventional practice.
3) Perforated ceiling panels provide a better supply system than grilles or ceiling diffusers in that the system design criteria are simpler and easier to apply, and precise adjustment of the fixtures is not required.

For the reasons described, an increased hood face velocity may be self-defeating because the increased air volume handled through the room makes the low-velocity distribution of supply air more difficult.

Selection of Hood Face Velocity: The interaction of supply air distribution and hood face velocity makes any blanket specification of hood face velocity inappropriate. Higher hood face velocities will be wasteful of energy and may provide no better or even poorer worker protection. The ANSI/ASHRAE Hood Performance Test[13.35.3] may be used as a specification. The specified performance should be required of both the hood manufacturer and the designer of the room air supply system.

The specification takes the form: AU_{yyy}, AI_{yyy}, or AM_{yyy}

where: AU identifies an "as used" test,
AI identifies an "as installed" test,
AM identifies an "as manufactured" test, and
yyy = control level, ppm, at the breathing zone of the worker.

Any well-designed airfoil hood, properly balanced, can achieve < 0.10 ppm control level when the supply air distribution is good. Therefore, it would seem appropriate that the "AM" requirements would be < 0.10 ppm. The "AU" requirement involves the design of the room supply system and the toxicity of the materials handled in the hood. The "AU" specification would be tailored to suit the needs of the laboratory room location.

For projected new buildings, it is frequently necessary to estimate the cost of air conditioning early — before the detailed design and equipment specifications are available. For that early estimating, the guidelines listed in Table 13-35-1 can be used.

13.35.2 Biological Safety Cabinets. Biological safety cabinets (BSCs) are classified as Class I, Class II; Types A, B1, B2, and B3; and Class III.

Class I BSC provides personnel and environmental protection but does not protect the product. The front panel can be open, allowing room air to enter the cabinet, sweep the inner surfaces, and exhaust out the duct. A front closure panel with glove ports may be installed. If gloves are installed, air is drawn through a secondary opening equipped with a roughing filter. A laboratory hood, as shown in VS-35-20, could be considered a Class I BSC if the exhausted air is passed through HEPA filters prior to release to the atmosphere.

Class II BSCs provide personnel, product and environmental protection. Class II cabinets differ in the proportion of air recirculated within the cabinet; velocity of airflow to the work

TABLE 13-35-1. Laboratory Hood Ventilation Rates

	Condition	acfm/ft² [am³/s/m²] Open Hood Face
1.	Ceiling panels properly located with average panel face velocity < 40 fpm [0.20 m³/s].(13.35.2) Horizontal sliding sash hoods. No equipment in hood closer than 12 inches [300 mm] to face of hood. Hoods located away from doors and trafficways.*	60 [0.30]
2.	Same as 1 above; some traffic past hoods. No equipment in hoods closer than 6 inches to face of hood. Hoods located away from doors and trafficways.*	80 [0.40]
3.	Ceiling panels properly located with average panel face velocity < 60 fpm [0.30 m/s](13.35.2) or ceiling diffusers properly located; no diffuser immediately in front of hoods; quadrant facing hood blocked; terminal throw velocity < 60 fpm [0.30 m/s]. No equipment in hood closer than 6 inches [150 mm] to face of hood. Hoods located away from doors or trafficways.*	80 [0.40]
4.	Same as 3 above; some traffic past hood. No equipment in hood closer than 6 inches [150 mm] to face of hood.	100 [0.50]
5.	Wall grilles are possible but not recommended for advance planning of new facilities.	

*Hoods near doors are acceptable if: 1) there is a second safe egress from the room; 2) traffic past hood is low; and 3) door is normally closed.

surface; where the exhausted air is discharged; and whether the contaminated air plenum is under positive pressure. A Type A cabinet (VS-35-10) may discharge the exhausted air, after HEPA filtration, directly into the room. Type A cabinets that discharge into the work area are not recommended for use with gases or vapors. A primary application is for sterile packaging. Care is required while decontaminating the cabinet.

Type B hoods (VS-35-11) discharge the exhaust but may recirculate within the cabinet. Type B1 cabinets recirculate about 30% of the air within the BSC and typically exhaust the remainder outside the laboratory (i.e., exhaust air is not discharged back into the room). The contaminated plenum is under negative pressure. Type B2 cabinets are referred to as "total exhaust" cabinets as the contaminated air is exhausted to the atmosphere after HEPA filtration without recirculation in the cabinet or return to the laboratory room air. Type B3 BSCs have HEPA filtered downflow air that is a portion of the mixed downflow and inflow air from a common exhaust plenum.

Class III BSCs (VS-35-20) provide the highest level of protection to personnel and the environment. The cabinet is totally enclosed with operations conducted through attached gloves. See "National Sanitation Foundation Standard No. 49"(13.35.4) for descriptions and requirements of the various classes of BSCs.

13.35.3 Clean Benches. Clean benches can be divided into laminar flow and exhausted clean benches.

Laminar flow clean benches provide product protection only. In a laminar flow clean bench, room air is HEPA-filtered, directed across the work area, and discharged back to the room. Air may be directed horizontally as depicted in VS-35-30 or vertically as in VS-35-31. Neither of these hoods provide worker protection. Workers using the Horizontal Laminar Flow Clean Bench are exposed to the product as the air sweeps across the product into the worker's face. Workers's arms or other objects protruding into the Vertical Laminar Flow Clean Bench opening may cause contaminated air to spill into the room. Personal protective equipment or general ventilation should be provided as needed.

Other types of clean benches incorporate the same general principles of biological safety cabinets and utilize HEPA filtered laminar flow within the hood to provide product protection and exhaust sufficient air to ensure flow into the hood at the face to provide operator protection.

13.35.4 Laboratory Equipment. Some laboratory equipment such as evaporation hoods (VS-35-40), discharge from instruments such as ICP or AA, and some ovens (VS-35-41) require local exhaust ventilation to adequately control contaminant releases. Often, especially designed ventilation specific to the operation provides better control than using a laboratory hood to control these releases.

REFERENCES

13.35.1 Knutson, G.W.: Effect of Slot Position on Laboratory Fume Hood Performance. Heating, Piping and Air Conditioning. (February 1984).

13.35.2 Caplan, K.J.; Knutson, G.W.: Influence of Room Air Supply on Laboratory Hoods. American Industrial Hygiene Association Journal. Vol. 43, pp 738-746 (October 1982).

13.35.3 American Society of Heating, Refrigerating and Air Conditioning Engineers: ANSI/ASHRAE Standard 110-1995, Method of Testing the Performance of Laboratory Fume Hoods. ASHRAE, Atlanta, GA (1995).

13.35.4 National Sanitation Foundation: Standard 49, Class II (Laminar Flow) Biohazard Cabinetry. NSF, Ann Arbor, MI (1987).

13.35.5 U.S. Air Force Technical Order 00-25-203: Standards and Guidelines for Design and Operation of Clean Rooms and Clean Work Stations. Office of Technical Services, Department of Commerce, Washington, DC (July 1963).

13.35.6 Harris, W.P.; Christofano, E.E.; Lippman, M.: Combination Hot Plate and Hood for Multiple Beaker Evaporation. American Industrial Hygiene Association Journal, 22(4) (August 1961).

FIGURE VS-35-01

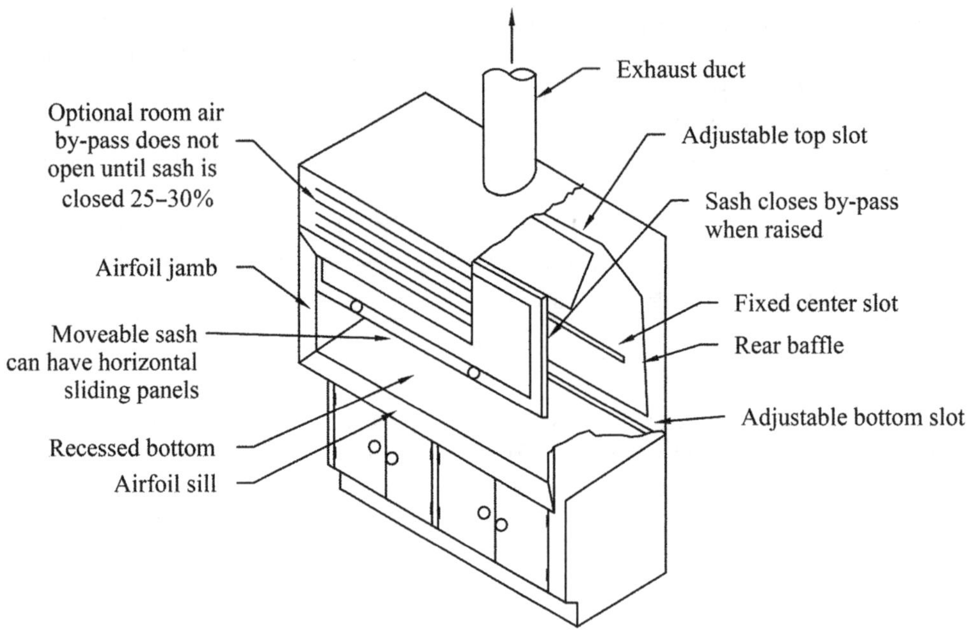

VERTICAL SASH AIRFOIL HOOD

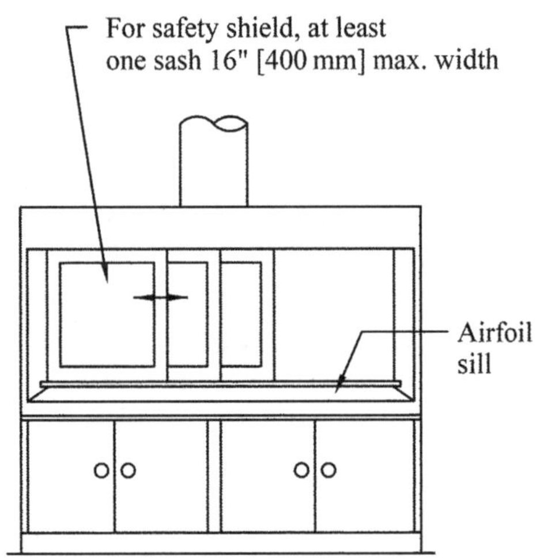

HORIZONTAL SASH AIRFOIL HOOD

Q = 80–100 acfm/ft^2 [0.4–0.5 am^3/s/m^2] full open area depending on quality of supply air distribution and uniformity of face velocity

$h_e = 0.56\ VP_d$

Duct velocity = 1000–2000 fpm [5–10 m/s] to suit conditions

Design specifications:

General use laboratory hoods – See VS-35-02

Perchloric acid – See VS-35-03

"Auxiliary Air" or "Compensating" hoods furnish some replacement air at hood face, design varies with vendor.

Work practices – See VS-35-04

ACGIH®	TITLE: TYPICAL LABORATORY HOOD	FIGURE: VS-35-01
		DATE: 1-12

CHECK CODES, REGULATIONS, AND LAWS (LOCAL, STATE, AND NATIONAL) TO ENSURE THAT DESIGN IS COMPLIANT.

FIGURE VS-35-02

GENERAL USE LABORATORY HOODS

A. Provide uniform exhaust air distribution in hood. Adjust baffles and airflow for $\leq$10% variation in point-to-point face velocity with sash in maximum open position.

B. Locate hood away from heavy traffic aisles and doorways. Hoods near doors are acceptable if: 1) there is a second safe means of egress from room, 2) traffic past hood is low, and 3) door is normally closed.

C. Use corrosion-resistant materials suitable for expected use.

D. Provide air cleaning on exhaust air if necessary and adequate stack height to minimize re-entry of contaminants or to comply with air pollution regulations.

E. Avoid sharp corners at jambs and sill. Tapered or round hood inlets are desirable; an airfoil shroud at sill is important.

F. Provide filters for radioactive materials in greater than "exempt" quantities.

G. By-pass opening in hood is desirable to avoid excessive indraft under partially closed sash condition. Opening to be baffled to prevent splash from eruption in hood as shown in VS-35-01.

H. Provide tempered or conditioned replacement air to laboratory. Replacement air volume to be selected for desired air balance with adjoining spaces.

I. In order to reduce airflow volumes, local exhaust hoods should be considered instead of laboratory bench hoods for fixed setups.

J. For air conservation, use horizontal sliding sash with airfoil sill.

K. All bench hoods should have a recessed work surface and airfoil sill.

ACGIH®	TITLE GENERAL USE LABORATORY HOODS	FIGURE VS-35-02 DATE 1-12

CHECK CODES, REGULATIONS, AND LAWS (LOCAL, STATE, AND NATIONAL) TO ENSURE THAT DESIGN IS COMPLIANT.

FIGURE VS-35-03

PERCHLORIC ACID HOODS

Perchloric acid is extremely dangerous because it is a very strong oxidizer. When the acid reacts with organic material, an explosive reaction product may be formed.

1. Do not use perchloric acid in a hood designed for other purposes. Identify perchloric acid hoods with large warning signs.

2. Provide exhaust ventilation and room supply air with minimal challenge to the hood.

3. Utilize local exhaust ventilation within the hood to minimize condensation of vapors inside the hood.

4. Locate all utility controls outside the hood.

5. Materials of construction for this type of hood and duct must be nonreactive, and acid resistant, and relatively impervious. AVOID ORGANIC MATERIALS unless known to be safe. Stainless steel type 316 with welded joints is preferred. Unplasticized polyvinyl chloride or an inorganic ceramic coating, such as porcelain, is acceptable.

6. Ease of cleanliness is paramount. Use stainless steel within accessible rounded corners and all-welded construction.

7. The work surface should be watertight with a minimum of 0.5 inch [13 mm] dished front and sides and an integral trough at the rear to collect the washdown water.

8. Design washdown facilities into the hood and duct. Use daily or more often to thoroughly clean perchloric acid from the exhaust system surfaces.

9. Each perchloric acid hood should have an individual exhuast system. Slope horizontal runs to drain. Avoid sharp turns.

10. Construct the hood and duct to allow easy visual inspection.

11. Where required, use a high-efficiency (greater than 80%) wet collector constructed for perchloric acid service. Locate as close to the hood as possible to minimize the accumulation of perchloric acid in the exhaust duct.

12. Use only an acid-resistant metallic fan protected by an inorganic coating or an air injector.

13. Lubricate the fan with fluorocarbon-type grease.

14. Locate the fan outside the building.

15. The exhaust discharge must terminate out-of-doors, preferably using a vertical discharge stack that extends well above the roof eddy zone.

ACGIH®	TITLE: PERCHLORIC ACID HOOD DATA	FIGURE: VS-35-03 / DATE: 1-12

CHECK CODES, REGULATIONS, AND LAWS (LOCAL, STATE, AND NATIONAL) TO ENSURE THAT DESIGN IS COMPLIANT.

FIGURE VS-35-04

WORK PRACTICES FOR LABORATORY HOODS

No large, open-face hood with a low face velocity can provide complete safety for a worker standing at the face against all events that may occur in the hood. The hood may not adequately protect the worker from volatile or otherwise airborne contaminants with a TLV in the low part-per-billion range. For more ordinary exposures, a properly designed hood in a properly ventilated room can provide adequate protection. However, certain work practices are necessary in order for the hood to perform capably. The following work practices are generally required; more stringent practices may be necessary in some circumstances.

1. Conduct all operations that may generate air contaminants at or above the appropriate TLV inside a hood.
2. Keep all apparatus at least 6 inches [150 mm] back from the face of the hood. A stripe on the bench surface is a good reminder.
3. Do not put your head in the hood when contaminants are being generated.
4. Do not use the hood as a waste disposal mechanism except for very small quantities of volatile materials.
5. Do not store chemicals or apparatus in the hood. Store hazardous chemicals in an approved safety cabinet.
6. Keep the hood sash closed as much as possible.
7. Keep the slots in the hood baffle free of obstruction by apparatus or containers.
8. Minimize foot traffic past the face of the hood.
9. Keep laboratory doors closed (exception: some laboratory designs require lab doors to be open).
10. Do not remove hood sash or panels except when necessary for apparatus set-up; replace sash or panels before operating.
11. Do not place electrical receptacles or other spark sources inside the hood when flammable liquids or gases are present. No permanent electrical receptacles are permited in the hood.
12. Use an appropriate barricade if there is a chance of explosion or eruption.
13. Provide adequate maintenance for the hood exhaust system and the building supply system. Use static pressure gauges on the hood throat, across any filters in the exhaust system, or other appropriate indicators to ensure flow is appropriate.
14. If hood sash is supposed to be partially closed for the operation, the hood should be so labeled and the appropriate closure point clearly indicated.

ACGIH®	TITLE WORK PRACTICES FOR LABORATORY HOODS	FIGURE VS-35-04 DATE 1-12

CHECK CODES, REGULATIONS, AND LAWS (LOCAL, STATE, AND NATIONAL) TO ENSURE THAT DESIGN IS COMPLIANT.

FIGURE VS-35-10

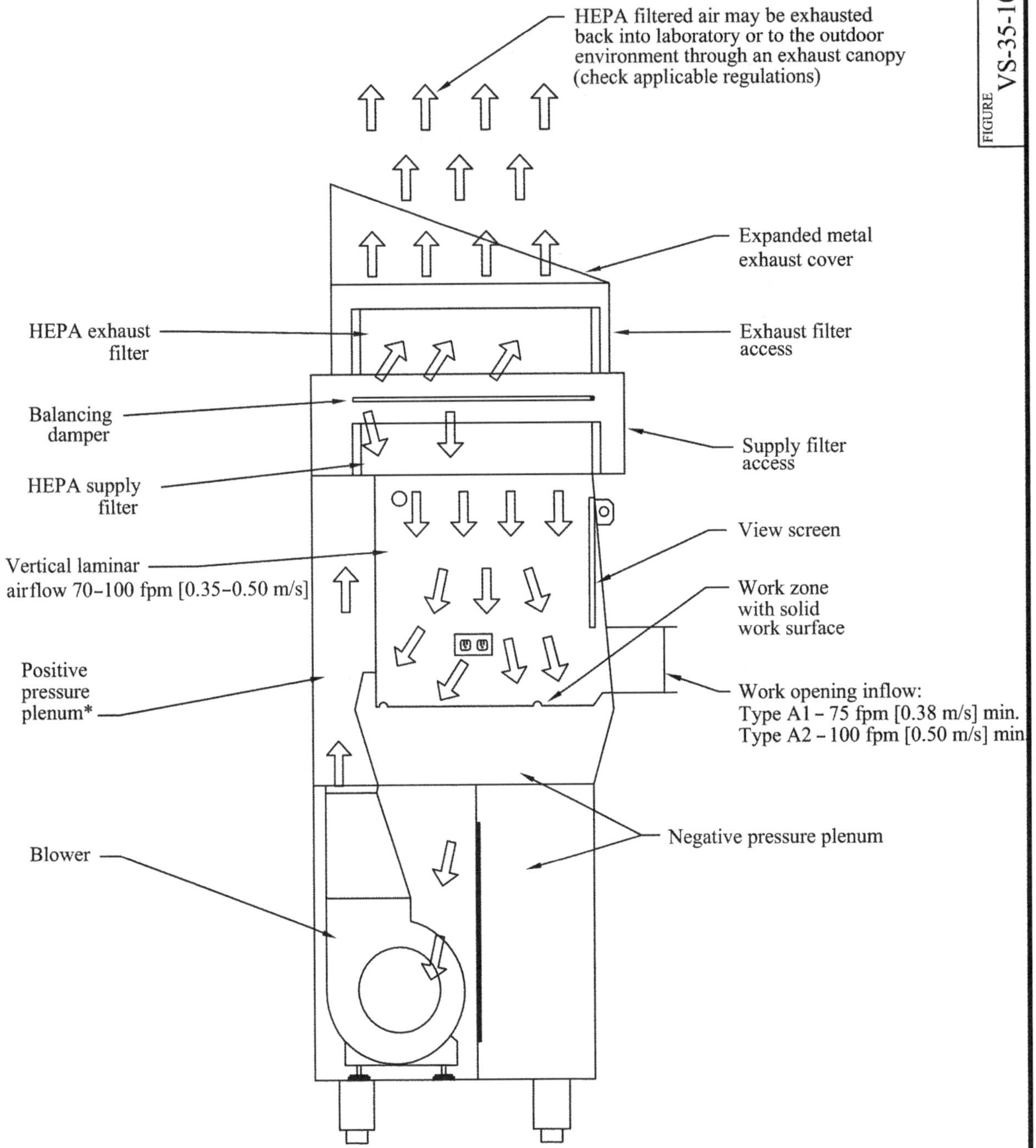

CLASS II TYPE A

* Type A2 cabinets require all biologically contaminated ducts and plenums under negative pressure or surrounded by negative pressure ducts and plenums.

ACGIH®	TITLE BIOLOGICAL SAFETY CABINET CLASS II, TYPE A	FIGURE VS-35-10 / DATE 1-12

CHECK CODES, REGULATIONS, AND LAWS (LOCAL, STATE, AND NATIONAL) TO ENSURE THAT DESIGN IS COMPLIANT.

FIGURE VS-35-11

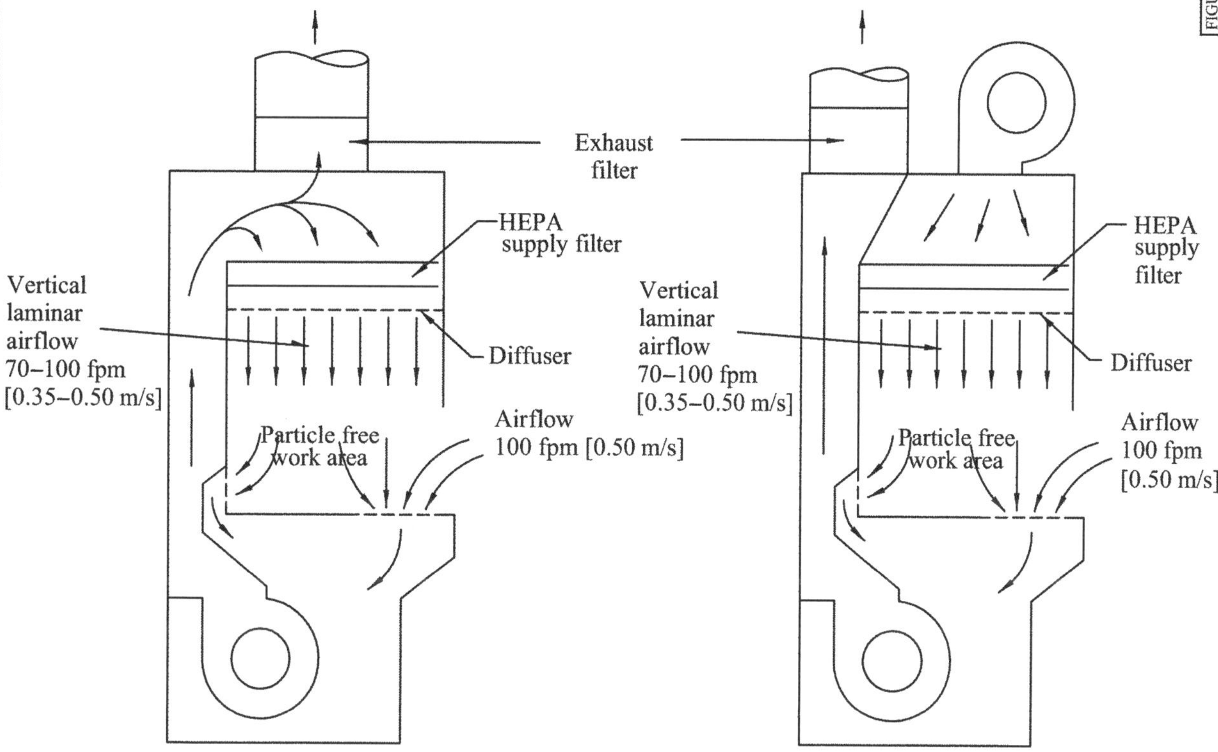

RECIRCULATING AIR HOOD *
TYPE B1

100% EXHAUST AIR HOOD
TYPE B2

* Recirculating Air Hoods are not recommended for use with gases or vapors.

Note: See "National Sanitation Foundation Standard 49" (Ref. 13.35.4) for requirements and definitions of types and classes.

For product protection only, see VS-35-30 and VS-35-31.

All Type B cabinets require biologically contaminated ducts and plenums to be under negative pressure or surrounded by negative pressure ducts and plenums.

ACGIH®	TITLE BIOLOGICAL SAFETY CABINET CLASS II, TYPE B	FIGURE VS-35-11 DATE 1-12

CHECK CODES, REGULATIONS, AND LAWS (LOCAL, STATE, AND NATIONAL) TO ENSURE THAT DESIGN IS COMPLIANT.

FIGURE VS-35-20

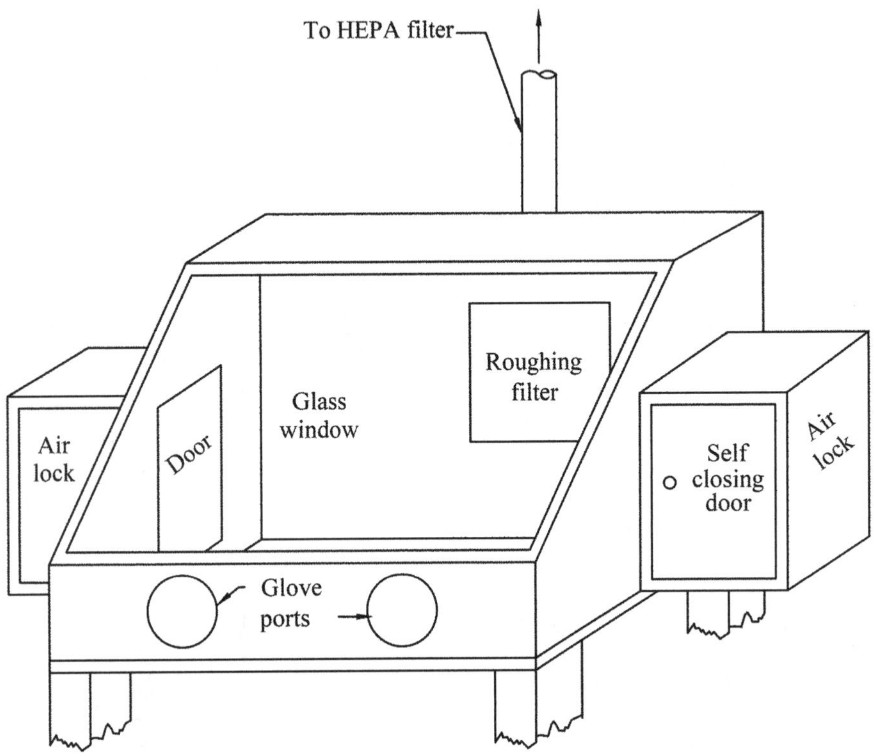

Q = 50 acfm/ft^2 [0.03 am^3/s/m^2] of open door area and 0.25" wg [60 Pa] SP on a closed system.
$h_e = 0.56\ VP_d$
Duct velocity = 2000–4000 fpm [10–20 m/s]
Filters : 1. Inlet air filters in doors
2. Roughing filter at exhaust connection to hood
3. HEPA filter
All facilities totally enclosed in hood. Exterior controls may be advisable.
Arm length rubber gloves are sealed to glove port rings.
Strippable plastic on interior and air cleaner on exhaust outlet may be used to facilitate decontamination of the system.
Filter units may be installed in the doors to allow the airflow necessary for burners etc.
For filters, see Chapter 6.

ACGIH®	TITLE DRY BOX OR GLOVE HOOD FOR HIGH TOXICITY AND RADIOACTIVE MATERIALS	FIGURE VS-35-20
		DATE 1-12

CHECK CODES, REGULATIONS, AND LAWS (LOCAL, STATE, AND NATIONAL) TO ENSURE THAT DESIGN IS COMPLIANT.

FIGURE VS-35-30

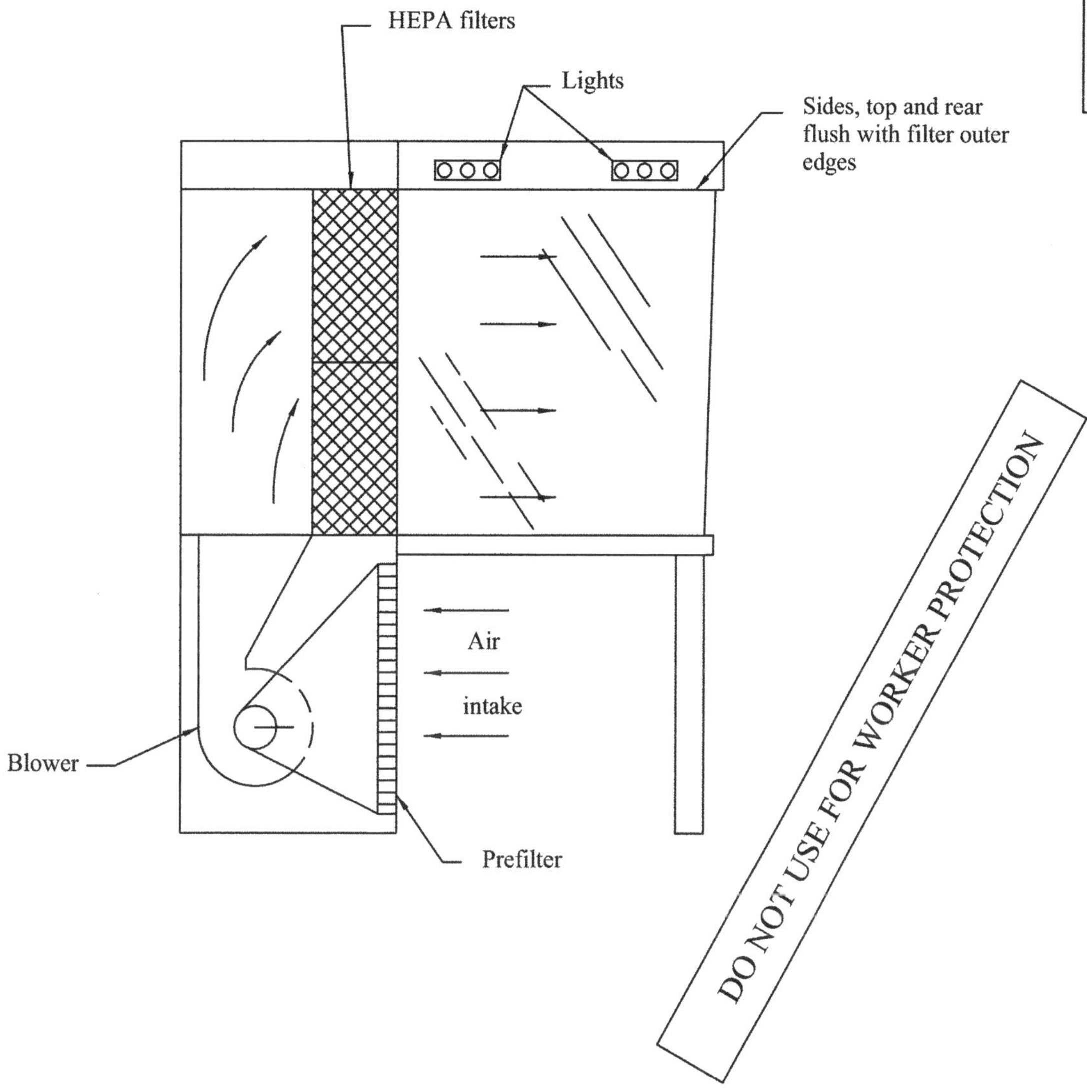

Face velocity = 90 fpm [0.45 m/s] ± 20 fpm [0.10 m/s]

Note : Total power input must be considered as part of air conditioning load.

This hood does not provide protection for the operator.

Do not use with toxic material.

Reference 13.35.5

ACGIH®	TITLE HORIZONTAL LAMINAR FLOW CLEAN BENCH (PRODUCT PROTECTION ONLY)	FIGURE VS-35-30 DATE 1-12

CHECK CODES, REGULATIONS, AND LAWS (LOCAL, STATE, AND NATIONAL) TO ENSURE THAT DESIGN IS COMPLIANT.

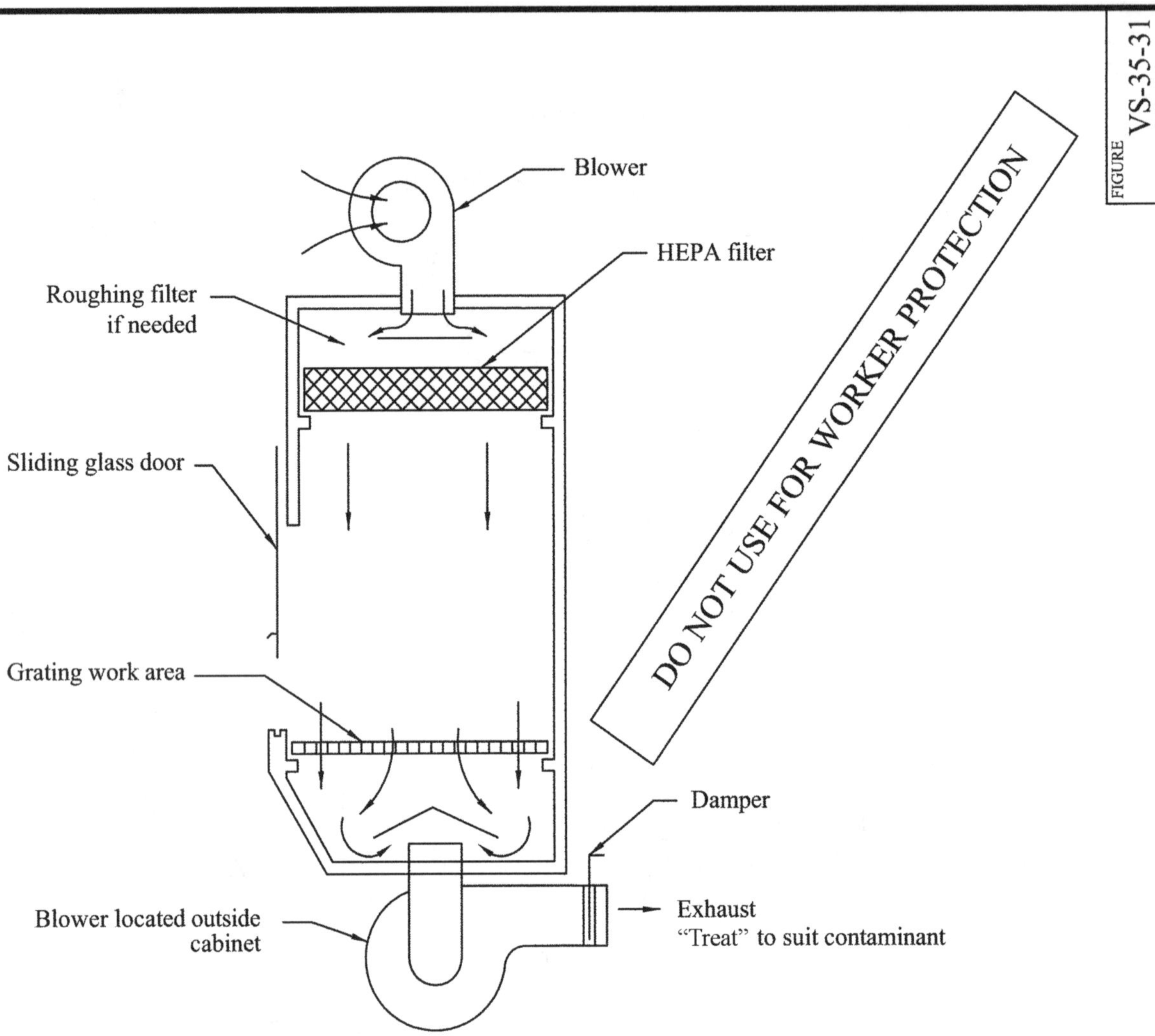

Vertical velocity = 90 fpm [0.45 m/s] with average minimum uniformity ± 20 fpm [0.10 m/s]

Duct velocity = 2000 – 4000 fpm [10–20 m/s] to suit conditions

Clean station for control of airborne particulates

Notes: Supply and exhaust should be maintained equal by flow meter control techniques.

This hood does not provide protection for the operator.

Do not use with toxic material.

ACGIH®	TITLE VERTICAL LAMINAR FLOW CLEAN BENCH (PRODUCT PROTECTION ONLY)	FIGURE VS-35-31 DATE 1-12

CHECK CODES, REGULATIONS, AND LAWS (LOCAL, STATE, AND NATIONAL) TO ENSURE THAT DESIGN IS COMPLIANT.

FIGURE VS-35-40

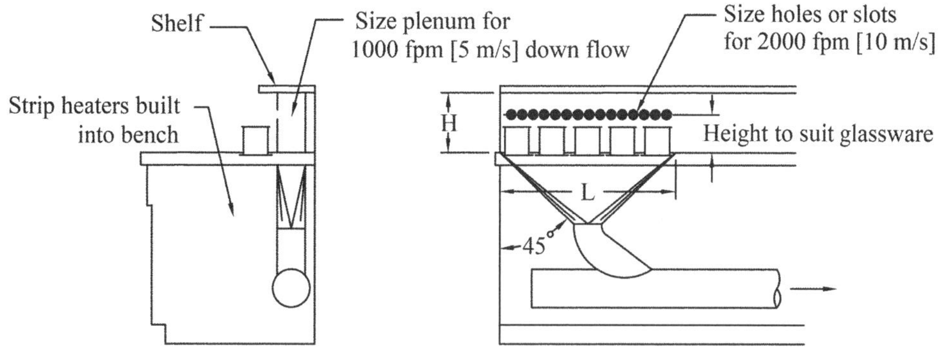

EVAPORATION BENCH

Q = 20 acfm/foot of hood [0.03 am^3/s/meter of hood] or 50 HL [0.25 HL]

Minimum duct velocity = 2000 fpm [10.00 m/s]

$h_e = 1.78VP_s + 0.25VP_d$

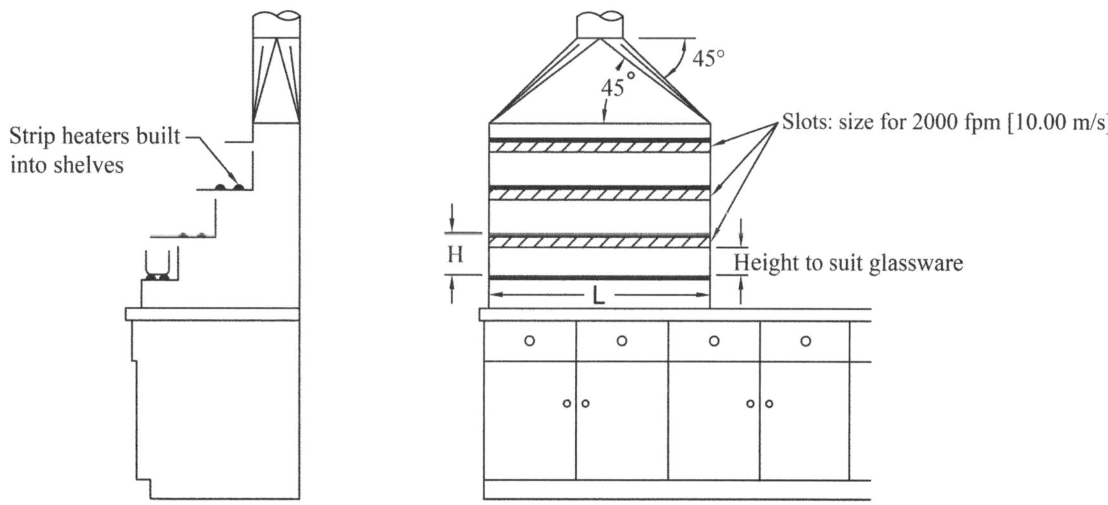

EVAPORATION HOOD

Q = 20 acfm/foot of hood [0.03 am^3/s/meter of hood] or 50 HL [0.25 HL] for each shelf

Minimum duct velocity = 2000 fpm [10.00 m/s]

$h_e = 1.78VP_s + 0.25VP_d$

Reference 13.35.6

ACGIH®	TITLE SPECIALIZED LABORATORY HOOD DESIGNS	FIGURE VS-35-40 DATE 1-12

CHECK CODES, REGULATIONS, AND LAWS (LOCAL, STATE, AND NATIONAL) TO ENSURE THAT DESIGN IS COMPLIANT.

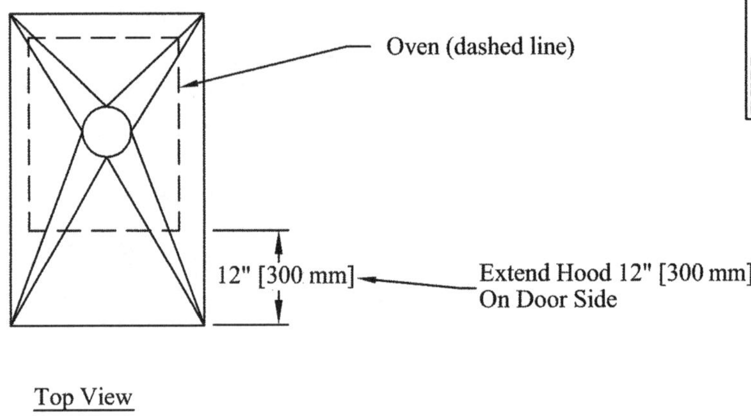

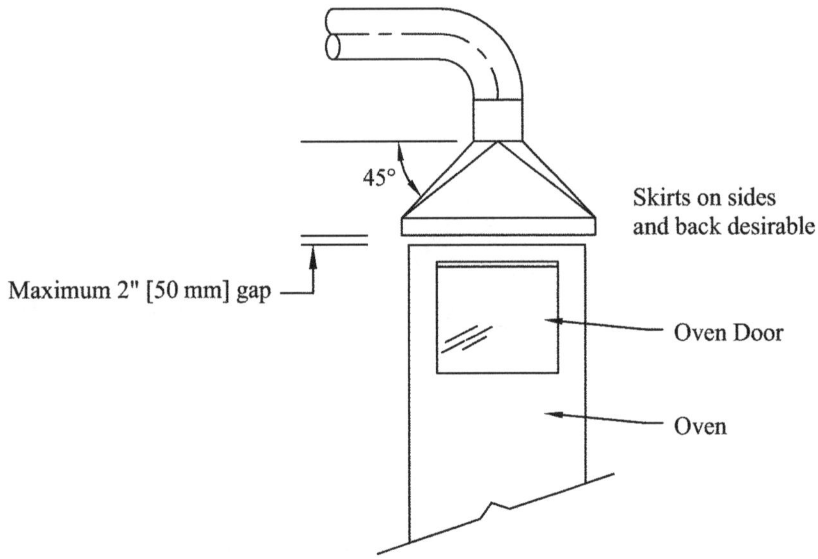

$Q = 200 - 400$ acfm [0.10–0.20 am^3/s]

Recommended duct velocity = 1000 – 3000 fpm [5–15 m/s]

$h_e = 0.25\ VP_d$ (based on transition angle)

Note: Consider heat from oven when sizing exhaust fan

ACGIH®	TITLE SMALL LABORATORY OVEN EXHAUST	FIGURE VS-35-41 DATE 1-12

CHECK CODES, REGULATIONS, AND LAWS (LOCAL, STATE, AND NATIONAL) TO ENSURE THAT DESIGN IS COMPLIANT.

FIGURE VS-35-41

13.40 LOW VOLUME-HIGH VELOCITY EXHAUST SYSTEMS

The low volume-high velocity (LVHV) exhaust system is a unique application of exhaust that uses small volumes of air at high velocities to control dust from portable hand tools and machining operations. Control is achieved by exhausting the air directly at the point of dust generation using close-fitting, custom-made hoods. Capture velocities are relatively high but the exhaust volume is low due to the small distance required. For flexibility, small diameter, light-weight plastic hoses are used with portable tools. This results in relatively high duct velocities, but allows the application of local exhaust ventilation to portable tools that otherwise would require larger flow rates and large duct sizes when controlled by conventional exhaust methods. The resulting additional benefit is the reduction of replacement air requirements.

This technique has found a variety of applications. Rock drilling dust has been controlled by using hollow core drill steel with suitable exhaust holes in the drill bits. Air is exhausted either by a multi-stage turbine, positive-displacement blowers or regenerative blowers (ring compressors), of the size generally used in industrial vacuum cleaners or, in the case of one manufacturer,[13.40.1] by the exhaust air from the pneumatic tool that operates a Venturi to withdraw air from the drill. Some applications use flexible connections to a central vacuum system to aid in the control of graphite dust at conventional machining operations. Hood static pressures (SP_h) are typically chosen in excess of 2 "Hg (27.2 "wg) [6.8 kPa] and minimum duct design velocities are typically greater than 5000 fpm [25 m/s]. One- to two-inch diameter flexible hoses are used with simple exhaust hoods mounted directly at the cutting tool. In a similar application for the machining of beryllium,[13.40.2] a central vacuum system utilizing 1.5-inch [38 mm] I.D. flexible hoses was employed. The exhaust hoods were made of lucite or transparent material and were tailor-made to surround the cutting tools and much of the work. Exhaust flow rates vary from 120 to 150 acfm [0.60 to 0.75 am^3/s] with inlet velocities of 11,000 to 14,000 fpm [55.00 to 70.00 m/s]. In another application, a portable orbital sanding machine has been fitted with a small exhaust duct surrounding the edge of the plate. A fitting has been provided to connect this to the flexible hose of a standard domestic vacuum cleaner.

VS-40-01 to VS-40-06 illustrate a custom-made line of exhaust hoods.[13.40.3] The required airflow rates range from 60 acfm for pneumatic chisels to 380 acfm for swing grinders. Due to the high entering velocities involved, system static pressures are in the range of 7" to 14" of mercury (95 to 190 "wg) [24 kPa to 48 kPa]. This high pressure is necessary to create the high capture velocities at the source to control the dust because the dust is typically generated at very high velocities. In addition, small flex lines can be significantly less cumbersome and will fit in tight spaces.. However, there are disadvantages associated with high velocities: 1) small metal parts can be sucked into the hood; 2) coolants may be disturbed; and 3) very high noise levels may be produced.

13.40.1 Design—Calculations. With the exception of the proprietary system mentioned, which was developed as a "package," the design calculations for these systems are largely empirical and little performance data are available for the user. In normal ventilation practice, air is considered to be incompressible since static pressures vary only slightly from atmospheric pressure. However, in LVHV systems, the extreme pressures required introduce problems of air density, compressibility, and viscosity, which are not easily solved. Also, pressure drop data for small diameter pipe, especially flexible tubing, is not commonly available. For practical purposes, the turbine exhauster should be selected for the maximum simultaneous exhaust flow rate required. Resistance in the pipe should be kept as low as possible; flexible tubing of less than 1 to 1.5-inch [25 to 38 mm] diameter should be limited to 10 feet [0.3 m] or less. In most applications, this is not a severe problem. To aid in static pressure assessment, see Figure 13-40-1.

The main consideration in piping for such systems is to provide smooth internal configuration so as to reduce pressure loss at the high velocities involved and to minimize abrasion. Ordinary pipe with threaded fittings is to be avoided because the lip of the pipe or male fitting, being of smaller diameter than the female thread, presents a discontinuity that increases pressure loss and may be a point of rapid abrasion. Duct velocities are the same as found in Chapter 5 except in cases where the mass of the contaminant contributes to a change in density of the airflow (+5%). Surges in material flow must be considered and an instantaneous "worse case density" should dictate power regiments.

For dust exhaust systems, a good dust collector and primary separator should be mounted ahead of the exhauster to minimize erosion of the precision blades and subsequent loss in performance. Final balance of the system can be achieved by varying the length and diameters of the small flexible hoses.

It must be emphasized that although data are empirical, LVHV systems require the same careful design as the more conventional ones. Abrupt changes of direction, expansions and contractions must be avoided and care must always be taken to minimize pressure losses.

The extra weight of the hood mounted on the tool may cause ergonomic issues and must also be considered.

REFERENCES

13.40.1 Clean Air Technologies, Inc., Concord, Ontario, Canada (2009).

13.40.2 Chamberlin, R.I.: The Control of Beryllium Machining Operations. A.M.A. Archives of Industrial Health, 19, No. 2 (February 1959).

13.40.3 Hoffman Air and Filtration Div., Hoffman Turbine, Division of Gardner Denver, Peachtree, GA (2004).

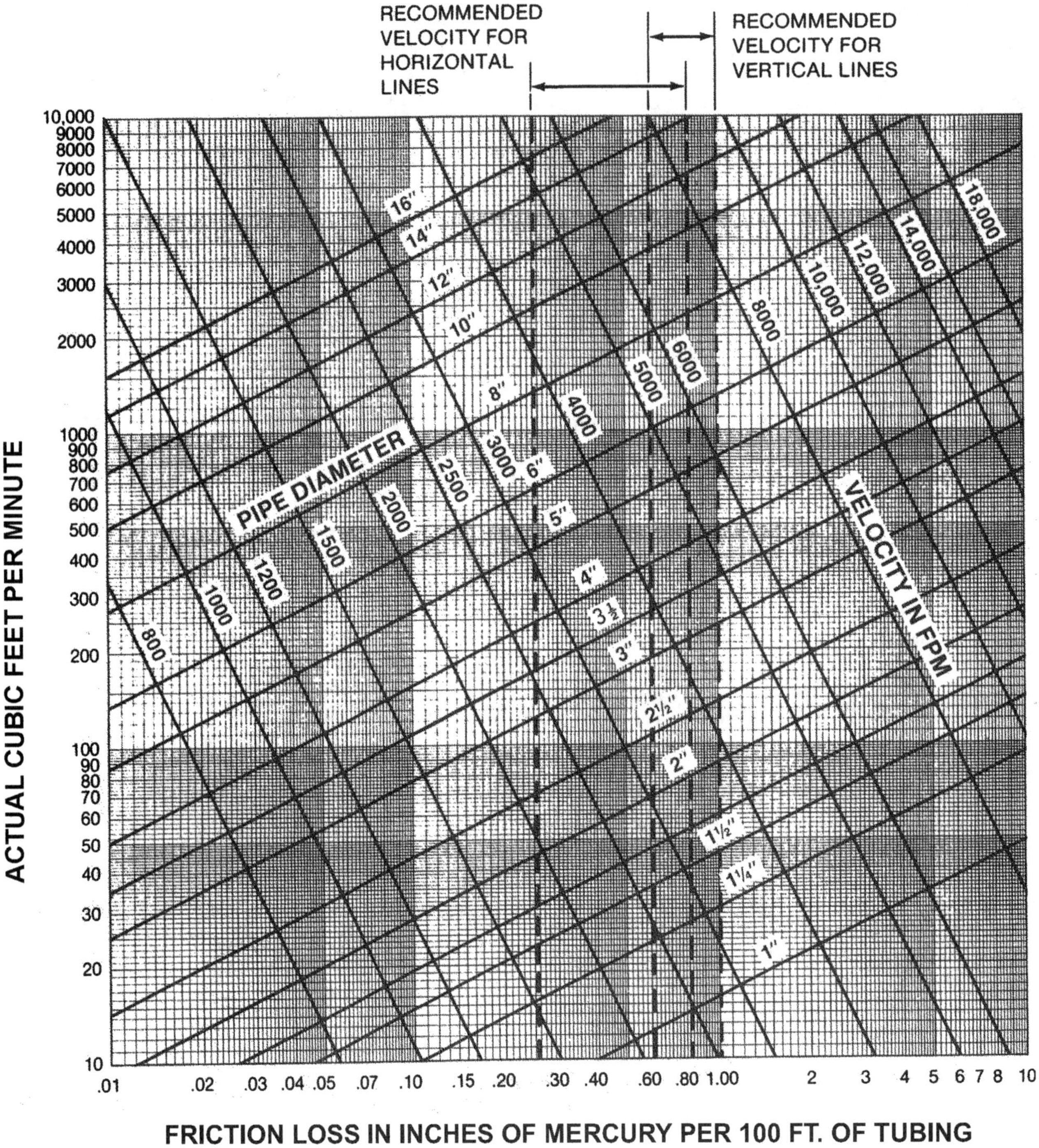

FIGURE 13-40-1 (IP). Friction loss chart — vacuum applications

FIGURE VS-40-01

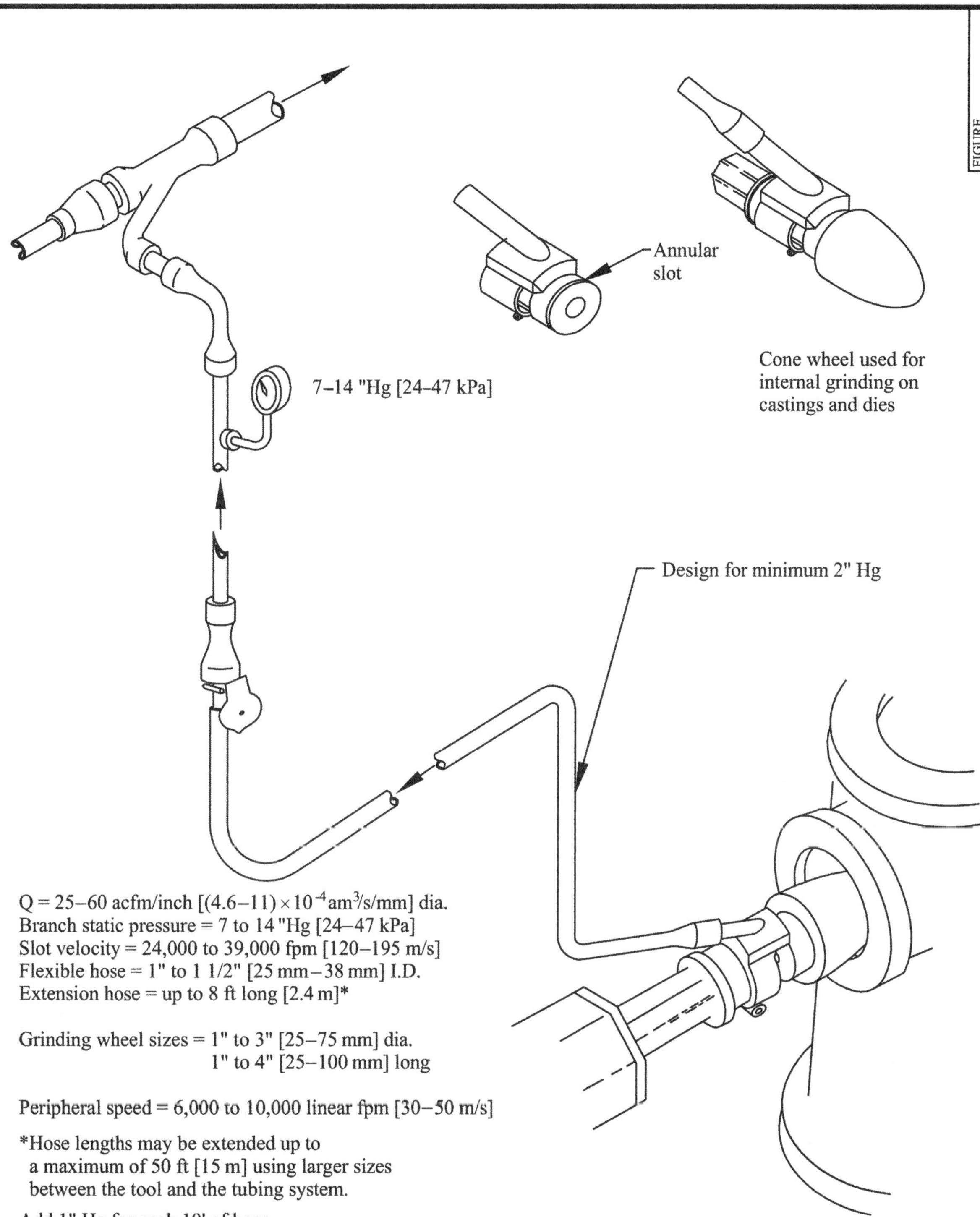

Q = 25–60 acfm/inch [(4.6–11) × 10^{-4} am^3/s/mm] dia.
Branch static pressure = 7 to 14 "Hg [24–47 kPa]
Slot velocity = 24,000 to 39,000 fpm [120–195 m/s]
Flexible hose = 1" to 1 1/2" [25 mm–38 mm] I.D.
Extension hose = up to 8 ft long [2.4 m]*

Grinding wheel sizes = 1" to 3" [25–75 mm] dia.
1" to 4" [25–100 mm] long

Peripheral speed = 6,000 to 10,000 linear fpm [30–50 m/s]

*Hose lengths may be extended up to a maximum of 50 ft [15 m] using larger sizes between the tool and the tubing system.

Add 1" Hg for each 10' of hose.

Reference 13.40.4

ACGIH®	TITLE EXTRACTOR HEAD FOR CONE WHEELS AND MOUNTED POINTS	FIGURE VS-40-01 DATE 1-12

CHECK CODES, REGULATIONS, AND LAWS (LOCAL, STATE, AND NATIONAL) TO ENSURE THAT DESIGN IS COMPLIANT.

FIGURE VS-40-02

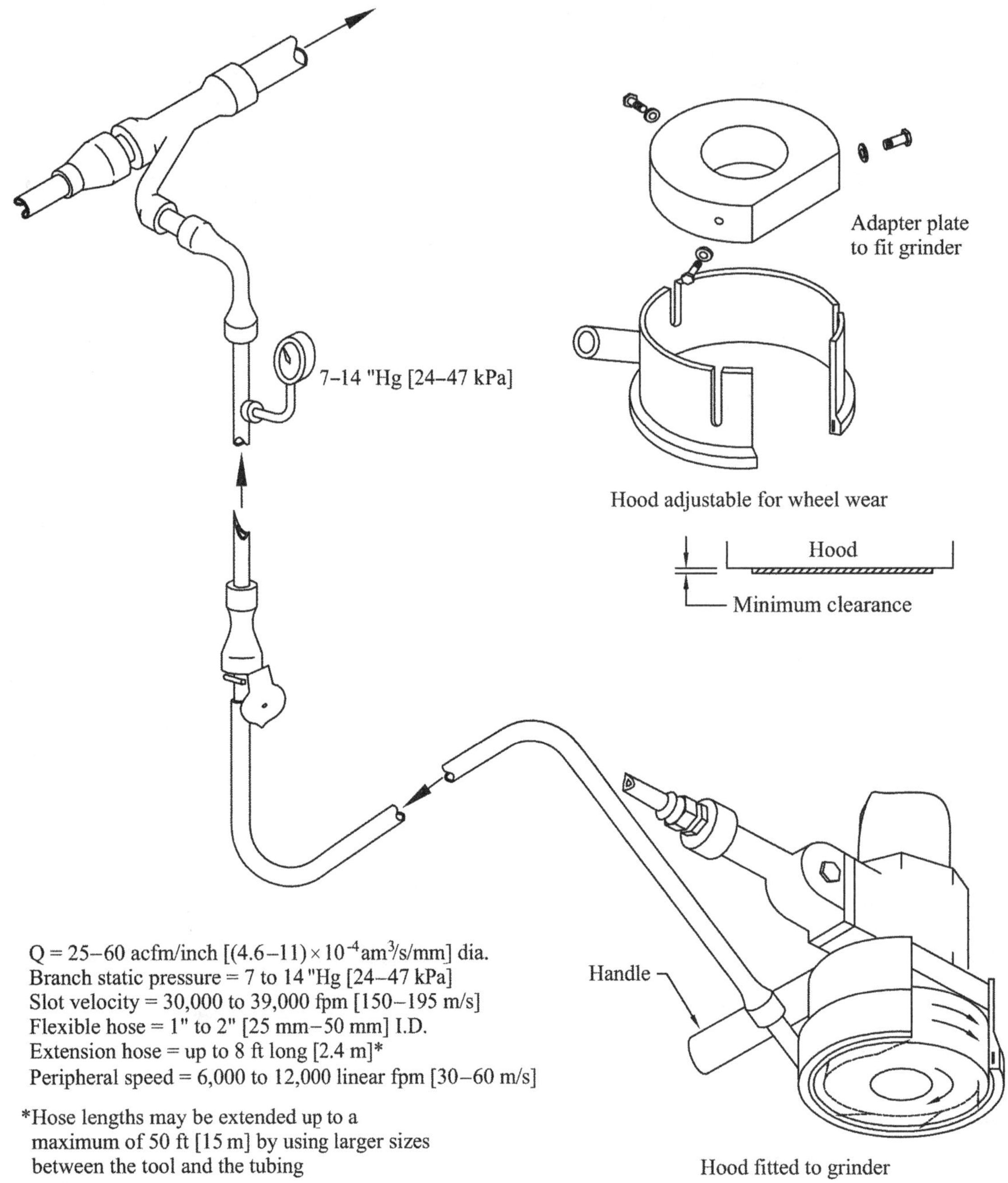

Q = 25–60 acfm/inch [(4.6–11) × 10^{-4} am^3/s/mm] dia.
Branch static pressure = 7 to 14 "Hg [24–47 kPa]
Slot velocity = 30,000 to 39,000 fpm [150–195 m/s]
Flexible hose = 1" to 2" [25 mm–50 mm] I.D.
Extension hose = up to 8 ft long [2.4 m]*
Peripheral speed = 6,000 to 12,000 linear fpm [30–60 m/s]

*Hose lengths may be extended up to a maximum of 50 ft [15 m] by using larger sizes between the tool and the tubing system.

Reference 13.40.4

ACGIH®	TITLE: HOOD FOR CUP TYPE SURFACE GRINDER AND WIRE BRUSHES	FIGURE: VS-40-02 / DATE: 1-12

CHECK CODES, REGULATIONS, AND LAWS (LOCAL, STATE, AND NATIONAL)
TO ENSURE THAT DESIGN IS COMPLIANT.

FIGURE VS-40-03

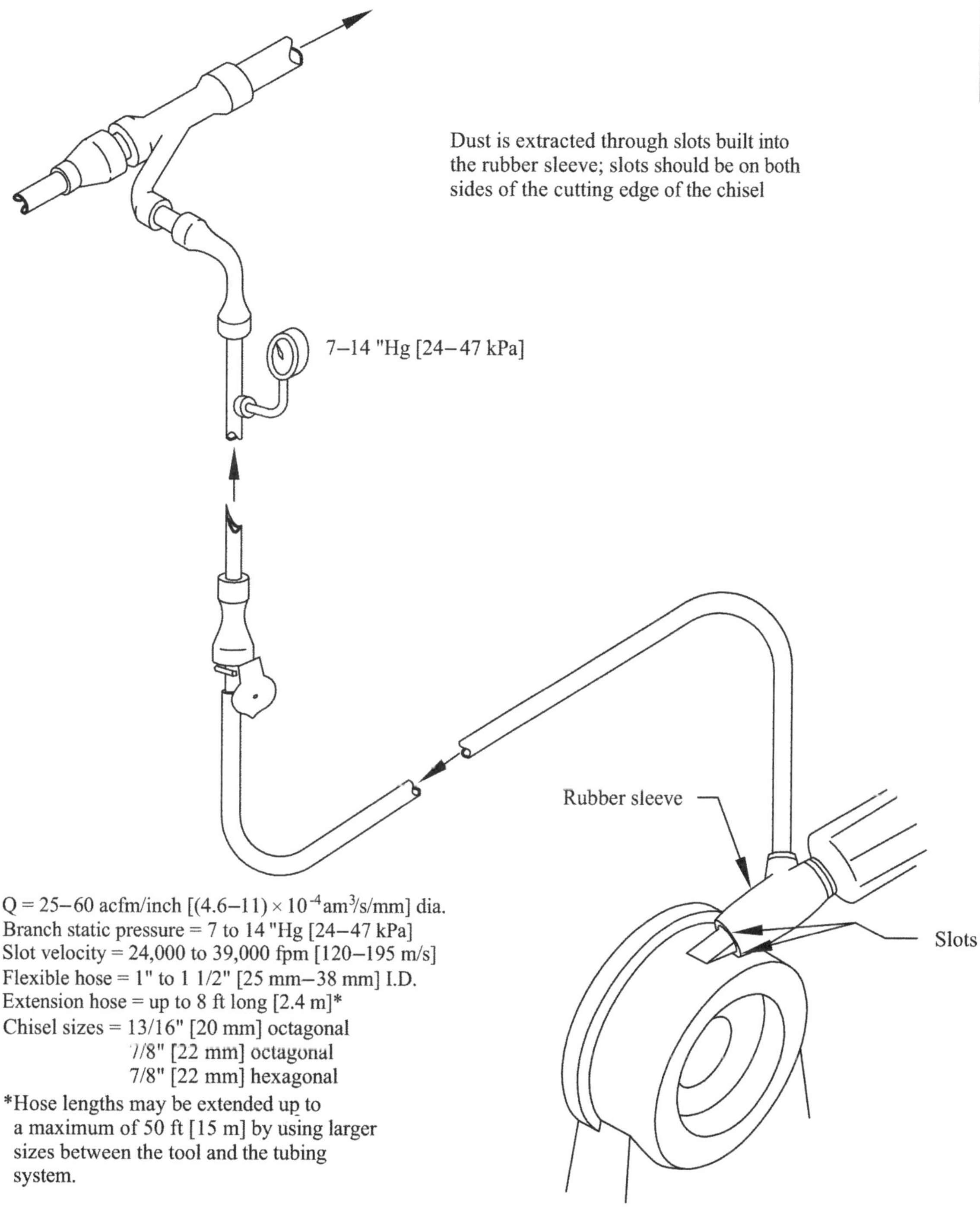

Q = 25–60 acfm/inch [(4.6–11) × 10^{-4} am^3/s/mm] dia.
Branch static pressure = 7 to 14 "Hg [24–47 kPa]
Slot velocity = 24,000 to 39,000 fpm [120–195 m/s]
Flexible hose = 1" to 1 1/2" [25 mm–38 mm] I.D.
Extension hose = up to 8 ft long [2.4 m]*
Chisel sizes = 13/16" [20 mm] octagonal
7/8" [22 mm] octagonal
7/8" [22 mm] hexagonal

*Hose lengths may be extended up to a maximum of 50 ft [15 m] by using larger sizes between the tool and the tubing system.

Reference 13.40.4

ACGIH®	TITLE CHISEL SLEEVE, PNEUMATIC	FIGURE VS-40-03 DATE 1-12

CHECK CODES, REGULATIONS, AND LAWS (LOCAL, STATE, AND NATIONAL) TO ENSURE THAT DESIGN IS COMPLIANT.

FIGURE VS-40-04

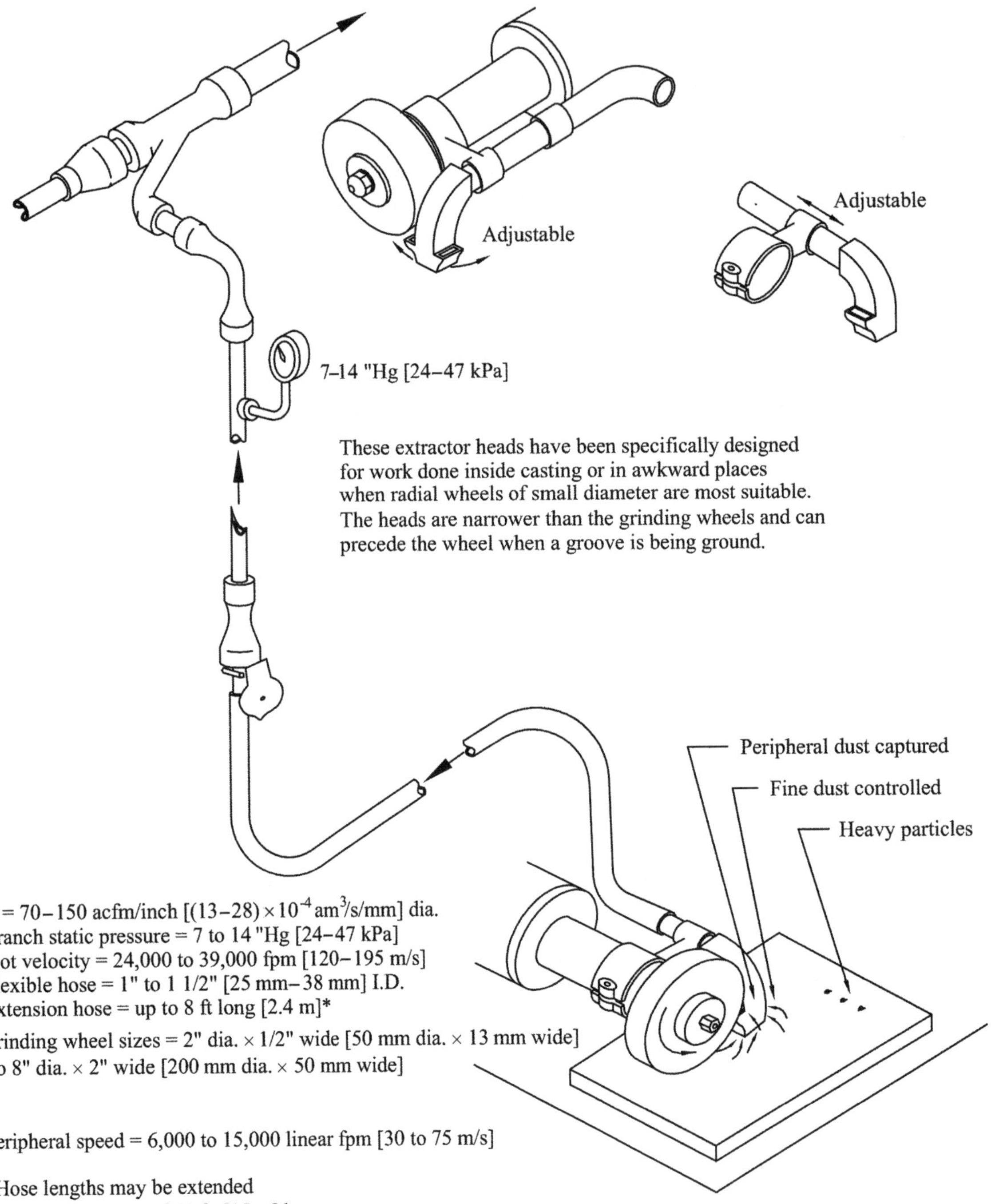

These extractor heads have been specifically designed for work done inside casting or in awkward places when radial wheels of small diameter are most suitable. The heads are narrower than the grinding wheels and can precede the wheel when a groove is being ground.

Q = 70–150 acfm/inch [(13–28) × 10^{-4} am^3/s/mm] dia.
Branch static pressure = 7 to 14 "Hg [24–47 kPa]
Slot velocity = 24,000 to 39,000 fpm [120–195 m/s]
Flexible hose = 1" to 1 1/2" [25 mm–38 mm] I.D.
Extension hose = up to 8 ft long [2.4 m]*
Grinding wheel sizes = 2" dia. × 1/2" wide [50 mm dia. × 13 mm wide]
To 8" dia. × 2" wide [200 mm dia. × 50 mm wide]

Peripheral speed = 6,000 to 15,000 linear fpm [30 to 75 m/s]

*Hose lengths may be extended up to a maximum of 50 ft [15 m] by using larger sizes between the tool and the tubing system.

Reference 13.40.4

ACGIH®	TITLE EXTRACTOR HEAD FOR GRINDERS, SMALL RADIAL	FIGURE VS-40-04
		DATE 1-12

CHECK CODES, REGULATIONS, AND LAWS (LOCAL, STATE, AND NATIONAL) TO ENSURE THAT DESIGN IS COMPLIANT.

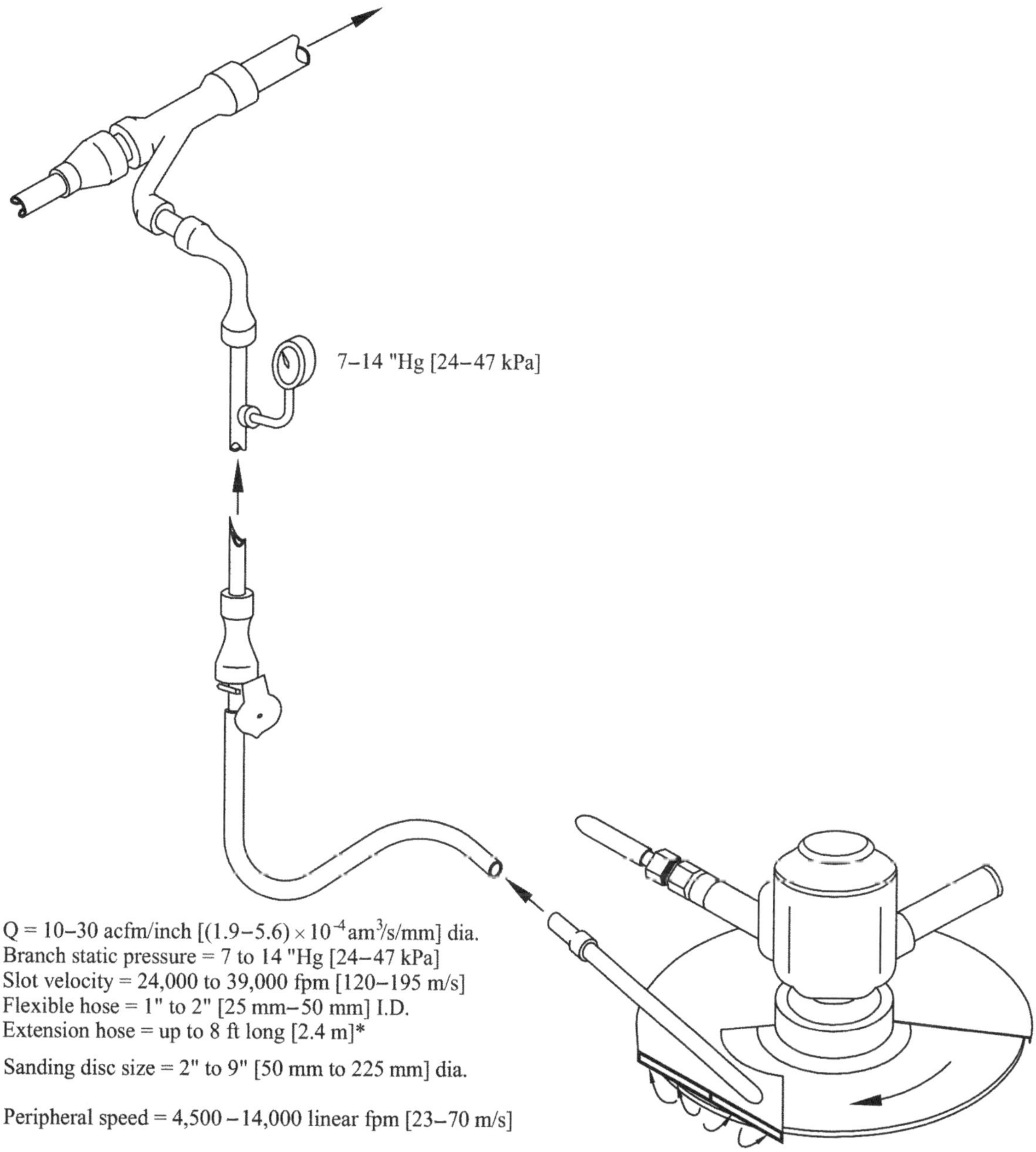

Q = 10–30 acfm/inch [(1.9–5.6) × 10^{-4} am^3/s/mm] dia.
Branch static pressure = 7 to 14 "Hg [24–47 kPa]
Slot velocity = 24,000 to 39,000 fpm [120–195 m/s]
Flexible hose = 1" to 2" [25 mm–50 mm] I.D.
Extension hose = up to 8 ft long [2.4 m]*

Sanding disc size = 2" to 9" [50 mm to 225 mm] dia.

Peripheral speed = 4,500 – 14,000 linear fpm [23–70 m/s]

*Hose lengths may be extended up to a maximum of 50 ft [15 m] by using larger sizes between the tool strike and the tubing system.

Reference 13.40.4

ACGIH®	TITLE EXTRACTOR HOOD FOR DISC SANDER	FIGURE VS-40-05 DATE 1-12

CHECK CODES, REGULATIONS, AND LAWS (LOCAL, STATE, AND NATIONAL) TO ENSURE THAT DESIGN IS COMPLIANT.

FIGURE VS-40-05

FIGURE VS-40-06

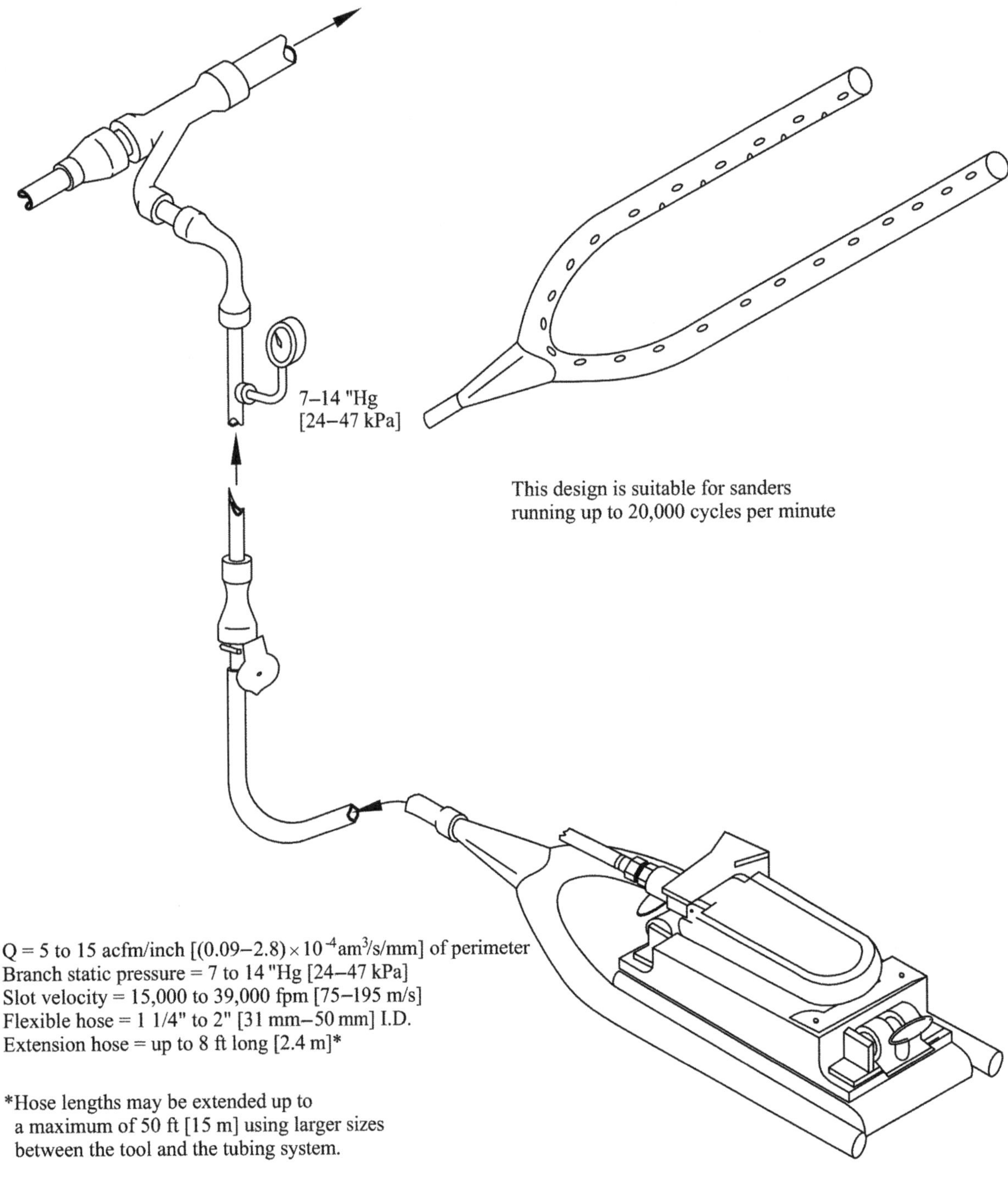

This design is suitable for sanders running up to 20,000 cycles per minute

Q = 5 to 15 acfm/inch [(0.09–2.8) $\times$ 10^{-4} $am^3/s/mm$] of perimeter
Branch static pressure = 7 to 14 "Hg [24–47 kPa]
Slot velocity = 15,000 to 39,000 fpm [75–195 m/s]
Flexible hose = 1 1/4" to 2" [31 mm–50 mm] I.D.
Extension hose = up to 8 ft long [2.4 m]*

*Hose lengths may be extended up to a maximum of 50 ft [15 m] using larger sizes between the tool and the tubing system.

Reference 13.40.4

ACGIH®	TITLE EXTRACTOR TOOL FOR VIBRATORY SANDER, HAND	FIGURE VS-40-06
		DATE 1-12

CHECK CODES, REGULATIONS, AND LAWS (LOCAL, STATE, AND NATIONAL) TO ENSURE THAT DESIGN IS COMPLIANT.

FIGURE VS-40-20 (IP)

System Notes:

Bell and socket, smooth-flow type tubing and fittings should be used throughout the system.

When system is used for vacuum cleaning of abrasive materials, Schedule No. 40 pipe and cast iron drainage fittings, or heavier, should be used in place of tubing.

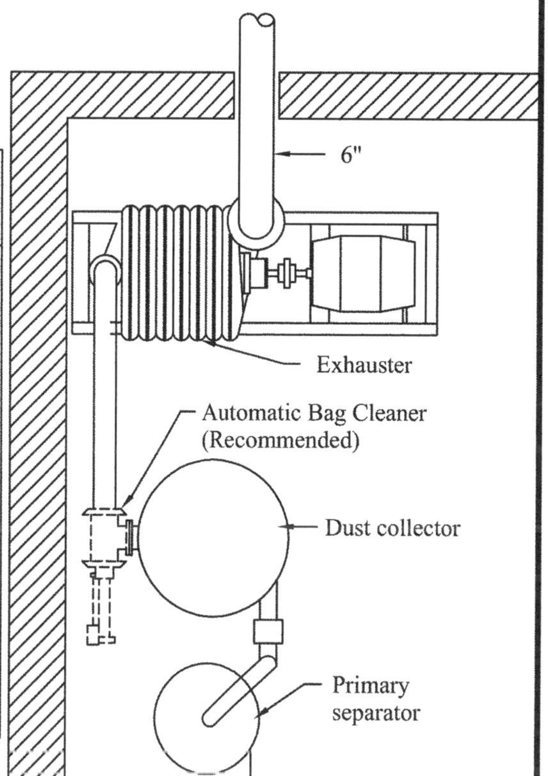

	acfm	I.D. Plastic Hose Size (inches)
Disk sanders 3–9 inch diam.	60–175	1–1.5
Vibratory pad sander – 4" × 9"	100	1.25
Router, 1/8"–1"	80–100	1–1.25
Belt sander 3"–4000 fpm	70	1
Pneumatic chisel	60	1
Radial wheel grinder	70	1
Surface die grinder, 1/4"	60	1
Cone wheel grinder	90	1.25
Cup stone grinder, 4"	100	1.25
Cup type brush, 6"	150	1.5
Radial wire brush, 6"	90	1.25
Hand wire brush 3" × 7"	60	1
Rip out knife	175	1.5
Rip out cast cutter	150	1.5
Saber saw	120	1.5
Swing frame grinder 2" × 18"	380	2.5
Saw abrasive 3"	100	1.25

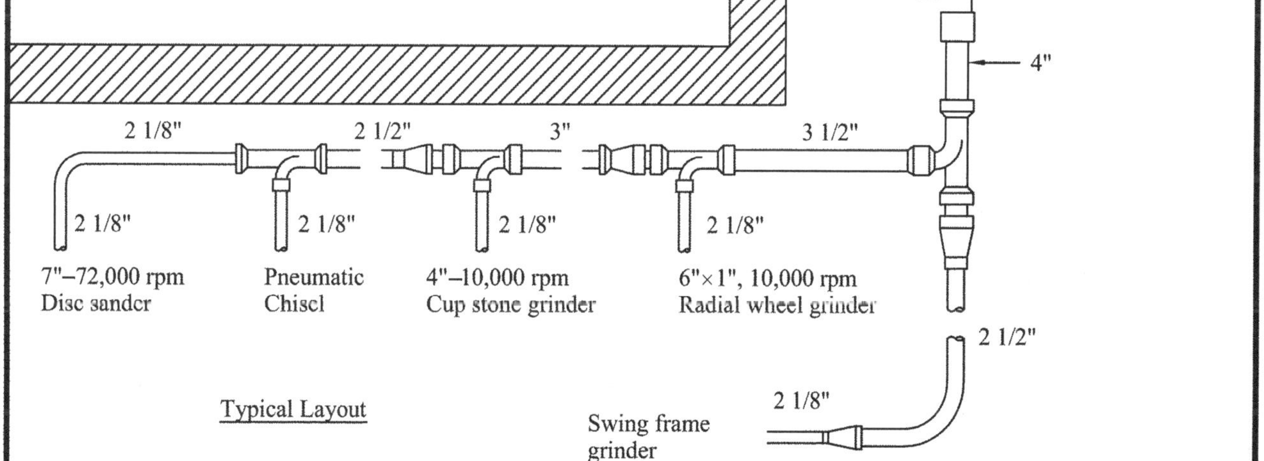

Typical Layout

Reference 13.40.4

ACGIH®

TITLE

TYPICAL SYSTEM
LOW VOLUME-
HIGH VELOCITY

FIGURE VS-40-20 (IP)

DATE 1-12

CHECK CODES, REGULATIONS, AND LAWS (LOCAL, STATE, AND NATIONAL) TO ENSURE THAT DESIGN IS COMPLIANT.

FIGURE VS-40-20 (SI)

System Notes:

Bell and socket, smooth-flow type tubing and fittings should be used throughout the system.

When system is used for vacuum cleaning of abrasive materials, Schedule DN 1000 pipe and cast iron drainage fittings, or heavier, should be used in place of tubing.

	m^3/s	I.D. Plastic Hose Size (mm)
Disk sanders 75–225 mm diam.	0.03–0.088	25–38
Vibratory pad sander – 100–225 mm	0.05	31
Router, 3–25 mm	0.04–0.05	25–31
Belt sander 75 mm–20 m/s	0.035	25
Pneumatic chisel	0.03	25
Radial wheel grinder	0.035	25
Surface die grinder, 6 mm	0.03	25
Cone wheel grinder	0.045	31
Cup stone grinder, 100 mm	0.05	31
Cup type brush, 150 mm	0.075	38
Radial wire brush, 150 mm	0.045	31
Hand wire brush 75 mm × 175 mm	0.03	25
Rip out knife	0.088	38
Rip out cast cutter	0.075	38
Saber saw	0.06	38
Swing frame grinder 50 mm–450 mm	0.19	63
Saw abrasive 75 mm	0.05	31

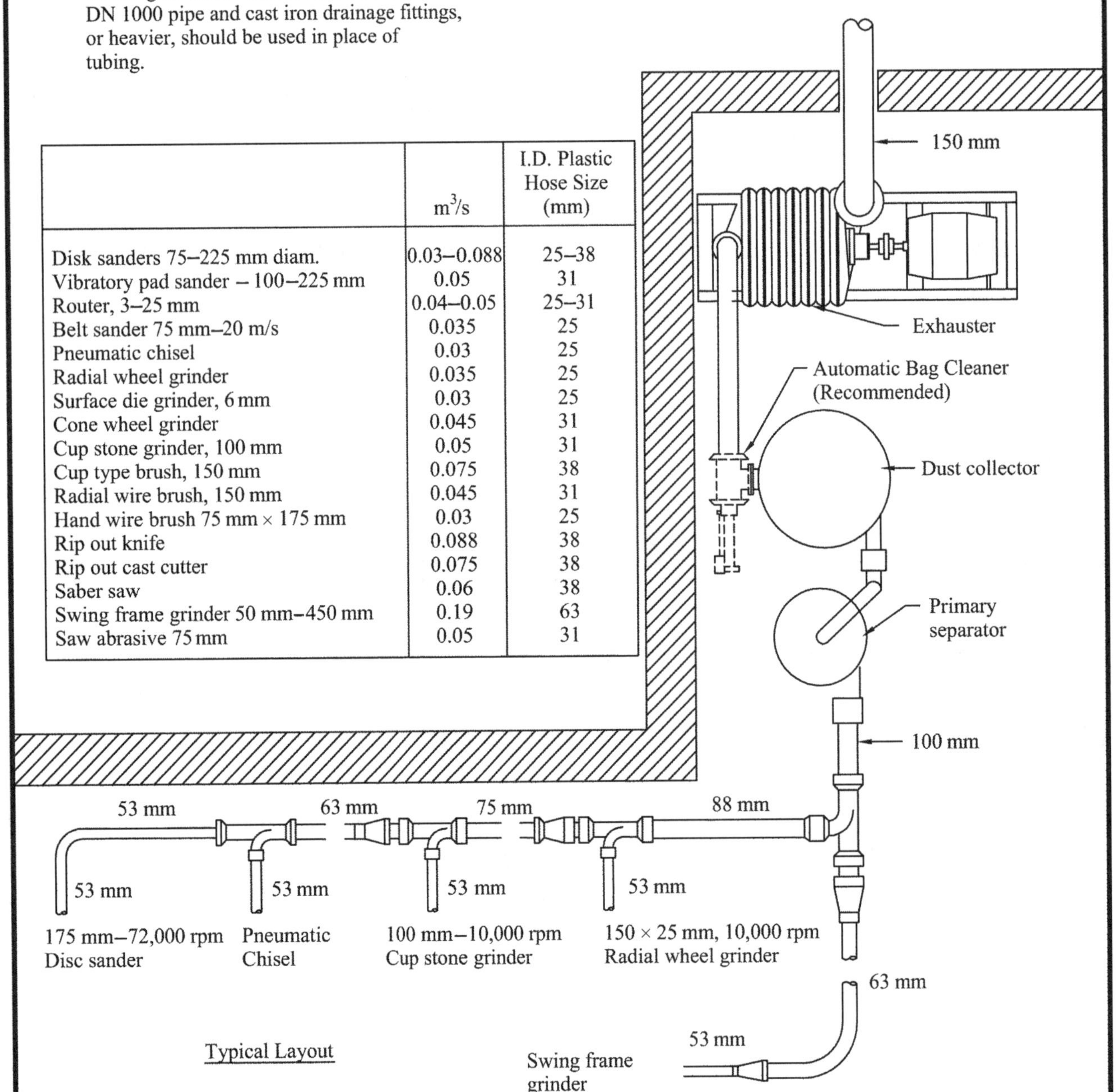

Typical Layout

Reference 13.40.4

ACGIH®

TITLE

TYPICAL SYSTEM
LOW VOLUME-
HIGH VELOCITY

FIGURE VS-40-20 (SI)

DATE 1-12

CHECK CODES, REGULATIONS, AND LAWS (LOCAL, STATE, AND NATIONAL) TO ENSURE THAT DESIGN IS COMPLIANT.

13.45 MACHINING

The primary purpose of metal cutting machines is to finish rough parts, formed by other processes, to specific dimensions. Finishing and shaping can be accomplished by a variety of cutting tools such as saws, broaches, and chisel-shaped tool bits held in fixtures with fixed or movable drives. Cutting is accomplished by creating relative motion between the tool edge and the material blank. Chips of varying sizes are produced; chip size depends on the material being cut, feed rate of the tool, and relative speed or feed between the tool and the metal being shaped.

Non-traditional methods of metal cutting and shaping include electrochemical, electrodischarge, wire electrodischarge, and laser beam machining. With the exception of the laser beam, each of the processes utilizes a circulating oil or water-base dielectric to facilitate molecular erosion as well as to remove process heat and particulate debris. The rate of metal removal is controlled by regulating the flow of electrical current between the shaped anode or wire and the work piece. The laser beam is used in a dry environment and metal cutting is accomplished by vaporizing the work piece along the cutting edge with a focused beam of high energy light. The process is flexible and a variety of metallic and non-metallic materials can be shaped by this means.[13.45.1]

It is estimated that up to 97% of the work involved in conventional metal cutting results in heat. The rate of heat removal must be controlled carefully in order to protect both the cutting tool and the metallurgy of the work being cut. Where convection or radiant cooling is insufficient, a cutting fluid can be used to reduce friction, carry away generated heat, and, more commonly, flush away metal chips produced by the cutting process. Cutting fluids include straight chained and synthetic mineral oils as well as soluble oil emulsions in water. A variety of water soluble lubri-coolants (1 to 5% mixture of lubricants, emulsifiers, rust inhibitors and other chemicals in water) are commonly used, particularly for high speed metal working machines. In some applications, the lubri-coolant mixture is applied as a mist by using a small volume of liquid in a high velocity air stream. In the more usual situation, liquid is applied by flooding the tool in the cutting zone to flush away cutting debris. The latter type of system requires a low pressure pump with valves, filters, settling chamber to separate the fluid from the chips, and a reservoir, which permits recirculation. Where liquids cannot be used, low temperature nitrogen or carbon dioxide gas can be used as a cooling media for both the tool and the cutting surface as well as a means of dispersing particulate debris.

The hazards created by skin exposure to the water lubri-coolant mixtures, particulates, and oil mist/vapor produced in the transfer of heat is best handled with engineering control – primarily ventilation. An additional health concern is the fact that soluble oil emulsions provide a breeding ground for bacteria and, therefore, it is common practice to add biocides to prevent odor generation and decomposition of the oil mixture. Biocides and other additives may be primary skin irritants or cause hypersensitive dermatitis. It is for these reasons that mist, vapors, and particulates must be controlled adequately. Mist and vapor from metalworking fluids can be reduced by:

1) Minimizing fluid delivery pressures.
2) Matching fluid composition to application.
3) Using low oil concentrations where possible.
4) Minimizing tramp oil contamination.
5) Maintaining control of metalworking fluid chemistry.
6) Covering fluid reservoirs and return systems as much as possible.
7) Maintaining a low metalworking fluid temperature.

Mist and vapors from machining operations can be controlled by a combination of machine enclosure and local exhaust ventilation. Exhaust hoods and enclosures should be designed so the machine can be serviced easily and the operation observed when required. Hood sides should act as splash guards since an indraft of air will not stop liquid directly thrown from rotating parts. All components should be robust and rigidly supported. To facilitate maintenance, service and tool adjustment portions of the hood enclosure that are not permanently fixed should be designed for easy removal. Thought should also be given to the use of sliding, hinged, or bellows-connected panels in locations where frequent access is required. All windowed openings must be shatter-proof with appropriate internal lighting. All non-fixed panels should be designed with overlapping, drip-proof edges. The use of gaskets or seals on abutting panels is not recommended. Ventilation rates vary; however, a minimum of 150 fpm [0.75 m/s] indraft is usually required to prevent vapor and mist from exiting the enclosure. Care must be taken to locate hood openings away from the direct path of coolant nozzled discharges or baffles used to prevent splashing at hood entrances. A typical machine enclosure may require a volumetric flow rate of from 400 to 500 acfm [0.20 to 0.25 am^3/s] depending on the hood and opening size. Additional air may be required to control heat generated by motors and other energy sources within the enclosure as well as to maintain adequate vision. Where coolant flumes are used for chip transport, additional exhaust ventilation is required to control air entrainment. Baffles above the liquid level are beneficial and flumes should be enclosed to the extent possible.

Local ventilation control is the preferred method – particularly in machine environments that are temperature controlled with refrigerated air conditioning systems. In more open workrooms, the use of dilution ventilation may be adequate to control air contaminants. For further information on dilution ventilation, see Chapter 2 of this manual.

13.45.1 General Design Considerations for Machine Enclosures.

1) Design enclosure with proper materials of construction that consider the metalworking fluid used, ambient

conditions or possible changes in metalworking fluid. Materials should be of sufficient strength for their own structural integrity and may be either mounted to the machine frame or supported separately from the floor, building steel, or adjacent equipment.

2) Design the enclosure with no sharp corners or extensions that may be a hazard to the operator or maintenance personnel.

3) Ergonomics should be considered in any enclosure design. Allow for access to maintain the machine and provide for tool changes. Enclosed wings that allow full tool retraction for service are recommended.

4) Enclosure should be fabricated with doors at access points. Hinged or sliding doors are recommended. Removable panels are not recommended because of the tendency not to replace them after removal.

5) Consideration should be given to the location of access doors used frequently versus other openings that are used sparingly. Safety interlocks are recommended to prevent machine operation if doors are open.

6) Design doors with drip edges on the inside to prevent metalworking fluids from dripping out of the enclosure.

7) Design the enclosure with consideration for external clearances such as cranes, other machines or building structure. This includes the removal of panels and the clearance required for major removal of the machine tool.

8) Include minimum 42" [1.1 m] clearance between the electrical enclosure and next grounded device. (See National Electrical Code requirements.)

9) The top of the enclosure should be slightly pitched to control internal metalworking fluid drippage from the roof.

10) Design machine utilities so that piping, wire ways, conduit and other utility accessories do not interfere with enclosure design, detract from efficiency of the enclosure or interfere with access doors or other opening.

11) Design machine tools so that chip conveyance is considered and the design of the enclosure does not impede the removal of chips or allow chips to accumulate.

12) The enclosure design should limit opening free area to minimize exhaust volumetric flow rate. This includes minimizing openings in areas that are not obvious such as chip chutes, flumes, or other entries into the machine.

13) Locate the enclosure duct take-offs so that metalworking fluid and chips are not directly removed. The purpose of the design is only to collect light contaminants and keep the enclosure under a negative pressure.

14) Openings should be limited only to those required for part entry and exit or utility penetration. Additional openings may be required to allow replacement air from the plant to enter the enclosure and mix with stagnant areas within the enclosure. It is important not to create a short circuit, which could direct the replacement air directly into the exhaust take-off without first mixing with the contaminants. Replacement air will help to condense metalworking fluid vapors.

15) Enclosures should be designed with tapered duct entry in order to minimize duct take-off velocity and entry losses.

16) High velocity duct entries or location of baffles that would restrict duct take-off openings and create the same effect as high velocity duct entries should be avoided to minimize entrainment of chips and fluid in the duct system. It is recommended that duct entry velocities be 2,000 fpm [10.00 m/s] or less for materials such as cast iron. Lower velocities may be required for less dense material such as aluminum or plastics. It is important to stay below the transport velocity of the chip/metalworking fluid, especially in the first vertical section leaving the hood (usually 1500 to 2000 fpm [7.50 to 10.00 m/s]).

17) If the machine contains electrical or laser, or other susceptible equipment, consideration must be given to protecting the environment they occupy (dust, mist, heat). It is preferred that the system be designed with this equipment outside the enclosure.

18) Part washers, pallet washers and blow-off station/booths are items that are mist–generating and need the same application of enclosure considerations as metal cutting equipment.

19) If the machine tool enclosure must be opened during operation, it will be necessary to increase the amount of air exhausted in order to purge the contaminants from the enclosure in a reasonable amount of time, prior to opening the door.

20) Some machine tools utilize oil mist lubricated bearings, which can be outside of the metal cutting machine enclosure. Special attention must be given to this mist producing lubrication method.

21) General safety considerations must be used in the design and all plant, state, and national codes must be incorporated. It is recommended that safety reviews and critiques be performed throughout the design and installation of the enclosure. Consideration must be given to confined space entry, fall protection, and safety lock-out regulations that may apply.

For additional information on control of metalworking fluids, see References 13.45.4, 13.45.5, 13.45.6 and 13.45.7.

REFERENCES

13.45.1 Rain, C.: Non-traditional Methods Advance Machining Industry. High Technology (November/ December 1957).

13.45.2 Schulte, H.F.; Hyatt, E.C.; Smith, Jr., F.S.: Exhaust Ventilation for Machine Tools Used on Materials of High Toxicity. American Medical Association Archives of Industrial Hygiene and Occupational Medicine, 5(21) (January 1952).

13.45.3 Mitchell, R.N.; Hyatt, E.C.: Beryllium – Hazard Evaluation and Control Covering a Five-Year Study. American Industrial Hygiene Quarterly, 18(3) (September 1957).

13.45.4 Johnston, W.J.: Oil Mist Control Systems, State-of-the-Art at Ford, Presented at the 50th Annual Industrial Ventilation Conference, Michigan State University (February 2001).

13.45.5 Adams, G.: Ventilation Control for Metalworking Fluids, GM Worldwide Facilities Group (2000).

13.45.6 ANSI B11 TR 2-1997, Mist Control Considerations for the Design, Installation and Use of Machine Tools Using Metalworking Fluids.

13.45.7 O'Brien, D.; Frede, J.C.: Guidelines for the Control of Exposure to Metal Working Fluids. National Institute for Occupational Safety and Health (February 1978).

FIGURE VS-45-01

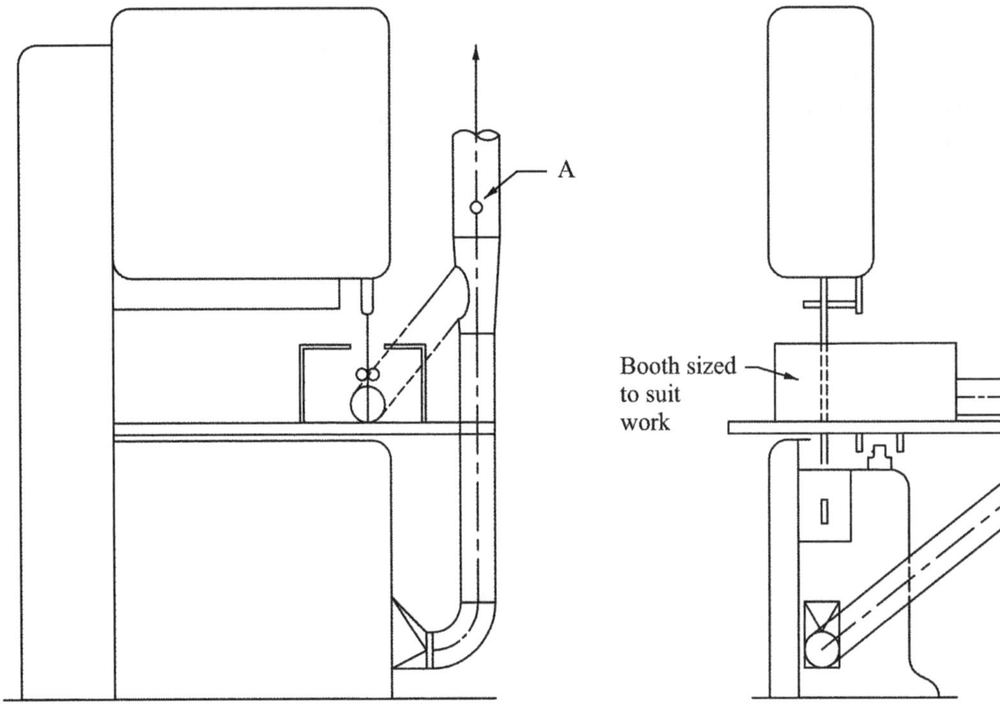

Q at booth = 225 acfm/ft^2 [1.13 am^3/s/m^2] open area
Q at bottom = 350 acfm [0.18 am^3/s]

Minimum duct velocity = 4000 fpm [20 m/s]

h_e = 1.75 VP_d, at point A

ACGIH®	TITLE METAL CUTTING BANDSAW	FIGURE VS-45-01 / DATE 1-12

CHECK CODES, REGULATIONS, AND LAWS (LOCAL, STATE, AND NATIONAL) TO ENSURE THAT DESIGN IS COMPLIANT.

FIGURE VS-45-02

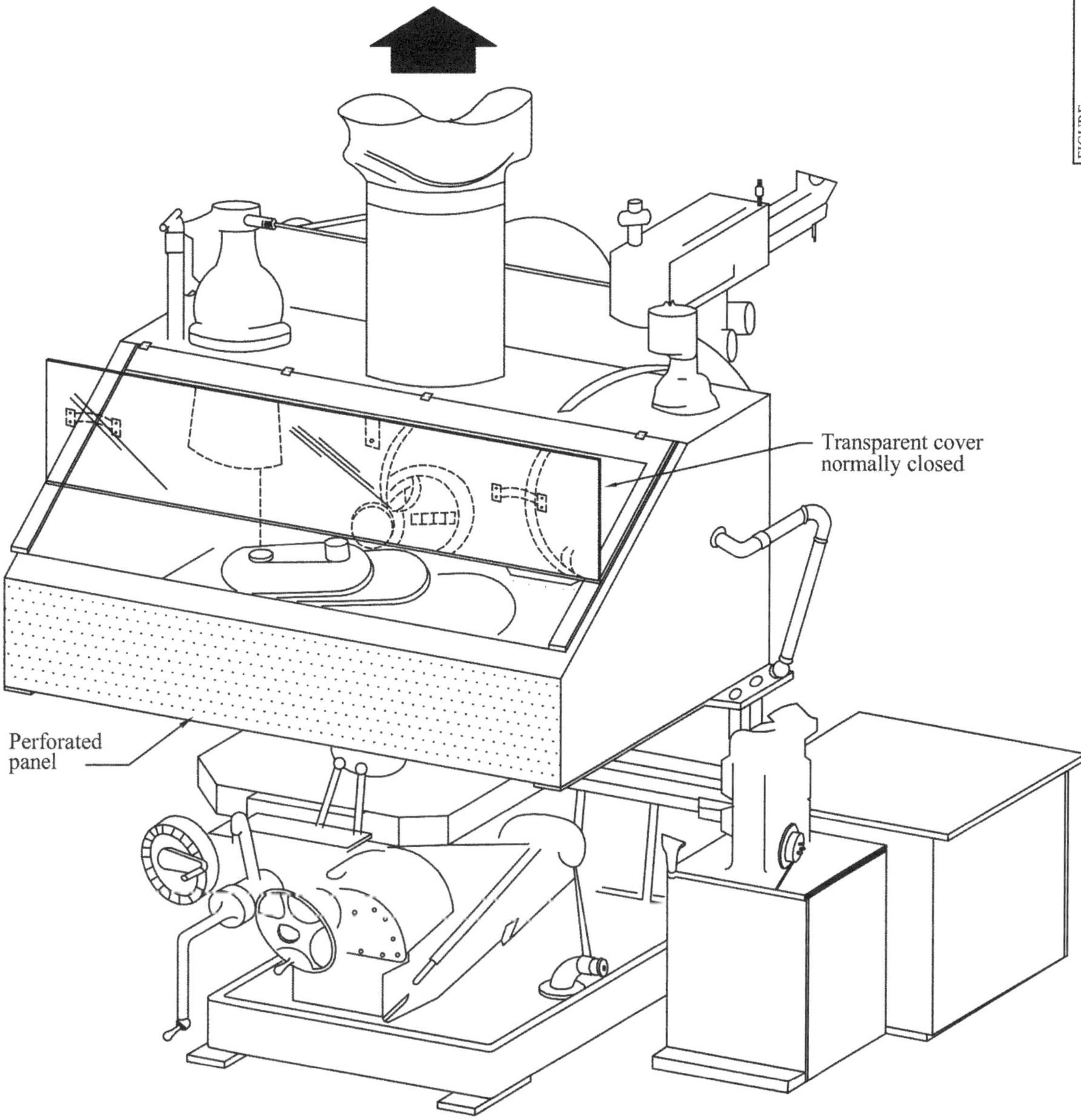

$Q = 300\ acfm/ft^2$ [$1.5\ am^3/s/m^2$] of open area
Minimum duct velocity = 3500 fpm [17.50 m/s]
$h_e = 0.35\ VP_d$

Reference 13.45.4

ACGIH®	TITLE HIGH TOXICITY MATERIALS MILLING MACHINE HOOD	FIGURE VS-45-02 DATE 1-12

CHECK CODES, REGULATIONS, AND LAWS (LOCAL, STATE, AND NATIONAL)
TO ENSURE THAT DESIGN IS COMPLIANT.

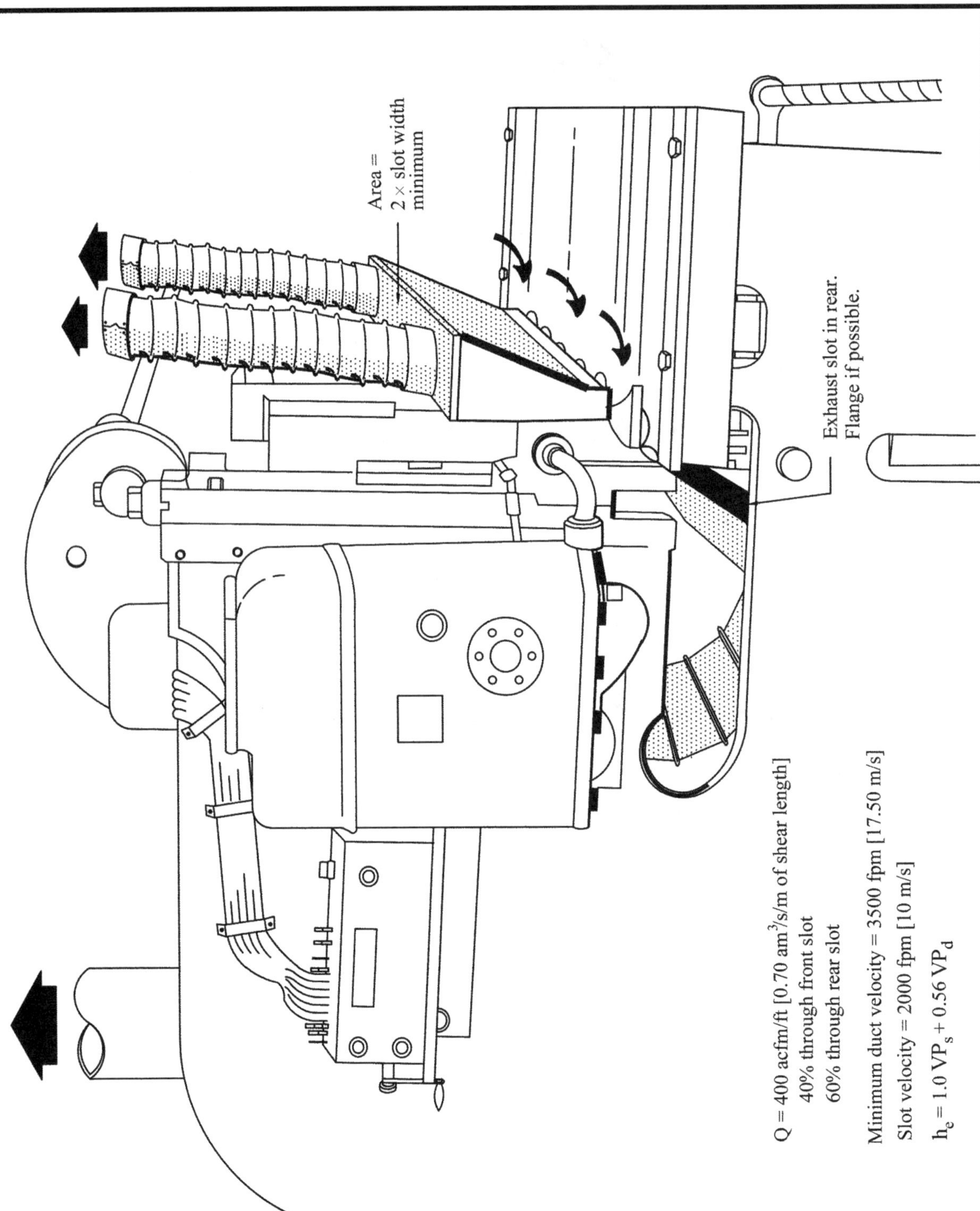

Reference 13.45.3

FIGURE VS-45-03

ACGIH®	TITLE METAL SHEARS HIGH TOXICITY MATERIALS	FIGURE VS-45-03
		DATE 1-12

CHECK CODES, REGULATIONS, AND LAWS (LOCAL, STATE, AND NATIONAL) TO ENSURE THAT DESIGN IS COMPLIANT.

FIGURE VS-45-04

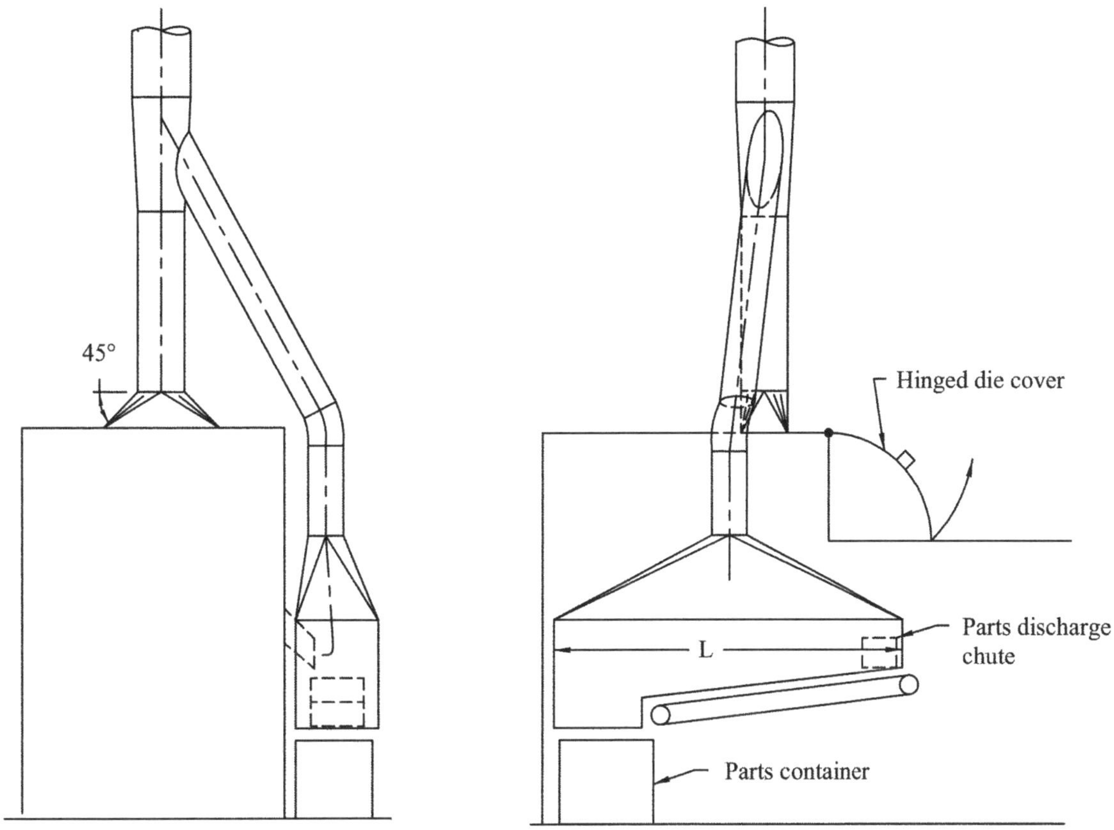

END VIEW

SIDE VIEW

Cold header:

$Q = 750\ \text{acfm/ft}^2\ [3.8\ \text{am}^3\text{/s/m}^2]$ of die opening
Minimum duct velocity = 3500 fpm [17.50 m/s]
$h_e = 1.78\ VP_s + 0.25\ VP_d$

Parts discharge and container:

$Q = 100\ \text{acfm/ft}^2\ [0.18\ \text{am}^3\text{/s/m}]$ of hood length
Minimum duct velocity = 3500 fpm [17.50 m/s]
$h_e = 0.25\ VP_d$

ACGIH®	TITLE COLD HEADING MACHINE VENTILATION	FIGURE VS-45-04 DATE 1-12

CHECK CODES, REGULATIONS, AND LAWS (LOCAL, STATE, AND NATIONAL) TO ENSURE THAT DESIGN IS COMPLIANT.

FIGURE VS-45-05

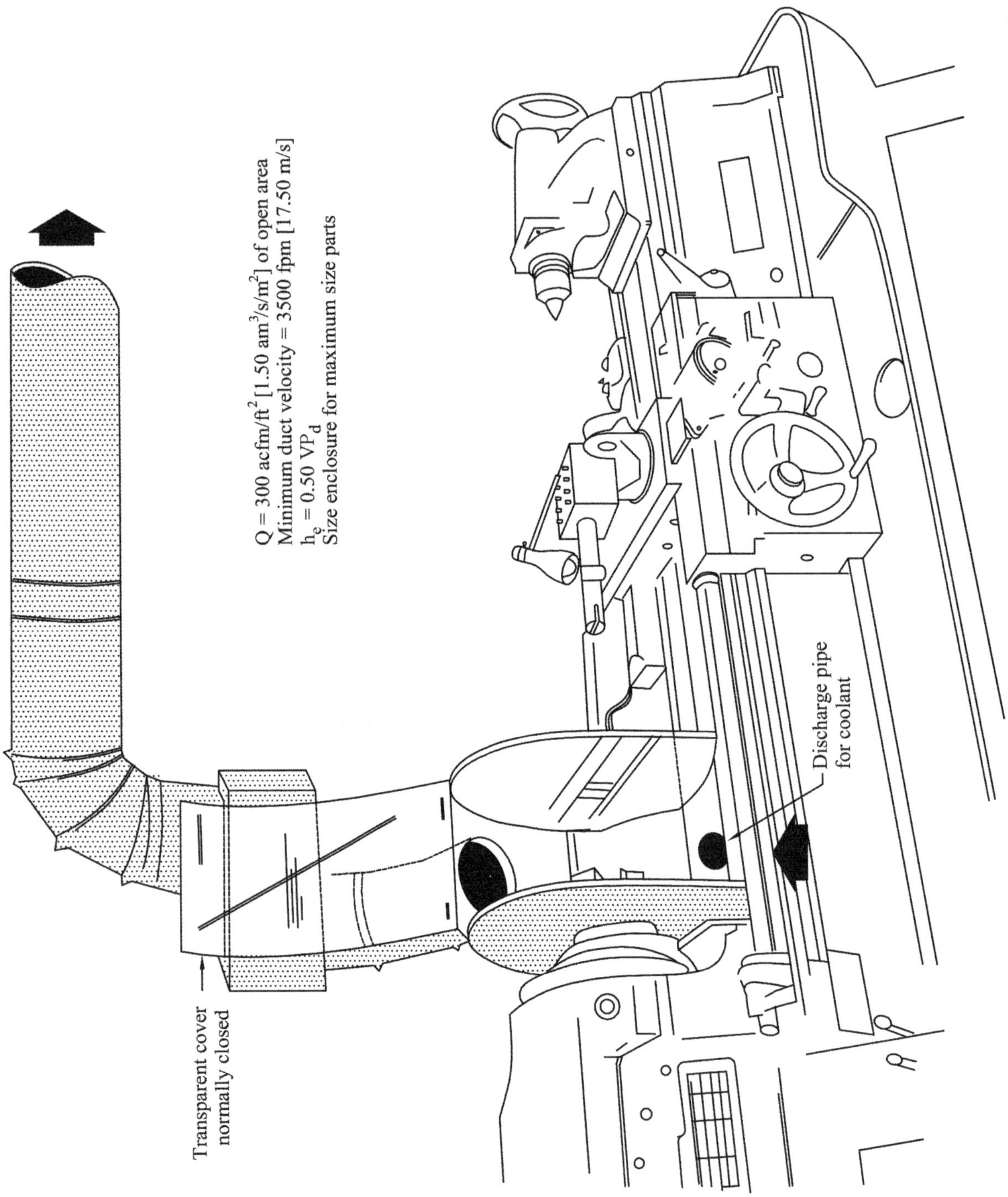

Reference 13.45.4

	TITLE	FIGURE VS-45-05
ACGIH®	LATHE HOOD	DATE 1-12

CHECK CODES, REGULATIONS, AND LAWS (LOCAL, STATE, AND NATIONAL) TO ENSURE THAT DESIGN IS COMPLIANT.

FIGURE VS-45-06

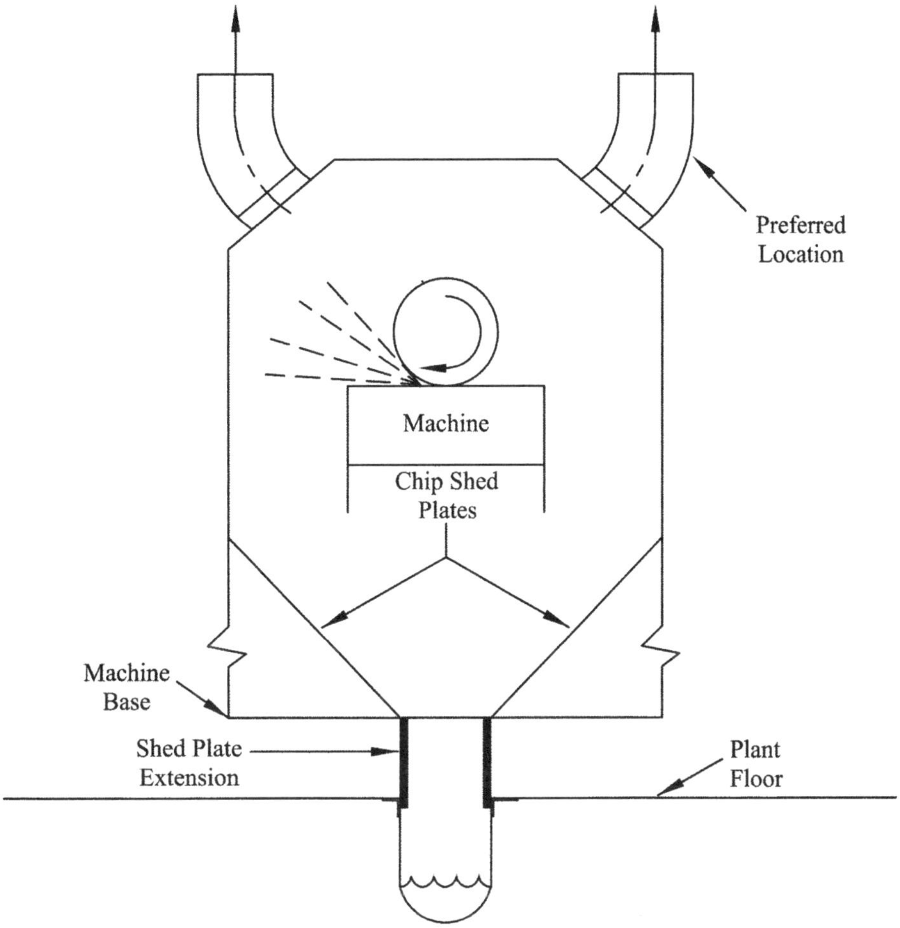

Q = 100 acfm/linear foot [0.16 am^3/s/ linear meter] of enclosure plus 100 acfm [0.05 am^3/s] for part exit and entrance opening.

Q = 150 acfm/linear foot [0.23 am^3/s/ linear meter] of enclosure for machines that handle large parts such as engine blocks and transmission cases, plus 100 acfm [0.05 am^3/s] for part exit and entrance opening.

Additional air at active machining stations for motors and other energy sources may be required.

ACGIH®	TITLE: IN-LINE TRANSFER MACHINES	FIGURE: VS-45-06 / DATE: 1-12

CHECK CODES, REGULATIONS, AND LAWS (LOCAL, STATE, AND NATIONAL) TO ENSURE THAT DESIGN IS COMPLIANT.

13.50 MATERIAL TRANSPORT

Ventilation of material transport systems generally requires the use of an exhausted enclosure because of the motion and quantity of material involved. If the enclosure were perfectly air tight there would be no need for exhaust. However, there usually are cracks and other leak points in addition to the openings necessary for personnel and material access.

For enclosures where there is little motion and low material quantity, the exhaust rate is the product of the total openings in square feet and some velocity between 50 and 200 fpm. However, in some cases, the inward flow of material and entrained air can overwhelm the exhaust flow rate calculated on the basis of enclosure openings. In such cases, the material flow rate, the dustiness of the material, and the height of fall in transferring from one surface to another must be considered in the system design.[13.50.1, 13.50.2] Other design factors include:

1) The rate of air induction into the space.
2) The location of cracks or other openings in relation to the "splash" or agitation of material during transfer.
3) The need to avoid excessive product withdrawal.
4) Adequate airflow for dilution of interior concentrations for visibility or safety from explosions.

13.50.1 Bucket Elevators. Air motion caused by the bucket moving within the elevator is not significant. The motion of buckets in one direction is offset by the opposite flow. Consequently, an exhaust rate of 100 acfm/ft^2 [0.5 am^3/s/m^2]of elevator cross-section is adequate for most elevator applications (see VS-50-01 for details). Additional ventilation is required as materials enter and leave the elevator (see VS-50-10, VS-50-20, and VS-50-21). Handling hot material often causes significant thermal bouyancy which requires increased exhaust ventilation to overcome this challenge.

13.50.2 Conveyors. Dust from the operation of belt conveyors originates mainly at the tail pulley where material is received and at the head pulley where material is discharged. The exhaust requirement at the head pulley is generally small because air is induced downward and away from this transfer point. An exhaust rate of 150 to 200 acfm/ft^2 [0.75 to 1.0 am^3/s/m^2] of opening often is adequate.

At the tail pulley, the exhaust requirements are determined by the amount of air induced by the delivery chute. An exhaust of 350 acfm/ft^2 [1.77 am^3/s/m^2] of belt width often is adequate where the material does not fall more than 3 feet. The exhaust point should be located at least twice the belt width away from the point where the material hits the belt. Where the material falls more than 3 feet, additional exhaust is required (see VS-50-20 for details). Note that very dry or dusty material may require flowrates 1.5 to 2.0 times these values.

Belt conveyors should be covered and exhausted at 30 foot [9.0 m] intervals at a rate of 350 acfm/ft [0.54 am^3/s/m] of belt width. Vibrating feeders should be exhausted at a rate of 500 acfm/ft [0.77 am^3/s/m] of feeder width. Rubber or canvas flexible seals should be provided from the feeder sides and end to the hopper sides and end.

The conveying of toxic material requires additional care in enclosure design to ensure that no air leaks out and that sufficient access is available for inspection and cleanout. The head pulley should be equipped with a scraper or brush (see VS-50-21).

13.50.3 Bin and Hopper Ventilation. For the mechanical loading of bins and hoppers, the exhaust rates previously listed for belt conveyors are appropriate. An exhaust rate of 150 acfm/ft^2 [0.75 am^3/s/m^2] of hopper cross-sectional area is adequate for manual loading operations. The enclosure should cover as much of the hopper opening as possible.

13.50.4 Loading and Unloading. For loading and unloading operations, a ventilation rate of 150 to 200 acfm/ft^2 [0.75 to 1.0 am^3/s/m^2] of enclosure opening is adequate *provided the enclosure is large enough to accommodate the "splash" effect.* The entrance to enclosure for truck dumps should be covered with flaps to minimize ventilation requirements. Rotary or bottom car dumps are generally exhausted at the rate of 50 to 100 acfm/ft^2 [0.08 to 0.15 am^3/s/m^2] of hopper area.

REFERENCES

13.50.1 DallaValle, J.M.: Exhaust Hoods, Industrial Press. New York (1946).

13.50.2 Hemeon, W.C.L.: Plant and Process Ventilation. Industrial Press (1963).

13.50.3 Rajhans, G.S.; Bragg, G.M.: Engineering Aspects of Asbestos Dust Control. Ann Arbor Science Publications, Inc., Ann Arbor, MI (1978).

13.50.4 National Grain and Feed Association: Dust Control for Grain Elevators. Washington, DC (1981).

FIGURE VS-50-01

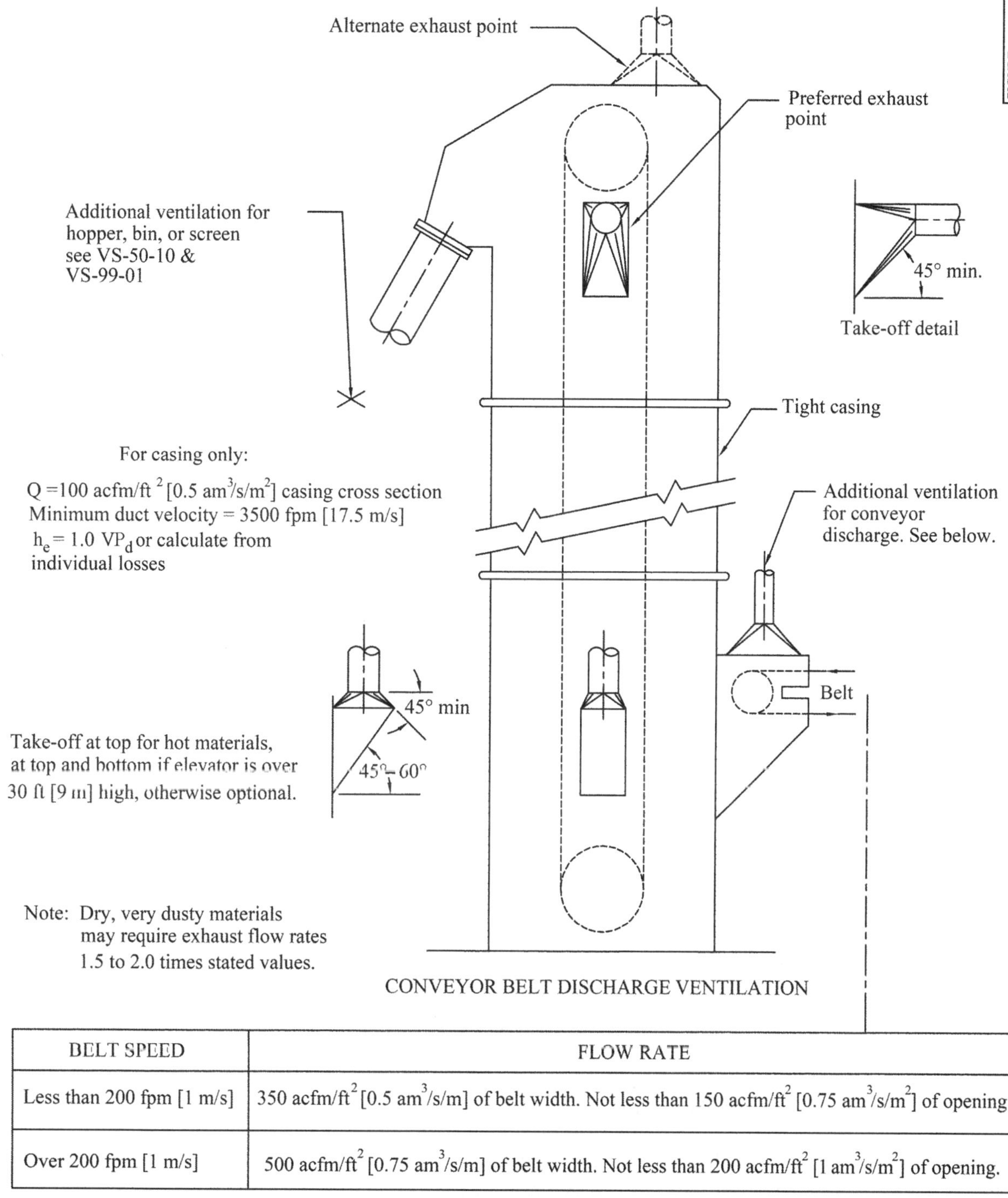

BELT SPEED	FLOW RATE
Less than 200 fpm [1 m/s]	350 acfm/ft^2 [0.5 am^3/s/m] of belt width. Not less than 150 acfm/ft^2 [0.75 am^3/s/m^2] of opening.
Over 200 fpm [1 m/s]	500 acfm/ft^2 [0.75 am^3/s/m] of belt width. Not less than 200 acfm/ft^2 [1 am^3/s/m^2] of opening.

ACGIH®	TITLE: BUCKET ELEVATOR VENTILATION	FIGURE: VS-50-01 DATE: 1-12

CHECK CODES, REGULATIONS, AND LAWS (LOCAL, STATE, AND NATIONAL) TO ENSURE THAT DESIGN IS COMPLIANT.

FIGURE VS-50-10

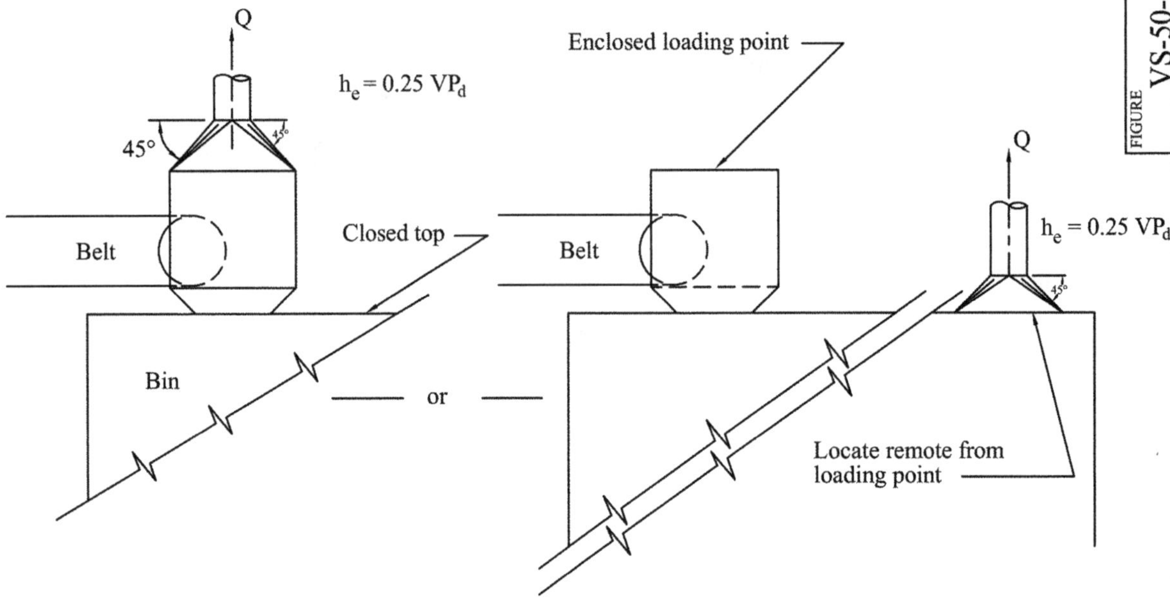

MECHANICAL LOADING

Belt speed		Flow rate (Q)
Less than 200 fpm [1 m/s]	—	350 acfm/ft [0.50 am^3/s/m] of belt width. Not less than 150 acfm/ft^2 [0.75 am^3/s/m^2] of opening.
Over 200 fpm [1 m/s]	—	500 acfm/ft [0.75 am^2/s/m] of belt width. Not less than 200 acfm/ft^2 [1 am^3/s/m^2] of opening.

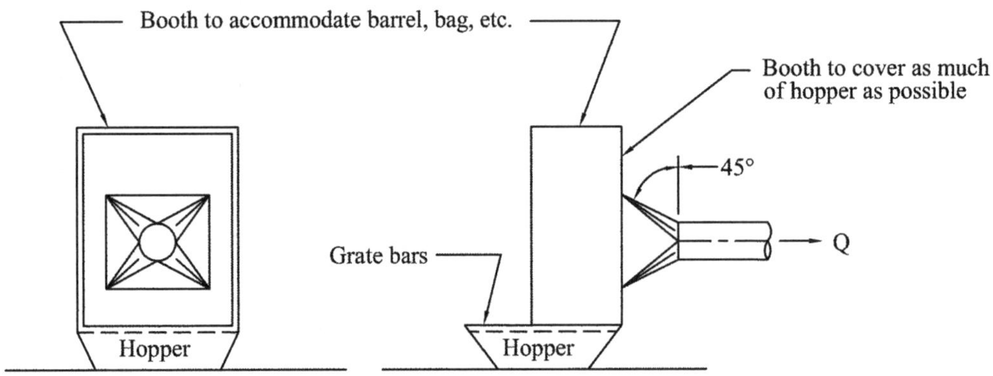

Minimum duct velocity = 3500 fpm [17.50 m/s]

Q = 150 acfm/ft^2 [0.75 am^3/s/m^2] face

$h_e = 0.25\ VP_d$

MANUAL LOADING

ACGIH®	TITLE BIN & HOPPER VENTILATION	FIGURE VS-50-10 DATE 1-12

CHECK CODES, REGULATIONS, AND LAWS (LOCAL, STATE, AND NATIONAL) TO ENSURE THAT DESIGN IS COMPLIANT.

FIGURE VS-50-20

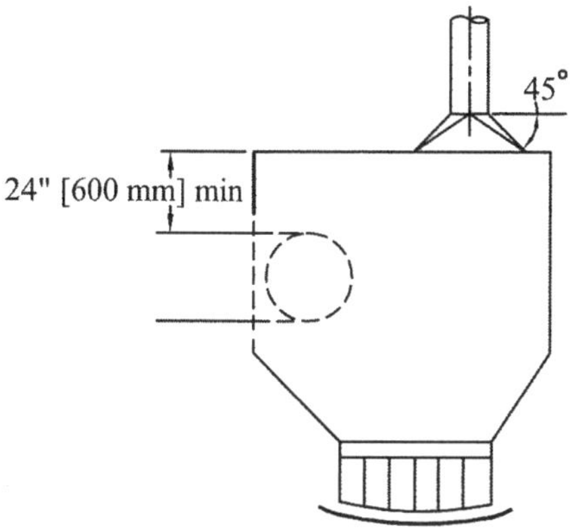

1. Conveyor transfer less than 3' [0.9 m] fall. For greater fall, provide additional exhaust at lower belt. See 3 below.
$h_e = 0.25\ VP_d$

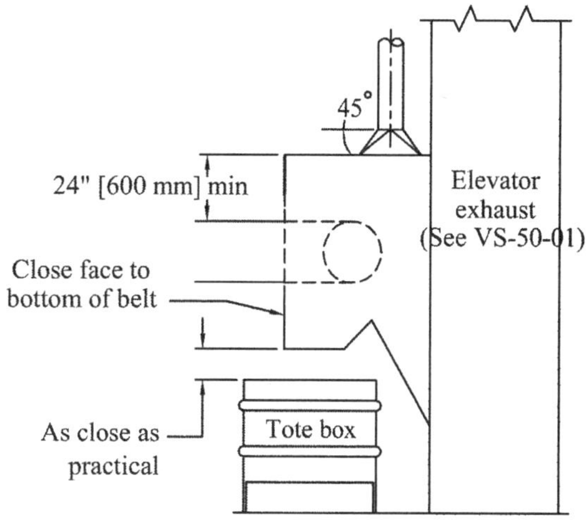

2. Conveyor to elevator with magnetic separator.
$h_e = 0.25\ VP_d$

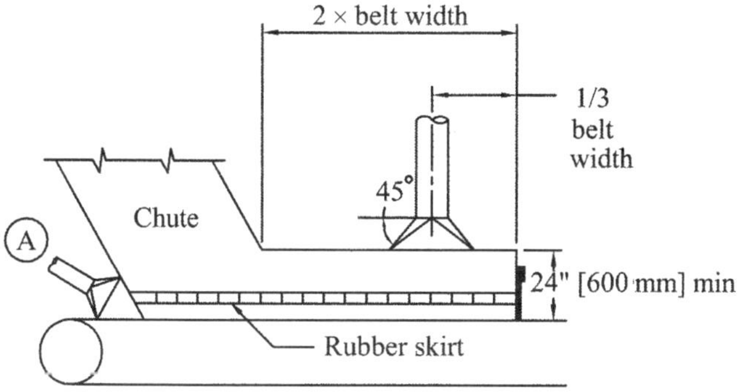

3. Chute to belt transfer and conveyor transfer, greater than 3' [0.9 m] fall. Use additional exhaust at (A) for dusty material as follows:
Belt width 12"–36" [300–900 mm], Q = 700 acfm [0.35 am^3/s]
Belt width above 36" [900 mm], Q = 1000 acfm [0.50 am^3/s]
$h_e = 0.25\ VP_d$

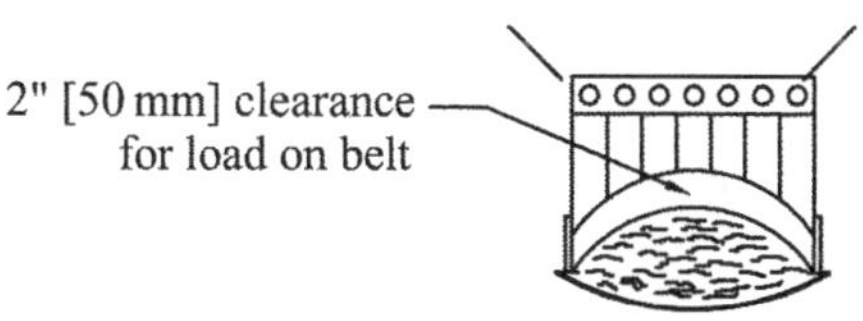

DETAIL OF BELT OPENING

DESIGN DATA

Transfer points:

Enclose to provide 150 – 200 fpm [0.75–1.00 m/s] indraft at all openings. (Underground mining tunnel ventilation will interfere with conveyor exhaust systems.)

Q = 350 acfm/ft [0.50 am^3/s/m] belt width for belt speeds under 200 fpm [1.00 m/s]. (minimum)
= 500 acfm/ft [0.75 am^3/s/m] belt width for belt speeds over 200 fpm [1.00 m/s] and for magnetic separators. (minimum)
Minimum duct velocity = 3500 fpm [17.50 m/s]
$h_e = 0.25\ VP_d$

Conveyor belts:

Cover belt between transfer points
Exhaust at transfer points
Exhaust additional 350 acfm/ft [0.50 am^3/s/m] of belt width at 30' [0.9 m] intervals. Use 45° minimum tapered connections.

Note: Dry, very dusty materials may require exhaust flow rates 1.5 to 2.0 times stated values.

ACGIH®	TITLE CONVEYOR BELT VENTILATION	FIGURE VS-50-20
		DATE 1-12

CHECK CODES, REGULATIONS, AND LAWS (LOCAL, STATE, AND NATIONAL) TO ENSURE THAT DESIGN IS COMPLIANT.

FIGURE VS-50-21

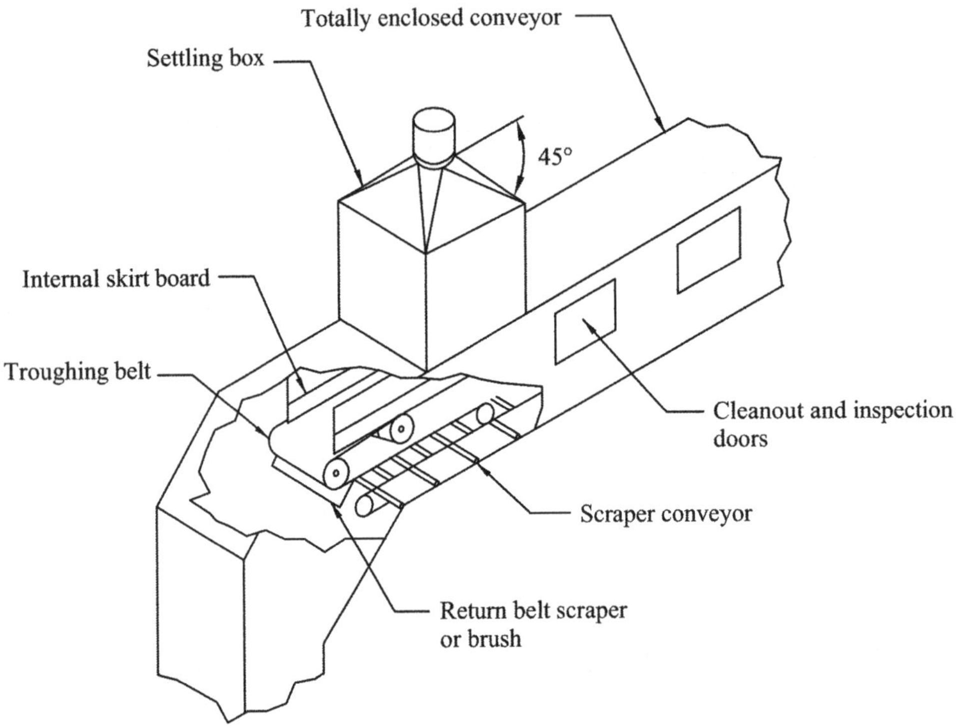

Q = 250 acfm/ft^2 [1.25 am^3/s/m^2] of open area
Minimum duct velocity = 3500 fpm [17.50 m/s]
$h_e = 0.4\ VP_d$

Reference: 13.50.3

ACGIH®	TITLE: TOXIC MATERIAL BELT CONVEYING HEAD PULLEY	FIGURE: VS-50-21 DATE: 1-12

CHECK CODES, REGULATIONS, AND LAWS (LOCAL, STATE, AND NATIONAL) TO ENSURE THAT DESIGN IS COMPLIANT.

FIGURE VS-50-22

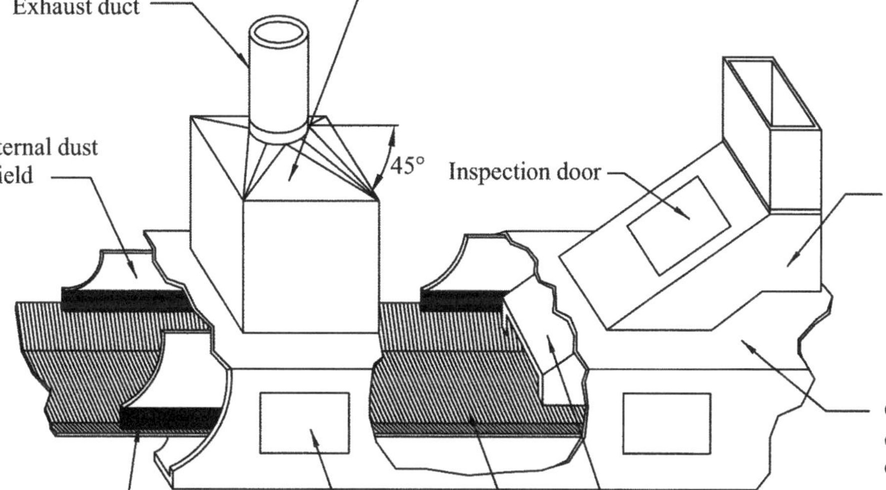

Q = 250 acfm/ft^2 [1.25 am^3/s/m^2] of open area

Minimum duct velocity = 3500 fpm [17.50 m/s]

h_e – 0.4 VP_d

Reference: 13.50.3

ACGIH®	TITLE: TOXIC MATERIAL CONVEYOR BELT LOADING	FIGURE: VS-50-22
		DATE: 1-12

CHECK CODES, REGULATIONS, AND LAWS (LOCAL, STATE, AND NATIONAL) TO ENSURE THAT DESIGN IS COMPLIANT.

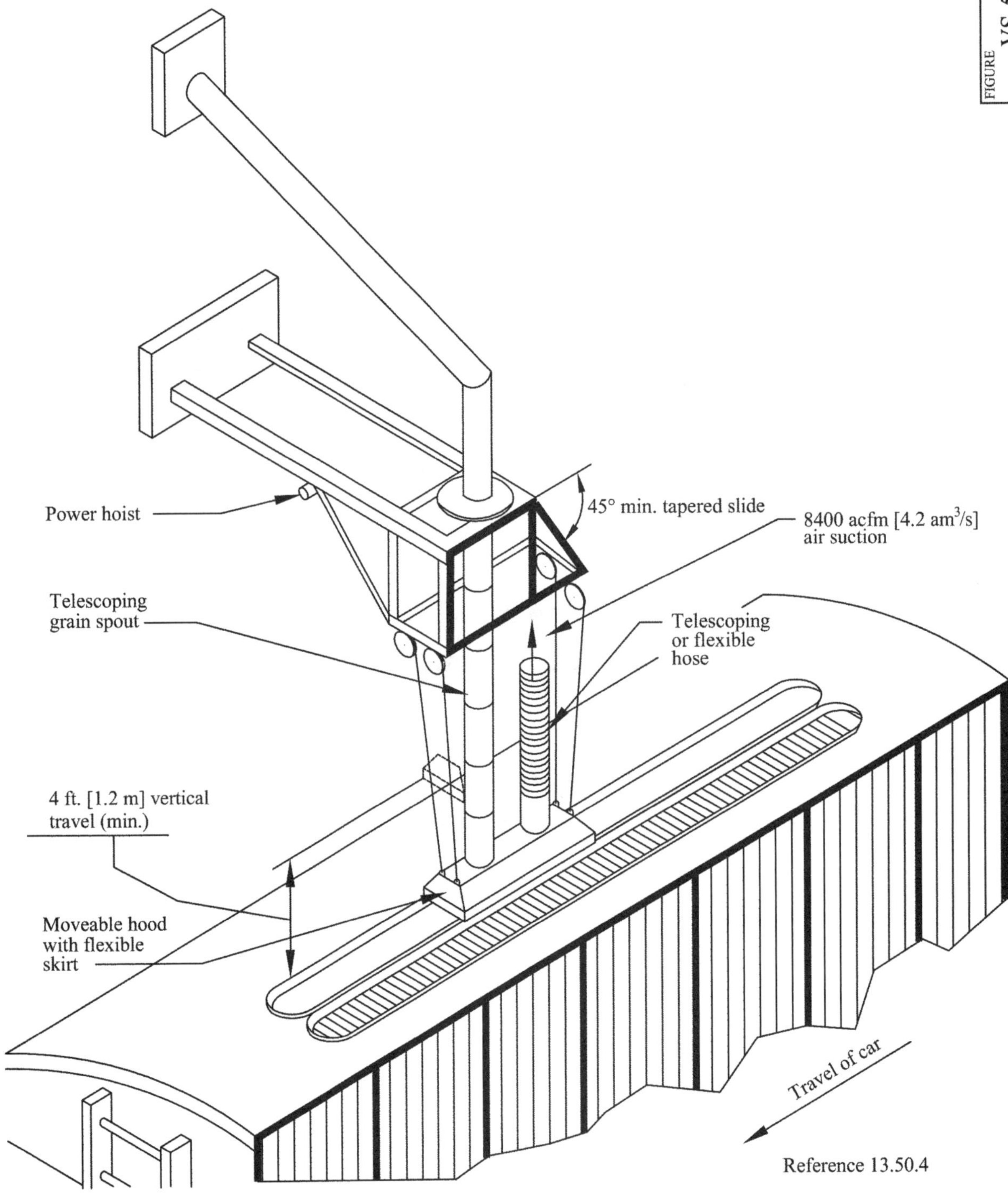

Reference 13.50.4

FIGURE VS-50-30

ACGIH®	TITLE: RAIL LOADING	FIGURE: VS-50-30
		DATE: 1-12

CHECK CODES, REGULATIONS, AND LAWS (LOCAL, STATE, AND NATIONAL) TO ENSURE THAT DESIGN IS COMPLIANT.

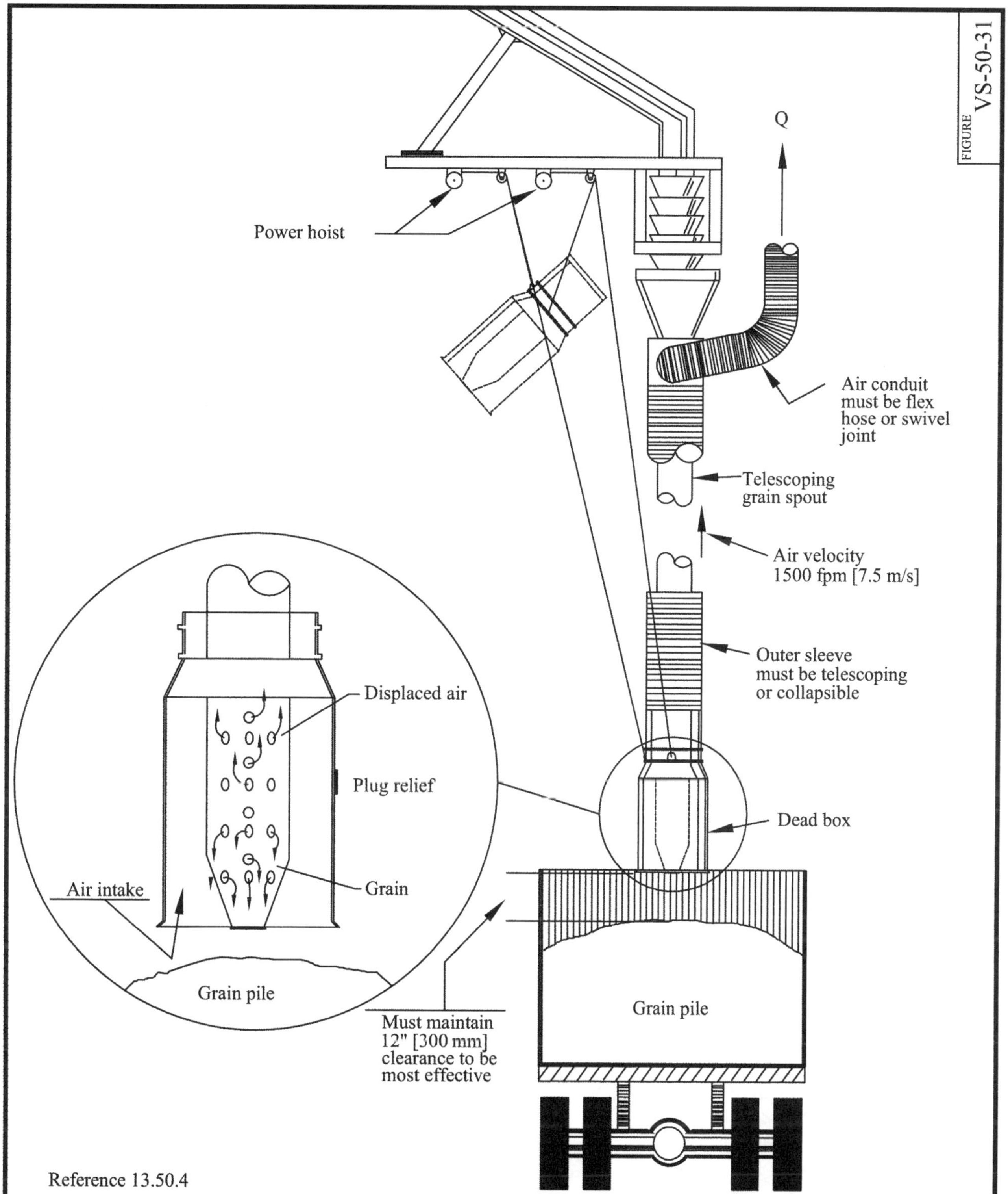

Reference 13.50.4

ACGIH®	TITLE TRUCK LOADING	FIGURE VS-50-31 DATE 1-12

CHECK CODES, REGULATIONS, AND LAWS (LOCAL, STATE, AND NATIONAL) TO ENSURE THAT DESIGN IS COMPLIANT.

13.55 METAL MELTING FURNACES

This set of VS-prints describes hood designs for a variety of metal melting furnaces, including electric induction, carbon arc, convection, and crucible, which use natural gas or electric resistance elements as the heat source. Exhaust ventilation is usually required to control fumes generated from the metal being oxidized during the melting as well as the combustion of contaminants carried with the scrap charge. In some cases, a single hood will suffice for charging, melting, and pouring (tapping). In other cases, a separate hood, remote from the primary melter, may be required for charging because of the nature of the charge or the large open area necessary at this phase of the operation. This is particularly true of electric arc furnaces which, in most cases, are completely open for charging. After charging, the direct exhaust can be used to achieve control during the remainder of the melting cycle. During tapping of the hot metal, either the remote (high canopy) hood or a side draft hood at the ladle lip can achieve control.

All metal will produce a slag that must be removed prior to tapping. This activity can produce a significant release of oxides and will require a separate exhaust system for the oxide/dross control. The remote or high canopy hood can often provide the control. Where metal purification is performed directly within the furnace or melting vessel, such as the addition of oxygen or chlorine, additional exhaust may be required to contain the rapidly generated plume. All systems must be designed to include the increase in air temperature under operating conditions to insure an adequate airflow into the hood. The air temperature rise is usually relatively low except where metal inoculants or oxidizers are added to the molten charge. During this phase of metal melting, a significant temperature rise will occur and it is customary to provide a large hood in which gas expansion can take place.

13.55.1 Electric Arc Furnace.[13.55.1] Small, low power (10 MVA or lower) Electric Arc Furnaces (EAFs) normally use custom designed side draft hoods. From the graph, either the melt rate or the oxygen lance rate can be used to determine the exhaust volume (Q) and the temperature of the exhaust gas. The duct velocity at the flange is a minimum of 3800 fpm [19.0 m/s]. A damper is placed in the duct leading from the slag door hood to the main duct. Eighty percent of Q is exhausted from around the electrodes and 20% from the slag door. A high canopy hood is required to capture charging and tapping fumes (VS-55-03B).

Most EAFs today are large Ultra High Powered (UHP) furnaces with melt rates of 100 TPH [Metric Tons per Hour] or more and oxygen lance rates well in excess of 1600 acfm [0.80 am^3/s]. The volume (Q) for the melt shop can be approximated from the geometry of the melt shop. VS-55-03B provides the information to approximate the overall 'Q' evacuated from the high canopy hood during charging and tapping. However, the heat load coming out of the EAF through the direct evacuation system (DES) must be considered. (The calculation of the DES heat load is beyond the scope of this Manual.) Approximately 20% of the energy entering the EAF leaves in the DES. The DES route generally removes 25% of the total Q during melting and contains approximately 80% of the dust and metallurgical fume. During melting and refining, the high canopy hood exhausts 75% of the total volume and contains only 20% of the dust and fume.

REFERENCES

13.55.1 Walli, R.: Private Communication to Gerry Lanham (2009).

13.55.2 American Foundrymen's Society, Inc.: Managing the Foundry Inplant Environment (2004).

FIGURE VS-55-01

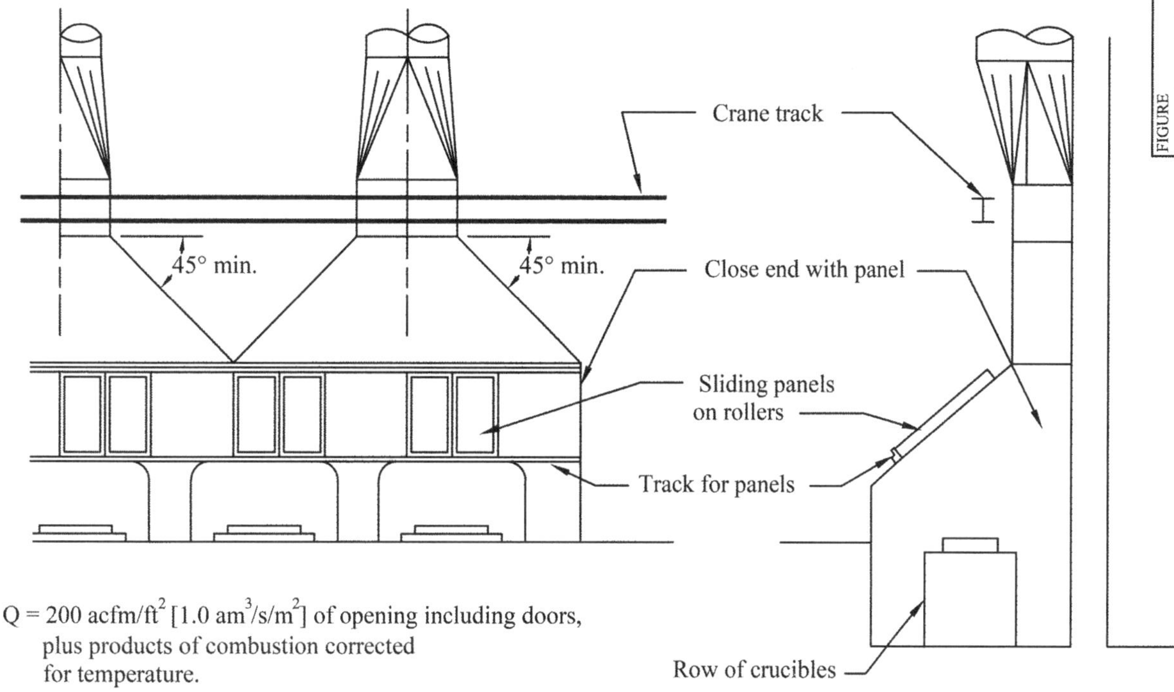

Q = 200 acfm/ft^2 [1.0 am^3/s/m^2] of opening including doors, plus products of combustion corrected for temperature.

Minimum duct velocity = 3500 fpm [17.50 m/s]

$h_e = 0.56\ VP_d$

Note: Same principle of sliding or swinging doors is applied to individual furnace enclosures.

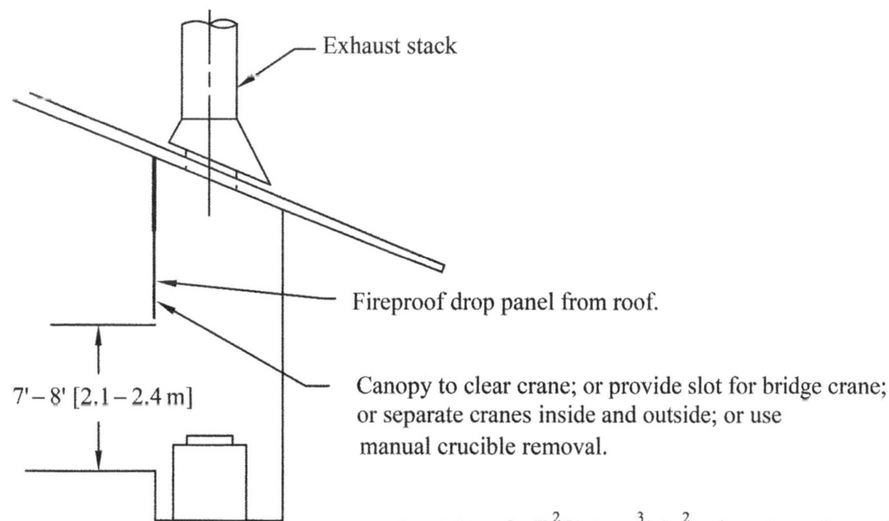

Q = 200 acfm/ft^2 [1.0 am^3/s/m^2] of total opening, minimum, plus products of combustion corrected for temperature.

ACGIH®

TITLE **MELTING FURNACE CRUCIBLE NON-TILT**

FIGURE VS-55-01

DATE 1-12

CHECK CODES, REGULATIONS, AND LAWS (LOCAL, STATE, AND NATIONAL) TO ENSURE THAT DESIGN IS COMPLIANT.

FIGURE VS-55-02

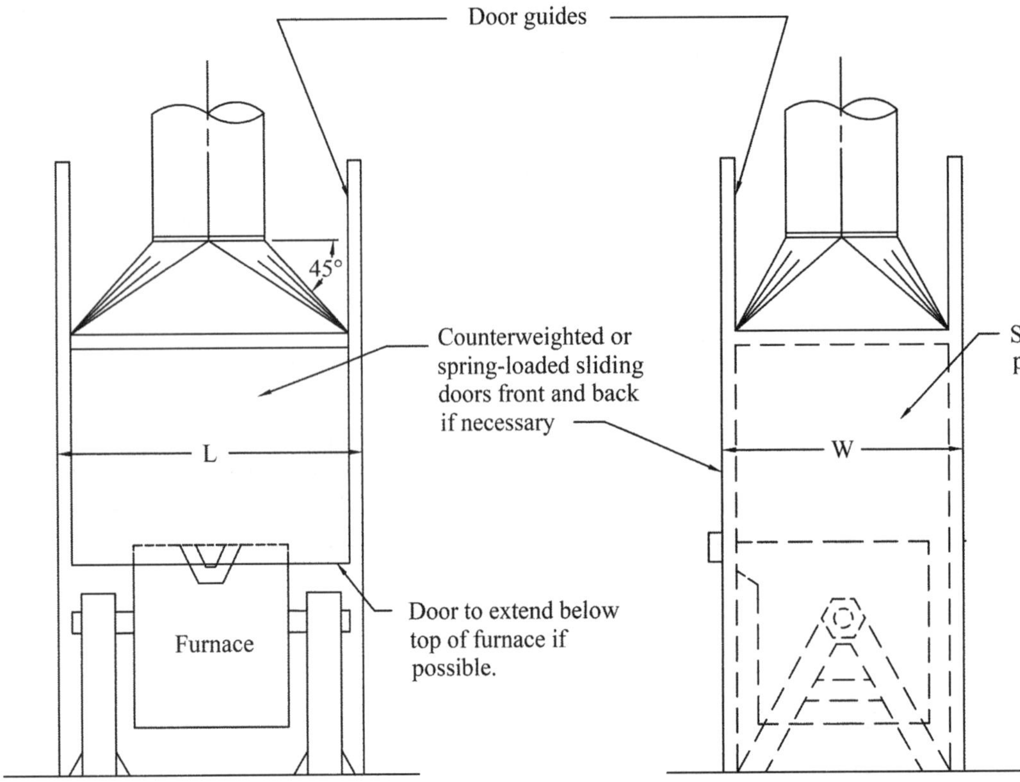

Q (acfm) = 200 LW [1.0 LW am^3/s]; but not less than 200 acfm/ft^2 [1.0 am^3/s/m^2] of all openings with doors open. Correct for products of combustion and temperature.

Minimum duct velocity = 3500 fpm [17.50 m/s]

$h_e = 0.25\ VP_d$

ACGIH®

TITLE: MELTING FURNACE TILTING

FIGURE: VS-55-02

DATE: 1-12

CHECK CODES, REGULATIONS, AND LAWS (LOCAL, STATE, AND NATIONAL) TO ENSURE THAT DESIGN IS COMPLIANT.

FIGURE VS-55-03

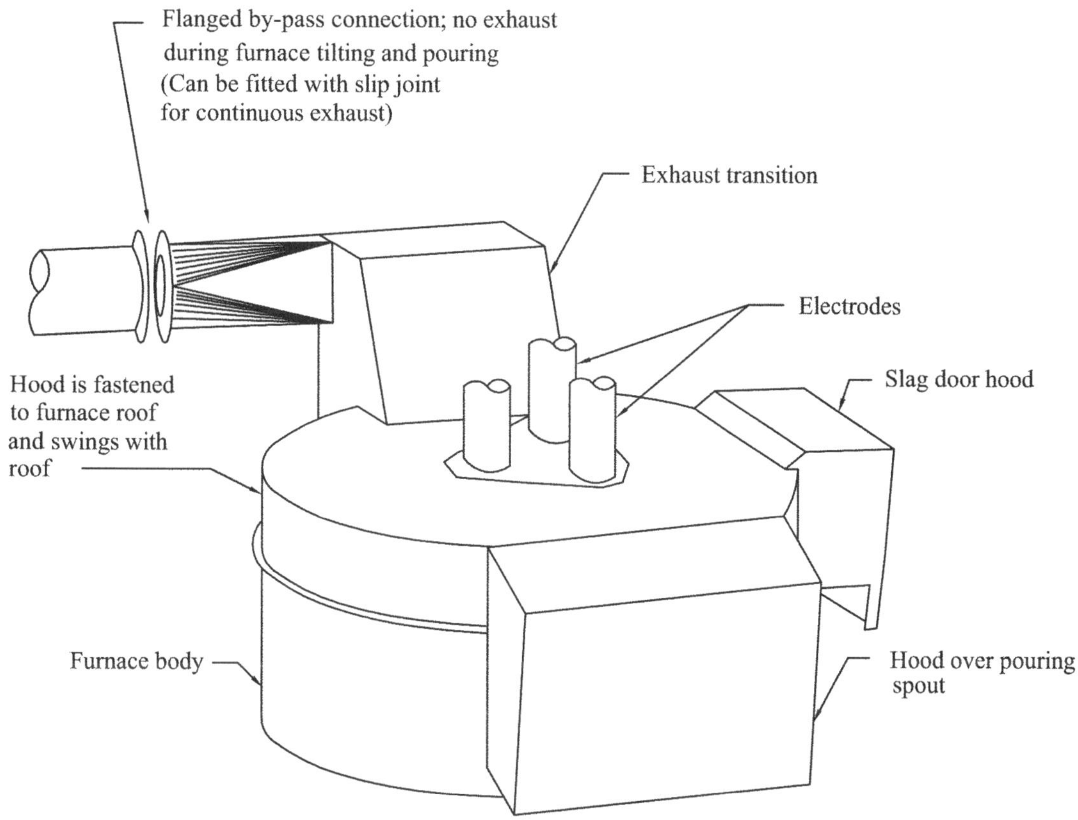

Close Capture:

For Q, SP and operating temperature, consult manufacturer
Approximate exhaust rate = 2500 scfm/ton [2.85 sm^3/s/metric ton] of charge

Alternate designs:

1. Some exhaust designs utilize direct furnace roof tap. For details consult manufacturer.
2. Canopy hoods require large exhaust and are not recommended. Canopy hoods can be used as secondary hoods to capture fugitive emissions and charging.
3. Close capture hoods can be mounted near electrodes.

References 13.55.1, 13.55.2

ACGIH®	TITLE MELTING FURNACE ELECTRIC, TOP ELECTRODE	FIGURE VS-55-03 DATE 1-12

CHECK CODES, REGULATIONS, AND LAWS (LOCAL, STATE, AND NATIONAL) TO ENSURE THAT DESIGN IS COMPLIANT.

FIGURE VS-55-04

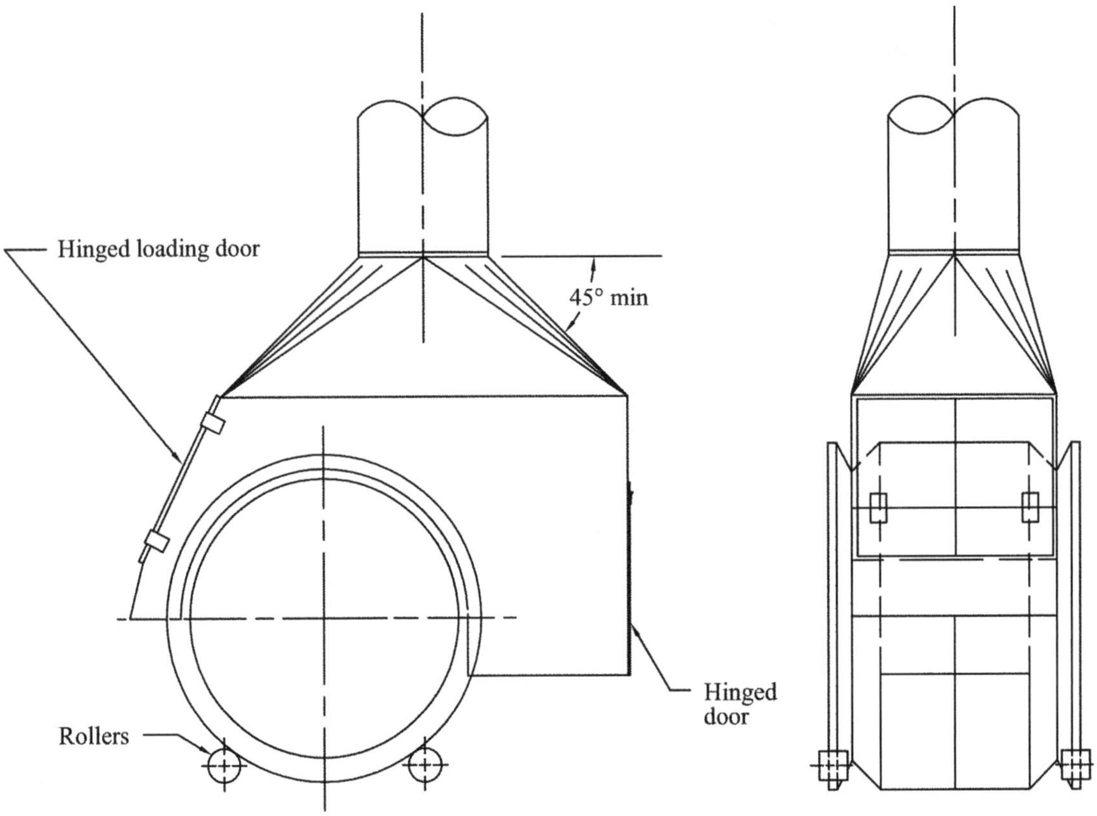

Q = 400 acfm/ft^2 [2.0 am^3/s/m^2] of opening
Minimum duct velocity = 3500 fpm [17.50 m/s]
$h_e = 1.78\ VP_s + 0.25\ VP_d$

ACGIH®

TITLE

MELTING FURNACE
ELECTRIC ROCKING

FIGURE VS-55-04

DATE 1-12

CHECK CODES, REGULATIONS, AND LAWS (LOCAL, STATE, AND NATIONAL) TO ENSURE THAT DESIGN IS COMPLIANT.

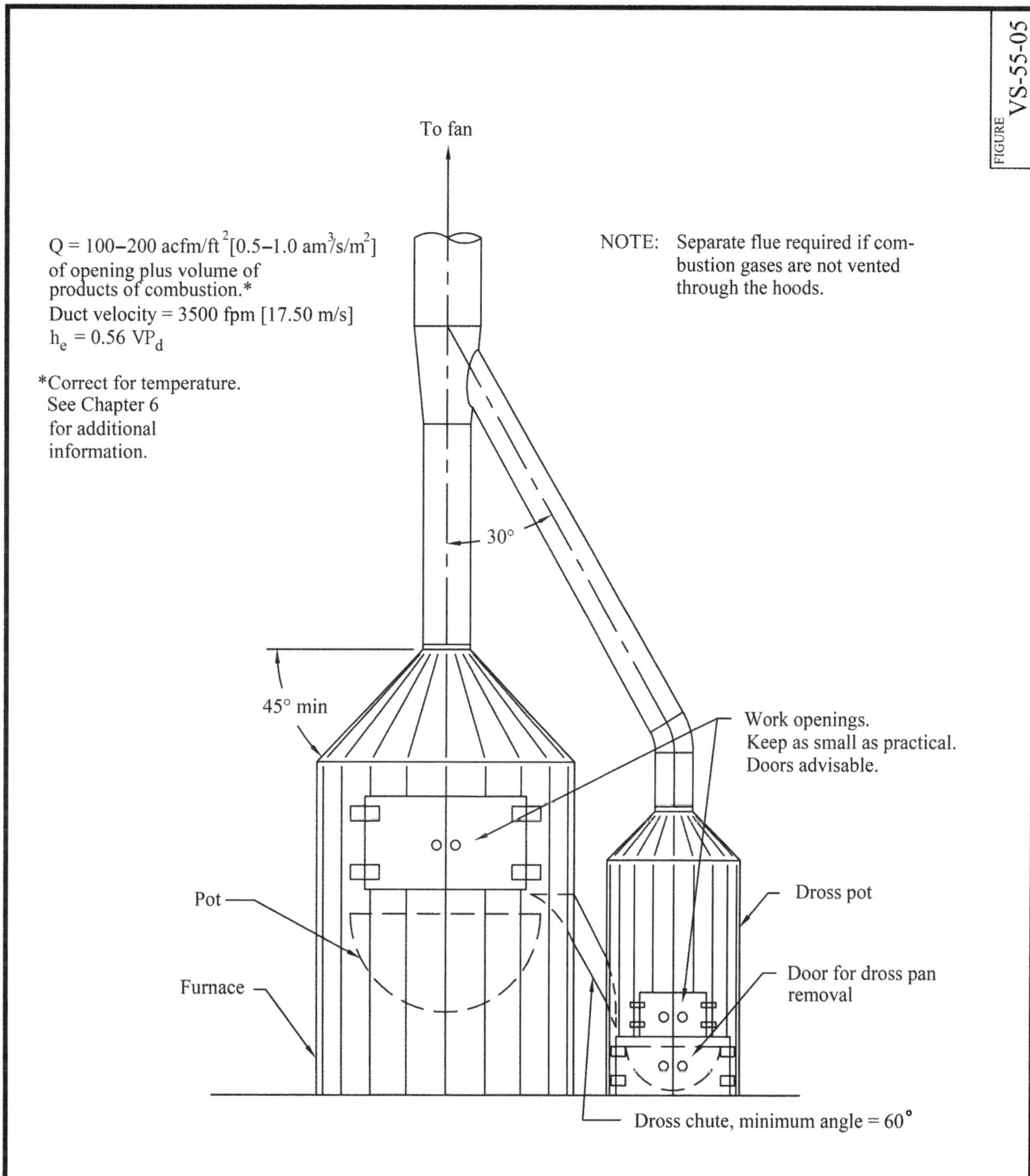

Stationary furnace or melting pot

ACGIH®	TITLE: MELTING POT AND FURNACE	FIGURE: VS-55-05
		DATE: 1-12

CHECK CODES, REGULATIONS, AND LAWS (LOCAL, STATE, AND NATIONAL) TO ENSURE THAT DESIGN IS COMPLIANT.

FIGURE VS-55-07

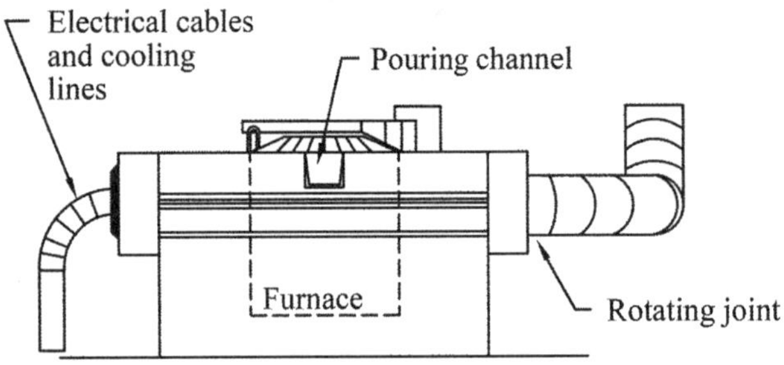

FRONT VIEW

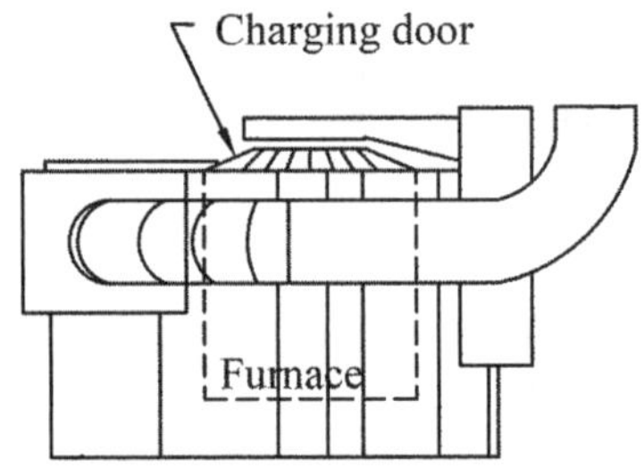

SIDE VIEW

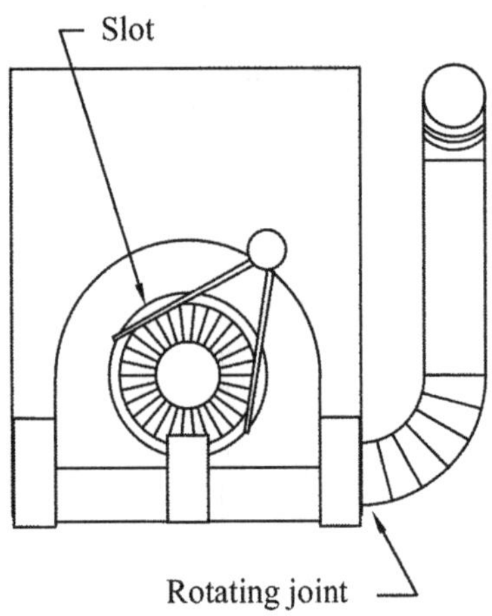

TOP VIEW

Q = 350 acfm/ft^2 [1.75 am^3/s/m^2] open area. Correct for temperature and combustion products.

Entry loss = 1.78 VP_s + 0.56 VP_d

Slot velocity = 2000 fpm [10 m/s]

Minimum duct velocity = 3500 fpm [17.50 m/s]

ACGIH®	TITLE INDUCTION MELTING FURNACE TILTING	FIGURE VS-55-07
		DATE 1-12

CHECK CODES, REGULATIONS, AND LAWS (LOCAL, STATE, AND NATIONAL) TO ENSURE THAT DESIGN IS COMPLIANT.

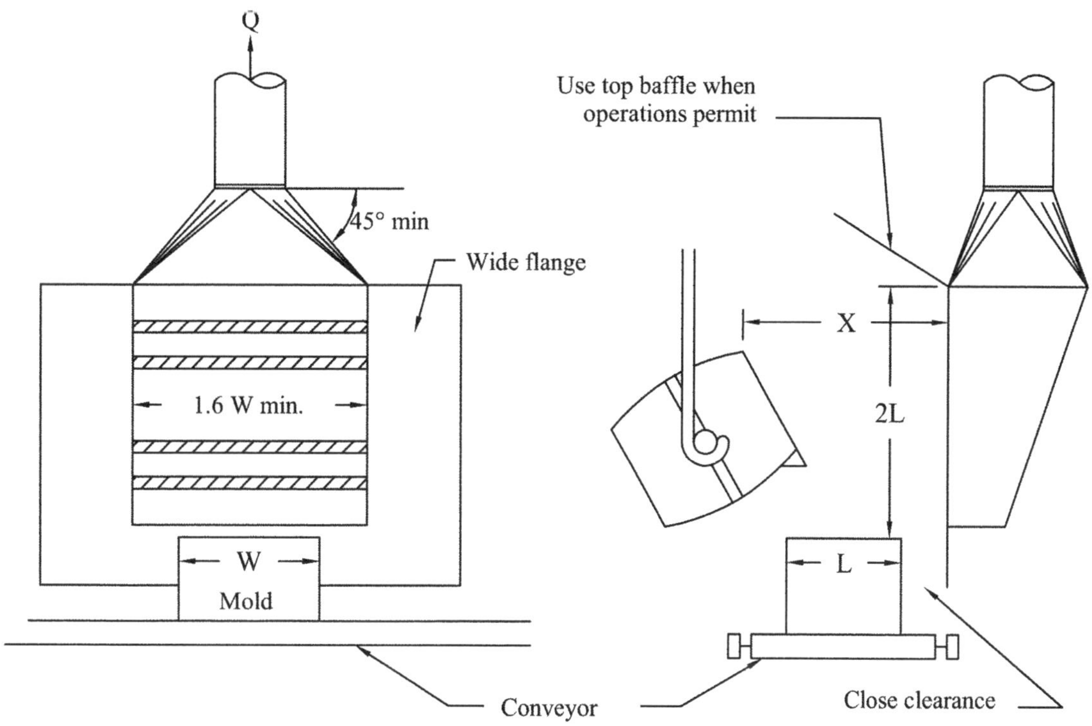

SMALL MOLDS

$Q_{HOOD} = 200\ (10\ X^2 + 3.2\ WL)$ acfm [$Q_{HOOD} = 1.0\ (10\ X^2 + 3.2\ WL)$ am^3/s, where A = hood area]
Minimum duct velocity = 3500 fpm [17.50 m/s]
$h_e = 1.78\ VP_s + 0.25\ VP_d$
Use slots for uniform distribution, size slots
for 2000 fpm [10 m/s]

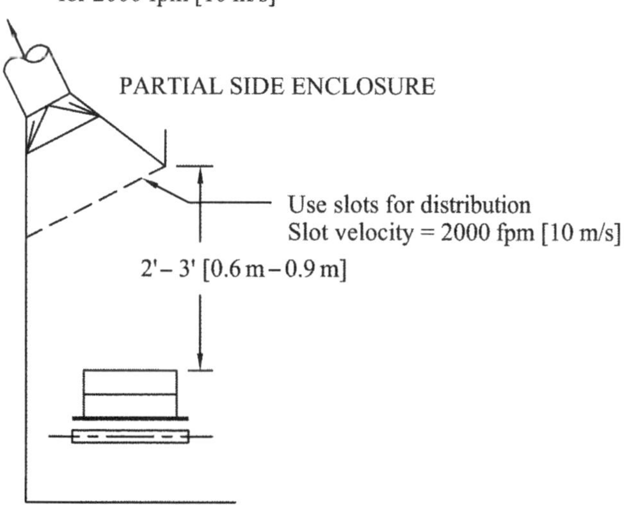

Note:
For large molds and ladles
provide large – draft hood
similar to shakeout.
Q = 400 acfm/ft^2 [2.0 am^3/s/m^2] working area

Q = 200 – 300 acfm/ft [0.3 – 0.45 m^3/s/m] of hood length

ACGIH®	TITLE: POURING STATION	FIGURE: VS-55-10 / DATE: 1-12

CHECK CODES, REGULATIONS, AND LAWS (LOCAL, STATE, AND NATIONAL)
TO ENSURE THAT DESIGN IS COMPLIANT.

FIGURE VS-55-20

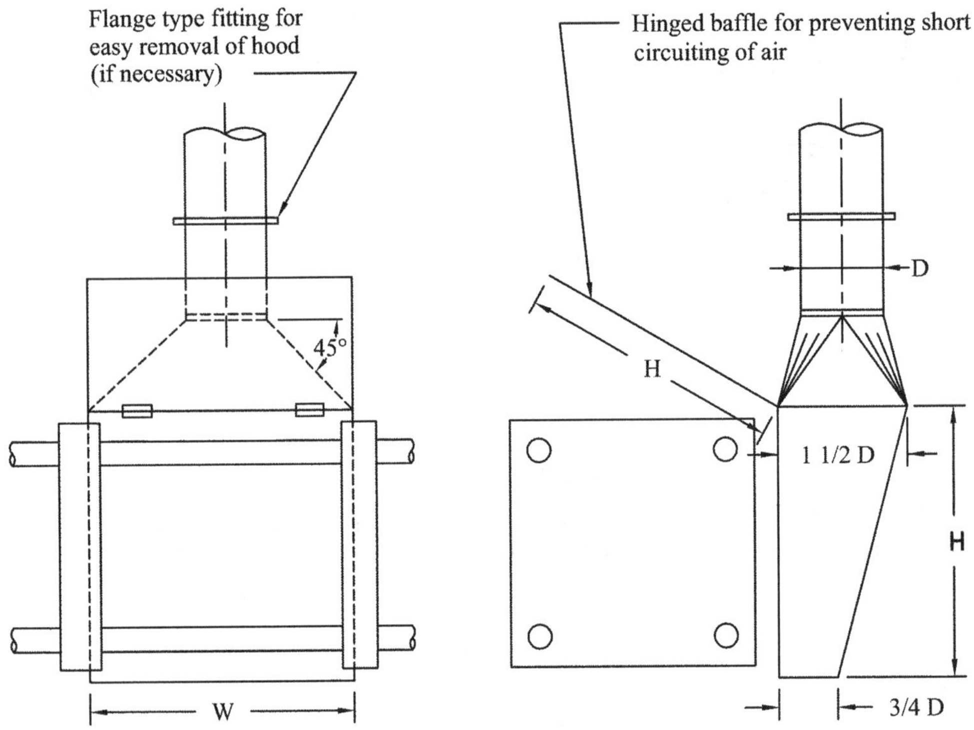

Note: Place hood as close to machine as possible. If more than 4 inches [100 mm] from back of machine, hinged side baffles should be used.

Note: Products of combustion require separate flue or may be vented into hood.

$Q = 300WH$ acfm $[Q = 1.5WH\ am^3/s]$
$h_e = 0.25\ VP_d$
Minimum duct velocity = 2000 fpm [10 m/s]

ACGIH®	TITLE FIXED POSITION DIE CASTING HOOD	FIGURE VS-55-20 DATE 1-12

CHECK CODES, REGULATIONS, AND LAWS (LOCAL, STATE, AND NATIONAL) TO ENSURE THAT DESIGN IS COMPLIANT.

FIGURE VS-55-21

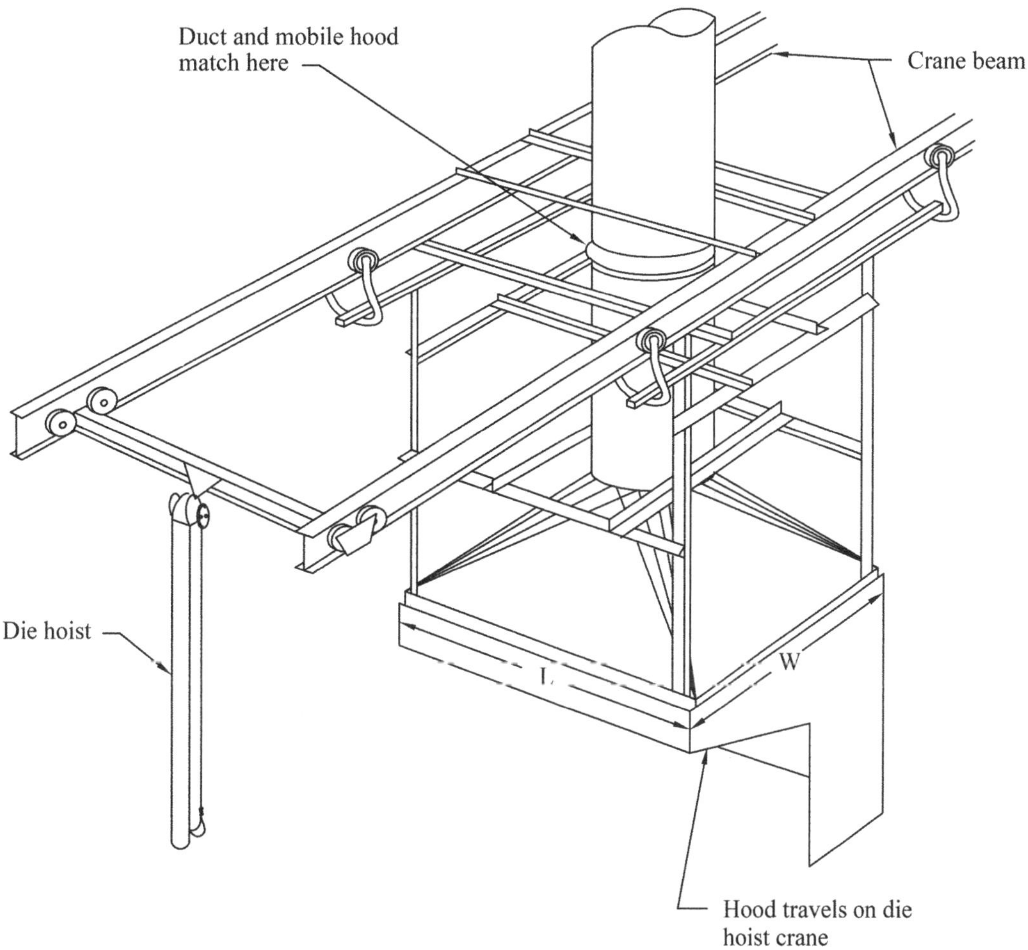

Q = 300 WL acfm [Q = 1.5 WL am^3/s]
Minimum duct velocity = 2000 fpm [10.00 m/s]
$h_e = 0.25\ VP_d$

ACGIH®	TITLE MOBILE HOOD DIE CASTING	FIGURE VS-55-21 DATE 1-12

CHECK CODES, REGULATIONS, AND LAWS (LOCAL, STATE, AND NATIONAL) TO ENSURE THAT DESIGN IS COMPLIANT.

13.60 MIXING

Mixing operations combine a large variety of materials, usually without significant chemical reactions. This section includes categories of mixing operations.

13.60.1 Mixing and Mulling. Mixers and mullers require exhaust ventilation to provide a minimum velocity of 150 fpm through all openings. Additional ventilation may be required when flammable solvents are used. The dilution ventilation rates should maintain concentrations within the muller below 25% of the Lower Explosive Limit (LEL). Some codes or standards may require ventilation rates that ensure the concentration of flammable vapor is maintained below 20% of the LEL.

13.60.2 Roll Mixing. The machinery shown in this subsection is used to mix and blend quantities of viscid materials, such as rubber and plastic, with additives that are dry powders or liquids. Emissions of gases and vapors may evolve due to chemical reactions or may be caused by the elevated temperature of the mixed materials. Particulate emissions can occur during additions as well as during mechanical blending.

The roller mill ventilation shown in VS-60-12 encloses the roller mill to the maximum extent possible except for a front opening of sufficient height and width to permit operator access and material entry/removal. An air curtain directed upward towards the top of the enclosure provides a barrier to contaminant escape but still permits operator access and material entry/removal. The design values provided are critical for operation.

REFERENCE

13.60.1 Hampl, V.; Johnston, O.E.; Murdock, D.M.: Application of an Air Curtain Exhaust System at a Milling Process. American Industrial Hygiene Association Journal, 49(4), pp 167-175 (1988).

FIGURE VS-60-01

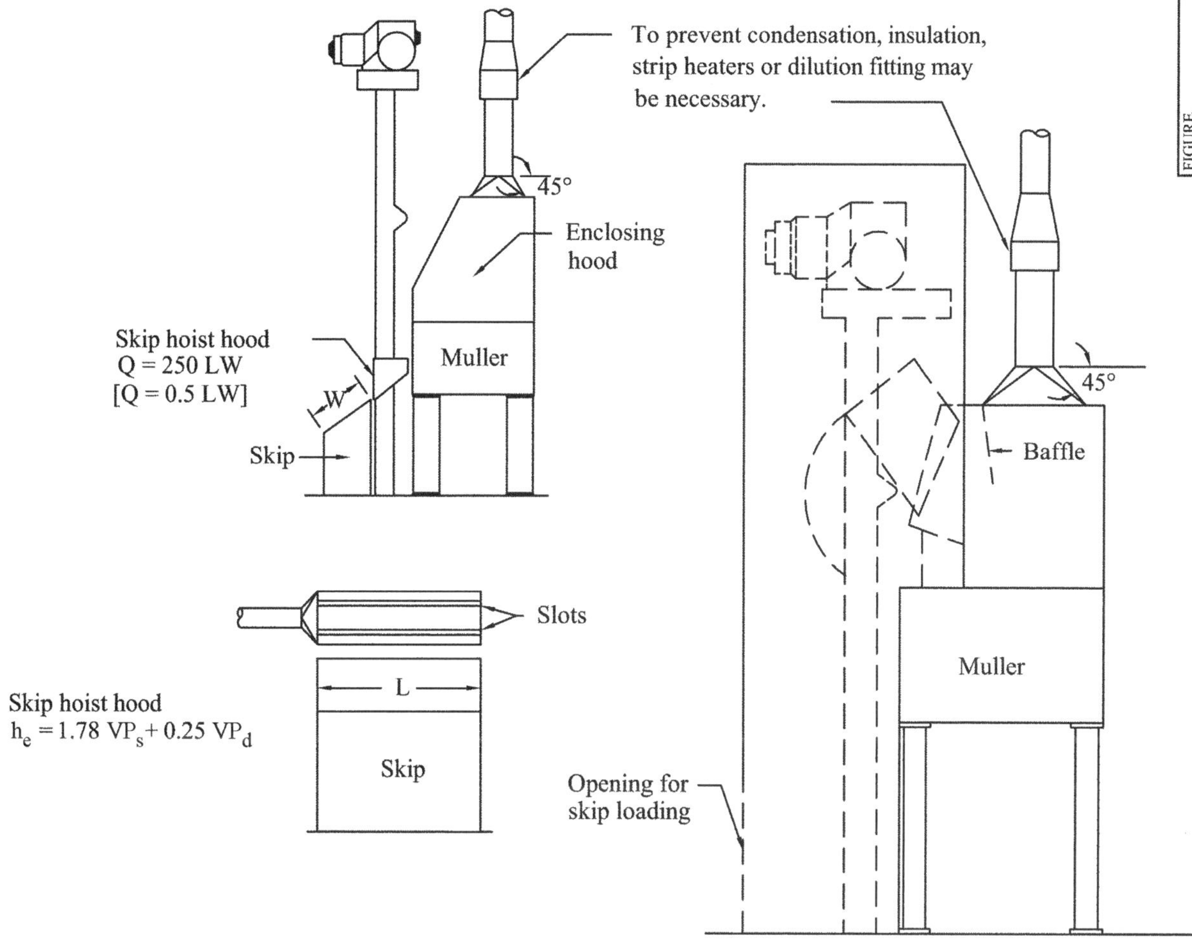

Q = 150 acfm/ft^2 [0.75 m^3/s/m^2] through all openings but not less than:

Muller diam. feet [m]	Exhaust, acfm [am^3/s]
4 [1.2]	750 [0.38]
6 [1.8]	900 [0.45]
7 [2.1]	1050 [0.53]
8 [2.4]	1200 [0.60]
10 [3.0]	1575 [0.79]

Minimum duct velocity = 4000 fpm [20 m/s]

$h_e = 0.25\ VP_d$

Notes:
1. Other types of mixers: enclose as much as possible and provide 150 acfm/ft^2 [0.75 am^3/s/m^2] through remaining openings
2. When flammable solvents are used in mixer, calculate minimum exhaust rate for dilution to 25% of the LEL. See Chapter 4.
3. For air cooled mullers see VS-60-02

ACGIH®	TITLE: MIXER AND MULLER HOOD	FIGURE: VS-60-01 / DATE: 1-12

CHECK CODES, REGULATIONS, AND LAWS (LOCAL, STATE, AND NATIONAL) TO ENSURE THAT DESIGN IS COMPLIANT.

FIGURE VS-60-02

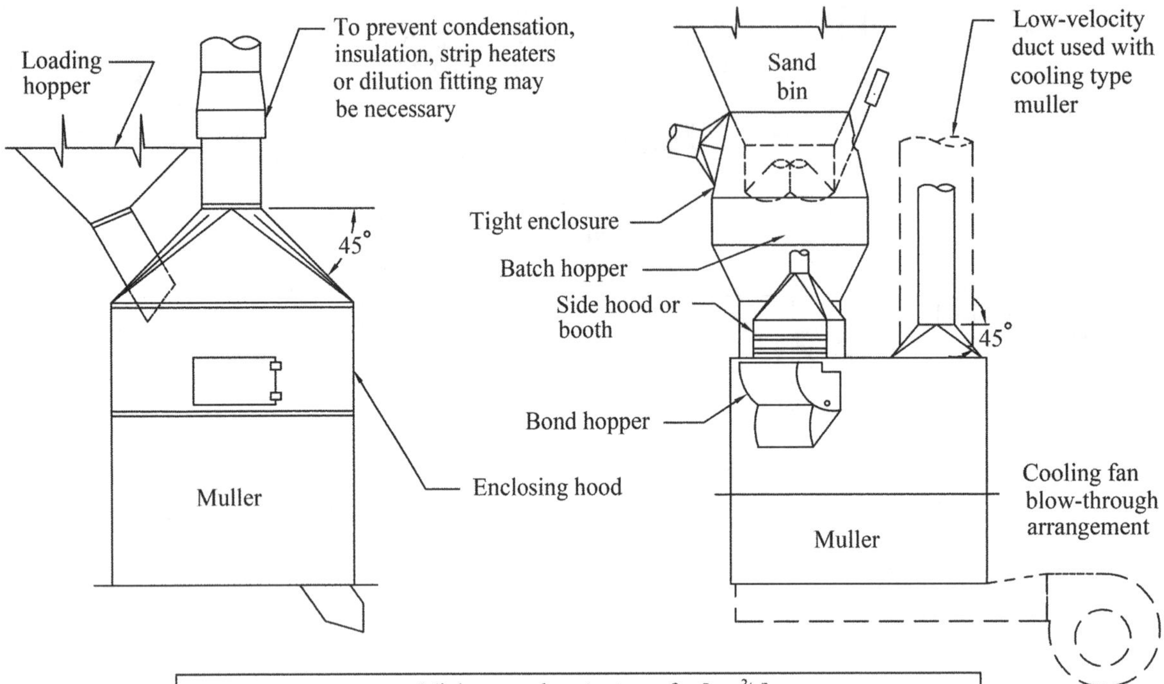

Minimum exhaust rate, acfm [am³/s]			
	Muller type		
Location	No cooling acfm [am³/s]	Blow-through cooling acfm [am³/s]	Draw-through cooling acfm [am³/s]
Batch hopper	Note 1	600 [0.30]	Note 1
Bond hopper	600 [0.30]	600 [0.30]	600 [0.30]
Muller:	Note 2	Note 3	Note 3
4' [1.2 m] diameter	750 [0.38]	"	"
6' [1.8 m] diameter	900 [0.45]	"	"
7' [2.1 m] diameter	1050 [0.53]	"	"
8' [2.4 m] diameter	1200 0.60]	"	"
10' [3.0 m] diameter	1575 [0.79]	"	"

Minimum duct velocity = 4500 fpm [23.00 m/s]

h_e = 0.25 VP

Notes:

1. Batch hopper requires separate exhaust with blow-through cooling. With other fan arrangement (muller under suction), separate exhaust may not be required. (If skip hoist is used, see VS-60-01)
2. Maintain 150 fpm [0.75 m/s] velocity through all openings in muller hood. Exhaust flow rates shown are the minimum for control.
3. Cooling mullers do not require additional exhaust if maintained in dust tight condition. Blow-through fan must be off during loading. If muller is not dust tight, exhaust as in note 2 plus cooling airflow rate.
4. When flammable solvents are used in mixer, calculate minimum exhaust flow rate for dilution to 25% of the LEL. See Chapter 4.

ACGIH®	TITLE AIR COOLED MIXER AND MULLER	FIGURE VS-60-02 DATE 1-12

CHECK CODES, REGULATIONS, AND LAWS (LOCAL, STATE, AND NATIONAL) TO ENSURE THAT DESIGN IS COMPLIANT.

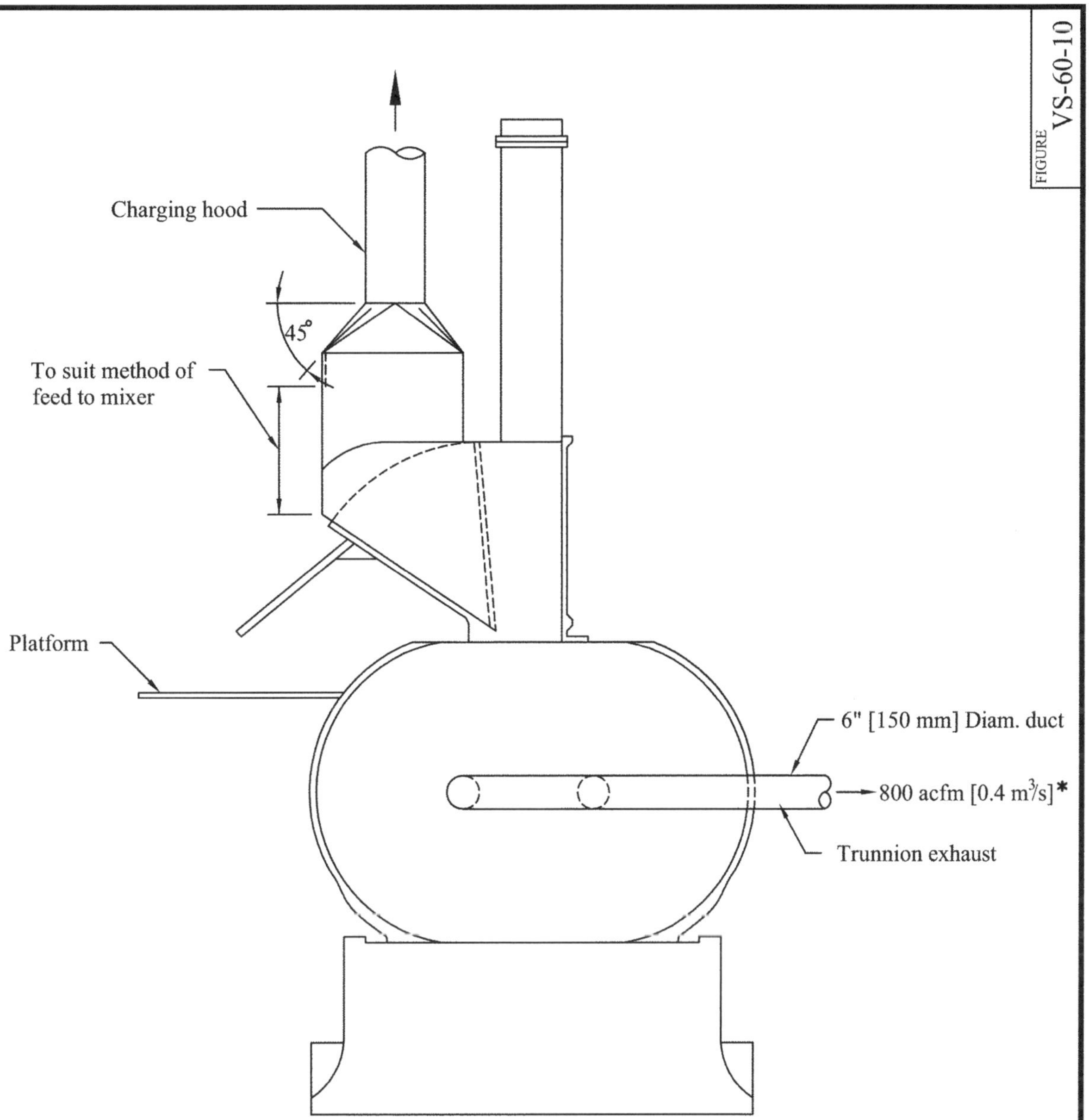

Q = 200–300 acfm/ft^2 [1.0–1.5 am^3/s/m^2] open face area.
500 acfm/ft [0.78 am^3/s/m] of belt width if belt feeder used.
Duct velocity = 3500 fpm [17.50 m/s] minimum.
h_e = 0.25 VP_d at hood
h_e = 1.0 VP_d at trunnion

* Consult mixer manufacturer for specific recommendations.

ACGIH®

TITLE: BANBURY MIXER

FIGURE: VS-60-10

DATE: 1-12

CHECK CODES, REGULATIONS, AND LAWS (LOCAL, STATE, AND NATIONAL) TO ENSURE THAT DESIGN IS COMPLIANT.

FIGURE VS-60-11

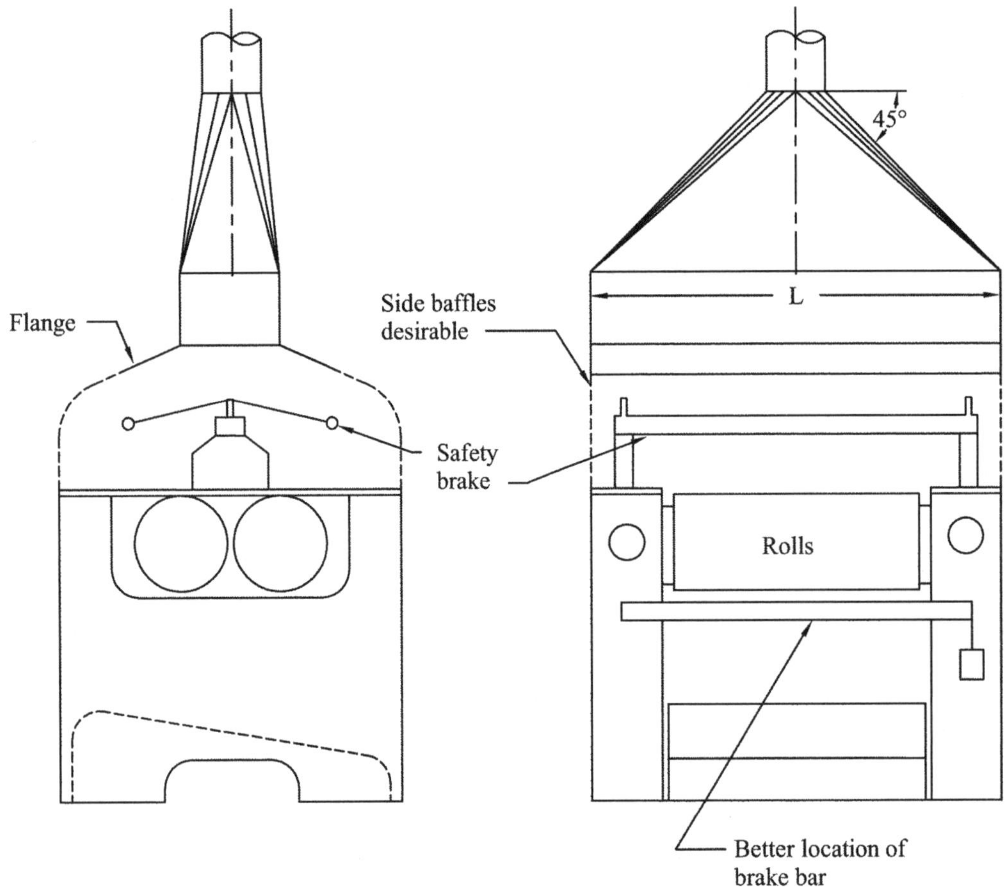

Q = 125 acfm/ft^2 [0.63 am^3/s/m^2] open face area
Minimum duct velocity = 2000 fpm [10 m/s]
h_e = 0.25 VP_d

NOTE: Both sides may be open

ACGIH®	TITLE RUBBER CALENDER ROLLS	FIGURE VS-60-11 DATE 1-12

CHECK CODES, REGULATIONS, AND LAWS (LOCAL, STATE, AND NATIONAL)
TO ENSURE THAT DESIGN IS COMPLIANT.

FIGURE VS-60-12

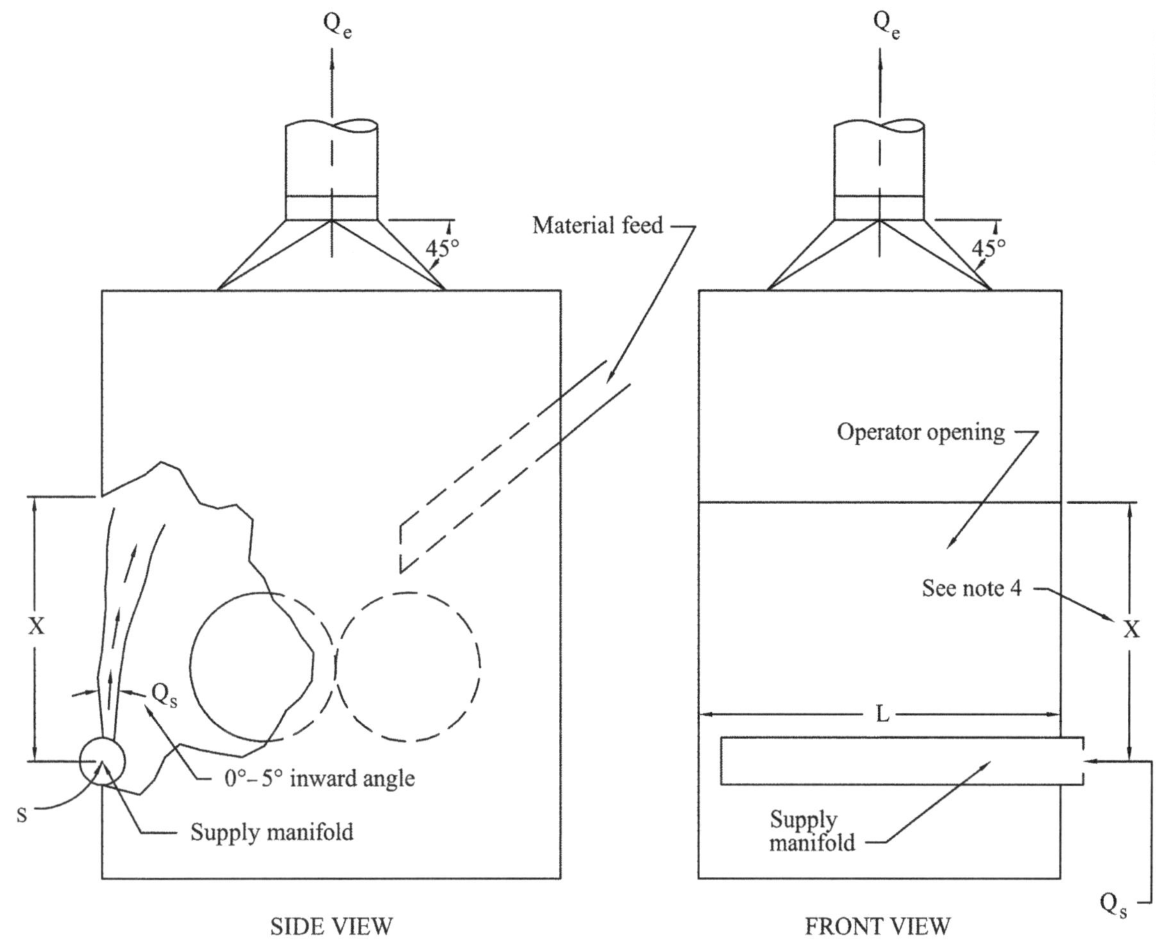

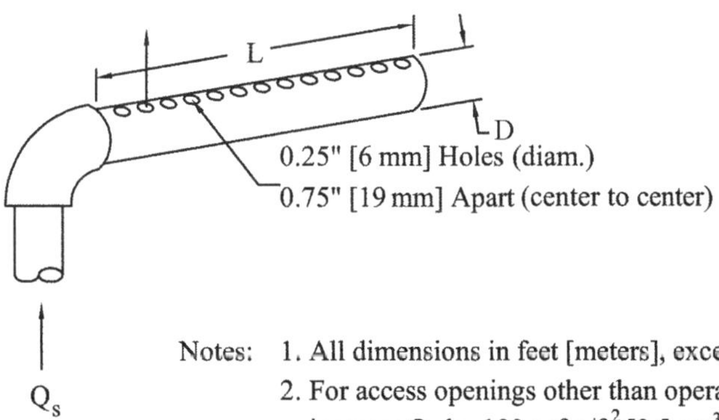

$D \geq 0.145\sqrt{L}$ ft (Manifold diameter)
[$D \geq 0.080\sqrt{L}$ m (Manifold diameter)]
$Q_s = 29.6$ L acfm (Supply flow)
[$Q_s = 0.05$ L am³/s (Supply flow)]
$Q_e = 320\sqrt{X}$ L acfm (Exhaust flow)
[$Q_e = 0.90\sqrt{X}$ L am³/s (Exhaust flow)]
SP_s = Manifold pressure = +1.2 "wg
[SP_s = Manifold pressure = +300 Pa]

Notes:
1. All dimensions in feet [meters], except as otherwise noted.
2. For access openings other than operator opening, increase Q_e by 100 acfm/ft^2 [0.5 am^3/s/m^2] of opening area.
3. If operator opening is required on both sides of mill, total Q_e will be sum of Q_e for both sides.
4. X not to exceed 6 ft [1.8 m].

Reference 13.60.1

ACGIH®	TITLE: ROLLER MILL VENTILATION	FIGURE: VS-60-12
		DATE: 1-12

CHECK CODES, REGULATIONS, AND LAWS (LOCAL, STATE, AND NATIONAL) TO ENSURE THAT DESIGN IS COMPLIANT.

13.65 MOVABLE EXHAUST HOODS

Movable exhaust hoods provide control for moving contaminant sources. In general, movable hoods are associated with flexible exhaust ducts, traveling exhaust hoods, swivel, slip or telescoping joints in duct sections, or systems that separate the hood from the duct for access to the process.

Flexible exhaust duct is possibly the most common way of providing a movable exhaust hood. A section of flexible duct connects to a relatively small exhaust hood. The duct section and hood may be supported by a counter-weighted or spring-loaded, hinged arm that allows the positioning of the exhaust hood near the source of contaminant generation. This type of device is known as "snorkel," "elephant trunk," or "flex-arm" exhaust. Illustrations of this type of exhaust in Chapter 13 include Welding Exhaust (VS-65-01) and Granite Cutting and Finishing (VS-65-02). Flexible exhaust duct use is also illustrated in Barrel Filling (VS-15-01), Metal Spraying (VS-90-30) Service Garage Ventilation (VS-85-01 and VS-85-02) and low volume/high velocity systems (Section 13.40). Frictional resistance can be very high in the flexible duct section as can the negative pressure or suction. Materials used in the construction of the duct may be metal or non-metal and the losses vary over a wide range and depend on the type and use. The application data provided by the manufacturer must be included in the design development of the system. When used, flexible duct should be non-collapsible with minimal length to reduce undesirable bends, which result in excessive static pressure losses.

Traveling exhaust hoods may be used for a variety of operations where the contaminant source moves from one point to another. This type is more suited to heavy duty requirements than the flexible exhaust duct. Examples of these operations include flame and plasma cutting, foundry pouring, heavy abrasive cutting, and similar operations. An illustration of a traveling exhaust hood is the Trav-L-Vent (VS-65-03). (See Reference 13.65.1.)

Telescoping or slip joints are duct sections that overlap, slide, or rotate to allow a section of one exhaust duct to slide into or rotate around another section of duct. This arrangement allows an exhaust hood to be moved from one position to another without disconnecting the duct or hood. Illustrations of the slip joint include the Core Grinder (VS-80-13) and Granite Cutting and Finishing (VS-65-02). This joint or duct section may include a swivel feature to permit rotation of the hood away from the process equipment being exhausted and may be used in the horizontal or vertical plane. Guide rails may be required for the horizontal application while pulleys and counter-weights may be required for vertical applications.

Separating exhaust duct sections is another method of providing a movable exhaust hood. This concept requires that the exhaust duct separate at or near the exhaust hood when the hood is moved for access to the process equipment or the process equipment moves. Illustrations of this method include hoods for Top Electrode Melting Furnace (VS-55-03), Core Making Machine—Small Roll-Over Type (VS-20-11) and Mobile Hood, Die Casting (VS-55-21). Alignment of the exhaust duct when the hood is in place is critical and any opening at this point should be included in the total exhaust calculations. Also, during the separation period, little or no exhaust control will be available at the contaminant source.

REFERENCE

13.65.1 Kirk & Blum Mfg.: Travel Vent Equipment Specifications and Layout, Indianapolis, IN (2005).

FIGURE VS-65-01

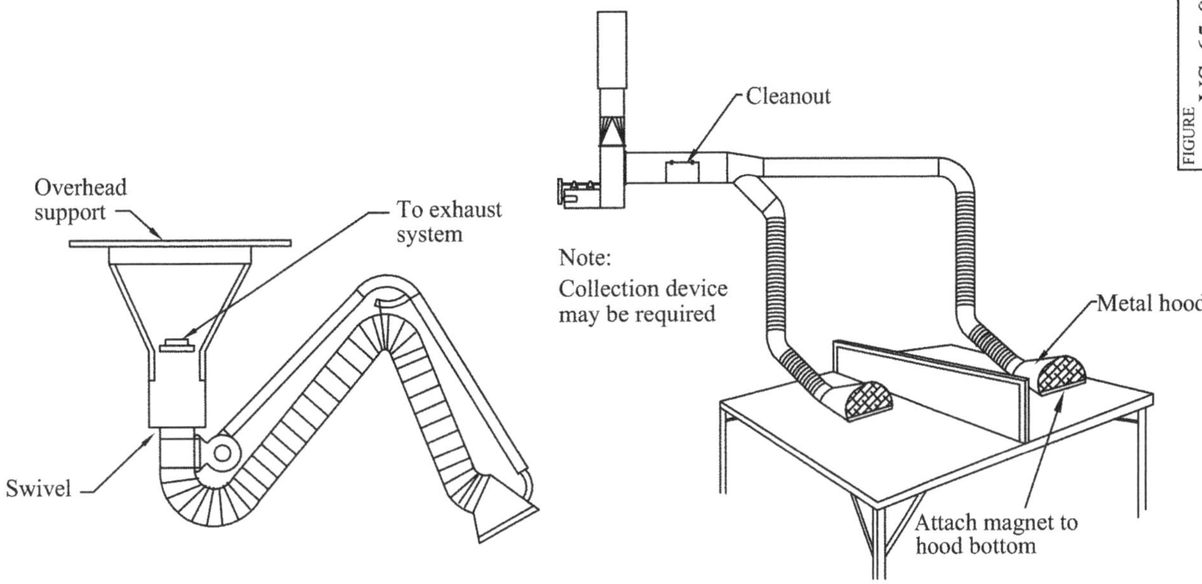

FLEXIBLE EXHAUST CONNECTIONS

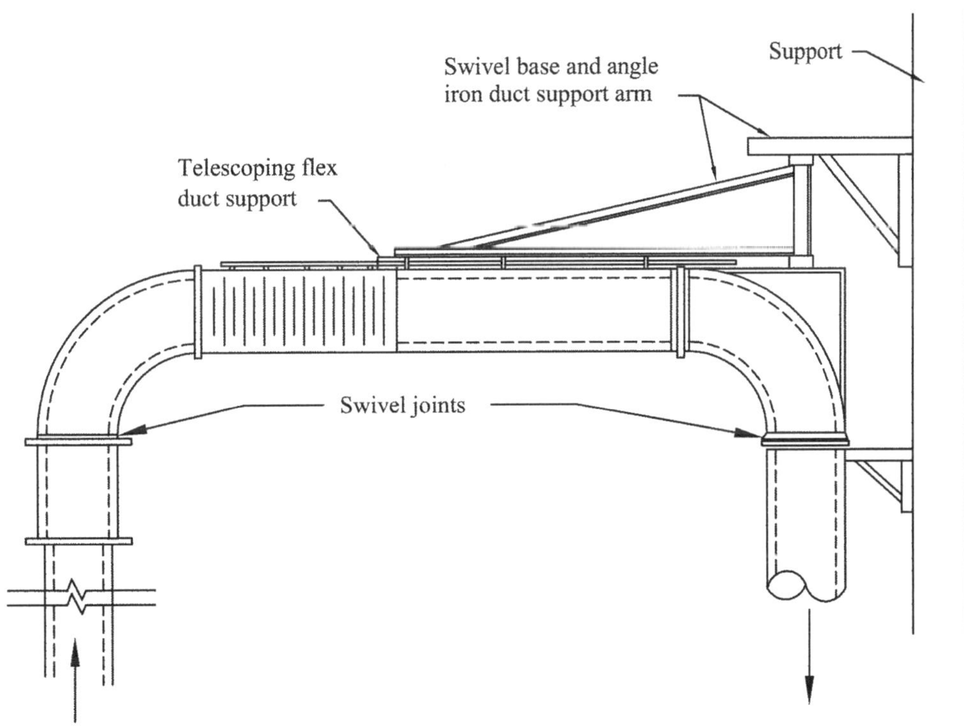

TITLE	FIGURE
MOVABLE EXHAUST HOODS	VS-65-01
	DATE 1-12

CHECK CODES, REGULATIONS, AND LAWS (LOCAL, STATE, AND NATIONAL) TO ENSURE THAT DESIGN IS COMPLIANT.

FIGURE VS-65-02

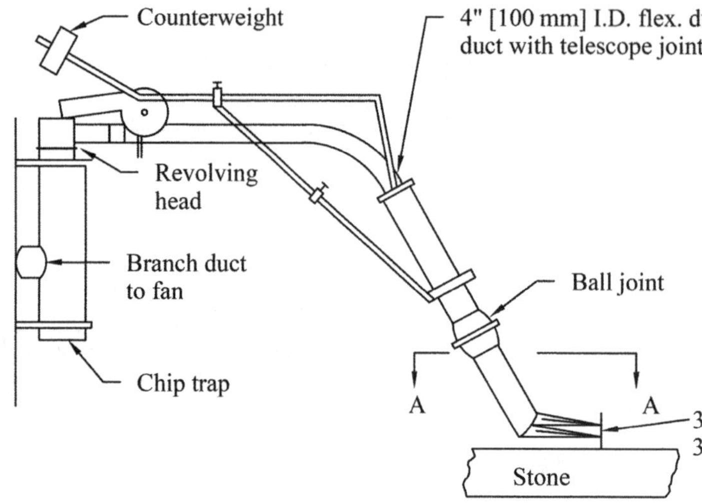

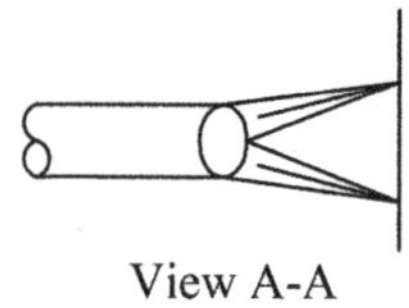

PNEUMATIC HAND TOOLS

Q = 400 acfm [0.20 am^3/s] minimum, tool 10" [250 mm] max distance from hood
Minimum duct velocity = 4000 fpm [20 m/s]

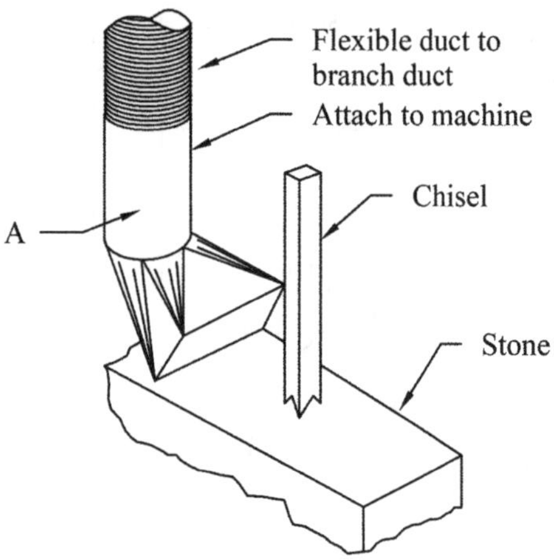

NOT RECOMMENDED

Abrasive blasting to be done in a room or cabinet; 500 fpm [2.5 m/s] at all openings. See "Abrasive Blasting", VS-80-01

SURFACE MACHINE HOODS

Hood	acfm [am^3/s]	Branch diam.
Baby surfacer	400 [0.20]	4" [100 mm]
Medium surfacer	600 [0.30]	5" [125 mm]

$h_e = 1.0\ VP_d$ (at point A)

ACGIH®

TITLE

GRANITE CUTTING AND FINISHING

FIGURE VS-65-02

DATE 1-12

CHECK CODES, REGULATIONS, AND LAWS (LOCAL, STATE, AND NATIONAL) TO ENSURE THAT DESIGN IS COMPLIANT.

FIGURE VS-65-03

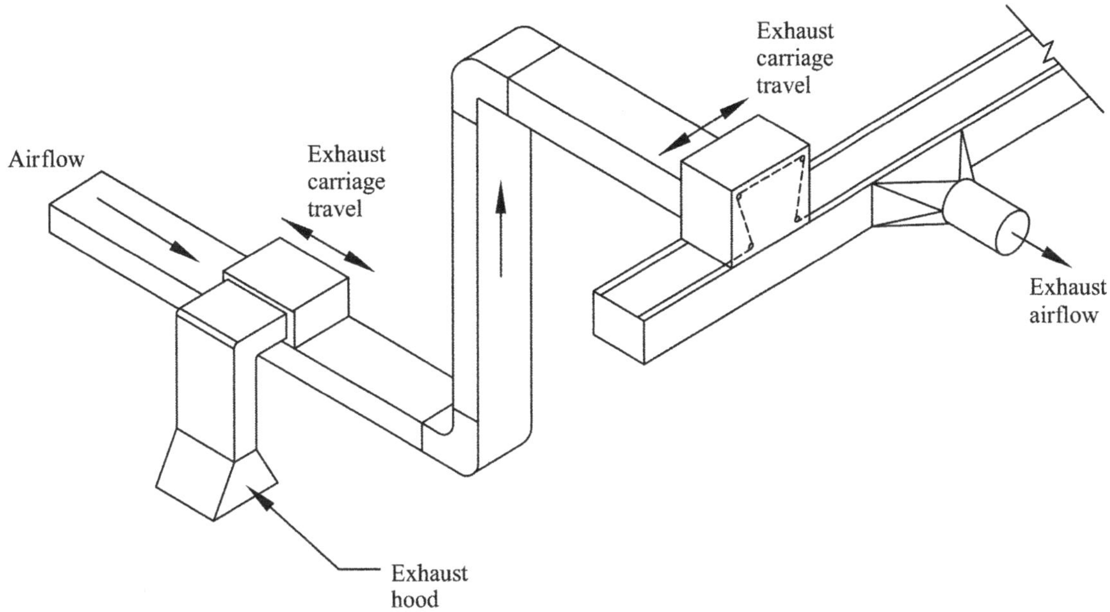

Typical Design Ranges:

Q = 2500 – 10,000 acfm [1.3–5.0 am^3/s]
V = 3500 – 4000 fpm [17.50–20.00 m/s]
h_e = 6" wg [1500 Pa] approx. total
SP_h = calculate separately

See Reference 13.65.1

ACGIH®	TITLE TRAV-L-VENT PERSPECTIVE LAYOUT	FIGURE VS-65-03 DATE 1-12

CHECK CODES, REGULATIONS, AND LAWS (LOCAL, STATE, AND NATIONAL)
TO ENSURE THAT DESIGN IS COMPLIANT.

13.67 VENTILATION REQUIREMENTS FOR ENGINEERED NANOMATERIALS

Engineered nanoparticles are materials purposefully produced with at least one dimension between 1 and 100 nanometers (nm, 10^{-9} m).[13.67.1] Nanoparticles often exhibit unique physical and chemical properties that impart specific characteristics essential in making engineered materials, but little is known about what effect these properties may have on human health. Research has shown that the physicochemical characteristics of particles can influence their effects in biological systems. These characteristics include particle size, shape, surface area, charge, chemical properties, solubility, oxidant generation potential, and degree of agglomeration. Until the results from research studies help us to more fully understand the characteristics of nanoparticles that may pose a health risk, precautionary measures are warranted.

Nanomaterial-enabled products such as nanocomposites, surface-coated materials, and materials comprised of nanostructures, such as integrated circuits, are unlikely to pose a risk of exposure during their handling and use as materials. However, some of the processes used in their production (e.g., formulating and applying nanoscale coatings) may lead to exposure to nanomaterials, and the cutting or grinding of such products could release respirable-sized nanoparticles.

Maintenance on production systems (including cleaning and disposal of materials from dust collection systems) is likely to result in exposure to nanoparticles if deposited nanomaterials are disturbed. The following workplace tasks can increase the risk of exposure to nanoparticles:

- Working with nanomaterials in liquid media without adequate protection
- Working with nanomaterials in liquid during pouring or mixing operations, or where a high degree of agitation is involved
- Generating nanoparticles in non-enclosed systems
- Handling (e.g., weighing, blending, spraying) powders of nanomaterials
- Maintenance on equipment and processes used to produce or fabricate nanomaterials and the cleaning-up of spills and waste material containing nanomaterials
- Cleaning of dust collection systems used to capture nanoparticles
- Machining, sanding, drilling, or other mechanical disruptions of materials containing nanoparticles

13.67.1 Precautionary Measures. Given the limited amount of information about health risks that may be associated with nanomaterials, taking measures to minimize worker exposures is prudent. For most processes and job tasks, the control of airborne exposure to nanoaerosols can be accomplished using a variety of engineering control techniques similar to those used in reducing exposure to general aerosols (see Chapters 4, 5, 6, and 8). The implementation of a risk management program in workplaces where exposure to nanomaterials exists can help to minimize the potential for exposure to nanoparticles. Elements of such a program should include the following:

- Evaluating the hazard posed by the nanomaterial based on available physical and chemical property data, toxicology, or health-effects data
- Assessing the worker's job tasks to determine the potential for exposure (see Chapter 1)
- Educating and training workers in the proper handling of nanomaterials (e.g., good work practices)
- Establishing criteria and procedures for installing and evaluating engineering controls (e.g., exhaust ventilation) at locations where exposure to nanomaterials might occur (see Chapters 5 and 6)
- Systematically evaluating exposures to ensure that control measures are working properly and that workers are being provided the appropriate personal protective equipment
- Developing procedures for determining the need for and selecting proper personal protective equipment (e.g., clothing, gloves, respirators)

13.67.2 Controlling Potential Worker Exposures. When controlling potential exposures within a workplace, it is recommended to use the hierarchy of controls to reduce worker exposures (see Chapter 1). The philosophy of the hierarchy of controls is to eliminate the hazard when possible (i.e., substitute with a less hazardous material) or, if not feasible, control the hazard at or as close to the source as possible.

13.67.2.1 Applicable Engineering Controls. If the potential hazard cannot be eliminated or substituted with a less hazardous or non-hazardous substance, then engineering controls should be installed and tailored to the process (see Chapters 5 and 6). Engineering control techniques such as source enclosure (i.e., isolating the generation source from the worker) and local exhaust ventilation systems should be effective for capturing airborne nanoparticles. However, the type of engineering control used should take into account information on the potential hazardous properties of the precursor materials and intermediates as well as those of the resulting nanomaterial. In light of current scientific knowledge about the generation, transport, and capture of aerosols,[13.67.2, 13.67.3] airborne exposure to nanomaterials can most likely be controlled using a variety of engineering control techniques similar to those used in reducing exposures to general aerosols.[13.67.4, 13.67.5] Current knowledge indicates that a well-designed exhaust ventilation system with a high-efficiency particulate air (HEPA) filter should effectively remove nanomaterials..[13.67.6]

Other factors that influence selection of controls include the physical form of the nanomaterial as well as task duration and frequency. For instance, working with nanomaterial in the slurry form in low quantities would require a less rigorous control system than those that would be required for large quantities of nanomaterials in a free or fine powder form. Unless cutting or grinding occurs, nanomaterials that are encapsulated in a solid, nanocomposites, and surface coated materials typically would not require engineering controls. Based on what is known regarding nanomaterial motion and behavior in air, traditional engineering control techniques should be effective to reduce the potential for worker exposure (see Chapters 5 and 6 and Figure 13-67-1). Given the state of the science and uncertainty in health risks from exposure to nanoparticles, a precautionary control banding approach would be a reasonable way to assess the potential risk, or potency, from airborne nanoparticles (see Chapter 1, Table 1-1). If little is known about the properties or hazards of the nanomaterial, the potency band should be judged to be 'high'. This is precautionary and is an appropriate position for new materials with unknown hazards. The quantity of the bulk nanomaterial that is synthesized or handled in the manufacture of a product will influence the selection of engineering controls as well. Once a potency band is determined, Table 13-67-1 can be used to select the appropriate ventilation control for the application, with consideration of the process size.

It may be helpful to walk through an example to illustrate how to select a suitable control device for a nanotechnology process. In this example, the control approach outlined above will be combined with process specific information that would be required to make a well-informed decision in selecting the best control device; weighing the factors of worker safety, the properties and behavior of nanoparticles, and nanomaterial process considerations. In this example, it is assumed that research quantities of nanoparticles are typically handled in laboratories with ventilation controls. The second assumption will be that carbon nanotubes are being handled in small quantities to conduct research. Since the quantities are small, local containment and control can be applied, such as low-flow vented work stations and small glove box chambers. The properties of nanomaterials, along with the unique methods that may be employed for producing them, may mean that traditional exhaust ventilation may be more energetic than necessary for removing incidentally released nanoparticles. When carbon nanotubes are being handled, the ventilation engineer should determine the aerodynamic diameter of the tubes, since their density is ~ 1.4 grams per cubic centimeter (g/cc; for comparison, the density of water is 1 g/cc). This information allows the engineer to determine the aerosolization potential, and airborne behavior of the nanomaterial. Utilizing this information, an engineering control should be selected that ensures protection of the worker, as well as minimizing nanomaterial aerosolization and losses. For low-density materials, like carbon nanotubes, it is possible to use low-flow (not low face velocity) hoods to reduce the turbulence within the hood, therefore reducing the potential for aerosolization and loss of the nanotubes in the hood. As the quantities of nanomaterials used are increased, care must be taken to ensure the selected ventilation control is adequate to prevent potential worker exposure and the migration of nanomaterials into the work room or adjacent areas. Ventilation systems to control nanoparticles should be designed, tested, and maintained using recommended approaches (see Chapters 5 and 6).

13.67.2.2 Collection Efficiency of Filters for Engineered Nanomaterials. HEPA filtration and a well-designed ventilation system should be sufficient to remove nanoparticles.[13.67.4, 13.67.5] Limited studies have reported the efficacy of filter media typically found in control systems in capturing nanoparticles. The lack of data on filtration performance against nanoparticles (in particular nanoparticles smaller than 20 nm) is primarily due to the challenges in generating and quantifying particles in those size ranges. Despite these limitations, the results of some studies using different filter media challenged with monodispersed aerosols (silver 4 to 10 nm and dioctylphthalate 32 to 420 nm) were in agreement with classical single-fiber theory showing an increase in filtration efficacy for smaller size particles.[13.67.7] No evidence for particle thermal rebound was observed. Similar results have been recently reported by Kim et al. using different filter media challenged with particles ranging in size from 2.5 to 20 nm, indicating that other filter media would behave similarly.[13.67.8] HEPA filters used in dust collection systems should be coupled with well-designed filter housings. If the filter is improperly seated within the housing, nanoparticles have the potential to bypass the filter, leading to lower than predicted filter efficiencies.[13.67.9]

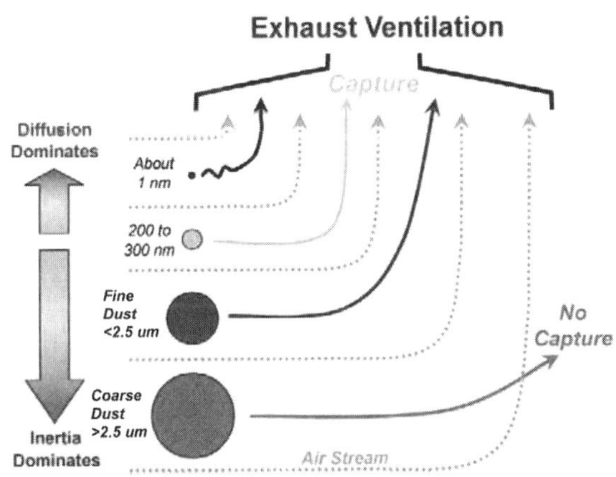

FIGURE 13-67-1. Exposure control of particles (illustration of how particle diameter-related diffusion and inertia influence particle capture efficiency in a ventilation system). Particles with a diameter of 200–300 nm have minimal diffusion and inertial properties and are easily transported by moving air and captured. Particle motion by diffusion increasingly dominates as particle diameter decreases below 200 nm. The inertial behavior of larger particles, especially those ejected from energetic processes such as grinding, increases significantly with particle diameter, enabling them to cross the streamlines of moving air and avoid capture.[13.67.10]

TABLE 13-67-1. Nanomaterial Process Control Recommendations Corresponding to Potency Bands from Chapter 1, Table 1-1

Potency Band	Process Size		
	Large	Medium	Small
1	Dilution ventilation	Dilution ventilation	Dilution ventilation
2	LEV–fixed position hoods Ventilated work benches/tables	LEV–fixed position hoods Ventilated work benches/tables	LEV–adjustable position hoods Ventilated work benches/tables
3	Clean rooms	Chemical fume hood Custom enclosures	Glove box Chemical fume hood Custom enclosures
4	Clean rooms	Chemical fume hood Custom enclosures	Glove box Chemical fume hood Custom enclosures

LEV–Local exhaust ventilation

Do not directly exhaust effluent air that is reasonably suspected to contain engineered nanoparticles.
Exhaust air must be passed through a HEPA filter, since at this time it is the only air pollution control device known to control nanoparticles with high efficiency.

REFERENCES

13.67.1 NIOSH: Approaches to Safe Nanotechnology: Managing the Health and Safety Concerns Associated with Engineered Nanomaterials. Cincinnati, OH: U.S. Department of Health and Human Services, Pub. # 2009-125 (2009).

13.67.2 Seinfeld J.A.; Pandis, S.N. Atmospheric chemistry and physics. New York: John Wiley and Sons (1998).

13.67.3 Hinds, W.C.: Aerosol technology: properties, behavior, and measurement of airborne particles, 2nd ed. New York: Wiley-Interscience (1999).

13.67.4 Ratherman, S.: Methods of control. In: Fundamentals of industrial hygiene, B. Plog, Ed. Itasca, IL: National Safety Council (1996).

13.67.5 Burton, J.: General methods for the control of airborne hazards. In: The occupational environment — its evaluation and control, S.R. DiNardi, Ed. Fairfax, VA: American Industrial Hygiene Association (1997).

13.67.6 U.S. DOE: Approach to Nanomaterial. ES&H, U.S. Department of Energy's Nanoscale Science Research Centers. Washington, DC: U.S. Department of Energy (2007).

13.67.7 Van Osdell, D.W.; Liu, B.Y.H.; Rubow, K.L.; Pui, D.Y.H.: Experimental-study of submicrometer and ultrafine particle penetration and pressure-drop for high-efficiency filters. Aerosol Sci Technol 12(4):911–925 (1990).

13.67.8 Kim, S.E.; Harrington, M.S.; Pui, D.Y.H.: Experimental study of nanoparticles penetration through commercial filter media. J Nanopart Res 9:117–125 (2007).

13.67.9 NIOSH: Filtration and air-cleaning systems to protect building environments. Cincinnati, OH: U.S. Department of Health and Human Services, Centers for Disease Control and Prevention, National Institute for Occupational Safety and Health, DHHS (NIOSH) Publication No. 2003–136 (2003).

13.67.10 Schulte, P.; Geraci, C.; Zumwalde, R.; Hoover, M.; Kuempel, E.: Occupational risk management of engineered nanoparticles. J Occup Environ Hyg 5: 239–249 (2008a).

13.70 OPEN SURFACE TANKS

Ventilation rates for plating, cleaning, and other open surface tank operations will depend on a number of parameters that include materials, tank configuration and location, and type of ventilation system. This section describes four hood/ventilation types: enclosing hoods, canopy hoods, lateral exhaust, and push-pull.

Enclosing hoods usually consist of a lateral hood with one end panel (two sides open) or panels at both tank ends (one side open). This hood configuration can provide increased efficiency by reducing the effects of cross-drafts and by directing more of the hood airflow over the tank open surface.

Canopy hoods may be open on four sides (free standing) or on three sides (such as against a wall). Control is achieved by airflow into the hood. It is, however, difficult in many cases to achieve sufficient control velocity without excessive airflow rates. Canopy hoods should not be used with highly toxic materials, in locations where high cross-drafts are unavoidable, or where the worker must bend over the tank.

Lateral exhaust consists of a slot hood which controls emissions by pulling air across the tank. A single slot may be used on one side of the tank where the tank width is 36 inches [0.9 m] or less. For widths greater than 36 inches [0.9 m] and where the process configuration will allow, two slot hoods on opposite sides of the tank or a slot hood along the tank centerline may be used. A single slot may be used up to a tank width of 48 inches [1.2 m], but only if the material hazard class is low and if cross-drafts are not present. (See Section 13.70.1.)

The airflow required will be that necessary to achieve a minimum control velocity determined by the hazard class of the material used for operation and the particular tank/ventilation system configuration. The procedure for determining the class and minimum control velocity for the three preceding hood types is provided in Tables 13-70-1 through 13-70-7 and the accompanying text. Exhaust flow for a canopy hood is determined from VS-99-03 and for a booth hood from Chapter 6, Figure 6-11 where W is the total opening width. The exhaust flow for a lateral hood is determined from Table 13-70-4.

Air and/or mechanical agitation of the tank solution may be used as an aid to the plating or cleaning process. Mechanical agitation creates a rolling motion and usually will not affect tank emissions. However, air agitation creates a boiling-like condition and may significantly increase tank emissions, thus creating need for increased exhaust flow to provide effective control.

Push-pull ventilation consists of a push jet located on one side of a tank with a lateral exhaust hood on the other side. Tank emissions are controlled by the jet formed over the tank surface that captures the emissions and carries them into the hood. Push-pull design criteria is provided in Section 13.72.

13.70.1 Tank Design Considerations.

1) Duct velocity = any desired velocity (see Chapters 6 and 9).
2) Entry loss = 1.78 slot VP plus duct entry loss for slot hoods. For canopy or enclosure hoods, entry loss = duct entry loss.
3) Maximum slot hood plenum velocity = 1/2 slot velocity (see Chapter 6).
4) Slot velocity = 2,000 fpm [10.00 m/s] unless distribution is provided by well-designed, tapered takeoff.
5) Provide ample area at the small end of the plenum.
6) If L = 6' [1.8 m] or greater, multiple takeoffs are desirable. If L = 10' [3.0 m] or greater, multiple takeoffs are necessary.
7) Tank width (W) means the effective width over which the hood must pull air to operate (e.g., where the hood face is set back from the edge of the tank, this setback must be added in measuring tank width).

   If W = 20" [0.5 m], slot on one side is suitable.

   If W = 20" to 36" [0.5 to 0.9 m], slots on both sides are desirable.

   If W = 36" to 48" [0.9 to 1.2 m], use slots on both sides or along tank centerline or use push-pull. A single slot along one side should not be used unless all other conditions are optimum.

   If W = 48" or greater, local exhaust usually is not practical. Consider using push-pull.

   Enclosure can be used for any width tank if process will permit.

   It is not practicable to ventilate across the long dimension of a tank whose ratio W/L exceeds 2.0. It is undesirable to do so when W/L exceeds 1.0
8) Liquid level should be 6" to 8" [150 to 200 mm] below top of tank with parts immersed.
9) Lateral hood types A, C, D, and E (VS-70-01 and -02) are preferred. Plenum acts as baffle to room air currents.
10) Provide removable covers on tank if possible.
11) Provide duct with cleanouts, drains, and corrosion-resistant coating if necessary. Use flexible connection at fan inlet.
12) Install baffles to reduce cross-drafts. A baffle is a vertical plate the same length as the tank and with the top of the plate as high as the tank is wide. If the exhaust hood is on the side of the tank against a building wall or close to it, it is perfectly baffled.
13) Replacement air to the tank area must be supplied evenly and directed toward the tank from above or in front of the tank so that cross-drafts do not occur.

13.70.2 Flow Rate Calculation for Good Conditions. (No cross-drafts, adequate and well-distributed replacement air.)

1) Establish process class by determining hazard potential from Tables 13-70-1 and 13-70-2; information from Threshold Limit Values, Solvent Flash Point, Solvent Drying Time Tables in Appendices A and B.
2) From Table 13-70-3, choose minimum control velocity according to hazard potential; evolution rate (process class); and hood design (see Table 13-70-5 for typical processes).
3) From Table 13-70-4, select the acfm/ft^2 for tank dimensions and tank location.
4) Multiply tank area by value obtained from Table 13-70-4 to calculate required air volume.
5) Process class can also be established directly from Tables 13-70-5 through 13-70-8 if process parameters are known. The class shown in these tables is based on the same information as Step 1 above for usual operating conditions. Higher temperatures, agitation or other conditions may result in a higher rate of evolution. *Note: There may be differences in the class shown in the table and the class calculated in Step 1 due to changes in the TLV® occurring between issues of this Manual. In such cases, the method providing the more stringent class should be used.*

13.70.3 Aspect Ratio Determination. The minimum flow rate requirements in Table 13-70-4 are for individual tanks and are a function of that tank's specific operation, contents, and aspect ratio. Where a line contains two or more ventilated tanks placed next to each other with no gap between them, the exhaust from each tank increases the effectiveness of the ventilation on the adjacent tank in the same manner as an extended length single tank. It would be appropriate to treat the group of two or more tanks as if they were combined while determining the aspect ratio. The aspect ratio for each group is determined from the total length of the group and is used to determine the flow rate for each tank. If the line contains non-ventilated tanks, the aspect ratio and flow rate should be determined using each individual and/or ventilated group of tanks only. Non-ventilated tanks should not be included in the aspect ratio determination.

EXAMPLE PROBLEM 1 (IP)

Given: Chrome Plating Tank 6' × 2.5'
High production decorative chrome
Free standing in room
No cross-drafts

a. Tank Hood. See VS-70-01. Use hood "A" along 6' side. Hood acts as baffle.
b. Component – chromic acid

Hazard potential: A (from Table 13-70-1; from Appendix A: TLV = 0.05 mg/m^3; from Appendix A: Flash point = Negligible).

Rate of evolution: 1 (from Table 13-70-2; from Table 13-70-7: Gassing rate = high).

Class: A-1

Control velocity = 150 fpm (from Table 13-70-3)

Minimum exhaust rate = 225 acfm/ft^2 (from Table 13-70-4; Baffled tank, W/L = 0.42)

Minimum exhaust flow rate = 225 × 15 = 3375 acfm

c. Hood Design

Design slot velocity = 2000 fpm

Slot area = Q/V = 3375 acfm/2000 fpm = 1.69 ft^2

Slot width = A/L = 1.69 ft^2/6 ft = 0.281' = 3.375"

Plenum depth = (2)(slot width) = (2)(3.375) = 6.75"

Duct area = Q/V = 3375 acfm/2500 fpm = 1.35 ft^2

Use 16" duct, area = 1.396 ft^2

Final duct velocity = Q/A = 3375/1.396 = 2420 fpm

Hood SP = Entry loss + Acceleration

= 1.78 VP_s + 0.25 VP_d + 1.0 VP_d (see Chapter 6)

= (1.78 × 0.25") + (0.25 × 0.37") + 0.37"

= 0.45 + 0.09 + 0.37

Hood SP = 0.91"

EXAMPLE PROBLEM 1 (SI) Chrome Plating Tank

Given: Chrome Plating Tank 1.8 m × 0.75
High production decorative chrome
Free standing in room
No cross-drafts

a. Tank Hood. See VS-70-01. Use hood "A" along 1.8 m side. Hood acts as baffle.
b. Component – chromic acid

Hazard potential: A (from Table 13-70-01; from Appendix A: TLV® = 0.05 mg/m^3; from Appendix A: Flashpoint = negligible)

Rate of evolution: 1 (from Table 13-70-2; from Table 13-70-7: gassing rate = high)

Class: A-1

Control velocity = 0.75 m/s (from Table 13-70-3)

Minimum exhaust rate = 1.3 m^3/s/m^2 (from Table 13-70-4; Baffled tank, W/L = 0.42)

Minimum exhaust flow rate = 1.13 × (.75 * 1.8) = 1.53 am^3/s

c. Hood Design

Design slot velocity = 10 m/s

Slot area = Q/V = 1.53/10 = 0.15 m^2

Slot width = A/L = 0.15/1.8 = 0.088 m^2 = 83 mm

Duct area = Q/V = 1.53/12.5 = 0.12 m^2

Use 380 mm duct, area = 0.12 m^2

Final duct velocity = Q/A = 1.53/0.12 = 12.75 m/s

Hood SP = Entry loss + Acceleration

$= 1.78\ VP_s + 0.25\ VP_d + Vp_d$

$= 1.78 \times 60 + 0.25 \times 97 + 97$

$= 106 + 24 + 97 = 227$

Hood SP = 227 Pa

TABLE 13-70-1. Determination of Hazard Potential

	HYGIENIC STANDARDS		
Hazard Potential	Gas and Vapor (see Appendix A)	Mist (see Appendix A)	Flash Point, F [C] (see Appendix B)
A	0–10 ppm	0–0.1 mg/m^3	—
B	11–100 ppm	0.11–1.0 mg/m^3	Under 100 F [< 38 C]
C	101–500 ppm	1.1–10 mg/m^3	100–200 F [38–93 C]
D	Over 500 ppm	Over 10 mg/m^3	Over 200 F [> 93 C]

TABLE 13-70-2. Determination of Rate of Gas, Vapor, or Mist Evolution

Rate	Liquid Temperature, F [C]	Degrees Below Boiling Point, F [C]	Relative Evaporation* (Time for 100% Evaporation)	Gassing**
1	Over 200 [> 93]	0–20 [0–11]	Fast (0–3 hours)	High
2	150–200 [66–93]	21–50 [11–28]	Medium (3–12 hours)	Medium
3	94–149 [34–65]	51–100 [28–55]	Slow (12–50 hours)	Low
4	Under 94 [< 34]	Over 100 [> 55]	Nil (Over 50 hours)	Nil

*Dry time relation (see Appendix B), Below 5–Fast; 5-15–Medium; 15-75–Slow; 75-over–Nil.

**Rate of gassing depends on rate of chemical or electrochemical action and therefore depends on the material treated and the solution used in the tank and tends to increase with: 1. amount of work in the tank at any one time, 2. strength of the solution in the tank, 3. temperature of the solution in the tank, and 4. current density applied to the work in electrochemical tanks.

TABLE 13-70-3. Minimum Control Velocity (fpm) [m/s] for Undisturbed Locations

Class (see Section 13.70.2)	Enclosing Hood One Open Side	Enclosing Hood Two Open Sides	Lateral Exhaust (see VS-70-01 and VS-70-02) (Note 1)	Canopy Hoods (see Chapter 6, Figure 6-11 and VS-99-03) Three Open Sides	Canopy Hoods Four Open Sides
A-1 and A-2 (Note 2)	100 [0.50]	150 [0.75]	150 [0.75]	Do not use	Do not use
A-3 (Note 2), B-1, B-2, and C-1	75 [0.38]	100 [0.50]	100 [0.50]	125 [0.63]	175 [0.88]
B-3, C-2, and D-1 (Note 3)	65 [0.33]	90 [0.45]	75 [0.38]	100 [0.50]	150 [0.75]
A-4 (Note 1), C-3, and D-2 (Note 3)	50 [0.25]	75 [0.38]	50 [0.25]	75 [0.38]	125 [0.63]
B-4, C-4, D-3 (Note 3), and D-4—Adequate General Room Ventilation Required (see Chapter 4)					

Notes: 1. Use aspect ratio to determine air volume; see Table 13-70-4.
2. Do not use canopy hood for Hazard Potential A processes.
3. Where complete control of hot water is desired, design as next highest class.

TABLE 13-70-4. Minimum Rate, acfm/ft^2 [m^3/s/m^2] of Tank Area for Lateral Exhaust

Required Minimum Control Velocity, fpm [m/s] (from Table 13-70-3)	acfm/ft^2 to maintain required minimum control velocities at following $\frac{\text{tank width}}{\text{tank length}}\left(\frac{W}{L}\right)$ ratios (see Section 13.70.3) 0.0–0.09	0.1–0.24	0.25–0.49	0.50–0.99	1.0–2.0 (Note 2)
Hood against wall or flanged (see Note 1 below and Section 13.70.1, Note 12). See VS-70-01 A and VS-70-02 D and E.					
50 [0.25]	50 [0.25]	60 [0.30]	75 [0.38]	90 [0.45]	100 [0.50]
75 [0.38]	75 [0.38]	90 [0.45]	110 [0.55]	130 [0.65]	150 [0.75]
100 [0.50]	100 [0.50]	125 [0.63]	150 [0.75]	175 [0.88]	200 [1.00]
150 [0.75]	150 [0.75]	190 [0.95]	225 [1.13]	250 Note 3 [1.25]	250 Note 3 [1.25]
Hood on free standing tank (see Note 1). See VS-70-01 B and VS-70-02 F.					
50 [0.25]	75 [0.38]	90 [0.45]	100 [0.50]	110 [0.55]	125 [0.63]
75 [0.38]	110 [0.55]	130 [0.65]	150 [0.75]	170 [0.85]	190 [0.95]
100 [0.50]	150 [0.75]	175 [0.88]	200 [1.00]	225 [1.13]	250 [1.25]
150 [0.75]	225 [1.13]	250 Note 3 [1.25]	250 Note 3 [1.25]	250 Note 3 [1.25]	250 Note 3 [1.25]

Note 1. Use W/2 as tank width in computing W/L ratio for hood along centerline or two parallel sides of tank. See VS-70-01 B and C and VS-70-02 F.
2. See Section 13.70.1, items 6 and 7.
3. While bracketed values may not produce 150 fpm [0.75 m/s] control velocity at all aspect ratios, the 250 acfm/ft^2 [1.25 m^3/s/m^2] is considered adequate for control.

TABLE 13-70-5. Typical Processes Minimum Control Velocity (fpm) [m/s] for Undisturbed Locations

Operation	Contaminant	Hazard See Section 13.70.2	Contaminant Evolution See Section 13.70.2	Lateral Exhaust Control Velocity (See VS-70-01 & VS-70-02)	Collector Recommended
Anodizing Aluminum	Chromic-Sulfuric Acids	A	1	150 [0.75]	X
Aluminum Bright Dip	Nitric & Sulfuric Acids	A	1	150 [0.75]	X
	Nitric & Phosphoric Acids	A	1	150 [0.75]	X
Plating–Chromium	Chromic Acid	A	1	150 [0.75]	X
Copper Strike	Cyanide Mist	C	2	75 [0.38]	X
Metal Cleaning (Boiling)	Alkaline Mist	C	1	100 [0.50]	X
Hot Water (if vent desired)					
Not boiling	Water Vapor	D	2	50 (Note 1) [0.25]	
Boiling		D	1	75 (Note 1) [0.38]	
Stripping–Copper	Alkaline-Cyanide Mists	C	2	75 [0.38]	X
Nickel	Nitrogen Oxide Gases	A	1	150 [0.75]	X
Pickling–Steel	Hydrochloric Acid	A	2	150 [0.75]	X
	Sulfuric Acid	B	1	100 [0.50]	X
Salt Solution					
Bonderizing & Parkerizing	Water Vapor	D	2	50* [0.25]	
Not Boiling	Water Vapor	D	2	50* [0.25]	
Boiling		D	1	75* [0.38]	
Salt Baths (Molten)	Alkaline Mist	C	1	100 [0.50]	X

*Note 1: Where complete control of water vapor is desired, design as next highest class.

TABLE 13-70-6. Airborne Contaminants Released by Metallic Surfaced Treatment, Etching, Pickling, Acid Dipping and Metal Cleaning Operations

Process	Type	Notes	Component of Bath That May Be Released to Atmosphere (13)	Physical and Chemical Nature of Major Atmospheric Contaminant	Class (12)	Usual Temp. Range, F [C]
Surface Treatment	Anodizing Aluminum		Chromic-Sulfuric Acids	Chromic Acid Mist	A-1	95 [35]
	Anodizing Aluminum		Sulfuric Acid	Sulfuric Acid Mist	B-1	60–80 [16–27]
	Black Magic		Conc. Sol. Alkaline Oxidizing Agents	Alkaline Mist, Steam	C-1	260–350 [127–177]
	Bonderizing	1	Boiling Water	Steam	D-2,1(14,15)	140–212 [60–100]
	Chemical Coloring		None	None	D-4	70–90 [21–32]
	Descaling	2	Nitric-Sulfuric, Hydrofluoric Acids	Acid Mist, Hydrogen Fluoride Gas, Steam	B-2,1(15)	70–150 [21–66]
	Ebonol		Conc. Sol. Alkaline Oxidizing Agents	Alkaline Mist, Steam	C-1	260–350 [127–177]
	Galvanic-Anodize	3	Ammonium Hydroxide	Ammonia Gas, Steam	B-3	140 [60]
	Hard-Coating Aluminum		Chromic-Sulfuric Acid	Chromic Acid Mist	A-1	120–180 [49–82]
	Hard-Coating Aluminum		Sulfuric Acid	Sulfuric Acid Mist	B-1	120–180 [49–82]
	Jetal		Conc. Sol. Alkaline Oxidizing Agents	Alkaline Mist, Steam	C-1	260–350 [127–177]
	Magcote	4	Sodium Hydroxide	Alkaline Mist, Steam	C-3,2(15)	105–212 [41–100]
	Magnesium Pre-Dye Dip		Ammonium Hydroxide-Ammonium Acetate	Ammonia Gas, Steam	B-3	90–180 [32–82]
	Parkerizing	1	Boiling Water	Steam	D-2,1(14,15)	140–213 [60–101]
	Zincete Immersion	5	None	None	D-4	70–90 [21–32]
Etching	Aluminum		Sodium Hydroxide-Soda Ash-Trisodium Phosphate	Alkaline Mist, Steam	C-1	160–180 [71–82]
	Copper	6	Hydrochloric Acid	Hydrogen Chloride Gas	A-2	79–90 [26–32]
	Copper	7	None	None	D-4	70 [21]
Pickling	Aluminum		Nitric Acid	Nitrogen Oxide Gases	A-2	70–90 [21–32]
	Aluminum		Chromic, Sulfuric Acids	Acid Mists	A-3	140 [60]
	Aluminum		Sodium Hydroxide	Alkaline Mist	C-1	140 [60]
	Cast Iron		Hydrofluoric-Nitric Acids	Hydrogen Fluoride-Nitrogen Oxide Gases	A-2,1(15)	70–90 [21–32]
	Copper		Sulfuric Acid	Acid Mist, Steam	B-3,2(15)	125–175 [52–79]
	Copper	8	None	None	D-4	70–175 [21–79]
	Duralumin		Sodium Fluoride, Sulfuric Acid	Hydrogen Fluoride Gas, Acid Mist	A-3	70 [21]
	Inconel		Nitric, Hydrofluoric Acids	Nitrogen Oxide, HF Gases, Steam	A-1	150–165 [66–74]
	Inconel		Sulfuric Acid	Sulfuric Acid Mist, Steam	B-2	160–180 [71–82]
	Iron and Steel		Hydrochloric Acid	Hydrogen Chloride Gas	A-2	70 [21]
	Iron and Steel		Sulfuric Acid	Sulfuric Acid Mist, Steam	B-1	70–175 [21–79]
	Magnesium		Chromic-Sulfuric, Nitric Acids	Nitrogen Oxide Gases, Acid Mist, Steam	A-2	70–160 [21–71]
	Monel and Nickel		Hydrochloric Acid	Hydrogen Chloride Gas, Steam	A-2	180 [82]
	Monel and Nickel		Sulfuric Acid	Sulfuric Acid Mist, Steam	B-1	160–190 [71–88]
	Nickel Silver		Sulfuric Acid	Acid Mist, Steam	B-3,2(15)	70–140 [21–60]
	Silver		Sodium Cyanide	Cyanide Mist, Steam	C-3	70–210 [21–99]
	Stainless Steel	9	Nitric, Hydrofluoric Acid	Nitrogen Oxide, Hydrogen Fluoride Gases	A-2	125–180 [52–82]
	Stainless Steel	9,10	Hydrochloric Acid	Hydrogen Chloride Gas	A-2	130–140 [54–60]
	Stainless Steel	9,10	Sulfuric Acid	Sulfuric Acid Mist, Steam	B-1	180 [82]
	Stainless Steel Immunization		Nitric Acid	Nitrogen Oxide Gases	A-2	70–120 [21–49]
	Stainless Steel Passivation		Nitric Acid	Nitrogen Oxide Gases	A-2	70–120 [21–49]

TABLE 13-70-6 (Cont.). Airborne Contaminants Released by Metallic Surfaced Treatment, Etching, Pickling, Acid Dipping and Metal Cleaning Operations

Process	Type	Notes	Component of Bath That May Be Released to Atmosphere (13)	Physical and Chemical Nature of Major Atmospheric Contaminant	Class (12)	Usual Temp. Range, F [C]
Acid Dipping	Aluminum Bright Dip		Phosphoric, Nitric Acids	Nitrogen Oxide Gases	A-1	200 [93]
	Aluminum Bright Dip		Nitric, Sulfuric Acids	Nitrogen Oxide Gases, Acid Mist	A-2,1(15)	70–90 [21–32]
	Cadmium Bright Dip		None	None	D-4	70 [21]
	Copper Bright Dip		Nitric, Sulfuric Acids	Nitrogen Oxide Gases, Acid Mist	A-2,1(15)	70–90 [21–32]
	Copper Semi-Bright Dip		Sulfuric Acid	Acid Mist	B-2	70 [21]
	Copper Alloys Bright Dip		Nitric, Sulfuric Acids	Nitrogen Oxide Gases, Acid Mist	A-2,1(15)	70–90 [21–32]
	Copper Matte Dip		Nitric, Sulfuric Acids	Nitrogen Oxide Gases, Acid Mist	A-2,1(15)	70–90 [21–32]
	Magnesium Dip		Chromic Acid	Acid Mist, Steam	A-2	190–212 [88–100]
	Magnesium Dip		Nitric, Sulfuric Acids	Nitrogen Oxide Gases, Acid Mist	A-2,1(15)	70–90 [21–32]
	Monel Dip		Nitric, Sulfuric Acids	Nitrogen Oxide Gases, Acid Mist	A-2,1(15)	70–90 [21–32]
	Nickel and Nickel Alloys Dip		Nitric, Sulfuric Acids	Nitrogen Oxide Gases, Acid Mist	A-2,1(15)	70–90 [21–32]
	Silver Dip		Nitric Acid	Nitrogen Oxide Gases	A-1	70–90 [21–32]
	Silver Dip		Sulfuric Acid	Sulfuric Acid Mist	B-2	70–90 [21–32]
	Zinc and Zinc Alloys Dip		Chromic, Hydrochloric	Hydrogen Chloride Gas (If HCl attacks Zn)	A-4,3 (15)	70–90 [21–32]
Metal Cleaning	Alkaline Cleaning	11	Alkaline Sodium Salts	Alkaline Mist, Steam	C-2,1(15)	160–210 [71–99]
	Degreasing		Trichloroethylene-Perchloroethylene	Trichlorethylene-Perchloroethylene Vapors	B (16)	188–250 [87–121]
	Emulsion Cleaning		Petroleum-Coal Tar Solvents	Petroleum-Coal Tar Vapors	B-3,2 (15) (17)	70–140 [21–60] 70–140 [21–60]
	Emulsion Cleaning		Chlorinated Hydrocarbons	Chlorinated Hydrocarbon Vapors	(17)	70–140 [21–60]

Notes:
1 Also Aluminum Seal, Magnesium Seal, Magnesium Dye Set, Dyeing Anodized Magnesium, Magnesium Alkaline Dichromate Soak, Coloring Anodized Aluminum.
2 Stainless Steel before Electropolishing.
3 On Magnesium.
4 Also Manodyz, Dow-12.
5 On Aluminum.
6 Dull Finish.
7 Ferric Chloride Bath.
8 Sodium Dichromate, Sulfuric Acid Bath and Ferrous Sulfate, Sulfuric Acid Bath.
9 Scale Removal.
10 Scale Loosening.
11 Soak and Electrocleaning.
12 See Section 13.70.2
13 Hydrogen gas also released by many of these operations.
14 Rate where essentially complete control of steam is required. Otherwise, adequate dilution ventilation may be sufficient.
15 The higher rate is associated with the higher value in the temperature range.
16 For vapor degreasers, rate is determined by operating procedure. See VS-70-21.
17 Class of operation is determined by nature of the hydrocarbon. Refer to Appendix A.

TABLE 13-70-7. Airborne Contaminants Released by Electropolishing, Electroplating and Electroless Plating Operations

Process	Type	Notes	Component of Bath That May Be Released to Atmosphere (19)	Physical and Chemical Nature of Major Atmospheric Contaminant	Class (18)	Usual Temp. Range, F [C]
Electropolishing	Aluminum	1	Sulfuric, Hydrofluoric Acids	Acid Mist, Hydrogen Fluoride Gas, Steam	A-2	140–200 [60–93]
	Brass, Bronze	1	Phosphoric Acid	Acid Mist	B-3	68 [20]
	Copper	1	Phosphoric Acid	Acid Mist	B-3	68 [20]
	Iron	1	Sulfuric, Hydrochloric, Perchloric Acids	Acid Mist, Hydrogen Chloride Gas, Steam	A-2	68–175 [20–79]
	Monel	1	Sulfuric Acid	Acid Mist, Steam	B-2	86–160 [30–71]
	Nickel	1	Sulfuric Acid	Acid Mist, Steam	B-2	86–160 [30–71]
	Stainless Steel	1	Sulfuric, Hydrofluoric, Chromic Acids	Acid Mist, Hydrogen Fluoride Gas, Steam	A-2,1 (20)	70–300 [21–149]
	Steel	1	Sulfuric, Hydrochloric, Perchloric Acids	Acid Mist, Hydrogen Gas, Steam	A-2	68–175 [20–79]
Strike Solutions	Copper		Cyanide Salts	Cyanide Mist	C-2	70–90 [21–32]
	Silver		Cyanide Salts	Cyanide Mists	C-2	70–90 [21–32]
	Wood's Nickel		Nickel Chloride, Hydrochloric Acid	Hydrogen Chloride Gas, Chloride Mist	A-2	70–90 [21–32]
Electroless Plating	Copper		Formaldehyde	Formaldehyde Gas	A-1	75 [24]
	Nickel	2	Ammonium Hydroxide	Ammonia Gas	B-1	190 [88]
Electroplating Alkaline	Platinum		Ammonium Phosphate, Ammonia Gas	Ammonia Gas	B-2	158–203 70–95]
	Tin		Sodium Stannate	Tin Salt Mist, Steam	C-3	140–170 [60–77]
	Zinc	3	None	None	D-4	170–180 [77–82]
Electroplating Fluoborate	Cadmium		Fluoborate Salts	Fluoborate Mist, Steam	C-3,2 (20)	70–170 [21–77]
	Copper		Copper Fluoborate	Fluoborate Mist, Steam	C-3,2 (20)	70–170 [21–77]
	Indium		Fluoborate Salts	Fluoborate Mist, Steam	C-3,2 (20)	70–170 [21–77]
	Lead		Lead Fluoborate-Fluoboric Acid	Fluoborate Mist, Hydrogen Fluoride Gas	A-3	70–90 [21–32]
	Lead-Tin Alloy		Lead Fluoborate-Fluoboric Acid	Fluoborate Mist	C-3,2 (20)	70–100 [21–38]
	Nickel		Nickel Fluoborate	Fluoborate Mist	C-3,2 (20)	100–170 [38–77]
	Tin		Stannous Fluoborate, Fluoboric Acid	Fluoborate Mist	C-3,2 (20)	70–100 [21–38]
	Zinc		Fluoborate Salts	Fluoborate Mist, Steam	C-3,2 (20)	70–170 [21–77]
Electroplating Cyanide	Brass, Bronze	4,5	Cyanide Salts, Ammonium Hydroxide	Cyanide Mist, Ammonia Gas	B-4,3 (20)	60–100 [16–38]
	Bright Zinc	5	Cyanide Salts, Sodium Hydroxide	Cyanide, Akaline Mists	C-3	70–120 [21–49]
	Cadmium	5	None	None	D-4	70–100 [21–38]
	Copper	5,6	None	None	D-4	70–160 [21–71]
	Copper	5,7	Cyanide Salts, Sodium Hydroxide	Cyanide, Alkaline Mists, Steam	C-2	110–160 [43–71]
	Indium	5	Cyanide Salts, Sodium Hydroxide	Cyanide, Alkaline Mists	C-3	70–120 [21–49]
	Silver	5	None	None	D-4	72–120 [22–49]
	Tin-Zinc Alloy	5	Cyanide Salts, Potassium Hydroxide	Cyanide, Alkaline Mists, Steam	C-3,2 (20)	120–140 [49–60]
	White Alloy	5,8	Cyanide Salts, Sodium Stannate	Cyanide, Alkaline Mists	C-3	120–150 [49–66]
	Zinc	5,9	Cyanide Salts, Sodium Hydroxide	Cyanide, Alkaline Mists	C-3,2 (7)	70–120 [21–49]

TABLE 13-70-7 (Cont.). Airborne Contaminants Released by Electropolishing, Electroplating and Electroless Plating Operations

Process	Type	Notes	Component of Bath That May Be Released to Atmosphere (19)	Physical and Chemical Nature of Major Atmospheric Contaminant	Class (18)	Usual Temp. Range, F [C]
Electroplating Acid	Chromium		Chromic Acid	Chromic Acid Mists	A-1	90–140 [32–60]
	Copper	10	Copper Sulfate, Sulfuric Acid	Sulfuric Acid Mist	B-4,3 (20,21)	75–120 [24–49]
	Indium	12	None	None	D-4	70–120 [21–49]
	Indium	13,14	Sulfamic Acid, Sulfamate Salts	Sulfamate Mist	C-3	70–90 [21–32]
	Iron		Chloride Salts, Hydrochloric Acid	Hydrochloric Acid Mist, Steam	A-2	190–210 [88–99]
	Iron	12	None	None	D-4	70–120 [21–49]
	Nickel	3	Ammonium Fluoride, Hydrofluoric Acid	Hydrofluoric Acid Mist	A-3	102 [39]
	Nickel and Black Nickel	12,15	None	None	C-4 (22)	70–150 [21–66]
	Nickel	9,12	Nickel Sulfate	Nickel Sulfate Mist	A-4	70–90 [21–32]
	Nickel	13,14	Nickel Sulfamate	Sulfamate Mist	A-2	75–160 [24–71]
	Palladium	15	None	None	D-4	70–120 [21–49]
	Rhodium	12,17	None	None	D-4	70–120 [21–49]
	Tin		Tin Halide	Halide Mist	C-2	70–90 [21–32]
	Tin	12	None	None	D-4	70–120 [21–49]
	Zinc		Zinc Chloride	Zinc Chloride Mist	B-3	75–120 [24–49]
	Zinc	12	None	None	D-4	70–120 [21–49]

Notes:
1 Arsine may be produced due to the presence of arsenic in the metal or polishing bath.
2 Alkaline Bath
3 On Magnesium
4 Also Copper-Cadmium Bronze
5 HCN gas may be evolved due to the acidic action of CO_2 in the air at the surface of the bath.
6 Conventional Cyanide Bath
7 Except Conventional Cyanide Bath
8 Albaloy, Spekwhite, Bonwhite (Alloys of Copper, Tin, Zinc)
9 Using Insoluble Anodes
10 Over 90 F
11 Mild Organic Acid Bath
12 Sulfate Bath
13 Sulfamate Bath
14 Air Agitated
15 Chloride Bath
16 Nitrite Bath
17 Phosphate Bath
18 See Section 13.70.2
19 Hydrogen gas also released by many of these operations.
20 The higher rate is associated with the higher value in the temperature range.
21 Baths operated at a temperature of over 140 F with a current density of over 45 amps/ft^2 and with air agitation will have a higher rate of evolution.
22 Local exhaust ventilation may be desired to control steam and water vapor.

TABLE 13-70-8. Airborne Contaminants Released by Stripping Operations

Coating to be Stripped	Base Metal (Footnote)	Component of Bath That May be Released to Atmosphere (1)	Physical and Chemical Nature of Major Atmospheric Contaminant	Class (e)	Usual Temp Range F [C]
Anodized Coatings	1,7	Chromic Acid	Acid Mist, Steam	A-2	120–200 [49–93]
Black Oxide Coatings	14	Hydrochloric Acid	Hydrogen Chloride Gas	A-3,2 (g)	70–125 [21–52]
Brass and Bronze	8,14 (a)	Sodium Hydroxide, Sodium Cyanide	Alkaline, Cyanide Mists	C-3,2 (g)	70–90 [21–32]
Cadmium	8,14 (a)	Sodium Hydroxide, Sodium Cyanide	Alkaline, Cyanide Mists	C-3,2 (g)	70–90 [21–32]
	2,4,14	Hydrochloric Acid	Acid Mist, Hydrogen Chloride Gas	A-3,2 (g)	70–90 [21–32]
Chromium	7,8,14 (a)	Sodium Hydroxide	Alkaline Mist, Steam	C-3	70–150 [21–66]
	2,4,8,14	Hydrochloric Acid	Hydrogen Chloride Gas	A-2	70–125 [21–52]
	2,4,8,18 (a)	Sulfuric Acid	Acid Mist	B-2	70–90 [21–32]
Copper	8,14	Sodium Hydroxide, Sodium Cyanide	Alkaline, Cyanide Mists	C-3,2 (g)	70–90 [21–32]
	7,12,14 (b)	None	None	D-4	70–90 [21–32]
	14 (a)	Alkaline Cyanide	Cyanide Mist	C-3,2 (g)	70–160 [21–71]
	1	Nitric Acid	Nitrogen Oxide Gases	A-1	70–120 [21–49]
	18 (a)	Sodium Hydroxide-Sodium Sulfide	Alkaline Mist, Steam	C-2	185–195 [85–91]
Gold	4,5,6,8,9,14 (a)	Sodium Hydroxide, Sodium Cyanide	Alkaline, Cyanide Mists	C-3,2 (g)	70–90 [21–32]
	4,5,18 (a)	Sulfuric Acid	Acid Mist	B-3,2 (g)	70–100 [21–38]
Lead	13 (c)	Acetic Acid, Hydrogen Peroxide	Oxygen Mist	D-3	70–90 [21–32]
	14 (a),(c)	Sodium Hydroxide	Alakaline Mist, Steam	C-3,2 (g)	70–140 [21–60]
Nickel	2,4	Sulfuric, Nitric Acids	Nitrogen Oxide Gases	A-2,1 (g)	70–90 [21–32]
	2,4 (a)	Hydrochloric Acid	Hydrogen Chloride Gas	A-3	70–90 [21–32]
	2,4,14 (a)	Sulfuric Acid	Acid Mist	B-3	70–90 [21–32]
	7	Hydrofluoric Acid	Hydrogen Fluoride Gas	A-3,2 (g)	70–90 [21–32]
	14	Fuming Nitric Acid	Nitrogen Oxide Gases	A-1	70–90 [21–32]
	(a),(d)	Hot Water	Steam	D-2 (h)	200 [93]
	1,18,19 (a)	Sulfuric Acid	Acid Mist, Steam	B-3,2 (g)	70–150 [21–66]
Phosphate Coatings	15	Chromic Acid	Acid Mist, Steam	A-3	165 [74]
	16	Ammonium Hydroxide	Ammonia Gas	B-3,2 (g)	70–90 [21–32]
Rhodium	10	Sulfuric, Hydrochloric Acids	Acid Mist, Hydrogen Chloride Gas	A-3,2 (g)	70–100 [21–38]
Silver	1	Nitric Acid	Nitrogen Oxide Gases	A-1	70–90 [21–32]
	2,11	Sulfuric, Nitric Acids	Nitrogen Oxide Gases, Steam	A-1	180 [82]
	8,14 (a)	Sodium Hydroxide, Sodium Cyanide	Alkaline, Cyanide Mists	C-3	70–90 [21–32]
	17 (a)	Sodium Cyanide	Cyanide Mist	C-3	70–90 [21–32]
Tin	2,3,4	Ferric Chloride, Copper Sulfate Acetic Acid	Acid Mist	B-4,3 (g)	70–90 [21–32]
	(a)	Sodium Hydroxide	Alkaline Mist	C-3	70–90 [21–32]
	2,14,14	Hydrochloric Acid	Hydrogen Chloride Gas	A-3,2 (g)	70–90 [21–32]
	14 (a)	Sodium Hydroxide	Alkaline Mist, Steam	C-2	70–200 [21–93]

TABLE 13-70-8 (Cont.). Airborne Contaminants Released by Stripping Operations

Coating to be Stripped	Base Metal Footnote	Component of Bath That May be Released to Atmosphere (1)	Physical and Chemical Nature of Major Atmospheric Contaminant	Class (e)	Usual Temp Range, F [C]
Zinc	1	Nitric Acid	Nitrogen Oxide Gases	A-1	70–90 [21–32]
	8,14	Sodium Hydroxide, Sodium	Alkaline, Cyanide Mists	C-3	70–90 [21–32]

Base Metal:
1. Aluminum
2. Brass
3. Bronze
4. Copper
5. Copper Alloys
6. Ferrous Metals
7. Magnesium
8. Nickel
9. Nickel Alloys
10. Nickel Plated Brass
11. Nickel Silver
12. Non-Ferrous Metals
13. Silver
14. Steel
15. Steel (Manganese Type Coatings)
16. Steel (Zinc Type Coatings)
17. White Metal
18. Zinc
19. Zinc Base Die Castings

Notes:
(a) Electrolytic Process
(b) Refers only to steel (14) when Chromic, Sulfuric Acids Bath is used.
(c) Also Lead Alloys
(d) Sodium Nitrate Bath
(e) See Section 13.70.2
(f) Hydrogen gas also released by some of these operations.
(g) The higher rate is associated with the higher value in the temperature range.
(h) Rate where essentially complete control of steam is required. Otherwise, adequate dilution ventilation may be sufficient.

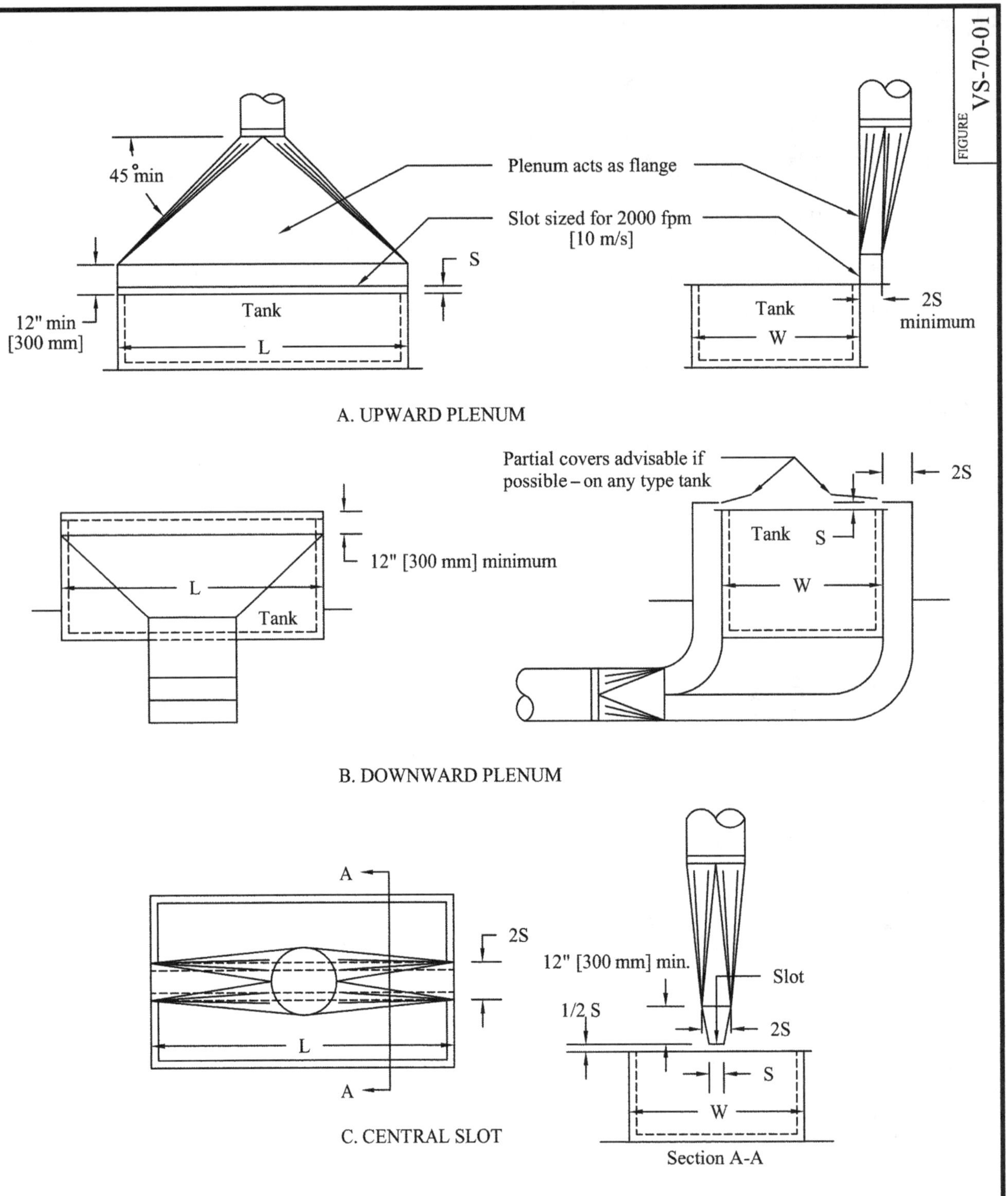

	TITLE	FIGURE VS-70-01
ACGIH®	OPEN SURFACE TANKS	DATE 1-12

CHECK CODES, REGULATIONS, AND LAWS (LOCAL, STATE, AND NATIONAL)
TO ENSURE THAT DESIGN IS COMPLIANT.

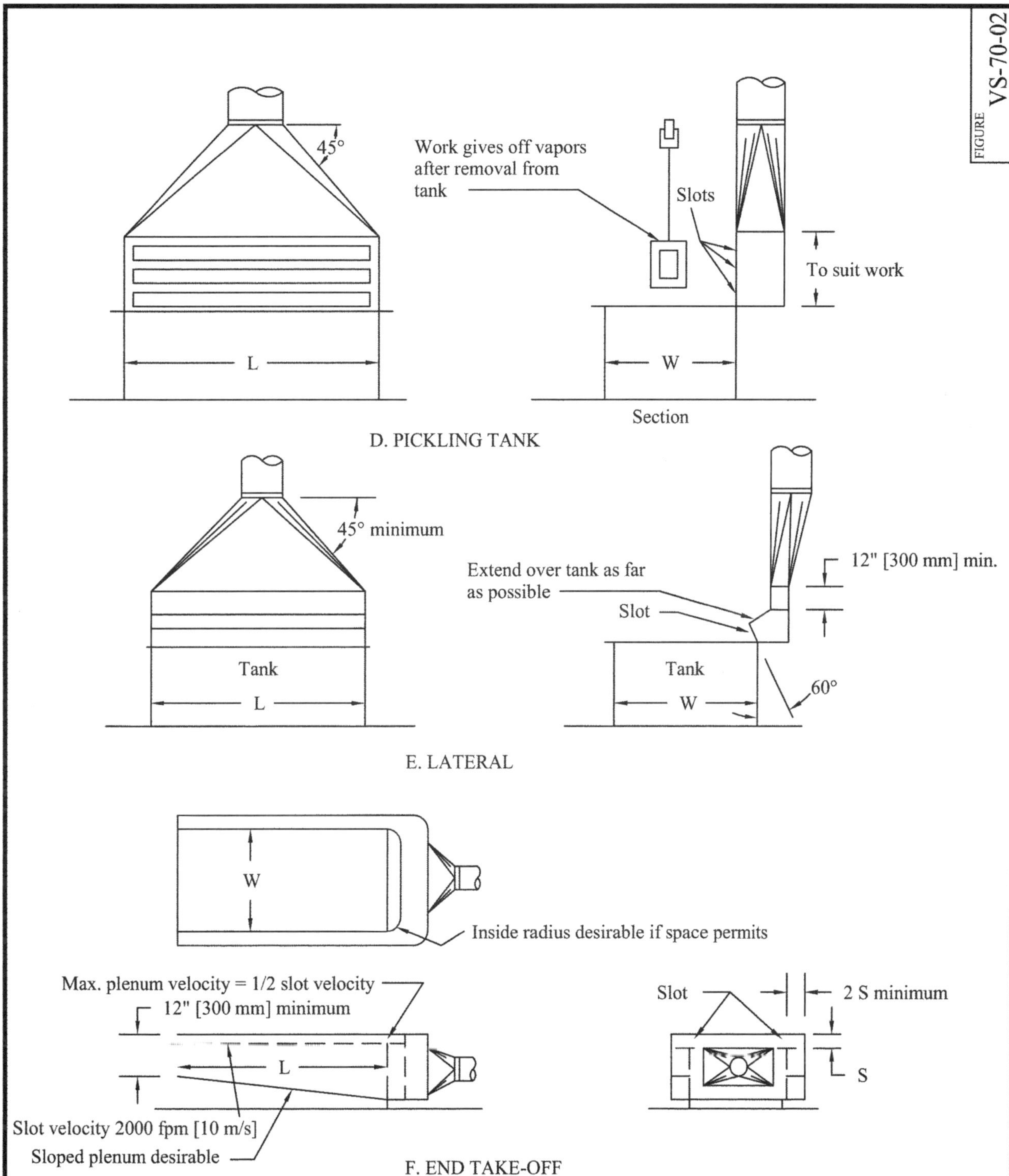

See Section 13.70

ACGIH®

TITLE: OPEN SURFACE TANKS

FIGURE: VS-70-02

DATE: 1-12

CHECK CODES, REGULATIONS, AND LAWS (LOCAL, STATE, AND NATIONAL) TO ENSURE THAT DESIGN IS COMPLIANT.

FIGURE VS-70-20

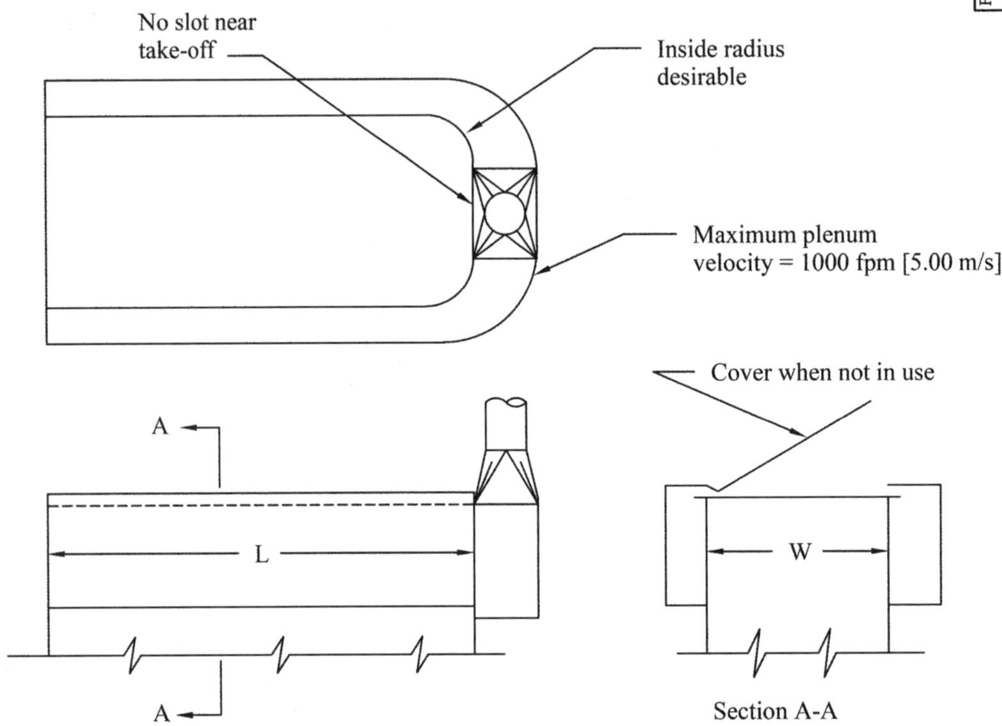

Q = 50 LW acfm [0.3 LW am^3/s]

Slot velocity = 2000 fpm [10.00 m/s]

h_e = 1.78 VP_s+ 0.25 VP_d

Duct velocity = 2000 fpm [10.00 m/s] minimum

Also provide:

1. Separate flue for combustion products.
2. For cleaning operation, appropriate respiratory protection is necessary.
3. For pit units, the pit should be mechanically ventilated.
4. For further safeguards, see VS-70-21.

NOTE: Provide downdraft grille for parts that cannot be removed dry; Q = 50 acfm/ft^2 [0.25 am^3/s/m^2] grille area.

See VS-70-21

ACGIH®	TITLE SOLVENT DEGREASING TANKS	FIGURE VS-70-20 DATE 1-12

CHECK CODES, REGULATIONS, AND LAWS (LOCAL, STATE, AND NATIONAL) TO ENSURE THAT DESIGN IS COMPLIANT.

FIGURE VS-70-21

Solvent vapor degreasing refers to boiling liquid cleaning systems utilizing trichloroethylene, perchloroethylene, methylene chloride, freons® or other halogenated hydrocarbons. Cleaning action is accomplished by the condensation of the solvent vapors in contact with the work surface producing a continuous liquid rinsing action. Cleaning ceases when the temperature of the work reaches the temperature of the surrounding solvent vapors. Since halogenated hydrocarbons are somewhat similar in their physical, chemical and toxic characteristics, the following safeguards should be provided to prevent the creation of a health or life hazard:

1. Vapor degreasing tanks should be equipped with a condenser or vapor level thermostat to keep the vapor level below the top edge of the tank by a distance equal to one-half the tank width or 36 inches [0.9 m], whichever is shorter.
2. Where water type condensers are used, inlet water temperature should not exceed 80 F [27 C] and the outlet temperature should not exceed 110 F [43 C]. For some solvent, lower water temperatures may be required.
3. Degreasers should be equipped with a boiling liquid thermostat to regulate the rate of vapor generation, and with a safety control at an appropriate height above the vapor line to prevent the escape of solvent in case of malfunction.
4. Tanks or machines of more than 4 ft^2 [0.36 m^2] of vapor area should be equipped with suitable gasketed cleanout or sludge doors, located near the bottom, to facilitate cleaning.
5. Work should be placed in and removed slowly from the degreaser, at a rate of no greater than 11 fpm [0.06 m/s], to prevent sudden disturbances of the vapor level.
6. CARE MUST BE TAKEN TO PREVENT DIRECT SOLVENT CARRYOUT DUE TO THE SHAPE OF THE PART.
7. Maximum rated workloads as determined by the rate of heat transfer (surface area and specific heat) should not be exceeded.
8. Special precautions should be taken where natural gas or other open flames are used to heat the solvent to prevent vapors* from entering the combustion air supply.
9. Heating elements should be designed and maintained so that their surface temperature will not cause the solvent or mixture to breakdown* or produce excessive vapors.
10. Degreasers should be located in such a manner that vapors* will not reach or be drawn into atmospheres used for gas or electrical arc welding, high temperature heat treating, combustion air or open electric motors.
11. Whenever spray or other mechanical means are used to disperse solvent liquids, sufficient enclosure or baffling should be provided to prevent direct release of airborne vapor to the top of the tank.
12. An emergency quick-drenching facility should be located in near proximity to the degreaser for use in the event of accidental eye contact with the degreasing liquid.

*Electric arcs, open flames and hot surfaces will thermally decompose halogenated hydrocarbons to toxic and corrosive substances (such as hydrochloric and/or hydrofluoric acid). Under some circumstances, phosgene may be formed.

ACGIH®	TITLE: SOLVENT VAPOR DEGREASING	FIGURE: VS-70-21
		DATE: 1-12

CHECK CODES, REGULATIONS, AND LAWS (LOCAL, STATE, AND NATIONAL) TO ENSURE THAT DESIGN IS COMPLIANT.

13.72 PUSH-PULL VENTILATION

Local exhaust ventilation consists of an exhaust hood which in effect creates a flow of air over a contaminant source, capturing the contaminant and removing it from the workplace. Local exhaust can be very effective where the contaminant is emitted over a narrow area and where the exhaust hood can be placed in reasonably close proximity to the emission source. The capture velocity of the hood decreases inversely with the square of the distance from the hood limiting the effectiveness of the hood to 3 to 4 ft [0.9 to 1.2 m]. Conversely, a jet of air can be blown up to 30 or more feet [9.0 m] in a coherent fashion. As such, local exhaust ventilation options may be limited for large open surface vessels, areas where contaminant generation occurs across a large area or where the process configuration makes local exhaust ventilation difficult. In such cases, push-pull ventilation may be appropriate.

The jet coupled with an exhaust hood forms a push-pull system. The push-pull system consists of a manifold, usually a pipe with drilled holes, which directs the jet across the contaminant emission area. The jet captures the contaminant and carries it into the exhaust hood. The primary purpose of the exhaust hood is to capture and remove the jet, not to provide capture velocity.

A jet is essentially a constant momentum process. The jet leaves the nozzle at a flow rate (Q_0) and high velocity (V_0). Downstream the velocity decreases and the flow rate increases due to the entrainment of the surrounding air. The product of the nozzle velocity and the nozzle flow per foot is referred to as the initial kinetic momentum (KM_0). Theoretically, this value remains essentially constant throughout the jet length. The selection of the initial jet flow and velocity values to achieve a momentum necessary for control is not arbitrary but is a function of the push distance.

13.72.1 Open Surface Tanks. Careful consideration must be given to the application of push-pull systems to open surface tank operations. When applied to plating or cleaning tanks, a properly applied jet will "attach" to the fluid surface and take on the characteristics of a wall jet.[13.72.1, 13.72.3] This tends to hold the jet within the tank and near the tank liquid surface. Ideally, the push manifold should be mounted on the tank such that there is no gap between the nozzle and the tank edge. This will optimize attachment of the jet. If the push manifold cannot be mounted without a gap, it should be located as close to the tank edge as possible and be baffled, if possible, to close the gap. Provisions must be incorporated to angle the jet downward into the tank and under tank obstructions. See VS-72-01.

The jet will usually flow around small obstructions such as parts hangers, parts, and parts baskets during placing and removal. However, large objects that may not cause a problem during the plating or cleaning operation may result in jet deflection and spillage when removing or placing the part in the tank. The flow of entrained air into the jet will offer some capture and removal of these spillage and dragout emissions. Possible solutions may be to shut the jet off during this part of the operation, or if the system configuration permits, configuring the jet to be parallel to the part or tank obstructions rather than perpendicular. In some instances, a second jet located above the tank may be beneficial in capturing dragout and jet spillage.

There are many situations such as process automation, high speed production, cables, bus bars, heaters, monitors, hoists, barrels, drives, racks, etc., which create obstructions that may hinder the performance of a push-pull system. Elevated process temperatures and automated finishing processes such as those discussed above may negate a reduction in exhaust volumes and, in severe cases, may preclude the use of push pull.

13.72.2 Exhaust Hood and Exhaust Flow. The exhaust hood should always be mounted or baffled in a manner to prevent pulling air from behind the hood rather than from the tank. The tank hood opening should extend the full width of the tank including flanges. As stated above, the primary purpose of the exhaust hood is to receive and remove the contaminant laden air flow supplied by the jet. However, where the process configuration causes jet deflection or spillage, as previously described, or emits contaminants at a high rate such as air agitation, high plating currents or utilizes highly volatile process materials, higher exhaust flows may be necessary for contaminant capture. Past editions of this manual have recommended 75 to 90 acfm/ft² [0.04 to 0.05 am³/s/m²] tank area. This value was based on laboratory testing of a 4 to 6 ft [1.2 to 1.8 m] wide tank with minimal obstructions and a simulated "low activity" plating process.[13.72.2] Field evaluation under similar conditions has produced satisfactory emission control.[13.72.2, 13.72.4]

A review of current field practice indicates that systems are being used with exhaust flows ranging from the 75 to 90 acfm/ft² [0.38 to 0.46 am³/s/m²] value for "low activity" processes and to over 200 acfm/ft² [1.0 am³/s/m²] for the more difficult higher emission processes. There is, however, a lack of documented case studies of operating push pull systems that can establish a minimum exhaust flow for the various processes encountered in open surface tank operations. While the 75 to 90 acfm/ft² [0.38 to 0.46 am³/s/m²] exhaust flow rate has been successful, it is believed that a minimum exhaust flow rate of 100 acfm/ft² [0.5 am³/s/m²] tank area for simple low activity processes should be considered. For high emission processes (air agitation, high current, etc.), higher flow rates based on similar process systems or designer experience may be required. Table 13-72-1 provides recommended jet flow rates and velocities for tank widths of 4 to 28 feet [2 to 8 meters].

13.72.3 Pilot System. Due to the wide variation in open surface tank operations and configurations, an evaluation using a pilot system is strongly recommended. Ideally, the pilot system would be applied to the existing process for which it is intended. The push system can be simple, a short section of PVC pipe with holes and a portable blower blown into the existing process exhaust hood. If an exhaust hood is not practical, use of test smoke and a velocity characterization of the jet can provide valuable design information.

13.72.4 Non Open Surface Tank Processes. Push-pull can be applied to processes other than open surface operations. Although there is usually less specific information to guide their design, the open surface information can be used as a starting point. Careful evaluation of the specific process must be made. If the jet is in the open, it will not be "contained" as completely as the open surface jet making capture more difficult. Combining the push pull with work practices can be effective. For example, a jet of air over a fiberglass boat hull buildup (from the rear) will carry emissions to the front. If the work is done from front to rear, the workers will be in a relatively clean stream of air. As with open surface tanks, a pilot system evaluation is recommended.

REFERENCES

13.72.1 Robinson, M.; Ingham, D.B.: Recommendations for the Design of Push-Pull Ventilation System for Open Surface Tanks. Ann. Occup. Hyg., Vol. 40, No. 6, pp. 693-704 (1996).

13.72.2 Klein, M.K.: A Demonstration of NIOSH Push Pull Ventilation Criteria. Am. Ind. Hyg. Assoc. J., Vol. 48, pp. 236-246 (1987).

13.72.3 Huebener, D.J.; Hughes, R.T.: Development of Push-Pull Ventilation. Am. Ind. Hyg. Assoc. J., Vol. 46, pp. 262-267 (1985).

13.72.4 Sciola, V.: Private Communication, Hamilton Standard, (1998).

TABLE 13-72-1 (IP). Push Nozzle Design Data

	x (ft)	D_{hole} (in) W_{slot} (in)	A_{hole} (ft^2)	$A_0/(ft^2/ft)$*	KM_0	V_0 (fpm)	Q_0 (acfm/ft)*	Q_e(acfm/ft)** Recommended
Holes on 2 in. centers	4	0.25	0.0003409	0.002045	49161	4903	10.0	400
	8	0.25	0.0003409	0.002045	98322	6933	14.2	800
	12	0.25	0.0003409	0.002045	147484	8492	17.4	1200
	16	0.25	0.0003409	0.002045	196645	9805	20.1	1600
	20	0.25	0.0003409	0.002045	245806	10963	22.4	2000
	24	0.25	0.0003409	0.002045	294967	12009	24.6	2400
	28	0.25	0.0003409	0.002045	344128	12971	26.5	2800
	4	0.2188	0.0002611	0.001567	49161	5602	8.8	400
	8	0.2188	0.0002611	0.001567	98322	7922	12.4	800
	12	0.2188	0.0002611	0.001567	147484	9703	15.2	1200
	16	0.2188	0.0002611	0.001567	196645	11204	17.6	1600
	20	0.2188	0.0002611	0.001567	245806	12526	19.6	2000
	24	0.2188	0.0002611	0.001567	294967	13721	21.5	2400
	28	0.2188	0.0002611	0.001567	344128	14821	23.2	2800
	4	0.1875	0.0001917	0.001150	49161	6537	7.5	400
	8	0.1875	0.0001917	0.001150	98322	9245	10.6	800
	12	0.1875	0.0001917	0.001150	147484	11322	13.0	1200
	16	0.1875	0.0001917	0.001150	196645	13074	15.0	1600
	20	0.1875	0.0001917	0.001150	245806	14617	16.8	2000
	24	0.1875	0.0001917	0.001150	294967	16012	18.4	2400
	28	0.1875	0.0001917	0.001150	344128	17295	19.9	2800
Slot	4	0.25		0.020833	49161	1536	32.0	400
	8	0.25		0.020833	98322	2172	45.3	800
	12	0.25		0.020833	147484	2661	55.4	1200
	16	0.25		0.020833	196645	3072	64.0	1600
	20	0.25		0.020833	245806	3435	71.6	2000
	24	0.25		0.020833	294967	3763	78.4	2400
	28	0.25		0.020833	344128	4064	84.7	2800
	4	0.1875		0.015625	49161	1774	27.7	400
	8	0.1875		0.015625	98322	2509	39.2	800
	12	0.1875		0.015625	147484	3072	48.0	1200
	16	0.1875		0.015625	196645	3548	55.4	1600
	20	0.1875		0.015625	245806	3966	62.0	2000
	24	0.1875		0.015625	294967	4345	67.9	2400
	28	0.1875		0.015625	344128	4693	73.3	2800
	4	0.125		0.010417	49161	2172	22.6	400
	8	0.125		0.010417	98322	3072	32.0	800
	12	0.125		0.010417	147484	3763	39.2	1200
	16	0.125		0.010417	196645	4345	45.3	1600
	20	0.125		0.010417	245806	4858	50.6	2000
	24	0.125		0.010417	294967	5321	55.4	2400
	28	0.125		0.010417	344128	5748	59.9	2800

* per ft push nozzle length

** per ft tank length

TABLE 13-72-1 (SI). Push Nozzle Design Data

	x (m)	D_{hole} (mm)	A_{hole} (m^2)	A_o(m^2/m)*	KM_0	V_0 (m/s)	Q_e(m^3/s/m)**
		W_{slot} (mm)					Recommended
Holes on 50 mm centers	2	6	0.00002827	0.0005372	0.0588	10.46	1.02
	3	6	0.00002827	0.0005372	0.0882	12.81	1.52
	4	6	0.00002827	0.0005372	0.1176	14.80	2.03
	5	6	0.00002827	0.0005372	0.1470	16.54	2.54
	6	6	0.00002827	0.0005372	0.1764	18.12	3.05
	7	6	0.00002827	0.0005372	0.2058	19.57	3.55
	8	6	0.00002827	0.0005372	0.2352	20.92	4.06
	2	5	0.00001964	0.0003731	0.0588	12.55	1.02
	3	5	0.00001964	0.0003731	0.0882	15.38	1.52
	4	5	0.00001964	0.0003731	0.1176	17.75	2.03
	5	5	0.00001964	0.0003731	0.1470	19.85	2.54
	6	5	0.00001964	0.0003731	0.1764	21.74	3.05
	7	5	0.00001964	0.0003731	0.2058	23.49	3.55
	8	5	0.00001964	0.0003731	0.2352	25.11	4.06
	2	4	0.00001257	0.0002388	0.0588	15.69	1.02
	3	4	0.00001257	0.0002388	0.0882	19.22	1.52
	4	4	0.00001257	0.0002388	0.1176	22.19	2.03
	5	4	0.00001257	0.0002388	0.1470	24.81	2.54
	6	4	0.00001257	0.0002388	0.1764	27.18	3.05
	7	4	0.00001257	0.0002388	0.2058	29.36	3.55
	8	4	0.00001257	0.0002388	0.2352	31.39	4.06
Slot	2	6		0.0060	0.0588000	3.13	1.02
	3	6		0.0060	0.0882	3.83	1.52
	4	6		0.0060	0.1176	4.43	2.03
	5	6		0.0060	0.1470	4.95	2.54
	6	6		0.0060	0.1764	5.42	3.05
	7	6		0.0060	0.2058	5.86	3.55
	8	6		0.0060	0.2352	6.26	4.06
	2	5		0.0050	0.0588	3.43	1.02
	3	5		0.0050	0.0882	4.20	1.52
	4	5		0.0050	0.1176	4.85	2.03
	5	5		0.0050	0.1470	5.42	2.54
	6	5		0.0050	0.1764	5.94	3.05
	7	5		0.0050	0.2058	6.42	3.55
	8	5		0.0050	0.2352	6.86	4.06
	2	4		0.0040	0.0588	3.83	1.02
	3	4		0.0040	0.0882	4.70	1.52
	4	4		0.0040	0.1176	5.42	2.03
	5	4		0.0040	0.1470	6.06	2.54
	6	4		0.0040	0.1764	6.64	3.05
	7	4		0.0040	0.2058	7.17	3.55
	8	4		0.0040	0.2352	7.67	4.06

* per meter push nozzle length

** per meter tank length

FIGURE
VS-72-01 (IP & SI)

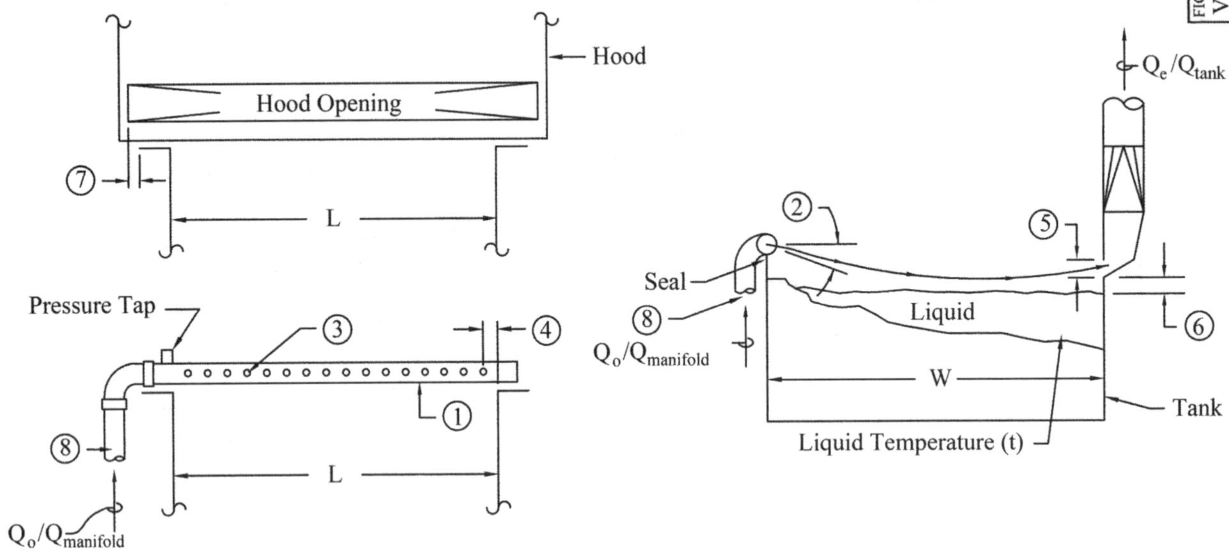

Push nozzle manifold	(1) – Circular, rectangular or square. Manifold cross-sectional area must be at least 2.5 times the total nozzle flow area. The manifold is mounted to the tank so that there is no gap between it and the tank.
Push nozzle angle	(2) – 0° to 20° down.
Nozzle openings	(3) – 1/8" to 1/4" [4 mm to 6 mm] dia. holes on 2" [50 mm] centers recommended. Other sizes and spacings may be used. See 13-72 text. Outer holes (4) must be 1/2" to 1" [12 mm to 25 mm] inside tank edge.
Exhaust opening	(5) – Size to achieve 2000 fpm [10 m/s] slot velocity. Outer edges of opening (7) must extend to edge of tank including flanges.
Liquid surface	(6) – Tank freeboard should be maintained at the minimum consistent with the process.
Push nozzle supply control valve	(8) – See Section 13.72.
Push nozzle supply flow	Q_o = manifold flow per unit length (L). See Section 13.72. $Q_{manifold}$ = total manifold flow ($Q_o \times L$). See Section 13.72.
Exhaust flow	Q_e = exhaust flow per unit length (L). See Section 13.72.

Tank surface area = L (length of tank) × W (width of tank)

ACGIH®	TITLE PUSH-PULL HOOD DESIGN DATA	FIGURE VS-72-01 (IP & SI) DATE 1-12

CHECK CODES, REGULATIONS, AND LAWS (LOCAL, STATE, AND NATIONAL) TO ENSURE THAT DESIGN IS COMPLIANT.

13.75 PAINTING OPERATIONS

Application of industrial paints and coatings is usually accomplished by one of four techniques: air atomization, electrostatic, airless, and high velocity low pressure methods. An important characteristic of the process is the transfer efficiency of the paint gun. Transfer efficiency is an expression of the percentage of nonvolatile materials contained in the paint that are actually deposited on the surface of the substrate. Paint components not transferred to the object appear as overspray on the object or in the air. Typical paint transfer efficiencies are shown below. Using an improper spray technique will also significantly reduce the transfer efficiency.

Spray System	% Efficiency
Air	15–20
Airless	25–45
Air-assisted airless	35–65
Electrostatic air	55–75
Electrostatic air-assisted airless	65–85
HVLP (high volume low pressure)	65–80
LVLP (low volume low pressure)	65–80
Electrostatic discs and bells	80–90

Potential health hazards exist from exposure to solid and liquid aerosols as well as to solvent vapors. In addition to the airborne exposures, hazards include the use of flammable and combustible liquids and the accumulation of flammable paint residues. Fire safety and proper electrical wiring are important concerns in most paint applications.

Control of airborne pollutants by ventilation may be accomplished through the use of spray booths such as shown on VS-75-01, VS-75-02, and VS-75-05. More specialized ventilation arrangements are shown on VS-75-03 and VS-75-04. The typical booth is a partial enclosure of sheet metal construction with openings for conveying the work piece into and out of the booth. Several factors are important in the performance of these booths. Booth depth is critical; spray rebound may escape from shallow booths and increase exposures. The size of the booth is governed principally by the size of the object being coated. Sufficient space must be provided to permit airflow on all sides of the object, to provide room to work, and to enable the air to enter the booth in a smooth, controlled manner without excessive wrap-around. While there is no definitive separation between the dimensions of a large and a small booth, small booths are typically bench- or table-mounted, and a person is able to stand in a typical large booth. In some cases, downdraft booths may be employed when large objects are painted.

VS-75-07 shows a cross-section of a typical High Production Downdraft Water Wash Paint Booth used to paint large objects such as automobiles, trucks and appliances. The booths may be several hundred feet long divided into zones for sequential coating applications. The zones are typically separated by vestibules to prevent coating drift between zones.

In air-atomization applications, the most common spray technique, it is important to use the minimum air pressure needed to accomplish the task. Excess air pressure results in increased dispersion of the paint and overspray as well as poor work quality.

Airless application results in aerosols with fewer particles in the respirable range. One study[13.75.1] suggests that approximately 20% of the particles in air-atomization applications are less than 12 microns while airless methods produce aerosols with only 2% less than this value. The larger aerosols produced by the airless technique will deposit more efficiently on the work piece, due to impaction, than the smaller particles produced by the compressed air method.

Electrostatic applications result in more efficient deposition of paint aerosols due to electrostatic forces. As a result, ventilation airflow requirements for control of electrostatic applications tend to be lower than for compressed air methods.

Many spray booths are equipped with disposable particulate filters that become “loaded” over time and result in increased system pressure loss. This loss can eventually reduce airflow to unacceptable levels and, hence, system performance must be monitored to determine when the disposable filters are to be replaced. Some regulators also require a carbon filtration or other vapor adsorption system top remove the Volatile Organic Compounds (VOCs). These filters increase the system pressure drop and typically require their own intrinsic booster fan. Water wash systems are available for cleaning particulate matter from the exhausted air but do little for solvent vapors. Due to environmental disposal issues many facilities are moving away from water wash booths.

Fan selection is an important component of a spray booth installation. Often the fan is an integral part of the system when purchased and may be installed in a different configuration than originally designed. This can result in reduced airflow, particularly if additional system resistance is encountered in the actual installation.[13.75.2]

Work practices remain an important aspect of controlling exposure to paint aerosols and solvent vapors. The worker should not stand downstream of the object being sprayed. See Chapter 6, Section 6.4.8, Worker Position Effect. A turntable can help to facilitate easy access to all sides of the object without the worker having to move. Extension arms on spray guns should be employed for hard to reach cavities. Proper location of the booth with respect to replacement air and obstructions is essential. Locating booths in corners or near disruptive air currents can defeat the protection of these hoods. Poor location of the booth may result in turbulent airflow which may reduce the protection provided by the booth.

The method of introducing supply air is critical for uniform air distribution and overspray control in a paint booth. Supply air may be mechanically introduced with a fan or by an air-handling unit if tempered air is required. To reduce turbulence in the booth, uniformly distribute the supply air over as large an area as possible, i.e., wall opposite the exhaust for lateral flow paint booths or a large portion of the ceiling for downdraft paint booths. In fully enclosed booths, inlet filters typically cover the inlet air openings to prevent foreign material from entering the booth and settling on the painted object. Alternatively, the supply air, especially when supplied via a mechanical system, may be filtered at the fan inlet. When an enclosed paint booth does not employ a mechanically supplied make-up air fan, the pressure loss across the inlet filters will establish the booth pressure and must be added to the system loss calculations.

If no mechanical air is supplied, the room or booth will be negatively pressurized. When a supply air fan is provided, the design and production team must decide on a room or booth pressure strategy. Typically, rooms with industrial ventilation systems are designed to maintain a negative pressure relative to the outdoors and adjacent rooms. Due to the nature of paint operations, a neutral or positive pressure in exceptional conditions may be specified. Listed below are some advantages and disadvantages:

Relative Room Pressure Decision		
Consideration	Negative Pressure	Positive Pressure
Material toxicity	Recommended for highly toxic materials	Toxic components may migrate into unprotected areas of the plant
Criticality of finish	May draw foreign material into booth through cracks and poorly seated inlet filters	Consider using higher quality inlet filters and filter housing or as a minimum used neutral room pressure
Booth location relative to other indoor operations	Reduces the probability of materials and odors migrating to other parts of the plant	Consider a stand alone booth (not in plant) May be used if environmental emissions are contained
Adjacent room		Nearby dirty processes, e.g., abrasive blasting may effect paint quality

Respiratory protection may be required in applications using toxic materials. This includes heavy metal pigments and organics such as isocyanates in urethane paints and amines in epoxy paints.

REFERENCES

13.75.1 National Institute for Occupational Safety and Health: An Evaluation of Engineering Control Technology for Spray Painting. DHHS (NIOSH) Pub. No. 81-121; NTIS Pub. No. PB-82-162-264. National Technical Information Service, Springfield, VA (1981).

13.75.2 Burgess, W.A.; Ellenbecker, M.J.; Treitman, R.D.: Ventilation for Control of the Work Environment. John Wiley and Sons, NY (1989).

13.75.3 National Fire Protection Association: Flammable and Combustible Liquids Code No. 33. NFPA, Boston, MA (1990).

13.75.4 National Fire Protection Association: National Electric Code No. 70. NFPA, Boston, MA (1990).

FIGURE VS-75-01

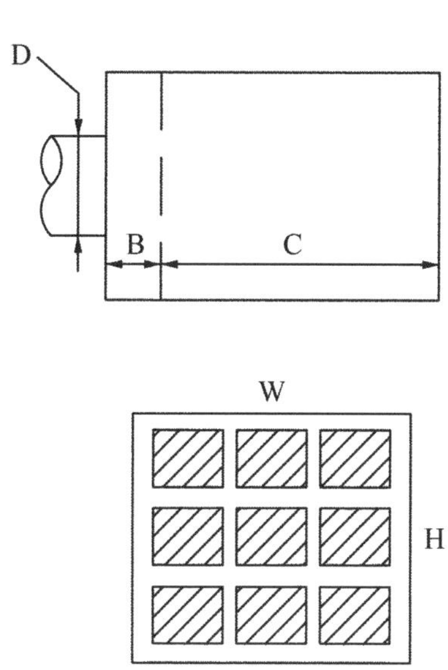

1. Split baffle or filters
$B = 0.75\ D$
Baffle area = 0.75 WH
For filter area, see note 2

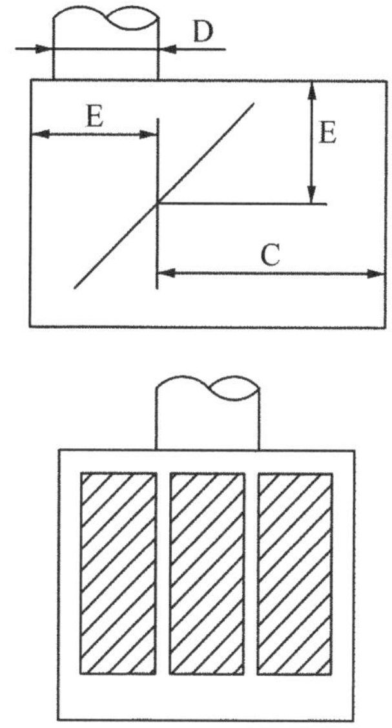

2. Angular baffle
$E = D + 6"$ [150 mm]
Baffle area = 0.40 WH
For filter area, see note 2

Air spray paint design data.

Any combination of duct connections and baffles may be used. Large, deep booths do not require baffles. Consult manufacturers for water-curtain designs. Use explosion proof fixtures and a non-sparking fan. Electrostatic spray booth requires automatic high-voltage disconnects for conveyor failure, fan failure or grounding.

Walk-in booth

- W = work size + 6' [1.8 m]
- H = work size + 3' [0.9 m] (minimum = 7' [2.1 m])
- C = work size + 6' [1.8 m]
- Q = 100 acfm/ft^2 [0.5 am^3/s/m^2] booth cross section
  May be 75 acfm/ft^2 [0.4 am^3/s/m^2] for very large, deep booth. Operator may require a NIOSH cerified respirator.
  $h_e = 1.78\ VP_s + 0.50\ VP_d$ (baffles)
  h_e = Dirty filter resistance + 0.50 VP_d (filters)
  Duct velocity = 2000 fpm [10.00 m/s]

Airless electrostatic and HVLP spray paint design

- Q = 60 acfm/ft^2 [0.30 am^3/s/m^2] booth cross section, walk-in booth
- Q = 60 – 100 acfm/ft^2 [0.30–0.50 am^3/s/m^2] of total open area, operator outside of booth

Operator outside booth

- W = work size + 2' [0.6 m]
- H = work size + 2' [0.6 m]
- C = 0.75 × larger front dimension
- Q = 100 – 150 acfm/ft^2 [0.50–0.75 am^3/s/m^2] of open area, including conveyor openings.

Notes :

1. Painter can walk into a large booth.
2. Baffle arrangements shown are for air distribution only.
3. Paint arresting filters usually selected for 100 – 500 fpm [0.5 – 2.5 m/s]. Consult filter manufacturer for specific details.
4. For construction and safety, consult NFPA.[13.75.3, 13.75.4]
5. Water curtain systems have unique environmental liquid waste requirements. Before choosing, perform a life cost analysis.
6. Object being painted may affect control flow. CFD or other analysis may be required.
7. Toxic materials in paint require higher inlet velocity.

TITLE	FIGURE
LARGE PAINT BOOTH	VS-75-01
	DATE 1-12

CHECK CODES, REGULATIONS, AND LAWS (LOCAL, STATE, AND NATIONAL) TO ENSURE THAT DESIGN IS COMPLIANT.

FIGURE VS-75-02

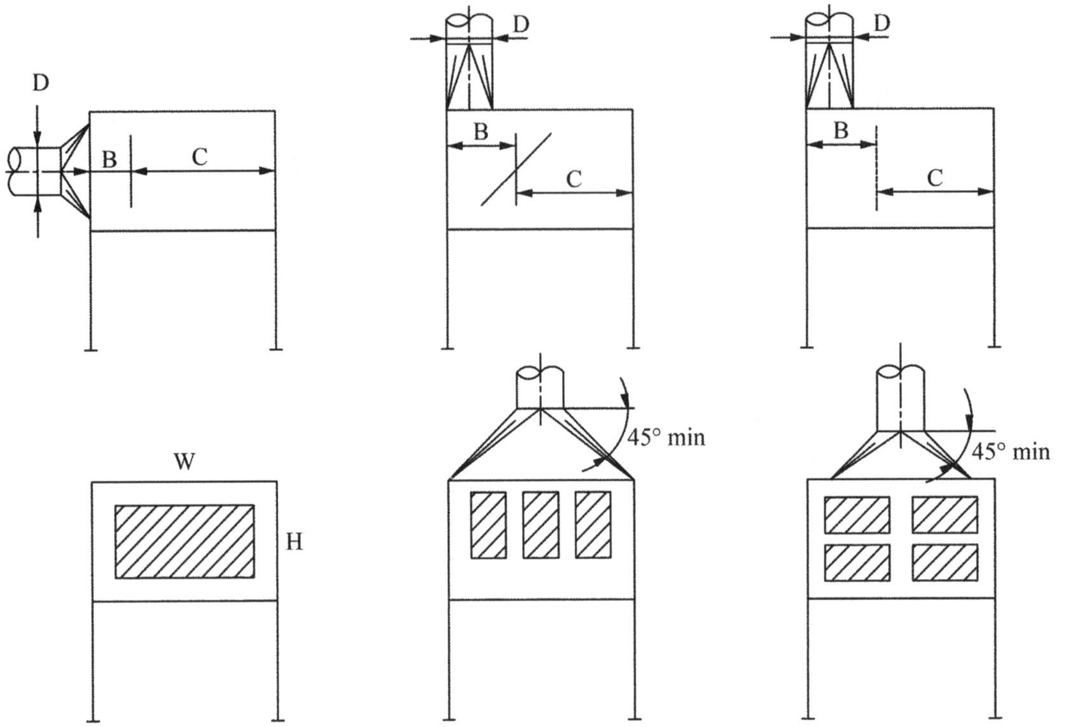

1. Solid baffle
   B = 0.75 D
   Baffle area = 0.60 WH

2. Angular baffle
   B = D + 6" [150 mm]
   Baffle area = 0.60 WH

3. Split baffle or filters
   B = D + 6" [150 mm]
   Baffle area = 0.75 WH

Air spray design data

Any combination of branch ducts and baffles may be used.

W = work size + 12" [300 mm]

H = work size + 12" [300 mm]

C = 0.75 W or H, whichever is larger

Q = 200 acfm/ft^2 [1.00 am^3/s/m^2] – for face area up to 4 ft^2 [0.36 m^2]

= 150 acfm/ft^2 [0.75 am^3/s/m^2] – for face area over 4 ft^2 [0.36 m^2]

h_e = 1.78 VP_s + 0.25 VP_d (baffles)

= dirty filter resistance + 0.25 VP_d (filters)

Duct velocity = 2000 fpm [10.00 m/s]

Airless spray paint design data

Q = 125 acfm/ft^2 [0.63 am^3/s/m^2] – for face area up to 4 ft^2 [0.36 m^2]

Q = 100 acfm/ft^2 [0.50 am^3/s/m^2] – for face area over 4 ft^2 [0.36 m^2]

Notes:

1. Baffle arrangements shown are for air distribution only.
2. Paint arresting filters usually selected for 100 – 500 fpm [0.50–2.50 m/s], consult manufacturer for specific details.
3. For construction and safety, consult NFPA.(13.75.3, 13.75.4)

ACGIH®	TITLE SMALL PAINT BOOTH	FIGURE VS-75-02 DATE 1-12

CHECK CODES, REGULATIONS, AND LAWS (LOCAL, STATE, AND NATIONAL) TO ENSURE THAT DESIGN IS COMPLIANT.

FIGURE VS-75-03

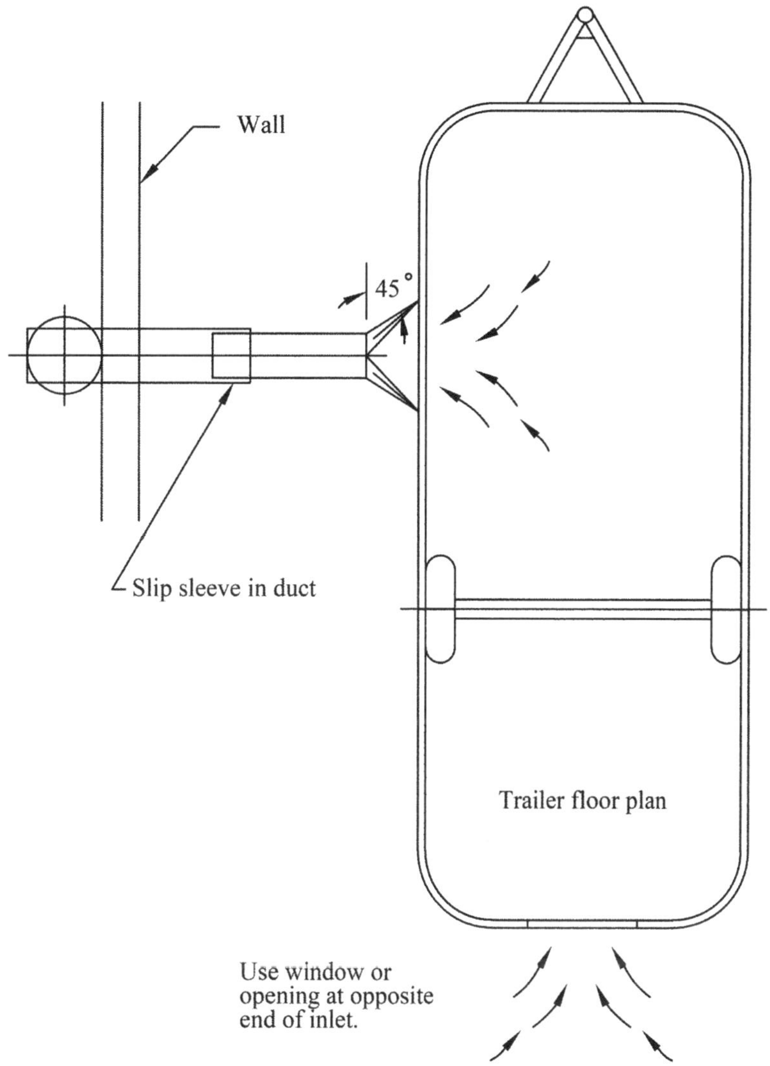

Q = 50 acfm/ft^2 [0.25 am^3/s/m^2] of cross-sectional trailer area
h_e = 0.25 VP_d
Minimum duct velocity = 2000 fpm [10.00 m/s]

Notes.

1. Paint arresting filters usually selected for 100–500 fpm [0.50–2.50 m/s], consult manufacturer for specific details.
2. For construction and safety, consult NFPA.[13.75.3, 13.75.4]
3. Operator must wear an appropriate, NIOSH certified respirator.
4. Consider interlock of exhaust fan with compressed air source to spray gun.

ACGIH®	TITLE TRAILER INTERIOR SPRAY PAINTING	FIGURE VS-75-03 DATE 1-12

CHECK CODES, REGULATIONS, AND LAWS (LOCAL, STATE, AND NATIONAL) TO ENSURE THAT DESIGN IS COMPLIANT.

FIGURE VS-75-04

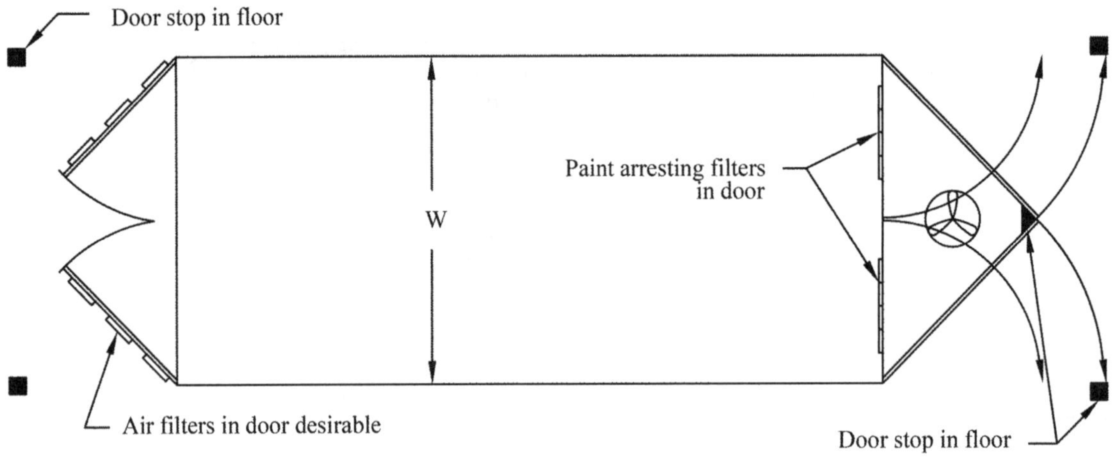

PLAN VIEW

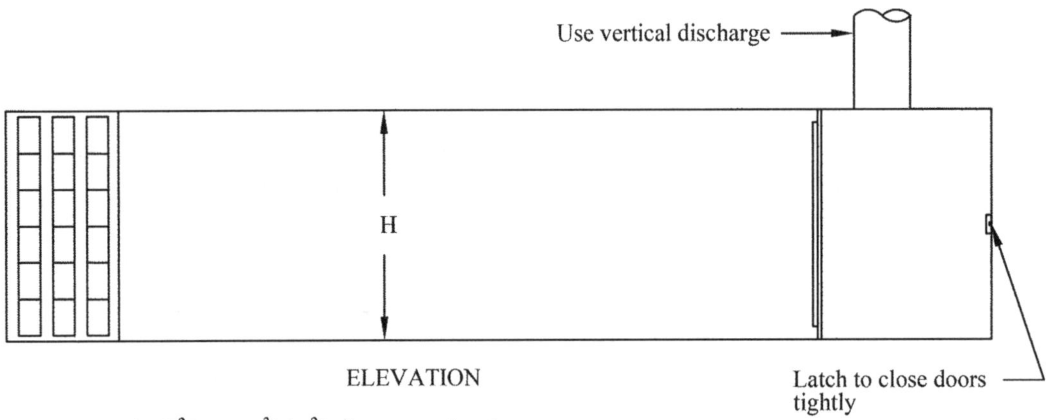

ELEVATION

Q = 100 acfm/ft^2 [0.5 am^3/s/m^2] of cross sectional area
(When W × H is greater than 150 ft^2 [13.5 m^2], Q = 50 acfm/ft^2 [0.25 am^3/s/m^2])

h_e = 0.56 VP_d plus resistance of each filter bank when dirty

Minimum duct velocity = 2000 fpm [10.00 m/s]

Notes:

1. Exhaust fan interlock with make-up air and compressed air to spray gun is desirable.
2. Paint arresting filters usually selected for 100–500 fpm [0.50–2.50 m/s]. Consult manufacturer for specific details.
3. For construction and safety, consult NFPA.(13.75.3, 13.75.4)
4. For airless electrostatic or HVLP spray painting use Q = 60 acfm/ft^2 [0.3 am^3/s/m^2] of cross section area.
5. Operators must wear appropriate NIOSH certified respirator.

ACGIH®	TITLE LARGE DRIVE-THROUGH SPRAY PAINT BOOTH	FIGURE VS-75-04 DATE 1-12

CHECK CODES, REGULATIONS, AND LAWS (LOCAL, STATE, AND NATIONAL) TO ENSURE THAT DESIGN IS COMPLIANT.

FIGURE VS-75-05

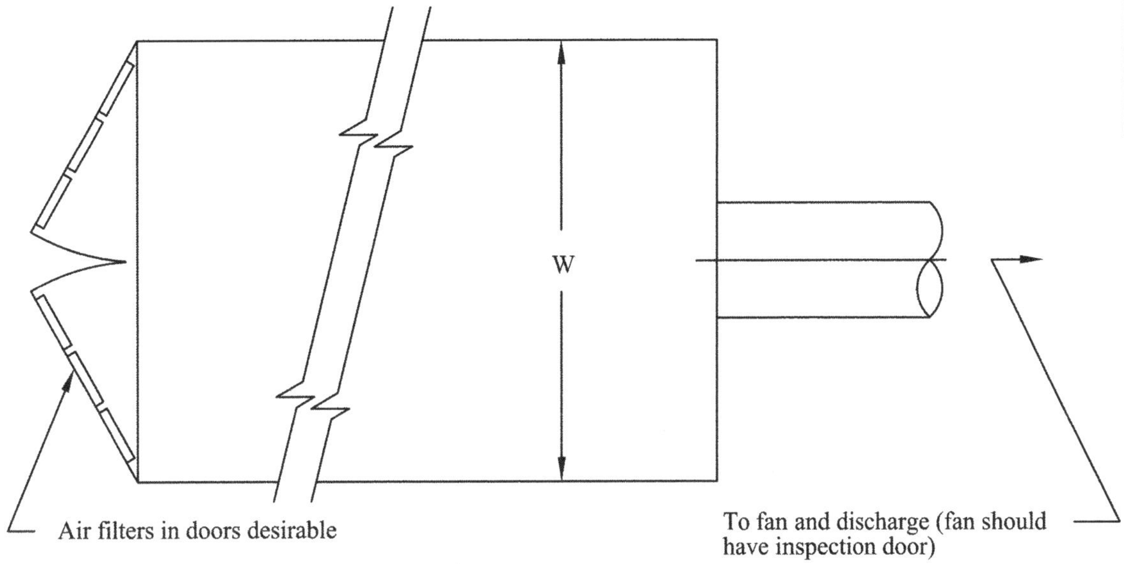

PLAN VIEW

$Q = 100$ acfm/ft^2 [0.50 am^3/s/m^2] of cross sectional area
(when $W \times H$ is greater than 150 ft [13.5 m^2], $Q = 50$ acfm/ft^2 [0.25 am^3/s/m^2])
$h_e = 0.56\ VP_d$ plus resistance of each filter bank when dirty
Minimum duct velocity = 2000 fpm [10.00 m/s]
Paint arresting filters to be sized for 100–500 acfm/ft^2 [0.50–2.50 am^3/s/m^2] of filter. Consult manufacturer for specific details.

Consider interlock of exhaust fan with replacement air fan and compressed air to spray gun.

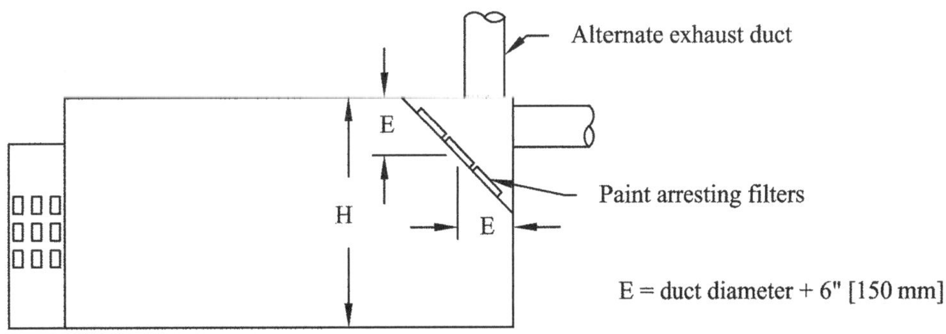

ELEVATION
Typical filter installation

Note: For airless electrostatic or HVLP spray painting use
$Q = 60$ acfm/ft^2 [0.30 am^3/s/m^2] of cross-sectional area may be permitted

For construction and safety consult NFPA(13.75.3, 13.75.4)

Note: Operators must wear appropriate NIOSH certified respirator

ACGIH®	TITLE: PAINT BOOTH VEHICLE SPRAY	FIGURE: VS-75-05 DATE: 1-12

CHECK CODES, REGULATIONS, AND LAWS (LOCAL, STATE, AND NATIONAL) TO ENSURE THAT DESIGN IS COMPLIANT.

FIGURE VS-75-06

Locate takeoffs a maximum of 10' [3 m] on center
Q = 50 acfm/ft^2 [0.25 am^3/s/m^2] drain board area, but not less than 100 fpm [0.50 m/s] indraft through openings
$h_e = 0.25\ VP_d$
Minimum duct velocity = 2000 fpm [10.00 m/s]

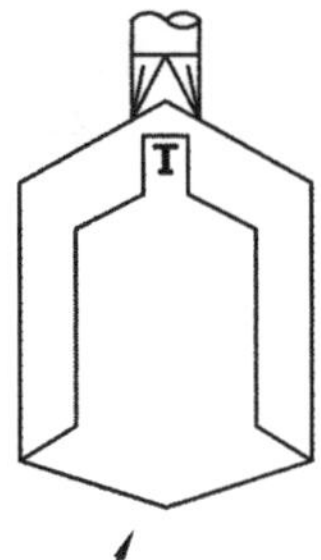

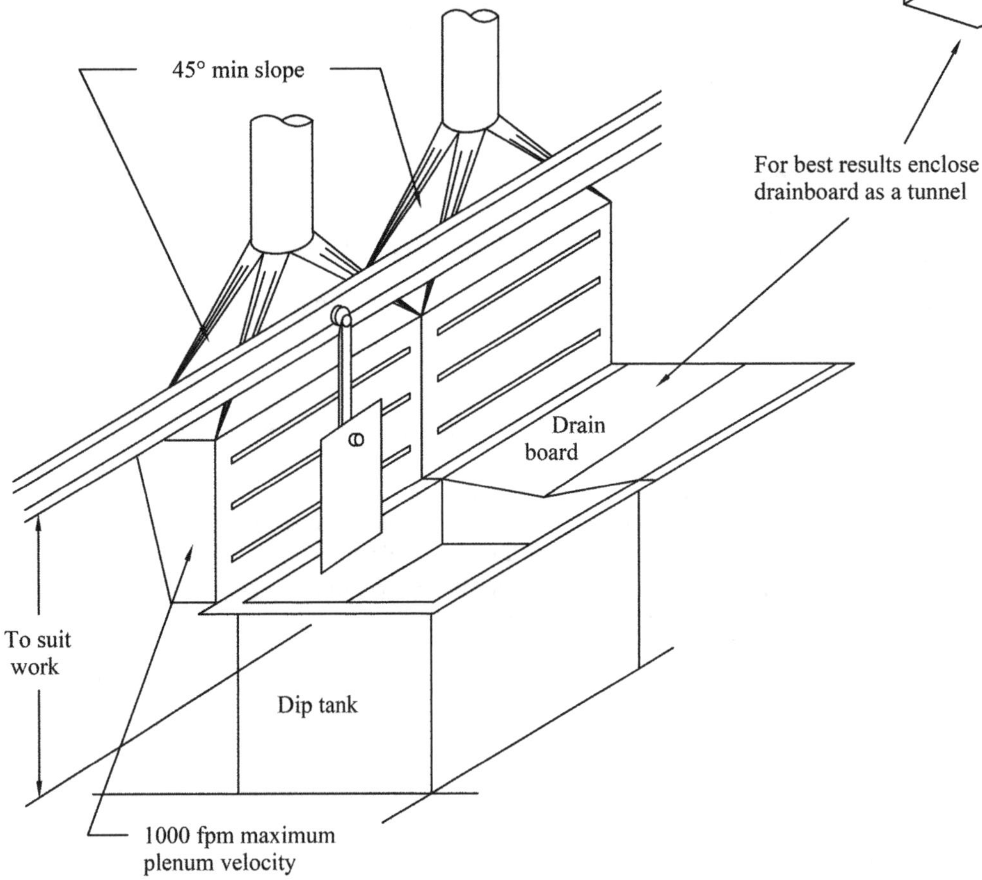

Q = 125 acfm/ft^2 [0.63 am^3/s/m^2] of tank and drainboard area
Slot velocity = 2000 fpm [10.00 m/s]
$h_e = 1.78\ VP_s + 0.25\ VP_d$
Minimum duct velocity = 2000 fpm [10.00 m/s]

For air drying in a room or enclosure, see Chapter 4 for dilution ventilation required.

For construction and safety, consult NFPA[13.75.3, 13.75.4]

ACGIH®	TITLE DIP TANK	FIGURE VS-75-06 DATE 1-12

CHECK CODES, REGULATIONS, AND LAWS (LOCAL, STATE, AND NATIONAL) TO ENSURE THAT DESIGN IS COMPLIANT.

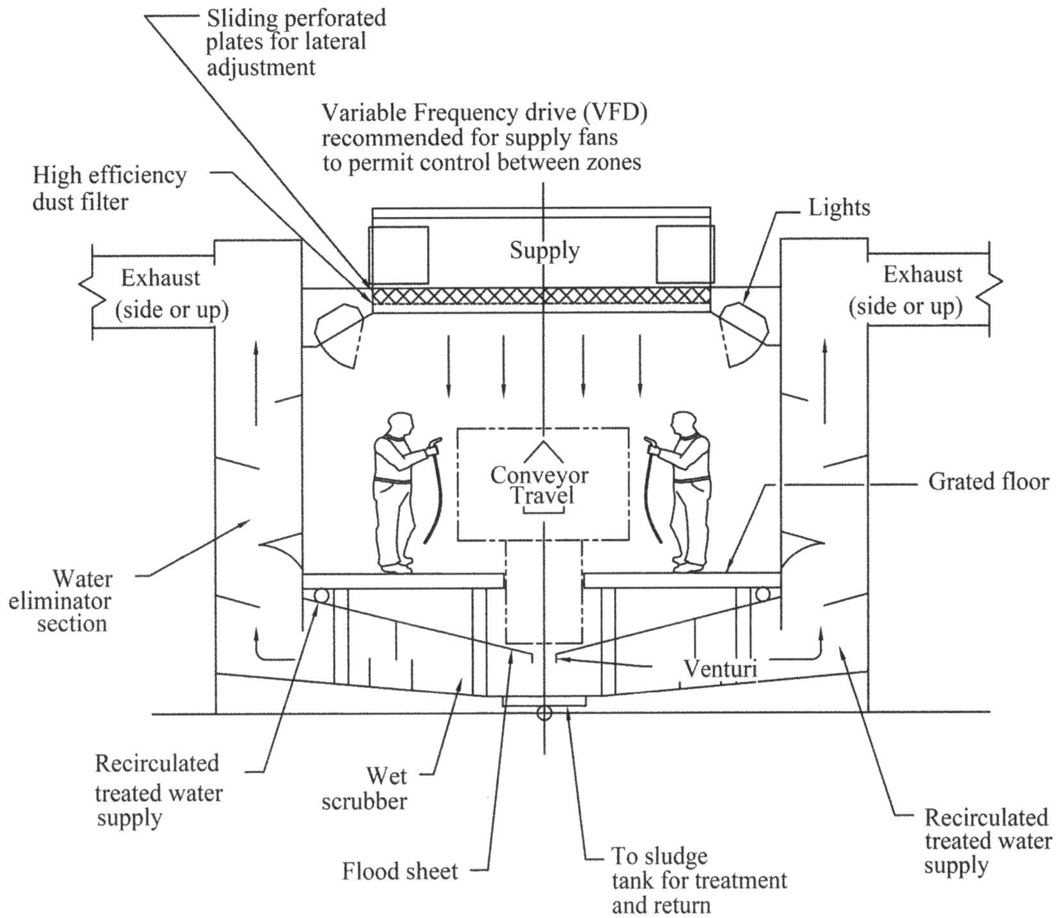

DOWN DRAFT AIRFLOW: 5' [1.5 m] ABOVE FLOOR		
50 – 70 fpm	[0.25–0.35 m/s]	Electrostatic and stationary bells
70 – 90 fpm	[0.35–0.45 m/s]	Robots and reciprocating bells
100 – 125 fpm	[0.50–0.63 m/s]	Manual painting zones (as shown)

NOTE: Above design values assume empty booth. Adjustments may be required to accomodate product shape and to control zone drift.

NOTES:

1. For construction and safety consult NFPA.[13.75.3, 13.75.4]
2. Pressure drop through scrubber depends on Venturi design. Typically 6.5" to 7.5" [1625 – 1875 Pa] S.P.
3. An appropriate NOISH certified respirator may be required for the operator.

ACGIH®

TITLE AUTOMATED/HIGH PROD. WATER WASH DOWNDRAFT PAINT BOOTH

FIGURE VS-75-07

DATE 1-12

CHECK CODES, REGULATIONS, AND LAWS (LOCAL, STATE, AND NATIONAL) TO ENSURE THAT DESIGN IS COMPLIANT.

FIGURE VS-75-07

FIGURE VS-75-20

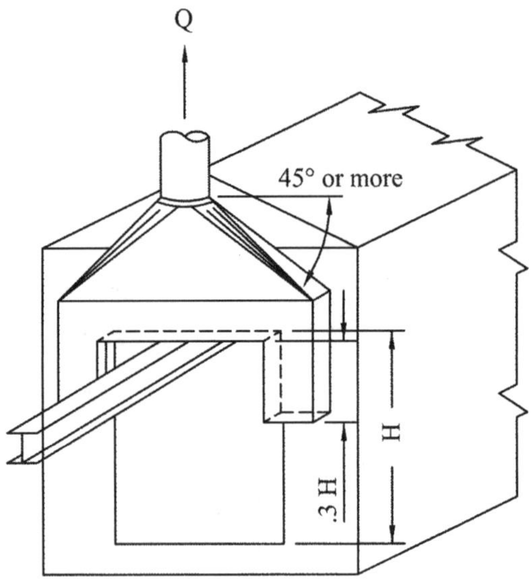

SLOT TYPE

$Q = 100$ acfm/ft^2 [0.5 am^3/s/m^2] door plus 1/2 products of combustion

Minimum duct velocity = 2000 fpm [10.00 m/s]

$h_e = 1.0\ VP_s + 0.25\ VP_d$

Size plenum for 1000 fpm [5.00 m/s] maximum

Slot on three sides with $V_s = 2000$ fpm [10.00 m/s]

Locate on inside or outside of door.

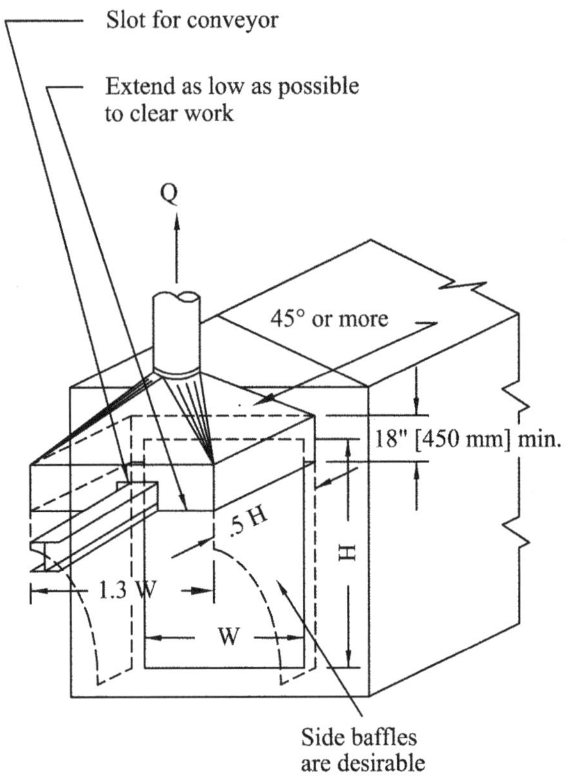

CANOPY TYPE

$Q = 200$ acfm/ft^2 [1 am^3/s/m^2] of hood face plus 1/2 products of combustion

$h_e = 0.25\ VP_d$

Duct velocity = 2000 fpm [10.00 m/s]

Notes: (APPLY TO BOTH DRAWINGS)

1. For dryers, include rate of water vapor liberated.
2. For flammable solvent drying refer to Chapter 4, "General Industrial Ventilation".
3. Hoods at each end of oven. Reduce size of doors as much as possible. Separate vent must be added for products of combustion.
4. For construction and safety, consult NFPA code (Reference 13.75.3).

ACGIH®	TITLE: DRYING OVEN VENTILATION	FIGURE: VS-75-20 — DATE: 1-12

CHECK CODES, REGULATIONS, AND LAWS (LOCAL, STATE, AND NATIONAL) TO ENSURE THAT DESIGN IS COMPLIANT.

FIGURE VS-75-30

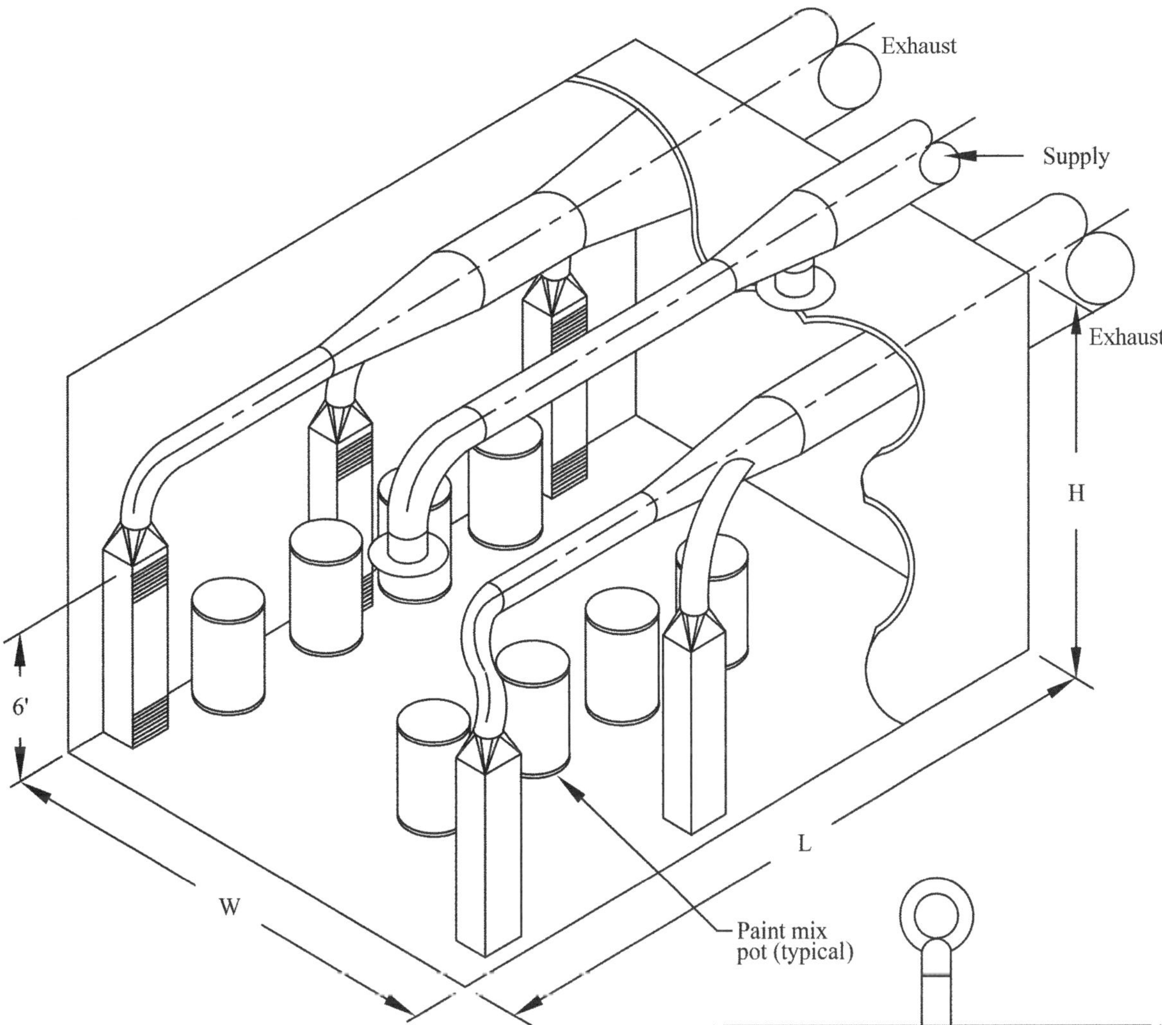

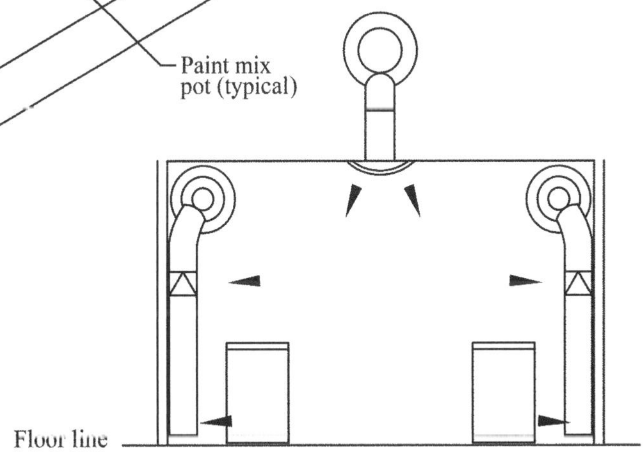

$Q_{(exhaust)}$ = 10 to 12 air changes/hour
$Q_{(supply)}$ = 10 to 12 air changes/hour
Balance room slightly negative
$Q_{(exhaust)} = 1.05\ Q_{(supply)}$
Entry loss = 1.78 VP_s + 0.56 VP_d
Minimum duct velocity = 2000 fpm [10.00 m/s]
Stack velocity = 3500 fpm [17.50 m/s]

For construction and safety consult appropriate NFPA Code.

ACGIH®	TITLE PAINT MIX STORAGE ROOM	FIGURE VS-75-30 DATE 1-12

CHECK CODES, REGULATIONS, AND LAWS (LOCAL, STATE, AND NATIONAL) TO ENSURE THAT DESIGN IS COMPLIANT.

13.80 MECHANICAL SURFACE CLEANING AND FINISHING

Mechanical surface cleaning is generally used to prepare a surface for painting, welding, or other operations. The surfaces may be coated with paint, rust, or oxidation; plated with other metals; or covered with molding sand, inorganic, organic, or biological matter. Mechanical cleaning may be accomplished by abrasive blasting, wire wheels, sand paper/sanding belts, grinding wheels, or use of abrasive chips in tumbling mills. The capture velocity needed to entrain large particles is often very high and the collection hood must be positioned so the materials are directed toward the hood. A minimum duct transport velocity of 3500 fpm [17.50 m/s] is needed but 4000 to 5000 fpm [20.00 to 25.00 m/s] is recommended. A hood that encloses as much of the operation as practical is desired. The toxicity of the material removed must be considered when cleaning mechanically. Complete enclosures may be used or the worker may need to wear a respirator in addition to using local exhaust ventilation.

For many mechanical cleaning and finishing operations, regulations from the Occupational Safety and Health Administration (OSHA)[13.80.1] and National Fire Protection Association (NFPA)[13.80.2] may apply.

13.80.1 Abrasive Cleaning. Abrasive blasting is one of the most widely used mechanical methods to remove surface organic coatings, scale, and rust from surface and parts in preparation for subsequent finishing operations. Sand was the original abrasive material used when the process was invented, but other types of abrasives have largely superseded its use, including plastic media, walnut shells, wheat starch, sodium bicarbonate, aluminum oxide, steel shot, and glass beads. Silica, a major component in sand, has been implicated in serious health problems. Therefore, do not use sand.

Abrasive blasting uses compressed air to direct abrasive media toward the desired surface with a blast nozzle. To comply with safety and environmental regulations that require strict control over the abrasive media and generated dusts, blasting is usually conducted in an enclosed facility. The blasting enclosures are of two types: blasting cabinets, where the operator stands outside the enclosure, and blasting rooms, where the operator stands inside the enclosure. Abrasive blasting enclosures are manufactured in many different sizes, from blasting cabinets that enclose small parts to walk-in blasting rooms the size of an aircraft hangar. Since workers must use respirators, the ventilation system is designed to improve visibly and reduce exposure hazards.

In general, blasting enclosures consist of a storage hopper, a media recovery system, a media separator, and a ventilation/emission control system. Most blasting procedures work in the following manner. Media is directed from the storage hopper through a compressed air blasting nozzle held by the worker. The media abrades the surface or parts. The media, any coatings, and the substrate partially break down. In blasting cabinets, spent media falls through a grate that the parts sit on and is recycled through the media recovery system. In blasting rooms, spent media falls to the floor. Some automated blasting rooms have grates in the floor; conveyors or augers transport the recovered media back to the hopper. Conveyors can be either mechanical or pneumatic. Other walk-in enclosures are not automated and workers must periodically shovel existing media into the recovery system during the blasting operation. The ventilation system collects the fine particles from the abraded coatings and blast media. All the spent media and coatings from both the floor and ventilation system usually pass through a cyclone or sieve system. The cyclone/sieve separates the coatings from any reusable blast media. Depending on its size, separated media is either sent back to the storage hopper for reuse or collected for disposal. The coatings and fine particulates are sent to an emission control system, such as an additional cyclone or, where most of the waste particulates fall into a drum for later disposal. The fine particulate airstream is then filtered before being exhausted to the atmosphere.

Various media types, blast nozzle pressures, blasting angles, standoff distances, and dwell time are used, depending on the substrate blasted and the media used. These factors influence dust generation rates and worker breathing zone concentrations. Typical stand-off distances in blasting rooms are 12 to 18 inches [300 to 450 mm] from the part. More visible dust is generated when workers use higher blast nozzle pressures (40 to 60 psi) [275 to 414 kPa] than lower nozzle pressures (20 to 30 psi) [138 to 207 kPa].

Abrasive blasting operations must be adequately ventilated. The primary purpose of the ventilation is to prevent the build-up of explosive dust concentrations. A blasting booth is a Class II, Division 1 environment, so precautions must be taken to ensure dust concentrations are controlled both inside and outside the booth. A secondary benefit is the capture of abraded coatings and control of contaminant levels.

Abrasive blasting cabinets have air intake vents located on top of the cabinet. These intakes are either baffled or have some sort of filter over them to prevent blast media being projected out of them. Air enters the cabinet and then is exhausted through an opening on the side or back of the cabinet. The exhaust duct leads to a media separator such as a cyclone, which separates the media from the abraded coatings. Air from the cyclone then passes through a filter before being exhausted into either the room or outside the facility.

Abrasive blasting rooms are either downdraft or crossdraft design (see Section 13.80.3). In downdraft ventilation design, air enters through inlets in the ceiling of the room and exit through the floor grille. In crossdraft design, air enters through one side of the room (usually the door) and exists through either a plenum or opening on the other side or sometimes along the bottom sides of the booth. Unlike paint booths, make-up air does not have to enter the booth in a laminar like fashion but it should not cause additional air currents.

Blasting cabinets and rooms should be under negative pressure relative to the facility in which they are located. Negative pressure helps contain particulates generated during the blasting procedure inside the enclosure. A manometer installed on the enclosure can indicate whether the booth is under negative pressure.

VS-80-01, -02, and -03 show suggested designs for abrasive blasting and tumbling mills. A supplied air respirator must be used in abrasive blasting rooms.

13.80.2 Grinding. Mechanical surface finishing uses organic bonded wheels, cones, saws, or other shapes rotating at a high rate of speed to smooth a surface; reduce an object or part in size; or perform other operations. As the object is being surfaced or finished, metallic particles are removed and leave the object at a high speed. In addition, the abrasive wheel is reduced in size and generates particles that must be controlled. Frequently, grinding is accomplished using fluids to keep the parts cool. This cooling fluid will be emitted as an aerosol or mist and needs to be controlled and provisions must be made in the duct to drain off the liquids that accumulate.

The hood used to capture the particles should enclose the operation as much as possible and be positioned to take advantage of the velocity and direction of the particles as they are generated. Design specifications for grinding and surfacing operations are shown in VS-80-10 through VS-80-19.

13.80.3 Buffing and Polishing. The same principles apply for buffing and polishing as for grinding and surfacing. The buffing wheel or belt should be enclosed as much as practical and positioned to take advantage of the centrifugal force of the particles as they leave the wheel or belt. The minimum duct velocity for the generated particles is 3500 fpm and 4500 fpm [17.50 to 22.50 m/s] if the material is wet or sticky. Since many varieties of metals and alloys are buffed and polished, it is extremely important not to mix ferrous and non-ferrous metals in the same exhaust systems (see NFPA codes).[13.80.2] VS-80-30 through VS-80-35 show suggested designs for buffing and polishing.

REFERENCES

13.80.1 U.S. Department of Labor, Occupational Safety and Health Administration: 29 CFR. 1910.

13.80.2 National Fire Protection Association Codes NFPA 61 (Standard for the protection of Fires and Dust Explosions in Agricultural and Food Processing Facilities – 2002 Edition), NFPA 68 (Guide for Venting of Deflagrations – 2002 Edition), NFPA 69 (Standard on Explosion Prevention Systems – 2002 Edition), NFPA 77 (Recommended Practice on Static Electricity – 2000 Edition), NFPA 91 (Standards for Exhaust Systems for Air Conveying of Vapors, Gases, Mists and Noncombustible Particulate Solids – 2004 Edition), NFPA 654 (Standard for the protection of Fires and Dust Explosions for Manufacturing, Processing, and Handling of Combustible Particulate Solids – 2006 Edition), NFPA, Quincy, MA.

13.80.3 Hogopian and Bastress: Recommended Ventilation Guidelines for Abrasive-Blasting Operations. CDC-99-74-33 (1975).

13.80.4 American Foundrymen's Society, Inc.: Foundry Ventilation Manual AFS, Des Plaines, IL (1985).

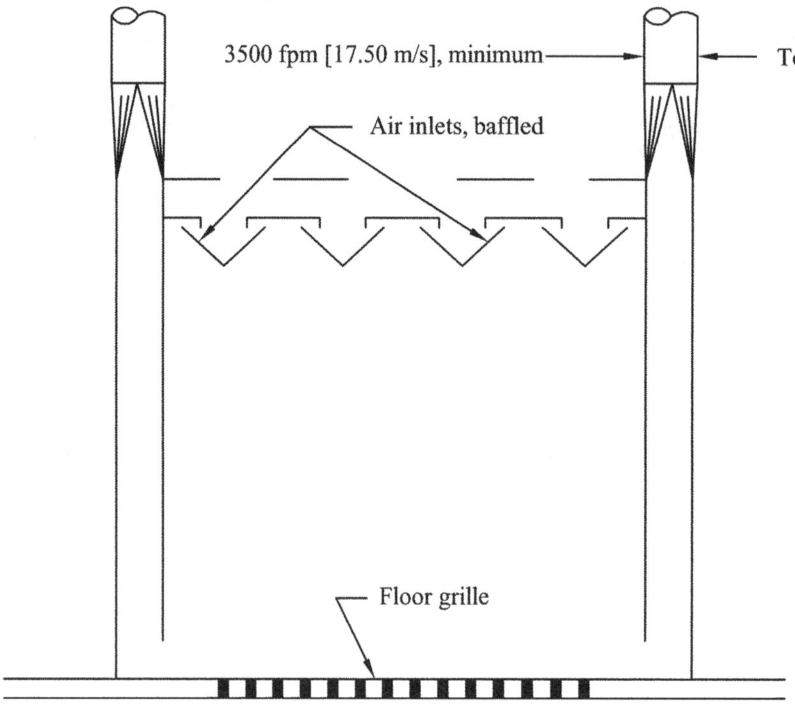

SECTION THROUGH TYPICAL ROOM

Q = 60–100 acfm/ft^2 [0.30–0.50 am^3/s/m^2] of floor for downdraft with typical choice 80 acfm/ft^2 [0.4 am^3/s/m^2]
Q = 100 acfm/ft^2 [0.5 am^3/s/m^2] of wall for crossdraft

Lower control velocities may be used depending on toxicity of the contaminant, object and blasting media and the size of the blasting room.

Notes:

1. The above ventilation is for operator visibility and to control escape of contaminants into adjacent work areas.
2. Operator in an abrasive blasting room is required to wear appropriate NIOSH certified respiratory protection.
3. For rotary tables use 200 acfm/ft^2 [1.0 am^3/s/m^2] of total opening (taken without curtains).
4. For blasting cabinets see VS-80-02.

Reference 13.80.3 and 13.80.4

ACGIH®	TITLE ABRASIVE BLASTING ROOM	FIGURE VS-80-01 DATE 1-12

CHECK CODES, REGULATIONS, AND LAWS (LOCAL, STATE, AND NATIONAL) TO ENSURE THAT DESIGN IS COMPLIANT.

FIGURE VS-80-02

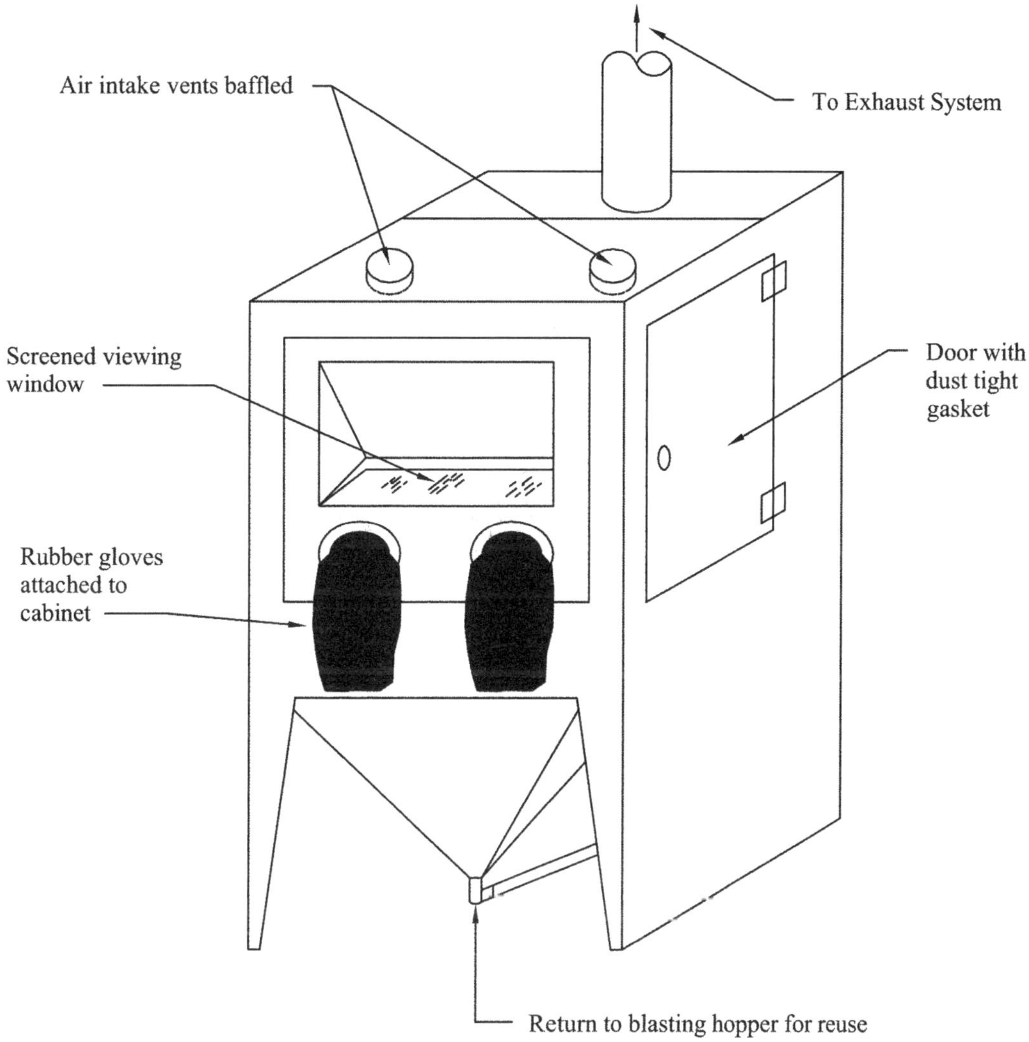

Notes: 20 air changes per minute
At least 500 fpm [2.50 m/s] inward velocity at all openings
Minimum duct velocity = 4000 fpm [20.00 m/s]
If cabinet has self-contained dust collector, consult manufacturer for losses.
$h_e = 1.0\ VP_d$

ACGIH®	TITLE ABRASIVE BLASTING CABINET	FIGURE VS-80-02 DATE 1-12

CHECK CODES, REGULATIONS, AND LAWS (LOCAL, STATE, AND NATIONAL) TO ENSURE THAT DESIGN IS COMPLIANT.

FIGURE VS-80-03

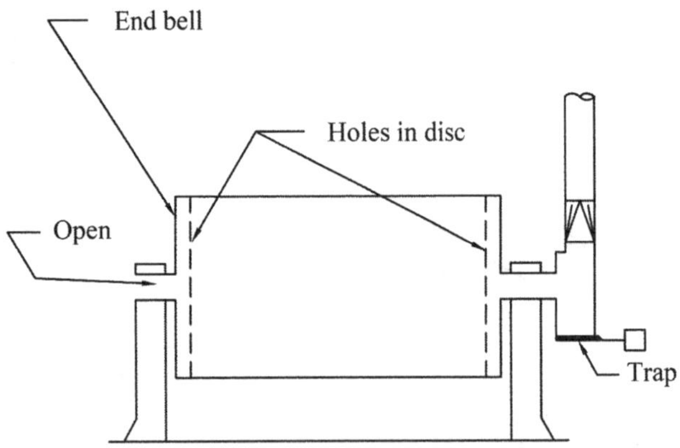

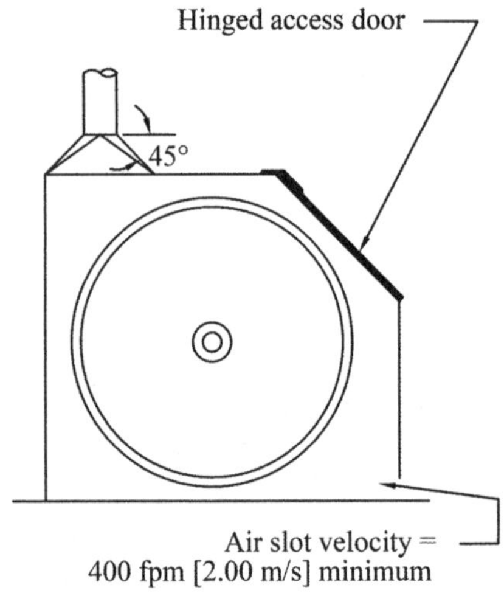

SECTION THROUGH HOLLOW
TRUNNION TUMBLER

Minimum duct velocity = 5000 fpm [25.00 m/s]
Entry loss (h_e) depends on design and
typically ranges from 3" to 9" wg [750 to 2250 Pa].

STAVE MILL
(END SECTION)

Minimum duct velocity = 3500 fpm [17.50 m/s]
h_e = 0.25 VP_d

EXHAUST RATES

Square mill side dia. in inches [mm]	Round mill I.D. in inches [mm]	Exhaust acfm/ft² [am³/s/m²] **	
		Trunnion	Stave
	Up to 24 incl. [600]	430 [2.15]	800 [4.00]
Up to 24 incl. [600]	25–30 [625–750]	680 [3.40]	900 [4.50]
25–30 [625–750]	31–36 [775–900]	980 [4.90]	980 [4.90]
31–36 [775–900]	37–42 [925–1050]	1330 [6.65]	1330 [6.65]
37–42 [925–1050]	43–48 [1075–1200]	1750 [8.75]	1750 [8.75]
43–48 [1075–1200]	49–54 [1225–1350]	2200 [11.00]	2200 [11.00]
49–54 [1225–1350]	55–60 [1375–1500]	2730 [13.65]	2730 [13.65]
55–60 [1375–1500]	61–66 [1525–1650]	3300 [16.50]	3300 [16.50]
61–66 [1525–1650]	67–72 [1675–1800]	3920 [19.61]	3920 [19.61]
67–72 [1675–1800]		4600 [23.00]	4600 [23.00]

* Low-loss designs have large air inlet openings in end bell.
Holes in end discs are sized for velocities of 1250 to 1800 fpm [6.30–9.00 m/s].

** For lengths over 72" [1.8 m], increase exhaust rate proportionately.

ACGIH®

TITLE: TUMBLING MILLS

FIGURE: VS-80-03

DATE: 1-12

CHECK CODES, REGULATIONS, AND LAWS (LOCAL, STATE, AND NATIONAL)
TO ENSURE THAT DESIGN IS COMPLIANT.

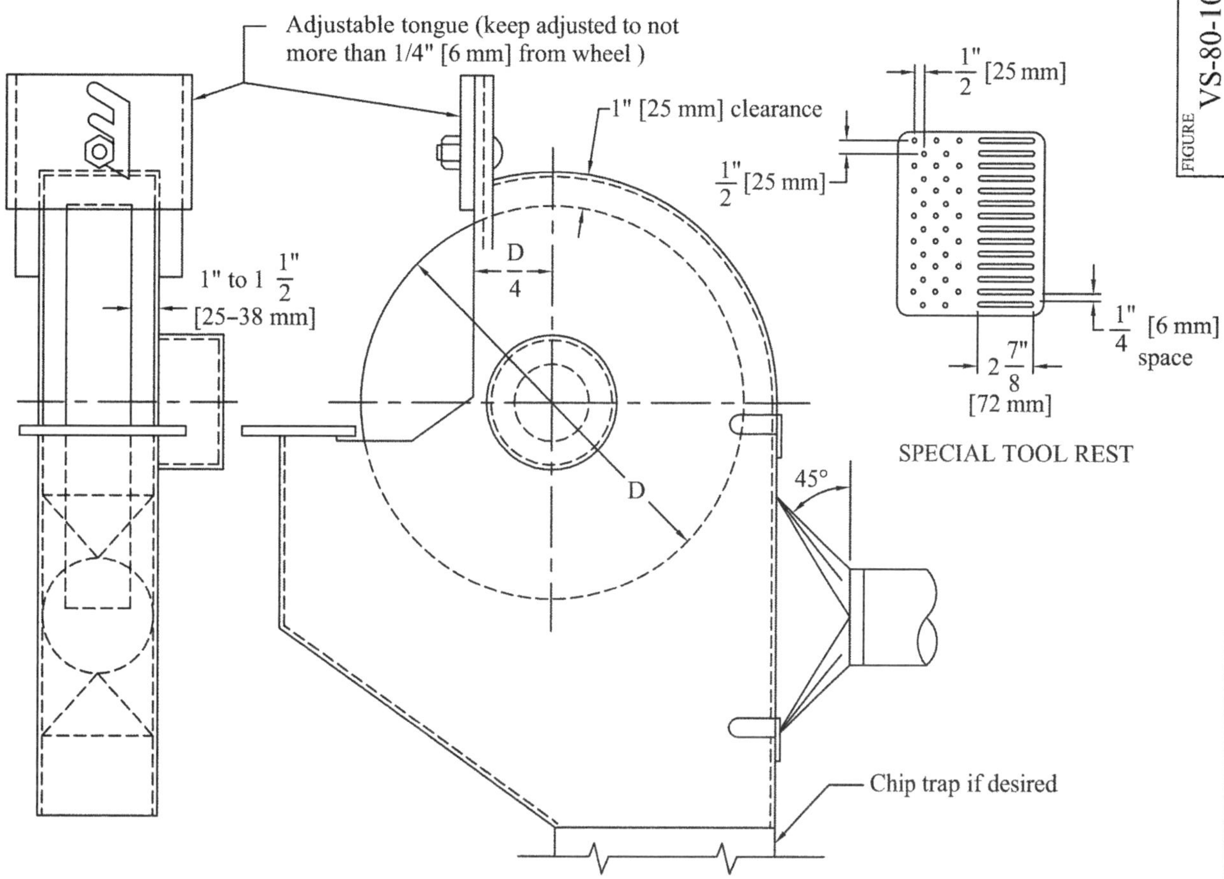

EXHAUST FLOW RATES, acfm [am^3/s]

Wheel dia. inches [mm]		Wheel width inches [mm]		Good enclosure*		Poor enclosure	
Up to 5	[Up to 125]	1	[25]	220	[0.11]	390	[0.20]
5 to 10	[125 to 250]	1.5	[38]	390	[0.20]	610	[0.31]
10 to 14	[250 to 350]	2	[50]	500	[0.25]	740	[0.37]
14 to 16	[350 to 400]	2	[50]	610	[0.31]	880	[0.44]
16 to 20	[400 to 500]	3	[75]	740	[0.37]	1000	[0.50]
20 to 24	[500 to 600]	4	[100]	880	[0.44]	1200	[0.60]
24 to 30	[600 to 750]	5	[125]	1200	[0.60]	1600	[0.80]
30 to 36	[750 to 900]	6	[150]	1600	[0.80]	2000	[1.00]

*Special hood and tool rest as shown, no more than 25% of the wheel exposed.
Minimum duct velocity = 4000 fpm [20.00 m/s]

$h_e = 0.65\ VP_d$ for straight take-off

$h_e = 0.40\ VP_d$ for tapered take-off

ACGIH®	TITLE: GRINDING WHEEL HOOD SURFACE SPEEDS ABOVE 6500 sfpm [32.50 sm/s]	FIGURE: VS-80-10 / DATE: 1-12

CHECK CODES, REGULATIONS, AND LAWS (LOCAL, STATE, AND NATIONAL) TO ENSURE THAT DESIGN IS COMPLIANT.

FIGURE VS-80-11

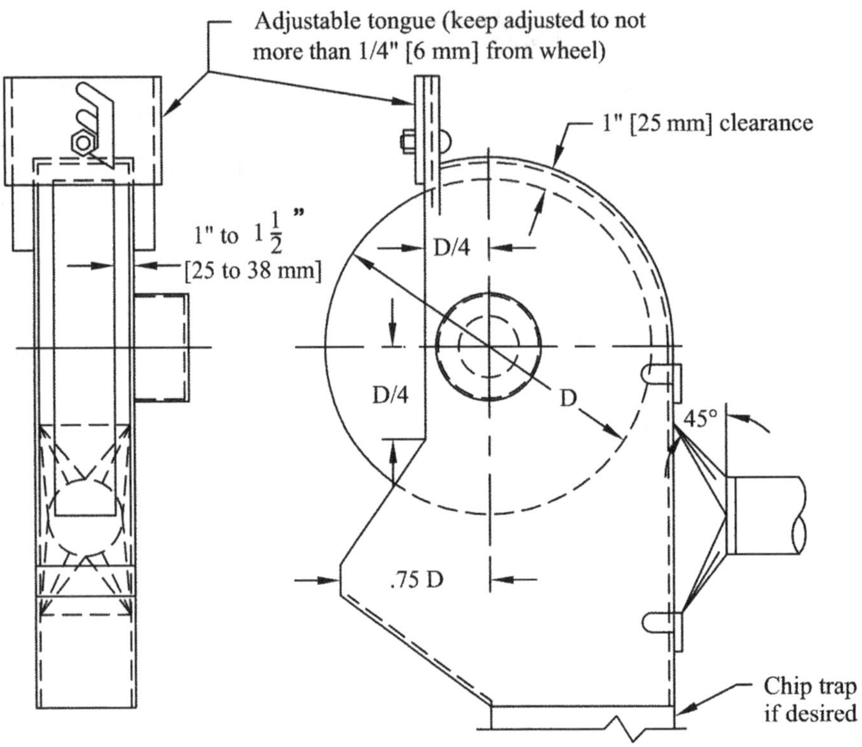

EXHAUST FLOW RATES, acfm [am³/s]

Wheel dia. inches [mm]	Wheel width inches [mm]	Good enclosure*	Poor enclosure
Up to 5 [Up to 125]	1 [25]	220 [0.11]	220 [0.11]
5 to 10 [125 to 250]	1.5 [38]	220 [0.11]	300 [0.15]
10 to 14 [250 to 350]	2 [50]	300 [0.15]	500 [0.25]
14 to 16 [350 to 400]	2 [50]	390 [0.20]	610 [0.31]
16 to 20 [400 to 500]	3 [75]	500 [0.25]	740 [0.37]
20 to 24 [500 to 600]	4 [100]	610 [0.31]	880 [0.44]
24 to 30 [600 to 750]	5 [125]	880 [0.44]	1200 [0.60]
30 to 36 [750 to 900]	6 [150]	1200 [0.60]	1600 [0.80]

* No more than 25% of wheel exposed.

Minimum duct velocity = 4000 fpm [20.00 m/s]

$h_e = 0.65\ VP_d$ for straight takeoff

$h_e = 0.40\ VP_d$ for tapered takeoff

ACGIH®

TITLE GRINDING WHEEL HOOD SURFACE SPEEDS BELOW 6500 sfpm [35.50 sm/s]

FIGURE VS-80-11

DATE 1-12

CHECK CODES, REGULATIONS, AND LAWS (LOCAL, STATE, AND NATIONAL) TO ENSURE THAT DESIGN IS COMPLIANT.

FIGURE VS-80-12

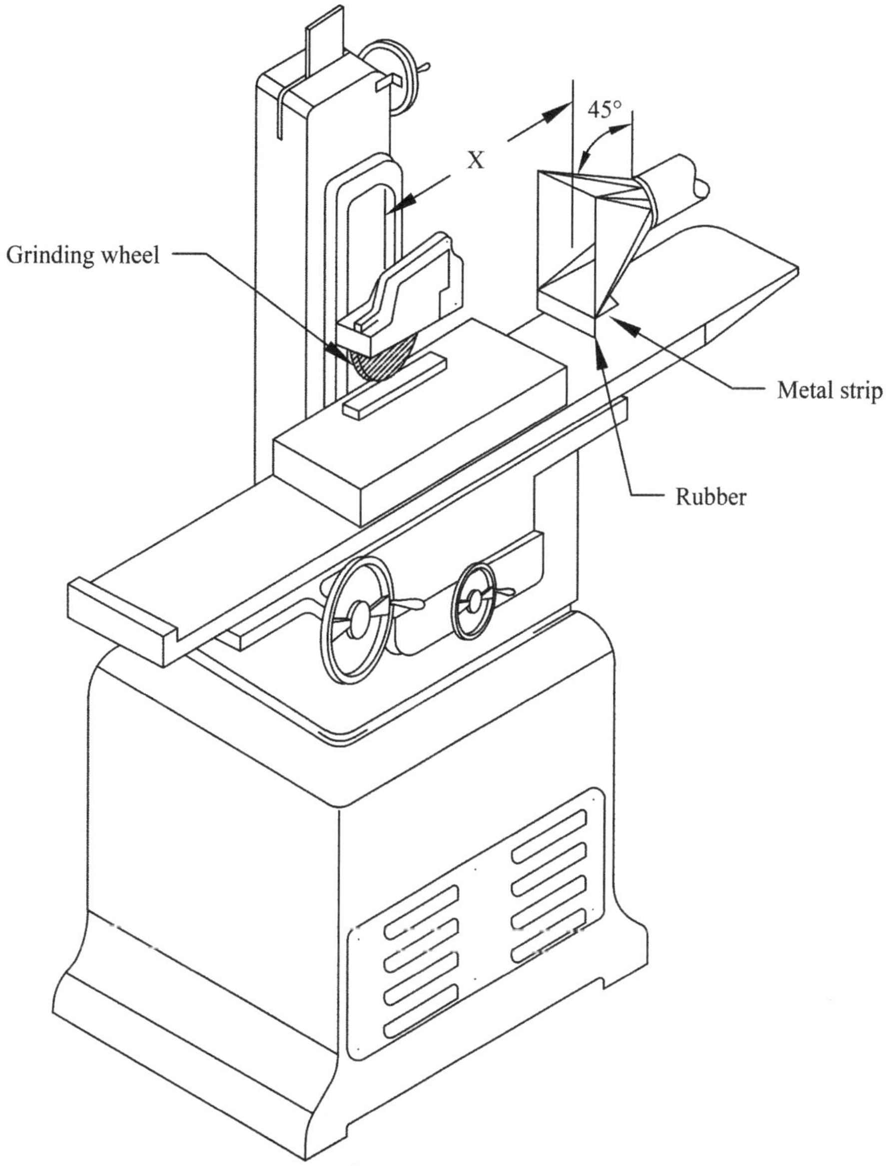

$Q = 0.043\ V_S\ (10X^2 + A)$ acfm [am^3/s]
Minimum duct velocity = 3500 fpm [17.50 m/s]
$h_e = 0.25\ VP_d$
X = distance from hood face to center of wheel, ft [m]
A = hood face area, ft^2 [m^2]
V_s = Wheel Speed, surface feet per min. (SFM or SFPM [sm/s])
= $\pi(D/12)\ R$ [= $\pi(D/1000)\ R$]
D = diameter in inches [mm]
R = rpm of grinding wheel

ACGIH®

TITLE	FIGURE VS-80-12
SURFACE GRINDER	DATE 1-12

CHECK CODES, REGULATIONS, AND LAWS (LOCAL, STATE, AND NATIONAL) TO ENSURE THAT DESIGN IS COMPLIANT.

FIGURE VS-80-13

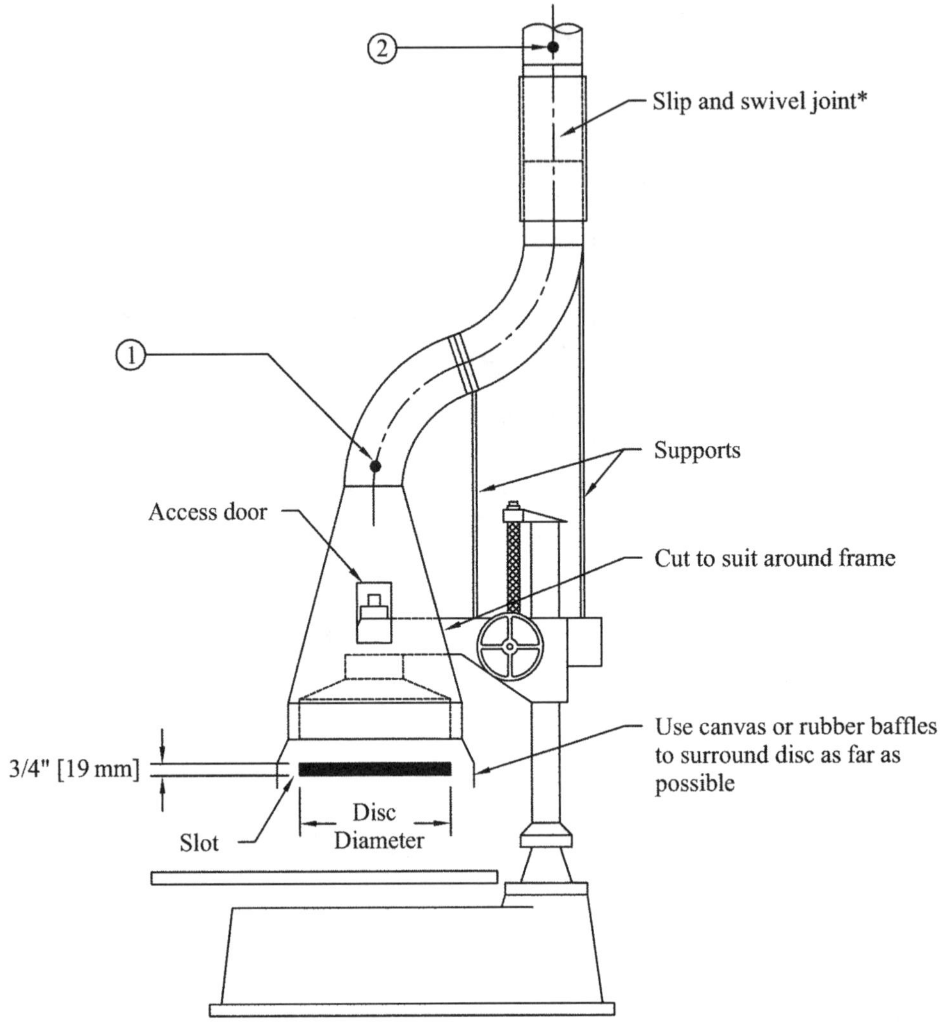

Disc diameter		Duct diameter		acfm	[am^3/s]
Up to 20"	Up to [500 mm]	6"	[150 mm]	900	[0.45]
20" to 30"	[500 to 750 mm]	8"	[200 mm]	1600	[0.80]
30" to 53"	[750 to 1325 mm]	12"	[300 mm]	3500	[1.75]
53" to 72"	[1325 to 1800 mm]	16"	[400 mm]	6300	[3.15]

Minimum duct velocity = 4000 fpm [20.00 m/s]
Minimum slot velocity = 2000 fpm [10.00 m/s]

① $h_e = 1.0\ VP_s + 0.40\ VP_d$

② h_e + elbow losses + swivel joint losses*

ACGIH®

TITLE: CORE GRINDER

FIGURE: VS-80-13

DATE: 1-12

CHECK CODES, REGULATIONS, AND LAWS (LOCAL, STATE, AND NATIONAL) TO ENSURE THAT DESIGN IS COMPLIANT.

FIGURE VS-80-14

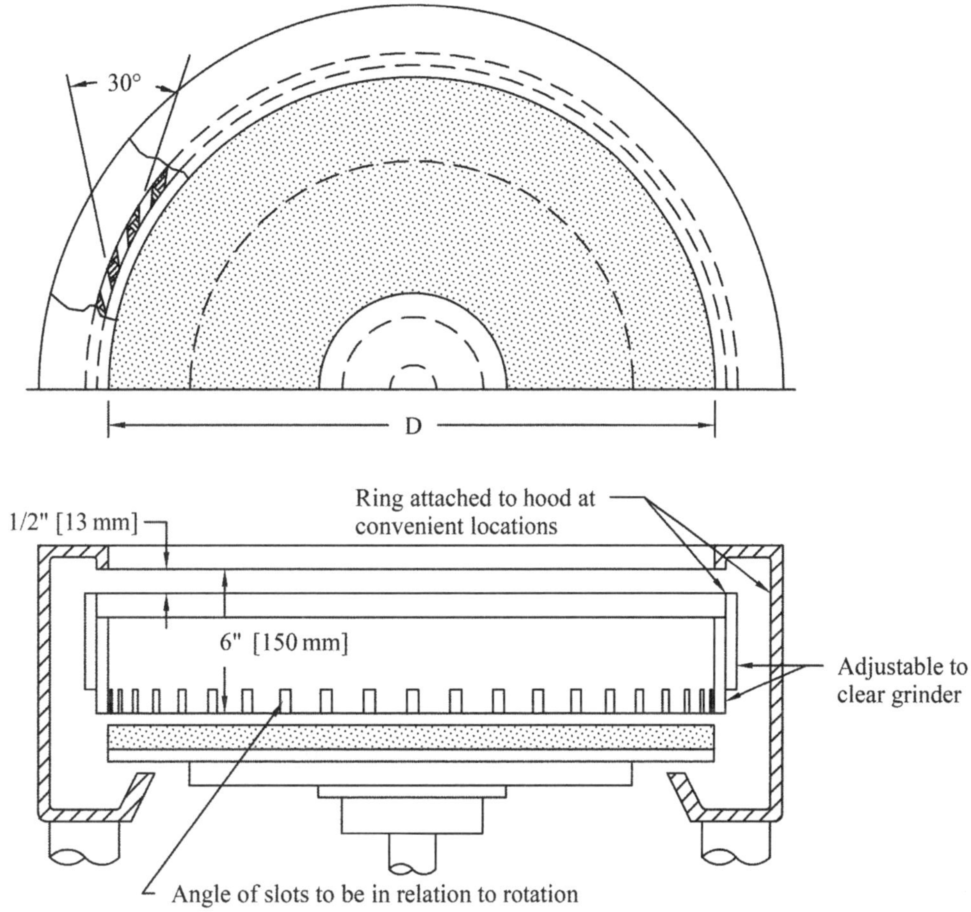

EXHAUST FLOW RATE, acfm [am³/s]

Disc diameter inches [mm]	1/2 or more of disc covered			Disc not covered		
	No.*	Exhaust flow rate, acfm	[am³/s]	No.*	Exhaust flow rate, acfm	[am³/s]
Up to 20 [Up to 500]	1	500	[0.25]	2	780	[0.39]
20 to 30 [500 to 750]	2	780	[0.39]	2	1500	[0.75]
30 to 53 [750 to 1325]	2	1800	[0.90]	4	3500	[1.75]
53 to 72 [1325 to 1800]	2	3100	[1.55]	5	6000	[3.00]

* Number of exhaust outlets around periphery of hood or equal distribution provided by other means.

Minimum slot velocity = 2000 fpm [10.00 m/s]
Minimum duct velocity = 4000 fpm [20.00 m/s]
$h_e = 1.0\ VP_s + 0.56\ VP_d$

ACGIH®

TITLE: VERTICAL SPINDLE DISC GRINDER

FIGURE: VS-80-14

DATE: 1-12

CHECK CODES, REGULATIONS, AND LAWS (LOCAL, STATE, AND NATIONAL) TO ENSURE THAT DESIGN IS COMPLIANT.

FIGURE VS-80-15

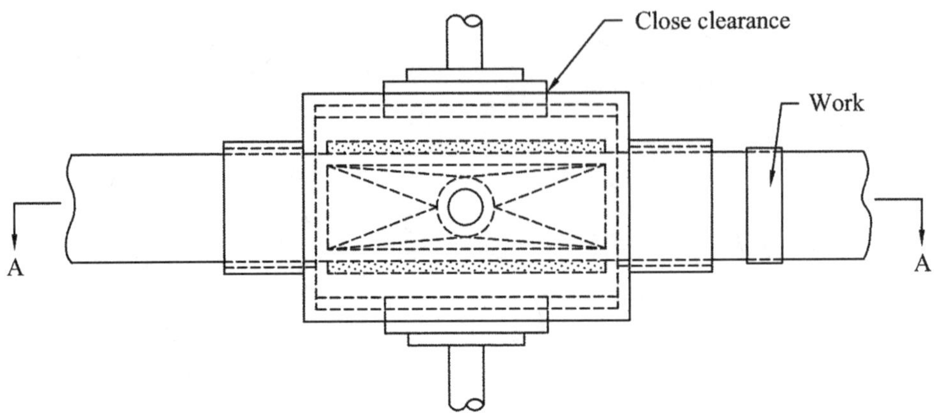

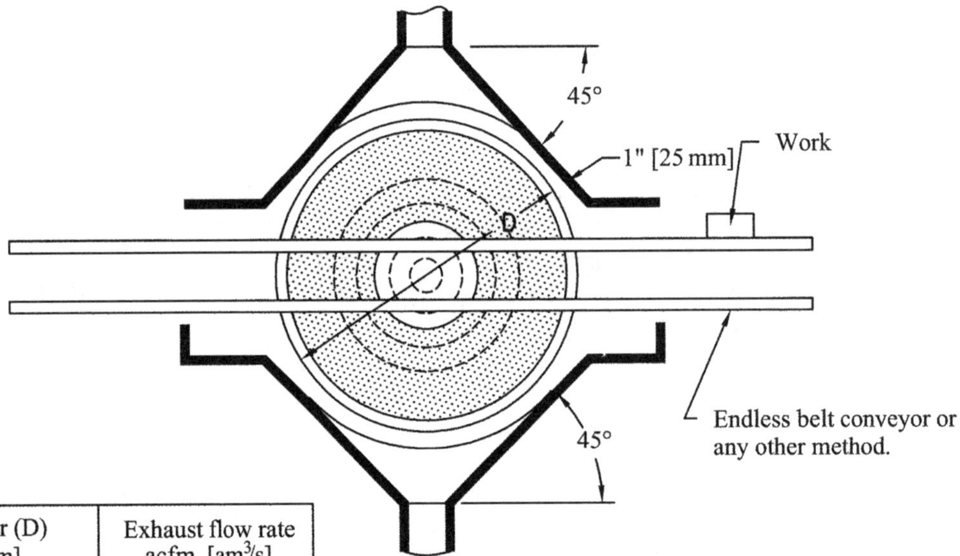

Disc diameter (D) inches [mm]		Exhaust flow rate acfm [am³/s]	
Up to 19	[Up to 475]	610	[0.31]
19 to 25	[475 to 625]	880	[0.44]
25 to 30	[625 to 750]	1200	[0.60]
30 to 53	[750 to 1325]	2000	[1.00]
53 to 72	[1325 to 1800]	6300	[3.15]

Section A-A

Note: If the disc is tightly enclosed by machine housing, then exhaust from the housing is acceptable.

Minimum duct velocity = 4000 fpm [20.00 m/s]
$h_e = 0.65\ VP_d$ straight take-off
$h_e = 0.45\ VP_d$ tapered take-off

ACGIH®

TITLE: HORIZONTAL DOUBLE-SPINDLE DISC GRINDER

FIGURE: VS-80-15

DATE: 1-12

CHECK CODES, REGULATIONS, AND LAWS (LOCAL, STATE, AND NATIONAL) TO ENSURE THAT DESIGN IS COMPLIANT.

FIGURE VS-80-16

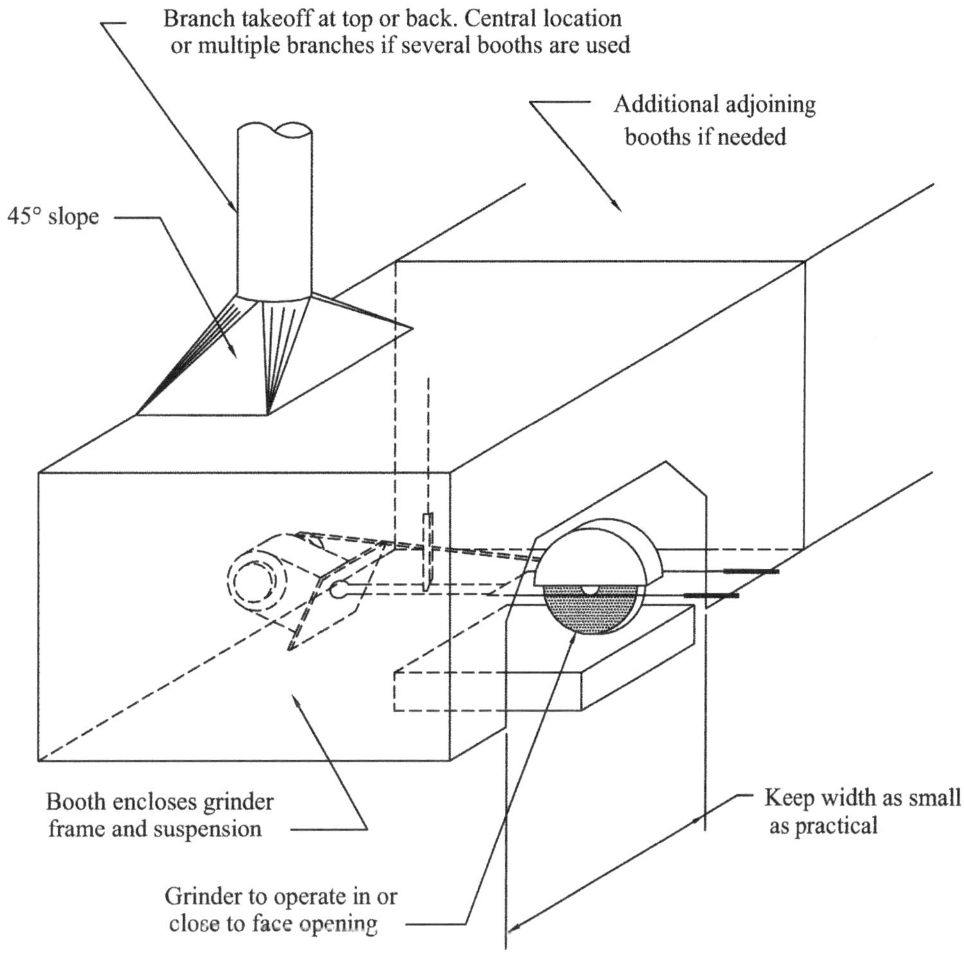

For a large opening, 4' to 6' [1.2 to 1.8 m] wide
Q = 150 acfm/ft^2 [0.75 am^3/s/m^2] of opening

For a small opening, 2' to 4' [0.6 to 1.2 m] with grinder in front
Q = 200 acfm/ft^2 [1.00 am^3/s/m^2] of opening

Minimum duct velocity = 3500 fpm [17.50 m/s]
$h_e = 0.25\ VP_d$

NOTE: Small local exhaust hoods mounted behind grinder wheel may trap the stream of sparks, but are usually not effective in control of airborne dust.

ACGIH®	TITLE: SWING GRINDER	FIGURE: VS-80-16
		DATE: 1-12

CHECK CODES, REGULATIONS, AND LAWS (LOCAL, STATE, AND NATIONAL) TO ENSURE THAT DESIGN IS COMPLIANT.

FIGURE VS-80-17

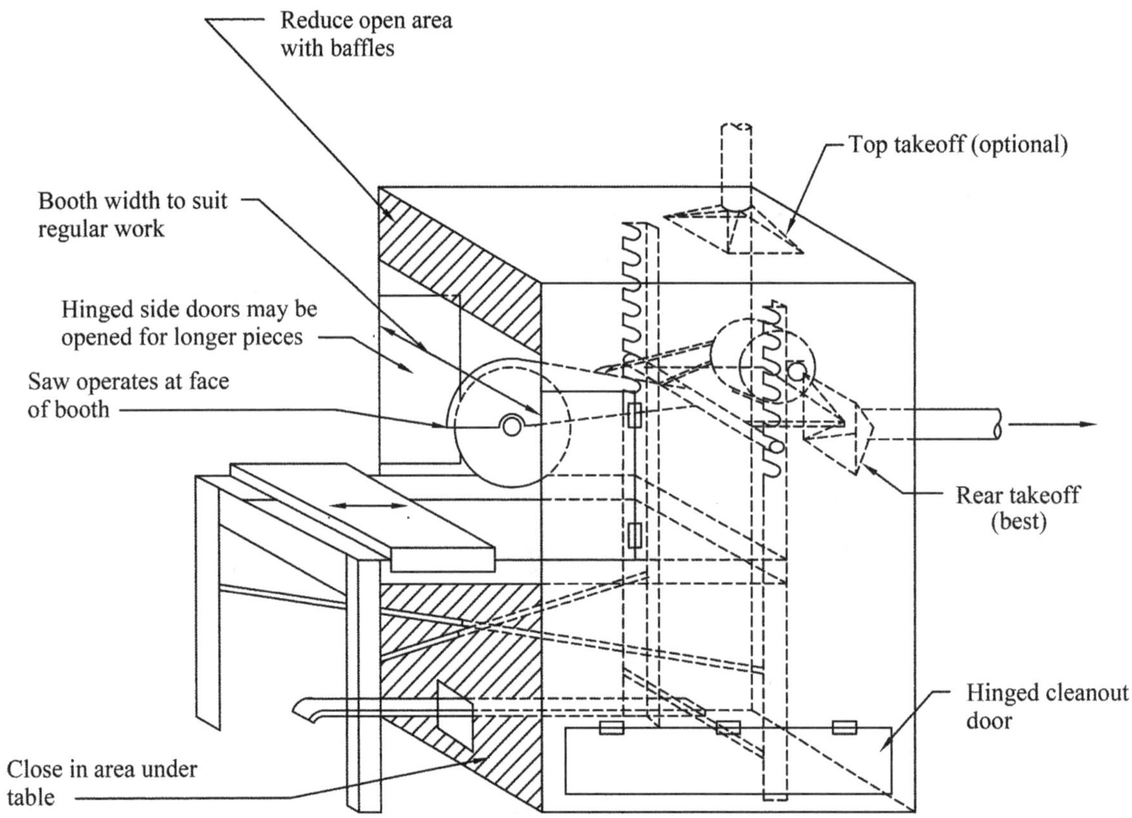

Q = 250 acfm/ft^2 [1.25 am^3/s/m^2] of open face area
Minimum duct velocity = 4000 fpm [20.00 m/s]
$h_e = 0.56\ VP_d$ (no taper)
$h_e = 0.25\ VP_d$ (with 45° taper)

ACGIH®	TITLE ABRASIVE CUT-OFF SAW	FIGURE VS-80-17
		DATE 1-12

CHECK CODES, REGULATIONS, AND LAWS (LOCAL, STATE, AND NATIONAL) TO ENSURE THAT DESIGN IS COMPLIANT.

FIGURE VS-80-18

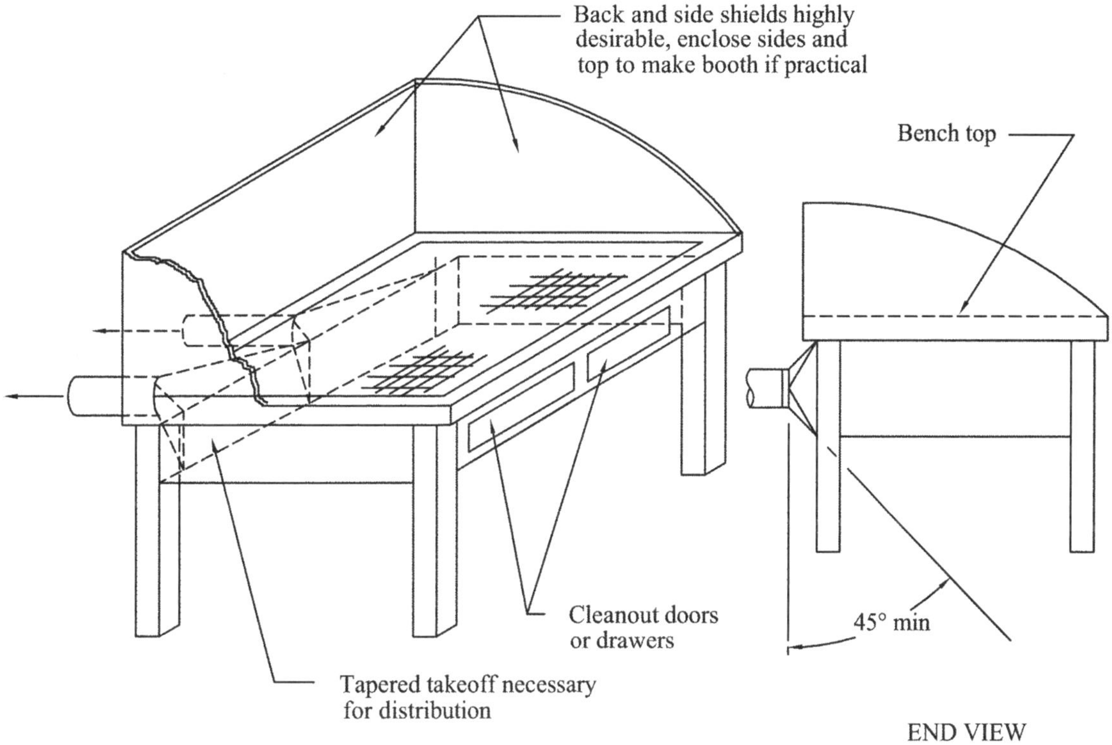

$Q = 150–250$ acfm/ft^2 [0.75–1.25 am^3/s/m^2] of bench area.

Minimum duct velocity = 3500 fpm [17.50 m/s]

$h_e = 0.25\ VP_d$

If slots are used for distribution

$h_e = 1.78\ VP_s + 0.25\ VP_d$

Notes:
1. If grinding in a booth, use 100 fpm [0.50 m/s] face velocity.
2. For downdraft grilles in floor: $Q = 100$ acfm/ft^2 [0.50 am^3/s/m^2] of working area.
3. Provide equal distribution.
4. Provide for cleanout.

ACGIH®	TITLE: HAND GRINDING BENCH	FIGURE: VS-80-18 / DATE: 1-12

CHECK CODES, REGULATIONS, AND LAWS (LOCAL, STATE, AND NATIONAL) TO ENSURE THAT DESIGN IS COMPLIANT.

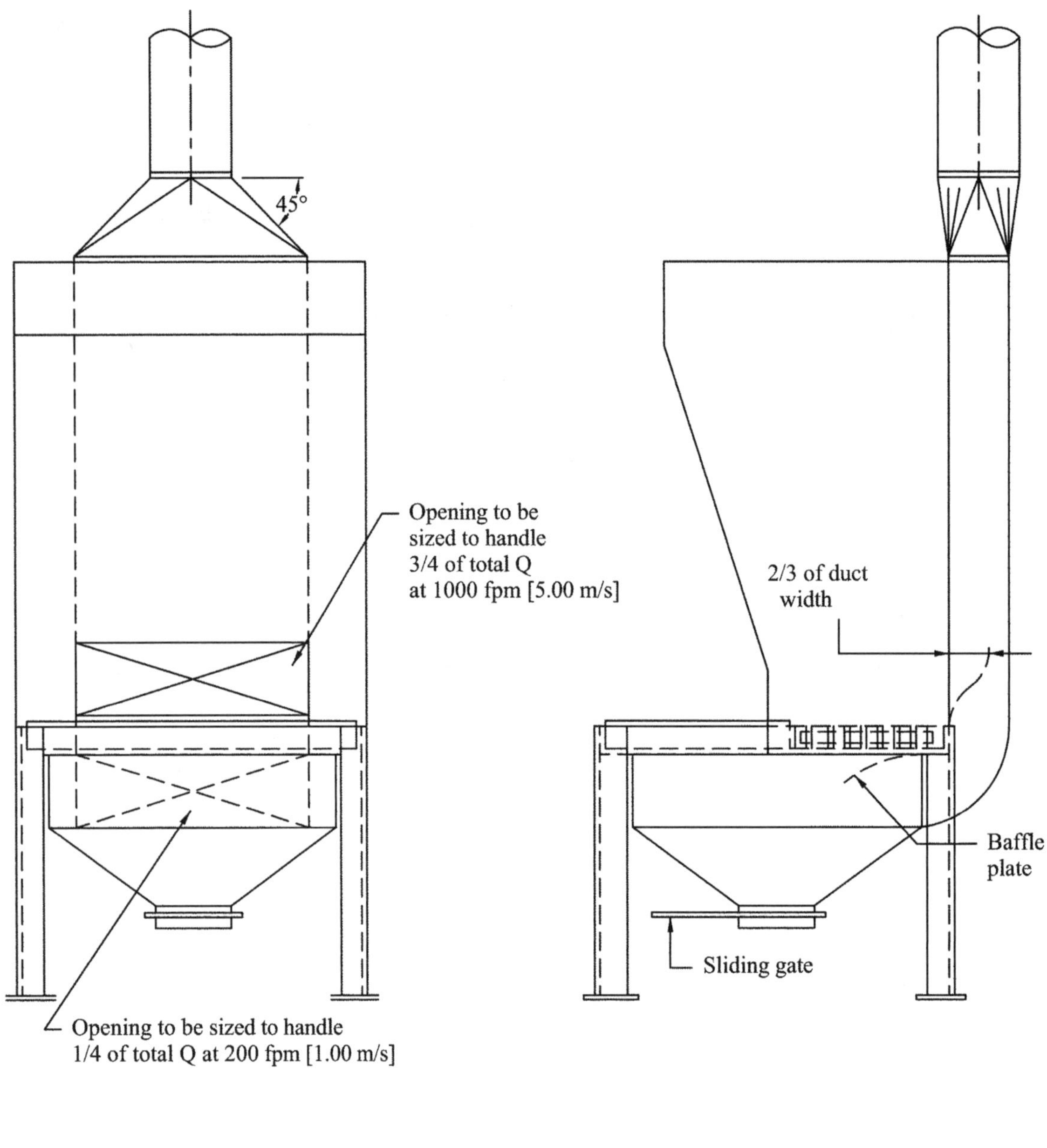

$Q = 150$ acfm/ft^2 [0.75 am^3/s/m^2] opening
Minimum duct velocity = 3500 fpm [17.50 m/s]
$h_e = 0.25\ VP_d$

ACGIH®	TITLE PORTABLE CHIPPING AND GRINDING TABLE	FIGURE VS-80-19
		DATE 1-12

CHECK CODES, REGULATIONS, AND LAWS (LOCAL, STATE, AND NATIONAL) TO ENSURE THAT DESIGN IS COMPLIANT.

FIGURE VS-80-30

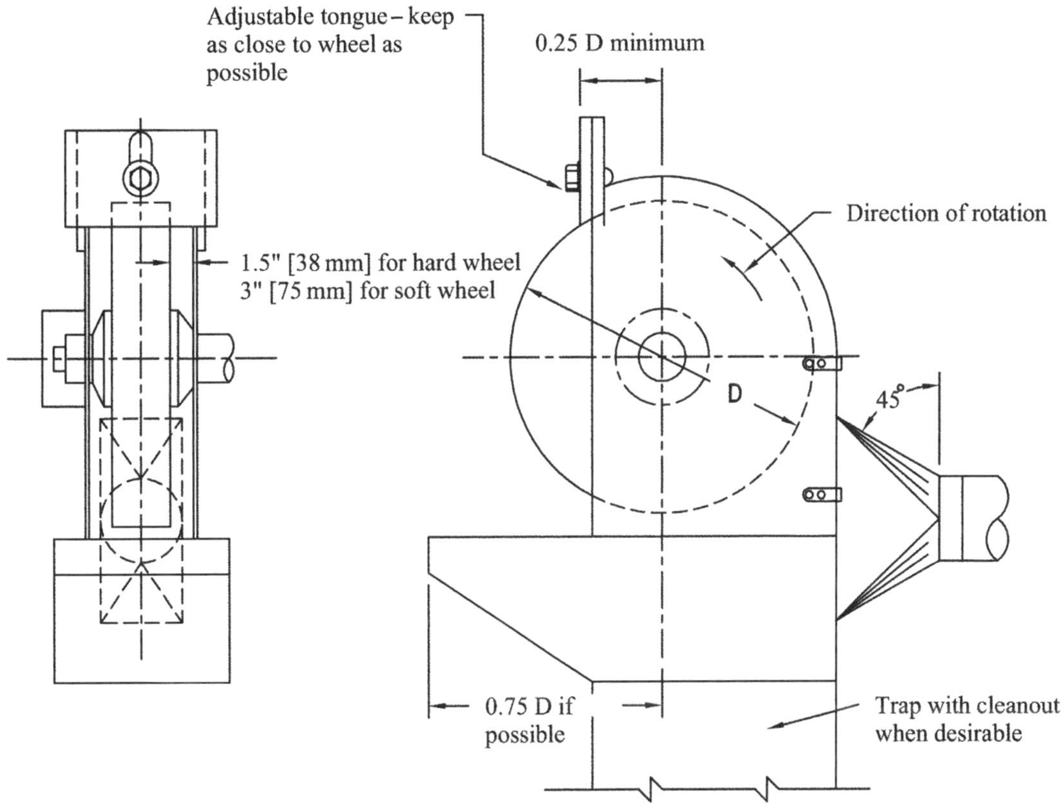

Minimum duct velocity: 3500 fpm [17.50 m/s], 4500 fpm [22.50 m/s] if material is wet or sticky

$h_e = 0.65\ VP_d$ for straight take-off
$h_e = 0.40\ VP_d$ for tapered take-off

Wheel dia. inches [mm]		Wheel width inches [mm]		Exhaust flow rate acfm [am³/s]		Exhaust flow rate acfm [am³/s]	
				Good enclosure *		Poor enclosure	
Up to 9	[Up to 225]	2	[50]	300	[0.15]	400	[0.20]
over 9 to 16	[over 225 to 400]	3	[75]	500	[0.25]	610	[0.31]
over 16 to 19	[over 400 to 475]	4	[100]	610	[0.31]	740	[0.37]
over 19 to 24	[over 475 to 600]	5	[125]	740	[0.37]	1200	[0.60]
over 24 to 30	[over 600 to 750]	6	[150]	1040	[0.52]	1500	[0.75]
over 30 to 36	[over 750 to 900]	6	[150]	1200	[0.60]	2000	[1.00]

* not more than 25% of the wheel is exposed

Note : Consult applicable NFPA codes

Caution : Do not mix ferrous and non-ferrous metals in same exhaust system.

ACGIH®

TITLE: MANUAL BUFFING AND POLISHING

FIGURE: VS-80-30

DATE: 1-12

CHECK CODES, REGULATIONS, AND LAWS (LOCAL, STATE, AND NATIONAL) TO ENSURE THAT DESIGN IS COMPLIANT.

FIGURE VS-80-31

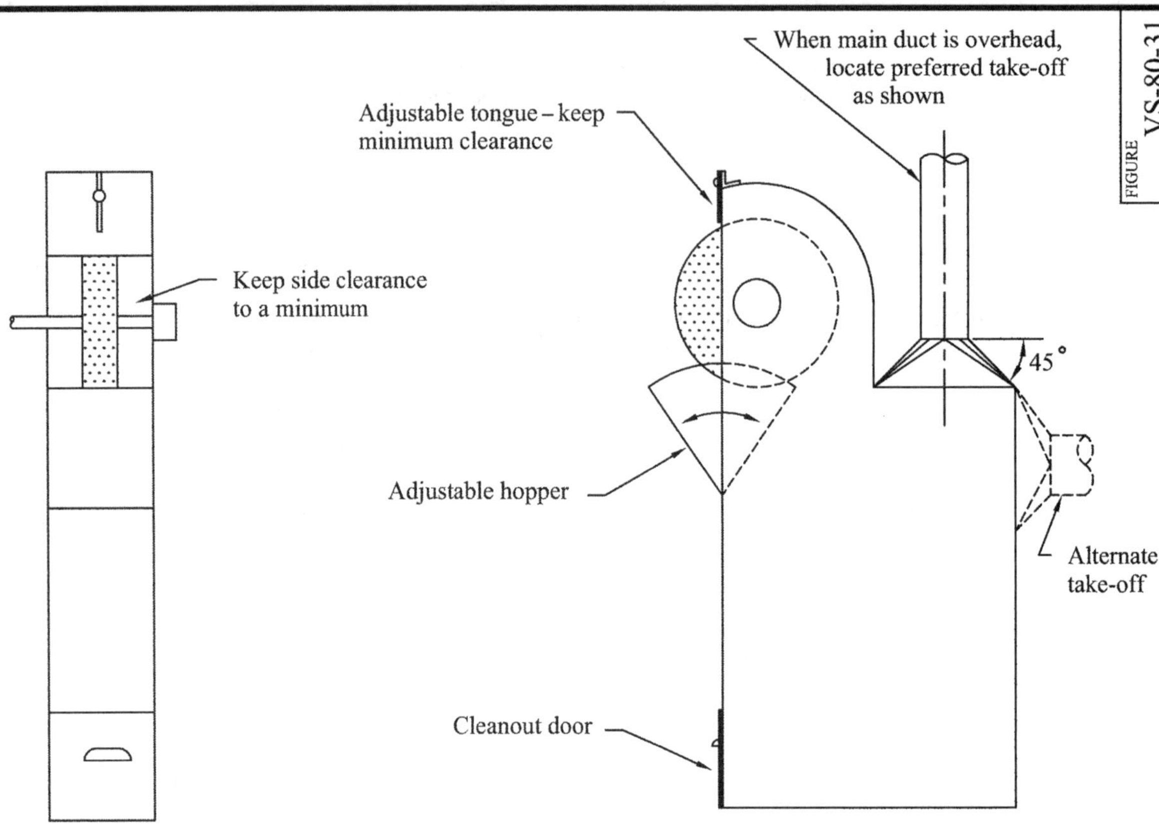

Wheel dia. inches [mm]		Wheel width inches [mm]		Exhaust flow rate acfm [am³/s]	
Up to 9	[Up to 225]	2	[50]	400	[0.20]
9 to 16	[225 to 400]	3	[75]	610	[0.31]
16 to 19	[400 to 475]	4	[100]	740	[0.37]
19 to 24	[475 to 600]	5	[125]	1200	[0.60]
24 to 30	[600 to 750]	6	[150]	1500	[0.75]
30 to 36	[750 to 900]	6	[150]	1900	[0.95]

Note: For wider wheels than listed, increase acfm [am³/s] with width
Minimum duct velocity = 3500 fpm [17.50 m/s]
= 4500 fpm [22.50 m/s] if material is wet or sticky
$h_e = 0.40\ VP_d$

Notes:
1. Consult applicable NFPA codes.
2. For titanium, aluminum, and magnesium, eliminate hopper. Use 5000 fpm [25.00 m/s] through hood cross-section.
3. Caution: Do not mix ferrous and non-ferrous metals in same exhaust system.

ACGIH®	TITLE: BUFFING LATHE	FIGURE: VS-80-31 / DATE: 1-12

CHECK CODES, REGULATIONS, AND LAWS (LOCAL, STATE, AND NATIONAL) TO ENSURE THAT DESIGN IS COMPLIANT.

FIGURE VS-80-32

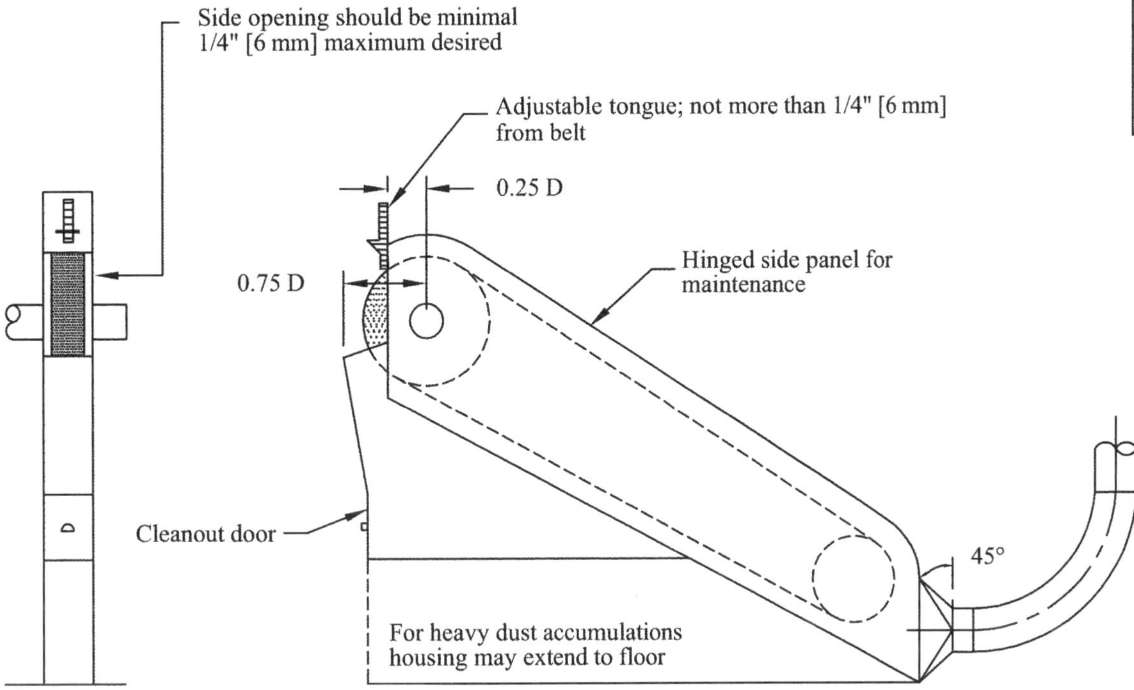

Belt width inches [mm]	Exhaust flow rate acfm [am³/s] Good enclosure *		Exhaust flow rate acfm [am³/s] Enclosure	
1 1/2 [38]	220	[0.11]	300	[0.15]
2 [50]	390	[0.20]	610	[0.31]
3 [75]	500	[0.25]	740	[0.37]
4 [100]	610	[0.31]	880	[0.44]
5 [125]	880	[0.44]	1200	[0.60]
6 [150]	1200	[0.60]	1570	[0.79]

* Hood as shown; no more than 25% of wheel exposed.

$h_e = 0.40\ VP_d$

Minimum duct velocity = 3500 fpm [17.50 m/s], 4500 fpm [22.50 m/s] if wet or sticky.

Notes:

1. Consult applicable NFPA Standards.
2. For titanium, aluminum and magnesium eliminate hopper and use 5000 fpm [25.00 m/s] through hood cross section.
3. Caution: Do not mix ferrous and non-ferrous metals in same exhaust system.

ACGIH®	TITLE: BACKSTAND IDLER POLISHING MACHINE	FIGURE: VS-80-32 / DATE: 1-12

CHECK CODES, REGULATIONS, AND LAWS (LOCAL, STATE, AND NATIONAL) TO ENSURE THAT DESIGN IS COMPLIANT.

FIGURE VS-80-33

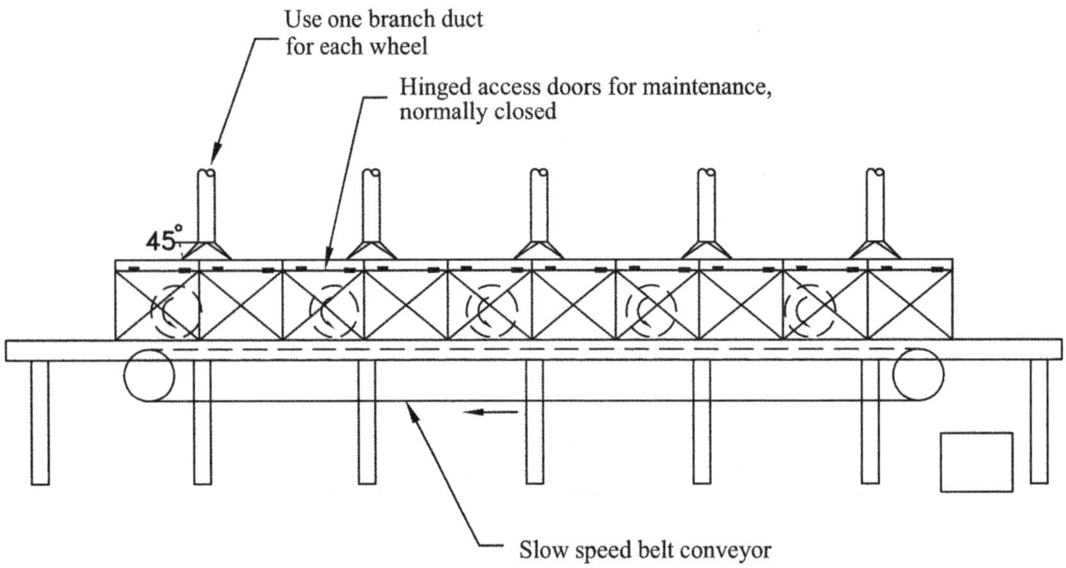

Q = 500 acfm/wheel [0.25 m^3/s/wheel], minimum
Not less than 250 acfm/ft^2 [1.25 $am^3/s/m^2$] total open area
Minimum duct velocity = 3500 fpm [17.50 m/s], 4500 fpm [22.50 m/s] if material is wet or sticky

$h_e = 1.78\ VP_s + 0.25\ VP_d$

Note:

1. Consult applicable NFPA standards. Reference 13.80.2
2. Caution: Do not mix ferrous and non-ferrous metals in same exhaust system.
3. Wheel adjustments on outside of enclosure.
4. For highly toxic material enclose the return strand of the belt conveyor.

ACGIH®	TITLE STRAIGHT LINE AUTOMATIC BUFFING	FIGURE VS-80-33 DATE 1-12

CHECK CODES, REGULATIONS, AND LAWS (LOCAL, STATE, AND NATIONAL) TO ENSURE THAT DESIGN IS COMPLIANT.

FIGURE VS-80-34

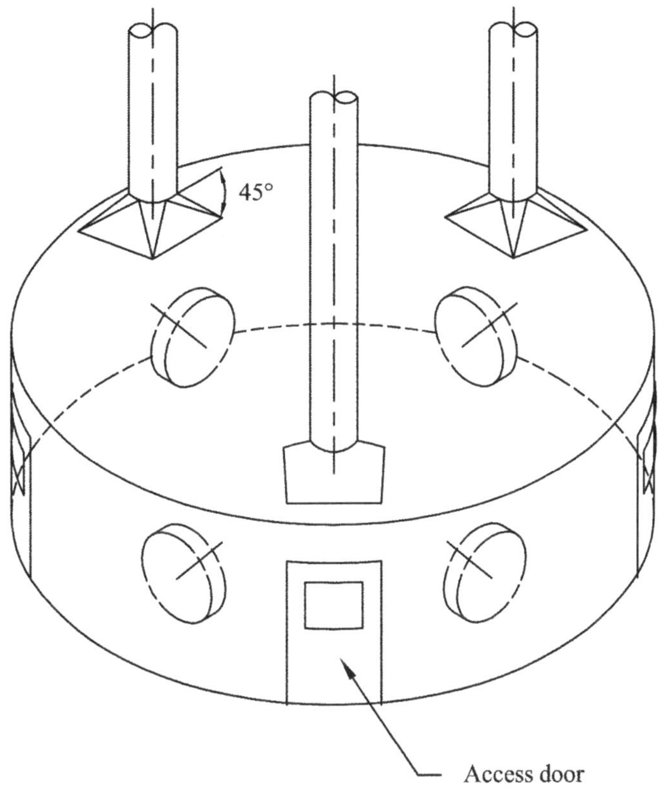

Q = 500 acfm/wheel [0.25 m^3/s/wheel], minimum
Not less than 250 acfm/ft^2 [1.25 $am^3/s/m^2$] total open area
Minimum duct velocity = 3500 fpm [17.50 m/s], 4500 fpm [22.50 m/s] if material is wet or sticky

$h_e = 1.78\ VP_s + 0.25\ VP_d$

On small, (2 or 3 spindle) machines one take-off may be used.
Multiple take-offs desirable.

Note:
1. Consult applicable NFPA standards, Reference 13.80.2
2. Caution: Do not mix ferrous and non-ferrous metals in same exhaust system.

ACGIH®	TITLE CIRCULAR AUTOMATIC BUFFING	FIGURE VS-80-34
		DATE 1-12

CHECK CODES, REGULATIONS, AND LAWS (LOCAL, STATE, AND NATIONAL) TO ENSURE THAT DESIGN IS COMPLIANT.

FIGURE VS-80-35

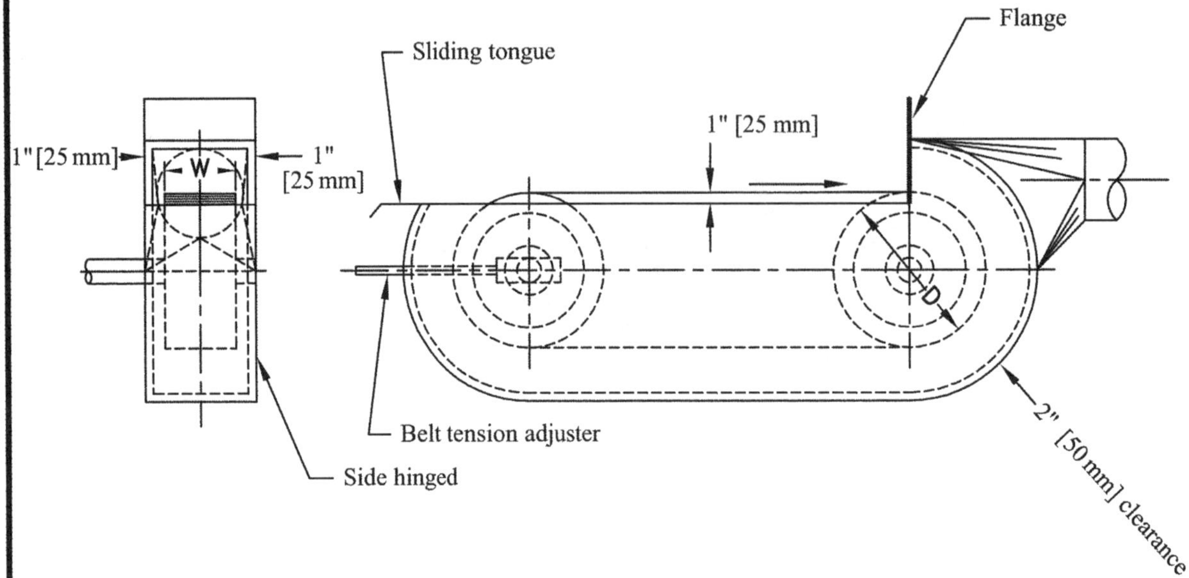

Belt width, inches [mm]	Exhaust flow rate, acfm [am³/s]
Up to 3 [Up to 75]	220 [0.11]
3 to 5 [75 to 125]	300 [0.15]
5 to 7 [125 to 175]	390 [0.20]
7 to 9 [175 to 225]	500 [0.25]
9 to 11 [225 to 275]	610 [0.31]
11 to 13 [275 to 325]	740 [0.37]

Minimum duct velocity = 3500 fpm [17.50 m/s], 4500 fpm [22.50 m/s] if material is wet or sticky
$h_e = 0.65\ VP_d$ for straight take-off
$h_e = 0.45\ VP_d$ for tapered take-off

Notes:
1. Consult applicable NFPA standards Reference 13.80.2
2. Caution : Do not mix ferrous and non-ferrous metals in same exhaust system.

ACGIH®	TITLE METAL POLISHING BELT	FIGURE VS-80-35 / DATE 1-12

CHECK CODES, REGULATIONS, AND LAWS (LOCAL, STATE, AND NATIONAL) TO ENSURE THAT DESIGN IS COMPLIANT.

13.85 VEHICLE VENTILATION

The objective of providing ventilation for vehicles in an environment is to keep a worker's exposure to toxic exhaust fumes and gases below the TLV®, the TWA and STEL, or other appropriate guidelines or standards. This can be achieved either by dilution or local exhaust ventilation.

It is difficult to establish dilution ventilation requirements accurately for the operating vehicles in a plant. For an existing facility, the designer has the opportunity to measure the emission in the field. Standard techniques can be used to measure gas flow rates, composition, temperatures, and contaminant levels. Using the equations in Chapter 4 and the measurements, the dilution rates can be calculated. However, it is not always possible to accurately determine the contaminant generation rate because generation is not uniform. Moreover, no such data are available to the designer for new vehicles.

The use of dilution ventilation is usually considered only after rejection of the source capture concept. Common reasons for rejecting source capture (local exhaust) are operating interference problems or layout constraints. For lift trucks or cars in motion or idling outside of stalls, local exhaust is not feasible. Hence the only method for control of health hazards is dilution ventilation.

Over the years, some empirical rates have been developed and applied successfully to achieve contaminant control. The recommended dilution rates based on average operating conditions are:

10,000 acfm [5.00 am^3/s] per propane fueled lift truck

16,000 acfm [8.00 am^3/s] per gasoline fueled lift truck

10,000 acfm [5.00 am^3/s] per operating automobile

20,000 acfm [10.00 am^3/s] (or more) per operating truck

100 acfm [0.05 am^3/s] per horsepower for diesel fueled vehicle

The above dilution rates for lift trucks apply under the following conditions:[(13.85.1)]

1) A regular maintenance program incorporating final engine tuning through carbon monoxide analysis of exhaust gas must be provided. CO concentration of gases should be limited to 1% for propane fueled trucks; 2% for gasoline fueled trucks. If no regularly scheduled maintenance program, increase the design ventilation rates by a factor of three.
2) The periods of lift truck engine operation do not exceed 50% of the working day (total engine operation of lift truck equal to or less than 4 hours in an 8-hour shift). If the operating time is greater than 50%, multiply the design flow rate by the actual operating percentage divided by 50%.
3) A reasonably good distribution of airflow must be provided. If there is poor air distribution, lift truck operation is not recommended.
4) The volume of space must amount to 150,000 ft^3[13,500 m^3] per lift truck or more. If the building volume is less than 150,000 ft^3 [13,500 m^3], apply the following: 75,000 ft^3 [6,750 m^3] use 150% of design flow rate; 30,000 ft^3 [2,700 m^3] use double the design flow rate. If the space is less than 25,000 ft^3 [2250 m^3], lift truck operation is not recommended.
5) The lift truck is powered by an engine of less than 60 HP [745 watts].

Where actual operating conditions vary from the above, the ventilation rate should be increased. On the other hand, mechanical ventilation may not be required in large buildings where lift truck operation is intermittent and where natural infiltration based on a maximum of one air change/hour for the net building volume exceeds the recommended dilution ventilation rate.

The alternative to dilution ventilation is to capture the contaminant at the source by installing local exhaust ventilation. For stationary vehicles in service garages, effective systems are shown in VS-85-01 (overhead and under floor). Exhaust volumes are shown on VS-85-02. The systems should be connected directly to the vehicle exhaust and should terminate outdoors above the roof. The design procedure outlined in Chapter 5 must be followed. For friction loss data of flexible ducts, manufacturers should be contacted. As with all flexible systems, the length of flexible duct must be minimized, and non-collapsible duct should be used. Unnecessary and/or sharp bends should be avoided.

REFERENCE

13.85.1 Hama, G.M.; Butler, K.E.: Ventilation for Lift Truck Operation. Heating, Piping and Air Conditioning (January 1970).

FIGURE VS-85-01

Note : When ventilating a garage use either the overhead or under floor system.
Exhaust to be discharged above roof. Exhaust volumes as noted on VS-85-02.

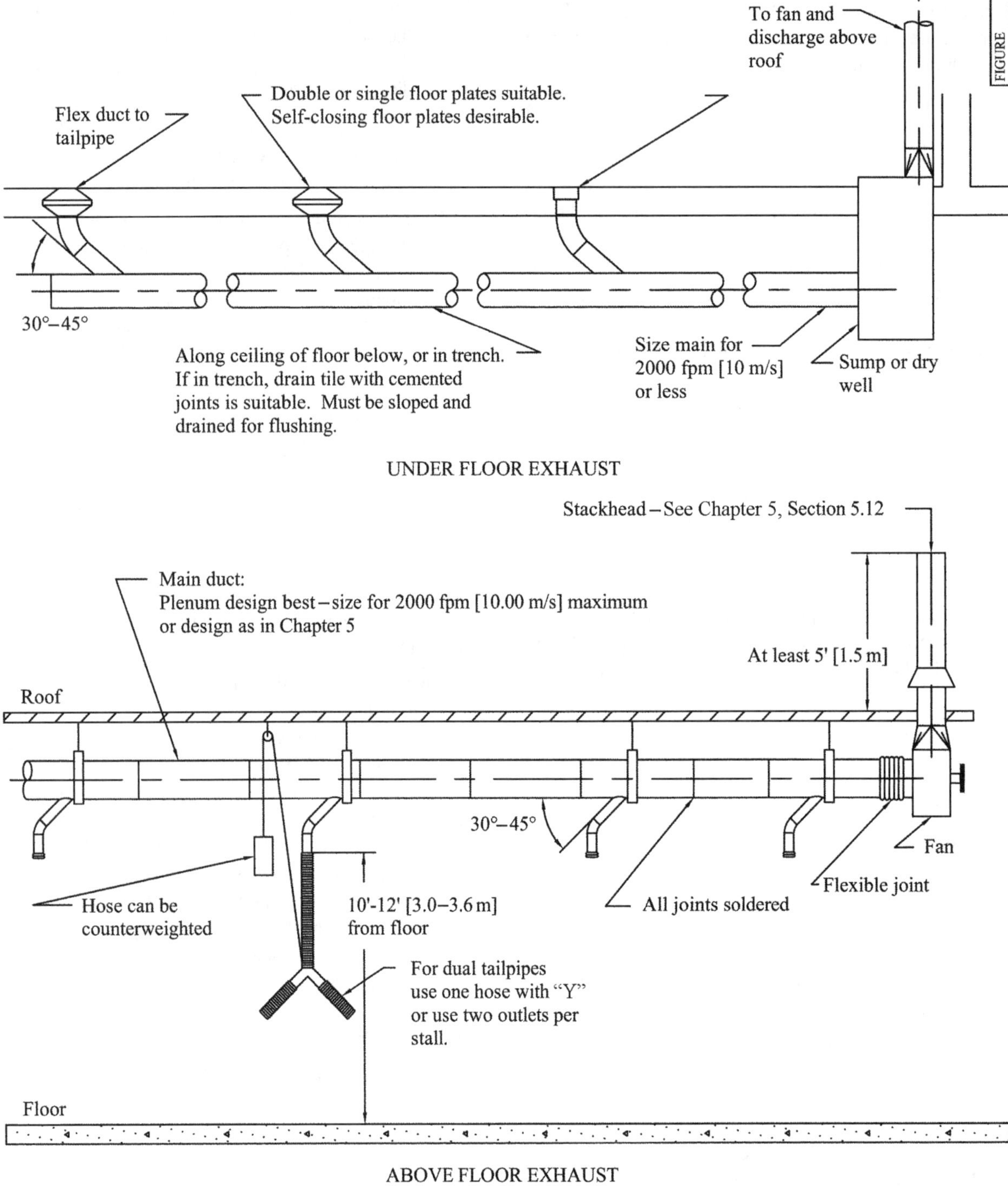

ACGIH®	TITLE: SERVICE GARAGE EXHAUST VENTILATION	FIGURE: VS-85-01 / DATE: 1-12

CHECK CODES, REGULATIONS, AND LAWS (LOCAL, STATE, AND NATIONAL) TO ENSURE THAT DESIGN IS COMPLIANT.

The exhaust flow rate for operating engines connected directly to a tail pipe exhaust system is a function of engine displacement, crankshaft revolutions, tailpipe exhaust temperature and the appropriate conversion factors. The exhaust volume is determined using the following equations, which also include a 20% safety factor.

4-Cycle Engine Exhaust Flow Rate Equations

The engine exhaust flow rate equation may take one of three forms depending on the units of the variables.

		Condition/Units of Measure		
Parameters	**Abbreviation**	**1**	**2**	**3**
Exhaust Flow Rate	Q_e	acfm	acfm	am^3/s
Engine Displacement	D_{eng}	in^3/cycle	liters/cycle	liters/cycle
Crankshaft Revolutions per Unit Time	N	rpm	rpm	rpm
Tailpipe Exhaust Temperature	T_{ex}	F	F	C
Standard Temperature	T_s	70 F	70 F	21 C

Where:

- Condition 1 – all parameters in IP units:

$$Q_e = (1.2)(D_{eng})(N)(0.5)(1\ ft^3/1728\ in^3)((460\ F + T_{ex})\ /\ 530\ F)$$

- Condition 2 – engine displacement in SI units; all other parameters in IP units:

$$Q_e = (1.2)(D_{eng})(N)(0.5)(0.0353\ ft^3/liter)((460\ F + T_{ex})\ /\ 530\ F)$$

- Condition 3 – all parameters in SI units:

$$[Q_e = (1.2)(D_{eng})(N)\left(\frac{1\ min}{60\ s}\right)(0.5)(.001\ m^3/liter)((273\ C + T_{ex})\ /\ 294\ C)]$$

Note 1: The above equations do not take into account the engine's volumetric efficiency (VE). VE is a ratio of the volume of fuel and air that actually enters the cylinder during the induction phase to the actual capacity of the cylinder under static conditions. Use of larger valves, multiple valves, cylinder head porting, tuned exhaust systems, variable valve timing, and forced air induction all can improve the VE above that which can be obtained from a normally tuned, naturally aspirated engine. If the VE is known, Q_e can be multiplied by VE to yield a more accurate representation of the exhaust flow.

If VE is unknown, it is assumed to be 100% (or 1.00, as is the case in the above equations). This is a conservative assumption when working with naturally aspirated engines, as VE is less than 100% is such cases.

Note 2: When calculating the exhaust flow rate for a dual exhaust system, divide the engine's exhaust flow rate (acfm) by two (2) to yield the flow rate for each exhaust pipe.

EXAMPLE PROBLEM 1 (5.7 Liter Engine)

A 5.7 liter, normally aspirated gas engine is operated at 1200 rpm. Its tailpipe exhaust gas temperature is 1000 F. What is the exhaust flow rate in acfm? (Condition 2)

$$Q_e = (1.2)(D_{eng})(N)(0.5)(0.0353\ ft^3/liter)((460\ F + T_{ex})\ /\ 530\ F)$$

$$= (1.2)(5.7\ L)(1200\ rpm)(0.5)(0.0353\ ft^3/L)(1460\ F\ /\ 530\ F)$$

$$= 400\ acfm$$

FIGURE VS-85-10

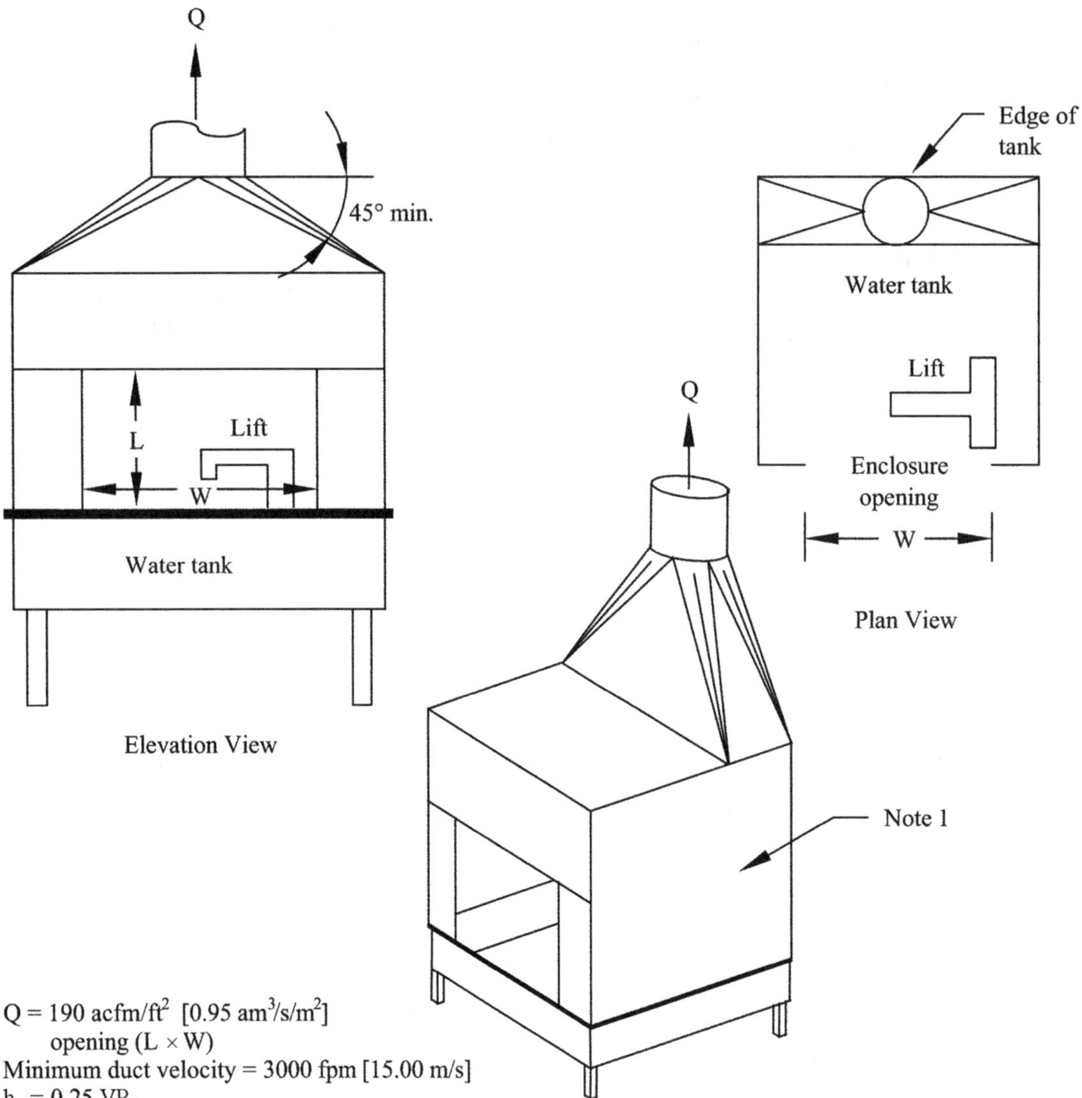

Q = 190 acfm/ft^2 [0.95 am^3/s/m^2]
opening (L × W)
Minimum duct velocity = 3000 fpm [15.00 m/s]
$h_e = 0.25\ VP_d$
L = Height of ventilated opening in inches [mm].
W = Width of ventilated opening in inches [mm].

Notes:
Enclosure can be made of metal or fire resistant curtain material

Reference: 13.85.3

TITLE	FIGURE VS-85-10
VENTILATED BOOTH FOR RADIATOR REPAIR SOLDERING	DATE 1-12

CHECK CODES, REGULATIONS, AND LAWS (LOCAL, STATE, AND NATIONAL) TO ENSURE THAT DESIGN IS COMPLIANT.

13.90 WELDING AND CUTTING

The purpose of welding and cutting ventilation is to control gases, fumes, and particulate generated during the welding and cutting operations.

13.90.1 Hazards. The generation rate of fumes and gases varies with the composition of the base metal, fluxes, and fillers, and with the rate and depth of welding. Exposure to the welder varies with the generation rate, duration and frequency of operations, work practices (particularly distance of the plume from the breathing zone), and the effectiveness of ventilation.

Contaminants from welding may include:

1) Fume from the base metals and filler or electrode metals
2) Fume from coatings (e.g., zinc oxide from galvanized surfaces, thoria from T.I.G. welding, and fluorides and NO_2 from electrode coatings)
3) Ozone due to ionization of oxygen by the ultraviolet light from arc welding
4) Carbon monoxide from ultraviolet effects on carbon dioxide in shield gas
5) Shield gases such as carbon dioxide, helium, and argon
6) Fluoride gases and other thermal decomposition products of fluxes and electrode coatings and
7) Flammable gases such as acetylene.

There are welding tasks that present enhanced hazards such as welding on materials containing or contaminated with heavy metals or welding in the presence of flammable vapors or halogenated hydrocarbons. If such welding is required, extraordinary precautions must be taken on a case-by-case basis. Even in the absence of such hazard materials, any welding operation in a confined space is potentially lethal and requires continuous and copious dilution ventilation.

13.90.2 General Recommendations.

1) Choose hood designs in the following descending order of effectiveness: enclosing hoods, vacuum nozzles, fixed slot/plenum hood on a worktable or rectangular hood fixed above a worktable, moveable hood above a worktable, moveable hood hanging freely or overhead canopy, dilution ventilation.
2) Integrate planning for ventilation systems with planning for materials handling.
3) Place welding curtains or other barriers to block cross-drafts.
4) Install turntables, work rests, and other aids to improve utilization of the hoods.
5) Avoid recirculating filtered air from welding hoods back into occupied spaces unless the welding is low-hazard and produces low quantities of gaseous contaminants.
6) Face velocity for enclosing hoods should be 100 to 130 fpm [0.5 to 0.65 m/s] with the higher values used for poor conditions such as high cross-draft velocities.
7) Capture velocity for non-enclosing hoods should be 100 to 170 fpm [0.5 to 0.85 m/s] with the higher values used for poor conditions such as high cross-draft velocities and with higher hazard levels.

Enclosing hoods are by far the most effective in controlling welding contaminants; however, they restrict access and force reconsideration of material and product handling. Capturing hoods are less effective than enclosures but for low hazard conditions can be adequate if properly used.

13.90.3 Dilution (General) Ventilation. Dilution ventilation should be used to complement local exhaust hoods. However, dilution ventilation may be adequate without local exhaust ventilation under the following conditions:

1) It is not a confined space,
2) The contaminants are relatively low toxicity (e.g., welding on mild steel or iron without coatings on the steel or unusual rod coatings),
3) Any impediments to air movement are at least 7 feet [2.1 m] away (e.g., welding curtains),
4) The welding is interrupted by long periods of other activities (e.g., setup, material movement, etc.) so that the total generation rate is much less than would occur in "production" welding,
5) For very short periods of welding with moderately toxic contaminants if an appropriate NIOSH-approved respirator is properly worn and maintained.

General ventilation may not be sufficient for toxic materials. Local ventilation and respiratory protection equipment will probably be necessary. Even if low toxicity welding, it is best to use local exhaust ventilation if at all possible rather than rely on general ventilation.

Airflow requirements for low-toxicity, non-production, non-confined space dilution ventilation depend on 1) the rate of generation, and 2) the effectiveness of cross-drafts in dispersing the welding effluent. For such welding in open areas where welding fume can easily rise away from the breathing zone:

Q_t = 800 acfm × lb/hour [0.40 am^3/s × 0.45 kg/hour] of the rod used

For such welding in open areas where welding fume is moderately blocked by welding curtains and other obstructions to cross-drafts that are not closer than about 7 feet [2.1 m] the airflow requirements should be roughly twice as much:

Q_t = 1600 acfm × lb/hour [0.80 am^3/s × 0.45 kg/hour] of the rod used

13.90.4 Movable Capturing Hoods. Use mobile hood system (see VS-90-02) if it is necessary to move the hood to keep it close enough to the point of weld as the welder moves about his or her work station.

Welders may forget or choose not to move the hood to keep it close enough to be effective. This is especially likely if the capture distance (X) is relatively short (e.g., less than 12"). For low values of X the velocity gradient is extreme. Therefore, moving the hood an inch or two closer has substantial effects on shield gas. Moving it an inch or two further out substantially reduces its effectiveness.

The need for hood movement can be reduced by using relatively large values of X in selecting airflows and by making the hood wide. The former reduces required up and back movements and the latter reduces the need for lateral movements. Increasing X greatly increases airflow requirements. Increasing the width moderately increases airflow requirements.

If the hood can be fixed in place, consider doing it since it would reduce the likelihood that the hood is left at an excessive distance from the weld.

The hood in VS-90-02 can be positioned within the range of the flexible ducts connecting it to a fan. There are commercially available portable hoods that have the hood and duct system both on a wheeled platform, typically with an air-cleaning device as well. The airflow requirements in VS-90-02 should apply to these hoods. The advantage of these hoods is the ability to provide temporary ventilation almost anywhere an electrical cord can reach.

Some concerns in using a portable hood include:

1) The portable hood "scoop" is typically small and the airflows are typically relatively low. For that reason, the hood must be kept very close to the weld to be effective, producing the problems described above
2) The air cleaner may be ineffective in some models even when new and the air cleaner in all of them must be properly maintained to remain effective.

13.90.5 Slot Hoods. The side panels on the slot "welding" (it could be used for many other tasks) hood (see VS-90-1) are intended to block cross-drafts. If recommended airflows are provided for a given slot length and depth of the work bench, then the design capture velocity will be provided or exceeded at any location on the bench. Thus, a welder can weld at any location on the bench, a great advantage over small capturing hoods.

Some concerns in using a slot hood include:

1) If welding the front of a bulky object, the object itself may block the flow such that the control velocity may be low. In addition, the plume could rise directly into the welder's face before flowing back to the slots.
2) The top slot must be located well above the height of the tallest object that will be welded to prevent the rising plume from escaping. It is desirable to have at least two slots, one low (e.g., 12" high) and one higher (e.g., 24" high).
3) The inward velocity close to the slots may be high enough to disrupt shield gases. The velocity gradient is close to linear (as opposed to quadratic for the non-slot capturing hoods), so welding can be done perhaps halfway between the front edge of the table and the slots.

13.90.6 Bench Top Enclosing Hoods. It is generally assumed that enclosing hoods are more effective than capturing hoods. The disadvantage of enclosing hoods is reduced access, making it more difficult to move materials to be welded and more difficult to access more than one side of an object to be welded. Some objects may be too large to fit into the hoods. If the user finds that his or her job is made more difficult by the hood, he or she may sabotage, jerry-rig "improvements," or fail to use it when welding.

For those reasons, the design and use of an enclosing hood requires careful thought and consideration. The hood should be designed for a large fraction of jobs, but it is often a mistake to try to make one hood fit every possible welding task. Very large objects should be welded elsewhere and ventilated using a different strategy. However, most objects that are smaller than the dimensions of the hood can be welded in an enclosing hood.

Likewise, problems of inconvenience often can be overcome by modifying the materials handling systems and by designing the hood to enhance usability. For example, convenient access to all sides of many objects to be welded can be obtained by installing a rotating work platform on the floor of the hood. Hoisted items can be set on to a heavy-duty roll-out platform. Some long, narrow objects can be passed through small holes in each side of the hood to allow welding at different points along their width.

The hoods also can be made more convenient for the welder by installing holders for welding rods and rests for the electrical cables. In some cases, the electrical cables can be passed through an eye-bolt that is screwed into the hood ceiling, reducing the dead weight of the cables.

Although it is generally true that enclosing hoods are more reliably successful than any other type of hood, it is possible that capturing hoods can be more effective than the bench top hood under some conditions. That is most likely when the welding operation requires the welder to put his head even partially into a concave object. On the other hand, if the welder is likely to fail to keep the capturing hood close enough to the source, then an enclosing hood is likely to work better in practice.

Face velocities for enclosing hoods should be 100 to 130 fpm through the enclosure face area.

13.90.7 Specialty Welding Hoods. VS-90-03 to VS-90-30 show designs for hoods for specific types of welding and welding-related activities.

REFERENCE

13.90.1 American Welding Society (AWS D. 1-72), P.O. Box 351040, Miami, Florida 33135.

FIGURE VS-90-01

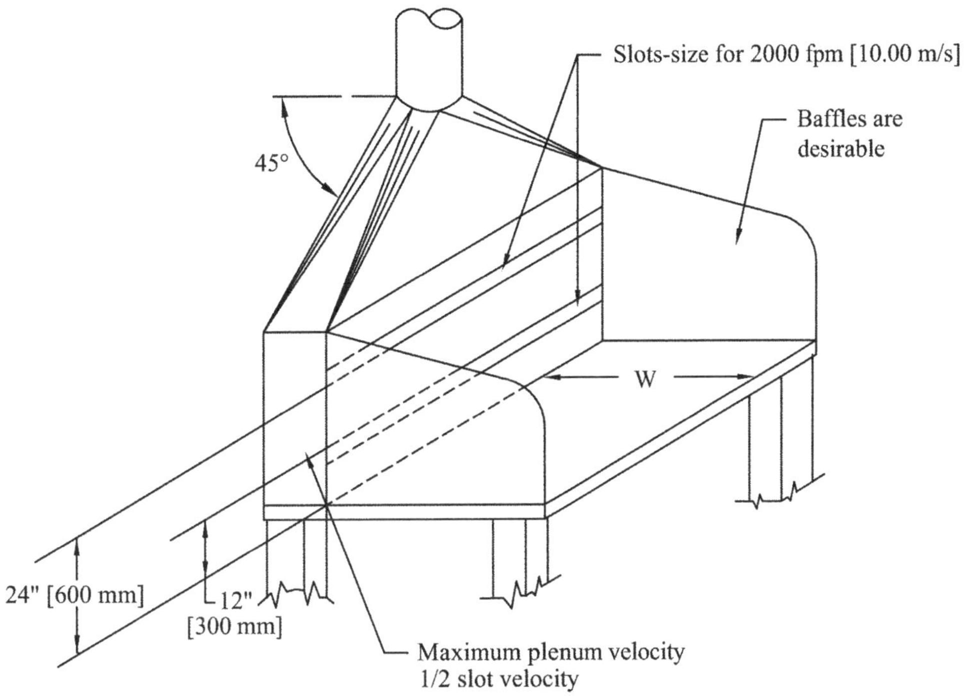

Q = 350 acfm/ft [0.54 am^3/s/m] of hood length
Hood length = required working space
W = 24" [600 mm] maximum, if W greater than 24" [600 mm] see Chapter 3
Minimum duct velocity = 2000 fpm [10.00 m/s]
$h_e = 1.78\ VP_s + 0.25\ VP_d$

General ventilation, where local exhaust cannot be used:

Rod, diam., in. [mm]		acfm/welder [am³/s/welder]	
5/32	[4]	1000	[0.50]
3/16	[5]	1500	[0.75]
1/4	[6]	3500	[1.75]
3/8	[9]	4500	[2.25]

or

A. For open areas, where welding fume can rise away from the breathing zone:
Q required (acfm) = 800 × lb/hour rod used
[Q required = 0.83 m^3/s/kg/hour rod used]

B. For enclosed areas or positions where fume does not readily escape breathing zone:
Q required (acfm) = 1600 × lb/hour rod used
[Q required = 1.7 m^3/s/kg/hour rod used]

For toxic materials higher airflows are necessary and operator may require respiratory protection equipment.

Other types of hoods
Local exhaust: See VS-90-02
Booth: For design see VS-90-30
Q = 100 acfm/ft^2 [0.50 am^3/s/m^2] of face opening
MIG welding may require precise airflow control

ACGIH®

TITLE
WELDING VENTILATION
BENCH HOOD

FIGURE VS-90-01

DATE 1-12

CHECK CODES, REGULATIONS, AND LAWS (LOCAL, STATE, AND NATIONAL) TO ENSURE THAT DESIGN IS COMPLIANT.

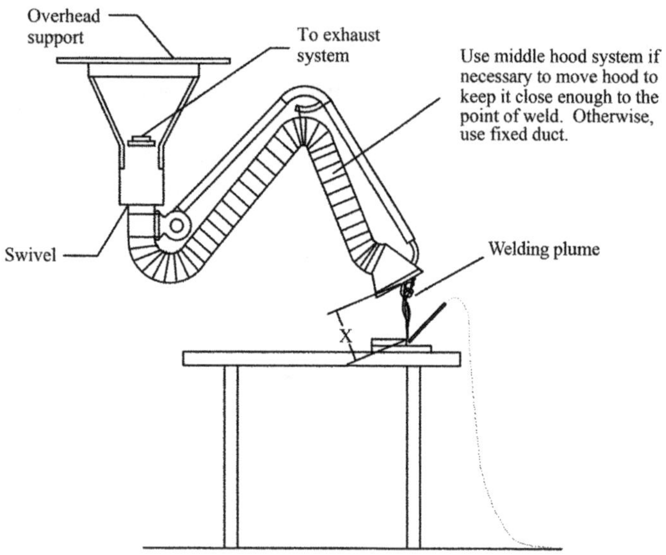

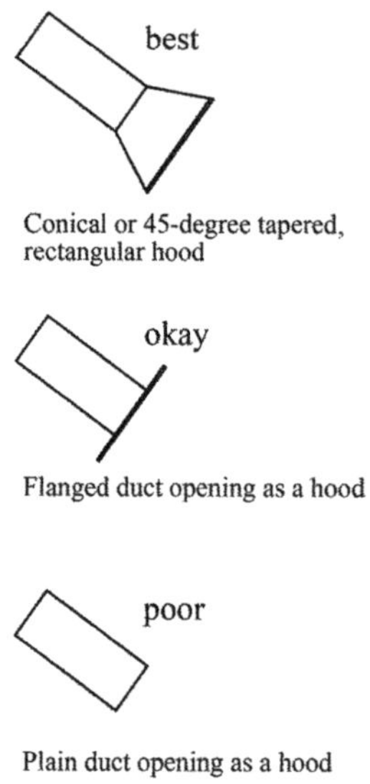

FIGURE VS-90-02

APPLICATION OF MOVEABLE CAPTURING HOODS

1. Not suitable for confined spaces unless coupled with appropriate use of respiratory protection and copious dilution ventilation with good mixing
2. Not suitable for highly hazardous contaminants (use enclosing hood instead)
3. The distance of the center of the hood face to the weld (X) is critical. The welder must be diligent in keeping the hood close to the weld for these hoods to be effective.

RECOMMENDED EXHAUST RATES (Q_t) AND DETERMINATION OF HOOD DIMENSIONS

$Q_t = 125\ C_f\ (10x^2 + \text{Area}_{face})$ acfm $[0.63\ C_f (10x^2 + A_{face})\ \text{am}^3/\text{s}]$

$C_f = 0.75$ for flanged or tapered hood

$C_f = 1.0$ for plain duct

X, inches [mm]		Plain duct acfm	[am³/s]	Flange or cone acfm	[am³/s]
6"	[150]	380	[0.19]	280	[0.14]
9"	[225]	770	[0.39]	580	[0.29]
12"	[300]	1350	[0.68]	1000	[0.50]

1. Q_t values above assume that the hood is elevated above the welding source to account for the rising plume. If the hood is level with the source, Q_t must be much higher.
2. Q_t values above are based on low toxicity welding. For moderate toxicity welding, increasing airflows by 20–50% (see also VS-90-01) may be sufficient. For still higher toxicity welding, an enclosing hood (VS-90-30) and respiratory protection may be necessary.
3. Velocities above 150 fpm [0.75 m/s] at the point of the weld may disturb shield gas.
4. The greater the width of the hood the less the hood must be moved left to right, which is often awkward. The hood width should be greater than the distance, X, from the face of the hood to the source.
5. The velocity gradient becomes extreme for X-distances less than 10" [250 mm]. It is best to design for X greater than 16" [400 mm].

DUCT SIZE SELECTION AND COMPUTATION OF SP_h

1. Minimum design velocity for ducts should be about 2,500 – 3,000 fpm [12.50 – 15.00 m/s].
2. The hood static pressure is computed from the entry loss coefficient (F_h) and the duct velocity pressure (VP):

$$SP_h = (1 + F_h)\ VP$$

Hood	F_h
Plain duct	0.93
Flanged duct	0.56
Conical	0.15
45-rectangular	0.25

ACGIH®

TITLE CAPTURING HOOD FOR LOW TOXICITY WELDING

FIGURE VS-90-02

DATE 1-12

CHECK CODES, REGULATIONS, AND LAWS (LOCAL, STATE, AND NATIONAL) TO ENSURE THAT DESIGN IS COMPLIANT.

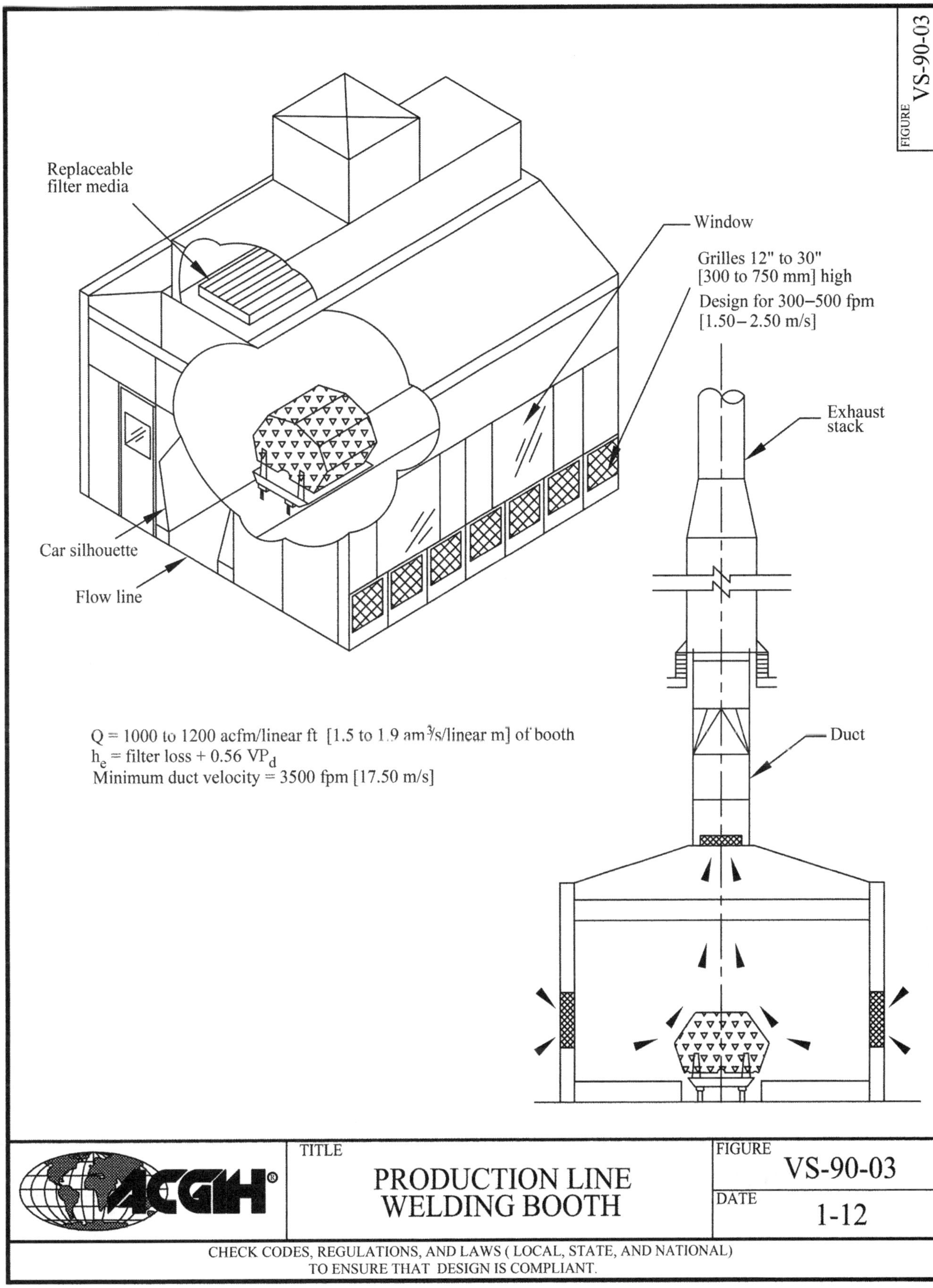

ACGIH®	TITLE: PRODUCTION LINE WELDING BOOTH	FIGURE: VS-90-03 / DATE: 1-12

CHECK CODES, REGULATIONS, AND LAWS (LOCAL, STATE, AND NATIONAL) TO ENSURE THAT DESIGN IS COMPLIANT.

FIGURE VS-90-10

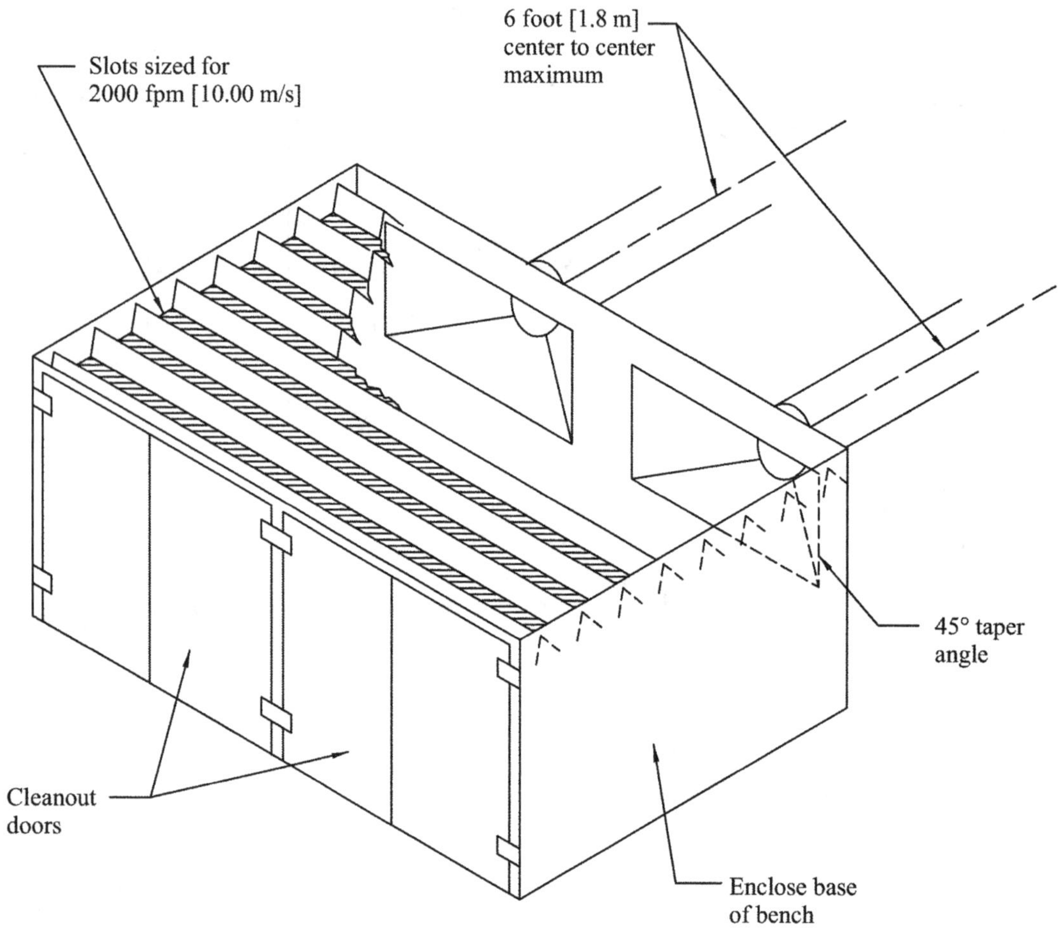

Q = 150 acfm/ft^2 [0.75 am^3/s/m^2] of gross bench area
Minimum duct velocity = 4000 fpm [20.00 m/s]
$h_e = 1.78\ VP_s + 0.25\ VP_d$

ACGIH®	TITLE TORCH CUTTING VENTILATION	FIGURE VS-90-10 DATE 1-12

CHECK CODES, REGULATIONS, AND LAWS (LOCAL, STATE, AND NATIONAL) TO ENSURE THAT DESIGN IS COMPLIANT.

FIGURE VS-90-20

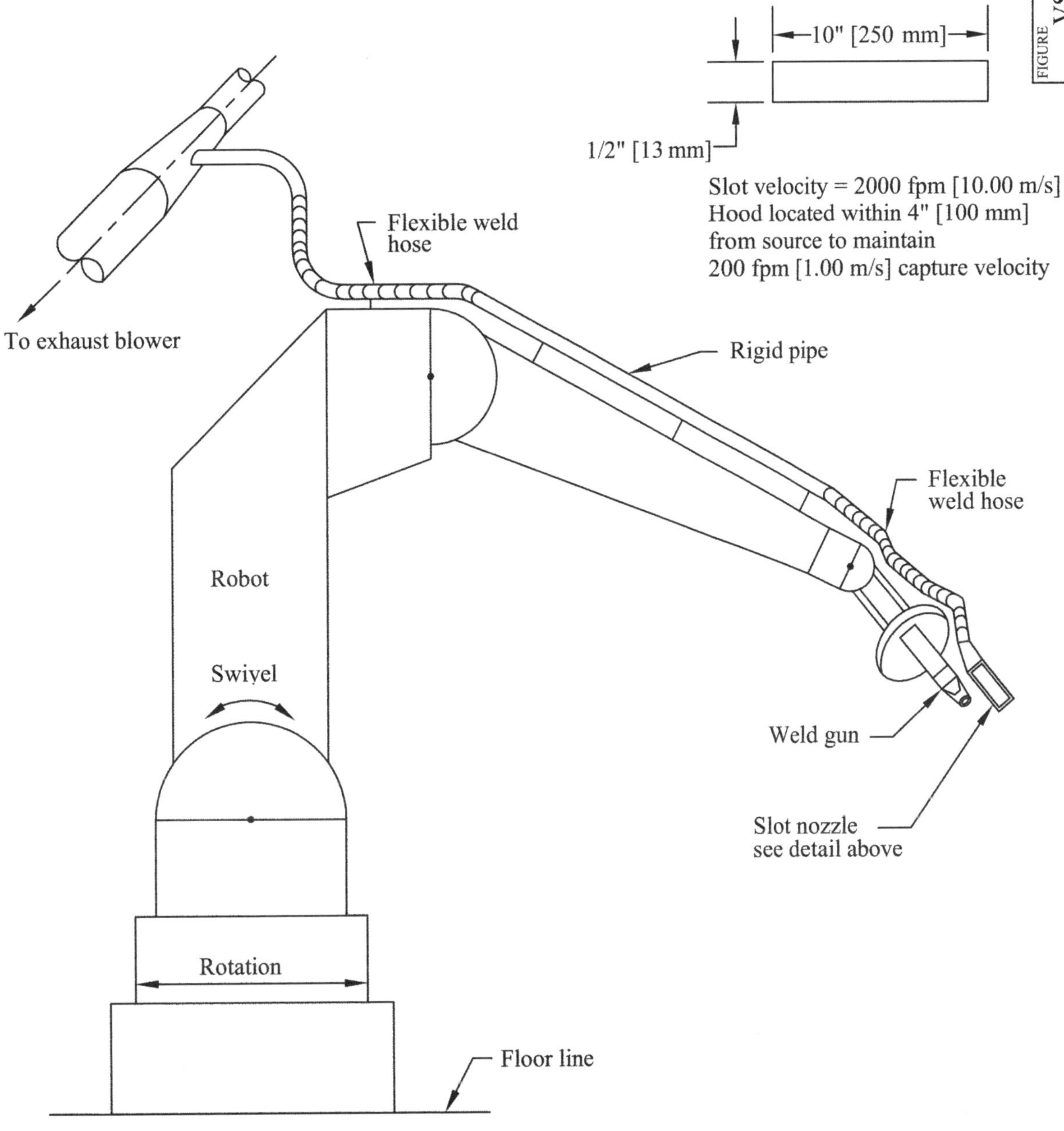

$h_e = 1.78\ VP_s + 0.25\ VP_d$

Minimum duct velocity = 3500 fpm [17.50 m/s]

ACGIH®	TITLE: ROBOTIC WELDING	FIGURE: VS-90-20
		DATE: 1-12

CHECK CODES, REGULATIONS, AND LAWS (LOCAL, STATE, AND NATIONAL) TO ENSURE THAT DESIGN IS COMPLIANT.

FIGURE VS-90-30

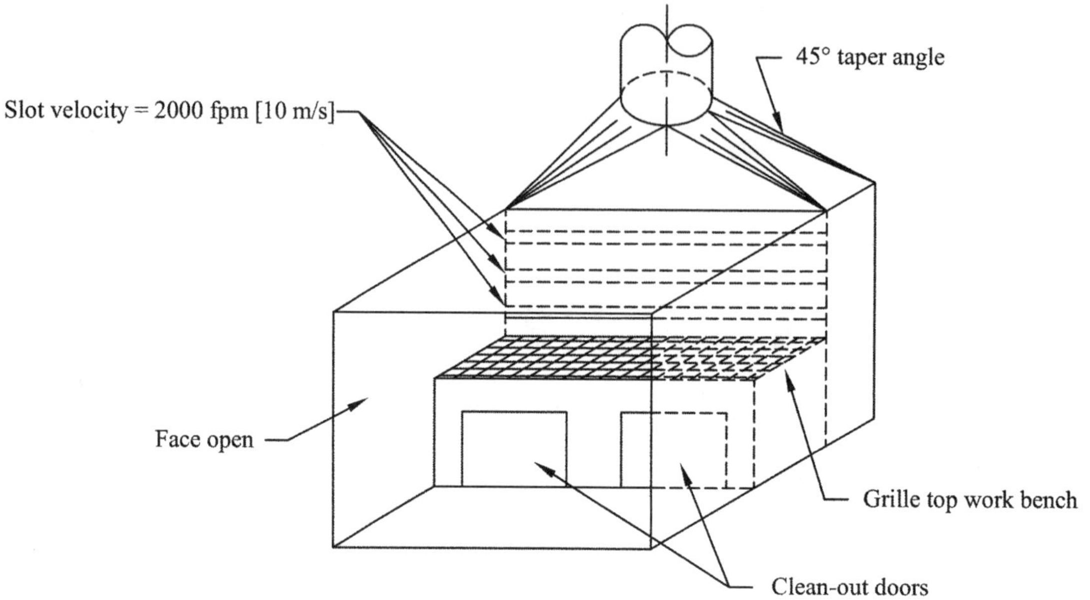

METALLIZING BOOTH

Non-toxic: $Q = 150$ acfm/ft^2 [0.75 am^3/s/m^2] face area

Toxic: Provide appropriate NIOSH certified respirator
$Q = 200$ acfm/ft^2 [1.0 am^3/s/m^2] face area

Minimum duct velocity = 3500 fpm [17.50 m/s]
$h_e = 1.78\ VP_s + 0.25\ VP_d$
Small lathe, etc., may be mounted in booth

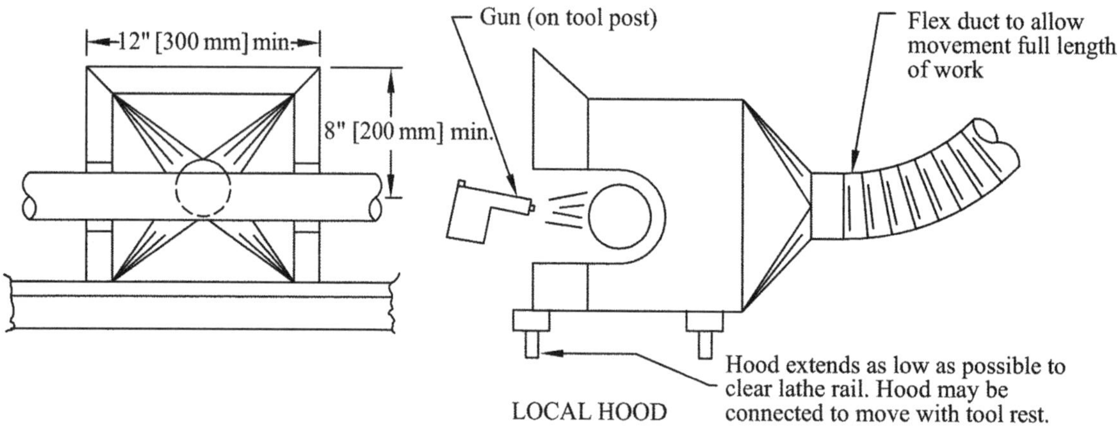

LOCAL HOOD

Note: Local hood may not be satisfactory for spraying toxic metals.
$Q = 200$ acfm/ft^2 [1.0 am^3/s/m^2] face openings
Minimum duct velocity = 3500 fpm [17.50 m/s]
$h_e = 0.25\ VP_d$

ACGIH®	TITLE: METAL SPRAYING	FIGURE: VS-90-30 / DATE: 1-12

CHECK CODES, REGULATIONS, AND LAWS (LOCAL, STATE, AND NATIONAL) TO ENSURE THAT DESIGN IS COMPLIANT.

13.95 WOODWORKING

Woodworking equipment generates large amounts of wood dust by abrasive or cutting action. It is important to provide good ventilation for all equipment as the broad particle size distribution of wood dust creates the potential for health, housekeeping problems, and fire hazards. Excessive amounts of dust, if allowed to accumulate inside equipment and in shop areas, can create fire or explosion hazards. An additional consideration should be the toxicity of the wood species used.

In many instances, woodworking equipment, such as saws and sanders, generates airflow patterns that make dust control difficult. Exhaust hoods should enclose the operation as much as possible. Where the equipment tends to eject wood dust (e.g., at sanding belt pulleys), the exhaust hood should be placed in the ejection path.

Enclosures must incorporate cleanout doors to prevent dust build-up. Duct velocities should be maintained at a minimum of 3500 fpm to prevent settling and subsequent clogging of the duct. Do not use auto-closing dampers or valves to turn off branches in tapered main systems unless minimum transport velocity can be maintained in all ducts at all times. (No purge cleaning.)

Exhaust flow rates will vary with equipment type and size. Design data are provided for a number of operations in VS-95-01 through VS-95-20 and in Table 13-95-1. The exhaust flow rates for many hoods shown in this section were developed based on the specific configuration shown in the drawings. The drawings show well-designed hoods that may not be found in woodworking equipment purchased "off-the-shelf." Readers are cautioned to evaluate the configuration of the specific equipment intended for use in their shop. In many cases, the recommended exhaust flow rates or hood static pressure must be increased to accommodate single exhaust ports, smaller duct connections, non-tapered entries, openings in equipment bases, etc. Additional information for hand held sanders using Low Volume-High Velocity (LV-HV) can be found in sub-section 13.40. Where information for a specific operation is not provided, data for similar listed operations can be used.

REFERENCES

13.95.1 Hampl, V.; Johnston, O.: Control of Wood Dust from Horizontal Belt Sanding. American Industrial Hygiene Association Journal, 46, 10, pp. 567-577 (1985).

13.95.2 Hampl, V.; Johnston, O.: Control of Wood Dust from Disc Sanders, Applied Occupational Hygiene, 6(11): 938-944 (November 1991).

13.95.3 Topmiller, J.L.; Watkins, D.A.; Schulman, S.A.; Murdock, D.J.: Controlling Wood Dust from Orbital Hand Sanders. Applied Occupational Hygiene, 11(9): 1131-1138 (September 1996).

13.95.4 Hampl, V.; Toppmiller, J.L.; Watkins, D.S.; Murdock, D.J.: Control of Wood Dust from Rotational Hand Sanders. Applied Occupational and Environmental Hygiene, 7(4): 263-270 (April 1992).

TABLE 13-95-1 (IP). Miscellaneous Woodworking Machinery Not Given in VS Prints

The following list of recommended exhaust volumes is for average-sized woodworking machines and is based on many years of experience. It must be noted that some modern, high-speed or extra-large machines may produce such a large volume of waste that greater exhaust must be used. Similarly, some small machines of the home workshop or bench-type may use less exhaust air than listed.

Self-Feed Table Rip Saw

	Exhaust Flow Rate, acfm		
Saw Diameter, Inches	**Bottom**	**Top**	**Total**
up to 16	440	350	790
over 16	550	350	900
self-feed not on table	800	500	1350

Gang Rip Saws

	Exhaust Flow Rate, acfm		
Saw Diameter, inches	**Bottom**	**Top**	**Total**
up to 24	550	354	900
over 24 to 36	800	440	1240
over 36 to 48	1100	550	1550
over 48	1400	550	2060

Vertical Belt Sanders

rear belt & both pulleys enclosed and

Top Run Horizontal Sanders

Belt Width, inches	**Exhaust Flow Rate, acfm**
up to 6	440
over 6 to 9	550
over 9 to 14	800
over 14	1100
Swing Arm Sander	2200

Single Planers or Surfacers

Planer Size	**Exhaust Flow Rate, acfm**
up to 20 in knives	785
over 20 to 26 in knives	1100
over 26 to 32 in knives	1400
over 32 to 38 in knives	1765
over 38 in knives	2200

Double Planers or Surfacers

	Exhaust Flow Rate, acfm		
Planer Size, inches	**Bottom**	**Top**	**Total**
up to 20 in knives	550	785	1335
over 20 to 26 in knives	785	1100	1885
over 26 to 32 in knives	1100	1400	2500
over 32 to 38 in knives	1400	1800	3200
over 38 in knives	1400	2200	3600

Moulders, Matchers, & Sizers

	Exhaust Flow Rate, acfm			
Size, inches	**Bottom**	**Top**	**Right**	**Left**
up to 7	440	550	350	350
over 7 to 12	550	800	440	440
over 12 to 18	800	1100	550	550
over 18 to 24	1100	1400	800	800
over 24	1400	1770	1100	1100

	Exhaust Flow Rate, acfm
Sash Stickers	500
Wood Shapers	440 to 1400
Tenoner	Same as moulder
Automatic Lathe	800 to 5000
Forming Lathe	350 to1400
Chain Mortise	350
Dowel Machine	350 to 500
Panel Raiser	550
Dovetail and Lock Corner	550 to 800
Pulley Sockets	550
Pulley Stile	550
Glue Jointer	800
Gainer	350 to1400
Router	350 to 800
Hogs	
up to 12" wide	1400
over 12" wide	3100
Floor Sweep	
6" to 8" diameter	800 to 1400

TABLE 13-95-1 (SI). Miscellaneous Woodworking Machinery Not Given in VS Prints

The following list of recommended exhaust volumes is for average-sized woodworking machines and is based on many years of experience. It must be noted that some modern, high-speed or extra-large machines may produce such a large volume of waste that greater exhaust must be used. Similarly, some small machines of the home workshop or bench-type may use less exhaust air than listed.

Self-Feed Table Rip Saw

	Exhaust Flow Rate, am^3/s		
Saw Diameter, mm	Bottom	Top	Total
up to 400	0.22	0.18	0.40
over 400	0.28	0.18	0.45
self-feed not on table	0.40	0.28	0.68

Gang Rip Saws

	Exhaust Flow Rate, am^3/s		
Saw Diameter, mm	Bottom	Top	Total
up to 600	0.28	0.18	0.45
over 600 to 900	0.40	0.22	0.62
over 900 to 1200	0.55	0.28	0.83
over 1200	0.70	0.43	1.00

Vertical Belt Sanders

rear belt & both pulleys enclosed and

Top Run Horizontal Sanders

Belt Width, mm	Exhaust Flow Rate, am^3/s
up to 150	0.22
over 150 to 225	0.28
over 225 to 350	0.40
over 350	0.55
Swing Arm Sander	1.10

Single Planers or Surfacers

Planer Size, mm	Exhaust Flow Rate, am^3/s
up to 500 mm knives	0.40
over 500 to 650 mm knives	0.55
over 650 to 800 mm knives	0.70
over 800 to 950 mm knives	0.88
over 950 mm knives	1.10

Double Planers or Surfacers

	Exhaust Flow Rate, am^3/s		
Planer size, mm	Bottom	Top	Total
up to 500 mm knives	0.28	0.39	0.67
over 500 to 650 mm knives	0.39	0.55	0.94
over 650 to 800 mm knives	0.55	0.70	1.25
over 800 to 950 mm knives	0.70	0.90	1.60
over 950 mm	0.70	1.10	1.80

Moulders, Matchers, & Sizers

	Exhaust Flow Rate, am^3/s			
Size, mm	Bottom	Top	Right	Left
up to 180	0.22	0.28	0.18	0.18
over 189 to 300	0.28	0.40	0.22	0.22
over 300 to 450	0.40	0.55	0.28	0.28
over 450 to 600	0.55	0.70	0.40	0.40
over 600	0.70	0.89	0.55	0.55

	Exhaust Flow Rate, am^3/s
Sash Stickers	0.25
Wood Shapers	0.22 to 0.70
Tenoner	Same as moulder
Automatic Lathe	0.40 to 2.5
Forming Lathe	0.18 to 0.70
Chain Mortise	0.18
Dowel Machine	0.18 to 0.25
Panel Raiser	0.28
Dovetail and Lock Corner	0.28 to 0.40
Pulley Sockets	0.28
Pulley Stile	0.28
Glue Jointer	0.4
Gainer	0.18 to 0.70
Router	0.18 to 0.40
Hogs	
up to 12" wide	0.07
over 12" wide	1.55
Floor Sweep	
6" to 8" diameter	0.40 to 0.70

FIGURE VS-95-01

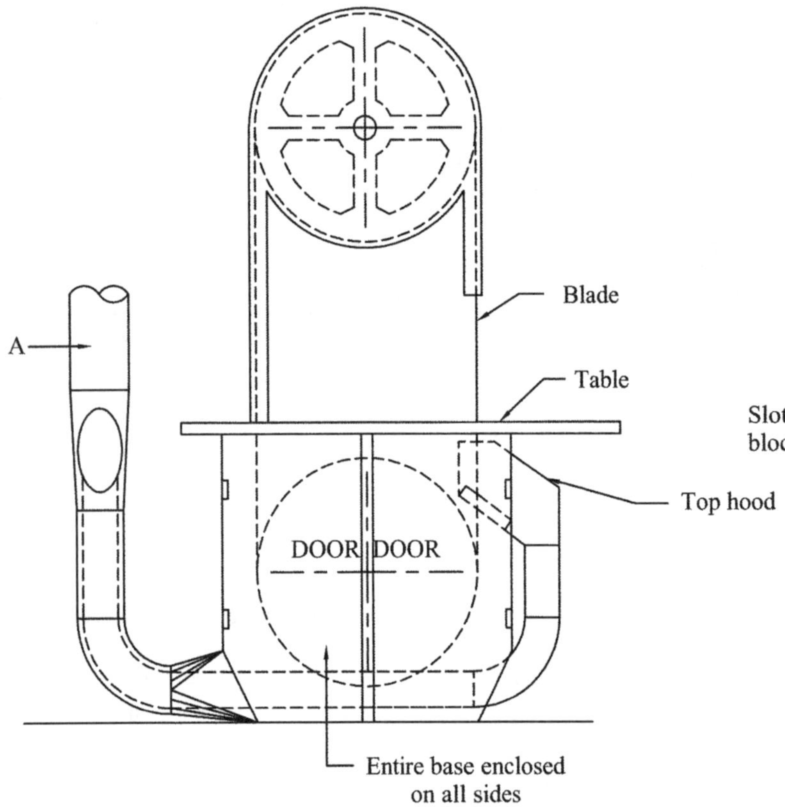

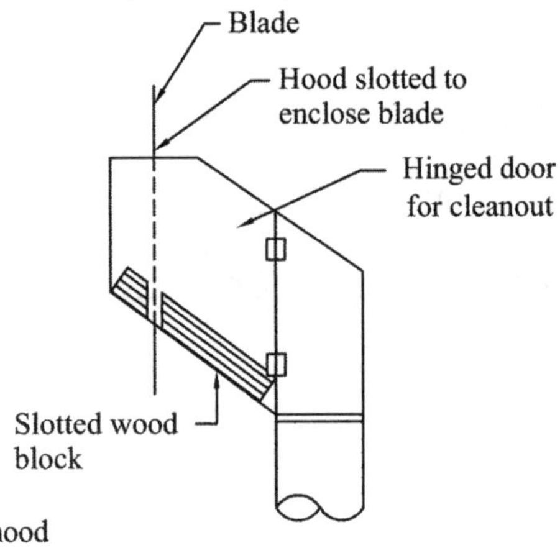

TOP HOOD DETAIL

Blade width inches [mm]	Exhaust flow rate, acfm [am³/s]		
	Bottom	Top	Total
Up to 2 [Up to 50]	350 [0.18]	350 [0.18]	700 [0.35]
2 to 3 [50 to 75]	350 [0.18]	550 [0.28]	900 [0.45]
3 to 4 [75 to 100]	550 [0.28]	800 [0.40]	1350 [0.68]
4 to 6 [100 to 150]	550 [0.28]	1100 [0.55]	1650 [0.83]
6 to 8 [150 to 200]	550 [0.28]	1400 [0.70]	1950 [0.98]

Minimum duct velocity = 3500 fpm [17.50 m/s]
$h_e = 1.75\ VP_d$ (Point Ⓐ in duct riser)

ACGIH®	TITLE	FIGURE VS-95-01
	BAND SAW	DATE 1-12

CHECK CODES, REGULATIONS, AND LAWS (LOCAL, STATE, AND NATIONAL) TO ENSURE THAT DESIGN IS COMPLIANT.

FIGURE VS-95-02

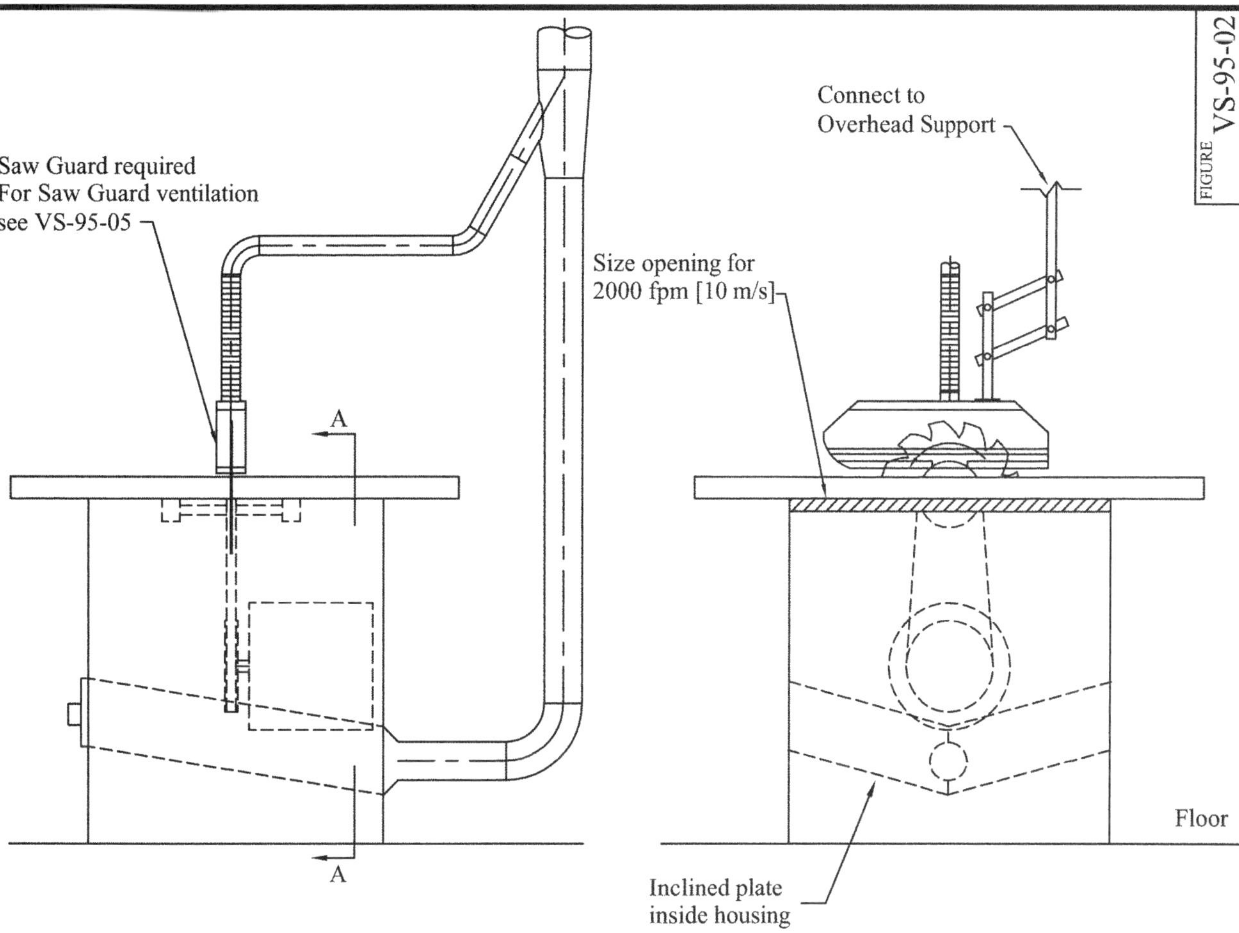

Saw blade diameter inches [mm]		Exhaust flow rate acfm [am³/s]	
		Base	Guard
up to 16	[up to 400]	545 [0.28]	100 [0.05]
over 16	[over 400]	785 [0.39]	100 [0.05]
Saw with dado blade		785 [0.39]	100 [0.05]

Minimum duct velocity = 4000 fpm [20.00 m/s]

$h_e = F_s\,VP_s + 0.25\,VP_d$

Shape	F_s
	1.78
	1.00
	0.49

ACGIH®

TITLE: FLOOR TABLE SAW

FIGURE: VS-95-02

DATE: 1-12

CHECK CODES, REGULATIONS, AND LAWS (LOCAL, STATE, AND NATIONAL) TO ENSURE THAT DESIGN IS COMPLIANT.

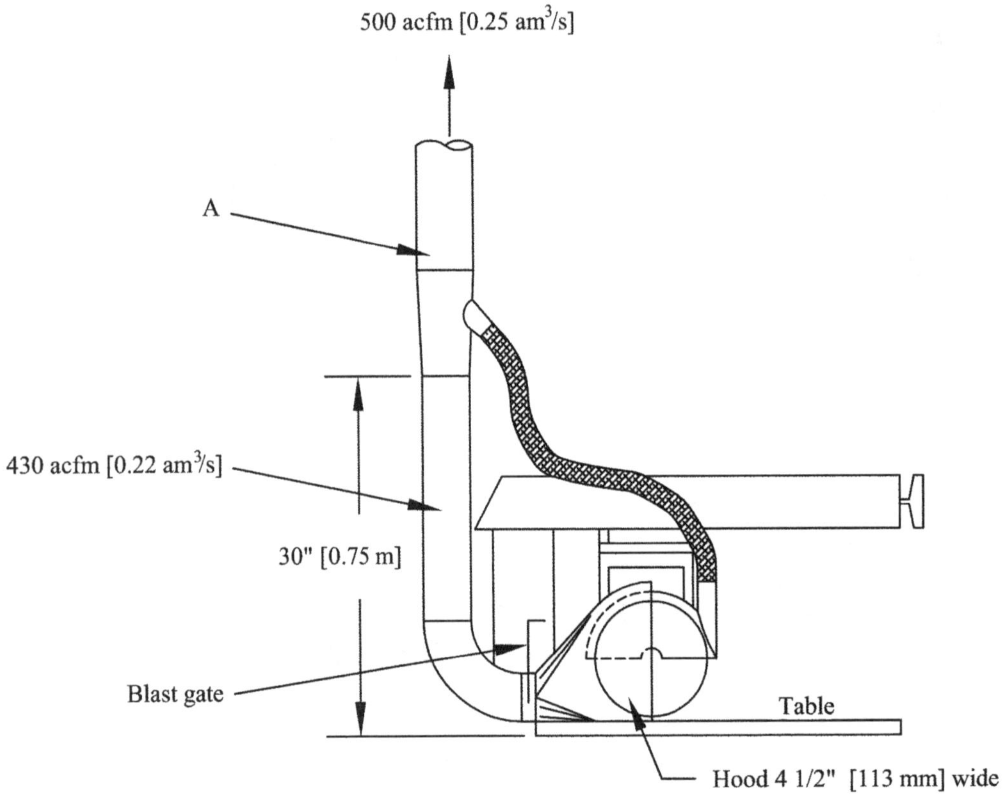

Minimum duct velocity = 4000 fpm [20.00 m/s]
$h_e = 3.5\ VP_d$ (point A in duct riser)

For booth enclosure, see VS-80-17

Note: Design cannot be used for diagonal or longitudinal cuts

ACGIH®	TITLE RADIAL ARM SAW	FIGURE VS-95-03 DATE 1-12

CHECK CODES, REGULATIONS, AND LAWS (LOCAL, STATE, AND NATIONAL) TO ENSURE THAT DESIGN IS COMPLIANT.

FIGURE VS-95-04

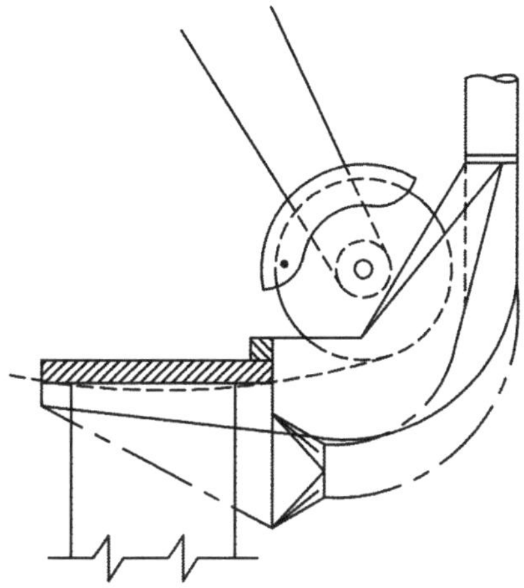

Type of hood where table is cut through

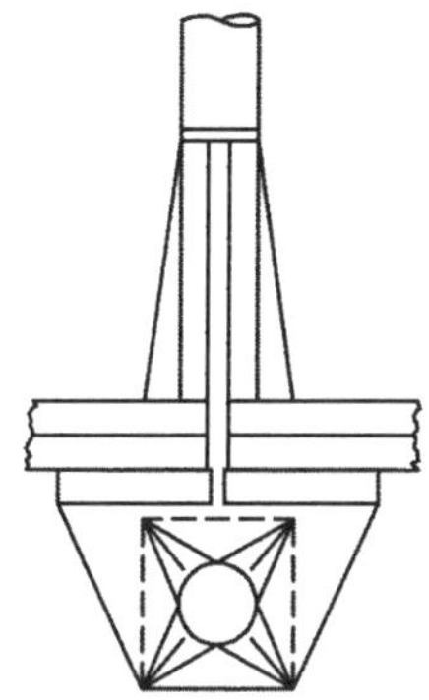

Front view of hood

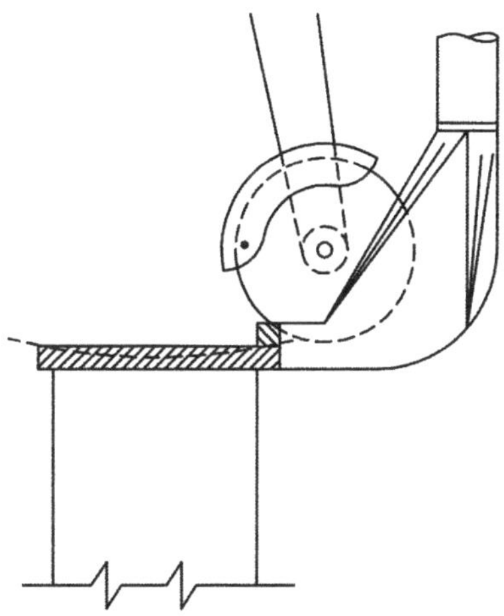

Type hood where table is not cut through

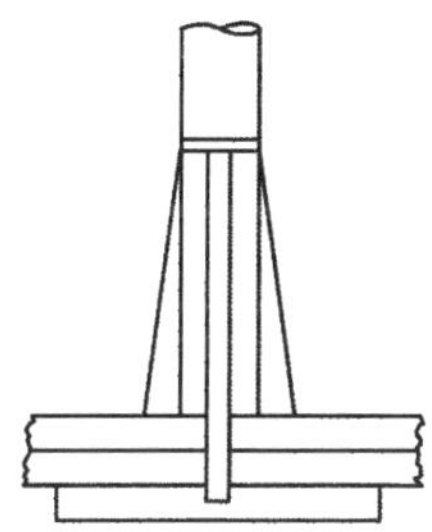

Front view of hood

Saw diameter inches	[mm]	Exhaust flow rate acfm	$[am^3/s]$
Up to 20 incl.	[Up to 500 incl.]	350	[0.18]
over 20	[over 500]	440	[0.22]

Minimum duct velocity = 4000 fpm [20.00 m/s]

$h_e = 1.78\ VP_s + 0.25\ VP_d$

ACGIH®

TITLE: SWING SAW

FIGURE: VS-95-04

DATE: 1-12

CHECK CODES, REGULATIONS, AND LAWS (LOCAL, STATE, AND NATIONAL) TO ENSURE THAT DESIGN IS COMPLIANT.

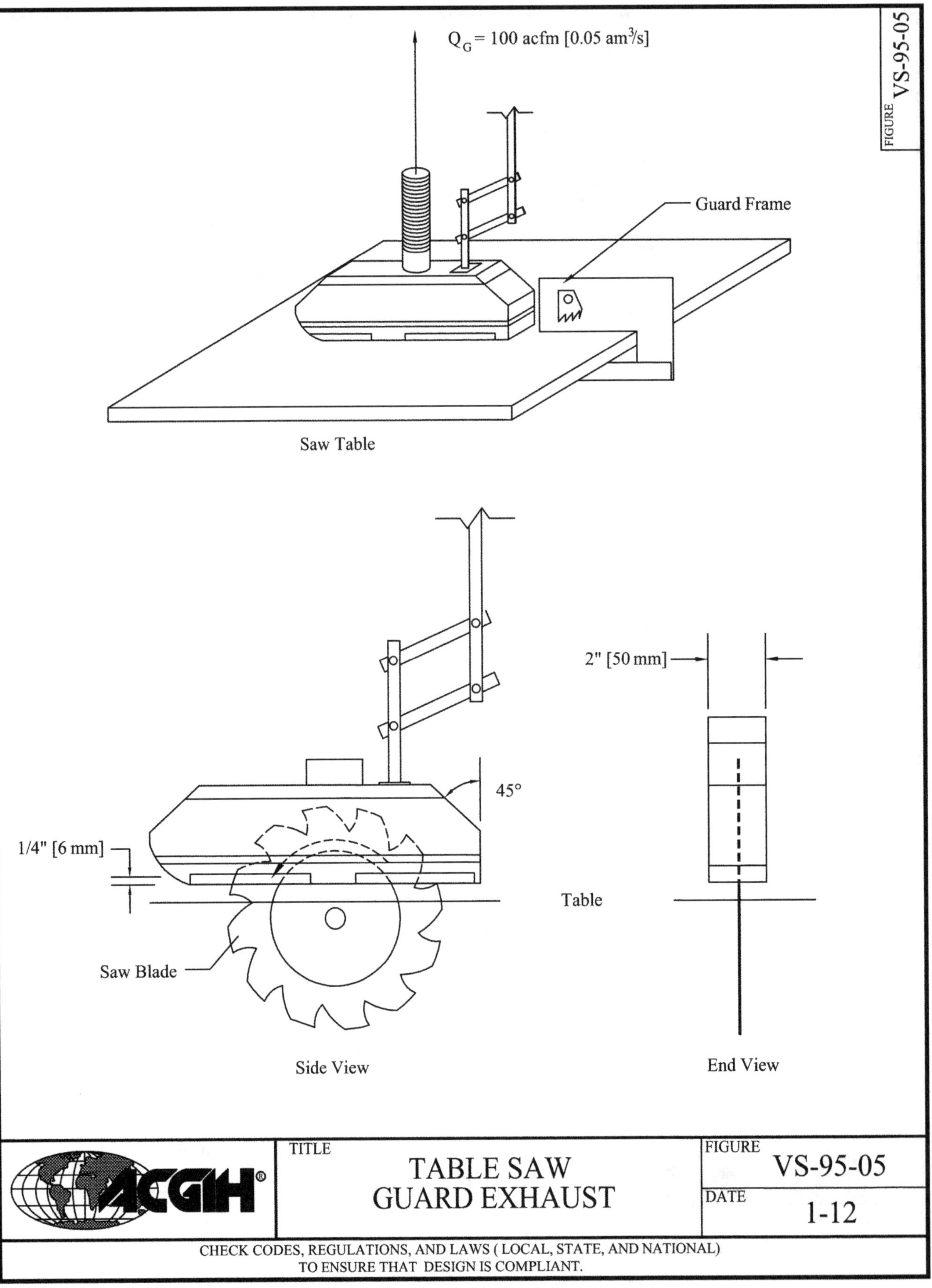

TITLE	FIGURE
TABLE SAW GUARD EXHAUST	VS-95-05
	DATE 1-12

CHECK CODES, REGULATIONS, AND LAWS (LOCAL, STATE, AND NATIONAL) TO ENSURE THAT DESIGN IS COMPLIANT.

FIGURE VS-95-07

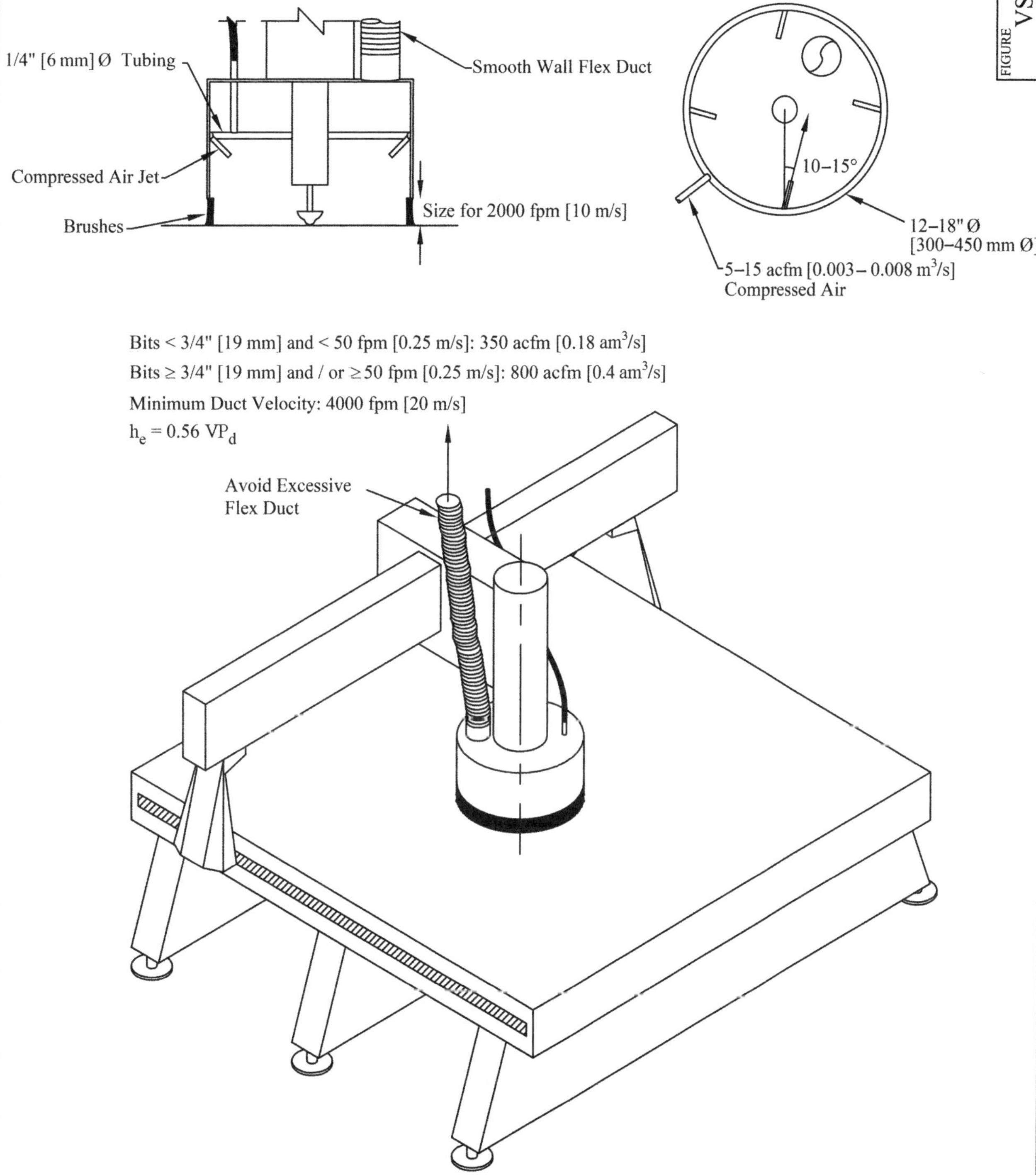

ACGIH®

TITLE	FIGURE
CNC ROUTER	VS-95-07
	DATE 1-12

CHECK CODES, REGULATIONS, AND LAWS (LOCAL, STATE, AND NATIONAL)
TO ENSURE THAT DESIGN IS COMPLIANT.

FIGURE VS-95-10

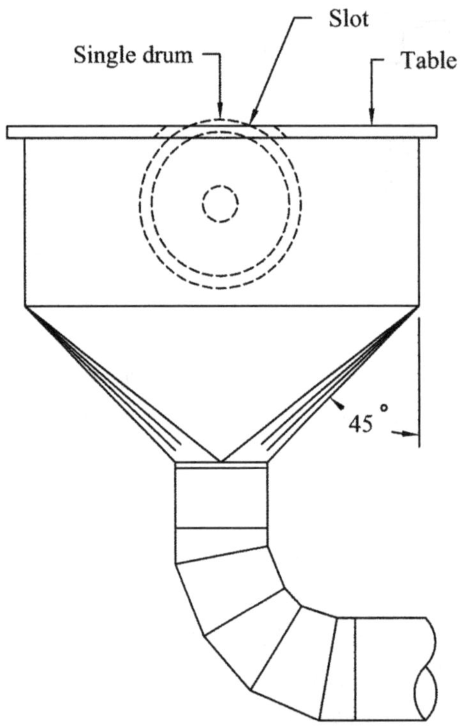

Drum surface square inches [m^2]		Exhaust flow rate acfm [am^3/s]	
Up to 200 (and less than 10" diam.)	[Up to 0.14] [and less than 250 mm diam.]	350	[0.18]
200 to 400	[0.14 to 0.28]	550	[0.28]
400 to 700	[0.28 to 0.49]	790	[0.40]
700 to 1400	[0.49 to 0.98]	1100	[0.55]
1400 to 2400	[0.98 to 1.68]	1400	[0.70]

Minimum duct velocity = 3500 fpm [17.50 m/s]
Entry loss depends on hood design.
$h_e = 1.78\ VP_s + 0.25\ VP_d$ as illustrated

ACGIH®

TITLE
SINGLE DRUM SANDER

FIGURE VS-95-10

DATE 1-12

CHECK CODES, REGULATIONS, AND LAWS (LOCAL, STATE, AND NATIONAL) TO ENSURE THAT DESIGN IS COMPLIANT.

FIGURE VS-95-11

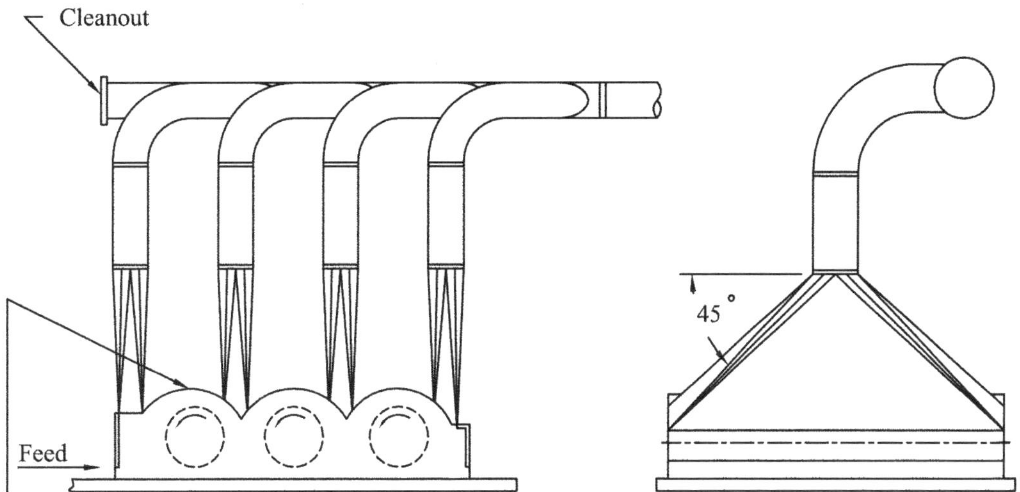

Exhaust flow rates

Drum length, in.	[mm]	Total exhaust for machine acfm/drum	[am³/s] *
Up to 31	[Up to 775]	550	[0.28]
31 to 49	[775 to 1225]	790	[0.40]
49 to 67	[1225 to 1675]	1100	[0.55]
over 67	[over 1675]	1400	[0.70]
Brush rolls	Brush rolls	350 acfm at brush	[0.18 am³/s at brush]

*Note: Provide one more take-off than the number of drums.

Minimum duct velocity = 3500 fpm [17.50 m/s]
$h_e = 0.25\ VP_d$

ACGIH®

TITLE: MULTIPLE DRUM SANDER

FIGURE: VS-95-11

DATE: 1-12

CHECK CODES, REGULATIONS, AND LAWS (LOCAL, STATE, AND NATIONAL) TO ENSURE THAT DESIGN IS COMPLIANT.

FIGURE VS-95-12

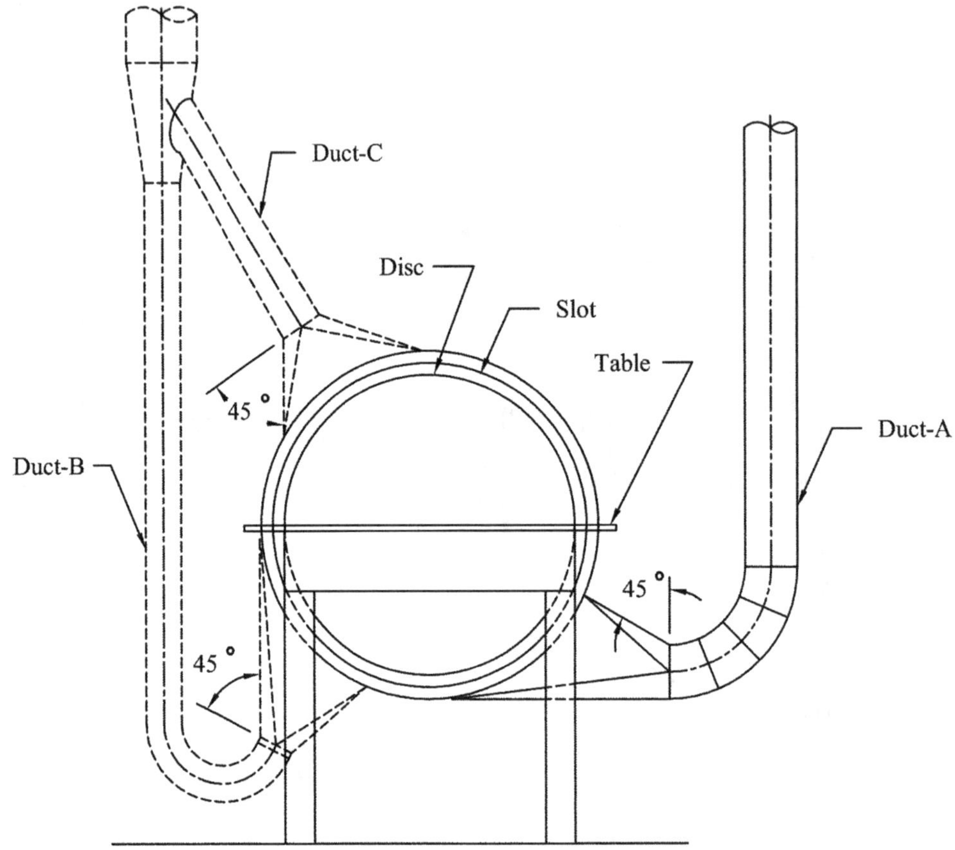

Disc diameter inches	[mm]	Total exhaust flow rate acfm	[am³/s]	Applies to duct
Up to 12	[Up to 300]	350	[0.18]	A
12 to 18	[300 to 450]	440	[0.22]	A
18 to 26	[450 to 650]	550	[0.28]	A
26 to 32	[650 to 800]	700*	[0.35]*	A-B
32 to 38	[800 to 950]	900*	[0.45]*	A-B
38 to 48	[950 to1200]	1300**	[0.65]**	A-B-C

* Two bottom branches.
** One top and two bottom branches.

Minimum duct velocity = 3500 fpm [17.50 m/s]
Entry loss depends on hood design.
$h_e = 1.0\ VP_s + 0.25\ VP_d$ as illustrated

ACGIH®

TITLE: DISC SANDERS

FIGURE: VS-95-12

DATE: 1-12

CHECK CODES, REGULATIONS, AND LAWS (LOCAL, STATE, AND NATIONAL) TO ENSURE THAT DESIGN IS COMPLIANT.

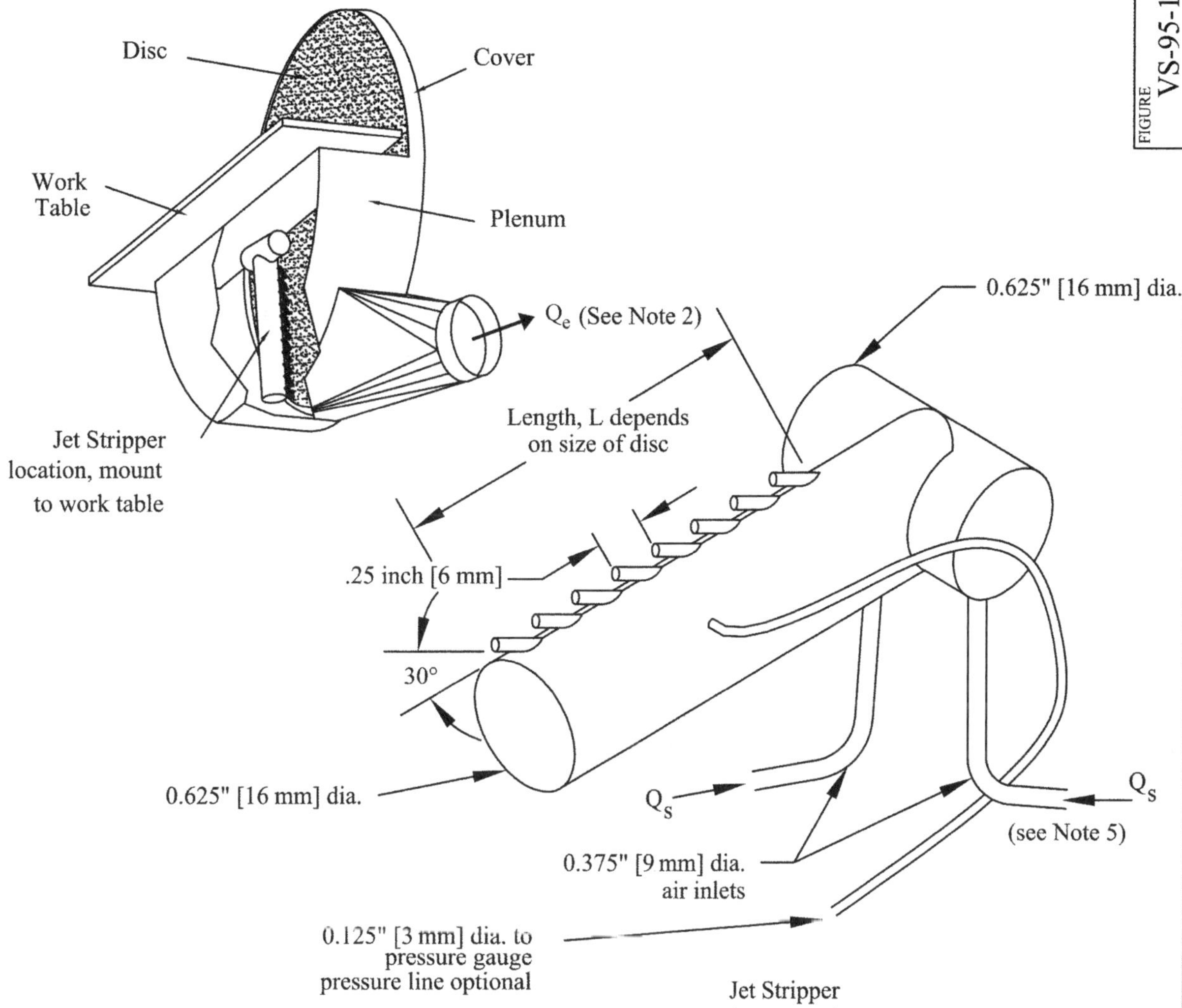

NOTES:

1. Jet stripper may be added as shown. It is used in addition to the ventilation specified in VS-95-12.
2. Q_e is same as shown in VS-95-12.
3. Clearance between stripper nozzle and disc is 0.25" [6 mm].
4. Nozzle opening diameter is 0.035" [1 mm].
5. Total airflow (Q_s) to stripper = 1.8 L (L in inches) acfm [0.000033 L (L in mm) am^3/s]
6. Stripper inlet pressure = 15 psi [100 kPa].

The Jet Stripper is patented (#5,099,616). Use of the device on an individual basis (not for sale) is permissible. For commercial production of the device contact NOISH, 1-800-35 NOISH to obtain information regarding licensing.

Reference 13.95.2

ACGIH®	TITLE JET STRIPPER FOR DISC SANDER	FIGURE VS-95-12a DATE 1-12

CHECK CODES, REGULATIONS, AND LAWS (LOCAL, STATE, AND NATIONAL) TO ENSURE THAT DESIGN IS COMPLIANT.

FIGURE VS-95-12a

FIGURE VS-95-13

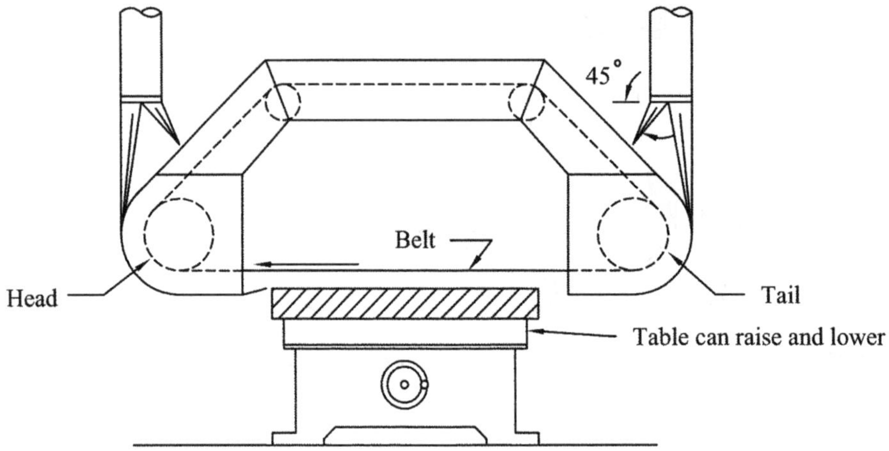

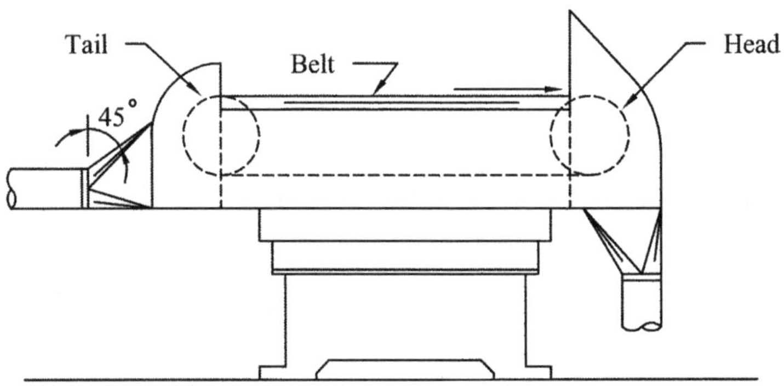

HORIZONTAL BELT SANDERS

Belt width inches [mm]	Exhaust flow rate acfm [am³/s]		
	Head end	Tail end	Total
up to 6 [up to 150]	440 [0.22]	350 [0.18]	790 [0.40]
6 to 9 [150 to 225]	550 [0.28]	350 [0.18]	900 [0.45]
9 to 14 [225 to 350]	800 [0.40]	440 [0.22]	1200 [0.60]
over 14 [over 350]	1100 [0.55]	550 [0.28]	1700 [0.85]

Minimum duct velocity = 3500 fpm [17.50 m/s]
$h_e = 0.40\ VP_d$

ACGIH®

TITLE: HORIZONTAL BELT SANDERS

FIGURE: VS-95-13

DATE: 1-12

CHECK CODES, REGULATIONS, AND LAWS (LOCAL, STATE, AND NATIONAL) TO ENSURE THAT DESIGN IS COMPLIANT.

FIGURE VS-95-14

Main Hood:
W = belt width, in.
Q_e = 75 acfm per inch [0.0014 am^3/s/mm] of belt width

Auxiliary Hood:
L_a = auxiliary hood length to extend 3" – 6" [75 – 150 mm] into end of sanding area
Q_a = 135 acfm per ft [0.21 am^3/s/mm] of hood length

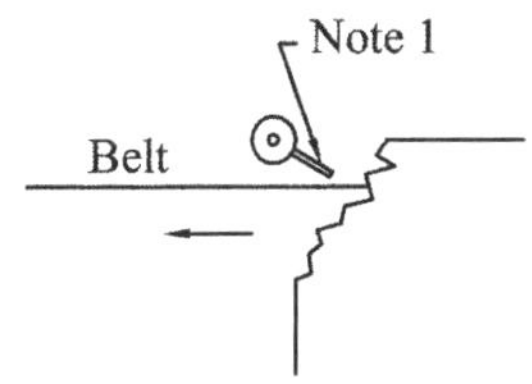

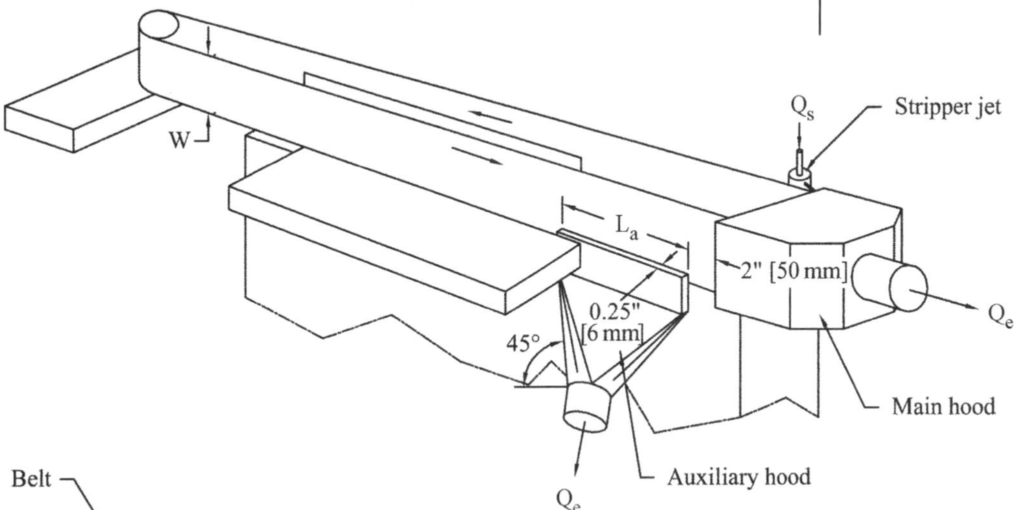

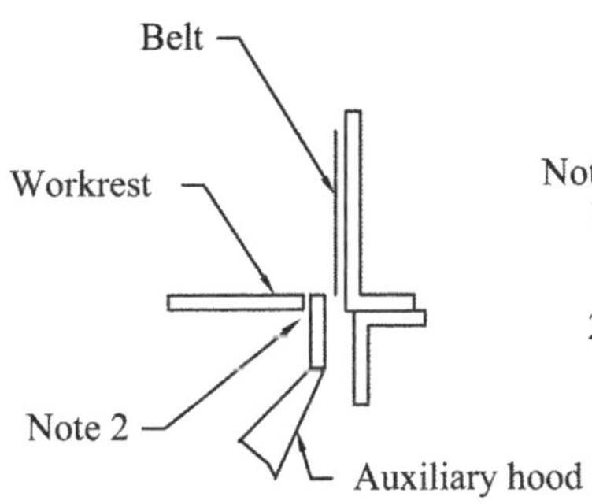

Notes:

1. Tips of stripper jets must be positioned 0.25" [6 mm] from belt and inside hood face.
2. Keep clearances between hood, belt and workrest to a minimum.

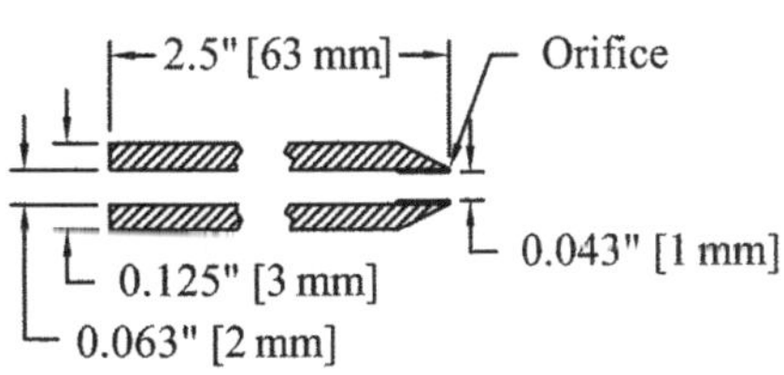

Tube and Orifice Detail

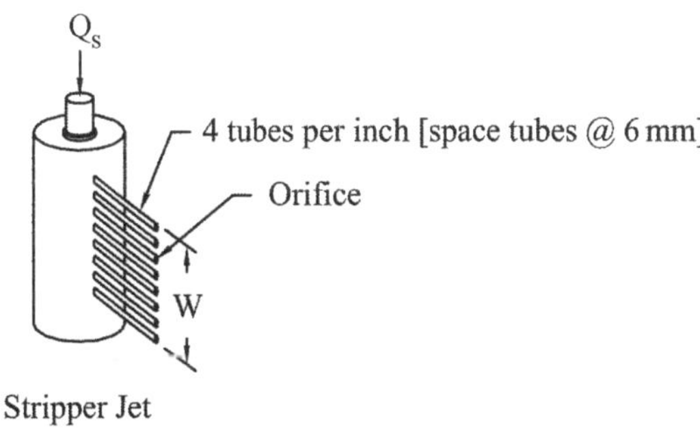

Stripper Jet

W = belt width, in. [mm]
Q_s = 0.5 acfm [0.00025 am^3/s] per tube, inlet pressure 10 – 12 psig [60 – 70 kPa]

Reference 13.95.1

ACGIH®

TITLE	FIGURE VS-95-14
HORIZONTAL BELT SANDER, STRIPPER SYSTEM	DATE 1-12

CHECK CODES, REGULATIONS, AND LAWS (LOCAL, STATE, AND NATIONAL) TO ENSURE THAT DESIGN IS COMPLIANT.

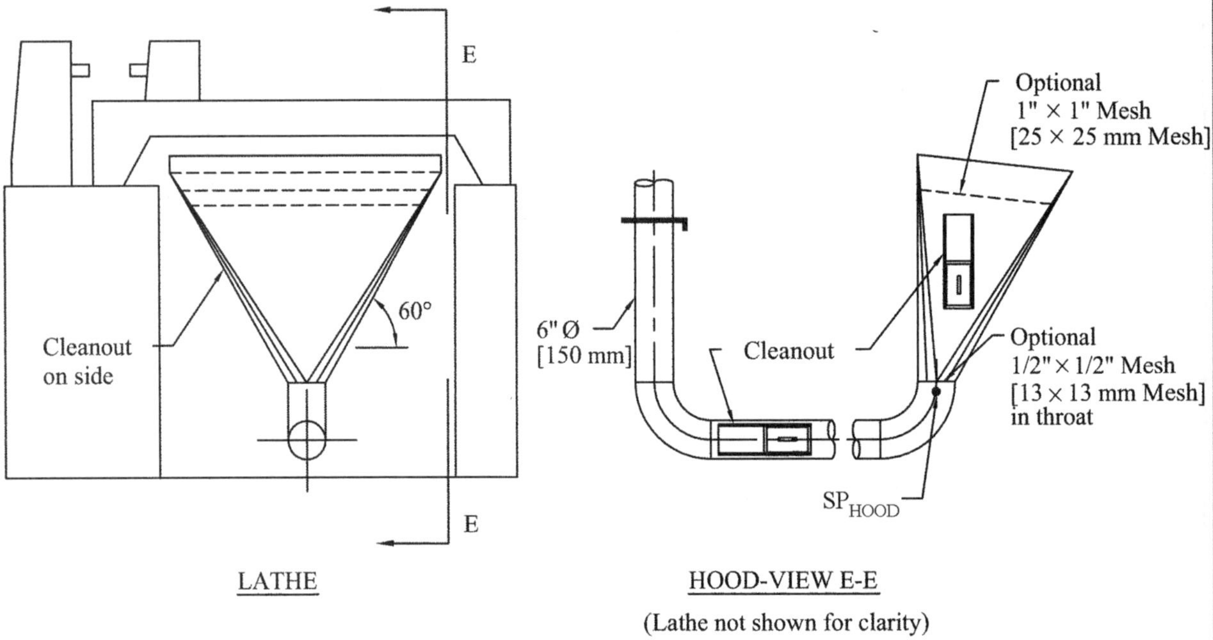

Q = 880 acfm [0.44 am^3/s]

Minimum duct velocity = 4000 fpm [20.00 m/s]

$h_e = .25\ VP_d$

Consider losses through screens, if used.

TITLE	FIGURE VS-95-15
WOODWORKING LATHE	DATE 1-12

CHECK CODES, REGULATIONS, AND LAWS (LOCAL, STATE, AND NATIONAL) TO ENSURE THAT DESIGN IS COMPLIANT.

FIGURE VS-95-15

FIGURE VS-95-20

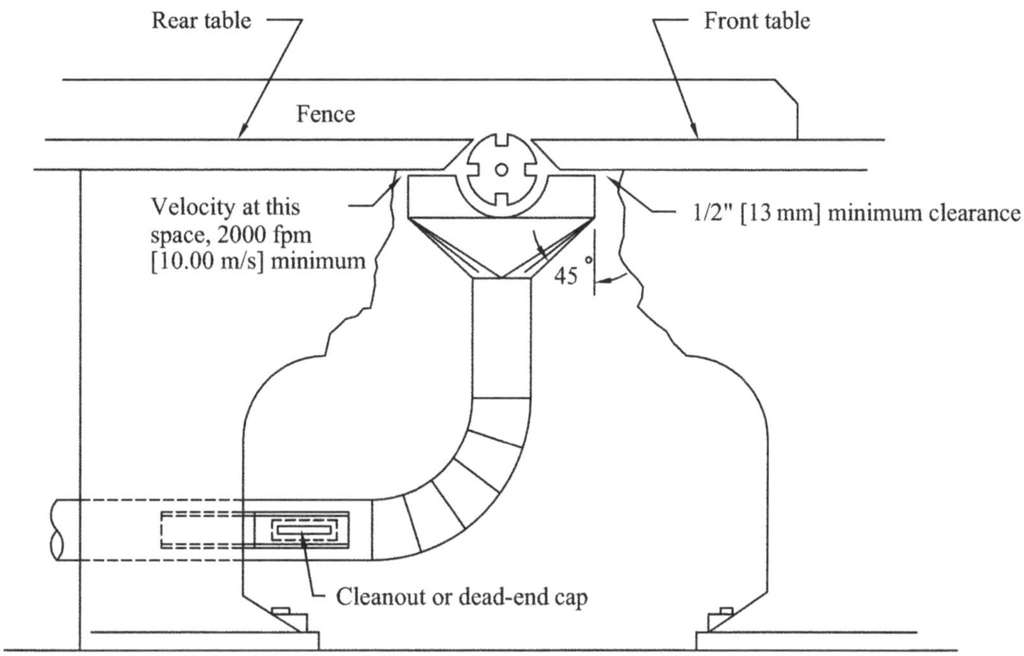

Knife length inches	[mm]	Exhaust flow rate acfm	[am^3/s]
Up to 6	[Up to 150]	350	[0.18]
6 to 12	[150 to 300]	440	[0.22]
12 to 20	[300 to 400]	550	[0.28]
over 20	[over 400]	800	[0.40]

Minimum duct velocity = 4000 fpm [20.00 m/s]
$h_e = 1.0\ VP_s + 0.25\ VP_d$

ACGIH®

TITLE
JOINTERS

FIGURE VS-95-20

DATE 1-12

CHECK CODES, REGULATIONS, AND LAWS (LOCAL, STATE, AND NATIONAL) TO ENSURE THAT DESIGN IS COMPLIANT.

FIGURE VS-95-30

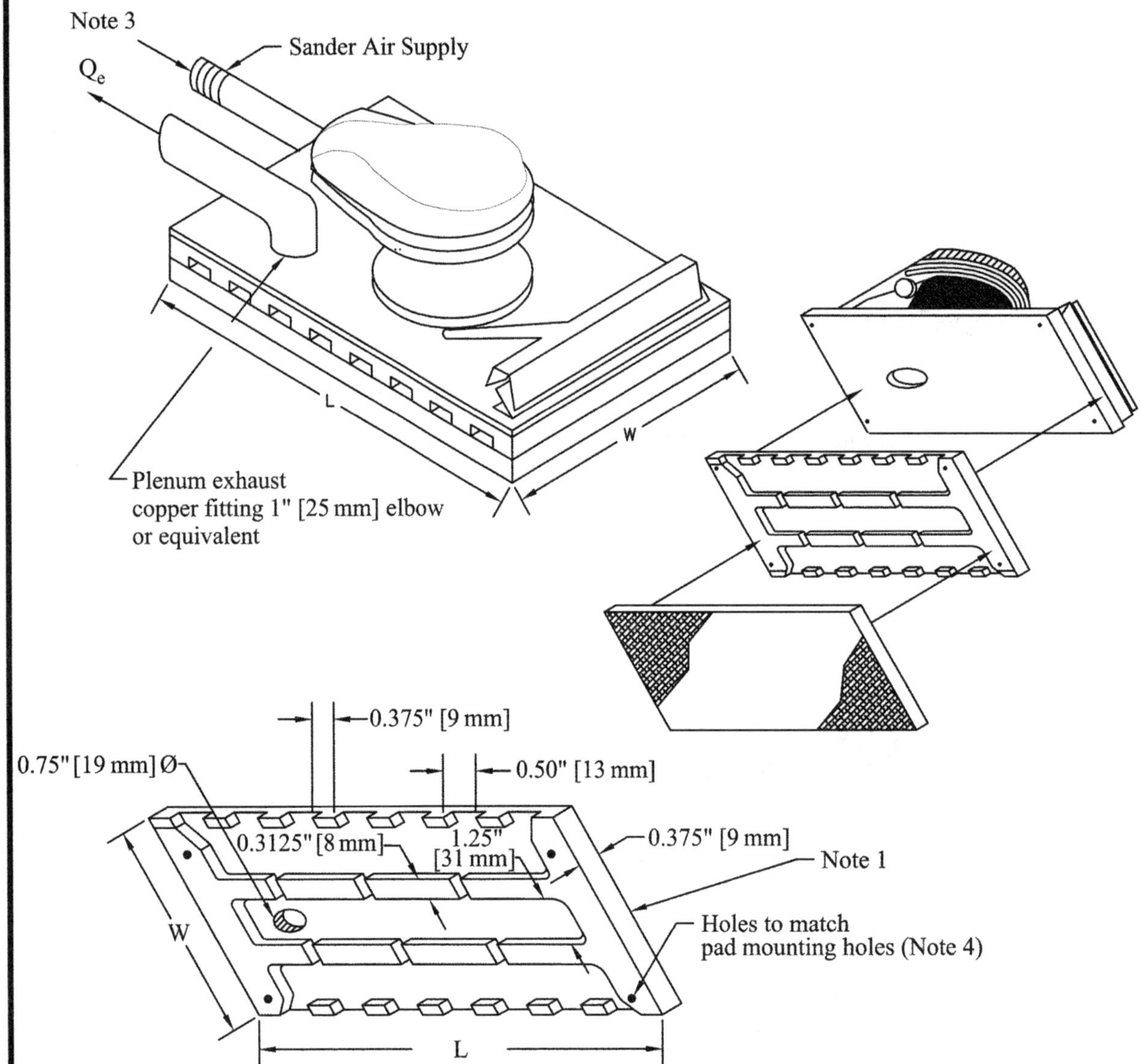

1. Plenum machined from metal or plastic.
2. Plenum can be used with air or electrically powered sander.
3. Q_e = 50 acfm [0.025 am^3/s] at approximately 25" wg [6.3 kPa].
4. Mounting holes located to match sander pad mounting holes.

The Exhaust Plenum Retrofit has a patent pending. Use of the device on an individual basis (not for sale) is permissible. For commercial production of the device contact NOISH, 1-800-35 NIOSH to obtain information regarding licensing.

Reference: 13.95.3

ACGIH®	TITLE EXHAUST PLENUM RETROFIT FOR ORBITAL HAND SANDER	FIGURE VS-95-30 DATE 1-12

CHECK CODES, REGULATIONS, AND LAWS (LOCAL, STATE, AND NATIONAL) TO ENSURE THAT DESIGN IS COMPLIANT.

FIGURE VS-95-31

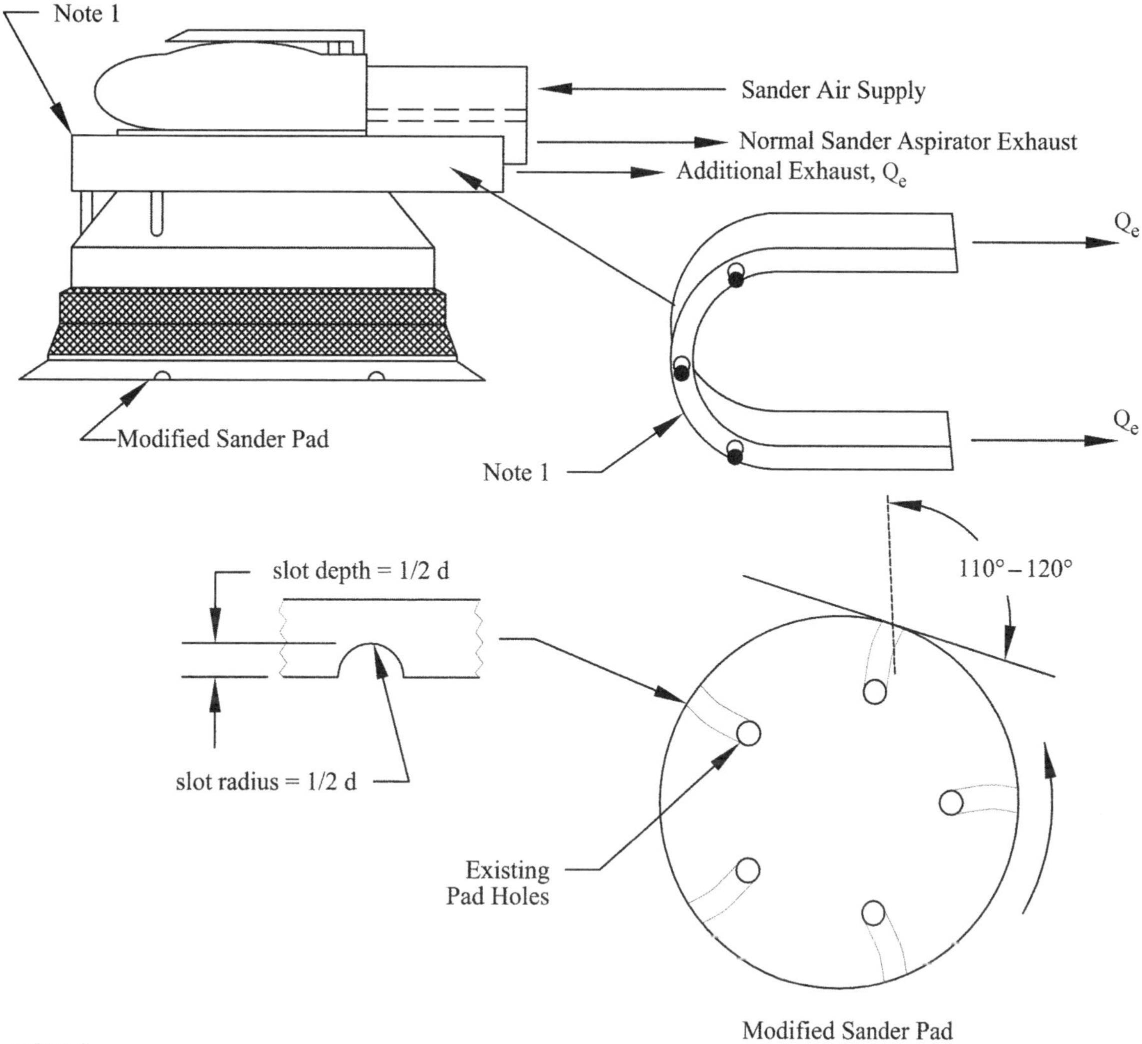

NOTES:
1. Additional exhaust plenum to fit sander.
2. Q_e = 50 acfm at 25" wg [0.25 am^3/s at 6.3 kPa]
3. Q_e is in addition to normal sander aspirator exhaust.
4. Additional exhaust may be supplied by standard shop vacuum cleaner.

The Auxiliary Exhaust Modification Configuration is patented (#5,105,858). Use of the device on an individual basis (not for sale) is permissible. For commercial production of the device on an individual basis contact NIOSH, 1-800-35 NIOSH for information regarding licensing.

Reference: 13.95.4

ACGIH®	TITLE AUXILLIARY EXHAUST RETROFIT FOR AIR POWERED RANDOM ORBITAL HAND SANDER	FIGURE VS-95-31 DATE 1-12

CHECK CODES, REGULATIONS, AND LAWS (LOCAL, STATE, AND NATIONAL) TO ENSURE THAT DESIGN IS COMPLIANT.

13.99 MISCELLANEOUS OPERATIONS

In the previous sections of the chapter, hood ventilation sketches were grouped together because they provided ventilation concepts for similar operations, used the same ventilation approach, or were applicable within the same industry. However, not all hood ventilation sketches are so easily categorized.

This section provides a location for those hood ventilation sketches that do not fit in other sections. Some have a unique application such as VS-99-04 for the Pistol Range. Others have such broad application that they could appear in many sections (e.g., the canopy hood in VS-99-03). In other cases, this section will be used for new ventilation sketches for a particular application or industry. Such sketches will reside in this section until other hood ventilation sketches are developed and a new section formed. Finally, this section will be used for tabular presentation of specific design parameters for a variety of operations that could not be adequately described in previous sections.

REFERENCES

13.99.1 Pennsylvania Department of Labor and Industry: Abrasive Wheel Manufacture. Safe Practices Bulletin No. 13.

13.99.2 Hartzell Propeller Fan Company, Bulletin 1001.

13.99.3 Goldfield, J.; Brandt. F.E.: Dust Control Techniques in the Asbestos Industry. A paper presented at the American Industrial Hygiene Conference, Miami Beach, FL (May 12-17, 1974).

13.99.4 Hama, G.M.: Ventilation Control of Dust from Bagging Operations. Heating and Ventilating (April, 1948).

13.99.5 Hutcheson, J.R.M.: Environmental Control in the Asbestos Industry of Quebec. C I MM Bulletin, Vol. 64, No. 712, pp. 83-89 (August 1971).

13.99.6 Private Communications, Occupational Health Protection Branch, Ontario Ministry of Labour, Ontario, Canada (October 1976).

13.99.7 Hama, G; Frederick, W.; Monteith, H.: Air Flow Requirements for Underground Parking Garages. American Industrial Hygiene Association Journal, Vol. 22, No. 6 (December 1961).

13.99.8 Kane, J.M.: Design of Exhaust Systems. Heating and Ventilating, 42, 68 (November 1945).

13.99.9 Oddie, W.M.: Pottery Dusts: Their Collection and Removal. Pottery Gazette, 53, 1280 (1928).

13.99.10 B. F. Sturtevant Company: What We Make. Catalog No. 500.

13.99.11 Brandt, A.D.: Industrial Health Engineering. John Wiley and Sons (New York, 1947).

13.99.12 Kane, J.M.: Foundry Ventilation. The Foundry (February and March 1946).

13.99.13 Kane, J.M.: Foundry Ventilation. University of Michigan Inservice Training Course (October 1945).

13.99.14 Fen, O.E.: The Collection and Control of Dust and Fumes from Magnesium Alloy Processing. Peters-Dalton, Inc. (January 1945).

13.99.15 Postman, B.F.: Practical Application of Industrial Exhaust Ventilation for the Control of Occupational Exposures. American Journal of Public Health, 30, 149 (1940).

13.99.16 New York Department of Labor: Rules Relating to the Control of Silica Dust in Stone Crushing Operations. Industrial Code Rule No. 34 (July 1942).

13.99.17 DallaValle, J.M.: Exhaust Hoods. Industrial Press, New York (1946).

13.99.18 Hatch, T., et al.: Control of the Silicosis Hazard in Hard Rock Industries. II. An Investigation of the Kelley Dust Trap for Use with Pneumatic Rock Drills of the Jackhammer Type. Journal of Industrial Hygiene, 14, 69 (February 1932).

13.99.19 Hay, P.S.: Modified Design of Hay Dust Trap. Journal of Industrial Hygiene, 12, 18 (January 1930).

13.99.20 Riley, E.C., et al.: How to Design Exhaust Hoods for Quartz-Fusing Operations. Heating and Ventilating, 37, 23 (April 1940).

13.99.21 Riley, E.C.; DallaValle, J.M.: A Study of Quartz-Fusing Operations with Reference to Measurement and Control of Silica Fumes. Public Health Reports, 54, 532 (1939).

13.99.22 Yaglou, C.P.: Ventilation of Wire Impregnating Tanks Using Chlorinated Hydrocarbons. Journal of Industrial Hygiene and Toxicology, 20, 401 (June 1938).

13.99.23 American Air Filter Co., Inc.: Usual Exhaust Requirements (for) Grain Elevators, Feed and Flour Mills, Louisville, KY (April 1956).

13.99.24 Gressel, M.G.: Hughes, R.T.: Effective Local Exhaust Ventilation for Controlling Formaldehyde Exposure During Embalming. Applied Occupational and Environmental Hygiene, 7(12): 840-845 (December 1992).

13.99.25 Estell, C.F.; Spencer, A.B.: Case Study: Control of Methylene Chloride Exposures During Commercial Furniture Stripping. American Industrial Hygiene Association Journal, 57(1):43-49 (1996).

13.99.26 Lauger, R.R.: How to Control Carcinogens in Chemical Production. Occupational Health and Safety (March-April 1977).

TABLE 13-99-1 (IP). Grain Elevators, Feed Mills, Flour Mills

The following data are offered as guides. Airflow rates can vary considerably depending on degree of enclosure, flow rate of material and dustiness of the grain. Minimum duct velocity = 3500 fpm. Ventilation control is desirable for the operations to minimize the explosive characteristics of the grain dusts and to preserve plant housekeeping standards.

<table>
<tr><th>Operation</th><th>Hood Design</th><th colspan="6">Air Volume</th></tr>
<tr><td rowspan="2">Bag Loading</td><td>VS-15-02, VS-15-03</td><td colspan="6">As Shown</td></tr>
<tr><td>Booth VS-15-01</td><td colspan="6">100 acfm</td></tr>
<tr><td rowspan="3">Belt Discharge</td><td>To belt—VS-50-02</td><td colspan="6">100 acfm/ft belt width up to 200 fpm belt speed</td></tr>
<tr><td>To bin—VS-50-10</td><td colspan="6">500 acfm/ft of belt width over 200 fpm belt speed</td></tr>
<tr><td>To elevator—VS-50-1, VS-50-20</td><td colspan="6">Increase 1/3 if material drop is over 10 ft</td></tr>
<tr><td>Bins</td><td>Direct Exhaust. Use Taper</td><td colspan="6">500 acfm/bin</td></tr>
<tr><td>Bucket elevator</td><td>VS-50-01</td><td colspan="6">100 acfm/ft^2 cross section</td></tr>
<tr><td>Cleaning machines</td><td>Consult Manufacturer</td><td colspan="6"></td></tr>
<tr><td rowspan="5">Distributors</td><td rowspan="5">Enclose discharge. 200 fpm in-draft through enclosure openings</td><td rowspan="2">No of Spouts</td><td colspan="3">Diameter of Spouts</td><td></td><td></td></tr>
<tr><td></td><td>6"</td><td>7"</td><td>8"</td><td>9"</td></tr>
<tr><td>0–6</td><td rowspan="3">Exhaust acfm</td><td>550</td><td>675</td><td>950</td><td>1250</td></tr>
<tr><td>6–12</td><td>950</td><td>1250</td><td>1500</td><td>1900</td></tr>
<tr><td>12–24</td><td>1500</td><td>1900</td><td>2250</td><td>2750</td></tr>
<tr><td>Feed Grinders</td><td>Consult Manufacturer</td><td colspan="6"></td></tr>
<tr><td>Floor Dump</td><td>Booth</td><td colspan="6">200 afcm/ft^2 open face area</td></tr>
<tr><td>Floor Sweep</td><td></td><td colspan="6">950 acfm in 4' × 8' opening</td></tr>
<tr><td>Garner Bin</td><td>Direct Exhaust. Use Taper</td><td colspan="6">1.25 acfm per bushel/min</td></tr>
<tr><td rowspan="4">Mixers</td><td rowspan="4">Ventilated Cover</td><td colspan="3">Mixer Capacity</td><td colspan="3">Exhaust acfm</td></tr>
<tr><td colspan="3">up to 0.05 ton</td><td colspan="3">300</td></tr>
<tr><td colspan="3">0.05 to 1.5 tons</td><td colspan="3">675</td></tr>
<tr><td colspan="3">over 1.5 tons</td><td colspan="3">950</td></tr>
<tr><td>Percentage Feeders</td><td>Enclosed Conveyor</td><td colspan="6">200 acfm at each feeder</td></tr>
<tr><td>Purifiers</td><td>Enclosure</td><td colspan="6">30–40 acfm /ft^2 screen area</td></tr>
<tr><td>Roll Stands</td><td>Enclosure</td><td colspan="6">60 acfm/linear ft</td></tr>
<tr><td rowspan="4">Scales</td><td rowspan="4">Enclosure</td><td colspan="3">Scale Capacity, Bushels</td><td colspan="3">Exhaust, acfm</td></tr>
<tr><td colspan="3">Up to 5</td><td colspan="3">250</td></tr>
<tr><td colspan="3">6 to 10</td><td colspan="3">400</td></tr>
<tr><td colspan="3">Over 11</td><td colspan="3">600</td></tr>
<tr><td>Scale Hopper</td><td>Direct Exhaust. Use Taper</td><td colspan="6">1.25 acfm per bushel/min</td></tr>
<tr><td>Screw Conveyor</td><td>Direct Exhaust. Use Taper</td><td colspan="6">200 acfm—ducts on 20 ft centers</td></tr>
<tr><td>Sifters</td><td>Enclosure</td><td colspan="6">200 acfm compartment</td></tr>
<tr><td>Track Sink</td><td>Direct exhaust from hopper. Use Taper</td><td colspan="6">100 acfm/ft^2 grate area</td></tr>
<tr><td>Tipper Car</td><td>Belt Discharge. VS-50-1, VS-50-20
Spout Ends–Tapered Connection. Spillage–exhaust under head pulley</td><td colspan="6">See "Belt Discharge" above. 200 acfm/ft^2 spout cross-section.
90 acfm/ft belt width.</td></tr>
</table>

TABLE 13-99-1 (SI). Grain Elevators, Feed Mills, Flour Mills

The following data are offered as guides. Airflow rates can vary considerably depending on degree of enclosure, flow rate of material andustiness of the grain. Minimum duct velocity = 3500 fpm. Ventilation control is desirable for the operations to minimize the explosive characteristics of the grain dusts and to preserve plant housekeeping standards.

<table>
<tr><th>Operation</th><th>Hood Design</th><th colspan="6">Air Volume</th></tr>
<tr><td>Bag Loading</td><td>VS-15-02, VS-15-03
Booth VS-15-01</td><td colspan="6">As Shown
0.5 am^3/s/m^2</td></tr>
<tr><td>Belt Discharge</td><td>To belt—VS-50-02
To bin—VS-50-10
To elevator—VS-50-1, VS-50-20</td><td colspan="6">0.5 am^3/s/m belt width up to 1 m/s belt speed
0.8 am^3/s/m of belt width over 1 m/s belt speed
Increase 1/3 if material drop is over 3 meters</td></tr>
<tr><td>Bins</td><td>Direct Exhaust. Use Taper</td><td colspan="6">0.25 am^3/s/bin</td></tr>
<tr><td>Bucket elevator</td><td>VS-50-01</td><td colspan="6">0.5 am^3/s/m^2 cross section</td></tr>
<tr><td>Cleaning machines</td><td>Consult Manufacturer</td><td colspan="6"></td></tr>
<tr><td rowspan="5">Distributors</td><td rowspan="5">Enclose discharge. 1 m/s in-draft through enclosure openings</td><td>No of</td><td colspan="4">Diameter of Spouts, mm</td><td></td></tr>
<tr><td>Spouts</td><td></td><td>15</td><td>18</td><td>20</td><td>22.5</td></tr>
<tr><td>0–6</td><td>Exhaust</td><td>0.25</td><td>0.34</td><td>0.48</td><td>0.62</td></tr>
<tr><td>6–12</td><td>am3/s</td><td>0.48</td><td>0.62</td><td>0.75</td><td>0.95</td></tr>
<tr><td>12–24</td><td></td><td>0.75</td><td>0.95</td><td>1.1</td><td>1.4</td></tr>
<tr><td>Feed Grinders</td><td>Consult Manufacturer</td><td colspan="6"></td></tr>
<tr><td>Floor Dump</td><td>Booth</td><td colspan="6">am^3/s/m^2 open face area</td></tr>
<tr><td>Floor Sweep</td><td></td><td colspan="6">0.5 am^3/s in 100 mm × 200 mm opening</td></tr>
<tr><td>Garner Bin</td><td>Direct Exhaust. Use Taper</td><td colspan="6">0.0015 am^3/s per m^3/s</td></tr>
<tr><td rowspan="4">Mixers</td><td rowspan="4">Ventilated Cover</td><td colspan="3">Mixer Capacity</td><td colspan="3">Exhaust am^3/s</td></tr>
<tr><td colspan="3">up to 0.05 metric ton</td><td colspan="3">0.15</td></tr>
<tr><td colspan="3">0.05 to 1.5 metric tons</td><td colspan="3">0.35</td></tr>
<tr><td colspan="3">over 1.5 metric tons</td><td colspan="3">0.5</td></tr>
<tr><td>Percentage Feeders</td><td>Enclosed Conveyor</td><td colspan="6">1 am^3/s at each feeder</td></tr>
<tr><td>Purifiers</td><td>Enclosure</td><td colspan="6">0.15–0.20 am^3/s/m^2 screen area</td></tr>
<tr><td>Roll Stands</td><td>Enclosure</td><td colspan="6">0.1 am^3/s/linear meter</td></tr>
<tr><td rowspan="4">Scales</td><td rowspan="4">Enclosure</td><td colspan="3">Scale Capacity, liters</td><td colspan="3">Exhaust, am^3/s</td></tr>
<tr><td colspan="3">Up to 175</td><td colspan="3">0.13</td></tr>
<tr><td colspan="3">175 to 350</td><td colspan="3">0.2</td></tr>
<tr><td colspan="3">Over 350</td><td colspan="3">0.3</td></tr>
<tr><td>Scale Hopper</td><td>Direct Exhaust. Use Taper</td><td colspan="6">5.9×10^{-4} am^3/s per liter per second</td></tr>
<tr><td>Screw Conveyor</td><td>Direct Exhaust. Use Taper</td><td colspan="6">1 am^3/s – ducts on 6 meter centers</td></tr>
<tr><td>Sifters</td><td>Enclosure</td><td colspan="6">1 am^3/s compartment</td></tr>
<tr><td>Track Sink</td><td>Direct exhaust from hopper. Use Taper</td><td colspan="6">0.05 am^3/s/m^2 grate area</td></tr>
<tr><td>Tipper Car</td><td>Belt Discharge. VS-50-1, VS-50-20
Spout Ends–Tapered Connection. Spillage–exhaust under head pulley</td><td colspan="6">See “Belt Discharge” above. 1 am^3/m^2 spout cross-section.
0.15 am^3/m belt width.</td></tr>
</table>

TABLE 13-99-2 (IP). Miscellaneous Specific Operations Standards

Minimum Design Operation or Industry	Ventilation: Type of Hood	Ventilation: Airflow or Capture Velocity	Duct Velocity fpm	Remarks
Abrasive Wheel Mfg				13.99.1
Grading Screen	Enclosure–booth	50 fpm at face	4000	
Barrels	Close canopy	400 fpm at face	4000	Barrels receive dust from cyclones
Grinding wheel dressing	Enclosure–booth	400 fpm at face	4000	
Aluminum Furnaces	Enclosure	150–200 fpm through opening	2000	13.99.2
Asbestos				
Bagging	Enclosure–booth	250 fpm through all openings	3500	13.99.3, 13.99.4
Carding	Enclosure	1600 fpm per card	3500	
Crushing	Enclosure	150 fpm through all openings	3500	13.99.5
Drilling of panels containing asbestos	Movable hood	400 fpm capture velocity	4500	13.99.6
Dumping	Booth	250 fpm through all openings	3500	13.99.6
Grinding of brake shoes	Enclosure	400 fpm minimum capture velocity at tool rest	3500	13.99.6
Hot press for brake shoes	Enclosure	250 fpm through all openings	3500	13.99.6
Mixing	Booth	250 fpm face velocity	3500	13.99.6
Preform press	Enclosure	250 fpm through all openings	3500	13.99.6
Screening	Enclosure	200 fpm through all openings but not less than 25 acfm/ft^2 screen areas	3500	13.99.3
Spool winding	Local hoods	50 acfm/spool	3500	
Spinning and twisting	Partial	50 acfm/spool	3500	Hinged front panels and skirt Wet twisting preferred
Weaving	Canopy with baffles	50 fpm through openings	3500	Wet twisting preferred
Auto parking garage	2-level	500 acfm/parking space		13.99.7
Ceramic	Enclosure	200 fpm through all openings	3500	13.99.8, .9, .10
Dry Pan	Local at die	500 acfm	3500	Automatic feed
Dry Press	Local at die	500 acfm	3500	Manual feed
	At supply bin	500 acfm	3500	Manual feed
Aerographing	Booth	100 fpm face velocity		
Spraying (lead glaze)	Booth	400 fpm face velocity	2000	
Coating Pans (pharmaceutical)	Airflow into opening of pan	100–150 fpm through opening	3000	13.99.8, 13,99.11 If heated air is supplied to pan, add volume of heated air to exhaust
Cooling Tunnels (foundry)	Enclosure	75–100 acfm/running foot of enclosure		13.99.12, 13.99.13
Core Knockout (manual)	Large side-draft or semi-booth–exhaust near floor	200–250 acfm/ft^2 dust producing area	3500	13.99.12, 13.99.13 13.99.14
Core Sanding (on lathe)	Downdraft under work	100 fpm at source	3500	13.99.15
Crushers and Grinders	Enclosure	200 fpm through openings	3500	13.99.16

TABLE 13-99-2 (IP) (Cont.). Miscellaneous Specific Operations Standards

Minimum Design Operation or Industry	Ventilation: Type of Hood	Ventilation: Airflow or Capture Velocity	Duct Velocity fpm	Remarks
Drilling (rocks)	Special trap (see References)	60 acfm–vertical (downward) work 200 acfm–horizontal work		13.99.17,13.99.18 13.99.19. may vary with size and speed of drill
Forge (hand)	Booth	200 fpm face velocity	1500	13.99.2
Outboard Test Tank	Side draft	200 acfm/ft^2 of tank opening		
Packaging Machines	Booth Downdraft Complete enclosure	50–100 fpm face velocity 95–100 fpm 100–400 fpm/opening	3000 to 4000	
Paper Machine	Canopy	200–300 fpm face velocity	1500	13.99.20, 13.99.21.
Quartz Fusing	Booth on bench	150–200 fpm face velocity		13.99.20, 13.99.21
Rotary Blasting Table	Enclosure	500 fpm through openings when in operation	3500	
Silver Soldering	Free hanging	100 fpm at source	2000	
Steam Kettles	Canopy	150 fpm face velocity	2000	
Varnish Kettles	Canopy	200–500 fpm face velocity	1500	
Wire Impregnating	covered tanks	200 acfm/ft^2 of opening		

TABLE 13-99-2 (SI). Miscellaneous Specific Operations Standards

Minimum Design Operation or Industry	Ventilation: Type of Hood	Ventilation: Airflow or Capture Velocity	Duct Velocity m/s	Remarks
Abrasive Wheel Mfg				13.99.1
Grading Screen	Enclosure–booth	0.25 m/s at face	20	
Barrels	Close canopy	2.0 m/s at face	20	Barrels receive dust from cyclones
Grinding Wheel Dressing	Enclosure–booth	2.0 m/s at face	20	
Aluminum Furnaces	Enclosure	0.75–1 m/s through opening	10	13.99.2
Asbestos				
Bagging	Enclosure–booth	0.13 m/s through all openings	18	13.99.3, 13.99.4
Carding	Enclosure	0.8 m/s per card	18	
Crushing	Enclosure	0.75 m/s through all openings	18	13.99.5
Drilling of panels containing asbestos	Movable hood	2.0 m/s capture velocity	23	13.99.6
Dumping	Booth	0.13 m/s through all openings	18	13.99.6
Grinding of brake shoes	Enclosure	2.0 m/s minimum capture velocity at tool rest	18	13.99.6
Hot press for brake shoes	Enclosure	0.13 m/s through all openings	18	13.99.6
Mixing	Booth	0.13 m/s face velocity	18	13.99.6
Preform press	Enclosure	0.13 m/s through all openings	18	13.99.6
Screening	Enclosure	0.10 m/s through all openings but not less than .13 am^3/s/m^2 screen areas	18	13.99.3
Spool winding	Local hoods	0.025 am^3/s/spool	18	
Spinning and twisting	Partial	0.025 am^3/s/spool	18	Hinged front panels and skirt Wet twisting preferred
Weaving	Canopy with baffles	0.25 m/s through openings	18	Wet twisting preferred
Auto parking garage	2-level	0.25 am^3/s/parking space		13.99.7
Ceramic	Enclosure	1.0 m/s through all openings	18	13.99.8, .9, .10
Dry Pan	Local at die	0.25 am^3/s	18	Automatic feed
Dry Press	Local at die	0.25 am^3/s	18	Manual feed
	At supply bin	0.25 am^3/s	18	Manual feed
Aerographing	Booth	0.5 m/s face velocity		
Spraying (lead glaze)	Booth	2.0 m/s face velocity	10	
Coating Pans (pharmaceutical)	Airflow into opening of pan	0.25–0.75 m/s through opening	15	13.99.8, 13,99.11 If heated air is supplied to pan, add volume of heated air to exhaust
Cooling Tunnels (foundry)	Enclosure	0.38–0.5 am^3/s/running meter of enclosure		13.99.12, 13.99.13
Core Knockout (manual)	Large side-draft or semi-booth–exhaust near floor	1.0–1.3 am^3/s/m^2 dust producing area	18	13.99.12, 13.99.13 13.99.14
Core Sanding (on lathe)	Downdraft under work	0.5 am/s at source	18	13.99.15
Crushers and Grinders	Enclosure	1.0 am/s through openings	18	13.99.16

TABLE 13-99-2 (SI) (Cont.). Miscellaneous Specific Operations Standards

Minimum Design Operation or Industry	Ventilation: Type of Hood	Ventilation: Airflow or Capture Velocity	Duct Velocity m/s	Remarks
Drilling (rocks)	Special trap (see References)	0.03 am^3/s – vertical (downward) work 0.10 am^3/s – horizontal work		13.99.17, 13.99.18 13.99.19. may vary with size and speed of drill
Forge (hand)	Booth	1.0 m/s face velocity	8	13.99.2
Outboard Test Tank	Side draft	1.0 am^3/s per m^2 tank opening		
Packaging Machines	Booth	0.25–0.5 m/s face velocity	15	
	Downdraft	0.5–0.75 m/s	to	
	Complete enclosure	0.5–2.0 m/s/opening	20	
Paper Machine	Canopy	1.0–1.5 m/s face velocity	8	13.99.20, 13.99.21.
Quartz Fusing	Booth on bench	0.75–1.0 m/s face velocity		13.99.20, 13.99.21
Rotary Blasting Table	Enclosure	2.5 m/s through openings when in operation	15	
Silver Soldering	Free hanging	0.5 m/s at source	10	
Steam Kettles	Canopy	0.75 m/s face velocity	10	
Varnish Kettles	Canopy	1.0–1.3 m/s face velocity	8	
Wire Impregnating	covered tanks	1.0 am^3/s/m^2 of opening		

FIGURE VS-99-01

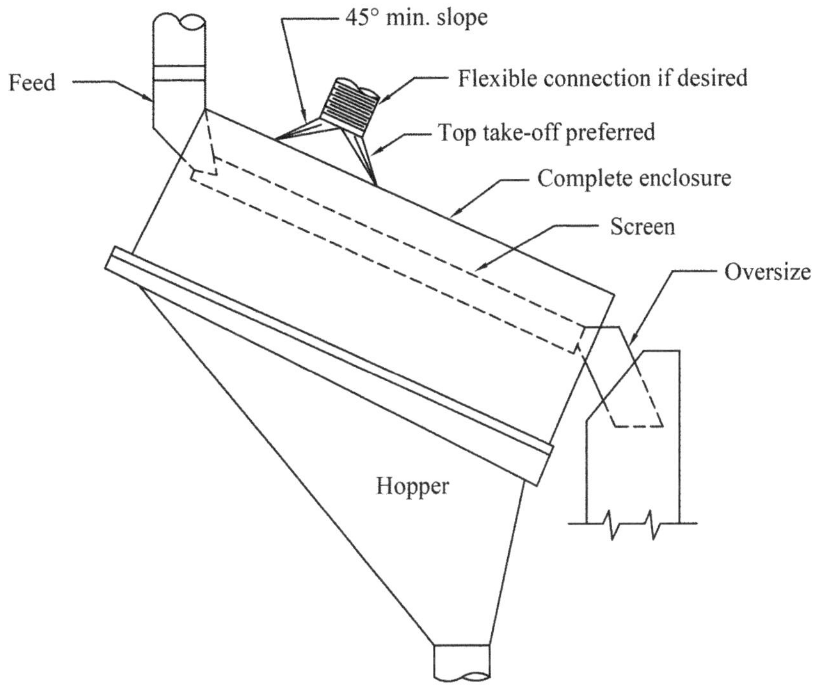

FLAT DECK SCREEN

Q = 200 acfm/ft^2 [1 am^3/s/m^2] through hood openings, but not less than 50 acfm/ft^2 [0.25 am^3/s/m^2] screen area. No increase for multiple decks

Minimum duct velocity = 3500 fpm [17.5 m/s]

$h_e = 0.56\ VP_d$

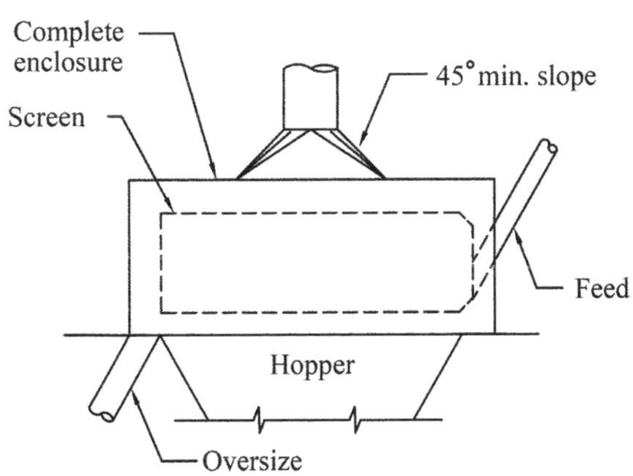

CYLINDRICAL SCREEN

Q = 100 acfm/ft^2 [0.5 am^3/s/m^2] circular cross section of screen; at least 400 acfm/ft^2 [2 am^3/s/m^2] of enclosure opening

Minimum duct velocity = 3500 fpm [17.5 m/s]

$h_e = 0.56\ VP_d$

Note: Dry, very dusty materials may require exhaust flow rates 1.5 to 2.0 times stated values.

ACGIH®	TITLE SCREENS	FIGURE VS-99-01 DATE 1-12

CHECK CODES, REGULATIONS, AND LAWS (LOCAL, STATE, AND NATIONAL) TO ENSURE THAT DESIGN IS COMPLIANT.

FIGURE VS-99-02

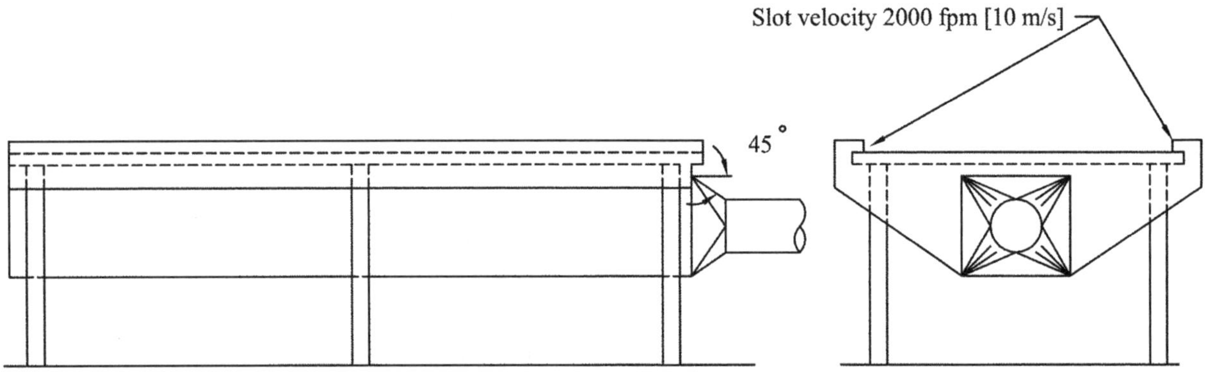

$Q = 50–100$ acfm/ft^2 [0.25–0.50 am^3/s/m^2] of table top.
Minimum duct velocity = 2000 fpm [10 m/s]
$h_e = 1.78\ VP_s + 0.25\ VP_d$

Note: See "Open Surface Tanks", VS-70-01 and VS-70-02 for other suitable slot types. Airflow rate may be calculated on dilution basis if data is available. Maximum plenum velocity = 1/2 slot velocity. Large plenum essential for good distribution.

ACGIH®	TITLE: TABLE SLOT	FIGURE: VS-99-02 / DATE: 1-12

CHECK CODES, REGULATIONS, AND LAWS (LOCAL, STATE, AND NATIONAL) TO ENSURE THAT DESIGN IS COMPLIANT.

FIGURE VS-99-03

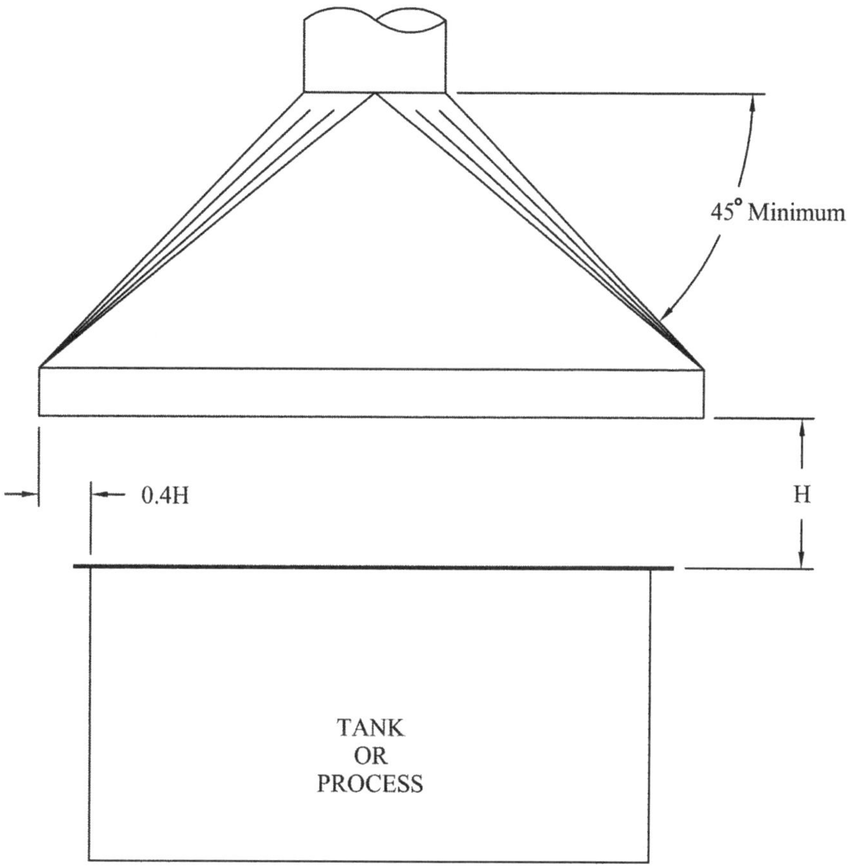

Not to be used where material is toxic and worker must bend over tank or process.
Side curtains are necessary when cross-drafts are present.

Q = 1.4PHV acfm [am³/s]	for open type canopy P = perimeter of tank, feet V = 50–500 fpm [0.25 – 2.50 m/s]; See Chapter 6
Q = (W + L)HV acfm [am³/s]	for two sides adjacent enclosed W & L are open sides of hood V = 50–500 fpm [0.25 – 2.50 m/s]; See Chapter 6
Q = WHV acfm [am³/s] or LHV acfm [am³/s]	for three sides enclosed (booth) V = 50–500 fpm [0.25 – 2.50 m/s]; See Chapter 6

$h_e = 0.25\ VP_d$
Duct velocity = 1000–3000 fpm [5.00–15.00 m/s]

ACGIH®

TITLE
CANOPY HOOD

FIGURE
VS-99-03

DATE
1-12

CHECK CODES, REGULATIONS, AND LAWS (LOCAL, STATE, AND NATIONAL)
TO ENSURE THAT DESIGN IS COMPLIANT.

FIGURE VS-99-04

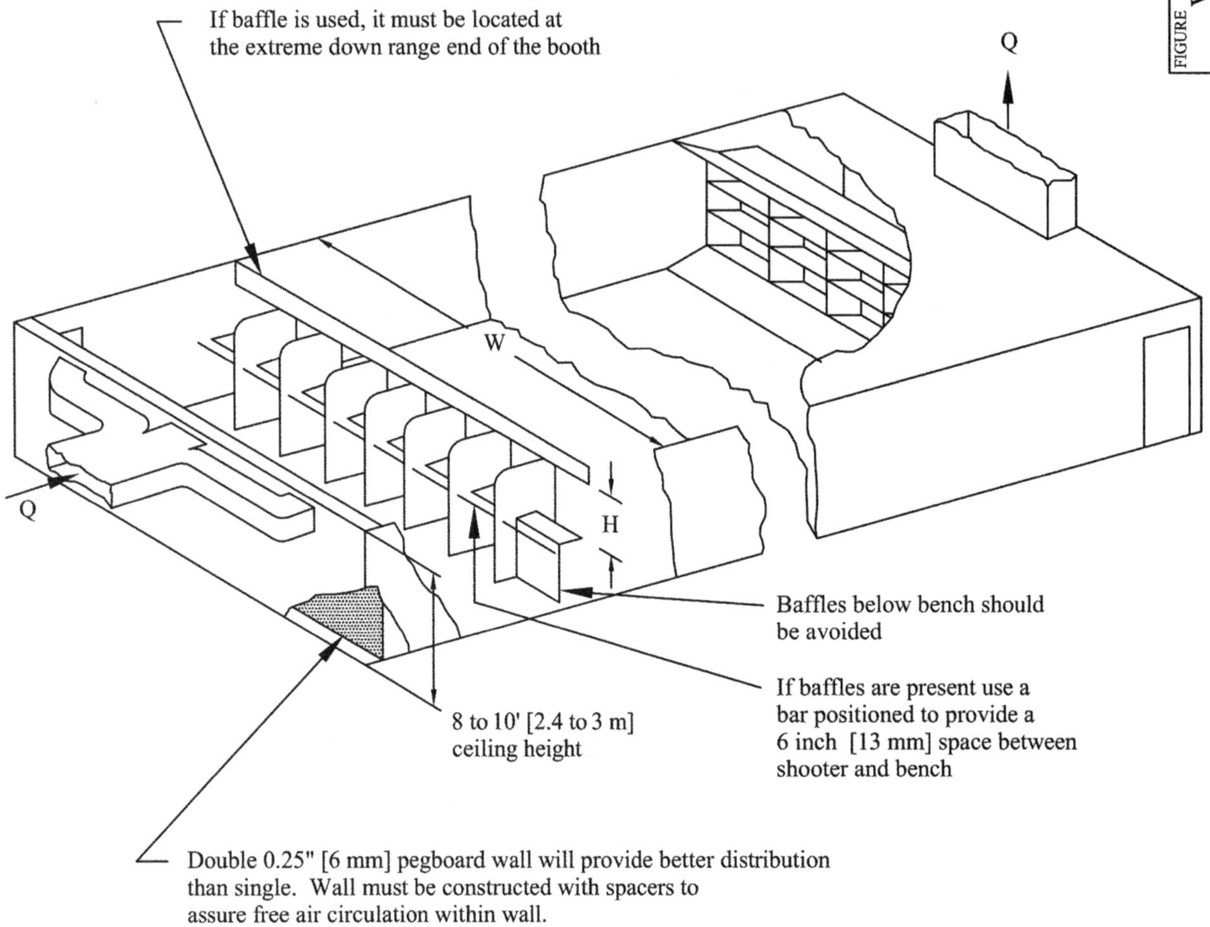

$Q_{(minimum)}$ = 50 HW, but not less than 20 acfm/ft^2 [0.1 am^3/s/m^2]
of room cross sectional area

Notes:

Replacement air distribution:
Uniform air distribution necessary.
Perforated rear wall or ceiling plenum system preferred. Grilles and diffusers are not recommended.

NIOSH certified dust respirator for lead is necessary during clean-up and lead removal from bullet trap.

Acoustical material on walls, ceiling and thick fabric on bench top are recommended.

TITLE	FIGURE
HAND GUN AND SMALL BORE RIFLE RANGE VENTILATION	VS-99-04
	DATE 1-12

CHECK CODES, REGULATIONS, AND LAWS (LOCAL, STATE, AND NATIONAL) TO ENSURE THAT DESIGN IS COMPLIANT.

FIGURE VS-99-04a

The range master, instructors, cleaning and maintenance crews are the primary personnel of concern when designing a firing range. The day-to-day contact with lead is usually at a higher dose than the occasional shooters. On the other hand, the occasional shooter may not be aware of their lead exposure that can be carried home on their clothing and hands.

1. Ensure that there are no obstructions to airflow (furniture, range master's station, coat racks, supplies, waiting personnel, etc.) located between the supply air plenum and the firing line. Provide coat racks, lockers, benches, etc. outside the range.

2. Do not allow space within the supply air plenum to be used as an observation or storage area. Provide ventilation storage areas for cleaning equipment, e.g., HEPA vacuum cleaners, wet mops and other supplies.

3. Each firing position should be designed to minimize surface areas that may obstruct airflow. Surface planes of booths, partitions, shelves, gates, etc. should be oriented parallel to the down-range flow direction to minimize turbulence at the firing line.

4. Provide restrooms or wash stations for hand cleaning. Consider providing employee lockers to separate street (clean) clothes and clothes (dirty) worn in the range. Showers may be needed if there are lead overexposures.

5. Provide ventilated weapons cleaning stations or workbenches that are not located in offices, classrooms or the lunchroom. If feasible, provide a separate room with ventilated workbenches.

6. Locate and design the range entrance doors to have minimum effect on the supply air plenum and airflow to the firing line.

7. Determine all shooter positions and design airflow distribution to accommodate all such positions. Typical positions are standing, kneeling, and prone.

8. Consider installing a target carrier system to reduce foot traffic downrange of the firing line, thus reducing lead dust re-entrainment.

9. Do not permit personnel to use a standard range for SWAT-type training unless the range is specifically designed for those operations.

   A. Higher air velocities with ceiling-to-floor airflow throughout the range are the preferred design for SWAT-type ranges because shooting is non-directional and frequently more intense.

10. Commission the range, to include performance requirements and acceptable testing, paying close attention to airflow along the outer walls. Contaminated air can travel back to the shooter position if the supply air is not properly balanced.

11. Consider using bullet traps or other dust reducing mechanisms to reduce lead dust generation from bullet impact on the backstop.

ACGIH®	TITLE HAND GUN AND SMALL BORE RIFLE RANGE DESIGN	FIGURE VS-99-04a DATE 1-12

CHECK CODES, REGULATIONS, AND LAWS (LOCAL, STATE, AND NATIONAL) TO ENSURE THAT DESIGN IS COMPLIANT.

FIGURE VS-99-04b

12. Where possible, provide maintenance access to the filter system to minimize worker exposure to excess lead dust when changing the filters. Consider using bag-in/bag-out (or similar) technology.

13. If air from exhaust ventilation is re-circulated into the range, ensure that (A) the system has a high efficiency filter followed by a reliable back-up (minimally a HEPA) filter, mounted in series; (B) controls monitor the concentration of lead in the return air; and (C) if the re-circulation system fails, airflow bypasses the re-circulation system and automatically exhausts outdoors. *All* of the three system components must be installed, operated, and maintained. [NOTE: Pressure drop monitoring across the HEPA filter serves as an effective monitoring approach.

14. Recent development of alternative non-lead bullets may permit lower flow rates or minimal industrial ventilation. Conduct a risk analysis to determine if materials (frequently other metals) in the alternate-bullets require industrial ventilation. At a minimum, comply with indoor air quality requirements.

ACGIH®	TITLE HAND GUN AND SMALL BORE RIFLE RANGE DESIGN	FIGURE VS-99-04b DATE 1-12

CHECK CODES, REGULATIONS, AND LAWS (LOCAL, STATE, AND NATIONAL) TO ENSURE THAT DESIGN IS COMPLIANT.

FIGURE VS-99-05

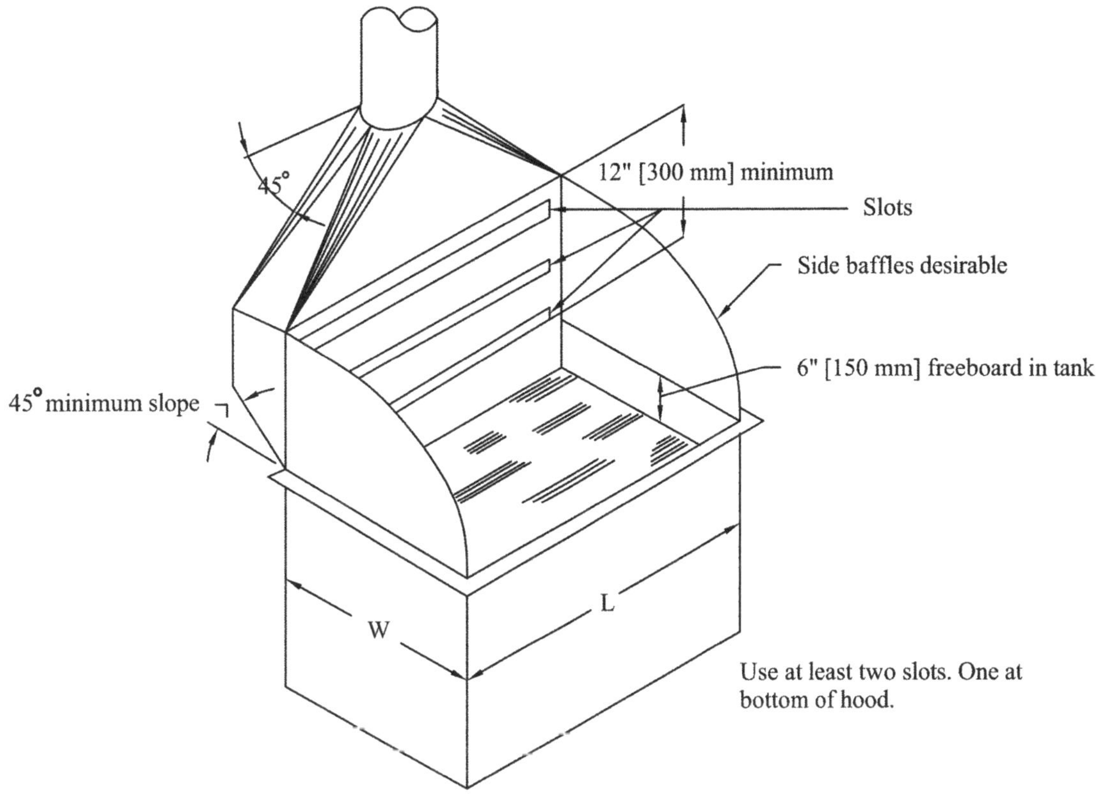

Q = 150 acfm/ft^2 [0.75 am^3/s/m^2] of bed
Slot velocity = 2000 fpm [10.00 m/s]
$h_e = 1.78\ VP_s + 0.25\ VP_d$
Minimum duct velocity = 3500 fpm [17.50 m/s]
W not to exceed 36" [0.9 m]

Free board must be maintained to prevent material carryout.

ACGIH®	TITLE FLUIDIZED BEDS	FIGURE VS-99-05 DATE 1-12

CHECK CODES, REGULATIONS, AND LAWS (LOCAL, STATE, AND NATIONAL) TO ENSURE THAT DESIGN IS COMPLIANT.

Exhaust ventilation volumes for operating engines are determined by the engine displacement, the engine RPM, and tailpipe exhaust temperature, plus a 20% safety factor. This is specified in actual cubic feet per minute (acfm) or [actual cubic meters per second (am^3/s)].

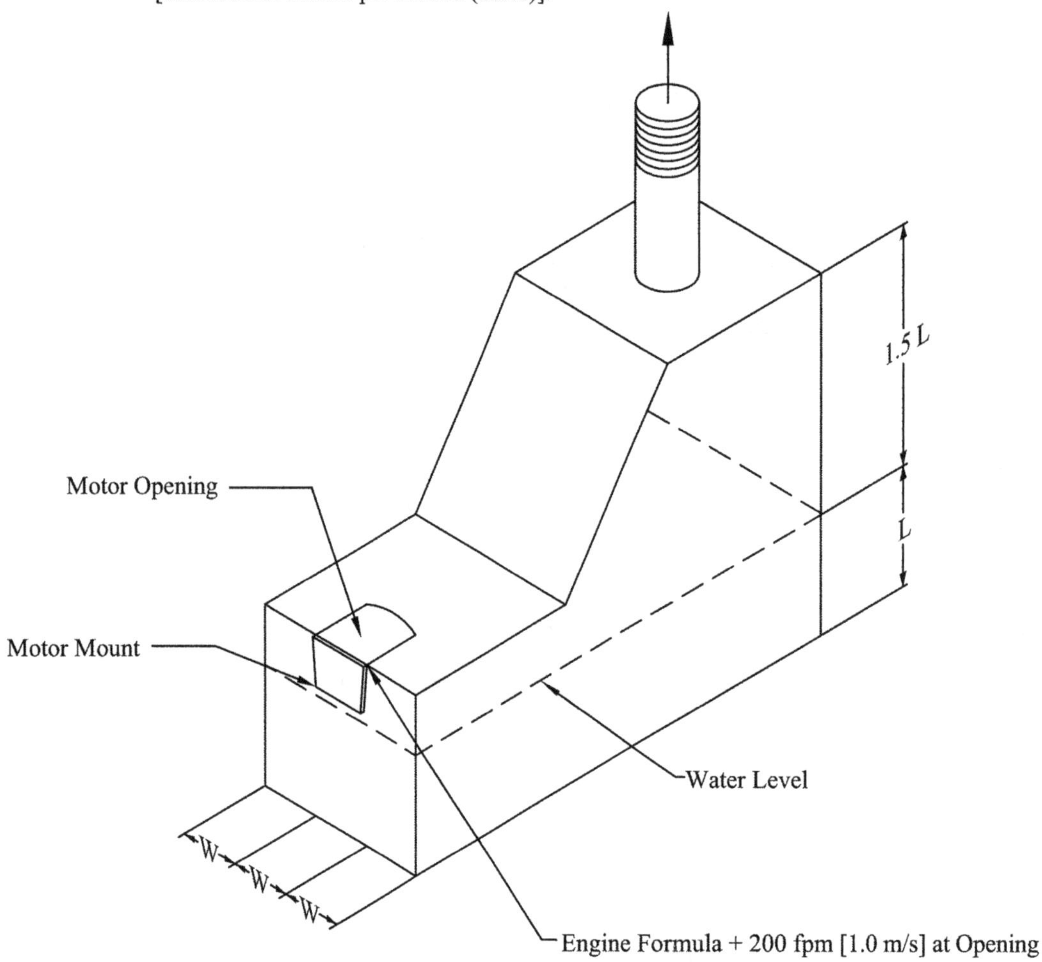

Engine Formula:

$$Q_e = (1.2)(D_{eng} \times N) \left(\frac{460\text{ F} + T_{eng}}{530\text{ F}}\right) \text{acfm}$$

$$\left[Q_e = (1.2)(D_{eng} \times N) \left(\frac{273\text{ C} + T_{eng}}{294\text{ C}}\right) \text{am}^3\text{/s} \right]$$

Where: Q_e = Exhaust flow (acfm) [am^3/s]

T_{eng} = Engine tail pipe temperature, (F) [C]

D_{eng} = Engine displacement, (ft^3) [m^3]

N = Engine revolutions per minute (IP and SI)

ACGIH®	TITLE OUTBOARD MOTOR TEST	FIGURE VS-99-06 DATE 1-12

CHECK CODES, REGULATIONS, AND LAWS (LOCAL, STATE, AND NATIONAL) TO ENSURE THAT DESIGN IS COMPLIANT.

FIGURE VS-99-07

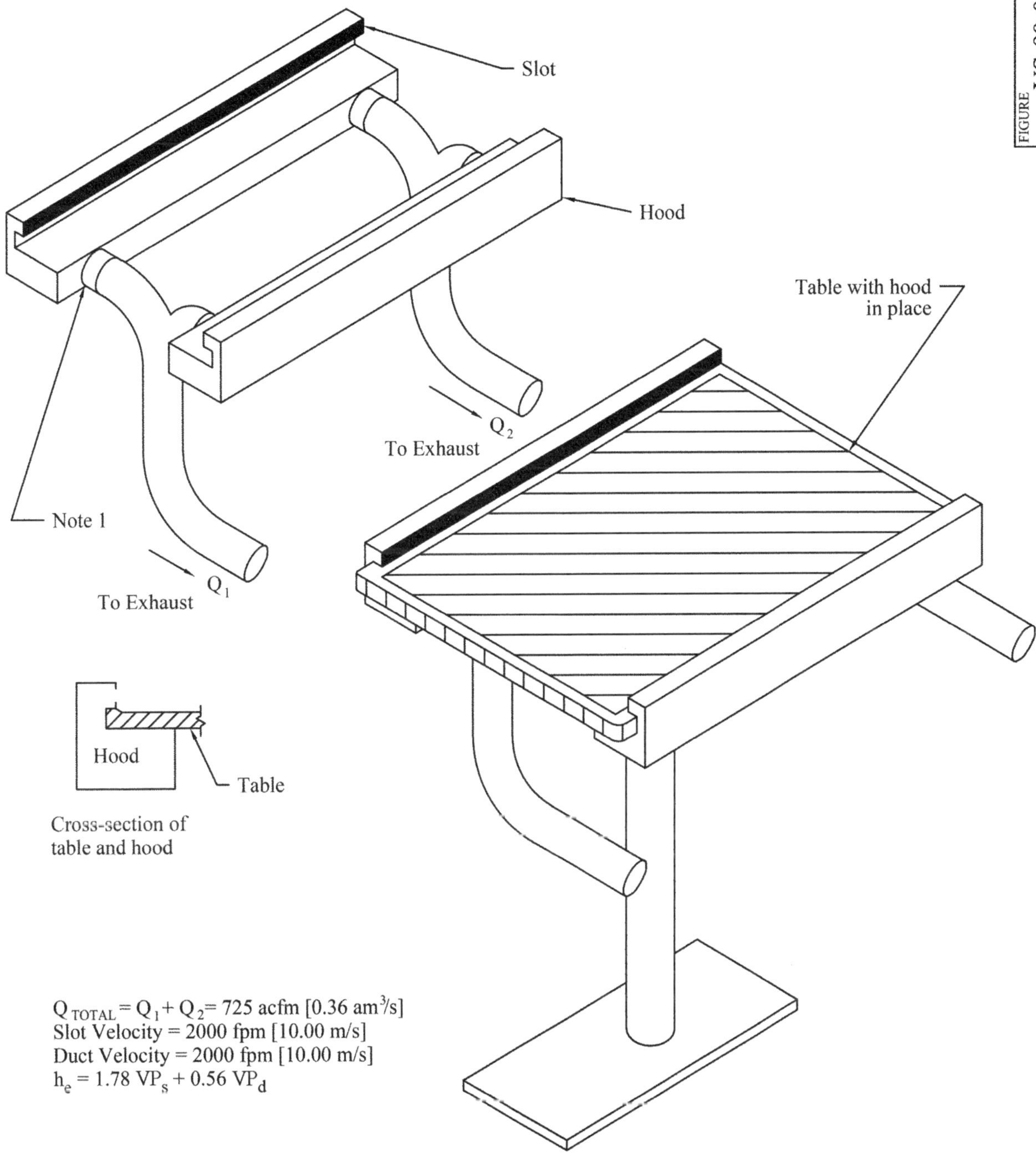

$Q_{TOTAL} = Q_1 + Q_2 = 725$ acfm [0.36 am³/s]
Slot Velocity = 2000 fpm [10.00 m/s]
Duct Velocity = 2000 fpm [10.00 m/s]
$h_e = 1.78\ VP_s + 0.56\ VP_d$

Note 1: Position of exhaust take-offs may vary due to table configuration. Place to assure same airflow distribution into slot.

Reference 13.99.24

ACGIH®	TITLE: MORTUARY TABLE	FIGURE: VS-99-07 DATE: 1-12

CHECK CODES, REGULATIONS, AND LAWS (LOCAL, STATE, AND NATIONAL) TO ENSURE THAT DESIGN IS COMPLIANT.

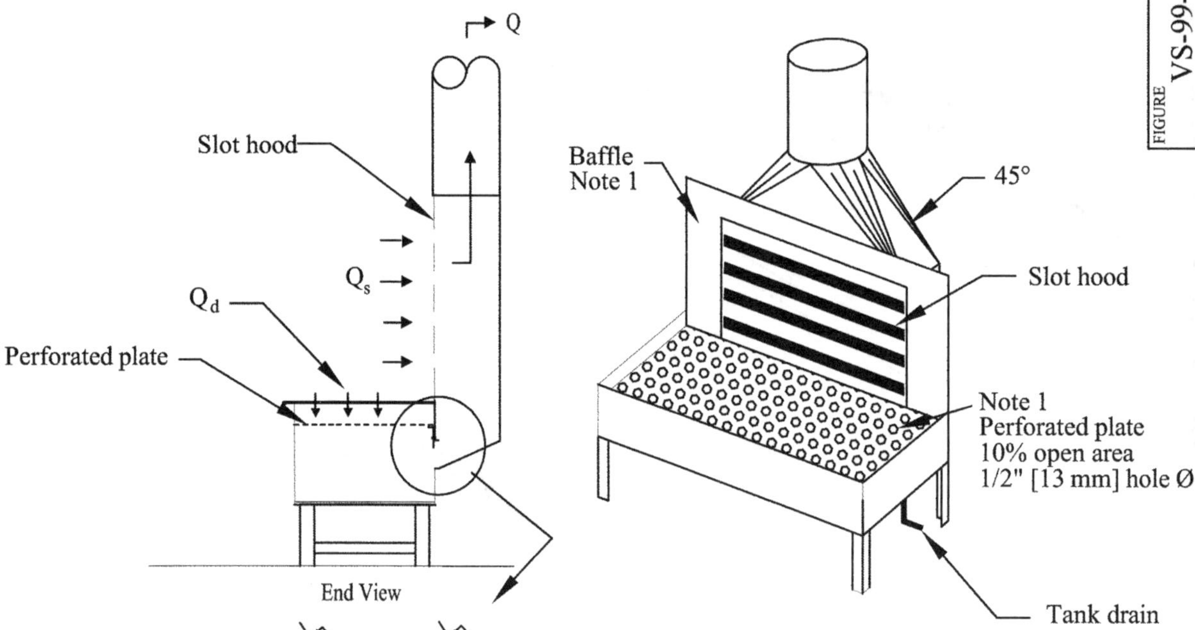

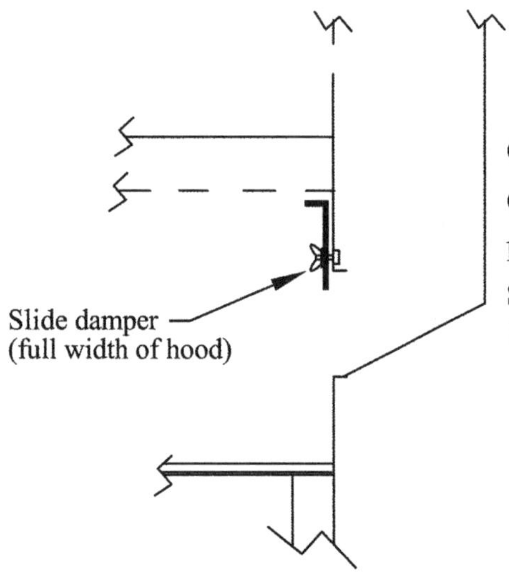

Q_s = 45 acfm/ft^2 [0.23 am^3/s/m^2] of tank area

Q_d = 65 acfm/ft^2 [0.33 am^3/s/m^2] of tank area

Plenum velocity less than 1/2 slot velocity

Slot velocity = 2000 fpm [10.00 m/s]

$h_e = 1.78\ VP_s + .25\ VP_d$

Note 1: Build perforated plate in sections for easy cleaning
Allow freeboard height at least 9" [225 mm]

Note 2: Baffle to extend to ends of tank
6" [150 mm] (minimum) above top of workpiece

Note 3: Adjust slot to achieve Q_s/Q_d flow distribution

Reference 13.99.25

TITLE	FIGURE
FURNITURE STRIPPING TANK	VS-99-08
	DATE 1-12

CHECK CODES, REGULATIONS, AND LAWS (LOCAL, STATE, AND NATIONAL) TO ENSURE THAT DESIGN IS COMPLIANT.

APPENDICES

APPENDIX A

Threshold Limit Values for Chemical Substances in the Work Environment with Intended Changes for 2013

Attached information in effect at time of publication of this Manual. Please refer to the latest edition of "Threshold Limit Values for Chemical Substances in Work Environment with Intended Changes" in ACGIH®'s *TLVs® and BEIs®* Book.

APPENDIX A

2013
Threshold Limit Values for Chemical Substances in the Work Environment

Adopted by ACGIH®
with Intended Changes

INTRODUCTION TO THE CHEMICAL SUBSTANCES

General Information

The TLVs® are guidelines to be used by professional industrial hygienists. The values presented in this book are intended for use only as guidelines or recommendations to assist in the evaluation and control of potential workplace health hazards and for no other use (e.g., neither for evaluating or controlling community air pollution; nor for estimating the toxic potential of continuous, uninterrupted exposures or other extended work periods; nor for proving or disproving an existing disease or physical condition in an individual). Further, these values are not fine lines between safe and dangerous conditions and should not be used by anyone who is not trained in the discipline of industrial hygiene. TLVs® are not regulatory or consensus standards.

Editor's note: The approximate year that the current *Documentation* was last substantially reviewed and, where necessary, updated may be found following the CAS number for each of the adopted entries in the alphabetical listing, e.g., Aldrin [309-00-2] (2006). The reader is advised to refer to the "TLV® Chronology" section in each *Documentation* for a brief history of the TLV® recommendations and notations.

Definition of the TLVs®

Threshold Limit Values (TLVs®) refer to airborne concentrations of chemical substances and represent conditions under which it is believed that *nearly all* workers may be repeatedly exposed, day after day, over a working lifetime, without adverse health effects.

Those who use the TLVs® **MUST** consult the latest *Documentation* to ensure that they understand the basis for the TLV® and the information used in its development. The amount and quality of the information that is available for each chemical substance varies over time.

Chemical substances with equivalent TLVs® (i.e., same numerical values) cannot be assumed to have similar toxicologic effects or similar biologic potency. In this book, there are columns listing the TLVs® for each chemical substance (that is, airborne concentrations in parts per million [ppm] or milligrams per cubic meter [mg/m^3]) and critical effects produced by the chemical substance. These critical effects form the basis of the TLV®.

ACGIH® recognizes that there will be considerable variation in the level of biological response to a particular chemical substance, regardless of the airborne concentration. Indeed, TLVs® do not represent a fine line between a healthy versus an unhealthy work environment or the point at which material impairment of health will occur. TLVs® will not adequately protect all workers. Some individuals may experience discomfort or even more serious adverse health effects when exposed to a chemical substance at the TLV® or even at concentrations below the TLV®. There are numerous possible reasons for increased susceptibility to a chemical substance, including age, gender, ethnicity, genetic factors (predisposition), lifestyle choices (e.g., diet, smoking, abuse of alcohol and other drugs), medications, and pre-existing medical conditions (e.g., aggravation of asthma or cardiovascular disease). Some individuals may become more responsive to one or more chemical substances following previous exposures (e.g., sensitized workers). Susceptibility to the effects of chemical substances may be altered during different periods of fetal development and throughout an individual's reproductive lifetime. Some changes in susceptibility may also occur at different work levels (e.g., light versus heavy work) or at exercise — situations in which there is increased cardiopulmonary demand. Additionally, variations in temperature (e.g., extreme heat or cold) and relative humidity may alter an individual's response to a toxicant. The *Documentation* for any given TLV® must be reviewed, keeping in mind that other factors may modify biological responses.

Although TLVs® refer to airborne levels of chemical exposure, dermal exposures may possibly occur in the workplace (*see* "Skin" on page 75 of the ***Definitions and Notations*** section).

Three categories of TLVs® are specified: time-weighted average (TWA); short-term exposure limit (STEL); and a ceiling (C). For most substances, a TWA alone or with a STEL is relevant. For some substances (e.g., irritant gases), only the TLV–C is applicable. If any of these TLV® types are exceeded, a potential hazard from that substance is presumed to exist.

Threshold Limit Value–Time-Weighted Average (TLV–TWA): The TWA concentration for a conventional 8-hour workday and a 40-hour workweek, to which it is believed that nearly all workers may be repeatedly exposed, day after day, for a working lifetime without adverse effect. Although calculating the average concentration for a workweek, rather than a workday, may be appropriate in some instances, ACGIH® does not offer guidance regarding such exposures.

Threshold Limit Value–Short-Term Exposure Limit (TLV–STEL): A 15-minute TWA exposure that should not be exceeded at any time during a workday, even if the 8-hour TWA is within the TLV–TWA. The TLV–STEL is the concentration to which it is believed that workers can be exposed continuously for a short period of time without suffering from 1) irritation, 2) chronic or irreversible tissue damage, 3) dose-rate-dependent toxic effects, or 4) narcosis of sufficient degree to increase the likelihood of accidental injury, impaired self-rescue, or materially reduced work efficiency. The TLV–STEL will not necessarily protect against these effects if the daily TLV–TWA is exceeded. The TLV–STEL usually supplements the TLV–TWA where there are recognized acute effects from a substance whose toxic effects are primarily of a chronic nature; however, the TLV–STEL may be a separate, independent exposure guideline. Exposures above the TLV–TWA up to the TLV–STEL should be less than

15 minutes, should occur no more than four times per day, and there should be at least 60 minutes between successive exposures in this range. An averaging period other than 15 minutes may be recommended when this is warranted by observed biological effects.

Threshold Limit Value–Ceiling (TLV–C): The concentration that should not be exceeded during any part of the working exposure. If instantaneous measurements are not available, sampling should be conducted for the minimum period of time sufficient to detect exposures at or above the ceiling value.

ACGIH® believes that TLVs® based on physical irritation should be considered no less binding than those based on physical impairment. There is increasing evidence that physical irritation may initiate, promote, or accelerate adverse health effects through interaction with other chemical or biologic agents or through other mechanisms.

Excursion Limits

For many substances with a TLV–TWA, there is no TLV–STEL. Nevertheless, excursions above the TLV–TWA should be controlled, even where the 8-hour TLV–TWA is within recommended limits. Excursion limits apply to those TLV–TWAs that do not have TLV–STELs.

> *Excursions in worker exposure levels may exceed 3 times the TLV–TWA for no more than a total of 30 minutes during a workday, and under no circumstances should they exceed 5 times the TLV–TWA, provided that the TLV–TWA is not exceeded.*

The approach here is that the maximum recommended excursion should be related to the variability generally observed in actual industrial processes. In reviewing large numbers of industrial hygiene surveys conducted by the U.S. National Institute for Occupational Safety and Health, Leidel et al. (1975) found that short-term exposure measurements were generally lognormally distributed.

While a complete discussion of the theory and properties of the lognormal distribution is beyond the scope of this section, a brief description of some important terms is presented. The measure of central tendency in a lognormal distribution is the antilog of the mean logarithm of the sample values. The distribution is skewed, and the geometric mean (m_g) is always smaller than the arithmetic mean by an amount that depends on the geometric standard deviation. In the lognormal distribution, the geometric standard deviation (sd_g) is the antilog of the standard deviation of the sample value logarithms, and 68.26% of all values lie between m_g/sd_g and $m_g \times sd_g$.

If the short-term exposure values in a given situation have a geometric standard deviation of 2.0, 5% of all values will exceed 3.13 times the geometric mean. If a process displays variability greater than this, it is not under good control, and efforts should be made to restore control.

The approach is a considerable simplification of the lognormal concentration distribution concept but is considered more convenient. If exposure excursions are maintained within the recommended limits, the geometric standard deviation of the concentration measurements will be near 2.0, and the goal of the recommendations will be accomplished. It is recognized that the geometric standard deviations of some common workplace exposures may exceed 2.0 (Buringh and Lanting, 1991). If such distributions are known and workers are not at increased risk of adverse health effects, recommended excursion limits should be modified based upon workplace-specific data. When the toxicologic data for a specific substance are available to establish a TLV–STEL or a TLV–C, these values take precedence over the excursion limit.

TWA and STEL versus Ceiling (C)

A substance may have certain toxicological properties that require the use of a TLV–C rather than a TLV–TWA excursion limit or a TLV–STEL. The amount by which the TLVs® may be exceeded for short periods without injury to health depends upon a number of factors such as the nature of the contaminant, whether very high concentrations — even for short periods — produce acute poisoning, whether the effects are cumulative, the frequency with which high concentrations occur, and the duration of such periods. All factors must be taken into consideration in arriving at a decision as to whether a hazardous condition exists.

Although the TWA concentration provides the most satisfactory, practical way of monitoring airborne agents for compliance with the TLVs®, there are certain substances for which it is inappropriate. In the latter group are substances that are predominantly fast-acting and whose TLV® is more appropriately based on this particular response. Substances with this type of response are best controlled by a TLV–C that should not be exceeded. It is implicit in these definitions that the manner of sampling to determine noncompliance with the TLVs® for each group must differ. Consequently, a single, brief sample that is applicable to a TLV–C is not appropriate to the TLV–TWA; here, a sufficient number of samples are needed to permit determination of a TWA concentration throughout a complete cycle of operation or throughout the workshift.

Whereas the TLV–C places a definite boundary that exposure concentrations should not be permitted to exceed, the TLV–TWA requires an explicit limit to the excursions which are acceptable above the recommended TLV–TWAs.

Mixtures

Special consideration should also be given to the application of the TLVs® in assessing the health hazards that may be associated with exposure to a mixture of two or more substances. A brief discussion of basic considerations involved in developing TLVs® for mixtures and methods for their development, amplified by specific examples, is given in Appendix E.

Deviations in Work Conditions and Work Schedules

Application of TLVs® to Unusual Ambient Conditions

When workers are exposed to air contaminants at temperatures and pressures substantially different than those at normal temperature and pressure (NTP) conditions (25°C and 760 torr), care should be taken in comparing sampling results to the applicable TLVs®. For aerosols, the TWA exposure concentration (calculated using sample volumes not adjusted to NTP conditions) should be compared directly to the applicable TLVs® published in the *TLVs® and BEIs®* book. For gases and vapors, there are a number of options for comparing air-sampling results to the TLV®, and these are discussed in detail by Stephenson and Lillquist (2001). One method that is simple in its conceptual approach is 1) to determine the exposure concentration, expressed in terms of mass per volume, at the sampling site using the sample volume not adjusted to NTP conditions, 2) if required, to convert the TLV® to mg/m^3 (or other mass per volume measure) using a molar volume of 24.4 L/mole, and 3) to compare the exposure concentration to the TLV®, both in units of mass per volume.

A number of assumptions are made when comparing sampling results obtained under unusual atmospheric conditions to the TLVs®. One such assumption is that the volume of air inspired by the worker per workday is not appreciably different under moderate conditions of temperature and pressure as compared to NTP (Stephenson and Lillquist, 2001). An additional assumption for gases and vapors is that absorbed dose is correlated to the partial pressure of the inhaled compound. Sampling results obtained under unusual conditions cannot easily be compared to the published TLVs®, and extreme care should be exercised if workers are exposed to very high or low ambient pressures.

Unusual Work Schedules

Application of TLVs® to work schedules markedly different from the conventional 8-hour day, 40-hour workweek requires particular judgment to provide protection for these workers equal to that provided to workers on conventional work shifts. Short workweeks can allow workers to have more than one job, perhaps with similar exposures, and may result in overexposure, even if neither job by itself entails overexposure.

Numerous mathematical models to adjust for unusual work schedules have been described. In terms of toxicologic principles, their general objective is to identify a dose that ensures that the daily peak body burden or weekly peak body burden does not exceed that which occurs during a normal 8-hour/day, 5-day/week shift. A comprehensive review of the approaches to adjusting occupational exposure limits for unusual work schedules is provided in *Patty's Industrial Hygiene* (Paustenbach, 2000). Other selected readings on this topic include Lapare et al. (2003), Brodeur et al. (2001), Caldwell et al. (2001), Eide (2000), Verma (2000), Roach (1978), and Hickey and Reist (1977).

Another model that addresses unusual work schedules is the Brief and Scala model (1986), which is explained in detail in *Patty's Industrial Hygiene* (Paustenbach, 2000). This model reduces the TLV® proportionately for both increased exposure time and reduced recovery (i.e., non-exposure) time, and is generally intended to apply to work schedules longer than 8 hours/day or 40 hours/week. The model should not be used to justify very high exposures as "allowable" where the exposure periods are short (e.g., exposure to 8 times the TLV–TWA for 1 hour and zero exposure during the remainder of the shift). In this respect, the general limitations on TLV–TWA excursions and TLV–STELs should be applied to avoid inappropriate use of the model with very short exposure periods or shifts.

The Brief and Scala model is easier to use than some of the more complex models based on pharmacokinetic actions. The application of such models usually requires knowledge of the biological half-life of each substance, and some models require additional data. Another model developed by the University of Montreal and the Institute de Recherche en Sante et en Securite du Travail (IRSST) uses the Haber method to calculate adjusted exposure limits (Brodeur et al., 2001). This method generates values close to those obtained from physiologically based pharmacokinetic (PBPK) models.

Because adjusted TLVs® do not have the benefit of historical use and long-time observation, medical supervision during initial use of adjusted TLVs® is advised. Unnecessary exposure of workers should be avoided, even if a model shows such exposures to be "allowable." Mathematical models should not be used to justify higher-than-necessary exposures.

TLV® Units

TLVs® are expressed in ppm or mg/m^3. An inhaled chemical substance may exist as a gas, vapor, or aerosol.

- A gas is a chemical substance whose molecules are moving freely within a space in which they are confined (e.g., cylinder/tank) at normal temperature and pressure (NTP). Gases assume no shape or volume.
- A vapor is the gaseous phase of a chemical substance that exists as a liquid or a solid at NTP. The amount of vapor given off by a chemical substance is expressed as the vapor pressure and is a function of temperature and pressure.
- An aerosol is a suspension of solid particles or liquid droplets in a gaseous medium. Other terms used to describe an aerosol include dust, mist, fume, fog, fiber, smoke, and smog. Aerosols may be characterized by their aerodynamic behavior and the site(s) of deposition in the human respiratory tract.

TLVs® for aerosols are usually established in terms of mass of the chemical substance in air by volume. These TLVs® are expressed in mg/m^3.

TLVs® for gases and vapors are established in terms of parts of vapor or gas per million parts of contaminated air by volume (ppm), but may also be expressed in mg/m^3. For convenience to the user, these TLVs® also reference molecular weights. Where 24.45 = molar volume of air in liters at NTP conditions (25°C and 760 torr), the conversion equations for gases and vapors [ppm ↔ mg/m^3] are as follows:

$$\text{TLV in ppm} = \frac{(\text{TLV in mg/m}^3)\,(24.45)}{(\text{gram molecular weight of substance})}$$

OR

$$\text{TLV in mg/m}^3 = \frac{(\text{TLV in ppm})\,(\text{gram molecular weight of substance})}{24.45}$$

When converting values expressed as an element (e.g., as Fe, as Ni), the molecular weight of the element should be used, not that of the entire compound.

In making conversions for substances with variable molecular weights, appropriate molecular weights should be estimated or assumed (*see* the TLV® *Documentation*).

User Information

Each TLV® is supported by a comprehensive *Documentation*. It is imperative to consult the latest *Documentation* when applying the TLV®.

Additional copies of the *TLVs® and BEIs®* book and the multi-volume *Documentation of the Threshold Limit Values and Biological Exposure Indices*, upon which this book is based, are available from ACGIH®. *Documentation* of individual TLVs® is also available. Consult the ACGIH® website (www.acgih.org/store) for additional information and availability concerning these publications.

Endnotes and Abbreviations

- * 2013 Adoption.
- ‡ See Notice of Intended Changes (NIC).
- ( ) Adopted values or notations enclosed are those for which changes are proposed in the NIC.
- † 2013 Revision or Addition to the Notice of Intended Changes.
- A Refers to Appendix A: Carcinogenicity.
- C Ceiling limit; *see* definition in the "Introduction to the Chemical Substances."
- (D) Simple asphyxiant; *see* discussion covering *Minimal Oxygen Content* found in the "Definitions and Notations" section following the NIC tables.
- (E) The value is for particulate matter containing no asbestos and < 1% crystalline silica.
- (F) Respirable fibers: length > 5 µm; aspect ratio $\geq$ 3:1, as determined by the membrane filter method at 400–450X magnification (4-mm objective), using phase-contrast illumination.
- (G) As measured by the vertical elutriator, cotton-dust sampler; *see* the TLV® *Documentation*.
- (H) Aerosol only.
- (I) Inhalable fraction; *see* Appendix C, paragraph A.
- (IFV) Inhalable fraction and vapor; *see* Notations/Endnotes section, p. 74.
- (J) Does not include stearates of toxic metals.
- (K) Should not exceed 2 mg/m^3 respirable particulate mass.
- (L) Exposure by all routes should be carefully controlled to levels as low as possible.
- (M) Classification refers to sulfuric acid contained in strong inorganic acid mists.
- (O) Sampled by method that does not collect vapor.
- (P) Application restricted to conditions in which there are negligible aerosol exposures.
- (R) Respirable fraction; *see* Appendix C, paragraph C.
- (T) Thoracic fraction; *see* Appendix C, paragraph B.
- (V) Vapor and aerosol.

- B = Background; *see* BEI Intro.
- BEI = Substances for which there is a Biological Exposure Index or Indices (*see* BEI® section).
  - BEI_A: *see* BEI® for Acetylcholinesterase Inhibiting Pesticides
  - BEI_M: *see* BEI® for Methemoglobin Inducers
  - BEI_P: *see* BEI® for Polycyclic Aromatic Hydrocarbons (PAHs)
- DSEN = Dermal Sensitization; *see* definition in the "Definitions and Notations" section.
- MW = Molecular weight.
- NOS = Not otherwise specified.
- Nq = Nonquantitative; *see* BEI Intro.
- Ns = Nonspecific; *see* BEI Intro.
- RSEN = Respiratory Sensitization; *see* definition in the "Definitions and Notations" section.
- SEN = Sensitization; *see* definition in the "Definitions and Notations" section.
- Skin = Danger of cutaneous absorption; *see* discussion under Skin in the "Definitions and Notations" section.
- Sq = Semi-quantitative; *see* BEI Intro.
- STEL = Short-term exposure limit; *see* definition in the "Introduction to the Chemical Substances."
- TWA = 8-hour, time-weighted average; *see* definition in the "Introduction to the Chemical Substances."
- ppm = Parts of vapor or gas per million parts of contaminated air by volume at NTP conditions (25°C; 760 torr).
- mg/m^3 = Milligrams of substance per cubic meter of air.

ADOPTED VALUES

Substance [CAS No.]	TWA (ppm/mg/m^3)	STEL/C (ppm/mg/m^3)	Notations
‡ Acetaldehyde [75-07-0] (1992)	—	(C 25 ppm)	(A3)
Acetic acid [64-19-7] (2003)	10 ppm	15 ppm	—
Acetic anhydride [108-24-7] (2010)	1 ppm	3 ppm	A4
‡ Acetone [67-64-1] (1996)	(500 ppm)	(750 ppm)	(A4); BEI
Acetone cyanohydrin [75-86-5], as CN (1991)	—	C 5 mg/m^3	Skin
Acetonitrile [75-05-8] (1996)	20 ppm	—	Skin; A4
Acetophenone [98-86-2] (2008)	10 ppm	—	—
Acetylene [74-86-2] (1990)	Simple asphyxiant (D)		
Acetylsalicylic acid (Aspirin) [50-78-2] (1977)	5 mg/m^3	—	—
Acrolein [107-02-8] (1995)	—	C 0.1 ppm	Skin; A4
Acrylamide [79-06-1] (2004)	0.03 mg/m^3 (IFV)	—	Skin; A3
Acrylic acid [79-10-7] (1986)	2 ppm	—	Skin; A4
Acrylonitrile [107-13-1] (1997)	2 ppm	—	Skin; A3
Adipic acid [124-04-9] (1990)	5 mg/m^3	—	—
Adiponitrile [111-69-3] (1990)	2 ppm	—	Skin
‡ Alachlor [15972-60-8] (2006)	1 mg/m^3 (IFV)	—	(SEN); A3
Aldrin [309-00-2] (2006)	0.05 mg/m^3 (IFV)	—	Skin; A3
Allyl alcohol [107-18-6] (1996)	0.5 ppm	—	Skin; A4
Allyl bromide [106-95-6] (2011)	0.1 ppm	0.2 ppm	Skin; A4
Allyl chloride [107-05-1] (2010)	1 ppm	2 ppm	Skin; A3
Allyl glycidyl ether (AGE) [106-92-3] (1995)	1 ppm	—	A4
‡ Allyl propyl disulfide [2179-59-1] (2001)	0.5 ppm	—	(SEN)
Aluminum metal [7429-90-5] and insoluble compounds (2007)	1 mg/m^3 (R)	—	A4
4-Aminodiphenyl [92-67-1] (1968)	— (L)	—	Skin; A1
2-Aminopyridine [504-29-0] (1966)	0.5 ppm	—	—
Amitrole [61-82-5] (1983)	0.2 mg/m^3	—	A3
Ammonia [7664-41-7] (1970)	25 ppm	35 ppm	—
Ammonium chloride fume [12125-02-9] (1970)	10 mg/m^3	20 mg/m^3	—
Ammonium perfluorooctanoate [3825-26-1] (1992)	0.01 mg/m^3	—	Skin; A3
Ammonium sulfamate [7773-06-0] (1956)	10 mg/m^3	—	—
tert-Amyl methyl ether (TAME) [994-05-8] (1999)	20 ppm	—	—
Aniline [62-53-3] (1979)	2 ppm	—	Skin; A3; BEI
o-Anisidine [90-04-0] (1979)	0.5 mg/m^3	—	Skin; A3; BEI_M
p-Anisidine [104-94-9] (1979)	0.5 mg/m^3	—	Skin; A4; BEI_M
Antimony [7440-36-0] and compounds, as Sb (1979)	0.5 mg/m^3	—	—
Antimony hydride [7803-52-3] (1990)	0.1 ppm	—	—
Antimony trioxide [1309-64-4], production (1977)	— (L)	—	A2
ANTU [86-88-4] (1990)	0.3 mg/m^3	—	A4; Skin
‡ Argon [7440-37-1] (1990)	(Simple asphyxiant (D))		

ACGIH® disclaims liability with respect to the use of TLVs®.

Brief RS; Scala RA: Occupational health aspects of unusual work schedules: a review of Exxon's experiences. Am Ind Hyg Assoc J 47(4):199-202 (1986).

Brodeur J; Vyskocil A; Tardif R; et al.: Adjustment of permissible exposure values to unusual work schedules. Am Ind Hyg Assoc J 62:584-594 (2001).

Buringh E; Lanting R: Exposure variability in the workplace: its implications for the assessment of compliance. Am Ind Hyg Assoc J 52:6-13 (1991).

Caldwell DJ; Armstrong TW; Barone NJ; et al.: Lessons learned while compiling a quantitative exposure database from the published literature. Appl Occup Environ Hyg 16(2):174-177 (2001).

Eide I: The application of 8-hour occupational exposure limits to non-standard work schedules offshore. Ann Occup Hyg 34(1):13-17 (1990).

Hickey JL; Reist PC: Application of occupational exposure limits to unusual work schedules. Am Ind Hyg Assoc J 38(11):613-621 (1977).

Lapare S; Brodeur J; Tardif R: Contribution of toxicokinetic modeling to the adjustment of exposure limits to unusual work schedules. Am Ind Hyg Assoc J 64(1):17-23 (2003).

Leidel NA; Busch KA; Crouse WE: Exposure measurement action level and occupational environmental variability. DHEW (NIOSH) Pub. No. 76-131; NTIS Pub. No. PB- 267-509. U.S. National Technical Information Service, Springfield, VA (December 1975).

Paustenbach DJ: Pharmacokinetics and Unusual Work Schedules. In: Patty's Industrial Hygiene, 5th ed., Vol. 3, Part VI, Law, Regulation, and Management, Chap. 40, pp. 1787-1901. RL Harris, Ed. John Wiley & Sons, Inc., New York (2000).

Roach SA: Threshold limit values for extraordinary work schedules. Am Ind Hyg Assoc J 39(4):345-348 (1978).

Stephenson DJ; Lillquist DR: The effects of temperature and pressure on airborne exposure concentrations when performing compliance evaluations using ACGIH TLVs and OSHA PELs. Appl Occup Environ Hyg 16(4):482-486 (2001).

Verma DK: Adjustment of occupational exposure limits for unusual work schedules. Am Ind Hyg Assoc J 61(3):367-374 (2000).

Substance [CAS No.]	TWA (ppm/mg/m^3)	STEL/C (ppm/mg/m^3)	Notations
	ADOPTED VALUES		
Arsenic [7440-38-2] and inorganic compounds, as As (1990)	0.01 mg/m^3	—	A1; BEI
Arsine [7784-42-1] (2006)	0.005 ppm	—	—
Asbestos [1332-21-4], all forms (1994)	0.1 f/cc (F)	—	A1
Asphalt (Bitumen) fume [8052-42-4], as benzene-soluble aerosol (1999)	0.5 mg/m^3 (I)	—	A4; BEI$_P$
‡ Atrazine [1912-24-9] (and related symmetrical triazines) (1985)	(5 mg/m^3)	—	(A4)
‡ Azinphos-methyl [86-50-0] (1999)	0.2 mg/m^3 (IFV)	—	Skin; (SEN); A4; BEI$_A$
Barium [7440-39-3] and soluble compounds, as Ba (1990)	0.5 mg/m^3	—	A4
‡ Barium sulfate [7727-43-7] (1983)	(10 mg/m^3) (E)	—	—
‡ Benomyl [17804-35-2] (2007)	1 mg/m^3 (I)	—	(SEN); A3
Benz[a]anthracene [56-55-3] (1990)	— (L)	—	A2; BEI$_P$
Benzene [71-43-2] (1996)	0.5 ppm	2.5 ppm	Skin; A1; BEI
Benzidine [92-87-5] (1979)	— (L)	—	Skin; A1
Benzo[b]fluoranthene [205-99-2] (1990)	— (L)	—	A2; BEI$_P$
Benzo[a]pyrene [50-32-8] (1990)	— (L)	—	A2; BEI$_P$
Benzotrichloride [98-07-7] (1994)	—	C 0.1 ppm	Skin; A2
Benzoyl chloride [98-88-4] (1992)	—	C 0.5 ppm	A4
Benzoyl peroxide [94-36-0] (1990)	5 mg/m^3	—	A4
Benzyl acetate [140-11-4] (1990)	10 ppm	—	A4
Benzyl chloride [100-44-7] (1990)	1 ppm	—	A3
‡ Beryllium [7440-41-7] and compounds, as Be (2008)	0.00005 mg/m^3 (I)	—	Skin; (SEN); A1
Biphenyl [92-52-4] (1979)	0.2 ppm	—	—
Bismuth telluride [1304-82-1] (1970)			
Undoped, as Bi_2Te_3	10 mg/m^3	—	A4
Se-doped, as Bi_2Te_3	5 mg/m^3	—	A4
Borate compounds, inorganic [1303-96-4; 1330-43-4; 10043-35-3; 12179-04-3] (2004)	2 mg/m^3 (I)	6 mg/m^3 (I)	A4
Boron oxide [1303-86-2] (1985)	10 mg/m^3	—	—
Boron tribromide [10294-33-4] (1990)	—	C 1 ppm	—
Boron trifluoride [7637-07-2] (1962)	—	C 1 ppm	—
Bromacil [314-40-9] (1976)	10 mg/m^3	—	A3
Bromine [7726-95-6] (1991)	0.1 ppm	0.2 ppm	—
Bromine pentafluoride [7789-30-2] (1979)	0.1 ppm	—	—
Bromoform [75-25-2] (2008)	0.5 ppm	—	A3
‡ 1-Bromopropane [106-94-5] (2003)	(10 ppm)	—	(—)
1,3-Butadiene [106-99-0] (1994)	2 ppm	—	A2
* Butane, all isomers [75-28-5; 106-97-8] (2012)		1000 ppm	—
n-Butanol [71-36-3] (1998)	20 ppm	—	—
sec-Butanol [78-92-2] (2001)	100 ppm	—	—
tert-Butanol [75-65-0] (1992)	100 ppm	—	A4
Butenes, all isomers [106-98-9; 107-01-7; 590-18-1; 624-64-6; 25167-67-3]	250 ppm	—	—
Isobutene [115-11-7] (2007)	250 ppm	—	A4
2-Butoxyethanol (EGBE) [111-76-2] (1996)	20 ppm	—	A3; BEI
2-Butoxyethyl acetate (EGBEA) [112-07-2] (2000)	20 ppm	—	A3
n-Butyl acetate [123-86-4] (1995)	150 ppm	200 ppm	—
sec-Butyl acetate [105-46-4] (1965)	200 ppm	—	—
tert-Butyl acetate [540-88-5] (1965)	200 ppm	—	—
‡ n-Butyl acrylate [141-32-2] (1996)	2 ppm	—	(SEN); A4
n-Butylamine [109-73-9] (1985)	—	C 5 ppm	Skin
Butylated hydroxytoluene (BHT) [128-37-0] (2001)	2 mg/m^3 (IFV)	—	A4
tert-Butyl chromate, as CrO_3 [1189-85-1] (1960)	—	C 0.1 mg/m^3	Skin

Substance [CAS No.]	TWA (ppm/mg/m^3)	STEL/C (ppm/mg/m^3)	Notations
	ADOPTED VALUES		
‡ n-Butyl glycidyl ether (BGE) [2426-08-6] (2002)	3 ppm	—	Skin; (SEN)
n-Butyl lactate [138-22-7] (1973)	5 ppm	—	—
n-Butyl mercaptan [109-79-5] (1968)	0.5 ppm	—	—
o-sec-Butylphenol [89-72-5] (1977)	5 ppm	—	Skin
p-tert-Butyl toluene [98-51-1] (1990)	1 ppm	—	—
Cadmium [7440-43-9] and	0.01 mg/m^3	—	A2; BEI
compounds, as Cd (1990)	0.002 mg/m^3 (R)	—	A2; BEI
Calcium chromate [13765-19-0], as Cr (1988)	0.001 mg/m^3	—	A2
Calcium cyanamide [156-62-7] (1973)	0.5 mg/m^3	—	A4
Calcium hydroxide [1305-62-0] (1979)	5 mg/m^3	—	—
Calcium oxide [1305-78-8] (1990)	2 mg/m^3	—	—
Calcium silicate, synthetic nonfibrous [1344-95-2] (1988)	10 mg/m^3 (E)	—	A4
Calcium sulfate [7778-18-9; 10034-76-1; 10101-41-4; 13397-24-5] (2005)	10 mg/m^3 (I)	—	—
Camphor, synthetic [76-22-2] (1990)	2 ppm	3 ppm	A4
Caprolactam [105-60-2] (1997)	5 mg/m^3 (IFV)	—	A5
Captafol [2425-06-1] (1990)	0.1 mg/m^3	—	Skin; A4
‡ Captan [133-06-2] (1999)	5 mg/m^3 (I)	—	(SEN); A3
Carbaryl [63-25-2] (2007)	0.5 mg/m^3 (IFV)	—	Skin; A4; BEI$_A$
Carbofuran [1563-66-2] (2001)	0.1 mg/m^3 (IFV)	—	A4; BEI$_A$
Carbon black [1333-86-4] (2010)	3 mg/m^3 (I)	—	A3
Carbon dioxide [124-38-9] (1983)	5000 ppm	30,000 ppm	—
Carbon disulfide [75-15-0] (2005)	1 ppm	—	Skin; A4; BEI
Carbon monoxide [630-08-0] (1989)	25 ppm	—	BEI
Carbon tetrabromide [558-13-4] (1972)	0.1 ppm	0.3 ppm	—
Carbon tetrachloride [56-23-5] (1990)	5 ppm	10 ppm	Skin; A2
Carbonyl fluoride [353-50-4] (1990)	2 ppm	5 ppm	—
Carbonyl sulfide [463-58-1] (2011)	5 ppm	—	—
Catechol [120-80-9] (1985)	5 ppm	—	Skin; A3
Cellulose [9004-34-6] (1985)	10 mg/m^3	—	—
Cesium hydroxide [21351-79-1] (1990)	2 mg/m^3	—	—
Chlordane [57-74-9] (1985)	0.5 mg/m^3	—	Skin; A3
Chlorinated camphene [8001-35-2] (1990)	0.5 mg/m^3	1 mg/m^3	Skin; A3
o-Chlorinated diphenyl oxide [31242-93-0] (1979)	0.5 mg/m^3	—	—
Chlorine [7782-50-5] (1986)	0.5 ppm	1 ppm	A4
Chlorine dioxide [10049-04-4] (1991)	0.1 ppm	0.3 ppm	—
Chlorine trifluoride [7790-91-2] (1979)	—	C 0.1 ppm	—
Chloroacetaldehyde [107-20-0] (1990)	—	C 1 ppm	—
Chloroacetone [78-95-5] (1986)	—	C 1 ppm	Skin
2-Chloroacetophenone [532-27-4] (1990)	0.05 ppm	—	A4
Chloroacetyl chloride [79-04-9] (1988)	0.05 ppm	0.15 ppm	Skin
Chlorobenzene [108-90-7] (1988)	10 ppm	—	A3; BEI
o-Chlorobenzylidene malononitrile [2698-41-1] (1990)	—	C 0.05 ppm	Skin; A4
Chlorobromomethane [74-97-5] (2008)	200 ppm	—	—
Chlorodifluoromethane [75-45-6] (1990)	1000 ppm	—	A4
Chlorodiphenyl (42% chlorine) [53469-21-9] (1979)	1 mg/m^3	—	Skin
Chlorodiphenyl (54% chlorine) [11097-69-1] (1990)	0.5 mg/m^3	—	Skin; A3
Chloroform [67-66-3] (1990)	10 ppm	—	A3
bis(Chloromethyl) ether [542-88-1] (1979)	0.001 ppm	—	A1
Chloromethyl methyl ether [107-30-2] (1979)	— (L)	—	A2
1-Chloro-1-nitropropane [600-25-9] (1971)	2 ppm	—	—
Chloropentafluoroethane [76-15-3] (1978)	1000 ppm	—	—
Chloropicrin [76-06-2] (1990)	0.1 ppm	—	A4
1-Chloro-2-propanol [127-00-4] and 2-Chloro-1-propanol [78-89-7] (1999)	1 ppm	—	Skin; A4
β-Chloroprene [126-99-8] (1990)	10 ppm	—	Skin

Substance [CAS No.]	ADOPTED VALUES TWA (ppm/mg/m^3)	STEL/C (ppm/mg/m^3)	Notations
2-Chloropropionic acid [598-78-7] (1988)	0.1 ppm	—	Skin
o-Chlorostyrene [2039-87-4] (1972)	50 ppm	75 ppm	—
o-Chlorotoluene [95-49-8] (1971)	50 ppm	—	—
Chlorpyrifos [2921-88-2] (2000)	0.1 mg/m^3 (IFV)	—	Skin; A4; BEI$_A$
Chromite ore processing (Chromate), as Cr	0.05 mg/m^3	—	A1
Chromium, [7440-47-3] and inorganic compounds, as Cr (1991)			
Metal and Cr III compounds	0.5 mg/m^3	—	A4
Water-soluble Cr VI compounds	0.05 mg/m^3	—	A1; BEI
Insoluble Cr VI compounds	0.01 mg/m^3	—	A1
Chromyl chloride [14977-61-8] (1990)	0.025 ppm	—	—
Chrysene [218-01-9] (1990)	— (L)	—	A3; BEI$_P$
‡ Citral [5392-40-5] (2009)	5 ppm (IFV)	—	Skin; (SEN); A4
* Clopidol [2971-90-6] (2012)	3 mg/m^3 (IFV)	—	A4
Coal dust (1995)			
Anthracite	0.4 mg/m^3 (R)	—	A4
Bituminous or Lignite	0.9 mg/m^3 (R)	—	A4
Coal tar pitch volatiles [65996-93-2], as benzene soluble aerosol (1984)	0.2 mg/m^3	—	A1; BEI$_P$
Cobalt [7440-48-4] and inorganic compounds, as Co (1993)	0.02 mg/m^3		A3; BEI
Cobalt carbonyl [10210-68-1], as Co (1980)	0.1 mg/m^3	—	—
Cobalt hydrocarbonyl [16842-03-8], as Co (1980)	0.1 mg/m^3	—	—
Copper [7440-50-8] (1990)			
Fume, as Cu	0.2 mg/m^3	—	—
Dusts and mists, as Cu	1 mg/m^3	—	—
Cotton dust, raw, untreated (2009)	0.1 mg/m^3 (T)	—	A4
Coumaphos [56-72-4] (2005)	0.05 mg/m^3 (IFV)	—	Skin; A4; BEI$_A$
Cresol, all isomers (2009) [95-48-7; 106-44-5; 108-39-4; 1319-77-3]	20 mg/m^3 (IFV)	—	Skin; A4
Crotonaldehyde [4170-30-3] (1995)	—	C 0.3 ppm	Skin; A3
Crufomate [299-86-5] (1971)	5 mg/m^3	—	A4; BEI$_A$
Cumene [98-82-8] (1997)	50 ppm	—	—
Cyanamide [420-04-2] (1974)	2 mg/m^3	—	—
Cyanogen [460-19-5] (1966)	10 ppm	—	—
Cyanogen chloride [506-77-4] (1977)	—	C 0.3 ppm	—
Cyclohexane [110-82-7] (1964)	100 ppm	—	—
Cyclohexanol [108-93-0] (1979)	50 ppm	—	Skin
Cyclohexanone [108-94-1] (1990)	20 ppm	50 ppm	Skin; A3
Cyclohexene [110-83-8] (1964)	300 ppm	—	—
Cyclohexylamine [108-91-8] (1990)	10 ppm	—	A4
Cyclonite [121-82-4] (1994)	0.5 mg/m^3	—	Skin; A4
Cyclopentadiene [542-92-7] (1963)	75 ppm	—	—
Cyclopentane [287-92-3] (1978)	600 ppm	—	—
Cyhexatin [13121-70-5] (1990)	5 mg/m^3	—	A4
* 2,4-D [94-75-7] (2012)	10 mg/m^3 (I)	—	Skin; A4
DDT [50-29-3] (1979)	1 mg/m^3	—	A3
Decaborane [17702-41-9] (1979)	0.05 ppm	0.15 ppm	Skin
Demeton [8065-48-3] (1998)	0.05 mg/m^3 (IFV)	—	Skin; BEI$_A$
‡ Demeton-S-methyl [919-86-8] (1998)	0.05 mg/m^3 (IFV)	—	Skin; (SEN); A4; BEI$_A$
Diacetone alcohol [123-42-2] (1979)	50 ppm	—	—
Diacetyl [431-03-8] (2011)	0.01 ppm	0.02 ppm	A4
Diazinon [333-41-5] (2000)	0.01 mg/m^3 (IFV)	—	Skin; A4; BEI$_A$
Diazomethane [334-88-3] (1970)	0.2 ppm	—	A2
Diborane [19287-45-7] (1990)	0.1 ppm	—	—
2-N-Dibutylaminoethanol [102-81-8] (1980)	0.5 ppm	—	Skin; BEI$_A$
Dibutyl phenyl phosphate [2528-36-1] (1987)	0.3 ppm	—	Skin; BEI$_A$

Substance [CAS No.]	ADOPTED VALUES TWA (ppm/mg/m^3)	STEL/C (ppm/mg/m^3)	Notations
Dibutyl phosphate [107-66-4] (2008)	5 mg/m^3 (IFV)	—	Skin
Dibutyl phthalate [84-74-2] (1990)	5 mg/m^3	—	—
Dichloroacetic acid [79-43-6] (2002)	0.5 ppm	—	Skin; A3
Dichloroacetylene [7572-29-4] (1992)	—	C 0.1 ppm	A3
o-Dichlorobenzene [95-50-1] (1990)	25 ppm	50 ppm	A4
p-Dichlorobenzene [106-46-7] (1990)	10 ppm	—	A3
3,3'-Dichlorobenzidine [91-94-1] (1990)	— (L)	—	Skin; A3
1,4-Dichloro-2-butene [764-41-0] (1990)	0.005 ppm	—	Skin; A2
Dichlorodifluoromethane [75-71-8] (1979)	1000 ppm	—	A4
1,3-Dichloro-5,5-dimethyl hydantoin [118-52-5] (1979)	0.2 mg/m^3	0.4 mg/m^3	—
1,1-Dichloroethane [75-34-3] (1990)	100 ppm	—	A4
1,2-Dichloroethylene, all isomers [156-59-2; 156-60-5; 540-59-0] (1990)	200 ppm	—	—
Dichloroethyl ether [111-44-4] (1985)	5 ppm	10 ppm	Skin; A4
Dichlorofluoromethane [75-43-4] (1977)	10 ppm	—	—
Dichloromethane [75-09-2] (1997)	50 ppm	—	A3; BEI
1,1-Dichloro-1-nitroethane [594-72-9] (1978)	2 ppm	—	—
1,3-Dichloropropene [542-75-6] (2003)	1 ppm	—	Skin; A3
2,2-Dichloropropionic acid [75-99-0] (1997)	5 mg/m^3 (I)	—	A4
Dichlorotetrafluoroethane [76-14-2] (1979)	1000 ppm	—	A4
‡ Dichlorvos (DDVP) [62-73-7] (1998)	0.1 mg/m^3 (IFV)	—	Skin; (SEN); A4; BEI$_A$
Dicrotophos [141-66-2] (1998)	0.05 mg/m^3 (IFV)	—	Skin; A4; BEI$_A$
Dicyclopentadiene [77-73-6] (1973)	5 ppm	—	—
Dicyclopentadienyl iron, as Fe [102-54-5] (1990)	10 mg/m^3	—	—
Dieldrin [60-57-1] (2009)	0.1 mg/m^3 (IFV)	—	Skin; A3
Diesel fuel [68334-30-5; 68476-30-2; 68476-31-3; 68476-34-6; 77650-28-3], as total hydrocarbons (2007)	100 mg/m^3 (IFV)	—	Skin; A3
Diethanolamine [111-42-2] (2008)	1 mg/m^3 (IFV)	—	Skin; A3
Diethylamine [109-89-7] (1992)	5 ppm	15 ppm	Skin; A4
2-Diethylaminoethanol [100-37-8] (1991)	2 ppm	—	Skin
* Diethylene glycol monobutyl ether [112-34-5] (2012)	10 ppm (IFV)	—	—
Diethylene triamine [111-40-0] (1985)	1 ppm	—	Skin
Di(2-ethylhexyl)phthalate (DEHP) [117-81-7] (1996)	5 mg/m^3	–	A3
* N,N-Diethylhydroxylamine [3710-84-7] (2012)	2 ppm	—	—
Diethyl ketone [96-22-0] (1995)	200 ppm	300 ppm	—
Diethyl phthalate [84-66-2] (1996)	5 mg/m^3	—	A4
Difluorodibromomethane [75-61-6] (1962)	100 ppm	—	—
Diglycidyl ether (DGE) [2238-07-5] (2006)	0.01 ppm	—	A4
Diisobutyl ketone [108-83-8] (1979)	25 ppm	—	—
Diisopropylamine [108-18-9] (1979)	5 ppm	—	Skin
N,N-Dimethyl acetamide [127-19-5] (1990)	10 ppm	—	Skin; A4; BEI
‡ Dimethylamine [124-40-3] (1989)	5 ppm	15 ppm	( ); A4
bis(2-Dimethylaminoethyl)ether (DMAEE) [3033-62-3] (1997)	0.05 ppm	0.15 ppm	Skin
Dimethylaniline [121-69-7] (1990)	5 ppm	10 ppm	Skin; A4; BEI$_M$
Dimethyl carbamoyl chloride [79-44-7] (2006)	0.005 ppm	—	Skin; A2
Dimethyl disulfide [624-92-0] (2006)	0.5 ppm	—	Skin
Dimethylethoxysilane [14857-34-2] (1991)	0.5 ppm	1.5 ppm	—
Dimethylformamide [68-12-2] (1979)	10 ppm	—	Skin; A4; BEI
1,1-Dimethylhydrazine [57-14-7] (1993)	0.01 ppm	—	Skin; A3
Dimethyl phthalate [131-11-3] (2005)	5 mg/m^3	—	—
Dimethyl sulfate [77-78-1] (1985)	0.1 ppm	—	Skin; A3
Dimethyl sulfide [75-18-3] (2001)	10 ppm	—	—

ADOPTED VALUES			
Substance [CAS No.]	TWA (ppm/mg/m^3)	STEL/C (ppm/mg/m^3)	Notations
Dinitrobenzene, all isomers [99-65-0; 100-25-4; 528-29-0; 25154-54-5] (1979)	0.15 ppm	—	Skin; BEI$_M$
Dinitro-o-cresol [534-52-1] (1979)	0.2 mg/m^3	—	Skin
3,5-Dinitro-o-toluamide [148-01-6] (2006)	1 mg/m^3	—	A4
Dinitrotoluene [25321-14-6] (1993)	0.2 mg/m^3	—	Skin; A3; BEI$_M$
1,4-Dioxane [123-91-1] (1996)	20 ppm	—	Skin; A3
Dioxathion [78-34-2] (2001)	0.1 mg/m^3 (IFV)	—	Skin; A4; BEI$_A$
1,3-Dioxolane [646-06-0] (1997)	20 ppm	—	—
Diphenylamine [122-39-4] (1990)	10 mg/m^3	—	A4
Dipropyl ketone [123-19-3] (1978)	50 ppm	—	—
Diquat [85-00-7; 2764-72-9; 6385-62-2]	0.5 mg/m^3 (I)	—	Skin; A4
(1990)	0.1 mg/m^3 (R)	—	Skin; A4
Disulfiram [97-77-8] (1979)	2 mg/m^3	—	A4
Disulfoton [298-04-4] (2000)	0.05 mg/m^3 (IFV)	—	Skin; A4; BEI$_A$
Diuron [330-54-1] (1974)	10 mg/m^3	—	A4
Divinyl benzene [1321-74-0] (1990)	10 ppm	—	—
‡ Dodecyl mercaptan [112-55-0] (2001)	0.1 ppm	—	(SEN)
Endosulfan [115-29-7] (2008)	0.1 mg/m^3 (IFV)	—	Skin; A4
Endrin [72-20-8] (1979)	0.1 mg/m^3	—	Skin; A4
Enflurane [13838-16-9] (1979)	75 ppm	—	A4
Epichlorohydrin [106-89-8] (1994)	0.5 ppm	—	Skin; A3
EPN [2104-64-5] (2000)	0.1 mg/m^3 (I)	—	Skin; A4; BEI$_A$
* Ethane [74-84-0]	*See* Appendix F: Minimal Oxygen Content		
Ethanol [64-17-5] (2008)	—	1000 ppm	A3
Ethanolamine [141-43-5] (1985)	3 ppm	6 ppm	—
Ethion [563-12-2] (2000)	0.05 mg/m^3 (IFV)	—	Skin; A4; BEI$_A$
2-Ethoxyethanol (EGEE) [110-80-5] (1981)	5 ppm	—	Skin; BEI
2-Ethoxyethyl acetate (EGEEA) [111-15-9] (1981)	5 ppm	—	Skin; BEI
Ethyl acetate [141-78-6] (1979)	400 ppm	—	—
Ethyl acrylate [140-88-5] (1986)	5 ppm	15 ppm	A4
Ethylamine [75-04-7] (1991)	5 ppm	15 ppm	Skin
Ethyl amyl ketone [541-85-5] (2006)	10 ppm	—	—
Ethyl benzene [100-41-4] (2010)	20 ppm	—	A3; BEI
Ethyl bromide [74-96-4] (1990)	5 ppm	—	Skin; A3
* Ethyl tert-butyl ether [637-92-3] (2012)	25 ppm	—	A4
Ethyl butyl ketone [106-35-4] (1995)	50 ppm	75 ppm	—
Ethyl chloride [75-00-3] (1992)	100 ppm	—	Skin; A3
Ethyl cyanoacrylate [7085-85-0] (1995)	0.2 ppm	—	—
Ethylene [74-85-1] (2001)	200 ppm	—	A4
Ethylene chlorohydrin [107-07-3] (1985)	—	C 1 ppm	Skin; A4
Ethylenediamine [107-15-3] (1990)	10 ppm	—	Skin; A4
Ethylene dibromide [106-93-4] (1980)	—	—	Skin; A3
Ethylene dichloride [107-06-2] (1977)	10 ppm	—	A4
Ethylene glycol [107-21-1] (1992)	—	C 100 mg/m^3 (H)	A4
Ethylene glycol dinitrate (EGDN) [628-96-6] (1980)	0.05 ppm	—	Skin
Ethylene oxide [75-21-8] (1990)	1 ppm	—	A2
Ethyleneimine [151-56-4] (2008)	0.05 ppm	0.1 ppm	Skin; A3
Ethyl ether [60-29-7] (1966)	400 ppm	500 ppm	—
Ethyl formate [109-94-4] (2011)	—	100 ppm	A4
2-Ethylhexanoic acid [149-57-5] (2006)	5 mg/m^3 (IFV)	—	—
‡ Ethylidene norbornene [16219-75-3] (1971)	(—)	(C 5 ppm)	—
Ethyl mercaptan [75-08-1] (2003)	0.5 ppm	—	—
N-Ethylmorpholine [100-74-3] (1985)	5 ppm	—	Skin
Ethyl silicate [78-10-4] (1979)	10 ppm	—	—
Fenamiphos [22224-92-6] (2005)	0.05 mg/m^3 (IFV)	—	Skin; A4; BEI$_A$
Fensulfothion [115-90-2] (2004)	0.01 mg/m^3 (IFV)	—	Skin; A4; BEI$_A$
Fenthion [55-38-9] (2005)	0.05 mg/m^3 (IFV)	—	Skin; A4; BEI$_A$
Ferbam [14484-64-1] (2008)	5 mg/m^3 (I)	—	A4

ADOPTED VALUES			
Substance [CAS No.]	TWA (ppm/mg/m^3)	STEL/C (ppm/mg/m^3)	Notations
Ferrovanadium dust [12604-58-9] (1990)	1 mg/m^3	3 mg/m^3	—
‡ Flour dust (2001)	0.5 mg/m^3 (I)	—	(SEN)
Fluorides, as F (1979)	2.5 mg/m^3	—	A4; BEI
Fluorine [7782-41-4] (1970)	1 ppm	2 ppm	—
Fonofos [944-22-9] (2005)	0.1 mg/m^3 (IFV)	—	Skin; A4; BEI$_A$
Formaldehyde [50-00-0] (1987)	—	C 0.3 ppm	SEN; A2
Formamide [75-12-7] (1985)	10 ppm	—	Skin
Formic acid [64-18-6] (1965)	5 ppm	10 ppm	—
Furfural [98-01-1] (1978)	2 ppm	—	Skin; A3; BEI
Furfuryl alcohol [98-00-0] (1979)	10 ppm	15 ppm	Skin
Gallium arsenide [1303-00-0] (2004)	0.0003 mg/m^3 (R)	—	A3
Gasoline [86290-81-5] (1990)	300 ppm	500 ppm	A3
Germanium tetrahydride [7782-65-2] (1970)	0.2 ppm	—	—
Glutaraldehyde [111-30-8], activated or inactivated (1998)	—	C 0.05 ppm	SEN; A4
Glycidol [556-52-5] (1993)	2 ppm	—	A3
‡ Glyoxal [107-22-2] (1999)	0.1 mg/m^3 (IFV)	—	(SEN); A4
Grain dust (oat, wheat, barley) (1985)	4 mg/m^3	—	—
Graphite (all forms except graphite fibers) [7782-42-5] (1988)	2 mg/m^3 (R)	—	—
Hafnium [7440-58-6] and compounds, as Hf (1990)	0.5 mg/m^3	—	—
Halothane [151-67-7] (1979)	50 ppm	—	A4
‡ Helium [7440-59-7] (1990)	(Simple asphyxiant (D))		
Heptachlor [76-44-8] and Heptachlor epoxide [1024-57-3] (1990)	0.05 mg/m^3	—	Skin; A3
Heptane, all isomers [108-08-7; 142-82-5; 565-59-3; 589-34-4; 590-35-2; 591-76-4] (1979)	400 ppm	500 ppm	—
Hexachlorobenzene [118-74-1] (1994)	0.002 mg/m^3	—	Skin; A3
Hexachlorobutadiene [87-68-3] (1979)	0.02 ppm	—	Skin; A3
Hexachlorocyclopentadiene [77-47-4] (1990)	0.01 ppm	—	A4
Hexachloroethane [67-72-1] (1990)	1 ppm	—	Skin; A3
Hexachloronaphthalene [1335-87-1] (1965)	0.2 mg/m^3	—	Skin
Hexafluoroacetone [684-16-2] (1986)	0.1 ppm	—	Skin
Hexafluoropropylene [116-15-4] (2009)	0.1 ppm	—	—
‡ Hexahydrophthalic anhydride, all isomers [85-42-7; 13149-00-3; 14166-21-3] (2002)	—	C 0.005 mg/m^3 (IFV)	(SEN)
Hexamethylene diisocyanate [822-06-0] (1985)	0.005 ppm	—	—
Hexamethyl phosphoramide [680-31-9] (1990)	—	—	Skin; A3
n-Hexane [110-54-3] (1996)	50 ppm	—	Skin; BEI
Hexane isomers, other than n-Hexane [75-83-2; 79-29-8; 96-14-0; 107-83-5] (1979)	500 ppm	1000 ppm	—
1,6-Hexanediamine [124-09-4] (1990)	0.5 ppm	—	—
1-Hexene [592-41-6] (1999)	50 ppm	—	—
sec-Hexyl acetate [108-84-9] (1963)	50 ppm	—	—
Hexylene glycol [107-41-5] (1974)	—	C 25 ppm	—
Hydrazine [302-01-2] (1988)	0.01 ppm	—	Skin; A3
‡ Hydrogen [1333-74-0] (1990)	(Simple asphyxiant (D))		
Hydrogenated terphenyls (nonirradiated) [61788-32-7] (1990)	0.5 ppm	—	—
Hydrogen bromide [10035-10-6] (2001)	—	C 2 ppm	—
Hydrogen chloride [7647-01-0] (2000)	—	C 2 ppm	A4
Hydrogen cyanide and cyanide salts, as CN (1991)			
Hydrogen cyanide [74-90-8]	—	C 4.7 ppm	Skin
Cyanide salts [143-33-9; 151-50-8; 592-01-8]	—	C 5 mg/m^3	Skin
Hydrogen fluoride [7664-39-3], as F (2004)	0.5 ppm	C 2 ppm	Skin; BEI
Hydrogen peroxide [7722-84-1] (1990)	1 ppm	—	A3

ADOPTED VALUES			
Substance [CAS No.]	TWA ($ppm/mg/m^3$)	STEL/C ($ppm/mg/m^3$)	Notations
Hydrogen selenide [7783-07-5], as Se (1990)	0.05 ppm	—	—
Hydrogen sulfide [7783-06-4] (2009)	1 ppm	5 ppm	—
‡ Hydroquinone [123-31-9] (2007)	1 mg/m^3	—	(SEN); A3
‡ 2-Hydroxypropyl acrylate [999-61-1] (1997)	0.5 ppm	—	Skin; (SEN)
Indene [95-13-6] (2007)	5 ppm	—	—
Indium [7440-74-6] and compounds, as In (1990)	0.1 mg/m^3	—	—
Iodine and iodides (2007)			
Iodine [7553-56-2]	0.01 ppm$^{(IFV)}$	0.1 ppm$^{(V)}$	A4
Iodides	0.01 ppm$^{(IFV)}$	—	A4
Iodoform [75-47-8] (1979)	0.6 ppm	—	—
Iron oxide (Fe_2O_3) [1309-37-1] (2005)	5 $mg/m^{3\,(R)}$	—	A4
Iron pentacarbonyl [13463-40-6], as Fe (1979)	0.1 ppm	0.2 ppm	—
Iron salts, soluble, as Fe (1990)	1 mg/m^3	—	—
Isoamyl alcohol [123-51-3] (1990)	100 ppm	125 ppm	—
Isobutanol [78-83-1] (1973)	50 ppm	—	—
Isobutyl acetate [110-19-0] (1966)	150 ppm	—	—
Isobutyl nitrite [542-56-3] (2000)	—	C 1 ppm$^{(IFV)}$	A3; BEI$_M$
Isooctyl alcohol [26952-21-6] (1990)	50 ppm	—	Skin
Isophorone [78-59-1] (1992)	—	C 5 ppm	A3
Isophorone diisocyanate [4098-71-9] (1985)	0.005 ppm	—	—
2-Isopropoxyethanol [109-59-1] (1990)	25 ppm	—	Skin
Isopropyl acetate [108-21-4] (2001)	100 ppm	200 ppm	—
Isopropylamine [75-31-0] (1962)	5 ppm	10 ppm	—
N-Isopropylaniline [768-52-5] (1990)	2 ppm	—	Skin; BEI$_M$
Isopropyl ether [108-20-3] (1979)	250 ppm	310 ppm	—
Isopropyl glycidyl ether (IGE) [4016-14-2] (1979)	50 ppm	75 ppm	—
Kaolin [1332-58-7] (1990)	2 $mg/m^{3\,(E,R)}$	—	A4
Kerosene [8008-20-6; 64742-81-0]/ Jet fuels, as total hydrocarbon vapor (2003)	200 $mg/m^{3\,(P)}$	—	Skin; A3
Ketene [463-51-4] (1962)	0.5 ppm	1.5 ppm	—
Lead [7439-92-1] and inorganic compounds, as Pb (1991)	0.05 mg/m^3	—	A3; BEI
Lead chromate [7758-97-6], as Pb	0.05 mg/m^3	—	A2; BEI
(1990) as Cr	0.012 mg/m^3	—	A2
Lindane [58-89-9] (1990)	0.5 mg/m^3	—	Skin; A3
Lithium hydride [7580-67-8] (1990)	0.025 mg/m^3	—	—
* L.P.G. (Liquefied petroleum gas) [68476-85-7]	*See* Appendix F: Minimal Oxygen Content		
Magnesium oxide [1309-48-4] (2000)	10 $mg/m^{3\,(I)}$	—	A4
Malathion [121-75-5] (2000)	1 $mg/m^{3\,(IFV)}$	—	Skin; A4; BEI$_A$
‡ Maleic anhydride [108-31-6] (2010)	0.01 $mg/m^{3\,(IFV)}$	—	(SEN); A4
* Manganese [7439-96-5], elemental and inorganic compounds, as Mn (2012)	0.02 $mg/m^{3\,(R)}$ 0.1 $mg/m^{3\,(I)}$	—	A4
Manganese cyclopentadienyl tricarbonyl [12079-65-1], as Mn (1992)	0.1 mg/m^3	—	Skin
Mercury [7439-97-6], alkyl compounds, as Hg (1992)	0.01 mg/m^3	0.03 mg/m^3	Skin
Mercury [7439-97-6], all forms except alkyl, as Hg (1991)			
Aryl compounds	0.1 mg/m^3	—	Skin
Elemental and inorganic forms	0.025 mg/m^3	—	Skin; A4; BEI
Mesityl oxide [141-79-7] (1992)	15 ppm	25 ppm	—
Methacrylic acid [79-41-4] (1992)	20 ppm	—	—
* Methane [74-82-8]	*See* Appendix F: Minimal Oxygen Content		
Methanol [67-56-1] (2008)	200 ppm	250 ppm	Skin; BEI
‡ Methomyl [16752-77-5] (1992)	(2.5 mg/m^3)	—	(A4); BEI$_A$
Methoxychlor [72-43-5] (1992)	10 mg/m^3	—	A4
2-Methoxyethanol (EGME) [109-86-4] (2005)	0.1 ppm	—	Skin; BEI
2-Methoxyethyl acetate (EGMEA) [110-49-6] (2005)	0.1 ppm	—	Skin; BEI
(2-Methoxymethylethoxy)propanol (DPGME) [34590-94-8] (1979)	100 ppm	150 ppm	Skin
4-Methoxyphenol [150-76-5] (1992)	5 mg/m^3	—	—
* 1-Methoxy-2-propanol [107-98-2] (2012)	50 ppm	100 ppm	A4
Methyl acetate [79-20-9] (1992)	200 ppm	250 ppm	—
Methyl acetylene [74-99-7] (1956)	1000 ppm	—	—
Methyl acetylene-propadiene mixture (MAPP) [59355-75-8] (1964)	1000 ppm	1250 ppm	—
‡ Methyl acrylate [96-33-3] (1997)	2 ppm	—	Skin; (SEN); A4
Methylacrylonitrile [126-98-7] (2010)	1 ppm	—	Skin; A4
Methylal [109-87-5] (1970)	1000 ppm	—	—
Methylamine [74-89-5] (1990)	5 ppm	15 ppm	—
Methyl n-amyl ketone [110-43-0] (1978)	50 ppm	—	—
N-Methyl aniline [100-61-8] (1992)	0.5 ppm	—	Skin; BEI$_M$
Methyl bromide [74-83-9] (1994)	1 ppm	—	Skin; A4
Methyl tert-butyl ether (MTBE) [1634-04-4] (1999)	50 ppm	—	A3
Methyl n-butyl ketone [591-78-6] (1995)	5 ppm	10 ppm	Skin; BEI
Methyl chloride [74-87-3] (1992)	50 ppm	100 ppm	Skin; A4
Methyl chloroform [71-55-6] (1992)	350 ppm	450 ppm	A4; BEI
Methyl 2-cyanoacrylate [137-05-3] (1995)	0.2 ppm	—	—
Methyl cyclohexane [108-87-2] (1962)	400 ppm	—	—
Methylcyclohexanol [25639-42-3] (2005)	50 ppm	—	—
o-Methylcyclohexanone [583-60-8] (1970)	50 ppm	75 ppm	Skin
2-Methylcyclopentadienyl manganese tricarbonyl [12108-13-3], as Mn (1970)	0.2 mg/m^3	—	Skin
Methyl demeton [8022-00-2] (2006)	0.05 $mg/m^{3\,(IFV)}$	—	Skin; BEI$_A$
Methylene bisphenyl isocyanate (MDI) [101-68-8] (1985)	0.005 ppm	—	—
4,4'-Methylene bis(2-chloroaniline) (MBOCA) [101-14-4] (1991)	0.01 ppm	—	Skin; A2; BEI
Methylene bis(4-cyclohexylisocyanate) [5124-30-1] (1985)	0.005 ppm	—	—
4,4'-Methylene dianiline [101-77-9] (1992)	0.1 ppm	—	Skin; A3
Methyl ethyl ketone (MEK) [78-93-3] (1992)	200 ppm	300 ppm	BEI
Methyl ethyl ketone peroxide [1338-23-4] (1992)	—	C 0.2 ppm	—
‡ Methyl formate [107-31-3] (1962)	(100 ppm)	150 ppm	(—)
Methyl hydrazine [60-34-4] (1991)	0.01 ppm	—	Skin; A3
Methyl iodide [74-88-4] (1978)	2 ppm	—	Skin
* Methyl isoamyl ketone [110-12-3] (2012)	20 ppm	50 ppm	—
Methyl isobutyl carbinol [108-11-2] (1966)	25 ppm	40 ppm	Skin
Methyl isobutyl ketone [108-10-1] (2009)	20 ppm	75 ppm	A3; BEI
‡ Methyl isocyanate [624-83-9] (1986)	0.02 ppm	(—)	Skin; ( )
Methyl isopropyl ketone [563-80-4] (2010)	20 ppm	—	—
Methyl mercaptan [74-93-1] (2003)	0.5 ppm	—	—
‡ Methyl methacrylate [80-62-6] (1992)	50 ppm	100 ppm	(SEN); A4
1-Methyl naphthalene [90-12-0] and 2-Methyl naphthalene [91-57-6] (2006)	0.5 ppm	—	Skin; A4
Methyl parathion [298-00-0] (2008)	0.02 $mg/m^{3\,(IFV)}$	—	Skin; A4; BEI$_A$
Methyl propyl ketone [107-87-9] (2006)	—	150 ppm	—
Methyl silicate [681-84-5] (1978)	1 ppm	—	—
α-Methyl styrene [98-83-9] (2009)	10 ppm	—	A3
Methyl vinyl ketone [78-94-4] (1994)	—	C 0.2 ppm	Skin; SEN
Metribuzin [21087-64-9] (1981)	5 mg/m^3	—	A4
Mevinphos [7786-34-7] (1998)	0.01 $mg/m^{3\,(IFV)}$	—	Skin; A4; BEI$_A$
Mica [12001-26-2] (1962)	3 $mg/m^{3\,(R)}$	—	—

ADOPTED VALUES			
Substance [CAS No.]	**TWA (ppm/mg/m^3)**	**STEL/C (ppm/mg/m^3)**	**Notations**
Mineral oil, excluding metal working fluids (2009)			
Pure, highly and severely refined	5 mg/m^3 $^{(I)}$	—	A4
Poorly and mildly refined	— $^{(L)}$	—	A2
Molybdenum [7439-98-7], as Mo (1999)			
Soluble compounds	0.5 mg/m^3 $^{(R)}$	—	A3
Metal and insoluble compounds	10 mg/m^3 $^{(I)}$	—	—
	3 mg/m^3 $^{(R)}$	—	—
Monochloroacetic acid [79-11-8] (2005)	0.5 ppm $^{(IFV)}$	—	Skin; A4
Monocrotophos [6923-22-4] (2002)	0.05 mg/m^3 $^{(IFV)}$	—	Skin; A4; BEI$_A$
Morpholine [110-91-8] (1992)	20 ppm	—	Skin; A4
‡ Naled [300-76-5] (2002)	0.1 mg/m^3 $^{(IFV)}$	—	Skin; (SEN); A4; BEI$_A$
‡ Naphthalene [91-20-3] (1992)	10 ppm	(15 ppm)	Skin; (A4)
β-Naphthylamine [91-59-8] (1979)	— $^{(L)}$	—	A1
* Natural gas [8006-14-2]	*See* Appendix F: Minimal Oxygen Content		
‡ Natural rubber latex [9006-04-6], as inhalable allergenic proteins (2007)	0.0001 mg/m^3 $^{(I)}$	—	Skin; (SEN)
‡ Neon [7440-01-9] (1992)	(Simple asphyxiant $^{(D)}$)		
Nickel [7440-02-0] and inorganic compounds including Nickel subsulfide, as Ni (1996)			
Elemental [7440-02-0]	1.5 mg/m^3 $^{(I)}$	—	A5
Soluble inorganic compounds (NOS)	0.1 mg/m^3 $^{(I)}$	—	A4
Insoluble inorganic compounds (NOS)	0.2 mg/m^3 $^{(I)}$	—	A1
Nickel subsulfide [12035-72-2], as Ni	0.1 mg/m^3 $^{(I)}$	—	A1
‡ Nickel carbonyl [13463-39-3], as Ni (1980)	(0.05 ppm)	(—)	(—)
Nicotine [54-11-5] (1992)	0.5 mg/m^3	—	Skin
Nitrapyrin [1929-82-4] (1992)	10 mg/m^3	20 mg/m^3	A4
Nitric acid [7697-37-2] (1992)	2 ppm	4 ppm	—
Nitric oxide [10102-43-9] (1992)	25 ppm	—	BEI$_M$
p-Nitroaniline [100-01-6] (1992)	3 mg/m^3	—	Skin; A4; BEI$_M$
Nitrobenzene [98-95-3] (1992)	1 ppm	—	Skin; A3; BEI
p-Nitrochlorobenzene [100-00-5] (1985)	0.1 ppm	—	Skin; A3; BEI$_M$
4-Nitrodiphenyl [92-93-3] (1992)	— $^{(L)}$	—	Skin; A2
Nitroethane [79-24-3] (1979)	100 ppm	—	—
‡ Nitrogen [7727-37-9] (1992)	(Simple asphyxiant $^{(D)}$)		
Nitrogen dioxide [10102-44-0] (2011)	0.2 ppm	—	A4
Nitrogen trifluoride [7783-54-2] (1992)	10 ppm	—	BEI$_M$
Nitroglycerin (NG) [55-63-0] (1980)	0.05 ppm	—	Skin
Nitromethane [75-52-5] (1997)	20 ppm	—	A3
1-Nitropropane [108-03-2] (1992)	25 ppm	—	A4
2-Nitropropane [79-46-9] (1992)	10 ppm	—	A3
N-Nitrosodimethylamine [62-75-9] (1992)	— $^{(L)}$	—	Skin; A3
Nitrotoluene, all isomers (1992) [88-72-2; 99-08-1; 99-99-0]	2 ppm	—	Skin; BEI$_M$
3,5-Nitro-o-toluidine [99-55-8] (2006)	1 mg/m^3 $^{(I)}$	—	A3
Nitrous oxide [10024-97-2] (1986)	50 ppm	—	A4
Nonane [111-84-2] (2011)	200 ppm	—	—
Octachloronaphthalene [2234-13-1] (1970)	0.1 mg/m^3	0.3 mg/m^3	Skin
Octane [111-65-9], all isomers (1979)	300 ppm	—	—
Osmium tetroxide [20816-12-0], as Os (1979)	0.0002 ppm	0.0006 ppm	—
Oxalic acid [144-62-7] (1992)	1 mg/m^3	2 mg/m^3	—
p,p'-Oxybis(benzenesulfonyl hydrazide) [80-51-3] (1997)	0.1 mg/m^3 $^{(I)}$	—	—
Oxygen difluoride [7783-41-7] (1983)	—	C 0.05 ppm	—
Ozone [10028-15-6] (1995)			
Heavy work	0.05 ppm	—	A4
Moderate work	0.08 ppm	—	A4
Light work	0.10 ppm	—	A4

ADOPTED VALUES			
Substance [CAS No.]	**TWA (ppm/mg/m^3)**	**STEL/C (ppm/mg/m^3)**	**Notations**
Heavy, moderate, or light workloads (≤ 2 hours)	0.20 ppm	—	A4
Paraffin wax fume [8002-74-2] (1972)	2 mg/m^3	—	—
Paraquat [4685-14-7], as the cation (1979)	0.5 mg/m^3	—	—
	0.1 mg/m^3 $^{(R)}$	—	—
Parathion [56-38-2] (2000)	0.05 mg/m^3 $^{(IFV)}$	—	Skin; A4; BEI
Particles (insoluble or poorly soluble) not otherwise specified	*See* Appendix B		
Pentaborane [19624-22-7] (1970)	0.005 ppm	0.015 ppm	—
Pentachloronaphthalene [1321-64-8] (1970)	0.5 mg/m^3	—	Skin
Pentachloronitrobenzene [82-68-8] (1988)	0.5 mg/m^3	—	A4
‡ Pentachlorophenol [87-86-5] (1992)	(0.5 mg/m^3)	(—)	Skin; A3; BEI
Pentaerythritol [115-77-5] (1970)	10 mg/m^3	—	—
‡ Pentane, all isomers [78-78-4; 109-66-0; 463-82-1] (1989)	(600 ppm)	—	
2,4-Pentanedione [123-54-6] (2010)	25 ppm	—	Skin
Pentyl acetate, all isomers [123-92-2; 620-11-1; 624-41-9; 625-16-1; 626-38-0; 628-63-7] (1997)	50 ppm	100 ppm	—
Perchloromethyl mercaptan [594-42-3] (1988)	0.1 ppm	—	—
Perchloryl fluoride [7616-94-6] (1962)	3 ppm	6 ppm	—
Perfluorobutyl ethylene [19430-93-4] (2001)	100 ppm	—	—
Perfluoroisobutylene [382-21-8] (1989)	—	C 0.01 ppm	—
Persulfates, as persulfate (1993)	0.1 mg/m^3	—	—
Phenol [108-95-2] (1992)	5 ppm	—	Skin; A4; BEI
Phenothiazine [92-84-2] (1968)	5 mg/m^3	—	Skin
N-Phenyl-β-naphthylamine [135-88-6] (1992)	— $^{(L)}$	—	A4
o-Phenylenediamine [95-54-5] (1988)	0.1 mg/m^3	—	A3
m-Phenylenediamine [108-45-2] (1988)	0.1 mg/m^3	—	A4
p-Phenylenediamine [106-50-3] (1988)	0.1 mg/m^3	—	A4
Phenyl ether [101-84-8], vapor (1979)	1 ppm	2 ppm	—
‡ Phenyl glycidyl ether (PGE) [122-60-1] (1992)	0.1 ppm	—	Skin; (SEN); A3
Phenylhydrazine [100-63-0] (1988)	0.1 ppm	—	Skin; A3
Phenyl mercaptan [108-98-5] (2001)	0.1 ppm	—	Skin
Phenylphosphine [638-21-1] (1992)	—	C 0.05 ppm	—
Phorate [298-02-2] (2002)	0.05 mg/m^3 $^{(IFV)}$	—	Skin; A4; BEI$_A$
Phosgene [75-44-5] (1992)	0.1 ppm	—	—
Phosphine [7803-51-2] (1992)	0.3 ppm	1 ppm	—
Phosphoric acid [7664-38-2] (1992)	1 mg/m^3	3 mg/m^3	—
Phosphorus (yellow) [12185-10-3] (1992)	0.1 mg/m^3	—	—
Phosphorus oxychloride [10025-87-3] (1979)	0.1 ppm	—	—
Phosphorus pentachloride [10026-13-8] (1985)	0.1 ppm	—	—
Phosphorus pentasulfide [1314-80-3] (1992)	1 mg/m^3	3 mg/m^3	—
Phosphorus trichloride [7719-12-2] (1992)	0.2 ppm	0.5 ppm	—
‡ Phthalic anhydride [85-44-9] (1992)	1 ppm	—	(SEN); A4
m-Phthalodinitrile [626-17-5] (2008)	5 mg/m^3 $^{(IFV)}$	—	—
o-Phthalodinitrile [91-15-6] (2011)	1 mg/m^3 $^{(IFV)}$	—	—
Picloram [1918-02-1] (1992)	10 mg/m^3	—	A4
Picric acid [88-89-1] (1992)	0.1 mg/m^3	—	—
Pindone [83-26-1] (1992)	0.1 mg/m^3	—	—
‡ Piperazine and salts [110-85-0], as piperazine (2011)	0.03 ppm $^{(IFV)}$	—	(SEN); A4
Platinum [7440-06-4], and soluble salts (1979)			
Metal	1 mg/m^3	—	—
Soluble salts, as Pt	0.002 mg/m^3	—	—
Polyvinyl chloride (PVC) [9002-86-2] (2007)	1 mg/m^3 $^{(R)}$	—	A4
Portland cement [65997-15-1] (2009)	1 mg/m^3 $^{(E,R)}$	—	A4
Potassium hydroxide [1310-58-3] (1992)	—	C 2 mg/m^3	—
* Propane [74-98-6]	*See* Appendix F: Minimal Oxygen Content		

ADOPTED VALUES			
Substance [CAS No.]	TWA (ppm/mg/m^3)	STEL/C (ppm/mg/m^3)	Notations
Propane sultone [1120-71-4] (1976)	— (L)	—	A3
n-Propanol (n-Propyl alcohol) [71-23-8] (2006)	100 ppm	—	A4
2-Propanol [67-63-0] (2001)	200 ppm	400 ppm	A4; BEI
Propargyl alcohol [107-19-7] (1992)	1 ppm	—	Skin
β-Propiolactone [57-57-8] (1992)	0.5 ppm	—	A3
Propionaldehyde [123-38-6] (1998)	20 ppm	—	—
Propionic acid [79-09-4] (1977)	10 ppm	—	—
Propoxur [114-26-1] (1992)	0.5 mg/m^3	—	A3; BEI$_A$
n-Propyl acetate [109-60-4] (1962)	200 ppm	250 ppm	—
Propylene [115-07-1] (2005)	500 ppm	—	A4
‡ Propylene dichloride [78-87-5] (2005)	10 ppm	—	(SEN); A4
Propylene glycol dinitrate [6423-43-4] (1980)	0.05 ppm	—	Skin; BEI$_M$
‡ Propylene oxide [75-56-9] (2000)	2 ppm	—	(SEN); A3
Propyleneimine [75-55-8] (2008)	0.2 ppm	0.4 ppm	Skin; A3
n-Propyl nitrate [627-13-4] (1962)	25 ppm	40 ppm	BEI$_M$
Pyrethrum [8003-34-7] (1992)	5 mg/m^3	—	A4
Pyridine [110-86-1] (1992)	1 ppm	—	A3
Quinone [106-51-4] (1970)	0.1 ppm	—	—
Resorcinol [108-46-3] (1992)	10 ppm	20 ppm	A4
Rhodium [7440-16-6], as Rh (1981)			
Metal and Insoluble compounds	1 mg/m^3	—	A4
Soluble compounds	0.01 mg/m^3	—	A4
Ronnel [299-84-3] (2005)	5 mg/m^3 (IFV)	—	A4; BEI$_A$
‡ Rosin core solder thermal decomposition products (colophony) [8050-09-7] (1992)	— (L)	—	(SEN)
Rotenone (commercial) [83-79-4] (1992)	5 mg/m^3	—	A4
Selenium [7782-49-2] and compounds, as Se (1992)	0.2 mg/m^3	—	—
Selenium hexafluoride [7783-79-1], as Se (1992)	0.05 ppm	—	—
Sesone [136-78-7] (1992)	10 mg/m^3	—	A4
Silica, crystalline — α-quartz [1317-95-9; 14808-60-7] and cristobalite [14464-46-1] (2009)	0.025 mg/m^3 (R)	—	A2
Silicon carbide [409-21-2] (2002)			
Nonfibrous	10 mg/m^3 (I,E)	—	—
	3 mg/m^3 (R,E)	—	—
Fibrous (including whiskers)	0.1 f/cc (F)	—	A2
Silicon tetrahydride [7803-62-5] (1992)	5 ppm	—	—
Silver [7440-22-4], and compounds (1992)			
Metal, dust and fume	0.1 mg/m^3	—	—
Soluble compounds, as Ag	0.01 mg/m^3	—	—
Sodium azide [26628-22-8] (1992)			
as Sodium azide	—	C 0.29 mg/m^3	A4
as Hydrazoic acid vapor	—	C 0.11 ppm	A4
Sodium bisulfite [7631-90-5] (1992)	5 mg/m^3	—	A4
Sodium fluoroacetate [62-74-8] (1992)	0.05 mg/m^3	—	Skin
Sodium hydroxide [1310-73-2] (1992)	—	C 2 mg/m^3	—
Sodium metabisulfite [7681-57-4] (1992)	5 mg/m^3	—	A4
Starch [9005-25-8] (1992)	10 mg/m^3	—	A4
Stearates (J) (1985)	10 mg/m^3	—	A4
Stoddard solvent [8052-41-3] (1980)	100 ppm	—	—
Strontium chromate [7789-06-2], as Cr (1989)	0.0005 mg/m^3	—	A2
Strychnine [57-24-9] (1992)	0.15 mg/m^3	—	—
Styrene, monomer [100-42-5] (1996)	20 ppm	40 ppm	A4; BEI
Subtilisins [1395-21-7; 9014-01-1], as 100% crystalline active pure enzyme (1972)	—	C 0.00006 mg/m^3	—
Sucrose [57-50-1] (1992)	10 mg/m^3	—	A4
Sulfometuron methyl [74222-97-2] (1991)	5 mg/m^3	—	A4
Sulfotepp (TEDP) [3689-24-5] (1993)	0.1 mg/m^3 (IFV)	—	Skin; A4; BEI$_A$
Sulfur dioxide [7446-09-5] (2008)	—	0.25 ppm	A4
Sulfur hexafluoride [2551-62-4] (1985)	1000 ppm	—	—
Sulfuric acid [7664-93-9] (2000)	0.2 mg/m^3 (T)	—	A2 (M)
Sulfur monochloride [10025-67-9] (1986)	—	C 1 ppm	—
Sulfur pentafluoride [5714-22-7] (1962)	—	C 0.01 ppm	—
Sulfur tetrafluoride [7783-60-0] (1992)	—	C 0.1 ppm	—
Sulfuryl fluoride [2699-79-8] (1992)	5 ppm	10 ppm	—
Sulprofos [35400-43-2] (2008)	0.1 mg/m^3 (IFV)	—	Skin; A4; BEI$_A$
Synthetic vitreous fibers (1999)			
Continuous filament glass fibers	1 f/cc (F)	—	A4
Continuous filament glass fibers	5 mg/m^3 (I)	—	A4
Glass wool fibers	1 f/cc (F)	—	A3
Rock wool fibers	1 f/cc (F)	—	A3
Slag wool fibers	1 f/cc (F)	—	A3
Special purpose glass fibers	1 f/cc (F)	—	A3
Refractory ceramic fibers	0.2 f/cc (F)	—	A2
2,4,5-T [93-76-5] (1992)	10 mg/m^3	—	A4
Talc [14807-96-6] (2009)			
Containing no asbestos fibers	2 mg/m^3 (E,R)	—	A4
Containing asbestos fibers	Use Asbestos TLV® (K)	—	A1
Tellurium [13494-80-9] and compounds (NOS), as Te, excluding hydrogen telluride (1992)	0.1 mg/m^3	—	—
Tellurium hexafluoride [7783-80-4], as Te (1992)	0.02 ppm	—	—
Temephos [3383-96-8] (2002)	1 mg/m^3 (IFV)	—	Skin; A4; BEI$_A$
Terbufos [13071-79-9] (1999)	0.01 mg/m^3 (IFV)	—	Skin; A4; BEI$_A$
Terephthalic acid [100-21-0] (1990)	10 mg/m^3	—	—
Terphenyls [26140-60-3] (1977)	—	C 5 mg/m^3	—
1,1,2,2-Tetrabromoethane [79-27-6] (2005)	0.1 ppm (IFV)	—	—
1,1,1,2-Tetrachloro-2,2-difluoroethane [76-11-9] (2007)	100 ppm	—	—
1,1,2,2-Tetrachloro-1,2-difluoroethane [76-12-0] (2007)	50 ppm	—	—
1,1,2,2-Tetrachloroethane [79-34-5] (1995)	1 ppm	—	Skin; A3
Tetrachloroethylene [127-18-4] (1990)	25 ppm	100 ppm	A3; BEI
Tetrachloronaphthalene [1335-88-2] (1992)	2 mg/m^3	—	—
Tetraethyl lead [78-00-2], as Pb (1992)	0.1 mg/m^3	—	Skin; A4
Tetraethyl pyrophosphate (TEPP) [107-49-3] (2006)	0.01 mg/m^3 (IFV)	—	Skin; BEI$_A$
Tetrafluoroethylene [116-14-3] (1997)	2 ppm	—	A3
Tetrahydrofuran [109-99-9] (2002)	50 ppm	100 ppm	Skin; A3
‡ Tetrakis (hydroxymethyl) phosphonium salts (2002)			
Tetrakis (hydroxymethyl) phosphonium chloride [124-64-1]	2 mg/m^3	—	(SEN); A4
Tetrakis (hydroxymethyl) phosphonium sulfate [55566-30-8]	2 mg/m^3	—	(SEN); A4
Tetramethyl lead [75-74-1], as Pb (1992)	0.15 mg/m^3	—	Skin
Tetramethyl succinonitrile [3333-52-6] (1992)	0.5 ppm	—	Skin
Tetranitromethane [509-14-8] (1992)	0.005 ppm	—	A3
Tetryl [479-45-8] (1984)	1.5 mg/m^3	—	—
Thallium [7440-28-0] and compounds, as Tl (2009)	0.02 mg/m^3 (I)	—	Skin
4,4'-Thiobis(6-tert-butyl-m-cresol) [96-69-5] (2010)	1 mg/m^3 (I)	—	A4
Thioglycolic acid [68-11-1] (1992)	1 ppm	—	Skin
Thionyl chloride [7719-09-7] (2009)	—	C 0.2 ppm	—
‡ Thiram [137-26-8] (2007)	0.05 mg/m^3 (IFV)	—	(SEN); A4
Tin [7440-31-5], and inorganic compounds, excluding Tin hydride, as Sn (1992)			
Metal	2 mg/m^3	—	—
Oxide and inorganic compounds	2 mg/m^3	—	—

ADOPTED VALUES			
Substance [CAS No.]	**TWA (ppm/mg/m^3)**	**STEL/C (ppm/mg/m^3)**	**Notations**
Tin [7440-31-5], organic compounds, as Sn	0.1 mg/m^3	0.2 mg/m^3	Skin; A4
‡ Titanium dioxide [13463-67-7] (1992)	(10 mg/m^3)	—	(A4)
o-Tolidine [119-93-7] (1992)	—	—	Skin; A3
Toluene [108-88-3] (2006)	20 ppm	—	A4; BEI
‡ Toluene-2,4- or 2,6-diisocyanate (or as a mixture) [584-84-9; 91-08-7] (1992)	(0.005 ppm)	(0.02 ppm)	( ); (SEN); (A4)
o-Toluidine [95-53-4] (1984)	2 ppm	—	Skin; A3; BEI$_M$
m-Toluidine [108-44-1] (1984)	2 ppm	—	Skin; A4; BEI$_M$
p-Toluidine [106-49-0] (1984)	2 ppm	—	Skin; A3; BEI$_M$
* Tributyl phosphate [126-73-8] (2012)	5 mg/m$^{3\,(IFV)}$	—	A3; BEI$_A$
‡ Trichloroacetic acid [76-03-9] (1992)	(1 ppm)	—	A3
1,2,4-Trichlorobenzene [120-82-1] (1975)	—	C 5 ppm	—
1,1,2-Trichloroethane [79-00-5] (1992)	10 ppm	—	Skin; A3
Trichloroethylene [79-01-6] (2006)	10 ppm	25 ppm	A2; BEI
Trichlorofluoromethane [75-69-4] (1992)	—	C 1000 ppm	A4
Trichloronaphthalene [1321-65-9] (1970)	5 mg/m^3	—	Skin
‡ 1,2,3-Trichloropropane [96-18-4] (1992)	(10 ppm)	—	(Skin; A3)
1,1,2-Trichloro-1,2,2-trifluoroethane [76-13-1] (1992)	1000 ppm	1250 ppm	A4
Trichlorphon [52-68-6] (1998)	1 mg/m$^{3\,(I)}$	—	A4; BEI$_A$
Triethanolamine [102-71-6] (1990)	5 mg/m^3	—	—
‡ Triethylamine [121-44-8] (1991)	1 ppm	(3 ppm)	Skin; A4
Trifluorobromomethane [75-63-8] (1979)	1000 ppm	—	—
1,3,5-Triglycidyl-s-triazinetrione [2451-62-9] (1994)	0.05 mg/m^3	—	—
‡ Trimellitic anhydride [552-30-7] (2007)	0.0005 mg/m$^{3\,(IFV)}$	0.002 mg/m$^{3\,(IFV)}$	Skin; (SEN)
Trimethylamine [75-50-3] (1990)	5 ppm	15 ppm	—
Trimethyl benzene (mixed isomers) [25551-13-7] (1970)	25 ppm	—	—
Trimethyl phosphite [121-45-9] (1980)	2 ppm	—	—
2,4,6-Trinitrotoluene (TNT) [118-96-7] (1984)	0.1 mg/m^3	—	Skin; BEI$_M$
Triorthocresyl phosphate [78-30-8] (1992)	0.1 mg/m^3	—	Skin; A4; BEI$_A$
Triphenyl phosphate [115-86-6] (1992)	3 mg/m^3	—	A4
Tungsten [7440-33-7], as W (1979)			
Metal and insoluble compounds	5 mg/m^3	10 mg/m^3	—
Soluble compounds	1 mg/m^3	3 mg/m^3	—
‡ Turpentine [8006-64-2] and selected monoterpenes [80-56-8; 127-91-3; 13466-78-9] (2001)	20 ppm	—	(SEN); A4
Uranium (natural) [7440-61-1] (1992)			
Soluble and insoluble compounds, as U	0.2 mg/m^3	0.6 mg/m^3	A1; BEI
n-Valeraldehyde [110-62-3] (1984)	50 ppm	—	—
Vanadium pentoxide [1314-62-1], as V (2008)	0.05 mg/m$^{3\,(I)}$	—	A3
Vinyl acetate [108-05-4] (1992)	10 ppm	15 ppm	A3
Vinyl bromide [593-60-2] (1996)	0.5 ppm	—	A2
Vinyl chloride [75-01-4] (1997)	1 ppm	—	A1
4-Vinyl cyclohexene [100-40-3] (1994)	0.1 ppm	—	A3
Vinyl cyclohexene dioxide [106-87-6] (1994)	0.1 ppm	—	Skin; A3
Vinyl fluoride [75-02-5] (1996)	1 ppm	—	A2
N-Vinyl-2-pyrrolidone [88-12-0] (2000)	0.05 ppm	—	A3
Vinylidene chloride [75-35-4] (1992)	5 ppm	—	A4
Vinylidene fluoride [75-38-7] (1996)	500 ppm	—	A4
Vinyl toluene [25013-15-4] (1992)	50 ppm	100 ppm	A4
Warfarin [81-81-2] (1992)	0.1 mg/m^3	—	—
‡ Wood dusts (2011)			
Western red cedar	0.5 mg/m$^{3\,(I)}$	—	(SEN); A4
All other species	1 mg/m$^{3\,(I)}$	—	—
Carcinogenicity			
Oak and beech	—	—	A1
Birch, mahogany, teak, walnut	—	—	A2
All other wood dusts	—	—	A4
Xylene [1330-20-7] (o, m & p isomers) [95-47-6; 108-38-3; 106-42-3] (1992)	100 ppm	150 ppm	A4; BEI
m-Xylene α,α'-diamine [1477-55-0] (1992)	—	C 0.1 mg/m^3	Skin
Xylidine (mixed isomers) [1300-73-8] (1999)	0.5 ppm $^{(IFV)}$	—	Skin; A3; BEI$_M$
Yttrium [7440-65-5] and compounds, as Y (1986)	1 mg/m^3	—	—
Zinc chloride fume [7646-85-7] (1992)	1 mg/m^3	2 mg/m^3	—
Zinc chromates [11103-86-9; 13530-65-9; 37300-23-5], as Cr (1992)	0.01 mg/m^3	—	A1
Zinc oxide [1314-13-2] (2001)	2 mg/m$^{3\,(R)}$	10 mg/m$^{3\,(R)}$	—
Zirconium [7440-67-7] and compounds, as Zr (1992)	5 mg/m^3	10 mg/m^3	A4

NOTICE OF INTENDED CHANGES FOR 2013

These substances, with their corresponding values and notations, comprise those for which 1) a limit is proposed for the first time, 2) a change in the Adopted value is proposed, 3) retention as an NIC is proposed, or 4) withdrawal of the *Documentation* and adopted TLV® is proposed. In each case, the proposals should be considered trial values during the period they are on the NIC. These proposals were ratified by the ACGIH® Board of Directors and will remain on the NIC for approximately one year following this ratification. If the Committee neither finds nor receives any substantive data that change its scientific opinion regarding an NIC TLV®, the Committee may then approve its recommendation to the ACGIH® Board of Directors for adoption. If the Committee finds or receives substantive data that change its scientific opinion regarding an NIC TLV®, the Committee may change its recommendation to the ACGIH® Board of Directors for the matter to be either retained on or withdrawn from the NIC.

Documentation is available for each of these substances and their proposed values.

This notice provides an opportunity for comment on these proposals. Comments or suggestions should be accompanied by substantiating evidence in the form of peer-reviewed literature and forwarded in electronic format to The Science Group, ACGIH®, at science@acgih.org. Please refer to the ACGIH® TLV®/BEI® Development Process on the ACGIH® website (http://www.acgih.org/TLV/DevProcess.htm) for a detailed discussion covering this procedure, methods for input to ACGIH®, and deadline date for receiving comments.

NOTICE OF INTENDED CHANGES (for 2013)			
Substance [CAS No.]	**TWA (ppm/mg/m^3)**	**STEL/C (ppm/mg/m^2)**	**Notations**
† Acetaldehyde [75-07-0]	—	C 25 ppm	A2
Acetone [67-64-1]	200 ppm	500 ppm	BEI
† Argon [7440-37-1]	WITHDRAW ADOPTED *DOCUMENTATION*; REFER TO APPENDIX F: MINIMAL OXYGEN CONTENT		
† Atrazine [1912-24-9]	2 mg/m$^{3\,(I)}$	—	A3
† Barium sulfate [7727-43-7]	5 mg/m$^{3\,(I)\,(E)}$	—	—
1-Bromopropane [106-94-5]	0.1 ppm	—	A3
† Dimethylamine [124-40-3]	5 ppm	15 ppm	DSEN; A4
† Ethyl isocyanate [109-90-0]	0.02 ppm	0.06 ppm	Skin; DSEN
† Ethylidene norbornene [16219-75-3]	2 ppm	4 ppm	—
† Helium [7440-59-7]	WITHDRAW ADOPTED *DOCUMENTATION*; REFER TO APPENDIX F: MINIMAL OXYGEN CONTENT		
† Hydrogen [1333-74-0]	WITHDRAW ADOPTED *DOCUMENTATION*; REFER TO APPENDIX F: MINIMAL OXYGEN CONTENT		
† Methomyl [16752-77-5]	0.2 mg/m$^{3\,(IFV)}$	—	Skin; BEI$_A$

Substance [CAS No.]	TWA	STEL/C	Notations
† Methyl formate [107-31-3]	50 ppm	150 ppm	Skin
† Methyl isocyanate [624-83-9]	0.02 ppm	0.06 ppm	Skin; DSEN
† Naphthalene [91-20-3]	10 ppm	—	Skin; A3
† Neon [7440-01-9]	WITHDRAW ADOPTED *DOCUMENTATION*; REFER TO APPENDIX F: MINIMAL OXYGEN CONTENT		
† Nickel carbonyl [13463-39-3], as Ni	—	C 0.05 ppm	A3
† Nitrogen [7727-37-9]	WITHDRAW ADOPTED *DOCUMENTATION*; REFER TO APPENDIX F: MINIMAL OXYGEN CONTENT		
† Pentachlorophenol [87-86-5]	0.5 mg/m^3 (IFV)	1 mg/m^3 (IFV)	Skin; A3; BEI
† Pentane, all isomers [78-78-4; 109-66-0; 463-82-1]	1000 ppm	—	—
† Peracetic acid [79-21-0]	—	0.4 ppm (IFV)	A4
† Phenyl isocyanate [103-71-9]	0.005 ppm	0.02 ppm	DSEN; RSEN
† Simazine [122-34-9]	0.5 mg/m^3 (I)	—	A4
† Titanium dioxide [13463-67-7]	1 mg/m^3 (R)	—	A3s
† Toluene-2,4- or 2,6-diisocyanate (or as a mixture) [584-84-9; 91-08-7]	0.001 ppm (IFV)	0.003 ppm (IFV)	Skin; DSEN; RSEN; A3; BEI
Trichloroacetic acid [76-03-9]	0.5 ppm	—	A3
1,2,3-Trichloropropane [96-18-4]	0.05 ppm	—	A2
† Triethylamine [121-44-8]	1 ppm	2 ppm	Skin; A4

NOTICE OF INTENDED CHANGES FOR 2013

2013 Notice of Intended Changes — Updates of Sensitizer (SEN) Notations to Dermal Sensitizer (DSEN) and/or Respiratory Sensitizer (RSEN). The following substances are placed on the NIC for review of this notation only.

NOTICE OF INTENDED CHANGES (for 2013)

Substance [CAS No.]	TWA (ppm/mg/m^3)	STEL/C (ppm/mg/m^3)	Notations
† Alachlor [15972-60-8]	1 mg/m^3 (IFV)	—	DSEN; A3
† Allyl propyl disulfide [2179-59-1]	0.5 ppm	—	DSEN
† Azinphos-methyl [86-50-0]	0.2 mg/m^3 (IFV)	—	Skin; DSEN; A4; BEI$_A$
† Benomyl [17804-35-2]	1 mg/m^3 (I)	—	DSEN; A3
† Beryllium [7440-41-7] and compounds, as Be	0.00005 mg/m^3 (I)	—	Skin; DSEN; RSEN; A1
† n-Butyl acrylate [141-32-2]	2 ppm	—	DSEN; A4
† n-Butyl glycidyl ether [2426-08-6]	3 ppm	—	Skin; DSEN
† Captan [133-06-2]	5 mg/m^3 (I)	—	DSEN; A3
† Citral [5392-40-5]	5 ppm (IFV)	—	Skin; DSEN; A4
† Demeton-S-methyl [919-86-8]	0.05 mg/m^3 (IFV)	—	Skin; DSEN; A4; BEI$_A$
† Dichlorvos [62-73-7]	0.1 mg/m^3 (IFV)	—	Skin; DSEN; A4; BEI$_A$
† Dodecyl mercaptan [112-55-0]	0.1 ppm	—	DSEN
† Flour dust	0.5 mg/m^3 (I)	—	RSEN
† Glyoxal [107-22-2]	0.1 mg/m^3 (IFV)	—	DSEN; A4
† Hexahydrophthalic anhydride, all isomers [85-42-7; 13149-00-3; 14166-21-3]	—	C 0.005 mg/m^3 (IFV)	DSEN; RSEN
† Hydroquinone [123-31-9]	1 mg/m^3	—	DSEN; A3
† 2-Hydroxypropyl acrylate [999-61-1]	0.5 ppm	—	Skin; DSEN
† Maleic anhydride [108-31-6]	0.01 mg/m^3 (IFV)	—	DSEN; RSEN; A4
† Methyl acrylate [96-33-3]	2 ppm	—	Skin; DSEN; A4
† Methyl methacrylate [80-62-6]	50 ppm	100 ppm	DSEN; RSEN; A4
† Naled [300-76-5]	0.1 mg/m^3 (IFV)	—	Skin; DSEN; A4; BEI$_A$
† Natural rubber latex [9006-04-6], as inhalable allergenic proteins	0.0001 mg/m^3 (I)	—	Skin; DSEN; RSEN
† Phenyl glycidyl ether [122-60-1]	0.1 ppm	—	Skin; DSEN; A3
† Phthalic anhydride [85-44-9]	1 ppm	—	DSEN; RSEN; A4
† Piperazine and salts [110-85-0], as piperazine	0.03 ppm (IFV)	—	DSEN; RSEN; A4
† Propylene dichloride [78-87-5]	10 ppm	—	DSEN; A4
† Propylene oxide [75-56-9]	2 ppm	—	DSEN; A3
† Rosin core solder thermal decomposition products (colophony) [8050-09-7]	— (L)	—	DSEN; RSEN
† Tetrakis (hydroxymethyl) phosphonium salts			
Tetrakis (hydroxymethyl) phosphonium chloride [124-64-1]	2 mg/m^3	—	DSEN; A4
Tetrakis (hydroxymethyl) phosphonium sulfate [55566-30-8]	2 mg/m^3	—	DSEN; A4
† Thiram [137-26-8]	0.05 mg/m^3 (IFV)	—	DSEN; A4
† Trimellitic anhydride [552-30-7]	0.0005 mg/m^3 (IFV)	0.002 mg/m^3 (IFV)	Skin; DSEN; RSEN
† Turpentine [8006-64-2] and selected monoterpenes [80-56-8; 127-91-3; 13466-78-9]	20 ppm	—	DSEN; A4
† Wood dusts			
Western red cedar	0.5 mg/m^3 (I)	—	DSEN; RSEN; A4
All other species	1 mg/m^3 (I)	—	—
Carcinogenicity			
Oak and beech	—	—	A1
Birch, mahogany, teak, walnut	—	—	A2
All other wood dusts	—	—	A4

Note: *See* Revisions or Additions (page xviii) for proposed revisions to the TLV®/BEI® Development Process: An Overview section. Proposed revisions include a revised submission deadline for comments.

ADOPTED APPENDICES

APPENDIX A: Carcinogenicity

ACGIH® has been aware of the increasing public concern over chemicals or industrial processes that cause or contribute to increased risk of cancer in workers. More sophisticated methods of bioassay, as well as the use of sophisticated mathematical models that extrapolate the levels of risk among workers, have led to differing interpretations as to which chemicals or processes should be categorized as human carcinogens and what the maximum exposure levels should be. The goal of the Chemical Substances TLV® Committee has been to synthesize the available information in a manner that will be useful to practicing industrial hygienists, without overburdening them with needless details. The categories for carcinogenicity are:

A1 — *Confirmed Human Carcinogen:* The agent is carcinogenic to humans based on the weight of evidence from epidemiologic studies.

A2 — *Suspected Human Carcinogen:* Human data are accepted as adequate in quality but are conflicting or insufficient to classify the agent as a confirmed human carcinogen; OR, the agent is carcinogenic in experimental animals at dose(s), by route(s) of exposure, at site(s), of histologic type(s), or by mechanism(s) considered relevant to worker exposure. The A2 is used primarily when there is limited evidence of carcinogenicity in humans and sufficient evidence of carcinogenicity in experimental animals with relevance to humans.

A3 — *Confirmed Animal Carcinogen with Unknown Relevance to*

Humans: The agent is carcinogenic in experimental animals at a relatively high dose, by route(s) of administration, at site(s), of histologic type(s), or by mechanism(s) that may not be relevant to worker exposure. Available epidemiologic studies do not confirm an increased risk of cancer in exposed humans. Available evidence does not suggest that the agent is likely to cause cancer in humans except under uncommon or unlikely routes or levels of exposure.

A4 — *Not Classifiable as a Human Carcinogen:* Agents which cause concern that they could be carcinogenic for humans but which cannot be assessed conclusively because of a lack of data. *In vitro* or animal studies do not provide indications of carcinogenicity which are sufficient to classify the agent into one of the other categories.

A5 — *Not Suspected as a Human Carcinogen:* The agent is not suspected to be a human carcinogen on the basis of properly conducted epidemiologic studies in humans. These studies have sufficiently long follow-up, reliable exposure histories, sufficiently high dose, and adequate statistical power to conclude that exposure to the agent does not convey a significant risk of cancer to humans; OR, the evidence suggesting a lack of carcinogenicity in experimental animals is supported by mechanistic data.

Substances for which no human or experimental animal carcinogenic data have been reported are assigned no carcinogenicity designation.

Exposures to carcinogens must be kept to a minimum. Workers exposed to A1 carcinogens without a TLV® should be properly equipped to eliminate to the fullest extent possible all exposure to the carcinogen. For A1 carcinogens with a TLV® and for A2 and A3 carcinogens, worker exposure by all routes should be carefully controlled to levels as low as possible below the TLV®. Refer to the "Guidelines for the Classification of Occupational Carcinogens" in the Introduction to the Chemical Substances in the *Documentation of the Threshold Limit Values and Biological Exposure Indices* for a complete description and derivation of these designations.

APPENDIX B: Particles (insoluble or poorly soluble) Not Otherwise Specified [PNOS]

The goal of the TLV®-CS Committee is to recommend TLVs® for all substances for which there is evidence of health effects at airborne concentrations encountered in the workplace. When a sufficient body of evidence exists for a particular substance, a TLV® is established. Thus, by definition the substances covered by this recommendation are those for which little data exist. The recommendation at the end of this Appendix is supplied as a guideline rather than a TLV® because it is not possible to meet the standard level of evidence used to assign a TLV®. In addition, the PNOS TLV® and its predecessors have been misused in the past and applied to any unlisted particles rather than those meeting the criteria listed below. The recommendations in this Appendix apply to particles that:

- Do not have an applicable TLV®;
- Are insoluble or poorly soluble in water (or, preferably, in aqueous lung fluid if data are available); and
- Have low toxicity (i.e., are not cytotoxic, genotoxic, or otherwise chemically reactive with lung tissue, and do not emit ionizing radiation, cause immune sensitization, or cause toxic effects other than by inflammation or the mechanism of "lung overload").

ACGIH® believes that even biologically inert, insoluble, or poorly soluble particles may have adverse effects and recommends that airborne concentrations should be kept below 3 mg/m^3, respirable particles, and 10 mg/m^3, inhalable particles, until such time as a TLV® is set for a particular substance.

APPENDIX C: Particle Size-Selective Sampling Criteria for Airborne Particulate Matter

For chemical substances present in inhaled air as suspensions of solid particles or droplets, the potential hazard depends on particle size as well as mass concentration because of 1) effects of particle size on the deposition site within the respiratory tract and 2) the tendency for many occupational diseases to be associated with material deposited in particular regions of the respiratory tract.

ACGIH® has recommended particle size-selective TLVs® for crystalline silica for many years in recognition of the well-established association between silicosis and respirable mass concentrations. The TLV®-CS Committee is now re-examining other chemical substances encountered in particle form in occupational environments with the objective of defining: 1) the size-fraction most closely associated for each substance with the health effect of concern and 2) the mass concentration within that size fraction which should represent the TLV®.

The Particle Size-Selective TLVs® (PSS–TLVs) are expressed in three forms:

1. *Inhalable Particulate Matter TLVs®* (IPM–TLVs) for those materials that are hazardous when deposited anywhere in the respiratory tract.
2. *Thoracic Particulate Matter TLVs®* (TPM–TLVs) for those materials that are hazardous when deposited anywhere within the lung airways and the gas-exchange region.
3. *Respirable Particulate Matter TLVs®* (RPM–TLVs) for those materials that are hazardous when deposited in the gas-exchange region.

The three particulate matter fractions described above are defined in quantitative terms in accordance with the following equations:[1–3]

A. IPM fraction consists of those particles that are captured according to the following collection efficiency regardless of sampler orientation with respect to wind direction:

$$\text{IPM}(d_{ae}) = 0.5\,[1 + \exp(-0.06\, d_{ae})] \quad \text{for } 0 < d_{ae} \le 100\ \mu\text{m}$$

where: $\text{IPM}(d_{ae})$ = the collection efficiency
d_{ae} = aerodynamic diameter of particle in μm

B. TPM fraction consists of those particles that are captured according to the following collection efficiency:

$$\text{TPM}(d_{ae}) = \text{IPM}(d_{ae})\,[1 - F(x)]$$

where: $F(x)$ = cumulative probability function of the standardized normal variable, x

$$x = \frac{\ln(d_{ae}/\Gamma)}{\ln(\Sigma)}$$

ln = natural logarithm
Γ = 11.64 μm
Σ = 1.5

C. RPM fraction consists of those particles that are captured according to the following collection efficiency:

$$RPM\ (d_{ae}) = IPM\ (d_{ae})\ [1 - F(x)]$$

where F(x) = same as above, but with Γ = 4.25 μm and Σ = 1.5

The most significant difference from previous definitions is the increase in the median cut point for a respirable particulate matter sampler from 3.5 μm to 4.0 μm; this is in accord with the International Organization for Standardization/ European Standardization Committee (ISO/CEN) protocol.[4,5] At this time, no change is recommended for the measurement of respirable particles using a 10-mm nylon cyclone at a flow rate of 1.7 liters per minute. Two analyses of available data indicate that the flow rate of 1.7 liters per minute allows the 10-mm nylon cyclone to approximate the particulate matter concentration which would be measured by an ideal respirable particulate sampler as defined herein.[6,7]

Collection efficiencies representative of several sizes of particles in each of the respective mass fractions are shown in Tables 1, 2, and 3. *Documentation* for the respective algorithms representative of the three mass fractions is found in the literature.[2–4]

TABLE 1. Inhalable Fraction

Particle Aerodynamic Diameter (μm)	Inhalable Particulate Matter (IPM) Fraction Collected (%)
0	100
1	97
2	94
5	87
10	77
20	65
30	58
40	54.5
50	52.5
100	50

TABLE 2. Thoracic Fraction

Particle Aerodynamic Diameter (μm)	Thoracic Particulate Matter (TPM) Fraction Collected (%)
0	100
2	94
4	89
6	80.5
8	67
10	50
12	35
14	23
16	15
18	9.5
20	6
25	2

TABLE 3. Respirable Fraction

Particle Aerodynamic Diameter (μm)	Respirable Particulate Matter (RPM) Fraction Collected (%)
0	100
1	97
2	91
3	74
4	50
5	30
6	17
7	9
8	5
10	1

References

1. American Conference of Governmental Industrial Hygienists: Particle Size-Selective Sampling in the Workplace. ACGIH®, Cincinnati, OH (1985).
2. American Conference of Governmental Industrial Hygienists: Particle Size-Selective Sampling for Particulate Air Contaminants. JH Vincent, Ed. ACGIH®, Cincinnati, OH (1999).
3. Soderholm, SC: Proposed International Conventions for Particle Size-Selective Sampling. Ann. Occup. Hyg. 33:301–320 (1989).
4. International Organization for Standardization (ISO): Air Quality—Particle Size Fraction Definitions for Health-Related Sampling. ISO 7708:1995. ISO, Geneva (1995).
5. European Standardization Committee (CEN): Size Fraction Definitions for Measurement of Airborne Particles. CEN EN481:1993. CEN, Brussels (1993).
6. Bartley, DL: Letter to J. Doull, TLV® Committee, July 9, 1991.
7. Lidén, G; Kenny, LC: Optimization of the Performance of Existing Respirable Dust Samplers. Appl. Occup. Environ. Hyg. 8(4):386–391 (1993).

APPENDIX D: Commercially Important Tree Species Suspected of Inducing Sensitization

Common	Latin
SOFTWOODS	
California redwood	*Sequoia sempervirens*
Eastern white cedar	*Thuja occidentalis*
Pine	*Pinus*
Western red cedar	*Thuja plicata*
HARDWOOD	
Ash	*Fraxinus spp.*
Aspen/Poplar/Cottonwood	*Populus*
Beech	*Fagus*
Oak	*Quercus*
TROPICAL WOODS	
Abirucana	*Pouteria*
African zebra	*Microberlinia*
Antiaris	*Antiaris africana, Antiaris toxicara*
Cabreuva	*Myrocarpus fastigiatus*
Cedar of Lebanon	*Cedra libani*
Central American walnut	*Juglans olanchana*
Cocabolla	*Dalbergia retusa*
African ebony	*Diospryos crassiflora*
Fernam bouc	*Caesalpinia*
Honduras rosewood	*Dalbergia stevensonii*
Iroko or kambala	*Chlorophora excelsa*
Kejaat	*Pterocarpus angolensis*
Kotibe	*Nesorgordonia papaverifera*
Limba	*Terminalia superba*
Mahogany (African)	*Khaya spp.*
Makore	*Tieghemella heckelii*
Mansonia/Beté	*Mansonia altissima*
Nara	*Pterocarpus indicus*
Obeche/African maple/Samba	*Triplochiton scleroxylon*
Okume	*Aucoumea klaineana*
Palisander/Brazilian rosewood/ Tulip wood/Jakaranda	*Dalbergia nigra*
Pau marfim	*Balfourodendron riedelianum*
Ramin	*Gonystylus bancanus*
Soapbark dust	*Quillaja saponaria*
Spindle tree wood	*Euonymus europaeus*
Tanganyike aningre	

APPENDIX E: Threshold Limit Values for Mixtures

Most threshold limit values are developed for a single chemical substance. However, the work environment is often composed of multiple chemical exposures both simultaneously and sequentially. It is recommended that multiple exposures that comprise such work environments be examined to assure that workers do not experience harmful effects.

There are several possible modes of chemical mixture interaction. Additivity occurs when the combined biological effect of the components is equal to the sum of each of the agents given alone. Synergy occurs where the combined effect is greater than the sum of each agent. Antagonism occurs when the combined effect is less.

The general ACGIH® mixture formula applies to the additive model. It is utilized when additional protection is needed to account for this combined effect.

The guidance contained in this Appendix does not apply to substances in mixed phases.

Application of the Additive Mixture Formula

The "TLV® Basis" column found in the table of Adopted Values lists the adverse effect(s) upon which the TLV® is based. This column is a resource that may help alert the reader to the additive possibilities in a chemical mixture and the need to reduce the combined TLV® of the individual components. Note that the column does not list the deleterious effects of the agent, but rather, lists only the adverse effect(s) upon which the threshold limit was based. The current *Documentation of the TLVs® and BEIs®* should be consulted for toxic effects information, which may be of use when assessing mixture exposures.

When two or more hazardous substances have a similar toxicological effect on the same target organ or system, their combined effect, rather than that of either individually, should be given primary consideration. In the absence of information to the contrary, different substances should be considered as additive where the health effect and target organ or system is the same.

That is, if the sum of

$$\frac{C_1}{T_1} + \frac{C_2}{T_2} + \cdots \frac{C_n}{T_n}$$

exceeds unity, the threshold limit of the mixture should be considered as being exceeded (where C_1 indicates the observed atmospheric concentration and T_1 is the corresponding threshold limit; see example). It is essential that the atmosphere is analyzed both qualitatively and quantitatively for each component present in order to evaluate the threshold limit of the mixture.

The additive formula applies to simultaneous exposure for hazardous agents with TWA, STEL, and Ceiling values. The threshold limit value time interval base (TWA, STEL, and Ceiling) should be consistent where possible. When agents with the same toxicological effect do not have a corresponding TLV® type, use of mixed threshold limit value types may be warranted. Table E-1 lists possible combinations of threshold limits for the additive mixture formula. Multiple calculations may be necessary.

Where a substance with a STEL or Ceiling limit is mixed with a substance with a TLV–TWA but no STEL, comparison of the short-term limit with the applicable excursion limit may be appropriate. Excursion limits are defined as a value five times the TLV–TWA limit. The amended formula would be:

$$\frac{C_1}{T_{1STEL}} + \frac{C_2}{(T_2)(5)} \leq 1$$

where: T_{1STEL} = the TLV–STEL
T_2 = the TLV–TWA of the agent with no STEL.

The additive model also applies to consecutive exposures of agents that occur during a single work shift. Those substances that have TLV–TWAs (and STELs or excursion limits) should generally be handled the same as if they were the same substance, including attention to the recovery periods for STELs and excursion limits as indicated in the "Introduction to Chemical Substances." The formula does not apply to consecutive exposures of TLV–Ceilings.

Limitations and Special Cases

Exceptions to the above rule may be made when there is a good reason to believe that the chief effects of the different harmful agents are not additive. This can occur when neither the toxicological effect is similar nor the target organ is the same for the components. This can also occur when the mixture interaction causes inhibition of the toxic effect. In such cases, the threshold limit ordinarily is exceeded only when at least one member of the series (C_1/T_1 or C_2/T_2, etc.) itself has a value exceeding unity.

Another exception occurs when mixtures are suspected to have a synergistic effect. The use of the general additive formula may not provide sufficient protection. Such cases at present must be determined individually. Potentiating effects of exposure to such agents by routes other than that of inhalation are also possible. Potentiation is characteristically exhibited at high concentrations, less probably at low. For situations involving synergistic effects, it may be possible to use a modified additive formula that provides additional protection by incorporating a synergy factor. Such treatment of the TLVs® should be used with caution, as the quantitative information concerning synergistic effects is sparse.

Care must be considered for mixtures containing carcinogens in categories A1, A2, or A3. Regardless of application of the mixture formula, exposure to mixtures containing carcinogens should be avoided or maintained as low as possible. See Appendix A.

The additive formula applies to mixtures with a reasonable number of agents. It is not applicable to complex mixtures with many components (e.g., gasoline, diesel exhaust, thermal decomposition products, fly ash, etc.).

Example

A worker's airborne exposure to solvents was monitored for a full shift as well as one short-term exposure. The results are presented in Table E-2.

According to the *Documentation of the TLVs® and BEIs®*, all three substances indicate irritation effects on the respiratory system and thus would be considered additive. Acetone and methyl ethyl ketone exhibit central nervous system effects.

Full shift analysis would utilize the formula:

$$\frac{C_1}{T_1} + \frac{C_2}{T_2} + \frac{C_3}{T_3} \leq 1$$

thus,

$$\frac{160}{500} + \frac{20}{200} + \frac{90}{200} = 0.32 + 0.10 + 0.45 = 0.87$$

The full-shift mixture limit is not exceeded.

Short-term analysis would utilize the formula:

$$\frac{C_1}{T_{1STEL}} + \frac{C_2}{(T_2)(5)} + \frac{C_3}{T_{3STEL}} \leq 1$$

thus,

$$\frac{490}{750} + \frac{150}{1000} + \frac{220}{300} = 0.65 + 0.15 + 0.73 = 1.53$$

The short-term mixture limit is exceeded.

TABLE E-2. Example Results

Agent	Full-Shift Results (*TLV–TWA*)	Short-Term Results (*TLV–STEL*)
1) Acetone1	60 ppm	490 ppm
	(500 ppm)	*(750 ppm)*
2) sec-Butyl acetate	20 ppm	150 ppm
	(200 ppm)	(N/A)
3) Methyl ethyl ketone	90 ppm	220 ppm
	(200 ppm)	*(300 ppm)*

APPENDIX F: Minimal Oxygen Content

Adequate oxygen delivery to the tissues is necessary for sustaining life and depends on 1) the level of oxygen in inspired air, 2) the presence or absence of lung disease, 3) the level of hemoglobin in the blood, 4) the kinetics of oxygen binding to hemoglobin (oxy-hemoglobin dissociation curve), 5) the cardiac output, and 6) local tissue blood flow. For the purpose of the present discussion, only the effects of decreasing the amount of oxygen in inspired air is considered.

The brain and myocardium are the most sensitive tissues to oxygen deficiency. The initial symptoms of oxygen deficiency are increased ventilation, increased cardiac output, and fatigue. Other symptoms that may develop include headache, impaired attention and thought processes, decreased coordination, impaired vision, nausea, unconsciousness, seizures, and death. However, there may be no apparent symptoms prior to unconsciousness. The onset and severity of symptoms depend on many factors such as the magnitude of the oxygen deficiency, duration of exposure, work rate, breathing rate, temperature, health status, age, and pulmonary acclimatization. The initial symptoms of increased breathing and increased heart rate become evident when hemoglobin oxygen saturation is reduced below 90%. At hemoglobin oxygen saturations between 80% and 90%, physiological adjustments occur in healthy adults to resist hypoxia, but in compromised individuals, such as emphysema patients, oxygen therapy would be prescribed for hemoglobin oxygen saturations below 90%. As long as the partial pressure of oxygen (ρO_2) in pulmonary capillaries stays above 60 torr, hemoglobin will be more than 90% saturated and normal levels of oxygen transport will be maintained in healthy adults. The alveolar ρO_2 level of 60 torr corresponds to 120 torr ρO_2 in the ambient air, due to anatomic dead space, carbon dioxide, and water vapor. For additional information on gas exchange and pulmonary physiology see Silverthorn(1) and Guyton.(2)

The U.S. National Institute for Occupational Safety and Health(3) used 60 torr alveolar ρO_2 as the physiological limit that establishes an oxygen-deficient atmosphere and has defined an oxygen-deficient atmosphere as one with an ambient ρO_2 less than 132 torr.(4) The minimum requirement of 19.5% oxygen at sea level (148 torr ρO_2, dry air) provides an adequate amount of oxygen for most work assignments and includes a margin of safety.(5) However, the margin of safety significantly diminishes as the O_2 partial pressure of the atmosphere decreases with increasing altitude, decreases with the passage of low pressure weather events, and decreases with increasing water vapor,(6) such that, at 5000 feet, the ρO_2 of the atmosphere may approach 120 torr because of water vapor and the passage of fronts and at elevations greater than 8000 feet, the ρO_2 of the atmosphere may be expected to be less than 120 torr.

The physiological effects of oxygen deficiency and oxygen partial pressure variation with altitude for dry air containing 20.948% oxygen are given in Table F-1. No physiological effects due to oxygen deficiency are expected in healthy adults at oxygen partial pressures greater than 132 torr or at elevations less than 5000 feet. Some loss of dark adaptation is reported to occur at elevations greater than 5000 feet. At oxygen partial pressures less than 120 torr (equivalent to an elevation of about 7000 feet or about 5000 feet accounting for water vapor and the passage of low pressure weather events) symptoms in unacclimatized workers include increased pulmonary ventilation and cardiac output, incoordination, and impaired attention and thinking. These symptoms are recognized as being incompatible with safe performance of duties.

Accordingly, ACGIH® recommends a minimal ambient oxygen partial pressure of 132 torr, which is protective against inert oxygen-displacing gases and oxygen-consuming processes for altitudes up to 5000 feet. Figure F-1 is a plot of ρO_2 with increasing altitude, showing the recommended minimal value of 132 torr. If the partial pressure of oxygen is less than 132 torr or if it is less than the expected value for that altitude, given in Table F-1, then additional work practices are recommended such as thorough evaluation of the confined space to identify the cause of the low oxygen concentration; use of continuous monitors integrated with warning devices; acclimating workers to the altitude of the work, as adaptation to altitude can increase an individuals work capacity by 70%; use of rest–work cycles with reduced work rates and increased rest periods; training, observation, and monitoring of workers; and easy, rapid access to oxygen-supplying respirators that are properly maintained.

Oxygen-displacing gases may have flammable properties or may produce physiological effects, so that their identity and source should be thoroughly investigated. Some gases and vapors, when present in high concentrations in air, act primarily as simple asphyxiants without other significant physiologic effects. A TLV® may not be recommended for each simple asphyxiant because the limiting factor is the available oxygen. Atmospheres deficient in O_2 do not provide adequate warning and most simple asphyxiants are odorless. Account should be taken of this factor in limiting the concentration of the asphyxiant particularly at elevations greater than 5000 feet where the ρO_2 of the atmosphere may be less than 120 torr.

References

1. Silverthorn DE: Human Physiology: An Integrated Approach, 2nd ed. Prentice-Hall, New Jersey (2001).
2. Guyton AC: Textbook of Medical Physiology, 8th ed. W.B. Saunders Co., Philadelphia (1991).
3. U.S. National Institute for Occupational Safety and Health: A Guide to Industrial Respiratory Protection, DHEW (NIOSH) Pub. No. 76-198. NIOSH, Cincinnati, OH (1976).
4. U.S. National Institute for Occupational Safety and Health: Working in Confined Spaces. DHHS (NIOSH) Pub. No. 80-106. NIOSH, Cincinnati, OH (1979).
5. NIOSH U.S. National Institute for Occupational Safety and Health: NIOSH Respirator Decision Logic. DHHS Pub. No. 87-108. NIOSH, Cincinnati, OH (1987).
6. McManus N: Safety and Health in Confined Spaces. Lewis Publishers, Boca Raton, FL (1999).

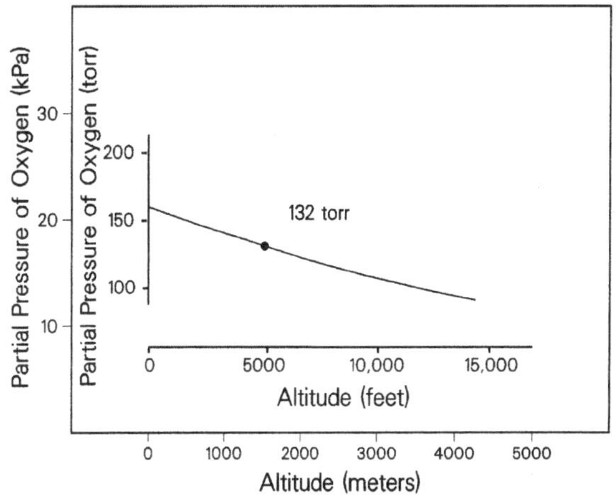

FIGURE F-1. Plot of oxygen partial pressure (ρO_2) (expressed in torr and kPa) with increasing altitude (expressed in feet and meters), showing the recommended oxygen partial pressure of 132 torr.

TABLE F-1. Barometric Pressure, Oxygen Partial Pressure, and Percent Oxygen Concentration Variation with Altitude and Physiological Effect [adapted from McManus[(6)]]

Altitude Feet (meters)	Barometric Pressure torr, Dry Air[A] (kilopascals)	ρO_2 Equivalent, torr dry air at 20.948% O_2[B] (kilopascals)	$\%O_2$ Equivalent, Dry Air at Sea Level[C] (percent)	Physiological Effect of ρO_2 Levels[D]
0 (0)	760 (101)	159 (21.2)	20.9	
1000 (305)	731 (97.4)	153 (20.4)	20.1	
2000 (610)	704 (93.8)	147 (19.6)	19.3	
3000 (914)	677 (90.3)	142 (18.9)	18.7	
4000 (1219)	652 (86.9)	137 (18.3)	18.0	
5000 (1524)	627 (83.6)	131 (17.5)	17.2	None in healthy adults
6000 (1829)	603 (80.4)	126 (16.8)	16.6	Loss of dark adaptation can occur at elevations above 5000 feet
7000 (2134)	580 (77.3)	121 (16.1)	16.0	Increased pulmonary ventilation and cardiac output, incoordination, and impaired attention and thinking

TABLE F-1 (con't.). Barometric Pressure, Oxygen Partial Pressure, and Percent Oxygen Concentration Variation with Altitude and Physiological Effect [adapted from McManus[6]]

Altitude Feet (meters)	Barometric Pressure torr, Dry Air^A (kilopascals)	ρO_2 Equivalent, torr dry air at 20.948% O_2^B (kilopascals)	$\%O_2$ Equivalent, Dry Air at Sea Level^C (percent)	Physiological Effect of ρO_2 Levels^D
8000 (2438)	559 (74.5)	117 (15.6)	15.4	Rapid exposure to altitudes over 8000 feet may cause high altitude sickness (respiratory alkalosis, headache, nausea, and vomiting) in unacclimatized individuals. Rapid ascent increases the risk of high altitude pulmonary edema and cerebral edema
9000 (2743)	537 (71.6)	112 (14.9)	14.7	
10000 (3048)	517 (68.9)	108 (14.4)	14.2	
11000 (3353)	498 (66.4)	104 (13.9)	13.7	Abnormal fatigue on exertion, faulty coordination, impaired judgment, emotional upset
12000 (3658)	479 (63.8)	100 (13.3)	13.2	
13000 (3962)	461 (61.5)	98 (12.9)	12.8	
14000 (4267)	443 (59.1)	93 (12.4)	12.2	Impaired respiration, very poor judgment and coordination, tunnel vision

^A Calculated from $P_{re:\ sea\ level} = 760 \times e^{-(\text{altitude in feet}/25970)}$

^B Calculated from $\rho O_2 = 0.20948 \times 760 \times e^{-(\text{altitude in feet}/25970)}$

^C Calculated from: $P_{\%O_2} = 20.948 \times e^{-(\text{altitude in feet}/25970)}$

^D The approximate physiological effect in healthy adults is influenced by duration of the oxygen deficiency, work rate, breathing rate, temperature, health status, age and pulmonary acclimatization.

APPENDIX G: Substances Whose Adopted *Documentation* and TLVs® Were Withdrawn For a Variety of Reasons, Including Insufficient Data, Regrouping, Etc.

[Individual entries will remain for a 10-year period, commencing with the year of withdrawal]

Substance [CRN]	Year Withdrawn	Reason
Acetylene tetrabromide	2006	Withdrawn in favor of its IUPAC name; *see* 1,1,2,2-Tetrabromoethane
Aliphatic hydrocarbon gases, Alkanes [C_1–C_4]	2013	Methane, Ethane, Propane, Liquefied petroleum gas (LPG) and Natural gas — refer to Appendix F: Minimal Oxygen Content. Butane and Isobutane — refer to Butane, all isomers
Aluminum [7429-90-5] and compounds, as Al	2008	Combined into Aluminum metal and insoluble compounds
Aluminum oxide [1344-28-1]	2008	Combined into Aluminum metal and insoluble compounds
Aluminum welding fumes	2004	TLV® withdrawn as a result of Substances of Variable Composition Appendix removal
APPENDIX B: Substances of Variable Composition B1: Polytetrafluoroethylene decomposition products B2: Welding fumes (not otherwise specified)	2004	Appendix withdrawn, insufficient data B1: *Documentation* withdrawn as a result of Appendix removal B2: *Documentation* and TLV® withdrawn as a result of Appendix removal
Borates, tetra, sodium salts	2005	Combined into Borate compounds, inorganic
Butane [106-97-8]	2004	Presently covered by Aliphatic hydrocarbon gases: Alkanes [C_1–C_4]
Calcium carbonate [471-34-1]	2007	Insufficient data
Dinitolmide	2007	Withdrawn in favor of its synonym 3,5-Dinitro-o-toluamide
Emery [1302-74-5]	2008	Combined into Aluminum metal and insoluble compounds
Ethane [74-84-0]	2004	Presently covered by Aliphatic hydrocarbon gases: Alkanes [C_1–C_4]
Glycerin mist [56-81-5]	2013	Insufficient data relevant to human occupational exposure
Iron oxide (Fe_2O_3) dust and fume, as Fe	2006	Combined into Iron oxide
Isopropanol	2006	Withdrawn in favor of its IUPAC name, 2-Propanol
Lead arsenate [3687-31-8], as $Pb_3(AsO_4)_2$	2009	Insufficient data
Liquefied petroleum gas (LPG) [68476-85-7]	2004	Presently covered by Aliphatic hydrocarbon gases: Alkanes [C_1–C_4]

Magnesite [546-93-0]	2006	Insufficient data
Methane [74-82-8]	2004	Presently covered by Aliphatic hydrocarbon gases: Alkanes [C_1–C_4]
Nonane [111-84-2], all isomers	2012	Refer to Nonane
Oil mist, mineral	2010	Refer to Mineral oil, excluding metal working fluids
Perlite [93763-70-3]	2006	Insufficient data
Piperazine dihydrochloride [142-64-3]	2012	Refer to Piperazine and salts
Propane [74-98-6]	2004	Presently covered by Aliphatic hydrocarbon gases: Alkanes [C_1–C_4]
Rouge	2006	Combined into Iron oxide
Rubber solvent (Naphtha) [8030-30-6]	2009	Refer to Appendix H: Reciprocal Calculation Method for Certain Refined Hydrocarbon Solvent Vapor Mixtures
Silica, amorphous — diatomaceous earth [61790-53-2]	2006	Insufficient data on single-substance exposure, most are co-exposures with crystalline silica
Silica, amorphous — fume [69012-64-2]	2006	Insufficient data
Silica, amorphous — fused [60676-86-0]	2006	Insufficient data
Silica amorphous — precipitated silica and silica gel [112926-00-8]	2006	Insufficient data
Silica, crystalline — cristobalite [14464-46-1]	2006	Combined into one TLV® and *Documentation*, i.e., Silica, crystalline
Silica, crystalline — quartz [14808-60-7]	2006	Combined into one TLV® and *Documentation*, i.e., Silica, crystalline
Silica, crystalline — tridymite [15468-32-3]	2005	Insufficient data
Silica, crystalline — tripoli [1317-95-9]	2006	Insufficient data and unlikely single-substance exposure. Combined into one TLV® and *Documentation*, i.e., Silica, crystalline
Silicon [7440-21-3]	2006	Insufficient data
Soapstone	2011	Refer to Talc
Tantalum [7440-25-7] and Tantalum oxide [1314-61-0] dusts, as Ta	2010	Insufficient data
Tetrasodium pyrophosphate [7722-88-5]	2006	Insufficient data
Triphenyl amine [603-34-9]	2008	Insufficient data
Vegetable oil mist	2006	Insufficient data
VM & P naphtha [8032-32-4]	2009	Refer to Appendix H: Reciprocal Calculation Method for Certain Refined Hydrocarbon Solvent Vapor Mixtures

APPENDIX H: Reciprocal Calculation Method for Certain Refined Hydrocarbon Solvent Vapors

The reciprocal calculation procedure (RCP) is a method for deriving occupational exposure limits (OEL) for refined hydrocarbon solvents. Refined hydrocarbon solvents often are found as mixtures created by distillation of petroleum oil over a particular boiling range. These mixtures may consist of up to 200 components consisting of aliphatic (alkane), cycloaliphatic (cycloalkane) and aromatic hydrocarbons ranging from 5 to 15 carbons.

The goal of the TLV®-CS Committee is to recommend TLVs® for all substances where there is evidence of health effects at airborne concentrations encountered in the workplace. When a sufficient body of evidence exists for a particular substance or mixture, a TLV® is established. However, hydrocarbon solvents are often complex and variable in composition. The use of the mixture formula, found in Appendix E: Threshold Limit Values for Mixtures, is difficult in such cases because these petroleum mixtures contain a large number of unique compounds, many of which do not have a TLV® recommendation.

There are two aspects of the RCP — the methodology and the group guidance values (GGVs). The methodology is based on the special case formula found in pre-2004 versions of the Mixture Appendix in *TLVs® and BEIs® Based on the Documentation of the Threshold Limit Values for Chemical Substances and Physical Agents and Biological Exposure Indices*. The RCP formula calculates a unique OEL based on the mass composition of the mixture, the GGVs and where applicable, substance-specific TLVs®.

Group guidance values are categorized based on similar chemical and toxicological concerns. Several entities (both trade groups and regulatory authorities) have adopted group guidance values to utilize with the reciprocal mixture formula (RMF) (Farmer, 1995; UK HSE, 2000; McKee et al., 2005). Two examples of published GGVs are found in Table 1. A mixture-specific time-weighted-average limit ($GGV–TWA_{mixture}$) is calculated based on the mass percent makeup of the designated groups utilizing the reciprocal mixture formula and the GGVs from column *B* or *C* and TLV® values in column *D* found in Table 1.

ACGIH® considers this method to be applicable for mixtures if the toxic effects of individual constituents are additive (i.e., similar toxicological effect on the same target organ or system). The principal toxicological effects of hydrocarbon solvent constituents are acute central nervous system (CNS) depression (characterised by effects ranging from dizziness and drowsiness to anaesthesia) and eye and respiratory tract irritation (McKee et al., 2005; ECETOC, 1997).

Application

The RCP is a special use application. It applies only to hydrocarbon solvents containing saturated aliphatics (normal, iso-alkanes and cycloalkanes) and aromatics with a carbon number of C_5 to C_{15} derived from petroleum and boiled in the range of 35–329°C. It does not apply to petroleum derived fuels, lubricating oils, or solvent mixtures for which there exists a unique TLV®. It does not apply to hydrocarbons with a toxicity that is significantly greater than the mixture at large, such as benzene (see limitations).

Where the mixture is comprised entirely of compounds with unique TLVs®, the mixture should be handled according to Appendix E. When the mixture contains an appreciable amount of a component for which there is a TLV® (i.e., when the use of the TLV® results in a lower $GGV\text{–}TWA_{mixture}$), those specific values should be entered into the RCP (see column *D*, Table 1). When the mixture itself has been assigned a unique TLV®, that value should be utilized rather than the procedures found in this appendix.

Exposure excursions above the calculated $GGV\text{–}TWA_{mixture}$ should be handled according to the procedures found in the Introduction to the TLVs® (see *Excursion Limits*).

The reciprocal calculation mixture formula is:

$$GGV_{mixture} = \frac{1}{\frac{F_a}{GGV_a} + \ldots + \frac{F_n}{GGV_n}}$$

where:

- $GGV_{mixture}$ = the calculated 8-hour TWA–OEL for the mixture
- GGV_a = the guidance value (or TLV®) for group (or component) *a*
- F_a = the liquid mass fraction of group (or component) *a* in the hydrocarbon mixture (value between 0–1)
- GGV_n = the guidance value (or TLV®) for the nth group (or component)
- F_n = the liquid mass fraction of the nth group (or component) in the hydrocarbon mixture (value between 0–1)

The resulting $GGV_{mixture}$ should identify the source of GGVs used in the calculation (i.e., column *B* or *C*).

The resulting calculated $GGV_{mixture}$ value should follow established recommendations regarding rounding. For calculated values < 100 mg/m³, round to the nearest 25. For calculated values between 100 and 600 mg/m³, round to the nearest 50, and for calculated values > 600 mg/m³, round to the nearest 200 mg/m³.

Limitations

1. The reciprocal formula requires that the composition of the mixture be characterized at least to the detail of mass percent of the groups found in Table 1.
2. The reciprocal formula does not apply to solvents containing benzene, or n-hexane, or methylnaphthalene, which have individual TLVs® significantly less than the GGV to which they would belong and have unique toxicological properties. Whenever present in the mixture, these components should be measured individually and evaluated using the methodology found in Appendix E, i.e., independent treatment or use of the additive formula depending on the TLV® basis.
3. Care in the use of GGV/RMF should be observed where the mixture in question is known to have significant toxicokinetic interactions of components that are manifested at or below GGV levels.
4. The use of the reciprocal formula should be restricted to applications where the boiling points of the solvents in the mixture are relatively narrow, within a range of less than 45°C (i.e., vapor pressure within approximately one order of magnitude). The procedure should not be used in situations where the liquid composition is significantly different from the vapor composition. If these conditions cannot be met, the reciprocal formula can be utilized by substituting $F_{(n)}$ in the equation with the vapor mass fraction for each group *(n)* in the hydrocarbon mixture, based on situation-specific airborne concentration measurements.
5. The group guidance values apply only to vapors and do not apply to mists or aerosols. The GGV/RMF procedure does not apply to mixtures containing olefins or other unsaturated compounds or polycyclic aromatic hydrocarbons (PAHs).

Example

A solvent containing the following mass composition is matched with the appropriate group guidance value:

Component	Percent by weight	Group Guidance Value (mg/m³)
C7–C8 alkanes cycloalkanes	45%	1500
C9–C10 alkanes cycloalkanes	40%	1200
C7–C8 aromatics	9%	200
Toluene	6%	75
Benzene	< 1%	-NA-

Based on Column *B*, Table 1 (McKee et al., 2005), the $GGV_{mixture}$ would be:

$$GGV_{mixture} = \frac{1}{\frac{.45}{1500} + \frac{.40}{1200} + \frac{.09}{200} + \frac{.06}{75}} = \frac{1}{.001884}$$

= 531 (rounded to 550 mg/m³)

Toluene (part of the aromatic C7, 8 fraction) is added as a TLV® rather than a GGV since it makes a difference in the resulting $GGV_{mixture}$. Benzene would be evaluated separately at the current TLV® for benzene.

References

European Centre for Ecotoxicology and Toxicology of Chemicals (ECETOC). Occupational exposure limits for hydrocarbon solvents. Special Report No. 13. Brussels, Belgium (1997).

Farmer TH: Occupational hygiene limits for hydrocarbon solvents. Annals of Occupational Hygiene 40: 237-242 (1995).

McKee RH; Medeiros AM; Daughtrey WC: A proposed methodology for setting occupational exposure limits for hydrocarbon solvents. J of Occ and Env Hygiene 2: 524-542 (2005).

UK Health and Safety Executive (UKHSE) EH40/2000. Occupational Exposure Limits (2000).

TABLE 1. Group Guidance Values

A Hydrocarbon Group	B McKee et al. (mg/m^3)	C UK-HSE 40/2000 (mg/m^3)	D ACGIH® Unique TLVs® (mg/m^3)
C_5–C_6 Alkanes	1500	1800	Pentane, all isomers (1770) Hexane isomers (1760)
C_7–C_8 Alkanes	1500	1200	Heptane, all isomers (1640) Octane, all isomers (1401)
C_5–C_6 Cycloalkanes	1500	1800	Cyclopentane (1720) Cyclohexane (350)
C_7–C_8 Cycloalkanes	1500	800	Methyl cyclohexane (1610)
C_7–C_8 Aromatics	200	500	Toluene (75) Xylene, all isomers (434) Ethyl benzene (434)
C_9–C_{15} Alkanes	1200	1200	Nonane (1050)
C_9–C_{15} Cycloalkanes	1200	800	
C_9–C_{15} Aromatics*	100	500	Trimethyl benzene, isomers (123) Naphthalene (52) Cumen (246)

*N-Hexane (TLV®-176 mg/m^3) and methylnaphthalenes (TLV®-3 mg/m^3) are significantly below the recommended GGV. Whenever present in the mixture, these components should be measured individually and evaluated using the methodology found in Appendix E, i.e., independent treatment or use of the additive formula depending on the critical effect.

APPENDIX B

Physical Constants/Conversion Factors

APPENDIX B

PHYSICAL CONSTANTS OF SELECTED MATERIALS

Substance	Formula	Molecular Weight	Specific Gravity	Flash Point F [C] Closed Cup	Open Cup	Explosion Limits (Volume Percent) Lower	Upper
Acetaldehyde	CH_3CHO	44.05	0.821	-17 [-27]	—	3.97	57.0
Acetic Acid	CH_3COOH	60.05	1.049	104 [41]	110 [43]	5.40	—
Acetic Anhydride	$(CH_3CO)_2O$	102.09	1.082	121 [49]	130 [54]	2.67	10.13
Acetone	CH_3COCH_3	58.08	0.792	0 [-18]	15 [-9]	2.55	12.80
Acrolein	$CH_2{:}CHCHO$	56.06	0.841	Gas		Unstable	
Acrylonitrile	$CH_2{:}CHCN$	53.06	0.806	—	32 [0]	3.05	17.0
Ammonia	NH_3	17.03	0.597	Gas		15.50	27.0
Amyl Acetate	$CH_3CO_2C_5H_{11}$	130.18	0.879	76 [24]	80 [27]	1.10	—
iso-Amyl Alcohol	$(CH_3)_2CHCH_2CH_2OH$	88.15	0.812	109 [43]	115 [46]	1.20	—
Aniline	$C_6H_5NH_2$	93.12	1.022	168 [76]	—	—	—
Arsine	AsH_3	77.93	2.695 (A)	Gas		—	—
Benzene	C_6H_6	78.11	0.879	12 [-11]	—	1.40	7.10
Bromine	Br_2	159.83	3.119	—	—	—	—
Butane	$CH_3(CH_2)_2CH_3$	58.12	2.085	Gas		1.86	8.41
1,3-Butadiene	$(CH_2{:}CH)_2$	54.09	0.621	Gas		2.00	11.50
n-Butanol	$C_2H_5CH_2CH_2OH$	74.12	0.810	84 [29]	110 [43]	1.45	11.25
2-Butanone (Methyl ethyl ketone)	$CH_3COC_2H_5$	72.10	0.805	30 [-1]	—	1.81	9.50
n-Butyl Acetate	$CH_3CO_2C_4H_9$	116.16	0.882	72 [22]	90 [32]	1.39	7.55
Butyl "Cellosolve"	$C_4H_9OCH_2CH_2OH$	118.17	0.903	141 [61]	165 [74]	—	—
Carbon Dioxide	CO_2	44.01	1.53	—	—	—	—
Carbon Disulphide	CS_2	76.13	1.263	-22 [-30]	—	1.25	50.0
Carbon Monoxide	CO	28.10	0.968	Gas		12.5	74.2
Carbon Tetrachloride	CCl_4	153.84	1.595	Nonflammable			
Celllosolve	$C_2H_5O(CH_2)_2OH$	90.12	0.931	104 [40]	120 [49]	2.6	15.7
Cellosolve Acetate	$CH_3CO_2C_4H_9O$	132.16	0.975	124 [51]	135 [57]	1.71	—
Chlorine	Cl_2	70.91	3.214	Gas		—	—
2-Chlorobutadiene	$CH_2{:}CClCHCH_2$	88.54	0.958	—	—	—	—
Chloroform	$CHCl_3$	119.39	1.478	Nonflammable			
1-Chloro-1-nitropropane	$NO_3ClC_3H_6$	139.54	1.209	144 [62]	—	—	—
Cyclohexane	C_6H_{12}	84.16	0.779	1 [-17]	—	1.26	7.75
Cyclohexanol	$CH_2(CH_2)_4CHOH$	100.16	0.962	154 [68]	—	—	—
Cyclohexanone	$CH_2(CH_2)_4CO$	98.14	0.948	147 [64]	—	—	—
Cyclohexene	$CH_2(CH_2)_3CH{:}CH$	82.14	0.810	—	—	—	—
Cyclopropane	$CH_2CH_2CH_2$	42.08	0.720	Gas		2.40	10.40
o-Dichlorobenzene	$Cl_2C_6H_4$	147.01	1.305	151 [66]	165 [74]	—	—
Dichlorodifluoromethane	CCl_2F_2	120.92	1.486	Nonflammable			
1,1-Dichlorethane	CH_2CHCl_2	98.97	1.175	—	—	—	—
1,2-Dichloroethane	$ClCH_2CH_2Cl$	98.97	1.257	56 [13]	65 [18]	6.2	15.9
1,2-Dichloroethylene (Ethylene Dichloride)	$ClCHCHCl$	96.95	1.291	43 [6]	—	9.7	12.8
Dichloroethylether	$ClCH_2CHClC_2H_5$	143.02	1.222	131 [55]	180 [82]	—	—
Dichloromethane	H_2CCl_2	84.94	1.336	—	—	—	—
Dichioromonofluoromethane	$HCCl_2F$	102.93	1.426	—	—	—	—
1,1-Dichloro-1-nitroethane	$H_3C_2Cl_2NO_3$	143.97	1.692	—	168 [76]	—	—
1,2-Dichloropropane	$CH_3CHClCH_2Cl$	112.99	1.159	59 [15]	65 [18]	3.4	14.5
Dichlorotetrafluoroethane	$CClF_2CClF_2$	170.93	1.433	Nonflammable			
Dimethylaniline	$(CH_3)_2NC_6H_5$	121.18	0.956	145 [63]	170 [77]	—	—
Dimethylsulfate	$(CH_3)_2SO_4$	126.13	1.332	182 [83]	240 [115]	—	—
Dioxane	$O(CH_2)_4O$	88.10	1.034	—	35 [2]	—	—
Ethyl Acetate	$CH_3CO_2C_2H_5$	88.10	0.901	24 [-4]	30 [-1]	2.18	11.4
Ethyl Alcohol	C_2H_5OH	46.07	0.789	55 [13]	—	3.28	18.95
Ethyl Benzene	$C_6H_5C_2H_5$	106.16	0.867	59 [15]	75 [24]	—	—

PHYSICAL CONSTANTS OF SELECTED MATERIALS (Cont.)

Substance	Formula	Molecular Weight	Specific Gravity	Flash Point F [C] Closed Cup	Flash Point F [C] Open Cup	Explosion Limits (Volume Percent) Lower	Explosion Limits (Volume Percent) Upper
Ethyl Bromide	C_2H_5Br	109.98	1.430	—	—	6.75	11.25
Ethyl Chloride	C_2H_5Cl	64.52	0.921	-58 [-50]	-45 [-43]	3.6	14.80
Ethylene Chlorohydrin	$ClCH_2CH_2OH$	80.52	1.213	—	140 [60]	—	—
Ethylenediamine	$NH_2CH_2CH_2NH_2$	60.10	0.899	—	—	—	—
Ethylene Oxide	CH_2CH_2O	44.05	0.887	—	—	3.0	80.0
Ethyl Ether	$(C_2H_5)_2O$	74.12	0.713	—	—	—	—
Ethyl Formate	$HCO_2C_2H_5$	74.08	0.917	—	—	2.75	16.40
Ethyl Silicate	$(C_2H_5)_4SiO_4$	208.30	0.933	—	125 [52]	—	—
Formaldehyde	HCHO	30.03	0.815	Gas		7.0	73.0
Gasoline	CHnH(2n + 2)	86.0	0.660	-50 [-46]	—	1.36.0	
Heptane	$CH_3(CH_2)_5CH_3$	100.20	0.684	25 [-4]	—	1.1	6.7
Hexane	$CH_3(CH_2)_4CH_3$	86.17	0.660	-7 [-22]	—	1.18	7.4
Hydrogen Chloride	HCl	36.47	1.268 (A)	—	—	—	—
Hydrogen Cyanide	HCN	27.03	0.688	Gas		5.6	40.0
Hydrogen Fluoride	HF	20.01	0.987	Gas		—	—
Hydrogen Selenide	H_2Se	80.98	2.12	Gas		—	—
Hydrogen Sulfide	H_2S	34.08	1.189 (A)	Gas		4.3	45.5
Iodine	I_2	253.82	4.93	—	—	—	—
Isophorone	$(CH_3)_3C(CH_2)_2CCHCO$	138.20	0.923	—	205 [96]	—	—
Mesityl Oxide	$(CH_3)_2{:}CHCOCH_3$	98.14	0.857	87 [31]	—	—	—
Methanol	CH_3OH	32.04	0.792	54 [12]	60 [16]	6.72	36.5
Methyl Acetate	$CH_3CO_2CH_3$	74.08	0.928	15 [-9]	20 [-7]	3.15	15.60
Methyl Bromide	CH_3Br	94.95	1.732	—	—	13.5	14.5
Methyl Butanone (Isopropyl butane)	$CH_3COCH(CH_3)_2$	86.13	0.803	—	—	—	—
Methyl Cellosolve	$HOCH_2CH_2OCH_3$	76.06	0.965	107 [42]	115 [46]	—	—
Methyl Cellosolve Acetate	$CH_3OCH_2CH_2OOCCH_3$	118.13	1.007	132 [56]	140 [60]	—	—
Methyl Chloride	CH_3Cl	50.49	1.785	Gas		8.25	18.70
Methyl Cyclohexane	$CH_3(CHC_5H_{10})$	98.18	0.769	25 [-4]	—	1.15	—
Methyl Cyclohexanol	$CH_3(CHC_4H_8CHOH)$	114.18	0.934	154 [68]	—	—	—
Methyl Cyclohexanone	$CH_5C_5H_9CO$	122.17	0.925	118 [48]	—	—	—
Methyl Formate	HCO_2CH_3	60.05	0.974	-2 [-19]	—	4.5	20.0
Methyl Isobutyl Ketone	$CH_3COC_4H_9$	100.16	0.801	73 [23]	—	—	—
Monochlorobenzene	C_6H_5Cl	112.56	1.107	90 [32]	—	—	—
Monofluorotrichloromethane	Cl_3CF	137.38	1.494	Nonflammable			
Mononitrotoluene	$CH_3C_6H_4NO_2$	137.13	1.163	223 [106]	—	—	—
Naphtha (coal tar)	$C_6H_4(CH_3)_2$	106.16	0.85	100-110 [38–43]	—	—	—
Nickel Carbonyl	$Ni(CO)_4$	170.73	1.31	—	—	—	—
Nitrobenzene	$C_6H_5NO_2$	123.11	1.205	190 [88]	—	1.8 (200 F) [93]	—
Nitroethane	$CH_3CH_2NO_2$	75.07	1.052	82 [28]	106 [41]	—	—
Nitrogen Oxides	NO	30.0	1.0367(A)	—	—	—	—
	N_2O	44.02	1.53	—	—	—	—
	N_2O_3	76.02	1.447	—	—	—	—
	NO_2	46.01	1.448	—	—	—	—
	N_2O_5	108.02	1.642	—	—	—	—
Nitroglycerine	$C_3H_5(ONO_2)_3$	227.09	1.601	—	—	—	—
Nitromethane	CH_3NO_2	61.04	1.130	95 [35]	112 [44]	—	—
2-Nitropropane	$CH_3CHNO_2CH_3$	89.09	1.003	—	103 [39]	—	—
Octane	$CH_3(CH_2)_6CH_3$	114.22	0.703	56 [13]	—	0.95	3.2
Ozone	O_3	48.0	1.658 (A)	—	—	—	—
Pentane	$CH_3(CH_2)_3CH_3$	72.15	0.626	-40 [-40]	—	1.4	7.8
Pentanone (Methylpropanone)	$CH_3COCH_2C_2H_5$	86.13	0.816	45 [7]	60 [16]	1.55	8.15
Phosgene	$O{:}C{:}Cl_2$	98.92	1.392	—	—	—	—
Phosphine	PH_3	34.0	1.146 (A)	—	205 [96]	—	—
Phosphorus Trichloride	PCl_3	137.35	1.574	—	—	—	—
iso-Propanol	$(CH_3)_2CHOH$	60.09	0.785	53 [12]	60 [16]	2.02	11.80

PHYSICAL CONSTANTS OF SELECTED MATERIALS (Cont.)

Substance	Formula	Molecular Weight	Specific Gravity	Flash Point F [C] Closed Cup	Flash Point F [C] Open Cup	Explosion Limits (Volume Percent) Lower	Explosion Limits (Volume Percent) Upper
Propane	$CH_3CH_2CH_3$	44.09	1.554	Gas		2.12	9.35
Propyl Acetate	$CH_3CO_2CH_2C_2H_5$	102.13	0.886	43 [4]	60 [16]	1.77	8.0
iso-Propyl Ether	$(CH_3)_4(CH)_2O$	102.17	0.725	-18 [-28]	-15 [-26]	—	—
Stibine	SbH_3	124.78	4.344 (A)	—	—	—	—
Styrene Monomer	$C_6H_5HC{:}CH_2$	104.14	0.903	90 [32]	—	1.1	6.1
Sulfur Chloride, Mono	S_2Cl_2	135.03	1.678	245 [118]	None	—	—
Di	SCl_2	102.97	1.621	—	—	—	—
Tetra	SCl_4	173.89	—	—	—		—
Sulfur Dioxide	SO_2	64.07	2.264 (A)	Gas		—	—
1,1,2,2, Tetrachloroethane	$Cl_2CHCHCl_2$	167.86	1.588	—	—	—	—
Tetrachloroethylene	$Cl_2C{:}CCl_2$	165.85	1.624	Nonflammable			
Toluene	$C_6H_5CH_3$	92.13	0.866	40 [4]	45 [7]	1.27	6.75
Toluidine	$CH_3C_6H_4NH_2$	107.15	0.999	188 [87]	205 [96]	—	—
Trichloroethylene	$ClCHCCl_2$	131.40	1.466	Nonflammable			
Turpentine (Turpene)	$C_{10}H_{16}$	136.23		95 [35]	—	0.8	—
Vinyl Chloride (Chloroethane)	C_2H_5Cl	62.50	0.908	Gas		4.0	21.70
Xylene	$C_6H_4(CH_3)_2$	106.16	0.881	63 [17]	75 [24]	1.0	6.0

SOLVENT DRYING TIME

Solvent	Dry Time Relation	Boiling Range Deg. C	Boiling Range Deg. F	Weight per Gal. Lbs.	Weight per m³ kg
Ethyl Ether, C.P.	1.0	34-35	93-95	5.98	716.58
Petrolene	1.8	61-96	142-205	5.83	698.61
Carbon Tetrachloride	1.9	76	169	13.30	1593.73
Acetone	2.0	55-58	133-136	6.35	760.92
Methyl Acetate	2.2	56-62	133-144	7.79	933.46
Ethyl Acetate 85-88%	2.5	74-77	165-171	7.37	883.15
Trichlorethylene	2.5	87	189	12.20	1461.93
Benzol (Industrial)	2.6	79-81	174-178	7.38	884.35
Methyl Ethyl Ketone	2.7	77-82	171-180	6.95	832.82
Isopropyl Acetate 8%	2.7	84-93	183-199	7.26	869.99
Ethylene Dichloride	3.0	84	183	10.49	1257.01
Solvsol 19/27	3.7	86-123	187-254	6.58	788.48
Ethylene Chloride	4.0	81-87	178-189	10.49	1257.02
Propylene Dichloride	4.1	92-97	199-207	9.64	1155.16
Troluoil	4.1	90-122	194-252	6.17	739.35
Methanol	5.0	64-65	147-149	6.63	794.47
Toluol (Industrial)	5.0	109-111	229-232	7.19	861.58
Methyl Propyl Ketone	5.2	101-107	214-225	6.77	811.25
V. M. & P.	5.8	95-141	203-286	6.23	746.54
Perchlorethylene	6.0	121	250	13.55	1623.70
Nor. Propyl Acetate	6.1	97-101	207-214	7.50	898.73
Sec. Butyl Acetate	6.5	106-135	223-275	7.13	854.39
Solox (Anhydrous)	6.5	71-78	160-172	6.80	814.84
Isobutyl Acetate 90%	7.0	106-117	223-243	7.28	872.36
Apocthinner	7.0	115-143	239-289	6.31	756.13
Ethyl Alcohol, Den. No. 1	7.7	78	172	6.64	795.67

SOLVENT DRYING TIME (Cont.)

Solvent	Dry Time Relation	Boiling Range Deg. C	Boiling Range Deg. F	Weight per Gal. Lbs.	Weight per m^3 kg
Solox	8.0	76-78	169-172	6.73	806.46
Isopropyl Alcohol 99%	8.6	79-82	174-180	6.75	808.85
Nor. Propyl Alcohol	9.1	96-98	205-208	6.73	806.46
Solvsol 24/34	9.4	101-168	214-334	6.80	814.84
Nor. Butyl Acetate	9.6	110-132	230-270	7.29	873.56
Diethyl Carbonate	9.6	100-130	212-266	8.14	975.41
Methyl Butyl Ketone	9.7	114-137	237-279	6.84	819.64
Xylol (Industrial)	9.7	127-144	261-291	7.17	859.18
Monochlor Benzol	10.0	130-132	266-270	9.20	1102.44
Tertiary Butyl Alcohol	11.9	82-83	180-181	6.55	784.87
Sec. Butyl Alcohol	14.0	99-100	210-212	6.85	820.84
Sec. Amyl Acetate	16.9	121-144	250-291	7.21	863.97
Amyl Acetate	17.4	105-142	221-288	7.24	867.57
Isobutyl Alcohol	17.7	107-111	225-232	6.70	802.86
Methyl Cellosolve	18.0	121-126	250-259	8.07	967.03
Butyl Propionate	18.0	124-171	255-340	7.31	875.96
Pentacetate	20.0	121-155	250-311	7.19	861.56
Turpentine	20.0	155-173	311-343	12.24	1466.72
Butanol	21.0	116-119	241-246	6.79	813.65
Sec. Amyl Alcohol	25.5	105-125	221-257	6.79	813.65
2-50-W Hi-Flash Naphtha	27.5	148-187	298-369	7.18	860.38
Amyl Alcohol (Fusel Oil)	32.1	126-130	259-266	6.76	810.05
Di Isopropyl Ketone	33.9	164-169	327-336	6.75	808.85
Ethyl Cellosolve	36.2	133-137	271-279	7.77	931.08
Odorless Mineral Spirits	38.6	150-210	302-394	6.52	781.29
Ethyl Lactate	40.0	119-176	246-349	8.59	1029.34
Sec. Hexyl Alcohol	41.7	157	315	6.97	835.22
Solvsol 30/40	43.2	142-199	288-390	7.06	845.00
Pentasol	45.0	112-140	234-284	6.76	810.05
Hi-Solvency Mineral Spirits	46.7	152-200	306-392	6.79	813.65
No. 380 Mineral Spirits	47.0	151-196	304-385	6.57	787.28
No. 10 Mineral Spirits	55.0	154-196	309-385	6.49	777.70
Distilled Water	60.0	100	212	8.32	996.99
Apco No. 125	60.5	162-200	324-392	6.52	781.29
Cellosolve Acetate	65.0	145-166	293-331	8.15	976.61
Sec. Butyl Lactate	73.5	172	342	8.14	975.42
Sec. Hexyl Acetate	76.5	129-159	264-316	7.19	861.58
Butyl Cellosolve	88.5	163-172	325-342	7.58	908.31
Dipentene	89.2	149-215	300-419	7.10	850.79
No. 140 Thinner	91.2	185-210	365-410	6.62	793.27
Octyl Acetate	152.5	195-203	383-397	7.20	862.78
Isobutyl Lactate	156.5	168-200	334-392	8.15	976.61
Hexalin	177.5	159-162	318-324	7.89	945.46
Solvsol 40/50	270.5	191-248	376-478	7.42	889.14
Methyl Hexalin	276.5	170-190	338-374	7.66	917.90
Butyl Lactate	339.0	185-195	365-383	8.14	975.42
Excellene	384.0	162-260	324-500	6.55	784.89
Special Heavy Naphtha	403.0	202-242	396-468	6.73	806.46
Dispersol	425.0	193-242	379-468	6.59	789.68
No. 50 Kerosene	626.7	178-256	352-493	6.76	810.05
Triethylene Glycol	Over 5200.0	276-310	529-590	9.30	1114.42
Dibutyl Phthalate	Over 5200.0	195-200	383-392	8.73	1046.12

Dry Time Relation: (IP & SI)
Below 5 — Fast
5-15 — Medium
15-75 — Slow
75 over — Nil

CONVERSION FACTORS*

to convert	into	multiply by
ampere-hours	coulombs	3600.0
angstrom units	inches	3.937×10^{-9}
angstrom units	microns	1×10^{-4}
atmospheres	centimeters of mercury	76.0
atmospheres	feet of water	33.96
atmospheres	inches of mercury	29.92
atmospheres	inches of water	407.52
atmospheres	millimeters of water	10351
atmospheres	millimeters of mercury	760
atmospheres	pascals	101513
atmospheres	pounds/square foot	2116.3
atmospheres	pounds/square inch	14.696
BTU	foot pounds	778
BTU	horsepower-hours	3.931×10^{-4}
BTU	joules	1.055×10^{-3}
BTU	kilowatt-hours	2.928×10^{-4}
BTU/hour	watts	0.2931
calories	BTU	3.9685×10^{-3}
centimeters	feet	3.281×10^{-2}
centimeters	inches	0.3937
centimeters	kilometers	1×10^{-5}
centimeters	meters	1×10^{-2}
centimeters	millimeters	10.0
coulombs	faradays	1.036×10^{-5}
cubic feet	cubic meters	0.02832
cubic feet	gallons	7.48
cubic feet	liters	28.3162
cubic inches	cubic centimeters	16.3872
cubic inches	liters	0.0164
cubic meters	cubic feet	35.3145
cubic centimeters	cubic inches	0.06102
cubic centimeters	pints (U.S. liquid)	2.113×10^{-3}
cubic feet/minute	pounds water/minute	62.43
cubic feet/minute	cubic meters/second	4.719×10^{-4}
cubic feet/minute/foot	cubic meters/second/ meter	1.548×10^{-3}
cubic feet/minute/ square foot	cubic meters/second/ square meter	5.085×10^{-3}
cubic feet/second	gallons/minute	448.83
days	seconds	86,400.0
degrees (angle)	radians	1.745×10^{-2}
degrees/second	revolutions/minute	0.1667
dynes	newtons	1×10^{-5}
ergs	BTU	9.480×10^{-11}
ergs	kilowatt-hours	2.778×10^{-14}
ergs	foot-pounds	7.3670×10^{-8}
faradays/second	amperes	96,500
feet	centimeters	30.48
feet	miles (nautical)	1.645×10^{-4}
feet	meters	0.3048
feet	miles (statute)	1.894×10^{-4}
feet/minute	centimeters/second	0.5080
feet/minute	meters/second	0.00508
feet/second	miles/hour	0.6818
feet/second	knots	0.5921
foot-pounds	BTU	1.286×10^{-3}
foot-pounds	kilowatt-hours	3.766×10^{-7}
gallons	cubic feet	0.1337
gallons	liters	3.785
gallons/minute	cubic feet/hour	8.0208
gallons/minute	cubic feet/second	0.00223
gallons of water	pounds of water	8.3453
grains	grams	0.0648
grams	grains	15.4324
grams	ounces (troy)	3.215×10^{-2}
grams	ounces (avoirdupois)	3.527×10^{-2}
grams	pounds	2.205×10^{-3}
grams/cubic foot	milligrams/cubic meter	2.2883
horsepower	BTU/minute	42.44
horsepower	foot-pounds/second	550
horsepower	foot-pounds/minute	33,000
horsepower	kilowatts	0.7457
horsepower	watts	745.7
inches	centimeters	2.540
inches	miles	1.578×10^{-5}
inches of mercury	atmospheres	0.03342
inches of mercury	feet of water	1.134
inches of mercury	inches of water	13.61
inches of mercury	millimeters of mercury	25.4
inches of mercury	millimeters of water	345.6
inches of mercury	pascals	3390
inches of mercury	pounds/square foot	76.70
inches of mercury	pounds/square inch	0.491
inches of water	atmospheres	0.002456
inches of water	feet of water	0.08333
inches of water	inches of mercury	0.0735
inches of water	millimeters of water	25.4
inches of water	millimeters of mercury	1.876
inches of water	pascals	249.1
inches of water	pounds/square inch	0.0361
inches of water	pounds/square foot	5.196
joules	BTU	9.480×10^{-4}
joules	ergs	1. 107
kilograms	pounds	2.205
kilograms	slugs	0.068522
kilometers	feet	3281.0
kilometers	miles	0.6214
kilometers	meters	1000.0
kilometers/hour	knots	0.5396
kilopascals	pounds/square inch	0.145
kilowatt-hours	BTU	3413.0
kilowatts	foot-pounds/second	737.6
kilowatts	horsepower	1.341
knots	feet/hour	6080.0
knots	statute miles/hour	1.151
knots	nautical miles/hour	1.0
light years	miles	5.9×10^{12}
liters	cubic inches	61.02

CONVERSION FACTORS* (Cont.)

to convert	into	multiply by	to convert	into	multiply by
liters	cubic centimeters	1000.0	pascals	newtons/square miles	1.0
liters	gallons (U.S. liquid)	0.2642	pascals	pounds/square foot	0.2089
liters	milliliters	1000.0	pascals	pounds/square inch	1.696×10^{-4}
liters	pints (U.S. liquid)	2.113	pints (liquid)	gallons	0.125
meters	centimeters	100.0 0	pints (liquid)	cubic centimeters	473.2
meters	feet	3.281	pints (liquid)	cubic inches	28.87
meters	kilometers	1×10^{-3}	pints (liquid)	quarts (liquid)	0.5
meters	miles (statute)	6.214×10^{-4}	pounds	ounces	16.0
meters	miles (nautical)	5.396×10^{-4}	pounds	ounces (troy)	14.5833
meters	millimeters	1000.0	pounds	pounds (troy)	1.21528
microns	meters	1×10^{-6}	pounds	kilograms	0.4536
miles (nautical)	miles (statute)	1.516	pounds/square in	inches of water	27.6807
miles (nautical)	kilometers	1.853	pounds/square in	millimeters of water	703.08
miles (nautical)	feet	6080.27	pounds/square in	Pascals	6.8952×10^{3}
miles (statute)	kilometers	1.609	quarts (dry)	cubic inches	67.20
miles (statute)	feet	5280.0	quarts (liquid)	gallons	0.25
miles (statute)	miles (nautical)	0.8684	quarts (liquid)	cubic inches	57.75
miles/hour	feet/minute	88.0	quarts (liquid)	liters	0.9463
miles/hour	feet/second	1.467	radians	minutes	3438.0
milligram/liter	parts/million	1.0	radians	degrees	57.30
milligrams/cubic meter	grains/cubic foot	4.37×10^{-4}	revolutions	degrees	360.0
milliliters	liters	1×10^{-3}	revolutions/minute	degrees/second	6.0
millimeters	inches	3.937×10^{-2}	seconds	minutes	1.667×10^{-2}
millimeters of mercury	atmospheres	0.001316	slugs	pounds	32.17
millimeters of mercury	feet of water	0.4464	square feet	square meters	0.0928
millimeters of mercury	inches of mercury	0.0394	tons (long)	pounds	2240.0
millimeters of mercury	inches of water	0.5357	tons (long)	tons metric	1.0160
millimeters of mercury	millimeters of water	13.61	tons (long)	tons (short)	1.120
millimeters of mercury	pascals	133	tons (long)	kilograms	1016.0
millimeters of mercury	pounds/square foot	2.789	tons (short)	tons (long)	0.89287
millimeters of mercury	pounds/square inch	0.01934	tons (short)	tons metric	0.9071
millimeters of water	inches of water	0.0394	tons (short)	kilograms	907.18
millimeters of water	Pascals	9.806	tons (short)	pounds	2000.0
newtons	dynes	1×10^{5}	watts	BTU/hour	3.4129
newtons	pounds	0.2248	watts	horsepower	1.341×10^{-3}
ounces	pounds	6.25×10^{-2}	yards	miles (nautical)	4.934×10^{-4}
ounces (troy)	ounces (avoirdupois)	1.09714	yards	meters	0.9144
pascals	atmospheres	9.872×10^{-4}	yards	miles (statute)	5.682×10^{-4}
pascals	inches of water	0.00401			
pascals	millimeters of water	0.102			

*These are soft conversion factors. See Definitions and Chapter 9, Appendix A9 for discussion of soft and hard conversions.

APPENDIX C

Testing and Measurement of Ventilation Systems

APPENDIX C

TESTING AND MEASUREMENT OF VENTILATION SYSTEMS

FINAL APRIL 03, 2006

C.1 INTRODUCTION

Airflow and static pressure measurements of a ventilation system are often required to determine the performance of the ventilation system. Some of the reasons include:

1) *Commissioning*: Recording the initial performance of the ventilation system and determining if it is functioning in accordance with specifications and whether it meets the design basis.
2) *Proof of Performance*: Determining the degree of compliance with applicable codes (e.g., OSHA, EPA) or trade association standards.
3) *Balancing Ventilation Systems*: Adjusting airflows to match the desired distribution. Balancing may be done during commissioning, after alterations are made, or to re-adjust systems where blast gates may have been tampered with.
4) *Baseline Maintenance*: Obtaining data through periodic checks and comparing to baseline or reference values to determine when maintenance or repairs are necessary.
5) *Troubleshooting Ventilation Systems*: Determining whether, where and why system components have changed causing undesirable changes in flows, such as partial blocking by obstructions or coatings, inadvertently disconnecting hoods, and addition or removal of branch ducts.
6) *Change Management*: Obtaining data to assist in the design of future systems or alterations such as adding or removing branches.

The most important indirect indicators of system functioning are hood static pressure (SP_h) and volumetric flow (Q) for each branch duct and the static pressure (SP_{end}) at the terminus of each run of duct – e.g., branch, submain, or main (Figure C-1).

No instrument measures volumetric flow directly in typical contaminant control systems. However, many instruments have sensors whose output can be used to measure air velocity at a point. Once the velocity is measured, it can be used to compute the airflow. The following sections provide cursory information required for measuring system performance in the field. Chapter 3 of *Industrial Ventilation: A Manual of Recommended Practice for Operation and Maintenance* (the O&M Manual) provides more comprehensive information and procedures for both instrumentation and accurate balancing and verification of performance.

C.2 VELOCITY AND STATIC PRESSURE MEASUREMENTS

It is possible to measure static pressure accurately using either a wall pressure tap or a Pitot tube probe. For both static pressure and velocity measurements, it is important when measuring static pressures that the probe is aligned correctly with airflow and that the measurement location is chosen to avoid conditions that are difficult to measure accurately. Three conditions produce error:

1) Flow disturbances at the opening of the probe;
2) Measurement error due to the probe not being perpendicular to the duct; and
3) Turbulence inside the duct.

To avoid the latter, take readings several diameters downstream of the source of the turbulence (e.g., elbow, hood, branch entries). <u>Static</u> pressure readings can be taken using a wall tap or with a Pitot tube inserted to the centerline of the duct. For a highly accurate estimate of the mean static pressure at a cross-section of duct, average three static pressure measurements taken at wall taps equally spaced around the perimeter of the duct. Pitot tubes can be used to measure the static pressure in a duct. Given the difficulties in creating and maintaining clean holes in the duct, taking measurements with a Pitot tube is often more convenient than wall taps — unless continuous monitoring is desired. Do not use a tube pushed into a hole for replacement of a Pitot tube.

C.2.1 Duct Conditions. Conditions inside ducts are important in considering the type of device to use and the corrections that should be made to its readings:

1) Air heavily contaminated with corrosive gases, dusts, fumes, mists, or products of combustion cannot be measured using some devices and presents problems in using all instruments. Drilling into those ducts can be hazardous, so it can be critical to know the properties of the chemicals and particulates that may be inside the duct. Fragile probes can be corroded, abraded, or otherwise damaged by airborne contaminants in the duct. Probes that contain sensors may report values incorrectly if coated with dust or other materials. Probes that act as a conduit for gases to reach the sensor may become blocked or allow these deleterious materials to reach the instrument where they block, coat, or damage it. Probes that conduct pressure only will pass along very little contaminant to the sensor, but they can

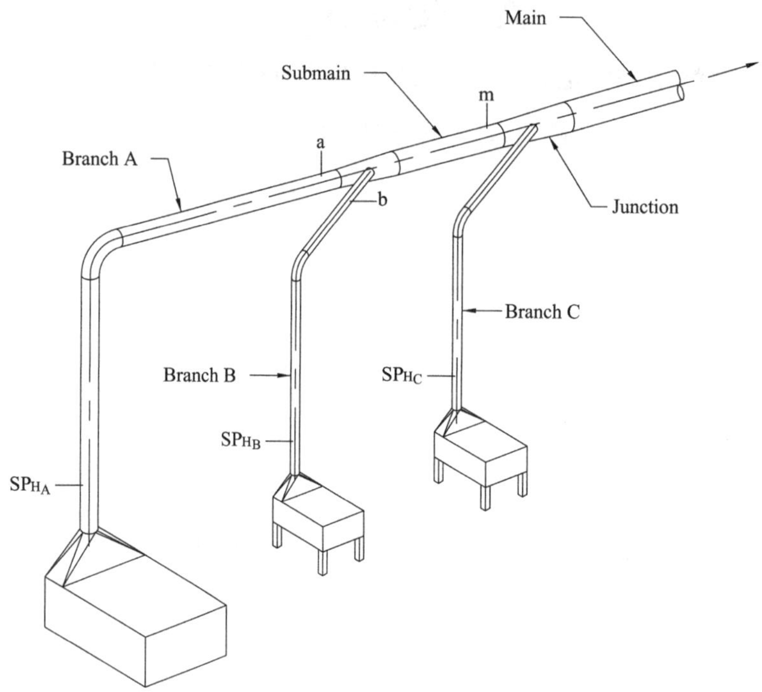

FIGURE C-1. Useful locations for measurements

become plugged.

2) High temperature conditions require density corrections for most devices and some sensors can be damaged if high temperature gases reach them.

3) High concentrations of water vapor and mist can plug probes so quickly that measurement becomes impracticable.

4) All field instruments have difficulty with very low velocities. Devices that measure pressure (including velocity pressure) for field use generally have degraded accuracy at static or velocity pressures below 0.1 "wg [25 Pa]. Thermal anemometers developed for the field become increasingly unreliable below 30 fpm [0.15 m/s].

5) It is difficult to measure accurately where airflows are non-uniform. Such locations include near the discharges of process equipment and fans and near elbows, sudden expansions or contractions, and less than seven duct diameters downstream of junction fittings. However, this is typical of many field conditions.

6) In some systems, measurements may vary substantially over relatively short periods. Systems with automatic control of dampers or fan speeds sometimes change so frequently that the measurements taken on the same day at different locations in the system may present a misleading picture of system behavior. Velocity sampling for isokinetic sampling of stacks is often plagued by variable air volume systems, particularly when one stack is used in common for several fans. It is therefore important to log the insertion depth or setting of all dampers before tests are done and confirm that they have not changed during the tests.

C.2.2 Instrument Use. It is important to use measurement instruments properly. Errors in measurement usually occur when any of the following "short cuts" are used:

1) Using a single reading at the center of the duct and then multiplying it by some factor such as '0.9' (called the Centerline Method).

2) Using more than the centerline but still taking fewer readings than recommended.

3) Using instruments not intended for measurement in industrial air streams. For example, one should avoid using thermal anemometers or similar devices in dust laden air streams or those with high moisture or heat.

4) Doing Pitot traverses in locations where the airflow has eddies, such as immediately downstream of elbows, hoods, etc.

5) Failure to consider heat, moisture or elevation corrections when calculating velocities and flows from measured data.

6) Using hood face velocities to determine branch airflows when it is possible to do a Pitot traverse in the duct instead.

7) Blocking hood airflows with the body or with bulky instruments when measuring hood face velocities.

8) Failing to calibrate measurement devices.

C.2.3 Sampling Locations. For sampling locations to be representative, every area of the duct should have an equal chance of being sampled. For that reason the traverse points are not evenly spaced, but instead are chosen so that each point represents close to the same amount of airflow. The measurement locations listed in Table C-1 for round ducts and Table C-2 for rectangular ducts provide the greatest accuracy for the least number of measurement locations for rectangular and round ducts, respectively.(C.1)

For round ducts, Table C-2 shows recommended insertion depths based on fractions of the duct diameter. In ideal conditions, the velocity profile is symmetric and it may be acceptable to traverse only one diameter. If the velocity is skewed in one direction (a typical situation in fieldwork), a single traverse will not represent the true average of the flow. In that case, or if accuracy is important, use two perpendicular diameters (Figure C-2). If unusually high accuracy is important, use three equally separated traverse diameters.

Taking more measurements on the same diameter is less useful than using a second or third diameter. For example, three 6-point traverses are much more accurate than one 20-point traverse. If only one traverse diameter is done, a 10-point traverse should be measured. For two perpendicular traverses, 8- or 10-point traverses should be used (Figure C-2). When traversing three diameters in the same cross-section, 6- or 8-point traverses are acceptable.

Determining the minimum number of measurement points for rectangular ducts is more complicated since the aspect ratio can vary from 1:1 to 5:1 or higher. ASHRAE(C.2) recommends a minimum of 25 points. The maximum distance between any two points should be 8 inches [200 mm]. Table C-2 shows traverse insertion depths for 5-, 6- and 7-point traverses for rectangular ducts (Figure C-3).

Elbows, junction fittings, dampers, fans, and other obstructions skew the airflow to one side of the duct. At seven or more diameters downstream for most upstream elements, the airflow often has reached reasonable symmetry. The exceptions are dampers and fans, which can disturb the flow for more than 10 duct diameters. Since airflow starts to change direction before encountering an elbow, damper, or other disturbances, it is important to take readings at least one duct diameter upstream of them. Table C-3 suggests some reasonable distances downstream and upstream from different disturbances.(C.3)

If it is not possible to find or use a location that meets those restrictions, there is no other choice than to use the best available location. For such relatively poor conditions, it is crucial to do two perpendicular traverses instead of relying on a single diameter traverse. A two-diameter traverse can be surprisingly accurate even when velocity contours are highly asymmetric. For example, in laboratory tests, the use of two diameter traverses should produce less than 6% error when taken only two diameters distance downstream from smooth elbows and less than 3% error when taken at least seven diameters distance downstream from obstructions.(C.3) Field measurements will have more variability.

It is likely that a hood airflow that is substantially lower than design values will be associated with inadequate performance for a given hood. For that reason, it is important to maintain hood exhaust volumes velocities at levels that meet or slightly exceed recommended levels.

In many cases, the most accurate way to determine the average velocity at an open hood face is to determine the airflow using Pitot traverses of the branch duct connected to the hood, and then divide the airflow by the open cross-sectional area of the hood face. Details for this method are included in Chapter 3 of the O&M Manual.

TABLE C-1. Recommended Traverse Insertion Depths for Round Ducts

Number of Insertions	Distance from wall in fractions of a duct diameter, log-linear rule Traverse Position									
	1	2	3	4	5	6	7	8	9	10
4	0.043	0.290	0.710	0.957						
6	0.032	0.135	0.321	0.679	0.865	0.968				
8	0.021	0.117	0.184	0.345	0.655	0.816	0.883	0.979		
10	0.019	0.077	0.153	0.217	0.361	0.639	0.783	0.847	0.923	0.981

TABLE C-2. Recommended Traverse Insertion Depths for Rectangular Ducts

Number of Insertions	Distance from wall in fractions of a duct width Traverse Position						
	1	2	3	4	5	6	7
5	0.074	0.288	0.500	0.712	0.926		
6	0.061	0.235	0.437	0.563	0.765	0.939	
7	0.053	0.203	0.366	0.500	0.634	0.797	0.947

*Use the lower end of the range if an elbow is between the damper or fan and the measurement point.

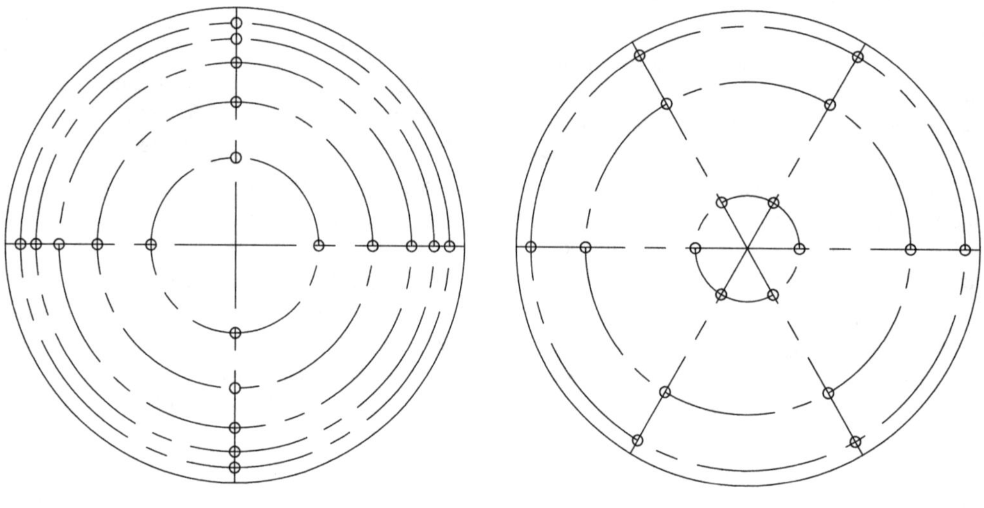

FIGURE C-2. Insertion depths for round ducts

C.3 COMPUTING AIRFLOW FROM ESTIMATED VELOCITY

One velocity measurement is seldom enough to determine the average velocity, because the velocities at a hood face or duct cross-section are seldom uniform. It is important to measure the velocities at several locations chosen to be representative then compute the average of those values. For the measured velocity to be a good estimate of the true value, measurements must be taken correctly, in the right locations, and there must be enough measurements to include most to the variability (both spatial and temporal).

As an example of computing the average velocity in a duct, consider the traverse values shown in Figure C-4. The average velocity in that duct would be computed as the simple average:

$$V = \frac{2950+3050+3170+3240+3290+3310+3260+3160+3060+2980}{10}$$

$$= 3150 \text{ fpm}$$

$$[V = \frac{14.99+15.49+16.10+16.46+16.71+16.81+16.56+16.05+15.54+15.14}{10}$$

$$= 15.99 \text{ m/s}]$$

The estimated airflow can be computed from the average velocity multiplied by the cross-sectional area normal to the flow:

$$Q = V * A$$

where: Q = volumetric flow rate, actual cubic feet per minute (acfm) [am^3/s]

V = average velocity normal to the cross-section, actual feet per minute (fpm) [m/s]

A = cross-sectional area of duct or hood at the measurement location, ft^2 [m^2]

EXAMPLE 1 (IP Units)

If the average velocity at the face of a 3 ft by 4 ft enclosing hood is 100 fpm, the estimated airflow into the hood can be determined as follows:

$$Q = (100 \text{ fpm}) \times (3 \text{ ft}) \times (4 \text{ ft}) = 1200 \text{ acfm}$$

EXAMPLE 2

If the average velocity in a 6-inch [152 mm] diameter duct (area = 0.1964 ft^2) [.018 m^2] is 3000 fpm [15.24 m/s], the estimated airflow would be:

$$Q = (3000 \text{ fpm}) \times (0.1964 \text{ ft}^2) = 588 \text{ acfm}$$

$$[Q = (15.24 \text{ m/s}) (0.018 \text{ m}^2) = 0.27 \text{ am}^3/\text{s}]$$

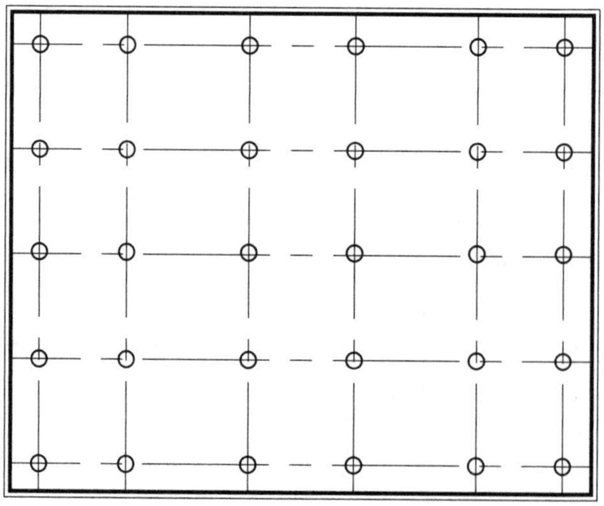

FIGURE C-3. Insertion depths for rectangular ducts

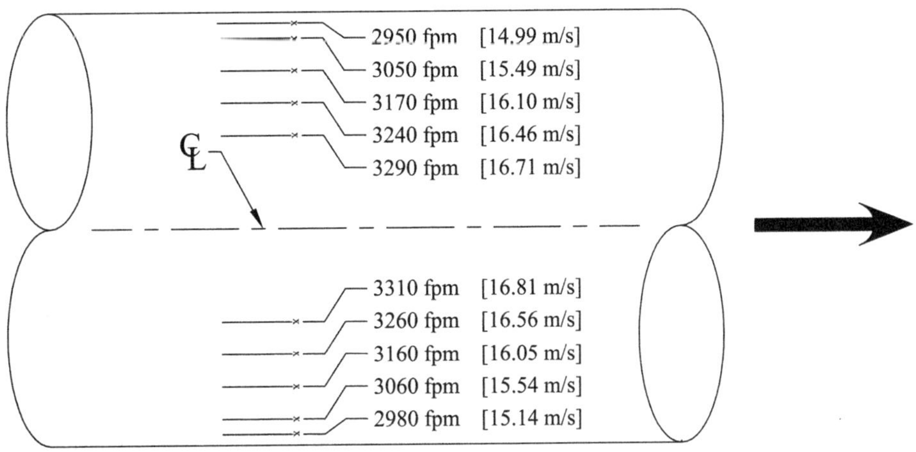

FIGURE C-4. Values measured in a duct cross-section

Most calculations are made on a calculator or computer. The readout of the device often implies greater accuracy than is warranted for ventilation calculations. In the above problem, the velocity was given as 3000 fpm [15.24 m/s]. The accuracy of the 3000 fpm [15.24 m/s] may be ± 3%. The answer should imply no greater accuracy. A scientifically correct answer to the above IP problem would be 590 acfm. For simplicity, this Manual will assume three places of accuracy. The calculations will be made without rounding until the final answer.

Ventilation practitioners sometimes convert the actual airflow to "standard" airflow by multiplying the true value by the density factor (df), the ratio of the actual density to standard density (when no moisture is present):

$$Q_{std} = df * Q_{act}$$

where: Q_{std} = airflow rate that would exist at normal temperature and pressure for dry air, scfm [nm³/s]

Q_{act} = actual airflow rate, acfm [am³/s]

df = Density Factor

C.4 COMPUTING VELOCITY FROM VELOCITY PRESSURE

In most ventilation systems, the actual airflow is more important than standard airflow. The hoods, ducts, air cleaning device, and fan design are based on actual airflow. Calculating airflow requires computation of the density factor. This is shown in Chapters 3 and 9 of this Manual.

Velocity can be computed from observed velocity pressures if the air density is known or can be determined. For the velocity at a single point:

$$V = 4005\sqrt{VP/df}$$

$$\left[V = 1.29\sqrt{VP/df}\right]$$

where: VP = Velocity pressure, "wg [Pa]

df = Ratio of actual density to standard density (Density Factor)

EXAMPLE 3

If the velocity pressure is 0.5 "wg [125 Pa] and the density factor is 0.95, the velocity would be computed as:

$$V = 4005\sqrt{0.5/0.95} = 2910\text{ fpm}$$

$$[V = 1.29\sqrt{125/0.95} = 14.78\text{ m/s}]$$

Although a velocity can be computed from a velocity pressure, the average velocity cannot be computed from the average velocity pressure. Instead, when determining the average velocity from velocity pressure measurements, compute the individual velocities from individual velocity pressures, then the average velocity can be computed from the mean of those velocities.

EXAMPLE 4

The average velocity for the ten velocity pressure measurements in Figure C-4 when the gas stream has a density factor of 0.88 is shown in Table C-4. As an example, the velocity at data point 1 is calculated:

$$V = 4005\sqrt{0.22/0.88} = 2003\text{ fpm}$$

$$[V = 1.29\sqrt{55/0.88} = 10.20\text{ m/s}]$$

EXAMPLE 5

Consider a system with a velocity pressure reading of 1.0 "wg [249 Pa] taken with a Pitot tube in a duct where the dry-bulb temperature is 300 F [149 C], the moisture content is 0.1 lbm water/lbm dry air $\left[0.1\frac{kg\ H_2O}{kg\ da}\right]$ and the static pressure is -23.5 "wg [-5.8 kPa]. The system is installed at an elevation of 5,000 ft [152.4 m]. What would the density factor and actual velocity be at that point?

$$df = (df_e)(df_P)(df_T)(df_m)$$

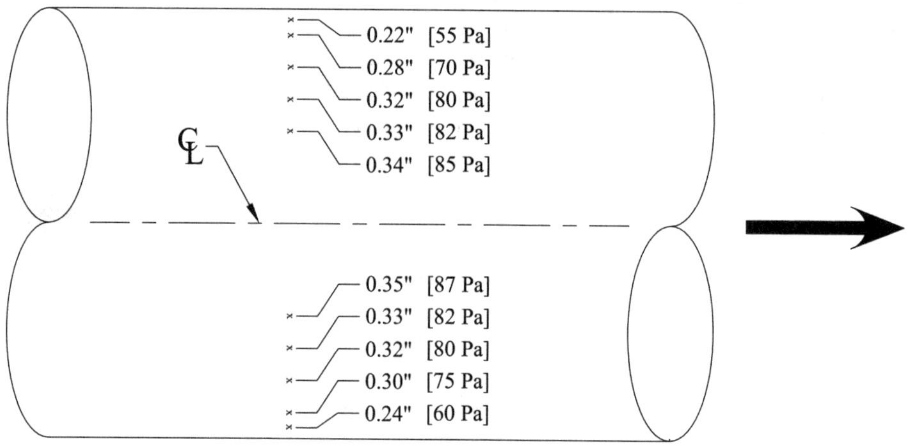

FIGURE C-5. Measured VP traverse values

$$df_e = (1-(6.73*10^{-6})(5000))^{5.258} = 0.84$$
$$[df_e = (1-(22.08 \times 10^{-6})(1520)^{5.258}) = 0.84]$$
$$df_P = (407 - 23.5)/407 = 0.94$$
$$[df_P = ((101384 - 5853)/101384) = 0.94]$$
$$df_T = 530/(300+460) = 0.70$$
$$[df_T = 294/(273 + 300) = 0.70]$$
$$df_m = (1 + 0.1)/(1 + 1.607*0.1) = 0.95$$
$$[df_m = [(1 + 0.1)/(1 + 1.607*.01) = 0.95]$$

$$df = (df_e)(df_P)(df_T)(df_m) = (0.835)(0.942)(0.697)(0.948) = 0.52$$ Same as IP

$V = 4005\sqrt{1.0/0.520} = 5550$ fpm $[V = 1.29\sqrt{249/0.52} = 28.23$ m/s$]$

Note that the value of velocity would have been 4005 fpm [20.35 m/s] — an error of 39% — if density effects had been ignored.

Measurement of air velocity at non-standard conditions requires calculation of the true air velocity, accounting for difference in air density due to air temperature, humidity, elevation and static pressure. The following examples illustrate the method of calculation and the effect of varying air density.

Two 10-Point traverse readings are shown in Tables C-5 and C-6.

EXAMPLE 6

Table C-5 shows the results of a 10-point traverse in the horizontal and vertical axis of a 24-inch [610 mm] duct. The following environmental conditions prevailed:

Conditions: SP = +2 "wg [488 Pa]
Dry-Bulb temperature = 79 F [26 C]
Wet-Bulb Temperature = 50 F [10 C]
Elevation = 1000 feet [308 m]
Duct diameter = 24" [610 mm]

From the psychrometric charts (Figures 9-g (IP) through 9-j (IP)) (with dry-bulb = 79 F and wet-bulb = 50 F), the water content ω is given by:

$$\omega = 0.0011 \text{ lbm water/lbm dry air}$$
$$[\omega = 0.0011 \text{ kg } H_2O/\text{kg} - \text{dry air}]$$
$$df = (df_e)(df_P)(df_T)(df_m)$$
$$df_e = (1-(6.73*10^{-6})(1000))^{5.258} = 0.97$$
$$[df_e = 1-((22.08)(10^{-6})(305))^{5.258} = 0.97]$$
$$df_P = (407 + 2.0)/407 = 1.001$$
$$[df_P = (101384 + 488)/101384 = 1.01]$$
$$df_T = 530/(460+79) = 0.98$$
$$[df_T = 294/(273 + 26) = 0.98]$$
$$df_m = (1 + 0.0011)/(1 + 1.607*0.0011) = 1.0\ [1.0]$$

$$df = (df_e)(df_P)(df_T)(df_m) = (0.97)(1.01)(0.98)(1.0) = 0.96\ [0.96]$$

The conversion from velocity pressure to velocity is given by:

$$V = 4005\sqrt{VP/0.96} \qquad \left[V = 1.29\sqrt{VP/0.96}\right]$$

EXAMPLE 7

Table C-6 shows the results of a 10-point traverse in the horizontal and vertical axis of a 12-inch [305 mm] duct. The following environmental conditions prevailed:

Conditions: SP = -2 "wg [-498 Pa]

Dry-Bulb temperature = 150 F [65.5 C]

Wet-Bulb Temperature = 100 F [38 C]

Elevation = 1500 feet [457.2 m]

Duct diameter = 12" [305 mm]

From the psychrometric chart (Figure 9-h), ω = 0.031 lbs [0.031 kg moisture/kg dry air] of moisture/lb of dry air

$$df = (df_e)\,(df_P)\,(df_T)\,(df_m)$$

$$df_e = (1\text{-}(6.73)(10^{-6})(1500))^{5.258} = 0.95$$

$$[df_e = (1\text{-}((22.08)(10^{-6}))(457.2))^{5.258} = 0.95]$$

$$df_P = (407 - 2.0)/407 = 1.00$$

$$[df_P = (101384 - 498)/101384 = 1.00]$$

$$df_T = 530/(460+150) = 0.87$$

$$[df_T = 294/(273 + 65.5) = 0.87]$$

$$df_m = (1 + 0.031)/(1 + 1.607)(0.031) = 0.98$$

$$[df_m = (1 + 0.031)/1 + 1.607)(0.031) = 0.98]$$

$$df = (df_e)\,(df_P)\,(df_T)\,(df_m)$$

$$= (0.95)\,(1.0)\,(0.87)\,(0.98)$$

$$= 0.81\ [0.81]$$

The conversion from velocity pressure to velocity is given by:

$$V = 4005\sqrt{VP/0.81} \qquad \left[V = 1.29\sqrt{VP/0.81}\right]$$

REFERENCES

C.1 American Society of Heating, Refrigerating and Air Conditioning Engineers: Practices for Measurement, Testing, Adjusting, and Balancing of Building Heating, Ventilation, Air Conditioning and Refrigeration Systems, ANSI/ASHRAE Standard 111-1988. ASHRAE.

C.2 American Society of Heating, Refrigerating and Air Conditioning Engineers: ASHRAE Handbook–1997 Fundamentals. ASHRAE.

C.3 Guffey, S.E.; Booth, D.W.: Comparison of Pitot Traverses Taken at Varying Distances Downstream of Obstructions. Am. Ind. Hyg. Assoc. J., Vol. 60, No. 2, pp. 165-174 (1999).

TABLE C-3. Minimum Distances for Velocity Measurement Locations

Location of Measurement Point in Multiples of the Duct Diameter, D			
Condition	**Downstream of Condition for VP or V Traverse**	**Downstream of Condition for SP**	**Upstream of Condition**
45° Tapered Hood Connection	6	2	N/A
Flanged Hood Connection or Takeoff	7	3	N/A
90° Radius Elbow	7	4	1
Junction Fitting	7	6	3
Damper or Fan	12–20*	6–20*	2–4*

TABLE C-4. 10-Point Velocity Average

First 5 VP Values			Second 5 VP Values		
Point	**VP "wg [Pa]**	**V fpm [m/s]**	**Point**	**VP "wg [Pa]**	**V fpm [m/s]**
1	0.22 [55]	2003 [10.18]	6	0.35 [87]	2526 [12.83]
2	0.28 [70]	2259 [11.48]	7	0.33 [82]	2453 [12.46]
3	0.32 [70]	2415 [12.28]	8	0.32 [80]	2415 [12.27]
4	0.33 [82.2]	2453 [12.46]	9	0.30 [80]	2338 [11.88]
5	0.34 [85]	2489 [12.64]	10	0.24 [60]	2092 [10.63]
Average velocity = 2340 fpm [11.89]					

TABLE C-5. 24-Inch [610 mm] Duct 10-Point Traverse

Vertical Traverse Pts			**Horizontal Traverse Pts**		
Traverse Pt	***VP, "wg [Pa]***	***V, fpm [m/s]***	***Traverse Pt***	***VP, "wg [Pa]***	***V, fpm [m/s]***
1	0.22 [55]	1924 [9.77]	1	0.23 [57]	1968 [9.99]
2	0.28 [69.8]	2171 [11.03]	2	0.27 [67]	2132 [10.83]
3	0.32 [80]	2321 [11.80]	3	0.33 [82]	2357 [11.97]
4	0.33 [82]	2357 [11.97]	4	0.34 [85]	2392 [12.15]
5	0.34 [85]	2392 [12.15]	5	0.34 [85]	2392 [12.15]
6	0.35 [87]	2427 [12.33]	6	0.35 [87]	2427 [12.32]
7	0.33 [82]	2357 [13.97]	7	0.34 [85]	2392 [12.15]
8	0.32 [80]	2321 [11.79]	8	0.32 [80]	2321 [11.80]
9	0.30 [75]	2247 [11.42]	9	0.32 [80]	2321 [11.80]
10	0.24 [60]	2010 [10.21]	10	0.25 [62]	2051 [10.42]
	Average	2253 [11.45]		Average	2275 [11.56]
Average Velocity = 2264 fpm [11.50 m/s]					
Volumetric flow = VA = (3.1416)*(2264) = 7110 acfm [VA = (0.29)*(11.50) = 3.34 am^3/s]					

TABLE C-6. 12-Inch [305 mm] Duct 10-Point Traverse

Vertical Traverse Pts			Horizontal Traverse Pts		
Traverse Pt	*VP, "wg [Pa]*	*V, fpm [m/s]*	*Traverse Pt*	*VP, "wg [Pa]*	*V, fpm [m/s]*
1	0.22 [55]	2093 [10.63]	1	0.23 [57]	2141 [10.87]
2	0.28 [70]	2362 [12.00]	2	0.27 [67]	2319 [11.78]
3	0.32 [70]	2525 [12.82]	3	0.33 [82]	2564 [13.03]
4	0.33 [82.2]	2564 [13.03]	4	0.34 [85]	2603 [13.22]
5	0.34 [85]	2603 [13.22]	5	0.34 [85]	2603 [13.22]
6	0.35 [87.2]	2641 [13.42]	6	0.35 [87.2]	2641 [13.45]
7	0.33 [82.2]	2564 [13.03]	7	0.34 [84.7]	2603 [13.22]
8	0.32 [80]	2525 [12.82]	8	0.32 [80]	2525 [12.83]
9	0.30 [75]	2445 [12.42]	9	0.32 [80]	2525 [12.83]
10	0.24 [60]	2187 [11.11]	10	0.25 [62.3]	2232 [13.34]
	Average	2451 [12.45]		Average	2476 [12.53]
Average Velocity = 2264 fpm [11.50 m/s]					
Volumetric flow = VA = (3.1416)*(2264) = 7110 acfm @ 150 F/100 F and 1500 ft ASL [VA = (0.29)(11.50) = 3.34 am^3/s @ 65 C/37 C and 457 m ASL]					

VELOCITY TRAVERSE (IP) DATA FORM

Process Location/Description

Duct/Traverse Schematic

Duct Description ____________ Shape: ☐ Round ☐ Rectangular ☐ ________

Orientation ________ Dimensions: D=____ L=____ W=____ Area=______

Density Factor

Elevation (ft) = ________ $df_e = [1-(6.73*10^{-6}(\text{elevation})]^{5.258}$ = ______

Static Pressure ("wg) = ______ $df_p = (407+SP)/407$ = ______

Temperature (F) = ________ $df_T = 530/(\text{temp}+460)$ = ______

Moisture (# H_2O/#da) = ______ $df_m = (1+\omega)/(1+1.607\omega)$ = ______

$df = (df_e)(df_p)(df_T)(df_m)$ = ______

Traverse ☐ Vertical ☐ Horizontal				Traverse ☐ Vertical ☐ Horizontal			
Point	Distance	VP, "wg	V, fpm	Point	Distance	VP, "wg	V, fpm
1				1			
2				2			
3				3			
4				4			
5				5			
6				6			
7				7			
8				8			
9				9			
10				10			
Velocity Subtotal				Velocity Subtotal			

Average Velocity = ________

Area = ________

Volumetric flow = ________

Note: Calculate Velocity from $V = 4005 * \sqrt{VP/df}$

FIGURE C-6 (IP). Sample pitot traverse data sheet

VELOCITY TRAVERSE (SI) DATA FORM

Process Location/Description

Duct/Traverse Schematic

Duct Description ______________ Shape: ☐ Round ☐ Rectangular ☐ ________

Orientation ____________ Dimensions: D=______ L=_____ W=_____ Area=________

Density Factor

Elevation (m) = ____________ $df_e = [1-(22.08*10^{-6}(elevation)]^{5.258}$ = ________

Static Pressure (Pa) = _______ $df_p = (101384+SP)/101384$ = ________

Temperature (C) = __________ $df_T = 294/(temp+273)$ = ________

Moisture ($kgH_2O/kgda$) = _____ $df_m = (1+\omega)/(1+1.607\omega)$ = ________

$df = (df_e)(df_p)(df_T)(df_m)$ = ________

Traverse ☐ Vertical ☐ Horizontal			
Point	Distance	VP, Pa	V, m/s
1			
2			
3			
4			
5			
6			
7			
8			
9			
10			
Velocity Subtotal			

Traverse ☐ Vertical ☐ Horizontal			
Point	Distance	VP, Pa	V, m/s
1			
2			
3			
4			
5			
6			
7			
8			
9			
10			
Velocity Subtotal			

Average Velocity = ____________

Area = ____________

Volumetric flow = ____________

Note: Calculate Velocity from $V = 1.29 * \sqrt{VP/df}$

FIGURE C-6 (SI). Sample pitot traverse data sheet

INDEX